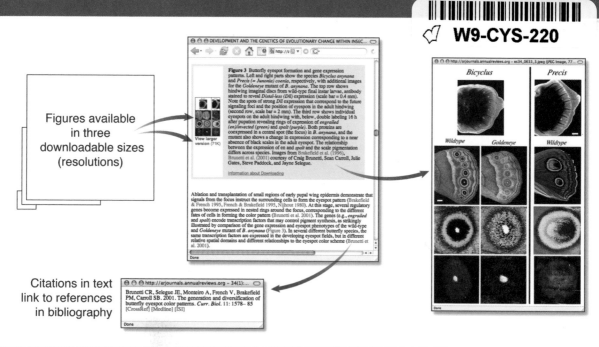

Figures available in three downloadable sizes (resolutions)

Citations in text link to references in bibliography

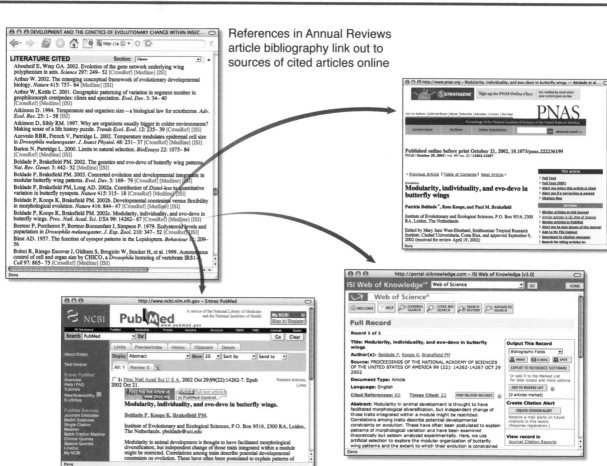

References in Annual Reviews article bibliography link out to sources of cited articles online

Annual Review of Ecology,
Evolution, and Systematics

Annual Review of Ecology, Evolution, and Systematics

Volume 38, 2007

Douglas J. Futuyma, *Editor*
State University of New York, Stony Brook

H. Bradley Shaffer, *Associate Editor*
University of California, Davis

Daniel Simberloff, *Associate Editor*
University of Tennessee

www.annualreviews.org • science@annualreviews.org • 650-493-4400

Annual Reviews
4139 El Camino Way • P.O. Box 10139 • Palo Alto, California 94303-0139

Annual Reviews
Palo Alto, California, USA

International Standard Serial Number: 1543-592X
International Standard Book Number: 978-0-8243-1438-5
Library of Congress Catalog Card Number: 71-135616

TYPESET BY APTARA, INC., FALLS CHURCH, VIRGINIA
PRINTED AND BOUND BY SHERIDAN BOOKS, INC., CHELSEA, MICHIGAN

Contents

Annual Review of
Ecology, Evolution,
and Systematics

Volume 38, 2007

Indexes

Errata

An online log of corrections to *Annual Review of Ecology, Evolution, and Systematics*
articles may be found at http://ecolsys.annualreviews.org/errata.shtml

Related Articles

Evolution of Animal Photoperiodism

William E. Bradshaw and Christina M. Holzapfel

Center for Ecology and Evolutionary Biology, University of Oregon, Eugene,
Oregon 97403; email: mosquito@uoregon.edu

Annu. Rev. Ecol. Evol. Syst. 2007. 38:1–25

The *Annual Review of Ecology, Evolution, and Systematics* is online at
http://ecolsys.annualreviews.org

This article's doi:
10.1146/annurev.ecolsys.37.091305.110115

Key Words

circannual rhythm, climatic adaptation, dormancy, migration,
seasonality

Abstract

Photoperiodism is the ability of organisms to assess and use the day
length as an anticipatory cue to time seasonal events in their life
histories. Photoperiodism is especially important in initiating physiological and developmental processes that are typically irrevocable
and that culminate at a future time or at a distant place; the further
away in space or time, the more likely a seasonal event is initiated
by photoperiod. The pervasiveness of photoperiodism across broad
taxa, from rotifers to rodents, and the predictable changes of photoperiodic response with geography identify it as a central component of fitness in temperate and polar seasonal environments. Consequently, the role of day length cannot be disregarded when evaluating
the mechanisms underlying life-historical events, range expansions,
invasions of novel species, and response to climate change among
animals in the temperate and polar regions of the world.

Of the four seasons, none lasts forever; of the days some are long and some are short.

Sun Tzu 6:31

INTRODUCTION

In seasonal environments, no life cycle can be complete without the means to exploit the favorable season, to avoid or mitigate the unfavorable season, and to switch between the two lifestyles in a timely manner. Animals exploit the favorable season through growth, development, and reproduction; many animals avoid or mitigate the unfavorable season through dormancy and migration. Successful individuals must be prepared for the appropriate seasonal activities when that season arrives. Reproducing too late in the fall exposes individuals to the exigencies of winter; entering dormancy too early misses the opportunity for continued reproduction and reduces nutritional reserves accumulated for overwintering and for reproduction the following spring. Fitness for animals in a seasonal environment then involves not only the abilities to cope with the changing seasons, but also the ability to express the appropriate phenotype so as not to miss out on opportunities and, at the same time, not to be exposed to lethal conditions. Fitness in seasonal environments is all about timing: the optimal time to migrate and reproduce, the optimal time to stop reproducing, and the optimal time to migrate again. Each of these activities requires preparation: acquiring resources or territories for reproduction, building up fat stores for dormancy, or molting old for new feathers for migration. For most animals, these go/no-go seasonal decisions are irrevocable, either for the lifetime of the individual or within the context of the normal progression of the seasons. Hence, fitness is dependent not only upon the optimal time for engaging in season-specific activities, but also upon the ability to forecast and prepare for the changing seasons in advance of their arrival.

A wide variety of animals from diverse taxa uses the day length or photoperiod as an anticipatory cue to make seasonal preparations. Photoperiod is most useful in predicting environmental conditions in the future or at distant localities; photoperiod provides a go/no-go signal that initiates a usually irrevocable cascade of physiological and developmental processes that culminate in reproduction, dormancy, or migration. Photoperiod, in addition to food, temperature, and other factors in the immediate environment, then affects the rates at which these processes proceed.

Experimental evaluation of photoperiodic response is more cumbersome than evaluating responses to other variables, most notably temperature; however, a correct photoperiodic response is a more important component of fitness than temperature in temperate and polar environments, where the predictability of seasonal change and its strong correlation with day length enable animals to exploit favorable temperatures and to avoid or mitigate unfavorable temperatures.

In this review, we show that photoperiodism is widespread among animals and that its evolution among taxa reveals many consistencies and some inconsistencies yet to be resolved. Most importantly, we show that photoperiodism cannot be disregarded when evaluating the mechanisms underlying life-historical events in any animal living at temperate and polar latitudes. Although we have made a comprehensive study of the literature on photoperiodism in all major animal taxa, the role of day length is well understood in only a few of those groups. We, however, review all taxa in which credible studies have been undertaken, emphasize case studies that illustrate general principles, comment on commonalities and differences of photoperiodism among taxa, discuss implications for rapid climate change, and propose avenues for future research on the evolution of photoperiodism.

period: the time required for a rhythm to complete one full cycle, or time from peak to peak of a rhythm

DAY LENGTH, SEASONALITY, AND PHOTOPERIODIC RESPONSE

Why Use Day Length?

Timing is crucial for maximal exploitation of the favorable season and for minimal exposure to the unfavorable season. However, reproduction, migration, and dormancy require physiological and developmental preparations that must be made in advance of the actual seasonal event. Day length provides a highly reliable calendar that animals can use to anticipate and prepare for seasonal change. Unlike temperature and rainfall, day length at a given spot on Earth is the same today as it was on this date 10 or 10,000 years ago. Hence, day length provides a consistent predictor of future environmental conditions over evolutionary time, enabling animals to use day length to prepare for and to optimize the timing of reproduction, dormancy, and migration in their seasonal life histories. Importantly, both seasonality and day length vary with changing geography.

Geographic Variation in Seasonality, Day Length, and Photoperiodism

The seasonal environment determines the optimal time to reproduce, migrate, or go dormant. In a population of individuals, each with its own genetically determined response to day length, some individuals will reproduce, migrate, or go dormant at the optimal time and thereby achieve greater fitness than others that reproduce, migrate, or go dormant at earlier or later times (Bradshaw et al. 2004, Cooke 1977, Lambrechts et al. 1997, Quinn et al. 2000, Templeton 1986). Hence, the seasonal environment imposes optimizing (stabilizing) selection within a population on the day length individuals use to time events in their seasonal life histories. Below, we describe how seasonality (the selective force) and day length (the environmental cue) change with geography, and then how photoperiodism (the biotic response) is integrated into geographically variable seasonal and photic contexts.

How does seasonality change with geography? Because of the 23° tilt of Earth's axis of daily rotation relative to the plane of its annual rotation about the Sun, the Northern and Southern Hemispheres experience opposite periods of summer and winter. Also because of this tilt, not only is winter day length shorter at higher than at lower latitudes, but also the angle of incident winter sunlight is more acute and imparts less heat per hour of daylight. Consequently, the latitudinal gradient in climate is primarily one of winter cold rather than summer heat (**Figure 1a**) and, as one proceeds poleward, spring arrives later and fall arrives earlier. Hence, the length of the favorable season declines regularly with increasing latitude. For example, in **Figure 1a**, the

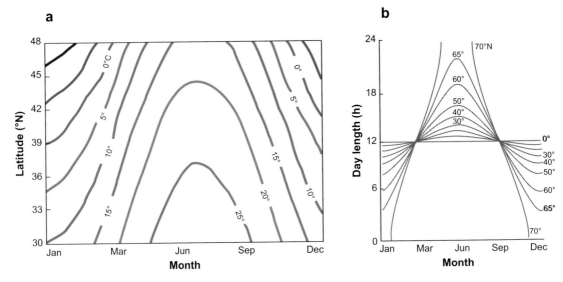

Figure 1

Geographic and seasonal variation in temperature and day length. (*a*) Isotherms for mean monthly temperature in central and eastern North America (Bradshaw et al. 2004). The latitudinal variation in climate is less a matter of summer warmth (June isotherms are far apart) than winter cold (January isotherms are close together), and northern populations experience a shorter growing season than southern populations. Hence, changes in season length and the timing of spring and fall activities have a greater effect on animal populations than do the direct effects of temperature. (*b*) Seasonal patterns in day length (sunrise to sunset) at different latitudes (°N) in the Northern Hemisphere (Danilevskii 1965). Day length at temperate and polar latitudes predicts future seasons more reliably than any other environmental cue.

15°C isotherm declines from 10.5 months at 30°N to 3 months at 48°N. When air masses encounter mountain ranges, they rise, expand, and cool so that the length of the favorable season also declines with increasing altitude (MacArthur 1972). In eastern North America, the number of freeze-free days decreases by approximately 9.3 days with every degree of increasing latitude, and by 94 days with every 1000 m of increasing elevation (Bradshaw 1976). Seasonal activities of temperate animals are therefore intimately related to the coming of spring and fall and to the length of the growing season. As the length of the growing season and the coming of spring and fall vary with geography, so do the optimal times to initiate growth, development, reproduction, dormancy, or migration.

How does day length vary with geography? Also because of Earth's tilt, day length varies with both time of year and latitude (**Figure 1*b***). At the equator, Earth's surface receives 12 hours of light per day (sunrise to sunset) all year long. As one proceeds north or south of the equator, the annual variation in day length becomes progressively more extreme, increasing from 0 h at the equator to 24 h in the summer at latitudes greater than 67°. There are several consequences of these patterns for animals using photoperiod to time their seasonal activities. First, at tropical latitudes below approximately 15°, the annual change in day length is not sufficient to provide a reliable seasonal cue. Even so, a few insects are responsive to day lengths at latitudes as low as 7°–9° (Norris 1965, Wolda & Denlinger 1984). Second, above 30° latitude, wherein lies the greatest proportion of Earth's landmass, day length provides a strong and highly reliable seasonal cue over evolutionary time. Third, at very high latitudes, animals encounter near-constant light in the summer and near-constant darkness during the winter. Consequently, animals must have mechanisms to cope with constant day length experienced by migratory animals in the tropics, rapidly changing day lengths encountered during spring and fall migrations, constant light at high latitudes, and constant darkness during the winter at high latitudes or subterranean hibernacula.

circannual: an internal, self-sustained rhythm with a period of approximately one year

refractoriness: nonresponsiveness to day length; may be spontaneous or induced by day length itself

How does photoperiodism integrate with seasonal and geographic variation in day length? So far in this discussion, we have treated absolute day length as the cue that times seasonal activities. This description is valid for the initiation of dormancy (diapause) in most arthropods and many short-lived vertebrates; for most deuterostomes, the timing of seasonal events represents an interaction between day length, circannual rhythmicity, and refractoriness. Circannual rhythms are endogenous (internal, self-sustained) physiological rhythms that persist under constant photoperiod and temperature. The period of this rhythm usually varies from 9–15 months and the rhythm can persist for several to as many as 10 years (Dawson 2002, Gwinner 1996). Under natural conditions, the circannual clock is set by seasonally changing day lengths. Short days in the late summer will lead to corrective advances of the rhythm because the circannual clock will be perceived as running behind real time; short days in the early summer will lead to corrective delays of the rhythm because the circannual clock will be perceived as running ahead of real time (Bromage et al. 2001). Finally, at specific times in their life cycles, animals may become refractory (nonresponsive) to day length. After reproduction, many birds and mammals become refractory to long days, thereby enabling them to ignore otherwise favorable conditions and to prepare for migration or dormancy in a timely manner. Responsiveness to long days is then usually restored by experiencing short days (Dawson 2002, Goldman et al. 2004). Most arthropods use day length to initiate dormancy (diapause), but, upon entering the diapause state, become refractory to photoperiod. Diapause is then terminated spontaneously or in response to prolonged cold temperatures (chilling) (Danks 1987, Tauber et al. 1986).

The interaction of photoperiodic, circannual rhythmicity and refractoriness not only enhances the ability of animals to time their seasonal development at intermediate latitudes, but also enables them to keep track of seasonal time with either unchanging or rapidly changing day lengths.

Does having the proper photoperiodic response actually affect fitness? Typical reciprocal transplant experiments cannot answer this question. Because of the covariation between the annual change in day length and seasonality (**Figure 1**), photoperiodic adaptation and seasonal adaptation will always be confounded in nature. Questions of potential effects of climate warming can be correctly assessed by

circadian: an internal, self-sustained rhythm with a period of approximately one day

amplitude: one-half of the maximum and minimum expression of a rhythm, or one-half the difference between a peak and a valley in a rhythm

transplants up and down a mountain or across longitudes at the same latitude. Transplants can also test for simple, local adaptations by asking if individuals of a species are more fit in one locality or another, but only so long as causality is not inferred. Our solution to this conundrum was to perform reciprocal transplants of the mosquito *Wyeomyia smithii* in controlled-environment rooms where we could program actual annual changes in day length while holding seasonality constant (Bradshaw et al. 2004). We measured fitness of southern, midlatitude, and northern populations (from 30°N–50°N) as the year-long cohort replacement rate, integrating the effects of all four seasons in a thermally benign midlatitude seasonal year. We either enforced the optimal seasonal timing of diapause and development with unambiguous long and short days (control) or programmed the natural midlatitude photoperiod. Relative to the control, southern populations entered diapause too late and lost 74% of fitness; northern populations entered diapause too early and lost 88% of fitness. The midlatitude populations suffered no loss in fitness. Hence, fitness was critically dependent on possessing the genetically determined correct response to day length. We know of no other study that has determined a composite index of fitness in a realistic seasonal setting where the annual change in day length has been varied while holding seasonality constant.

DOES PHOTOPERIODISM COEVOLVE WITH CIRCADIAN RHYTHMICITY?

In 1936, Bünning (1936) proposed that circadian rhythms regulating daily activities formed the basis (*Grundlage*) of the photoperiodic timer regulating seasonal activities. Bünning's hypothesis has had enduring appeal because, if true, it would mean that a single mechanism is responsible for the regulation of both daily and seasonal cycles in animals. If, however, the same suite of genes responsible for the expression of seasonal timing is also responsible for the expression of daily cycles, then adaptive modification of photoperiodic time measurement necessarily implies genetic modification of the circadian clock, that is, the two must evolve together because of the causal, pleiotropic relationship between them. If genetic variation in photoperiodic response were caused by genetic variation in the circadian clock, then there should be a correlation between photoperiodic responsiveness and genetically determined variation in (*a*) the period or amplitude of circadian clock expression, (*b*) the time of the onset of circadian activity each day, or (*c*) the duration of circadian activity each day (Majoy & Heideman 2000). We acknowledge that we do not cover the many studies on the relationship between circadian rhythms and photoperiodism conducted on single, often highly inbred, long-established lab stocks that have experienced only daily and not seasonal cycles for many generations. Rather, we focus on studies that have considered covariation of photoperiodism (the trait under selection) and circadian rhythmicity (the putative causal mechanism) over evolutionary time across a geographic gradient of seasonality (the selective force).

First, in the white-footed mouse, *Peromyscus leucopus*, southern populations are nonresponsive to short days and reproduce year round; northern populations are polymorphic in response to short days so that some mice undergo reproductive regression in the winter, whereas other mice in the same population may continue to reproduce throughout the year (Heideman et al. 1999, Lynch et al. 1981). Northern and southern populations do not differ in the period of the circadian activity rhythm in constant darkness or in the timing or duration of nocturnal activity under either long or short days (Carlson et al. 1989). Similar responses are obtained in comparison of lines from within a single population divergently selected for response and nonresponse to short days (Majoy & Heideman 2000). Within a single northern population, circadian entrainment maintained a stable pattern for 30 weeks through gonadal regression and spontaneous redevelopment, and mice did not differ in the timing of activity onset after lights off (Johnston & Zucker 1980). These results indicate that in *P. leucopus*, neither genetic variation within populations nor the genetic differences between populations in photoperiodic response is due to a corresponding genetic modification in formal properties of the circadian clock. Rather, genetic variation in *P. leucopus* and its congener, *P. maniculatus*, is due to the effects of melatonin on target organs downstream from the circadian clock (Desjardins et al. 1986, Heath & Lynch 1982, Ruf et al. 1997).

entrainment: ability of a zeitgeber to regulate or drive the period of an otherwise self-sustained rhythm

critical photoperiod: day length stimulating 50% development and 50% dormancy, or the day length at the inflection point of a photoperiodic response curve

Second, in the pitcher-plant mosquito *W. smithii*, the switching day length (critical photoperiod) regulating larval diapause increases regularly with latitude and altitude ($R^2 > 0.90$ repeatedly) and the critical photoperiod varies by 10 SD of mean phenotype from 30°N–50°N (Bradshaw & Holzapfel 2001a,b). However, neither the period nor the amplitude of the circadian rhythm is correlated with critical photoperiod over the same geographic range (Bradshaw et al. 2003, 2006). These results indicate that photoperiodic response and circadian rhythmicity have evolved independently of each other over the climatic gradient of eastern North America.

Third, in *Drosophila littoralis*, critical photoperiod is correlated with both the median timing of adult eclosion and the free-running eclosion rhythm over 30° latitude in Eastern Europe (Lankinen 1986). However, this covariation of photoperiodism with the eclosion rhythm is due to linkage and not pleiotropy, that is, closely linked but different genes (Lankinen & Forsman 2006).

These results do not mean that no circadian genes are involved in photoperiodism, incidental to their functional role in the circadian clock. Given the pervasiveness of circadian rhythmicity on cellular physiology in general (Claridge-Chang et al. 2001, McDonald & Rosbash 2001), we would be surprised if circadian rhythmicity had zero effect on the expression of photoperiodic response. In Syrian (golden) hamsters, a mutation in the circadian rhythm gene *casein kinase 1ε* (*tau*) changes both the period of the circadian wheel-running rhythm (Ralph & Menaker 1988) and the total period (light + dark) under which hamsters are photoperiodic (Shimomura et al. 1997). In various flies (Diptera), there is increasing evidence that the circadian rhythm gene *timeless* somehow affects photoperiodic control of diapause (Goto et al. 2006; Mathias et al. 2005, 2007; Pavelka et al. 2003). The noninvolvement of circadian rhythmicity in the evolution of photoperiodism in mice, mosquitoes, and flies does, however, mean that whatever the connection between the circadian clock and the photoperiodic timer within specific populations, this connection does not impede their independent evolution in nature. In sum, comparisons between photoperiodic time measurement and circadian expression across climatic gradients in mice, mosquitoes, and flies indicate that the photoperiodic timer and the circadian clock are capable of and, indeed, have undergone independent evolution over seasonal climatic gradients in both North America and Europe.

PREVALENCE OF PHOTOPERIODISM AMONG ANIMALS

Photoperiodism has been documented in rotifers, annelids, mollusks, arthropods, echinoderms, bony fish, frogs, turtles, lizards, birds, and mammals. In general, photoperiod provides the go/no-go cue for the direct timing of seasonal events, or for the initiation of physiological, endocrinological, and developmental cascades that, once started, are irrevocable or, at least under natural conditions, usually not reversed before the completion of the seasonal event under selection. In ectotherms, photoperiod can interact with temperature to modulate rates of conversion to sexual forms in an oligochaete annelid (Schierwater & Hauenschild 1990), rates of egg laying in a freshwater snail (Joose 1984), rates of vernal development in tree-hole mosquitoes (Holzapfel & Bradshaw 1981), rates of smoltification in Atlantic salmon (McCormick et al. 1998), accelerated feeding in frogs (Wright et al. 1988), thermal preferences in turtles (Grahm & Hutchison 1979, Hutchison & Maness 1979, Kosh & Hutchison 1968), and temperature-dependent rates of metabolism (Angilleta 2001), growth rate (Uller & Olsson 2003), and thermal homeostasis (Lashbrook & Livezey 1970) in lizards.

Photoperiod by itself is crucial for the go/no-go initiation of events that occur at future times or distant places. The optimal timing of these long-range transitions is determined by selection over many years. Interaction of photoperiod with temperature in ectotherms then fine-tunes the actual completion of those events in concert with annual variation in the thermal environment.

Invertebrates

Other than arthropods, the literature on photoperiodism in most invertebrate groups, including invertebrate chordates, is scant. Nonetheless, we did find examples of photoperiodism in rotifers, annelids, mollusks, and echinoderms.

Rotifers. In the monogonont rotifer *Notommata copeus*, long days provide the go/no-go signal for switching from parthenogenetic to mictic females. Mictic females lay haploid eggs that develop into either males or haploid females that, when fertilized, produce diapausing female embryos (Pourriot & Clément 1975).

Annelids. In polychaetes, increasing day lengths initiate the seasonal reproductive cycle of *Nereis (Neanthes) limnicola* (Fong & Pease 1992). Photoperiodic setting of the circannual clock in the semelparous *N. virens* determines the irrevocable switch from somatic to reproductive growth (Last & Olive 1999, 2004). In the oligochaete *Sylaria lacustris*, short days stimulate the irreversible switch from vegetative to sexual reproduction and formation of overwintering cocoons (Schierwater & Hauenschild 1990).

Mollusks. We could not find any clear studies of photoperiodism in bivalves, cephalopods, chitins, or marine gastropods. In the freshwater snail *Lymnaea stagnalis*, long days control the rate of egg laying and override inhibitory effects of both low temperature and starvation (Joose 1984). In the terrestrial snail *Helix aspersa*, short days trigger both dormancy and supercooling ability (Ansart et al. 2001), and in the terrestrial slug *Limax valentianus*, photoperiod is the essential factor regulating both the initiation of reproduction and the rate of egg production (Hommay et al. 2001).

Arthropods. The literature on arthropod photoperiodism is vast. Generalizations about the use of photoperiodism in insects are highly elusive and, indeed, "it is fair to say that for every life cycle that we consider potentially reasonable controlled by external cues, some insect can be found to illustrate it" (Danks 1994, p. 357). Several excellent and comprehensive reviews exist on photoperiodic and physiological control of seasonal development in arthropods (Danilevskii 1965, Danks 1987, Saunders 2002, Tauber et al. 1986, Nijhout 1994). From this literature, we make several generalizations about photoperiodism in arthropods.

1. Photoperiod provides the primary go/no-go signal for the initiation of neuro-endocrine cascades leading to diapause, migration, or reproduction. Photoperiod may also modulate the rates of completion of these events through its interaction with food, temperature, and moisture. When there is an interaction between photoperiod and temperature, high temperatures tend to reinforce long-day effects, and low temperatures short-day effects.
2. The switching day length (critical photoperiod) initiating diapause increases regularly with latitude and altitude, where the correlation between critical photoperiod and geography can exceed 0.95. To our knowledge, this generalization is the most robust of any ecogeographic rule.
3. The critical photoperiod usually occurs approximately a generation before the onset of adverse conditions. Indeed, the importance of photoperiod as a long-term, reliable, and predictive cue lies in its ability to induce diapause at the optimal time of year, even when temperature and food are otherwise favorable for growth, development, and reproduction.
4. In temperate environments with warm, moist summers and harsh winters, especially in midcontinental or eastern continental climates, long-day arthropods usually enter a hibernal diapause that is initiated by short or shortening days. In temperate environments with hot, dry summers and mild winters, especially on western continental slopes, short-day arthropods may enter an aestival diapause that is initiated by long days or enter both an aestival and hibernal diapause cued by opposite day lengths.
5. High-latitude or polar arthropods may extend development over two or more years, and individuals may enter a photoperiodically mediated diapause two or more times at two or more stages during their life cycles.
6. Photoperiodic response is a polygenic trait, generally with a high heritability and a complex underlying genetic architecture involving pleiotropy, dominance, and epistasis.

7. Photoperiod is generally important for timing the switch from continuous development to diapause, but diapause is usually terminated spontaneously or in response to prolonged exposure to cold temperatures (chilling).

8. Among arthropods, photoreception related to photoperiodism may involve the ocelli, the compound eyes, or the brain itself. After interpretation of day length, the go/no-go photoperiodic response is initiated from the brain via peptide hormones to the corpora alata (allatostatins and allatoropins) that regulate the release of juvenile hormone (a terpenoid), or to the prothoracic glands (prothoracicotropic hormone) that release ecdysteroid, or, in *Bombyx mori*, via neural connections to the suboesophageal ganglion that releases diapause hormone (a peptide). For most arthropods, photoperiodic control of seasonal polyphenisms and seasonal development, diapause, migration, and reproduction involves juvenile hormone, ecdysteroid, or their interaction.

Echinoderms. Gametogenesis in the sea urchin *Strongylocetrotus purpuratus* requires the shortening day lengths of fall and short days of winter; gametogenesis in the starfish *Pisaster ocraceus* depends upon the increasing day lengths of spring and long days of summer. In both species, the principal roles of photoperiod are in the initiation of physiological and developmental processes that ultimately culminate in reproduction and in the setting of the circannual clock that mediates seasonal windows for the timing of reproduction. In *P. ocraceus*, the effect of day length on gonadogenesis remains intact after removal of the terminal eyespots (ocelli), indicating that the terminal ocelli are not necessary for the go/no-go timing of seasonal reproduction (Halberg et al. 1987, Pearse et al. 1986).

Cephalochordates, Jawless Craniata, and Cartilaginous Fishes

We found no evidence for photoperiodic regulation of seasonal growth, development, or reproduction in amphioxus, hagfish, lampreys, or dogfish. The critical experiments for detecting photoperiodically timed life-history events have either not been run (amphioxus, hagfish, skates, or rays) or have considered only one life-cycle transition: metamorphosis in the sea lamprey *Pteromyzon marinus* (Holmes et al. 1994) or embryogenesis in the dogfish *Scyliorhinus canicula* (Thomason et al. 1996). We know of no studies that have examined the effect of day length on gonadal development, sexual maturation, or reproduction in any of these groups. The apparent presence of photoperiodism and circannual rhythms in echinoderms indicates that photoperiodism and its regulation of circannual rhythmicity either appeared early in deuterostome evolution and were lost (or were not looked for) in cephalochordates, hagfish, lampreys, and cartilaginous fish, or evolved separately in echinoderms and bony vertebrates, where photoperiod provides pivotal go/no-go signals for the seasonal timing of life-history events in teleost fish and tetrapods.

Bony Vertebrates

Photoreception. Vertebrate life cycles in general consist of a reproductive stage brought on by a combination of direct effects of absolute day length, increasing day lengths, entrainment (setting) of the circannual clock by photoperiod, and temperature. Usually, day length and the circannual clock are the most important for providing the go/no-go signal that initiates seasonal gonadal development or migration, and more proximal environmental factors such as temperature and food affect mating and the actual production of offspring. The reproductive condition may become refractory (nonresponsive) to long days either spontaneously or in response to long summer days. Refractoriness is generally terminated by short days, declining day lengths, cold temperatures (commonly in ectotherms), or a combination of these factors. The refractory period provides nonreproductive time to accumulate nutritional reserves before winter or to molt before migration, when days are warm and resources abundant.

Among vertebrates, photoreception related to photoperiodism is exclusively retinal in mammals and may involve retinal, pineal, or deep brain (mediobasal hypothalamus) photoreceptors in other vertebrates (Björnsson 1997, Borg et al. 2004, Bromage et al. 2001, Dawson et al. 2001, Dawson 2002, Tosini 1997, Tosini et al. 2001, Vígh et al. 2002). There are two general hypotheses concerning the adaptive significance of possessing extraretinal photoreception for biological timing. First, vision requires focused representation of the environment, and a point on the retina projects to a precise positional map in the brain. By contrast, an irradiance detector integrates light from the whole field of view, as would occur from pineal or deep-brain photoreception (Foster et al. 1994). Even in mammals where light enters via the retina, photoreception is mainly by the retinal ganglion cells (Menaker 2003), and neural projections to the brain lack spatial order. Second, Menaker & Tosini (1996) propose that when animals become nocturnal, they reduce the number of their photoreceptors, so that exclusive retinal photoreception for the photoperiodic timer in mammals is ultimately due to their furtive nocturnal habits when they coexisted with ruling reptiles. Nocturnal geckos, snakes, alligators, owls, and benthic hagfish lack parietal eyes. Apparently, it is not nocturnality, per se, that matters; rather it is the photic environment under which animals have evolved that is ultimately responsible for reliance on extraretinal photoreceptors used in photoperiodic regulation of seasonal activities in vertebrates.

After interpretation of day length, the go/no-go photoperiodic response is initiated from the brain via hypothalamic inhibitory or releasing factors (peptides) that are transported to the anterior pituitary via the hypothalamic-hypophyseal portal system (Turner & Bagnara 1971, McNamara 2003). Tropic hormones from the anterior pituitary then control the expression of growth, molting, color, pelage, development, and reproduction. In mammals, photoreception from the retinal ganglion cells is transmitted electrically via the suprachiasmatic nucleus to the pineal that synthesizes and releases melatonin (5-methoxy, N-acetyltryptamine). Melatonin production is inhibited by light and proceeds in darkness, suggesting that day length signals are encoded in the duration of nocturnal melatonin secretion and decoded in melatonin target cells to provide responses associated with day length (Goldman 2001). For other vertebrates, the pineal and melatonin can play a role in circadian rhythmicity (Cassone 1998), but melatonin does not appear to be a crucial component of photoperiodic response (Goldman 2001, Mayer et al. 1997, Underwood & Goldman 1987), except for photoperiodic control of seasonal variation in bird song (Bentley et al. 1999, MacDougall-Shackleton et al. 2001).

Chondrostean, holostean, and teleostean fish. We found no studies of photoperiodism in chondrostean (paddlefish, sturgeon, *Polypterus*) or holostean (gar pike, bowfin) fish. Photoperiodism is widespread among teleostean (higher bony) fish and has been studied in at least nine orders. In various species, photoperiod may provide the go/no-go signal for seasonal dormancy (Podrabsky & Hand 1999) as well as migration, sexual maturation, and associated physiology and behavior in migratory fish (Bromage et al. 2001). Fish with short gonadal maturation cycles generally respond positively to a single constant day length; fish with long gonadal maturation cycles usually require sequentially changing day lengths (Bromage et al. 2001). Particularly in long-lived, iteroparous fish, step-up or step-down transitions in photoperiod set the phase of the circannual rhythm (Davies & Bromage 2002, Duston & Bromage 1986, Holcombe et al. 2000, Randall et al. 1998). Reproduction is then controlled by an endogenous circannual rhythm that, under natural conditions, is entrained by seasonal changes in day length (Randall et al. 1998).

Photoperiodism has been studied primarily in game, commercial, or farmed fish, especially in salmonids (Bromage et al. 2001), where photoperiod "is regarded as the major proximate cue which adjusts the seasonal timing of reproduction" (Taranger et al. 1998, p. 403). In salmonids, photoperiod provides the go/no-go stimulus for smoltification and migration to sea and for the initiation of gonadogenesis and migration back to freshwater for spawning (Clarke et al. 1994, Quinn & Adams 1996). Smolting and sexual maturation are likely gated events in a circannual cycle so that if some size or physiological threshold is not reached during the circannual window, smolting or sexual maturation may be delayed for a year (Arnesen et al. 2003, Duston & Saunders 1990). Once a threshold size is reached, photoperiod determines the initiation of sexual maturation, which may take place at sea months before final spawning in freshwater; conversely, the timing of spawning itself is more affected by stream flow and temperature, which provide indicators of immediate environmental conditions at the spawning sites (Baras & Philippart 1999, Dabrowski et al. 1996, Davies & Bromage 2002, Huber & Bengston 1999). In general, the importance of photoperiod as the critical go/no-go determinant of sexual maturation increases with time and distance between the initiation of sexual maturation or migration and actual spawning (Clarke et al. 1994, Quinn & Adams 1996).

Sarcopterygian (lobe-fin) fish. We know of no studies on photoperiodism in coelacanths and only one among the Dipnoi (lungfish) where the onset of spawning in the Australian lungfish *Neoceratodus fosteri* is controlled by increasing day lengths of spring (Kemp 1984).

Amphibians. "The role of photoperiod in the control of amphibian reproduction is inconclusive due to the limited number of studies available" (Pancharatna & Patil 1997, p. 111). In addition, many of the available studies are inconclusive because they use tropical animals that, at their native latitudes, receive little change in annual day length to serve as a cue (Pancharatna & Patil 1997, Saidapur & Hoque 1995) because they use animals from a commercial source or from otherwise unknown localities (Delgado et al. 1987, de Vlaming & Bury 1970, Eichler & Gray 1976, Inai et al. 2003, Jacobs et al. 1988, Willis et al. 1956, Wright et al. 1988), or because they use only extreme light regimens unlikely to be found in the subject's native habitat (Delgado et al. 1987, Eichler & Gray 1976, Saidapur & Hoque 1995) or that produce deleterious effects (Rastogi et al. 1978). In two studies of the effects of photoperiod on weight gain in *Rana pipiens*, tadpoles were provided enough food "to last the entire daylength" (Wright et al. 1988, p. 316) so that it was not clear whether accelerated development and metamorphosis on longer than shorter days was due to day length as an environmental signal or simply light-dependent feeding behavior.

A good example of amphibian photoperiodism in an ecological context is provided by *Rana temporaria* in Scandinavia (Laurila et al. 2001). In northern populations that have a strictly limited growing season, photoperiod provides the go/no-go signal for the impending winter; in the south where winter comes later and there is greater developmental flexibility, photoperiod provides a modulating effect on temperature-dependent processes.

At ecologically relevant photoperiods in amphibians of known geographic origin, photoperiod often has a modulating effect on temperature-dependent processes (de Vlaming & Bury 1970, Rastogi et al. 1976). In at least one anuran (Rastogi et al. 1976) and one urodele (Werner 1969), the combined effects of warm temperatures and long days could prevent testicular regression, so it would appear that a refractory period is not obligatory among amphibians. We are therefore left with the questions of how frequently a true photorefractory state occurs in amphibians, whether a refractory state can generate a circannual rhythm, and whether photoperiod is as effective in setting that rhythm as it is in teleosts (Bromage et al. 2001), birds (Dawson 2002), and mammals (Goldman 2001).

Turtles, snakes, and lizards. Photoperiodism in these reptiles may affect seasonal development either directly by regulating the timing of reproduction or indirectly by modifying thermal behavior and thermal preferences, thereby modifying temperature-dependent physiological processes. However, all the studies we reviewed concerned animals collected directly from the field and were not run through two generations to remove maternal and other field effects.

In turtles, long days induce an increase in critical thermal maxima and a preference for warmer temperatures (Grahm & Hutchison 1979, Hutchison & Maness 1979, Kosh & Hutchison 1968). Long days therefore act as an anticipatory cue to pre-acclimate turtles for future summer heat in at least two ways. First, the direct effects of photoperiod on critical thermal maxima are equivalent to a 3°C–4°C increase in acclimation temperature. Second, by inducing a preference for warmer temperatures during early afternoon at close to the daily thermal maximum, long days are actually accelerating direct thermal acclimation (R. Huey, personal communication). To our knowledge, there are no studies determining the effect of photoperiod on the timing of reproduction or hibernation, or testing for circannual rhythms in turtles.

In snakes, we found no well-designed study of photoperiodism. The one study we did find (Hawley & Aleksiuk 1976) used an inappropriate long day for testing for a photoperiodic effect on reproduction.

In lizards, experimental biochronometry has focused largely on circadian (daily) rhythms and not photoperiodism. Photoperiodism in lizards has been studied primarily in the green anole *Anolis carolinensis*, in which photoperiod appears to play two roles. First, long days in summer sustain testicular development and short days in the fall induce regression of the testes, reducing or eliminating the ability of warm temperature to promote spermatogenesis. Second, long days in the spring initiate an increase in feeding and growth, leading to the onset of reproduction (Fox & Dessauer 1957; Licht 1966, 1967, 1973).

Photoperiod can serve as a seasonal modulator of temperature-dependent processes in lizards. With increasing latitude, photoperiod has an increasing influence on metabolic rate (Angilleta 2001), on growth rate (Uller & Olsson 2003), and on ability to maintain a constant body temperature with increasing temperature in the spring (Lashbrook & Livezey 1970). Given their tendency to hibernate in sites with minimal photic or thermal cues, temperate-zone lizards might be expected to rely on a circannual clock for the timing of seasonal activities (Gwinner 1986). Long days retard and short days advance the reproductive cycle of *Cnemidophorus* (Cuellar & Cuellar 1977), but otherwise the role of photoperiod in setting the circannual rhythm of lizards remains untested and unknown.

Birds. "In birds, the annual change in daylength is the most important environmental cue used for synchronizing breeding moult, and migration with recurrent seasonal fluctuation in environmental conditions" (Coppack & Pulido 2004, p. 131). Temperate birds must coordinate several key events into their phenology, principally reproduction, molting, and migration (Dawson et al. 2001, Dawson 2002, Gwinner 1996). These processes are mutually exclusive energetic activities and are sequentially orchestrated in the seasonal life history of birds, both in nature and as a circannual rhythm under constant conditions. In addition, birds are constrained by the energetics of flight, which places a premium on healthy flight feathers and low body mass. Consequently, the difference in mass between reproductive and nonreproductive gonads may differ by a factor of 100, and molting usually takes place after reproduction and before migration (Dawson et al. 2001, Dawson 2002).

In birds, gonadal development begins early in the season, progresses gradually, and ceases abruptly prior to molting. The late-summer molt then marks the termination of the breeding season and provides the seasonal link to migration. Prior to migration, birds become more active and exhibit directional preferences in orientation cages provided with an artificial sun or night sky. This migratory restlessness is referred to as Zugunruhe and can be quantified in controlled environments to estimate the onset, duration, intensity, and directionality of migration.

In passerines such as starlings and sparrows, increasing day lengths of spring initiate gonadal maturation and breeding (Dawson et al. 2001, Dawson 2002, Gwinner 1996). The longer days of summer induce a photorefractory state in which reproductive processes are no longer sustained by long days and the gonads regress rapidly to an essentially prepuberty condition. Following the summer molt, short days of autumn and winter lead to a dissolution of refractoriness and birds again become responsive to increasing or long days during the late fall or early winter.

Photoperiod provides "the most important zeitgeber [setting agent] of circannual rhythms" (Gwinner & Helm 2003, p. 83). Circannual clocks enable birds to keep track of time in unvarying tropical day lengths, and during fall and spring migration through zones of rapidly changing day length (Gwinner 1986, 1996). Under constant temperature and photoperiod, both the duration and direction of Zugunruhe in orientation cages have been observed in some birds to correspond to the mid-migration changes in migration direction of wild populations. Hence, circannual rhythmicity may account for the accurate migratory pathways of even naïve birds and provide an internal temporal reference for course changes (Helm & Gwinner 2006).

Zeitgeber: an external cue that sets the timing of or entrains an internal, otherwise self-sustained rhythm

Mammals. "Photoperiodic information has been shown to be the strongest synchronizer of seasonal functions in most species" of mammals (Hofman 2004, p. 63). In addition to regulating annual reproductive cycles, photoperiod can control the timing of seasonal shedding and change in color and thickness of fur (Farner 1961), tendency to enter torpor (Heideman et al. 1999, Lynch et al. 1981), investment in nest insulation (Heideman et al. 1999, Lynch et al. 1981), temporal niche partitioning in reproduction between sympatric congeners (Dickman 1982), and embryonic diapause in which implantation of the zygote is delayed for varying durations on the basis of photoperiod (Farner 1961, McConnell et al. 1986, McConnell & Tyndale-Biscoe 1985, Renfree et al. 1981, Thom et al. 2004, Tyndale-Biscoe 1980). The incidence of photoperiodically induced gonadal regression in mice (*Peromyscus*) (Heideman et al. 1999, Lowrey et al. 2000, Lynch et al. 1981, Sullivan & Lynch 1986) and embryonic diapause in mustelids (Thom et al. 2004) increases with latitude.

Temperate mammals may be short-day (early spring) or long-day (spring and summer) breeders. Both short-day and long-day breeders may go through a period of photorefractoriness (nonresponsiveness to day length) that may interact with circannual rhythmicity. Long-day breeders cease reproduction either through the action of decreasing day lengths or through the action of long days, themselves inducing a photorefractory state (inability of long days to sustain reproduction) followed by gonadal regression. In both types, prolonged exposure to short days renders them refractory to short days and allows for the initiation of gonadal maturation (Goldman 2001). "No mammalian species are known to become refractory exclusively to long days. Mammals become refractory either to short days, or to both short and long days" (Goldman et al. 2004, p. 132).

As in birds (above), photoperiod sets the timing of the circannual clock among a wide variety of mammals (Goldman 2001, Gwinner 1986). The circannual component of mammalian reproductive cycles may be especially important for high-latitude species where the season favorable for reproduction is highly restricted, where animals experience constant light in the summer, and where individuals may hibernate under relatively constant conditions for over six months of the year.

CONCLUSIONS

Being in the right physiological, developmental or reproductive condition at the right time and place is an essential component of fitness in seasonal environments. A wide variety of vertebrates and invertebrates in marine, freshwater, and terrestrial habitats use the day length to anticipate and prepare for seasonal transitions or events in their life histories. Several generalizations can be made about photoperiodism in animals:

1. Unlike temperature or rainfall, the annual change in day length is invariant from year to year, and day length therefore provides a highly reliable anticipatory cue for future or distant seasonal conditions.

2. A specific photoperiodic response is based on selection through evolutionary time for the optimal seasonal time to develop, migrate, reproduce, or go dormant.

3. Photoperiodism regulates a go/no-go response that initiates a cascade of physiological, developmental, or reproductive processes that are generally irrevocable within the lifetime of the individual or are not reversed before completion of the seasonal event under selection.

4. In ectotherms, photoperiodism may act in concert with temperature to regulate subsequent continuous rate processes and thereby fine-tune the actual timing of the seasonal event in a thermal environment that varies from year to year.

5. Photoperiod tends to provide the most important cue for events that are distant in time or space; temperature, food, and other ecological conditions become more important closer to the actual event itself.

6. Animals may respond to either absolute or changing day lengths; reliance on absolute day length is more prevalent in short- than long-lived animals.

7. Critical photoperiod, threshold day length, or the incidence of photoperiodism within and among species tends to increase with latitude or altitude in the temperate zone.

8. Photoperiodic response within populations may be affected by circadian rhythmicity, but the seasonal photoperiodic timer and the daily circadian clock can evolve independently over seasonal, geographic gradients.

Animals use day length in conjunction with circannual rhythmicity and refractory periods to keep track of seasonal time not only at temperate latitudes but also at tropical overwintering localities with constant day length, during migration through zones of rapidly changing day length, during polar summers with constant light, and in winter hibernacula or during polar winters with constant darkness.

Day length is perceived through either optic (compound eye, retina) or extraoptic (ocelli, pineal, brain) photoreceptors. In vertebrates, primary photoreception is by nonvisual irradiance detectors in the retina, pineal, or hypothalamus. In all animals, day length is assessed within the brain and transmission of the photoperiodic signal to target organs involves peptide hormones at some step in the pathway (with the possible exception of melatonin's effect on song centers of the avian brain). Melatonin serves as an interval signal for mammalian photoperiodism but does not appear to play a significant role in photoperiodic response of other vertebrates except, again, for avian song.

Most importantly, photoperiodic response is a crucial component of fitness that cannot be overlooked when considering the present distribution of animals, life-history evolution, range expansion or contraction, invasiveness of agricultural pests or vectors of disease, outcrossing of managed populations for the maintenance of genetic variability, the introduction or transplantation of species for agricultural or biological control or for biological conservation, and the potential persistence of populations confronted with rapid climate change.

PREDICTIONS FOR RAPID CLIMATE CHANGE

Recent, rapid climate change has resulted primarily in warmer winters rather than warmer summers, and the rate of winter warming has increased with latitude (IPCC 2001, 2007). Warmer winters have resulted in earlier springs, later onset of winters, and longer growing seasons. However, at any locality on Earth, climate warming does not alter day length. Animals from rotifers to rodents use this high reliability of day length to time the seasonal events in their life cycles that are crucial to fitness in temperate and polar environments: when to develop, when to reproduce, when to enter dormancy, and when to migrate. Climate warming is changing the optimal timing of these events and, consequently, is imposing selection on the photoperiodic response used to time them. For example, recent genetic shifts in photoperiodic response in the pitcher-plant mosquito have occurred over as short a time span as five years (Bradshaw & Holzapfel 2001a, 2006).

By contrast, there are, to our knowledge, no examples of genetic increases in thermal optima or heat tolerance associated with climate warming in any animal. We therefore predict that, when confronted with continued, rapid climate change, the differential ability of animals to track that change, and hence the composition of future biotic communities, will depend on the evolvability of their respective photoperiodic responses (Bradshaw & Holzapfel 2001a, 2006, 2007).

FUTURE ISSUES

1. How prevalent are photoperiodism and circannual rhythmicity among invertebrates other than arthropods?

2. How many times has photoperiodism evolved? For instance, did it arise independently in echinoderms and vertebrates or was there a common deuterostome origin?

3. What is the genetic basis of photoperiodic response within populations, how does this variation respond to selection along climatic gradients, and does it relate to the circadian clock?

4. Is there a central circannual clock or is circannual rhythmicity simply the concatenation of multiple, long-term physiological processes?

5. What are the relative contributions to fitness of the photoperiodic timer and the circadian clock in natural populations?

6. How well do rates of evolution (genetic change) in photoperiodic response track rapid climate change among diverse animal taxa?

DISCLOSURE STATEMENT

The authors are not aware of any biases that might be perceived as affecting the objectivity of this review.

ACKNOWLEDGMENTS

We thank Michael Menaker, Ray Huey, Paul Heidemann, Jeffrey Hard, Kevin Emerson, Herbert Underwood, Serge Daan, Gregory Ball, Vincent Cassone, Ruth Shaw, Barbara Helm, David Saunders, Peter Zani, Denis Réale, Andrew McAdam, and Stanley Boutin for useful discussions; Kevin Emerson and Douglas Futuyma for reviewing previous versions of the manuscript; and the John Simon Guggenheim Memorial Foundation, the Fulbright Commission, and the National Science Foundation programs in Population Biology and in Ecological and Evolutionary Physiology for their support of our research on the genetics, physiology, evolution, and ecology of seasonal adaptations.

LITERATURE CITED

Angilleta JJ Jr. 2001. Variation in metabolic rate between populations of a geographically widespread lizard. *Physiol. Biochem. Zool.* 74:11–21

Ansart A, Vernon P, Daguzan J. 2001. Photoperiod is the main cue that triggers supercooling ability in the land snail, *Helix aspersa* (Gastropoda: Helicidae). *Cryobiology* 42:266–73

Arnesen AM, Taften H, Agustsson T, Stefansson SO, Handeland SO, Björnsson BT. 2003. Osmoregulation, feed intake, growth and growth hormone levels in 0+ Atlantic salmon (*Salmo salar* L.) transferred to seawater at different stages of smolt development. *Aquaculture* 222:167–87

Baras E, Philippart JC. 1999. Adaptive and evolutionary significance of a reproductive thermal threshold in *Barbus barbus*. *J. Fish Biol.* 55:354–75

Bentley GE, Van't Hoff TJ, Ball GF. 1999. Seasonal neuroplasticity in the songbird telencephalon: a role for melatonin. *Proc. Natl. Acad. Sci. USA* 96:4674–79

Björnsson BT. 1997. The biology of salmon growth hormone: from daylight to dominance. *Fish Physiol. Biochem.* 17:9–24

Borg B, Bornestaf C, Hellqvist A, Schmitz M, Mayer I. 2004. Mechanisms in the photoperiodic control of reproduction in the stickleback. *Behaviour* 141:1521–30

Bradshaw WE. 1976. Geography of photoperiodic response in a diapausing mosquito. *Nature* 262:384–86

Bradshaw WE, Holzapfel CM. 2001a. Genetic shift in photoperiodic response correlated with global warming. *Proc. Natl. Acad. Sci. USA* 98:14509–11

Bradshaw WE, Holzapfel CM. 2001b. Phenotypic evolution and the genetic architecture underlying photoperiodic time measurement. *J. Insect. Physiol.* 47:809–20

Bradshaw WE, Holzapfel CM. 2006. Evolutionary response to rapid climate change. *Science* 312:1477–78

Bradshaw WE, Holzapfel CM. 2007. Genetic response to rapid climate change: It's seasonal timing that matters. *Mol. Ecol.* In press

Bradshaw WE, Holzapfel CM, Mathias D. 2006. Circadian rhythmicity and photoperiodism in the pitcher-plant mosquito: Can the seasonal timer evolve independently of the circadian clock? *Am. Nat.* 167:601–5

Bradshaw WE, Quebodeaux MC, Holzapfel CM. 2003. Circadian rhythmicity and photoperiodism in the pitcher-plant mosquito: adaptive response to the photic environment or correlated response to the seasonal environment? *Am. Nat.* 161:735–48

Bradshaw WE, Zani PA, Holzapfel CM. 2004. Adaptation to temperate climates. *Evolution* 58:1748–62

Bromage N, Porter M, Randall C. 2001. The environmental regulation of maturation in farmed finfish with special reference to the role of photoperiod and melatonin. *Aquaculture* 197:63–98

Bünning E. 1936. Die endogene Tagesrhythmik als Grundlage der photoperiodischen Reaktion. *Ber. Dtsch. Bot. Ges.* 54:590–607

Carlson LL, Zimmermann A, Lynch GR. 1989. Geographic differences for delay of sexual maturation in *Peromyscus leucopus*: effects of photoperiod, pinealectomy, and melatonin. *Biol. Reprod.* 41:1004–13

Cassone VM. 1998. Melatonin's role in vertebrate circadian rhythms. *Chronobiol. Int.* 15:457–73

Claridge-Chang A, Wijnen H, Nacef F, Boothroyd C, Rajewsky N, Young MW. 2001. Circadian regulation of gene expression systems in the *Drosophila* head. *Neuron* 37:657–71

Clarke WC, Withler RE, Shelbourn JW. 1994. Inheritance of smolting phenotypes in backcrosses and hybrid stream-type × ocean-type Chinook salmon (*Oncorhynchus tshawytscha*). *Estuaries* 17:13–25

Cooke BD. 1977. Factors limiting the distribution of the wild rabbit in Australia. *Proc. Ecol. Soc. Aust.* 10:113–20

Coppack T, Pulido F. 2004. Photoperiodic response and the adaptability of avian life cycles to environmental change. *Adv. Ecol. Res.* 35:131–50

Cuellar HS, Cuellar O. 1977. Evidence for endogenous rhythmicity in the reproductive cycle of the parthenogenetic lizard *Cnemidophorus uniparens* (Reptilia: Teiidae). *Copeia* 1977:554–57

Dabrowski K, Cieresko A, Toth GP, Christ SA, El-Saidy D, Ottobre JS. 1996. Reproductive physiology of yellow perch (*Perca flavescens*): environmental and endocrinological cues. *J. Appl. Ichthyol.* 12:139–48

Danilevskii AS. 1965. *Photoperiodism and Seasonal Development in Insects*. Edinburgh: Oliver & Boyd

A comprehensive review of fish photoperiodism.

An excellent review of insect photoperiodism in a climatic and geographic context.

An excellent review of photoperiodism and dormancy in an ecological context.

Danks HV. 1987. *Insect Dormancy: An Ecological Perspective*. **Ottawa: Biol. Surv. Can. (Terr. Arthropods)**

Danks HV. 1994. Insect life-cycle polymorphisms: current ideas and future prospects. In *Insect Life-Cycle Polymorphism: Theory, Evolution and Ecological Consequences for Seasonality and Diapause Control*, ed. HV Danks, pp. 349–65. Dordrecht: Kluwer Acad.

Davies B, Bromage N. 2002. The effects of fluctuating seasonal and constant water temperatures on the photoperiodic advancement of reproduction in female rainbow trout, *Oncorhynchus mykiss. Aquaculture* 205:183–200

Dawson A. 2002. Photoperiodic control of the annual cycle in birds and comparison with mammals. *Ardea* **90:355–67**

The best reviews on avian photoperiodism.

Dawson A, King VM, Bentley GE, Ball GF. 2001. Photoperiodic control of seasonality in birds. *J. Biol. Rhythms* **16:365–80**

de Vlaming VL, Bury RB. 1970. Thermal selection in tadpoles of the tailed-frog, *Ascaphus truei. J. Herpetol.* 4:179–89

Delgado MJ, Gutiérrez P, Alonso-Bedate M. 1987. Melatonin and photoperiod alter growth and larval development in *Xenopus laevis* tadpoles. *Comp. Biochem. Physiol.* 86A:417–21

Desjardins C, Bronson FH, Blank JL. 1986. Genetic selection for reproductive photoresponsiveness in deer mice. *Nature* 322:172–73

Dickman CR. 1982. Some ecological aspects of seasonal breeding in *Antechinus* (Dasyuridae, Marsupialia). In *Carnivorous Marsupials*, pp. 139–50, ed. M Archer. Sydney, NSW: Roy. Zool. Soc.

Duston J, Bromage N. 1986. Photoperiodic mechanisms and rhythms of reproduction in the female rainbow trout. *Fish Physiol. Biochem.* 2:35–51

Duston J, Saunders RL. 1990. The entrainment role of photoperiod on hypoosmoregulatory and growth-related aspects of smolting in Atlantic salmon (*Salmo salar*). *Can. J. Zool.* 68:707–15

Eichler VB, Gray LSJ. 1976. The influence of environmental lighting on the growth and prometamorphic development of larval *Rana pipiens. Dev. Growth Differ.* 18:177–82

Farner DS. 1961. Comparative physiology: photoperiodicity. *Annu. Rev. Physiol.* 23:71–96

Fong PP, Pearse JS. 1992. Evidence for a programmed circannual life cycle modulated by increasing daylengths in *Neanthes limnicola* (Polychaeta: Nereidae) from Central California. *Biol. Bull.* 182:289–97

Foster RG, Grace MS, Provencio I, Degrip WJ, Garcia-Fernandez JM. 1994. Identification of vertebrate deep brain photoreceptors. *Neurosci. Behav. Rev.* 18:541–46

Fox W, Dessauer HC. 1957. Photoperiodic stimulation of appetite and growth in the male lizard, *Anolis carolinensis. J. Exp. Zool.* 134:557–75

An excellent review of mammalian photoperiodism that includes ecological as well as mechanistic considerations.

Goldman BD. 2001. Mammalian photoperiodic system: formal properties and neuroendocrine mechanisms of photoperiodic time measurement. *J. Biol. Rhythms* **16:283–301**

Goldman B, Gwinner E, Karsch FJ, Saunders D, Zucker I, Gall GF. 2004. Circannual rhythms and photoperiodism. In *Chronobiology: Biological Timekeeping*, ed. JC Dunlap, JJ Loros, PJ DeCoursey, pp. 107–42. Sunderland, MA: Sinauer Assoc.

Goto SG, Han B, Denlinger DL. 2006. A nondiapausing variant of the flesh fly, *Sarcophaga bullata*, that shows arrhythmic adult eclosion and elevated expression of two circadian clock genes, *period* and *timeless*. *J. Insect Physiol.* 52:1213–18

Grahm TE, Hutchison VH. 1979. Effect of temperature and photoperiod acclimatization on thermal preferences of selected freshwater turtles. *Copeia* 1979:165–69

Gwinner E. 1986. *Circannual Clocks*. Berlin: Springer-Verlag

Gwinner E. 1996. Circannual clocks in avian reproduction and migration. *Ibis* 138:47–63

Gwinner E, Helm B. 2003. Circannual and circadian contributions to the timing of avian migration. In *Avian Migration*, ed. P Berthold, E Gwinner, E Sonnenschein, pp. 81–95. Berlin: Springer-Verlag

Halberg F, Shankaraiah K, Giese AC, Halberg F. 1987. The chronobiology of marine invertebrates: methods of analysis. In *Reproduction of Marine Invertebrates*, Vol. 9, ed. AC Giese, JS Pearse, VB Pearse, pp. 331–84. Palo Alto, CA: Blackwell

Hawley AWL, Aleksiuk M. 1976. The influence of photoperiod and temperature on seasonal testicular recrudescence in the red-sided garter snake (*Thamnophis sirtalis parietalis*). *Comp. Biochem. Physiol.* 53A:215–21

Heath HW, Lynch GR. 1982. Intraspecific differences for melatonin-induced reproductive regression and the seasonal molt in *Peromyscus leucopus*. *Gen. Comp. Endocrinol.* 48:289–95

Heideman PD, Bruno TB, Singley JW, Smedley JV. 1999. Genetic variation in photoperiodism in *Peromyscus leucopus*: geographic variation in an alternative life-history strategy. *J. Mammal.* 80:1232–42

Helm B, Gwinner E. 2006. Migratory restlessness in an equatorial nonmigratory bird. *PLoS Biol.* 4:611–14

Hofman MA. 2004. The brain's calendar: neural mechanisms of seasonal timing. *Biol. Rev.* 79:61–77

Holcombe GW, Pasha MW, Jensen KM, Tietge UE, Ankley GT. 2000. Effects of photoperiod manipulation on brook trout reproductive development, fecundity, and circulating sex steroid concentrations. *North Am. J. Aquacult.* 62:1–11

Holmes JA, Beamish FWH, Seelye JG, Sower SA, Youson JH. 1994. Long-term influence of water temperature, photoperiod, and food deprivation on metamorphosis of sea lamprey, *Pteromyzon marinus*. *Can. J. Fish. Aquat. Sci.* 51:2045–51

Holzapfel CM, Bradshaw WE. 1981. Geography of larval dormancy in the tree-hole mosquito, *Aedes triseriatus* (Say). *Can. J. Zool.* 59:1014–21

Hommay G, Kienlen JC, Gertz C, Hill A. 2001. Growth and reproduction of the slug *Limax valentianus* Férussac in experimental conditions. *J. Molluscan Stud.* 67:191–207

Huber M, Bengtson DA. 1999. Effects of photoperiod and temperature on the regulation of the onset of maturation in the estuarine fish *Menidia beryllina* (Cope) (Antherinidae). *J. Exp. Marine Biol. Ecol.* 240:285–302

Hutchison VH, Maness JD. 1979. The role of behavior in temperature acclimation and tolerance in ectotherms. *Am. Zool.* 19:367–84

Inai Y, Nagai K, Ukena K, Oishi T, Tsutsui K. 2003. Seasonal changes in neurosteroid concentrations in amphibian brain and environmental factors regulating their changes. *Brain Res.* 959:214–25

The definitive treatise on circannual rhythms (update in Gwinner 1996).

IPCC. 2001. *Climate Change 2001: The Scientific Basis. Contribution of Working Group I to the Third Assessment Report of the Intergovernmental Panel on Climate Change.* Cambridge, UK: Cambridge Univ. Press

IPCC. 2007. *Climate Change 2007: The Physical Science Basis. Summary for Policymakers. Contribution of Working Group I to the Fourth Assessment Report of the Intergovernmental Panel on Climate Change.* Geneva, Switz.: IPCC Secr.

Jacobs GFM, Goyvaerts MP, Vandorpe G, Quaghebeur AML, Kühn ER. 1988. Luteinizing hormone-releasing hormone as a potent stimulator of the thryroidal axis in ranid frogs. *Gen. Comp. Endocrinol.* 70:274–83

Johnston PG, Zucker I. 1980. Photoperiodic regulation of the testes of adult white-footed mice (*Peromyscus leucopus*). *Biol. Reprod.* 23:859–66

Joose J. 1984. Photoperiodicity, rhythmicity and endocrinology of reproduction in the snail *Lymnaea stagnalis*. In *Photoperiodic Regulation of Insect and Molluscan Hormones*, pp. 204–20, ed. R Porter, GM Collins. London: Pitman

Kemp A. 1984. Spawning of the Australian lungfish, *Neoceratodus fosteri* (Krefft) in the Brisbane River and Enoggera Reservoir, Queensland. *Mem. Queensl. Mus.* 21:391–99

Kosh RJ, Hutchison VH. 1968. Daily rhythmicity of temperature tolerance in eastern painted turtles, *Chrysemys picta*. *Copeia* 1968:244–46

Lambrechts MMBJ, Maistre M, Perret P. 1997. A single response mechanism is responsible for evolutionary adaptive variation in a bird's laying date. *Proc. Natl. Acad. Sci. USA* 94:5153–55

Lankinen P. 1986. Geographical variation in circadian eclosion rhythm and photoperiodic adult diapause in *Drosophila littoralis*. *J. Comp. Physiol. A* 159:123–42

Lankinen P, Forsman P. 2006. Independence of genetic geographical variation between photoperiodic diapause, circadian eclosion rhythm, and Thr-Gly repeat region of the *period* gene in *Drosophila littoralis*. *J. Biol. Rhythms* 21:3–12

Lashbrook MK, Livezey RL. 1970. Effects of photoperiod on heat tolerance in *Sceloporus occidentalis occidentalis*. *Physiol. Zool.* 43:38–46

Last KS, Olive PJW. 1999. Photoperiodic control of growth and segment proliferation by *Nereis* (*Neanthes*) *virens* in relation to state of maturity and season. *Mar. Biol.* 134:191–99

Last KS, Olive PJW. 2004. Interaction between photoperiod and an endogenous seasonal factor influencing the diel locomotor activity of the benthic polychaete *Nereis virens* Sars. *Biol. Bull.* 206:103–12

Laurila A, Pakkasmaa SMJ, Merilä J. 2001. Influence of seasonal time constraints on growth and development of common frog tadpoles: a photoperiod experiment. *Oikos* 95:451–60

Licht P. 1966. Reproduction in lizards: influence of temperature on photoperiodism in testicular recrudescence. *Science* 154:1668–70

Licht P. 1967. Environmental control of annual testicular cycles in the lizard *Anolis carolinensis* II. Seasonal variations in the effects of photoperiod and temperature on testicular recrudescence. *J. Exp. Zool.* 166:243–54

Licht P. 1973. Influence of temperature and photoperiod on the annual ovarian cycle in the lizard *Anolis carolinensis*. *Copeia* 1973:465–72

Lowrey PL, Shimomura K, Antoch MP, Yamazaki S, Zemenides PD, et al. 2000. Positional syntenic cloning and functional characterization of the mammalian circadian mutation *tau*. *Science* 288:483–91

Lynch GR, Heath HW, Johnston CM. 1981. Effect of geographic origin on the photoperiodic control of reproduction in the white-footed mouse, *Peromyscus leucopus. Biol. Reprod.* 25:475–80

MacArthur RH. 1972. *Geographical Ecology.* New York, NY: Harper & Row

MacDougall-Shackleton SA, Deviche PJ, Crain RD, Gall GF, Hahn TP. 2001. Seasonal changes in brain GnRH immunoreactivity and song-control nuclei volumes in an opportunistically breeding songbird. *Brain Behav. Evol.* 58:38–48

Majoy SB, Heideman PD. 2000. Tau differences between short-day responsive and short-day nonresponsive white-footed mice (*Peromyscus leucopus*) do not affect reproductive photoresponsiveness. *J. Biol. Rhythms* 15:500–12

Mathias D, Jacky L, Bradshaw WE, Holzapfel CM. 2005. Geographic and developmental variation in expression of the circadian rhythm gene, *timeless*, in the pitcher-plant mosquito, *Wyeomyia smithii. J. Insect Physiol.* 51:661–67

Mathias D, Jacky L, Bradshaw WE, Holzapfel CM. 2007. Quantitative trait loci associated with photoperiodic response and stage of diapause in the pitcher plant mosquito, *Wyeomyia smithii. Genetics* 176:391–402

Mayer I, Bornestaf C, Borg B. 1997. Melatonin in nonmammalian vertebrates: physiological role in reproduction? *Comp. Biochem. Physiol.* 118A:515–31

McConnell SJ, Tyndale-Biscoe CH. 1985. Response in peripheral plasma melatonin to photoperiod change and the effects of exogenous melatonin on seasonal quiescence in the tammar, *Macropus eugenii. J. Reprod. Fertil.* 73:529–38

McConnell SS, Tyndale-Biscoe CH, Hinds LA. 1986. Change in duration of elevated concentrations of melatonin is the major factor in photoperiod response of the tammar, *Macropus eugenii. J. Reprod. Fertil.* 77:623–32

McCormick SS, Hansen LP, Quinn TP, Saunders RL. 1998. Movement, migration, and smolting of Atlantic salmon (*Salmo salar*). *Can. J. Fish. Aquat. Sci.* 55(Suppl. 1):77–92

McDonald MJ, Rosbash M. 2001. Microarray analysis and organization of circadian gene expression in *Drosophila. Cell* 107:567–78

McNamara P. 2003. Hormonal rhythms. In *Molecular Biology of Circadian Rhythms*, ed. A Sehgal, pp. 231–53. Hoboken, NJ: John Wiley & Sons

Menaker M. 2003. Circadian photoreception. *Science* 299:213–14

Menaker M, Tosini G. 1996. The evolution of vertebrate circadian systems. In *Sixth Sapporo Symposium on Biological Rhythms: Circadian Organization and Oscillatory Coupling*, ed. KI Honma, S Honma, pp. 39–52. Sapporo, Japan: Hokkaido Univ. Press

Nijhout HF. 1994. *Insect Hormones*. Princeton, NJ: Princeton Univ. Press.

Norris MJ. 1965. The influence of constant and changing photoperiods on imaginal diapause in the red locust (*Nomadacris septemfasciata* Serv.). *J. Insect Physiol.* 50:600–3

Pancharatna K, Patil MM. 1997. Role of temperature and photoperiod in the onset of sexual maturity in female frogs, *Rana cyanophlyctis. J. Herpetol.* 31:111–14

A straightforward account of the functional roles of vertebrate hormones.

An excellent, clear, and concise treatise on insect hormones.

Pavelka J, Shimada K, Kostál V. 2003. TIMELESS: a link between fly's circadian and photoperiodic clocks? *Eur. J. Entomol.* 100:255–65

Pearse JS, Eernisse DJ, Pearse VB, Beauchamp KA. 1986. Photoperiodic regulation of gametogenesis in sea stars, with evidence for an annual calendar independent of fixed daylength. *Am. Zool.* 26:417–31

Podrabsky JE, Hand SC. 1999. The bioenergetics of embryonic diapause in an annual killifish, *Austrofundulus limnaeus*. *J. Exp. Biol.* 202:2567–80

Pourriot R, Clément P. 1975. Influence de la durée de l'éclairement quotidien sur le taux de femelles mictiques chez *Notommata copeus* Ehr. (Rotifère). *Oecologia (Berlin)* 22:67–77

Quinn TP, Adams DJ. 1996. Environmental changes affecting the migratory timing of American shad and sockeye salmon. *Ecology* 77:1151–62

Quinn TP, Unwin MJ, Kinnison MT. 2000. Evolution of temporal isolation in the wild: genetic divergence in timing of migration and breeding by introduced Chinook salmon populations. *Evolution* 54:1372–85

Ralph MR, Menaker M. 1988. A mutation of the circadian system in golden hamsters. *Science* 241:1225–27

Randall CF, Bromage NR, Duston J, Symes U. 1998. Photoperiod-induced phase-shifts of the endogenous clock controlling reproduction in the rainbow trout: a circannual phase-response curve. *J. Reprod. Fertil.* 112:399–405

Rastogi RK, Iela L, Delrio G, Di Meglio M, Russo A, Chieffi G. 1978. Environmental influence on testicular activity in the green frog, *Rana esculenta*. *J. Exp. Zool.* 206:49–64

Rastogi RK, Iela L, Saxena PK, Chieffi G. 1976. The control of spermatogenesis in the green frog, *Rana esculenta*. *J. Exp. Zool.* 196:151–66

Renfree MB, Lincoln DW, Almeida OFX, Short RV. 1981. Abolition of seasonal embryonic diapause in a wallaby by pineal denervation. *Nature* 293:138–39

Ruf T, Korytko AI, Stieglitz A, Lavenburg KR, Blank JL. 1997. Phenotypic variation in seasonal adjustments of testis size, body weight, and food intake in deer mice: role of pineal function and ambient temperature. *J. Comp. Physiol. B* 167:185–92

Saidapur SK, Hoque B. 1995. Effect of photoperiod and temperature on ovarian cycle of the frog *Rana tigrina*. *J. Biosci.* 20:445–52

Saunders DS. 2002. *Insect Clocks*. Amsterdam, Neth.: Elsevier

Schierwater B, Hauenschild C. 1990. A photoperiod determined life-cycle in an oligocheate worm. *Biol. Bull.* 178:111–17

Shimomura K, Nelson DE, Ihara NL, Menaker M. 1997. Photoperiodic time measurement in *tau* mutant hamsters. *J. Biol. Rhythms* 12:423–30

Sullivan JK, Lynch GR. 1986. Photoperiod time measurement for activity, torpor, molt and reproduction in mice. *Physiol. Behav.* 36:167–74

Taranger GL, Haux C, Stefansson SO, Björnsson BT, Walther BT, Hansen T. 1998. Abrupt changes in photoperiod affect age at maturity, timing of ovulation and plasma testosterone and estradiol-17β profiles in Atlantic salmon, *Salmo salar*. *Aquaculture* 162:85–98

Tauber MJ, Tauber CA, Masaki S. 1986. *Seasonal Adaptations of Insects*. New York, NY: Oxford Univ. Press

The definitive treatise on the physiology of insect circadian rhythms and photoperiodism, with a thought-provoking final chapter.

An excellent, exhaustive treatise on seasonal adaptations of insects.

Templeton AR. 1986. Coadaptation and outbreeding depression. In *Conservation Biology: The Science of Scarcity and Diversity*, ed. E Soulé, pp. 105–16. Sunderland, MA: Sinauer Assoc.

Thom MD, Johnson DDP, MacDonald DW. 2004. The evolution and maintenance of delayed implantation in the mustelidae (Mammalia:Carnivora). *Evolution* 58:175–83

Thomason JC, Conn W, LeComte E, Davenport J. 1996. Effect of temperature and photoperiod on the growth of the embryonic dogfish, *Scyliorhinus canicula*. *J. Fish Biol.* 49:739–42

Tosini G. 1997. The pineal complex of reptiles: physiological and behavioral roles. *Ethol. Ecol. Evol.* 9:313–33

Tosini G, Bertolucci C, Foà A. 2001. The circadian system of reptiles: a multioscillatory and multiphotoreceptive system. *Physiol. Behav.* 72:461–71

Turner CL, Bagnara JT. 1971. *General Endocrinology*. Philadelphia, PA: W.B. Saunders

Tyndale-Biscoe CH. 1980. Photoperiod and the control of seasonal reproduction in marsupials. In *Endocrinology 1980. Proceedings of the VI International Congress of Endocrinology, Melbourne, Australia, February 10–16, 1980*, ed. IA Cummin, JW Funder, FAO Mendelsohn, pp. 277–82. Amsterdam, Neth.: Elsevier/North Holl. Biomed.

Uller T, Olsson M. 2003. Life in the land of the midnight sun: Are northern lizards adapted to longer days? *Oikos* 101:317–22

Underwood H, Goldman BD. 1987. Vertebrate circadian and photoperiodic systems: role of the pineal gland and melatonin. *J. Biol. Rhythms* 2:279–315

Vígh BM, Manzano MJ, Zádori A, Frank CL, Lukáts A, et al. 2002. Nonvisual photoreceptors of the deep brain, pineal organs and retina. *Histol. Histopathol.* 17:555–90

Werner JK. 1969. Temperature-photoperiod effects on spermatogenesis in the salamander *Plethodon cinereus*. *Copeia* 1969:592–602

Willis YL, Moyle DL, Baskett TS. 1956. Emergence, breeding, hibernation, movements and transformation of the bullfrog, *Rana catesbeiana*, in Missouri. *Copeia* 1956:30–41

Wolda H, Denlinger DL. 1984. Diapause in a large aggregation of a tropical beetle. *Ecol. Entomol.* 9:217–30

Wright ML, Jorey ST, Myers YM, Fieldstad ML, Paquette CM, Clark MB. 1988. Influence of photoperiod, daylength, and feeding schedule on tadpole growth and development. *Dev. Growth Differ.* 30:315–23

RELATED RESOURCES

Møller AP, Fiedler W, Berthold P, eds. 2004. Birds and climate change. *Adv. Ecol. Res.* Vol. 35.

Helm B, Gwinner E, Trost L. 2005. Flexible seasonal timing and migratory behavior. Results from stonechat breeding programs. *Ann. N. Y. Acad. Sci.* 1046:216–227

J. Biol. Rhythms 16(4), 2001.

Virus Evolution: Insights from an Experimental Approach

Santiago F. Elena and Rafael Sanjuán

Instituto de Biología Molecular y Celular de Plantas, Consejo Superior de Investigaciones Científicas-Universidad Politécnica de Valencia, 46022 València, Spain; email: sfelena@ibmcp.upv.es

Annu. Rev. Ecol. Evol. Syst. 2007. 38:27–52

First published online as a Review in Advance on June 28, 2007

The *Annual Review of Ecology, Evolution, and Systematics* is online at
http://ecolsys.annualreviews.org

This article's doi:
10.1146/annurev.ecolsys.38.091206.095637

Key Words

adaptation, epistasis, error threshold, quasi-species, robustness, trade-offs

Abstract

Viruses represent a serious problem faced by human and veterinary medicine and agronomy. New viruses are constantly emerging while old ones evolve and challenge the latest advances in antiviral pharmaceutics, thus generating tremendous social alarm, sanitary problems, and economical losses. However, they constitute very powerful tools for experimental evolution. These two faces of virology are tightly related because future antiviral treatments shall be rationally designed by considering evolutionary principles. Evidence indicates that the evolution of viruses is determined mainly by key features such as their small genomes, enormous population sizes, and short generation times, and at least for RNA viruses, large selection coefficients, antagonistic epistasis, and high mutation rates. We summarize recent advances in the field of experimental virus evolution. Increasing our understanding of the roles of selection, mutation, chance, and historical contingency on the ecology and epidemiology of viral infections could determine our ability to combat them.

INTRODUCTION

RdRp: RNA-dependent RNA polymerase

VSV: vesicular stomatitis virus

Fitness: the number of descendants an individual generates per time unit, usually relative to that of a reference genotype

Complementation: interaction between two viral genomes within an infected cell such that the virus can function de<pite each genome carrying different mutated, nonfunctional genes

Viruses, and in particular those having RNA as genetic material, are the most abundant parasites infecting animals, plants, and bacteria (Domingo & Holland 1997). Despite tremendous economical efforts, the number of eradicated viruses is quite limited and the perspectives for future eradications would most likely be overbalanced by the emergence or reemergence of other viruses (Murphy & Nathanson 1994). The fact that few viruses can be effectively controlled with state-of-the-art pharmacology, as well as the pervasive emergence of new viruses, could be a consequence of both the intrinsic RNA virus ability to evolve and the human-induced alterations in viruses' natural ecosystems. Immune escape strains and strains resistant to antivirals may arise soon after challenged by the immune system or drugs (Althaus & Bonhoeffer 2005, Kalia et al. 2005), and viruses may jump the species barrier from their natural reservoir host to a naïve one (Daszak et al. 2000, Kuiken et al. 2006).

The interest in studying the evolution of RNA viruses is not only motivated by the need to develop new rational antiviral strategies. They also constitute useful tools for experimentally addressing fundamental evolutionary questions while still allowing one to pay attention to the molecular details. After decades of research, some seemingly general properties of RNA viruses have been established. (*a*) As a consequence of the lack of proofreading activity in their RNA-dependent RNA polymerase (RdRp), genomic mutation rates are, on average, in the range 0.13–1.15 (Drake & Holland 1999). (*b*) Genomes are small [e.g., 3569 nt for MS2 to 11,162 nt for vesicular stomatitis virus (VSV)], with many examples of overlapping reading frames and multifunctional proteins. (*c*) Replication rates are fast, reaching tremendous population sizes shortly after infection. (*d*) Variability is a key factor for pathogenicity (e.g., immune escape, antiviral resistant, or host-range mutants). (*e*) Abundant molecular, functional, and structural information makes it relatively easy to map genotypes into phenotypic space. (*f*) The average fitness effect of point mutations are large (~20%, with up to 40% lethals for VSV; Sanjuán et al. 2004a) and the average interaction between deleterious mutations antagonistic, a hallmark of nonrobust genomes (Elena et al. 2006).

In this review, we summarize recent advances in the field of experimental virus evolution, focusing not only on the experiments but also on the underlying evolutionary theories. Particular emphasis is on the dual role of mutation and its interplay with population size. At large population sizes, evolutionary dynamics are driven by competition between beneficial variants. At small population sizes, deleterious mutations have a chance to spread and may drive extinction or the evolution of robustness mechanisms. One such potential mechanism discussed, complementation during multiple infections, allows for the evolution of social behavior, which is briefly discussed at the end of the review.

The Molecular Quasi-species and the Classical Population Genetics Frameworks

Virologists usually, although not always in a precise way, employ the term quasi-species to refer to highly polymorphic viral populations. The quasi-species theory

(Eigen et al. 1989) describes the evolution of an infinite asexual population with short genomes and high error rates. Eigen and coworkers studied mutation-selection dynamics and found that populations reached an equilibrium composed by a polymorphic assemblage of mutant genomes with a rare wild type. Nonetheless, it is worth highlighting that most of the quasi-species formulation is fully equivalent to the classic mutation-selection balance for haploid asexuals developed long ago within the framework of population genetics (Wilke 2005). However, the quasi-species theory has made some specific and relevant predictions (Eigen et al. 1989): (*a*) There exists an error threshold beyond which selection cannot further maintain the genetic structure of the population, the wild type is lost, and the population randomly drifts in genotypic space; (*b*) selection does not act on individual genomes but on the cloud of closely related genomes, implying that at high mutation rates a slow-replicating quasi-species can outcompete a faster-replicating one if the first is more robust to deleterious mutational effects (van Nimwegen et al. 1999, Wilke 2001).

Error threshold: critical mutation rate beyond which selection cannot further maintain the information encoded in the genome

THE ADAPTIVE PROCESS

The Dynamics of Adaptation

Similar to other microbes, a common observation in long-term evolution experiments with viruses is fitness trajectory in which gains are initially rapid but tend to decelerate over time (Burch & Chao 1999, Elena et al. 1998, Novella et al. 1995, Wichman et al. 1999) (**Figure 1**). Such dynamics indicate that, after being placed in a new environment, populations are evolving from a region of low fitness toward an adaptive peak or plateau. What determines the rate of adaptation? What kind of molecular changes are associated with fitness increases? These questions are explored below.

In negative-stranded RNA viruses, the genome is complementary to the messenger RNA and hence is not directly translated. In this kind of virus, recombination is rare (Chare et al. 2003), probably owing to the tight binding between viral ribonucleoproteins and genomic RNA, and thus populations behave truly clonally. An important consequence of asexuality is clonal interference. Clones that carry different beneficial mutations compete with one another, thereby interfering with their spread and substitution in the population (Gerrish & Lenski 1998). In general, all but one lineage will be excluded by the one with the most beneficial combination of mutations. Clonal interference becomes increasingly intense for large populations and high mutation rates and thus should be relevant to the evolution of RNA viruses. Its evolutionary consequences include the following: (*a*) the probability of substitution of a given beneficial mutation declines with increasing population size or mutation rate, but the individual substitutions entail larger fitness gains, (*b*) the rate of fitness improvement shows diminishing returns with an increasing supply of beneficial mutations caused by large population size or high mutation rates, and (*c*) many beneficial mutations become transiently common only to be excluded later by interfering mutations, giving rise to a leapfrog event in which the most abundant genotype at a given moment is phylogenetically related to an earlier dominant time than to the immediately preceding one.

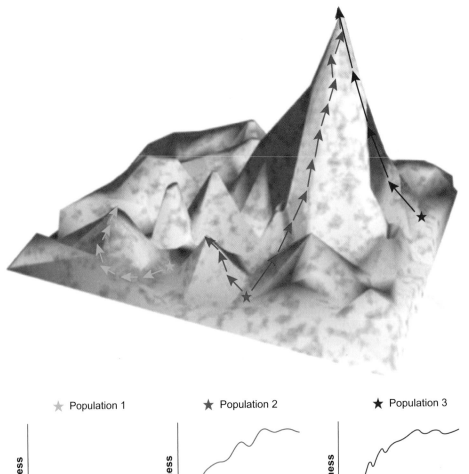

★ Population 1   ★ Population 2   ★ Population 3

Figure 1

Trajectories of virus fitness evolution. The evolutionary trajectory of three independent populations is represented in a hypothetical rugged fitness landscape by arrows and lines of different colors. Populations starting from different points can reach the same adaptive peak, in a case of convergent evolution (*population 3 and high population 2 dynamics*), and in this particular case, the global optimum. Starting from the same point, replicate populations can diverge into two different trajectories and reach completely different adaptive peaks (*population 2 dynamics*) depending upon the availability of beneficial mutations. A population can be trapped in a local adaptive peak (*population 1 and low population 2 dynamics*).

The first two predictions have been confirmed for RNA viruses. Clarke et al. (1994) and Quer et al. (1996) observed that when two equally fit clones of VSV competed, they coexisted for long periods of time until coexistence was suddenly broken and one of them was displaced by the other. Interestingly, clones of both competitors isolated right before the displacement showed an increase in fitness relative to their ancestors, suggesting the existence of an arms race between both competitors, although the reasons for coexistence breaking remained unclear. Later, Miralles et al. (1999, 2000) expanded this study. Two genotypes of VSV carrying distinguishable states of a neutral marker were mixed at equal frequency and allowed to compete at increasing population sizes. Increasing population size has the effect of increasing the number of available beneficial mutations and thus strengthens clonal interference. **Figure 2a** shows the results of the competition dynamics. After one of the two competitors got fixed in the population, the fitness effect of the mutation responsible for the fixation was measured. The magnitude of the fitness effect fixed increased with the intensity of clonal interference, as predicted above (**Figure 2b**). Furthermore, population size had the predicted diminishing-returns effect on the rate of adaptation (**Figure 2c**). With a similar experimental design, Burch & Chao (1999) showed that in φ6 the beneficial effect of the fixed mutations was proportional to population size (that is, the intensity of clonal interference). Finally, Wichman et al. (2005) showed that clonal interference among co-infecting genotypes of φX174 was coupled with a sustained molecular evolution after 13,000 generations.

All of these experiments create a picture in which adaptive evolution of RNA viruses occurs throughout the competition of large numbers of variants created continuously by the error-prone replication of their genomes, and only the fittest available genotype will succeed in the population. However, this competition process between multiple genotypes has a diminishing-returns effect on the rate of virus adaptation.

HIV-1: human immunodeficiency virus type 1

The Molecular Basis of Adaptation

One of the goals of modern evolutionary biology is to learn about the molecular basis of the adaptive process, something that can be easily achieved with the use of RNA viruses as model systems. Perhaps one of the most amazing realizations after sequencing virus lineages evolved in a common environment is the large amount of evolutionary parallelisms and convergences, both at synonymous and nonsynonymous sites (Bull et al. 1997, Cuevas et al. 2002, Novella & Ebendick-Corpus 2004, Wichman et al. 1999). Furthermore, this phenomenon is not exclusive of experimental evolution but also widespread and observed across human immunodeficiency virus type 1 (HIV-1)-infected patients treated with the same antiviral drug (Martínez-Picado et al. 2000), where not only do the same mutations arise, but they do so in a conserved order (Boucher et al. 1992). Although these results could in principle be explained from a neutralist point of view as resulting from mutational bias, it is more likely that parallel and convergent substitutions are adaptive. This pattern would result from organisms facing identical selective pressures, with few alternative adaptive pathways, as expected for viruses' simple and compacted genomes. In addition, the

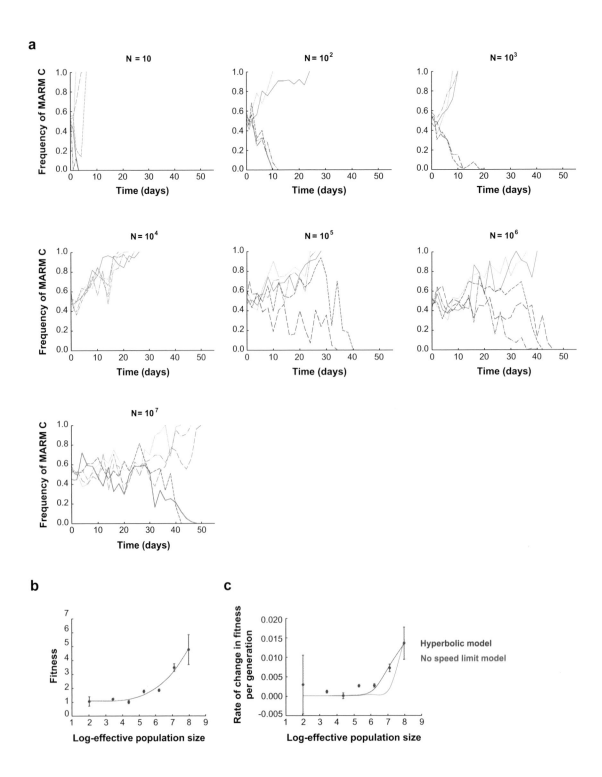

observation that beneficial mutations often become fixed in an ordered way supports clonal interference (Gerrish & Lenski 1998).

Convergence at nonsynonymous sites tells us that RNA itself can be the target of selection. Genomic RNA is involved in many RNA-RNA and RNA-protein interactions, and contains regulatory signals that affect its own replication, transcription, and encapsidation (reviewed in Novella 2003). Furthermore, RNA silencing, at least in plants, acts as an elaborate and adaptive antiviral response (Lecellier & Voinnet 2004), which depends upon sequence complementarities between the target RNA to be sliced and the small interfering RNAs (siRNAs) guiding the nuclease of the RNA-induced silencing complex. Under the selective pressure imposed by RNA silencing, RNA viruses have evolved strategies to avoid it. One such strategy is the evolution of proteins with the ability to suppress the silencing response (Voinnet et al. 1999). An alternative is to get involved in a run-away strategy in which changes in synonymous sites would be constantly selected to minimize homology with siRNAs.

Antagonistic epistasis: when the combined effect of mutations is weaker than expected from their individual effects

Mutational robustness: the constancy of phenotypic expression in the face of mutation

Synergistic epistasis: when the combined effect of mutations is stronger than expected from their individual effects

Epistasis and Pleiotropy Generate Trade-Offs on the Rate of Adaptation

Despite the adaptability of viral populations, theoretical considerations suggest that some sort of evolutionary constraint may exist for adaptation as a consequence of environmental, selective, genetic, or functional trade-offs. One source of genetic trade-offs is epistasis. The nature, frequency, and intensity of epistasis are barely known, despite their importance for many evolutionary theories seeking to understand the evolution of genetic systems. Viruses may shed light onto the epistasis problem owing to their easy-to-handle genomes. In particular, site-directed mutagenesis of infectious cDNAs allows creating large numbers of genomes carrying known numbers of mutations and analyzing the fitness effects of each mutation independently as well as in combination. Consequently, in recent years, multiple studies have converged to a common picture in which antagonistic epistasis among pairs of mutations is the rule for RNA genomes (Bonhoeffer et al. 2004, Burch & Chao 2004, Sanjuán et al. 2004b). Other implications of antagonistic epistasis are discussed below in the context of mutational robustness. Here, we focus only on their effect on the rate of viral evolution.

Synergistic epistasis should accelerate the rate of evolution, compared with the case of multiplicative fitness effects, because of the extra fitness differential recovered after the fixation of the two mutations. Conversely, antagonistic epistasis should slow

---

Figure 2

The effect of clonal interference on the evolutionary dynamics of VSV populations. (*a*) Temporal dynamics of the two competitors until one of them gets fixed. Each competitor is represented by a different color. (*b*) Correlation between the magnitude of the fitness effect fixed and the intensity of clonal interference. (*c*) Diminishing-returns effect of population size on the rate of viral adaptation. The lines represent the fit to a model with a limit in the rate of an adaptation (hyperbolic) model and the fit to a linear (no speed limit) model. (Panels b and c are taken from Miralles et al. 1999.)

Arboviruses: viruses that use arthropods as vectors of transmission

down adaptation. Sanjuán et al. (2005) tested these predictions by constructing VSV genotypes that carried pairs of deleterious mutations showing different sign of epistasis. In this experimental setting, adaptation consisted mainly of the compensation of the artificially introduced deleterious mutations. As predicted, a negative correlation was observed between the magnitude of fitness improvement and the sign of epistasis: On average, larger fitness increases were associated with pairs of mutations that interacted synergistically, whereas smaller fitness increases were associated with antagonistic epistasis.

In general, differences in cell types and tissues within a given host, differences in host species, or the presence/absence of antiviral responses all represent instances of environmental heterogeneity. Fluctuation between different hosts is an important component of the infectious cycle of arboviruses such as VSV or Easter equine encephalitis virus. Arboviruses are transmitted among vertebrate hosts by insect and tick vectors. Although some can persist by vertical transmission from female arthropods to their offspring, most replicate alternately in vertebrate and vectors during horizontal transmission cycles. Host radiation allows a virus to expand its ecological niche by adapting to one or more novel hosts. With trade-offs, constant environments promote the evolution of specialists, whereas changing environments favor generalists, even if these had suboptimal fitness compared with specialists. Without trade-offs, a single genotype would be expected to prevail in all cases. A simple mechanism for such trade-offs is antagonistic pleiotropy, in which a particular mutation that is beneficial in one environment is harmful others. Alternatively, the trade-off could be generated if mutations accumulated by drift in genes that are not necessary in some environment but useful in others. Owing to their extremely compacted genomes and the necessity of expressing all genes during the infectious cycle, the latter explanation is unlikely to operate for viruses.

Several studies have confirmed the hypothesis that adaptation to a novel host would decrease competitive ability in the original host and that adaptation is host specific (Crill et al. 2000, Turner & Elena 2000, Weaver et al. 1999). Furthermore, when environments were forced to fluctuate between two novel conditions, two general conclusions were drawn. First, the fitness cost in the ancestral host was as large as it was for the more costly of the two new environments (Turner & Elena 2000). Second, on average, populations evolved in fluctuating conditions improved fitness in each host as much as the populations evolved in each single host, suggesting that fluctuating environments would select for mutations with no pleiotropic effects (Novella et al. 1999, Turner & Elena 2000, Weaver et al. 1999). These results apparently contradict the existence of trade-offs and host specificity. To reconcile these results, it is possible to imagine that there are two classes of beneficial mutations. One class is beneficial only on a particular host and has antagonistic pleiotropic effects on other hosts; mutations in genes that affect interactions with host receptors and other host-specific molecules are good candidates. The other class produces beneficial effects in all hosts; mutations in genes involved in RNA processing and elongation are candidates. Even if mutations with host-specific benefits were more common than the generally beneficial mutations, the latter class would be differentially enriched in viral populations that evolved on alternating host types.

THE EVOLUTION OF HIGH MUTATION RATES IN RNA VIRUSES

The evolutionarily optimal mutation rate of viruses should be determined by the following factors (Sniegowski et al. 2000): (*a*) Because most mutations are deleterious, a selective pressure exists for reducing mutation rates toward whatever limit is imposed by biochemical restrictions; (*b*) the mechanisms of replication fidelity could come at a kinetic or energetic cost and thus be selected against (Dawson 1998, Kimura 1967); and (*c*) raising error rates provides more chances to generate beneficial mutations and to explore adaptive landscapes. On the basis of *c*, researchers have often argued that elevated mutation rates are maintained in RNA viruses because of the rapid adaptive capacity they bestow (Domingo & Holland 1997, Holland et al. 1982, Pfeiffer & Kirkegaard 2005, Vignuzzi et al. 2006). However, hypotheses about the high adaptability of RNA viruses should take into account the interplay between the three above-mentioned factors.

Biochemical experiments with avian myeloblastosis virus (Kunkel et al. 1986) and VSV (Steinhauer et al. 1992), and structural analyses of HIV-1 reverse transcriptase (RT), have established that RNA virus RdRp lacks 3′-exonuclease activity. This provides a basis for error-prone replication, but it remains to be elucidated whether this lack of activity is a consequence of fundamental biochemical restrictions or the product of natural selection. Variability in mutation rates within and between RNA virus species (Drake & Holland 1999) gives weak support to the latter possibility, as a large portion of this variation is most probably due to inaccurate estimates. However, recent experiments demonstrate that specific genotypic changes modify replication fidelity. For example, serial passage of poliovirus-1 in the presence of increasing concentrations of ribavirin resulted in the fixation of a single nonsynonymous nucleotide substitution in the polymerase gene that conferred a threefold increase in replication fidelity (Pfeiffer & Kirkegaard 2003). Similarly, experiments with HIV-1 RT variants isolated from patients undergoing antiretroviral therapy revealed the existence of substitutions that specifically confer mutator or antimutator phenotypes (Cases-González et al. 2000, Gutiérrez-Rivas & Menéndez-Arias 2001). Some of these substitutions have also been assayed in cell culture, suggesting a good correlation between in vitro and in vivo error rates (Furió et al. 2007, Mansky et al. 2003). Finally, the presence of 3′-exonuclease activity in eukaryotic RNA polymerases (Thomas et al. 1998) further suggests that the observed mutation rates may not be due merely to biochemical restrictions.

RT: reverse transcriptase

L: mutational load

Genomic deleterious mutation rate (U_d): the number of deleterious mutations produced per genome and replication round

Mutational Load

Mutations are more often deleterious than beneficial (Sanjuán et al. 2004a). Hence, high mutation rates must be detrimental in the short term. A well-known result for asexual species is that, in the absence of epistasis, the equilibrium mutational load is $L = 1 - \exp(-U_d)$, where L is the mutational load and U_d is the genomic deleterious mutation rate (Kimura & Maruyama 1966). Owing to RNA viruses high error rates, L should be especially elevated for them. Assuming a genomic mutation rate of 1.0 per replication event and that approximately 70% of random mutations are deleterious (Sanjuán et al. 2004a), the fitness load due to mutation should be $1 - \exp(-0.7) \approx 0.50$, which means that half the replication capacity would be lost. If the mutation rate was 0.1, L would drop to only 0.07. It is possible that RNA virus mutation rates are actually below 1.0 (Chao et al. 2002, Furió et al. 2005), but otherwise RNA virus populations harbor a considerably high L and, consequently, are under strong selection pressure for reducing mutation rates.

In addition to U_d, the selection coefficient (s) shall be important in determining L in two situations. First, if the population is out of the mutation-selection balance, the larger s is, the faster fitness declines toward the equilibrium (Johnson 1999). Second, for small genomes replicating at high mutation rates, the mean equilibrium fitness depends not only upon the rate of deleterious mutations but also upon their average s (Krakauer & Plotkin 2002, Schuster & Swetina 1988, van Nimwegen et al. 1999, Wilke 2001). In both scenarios, the larger s is, the lower average fitness. This further supports RNA viruses harboring high L.

The Cost of Replication Fidelity

Most polymerases show fidelities well above the substrate specificity predicted from differences in thermodynamic stability between cognate and noncognate base pairs (Showalter & Tsai 2002). However, it was soon realized that the maintenance of replication fidelity must have a replication efficiency cost that would prevent polymerases from evolving the biochemically lowest possible mutation rate (Kimura 1967). Theory has been developed that takes into account the balance between the cost of replication fidelity and L, two opposing forces that should determine a non-null evolutionarily stable mutation rate (Dawson 1998).

Only recently have researchers suggested that this cost may be relevant to the evolution of mutation rates in RNA viruses. Furió et al. (2005) estimated both mutation rate and fitness for VSV mutants carrying single amino acid substitutions in the RdRp gene. Changes leading to lower mutation rates also led to slower growth rates. In good agreement, evolution under rapid growth conditions increased the mutation rate toward the theoretically optimal rate. To shed some light on the biochemical basis of the fidelity cost, data from in vitro experiments with HIV-1 RT have been recently analyzed (Furió et al. 2007). A positive correlation between the in vitro mutation rate and the catalytic constant for cognate nucleotide incorporation has been observed, suggesting that an increased fidelity could negatively impact the rate of replication. A clue to the mechanism underlying the cost of fidelity in HIV-1 RT came from the observation that, if an incorrect nucleotide is incorporated to the nascent chain, its extension occurs at a much slower rate than for the correct pair (Kunkel 2004). Inefficient misspair extension would hence reduce error rates at the expense of decreasing polymerization speed.

Differences between mutation rates among taxa can be explained by the trade-off between the accuracy and the efficiency of replication because the cost of fidelity should be stronger for species that rely critically on fast replication for survival. This is the case for RNA viruses because their rapid infection cycles allow them to reach high titers before the onset of the host's defense mechanisms (Coffin 1995). This means that the cost of proofreading functions, in terms of replication rate, may be excessive for RNA viruses. This reasoning may also apply to DNA viruses because their parasitic lifestyle is similar to that of RNA viruses, but the fact is that DNA viruses show substantially lower mutation rates (Drake & Holland 1999). This is an unresolved problem, although it must be noted that genetic information is more compressed in RNA viruses than in DNA viruses because replication and transcription are biochemically equivalent and often catalyzed by a common molecular complex. Increased functional overlapping may impose more functional restrictions and hence more stressed fitness trade-offs.

Lethal mutagenesis: the deterministic extinction of a population due to an excessive mutation rate

Beneficial Mutations and Adaptation

In a changing environment, optimal replication occurs at a nonvanishing error rate to allow the organism to keep up with environmental changes. If deleterious mutations are neglected, the rate of adaptation increases monotonically with mutation rate until clonal interference becomes important (Gerrish & Lenski 1998). However, when deleterious mutations are taken into account, the optimal mutation rate must be high enough to supply adaptive variability but low enough to prevent the accumulation of deleterious mutations (Johnson & Barton 2002, Orr 2000). In asexual species, modifier alleles that increase mutation rate are more likely to be linked to beneficial mutations and hence have a chance to get hitchhiked to fixation. In sexual species, however, linkage is rapidly dissipated by recombination, hitchhiking is weak (Kimura 1967), and thus the advantage of mutator alleles is often not enough to overcome the short-term increase in L. Many RNA viruses, for example HIV-1, show high levels of recombination (Lemey et al. 2006). Hence, the argument that RNA viruses adapt fast owing to their high mutation rates encounters some conceptual flaws.

Mutator genotypes ought to be favored by selection if they often face novel environmental conditions (de Visser 2002), as is the case for rapidly changing environments. Immunity and, in general, variable environments, may favor high mutation rates in RNA viruses. The evolution of viral mutation rates in the presence of an adaptive immune system has been modeled (Kamp et al. 2002), and researchers concluded that the optimal genomic mutation rate per infection cycle should equal the rate at which the immune system adapts to each new viral antigen. However, no experimental work has addressed the influence of immunity in the evolution of RNA virus mutation rates. It is known, though, that the addition of chemical mutagens does not translate into a higher rate of adaptation (Lee et al. 1997). Increasing the mutation rate beyond the already high spontaneous values is counterproductive for RNA virus fitness and adaptation because it favors mutation accumulation and, in some cases, viral extinction through lethal mutagenesis (Anderson et al. 2004).

RNA viruses should be viewed as extant mutators, and thus the adaptive value of their high mutation rates is probably better tested by conducting experiments with increased-fidelity mutants. Given RNA virus' high L, it should a priori be expected that lowering the mutation rate could come about with little loss of adaptive capacity while conferring the benefit of slowing down the accumulation of deleterious mutations. This expectation is supported by the observation that lamivudine-resistant HIV-1 clones with increased RT fidelity do not pay any cost in terms of adaptability (Keulen et al. 1999). However, recent work with poliovirus-1 has challenged this view. A mutant genotype carrying a substitution at the polymerase gene that confers a threefold increase in replication fidelity was less pathogenic in mice than the wild type (Pfeiffer & Kirkegaard 2005, Vignuzzi et al. 2006). An RNA accumulation defect was observed in the high-fidelity genotype (Pfeiffer & Kirkegaard 2005), in agreement with the cost-of-fidelity hypothesis. However, this difference was apparently too slight to explain differences in pathogenesis. Infection of mice revealed that the high-fidelity mutant was less able to spread to the brain tissue, probably owing to its restricted variability and its consequently diminished capacity to invade different local microenvironments.

As a general remark, although immunity and changing environments could facilitate the evolution and maintenance of high mutation rates in RNA viruses, the same should be valid for DNA viruses. A priori, losing the ability to replicate with high fidelity should be evolutionarily easier than gaining it. Therefore, if RNA viruses owed their rapid adaptation to their elevated mutation rates, then there is apparently no reason why DNA viruses should have not evolved high mutation rates as well. Why DNA and RNA viruses, which apparently share similar lifestyles, show different mutation rates, is a major unsolved question.

GENETIC CONTAMINATION

Genetic Drift and Transmission Bottlenecks

Genetic drift can play an evolutionary role comparable with that of natural selection (Kimura 1983). The influence of drift depends upon effective population size (N_e), which in turn is determined by factors such as the reproduction mode, the historical population bottlenecks, the linkage between genes, or natural selection. These factors can reduce genetic variability and hence make N_e substantially lower than census sizes. The relevance of drift also depends on the magnitude of s. Selection should prevail over drift when $N_e s > 1$, whereas drift should prevail otherwise (Ohta 1992). Under drift, deleterious mutations behave as nearly neutral and hence accumulate in the population. This can lead to fitness declines and potentially jeopardize the survival of small populations. A mechanism for mutation accumulation is Muller's ratchet: In finite populations, mutation-free individuals will become rare at low population sizes, hence making it plausible that they get lost by drift (Haigh 1978). In the absence of compensatory and back mutations or recombination, the loss is irreversible and the ratchet clicks. If mutation accumulation is sustained, Muller's ratchet can result in extinction by mutational meltdown (Lynch et al. 1993).

Considering the fact that RNA viruses show large s and rapidly reach population sizes of several billion particles (Coffin 1995), it should be concluded that selection is the main factor determining their evolution and that drift plays only a minor role. However, data indicate that, in natural populations, N_e in RNA viruses is several orders of magnitude lower than particle counts (Brown 1997, García-Arenal et al.

Genetic drift: changes in genotypic frequencies due to random sampling between generations

Effective population size (N_e): the number of viral particles that effectively contribute to the next infectious cycle

2001). One reason is that viral populations experience strong bottlenecks upon transmission, especially between individual hosts (Edwards et al. 2006), but probably also between organ compartments within individuals (Itescu et al. 1994). For example, bottlenecks take place during plant virus systemic movement between leaves of the same plant (Hall et al. 2001, Li & Roossinck 2004), being the number of viral particles propagated on the order of tens for tobacco mosaic virus (Sacristán et al. 2003) or as low as 4 for wheat streak mosaic virus (Hall et al. 2001). Similarly, strong bottlenecks upon aphid-mediated plant-to-plant transmission have been reported (Ali et al. 2006). Furthermore, in HIV-1, N_e upon homosexual transmission has been estimated by coalescence methods to be as low as 1.6–2.0 particles (Edwards et al. 2006).

FMDV: foot-and-mouth disease virus

The evolutionary consequences of these transmission bottlenecks have been studied extensively in cell cultures by performing serial transfers of randomly chosen lytic plaques. The sampling of individual infection units at each passage dramatically reduces N_e to little more than one individual, which maximizes genetic drift and onsets Muller's ratchet. Sustained plaque-to-plaque passages of a variety of RNA viruses, including $\phi6$ (Chao 1990), MS2 (de la Peña et al. 2000), VSV (Duarte et al. 1992), foot-and-mouth disease virus (FMDV) (Escarmís et al. 1996), and HIV-1 (Yuste et al. 1999), typically resulted in significant fitness losses, ranging from 20%–90% relative to the ancestor after approximately 20–30 passages, and in a few cases achieved viral extinction. Several mutations were fixed in these low-fitness populations, albeit less than expected. For example, in FMVD, an average of two to three mutations was found in lines that had experienced 35% fitness losses. These results are not striking given the elevated s showed by RNA viruses. One or few changes are sufficient to produce dramatic fitness losses, and lines with more mutations went extinct.

The main difference between plaque-to-plaque experiments and natural infections is that in the latter, the number of generations between bottleneck events is much larger, thus providing more chances for beneficial or compensatory mutations to get fixed. Together with the fact that RNA viruses can recover from fixation of deleterious mutations even at small population sizes (Burch & Chao 1999), this difference should explain why natural RNA virus populations are not extinguished or even endangered by Muller's ratchet.

Lethal Mutagenesis versus Error Catastrophe

Genetic contamination can also take place at large population sizes. Population genetics predicts that in an arbitrarily large population, the frequency of genotypes carrying deleterious mutations at the mutation-selection balance is U_d/s and the frequency of the mutation-free genotype is $\exp(-U_d/s)$ (Kimura & Maruyama 1966). Therefore, mutagenesis will make the mutation-free class rarer and average fitness will rapidly decrease. However, this does not provide enough information about the probability of extinction by lethal mutagenesis. To address this point, it is necessary to express fitness in an absolute instead of relative scale (Bull et al. 2007). If the absolute number of progeny per individual is lower than one, population size will deterministically decline, leading to lethal mutagenesis. Once the population size is low enough to make drift important relative to selection, extinction would be boosted by the stochastic mutation-accumulation processes discussed in the previous section.

A concept related to mutation accumulation in large populations is the error catastrophe, a prediction of the quasi-species theory (Eigen et al. 1989). Beyond an error threshold, the mutation-selection balance is expected to be lost and the population becomes a pool of randomly drifting genotypes. In general, the existence of an error threshold depends critically upon the assumed fitness landscape, which in the original formulation consisted of a single fit wild-type sequence and all other genotypes with a constant, non-null, fitness, regardless of the number of mutations they carried. This is an unrealistic model because lethal mutations are highly frequent in RNA viruses (Sanjuán et al. 2004a) and average fitness generally declines with increasing numbers of mutations. Thus, despite error thresholds that arise in some generalizations of the quasi-species model (e.g., Tarazona 1992), it takes place only under some special set of conditions (Summers & Litwin 2006, Wagner & Krall 1993, Wiehe 1997).

The consequences of artificially increasing error rates in RNA viruses have been explored in several cell culture experiments. Chemical mutagens have been used in a variety of RNA viruses, including VSV (Lee et al. 1997), HIV-1 (Loeb et al. 1999), poliovirus-1 (Crotty et al. 2001), FMDV (Sierra et al. 2000), and lymphocytic choriomeningitis virus (Grande-Pérez et al. 2002), among others. All these studies have proven that mutagens are detrimental to viral fitness and, in some cases, extinction was observed. For example, in HIV-1, the addition of the base analog 5-hydroxydeoxycytidine resulted in a loss of infectivity after 9–24 serial passages. Similarly, in FMDV, 5-fluorouracil or 5-azacytidine caused occasional extinction after 11–21 passages. In many cases, extinction was accompanied by an increase in the average number of mutations per genome. Virologists have often relied on the notion of error threshold to explain these experimental observations. The very idea behind this interpretation is that the error catastrophe is a form of lethal mutagenesis. However, the two concepts are not equivalent (**Table 1**) and, ironically, the

Table 1 Differences between lethal mutagenesis and error threshold models

	Lethal mutagenesis	**Error catastrophe**
Nature of the process	Mutation accumulation and population extinction	Change in the genetic composition of the population
Name of the threshold	Extinction threshold	Error threshold
Key parameters	U_d, fecundity of the wild type, and s	U_d and s
Demography	Population size declines	No changes in population size are required
Fate of the wild type	Not necessarily extinguished	Extinguished
Dependence on mutation rate	Gradual; extinction is more likely and occurs faster at higher mutation rates	Phase transition; beyond the error threshold, further increases in mutation rate have no effect
Effect of mutations on fitness	The fitness of mutant genotypes decays with mutation number	The fitness of mutant genotypes does not decay with mutation number
Mutational pattern	Only that induced by the drug; no specific changes are required	The consensus sequence randomly drifts through time

crucial assumption leading to the prediction of an error catastrophe (that is, a lower bound to fitness) actually retards extinction (Bull et al. 2007). Making the distinction is important because error catastrophe has been proposed as a candidate therapeutic strategy (Anderson et al. 2004, Domingo et al. 2001).

MOI: multiplicity of infection

MUTATIONAL ROBUSTNESS

Mechanisms of Robustness

Both s and the fraction of mutations that are neutral are useful measures of mutational robustness. For any given protein, the latter fraction can be accurately predicted from thermodynamic parameters (Bloom et al. 2005). Thermostability is a form of environmental robustness and, on the ground of theoretical arguments and RNA-folding simulations, it has been proposed that environmental and mutational robustness ought to be correlated (Ancel & Fontana 2000). Although mutational robustness has now a well-founded biophysical ground, it may sound intuitive that the transition from a nonrobust to a robust state should be complex, involving many genetic changes and adjustments. However, using β-lactamase as a model, it has been shown that a single amino acid substitution is enough to increase thermostability and render a more robust protein (Bloom et al. 2005). Additionally, the virological literature provides many examples of thermosensitive phenotypes associated with one or few nucleotide substitutions (Dupraz & Spahr 1993, Marriott et al. 1999). Therefore, current data and theory support the notion that robustness does not constitute a difficult evolutionary transition.

Misfolded proteins, in addition to being nonfunctional, aggregate and are often toxic for the cell, imposing metabolic burdens (Goldberg 2003). Molecular chaperones, originally identified as heat-shock proteins, assist folding and tag unfolded proteins for degradation (Feldman & Frydman 2000). Chaperones ought to have an important role in buffering not only environmental noise, but also mutational effects (Fares et al. 2002). Although the fact that chaperones are overexpressed in response to viral infections may suggest that chaperones are antiviral factors, animal viruses and bacteriophages depend upon chaperones to complete many steps of their infection cycle, including endocytosis, capsid disassembling, early replication, enhancement of transcription and translation, and virion assembly (Mayer 2005). Interestingly, thermosensitive mutants of tobacco mosaic virus trigger a much stronger expression of heat-shock protein 70 than the wild type, suggesting that the host's chaperones also assist the folding of viral proteins (Jockusch et al. 2001). Therefore, it is possible that chaperones may provide a mechanism for buffering mutational effects in RNA viruses. However, this hypothesis remains to be tested.

An additional putative mechanism for mutational buffering could be genetic complementation during co-infections. It is well known that defective viral particles lacking a portion of the genome can be stably maintained in populations at high multiplicity of infection (MOI), that is, the average number of viral particles infecting a host cell, impacting the evolution of the full virus (Bangham & Kirkwood 1993). Tobacco aspermy virus genotypes carrying single lethal substitutions have also been reported to be at higher frequencies than expected, presumably maintained by complementation (Moreno et al. 1997). It has also been hypothesized that deleterious mutants of FMDV could be stably maintained by complementation at frequencies higher than the mutation-selection balance expectation (Wilke & Novella 2003). Similarly, high levels of co-infection lessen the effectiveness of selection at purging deleterious mutations (Froissart et al. 2004).

The Evolution of Robustness

Insofar as robustness has a heritable basis, shows variability among individuals, and affects the probability of survival, it is a potential target for selection and evolutionary optimization. However, for mutational robustness to provide a selective advantage, the genes involved in expression of the phenotype have to be mutated. Therefore, the selective advantage of robustness can be, at most, equal to the mutation rate, which is typically small (Wagner 2005). The difficulty is somehow lessened in RNA viruses because the per site mutation rate is orders of magnitude higher than for their cellular hosts. Also, this putative advantage would also be fuelled by the strong deleterious coefficients characteristic of RNA viruses.

Even though robustness could potentially evolve in RNA viruses, the fact is that they remain highly sensitive to mutation compared to more complex microorganisms (Elena et al. 2006). In general, a lack of robustness is expected in small, compact genomes that have no redundancy, no repair systems, and exhibit strong pleiotropy (Krakauer & Plotkin 2002). These systems usually exist as very large populations, thus making selection very efficient at purging deleterious mutations ($N_e s \gg 1$) and promoting the preservation of the unmutated genotype at high frequencies. Individual hypersensitivity seems to be the predominant survival strategy for RNA viruses, but, under some circumstances, the evolution of mutational robustness may be favored. These conditions were identified using digital organisms (Wilke et al. 2001). Digital organisms that had been replicating at low population sizes and high genomic mutation rates evolved increased mutational robustness but paid the cost of reduced replication rates. Using Wright's adaptive landscape metaphor, these organisms had evolved toward a flat peak, as opposed to organisms that had been replicating at high population sizes and low mutation rates (**Figure 3**). The latter were faster replicators, but were more sensitive to mutation and hence were located in a higher peak (**Figure 3**). The flatter population was readily outcompeted by the fitter at low mutation rates, but it benefited from a selective advantage at high mutation rates. Similarly, flat populations should be good competitors in small populations, where genetic drift favors mutation accumulation (Krakauer & Plotkin 2002, Schuster & Swetina 1988, van Nimwegen et al. 1999, Wilke 2001). Quasi-species theory provides a suitable theoretical framework for these results because neutral and back mutations are not ignored and, as a consequence, the average fitness of the population at the mutation-selection balance depends upon the geometry of the fitness landscape, that is, upon s (Schuster & Swetina 1988, van Nimwegen et al. 1999).

Experiments proving the evolution of mutational robustness in RNA viruses are scarce. Some clues first came from work with φ6 (Burch & Chao 2000) showing that

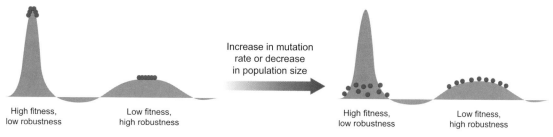

Increase in mutation
rate or decrease
in population size

High fitness,
low robustness

Low fitness,
high robustness

High fitness,
low robustness

Low fitness,
high robustness

Figure 3

Schematic representation of a landscape characterized by a peak of high fitness but low robustness and another one of low fitness but high neutrality (that is, robustness). As mutation rate increases or population size decreases, populations at the high peak suffer disproportionately larger fitness declines than those inhabiting the flatter surface.

the evolution of different genotypes depended on the topology of the neighboring adaptive landscape. Interestingly, one genotype repeatedly evolved decreased fitness despite passages done at high population sizes, whereas another genotype repeatedly increased in fitness owing to the availability of beneficial mutations in its mutational neighborhood. More recent work with φ6 has provided indirect evidence for the adaptive evolution of robustness (Montville et al. 2005). If robustness had a selective value, selection for robustness might be relaxed under co-infection regimes owing to genetic complementation. This prediction was tested by subjecting lineages that had previously evolved under high versus low MOI to mutation accumulation through plaque-to-plaque passages. In good agreement with the hypothesis, fitness declined faster in lineages that had evolved under high MOI.

A more direct approach was undertaken with viroids, plant pathogens constituted by small, noncoding, RNA molecules (Codoñer et al. 2006). Two viroid species belonging to different families and characterized by different secondary structures were competed in common host plants. One of the viroids was characterized by fast population growth and genetic homogeneity, whereas the other showed slow population growth and a high degree of variation. As expected, under standard nonmutagenic infection conditions, the faster replicator outcompeted the slower one. However, this advantage vanished when mutation rate was artificially increased, and the slower but more robust replicator was not outcompeted anymore.

SOCIAL CONFLICT AND COOPERATION IN THE VIRAL WORLD

Usually when talking about sociality, we tend to imagine the intricate relationships established between higher organisms (from social insects to the human society), but we hardly imagine viruses to be good candidates for testing theoretical predictions about cooperation, altruism, selfishness, or cheating. This appreciation is not correct because social interactions among viruses are quite common in nature, as, for example, synergistic symptoms among co-infecting plant virus (Malpica et al. 2006), predator-prey-like coevolutionary dynamics between some virus and their defective interfering particles (Bangham & Kirkwood 1993, DePolo et al. 1987), or the existence of coviruses, that is, genome segments encapsidated into separate particles, which are required for a successful infection (Nee 2000).

Viral sociality has also been studied in the laboratory. In a series of clever experiments, Turner & Chao (1999) explored the evolution of competitive interactions among φ6 at different MOIs. At high MOI, many viruses infected the same cell; at low MOI, only one virus infected each cell. Hence, they argued that different selective constraints were acting on each type of experiment. They found that the fitness of the virus evolved at high MOI relative to their ancestors generates a payoff matrix similar to the prisoner's dilemma strategy (that is, the cost of defecting is smaller than its advantage). In this strategy, selfishness evolves. How can cooperation and defection be defined in a viral system? Simply by manufacturing (cooperation) or sequestering (selfishness) diffusible intracellular products; this is a different side of the complementation phenomena discussed in the previous section. In single infections, either competition was absent or selection favored the evolution of cooperation among closely related individuals (resulting from the replication of the initially infecting phage). At high MOI, however, strong intracellular competition occurred between less-related individuals (originated from different infected hosts). As Turner & Chao postulated, the evolution of a prisoner's dilemma strategy can be attributed to the absence of clonal structure at high MOI, leading to the evolution of selfishness. Indeed, later on Turner & Chao (2003) showed that clonal selection allowed viruses to escape from the prisoner's dilemma. The ancestral φ6 clone was propagated under strict clonal conditions, creating the opportunity for cooperation to evolve. When the evolved cooperators were mixed with their selfish counterparts, both strategies coexisted in mixed polymorphism (Turner & Chao 2003). The transition from the prisoner's dilemma can occur either by selection for cheating and the associated cost or by selection for decreased sensitivity to cheaters. The Turner & Chao (2003) results supported the latter possibility. One plausible explanation for the evolution of selfish viruses relies on the fact that RNA virus genomes serve both as a template for replication and transcription. A virus with increased transcription activity, that is, contributing more proteins to the pool, would in fact act as a cooperator. By contrast, a virus that changes its schedule and spends more time on replication would act as a cheater and would encapsidate using the cooperator's coat proteins.

Sachs & Bull (2005) observed the evolution of cooperation between f1 and IKe. Both phages produce nonlytic infections in *Escherichia coli* and they were engineered to contain distinct antibiotic resistance markers. When both antibiotics are present in the media, only co-infected cells can survive and produce viral progeny. Thus, it is to the benefit of both viruses to remain together, despite the fact that cellular resources consumed by one virus are detrimental to the replication opportunities of the other. The evolutionary solution to this conflict was that both phages copackaged their genomes into f1 protein coats, ensuring cotransmission. In parallel, the IKe genome got smaller by deleting its own coat protein gene.

SUMMARY POINTS

1. Small genome size, high mutation rates, large selection coefficients, short generation times, and large population sizes are the key parameters for understanding RNA virus evolution.

2. Despite the remarkable evolutionary potential of RNA viruses, it is important to stress that factors such as clonal interference among coexisting beneficial mutations and negative epistasis among beneficial mutations and their pleiotropic effects can create trade-offs, limiting their evolution.

3. RNA viruses' high mutation rate may or may not have evolved as a strategy for accelerating adaptation. It is possible that the mechanisms of replication fidelity impose a fitness burden to systems that, as RNA viruses, rely critically on fast replication.

4. When the purifying selection is relaxed, for instance, during bottleneck transmission events, or when the mutation rate is artificially increased, deleterious mutations accumulate in viral populations, potentially triggering their extinction.

5. The general viral strategy to cope with deleterious mutational effects is individual hypersensitivity, which makes selection efficient at preserving the unmutated genotype. However, alternative strategies such as increased mutational neutrality, genetic complementation during co-infection, genome segmentation, or the use of cellular buffering mechanisms are also likely.

6. RNA viruses evolve social behaviors, including cooperative synergistic interspecies interactions or cheating.

FUTURE ISSUES

1. Researchers must adopt a unified theoretical framework for viral evolution from quasi-species and classical population genetics theories. The two are fundamentally equivalent but differ at some sets of assumptions, which give rise to sometimes different predictions.

2. Is there an error threshold beyond which genetic information is lost? And if so, is it equivalent to lethal mutagenesis? Can it be the basis for a new approach to antiviral therapies?

3. Is RNA viruses' high mutation rate beneficial per se or an unavoidable consequence of selection for fast replication?

4. From an evolutionary perspective, why do DNA viruses maintain mutation levels much lower than those of RNA viruses despite sharing the same lifestyle?

5. Can RNA viruses evolve mechanisms that buffer the deleterious effects of mutations and how may such mutations affect virus evolvability?

DISCLOSURE STATEMENT

The authors are not aware of any biases that might be perceived as affecting the objectivity of this review.

ACKNOWLEDGMENTS

Funding was provided by the Spanish Ministerio de Educación y Ciencia-FEDER (BFU2005-23720-E/BMC and BFU2006-14819-C02-01/BMC), the Generalitat Valenciana (ACOMP06/015 and GV06/031), and the EMBO Young Investigator Program.

LITERATURE CITED

Ali A, Li H, Schneider WL, Sherman DJ, Gray S, et al. 2006. Analysis of genetic bottlenecks during horizontal transmission of *Cucumber mosaic virus*. *J. Virol.* 80:8345–50

Althaus CL, Bonhoeffer S. 2005. Stochastic interplay between mutation and recombination during the acquisition of drug resistance mutations in human immunodeficiency virus type 1. *J. Virol.* 79:13572–78

Ancel LW, Fontana W. 2000. Plasticity, evolvability, and modularity in RNA. *J. Exp. Zool.* 288:242–83

Anderson JP, Daifuku R, Loeb LA. 2004. Viral error catastrophe by mutagenic nucleosides. *Annu. Rev. Microbiol.* 58:183–205

Bangham CR, Kirkwood TB. 1993. Defective interfering particles and virus evolution. *Trends Microbiol.* 1:260–64

Bloom JD, Silberg JJ, Wilke CO, Drummond DA, Adami C, Arnold FH. 2005. Thermodynamic prediction of protein neutrality. *Proc. Natl. Acad. Sci. USA* 102:606–11

Bonhoeffer S, Chappey C, Parkin NT, Whitcomb JM, Petropoulos CJ. 2004. Evidence for positive epistasis in HIV-1. *Science* 306:1547–50

Boucher CA, O'Sullivan E, Mulder JW, Ramautarsing C, Kellam P, et al. 1992. Ordered appearance of zidovudine resistance mutations during treatment of 18 human immunodeficiency virus-positive subjects. *J. Infect. Dis.* 165:105–10

Brown AJL. 1997. Analysis of HIV-1 *env* gene sequences reveals evidence for a low effective number in the viral population. *Proc. Natl. Acad. Sci. USA* 94:1862–65

Bull JJ, Badgett MR, Wichman HA, Huelsenbeck JP, Hillis DM, et al. 1997. Exceptional convergent evolution in a virus. *Genetics* 147:1497–507

Bull JJ, Sanjuán R, Wilke CO. 2007. Theory of lethal mutagenesis for viruses. *J. Virol.* 81:2930–39

Burch CL, Chao L. 1999. Evolution by small steps and rugged landscapes in the RNA virus φ6. *Genetics* 151:921–27

Burch CL, Chao L. 2000. Evolvability of an RNA virus is determined by its mutational neighbourhood. *Nature* 406:625–28

Burch CL, Chao L. 2004. Epistasis and its relationship to canalization in the RNA virus φ6. *Genetics* 167:559–67

Cases-González CE, Gutierrez-Rivas M, Menéndez-Arias L. 2000. Coupling ribose selection to fidelity of DNA synthesis. The role of Tyr-115 of human immunodeficiency virus type 1 reverse transcriptase. *J. Biol. Chem.* 275:19759–67

Chao L. 1990. Fitness of RNA virus decreased by Muller's ratchet. *Nature* 348:54–55

Chao L, Rang CU, Wong LE. 2002. Distribution of spontaneous mutants and inferences about the replication mode of the RNA bacteriophage φ6. *J. Virol.* 76:3276–81

Chare ER, Gould EA, Holmes EC. 2003. Phylogenetic analysis reveals a low rate of homologous recombination in negative-sense RNA viruses. *J. Gen. Virol.* 84:2691–703

Clarke DK, Duarte EA, Elena SF, Moya A, Domingo E, Holland JJ. 1994. The red queen reigns in the kingdom of RNA viruses. *Proc. Natl. Acad. Sci. USA* 91:4821–24

Codoñer FM, Daròs JA, Solé RV, Elena SF. 2006. The fittest versus the flattest: experimental confirmation of the quasispecies effect with subviral pathogens. *PLoS Pathog.* 2:e136

Coffin JM. 1995. HIV population dynamics in vivo: implications for genetic variation, pathogenesis, and therapy. *Science* 267:483–89

Crill WE, Wichman HA, Bull JJ. 2000. Evolutionary reversals during viral adaptation to alternating hosts. *Genetics* 154:27–37

Crotty S, Cameron CE, Andino R. 2001. RNA virus error catastrophe: direct molecular test by using ribavirin. *Proc. Natl. Acad. Sci. USA* 98:6895–900

Cuevas JM, Elena SF, Moya A. 2002. Molecular basis of adaptive convergence in experimental populations of RNA viruses. *Genetics* 162:533–42

Daszak P, Cunningham AA, Hyatt AD. 2000. Emerging infectious diseases of wildlife—threats to biodiversity and human health. *Science* 287:443–49

Dawson KJ. 1998. Evolutionarily stable mutation rates. *J. Theor. Biol.* 194:143–57

de la Peña M, Elena SF, Moya A. 2000. Effect of deleterious mutation-accumulation on the fitness of RNA bacteriophage MS2. *Evolution* 54:686–91

DePolo NJ, Giachetti C, Holland JJ. 1987. Continuing coevolution of virus and defective interfering particles and of viral genome sequences during undiluted passages: virus mutants exhibiting nearly complete resistance to formerly dominant defective interfering particles. *J. Virol.* 61:454–64

de Visser JA. 2002. The fate of microbial mutators. *Microbiology* 148:1247–52

Domingo E, Holland JJ. 1997. RNA virus mutations and fitness for survival. *Annu. Rev. Microbiol.* 51:151–78

Domingo E, Biebricher CK, Eigen M, Holland JJ. 2001. *Quasispecies and RNA virus evolution: principles and consequences*. Austin, Tex.: Landes Biosci.

Drake JW, Holland JJ. 1999. Mutation rates among RNA viruses. *Proc. Natl. Acad. Sci. USA* 96:13910–13

Duarte EA, Clarke DK, Moya A, Domingo E, Holland JJ. 1992. Rapid fitness losses in mammalian RNA virus clones due to Muller's ratchet. *Proc. Natl. Acad. Sci. USA* 89:6015–19

Dupraz P, Spahr PF. 1993. Analysis of deletions and thermosensitive mutations in Rous sarcoma virus gag protein p10. *J. Virol.* 67:3826–34

Edwards CTT, Holmes EC, Wilson DJ, Viscidi RP, Abrams EJ, et al. 2006. Population genetic estimation of the loss of genetic diversity during horizontal transmission of HIV-1. *BMC Evol. Biol.* 6:28

Eigen M, McCaskill J, Schuster P. 1989. The molecular quasi-species. *Adv. Chem. Phys.* 75:149–263

Elena SF, Carrasco P, Darós JA, Sanjuán R. 2006. Mechanisms of genetic robustness in RNA viruses. *EMBO Rep.* 7:168–73

Elena SF, Dávila M, Novella IS, Holland JJ, Domingo E, Moya A. 1998. Evolutionary dynamics of fitness recovery from the debilitating effects of Muller's ratchet. *Evolution* 52:309–14

Escarmís C, Dávila M, Charpentier N, Bracho A, Moya A, Domingo E. 1996. Genetic lesions associated with Muller's ratchet in an RNA virus. *J. Mol. Biol.* 264:255–67

Fares MA, Ruíz-González MX, Moya A, Elena SF, Barrio E. 2002. Endosymbiotic bacteria: GroEL buffers against deleterious mutations. *Nature* 417:398

Feldman DE, Frydman J. 2000. Protein folding in vivo: the importance of molecular chaperones. *Curr. Opin. Struct. Biol.* 10:26–33

Froissart R, Wilke CO, Montville R, Remold SK, Chao L, Turner PE. 2004. Co-infection weakens selection against epistatic mutations in RNA viruses. *Genetics* 168:9–19

Furió V, Moya A, Sanjuán R. 2005. The cost of replication fidelity in an RNA virus. *Proc. Natl. Acad. Sci. USA* 102:10233–37

Furió V, Moya A, Sanjuán R. 2007. The cost of replication fidelity in human immunodeficiency virus type 1. *Proc. R. Soc. B.* 274:225–30

García-Arenal F, Fraile A, Malpica JM. 2001. Variability and genetic structure of plant virus populations. *Annu. Rev. Phytopathol.* 39:157–86

Gerrish PJ, Lenski RE. 1998. The fate of competing beneficial mutations in an asexual population. *Genetica* 102/103:127–44

Goldberg AL. 2003. Protein degradation and protection against misfolded or damaged proteins. *Nature* 426:895–99

Grande-Pérez A, Sierra S, Castro MG, Domingo E, Lowenstein PR. 2002. Molecular indetermination in the transition to error catastrophe: systematic elimination of lymphocytic choriomeningitis virus through mutagenesis does not correlate linearly with large increases in mutant spectrum complexity. *Proc. Natl. Acad. Sci. USA* 99:12938–43

Gutierrez-Rivas M, Menéndez-Arias L. 2001. A mutation in the primer grip region of HIV-1 reverse transcriptase that confers reduced fidelity of DNA synthesis. *Nucl. Acids Res.* 29:4963–72

Hall JS, French R, Morris TJ, Stenger DC. 2001. Structure and temporal dynamics of populations within wheat streak mosaic virus isolates. *J. Virol.* 75:10231–43

Haigh J. 1978. The accumulation of deleterious genes in a population—Muller's Ratchet. *Theor. Pop. Biol.* 14:251–67

Holland JJ, Spindler K, Horodyski F, Grabau E, Nichol S, VandePol S. 1982. Rapid evolution of RNA genomes. *Science* 215:1577–85

Itescu S, Simonelli PF, Winchester RJ, Ginsberg HS. 1994. Human immunodeficiency virus type 1 strains in the lungs of infected individuals evolve independently from those in peripheral blood and are highly conserved in the C-terminal region of the envelope V3 loop. *Proc. Natl. Acad. Sci. USA* 91:11378–82

Jockusch H, Wiegand C, Mersch B, Rajes D. 2001. Mutants of tobacco mosaic virus with temperature-sensitive coat proteins induce heat shock response in tobacco leaves. *Mol. Plant Microbe Interact.* 14:914–17

Johnson T, Barton NH. 2002. The effect of deleterious alleles on adaptation in asexual populations. *Genetics* 162:395–411

Johnson T. 1999. The approach to mutation-selection balance in an infinite asexual population, and the evolution of mutation rates. *Proc. R. Soc. B.* 266:2389–97

Kalia V, Sarkar S, Gupta P, Montelaro RC. 2005. Antibody neutralization escape mediated by point mutations in the intracytoplasmic tail of human immunodeficiency virus type 1 gp41. *J. Virol.* 79:2097–107

Kamp C, Wilke CO, Adami C, Bornholdt S. 2002. Viral evolution under the pressure of an adaptive immune system: optimal mutation rates for viral escape. *Complexity* 8:28–33

Keulen W, van Wijk A, Schuurman R, Berkhout B, Boucher CA. 1999. Increased polymerase fidelity of lamivudine-resistant HIV-1 variants does not limit their evolutionary potential. *AIDS* 13:1343–49

Kimura M. 1967. On the evolutionary adjustment of spontaneous mutation rates. *Genet. Res. Camb.* 9:23–34

Kimura M. 1983. *The neutral theory of molecular evolution.* Cambridge, UK: Cambridge Univ. Press

Kimura M, Maruyama T. 1966. The mutational load with epistatic gene interactions in fitness. *Genetics* 54:1337–51

Krakauer DC, Plotkin JB. 2002. Redundancy, antiredundancy, and the robustness of genomes. *Proc. Natl. Acad. Sci. USA* 99:1405–9

Kuiken T, Holmes EC, McCauley J, Rimmetzwaan GF, Williams CS, Grenfell BT. 2006. Host species barriers to influenza virus infections. *Science* 312:394–97

Kunkel TA, Beckman RA, Loeb LA. 1986. On the fidelity of DNA synthesis. Pyrophosphate-induced misincorporation allows detection of two proofreading mechanisms. *J. Biol. Chem.* 261:13610–16

Kunkel TA. 2004. DNA replication fidelity. *J. Biol. Chem.* 279:16895–98

Lecellier CH, Voinnet O. 2004. RNA silencing: no mercy for viruses? *Immunol. Rev.* 198:285–303

Lee CH, Gilbertson DL, Novella IS, Huerta R, Domingo E, Holland JJ. 1997. Negative effects of chemical mutagenesis on the adaptive behavior of vesicular stomatitis virus. *J. Virol.* 71:3636–40

Lemey P, Rambaut A, Pybus OG. 2006. HIV evolutionary dynamics within and among hosts. *AIDS Rev.* 8:125–40

Li H, Roossinck MJ. 2004. Genetic bottlenecks reduce population variation in an experimental RNA virus population. *J. Virol.* 78:10582–87

Loeb LA, Essigmann JM, Kazazi F, Zhang J, Rose KD, Mullins JI. 1999. Lethal mutagenesis of HIV with mutagenic nucleoside analogs. *Proc. Natl. Acad. Sci. USA* 96:1492–97

Lynch M, Bürger R, Butcher D, Gabriel W. 1993. The mutational meltdown in asexual popualtions. *J. Heredity* 84:339–44

Malpica JM, Sacristán S, Fraile A, García-Arenal F. 2006. Association and host selectivity in multi-host pathogens. *PLoS ONE* 1:e41

Mansky LM, Le Rouzic E, Benichou S, Gajary LC. 2003. Influence of reverse transcriptase variants, drugs, and Vpr on human immunodeficiency virus type 1 mutant frequencies. *J. Virol.* 77:2071–80

Marriott AC, Wilson SD, Randhawa JS, Easton AJ. 1999. A single amino acid substitution in the phosphoprotein of respiratory syncytial virus confers thermosensitivity in a reconstituted RNA polymerase system. *J. Virol.* 73:5162–65

Martínez-Picado J, DePasquale MP, Kartsonis N, Hanna GJ, Wong J, et al. 2000. Antiretroviral resistance during successful therapy of HIV type 1 infection. *Proc. Natl. Acad. Sci. USA* 97:10948–53

Mayer MP. 2005. Recruitment of Hsp70 chaperones: a crucial part of viral survival strategies. *Rev. Physiol. Biochem. Pharmacol.* 153:1–46

Miralles R, Gerrish PJ, Moya A, Elena SF. 1999. Clonal interference and the evolution of RNA viruses. *Science* 285:1745–47

Miralles R, Moya A, Elena SF. 2000. Diminishing returns of population size in the rate of RNA virus adaptation. *J. Virol.* 74:3566–71

Montville R, Froissart R, Remold SK, Tenaillon O, Turner PE. 2005. Evolution of mutational robustness in an RNA virus. *PLoS Biol.* 3:e381

Moreno IM, Malpica JM, Rodriguez-Cerezo E, García-Arenal F. 1997. A mutation in tomato aspermy cucumovirus that abolishes cell-to-cell movement is maintained to high levels in the viral RNA population by complementation. *J. Virol.* 71:9157–62

Murphy FA, Nathanson N. 1994. The emergence of new virus diseases: an overview. *Semin. Virol.* 5:87–102

Nee S. 2000. Mutualism, parasitism and competition in the evolution of coviruses. *Philos. Trans. R. Soc. B.* 355:1607–13

Novella IS. 2003. Contributions of vesicular stomatitis virus to the understanding of RNA virus evolution. *Curr. Opin. Microbiol.* 6:399–405

Novella IS, Duarte EA, Elena SF, Moya A, Domingo E, Holland JJ. 1995. Exponential increases of RNA virus fitness during repeated transmission. *Proc. Natl. Acad. Sci. USA* 92:5841–44

Novella IS, Ebendick-Corpus BE. 2004. Molecular basis of fitness loss and fitness recovery in vesicular stomatitis virus. *J. Mol. Biol.* 342:1423–30

Novella IS, Hershey CL, Escarmís C, Domingo E, Holland JJ. 1999. Lack of evolutionary stasis during alternating replication of an arbovirus in insect and mammalian cells. *J. Mol. Biol.* 287:459–65

Ohta T. 1992. The nearly neutral theory of molecular evolution. *Annu. Rev. Ecol. Syst.* 23:263–86

Orr HA. 2000. The rate of adaptation in asexuals. *Genetics* 155:961–68

Pfeiffer JK, Kirkegaard K. 2003. A single mutation in poliovirus RNA-dependent RNA polymerase confers resistance to mutagenic nucleotide analogs via increased fidelity. *Proc. Natl. Acad. Sci. USA* 100:7289–94

Pfeiffer JK, Kirkegaard K. 2005. Increased fidelity reduces poliovirus fitness and virulence under selective pressure in mice. *PLoS Pathog.* 1:e11

Quer J, Huerta R, Novella IS, Tsimring LS, Domingo E, Holland JJ. 1996. Reproducible nonlinear population dynamics and critical points during replicate competitions of RNA virus quasispecies. *J. Mol. Biol.* 264:465–71

Sachs JL, Bull JJ. 2005. Experimental evolution of conflict mediation between genomes. *Proc. Natl. Acad. Sci. USA* 102:390–95

Sacristán S, Malpica JM, Fraile A, García-Arenal F. 2003. Estimation of population bottlenecks during systemic movement of tobacco mosaic virus in tobacco plants. *J. Virol.* 77:9906–11

Sanjuán R, Cuevas JM, Moya A, Elena SF. 2005. Epistasis and the adaptability of an RNA virus. *Genetics* 170:1001–8

Sanjuán R, Moya A, Elena SF. 2004a. The distribution of fitness effects caused by single-nucleotide substitutions in an RNA virus. *Proc. Natl. Acad. Sci. USA* 101:8396–401

Sanjuán R, Moya A, Elena SF. 2004b. The contribution of epistasis to the architecture of fitness in an RNA virus. *Proc. Natl. Acad. Sci. USA* 101:15376–79

Schuster P, Swetina J. 1988. Stationary mutant distributions and evolutionary optimization. *Bull. Math. Biol.* 50:635–60

Showalter AK, Tsai MD. 2002. A reexamination of the nucleotide incorporation fidelity of DNA polymerases. *Biochemistry* 41:10571–76

Sierra S, Dávila M, Lowenstein PR, Domingo E. 2000. Response of foot-and-mouth disease virus to increased mutagenesis: influence of viral load and fitness in loss of infectivity. *J. Virol.* 74:8316–23

Sniegowski PD, Gerrish PJ, Johnson T, Shaver A. 2000. The evolution of mutation rates: separating causes from consequences. *BioEssays* 22:1057–66

Steinhauer DA, Domingo E, Holland JJ. 1992. Lack of evidence for proofreading mechanisms associated with an RNA virus polymerase. *Gene* 122:281–88

Summers J, Litwin S. 2006. Examining the theory of error catastrophe. *J. Virol.* 80:20–26

Tarazona P. 1992. Error thresholds for molecular quasispecies as phase transitions: from simple landscapes to spin-glass models. *Phys. Rev. A* 45:6038–50

Thomas MJ, Platas AA, Hawley DK. 1998. Transcriptional fidelity and proofreading by RNA polymerase II. *Cell* 93:627–37

Turner PE, Chao L. 1999. Prisioner's dilemma in an RNA virus. *Nature* 398:441–43

Turner PE, Chao L. 2003. Escape from prisoner's dilemma in RNA phage φ6. *Am. Nat.* 161:497–505

Turner PE, Elena SF. 2000. Cost of host radiation in an RNA virus. *Genetics* 156:1465–70

van Nimwegen E, Crutchfield JP, Huynen M. 1999. Neutral evolution of mutational robustness. *Proc. Natl. Acad. Sci. USA* 96:9716–20

Vignuzzi M, Stone JK, Arnold JJ, Cameron CE, Andino R. 2006. Quasispecies diversity determines pathogenesis through cooperative interactions in a viral population. *Nature* 439:344–48

Voinnet O, Pinto YM, Baulcombe DC. 1999. Suppression of gene silencing: a general strategy used by diverse DNA and RNA viruses of plants. *Proc. Natl. Acad. Sci. USA* 96:14147–52

Wagner GP, Krall P. 1993. What is the difference between models of error thresholds and Muller's ratchet? *J. Math. Biol.* 32:33–44

Wagner A. 2005. *Robustness and Evolvability in Living Systems*. Princeton, NJ: Princeton Univ. Press

Weaver WC, Brault AC, Kang W, Holland JJ. 1999. Genetic and fitness changes accompanying adaptation of an arbovirus to vertebrate and invertebrate cells. *J. Virol.* 73:4316–26

Wichman HA, Badgett MR, Scott LA, Boulianne CM, Bull JJ. 1999. Different trajectories of parallel evolution during viral adaptation. *Science* 285:422–24

Wichman HA, Millstein J, Bull JJ. 2005. Adaptive molecular evolution for 13000 phage generations: a possible arms race. *Genetics* 170:19–31

Wiehe T. 1997. Model dependency of error thresholds: the role of fitness functions and contrasts between the finite and infinite sites models. *Genet. Res. Camb.* 69:127–36

Wilke CO. 2001. Selection for fitness versus selection for robustness in RNA secondary structure folding. *Evolution* 55:2412–20

Wilke CO. 2005. Quasispecies theory in the context of population genetics. *BMC Evol. Biol.* 5:44

Wilke CO, Novella IS. 2003. Phenotypic mixing and hiding may contribute to memory in viral quasispecies. *BMC Microbiol.* 3:11

Wilke CO, Wang JL, Ofria C, Lenski RE, Adami C. 2001. Evolution of digital organisms at high mutation rate leads to the survival of the flattest. *Nature* 412:331–33

Yuste E, Sánchez-Palomino S, Casado C, Domingo E, López-Galíndez C. 1999. Drastic fitness loss in human immunodeficiency virus type 1 upon serial bottleneck events. *J. Virol.* 73:2745–51

The Social Lives of Microbes

Stuart A. West,[1] Stephen P. Diggle,[2]
Angus Buckling,[3] Andy Gardner,[1,4]
and Ashleigh S. Griffin[1]

[1] Institute of Evolutionary Biology, University of Edinburgh, Edinburgh EH9 3JT, United Kingdom; email: Stu.West@ed.ac.uk, a.griffin@ed.ac.uk

[2] Institute of Infection, Immunity & Inflammation, Center for Biomolecular Sciences, University of Nottingham, Nottingham NG7 2RD, United Kingdom; email: steve.diggle@nottingham.ac.uk

[3] Department of Zoology, Oxford University, Oxford OX1 3PS, United Kingdom; email: angus.buckling@zoo.ox.ac.uk

[4] St. John's College, Oxford University, Oxford OX1 3JP, United Kingdom; email: andy.gardner@sjc.ox.ac.uk

Annu. Rev. Ecol. Evol. Syst. 2007. 38:53–77

The *Annual Review of Ecology, Evolution, and Systematics* is online at
http://ecolsys.annualreviews.org

This article's doi:
10.1146/annurev.ecolsys.38.091206.095740

Key Words

altruism, cooperation, kin selection, public goods, social evolution

Abstract

Our understanding of the social lives of microbes has been revolutionized over the past 20 years. It used to be assumed that bacteria and other microorganisms lived relatively independent unicellular lives, without the cooperative behaviors that have provoked so much interest in mammals, birds, and insects. However, a rapidly expanding body of research has completely overturned this idea, showing that microbes indulge in a variety of social behaviors involving complex systems of cooperation, communication, and synchronization. Work in this area has already provided some elegant experimental tests of social evolutionary theory, demonstrating the importance of factors such as relatedness, kin discrimination, competition between relatives, and enforcement of cooperation. Our aim here is to review these social behaviors, emphasizing the unique opportunities they offer for testing existing evolutionary theory as well as highlighting the novel theoretical problems that they pose.

1. INTRODUCTION

Microorganisms exhibit a stunning array of social behaviors (Crespi 2001). Individuals communicate and cooperate to perform activities such as dispersal, foraging, construction of biofilms, reproduction, chemical warfare, and signaling. Microbiologists are making amazing advances in our understanding of these behaviors from a mechanistic perspective, examining the molecular mechanisms involved and the underlying genetic regulation (Kolter & Greenberg 2006, Lazdunski et al. 2004, Parsek & Greenberg 2005, Webb et al. 2003, Williams et al. 2007).

Our aim in this review is to highlight the numerous opportunities microbes offer for evolutionary biologists and ecologists interested in social evolution. First, there is a huge amount of microbe biology requiring an evolutionary explanation. Cooperation and communication appear to be extremely important to microbes; for example, 6–10% of all genes in the bacterium *Pseudomonas aeruginosa* are controlled by cell-cell signaling systems (Schuster et al. 2003). Consequently, if explaining cooperation is one of the greatest problems for evolutionary biology, then explaining cooperation in microbes is one of the key aspects of this problem. Second, social evolution theory was largely developed to explain known behaviors in animals such as insects, mammals, and birds. The huge variety of social behaviors discovered in microbes offers a unique opportunity to test how generally that theory can be applied to other taxa, which did not play a major role in the development of that theory. Furthermore, microbes provide most of the diversity of life, and so previous focus on animals may have given a misleading impression of the importance of different factors for social evolution.

Third, microbe systems are uniquely amenable to experimental study: They have short generation times for selection experiments, and genetic mutants that do not cooperate (cheats) can be created relatively easily. Fourth, the biology of microbes means they offer some novel problems. For example, genes for social behaviors can be transferred horizontally between different bacterial lineages, by mobile genetic elements (Smith 2001). Fifth, the social lives of microbes have a significant impact on human life; cooperative behaviors play key roles in the ability of microbes to (*a*) infect and harm humans, livestock, and agricultural plants (André & Godelle 2005, Williams et al. 2000) and (*b*) provide beneficial services such as breaking down sewage (Valle et al. 2004) or symbiotically aiding plant growth (Kiers et al. 2002).

In the first part of this review, we provide a brief summary of social evolution theory relevant to microbes (Section 2). In the main parts of the paper, we give a tour of microbial social behaviors (Sections 3 and 4). To highlight the general issues, we discuss two categories of social behavior in some detail: the production of public goods and fruiting bodies (Section 3). In the final sections, we consider two topics that arise repeatedly: the implications for the evolution of parasite virulence (Section 5) and the application of the kin selection coefficient of relatedness to microorganisms (Section 6). Although this paper is aimed primarily at evolutionary biologists and ecologists, interdisciplinary research is key to this field, with microbiologists discovering behaviors that need evolutionary explanations and the evolutionary and ecological approaches providing problems for which there must be a mechanistic answer (West et al. 2006).

2. SOCIAL EVOLUTION THEORY

Theoretical explanations for any behavior can be broadly classified into two categories: direct and indirect fitness benefits (Grafen 1984, Lehmann & Keller 2006, Sachs et al. 2004, West et al. 2006). This follows from Hamilton's (1964) insight that individuals gain inclusive fitness through their impact on the reproduction of related individuals (indirect fitness effects) as well as directly through their impact on their own reproduction (direct fitness effects). Cooperative behaviors that benefit other individuals have posed particular problems for evolutionary biologists because although the cost of a cooperative behavior may be very obvious, the benefits are often obscure. We do not discuss the use and misuse of terms such as cooperation and altruism, as we have recently covered that in detail elsewhere (West et al. 2007).

The first explanation for the evolution of cooperation is that it provides a direct fitness benefit to the individual that performs the behavior, which outweighs the cost of performing the behavior. In this case, cooperation is mutually beneficial (West et al. 2007). One possibility is that individuals have a shared interest in cooperation. For example, in many cooperative breeding species, larger group size may benefit all the members of the group through factors such as greater survival or higher foraging success. In this case, individuals can be selected to help rear offspring that are not their own to increase group size (Kokko et al. 2001). Another possibility is that there is some mechanism for enforcing cooperation, by rewarding cooperators or punishing cheaters (Frank 2003, Trivers 1971). This could happen in a variety of ways, which have been termed punishment, policing, sanctions, reciprocal altruism, indirect (reputation-based) reciprocity, and strong reciprocity.

The second class of explanations for cooperation is that it provides an indirect benefit by improving the fitness of other individuals who carry the cooperative gene (Hamilton 1964). In this case, cooperation is altruistic. The simplest and most common reason for two individuals to share genes in common is for them to be genealogical relatives (kin), and so this is often termed kin selection (Maynard Smith 1964). By helping a close relative reproduce, an individual transmits genes to the next generation, albeit indirectly. Hamilton (1964) pointed out that kin selection could occur via two mechanisms: (*a*) kin discrimination, when cooperation is preferentially directed toward relatives, and (*b*) limited dispersal (population viscosity), which keeps relatives in spatial proximity to one another, allowing cooperation to be directed indiscriminately toward all neighbors (who tend to be relatives). The clonal nature of most microbes raises some complications for the concept of genetic relatedness, which we discuss in Section 6. The other way to obtain an indirect fitness benefit is if cooperation is directed toward nonkin who share the same cooperative gene, that is, genetic relatives who are not necessarily genealogical relatives. This assortment or greenbeard mechanism requires a single gene (or a number of tightly linked genes) that both causes the cooperative behavior and can be recognized by other individuals owing to a distinctive phenotypic marker, such as a green beard (Dawkins 1976, Hamilton 1964).

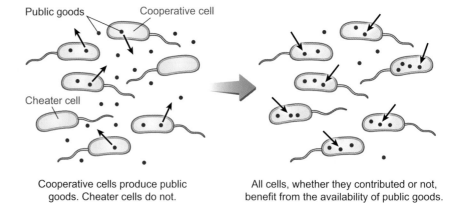

Public goods   Cooperative cell

Cheater cell

Cooperative cells produce public
goods. Cheater cells do not.

All cells, whether they contributed or not,
benefit from the availability of public goods.

Figure 1

The tragedy of the commons with public goods. Cheats who do not pay the cost of producing
public goods can still exploit the benefits of public goods produced by other cells.

3. MICROBIAL CASE STUDIES

3.1. Public Goods

Possibly the most common form of social behavior in microbes is the production
of public goods. Public goods are products manufactured by an individual that can
then be utilized by the individual or its neighbors (**Figure 1**). For example, bacteria
produce numerous factors that are released into the environment beyond the cell
membrane (**Table 1**). Public goods lead to the problem of cooperation because they
are metabolically costly to the individual to produce but provide a benefit to all the
individuals in the local group or population (West et al. 2006). Consequently, although
the production of public goods can have direct fitness benefits to the individuals
producing them, they can also have indirect fitness benefits to the individuals around
them. The specific problem in this case is what stops the spread of cheats that do not
produce public goods (or produce less) but benefit from those produced by others?

How can the social nature of a microbial behavior such as the production of
some postulated potential good be determined (West et al. 2006)? A useful feature
of working with microbes is that mutants who do not perform the behavior (termed
cheats) often already exist, or can be constructed, or will evolve relatively quickly if
they are selectively favored (Dugatkin et al. 2005, Foster et al. 2004, Griffin et al.
2004, Harrison & Buckling 2005, Rainey & Rainey 2003, Velicer et al. 2000). Once
these are acquired, a first step is to examine the relative fitness of the wild type that
performs the putative social behavior and the mutant that does not, when grown
alone (monoculture) or in a mixture. This examines the social costs and benefits of
the behavior. Such experiments require that the wild type and mutant can be distinguished when grown in a mixture. This can be done in a number of ways, including
phenotypic features associated with the behavior (Griffin et al. 2004) or by insertion
of a phenotypic or molecular marker (Foster et al. 2004, Mehdiabadi et al. 2006).

Table 1 Potential public goods

Public good	Role
Siderophores	Iron-scavenging molecules (West & Buckling 2003).
Invertase	An enzyme for digesting sucrose (Greig & Travisano 2004).
β-lactamase	Inactivates and therefore gives resistance to antibiotics (Ciofu et al. 2000)
Biosurfactants	Extracellular matrices for facilitating movement over surfaces, e.g., Rhamnolipid (*P. aeruginosa*) and Serrawettin (*Serratia marcescens*) (Daniels et al. 2004, Velicer & Yu 2003)
Exopolysaccharides such as alginate or adhesive polymers	Providing structure for growth, and the ability to colonize different habitats (Davies & Geesey 1995, Rainey & Rainey 2003)
Host-manipulation factors	Increasing host susceptibility to predation, immune suppression, host castration (Brown 1999)
Shiga toxins	Breaking down host tissue (O'Loughlin & Robins-Browne 2001)
Protein synthesis	Growth (Turner & Chao 1999)
Toxic and lytic secondary metabolites	To kill and degrade prey organisms
Adhesive polymer	Colonization of the air-liquid interface (Rainey & Rainey 2003)
Quorum-sensing molecules	Cell-cell signals (Williams et al. 2007), iron chelation (Diggle et al. 2007, Kaufmann et al. 2005), immune modulators (Pritchard et al. 2003), biosurfactants (Daniels et al. 2006), plant systemic resistance (Schuhegger et al. 2006)
Proteases	Extracellular protein digestion (Hase & Finkelstein 1993)
Extracellular DNA	Structural component of biofilms (Spoering & Gilmore 2006)
Antibiotics	To kill competitors (although better to conceptualize as spiteful rather than public good; see Section 4.6)
Membrane vesicles	Common biofilm component (Schooling & Beveridge 2006), transport of cell-to-cell signals (Mashburn & Whiteley 2005)
Rhamnolipids	Antiprotozoan defense mechanism (Cosson et al. 2002), mediate detachment from biofilms (Boles et al. 2005)
Microbial repellents	Repels competitors (Burgess et al. 2003)
Resources supplied by symbionts to their hosts, such as nitrogen fixation by rhizobia	Aids host growth and, in some cases, avoid enforcement by the host (Kiers et al. 2003, Kiers & van der Heijden 2006, West et al. 2002)

Recent research on siderophores provides an example of how this approach can be used to test the social nature of a potential public goods behavior. Siderophores are iron-scavenging molecules produced by many species of bacteria (Ratledge & Dover 2000, West & Buckling 2003). Iron is a major limiting factor for bacterial growth because most iron in the environment is in the insoluble Fe(III) form, and, in the context of bacterial parasites, is actively withheld by hosts. Recently, Griffin et al. (2004) investigated the social nature of the production of the siderophore pyoverdine in *P. aeruginosa*. Siderophore production is beneficial when iron is limiting, as shown by the fact that the wild type that produces siderophores outcompetes a mutant that does not, when the strains are grown in pure culture. However, siderophore production is also metabolically costly, as demonstrated by the fact that mutants outcompete wild-type strains in an iron-rich environment. Consequently, in mixed populations where both wild-type and mutant bacteria are present, the mutants can gain the benefit of siderophore production without paying the cost, and hence increase in frequency.

The production of public goods such as siderophores could provide both a direct and indirect fitness benefit because they benefit the individual who produces them and their neighbors. This is an example of what has been termed a whole-group trait (Pepper 2000). Consequently, we would expect public goods to be subject to kin selection, with selection for higher levels of public goods production when there is higher relatedness between interacting individuals. An experimental evolution approach with siderophore production in *P. aeruginosa* (Griffin et al. 2004) provided support for the predicted effect of relatedness. Relatedness was manipulated by allowing the bacteria to grow and interact in groups derived from a single clone (relatively high relatedness) or from two clones (relatively low relatedness). The cooperative wild-type strain outcompeted the selfish mutant strain only when cultured under conditions of relatively high relatedness.

A useful feature of microbes is that they have provided excellent opportunities for testing how the relative cost and benefit of cooperation varies with population demographics such as density or frequency and environmental factors such as resource availability. For example, (*a*) at lower population density, cells will be less able to use the public goods produced by other cells and so cheats will do worse (Brown & Johnstone 2001, Greig & Travisano 2004), (*b*) cheats do better when they are at lower frequencies in the population (frequency dependence) because they are then better able to exploit cooperators (Dugatkin et al. 2003, 2005; Harrison et al. 2006; Ross-Gillespie et al. 2007; Velicer et al. 2000), (*c*) individuals would be expected to change their production of public goods depending upon the availability of resources in the environment, which will also be influenced by the behavior of their neighbors, and (*d*) competition between relatives reduces the kin-selected benefit of cooperation and hence favors reduced levels of cooperation (Griffin et al. 2004). The work in these areas stresses the importance of developing specific theory that can be tested with specific systems (Ross-Gillespie et al. 2007).

A relatively unexplored possibility in microbes is that selection would favor mechanisms that allow public goods to be preferentially directed toward closer relatives, analogous to kin discrimination. For example, with the production of public goods, selection would favor the production of highly specific molecules that other lineages (clones) could not utilize. Consistent with this, in *P. aeruginosa* there is variation across strains in the form of pyoverdine produced and in the ability of strains to uptake iron chelated by pyoverdines produced by other strains (Meyer et al. 1997). Furthermore, sequence data suggest that the genes involved in pyoverdine production are under selection for novelty and specificity (diversifying selection) (Smith et al. 2005).

We have focused on siderophores in this section because they provide a specific example where there have been several studies from an evolutionary perspective. However, as illustrated by **Table 1**, microbes produce a fascinating diversity of molecules that may function as public goods. There is enormous potential for future work in this area because, in practically all of these cases, even the most basic social nature of these traits has yet to be examined.

3.2. Fruiting Bodies

One of the most striking forms of cooperation in microbes is the development of fruiting bodies, as found in the cellular slime molds, or social amoebae, in which there is a huge diversity in the way fruiting bodies are formed (Bonner 1967). The most studied slime mold from this perspective is *Dictyostelium discoideum*, a predator of bacteria that is common in the soil. When starving, the usually solitary single-celled amoebae aggregate and form a multicellular slug (pseudoplasmodium, or grex) that can contain 10^4–10^6 cells. This slug migrates to the soil surface, where it transforms into a fruiting body composed of a spherical sorus of spores and a stalk consisting of nonviable stalk cells that hold the sorus aloft. Roughly 20% of the cells die to form the stalk in what appears to be an altruistic act. This division between nonviable stalk cells and spores leads to clear potential for conflict. If the slug is composed of a single lineage, with a mass of genetically identical cells, then kin selection would lead to no conflict over cell fates. However, when multiple lineages occur in a slug, each lineage is selected to make a relatively larger contribution to the spore cells and a relatively smaller contribution to the stalk, at a cost to the other lineages (Strassmann et al. 2000).

In *D. discoideum*, cells from distinct lineages will come together to form slugs and fruiting bodies, suggesting that the conflict exists in natural populations (Fortunato et al. 2003, Strassmann et al. 2000). Evidence for a struggle resulting from this conflict comes from the observation that, when mixed, different lineages show a variable ability to exploit other lineages in the race to provide a higher proportion of the spore cells (Fortunato et al. 2003, Strassmann et al. 2000). Furthermore, this conflict is costly, as shown by the fact that slugs formed by multiple lineages (chimeras with conflict) are not as good at moving as slugs formed by single lineages (where no conflict occurs) (Foster et al. 2002). Consequently, they are less able to migrate to a position where fruiting body formation is possible (Castillo et al. 2005). However, this effect is complicated by the fact that larger slugs are better at moving (Foster et al. 2002). Consequently, under certain conditions, it can be better to be in a larger chimera slug than in a smaller slug consisting of only one lineage (Foster et al. 2002). This suggests that we may expect variation in the allocation of resources to conflict across species that typically occur at different densities, or even facultative adjustment in response to local population density.

Another possible way to avoid the costs of forming chimeras would be if amoebae preferentially formed slugs with relatives (Mehdiabadi et al. 2006). Such kin discrimination has been observed in the species *D. purpureum* (Mehdiabadi et al. 2006). When pairs of clones were placed together in equal proportions at high density, in the absence of food, the slugs that were formed tended to be dominated by one or another of the strains rather than an equal mix. This led to the mean relatedness within slugs being approximately 0.8 rather than the expected of 0.5. A more extreme case has been found in *D. discoideum*, where a greenbeard mechanism preferentially directs cooperation toward individuals who share the same cooperative gene. Individuals who share the *csa* cell adhesion gene adhere to each other in aggregation streams—excluding mutants that do not share the gene—allowing them to cooperatively form fruiting bodies (Queller et al. 2003).

A system of differentiation analogous to the slime molds occurs in soil-dwelling myxobacteria. In *Myxococcus xanthus*, a lack of amino acids triggers aggregation into local groups that exchange intercellular signals and construct spore-bearing fruiting bodies. As in the *Dictyostelium* species, only a fraction of cells can become spores in these fruiting bodies, leading to the same conflicts of interest (Fiegna & Velicer 2006, Fiegna et al. 2006, Kadam & Velicer 2006). Research on *M. xanthus* has provided some elegant experiments on the costs and benefits of making fruiting bodies (Fiegna et al. 2006, Velicer et al. 2000). An important feature of this research is that it has utilized a number of cheater mutants that contribute less to the production of the nonspore parts of the fruiting body. The advantage of replicating the experiments with multiple mutants is that it demonstrates the generality of any results and that those results are not just due to a pleiotropic effect or correlated character. In addition, the mutant strains used here were constructed in two different ways: defined single gene mutants and selection under controlled laboratory conditions (Velicer et al. 2000). Each type of mutant cheater has its advantage. Strains that have had the gene for a cooperative behavior artificially knocked out have well-defined, clear, and large effects, whereas spontaneous mutations from the laboratory or isolates from the field may be more natural and may give a better indication of the kind of variation upon which natural selection could act.

4. OTHER SOCIAL BEHAVIORS

4.1. Resource Use

There can be a trade-off between the rate at which a resource can be used and the efficiency of its use (Kreft 2004, Kreft & Bonhoeffer 2005, MacLean & Gudelj 2006, Pfeiffer et al. 2001). For example, respiration of ATP leads to a higher energy yield than fermentation. Consequently, individuals that use both respiration and fermentation can produce energy at a higher rate but use the resources less efficiently than individuals that employ respiration only (Pfeiffer et al. 2001). Fermentation can therefore provide a direct benefit to the individual at a cost that is shared among the local population. This allows respiro-fermenters to outcompete respirers, leading to the familiar tragedy of the commons problem where the selfish interest of individuals leads to less efficient resource use (Hardin 1968). A solution to this problem, that can be important in microbes, is if interacting individuals are relatives, as this also leads to an indirect fitness cost of less efficient resource use (fermentation) by reducing the resources available for relatives. Consequently, more efficient resource use can be thought of as a cooperative behavior, with more efficient use favored by kin selection.

4.2. Quorum Sensing

Perhaps the paradigm for bacterial cooperation and communication can be seen in the diverse quorum sensing (QS) systems, which occur widely across bacteria (Williams et al. 2007). QS describes the phenomenon whereby the accumulation of signaling molecules in the surrounding environment enables a single cell to assess the number of bacteria (cell density) so that the population as a whole can make a coordinated response. In many situations, a group venture is not worth taking on unless there is a sufficient number of collaborators to make it worthwhile. QS systems have been shown to regulate the expression of genes involved in plasmid transfer, bioluminescence, population mobility, biofilm maturation, and virulence (Williams et al. 2007). Many of the behaviors regulated by QS appear to be cooperative, for example, producing public goods such as exoenzymes, biosurfactants, antibiotics, and exopolysaccharides (see **Table 1**).

Little attention has been given to the evolutionary implications of QS (Brown & Johnstone 2001). Microbiologists frequently assume that QS is readily selected for because it benefits the local group or population as a whole (Henke & Bassler 2004, Shapiro 1998). In contrast, evolutionary theory suggests that cooperative communication is maintained only by selection under fairly restrictive conditions (Maynard Smith & Harper 2003). This raises the question as to whether QS in microbes is truly a cooperative behavior (Diggle et al. 2007, Keller & Surette 2006, Redfield 2002). Are QS molecules true signals? The fact that a compound produced by cell A elicits a response in cell B does not necessarily mean that there is signaling between the cells; it may represent cell B using the molecule as an environmental cue to guide future action, or it may represent cell A coercing cell B into a certain behavior. We suspect that within species, QS signaling between relatives will be favored by kin selection (Brown & Johnstone 2001, Diggle et al. 2007). In contrast, between-species QS (sometimes termed bacterial cross talk) is more problematic to explain and it could be that, in these cases, QS molecules are used by other species as cues or for coercion.

A fundamental first step is to determine the fitness consequences of producing and responding to a signal (Brown & Johnstone 2001). Although there is undoubtedly a metabolic cost for signal production (Keller & Surette 2006), it is likely that the cost of responding is far more expensive metabolically. For example, as mentioned above, 6%–10% of *P. aeruginosa* genes appear to be QS-regulated. Given high costs, QS signaling or response could be potentially exploitable by QS cheats (Diggle et al. 2007, Keller & Surette 2006). In theory, QS cheats could take the form of either (*a*) a signal-negative (mute) strain that does not make the molecule but can respond to it, or (*b*) a signal-blind (deaf) strain that may (or may not) make a signal but, more importantly, does not respond to it. Both types of mutants can be constructed, and signal-blind mutants are often isolated from clinical infections of *P. aeruginosa* (Smith et al. 2006).

Signaling in bacteria has a number of complexities that offer novel problems from an evolutionary perspective. First, the signal can be degraded. *N*-acyl homoserine lactone (AHL) signals are rendered biologically inactive in alkaline environments (Yates et al. 2002). Therefore, in certain environmental niches, signaling may be ineffective or the level and response to QS may vary. AHLs can also be degraded by enzymes produced by bacteria, a process known as quorum quenching (Dong & Zhang 2005, Sio et al. 2006). Can this behavior be considered coercive or spiteful, and are there indirect or direct fitness benefits for the AHL degrader?

Second, the genes required for signal generation (*luxI* homologs) and response (*luxR* homologs) are not always found on the bacterial chromosome. A number of these homologs have been identified on plasmids such as the *Agrobacterium* Ti plasmid (Zhang et al. 1993) and *Rhizobium* symbiotic plasmids (Smith 2001, Wisniewski-Dye & Downie 2002). Although this may just represent an easy way to obtain QS mechanisms, it could also be a mechanism by which signaling is forced onto a cell that does not contain the QS machinery, coercing it into cooperative behavior (Smith 2001). An issue here is that the interests of the bacteria and the plasmids may conflict. Third, a number of other roles have been assigned to QS molecules. QS molecules can function as public goods, for example, iron chelators (Diggle et al. 2007), immunomodulatory compounds (Pritchard et al. 2003), and biosurfactants (Daniels et al. 2006). QS compounds can also be harmful or spiteful; for example, the lantibiotics typified by lactococcal nisin and produced by *Lactococcus lactis* are potent bacteriocides against many bacteria (Dodd et al. 1996, Stein 2005). The consequences of QS signals having multiple functions must be explored theoretically, as it will alter the costs and benefits of signaling as well as how they vary with parameters such as relatedness (Diggle et al. 2007).

Remarkably, given the size of the literature on the subject, it remains to be conclusively demonstrated that QS systems are mechanisms for cell-to-cell communication. Redfield (2002) raised the possibility that QS molecules allow individual cells to determine how rapidly secreted molecules move away from the cell. This diffusion sensing could allow cells to regulate the secretion of public goods to minimize losses due to extracellular diffusion and mixing. A possibility here is that the production of these molecules may have initially evolved for one reason (e.g., diffusion sensing) but are now maintained for another (e.g., QS).

4.3. Biofilms

Traditionally, bacteria have been thought of as free-swimming planktonic organisms. However, most bacterial species are capable of forming structured multicellular communities known as biofilms (Kolter & Greenberg 2006). Biofilms are ubiquitous, being found in such diverse environments as dental plaques, wounds, rock surfaces, and at the bottom of rivers. They have a definite structure, including water channels, which may involve a number of different specialist cells, and they are often enclosed by an exopolysaccharide matrix, which can make them difficult to eradicate. For example, biofilm-growing cells are often significantly more resistant to antibiotics than free-living (planktonic) cells, and the matrix can also help protect bacterial cells against the host immune system during infection. The life cycle of a biofilm can be split up into distinct stages: (*a*) reversible attachment, often mediated by flagella and type IV pili, (*b*) irreversible attachment to a surface, (*c*) formation of microcolonies, (*d*) differentiation into intricate architectures, and (*e*) dispersal (**Figure 2**). Biofilms are of particular interest from an evolutionary perspective because the close proximity of individuals in a biofilm can make cooperation and communication particularly important.

Many forms of cooperation can be involved in the establishment and growth of a biofilm. First, there is the cooperative production of an extracellular matrix (ECM), which surrounds the biofilm and may be important in maintaining structure. There

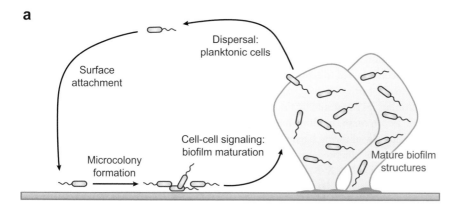

a

Dispersal:
planktonic cells

Surface
attachment

Cell-cell signaling:
biofilm maturation

Microcolony
formation

Mature biofilm
structures

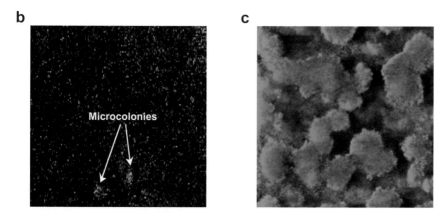

b

Microcolonies

c

Figure 2

Life cycle of a bacterial biofilm. (*a*) Planktonic cells are released from mature biofilms and, via motility mechanisms, settle on a new surface. Cells become irreversibly attached and begin to form microcolonies. Mechanisms such as cell-cell signaling systems lead to the differentiation of mature biofilm structures. Diagram adapted from Kolter & Losick (1998). (*b*) Scanning electron microscopy image of *Pseudomonas aeruginosa* attachment to a stainless steel coupon. The formation of microcolonies can be observed. Image taken with permission from Diggle et al. (2006). (*c*) Scanning confocal microscopy image of mature five-day-old *P. aeruginosa* biofilms grown in flow cell chambers. Image courtesy of S. Crusz.

are many components found in the ECM that may be involved in holding the biofilm structure together, including alginate, the lectin LecA, and carbohydrate polymers (Davies & Geesey 1995, Diggle et al. 2006, Friedman & Kolter 2004, Matsukawa & Greenberg 2004, Nivens et al. 2001). Second, numerous other public goods can be important in biofilms, such as rhamnolipid, a biosurfactant that aids in biofilm detachment (Boles et al. 2005), and microvesicles, which are a component of the ECM and can contain signal molecules and proteases (Schooling & Beveridge 2006). Third, cell death can occur in a subpopulation of cells. Researchers have hypothesized that this is a cooperative behavior to release extracellular DNA that aids in structuring the biofilm (Whitchurch et al. 2002). However, a complication here is that in *P. aeruginosa* cell death involves a genomic prophage (Webb et al. 2003). Consequently, this could be an altruistic behavior by the individual cells, or a selfish manipulation by a prophage. Fourth, dispersal from biofilms may be a social trait, if it has been selected to reduce competition with nondispersing relatives (Hamilton & May 1977). Fifth, QS may play an important role in the development and structuring of biofilms, as suggested by the poor biofilm formation of some QS mutants (Davies et al. 1998). Sixth, biofilms may also be important in protecting against protozoan predation, by allowing bacterial prey to reach a size beyond the maximum that can be handled by the predator, or through group-coordinated chemical warfare against the predator (Matz & Kjelleberg 2005).

When viewed under a confocal microscope, it is immediately obvious that biofilms are not simply random clusters of cells (**Figure 2**). Researchers have suggested that there may be some specialization within biofilms, analogous to how caste development in social insects allows individuals to specialize in certain behaviors. In *P. aeruginosa* biofilms, several phenotypically different cell variants that exhibited different behaviors have been isolated (Boles et al. 2004), in particular, a wrinkly variant that showed faster biofilm development, and greater resistance to stress, and a mini variant that showed greater dispersal from the biofilm. This variation is heritable, but variants can change back to the wild type or each other. Although it is tempting to think that this represents specialization favored by kin selection, there are alternatives. For example, a variant may spread selfishly within a biofilm at the expense of other types, even if this leads to a long-term decrease in the fitness of the biofilm or that lineage, analogous to how a cancer selfishly spreads within an individual, despite the long-term fitness consequences.

Kin selection would clearly be important in biofilms initiated by one or a small number of clonal lineages. However, naturally formed biofilms very rarely contain just one species of bacteria, let alone a single clonal lineage. For example, the colonization of human teeth and the oral mucosa can involve up to 500 species of bacteria (Kolenbrander et al. 2002). Nonetheless, kin selection may still be important in such cases if social interactions take place on a local scale (see also Section 6). For example, if the benefits of producing the materials that structure the biofilm, such as exopolysaccharides, or other public goods, are shared primarily with neighboring cells, then the clonal growth of bacteria means that these benefits can still be shared with close relatives (Xavier & Foster 2007). In this case, it may be useful to think of biofilms as consisting of a number of clonal lineages (groups of lineages), with cooperation primarily within lineages but competition primarily between lineages. Diversity may also promote cooperation in some situations because rarer and more specialized cooperators can be harder to exploit. Experiments on biofilm formation in *P. fluorescens*, where more diverse groups were found to be both more productive and less susceptible to invasion by cheats, have provided support for this (Brockhurst et al. 2006). A final complication is that cooperative interactions can occur between species within biofilms. For example, cooperation between *Streptococcus oralis* and *Actinomyces naeslundii* is important in the early colonization of human teeth, allowing these species to grow where neither can survive alone (Palmer et al. 2001).

4.4. Persisters

In populations of microbes, researchers have discovered that, at any point in time, a small proportion of cells do not grow at the normal rate but exist in a quiescent, nongrowing state (Lewis 2007). These cells are sometimes known as persister cells because they persist in the face of catastrophes such as antibiotic treatment (Cozens et al. 1986). Persister cells are resistant to antibiotics because antibiotics rely on disrupting the translation of the mRNA code to polypeptide chains, and this process does not occur in nongrowing cells. Researchers have suggested that persister cells are responsible for the recalcitrance of microbial populations in biofilms rather than, as is often assumed, the production of a protective ECM (Lewis 2007). Two important aspects of persister biology are that (*a*) the cells appear not to be genetically resistant to treatment by antibiotics—following antibiotic treatment, persisters give rise to new populations that have the same vulnerability to antibiotic treatment; and (*b*) persistence is controlled by a phenotypic switch—all cells have the potential to become persisters and persisters readily revert to normal growth rate (Balaban et al. 2004, Kussell et al. 2005, Lewis 2007).

In the microbiology literature, the general assumption has been that persistence is favored because it provides a benefit at the population level (reviewed by A. Gardner, S.A. West & A.S. Griffin, submitted). In contrast, a recent theoretical analysis suggests that persistence is better understood as a potentially social trait that can have both direct and indirect fitness consequences (A. Gardner, S.A. West & A.S. Griffin, submitted). Persistence provides a direct benefit by producing a phenotype that survives catastrophes, but it is also costly because it reduces the short-term growth rate. However, this reduced growth rate also reduces local competition for resources and hence provides an indirect benefit to relatives competing for the same resources. This means that higher levels of persistence are predicted with a higher relatedness between competing bacteria. Theory also predicts that a higher level of persistence will be favored when there is greater competition for resources. Data from clinical infections support this prediction, where there appears to be higher levels of persistence in populations that have entered the stationary phase owing to resource depletion (A. Gardner, S.A. West & A.S. Griffin, submitted). A novel feature of selection on persistence is that this last prediction and pattern are in a direction opposite to the usual prediction, that more local competition selects against cooperation (West et al. 2002).

4.5. Cell Death

Researchers have suggested that programmed cell death (PCD; also known as apoptosis or autolysis) is adaptive in several situations. PCD clearly provides no direct fitness benefit, and so explanations must rely on indirect benefits for relatives or cells being forced into cell death by others. Several altruistic possibilities have been suggested, such as providing resources that could be used by other cells for growth and survival in yeast (Fabrizio et al. 2006, Gourlay et al. 2006) or the formation of fruiting bodies in *M. xanthus* (Wireman & Dworkin 1977). In biofilms, PCD may also be useful for (*a*) creating channels within biofilms, which are responsible for the transport of nutrients and waste to and from cells deep in the biofilm or through which cells can disperse, and (*b*) releasing extracellular DNA that can be used for structuring biofilms (Allesen-Holm et al. 2006, Webb et al. 2003, Whitchurch et al. 2002). QS may be involved in controlling PCD (D'Argenio et al. 2002).

If PCD is occurring as a form of adaptive suicide, then this is clearly an area where microbes differ from animals, where there is a relative lack of such behaviors. Charlesworth (1978) noted that it is extremely hard for a gene causing suicide to spread because only relatives that do not share the gene would benefit. The possible solution to this problem in microbes is that selection could favor a low probability of PCD among a large population of cells, possibly depending upon individual condition, environmental conditions, or signaling. However, we stress that the adaptive nature of altruistic PCD remains to be demonstrated empirically, and alternative explanations exist. For example, in *P. aeruginosa*, cell death is mediated by a bacteriophage (Webb et al. 2003), and although this may reflect host control (D'Argenio et al. 2002), there is also the possibility that the bacteriophage may favor host death under the low growth conditions found in biofilms, to aid their own transmission.

4.6. Bacteriocins

Bacteriocins are the most abundant of a range of antimicrobial compounds facultatively produced by bacteria and are found in all major bacterial lineages (Riley & Wertz 2002). They are a diverse family of proteins with a variety of antimicrobial killing activities, many of which can be produced by a single bacterium, including enzyme inhibition, nuclease activity, and pore formation in cell membranes (Reeves 1972, Riley & Wertz 2002). Bacteriocin production is spiteful because it is costly to the individual cell that performs it and also costly to the recipient, which is killed (Gardner et al. 2004). The individual cost of bacteriocin production may be the diversion of resources from other cellular functions. However, in many gram-negative bacteria, such as *Escherichia coli*, cell death is required for the release of bacteriocins (Chao & Levin 1981, Riley & Wertz 2002). Individuals from the same lineage (clone mates) are protected from the toxic effects of bacteriocins as a result of genetic linkage between the bacteriocin gene and an immunity gene that encodes a factor that deactivates the bacteriocin (Riley & Wertz 2002). Bacteriocin production can therefore provide an indirect fitness benefit by reducing the level of competition experienced by relatives and hence can also be viewed as indirect altruism. Maximum bacteriocin production should be favored at intermediate levels of relatedness; when relatedness is extremely high, there are few susceptible competitors to attack, and when relatedness is extremely low, there are fewer relatives to enjoy the benefits of relaxed competition (Gardner et al. 2004).

Given the strong selective pressure imposed by bacteriocins, it is unsurprising that bacteriocin resistance readily evolves. As with bacteriocin production, such resistance carries pleiotropic fitness costs. Such costs are critical for coexistence (via negative-frequency-dependent selection) between bacteriocin-producing, sensitive, and resistant strains in spatially structured environments (Czaran & Hoekstra 2003, Czaran et al. 2002, Kerr et al. 2002), as producer beats sensitive, sensitive beats resistant, and resistant beats producer (Kerr et al. 2002). Furthermore, costly resistance is also likely to explain why there is considerable diversity in bacteriocin types maintained within microbial species. It is likely to be too costly to be resistant to all bacteriocins at once, and the probability of resistance will be lowest with respect to rare bacteriocins. As such, rare bacteriocin alleles will increase in frequency, allowing diversity to be maintained. Another interesting complication that can occur with bacteriocins is that they may help disfavor siderophore cheats: A pyoverdin (siderophore) receptor (type II) of *P. aeruginosa* also acts as a receptor for a *P. aeruginosa* bacteriocin (Pyocin S3), and so a cheat that evolved from a different lineage is likely to be susceptible to the toxins produced by the population being exploited (Tümmler & Cornelis 2005).

5. VIRULENCE

An appreciation of the widespread importance of cooperative social behaviors in microbes is leading to a change in how evolutionary biologists think about parasite virulence. In the 1990s, researchers saw a huge explosion of evolutionary theory designed to predict the damage that a parasite should inflict on its host, termed parasite virulence (reviewed by Frank 1996). The underlying idea in this theory was that higher growth rates enhance parasite transmission rates but incur greater mortality for the host and hence reduce the total duration of infection, and this results in a trade-off between the transmission benefit and the host-mortality cost. This is a social problem because the optimal use of a host's resources will depend upon the relatedness between the parasites infecting a host (Frank 1996). If only one lineage infects a host (high relatedness), this selects for a prudent use of host resources. As multiple lineages infect a host (low relatedness), parasites are selected to use host resources before their competitors, and this selects for higher growth rates and hence a higher virulence. Consequently, a tragedy of the commons arises, where higher growth rates provide a benefit to the individual and the cost of higher host mortality is shared over the group, analogous to the evolution of resources discussed in Section 4.1.

However, despite huge amounts of theoretical attention, there is a lack of empirical support for these predictions (Read & Taylor 2001). One possible explanation is that the diverse social lives of microbes can also lead to different predictions (West & Buckling 2003). The above prediction assumes that parasites can regulate their own growth rate, to whatever level they choose. However, an alternative possibility is that parasite growth rates are limited by host measures and that cooperative behaviors play a key role in increasing growth rates within hosts (West & Buckling 2003). Many public goods produced by microbes play important roles in acquiring resources from hosts or evading immune responses (**Table 1**). Indeed, we have found that when microbiologists refer to virulence factors, they are usually discussing a cooperative public good. Consequently, cheats that do not perform cooperative behaviors, such as siderophore production, cause a lower virulence than cooperators (Harrison et al. 2006). This leads to the prediction that growth rates and virulence should be higher when relatedness between parasites is greater (Brown et al. 2002, West & Buckling

2003). This prediction is in the opposite direction from that predicted in the previous paragraph. It is even possible to make more complicated predictions: With spiteful behaviors such as bacteriocin production, we predict the lowest growth rates and virulence at intermediate levels of relatedness because it is here that bacteriocin-mediated killing of bacterial cells is most favored (Gardner et al. 2004). The possible role of bacteriocins has been supported by experiments showing that infections of caterpillars with multiple species of bacteria that can kill each other lead to reduced virulence, compared with single-species infections (Massey et al. 2004).

6. RELATEDNESS

In the above sections we used the term relatedness rather informally. Social biologists familiar with diploid eukaryotes will be familiar with the classic results that relatedness between full sibs is $r = 0.5$ and between half-sibs is $r = 0.25$, in the absence of inbreeding (Hamilton 1964). Formally, relatedness describes the statistical association (regression) between genes in different individuals (Frank 1998, Hamilton 1970). In principle, the concept of relatedness is also clear in microbes. In the simplest case, if N unrelated lineages (clones) mix equally in a social arena, then average relatedness will be $r = 1/N$ (West & Buckling 2003). This comes from the average of individuals being related by $r = 1$ to their clonemates, and by $r = 0$ to individuals in the other lineages (assuming they are chosen randomly from the global population). However, several complications can occur, making the application of relatedness to microbial social behaviors less clear.

First, the relevant relatedness should be measured over the scale at which social interactions occur (West & Buckling 2003, West et al. 2006). This will vary between species and even between traits within species. For example, the relevant scale is (a) the distance over which public goods disperse when considering public goods such as siderophores (Section 3.1), (b) the whole fruiting body when considering cooperative stalk formation in *Dictyostelium* species (Section 3.2), (c) the whole host when considering symbiotic cooperation through partner fidelity feedback or how growth rates may be adjusted in response to their effect on host mortality (virulence), and (d) the scale at which enforcement occurs to maintain cooperation by symbionts (e.g., individual root nodules).

Consequently, when social interactions take place over a limited spatial scale, increased genetic variability within a patch will not necessarily lead to a decrease in relatedness. This is illustrated by considering biofilms, where clonal growth and low dispersal can lead to the biofilm consisting of a number of competing lineages (Xavier & Foster 2007). With respect to public goods such as exopolysaccharides, which provide a benefit over a scale smaller than the whole biofilm, the recipients of the social behavior can be strongly related to the actor even when the biofilm consists of lots of lineages or even multiple species. The scale at which interactions occur will vary in nature and could be manipulated experimentally, through factors such as shaken or unshaken liquid cultures, agar plates with a variable agar concentration, or in vivo. If selection can influence the scale of interaction, for example, by altering the diffusibility of public goods molecules, a number of questions arise, including whether harmful products such as bacteriocins tend to be dispersed over larger distances than helpful products such as siderophores or whether the dispersal distance of different QS signals correlates with the kind of behaviors they trigger.

Second, the rapid generation time of microbes means that mutations can have appreciable effects on relatedness. Consider a population of clones that perform a cooperative behavior favored by kin selection, in which a mutant that does not perform this behavior arises. The relatedness between these mutants and the ancestral strain would be $r = 0$ (for reasons described above). Over time, mutation could therefore lead to a reduction in relatedness, allowing selfish cheaters to spread, and to the breakdown of cooperation. This suggests that the stability of cooperation will be influenced by factors such as mutation rates, patch lifetimes, and dispersal rates (Michod & Roze 2001). West et al. (2006) have previously suggested that mutation in populations with low disperal provides a possible explanation for why cooperative traits such as QS signaling and protease production decline over time in long-term *P. aeruginosa* infections (Lee et al. 2005). If mutation rates are important, this also suggests the possibility for selection on mutation rates of social traits, as shown by the fact that cheat mutants that do not produce siderophores can arise and spread faster in populations with an elevated mutation rate (Harrison & Buckling 2005).

Third, genes for cooperative behaviors can sometimes be transmitted horizontally between different bacterial lineages, by mobile genetic elements such as conjugative plasmids or lysogenic phages (Smith 2001). Smith (2001) suggested that this can be a way of stopping the spread of cheats within a population by reinfecting them with cooperative behavior. This fascinating possibility requires further theoretical and empirical exploration. It would be useful to model the coevolutionary dynamics from the perspective of all the parties involved: donor, recipient, and the mobile genetic element. For example, the extent to which transfer could be counteracted or aided by selection to avoid or uptake such genetic elements is not clear. This emphasizes that, when determining the importance of kin selection, what matters is similarity at the gene controlling the social behavior (Dawkins 1979, Grafen 1985). In animals, we are accustomed to estimating this with pedigrees because here relatedness is (on average) the same throughout most of the genome, whereas in microorganisms we may have to get used to focusing on the actual gene.

A general point here is that the application of social evolution theory to microbes offers some novel problems compared with the traditional social evolution study of organisms such as insects and birds. In particular, although the same social evolution theory should apply to both macroscopic and microscopic organisms, the details can be different, with the important factors varying in very general ways. For example, considering kin selection, (*a*) limited dispersal may be the key mechanism in microorganisms (Griffin et al. 2004, West et al. 2006), whereas kin discrimination is the key mechanism in macroorganisms such as cooperative breeding vertebrates (Griffin & West 2003), (*b*) greenbeard effects may be more important in microorganisms because the required signaling and recognition can take place at the level of individual cells interacting with their neighbors (Queller et al. 2003), (*c*) spite may be more common in microorganisms owing to local competition for resources and extreme differences in relatedness (e.g., clonal versus unrelated, with nothing in between) (Gardner et al. 2004), and (*d*) adaptive suicide could be occurring in microorganisms.

7. CONCLUSIONS

To a large extent, the field of social evolution in microbes is still relatively wide open. There are a huge number of behaviors and traits that are likely to be social but that have not been examined theoretically or empirically from an evolutionary perspective. Much of the early evolutionary work in this area was aimed at the basics, such as showing that cheats or conflicts of interest could occur (e.g., Strassmann et al. 2000, Velicer et al. 2000). Although similar studies are still required for a huge range of traits, there are also possibilities for using microbes to test specific predictions of social evolution theory (Griffin et al. 2004) or for clarifying the complexities that can occur in microbes (Section 6). Microbiologists have given these evolutionary issues little attention because (*a*) their focus is on the underlying mechanisms and genetic control of behaviors, and (*b*) it is frequently assumed that cooperative behaviors can be explained by species or even community-level benefits (i.e., cooperation is not a problem and conflict does not occur). Future progress will be maximized by interdisciplinary exchange between microbiologists, evolutionary biologists, and ecologists.

DISCLOSURE STATEMENT

The authors are not aware of any biases that might be perceived as affecting the objectivity of this review.

ACKNOWLEDGMENTS

The authors thank Bernie Crespi for inviting the review and for his comments, and the Biotechnology and Biological Sciences Research Council, the Natural Environmental Research Council, and the Royal Society for funding.

LITERATURE CITED

Allesen-Holm M, Barken KB, Yang L, Klausen M, Webb JS, et al. 2006. A characterization of DNA release in *Pseudomonas aeruginosa* cultures and biofilms. *Mol. Microbiol.* 59:1114–28

André JB, Godelle B. 2005. Multicellular organization in bacteria as a target for drug therapy? *Ecol. Lett.* 8:800–10

Balaban NQ, Merrin J, Chait R, Kowalik L, Leibler S. 2004. Bacterial persistence as a phenotypic switch. *Science* 305:1622–25

Boles BR, Thoendel M, Singh PK. 2004. Self-generated diversity produces "insurance effects" in biofilm communities. *Proc. Natl. Acad. Sci. USA* 101:16630–35

Boles BR, Thoendel M, Singh PK. 2005. Rhamnolipids mediate detachment of *Pseudomonas aeruginosa* from biofilms. *Mol. Microbiol.* 57:1210–23

Bonner JT. 1967. *The Cellular Slime Molds*. Princeton: Princeton Univ. Press

Brockhurst MA, Hochberg ME, Bell T, Buckling A. 2006. Character displacement promotes cooperation in bacterial biofilms. *Curr. Biol.* 16:2030–34

Brown SP. 1999. Cooperation and conflict in host-manipulating parasites. *Proc. R. Soc. London Ser. B* 266:1899–904

Brown SP, Hochberg ME, Grenfell BT. 2002. Does multiple infection select for raised virulence? *Trends Microbiol.* 10:401–5

Brown SP, Johnstone RA. 2001. Cooperation in the dark: signaling and collective action in quorum-sensing bacteria. *Proc. R. Soc. London Ser. B* 268:961–65

Burgess JG, Boyd KG, Armstrong E, Jiang Z, Yan L, et al. 2003. The development of a marine natural product-based antifouling paint. *Biofouling* 19(Suppl.):197–205

Castillo D, Switz G, Foster KR, Strassman JS, Queller DC. 2005. A cost to chimerism in *Dictyostelium discoideum* on natural substrates. *Evol. Ecol. Res.* 7:263–71

Chao L, Levin BR. 1981. Structured habitats and the evolution of anticompetitor toxins in bacteria. *Proc. Natl. Acad. Sci. USA* 78:6324–28

Charlesworth B. 1978. Some models of evolution of altruistic behavior between siblings. *J. Theor. Biol.* 72:297–319

Ciofu O, Beveridge TJ, Kadurugamuwa J, Walther-Rasmussen J, Høiby N. 2000. Chromosomal β-lactamase is packaged into membrane vesicles and secreted from *Pseudomonas aeruginosa*. *J. Antimicrob. Chemother.* 45:9–13

Cosson P, Zulianello L, Join-Lambert O, Faurisson F, Gebbie L, et al. 2002. *Pseudomonas aeruginosa* virulence analyzed in a *Dictyostelium discoideum* host system. *J. Bacteriol.* 184:3027–33

Cozens RM, Tuomanen E, Tosch W, Zak O, Suter J, Tomasz A. 1986. Evaluation of the bactericidal activity of ß-lactam antibiotics on slowly growing bacteria cultured in the chemostat. *Animicrob. Agents Chemother.* 29:797–802

Crespi BJ. 2001. The evolution of social behavior in microorganisms. *Trends Ecol. Evol.* 16:178–83

Czaran TL, Hoekstra RF. 2003. Killer-sensitive coexistence in metapopulations of micro-organisms. *Proc. R. Soc. London Ser. B Biol. Sci.* 270:1373–78

Czaran TL, Hoekstra RF, Pagie L. 2002. Chemical warfare between microbes promotes biodiversity. *Proc. Natl. Acad. Sci. USA* 99:786–90

D'Argenio DA, Calfee MW, Rainey PB, Pesci EC. 2002. Autolysis and autoaggregation in *Pseudomonas aeruginosa* colony morphology mutants. *J. Bacteriol.* 184:6481–89

Daniels R, Reynaert S, Hoekstra H, Verreth C, Janssens J, et al. 2006. Quorum signal molecules as biosurfactants affecting swarming in *Rhizobium etli*. *Proc. Natl. Acad. Sci. USA* 103:14965–70

Daniels R, Vanderleyden J, Michiels J. 2004. Quorum sensing and swarming migration in bacteria. *FEMS Microbiol. Rev.* 28:261–89

Davies DG, Geesey GG. 1995. Regulation of the alginate biosynthesis gene *algC* in *Pseudomonas aeruginosa* during biofilm development in continuous culture. *Appl. Environ. Microbiol.* 61:860–67

Davies DG, Parsek MR, Pearson JP, Iglewski BH, Costerton JW, Greenberg EP. 1998. The involvement of cell-to-cell signals in the development of a bacterial biofilm. *Science* 280:295–98

Dawkins R. 1976. *The Selfish Gene*. Oxford: Oxford Univ. Press

Dawkins R. 1979. Twelve misunderstandings of kin selection. *Z. Tierpsychol.* 51:184–200

Diggle SP, Gardner A, West SA, Griffin AS. 2007. Evolutionary theory of bacterial quorum sensing: When is a signal not a signal? *Philos. Trans. R. Soc. London Ser. B.* 362:1241–49

Diggle SP, Matthijs S, Wright VJ, Fletcher MP, Chhabra SR, et al. 2007. The *Pseudomonas aeruginosa* 4-quinolone signal molecules HHQ and PQS play multifunctional roles in quorum sensing and iron entrapment. *Chem. Biol.* 14:87–96.

Diggle SP, Stacey RE, Dodd C, Cámara M, Williams P, Winzer K. 2006. The galactophilic lectin, LecA, contributes to biofilm development in *Pseudomonas aeruginosa*. *Environ. Microbiol.* 8:1095–104

Dodd HM, Horn N, Chan WC, Giffard CJ, Bycroft BW, et al. 1996. Molecular analysis of the regulation of nisin immunity. *Microbiology* 142:2385–92

Dong YH, Zhang LH. 2005. Quorum sensing and quorum-quenching enzymes. *J. Microbiol.* 43:101–9

Dugatkin LA, Perlin M, Atlas R. 2003. The evolution of group-beneficial traits in the absence of between-group selection. *J. Theor. Biol.* 220:67–74

Dugatkin LA, Perlin M, Lucas JS, Atlas R. 2005. Group-beneficial traits, frequency-dependent selection and genotypic diversity: an antibiotic resistance paradigm. *Proc. R. Soc. London Ser. B* 272:79–83

Fabrizio P, Battistella L, Vardavas R, Gattazzo C, Liou L, et al. 2006. Superoxide is a mediator of an altruistic aging program in *Saccharomyces cerevisiae*. *J. Cell Biol.* 166:1055–67

Fiegna F, Velicer GJ. 2006. Exploitative and hierarchical antagonism in a cooperative bacteria. *PLoS Biol.* 3:e370

Fiegna F, Yu YTN, Kadam SV, Velicer GJ. 2006. Evolution of an obligate social cheater to a superior cooperator. *Nature* 441:310–14

Fortunato A, Strassmann JE, Queller DC. 2003. A linear dominance hierarchy among clones in chimeras of the social amoeba *Dictyostelium discoideum*. *J. Evol. Biol.* 16:438–45

Fortunato A, Strassmann JE, Santorelli L, Queller DC. 2003. Co-occurence in nature of different clones of the social amoeba, *Dictyostelium discoideum*. *Mol. Ecol.* 12:1031–38

Foster KR, Fortunato A, Strassmann JE, Queller DC. 2002. The costs and benefits of being a chimera. *Proc. R. Soc. London Ser. B* 269:2357–62

Foster KR, Shaulsky G, Strassmann JE, Queller DC, Thompson CRL. 2004. Pleiotropy as a mechanism to stabilize cooperation. *Nature* 431:693–96

Frank SA. 1996. Models of parasite virulence. *Q. Rev. Biol.* 71:37–78

Frank SA. 1998. *Foundations of Social Evolution*. Princeton: Princeton Univ. Press

Frank SA. 2003. Repression of competition and the evolution of cooperation. *Evolution* 57:693–705

Friedman L, Kolter R. 2004. Genes involved in matrix formation in *Pseudomonas aeruginosa* PA14 biofilms. *Mol. Microbiol.* 51:675–90

Gardner A, West SA, Buckling A. 2004. Bacteriocins, spite and virulence. *Proc. R. Soc. London Ser. B* 271:1529–35

Gourlay CW, Du W, Ayscough KR. 2006. Apoptosis in yeast—mechanisms and benefits to a unicellular organism. *Mol. Microbiol.* 62:1515–21

Grafen A. 1984. Natural selection, kin selection and group selection. In *Behavioural Ecology: An Evolutionary Approach*, ed. JR Krebs, NB Davies, pp. 62–84. Oxford, UK: Blackwell

Grafen A. 1985. A geometric view of relatedness. *Oxford Surv. Evol. Biol.* 2:28–89

Greig D, Travisano M. 2004. The prisoner's dilemma and polymorphism in yeast *SUC* genes. *Biol. Lett.* 271:S25–26

Griffin AS, West SA. 2003. Kin discrimination and the benefit of helping in cooperatively breeding vertebrates. *Science* 302:634–36

Griffin AS, West SA, Buckling A. 2004. Cooperation and competition in pathogenic bacteria. *Nature* 430:1024–27

Hamilton WD. 1964. The genetical evolution of social behavior, I & II. *J. Theor. Biol.* 7:1–52

Hamilton WD. 1970. Selfish and spiteful behavior in an evolutionary model. *Nature* 228:1218–20

Hamilton WD, May R. 1977. Dispersal in stable habitats. *Nature* 269:578–81

Hardin G. 1968. The tragedy of the commons. *Science* 162:1243–48

Harrison F, Browning LE, Vos M, Buckling A. 2006. Cooperation and virulence in acute *Pseudomonas aeruginosa* infections. *BMC Biol.* 4:21

Harrison F, Buckling A. 2005. Hypermutability impedes cooperation in pathogenic bacteria. *Curr. Biol.* 15:1968–71

Hase C, Finkelstein RA. 1993. Bacterial extracellular zinc-containing metalloproteases. *Microbiol. Rev.* 57:823–37

Henke JM, Bassler BL. 2004. Bacterial social engagements. *Trends Cell Biol.* 14:648–56

Kadam SV, Velicer GJ. 2006. Variable patterns of density-dependent survival in social bacteria. *Behav. Ecol.* 17:833–38

Kaufmann GF, Sartorio R, Lee SH, Rogers CJ, Meijler MM, et al. 2005. Revisiting quorum sensing: discovery of additional chemical and biological functions for 3-oxo-*N*-acylhomoserine lactones. *Proc. Natl. Acad. Sci. USA* 102:309–14

Keller L, Surette MG. 2006. Communication in bacteria: an ecological and evolutionary perspective. *Nat. Rev. Microbiol.* 4:249–58

Kerr B, Riley MA, Feldman MW, Bohannan BJM. 2002. Local dispersal promotes biodiversity in a real-life game of rock-paper-scissors. *Nature* 418:171–74

Kiers ET, Rousseau RA, West SA, Denison RF. 2003. Host sanctions and the legume-Rhizobium mutualism. *Nature* 425:78–81

Kiers ET, van der Heijden MGA. 2006. Mutualistic stability in the arbuscular mycorrhizal symbiosis: exploring hypotheses of evolutionary cooperation. *Ecology* 87:1627–36

Kiers ET, West SA, Denison RF. 2002. Mediating mutualisms: farm management practices and evolutionary changes in symbiont cooperation. *J. Appl. Ecol.* 39:745–54

Kokko H, Johnstone RA, Clutton-Brock TH. 2001. The evolution of cooperative breeding through group augmentation. *Proc. R. Soc. London Ser. B* 268:187–96

Kolenbrander PE, Andersen RN, Blehert DS, Egland PG, Foster JS, Palmer RJ. 2002. Communication among oral bacteria. *Microbiol. Mol. Biol. Rev.* 66:486–505

Kolter R, Greenberg EP. 2006. The superficial life of microbes. *Nature* 441:300–2

Kolter R, Losick R. 1998. One for all and all for one. *Science* 280:226–27

Kreft JU. 2004. Biofilms promote altruism. *Microbiology* 150:2751–60

Kreft JU, Bonhoeffer S. 2005. The evolution of groups of cooperating bacteria and the growth rate versus yield trade-off. *Microbiology* 151:637–41

Kussell E, Kishony R, Balaban NQ, Leibler S. 2005. Bacterial persistence: a model of survival in changing environments. *Genetics* 169:1807–14

Lazdunski AM, Ventre I, Sturgis JN. 2004. Regulatory circuits and communication in gram-negative bacteria. *Nat. Rev. Microbiol.* 2:581–92

Lee B, Haagensen JAJ, Ciofu O, Andersen JB, Høiby N, Molin S. 2005. Heterogeneity of biofilms formed by nonmucoid *Pseudomonas aeruginosa* isolates from patients with cystic fibrosis. *J. Clin. Microbiol.* 43:5247–55

Lehmann L, Keller L. 2006. The evolution of cooperation and altruism. A general framework and classification of models. *J. Evol. Biol.* 19:1365–78

Lewis K. 2007. Persister cells, dormancy and infectious disease. *Nat. Rev. Microbiol.* 5:48–56

MacLean RC, Gudelj I. 2006. Resource competition and social conflict in experimental populations of yeast. *Nature* 441:498–501

Mashburn LM, Whiteley M. 2005. Membrane vesicles traffic signals and facilitate group activities in a prokaryote. *Nature* 437:422–25

Massey RC, Buckling A, ffrench-Constant R. 2004. Interference competition and parasite virulence. *Proc. R. Soc. London Ser. B* 271:785–88

Matsukawa M, Greenberg EP. 2004. Putative exopolysaccharide synthesis genes influence *Pseudomonas aeruginosa* biofilm development. *J. Bacteriol.* 186:4449–56

Matz C, Kjelleberg S. 2005. Off the hook—how bacteria survive protozoan grazing. *Trends Microbiol.* 13:302–7

Maynard Smith J. 1964. Group selection and kin selection. *Nature* 201:1145–47.

Maynard Smith J, Harper D. 2003. *Animal Signals.* Oxford: Oxford Univ. Press

Mehdiabadi NJ, Jack CN, Farnham TT, Platt TG, Kalla SE, et al. 2006. Kin preference in a social microbe. *Nature* 442:881–82

Meyer JM, Stintzi A, Vos DD, Cornellis P, Tappe R, et al. 1997. Use of siderophores to type pseudomonads: the three *Pseudomonas aeruginosa* pyoverdine systems. *Microbiology* 143:35–43

Michod RE, Roze D. 2001. Coopration and conflict in the evolution of multicellularity. *Heredity* 86:1–7

Nivens DE, Ohman DE, Williams J, Franklin MJ. 2001. Role of alginate and its O acetylation in formation of *Pseudomonas aeruginosa* microcolonies and biofilms. *J. Bacteriol.* 183:1047–57

O'Loughlin EV, Robins-Browne RM. 2001. Effect of Shiga toxin and Shiga-like toxins on eukaryotic cells. *Microbes Infect.* 3:493–507

Palmer RJJ, Kazmerzak K, Hansen MC, Kolenbrander PE. 2001. Mutualism versus independence: strategies of mixed-species oral biofilms in vitro using saliva as the sole nutrient source. *Infect. Immun.* 69:5794–804

Parsek MR, Greenberg EP. 2005. Sociomicrobiology: the connections between quorum sensing and biofilms. *Trends Microbiol.* 13:27–33

Pepper JW. 2000. Relatedness in trait group models of social evolution. *J. Theor. Biol.* 206:355–68

Pfeiffer T, Schuster S, Bonhoeffer S. 2001. Cooperation and competition in the evolution of ATP-producing pathways. *Science* 292:504–7

Pritchard D, Hooi DSW, Watson E, Chow S, Telford G, et al. 2003. Bacterial quorum sensing signaling molecules as immune modulators. In *Bacterial Evasion of Host Immune Responses*, ed. B Henderson, PCF Oyson, pp. 201–22. Cambridge: Cambridge Univ. Press

Queller DC, Ponte E, Bozzaro S, Strassmann JE. 2003. Single-gene greenbeard effects in the social amoeba *Dictostelium discoideum. Science* 299:105–6

Rainey PB, Rainey K. 2003. Evolution of cooperation and conflict in experimental bacterial populations. *Nature* 425:72–74

Ratledge C, Dover LG. 2000. Iron metabolism in pathogenic bacteria. *Annu. Rev. Microbiol.* 54:881–941

Read AF, Taylor LH. 2001. The ecology of genetically diverse infections. *Science* 292:1099–102

Redfield RJ. 2002. Is quorum sensing a side effect of diffusion sensing? *Trends Microbiol.* 10:365–70

Reeves P. 1972. *The Bacteriocins.* New York: Springer-Verlag

Riley MA, Wertz JE. 2002. Bacteriocins: evolution, ecology and application. *Annu. Rev. Microbiol.* 56:117–37

Ross-Gillespie A, Gardner A, West SA, Griffin AS. 2007. Frequency dependence and cooperation: theory and a test with bacteria. *Am. Nat.* 170:331–42

Sachs JL, Mueller UG, Wilcox TP, Bull JJ. 2004. The evolution of cooperation. *Q. Rev. Biol.* 79:135–60

Schooling SR, Beveridge TJ. 2006. Membrane vesicles: an overlooked component of the matrices of biofilms. *J. Bacteriol.* 188:5945–57

Schuhegger R, Ihring A, Gantner S, Bahnweg G, Knappe C, et al. 2006. Induction of systemic resistance in tomato by *N*-acyl-L-homoserine lactone-producing rhizosphere bacteria. *Plant Cell Environ.* 29:909–18

Schuster M, Lostroh CP, Ogi T, Greenberg EP. 2003. Identification, timing and signal specificity of *Pseudomonas aeruginosa* quorum-controlled genes: a transcriptome analysis. *J. Bacteriol.* 185:2066–79

Shapiro JA. 1998. Thinking about bacterial populations as multicellular organisms. *Annu. Rev. Microbiol.* 52:81–104

Sio CF, Otten LG, Cool RH, Diggle SP, Braun PG, et al. 2006. Quorum quenching by an *N*-acyl-homoserine lactone acylase from *Pseudomonas aeruginosa* PAO1. *Infect. Immun.* 74:1673–82

Smith EE, Buckley DG, Wu Z, Saenphimmachak C, Hoffman LR, et al. 2006. Genetic adaptation by *Pseudomonas aeruginosa* to the airways of cyctic fibrosis patients. *Proc. Natl. Acad. Sci. USA* 103:8487–92

Smith EE, Sims EH, Spencer DH, Kaul R, Olson MV. 2005. Evidence for diversifying selection at the pyoverdine locus of *Pseudomonas aeruginosa. J. Bacteriol.* 187:2138–47

Smith J. 2001. The social evolution of bacterial pathogenesis. *Proc. R. Soc. London Ser. B* 268:61–69

Spoering AL, Gilmore MS. 2006. Quorum sensing and DNA release in bacterial biofilms. *Curr. Opin. Microbiol.* 9:133–37

Stein T. 2005. *Bacillus subtilis* antibiotics: structures, syntheses and specific functions. *Mol. Microbiol.* 56:845–57

Strassmann JE, Zhu Y, Queller DC. 2000. Altruism and social cheating in the social amoeba *Dictyostelium discoideum*. *Nature* 408:965–67

Trivers RL. 1971. The evolution of reciprocal altruism. *Q. Rev. Biol.* 46:35–57

Turner PE, Chao L. 1999. Prisoner's dilemma in an RNA virus. *Nature* 398:441–43

Tümmler B, Cornelis P. 2005. Pyoverdine receptor: a case of positive Darwinian selection in *Pseudomonas aeruginosa*. *J. Bacteriol.* 187:3289–92

Valle A, Bailey MJ, Whiteley AS, Manefield M. 2004. N-acyl-L-homoserine lactones (AHLs) affect microbial community composition and function in activated sludge. *Environ. Microbiol.* 6:424–33

Velicer GJ, Kroos L, Lenski RE. 2000. Developmental cheating in the social bacterium *Myxococcus xanthus*. *Nature* 404:598–601

Velicer GJ, Yu YN. 2003. Evolution of novel cooperative swarming in the bacterium *Myxococcus xanthus*. *Nature* 425:75–78

Webb JS, Givskov M, Kjelleberg S. 2003. Bacterial biofilms: prokaryotic adventures in multicellularity. *Curr. Opin. Microbiol.* 6:578–85

Webb JS, Thompson LS, James S, Charlton T, Tolker-Nielsen T, et al. 2003. Cell death in *Pseudomonas aeruginosa* biofilm development. *J. Bacteriol.* 185:4582–92

West SA, Buckling A. 2003. Cooperation, virulence and siderophore production in bacterial parasites. *Proc. R. Soc. London Ser. B* 270:37–44

West SA, Griffin AS, Gardner A. 2007. Social semantics: altruism, cooperation, mutualism, strong reciprocity and group selection. *J. Evol. Biol.* 20:415–32

West SA, Griffin AS, Gardner A, Diggle SP. 2006. Social evolution theory for microbes. *Nat. Rev. Microbiol.* 4:597–607

West SA, Kiers ET, Simms EL, Denison RF. 2002. Sanctions and mutualism stability: Why do rhizobia fix nitrogen? *Proc. R. Soc. London Ser. B* 269:685–94

West SA, Pen I, Griffin AS. 2002. Cooperation and competition between relatives. *Science* 296:72–75

Whitchurch CB, Tolker-Nielsen T, Ragas PC, Mattick JS. 2002. Extracellular DNA required for bacterial biofilm formation. *Science* 295:1487

Williams P, Cámara M, Hardman A, Swift S, Milton D, et al. 2000. Quorum sensing and the population-dependent control of virulence. *Philos. Trans. R. Soc. London Ser. B* 355:667–80

Williams P, Winzer K, Chan W, Cámara M. 2007. Look who's talking: communication and quorum sensing in the bacterial world. *Philos. Trans. R. Soc. London Ser. B.* 362(1483):1119–34

Wireman JW, Dworkin M. 1977. Developmentally induced autolysis during fruiting body formation by *Myxococcus xanthus*. *J. Bacteriol.* 129:796–802

Wisniewski-Dye F, Downie JA. 2002. Quorum-sensing in Rhizobium. *Antonie Van Leeuwenhoek* 81:397–407

Xavier JB, Foster KR. 2007. Cooperation and conflict in microbial biofilms. *Proc. Natl. Acad. Sci. USA* 104:876–81

Yates EA, Philipp B, Buckley C, Atkinson S, Chhabra SR, et al. 2002. *N*-acylhomoserine lactones undergo lactonolysis in a pH-, temperature-, and acyl chain length-dependent manner during growth of *Yersinia pseudotuberculosis* and *Pseudomonas aeruginosa*. *Infect. Immun.* 70:5635–46

Zhang LH, Murphy PJ, Kerr A, Tate ME. 1993. *Agrobacterium* conjugation and gene-regulation by *N*-acyl-L-homoserine lactones. *Nature* 362:446–48

Sexual Selection and Speciation

Michael G. Ritchie

School of Biology, University of St. Andrews, Fife KY16 9TS, Scotland;
email: mgr@st-and.ac.uk

Annu. Rev. Ecol. Evol. Syst. 2007. 38:79–102

First published online as a Review in Advance on
June 28, 2007

The *Annual Review of Ecology, Evolution, and
Systematics* is online at
http://ecolsys.annualreviews.org

This article's doi:
10.1146/annurev.ecolsys.38.091206.095733

Key Words

reproductive isolation, sexual conflict, comparative method,
experimental evolution, genomics

Abstract

Sexual selection has a reputation as a major cause of speciation, one
of the most potent forces driving reproductive isolation. This reputation arises from observations that species differ most in traits
involved with mating success and from successful models of sexual
selection–driven speciation. But how well proven is the case? Models
confirm that the process can occur, but is strongest in conjunction
with ecological or niche specialization. Some models also show that
strong sexual selection can act against speciation. Studies using the
comparative method are equivocal and often inconclusive, but some
phylogeographic studies are more convincing. Experimental evolution and genetic or genomic analyses are in their infancy, but look
particularly promising for resolving the importance of sexual selection. The case for sexual selection is not as strongly supported as,
for example, allopatric speciation. Sexual selection probably contributes most effectively alongside ecological selection or selection
for species recognition than as a solitary process.

INTRODUCTION

Sexual selection: the component of natural selection arising owing to variation in mating or fertilization success

The study of speciation has been dominated by debates about the geographic context in which speciation occurs, for example, Mayr's (1942) "Modes of Speciation," and by analyses of the genetics of reproductive isolation between species. The majority of the latter involves studies of hybrid sterility or inviability, inspired by Haldane's rule (1922). Since Dobzhansky (1937) categorized the causes of reproductive isolation, it has been acknowledged that the most common cause of reproductive isolation in sexual animals are obstacles to fertilization between species. Postinsemination but prezygotic effects on reproductive isolation are important, but it is likely that sexual isolation—premating isolation due to courtship traits and associated preferences—is the most common cause of reproductive isolation in animal species.

A probable resolution to the debate over modes of speciation is that allopatric speciation predominates. However, sympatric speciation is possible, although rare. If we were to ask what the most common processes generating reproductive isolation are—genetic drift, ecological adaptation inadvertently or pleiotropically causing isolation, direct selection for isolation, or sexual selection—there would probably be less consensus. Recently, sexual selection has attained widespread recognition as an engine of speciation, perhaps the most important of the forces that generates new species. But how does sexual selection influence reproductive isolation? By acting only directly on behavior, or also indirectly? Does it act in conjunction with ecological divergence? Can sexual selection inhibit speciation? What are the sources of evidence that implicate sexual selection, and are they unambiguous? Is the case as well proven as allopatric speciation?

HISTORY

Darwin's other great book, published 12 years after *The Origin of Species*, introduced the concept of sexual selection as a major source of selection on males and females (Darwin 1871). This had even less to say about the process of speciation than *The Origin of Species*, although the idea that this form of selection would influence speciation is clearly implicit (for example, Darwin regularly points out that sexually selected male traits are species or racially specific). Explicit discussion of the role of sexual selection in speciation is much more recent. During the modern synthesis, the processes of mate choice and species recognition were usually seen as fundamentally different, whereas most modern treatments (with important exceptions) recognize that these are part of a continuum: Females may discriminate against some males because they are of low quality or less stimulating, but when that process leads to discrimination between geographic races then it contributes to species recognition (Ryan & Rand 1993). Species recognition predominated in early discussions of sexual behaviors, with the concomitant assumption that change in species recognition systems arose owing to direct selection to avoid deleterious hybridization (reinforcement) or to avoid confusion with other species. This raised the problem of how speciation could be completed in allopatry, and some authors argued that the process was finalized only following hybridization or subsequent interspecific encounters (Dobzhansky 1940).

Others solved this by arguing that the fertilization system was finely tuned to environmental conditions and diverged in response to altered environments encountered in allopatry (Paterson 1985). Environmental adaptation of signals and preferences is a potent source of selection on sexual communication and can indirectly cause sexual isolation. Sexual selection will contribute to this process, but in a less direct manner than selection from intra- and intersexual selection.

As recently as the 1980s, researchers began to emphasize the continuity of mate choice within populations, between population divergence and speciation. This led to the natural conclusion that sexual selection within populations could lead indirectly to sexual isolation between populations, without requiring a link to the environment or other species. Fisher's (1930) verbal model of his runaway process of coevolutionary divergence between male traits and female preferences inspired several more formal models (Fisher's own account of the runaway process is explicitly framed as an intraspecific process, although calling hybridization "the grossest blunder in sexual preference" suggests a continuity of process). Initially, the most successful and influential models were those of Lande (1981, 1982), which showed that so long as there is genetic variation for a male trait and a female preference for that trait, assortative mating would generate a positive genetic covariance between the two. Usually the system remains at an equilibrium, where viability selection counters indeterminate changes in traits and preferences. However, if the disequilibrium is particularly strong, the system has the potential to become unstable and evolve away from the equilibrium. Clearly, if trait and preference distributions become nonoverlapping between populations, then sexual isolation is expected to follow. Lande's models were explicitly interpreted as models of speciation.

West-Eberhard (1983) also presented a highly influential review, which argued that social evolution, including both inter- and intrasexual selection, could cause speciation [Coyne & Orr (2004) credit this with being the first review "to emphasize what now seems an obvious link between sexual selection and speciation"]. While other authors, notably Paterson (1985), criticized the typological concepts of species recognition inherent in the new synthesis, West-Eberhard's (1983) paper contained the most cogent discussions of how social competition for mates could cause speciation in allopatry without any recourse to interspecific interactions. Analysis of the frequency of papers published on sexual selection and speciation shows a clear and significant upturn as recently as the 1990s, which probably reflects the natural lag time from the seminal West-Eberhard and Lande papers.

Other stimulants to this upturn of interest include many detailed comparisons of sister species that emphasize divergence in relevant behaviors such as male morphology, courtship song or genitalia, the rise of the comparative method in evolutionary studies, and recent experimental and genomic analyses. Studies of sexual selection within behavioral ecology have increased exponentially through this period, and there have been several very important changes in basic paradigms. The covariance between traits and preferences predicted in the Lande models has sometimes been found within populations. However, expectations for between-population covariances (more important for speciation) are more complex. Alternative sources of selection on traits or preferences can act against covariance among populations, and

Coevolutionary divergence: traits (or species) that mutually influence each other's evolution, e.g., male mating signals and female mating preferences

Assortative mating: nonrandom mating, usually due to similar individuals mating more often than expected, which can contribute to sexual selection

Comparative method: analyses of the evolution of traits upon an explicit phylogenetic hypothesis of species relationships, e.g., whether a trait influences speciation

noncoevolutionary models predict that traits and signals may be substantially out of step during evolutionary change. Male traits may evolve to exploit arbitrary or biased female preferences that have evolved in another context (Ryan 1990), for example, feeding behavior (Macías Garcia & Ramirez 2005). Sometimes this can lead to females preferring attributes of heterospecific males (Ryan & Wagner 1987). There has also been an upsurge in the importance of sexual conflict underlying the coevolution between traits and preferences, which originates with Parker (1979), with males evolving traits to overcome female reluctance to mate and female counteradaptations to these male exploitatory traits. Predictions of sexual conflict for coevolution between populations are not so clear. Although we may expect a broad correspondence between traits and preferences, it is possible for males and females to be out of step while one sex (probably temporarily) has the upper hand in the antagonistic coevolution (Arnqvist & Rowe 2002). This move away from an expectation of simple stepwise coevolution perhaps calls into question one of the underlying principles behind sexual selection as a generative force in speciation. Here I review the theoretical background and the main sources of empirical evidence presented to support the idea of speciation by sexual selection (this is a representative rather than comprehensive set of examples). Throughout, emphasis is placed on what constitutes decisive rather than supporting evidence, as much of the evidence in this area is rather indirect.

THEORETICAL STUDIES OF SEXUAL SELECTION AND SPECIATION

Coevolution

Lande's quantitative genetic models (1981, 1982) [and Kirkpatrick's (1982) similar few-loci models] were of what may be called pure sexual selection in that there was only viability selection on the male trait and coevolution on the mate recognition system. Major changes since have included the incorporation of other forms of selection, emphasis on the possibility of sexual selection facilitating sympatric speciation, and sexual conflict. Simulation studies confirmed that the Fisher-Lande process was possible (Wu 1985), at least in large populations (Nichols & Butlin 1989). Initial models of the effects of direct selection on preferences (direct or indirect costs of being choosy, for example) suggested that runaway coevolution was less likely (Bulmer 1989, Kirkpatrick & Ryan 1991). More recent studies have clarified the restrictive conditions under which runaway coevolution may occur (Hall et al. 2000). An intriguing additional component to coevolutionary models of trait-preference evolution is the incorporation of learning into the dynamics of population divergence. Noting that songbirds are a particularly speciose group, Lachlan & Servedio (2004) modeled coevolution in which males were predisposed to learn songs. Learning favors common traits, leading to faster divergence between populations via drift. Learning therefore leads to more rapid allopatric speciation, although if female preferences are costly (or trait and preference unlinked), the system behaves more like a conventional Fisher-Lande system, or worse. However, most empirical studies of song dialects in birds find weak or no relationships between song and genetic variation (e.g., Nicholls et al. 2006, MacDougall-Shackleton & MacDougall-Shackleton 2001, Wright & Wilkinson 2001). Condition dependence of male traits can lead to accelerated and extended evolution of male signals (Lorch et al. 2003). It seems likely that this would also accelerate allopatric speciation, but this has not been modeled explicitly.

Sexual Selection Against Hybrids: Parapatric and Sympatric Speciation

The issue of defining species recognition and sexual selection as different processes is critical when discussing speciation by reinforcement (selection against deleterious hybridization) or the evolution of behaviors to avoid signal confusion or mating competition from heterospecifics. In these cases, behavior is evolving to be species specific; therefore, species recognition is clearly a valid description of the function of the behavior. However, the underlying process is selection on variation in mating success or, in the case of reinforcement, a form of sexual selection of good genes or genetic complementarity. Reinforcement has been extensively reviewed elsewhere (e.g., Butlin 1989, Servedio 2004), so this section is restricted to situations where sexual selection is explicitly incorporated into interspecific interactions.

Speciation by reinforcement can occur when sexual selection is the main source of selection against hybrids. Hybrids often have aberrant behaviors, in addition to low viability or gametic sterility, and there are cases where postmating isolation is entirely due to behavioral abnormalities. Kawata & Yoshimura (2000) modeled hybridization in which the interacting populations differed in trait and preference distributions in the absence of postmating isolation. Selection against hybrids was frequency-dependent sexual selection on males. Nonoverlapping distributions evolved under a range of choice models. The probability of speciation was, not surprisingly, related to trait overlap between species, but even large overlap could result in speciation. In real populations, drift may counter the frequency-dependent selection favoring extinction of one form, or a hybrid swarm may form. Servedio (2004) attempted to examine the interaction between multiple sources of selection that contribute to reinforcement, including selection against hybrids, assortative mating, and local adaptation of mating signals. The importance of sexual selection is distinguishable from indirect viability selection in these models. Sexual selection alone can contribute to isolation, but is much more powerful when the traits involved are under environmental selection (or there are also incompatibilities). This makes the important point that sexual selection on locally adapted signals may be an important component of many systems usually interpreted as reinforcement or reproductive character displacement (see also Liou & Price 1994).

Sympatric speciation has been extensively modeled because of its controversial nature. Early models implied that gene flow and recombination restricted the buildup of assortative mating between sympatric morphs (e.g., Felsenstein 1981). More recent models can be divided into those that incorporate sexual selection into disruptive selection by ecological niche divergence or those that examine if sexual selection alone can cause sympatric speciation. The former are much more convincing, although there is debate about whether speciation due to enhanced assortative mating constitutes speciation by sexual selection. For example, Kondrashov & Kondrashov (1999) showed that disruptive selection on traits involved in ecological adaptation can drive bimodality within a population. If a behavioral trait became involved, then it was effectively recruited to facilitate assortative mating between the morphs. Even if female preference for the trait was influenced by an independent set of loci, assortative mating deterministically built up isolation between the morphs under a broad range of conditions. Dieckmann & Doebeli (1999) reached similar conclusions with an adaptive dynamics model of resource competition driving sympatric divergence. Even if an indicator mating character was independent of the resource specialization, sympatric morphs could appear with alternative characters in disequilibrium with the specialization (see also Drossel & McKane 2000). However, Kondrashov & Kondrashov (1999) were insistent that such models, with sexual traits facilitating assortative mating, are not examples of speciation by sexual selection but of natural selection driving adaptation (perhaps this is another case in which behavior can be thought of as a species-recognition trait rather than one evolving primarily under sexual selection). This is potentially a critically important point in assessing the importance of sexual selection in speciation. If a trait that contributes to assortative mating is intimately associated with niche specialization (beak size in finches, body size in sticklebacks), then the trait may have facilitated speciation by ecological selection rather than sexual selection. However, distinguishing assortative-mating-facilitating adaptation from, say, sexual selection for good genes may seem a rather subtle and empirically fraught proposition (Schwartz & Hendry 2006), and other authors (e.g., Kirkpatrick & Ravigne 2002) do not make this distinction (it seems inevitable that frequency-dependent selection against individuals struggling to mate assortatively will generate sexual selection).

Models of sympatric speciation by assortative mating independent of ecology have been proposed, but are less convincing. Turner & Burrows (1995) modeled a situation where extremes of a unidimensional male trait were opposed by viability selection (visually driven predation, for example). A mutation in female preference that altered the direction of preference could spread until the alternative forms became fixed. This was inspired by the species flocks of African cichlid fishes, where closely related species differ by color and sometimes ecology. Incorporation of the peculiarities of cichlid sex determination into such models also facilitates sympatric speciation by assortative mating (Lande et al. 2001). These studies show that given the (exacting) conditions of the models, sympatric speciation can occur by sexual selection. However, Kondrashov & Kondrashov (1999) believe that their ecological-selection model with the secondary recruitment of behavior is "more economical and explains all the evidence" (see also Arnegard & Kondrashov 2004).

Higashi et al. (1999) produced a model of sympatric divergence in traits and preferences. In a large population starting with continuous normal distributions of male traits and preferences, assortative mating generates three possible outcomes: The population remains stable, the trait shifts, or trait and preference diverge to produce two species. They describe the latter as like two bouts of Fisher's runaway process for higher and lower values of the trait. However, most models have implied that pure sexual selection is unlikely to promote sympatric speciation. Sexual selection against rare forms will oppose sympatric speciation (Kirkpatrick & Nuismer 2004), and the circumstances favoring it are rare (see also Kondrashov & Shpak 1998). Numerous factors act against sympatric speciation by sexual selection alone, including polygenic determination of the traits involved and frequency-dependent selection that eliminates the rarer forms (requiring symmetrical distributions, disruptive selection, or some strong counteradvantage to favor the process) (Arnegard & Kondrashov 2004, Gourbiere 2004, van Doorn et al. 2004).

Sexual Conflict and Speciation

Sexual conflict arises when the sexes differ in the optimal outcome of an interaction and underlies most processes in sexual selection, from copulation to polyspermy and female postmating fecundity (Parker 1979). The first model to explicitly address the influence of sexual conflict on speciation concerned conflict over mating at a parapatric secondary contact (Parker & Partridge 1998, 1999). This proposed that the effect of conflict on speciation depended on which sex gained the upper hand in determining the outcome: If female preference predominated a mating system, speciation was more likely, but if male competition overcame female preference, then speciation would be less likely. This is one of a few models (see also Kirkpatrick & Nuismer 2004, Gavrilets & Waxman 2002) to argue that strong sexual selection can sometimes inhibit speciation, a logical conclusion that is underappreciated (some comparative studies employ one-tailed tests).

More recently, Gavrilets and colleagues (Gavrilets et al. 2001) modeled the influence of sexual conflict on speciation. Preference functions evolve to resist male traits and potentially give rise to isolation, although other possibilities include asymmetrical isolation or extinction. Another model illustrated the potential for allopatric speciation (Gavrilets 2000). The preference function determined the number of compatible males available to females, and conflict led to perpetual changes in traits and preferences, in a manner reminiscent of Fisher's runaway process, but with females perpetually evolving to decrease the mating rate while males evolved to increase it. Holland & Rice (1998) term such antagonistic sexual coevolution a chase-away process. This process was intimately related to population size, with larger populations facilitating conflict and consequently divergence rate. The dynamics of conflict coevolution in sympatry are particularly interesting (Gavrilets & Waxman 2002). If females evolve two differing resistance strategies, males can chase them and can themselves separate into two distinct assortative mating types. Alternatively, males can become immobilized in a trough of low mating success between them, like Buridan's ass (trapped by being equidistant between two equally attractive sites, like a Scotsman between a chip shop and a pub). The message of these models is that if female preference evolves as resistance against males, a strong evolutionary dynamic can be set up that can facilitate sexual isolation between populations, although one would imagine that conflict would lead to unpredictable or asymmetric isolation more readily than Fisher-Lande coevolution.

Looking across these models, the most obvious way in which sexual selection could accelerate speciation (independent of direct selection for speciation by reinforcement) is via increased coevolution of male traits and female preferences in allopatric populations or if traits involved in mate recognition were under direct environmental selection. Behavioral traits that facilitate assortative mating can act alongside ecological selection in parapatry or sympatry (although in this case there is debate about whether sexual selection is really the driving force or simply facilitates ecological divergence). Sympatric speciation by sexual selection alone is not impossible, but remains controversial (whether sexual selection between sperm- and egg-recognition proteins is a special case is an intriguing possibility; van Doorn et al. 2001). Sexual conflict can drive speciation in a manner similar to Fisher's runaway process, but can also lead to asymmetrical preferences. Models of sympatric speciation by sexual conflict seem more convincing than those relying on assortative mating or coevolution alone.

EMPIRICAL WORK

Comparative Studies

The most consistently quoted source of evidence that sexual selection causes speciation has been the observation of closely related species that differ primarily or solely in traits involved in sexual communication or sexual selection. However, many of these examples are largely anecdotal. Proving that sexual selection has caused speciation in such cases requires demonstrating that the traits have evolved by sexual selection and that they were the first traits to diverge and cause reproductive isolation (Panhuis et al. 2001). Detailed comparisons to assess the rate of evolution of sexual behaviors versus other traits, such as ecological specialization or incompatibilities, are usually lacking. In an influential study, Coyne & Orr (1989; see also Coyne & Orr 1997) surveyed the rate of evolution of pre- and postmating isolation in species pairs of *Drosophila* using genetic distance as a measure of divergence time. They found that these evolve at similar rates unless species are sympatric or parapatric, when sexual isolation accelerates over postmating isolation (the inferred average time to speciation is 2.7 My for allopatric species versus 0.2 My for sympatric, a remarkable difference). This pattern is liable to multiple interpretations. It suggests that interspecific encounters promote sexual isolation, which is compatible with reinforcing selection or reproductive character displacement. Coyne & Orr (1989) also speculated that sexual selection alone may lead to increased divergence in sympatry, but it would have to be accentuated by the presence of another species, as there is no reason to suppose that sexual selection should be less important in allopatry. Resource competition may be greater where closely related species meet, and ecological selection could favor behavioral separation by a Kondrashov-like (Kondrashov & Kondrashov 1999) process. Funk et al. (2006), in a similar but more taxonomically widespread study, found that ecological differentiation was related to sexual but not postmating isolation. Sperm proteins of marine invertebrates evolve more quickly in sympatry (van Doorn et al. 2001). Mendelson (2003) examined divergence rates of assortative mating and egg inviability in some species of darters (fish) and found that behavioral divergence occurred more quickly. Dealing with traits potentially under sexual selection, Gleason

& Ritchie (1998) found that courtship song diverged particularly quickly in the *willistoni* group of *Drosophila*, sometimes before any postmating isolation was detectable. More studies like these are sorely needed. However, even accurately quantifying that sexual isolation evolved most quickly in a group would not necessarily prove that sexual selection caused this, especially if we consider species recognition a distinct process.

Another way of asking if sexual selection influences speciation is to perform comparisons using the comparative method, which assesses if number of species or rates of speciation are greater in species that show signals of sexual selection while controlling for phylogenetic relatedness. The first study addressing this was done by Barraclough et al. (1995), who examined species richness in passerine birds and found that clades with more sexually dichromatic species contained more species. They discussed potential confounding factors such as species recognition and issues of species definition. Price (1998) reanalyzed these data and found that the pattern was mainly due to allopatric comparisons and argued that ecological equivalents were of uncertain importance. Owens et al. (1999) confirmed an effect of dichromatism in birds, although they found no effect of polyandry or sexual size dimorphism on species richness. Similar studies in birds include those by Møller & Cuervo (1998) and Mitra et al. (1996), both of which found marginal effects of feather ornamentation or mating system on species richness when making sister group comparisons in birds.

Arnqvist et al. (2000) compared species richness in insect clades differing in levels of polyandry; 25 phylogenetic contrasts were possible, and inferred speciation rates were approximately four times greater in the polyandrous clades (an even stronger pattern was seen when restricted to more closely related groups). This is an impressive and seemingly robust study. The mechanism is unknown, but candidates include sperm competition, cryptic female choice, or antagonistic coevolution, emphasizing the need for speciation biologists to study postinsemination effects (Howard 1999). Fast evolution of male genitalia is a well-characterized phenomenon (Eberhard 2004), and accelerated genital evolution is also seen in comparative studies of mono- and polyandrous insect clades (Arnqvist 1998). Although rapid genital evolution was originally interpreted in terms of species recognition, recent studies have suggested that male coercion is more important (more "male lock" than "lock and key") (e.g., Jagadeeshan & Singh 2006).

However, other apparently equally detailed comparative studies fail to support a role of sexual selection or sexual conflict in speciation. Gage et al. (2002) carried out a large independent-contrasts study of mammals, butterflies, and spiders. They found that species richness was unrelated to sexual size dimorphism (across more than 700 genera) or measures of polyandry (120 genera). They discuss a number of factors that might act against finding a result, but the contrast with Arnqvist et al. (2000) is difficult to understand, as the methods are similar. Gage et al.'s (2002) paper is more extensive taxonomically, although it has been suggested that it does not control for phylogeny so well (the butterflies are a geographic, not a phylogenetic, assemblage, although comparative approaches were used within that constraint). Morrow et al. (2003) carried out probably the most comprehensive study of the influence of sexual size dimorphism, sexual dichromatism, and testes mass on species richness in birds.

They identified 130–180 independent contrasts with which to test the hypothesis. None of the predictors were significant; in fact two had a negative effect on species richness (see also Morrow et al. 2003).

Some studies are difficult to interpret. Katzourakis et al. (2001) analyzed predictors of species richness including male testes size and female spermathecal width—indicators of sperm competition—in an extensive phylogeny of hoverflies. The results were equivocal, with contrasts in testes size significantly different from zero ($p = 0.014$) but no significant regression of these contrasts on species richness. Stuart-Fox & Owens (2003) analyzed a fairly complete lizard phylogeny with sexual size dimorphism, sexual dichromatism, and the possession of sexual ornaments as predictors of species richness. Ornaments did not predict species richness—dichromatism was marginal ($p = 0.056$)—but sexual size dimorphism did, but in the opposite direction to that predicted, with more species in groups with less sexual dimorphism. Stuart-Fox & Owens (2003) conclude the latter may reflect a process other than sexual selection, for example, resource partitioning. They also suggest that dichromatism may represent male competition or species recognition, considerable caveats to the previous studies supporting an association in birds.

Ritchie et al. (2005) examined inferred rates of speciation in a monophyletic group of Mexican fish in which body shape dimorphism was thought to reflect sexual conflict. In 10 out of 16 phylogenetically independent contrasts, speciation occurred more quickly in dimorphic lineages, but this was not statistically significant. However, Mank (2007) recently completed a sister group comparison across all fish and found consistent evidence of more species in groups containing a greater proportion of species with apparent sexually selected traits.

The comparative method has been proposed as a particularly appropriate way to assess the influence of traits on speciation, but is fraught with difficulties. Looking across the studies currently available, supporting evidence for a role of sexual selection is not overwhelming (although some authors find it to be "particularly convincing" regarding measures of sexual dichromatism in birds). Of 11 studies considered here, two were convincing in support of the hypotheses, two apparently equally convincing against the hypothesis, and the remainder either gave inconsistent results with different traits thought to correlate with the intensity of sexual selection or were of marginal statistical significance (at most, five or six provide positive evidence). Tests are often understandably statistically weak owing to a low number of independent contrasts, but we should expect a strong effect to be more consistently detectable. Furthermore, species richness depends on the net diversification rate; clades vary in richness owing to extinction as well as speciation. If both processes are potentially influenced by the traits of interest, it may not be possible to resolve their relative importance simply by analyses of the number of extant species. It does seem likely that extreme sexual selection may be associated with an increased risk of extinction (Gavrilets et al. 2001), and there is empirical evidence that, for example, dichromatic bird species have elevated extinction rates (Doherty et al. 2003). Morrow et al. (2003) actually concluded that the role of sexual selection in promoting speciation was so well established that increased extinction in sexually selected species must explain their inability to find such a relationship in their analysis. A rather more subtle potential bias of the comparative approach concerns deciding what to include. A species complex where different pheromone blends distinguish reproductively isolated insect species may have the number of described species underrepresented. It is perhaps not too surprising that more species of birds have been described in clades with elaborate sexual dichromatism.

There are other problems with cause and effect in correlations between species richness and apparently sexually selected traits: What if adaptive radiation favors polyandry (Gage et al. 2002), or what if ecological diversification increases the need for species recognition? Price (1998) has argued that sexually selected traits may be particularly prone to diverge during adaptive radiation (different traits may propagate better in different environments) but that coevolution of preferences may be unlikely. For example, in bird species, females may all prefer males with larger tails, but male tail length reflects varying costs in different environments. Price (1988) argued that imprinting in birds (and possibly cichlid fish) may mean that some striking morphologies evolve as recognition traits not under sexual selection (see also Lachlan & Servedio 2004). In many of the species groups where male traits are species specific (e.g., cichlids, *Anolis*, lacewings, Hawaiian *Drosophila*, or crickets), female preferences have been studied in detail in only one or two species, so covarying female preferences are usually an assumption. Price (1998) argues that this is unjustified, as preferences may be asymmetrical, costly, or have evolved under sensory exploitation. More empirical studies examining Fisher-Lande coevolution between preference and trait variation within and between populations are required. Although comparative approaches are likely to continue to be used extensively in this area in the future, the caveats associated with this approach perhaps mean that more decisive results are to be obtained from other types of study.

Case Studies

African cichlids and other fish. The cichlids from the crater lakes of central Africa (particularly Victoria and Malawi) represent a remarkable recent radiation. Inferred speciation rates rival or exceed any other, with speciation intervals of as little as a few thousand years for Lake Victoria (Turner 1999) and the fastest rates of all in Lake Nabugabo (Coyne & Orr 2004). Several authors have argued that sexual selection is an important driving force behind this, as many of the most closely related species are distinguished by body color, a target of female mate choice (Seehausen et al. 1999). These have become a flagship example of sexual selection and sympatric speciation and have inspired several models of this process (including those of Turner & Burrows 1995, van Doorn et al. 1998, and Lande et al. 2001). However, many species also differ in ecological niche and there is debate over the relative importance of ecologically driven selection or if the divergence of color between species is driving speciation or is recruited into ecological adaptation to facilitate assortative mating (Kondrashov & Shpak 1998, Arnegard & Kondrashov 2004). Unfortunately, the rapidity of the radiation means that attempts to obtain a molecular phylogeny within Lakes Victoria and Malawi have been of limited success, precluding careful comparisons of the divergence rates in ecology and behavior. Seehausen et al. (1999) examined the evolution

Quantitative trait loci: genes influencing a trait whose approximate genomic location, magnitude, and directionality have been inferred statistically

Phylogeography: analyses of patterns of geographic variation in traits or genetic markers among related populations or closely related species

of cichlid color patterns against a phylogeny of a wider assemblage of cichlids. Interestingly, they were able to detect the influence of ecology on the characteristic dark banding patterns of cichlids, but in contrast, nuptial hue coloration varied with mating system rather than ecology. Speciation in promiscuous species was associated with changes in hue, but it was not clear that their speciation rate was accelerated over other groups. Seehausen et al. (1999) compared the distribution of species within Lake Victoria and found that color heteromorphic pairs were more likely to be found in sympatry; if these pairs have large niche overlap (see also Haesler & Seehausen 2005), this would imply that behavior diverges more quickly. Color (and vision) also diverges under direct natural environmental selection for efficient propagation and detection as light quality changes in different regions within lakes (Terai et al. 2006).

Quantitative trait loci have been identified for morphologies involved in niche adaptation in cichlids, and their genetic architecture implies a role of directional selection in divergence between species (Albertson et al. 2003). Decisive genetic evidence for strong sexual selection on nuptial color could be found if similar quantitative trait loci analyses of color differences between sibling species occupying similar ecological niches found evidence for strong selection. Quantitative genetic studies suggest that relatively few genes may be involved (Haesler & Seehausen 2005, Barson et al. 2007). Most of the behavior genetic studies have concentrated on finding a match between the predictions of models of speciation derived for cichlids, but these architectures will probably be compatible with other interpretations (predictions of the number of loci and the likelihood of sympatric speciation vary between different models). Adaptive radiation, direct ecological selection involving behavior, and sexual selection have undoubtedly all contributed to the extremely rapid divergence of these cichlids, but the relative importance of these is uncertain and difficult to ascertain.

Wilson et al. (2000) attempted to distinguish the roles of assortative mating and ecological adaptation to genetic differentiation among morphs of a Midas cichlid from Nicaragua. Phylogeography was used to examine whether genetic variation was structured by trophic morph or nuptial color. Both mtDNA and microsatellites were partitioned more by color than jaw morph in at least one lake. Similar studies in the African cichlids would be most interesting. Studies of genetic differentiation between populations of four species of Mexican Goodeid fish found greater differentiation between populations within two dimorphic than two monomorphic species, and evidence for greater sex-biased gene flow in the dimorphic species (Ritchie et al. 2007).

Assortative mating by size contributes to some of the recent radiations of sticklebacks into different ecological niches in Canada and elsewhere (Nagel & Schluter 1998). Hybrids have low mating success, which would contribute to their isolation, but in this case ecological selection seems to be the main source of selection (Schluter 2000). Morphological traits such as body size (or divergence in nest structure in an Icelandic radiation; Ólafsdóttir et al. 2006) probably provide examples of traits intimately associated with the ecological selection facilitating speciation, as in some of the models of sympatric ecological speciation (e.g., Dieckmann & Doebeli 1999).

The guppy provides an intriguing potential counterexample of the role of sexual selection in fish speciation (Magurran 1999). Guppies show rapid evolution, and transplantation experiments have shown adaptation to local conditions within a few generations. Why, then, have divergent populations of the guppy from Trinidad not completed speciation? Female guppies are subject to numerous encounters with amorous males who can "sneak mate," bypassing female mating preferences. Magurran (1999) has argued that the guppy may provide an example of nonspeciation due to male-biased sexual conflict (Parker & Partridge 1998).

Hawaiian *Drosophila* and crickets. The iconic example of sexual selection and island speciation is the Hawaiian *Drosophila* (Templeton 1978, Boake 2005). Their radiation is similar to that of cichlids, with rapid evolution on these oceanic islands (around 25% of the world's species of *Drosophila* are endemic to Hawaii). There is also a great ecological radiation, with the flies occupying a much broader range of niches than are usually occupied by fruit flies and elaborate and unusual secondary sexual traits, including painted wings, unusual morphologies, and elaborate courtship songs (Hoikkala et al. 1989). Carson, Kaneshiro and colleagues (e.g., Kaneshiro & Boake 1987) have argued that sexual selection has been particularly important for this radiation. Genetic revolutions during founder-flush cycles are thought to have facilitated switches between different coadapted gene complexes and the evolution of unusual courtship behaviors (Carson & Templeton 1984). Sexual selection is an important component of the elaborate models of speciation inspired by these flies, including asymmetric isolation due to relaxation of preferences and simplification of courtship repertoires (Kaneshiro 1989). Barton & Charlesworth (1984) produced cogent criticism of these models of speciation, arguing that details of the genetic transilience model of speciation are unnecessary and direct ecological selection may suffice. The Kaneshiro hypothesis of asymmetrical sexual selection and speciation (Kaneshiro 1989) has also had a rough reception and lacks widespread acceptance. The Hawaiian *Drosophila* are difficult to breed, and detailed studies of their post-mating isolation are lacking. Sexual isolation is clearly important to their speciation, but its relative role, and even the contribution of the elaborate traits to isolation, is unclear (Boake 2005). The Hawaiian *Drosophila* have probably been the single most important radiation giving credence to the role of sexual selection in driving speciation, but the evidence is almost wholly indirect. Rather depressingly, the possible loss in the wild of iconic species such as *D. heteroneura* means we may never be in a position to resolve these issues.

More recently, explosive speciation of crickets of the genus *Laupala* on Hawaii has also been described (Otte 1989). Mendelson & Shaw (2005) generated a phylogeny of this genus and examined the timescales involved. The inferred rate of speciation on the Big Island exceeds that of *Drosophila* and is second only to the African cichlids. Each species of *Laupala* is characterized by a unique calling song, and studies of two species have indicated that female preferences are tuned to male song and that the species are reproductively isolated (Shaw & Parsons 2002). Mendelson & Shaw (2005) argue that ecological selection or reinforcing selection can be reasonably ruled out as an alternative to sexual selection as causes of speciation in this group. *Laupala* illustrates the difficulty of a conclusive broad comparison of a species group because it is difficult to know if sexual selection, drift, or other factors associated with speciation

Kaneshiro hypothesis: asymmetric sexual isolation arises during founder event speciation because sexual selection acts against fussy females in derived species

on islands is consistently most important across the group. However, the crickets are more ecological generalists than Hawaiian *Drosophila*.

Other examples. As we have seen, broad comparisons among species groups can be indecisive. With fewer species, it may be possible to examine gene flow directly and correlate this with variation in traits known to be evolving under sexual selection. An example of assortative mating driving divergence in a natural context is a recent study of the Amazonian frog *Physalaemus petersi* (Boul et al. 2007). Different populations within this species have independently evolved two call types important to mate choice. Phonotaxis experiments between distinct but geographically close populations demonstrated assortative female call preferences, and genetic differentiation follows divergence in calls. Furthermore, mtDNA divergence implies that directional selection has fixed call types between populations. Boul et al. (2007) argue that the appearance of complex calls in some populations has altered female preferences for normal calls, driving assortative mating and reducing gene flow between diverging populations. The association between call, female preference, and gene flow is clear, but why sexual selection has favored males with complex calls is not [although sensory biases for such songs have been seen in related species (Ryan et al. 1990)]. A similar study compared populations of a jumping spider from Arizona (Masta & Maddison 2002). Phylogeographic analyses implied a role of positive selection in the fixation of sexually dimorphic male color and morphological ornaments, presumed to be sexual selection. Females showed a reluctance to mate with males of other morphs in some crosses between allopatric types (there was also some postmating isolation), but there was no direct comparison of genetic differentiation and behavior. These and the previously described fish studies (Wilson et al. 2000; Ritchie et al. 2007) perhaps suggest that phylogeographic studies are a more appropriate scale than broad comparative studies at which to detect an influence of sexual selection on evolutionary differentiation.

Most other studies only address components of the story. Gray & Cade (2000) examined variation in song and female preferences in the cricket *Gryllus texensis*. Females prefer songs of their own species over the geographically overlapping *G. rubens*, yet there is a lack of character displacement and no apparent postmating isolation or ecological divergence. This, and numerous other examples of primary divergence in mating signals and preferences, is compatible with speciation by co-evolution of trait and preference (although song and preference were not positively correlated among populations of *G. texensis*) but does not satisfy all the criteria necessary to unambiguously demonstrate speciation by sexual selection (Panhuis et al. 2001).

Experimental Evolution

Ultimately the most decisive way of testing models of speciation is to examine their effectiveness during experimental evolution. There have been many studies of factors promoting reproductive isolation or assortative mating in the laboratory, and the most successful experiments have involved selection from a multiplicity of sources

(Rice & Hostert 1993). However, there have been few direct tests of the importance of sexual selection. Rice (1998) pioneered groundbreaking studies of responses to elevated sexual conflict and has argued that this could drive both pre- and post-mating reproductive isolation. Martin & Hosken (2003) performed one important experiment addressing this issue. They reared dung flies under three regimes: enforced monogamy and free mating, under high or low population density [following Gavrilets' model (2000) showing that sexual conflict would be especially important at high density because selection is more effective with more genetic variation and less drift]. After 35 generations, tests for assortative mating between replicates found increased positive assortative mating between lines within the free-mating regimes. Furthermore, this was greatest at high density and was driven by increased female reluctance to mate. This is consistent with sexual conflict driving allopatric divergence in assortative mating.

Synonoymous/nonsynonymous mutations: synonymous mutations do not change a protein (owing to code redundancy); diversifying selection will increase nonsynonymous but not synonymous replacements

Wigby & Chapman (2006) have produced the only attempt to replicate such an experiment so far. They reared *D. melanogaster* for a similar period on male-biased, female-biased, or equal-sex ratios, again manipulating sexual conflict. However, they did not find any consistent pattern of increased reluctance to mate under higher sexual conflict or between lines or regimes. Sexual conflict is strong in *D. melanogaster* so the contrast is hard to explain, although perhaps postinsemination responses to the toxicity or manipulative effects of semen components rather than reluctance to copulate are more important in this species. Female postcopulatory resistance can evolve in the laboratory (Lew et al. 2006) and is a component of natural between-species variation (Sakaluk et al. 2006). It is critically important to see further replication of such laboratory evolution experiments, as these are much more decisive tests of theory (Rice & Hostert 1993). Traits with the potential to influence sexual isolation such as male courtship song have shown rapid laboratory evolution under increased sexual selection in *D. pseudoobscura* (Snook et al. 2005), although differences in mating rate within lines seem slight (Crudgington et al. 2005).

Genetic Studies

There has been a long history of attempting to identify genes that cause reproductive isolation (Coyne & Orr 2004). Probably the ultimate way of demonstrating the cause of speciation in nature would be to identify the sources of selection responsible for the substitution of genes causing reproductive isolation between species. There are too few examples of potential so-called speciation genes that have been characterized to enable general conclusions to be drawn, but some are contributing to the evidence that sexual reproduction (and by implication sexual selection) is a pervasive source of selection driving gene substitution.

Most early studies of speciation genes attempted to identify genes causing hybrid male sterility or inviability. One gene found in the *simulans* clade of *melanogaster* is *Odysseus*; when crossed into the wrong genetic background, it is responsible for approximately 40% of the sterility typically seen between species (Perez & Wu 1995). It shows classic signs of strong directional selection—rapid sequence divergence and a greatly elevated rate of synonymous over nonsynonymous substitutions within the clade (Ting et al. 1998). But what caused this selection? The normal function of *Odysseus* is not clear, but it does influence sperm production and potentially sperm competition in *D. simulans* (Sun et al. 2004), so postcopulatory sexual selection may have driven its divergence and indirectly contributed to hybrid sterility. Genes that show male-biased expression show greater signatures of selection across species of *Drosophila* (the ratio of nonsynonymous to synonymous mutations is approximately double in male-biased genes; Zhang et al. 2004). The excess of nonsynonymous substitutions is positively related to the recombination rate in male-biased genes but not those in females-biased or unbiased genes, an important observation as it supports the interpretation that strong directional selection (rather than relaxed constraints) underlies this rapid evolution (Zhang & Parsch 2005). Male-biased genes are also more likely to be misexpressed in sterile hybrid males (Michalak & Noor 2003). Why are male-biased genes evolving quickly? The fast-male evolution explanations of Haldane's rule (which argued that males are under stronger selection) were thought not to be of great general importance because hybrid female sterility is seen in butterflies and birds with opposite chromosomal systems of sex determination, but fast-male evolution does occur in species with conventional sex determination (Wu et al. 1996, Presgraves & Orr 1998).

Genes involved in sexual or gametic recognition (for example, mating-type loci in *Chlamydomonas*, sperm-egg recognition genes) also show strong signals of rapid directional selection compared to other loci (e.g., Metz & Palumbi 1996, Ferris et al. 1997; reviewed in Swanson & Vacquier 2002). Genes classified as involved in sexual reproduction (that is, known to influence mating behavior, fertilization, spermatogenesis, or sex determination) also show faster evolution and an excess of nonsynonymous substitutions in a range of organisms (e.g., Civetta & Singh 1998; reviewed in Singh & Kulathinal 2000). Many authors think that sexual selection, particularly antagonistic selection, is likely to drive elevated evolutionary rates in sex-biased or sex-related genes. Zhang et al. (2004) make the important point that male-male competition may be a more likely explanation than intersexual antagonism for fast-male evolution. Antagonistic coevolution alone cannot explain fast-male evolution, unless the interacting female loci are systematically underrepresented [faster evolution of sperm than egg proteins is seen in marine invertebrates, a system where the female counterparts are well known; van Doorn et al. (2001) suggest that intrasexual competition for fertilization drives this]. If fast-male evolution is driven by spermatogenesis rather than sperm competition or other intrasexual selection, this intriguing and strong correlation between sex and fast evolution of genes may have a relatively simple explanation (at least 12% of the genes in *D. melanogaster* are expressed only in testes; Boutanaev et al. 2002). The bias in sexual-function-related genes (rather than sex-biased ones) is also seen in genes expressed in female reproductive tracts (Singh & Kulathinal 2000) and must indicate a stronger selection, perhaps reflecting antagonistic coevolution. However, the nature of selection on female-biased genes may differ from that on male-biased ones, with directional selection on male-biased genes and purifying or balancing selection being more likely to be seen on female-biased genes (Proschel et al. 2006). It remains to be seen if such fast-evolving genes are more likely to be identified as speciation genes, which seems likely.

Clearly, genomic studies provide a great opportunity to identify which genes evolve most rapidly between species. Identifying the sources of selection acting on these gene replacements will require correlational or manipulative studies of individual loci. Recent studies of selection on loci thought to influence behavioral reproductive isolation, such as opsin genes in cichlids (Terai et al. 2006) and genes involved in pheromone production in moths (Groot et al. 2006) and *Drosophila* (Greenberg et al. 2003; but see Coyne & Elwyn 2006), have implicated ecological selection through environmental adaptation, efficient signal detection, or directional selection from interspecific interactions in their divergence. None of these studies provides examples of loci under primarily sexual rather than ecological selection.

CONCLUSIONS

The evidence that sexual selection is an important cause of speciation comes from numerous sources. Of necessity, much of this is theoretical, correlational, or indirect. Decisive studies are surprisingly thin on the ground, but some can be found. The concept of speciation by sexual selection is currently undergoing a surge in popularity, and perhaps as a consequence supporting evidence tends to be accepted rather uncritically and competing or alternative explanations not so thoroughly considered (as seen with other hypotheses; Simmons et al. 1999). Currently, the case for speciation occurring primarily by sexual selection is certainly not as well made as, say, the case for the preponderance of allopatric speciation. It seems highly likely that sexual selection (or sexual conflict) contributes to divergence in traits that influence sexual isolation in allopatry. The role of sexual selection in sympatric speciation is much more contentious, with theoretical support but no wholly convincing empirical example. So far, much of the evidence has come from comparative studies, but smaller-scale phylogeographic studies, genetic and genomic studies, and experimental evolution are perhaps more likely to provide conclusive evidence in the future. The major challenge for empirical studies is distinguishing between cases where sexual selection has directly driven reproductive isolation rather than acted as a secondary force alongside ecological selection, or where selection was for species recognition. It seems unlikely that sexual selection often acts alone.

DISCLOSURE STATEMENT

The author is not aware of any biases that might be perceived as affecting the objectivity of this review.

ACKNOWLEDGMENTS

My work is funded mainly by the Natural Environment Research Council, United Kingdom. Many ideas in this review have benefited from discussions with colleagues, especially Roger Butlin, Tino Macias, and Anne Magurran. Roger Butlin, Mark Kirkpatrick, Tino Macias, and Mohamed Noor gave very helpful comments on the manuscript.

LITERATURE CITED

Albertson RC, Streelman JT, Kocher TD. 2003. Directional selection has shaped the oral jaws of Lake Malawi cichlid fishes. *Proc. Natl. Acad. Sci. USA* 100:5252–57

Arnegard ME, Kondrashov AS. 2004. Sympatric speciation by sexual selection alone is unlikely. *Evolution* 58:222–37

Arnqvist G. 1998. Comparative evidence for the evolution of genitalia by sexual selection. *Nature* 393:784–86

Arnqvist G, Edvarrdsson M, Friberg U, Nilsson T. 2000. Sexual conflict promotes speciation in insects. *Proc. Natl. Acad. Sci. USA* 97:10460–64

Arnqvist G, Rowe L. 2002. Antagonistic coevolution between the sexes in a group of insects. *Nature* 415:787–89

Barraclough TG, Harvey PH, Nee S. 1995. Sexual selection and taxonomic diversity in passerine birds. *Proc. R. Soc. London B Biol. Sci.* 259:211–15

Barson NJ, Knight ME, Turner GF. 2007. The genetic architecture of male color differences between a sympatric Lake Malawi cichlid species pair. *J. Evol. Biol.* 20:45–53

Barton NH, Charlesworth B. 1984. Genetic revolutions, founder effects and speciation. *Annu. Rev. Ecol. Syst.* 15:133–64

Boake CRB. 2005. Sexual selection and speciation in Hawaiian *Drosophila*. *Behav. Genet.* 35:297–303

Boul KE, Funk WC, Darst CR, Cannatella DC, Ryan MJ. 2007. Sexual selection drives speciation in an Amazonian frog. *Proc. R. Soc. London B Biol. Sci.* 274:399–406

Boutanaev AM, Kalmykova AI, Shevelyou YY, Nurminsky DI. 2002. Large clusters of coexpressed genes in the *Drosophila* genome. *Nature* 420:666–69

Bulmer M. 1989. Structural instability of models of sexual selection. *Theor. Popul. Biol.* 35:195–206

Butlin RK. 1989. Reinforcement of premating isolation. In *Speciation and Its Consequences*. ed. D Otte, JA Endler, pp. 158–79. Sunderland, MA: Sinauer

Carson HL, Templeton AR. 1984. Genetic revolutions in relation to speciation phenonema: the founding of new populations. *Annu. Rev. Ecol. Syst.* 15:97–131

Civetta A, Singh RS. 1998. Sex-related genes, directional sexual selection, and speciation. *Mol. Biol. Evol.* 15:901–9

Coyne JA, Elwyn S. 2006. Does the desaturase-2 locus in *Drosophila melanogaster* cause adaptation and sexual isolation? *Evolution* 60:279–91

Coyne JA, Orr HA. 1989. Patterns of speciation in *Drosophila*. *Evolution* 43:362–81

Coyne JA, Orr HA. 1997. "Patterns of speciation in *Drosophila*" revisited. *Evolution* 51:295–303

Coyne JA, Orr HA. 2004. *Speciation*. Sunderland, MA: Sinauer. 545 pp.

Crudgington HS, Beckerman AP, Brustle L, Green K, Snook RR. 2005. Experimental removal and elevation of sexual selection: Does sexual selection generate manipulative males and resistant females? *Am. Nat.* 165:S72–87

Darwin C. 1871. *The Descent of Man, and Selection in Relation to Sex.* London: J. Murray

Dieckmann U, Doebeli M. 1999. On the origin of species by sympatric speciation. *Nature* 400:354–57

Dobzhansky T. 1937. *Genetics and the Origin of Species*. New York: Columbia Univ. Press

Dobzhansky T. 1940. Speciation as a stage in evolutionary divergence. *Am. Nat.* 74:312–21

Doherty PF Jr, Sorci G, Royle JA, Hines JE, Nichols JD, Boulinier T. 2003. Sexual selection affects local extinction and turnover in bird communities. *Proc. Natl. Acad. Sci. USA* 100:5858–62

Drossel B, McKane A. 2000. Competitive speciation in quantitative genetic models. *J. Theor. Biol.* 204:467–78

Eberhard WG. 2004. Rapid divergent evolution of sexual morphology: comparative tests of antagonistic coevolution and traditional female choice. *Evolution* 58:1947–70

Felsenstein J. 1981. Skepticism towards Santa Rosalia, or why are there so few kinds of animals? *Evolution* 35:124–38

Ferris PJ, Pavlovic C, Fabry S, Goodenough UW. 1997. Rapid evolution of sex-related genes in *Chlamydomonas*. *Proc. Natl. Acad. Sci. USA* 94:8634–39

Fisher RA. 1930. *The Genetical Theory of Natural Selection*. Oxford, UK: Oxford Univ. Press

Funk DJ, Nosil P, Etges WJ. 2006. Ecological divergence exhibits consistently positive associations with reproductive isolation across disparate taxa. *Proc. Natl. Acad. Sci. USA* 103:3209–13

Gage MJG, Parker GA, Nylin S, Wiklund C. 2002. Sexual selection and speciation in mammals, butterflies and spiders. *Proc. R. Soc. London B Biol. Sci.* 269:2309–16

Gavrilets S. 2000. Rapid evolution of reproductive barriers driven by sexual selection. *Nature* 403:886–89

Gavrilets S, Arnqvist G, Friberg U. 2001. The evolution of female mate choice by sexual conflict. *Proc. R. Soc. London B Biol. Sci.* 268:531–39

Gavrilets S, Waxman D. 2002. Sympatric speciation by sexual conflict. *Proc. Natl. Acad. Sci. USA* 99:10533–38

Gleason JM, Ritchie MG. 1998. Evolution of courtship song and reproductive isolation in the *Drosophila willistoni* species complex: Do sexual signals diverge the most quickly? *Evolution* 52:1493–500

Gourbiere S. 2004. How do natural and sexual selection contribute to sympatric speciation? *J. Evol. Biol* 17:1297–309

Gray DA, Cade WH. 2000. Sexual selection and speciation in field crickets. *Proc. Natl. Acad. Sci. USA* 97:14449–54

Greenberg AJ, Moran JR, Coyne JA, Wu CI. 2003. Ecological adaptation during incipient speciation revealed by precise gene replacement. *Science* 302:1754–57

Groot AT, Horovitz JL, Hamilton J, Santangelo RG, Schal C, Gould F. 2006. Experimental evidence for interspecific directional selection on moth pheromone communication. *Proc. Natl. Acad. Sci. USA* 103:5858–63

Haesler MP, Seehausen O. 2005. Inheritance of female mating preference in a sympatric sibling species pair of Lake Victoria cichlids: implications for speciation. *Proc. Natl. Acad. Sci. USA* 272:237–45

Haldane JBS. 1922. Sex ratio and unisexual sterility in hybrid animals. *J. Genet.* 12:7–109

Hall DW, Kirkpatrick M, West B. 2000. Runaway sexual selection when female preferences are directly selected. *Evolution* 54:1862–69

Higashi M, Takimoto G, Yamamura N. 1999. Sympatric speciation by sexual selection. *Nature* 402:523–26

Hoikkala A, Hoy RR, Kaneshiro KY. 1989. High-frequency clicks of Hawaiian picture winged *Drosophila* species. *Anim. Behav.* 37:927–34

Holland B, Rice WR. 1998. Chase-away sexual selection: antagonistic seduction versus resistance. *Evolution* 52:1–7

Howard DJ. 1999. Conspecific sperm and pollen precedence and speciation. *Annu. Rev. Ecol. Syst.* 30:109–32

Jagadeeshan S, Singh RS. 2006. A time-sequence functional analysis of mating behavior and genital coupling in *Drosophila*: role of cryptic female choice and male sex-drive in the evolution of male genitalia. *J. Evol. Biol.* 19:1058–70

Kaneshiro KY. 1989. The dynamics of sexual selection and founder effects in species formation. In *Genetics, Speciation and the Founder Principle*, ed. LV Giddings, KY Kaneshiro, WW Anderson, pp. 279–96. New York: Oxford Univ. Press

Kaneshiro KY, Boake CRB. 1987. Sexual selection and speciation: issues raised by Hawaiian *Drosophila*. *Trends Ecol. Evol.* 2:207–12

Katzourakis A, Purvis A, Azmeh S, Rotheray G, Gilbert F. 2001. Macroevolution of hoverflies (Diptera: Syrphidae): The effect of using higher-level taxa in studies of biodiversity, and correlates of species richness. *J. Evol. Biol.* 14:219–27

Kawata M, Yoshimura J. 2000. Speciation by sexual selection in hybridizing populations without viability selection. *Evol. Ecol. Res.* 2:897–909

Kirkpatrick M. 1982. Sexual selection and the evolution of female choice. *Evolution* 36:1–12

Kirkpatrick M, Nuismer SL. 2004. Sexual selection can constrain sympatric speciation. *Proc. R. Soc. London B Biol. Sci.* 271:687–93

Kirkpatrick M, Ravigne V. 2002. Speciation by natural and sexual selection: models and experiments. *Am. Nat.* 159:S22–35

Kirkpatrick M, Ryan MJ. 1991. The evolution of mating preferences and the paradox of the lek. *Nature* 350:33–38

Kondrashov AS, Kondrashov FA. 1999. Interactions among quantitative traits in the course of sympatric speciation. *Nature* 400:351–54

Kondrashov AS, Shpak M. 1998. On the origin of species by means of assortative mating. *Proc. R. Soc. London B Biol. Sci.* 265:2273–78

Lachlan RF, Servedio MR. 2004. Song learning accelerates allopatric speciation. *Evolution* 58:2049–63

Lande R. 1981. Models of speciation by sexual selection on polygenic traits. *Proc. Natl. Acad. Sci. USA* 78:3721–25

Lande R. 1982. Rapid origin of sexual isolation and character divergence in a cline. *Evolution* 36:213–23

Lande R, Seehausen O, van Alphen JJM. 2001. Mechanisms of rapid sympatric speciation by sex reversal and sexual selection in cichlid fish. *Genetica* 112:435–43

Lew TA, Morrow EH, Rice WR. 2006. Standing genetic variance for female resistance to harm from males and its relationship to intralocus sexual conflict. *Evolution* 60:97–105

Liou LW, Price TD. 1994. Speciation by reinforcement of premating isolation. *Evolution* 48:1451–59

Lorch PD, Proulx S, Rowe L, Day T. 2003. Condition-dependent sexual selection can accelerate adaptation. *Evol. Ecol. Res.* 5:867–81

Macías Garcia C, Ramirez E. 2005. Evidence that sensory traps can evolve into honest signals. *Nature* 434:501–5

MacDougall-Shackleton EA, MacDougall-Shackleton SA. 2001. Cultural and genetic evolution in mountain white-crowned sparrows: Song dialects are associated with population structure. *Evolution* 55:2568–75

Magurran AE. 1999. Population differentiation without speciation. In *Evolution of Biological Diversity*, ed. AE Magurran, RM May, pp. 160–83. Oxford: Oxford Univ. Press

Mank JE. 2007. Mating preferences, sexual selection and patterns of cladogenesis in ray-finned fishes. *J. Evol. Biol*

Martin OY, Hosken DJ. 2003. The evolution of reproductive isolation through sexual conflict. *Nature* 423:979–82

Masta SE, Maddison WP. 2002. Sexual selection driving diversification in jumping spiders. *Proc. Natl. Acad. Sci. USA* 99:4442–47

Mayr E. 1942. *Systematics and the Origin of Species.* New York: Columbia Univ. Press. 334 pp.

Mendelson TC. 2003. Sexual isolation evolves faster than hybrid inviability in a diverse and sexually dimorphic genus of fish (Percidae: Etheostoma). *Evolution* 57:317–27

Mendelson TC, Shaw KL. 2005. Rapid speciation in an arthropod. *Nature* 433:375–76

Metz E, Palumbi S. 1996. Positive selection and sequence rearrangements generate extensive polymorphism in the gamete recognition protein bindin. *Mol. Biol. Evol.* 13:397–406

Michalak P, Noor MAF. 2003. Genome-wide patterns of expression in *Drosophila* pure species and hybrid males. *Mol. Biol. Evol.* 20:1070–76

Mitra S, Landel H, Pruett-Jones S. 1996. Species richness covaries with mating system in birds. *Auk* 113:544–51

Møller AP, Cuervo JJ. 1998. Speciation and feather ornamentation in brids. *Evolution* 52:859–69

Morrow EH, Pitcher TE, Arnqvist G. 2003. No evidence that sexual selection is an 'engine of speciation' in birds. *Ecol. Lett.* 6:228–34. Erratum. *Ecol. Lett.* 6:1038

Nagel L, Schluter D. 1998. Body size, natural selection, and speciation in sticklebacks. *Evolution* 52:209–18

Nicholls JA, Austin JJ, Moritz C, Goldizen AW. 2006. Genetic population structure and call variation in a passerine bird, the satin bowerbird, *Ptilonorhynchus violaceus*. *Evolution* 60:1279–90

Nichols RA, Butlin RK. 1989. Does runaway sexual selection work in finite populations? *J. Evol. Biol.* 2:299–313

Ólafsdóttir GÁ, Ritchie MG, Snorrason SS. 2006. Positive assortative mating between recently described sympatric morphs of Icelandic sticklebacks. *Biol. Lett.* 2:250–52

Otte D. 1989. Speciation in Hawaiian crickets. In *Speciation and Its Consequences*, ed. D Otte, JA Endler, pp. 482–526. Sunderland, MA: Sinauer

Owens IPF, Bennett PM, Harvey PH. 1999. Species richness among birds: body size, life history, sexual selection or ecology? *Proc. R. Soc. London B Biol. Sci.* 266:933–39

Panhuis TM, Butlin R, Zuk M, Tregenza T. 2001. Sexual selection and speciation. *Trends Ecol. Evol.* 16:364–71

Parker GA. 1979. Sexual selection and sexual conflict. In *Sexual Selection and Reproductive Competition in Insects*, ed. MS Blum, NA Blum, pp. 123–66. New York: Academic

Parker GA, Partridge L. 1998. Sexual conflict and speciation. *Philos. Trans. R. Soc. London B Biol. Sci.* 353:261–74

Partridge L, Parker GA. 1999. Sexual conflict and speciation. In *Evolution of Biological Diversity*, ed. AE Magurran, RM May, pp. 130–59. Oxford: Oxord Univ. Press

Paterson HEH. 1985. The recognition concept of species. In *Species and Speciation*, ed. S Vrba, pp. 21–29. Pretoria, S. Afr.: Transvsaal Mus.

Perez DE, Wu CI. 1995. Further characterization of the *Odysseus* locus of hybrid sterility in *Drosophila*. One gene is not enough. *Genetics* 140:201–106

Presgraves DC, Orr HA. 1998. Haldane's rule in taxa lacking a hemizygous X. *Science* 282:952–54

Price T. 1998. Sexual selection and natural selection in bird speciation. *Philos. Trans. R. Soc. London B Biol. Sci.* 353:251–60

Proschel M, Zhang Z, Parsch J. 2006. Widespread adaptive evolution of *Drosophila* genes with sex-biased expression. *Genetics* 174:893–900

Rice WR. 1998. Intergenomic conflict, interlocus antagonistic coevolution, and the evolution of reproductive isolation. In *Endless Forms: Species and Speciation.*, ed. DJ Howard, SH Berlocher, pp. 261–70. Oxford: Oxford Univ. Press

Rice WR, Hostert EE. 1993. Laboratory experiments on speciation: What have we learned in 40 years? *Evolution* 47:1637–53

Ritchie MG, Webb SA, Graves JA, Magurran AE, Macias Garcia C. 2005. Patterns of speciation in endemic Mexican Goodeid fish: sexual conflict or early radiation? *J. Evol. Biol.* 18:922–29

Ritchie MG, Hamill RM, Graves JA, Magurran AE, Webb SA, Macías Garcia C. 2007. Sex and differentiation; population genetic divergence and sexual dimorphism in Mexican goodeid fish. *J. Evol. Biol.* 20(5):2048–55

Ryan MJ. 1990. Sexual selection, sensory systems and sensory exploitation. *Oxf. Surv. Evol. Biol.* 7:157–95

Ryan MJ, Fox JH, Wilczynski W, Rand AS. 1990. Sexual selection for sensory exploitation in the frog *Physalaemus pustulosus*. *Nature* 343:66–67

Ryan MJ, Rand AS. 1993. Species recognition and sexual selection as a unitary problem in animal communication. *Evolution* 47:647–57

Ryan MJ, Wagner WEJ. 1987. Asymmetries in mating preferences between species: Female swordtails prefer heterospecific males. *Science* 236:595–97

Sakaluk SK, Avery RL, Weddle CB. 2006. Cryptic sexual conflict in gift-giving insects: chasing the chase-away. *Am. Nat.* 167:94–104

Schluter D. 2000. Ecological character displacement in adaptive radiation. *Am. Nat.* 156:S4–S16

Schwartz AK, Hendry AP. 2006. Sexual selection and the detection of ecological speciation. *Evol. Ecol. Res.* 8:399–413

Seehausen O, Mayhew PJ, Van Alphen JJM. 1999. Evolution of color patterns in East African cichlid fish. *J. Evol. Biol.* 12:514–34

Servedio MR. 2004. The evolution of premating isolation: local adaptation and natural and sexual selection against hybrids. *Evolution* 58:913–24

Servedio MR. 2004. The what and why of research on reinforcement. *PLoS Biol.* 2:2032–35

Shaw KS, Parsons YM. 2002. Divergence of mate recognition behavior and its consequences for genetic architectures of speciation. *Am. Nat.* 159:S61–75

Simmons LW, Tomkins JL, Kotiaho JS, Hunt J. 1999. Fluctuating paradigm. *Proc. R. Soc. London B Biol. Sci.* 266:593–95

Singh RS, Kulathinal RJ. 2000. Sex gene pool evolution and speciation: a new paradigm. *Genes Genet. Syst.* 75:119–30

Snook RR, Robertson A, Crudgington HS, Ritchie MG. 2005. Experimental manipulation of sexual selection and the evolution of courtship song in *Drosophila pseudoobscura*. *Behav. Genet.* 35:245–55

Stuart-Fox D, Owens IPF. 2003. Species richness in agamid lizards: chance, body size, sexual selection or ecology? *J. Evol. Biol* 16:659–69

Sun S, Ting CT, Wu CI. 2004. The normal function of a speciation gene, *Odysseus*, and its hybrid sterility effect. *Science* 305:81–83

Swanson WJ, Vacquier VD. 2002. The rapid evolution of reproductive proteins. *Nat. Rev. Genet.* 3:137–44

Templeton AR. 1978. Once again, why 300 species of Hawaiian *Drosophila*? *Evolution* 33:513–17

Terai Y, Seehausen O, Sasaki T, Takahashi K, Mizoiri S, et al. 2006. Divergent selection on opsins drives incipient speciation in Lake Victoria cichlids. *PLoS Biol.* 4:e433

Ting CT, Tsaur SC, Wu ML, Wu CI. 1998. A rapidly evolving homeobox at the site of a hybrid sterility gene. *Science* 282:1501–04

Turner GF. 1999. Explosive speciation of African cichlid fishes. In *Evolution of Biological Diversity*, ed. AE Magurran, RM May, pp. 113–29. Oxford: Oxford Univ. Press

Turner GF, Burrows MT. 1995. A model of sympatric speciation by sexual selection. *Proc. R. Soc. London B Biol. Sci.* 260:287–92

van Doorn GS, Dieckmann U, Weissing FJ. 2004. Sympatric speciation by sexual selection: a critical reevaluation. *Am. Nat.* 163:709–25

van Doorn GS, Luttikhuizen PC, Weissing FJ. 2001. Sexual selection at the protein level drives the extraordinary divergence of sex-related genes during sympatric speciation. *Proc. R. Soc. London B Biol. Sci.* 268:2155–61

van Doorn GS, Noest AJ, Hogeweg P. 1998. Sympatric speciation and extinction driven by environment dependent sexual selection. *Proc. R. Soc. London B Biol. Sci.* 265:1915–19

West-Eberhard MJ. 1983. Sexual selection, social competition and speciation. *Q. Rev. Biol.* 58:155–83

Wigby S, Chapman T. 2006. No evidence that experimental manipulation of sexual conflict drives premating reproductive isolation in *Drosophila melanogaster*. *J. Evol. Biol.* 19:1033–39

Wilson AB, Noack-Kunnmann K, Meyer A. 2000. Incipient speciation in sympatric Nicaraguan crater lake cichlid fishes: sexual selection versus ecological diversification. *Proc. R. Soc. London B Biol. Sci.* 267:2133–41

Wright TF, Wilkinson GS. 2001. Population genetic structure and vocal dialects in an amazon parrot. *Proc. R. Soc. London B Biol. Sci.* 268:609–16

Wu CI. 1985. A stochastic simulation study on speciation by sexual selection. *Evolution* 39:66–82

Wu CI, Johnson NA, Palopoli MF. 1996. Haldane's rule and its legacy: Why are there so many sterile males? *Trends Ecol. Evol.* 11:281–84

Zhang Z, Hambuch TM, Parsch J. 2004. Molecular evolution of sex-biased genes in *Drosophila*. *Mol. Biol. Evol.* 21:2130–39

Zhang Z, Parsch J. 2005. Positive correlation between evolutionary rate and recombination rate in *Drosophila* genes with male-biased expression. *Mol. Biol. Evol.* 22:1945–47

Kin Selection and the Evolutionary Theory of Aging

Andrew F.G. Bourke

School of Biological Sciences, University of East Anglia, Norwich, Norfolk NR4 7TJ, United Kingdom; email: a.bourke@uea.ac.uk

Annu. Rev. Ecol. Evol. Syst. 2007. 38:103–28

First published online as a Review in Advance on July 3, 2007

The *Annual Review of Ecology, Evolution, and Systematics* is online at http://ecolsys.annualreviews.org

This article's doi: 10.1146/annurev.ecolsys.38.091206.095528

Key Words

kin conflict, life history, longevity, senescence, social evolution

Abstract

Researchers are increasingly recognizing that social effects influence the evolution of aging. Kin selection theory provides a framework for analyzing such effects because an individual's longevity and mortality schedule may alter its inclusive fitness via effects on the fitness of relatives. Kin-selected effects on aging have been demonstrated both by models of intergenerational transfers of investment by caregivers and by spatially explicit population models with limited dispersal. They also underlie coevolution between the degree and form of sociality and patterns of aging. In this review I critically examine and synthesize theory and data concerning these processes. I propose a classification, stemming from kin selection theory, of social effects on aging and describe a hypothesis for kin-selected conflict over parental time of death in systems with resource inheritance. I conclude that systematically applying kin selection theory to the analysis of the evolution of aging adds considerably to our general understanding of aging.

INTRODUCTION

The evolutionary theory of aging provides an explanation for aging based on natural selection. The theory proposes that aging occurs because extrinsic mortality reduces the relative size of cohorts of older individuals, causing the potential genetic contribution of older cohorts to future generations to fall. Therefore, any gene increasing survival or fecundity is more strongly selected for when its phenotypic effects occur at younger ages, and conversely, any gene decreasing survival or fecundity is less strongly selected against when its phenotypic effects occur at greater ages. As a result, age-specific phenotypic effects tend to evolve whereby genes for increased survival or fecundity have an effect at younger ages and genes for decreased survival or fecundity have an effect at greater ages, so explaining the occurrence of aging. The evolutionary theory of aging was initially formulated in the mid-twentieth century as an amalgam of contributions from R.A. Fisher, J.B.S. Haldane, P.B. Medawar, G.C. Williams, and W.D. Hamilton, with the last three authors having made the major contributions (reviewed in Charlesworth 2000, Rose 1991). The early theorists also identified two principal genetic routes to aging. The first, antagonistic pleiotropy, proposes that aging stems from selection of pleiotropic genes having a positive effect on survival and fecundity early in life and a negative effect late in life. The second, mutation accumulation, proposes that aging stems from lack of selection against genes of purely negative effect, where these effects occur only late in life (e.g., Hughes & Reynolds 2005, Kirkwood & Austad 2000).

A major area of research has developed on the basis of the evolutionary theory of aging (e.g., Arking 2006, Austad 1997a, Carey 2003, Charlesworth 1980, Finch 1990, Rose 1991), with more derived versions of the original theory having been constructed to account for various complicating factors (e.g., for the case when extrinsic mortality reduces population density and so increases resource availability for older age classes; Abrams 1993). The theory has also been integrated with the general study of life history evolution because finite resources entail the occurrence of trade-offs between somatic maintenance and reproduction (Barnes & Partridge 2003, Kirkwood 1977), and because the age at first reproduction represents the critical period beyond which aging is predicted to occur (Charlesworth 1980, Rose 1997, Stearns 1992). Overall, empirical work has provided strong support for the evolutionary theory of aging from several sources. These include comparative analyses of life histories as a function of the strength of extrinsic mortality and experimental manipulations of schedules of reproduction (e.g., Hughes & Reynolds 2005, Kirkwood & Austad 2000, Rose 1991, Stearns 1992), although results from some studies have proved more consistent with derived versions of the theory than with the original theory (Reznick et al. 2004, Williams et al. 2006). Related studies have provided better support for antagonistic pleiotropy than for mutation accumulation as genetic routes to aging (e.g., Campisi 2005, Hughes & Reynolds 2005, Kirkwood & Austad 2000, Leroi et al. 2005, Partridge & Barton 1993, Partridge & Gems 2002a, Rose 1991), with some phenomena not readily reconcilable with either route (Mitteldorf 2004).

Another major body of evolutionary theory, kin-selection theory (Hamilton 1964, Lehmann & Keller 2006, Michod 1982, West-Eberhard 1975), has sought to explain the evolution of social behavior as a function of the genetic relatedness of interacting individuals. Kin selection theory is based on Hamilton's rule (Grafen 1985, Hamilton 1964), which states that a social action is naturally selected when the sum of rb and c exceeds zero, where r is the relatedness of the interactants, b is the change in offspring number of the recipient of the act, and c is the change in offspring number of the social actor. Although controversial in some of its applications (Alonso & Schuck-Paim 2002, Wilson & Hölldobler 2005), kin selection is widely accepted as a powerful and robust explanation for various forms of social behavior across many taxa (e.g., Bourke 2005, Emlen 1995, Foster et al. 2006, Trivers 1985). These forms include kin-selected conflict, whereby individuals within social groups differ in their relatedness to the group's progeny, and hence in their fitness optima, leading to potential within-group conflict (Ratnieks & Reeve 1992, Trivers 1974). A central tenet of the theory is that an individual's fitness (its inclusive fitness) depends upon its own offspring output plus its effects (through social actions) on the offspring output of relatives (Hamilton 1964). A key prediction is that altruism (entailing lifetime reproductive costs) cannot evolve unless interactants are related (Hamilton 1964, Lehmann & Keller 2006).

The evolutionary theory of aging, including its derived versions, has traditionally considered patterns of aging with reference to an individual's fitness evaluated in the absence of social effects (e.g., Hughes & Reynolds 2005, Rose 1991). However, kin selection theory predicts that social effects will influence patterns of aging because a focal individual's longevity and mortality schedule may affect the fitness of other individuals. If the interacting individuals are relatives, the focal individual's inclusive fitness will be altered, and hence patterns of aging in the focal individual will, in principle, be subject to kin selection. Recently, researchers have incorporated social and kin-selected effects into evolutionary models of aging (e.g., Lee 2003, Travis 2004), albeit implicitly in some cases. In a related development, researchers have argued that social organisms exhibit a complex, coevolutionary relationship between the trajectory of social evolution and patterns of aging (e.g., Alexander et al. 1991, Carey & Judge 2001). Nonetheless, patterns of aging stemming from social effects have not been systematically considered in the synthetic, overarching framework provided by kin selection theory. Applying this framework is an important exercise because it allows social hypotheses of aging to be classified and rigorously assessed. It also reveals hitherto underappreciated possibilities, such as the occurrence of potential kin-selected conflict over the timing of death and aging. In addition, explorations of coevolution between sociality and aging have tended to proceed independently of one another.

In this review I therefore aim to (*a*) critically review formal models that propose social effects on aging, (*b*) propose a framework provided by kin selection theory for classifying social effects on aging, while also collating examples of such effects, and (*c*) synthesize theory and data regarding coevolution between sociality and aging. I also discuss some limits to a social theory of aging by considering whether such a theory uniquely explains conserved, single-gene effects underpinning aging, programmed aging, and adaptive aging. Although some authors (e.g., Hayflick 2000) argue that aging and life span determination are distinct processes, I do not maintain a strict distinction between them. This is because the timing and rate of aging must

Relatedness: probability that two individuals share a gene as the result of kinship

Social action: an action by one individual (the actor) affecting the survivorship or offspring output of another individual (the recipient)

Inclusive fitness: in kin selection theory, the fitness of a focal individual incorporating effects of its social actions toward relatives

Altruism: social action in which the actor experiences a loss in survivorship or offspring output and the recipient a gain

affect average life span, rendering both aging and life span determination potentially subject to the same evolutionary forces. Overall, I seek to place the social evolutionary theory of aging within kin selection theory and to demonstrate that such a theory explains a variety of aging-related phenomena, generates novel predictions, and adds considerably to our general understanding of the evolution of aging.

REVIEW OF MODELS PROPOSING SOCIAL EFFECTS ON AGING

Formal models incorporating social effects on aging fall into two broad classes: those involving altruistic, generally postreproductive care (intergenerational transfers) and those involving limited dispersal in spatially explicit grids. Both classes of models, by considering how altruism interacts with aging, echo some early explanations of why organisms die, which from the late nineteenth century onward proposed population-level benefits of aging (reviewed in Austad 1997a, Rose 1991, Travis 2004, Mitteldorf 2006). The older models, however, frequently proposed that organismal death either benefited nonrelatives or brought long-term evolutionary benefits, thus invoking an unsustainable, naive group selection. (I remark in passing that much current theorizing in the nonevolutionary literature on possible causes of aging retains a naive group selectionist flavor.) As Travis (2004) noted, another feature of a social evolutionary theory of aging is that many of its predictions, in addition to many of those of the classic theory, are independent of the underlying genetic basis of aging and the proximate mechanisms of aging. For example, if individuals gain inclusive fitness from postreproductive life, the degree to which the force of natural selection is attenuated in later life would be reduced with respect to both late-acting effects of pleiotropic genes and the accumulation of late-acting deleterious mutations.

Models Based on Postreproductive Caregiving and Intergenerational Transfers

In line with suggestions from earlier authors (e.g., Fisher 1930, Hamilton 1966, Williams 1957), and stimulated in particular by an interest in explaining menopause in human females using kin selection theory, many researchers have argued that aging will be moderated in species in which care is provided by parents, grandparents, or helpers (reviewed in Arking 2006, Carey & Gruenfelder 1997, Lee 2003). In general, if older individuals enhance their inclusive fitness through care dispensed to relatives (e.g., offspring, grandoffspring, other relatives), selection should counteract aging later in life. Several researchers (e.g., Hawkes et al. 1998, Mace 2000, Roach 1992, Rogers 1993, Shanley & Kirkwood 2001) have modeled this process as it applies to human menopause, with particular reference to the grandmother hypothesis (that aid from grandmothers mediates selection for postreproductive life span extension). Overall, these models find that early reproductive cessation (menopause), and a corresponding extension of postreproductive life span in human females, can be accounted for by the risk of death from childbirth in older mothers combined with inclusive fitness gains from providing care to both offspring and grandoffspring.

Among nonhuman animals, reproductive cessation may arise through physiological changes (i.e., postreproductive individuals are physiologically unable to reproduce, as in menopausal human females) or for other reasons (i.e., postreproductive individuals are physiologically capable of reproducing but have ceased to do so) (Austad 1997b). In principle, kin-selection models for reproductive cessation could apply to both these cases.

Lee (2003) has provided the most general formal model of the effects of intergenerational transfers of investment on aging. Lee (2003) showed analytically that the age-specific strength of selection on mortality (where weaker selection implies higher mortality) in a focal individual is a weighted average of two effects. The first is the classic effect (i.e., that predicted by traditional aging theory as outlined in the previous section), which is proportional to remaining fecundity. The second is a transfer effect, which is proportional to the amount of remaining investments to be transferred to others. When investments are never transferred to others, the transfer effect is zero and the model reverts to the classic theory. When investments are transferred, Lee (2003) argued that selection would bring about an optimal allocation of resources between number of offspring produced (fecundity) and level of investment per offspring. In this case; fitness would not be affected by small changes in fecundity, reducing the weight (which depends on the relationship of fitness with resource consumption) upon the classic effect to zero. Hence the age-specific strength of selection on mortality would be influenced only by the transfer effect. In such circumstances, because juveniles receive investments (enhancing their survivorship) but do not transfer them (hence the size of the transfer effect has not started to fall), the model predicts that juvenile mortality should fall with increasing juvenile age. In addition, because postreproductive adults may still make investments, the transfer effect remains positive after reproduction has ceased, so the model predicts the extension of postreproductive life span. This way, the model predicts two features of aging in humans and other caregiving mammals not predicted by the classic theory. Because Lee's (2003) model incorporates both classic and social effects upon aging in a single framework, it represents a major advance in the field. However, although Lee (2003) recognized links with kin selection theory, kin selection did not enter the formal model. In fact, as Rogers (2003) pointed out, Lee (2003) implicitly assumed asexual reproduction (and hence a parent-offspring relatedness of 1). This precludes the possibility of parent-offspring conflict over the timing of intergenerational transfers of investment (see following section) and hence does not capture the range of evolutionary possibilities inherent in a model with sexual reproduction. Adding kin selection explicitly to Lee's (2003) model would therefore represent an important next step.

Models Based on Limited Dispersal in Spatially Explicit Grids

A second class of models has employed simulations of populations or metapopulations occupying spatially explicit grids to investigate aging as a function of dispersal (Dytham & Travis 2006, Mitteldorf 2006, Travis 2004). These models have shown that a shorter, deterministic life span, or an increased rate of aging, or both, can evolve when dispersal is limited. In two models, these effects were considered to occur through kin selection for altruistic early dying that benefits relatives, both when dispersal distance was fixed (Travis 2004) and when it coevolved with aging (Dytham & Travis 2006). In a third model (Mitteldorf 2006), they were considered to occur through interdemic group selection involving a balance between within-population selection for greater life span (selfishness) and between-population selection for shorter life span (altruism) because populations whose members had greater life spans went extinct more rapidly. Mitteldorf (2006) proposed that this result supported a novel demographic theory of aging distinct from all previous evolutionary theories of aging, including (implicitly) those invoking kin selection. Likewise, Longo et al. (2005) argued for a theory of programmed and altruistic aging based either on kin selection (where relatives benefited) or on group selection (where there was death for the benefit of unrelated organisms). However, it is not clear that kin selection can be excluded as the main factor underlying Mitteldorf's (2006) model. Like those of Travis (2004) and Dytham & Travis (2006), Mitteldorf's (2006) model assumed asexual reproduction and limited dispersal and found that increasing dispersal led to reduced aging. This suggests the presence of kin selection, as increased dispersal would cause local relatedness to fall. In addition, Mitteldorf's (2006) model assumed that density limits population growth at a local level, but with delayed effect, that is density dependence is proportional to past population size. Previous general models (e.g., Wilson et al. 1992) have shown that limited dispersal combined with local density dependence inhibits the evolution of altruism by kin selection because altruistic genotypes remain in competition with one another. However, altruism reemerges if density dependence is delayed (West et al. 2002) because the severity of density dependence becomes less correlated with population composition (Kelly 1992). Hence, a possible alternative interpretation of Mitteldorf's (2006) results is that they arose, under conditions of limited dispersal and delayed density dependence, from such an interaction of kin selection with kin competition. This issue would be clarified by the development of an analytical version of Mitteldorf's (2006) model.

All the limited-dispersal models assumed either declining individual fecundity (Dytham & Travis 2006, Travis 2004) or increasing mortality (Mitteldorf 2006) with increasing age. This means that, as Travis (2004) noted for his model, none provides an alternative to the classic theory of aging, which explains why fecundity should decrease or mortality increase with age from first principles. Instead, these models demonstrate that existing patterns of aging can be modified by social effects as a function of population structure. The limited-dispersal models resemble Lee's (2003) theory of intergenerational transfers in involving an investment or resource (care or space to live) transmitted to individuals that are, or are likely to be, relatives. They also resemble Lee's (2003) model in assuming asexual reproduction and hence in not allowing the possibility of kin-selected conflict over the timing of intergenerational transfers.

KIN SELECTION THEORY AND THE EVOLUTION OF LIFE SPAN AND RATE OF AGING

In this section, I systematically investigate the diversity of possible kin-selected effects on life span and rate of aging by classifying them on the basis of the consequences of

Table 1 Classification with examples of hypothesized social effects on life span and rate of aging

	(*i*) Recipient's fitness is increased	**(*ii*) Recipient's fitness is decreased**
(*a*) Focal individual increases life span	Kin-selected delayed aging in caregiving parent or grandparent	Longer-lived individual denies inherited resources to potential successor; corollary of case *b* × *i*.
(*b*) Focal individual decreases life span or dies	Kin-selected accelerated aging to release inherited resources to potential successor	Death of caregiver curtails care provided to dependent individuals; corollary of case *a* × *i*.

either an increase or a decrease in a focal individual's life span for the fitness of another individual, the recipient (**Table 1**). In each case, I discuss ways in which such an increase or decrease might affect the recipient's fitness and whether the phenomenon under discussion could cause an evolutionary change in the focal individual's life span or pattern of aging through a net increase in its own inclusive fitness. I also consider whether these phenomena provide predictions regarding whether changes in life span in a focal individual should be brought about by changing the date of onset of aging, changing the rate of aging, or changing both the age of onset and the rate of aging. Finally, I review possible evidence, taken from the literature, for the hypothesized effects. The proposed classification applies to kin-selected effects on aging in organisms living in groups of adults without being restricted to such organisms (e.g., they potentially apply to organisms that have parental care but are otherwise solitary). It resembles but cuts across Hamilton's (1964) classification of social actions based on whether effects on the fitness of actors and recipients are positive or negative. I seek reasons for positive or negative effects on life span, and, in this case, effects of the same sign may represent differing Hamiltonian social actions. For example, extending one's life span could be altruistic (if there is postreproductive care of others) or selfish (if resources are denied to others).

Increased Focal Individual's Life Span, Increased Recipient's Fitness

Theory. An important means by which an increase in a focal individual's life span might increase the fitness of a recipient is the provision of care to dependent individuals. Caregiving forms the background to Lee's (2003) model of intergenerational transfers and to the many models invoking kin-selected effects upon aging in the context of reproductive cessation in female mammals, particularly human menopause (see previous section). Briefly, parental care, care by grandparents, and care by helpers are all common phenomena (Clutton-Brock 1991, Wilson 1975). Given that the duration of care affects the amount of care received, increasing the caregiver's life span would, other things equal, raise the recipients' fitness. If caregivers and recipients are related, it would also, by allowing more relatives to be reared, raise the caregiver's inclusive fitness. The most probable mechanism for such an increase in life span would be a delayed onset of aging, as merely decreasing the rate of aging might reduce the quality of care provided. To the extent that bearing young oneself and providing care are mutually exclusive activities (Rogers 1993), such delayed aging should apply more to survival and performance than to fecundity. Indeed, Lee's (2003) intergenerational-transfer model explicitly predicted postreproductive survival of a caregiving parent. Some authors distinguish between selection for an extended postreproductive life span to deliver care to offspring and selection for the same trait to deliver care to grandoffspring or other related young (e.g., Packer et al. 1998). However, although the effects of these two processes may differ quantitatively, kin selection theory makes no fundamental distinction between them. This is because care directed at dependent young related in any degree can, in principle, increase inclusive fitness.

Despite an emphasis in the literature on the evolution of postreproductive life span via caregiving, there are other ways in which an increase in a focal individual's life span might increase the fitness of a recipient. In addition, despite the present review's emphasis on kin selection, other social evolutionary processes conceivably affect aging. Consider the possibility that the mere presence of individuals benefits others, with no specific behaviors being directed at recipients. Such passive benefits of living together were considered in Kokko et al.'s (2001) group augmentation model. A possible example comes from colonially breeding species in which no reproduction occurs below a threshold density (Allee effect, e.g., Stephens et al. 1999). Here, the presence of any given individual within the colony indirectly increases the fitness of others (up to the point where density-dependent decreases in reproductive success occur). Because the additional breeding success of others helps maintain the critical density, the focal individual might indirectly benefit, essentially through by-product mutualism (Connor 1995). If so, to the extent that a focal individual's longer life provides a mechanism for maintaining its presence in the colony, Allee effects could select for an extension to the focal individual's life span. There is no incentive for cheating in such a system because the benefit to others is conferred by an individual's presence alone and so is essentially cost free. The likeliest mechanism for an increase in life span brought about in this way would be delayed onset of aging (with respect to survival, performance, and fecundity, as there would be no incentive to remain in the colony as a nonbreeding individual).

Evidence. Strong evidence for kin selection causing an increase in postreproductive life span would be provided by a phylogenetically corrected comparative analysis to test for the predicted positive association between the relative duration of postreproductive life span and the duration and frequency of care shown to offspring or other relatives. Such an analysis has yet to be conducted. In a recent review, Cohen (2004) collated data on the postreproductive life span in female mammals, but he did not include data on duration of care or group kin structure. Nonetheless, within-taxon comparisons provide suggestive evidence for kin selection (through the provision of care to offspring, grandoffspring, and perhaps other related young) leading to extensions in caregivers' postreproductive life spans or overall life spans (**Table 2**). However, some cautionary notes are warranted. Other studies have suggested that the average inclusive fitness benefits of postreproductive life may sometimes be too small for reproductive cessation to be adaptive (e.g., Japanese macaques *Macaca fuscata*; Pavelka et al. 2002) or that a measurable postreproductive life span can occur in the absence of kin-selected benefits as a nonadaptive correlate of reproductive life span (e.g., guppies *Poecilia reticulata* lacking parental care, Reznick et al. 2006).

In his review, Cohen (2004) found postreproductive life span in females to be widespread but not universal across mammalian orders. He hypothesized that a physiological disassociation between somatic and reproductive aging in female mammals leads to independent selection upon them. This physiological disassociation was hypothesized to stem from the timing of reproductive cessation in mammals being a function of the initial (fixed) number of oocytes and the rate of loss of oocytes, whereas somatic aging is likely to stem from aging across many cell types. Furthermore, Cohen (2004) argued that selection had not synchronized reproductive and somatic aging because it would be costly to females to increase their initial stock of oocytes or to reduce the rate of their loss. However, such costliness has not been demonstrated. Overall, Cohen's (2004) hypothesis provides a possible proximate explanation for widespread reproductive cessation in female mammals. But this phenomenon is not evidence against kin-selection hypotheses for postreproductive life span in humans and other mammals. At the ultimate (evolutionary) level, a postreproductive life span in females might be expected to be widespread in a taxon whose hallmark is maternal care of young. In addition, it is still possible, as Cohen (2004) stated, that variation in the relative duration of postreproductive life span is determined by kin selection, as comparisons in mammals suggest (**Table 2**).

Finally, there are too few data to assess whether, other things being equal, species exhibiting Allee effects display extended life spans. Møller (2006) found no significant difference in life span between colonially breeding and solitarily breeding birds when correcting for phylogeny, sampling effort, body mass, and survival rate. However, although colonially breeding species are good candidates for Allee effects (e.g., Serrano et al. 2005), Møller's (2006) study was not designed to test for the influence of Allee effects.

Increased Focal Individual's Life Span, Decreased Recipient's Fitness

Theory. Merely existing deprives other individuals of the resources that a focal individual consumes. Hence, an increase in the life span of a focal individual will, other things equal, reduce the fitness of others. When nonrelatives suffer reduced fitness, this has no evolutionary consequences for focal individuals, whose inclusive fitness is unaffected. However, when relatives lose fitness, kin selection acts as a brake on the level of cost that one individual inflicts on another. For example, consider the case, common in social systems (Myles 1988, Ragsdale 1999), of resource inheritance. This occurs when, upon their deaths, parents transmit to offspring a resource required for breeding (e.g., nest, territory, rank). Parental death therefore benefits offspring, and a corollary is that an increase in parental life span may decrease offspring fitness by delaying the uptake of the inherited benefit. This situation may influence adult life span because, as touched on by previous authors (Alexander et al. 1991, Austad 1997b, Bonsall 2006, Bourke 1994, Hart & Ratnieks 2005, Reeve & Sherman 1991), it creates

Table 2 Comparative evidence for increased focal individual's life span leading to increased recipient's fitness via kin selection

	Comparison		Interpretation	References
	Group 1	**Group 2**		
Across species				
Whales	Female odontocete whales exhibit postreproductive life span extension and age-related declines in fecundity, live in kin groups, and have relatively prolonged maternal care	Female mysticete whales exhibit no substantial postreproductive life span extension or age-related declines in fecundity, do not live in kin groups, or have relatively prolonged maternal care	Supports hypothesis that maternal care (and possibly care of other relatives) selects for extended postreproductive life span in females; care could encompass intergenerational transmission of useful cultural information	Marsh & Kasuya (1986), Hoelzel (1994), Carey & Gruenfelder (1997), McAuliffe & Whitehead (2005)
African primates and carnivores	Olive baboons *Papio cynocephalus anubis* have a greater period of offspring dependency and greater female postreproductive life span; there was no effect of grandmothers on daughters' productivity or grandoffspring survival	African lions *Panthera leo* have a shorter period of offspring dependency and shorter female postreproductive life span; there was no effect of grandmothers on daughters' productivity or grandoffspring survival	Supports hypothesis that maternal care selects for extended female postreproductive life span; does not support grandmother hypothesis in these species	Packer et al. (1998)
Apes and humans (*Homo sapiens*)	Female humans from present-day foraging societies have postreproductive life span extension and higher lifetime productivity (number of daughters)	Female apes show no or limited postreproductive life span extension and lower lifetime productivity	Supports grandmother hypothesis for human menopause	Hawkes et al. (1998)
Within species				
Sockeye salmon *Onchorhynchus nerka*	Semelparous females that arrive early to breed in the natal stream have a longer reproductive life span, invest relatively less in egg production, and provide care to young in the form of defense of the nest against superimposition by later-arriving females	Semelparous females that arrive late to breed in the natal stream have a shorter reproductive life span, invest relatively more in egg production, and provide no care to young (since risk of their nests being superimposed is small)	Supports hypothesis that maternal care selects for extended female life span	Hendry et al. (2004)

(Continued)

Table 2 *(Continued)*

	Comparison		Interpretation	References
	Group 1	**Group 2**	**Interpretation**	**References**
Marsupial families Didelphidae and Dasyuridae (e.g., genus *Antechinus*)	Males are semelparous, provide no parental care, and show a marked die-off immediately after the rutting season	Females may be iteroparous,[a] provide parental care, and do not show marked die-off immediately after each rutting season	Supports hypothesis that maternal care selects for slower aging and extended female life span	Austad (1997a), Cockburn (1997)
African elephant (genus *Loxodonta*)	Normal group is led by a matriarch, usually the oldest female; group has a matrilineal kin structure; females exhibit postreproductive life span extension and age-related declines in fecundity and have prolonged maternal care	Group lacking matriarch loses organization and may disband or lose members to predators	Supports hypothesis that maternal care (and possibly care of other relatives) selects for extended postreproductive life span in females; as in odontocete whales, care could encompass intergenerational transmission of useful cultural information	Carey & Gruenfelder (1997)
Anthropoid primates	Within species with maternal care, females outlive males	Within species with paternal care, males outlive females	Supports hypothesis that parental care selects for extended life span in the caring sex	Allmann et al. (1998)
Humans *Homo sapiens* (Paraguayan Ache people and eighteenth-century Swedes)	Intergenerational-transfer model predicts observed depression in age-specific mortality rate at greater ages	Classic theory does not predict observed depression in age-specific mortality rate at greater ages	Supports intergenerational-transfer model in humans	Lee (2003)
Humans *Homo sapiens* (Pre-modern Finns and Canadians)	Females with longer postreproductive life spans have more grandchildren	Females with shorter postreproductive life spans have fewer grandchildren	Supports grandmother hypothesis for human menopause	Lahdenperä et al. (2004)

[a]Iteroparous: having more than one reproductive episode per lifetime.

potential kin conflict over the timing of parental death, resembling parent-offspring conflict over the level of parental investment (Trivers 1974).

To develop this hypothesis, say a parent produces c offspring and a nondispersing offspring needs to inherit the parental territory to produce b offspring. Assuming diploidy and sexual reproduction, the offspring is related to its own progeny by $r = 0.5$ and to its parent's progeny (siblings) by 0.5, and hence values each class of progeny equally. But the parent is related to its own progeny by 0.5 and to its offspring's progeny (grandoffspring) by 0.25, and so values each of its own progeny

Semelparous: having a single reproductive episode per lifetime

twice as highly as it values each progeny of its offspring. Hence, by Hamilton's rule, if parental fecundity (c) is declining, the offspring favors the handover of the territory when c falls below b, that is, when $c/b < 1$. But the parent favors it when $c < 0.5b$, that is, when $c/b < 0.5$. This creates a region ($0.5 < c/b < 1$) of potential kin-selected conflict in which the offspring favors the handover of the territory but the parent does not. In this region, an increase in parental life span would decrease the fitness of offspring. One might therefore expect the behavioral expression of conflict [i.e., actual conflict in the sense discussed by Ratnieks & Reeve (1992)] to occur, for example, offspring challenges (harassment) against parents and parental resistance to these challenges. From the parents' standpoint, such harassment might, within the parental lifetime, lead to further fecundity declines and so create greater incentive to offspring to harass. Over evolutionary time, it might constitute a further source of extrinsic mortality, increasing the intrinsic rate of decline in parental fecundity. This way, over both the timescale of one generation and over an evolutionary timescale, resource inheritance systems conceivably create positive feedback between offspring harassment of aging parents and the rate of fecundity decline in parents, leading to accelerated parental aging and death. When $c/b < 0.5$, both parent and offspring would favor territory handover and the parent should relinquish the territory. In this region, a decrease in parental life span would increase the fitness of offspring. This might create selection for immediate parental death by adaptive suicide. Note that the foregoing reasoning invokes a decline in parental fecundity. Therefore, parent-offspring conflict is likely to augment (by affecting its onset and rate) aging that is occurring for conventional reasons. This hypothesis, if correct, demonstrates an ineluctable connection between aging and kin-selected conflict in systems with resource inheritance and sexual reproduction. Ronce et al. (1998) showed that, when a single parent occupies a site and produces both dispersing and nondispersing offspring, the percentage of nondispersing offspring should rise as the parent ages because parental aging increases the chance of a vacant site arising. This model complements the hypothesis developed above because it demonstrates that parental aging increases the likelihood of offspring remaining in the natal area in order to inherit it.

Evidence. Evidence exists that, in social systems with resource inheritance, resource holders and resource inheritors indeed differ over the timing of resource handover and hence over the timing of the death of resource holders. For example, in bumble bees (*Bombus* spp), in which workers inherit the mother queen's nest upon her death and may then reproduce within it (by laying male eggs), would-be reproductive workers harass the queen and sometimes even kill her toward the end of the colony cycle (Bourke 1994). Worker matricide, preceded by worker aggression toward the mother queen, is likely in other social Hymenoptera with reproduction by queenless workers (Bourke 2005, Ratnieks et al. 2006). Honey bee (*Apis mellifera*) workers also oust colony queens that are injured, diseased, insufficiently fecund, or old, but in this case workers do not reproduce, instead replacing the colony queen with a sister or half-sister (Winston 1987). In some ants and in the social naked mole rat (*Heterocephalus glaber*), resource inheritance occurs but data on subordinates' aggression in relation to a breeder's condition are lacking (Hart & Ratnieks 2005). Hence, although offspring aggression against parents clearly constitutes a source of extrinsic mortality in some social species with resource inheritance, whether falling parental productivity generally attracts offspring harassment remains to be tested in a wider variety of species. Evidence for the predicted acceleration in parental aging is considered in the following subsection.

Decreased Focal Individual's Life Span, Increased Recipient's Fitness

Theory. There are at least two ways in which a decrease in a focal individual's life span could increase the fitness of another individual. First, in systems of resource inheritance, the death of a resource-holding focal individual releases benefits to its successor [this also resembles the case modeled by Travis (2004), Dytham & Travis (2006), and Mitteldorf (2006)]. As discussed in the previous subsection, if the resource holder passes a threshold beyond which its cost-to-benefit ratio for dying is sufficiently low, then its inclusive fitness is enhanced by its death and selection should act upon it to commit adaptive suicide, possibly by accelerated aging. One signature of kin-selected aging in social systems with resource inheritance might be sex-specific differences in aging patterns as a function of sex-specific differences in philopatry. When a member of the philopatric sex dies, its resources are inherited by a relative (and hence the focal individual potentially experiences the threshold beyond which adaptive suicide is selected). However, when a member of the dispersing sex dies, there are no benefits to relatives (and hence the focal individual never reaches the threshold for adaptive suicide). Therefore, the theory predicts that, other things being equal, within species the philopatric sex should show a greater rate of aging. Second, in some social contexts, individuals might be selected to develop specific behaviors or structures for the defense of group members. Defense at times of danger to group members is likely to enhance recipients' fitness while risking the defender's life. If the focal individual nonetheless benefits through increased inclusive fitness, the result could be selection for adaptive suicide. Note that in the first situation outlined above, adaptive suicide would be more likely to involve obligate (programmed) death, whereas in the second, being conditional on a behavioral response to danger, it would be more likely to involve facultative suicide and hence not necessarily aging per se. In both cases, an essential condition for adaptive suicide under kin selection theory is that the recipients should be relatives.

Evidence. Both obligate and facultative adaptive suicide benefiting relatives (particularly offspring) are known or suspected in a number of taxa. For example, as mentioned by previous authors (Finch 1990, Kirkwood 1985, Longo et al. 2005), in some invertebrates mothers obligately die or become moribund upon giving birth and their offspring then consume their bodies [e.g., mites (Elbadry & Tawfik 1966), spiders (Evans et al. 1995)]. Foster (1977) presented evidence that the neotropical tree *Tachigalia versicolor* dies after a single episode of reproduction to create a light gap for offspring. Facultative adaptive suicide to protect clone mates from parasites and predators has been reported in bacteria (Crespi 2001, Lewis 2000) and aphids (McAllister & Roitberg 1987). Facultative self-sacrificial defense by workers in favor of related nest mates is widespread in social Hymenoptera and termites (Wilson 1971). As regards parent-offspring conflict over resource inheritance leading to accelerated aging in the parent (see previous subsection), a possible case occurs in *Vespula* wasps, in which old queens appear to undergo physiological breakdown (Spradbery 1973) in favor of reproductive daughter workers (Bourke 1994). However, the prediction that this phenomenon occurs widely under the relevant conditions needs comprehensive testing. Likewise, although aging has been detected in the wild in social birds and mammals with differential dispersal by sex (e.g., McDonald et al. 1996, Nussey et al. 2006), at present there are insufficient data on sex-specific differences in aging within species to test the hypothesis of differential rates of aging as a function of the identity of the philopatric sex.

Not all cases of rapid, apparently suicidal death represent adaptive suicide benefiting relatives. Many animal and plant species exhibit a semelparous life history (monocarpic senescence in plants) in which adults die rapidly after their single bout of reproduction. This trait is believed to have evolved as a result of the comparative magnitude of juvenile and adult mortality reducing the benefits to individuals of attempting additional bouts of reproduction (Finch 1990, Rose 1991, Stearns 1992). There are also cases in which adaptive suicide has been postulated but where the suggested social benefit is problematic. First, Blest (1963) proposed that cryptic adult moths should have truncated postreproductive life spans to reduce opportunities for birds to learn how to find conspecifics (likewise he proposed that aposematic adult moths should have prolonged postreproductive life spans to reinforce the deterrent effects of aposematism on birds). Although there was evidence, which has since been held to support Blest's (1963) proposed mechanism (Carey 2001a), that moths showed the expected patterns of life span (Blest 1963), there is no evidence that the adult moths occurred in kin groups as required. Second, Fabrizio et al. (2004) and Longo et al. (2005) argued for suicide in populations of yeast (*Saccharomyces cerevisiae*) cells whose evolutionary basis is an enhancement of the mutation rate followed by the regrowth of a small mutant subpopulation adapted to the altered culture conditions. However, these authors discussed but could not exclude an adaptive aging program based on kin selection operating in wild populations (e.g., to benefit clone mates; Büttner et al. 2006) or, in their experimental conditions, selection for rare mutants adapted to the altered medium.

Decreased Focal Individual's Life Span, Decreased Recipient's Fitness

Theory. Selection for decreased life span in order to decrease a recipient's fitness seems unlikely because an increase in a focal individual's inclusive fitness would rarely follow. An exception might occur if conditions favored spite (whereby a social act brings about a reduction in survivorship or offspring for both parties; Hamilton 1964), but such conditions are probably relatively rare (Gardner & West 2004, Grafen 1985, Lehmann et al. 2006). A corollary of the occurrence of caregiving is that premature death in a caregiver would decrease the fitness of recipients. Such a decrease could be very severe (involving, for example, the death of young unable to defend or feed themselves), and the caregiver's loss of inclusive fitness would be correspondingly high. This scenario might therefore bring about selection for risk-averse behaviors in caregivers, ensuring their survival throughout the period of the recipients' dependency and as a result prolonging their own life spans.

Evidence. I know of no cases of individuals decreasing their life spans to reduce the fitness of recipients. As regards selection for risk aversion, the provision of care appears to promote risk-averse behavior in primates, as individuals not engaged in caregiving are more risk seeking than caregivers (Allman & Hasenstaub 1999). However, note that, if social living insures against the loss of investment in dependent young as a result of a focal individual's death (Gadagkar 1990), existing sociality might mask the occurrence of specific risk-averse behaviors in caregivers.

COEVOLUTION OF AGING PATTERNS WITH HELPING BEHAVIOR AND SOCIALITY

There are several contexts in which aging patterns have coevolved with helping behavior (incorporating altruism, parental care, and intergenerational transfers) or sociality. In this section, I review and synthesize them.

Coevolution of Postreproductive Life Span with Postreproductive Helping Behavior

Carey & Judge (2001) and Carey (2003) proposed a verbal model for self-reinforcing life span extension in humans. The starting point is selection for parental care and increased investment per offspring. In the model, this leads to greater offspring survival and reduced parental fecundity, followed by greater parental survival (through reduced costs of reproduction) and hence extended parental life span. In turn, this provides more opportunity for intergenerational transfers, including those to grand-offspring. This provides yet more investment to individual offspring, creating a causal loop of positive feedback that promotes extended life span associated with pronounced, intergenerational transfers. Kaplan & Robson (2002) also argued for self-reinforcing life span extension in humans, although these authors emphasized the effect of investment in somatic capital (in humans, a large brain) having a future payoff in promoting life span extension. Following Lee's (2003) general model of intergenerational transfers, a model by Chu & Lee (2006) also produced coevolution of life span and intergenerational transfers. In general, it appears that a web of positive feedback exists between group size, brain size, and individual life span in primates (e.g., Allman et al. 1993, Allman & Hasenstaub 1999, Kaplan & Robson 2002, Lindenfors 2005, Walker et al. 2006). Carey & Judge (2001) did not discuss what might set a limit on self-reinforcing life span extension in their model. Presumably, however, in the postreproductive phase, the influence of extrinsic mortality devaluing care delivered late compared to care delivered early must still apply (see Roach 1992). In addition, parent-offspring conflict over the timing of transfers (when these involve resource inheritance) may also prevent indefinite life span extension (see previous section). Carey & Judge (2001) also implied that their model had general applicability to social species. The model indeed appears relevant to vertebrates with postreproductive care, but for other social species (e.g., cooperative breeders in vertebrates, social insects) it appears less applicable because it does not account for the effects of the presence of helpers alongside reproductives. Specifically, in these species, the fecundity of reproductives often increases rather than decreases, as Carey & Judge (2001) assume, owing to the presence of helpers [e.g., social insects (Wilson 1971), social mammals (Russell 2004)]. Nonetheless, reproductives in these species generally exhibit extended life span, almost certainly because of disruptive selection on the life spans of helper and reproductive phenotypes (see below).

Coevolution of Life Span with Sociality at the Origin of Social Groups

Phylogenetic analyses suggest that, in birds, low adult mortality predisposes lineages to adopt cooperative breeding because low adult mortality reduces the turnover of breeding individuals and hence limits the number of opportunities for independent breeding (Arnold & Owens 1998). Ridley et al. (2005) proposed a model that provided a route by which low mortality might facilitate group living (and hence cooperative breeding) in territorial species. According to these authors, longevity favors the establishment of reciprocally altruistic, nonaggressive associations between neighbors, which increases the chance that nonbreeding subordinates eventually acquire breeding rank in their natal area (local dominance), which in turn promotes group living. Hence, in birds, a preexisting feature of the life history (low adult mortality) may promote the origin of sociality. This may itself further influence life span, given that cooperatively breeding birds have more prolonged periods of offspring dependency than noncooperatively breeding ones (Langen 2000; see also the following subsection). By contrast, evidence exists that insect species resembling those likely to have been ancestral to social species were already characterized by offspring with a long period of dependency, combined with short-lived adults (Field & Brace 2004). In these cases, group living may have evolved to insure against total reproductive failure as a consequence of mortality in adult caregivers (Field et al. 2000, Gadagkar 1990, Queller 1994).

Coevolution of Aging Patterns with Sociality in Established Social Groups

Once group living has evolved, there are likely to be further coevolutionary changes to life span (Alexander et al. 1991, Carey 2001b). For example, societies with small group sizes are typified by age-based queues for the top-ranked breeding position, a phenomenon likely to cause mutually dependent evolution of breeder and subordinate life spans (Mesterton-Gibbons et al. 2006, Shreeves & Field 2002). In societies characterized by caregiving, nonreproductive helpers, Alexander et al. (1991) proposed that disruptive selection should act to reduce the life span of helper phenotypes and increase the life span of reproductive phenotypes. Their argument was that in helpers, the age of first helping represents the critical period that triggers the onset of aging (corresponding to the age of first reproduction in reproductive phenotypes). Unlike reproduction, which requires sexual maturity, helping can be effectively performed by juveniles or young adults. Therefore, helping behavior induces earlier aging in helpers relative to reproductives. In addition, helpers perform relatively risky foraging tasks, increasing their level of extrinsic mortality and further promoting earlier aging. By contrast, reproductives avoid risky tasks and hence have diminished extrinsic mortality, thus delaying aging. These effects are self-reinforcing because relatively earlier aging in helpers decreases the value of any later reproduction by them, promoting further phenotypic specialization as helpers (Alexander et al. 1991). Eventually, such a process is likely to release a cascade of consequences propelling the lineage toward advanced sociality with short-lived, nonreproductive helpers and long-lived, highly fecund reproductives (Alexander et al. 1991; Bourke 1999; Crespi 2004, 2007). Evidence for disruptive selection on aging in social groups is of two types. First, within highly social taxa, helpers have shorter mean life spans than reproductives, examples being social mole rats (Dammann & Burda 2006, Jarvis et al. 1994) and social insects (Alexander et al. 1991, Finch 1990, Wilson 1971). Second, helpers in highly social species have relatively shorter life spans than individuals in related solitary species, and reproductives in highly social species have relatively longer life spans than individuals in related, solitary species [insects (Alexander et al. 1991, Carey 2001b, Keller 1998, Keller & Genoud 1997), rodents (Sherman & Jarvis 2002)]. Evidence exists that, once advanced sociality has arisen, aging patterns within the helper class may further evolve as a function of between-group selection for group-level traits (Amdam & Omholt 2002, O'Donnell & Jeanne 1995, Oster & Wilson 1978), differing patterns of intergenerational transfers (Amdam & Page 2005), or differences in extrinsic mortality between categories of helpers (Chapuisat & Keller 2002). Within-group variation in aging patterns demonstrates that the genetic basis of aging can be modulated by differential gene expression (Finch 1990, Keller & Jemielity 2006). For all these reasons, highly social species provide special opportunities for within-species investigations of both proximate and ultimate mechanisms underlying aging (Buffenstein 2005, Carey 2001b, Gräff et al. 2007, Keller & Genoud 1997, Jemielity et al. 2007, Keller & Jemielity 2006, Parker et al. 2004, Seehuus et al. 2006, Sherman & Jarvis 2002).

LIMITS TO A SOCIAL THEORY OF AGING

To conclude, I consider a set of interrelated issues that are potentially addressed by a social theory of aging but that remain controversial. First, recent discoveries demonstrate that some of the molecular pathways affecting aging are influenced by single genes and are conserved across a very broad phylogenetic span (Arking 2006, Guarente & Kenyon 2000). These findings appear inconsistent with the classic theory, which implies that aging should affect all cells, tissues, and organs in a haphazard way and that genetic influences upon aging should vary widely across lineages (Partridge & Gems 2002a,b, 2006). By contrast, they appear to follow from a social theory that invokes the ancient evolution of single genes underpinning deterministic life spans (Dytham & Travis 2006, Mitteldorf 2004, Travis 2004). However, it is not clear that such effects are inconsistent with the classic theory. Because molecular signaling pathways related to nutrition are strongly implicated in aging, it is possible, under the classic theory, that what has been conserved over evolutionary history is the manner (involving one or a few genes) in which these pathways are genetically controlled (Partridge & Gems 2002a). More generally, single genes might influence the molecular pathways underlying the trade-off between somatic maintenance (survival) and reproduction (Barnes & Partridge 2003). If so, conserved, single-gene effects on aging do not uniquely support a social theory of aging.

Second, many authors have argued that, under the classic theory, aging is not a coordinated process under direct genetic control like development and hence is not programmed (e.g., Hayflick 2000, Kirkwood 2005, Kirkwood & Austad 2000, Partridge & Gems 2002a). Some authors (e.g., Kirkwood 2005) hold this to be the case even in semelparous organisms in which death rapidly follows reproduction, but others do not (Rose 1991), with plant biologists even postulating a so-called death hormone to account for monocarpic senescence (Wilson 1997). Still other authors have argued that analogies with programmed cell death (apoptosis) show organismal aging is programmed (Longo et al. 2005, Skulachev 2001). By contrast, a social theory of aging predicts programmed aging less equivocally (e.g., deterministic life spans in the limited-dispersal models; Dytham & Travis 2006, Mitteldorf 2006, Travis 2004). However, the dispute over whether the classic theory entails programmed aging appears partly semantic. Partridge & Gems (2002a) argue that aging is under indirect genetic control in that there is genetic variation in the mechanisms regulating the rate at which damage from aging is repaired. Such a process closely resembles what Longo et al. (2005) regard as programming. This resemblance makes it less clear that programmed aging occurs only under the social theory and supports the case that its multiple meanings make programmed aging a term best avoided (Partridge & Gems 2006, Rose 1991).

Finally, some authors have suggested that, unlike the classic theory (e.g., Kirkwood 1985), the social theory shows aging to be adaptive (e.g., Longo et al. 2005; Mitteldorf 2004, 2006; Travis 2004). The social theory does propose adaptive reasons for aging (for example, in the sense that genes affecting aging are subject to kin selection), but the contrast with the classic theory is again partly semantic. Under antagonistic pleiotropy in the classic case, aging is adaptive in the sense that the causative gene spreads through a population by selection and so must be associated with higher individual fitness (Rose 1991). However, one of the gene's effects (aging in later life) is detrimental to individual survival. Hence, labeling aging via antagonistic pleiotropy as adaptive or nonadaptive depends upon whether one defines adaptation in terms of evolutionary fitness or individual welfare. More broadly, adaptation must always be defined relative to some entity's interests (Dawkins 1982), so it is uninformative to label aging as adaptive or nonadaptive without this qualification.

SUMMARY POINTS

1. The evolution of life span and rate of aging should in principle be subject to kin selection because a focal individual's longevity and mortality schedule may affect the fitness of relatives and hence the individual's inclusive fitness.

2. Two classes of formal models have demonstrated that aging may be influenced by social effects through kin selection. Intergenerational-transfer models have shown that reduced juvenile mortality, reproductive cessation, and postreproductive life span can arise from kin selection for care to offspring or grandoffspring. Spatially explicit population models have shown that shorter, deterministic life spans or increased rates of aging can evolve when limited dispersal allows relatives to benefit from the deaths of others.

3. In principle, a focal individual could increase its inclusive fitness by increasing or decreasing its life span and simultaneously increasing or decreasing the fitness of related individuals (recipients). Evidence exists for kin selection occurring for all four combinations of changes in a focal individual's life span and a recipient's fitness, except for a decrease in a focal individual's life span associated with a decrease in a recipient's fitness.

4. In sexual organisms with resource inheritance, kin selection theory predicts potential conflict between parents undergoing aging and their offspring over the timing of resource inheritance, conceivably leading to offspring harassment of parents and accelerated parental aging.

5. In social organisms, the degree and form of sociality and patterns of aging are likely to coevolve. Examples include positive feedback between selection for postreproductive life span extension and selection for postreproductive intergenerational transfers and, in social groups with nonreproductive helpers, disruptive selection for reduced life span in helper phenotypes and extended life span in reproductive phenotypes.

6. Overall, a social evolutionary theory of aging based on kin selection, although not uniquely explaining either conserved, single-gene effects underpinning aging, programmed aging, or adaptive aging, encompasses many previously proposed social theories of aging, explains a wide range of aging-related phenomena in a broad variety of taxa, generates novel predictions, and considerably enriches our general understanding of the evolution of life span and rate of aging.

DISCLOSURE STATEMENT

The author is not aware of any biases that might be perceived as affecting the objectivity of this review.

ACKNOWLEDGMENTS

I thank Bernie Crespi for inviting me to contribute a review. I also thank Peter Bennett, Mike Cant, Ian Owens, David Richardson, and Andy Russell for valuable discussions, and Peter Bennett, Tracey Chapman, Bernie Crespi, Laurent Keller, and Doug Yu for helpful comments on the manuscript.

LITERATURE CITED

Abrams PA. 1993. Does increased mortality favor the evolution of more rapid senescence? *Evolution* 47:877–87

Alexander RD, Noonan KM, Crespi BJ. 1991. The evolution of eusociality. In *The Biology of the Naked Mole-Rat*, ed. PW Sherman, JUM Jarvis, RD Alexander, pp. 3–44. Princeton: Princeton Univ. Press

Allman J, Hasenstaub A. 1999. Brains, maturation times, and parenting. *Neurobiol. Aging* 20:447–54

Allman J, McLaughlin T, Hakeem A. 1993. Brain weight and life-span in primate species. *Proc. Natl. Acad. Sci. USA* 90:118–22

Allman J, Rosin A, Kumar R, Hasenstaub A. 1998. Parenting and survival in anthropoid primates: caretakers live longer. *Proc. Natl. Acad. Sci. USA* 95:6866–69

Alonso WJ, Schuck-Paim C. 2002. Sex-ratio conflicts, kin selection, and the evolution of altruism. *Proc. Natl. Acad. Sci. USA* 99:6843–47

Amdam GV, Omholt SW. 2002. The regulatory anatomy of honeybee lifespan. *J. Theor. Biol.* 216:209–28

Amdam GV, Page RE. 2005. Intergenerational transfers may have decoupled physiological and chronological age in a eusocial insect. *Ageing Res. Rev.* 4:398–408

Arking R. 2006. *The Biology of Aging: Observations and Principles*. New York: Oxford Univ. Press. 3rd ed.

Arnold KE, Owens IPF. 1998. Cooperative breeding in birds: a comparative test of the life history hypothesis. *Proc. R. Soc. London Ser. B* 265:739–45

Austad SN. 1997a. *Why We Age: What Science is Discovering about the Body's Journey Through Life*. New York: Wiley & Sons

Austad SN. 1997b. Postreproductive survival. In *Between Zeus and the Salmon: The Biodemography of Longevity*, ed. KW Wachter, CE Finch, 161–74. Washington, DC: Natl. Acad.

Barnes AI, Partridge L. 2003. Costing reproduction. *Anim. Behav.* 66:199–204

Blest AD. 1963. Longevity, palatability and natural selection in five species of New World saturniid moth. *Nature* 197:1183–86

Bonsall MB. 2006. Longevity and ageing: appraising the evolutionary consequences of growing old. *Philos. Trans. R. Soc. London Ser. B* 361:119–35

Bourke AFG. 1994. Worker matricide in social bees and wasps. *J. Theor. Biol.* 167:283–92

Bourke AFG. 1999. Colony size, social complexity and reproductive conflict in social insects. *J. Evol. Biol.* 12:245–57

Bourke AFG. 2005. Genetics, relatedness and social behavior in insect societies. In *Insect Evolutionary Ecology*, ed. MDE Fellowes, GJ Holloway, J Rolff, pp. 1–30. Wallingford: CABI Publ.

Buffenstein R. 2005. The naked mole-rat? A new long-living model for human aging research. *J. Gerontol. Biol. Sci.* 60A:1369–77

Büttner S, Eisenberg T, Herker E, Carmona-Gutierrez D, Kroemer G, Madeo F. 2006. Why yeast cells can undergo apoptosis: death in times of peace, love, and war. *J. Cell Biol.* 175:521–25

Campisi J. 2005. Senescent cells, tumor suppression, and organismal aging: good citizens, bad neighbors. *Cell* 120:513–22

Carey J. 2001a. Insect biodemography. *Annu. Rev. Entomol.* 46:79–110

Carey JR. 2001b. Demographic mechanisms for the evolution of long life in social insects. *Exper. Gerontol.* 36:713–22

Carey JR. 2003. *Longevity: The Biology and Demography of Life Span.* Princeton: Princeton Univ. Press

Carey JR, Gruenfelder C. 1997. Population biology of the elderly. In *Between Zeus and the Salmon: The Biodemography of Longevity,* ed. KW Wachter, CE Finch, pp. 127–60. Washington, DC: Natl. Acad.

Carey JR, Judge DS. 2001. Life span extension in humans is self-reinforcing: a general theory of longevity. *Popul. Dev. Rev.* 27:411–36

Chapuisat M, Keller L. 2002. Division of labour influences the rate of ageing in weaver ant workers. *Proc. R. Soc. London Ser. B* 269:909–13

Charlesworth B. 1980. *Evolution in Age-structured Populations.* Cambridge, United Kingdom: Cambridge Univ. Press

Charlesworth B. 2000. Fisher, Medawar, Hamilton and the evolution of aging. *Genetics* 156:927–31

Chu CYC, Lee RD. 2006. The coevolution of intergenerational transfers and longevity: an optimal life history approach. *Theor. Pop. Biol.* 69:193–201

Clutton-Brock TH. 1991. *The Evolution of Parental Care.* Princeton, NJ: Princeton Univ. Press

Cockburn A. 1997. Living slow and dying young: senescence in marsupials. In *Marsupial Biology: Recent Research, New Perspectives,* ed. NR Saunders, LA Hinds, pp. 163–71. Sydney: Univ. NSW Press

Cohen AA. 2004. Female postreproductive lifespan: a general mammalian trait. *Biol. Rev.* 79:733–50

Connor RC. 1995. The benefits of mutualism: a conceptual framework. *Biol. Rev.* 70:427–57

Crespi BJ. 2001. The evolution of social behavior in microorganisms. *Trends Ecol. Evol.* 16:178–83

Crespi BJ. 2004. Vicious circles: positive feedback in major evolutionary and ecological transitions. *Trends Ecol. Evol.* 19:627–33

Crespi B. 2007. Comparative evolutionary ecology of social and sexual systems: water-breathing insects come of age. In *Evolutionary Ecology of Social and Sexual Systems: Crustaceans as Model Organisms,* ed. JE Duffy, M Thiel. New York: Oxford Univ. Press. In press

Dammann P, Burda H. 2006. Sexual activity and reproduction delay ageing in a mammal. *Curr. Biol.* 16:R117–18

Dawkins R. 1982. *The Extended Phenotype.* Oxford, United Kingdom: Freeman

Dytham C, Travis JMJ. 2006. Evolving dispersal and age at death. *Oikos* 113:530–38

Elbadry EA, Tawfik MSF. 1966. Life cycle of mite *Adactylidium* sp (Acarina: Pyemotidae) a predator of thrips eggs in United Arab Republic. *Ann. Entomol. Soc. Am.* 59:458–61

Emlen ST. 1995. An evolutionary theory of the family. *Proc. Natl. Acad. Sci. USA* 92:8092–99

Evans TA, Wallis EJ, Elgar MA. 1995. Making a meal of mother. *Nature* 376:299

Fabrizio P, Battistella L, Vardavas R, Gattazzo C, Liou LL, et al. 2004. Superoxide is a mediator of an altruistic aging program in *Saccharomyces cerevisiae*. *J. Cell Biol.* 166:1055–67

Field J, Brace S. 2004. Pre-social benefits of extended parental care. *Nature* 428:650–52

Field J, Shreeves G, Sumner S, Casiraghi M. 2000. Insurance-based advantage to helpers in a tropical hover wasp. *Nature* 404:869–71

Finch CE. 1990. *Longevity, Senescence, and the Genome*. Chicago: Univ. Chicago Press

Fisher RA. 1930. *The Genetical Theory of Natural Selection*. Oxford: Clarendon

Foster KR, Wenseleers T, Ratnieks FLW. 2006. Kin selection is the key to altruism. *Trends Ecol. Evol.* 21:57–60

Foster RB. 1977. *Tachigalia versicolor* is a suicidal neotropical tree. *Nature* 268:624–26

Gadagkar R. 1990. Evolution of eusociality: the advantage of assured fitness returns. *Philos. Trans. R. Soc. London Ser. B* 329:17–25

Gardner A, West SA. 2004. Spite and the scale of competition. *J. Evol. Biol.* 17:1195–203

Grafen A. 1985. A geometric view of relatedness. In *Oxford Surveys in Evolutionary Biology*, ed. R Dawkins, M Ridley, 2:28–89. Oxford: Oxford Univ. Press

Gräff J, Jemielity S, Parker JD, Parker KM, Keller L. 2007. Differential gene expression between adult queens and workers in the ant *Lasius niger*. *Mol. Ecol.* 16:675–83

Guarente L, Kenyon C. 2000. Genetic pathways that regulate ageing in model organisms. *Nature* 408:255–62

Hamilton WD. 1964. The genetical evolution of social behavior I, II. *J. Theor. Biol.* 7:1–52

Hamilton WD. 1966. The moulding of senescence by natural selection. *J. Theor. Biol.* 12:12–45

Hart AG, Ratnieks FLW. 2005. Crossing the taxonomic divide: conflict and its resolution in societies of reproductively totipotent individuals. *J. Evol. Biol.* 18:383–95

Hawkes K, O'Connell JF, Blurton Jones NG, Alvarez H, Charnov EL. 1998. Grandmothering, menopause, and the evolution of human life histories. *Proc. Natl. Acad. Sci. USA* 95:1336–39

Hayflick L. 2000. The future of ageing. *Nature* 408:267–69

Hendry AP, Morbey YE, Berg OK, Wenburg JK. 2004. Adaptive variation in senescence: reproductive lifespan in a wild salmon population. *Proc. R. Soc. London Ser. B* 271:259–66

Hoelzel AR. 1994. Genetics and ecology of whales and dolphins. *Annu. Rev. Ecol. Syst.* 25:377–99

Hughes KA, Reynolds RM. 2005. Evolutionary and mechanistic theories of aging. *Annu. Rev. Entomol.* 50:421–45

Jarvis JUM, O'Riain MJ, Bennett NC, Sherman PW. 1994. Mammalian eusociality: a family affair. *Trends Ecol. Evol.* 9:47–51

Jemielity S, Kimura M, Parker KM, Parker JD, Cao XJ, et al. 2007. Short telomeres in short-lived males: what are the molecular and evolutionary causes? *Aging Cell* 6:225–33

Kaplan HS, Robson AJ. 2002. The emergence of humans: the coevolution of intelligence and longevity with intergenerational transfers. *Proc. Natl. Acad. Sci. USA* 99:10221–26

Keller L. 1998. Queen lifespan and colony characteristics in ants and termites. *Insectes Soc.* 45:235–46

Keller L, Genoud M. 1997. Extraordinary lifespans in ants: a test of evolutionary theories of ageing. *Nature* 389:958–60

Keller L, Jemielity S. 2006. Social insects as a model to study the molecular basis of ageing. *Exper. Gerontol.* 41:553–56

Kelly JK. 1992. Kin selection in density regulated populations. *J. Theor. Biol.* 157:447–61

Kirkwood TBL. 1977. Evolution of ageing. *Nature* 270:301–4

Kirkwood TBL. 1985. Comparative and evolutionary aspects of longevity. In *Handbook of the Biology of Aging*, ed. CE Finch, EL Schneider, pp. 27–44. New York: Van Nostrand Reinhold

Kirkwood TBL. 2005. Understanding the odd science of aging. *Cell* 120:437–47

Kirkwood TBL, Austad SN. 2000. Why do we age? *Nature* 408:233–38

Kokko H, Johnstone RA, Clutton-Brock TH. 2001. The evolution of cooperative breeding through group augmentation. *Proc. R. Soc. London Ser. B* 268:187–96

Lahdenperä M, Lummaa V, Helle S, Tremblay M, Russell AF. 2004. Fitness benefits of prolonged postreproductive lifespan in women. *Nature* 428:178–81

Langen TK. 2000. Prolonged offspring dependence and cooperative breeding in birds. *Behav. Ecol.* 11:367–77

Lee RD. 2003. Rethinking the evolutionary theory of aging: transfers, not births, shape senescence in social species. *Proc. Natl. Acad. Sci. USA* 100:9637–42

Lehmann L, Bargum K, Reuter M. 2006. An evolutionary analysis of the relationship between spite and altruism. *J. Evol. Biol.* 19:1507–16

Lehmann L, Keller L. 2006. The evolution of co-operation and altruism—a general framework and a classification of models. *J. Evol. Biol.* 19:1365–76

Leroi AM, Bartke A, De Benedictis G, Franceschi C, Gartner A, et al. 2005. What evidence is there for the existence of individual genes with antagonistic pleiotropic effects? *Mech. Ageing Dev.* 126:421–29

Lewis K. 2000. Programmed death in bacteria. *Microbiol. Mol. Biol. Rev.* 64:503–14

Lindenfors P. 2005. Neocortex evolution in primates: the 'social brain' is for females. *Biol. Lett.* 1:407–10

Longo VD, Mitteldorf J, Skulachev VP. 2005. Programmed and altruistic ageing. *Nature Rev. Gen.* 6:866–72

Mace R. 2000. Evolutionary ecology of human life history. *Anim. Behav.* 59:1–10

Marsh H, Kasuya T. 1986. Evidence for reproductive senescence in female cetaceans. *Rep. Intl. Whal. Com. Spec. Iss.* 8:57–74

McAllister MK, Roitberg BD. 1987. Adaptive suicidal behavior in pea aphids. *Nature* 328:797–99

McAuliffe K, Whitehead H. 2005. Eusociality, menopause and information in matrilineal whales. *Trends Ecol. Evol.* 20:650

McDonald DB, Fitzpatrick JW, Woolfenden GE. 1996. Actuarial senescence and demographic heterogeneity in the Florida Scrub Jay. *Ecology* 77:2373–81

Mesterton-Gibbons M, Hardy ICW, Field J. 2006. The effect of differential survivorship on the stability of reproductive queueing. *J. Theor. Biol.* 242:699–712

Michod RE. 1982. The theory of kin selection. *Annu. Rev. Ecol. Syst.* 13:23–55

Mitteldorf J. 2004. Ageing selected for its own sake. *Evol. Ecol. Res.* 6:937–53

Mitteldorf J. 2006. Chaotic population dynamics and the evolution of ageing. *Evol. Ecol. Res.* 8:561–74

Møller AP. 2006. Sociality, age at first reproduction and senescence: comparative analyses of birds. *J. Evol. Biol.* 19:682–89

Myles TG. 1988. Resource inheritance in social evolution from termites to man. In *The Ecology of Social Behavior*, ed. CN Slobodchikoff, pp. 379–423. San Diego: Academic

Nussey DH, Kruuk LEB, Donald A, Fowlie M, Clutton-Brock TH. 2006. The rate of senescence in maternal performance increases with early-life fecundity in red deer. *Ecol. Lett.* 9:1342–50

O'Donnell S, Jeanne RL. 1995. Implications of senescence patterns for the evolution of age polyethism in eusocial insects. *Behav. Ecol.* 6:269–73

Oster GF, Wilson EO. 1978. *Caste and Ecology in the Social Insects*. Princeton: Princeton Univ. Press

Packer C, Tatar M, Collins A. 1998. Reproductive cessation in female mammals. *Nature* 392:807–11

Parker JD, Parker KM, Sohal BH, Sohal RS, Keller L. 2004. Decreased expression of Cu-Zn superoxide dismutase 1 in ants with extreme lifespan. *Proc. Natl. Acad. Sci. USA* 101:3486–89

Partridge L, Barton NH. 1993. Optimality, mutation and the evolution of ageing. *Nature* 362:305–11

Partridge L, Gems D. 2002a. Mechanisms of ageing: public or private? *Nature Rev. Gen.* 3:165–75

Partridge L, Gems D. 2002b. The evolution of longevity. *Curr. Biol.* 12:R544–46

Partridge L, Gems D. 2006. Beyond the evolutionary theory of ageing, from functional genomics to evo-gero. *Trends Ecol. Evol.* 21:334–40

Pavelka MSM, Fedigan LM, Zohar S. 2002. Availability and adaptive value of reproductive and postreproductive Japanese macaque mothers and grandmothers. *Anim. Behav.* 64:407–14

Queller DC. 1994. Extended parental care and the origin of eusociality. *Proc. R. Soc. London Ser. B* 256:105–11

Ragsdale JE. 1999. Reproductive skew theory extended: the effect of resource inheritance on social organization. *Evol. Ecol. Res.* 1:859–74

Ratnieks FLW, Foster KR, Wenseleers T. 2006. Conflict resolution in insect societies. *Annu. Rev. Entomol.* 51:581–608

Ratnieks FLW, Reeve HK. 1992. Conflict in single-queen Hymenopteran societies: the structure of conflict and processes that reduce conflict in advanced eusocial species. *J. Theor. Biol.* 158:33–65

Reeve HK, Sherman PW. 1991. Intracolonial aggression and nepotism by the breeding female naked mole-rat. In *The Biology of the Naked Mole-Rat*, ed. PW Sherman, JUM Jarvis, RD Alexander, pp. 337–57. Princeton: Princeton Univ. Press

Reznick D, Bryant M, Holmes D. 2006. The evolution of senescence and postreproductive lifespan in guppies (*Poecilia reticulata*). *PLoS Biol.* 4:136–43

Reznick DN, Bryant MJ, Roff D, Ghalambor CK, Ghalambor DE. 2004. Effect of extrinsic mortality on the evolution of senescence in guppies. *Nature* 431:1095–99

Ridley J, Yu DW, Sutherland WJ. 2005. Why long-lived species are more likely to be social: the role of local dominance. *Behav. Ecol.* 16:358–63

Roach DA. 1992. Parental care and the allocation of resources across generations. *Evol. Ecol.* 6:187–97

Rogers AR. 1993. Why menopause? *Evol. Ecol.* 7:406–20

Rogers AR. 2003. Economics and the evolution of life histories. *Proc. Natl. Acad. Sci. USA* 100:9114–15

Ronce O, Clobert J, Massot M. 1998. Natal dispersal and senescence. *Proc. Natl. Acad. Sci. USA* 95:600–5

Rose MR. 1991. *Evolutionary Biology of Aging*. New York: Oxford Univ. Press

Rose MR. 1997. Toward an evolutionary demography. In *Between Zeus and the Salmon: The Biodemography of Longevity*, ed. KW Wachter, CE Finch, pp. 96–107. Washington, DC: Natl. Acad.

Russell AF. 2004. Mammals: comparisons and contrasts. In *Ecology and Evolution of Cooperative Breeding in Birds*, ed. WD Koenig, JL Dickinson, 210–27. Cambridge, United Kingdom: Cambridge Univ. Press

Seehuus SC, Norberg K, Gimsa U, Krekling T, Amdam GV. 2006. Reproductive protein protects functionally sterile honey bee workers from oxidative stress. *Proc. Natl. Acad. Sci. USA* 103:962–67

Serrano D, Oro D, Esperanza U, Tella JL. 2005. Colony size selection determines adult survival and dispersal preferences: Allee effects in a colonial bird. *Am. Nat.* 166:E22–31

Shanley DP, Kirkwood TBL. 2001. Evolution of the human menopause. *BioEssays* 23:282–87

Sherman PW, Jarvis JUM. 2002. Extraordinary life spans of naked mole rats (*Heterocephalus glaber*). *J. Zool.* 258:307–11

Shreeves G, Field J. 2002. Group size and direct fitness in social queues. *Am. Nat.* 159:81–95

Skulachev VP. 2001. The programmed death phenomena, aging, and the Samurai law of biology. *Exper. Gerontol.* 36:995–1024

Spradbery JP. 1973. *Wasps.* London: Sidgwick Jackson

Stearns SC. 1992. *The Evolution of Life Histories*. Oxford: Oxford Univ. Press

Stephens PA, Sutherland WJ, Freckleton RP. 1999. What is the Allee effect? *Oikos* 87:185–90

Travis JMJ. 2004. The evolution of programmed death in a spatially structured population. *J. Gerontol. Biol. Sci.* 59A:301–5

Trivers R. 1985. *Social Evolution*. Menlo Park, CA: Benjamin/Cummings

Trivers RL. 1974. Parent-offspring conflict. *Am. Zool.* 14:249–64

Walker R, Burger O, Wagner J, Von Rueden CR. 2006. Evolution of brain size and juvenile periods in primates. *J. Hum. Evol.* 51:480–89

West-Eberhard MJ. 1975. The evolution of social behavior by kin selection. *Q. Rev. Biol.* 50:1–33

West SA, Pen I, Griffin AS. 2002. Cooperation and competition between relatives. *Science* 296:72–75

Williams GC. 1957. Pleiotropy, natural selection, and the evolution of senescence. *Evolution* 11:398–411

Williams PD, Day T, Fletcher Q, Rowe L. 2006. The shaping of senescence in the wild. *Trends Ecol. Evol.* 21:458–63

Wilson DS, Pollock GB, Dugatkin LA. 1992. Can altruism evolve in purely viscous populations? *Evol. Ecol.* 6:331–41

Wilson EO. 1971. *The Insect Societies*. Cambridge, MA: Belknap Press

Wilson EO. 1975. *Sociobiology: The New Synthesis*. Cambridge, MA: Belknap Press

Wilson EO, Hölldobler B. 2005. Eusociality: origin and consequences. *Proc. Natl. Acad. Sci. USA* 102:13367–71

Wilson JB. 1997. An evolutionary perspective on the 'death hormone' hypothesis in plants. *Physiol. Plant.* 99:511–16

Winston ML. 1987. *The Biology of the Honey Bee*. Cambridge, MA: Harvard Univ. Press

Climate Change and Invasibility of the Antarctic Benthos

Richard B. Aronson,[1] Sven Thatje,[2]
Andrew Clarke,[3] Lloyd S. Peck,[3] Daniel B. Blake,[4]
Cheryl D. Wilga,[5] and Brad A. Seibel[5]

[1]Dauphin Island Sea Lab, Dauphin Island, Alabama 36528; email: raronson@disl.org

[2]National Oceanography Centre, Southampton, School of Ocean and Earth Science, University of Southampton, Southampton SO14 3ZH, United Kingdom; email: svth@noc.soton.ac.uk

[3]British Antarctic Survey, NERC, Cambridge CB3 0ET, United Kingdom; email: andrew.clarke@bas.c.uk, l.peck@bas.ac.uk

[4]Department of Geology, University of Illinois, Urbana, Illinois 61801; email: dblake@uiuc.edu

[5]Department of Biological Sciences, University of Rhode Island, Kingston, Rhode Island 02881; email: cwilga@uri.edu, seibel@uri.edu

Annu. Rev. Ecol. Evol. Syst. 2007. 38:129–54

The *Annual Review of Ecology, Evolution, and Systematics* is online at http://ecolsys.annualreviews.org

This article's doi:
10.1146/annurev.ecolsys.38.091206.095525

Key Words

climate change, Decapoda, invasive species, physiology, polar, predation

Abstract

Benthic communities living in shallow-shelf habitats in Antarctica (<100-m depth) are archaic in structure and function compared to shallow-water communities elsewhere. Modern predators, including fast-moving, durophagous (skeleton-crushing) bony fish, sharks, and crabs, are rare or absent; slow-moving invertebrates are generally the top predators; and epifaunal suspension feeders dominate many soft-substratum communities. Cooling temperatures beginning in the late Eocene excluded durophagous predators, ultimately resulting in the endemic living fauna and its unique food-web structure. Although the Southern Ocean is oceanographically isolated, the barriers to biological invasion are primarily physiological rather than geographic. Cold temperatures impose limits to performance that exclude modern predators. Global warming is now removing those physiological barriers, and crabs are reinvading Antarctica. As sea temperatures continue to rise, the invasion of durophagous predators will modernize the shelf benthos and erode the indigenous character of marine life in Antarctica.

INTRODUCTION

Global climate change is altering the geographic ranges, behaviors, and phenologies of terrestrial, freshwater, and marine species. A warming climate, therefore, appears destined to change the composition and function of marine communities in ways that are complex and not entirely predictable (Clarke et al. 2007, Fields et al. 1993, Helmuth et al. 2006, Smetacek & Nicol 2005, Walther et al. 2002). Higher temperatures are expected to increase the introduction and establishment of exotic species, thereby changing trophic relationships and homogenizing biotas (Stachowitz et al. 2002). Because organisms in polar regions are adapted to the coldest temperatures and most intense seasonality of resource supply on Earth (Peck et al. 2006), polar species and the communities they comprise are especially at risk from global warming and the concomitant invasion of species from lower latitudes (Aronson & Blake 2001, Barnes et al. 2006, Thatje et al. 2005).

Shallow-water, benthic communities in Antarctica (<100-m depth) are unique. Nowhere else do giant pycnogonids, nemerteans, and isopods occur in shallow marine environments, cohabiting with fish that have antifreeze glycoproteins in their blood. An emphasis on brooding and lecithotrophic reproductive strategies (Pearse et al. 1991, Thorson 1950) and a trend toward gigantism (Chapelle & Peck 1999, Peck 2002) are among the unusual features of the invertebrate fauna. Ecological and evolutionary responses to cold temperature underlie these peculiarities, making the Antarctic bottom fauna particularly vulnerable to climate change. The Antarctic benthos, living at the lower thermal limit to marine life, serves as a natural laboratory for understanding the impacts of climate change on marine systems in general.

Recent advances in the physiology, ecology, and evolutionary paleobiology of marine life in Antarctica make it possible to predict the nature of biological invasions facilitated by global warming and the likely responses of benthic communities to such invasions. This review draws on paleontology, biogeography, oceanography, physiology, molecular ecology, and community ecology. We explore the climatically driven origin of the peculiar community structure of modern benthic communities in Antarctica and the macroecological consequences of present and future global warming.

THE MODERN FAUNA

The Southern Ocean

The Southern Ocean comprises all waters south of the Polar Front, a well-defined circum-Antarctic oceanographic feature that marks the northernmost extent of cold surface water. The total area of the Southern Ocean is about 34.8 million km², of which up to 21 million km² are covered by ice at the winter maximum and about 7 million km² are covered at the summer minimum (Zwally et al. 2002).

Much of the Southern Ocean overlies deep seafloor, the fauna of which is in general poorly known (Brandt et al. 2007). The continental shelf around Antarctica is unusually deep, partly from isostatic depression from the enormous mass of continental ice, but predominantly as a result of ice scour during previous extensions of the ice sheet (Anderson 1999, Huybrechts 2002). Continental shelves elsewhere in the world are typically 100–200 m deep and 75 km wide (Walsh 1988). Around Antarctica the shelf edges are closer inshore. They average over 450 m deep, and in places they extend to more than 1000-m depth. Conventionally the 1000-m isobath is taken to mark the edge of the Antarctic continental shelf (Clarke & Johnston 2003), and the 3000-m isobath is taken as the transition from the continental slope to the continental rise (Snelgrove 2001).

Decapods: crabs, hermit crabs, lobsters, shrimp, and related crustaceans

Teleosts: modern bony fish

Benthos of the Southern Ocean

The total species list for the Southern Ocean benthos currently exceeds 4100 (Clarke & Johnston 2003). Gutt et al. (2004) estimated that the total macrofaunal diversity of the Southern Ocean continental shelf could exceed 15,000, meaning that most of the fauna remains undescribed.

Most noteworthy in terms of high species richness are the pycnogonids and ascidians. Polychaetes are also diverse, but that is not surprising because they are speciose almost everywhere else as well. For a number of taxa, low species richness in the Southern Ocean is not typical of that group in warmer waters; examples include gastropods, bivalves, decapods, and teleosts, all of which are highly speciose in many oceans but poorly represented in the Southern Ocean. Amphipods and isopods are well represented, although they are not particularly diverse. Some amphipod and isopod lineages radiated in the Southern Ocean (Brandt 1992, 2000; Held 2000; Watling & Thurston 1989).

Antarctica contains ~11% of the world's continental-shelf area. With the caveat that the Antarctic shelf is configured differently from continental shelves elsewhere in the world, we can calculate the fraction of the world's continental-shelf fauna found in Antarctica for each group (**Figure 1**). We see that none of the taxa there achieve levels of representation greater than about 15% in the Southern Ocean, and most are well below. The data do, however, reinforce the conclusion that polychaetes, bryozoans, sponges, pycnogonids, amphipods, and ascidians are well-represented in the Southern Ocean benthic fauna, whereas gastropods and bivalves in particular are not.

Deep-Sea Affinities

Many researchers have commented on the similarities between the fauna of the Antarctic continental shelf and typical deep-sea (bathyal to abyssal) communities. Two aspects in particular have attracted attention. The first is the importance of echinoderms in the fauna, and the second is the evolutionary connection between the faunas of the Antarctic continental shelf and the adjacent deep sea.

The unusual depth of the Antarctic continental shelf means that many taxa, elsewhere considered to be typical of shallow shelves, are living at depths traditionally regarded as bathyal. On the deeper parts of the Southern Ocean shelves, organisms require physiological adaptations to pressure that parallel those found in truly deep-sea

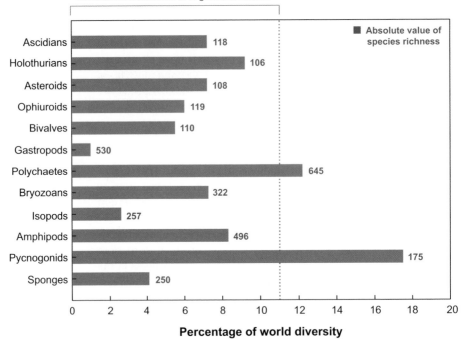

Antarctic shelf area as % of global shelf area

Taxon	Absolute value
Ascidians	118
Holothurians	106
Asteroids	108
Ophiuroids	119
Bivalves	110
Gastropods	530
Polychaetes	645
Bryozoans	322
Isopods	257
Amphipods	496
Pycnogonids	175
Sponges	250

Percentage of world diversity (x-axis: 0 2 4 6 8 10 12 14 16 18 20)

Legend: ■ Absolute value of species richness

Figure 1

The most speciose taxa of benthic invertebrates in Antarctica (species richness of more than 100), shown as percentages of the worldwide species diversities of those taxa (bars) and absolute values of species richness (adjacent numbers). Colloquial names are used because the groups represent different taxonomic levels (phylum, class, or order). Vertical dotted line shows the area of the Antarctic continental shelf as a percentage of the total, worldwide continental-shelf area. Benthic hydrozoans are omitted because, although there are 186 species known from Antarctica, no reliable data exist for worldwide species richness. Values are from Clarke & Johnston (2003).

organisms in other oceans (Clarke 2003). Echinoderms are an important component of the Antarctic shelf fauna, as they are in the deep sea (Gage & Tyler 1991). The importance of echinoderms within the benthic fauna of the Antarctic shelf, and in particular the notable diversity of crinoids, may be related in part to the depth of the shelf.

Isopod crustaceans are a second group that is well represented both in the deep sea and on the Antarctic shelf, and this has prompted much discussion about evolutionary links between the continental shelves at high southern latitudes and the deep-sea fauna. Menzies et al. (1973) analyzed the depth ranges of the various components of the Southern Ocean isopod fauna and demonstrated a complex evolutionary history. Some shallow-water taxa moved into deeper water (evolutionary polar submergence), whereas others colonized the continental shelf from deeper water (evolutionary polar emergence; see also Brandt 1991, 1992; Brandt et al. 2007; Zinsmeister & Feldmann 1984).

These patterns are clearly linked to the glacial history of Antarctica. Brey et al. (1996) showed that many shelf taxa in Antarctica have more extended bathymetric ranges than do comparable taxa on continental shelves elsewhere, suggesting that movement in and out of deeper water, driven by glacial cycles, may represent a general evolutionary history for the fauna. In summary, exchanges between the shelf and deep sea in Antarctica were aided by the generally deep nature of the shelves, requiring the shelf fauna to tolerate greater hydrostatic pressures than shelf faunas elsewhere, and by the absence of a strong thermal gradient between the deep sea and shallow water in the Southern Ocean.

Durophagy: skeleton-crushing predation

Epifaunal suspension feeder: an animal that lives on the sediment surface and meets its energetic requirements by filtering organic particles from the water column

Benthic Ecology of the Antarctic

Benthic work in the Antarctic has been concentrated at a relatively small number of sites, and the ecologies of those assemblages have influenced much of our thinking about benthic ecology in the Southern Ocean. Particularly important have been studies of soft substrata in McMurdo Sound and along the Antarctic Peninsula, and investigations of suspension-feeding communities in the Weddell and Ross Seas.

A critical feature of benthic ecology in Antarctica is that skeleton-crushing predation (durophagy) is very limited at present (Arntz et al. 1994, Dayton et al. 1974, Dell 1972, McClintock & Baker 1997). Evolutionarily modern (i.e., post-Paleozoic), durophagous predators—fish and decapods—are largely responsible for structuring food webs in subtidal marine communities at temperate, subtropical, and tropical latitudes. In contrast, there are no brachyuran crabs, lobsters, sharks, or rays in Antarctica; the diversity and abundance of skates (Rajidae) are low; and the teleostean fauna is dominated by two nondurophagous clades, namely the notothenioids and the liparids (Clarke & Johnston 1996, 2003; Dayton et al. 1994; Eastman & Clarke 1998; Long 1992, 1994). Weddell seals feed at depth, but their diet consists primarily of free-swimming prey such as fish and squid; although these seals take benthic prey occasionally, there are no marine mammals in Antarctica that are ecologically equivalent to the durophagous, bottom-feeding walruses or gray whales of the Arctic.

Lacking modern predators, benthic communities in Antarctica are functionally Paleozoic, with clear affinities to archaic, deep-sea biotas: slow-moving invertebrates, including asteroids and nemerteans, are generally the top predators of invertebrates, and dense populations of epifaunal suspension feeders, including ophiuroids, crinoids, bryozoans, and brachiopods, are widespread in shallow-water environments (Arntz et al. 2005, Aronson & Blake 2001, Clarke et al. 2004, Gili et al. 2006). As with all faunas, however, the benthic fauna is a mixture of relict species, taxa that arrived from elsewhere, and taxa that have evolved in situ (Clarke & Crame 1989). Thus, although the overall appearance of the fauna may be archaic, the taxa themselves are not necessarily ancient or primitive, and many are highly derived (Aronson et al. 1997).

In the Southern Ocean, suspension-feeding assemblages can cover large areas of the seafloor. Antarctic suspension-feeding communities can extend continuously for hundreds of kilometers and develop a complex, three-dimensional, tiered structure (Gili et al. 2006). Although these Antarctic communities are typically quite deep and thus subject to greatly reduced ice-scour, rich suspension-feeding communities also exist at depths less than 100 m (Raguá-Gil et al. 2004).

ACC: Antarctic Circumpolar Current

Predation in the Arctic

In the Arctic, durophagous teleosts feed in shallow-water, soft-substratum habitats. These teleosts, which include cod (Gadidae), flatfish (Pleuronectidae), sculpins (Cottidae), and eelpout (Zoarcidae), avoid mortality from cold sea temperatures by means of antifreeze proteins, other biochemical adaptations, and migratory behaviors (Helfman et al. 1997). The dogfish *Squalus acanthias* (Squalidae) is the most common shark in shallow Arctic waters. Although *S. acanthias* eats small crabs occasionally, its diet consists primarily of fish. Crabs do not occur naturally in shallow benthic communities in the Arctic. As mentioned above, walruses and gray whales are important sources of disturbance and durophagy in Arctic soft-substratum communities. The oceanographic and ecological differences between Arctic and Antarctic marine ecosystems are reviewed elsewhere (Dayton et al. 1994, Smetacek & Nicol 2005).

Origin and Isolation of the Antarctic Fauna

It is now generally accepted that the marine fauna of the Antarctic continental shelf has a long history of evolution in situ (Clarke & Crame 1989, Dell 1972, Knox & Lowry 1977, Lipps & Hickman 1982). Phylogenetic analyses (e.g., Brandt 1991, 1992) show that older groups have an evolutionary history extending back before the fragmentation of Gondwana. Biogeographic analyses also show that taxa migrated to and from what is now Antarctica along the Scotia arc (Dell 1972, Knox & Lowry 1977). A high degree of endemism in the Southern Ocean (Arntz et al. 1997) highlights the length of time during which the fauna has been isolated (Clarke & Crame 1997).

The Southern Ocean has long been regarded as one of the most clearly defined large marine ecosystems on earth, being bounded by the Antarctic Continent to the south and the Polar Front to the north. The Polar Front is the most intense of a series of eastward-flowing jets of the Antarctic Circumpolar Current (ACC) and represents a physical barrier to free north/south exchange of water. Visible as a sharp change in surface water temperatures and detectable to significant depths (>1000 m), it marks a distinct biogeographic discontinuity. With the exception of migratory seabirds and marine mammals, few epipelagic or benthic taxa have distributions both inside and outside the Southern Ocean (Dell 1972).

There are, however, strong biogeographic links between the benthic marine faunas of the Antarctic Peninsula and the Magellan Region of South America (Arntz et al. 2005). The traditional interpretation of these affinities is that the two regions were once contiguous and that only during the Cenozoic have they been separated by the deep waters of Drake Passage and development of the Polar Front. Although the possibility of a continued low level of faunal exchange has been recognized (Dell 1972), it is usually assumed that the strength of the ACC and the intensity of the associated Polar Front render this a rare event; however, there are observations of Antarctic plankton well to the north of the Polar Front (Antezana 1999, Hodgson et al. 1997), and it has long been recognized that pumice and driftwood can move both into and out of the Southern Ocean (Barber et al. 1959, Coombs & Landis 1966).

Mya: million years ago

Observations by Thatje & Fuentes (2003) highlight the permeability of the Polar Front. They reported zoeal stages of anomuran and brachyuran crabs associated with specimens of the Subantarctic copepod genus *Acartia* from the waters off King George Island, Antarctic Peninsula (62°S). Five early zoeal stages were assigned to the (anomuran) mole crab *Emerita* sp., which is known from shallow waters off sandy beaches, and two early and two advanced zoea were assigned to the brachyuran genus *Pinnotheres*.

These observations raise questions about how the organisms got to Antarctica and how often this might happen. Thatje & Fuentes (2003) suggested that the larvae might have crossed the Polar Front by means of eddies or more substantial intrusions of Subantarctic water masses through the ACC. Far from being a continuous, well-defined barrier, the ACC has an exceedingly complex mesoscale structure. The detail now available from satellite imagery reveals a highly variable and dynamic structure, including eddies over a wide range of scales (Glorioso et al. 2005, Olbers et al. 2004). Eddies are an important mechanism for transport of organisms across the ACC: warm-core rings can transport subantarctic plankton, including larvae, to Antarctica, and cold-core rings can carry Antarctic plankton into the warmer waters to the north.

How frequently these events transport biota across the Polar Front is difficult to estimate, but given that even low levels of current and historical sampling have revealed several such events in modern times, it is clear that the Southern Ocean is far from isolated. In fact, the porosity of oceanographic features has exerted a significant influence on the evolution of the Southern Ocean fauna. Using molecular techniques, Page & Linse (2002) described speciation and dispersal of limatulid bivalves across the Polar Front well after the formation of the ACC, suggesting that neither the ACC nor the Polar Front has been an absolute physical barrier to invasion.

There are, of course, other mechanisms that can transport marine organisms across the Polar Front. Particularly important for encrusting biota are pumice and drift-wood, and for a range of organisms, kelp rafts (Helmuth et al. 1994, Highsmith 1985). More recently, humans have increased the likelihood of introduction of new organisms to Antarctica through increases in anthropogenic flotsam (Barnes 2002) and in the frequency with which ships cross the Polar Front (Barnes et al. 2006).

PREDATION IN DEEP TIME

The reduced level of durophagy in Antarctica and the peculiar composition of the shallow bottom fauna can be traced to a global cooling trend that began late in the Eocene, about 40 Mya (million years ago). In this section we trace the history of predation and its effects on benthic faunas through the Phanerozoic, concentrating on post-Paleozoic patterns and processes. We then apply inferences about escalating predation in the Mesozoic to events in Antarctica during the Eocene.

Phanerozoic Patterns of Predation

Neoselachians: modern
sharks, skates, and rays

Durophagous predation increased several times during the Phanerozoic, but the fossil record suggests that some of the most significant changes occurred as part of the Mesozoic marine revolution (Aberhan et al. 2006, Vermeij 1977). Modern, skeleton-breaking predators, particularly teleosts, neoselachians, and decapod crustaceans, began to diversify in nearshore environments during the Jurassic (Thies & Reif 1985; Vermeij 1987). Radiations of durophagous taxa are thought to have driven the evolution of architectural defenses in a variety of marine invertebrates during the Mesozoic and Cenozoic. In gastropods, adaptations of the shell to increasing predation pressure included increased spination, ribbing, and other defensive sculpture (Vermeij 1977, 1987).

The coleoid cephalopods—those forms, including octopus, squid, and cuttlefish, that lack an external shell—radiated beginning in the Mesozoic, at least partially in response to the evolution of skeleton-crushing fish and marine reptiles. The slow-moving, externally shelled (ectocochleate) cephalopods—ammonoids and nautiloids—came under increasing predation pressure in nearshore environments. The faster-moving, fish-like coleoids radiated into a variety of niches in nearshore environments that had been vacated by extinction and ecological displacement of the ectocochleates (Aronson 1991; Packard 1972). Octopods are predators of skeletonized invertebrates in shallow benthic communities at temperate and tropical latitudes, and they can be significant consumers in some situations.

A direct consequence of increased predation, evident at multiple spatio-temporal scales, is a decline in the occurrence of dense populations of prey. Dense populations of epifaunal, suspension-feeding ophiuroids are excluded from most modern coastal environments by predatory fish and crabs (Aronson 1989). The primary predators of epifaunal ophiuroids living in dense populations are slow-moving invertebrates, including asteroids and polychaetes, and the low levels of predation the ophiuroids experience are reflected in low frequencies of sublethal arm damage (regenerating arms). On a macroecological-macroevolutionary scale, the diversification of durophagous predators during the Mesozoic caused a global decline in the occurrence of these dense, low-predation ophiuroid populations in shallow-water habitats (Aronson 1992).

The Mesozoic-Cenozoic escalation in predation resulted from radiations in nearshore environments of taxa comprising the "Modern evolutionary fauna" (sensu Sepkoski 1991). The Modern fauna, which is rich in mollusks, active bioturbators, and durophagous predators, replaced the "Paleozoic evolutionary fauna" in coastal environments. The Paleozoic fauna, rich in brachiopods, bryozoans, echinoderms, and other epifaunal, suspension-feeding taxa (as well as ectocochleate cephalopods), was progressively restricted to more offshore, deeper-water habitats as the Modern fauna spread from onshore to offshore environments. Durophagous predators originated onshore, largely eliminating epifaunal, suspension-feeding populations from soft-substratum habitats (Jablonski & Bottjer 1991, Sepkoski 1991). From the Jurassic onward, epifaunal suspension feeders on soft substrata were replaced by infaunal and more mobile epifaunal suspension feeders, giving onshore soft-substratum communities their modern, bivalve-dominated ecology (Aberhan et al. 2006, Bottjer

& Ausich 1986, Stanley 1977). As a broad generalization, predation is lower and community structure is archaic in offshore, deep-water habitats compared to nearshore, shallow-water habitats.

The decline of dense ophiuroid populations in coastal waters during the Mesozoic is one aspect of this onshore-offshore trend. Another is that stalked crinoids were abundant in shallow water in the Paleozoic and early Mesozoic (Bottjer & Jablonski 1988, Meyer & Macurda 1977). Living stalked crinoids, most of which are in the order Isocrinida, occur only offshore, in water deeper than ~100 m, where predation pressure is reduced (Oji 1996). The unstalked crinoids (order Comatulida), which are mobile, replaced the stalked crinoids in shallow-water environments (Meyer & Macurda 1977).

Predator-prey dynamics evolved within the context of macroevolutionary events and trends. The end-Permian mass extinction accelerated diversification of the Modern evolutionary fauna and increased the complexity of post-Paleozoic onshore communities, perhaps by creating an ecological vacuum of vacant niches (Sepkoski 1991, Wagner et al. 2006). The rapid escalation of predator-prey interactions during the Mesozoic also coincided with increased productivity (Vermeij 1987). More energy permitted the construction of more elaborate antipredatory shell architectures and enabled infauna to dig more actively and to greater depths (Bambach 1993). Thus, increased productivity helped drive the post-Paleozoic radiation of (infaunal) bivalves. These evolutionary innovations of predators and prey transcended the end-Cretaceous mass extinction, and escalation continued in the Cenozoic. Episodes of elevated extinction in the Cenozoic selectively exterminated well-defended prey taxa, temporarily reversing escalatory trends (Vermeij 1987).

Recent work on living communities in the Adriatic Sea corroborates the idea that both increasing durophagous predation and increasing productivity drove the post-Paleozoic infaunalization of suspension-feeding assemblages in shallow-water, soft-substratum environments. Given the background condition of low predation in the Adriatic, areas of low productivity support Paleozoic-type dense assemblages of epifaunal suspension feeders. Areas of high productivity are characterized by Modern-style assemblages of more energetic, infaunal suspension and deposit feeders (McKinney & Hageman 2006). The same pattern holds in Antarctica, with retrograde, epifaunally dominated benthic communities requiring oligotrophic conditions in addition to low levels of predation (Dayton & Oliver 1977).

Antarctica from the Eocene to the Present

In the late Eocene to early Oligocene, global cooling reduced sea temperatures in Antarctica, with a drop of 4–9°C in both open-ocean and coastal environments (Dutton et al. 2002, Ehrmann & Mackensen 1992, Kennett & Warnke 1992, Mackensen & Ehrmann 1992). The opening of the Drake Passage during the Eocene, as early as 41 Mya (Scher & Martin 2006), and the consequent establishment of circum-Antarctic circulation, drove the long-term shift from the cool-temperate climate of the Eocene to the glaciated, polar climate in Antarctica today (Ivany et al. 2006, Tripati et al. 2005, Zachos et al. 2001). Climatic cooling reduced predation pressure, causing a fundamental shift in the structure of benthic communities in Antarctica.

Most of what we know about this critical time in the history of the Antarctic benthos comes from studies of the shallow-water paleoenvironments represented in the La Meseta Formation at Seymour Island off the Antarctic Peninsula. La Meseta time extends from late in the early Eocene through the end of the late Eocene, or 52–33.5 Mya (Dutton et al. 2002). Sea temperatures at Seymour Island dropped from ~14°C in the middle Eocene, 45 Mya, to <10°C in the late Eocene, 35 Mya (Dutton et al. 2002, Ivany et al. 2004).

Prior to the late Eocene cooling event (i.e., lower in the formation), the benthic fauna was broadly typical of Cenozoic faunas elsewhere: it was dominated by bivalve and gastropod mollusks. Significant changes occurred in the composition of molluscan assemblages as cooling proceeded (Stilwell & Zinsmeister 1992). In functional terms, the gastropods display a notable decline upsection in the expression of defensive architectural features (Werner et al. 2004).

Twelve dense, autochthonous assemblages of suspension-feeding echinoderms, including ophiuroids (*Ophiura hendleri*) and stalked, isocrinid crinoids (*Metacrinus fossilis*), were recovered from the uppermost portion of the La Meseta Formation (representing cold conditions), whereas none were found in the rest of the section below (representing warmer conditions). The presence of dense concentrations of these taxa in a nearshore-shelf setting, and the low frequencies of sublethal injury (regenerating arms) in their populations, indicates that predation pressure from durophagous predators was reduced when temperatures were cooler (Aronson & Blake 2001, Aronson et al. 1997). Dense populations of other epifaunal suspension feeders, including the comatulid crinoid *Notocrinus rasmusseni* and the hiatellid bivalve *Hiatella tenuis*, also appear in the upper La Meseta Formation, supporting the hypothesized decline in importance of durophagous predators.

Dense clusters of tens to hundreds of terebratellid brachiopods, *Bouchardia antarctica*, occur throughout the La Meseta Formation (Bitner 1996, Wiedman & Feldmann 1988) and are reminiscent of brachiopod-rich Paleozoic faunas. Unlike the predation-sensitive echinoderms, brachiopods are only incidentally affected by durophagous predators (Kowalewski et al. 2005) and so could persist both before and after the cooling-mediated decline in predation pressure. Brachiopods are low-energy suspension feeders (James et al. 1992, Rhodes & Thompson 1993), making it possible for dense populations to persist during times of both low or fluctuating productivity prior to the cooling trend, and during periods of sporadically high productivity as sea temperatures dropped (see Diester-Haass & Zahn 1996, Scher & Martin 2006).

Isolated bones of teleosts are infrequent and scattered throughout the La Meseta Formation. Shark remains are concentrated in the lower portions, deposited before the cooling began, and there is only one record of skate teeth (Long 1992, 1994). It is not known when the early Tertiary fish fauna became extinct, but molecular evidence puts the endemic notothenioid radiation approximately in the middle Miocene, coincident with a decline in temperature (Cheng & Chen 1999).

Crab fossils occur throughout the La Meseta Formation, but again they are concentrated in the lower portions (Feldmann & Wilson 1988). Their subsequent decline is poorly constrained, because there are only three species of post-Eocene fossil decapods known from Antarctica: a brachyuran crab and a lobster from the early Miocene, and a palinurid (spiny) lobster from the Pliocene (Feldmann & Schweitzer 2006). Sea temperatures in the Southern Ocean may have been warm during both the early Miocene and the Pliocene (Hillenbrand & Ehrmann 2005, Quilty 1990, Zachos et al. 2001), possibly allowing incursions of reptant decapods. In contrast, isopods probably radiated during the cooling trends of the Oligocene and Miocene (Brandt 1991, 2000; Brandt et al. 1998), and they appear to have evolved to occupy many of the benthic niches vacated by the extinctions of decapods.

Reptant: formerly a taxonomic designation and now a functional term, referring to bottom-dwelling, walking decapods other than benthic shrimp

The few available fossils of durophagous predators do not by themselves yield much insight into temporal trends in predation. What we do know is that (a) durophagous predators currently are absent from marine communities in Antarctica; (b) the porous ACC and the Scotia arc provide avenues for benthic faunal elements to invade Antarctica; and therefore, (c) physiological constraints rather than geographic barriers are keeping the durophagous predators out.

PHYSIOLOGICAL LIMITS TO DUROPHAGOUS PREDATION

General Considerations

Variation in the solubility of calcium carbonate ($CaCO_3$) with latitude is often cited as an important factor shaping the polar marine fauna. Durophagous fish and crustaceans utilize heavily calcified teeth and chelae, respectively, and heavy calcification is a mainstay of the antipredatory defenses of skeletonized prey. The solubility of $CaCO_3$ varies inversely with temperature, being greater in polar waters than in the tropics (Revelle & Fairbridge 1957). The added cost of depositing $CaCO_3$ from seawater has been used to explain the lack of durophagous predators and the thin shells of mollusks at high latitudes (Arnaud 1974, Vermeij 1978). As a related matter, the slow growth rates of ectotherms living in Antarctic marine environments (Arntz et al. 1994, Peck 2002) could be an impediment to durophagy, because of the time required to produce effective crushing structures or viable architectural defenses.

Some patterns are, however, difficult to explain if $CaCO_3$ deposition is a major constraint at low temperature. Many successful taxa in Antarctica have high skeleton-to-tissue ratios. For example, brachiopods both inside Antarctica and at lower latitudes are 93–97.5% $CaCO_3$ (Peck 1993). Also, groups of both heavily calcified invertebrates (e.g., solitary corals, crinoids, ophiuroids, and calcareous sponges) and lightly calcified taxa (e.g., nemerteans, anthozoans, and amphipods) have species that attain large size in Antarctica (Chapelle & Peck 1999, 2004). Animals with high skeletal content and trends of increasing size toward the poles would not be expected if the increased cost of calcification at low temperature limited skeletal construction.

Teleosts

Durophagous predators eat echinoderms, polychaetes, crustaceans, and mollusks, the last requiring the greatest force generation by the predator. Antarctic benthic-feeding fish take polychaetes, crustaceans, and echinoderms. There are, however, no fish with feeding habits analogous to the subtropical black drum *Pogonias cromis* (Sciaenidae), or tropical triggerfish (Balistidae), wrasses (Labridae), or balloonfish (Diodontidae), which feed primarily on hard-shelled prey and often include mollusks in their diets.

Notothenioids, the dominant teleosts in shallow benthic communities in Antarctica, are lightly calcified and as a group possess simple conical teeth. In most species the teeth are not ankylosed to the jaw; they are attached via unossified ligaments (Eastman 1993). This arrangement is not suitable for crushing skeletonized prey, because the attachment is too weak to transfer the necessary forces. Species of benthic-feeding notothenioids in Antarctica either have a catholic diet or are ambush predators taking mainly peracarids (Gon & Heemstra 1990). The liparids, which generally live in deeper water, are not heavily calcified either, nor are they durophagous. The only fish in Antarctica with mouthparts suitable for durophagy are the skates. *Raja georgiana* consumes a wide range of prey including crustaceans, echinoderms, cnidarians, polychaetes, and fish, but not mollusks or brachiopods (Long 1994). Thus, following the disappearance of durophagous predators after the Eocene, groups radiating in or colonizing the Southern Ocean have not included a taxon with strong crushing mouthparts.

The power output from a given mass of fish muscle at 0°C is about one tenth that at 25°C (Wakeling & Johnston 1998). In the absence of other compensating mechanisms, a predatory fish would require as much as ten times the muscle mass to crush heavily skeletonized prey in Antarctica as compared to the tropics. Other physiological attributes of Antarctic teleosts, including the production of antifreeze compounds (Cheng and DeVries 1991, De Vries 1971), the lack of a standard heat-shock response (Hofman et al. 2000), and a marked sensitivity to raised temperature (Peck 2005, Peck et al. 2006, Somero & De Vries 1967), do not impinge directly on the question of whether or not durophagy is viable in benthic communities in Antarctica.

Reptant Decapods and Magnesium

The physiological considerations elaborated above also apply to the skeleton-crushing reptant decapods. The primary reason they are excluded from Antarctica, however, appears to be their inability to regulate the concentration of magnesium ions in their hemolymph at low temperatures. Ventilation and circulation are inhibited in brachyuran and anomuran crabs by their poor ability to down-regulate $[Mg^{2+}]$ below that of seawater (Frederich et al. 2001). High concentrations of magnesium in the hemolymph have relaxant or anesthetic properties, and the narcotic effect is stronger at low temperature. The reductions in heart rate and metabolism caused by high $[Mg^{2+}]$ reduce aerobic scope (Frederich et al. 2001). Aerobic scope is the crucial factor setting temperature tolerance limits (Pörtner 2002, Pörtner et al. 2006) and activity limits (Peck et al. 2004) in marine animals.

Elevated hemolymph $[Mg^{2+}]$, therefore, explains the absence of reptant decapods in low-temperature environments, setting a lower temperature limit of 0–1°C for

Hemolymph: the blood of arthropods, mollusks, and other invertebrates with an open circulatory system

crabs. Among the crabs, the anomuran king crabs (Lithodidae) are the most cold-tolerant, especially at the larval stage, within the limits imposed by their inability to regulate magnesium (Anger et al. 2003, Thatje et al. 2005). Caridean shrimp, and peracarids such as isopods and amphipods, are able to down-regulate $[Mg^{2+}]$ in their hemolymph below levels found in seawater, which accounts for the predominance of isopods and amphipods in the shallow Antarctic benthos (Frederich et al. 2001).

Neoselachian Sharks

The absence of sharks from Antarctic waters may be due to interactions of geography, temperature, and physiology. Most sharks are denser than seawater and must swim to remain in the water column, even though large fat stores and accumulation of trimethylamine oxide (TMAO) aid in reducing density (Carrier et al. 2004). Furthermore, fast-moving, pelagic sharks must swim to aerate their gills, incurring a cost in increased metabolism.

The active lifestyles of fast-moving pelagic sharks are hypothesized to be prohibitively expensive in extremely cold, oligotrophic polar environments (Priede et al. 2006). Pelagic cold-water sharks, like other pelagic predators, have metabolic rates five to ten times higher than related benthic species; however, corrected for temperature, these rates are not dramatically different from those of Antarctic teleosts and invertebrates with similar lifestyles at lower temperatures (Carrier et al. 2004, Seibel & Drazen 2007). All durophagous sharks are benthic and feed on benthic organisms, whereas pelagic sharks feed on fish and squid. Benthic sharks spend most of their time on the substratum, have small home ranges and have a body shape that is not as streamlined as that of pelagic sharks (Carrier et al. 2004).

Accumulating a lipid-rich liver requires a great investment of energy. A lipid store of 20% mass could support a benthic shark for more than three years. Although it seems counterintuitive, many polar organisms have larger fat stores than their temperate counterparts, perhaps as an adaptation to a seasonally variable food supply. Thus, energetically expensive lifestyles are not precluded by a seasonal food supply, but rather they may be limited by temperature effects on aerobic metabolism (Clarke & Johnston 1996, Seibel & Drazen 2007).

Benthic sharks lack the metabolism either to swim continuously in the water column or intermittently traverse the substratum to cross the Southern Ocean to Antarctica. Furthermore, there may be a physiological barrier to reinvasion. Although TMAO, which counters the harmful effects of high pressure and lowers the body freezing point, may be beneficial to sharks in deep-water environments, no sharks have been collected from depths greater than 3700 m (Priede et al. 2006, Samerotte et al. 2007). It may be that in cold, very deep waters, TMAO amplifies the inwardly directed osmotic gradient, and the kidneys are unable to compensate (Samerotte et al. 2007).

Octopods

Pareledone is the most abundant and diverse genus of benthic octopods in Antarctica (Allcock et al. 2001). Although present in shallow-water benthic communities,

Pareledone spp. appear to exert only a minor influence on food-web dynamics. Benthic octopods in Antarctica eat a variety of prey, including amphipods, polychaetes, gastropods, bivalves, ophiuroids, and fish (Daly 1996, Piatkowski et al. 2003).

The metabolic rates of benthic octopods across latitudes can be described by a single relationship incorporating mass and temperature (Seibel 2007). Polar octopods are slower in terms of metabolism and activity than their temperate and tropical counterparts. They do not possess adaptations to compensate for the depression of metabolic rate in their cold environment (Daly & Peck 2000).

THE RETURN OF DUROPHAGOUS PREDATORS

Recent Invasions

The first indication that alien predators had invaded Antarctica came from records of adult male and female brachyuran crabs, *Hyas araneus*, occurring off King George Island in 1986 (Tavares & De Melo 2004). *H. araneus* is a spider crab (Majidae) from the Northern Hemisphere, which ranges from boreal to Subarctic waters. It could have been introduced to Antarctica via the transport of adults on ships' hulls or larvae in ballast water (Barnes et al. 2006, Tavares & De Melo 2004). This first record of *H. araneus*, which is physiologically preadapted to conditions in the Southern Ocean, serves as a warning of their potential threat to Antarctic benthic communities.

Large populations of predatory king crabs (Lithodidae) native to the Southern Hemisphere were recently discovered in deep-water, continental-slope environments off the Antarctic Peninsula (Thatje & Arntz 2004, Thatje & Lörz 2005). Lithodids are common at high latitudes in both hemispheres and are frequent in the deep sea. In the Southern Hemisphere they were, however, believed to be entirely Subantarctic (Dawson 1989). In recent years the number of lithodid species known from the Southern Ocean has increased to about 14 species (Ahyong & Dawson 2006, Thatje & Lörz 2005, Thatje et al. 2005), including records of *Neolithodes capensis* and *Paralomis birsteini* described from the Antarctic continental slope in the Bellingshausen Sea (García Raso et al. 2005). It may be more than coincidental that a northern lithodid, *Lithodes maja*, was recently recorded for the first time off southeast Greenland (Woll & Burmeister 2002).

The timescale over which lithodids would be capable of colonizing Antarctica remains unclear (Thatje et al. 2005). They possess demersally drifting larvae, making range expansions over long distances improbable (Thatje et al. 2005). On the other hand, their larvae are lecithotrophic, with embryonic development lasting as long as two years. Adults are long-lived and grow very slowly (Lovrich & Vinuesa 1999). Low metabolism and adjustment of all life history stages to low temperatures may be the keys to ecological success for this group in polar seas (Kattner et al. 2003, Pörtner 2002, Thatje et al. 2005).

Limits to Invasibility

New discoveries of lithodids in Antarctica could represent recent invasions, but at least some of these new records simply reflect the inadequacy of earlier observations

(García Raso et al. 2005). Our understanding of the history of lithodids in Antarctica is further obscured by the fact that some species could be endemic. Ambient temperatures in deep-water, slope habitats are high enough to support lithodids in a torpid, hypometabolic state (Thatje et al. 2005). Regardless of how long they have been in Antarctica, however, they remain absent from shallow-shelf habitats, where water temperatures remain too low.

The lower temperature threshold of two Subantarctic lithodid species is ~1°C, a direct consequence of the problems of magnesium regulation (Thatje et al. 2005). Although physiological data are lacking for lithodids found south of the Polar Front, the 1°C threshold corresponds to the temperature of the Upper Circumpolar Deep Water, which characterizes the deep-water locations of, for example, *Lithodes murrayi* in the Bellingshausen Sea (Klages et al. 1995). Typical sea temperatures on the Antarctic continental shelf are lower, particularly on the Weddell and Ross Sea shelves where they routinely fall below 0°C. This suggests that at present most of the Antarctic shelf is unsuitable for lithodids to establish viable populations.

Should the sea-surface warming demonstrated for the western Antarctic Peninsula—a rise of more than 1°C in the past 50 years (Meredith & King 2005)—extend deeper into the water column, lithodids from deep water could colonize shallow waters along the Peninsula. Lithodids are generalist predators, feeding on echinoderms, benthic mollusks, and bryozoans (Comoglio & Amin 1999). Their invasion of shallow-water habitats off the Antarctic Peninsula would have significant consequences for benthic community structure.

Other anomuran and brachyuran crabs from southern South America would be likely candidates for invasion if sea temperatures increased significantly. At this point a severe impediment to invasion is the mismatch between the short Antarctic growing season and the longer development times of planktotrophic larvae (Thatje et al. 2003). Rising sea temperatures and continued breakup of the ice shelves should both increase phytoplankton productivity and prolong the growing season (Doney 2006, Smetacek & Nicol 2005). These effects should better synchronize development times with the annual period during which phytoplankton are readily available. Rising temperatures should also speed larval development (O'Connor et al. 2007, Peck et al. 2006), which could either reduce or increase the mismatch.

The diversity of sharks is high on outer shelves and slopes in cool-temperate environments (Carrier et al. 2004). There is, therefore, a high probability that fast-moving, pelagic shark species would reinvade Antarctic waters if sea temperatures continued to increase. For benthic, durophagous sharks, however, the problem remains of crossing deep ocean to reach shallow Antarctic waters. Demersal, durophagous teleosts, which produce planktonic larvae, could invade more easily than benthic sharks if temperatures in shallow water increased to the point that antifreeze proteins became unnecessary.

Lack of suitable habitat in Antarctica would prevent some invasions. For example, larvae of the mole crab *Emerita* sp. would not be able to settle and metamorphose without sandy-beach habitat, which is virtually absent from Antarctica. On the other hand, warming sea temperatures might cause physiological problems for shallow-water, Antarctic cold-stenotherms, which could soon be exposed to conditions

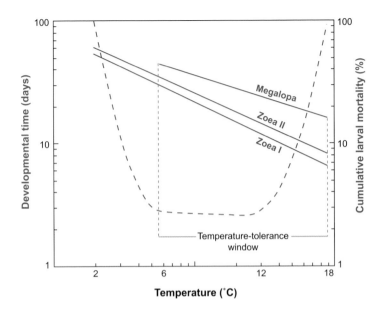

Figure 2

Physiological temperature-tolerance window and cumulative mortality in all larval stages of the spider crab *Hyas araneus*. In nature, the benthic megalopa stage usually encounters temperatures ranging from 5 to 13°C; successful larval development and, thus, survival at low temperatures typical of shallow Antarctic seas (<1°C) is unlikely. Redrawn from Thatje (2005).

approaching their upper temperature thresholds (Clarke et al. 2007, Peck 2005, Pörtner 2002). Physiological stress, increased mortality, and lowered recruitment success of native species could open ecological niche space for invading predators, which would be at a competitive advantage.

Under present climatic conditions, the introduction of Subarctic and Arctic species in ballast water or as fouling organisms on ships' hulls continues to be a cause for serious concern. Because they are cold-adapted, these species have the potential to establish viable populations rapidly in Antarctica. The record of two adult *Hyas araneus* from Antarctica raises the question of how these Subarctic brachyurans managed, as larvae or adults, to survive the transit through tropical seas at temperatures lethal to this species (**Figure 2**). Development of double-hulled ships in the 1980s has increased the thermal insulation of ballast water, potentially increasing the survival of alien organisms as they are transported across the oceans. Increased ship traffic into and out of Antarctica is increasing the risk of faunal exchange and the establishment of alien species. Such invaders will themselves be faced with changing ambient conditions fostered by global warming.

CONCLUSION

Beginning in the late Eocene, global cooling reduced durophagous predation in Antarctica. Despite some climatic reversals, the post-Eocene cooling trend drove shallow-water, benthic communities to the retrograde, Paleozoic-type structure and function we see today. Now, global warming is facilitating the return of durophagous predators, which are poised to eliminate that anachronistic character and remodernize the Antarctic benthos in shallow-water habitats. Rising sea temperatures should in general act to reduce the mismatch between the development times of invasive larvae and the length of the growing season. Increased survivability of planktotrophic larvae will decrease the selective advantage of brooding and lecithotrophy, increasing the pool of potentially invasive species. Warming temperatures will also increase the scope for more rapid metabolism and should ultimately obviate the adaptive value of gigantism. All of these effects appear destined to amplify the ongoing, worldwide homogenization of marine biotas by reducing the endemic character of the Antarctic fauna.

The fact that benthic predators are already beginning to invade the Antarctic Peninsula should be taken as an urgent warning. Controlling the discharge of ballast water from ships will be difficult but not impossible. Whether or not humans are the proximal vectors, however, the long-term threat of invasion in Antarctica has its roots in climate change. The Antarctic Treaty cannot control global warming. Global environmental policy must immediately be directed to reducing and reversing anthropogenic emissions of greenhouse gases into the atmosphere if marine life in Antarctica is to survive in something resembling its present form.

SUMMARY POINTS

1. Shallow-water, benthic communities in Antarctica are functionally similar to communities of the Paleozoic and the modern deep sea.

2. The historical absence of skeleton-crushing (durophagous) predators is an important reason for the archaic character of the Antarctic bottom fauna.

3. Durophagous predators disappeared as sea temperatures declined, beginning in the late Eocene (approximately 40 Mya).

4. The barriers to reinvasion by durophagous predators are physiological (intolerance of cold temperatures) rather than geographic or oceanographic.

5. Durophagous king crabs have recently been discovered on the continental slope off the Antarctic Peninsula.

6. Warming temperatures could facilitate the return of king crabs and other predators to shallow-shelf habitats.

7. Viable populations of durophagous predators would radically alter the structure of benthic food webs in Antarctica.

FUTURE ISSUES

1. Document ongoing invasions through continued long-term monitoring of the plankton and the benthos.

2. Determine experimentally whether physiologies and life histories of actual and potential invaders are amenable to warming sea temperatures in Antarctica.

3. Monitor and experiment on shallow benthic communities to detect and predict incipient changes to food-web structure.

4. Expand the scope of ecological observations beyond a few intensively studied areas.

DISCLOSURE STATEMENT

The authors are not aware of any biases that might be perceived as affecting the objectivity of this review.

ACKNOWLEDGMENTS

We thank Bill Baker, Alistair Crame, Rodney Feldmann, Alexander Glass, Drew Harvell, Linda Ivany, David Jablonski, Susan Kidwell, James McClintock, Ryan Moody, Sean Powers, Victor Smetacek, Jan Strugnell, Simon Thrush, John Werner, and William Zinsmeister for helpful advice and discussion. The impetus to write this review came from a workshop on *The Role of the Southern Ocean in Global Processes*, organized by the British Antarctic Survey and held at The Royal Society, London, in July 2003. Many of the ideas expressed here grew out of research funded by the U.S. National Science Foundation's Office of Polar Programs (grants OPP-9908828 and ANT-0245563 to R.B.A., and OPP-9908856 to D.B.B.). This is Contribution No. 385 from the Dauphin Island Sea Lab.

LITERATURE CITED

Aberhan M, Kiessling W, Fürisch FT. 2006. Testing the role of biological interactions in the evolution of mid-Mesozoic marine benthic ecosystems. *Paleobiology* 32:259–77

Ahyong ST, Dawson EW. 2006. Lithodidae from the Ross Sea, Antarctica, with descriptions of two new species (Crustacea: Decapoda: Anomura). *Zootaxa* 1303:45–68

Allcock AL, Piatkowski U, Rodhouse PGK, Thorpe JP. 2001. A study on octopodids from the eastern Weddell Sea, Antarctica. *Polar Biol.* 24:832–38

Anderson JB. 1999. *Antarctic Marine Geology*. Cambridge: Cambridge Univ. Press. 289 pp.

Anger K, Thatje S, Lovrich G, Calcagno J. 2003. Larval and early juvenile development of *Paralomis granulosa* reared at different temperatures: tolerance of cold and food limitation in a lithodid crab from high latitudes. *Mar. Ecol. Prog. Ser.* 253:243–51

Antezana T. 1999. Plankton of southern Chilean fjords: trends and linkages. *Sci. Mar.* 63(Suppl. 1):69–80

Arnaud PM. 1974. Contribution à la bionomie marine benthique des régions antarctiques et subantarctiques. *Téthys* 6:567–653

Arntz WE, Brey T, Gallardo VA. 1994. Antarctic zoobenthos. *Oceanogr. Mar. Biol. Ann. Rev.* 32:241–304

Arntz WE, Gutt J, Klages M. 1997. Antarctic marine biodiversity: an overview. In *Antarctic Communities: Species, Structure and Survival*, ed. B Battaglia, J Valencia, DWH Walton, pp. 3–14. Cambridge: Cambridge Univ. Press

Arntz WE, Thatje S, Gerdes D, Gili J-M, Gutt J, et al. 2005. The Antarctic-Magellan connection: macrobenthos ecology on the shelf and upper slope, a progress report. *Sci. Mar.* 69(Suppl. 2):237–69

Aronson RB. 1989. Brittlestar beds: low-predation anachronisms in the British Isles. *Ecology* 70:856–65

Aronson RB. 1991. Ecology, paleobiology and evolutionary constraint in the octopus. *Bull. Mar. Sci.* 49:245–55

Aronson RB. 1992. Biology of a scale-independent predator-prey interaction. *Mar. Ecol. Prog. Ser.* 89:1–13

Aronson RB, Blake DB. 2001. Global climate change and the origin of modern benthic communities in Antarctica. *Am. Zool.* 41:27–39

Aronson RB, Blake DB, Oji T. 1997. Retrograde community structure in the late Eocene of Antarctica. *Geology* 25:903–6

Bambach RK. 1993. Seafood through time: changes in biomass, energetics, and productivity in the marine ecosystem. *Paleobiology* 19:372–97

Barber HN, Dadswell HE, Ingle HD. 1959. Transport of driftwood from South America to Tasmania and Macquarie Island. *Nature* 184:203–4

Barnes DKA. 2002. Invasions by marine life on plastic debris. *Nature* 416:808–9

Barnes DKA, Hodgson DA, Convey P, Allen CS, Clarke A. 2006. Incursion and excursion of Antarctic biota: past, present and future. *Glob. Ecol. Biogeogr.* 15:121–42

Bitner MA. 1996. Brachiopods from the Eocene La Meseta Formation of Seymour Island, Antarctic Peninsula. *Palaeontol. Polon.* 55:65–100

Bottjer DJ, Ausich WI. 1986. Phanerozoic development of tiering in soft substrata suspension-feeding communities. *Paleobiology* 12:400–20

Bottjer DJ, Jablonski D. 1988. Paleoenvironmental patterns in the evolution of post-Paleozoic benthic marine invertebrates. *Palaios* 3:540–60

Brandt A. 1991. Zur Besiedlungsgeschichte des antarktischen Schelfes am Beispiel der Isopoda (Crustacea, Malacostraca). *Ber. Polarforsch.* 98:1–240

Brandt A. 1992. Origin of Antarctic Isopoda (Crustacea, Malacostraca). *Mar. Biol.* 113:15–23

Brandt A. 2000. Hypotheses on Southern Ocean peracarid evolution and radiation (Crustacea, Malacostraca). *Antarct. Sci.* 12:269–75

Brandt A, Crame JA, Polz H, Thomson MRA. 1998. Late Jurassic Tethyan ancestry of Recent southern high-latitude marine isopods (Crustacea, Malacostraca). *Palaeontology* 42:663–75

Brandt A, Gooday AJ, Brandão SN, Brix S, Brökeland W, et al. 2007. First insights into the biodiversity and biogeography of the Southern Ocean deep sea. *Nature* 447:307–11

Brey T, Dahm C, Gorny M, Klages M, Stiller M, Arntz WE. 1996. Do Antarctic benthic invertebrates show an extended level of eurybathy? *Antarct. Sci.* 8:3–6

Carrier JC, Musick J, Heithau M, eds. 2004. *Biology of Sharks and their Relatives*. Boca Raton: CRC Press. 616 pp.

Chapelle G, Peck LS. 1999. Polar gigantism dictated by oxygen availability. *Nature* 399:114–15

Chapelle G, Peck LS. 2004. Amphipod crustacean size spectra: new insights in the relationship between size and oxygen. *Oikos* 106:167–75

Cheng C-HC, Chen LB. 1999. Evolution of an antifreeze glycoprotein. *Nature* 401:443–44

Cheng C-HC, De Vries AL. 1991. The role of antifreeze glycopeptides and peptides in the freezing avoidance of cold-water fish. In *Life under Extreme Conditions: Biochemical Adaptation*, ed. G. di Prisco, pp. 1–15. Berlin: Springer-Verlag

Clarke A. 2003. The polar deep seas. In *Ecosystems of the Deep Oceans*, ed. PA Tyler, pp. 239–60. Amsterdam: Elsevier

Clarke A, Aronson RB, Crame JA, Gili J-M, Blake DB. 2004. Evolution and diversity of the benthic fauna of the Southern Ocean continental shelf. *Antarct. Sci.* 16:559–68

Clarke A, Crame JA. 1989. The origin of the Southern Ocean marine fauna. In *Origins and Evolution of the Antarctic Biota*, ed. JA Crame, pp. 253–68. London: Geol. Soc., Spec. Publ. 47

Clarke A, Crame JA. 1997. Diversity, latitude and time: patterns in the shallow sea. In *Marine Biodiversity: Causes and Consequences*, ed. RFG Ormond, JD Gage, MV Angel, pp. 122–47. Cambridge: Cambridge Univ. Press

Clarke A, Johnston IA. 1996. Evolution and adaptive radiation of Antarctic fishes. *TREE* 11:212–18

Clarke A, Johnston NM. 2003. Antarctic marine benthic diversity. *Oceanogr. Mar. Biol. Ann. Rev.* 41:47–114

Clarke A, Murphy EJ, Meredith MP, King JC, Peck LS, et al. 2007. Climate change and the marine ecosystem of the western Antarctic Peninsula. *Philos. Trans. R. Soc. London Ser. B* 362:149–66

Comoglio LI, Amin OA. 1999. Feeding habits of the false southern king crab *Paralomis granulosa* (Lithodidae) in the Beagle Channel, Tierra del Fuego, Argentina. *Sci. Mar.* 63(Suppl. 1):361–66

Coombs DS, Landis CA. 1966. Pumice from the South Sandwich eruption of March 1962 reaches New Zealand. *Nature* 209:289–90

Daly HI. 1996. *Ecology of the Antarctic Octopus* Pareledone *from the Scotia Sea*. Ph.D. Dissertation. Aberdeen: Univ. Aberdeen. 162 pp.

Daly HI, Peck LS. 2000. Energy balance and cold adaptation in the octopus *Pareledone charcoti*. *J. Exp. Mar. Biol. Ecol.* 245:197–214

Dawson EW. 1989. King crabs of the world (Crustacea: Lithodidae) and their fisheries: a comprehensive bibliography. *Misc. Publ. N. Z. Oceanogr. Inst.* 101:1–338

Dayton PK, Mordida BJ, Bacon F. 1994. Polar marine communties. *Am. Zool.* 34:90–99

Dayton PK, Oliver JS. 1977. Antarctic soft-bottom benthos in oligotrophic and eutrophic environments. *Science* 197:55–58

Dayton PK, Robilliard GA, Paine RT, Dayton LB. 1974. Biological accommodation in the benthic community at McMurdo Sound, Antarctica. *Ecol. Monogr.* 44:105–28

Dell RK. 1972. Antarctic benthos. *Adv. Mar. Biol.* 10:1–216

De Vries AL. 1971. Glycoproteins as biological antifreeze agents in Antarctic fishes. *Science* 172:1152–55

Diester-Haass L, Zahn R. 1996. Eocene-Oligocene transition in the Southern Ocean: history of water mass circulation and biological productivity. *Geology* 24:163–66

Doney SC. 2006. Plankton in a warmer world. *Nature* 444:695–96

Dutton AL, Lohmann KC, Zinsmeister WJ. 2002. Stable isotope and minor element proxies for Eocene climate of Seymour Island, Antarctica. *Paleoceanography* 17:1016–28

Eastman JT. 1993. *Antarctic Fish Biology: Evolution in a Unique Environment.* New York: Academic. 322 pp.

Eastman JT, Clarke A. 1998. A comparison of adaptive radiations of Antarctic fish with those of nonantarctic fish. In *Fishes of Antarctica: A Biological Overview*, ed. G di Prisco, E Pisano, A Clarke, pp. 3–26. Berlin: Springer-Verlag

Ehrmann WU, Mackensen A. 1992. Sedimentological evidence for the formation of an East Antarctic ice sheet in Eocene/Oligocene time. *Palaeogeogr. Palaeoclimatol. Palaeoecol.* 93:85–112

Feldmann RM, Schweitzer CE. 2006. Paleobiogeography of Southern Hemisphere decapod Crustacea. *J. Paleont.* 80:83–103

Feldmann RM, Wilson MT. 1988. Eocene decapod crustaceans from Antarctica. See Feldman & Woodburne 1988, pp. 465–88

Feldmann RM, Woodburne MO, eds. 1988. *Geology and Paleontology of Seymour Island, Antarctic Peninsula.* Boulder: Geol. Soc. Am., Mem. 169

Fields PA, Graham JB, Rosenblatt RH, Somero GN. 1993. Effects of expected global climate change on marine faunas. *TREE* 8:361–67

Frederich M, Sartoris FJ, Pörtner H-O. 2001. Distribution patterns of decapod crustaceans in polar areas: a result of magnesium regulation? *Polar Biol.* 24:719–23

Gage JD, Tyler PA. 1991. *Deep-Sea Biology: A Natural History of Organisms at the Deep-Sea Floor.* Cambridge: Cambridge Univ. Press. 504 pp.

García Raso JE, Manjón-Cabeza ME, Ramos A, Olasi I. 2005. New record of Lithodidae (Crustacea, Decapoda, Anomura) from the Antarctic (Bellingshausen Sea). *Polar Biol.* 28:642–46

Gili J-M, Arntz WE, Palanques A, Orejas C, Clarke A, et al. 2006. A unique assemblage of epibenthic sessile suspension feeders with archaic features in the high-Antarctic. *Deep-Sea Res. II* 53:1029–52

Glorioso PD, Piola AR, Leben RR. 2005. Mesoscale eddies in the Subantarctic Front, southwest Atlantic. *Sci. Mar.* 69(Suppl. 2):7–15

Gon O, Heemstra PC. 1990. *Fishes of the Southern Ocean.* Grahamstown: J.L.B. Smith Inst. Ichthyol. 462 pp.

Gutt J, Sirenko BI, Smirnov IS, Arntz WE. 2004. How many macrozoobenthic species might inhabit the Antarctic shelf? *Antarct. Sci.* 16:11–16

Held C. 2000. Phylogeny and biogeography of serolid isopods (Crustacea, Isopoda, Serolidae) and the use of ribosomal expansion segments in molecular systematics. *Mol. Phylogenet. Evol.* 15:165–78

Helfman GS, Colette BB, Facey DE. 1997. *The Diversity of Fishes.* Oxford: Blackwell. 528 pp.

Helmuth B, Mieskowska N, Moore P, Hawkins SJ. 2006. Living on the edge of two changing worlds: forecasting the responses of rocky intertidal ecosystems to climate change. *Annu. Rev. Ecol. Evol. Syst.* 37:373–404

Helmuth B, Veit RR, Holberton R. 1994. Long distance dispersal of a subantarctic brooding bivalve (*Gaimarida trapesina*) by kelp-rafting. *Mar. Biol.* 120:421–26

Highsmith RC. 1985. Floating and algal rafting as potential dispersal mechanisms in brooding invertebrates. *Mar. Ecol. Prog. Ser.* 25:169–79

Hillenbrand C-D, Ehrmann W. 2005. Late Neogene to Quaternary environmental changes in the Antarctic Peninsula region: evidence from drift sediments. *Global Planet. Change* 45:165–91

Hodgson DA, Vyverman W, Tyler P. 1997. Diatoms of meromictic lakes adjacent to the Gordon River, and of the Gordon River estuary in south-west Tasmania. *Bibl. Diatomol.* 35:1–172

Hofmann GE, Buckley BA, Airaksinen S, Keen JE, Somero GN. 2000. Heat-shock protein expression is absent in the Antarctic fish *Trematomus bernacchii* (Family Nototheniidae). *J. Exp. Biol.* 203:2331–39

Huybrechts P. 2002. Sea-level changes at the LGM from ice-dynamic reconstructions of the Greenland and Antarctic ice sheets during the glacial cycles. *Quat. Sci. Rev.* 22:203–31

Ivany LC, Blake DB, Lohmann KC, Aronson RB. 2004. Eocene cooling recorded in the chemistry of La Meseta Formation mollusks, Seymour Island, Antarctic Peninsula. *Boll. Geofis.* 45(Suppl. 2):242–45 (Abstr.)

Ivany LC, Van Simaeys S, Domack EW, Samson SD. 2006. Evidence for an earliest Oligocene ice sheet on the Antarctic Peninsula. *Geology* 34:377–80

Jablonski D, Bottjer DJ. 1991. Environmental patterns in the origins of higher taxa: the post-Paleozoic fossil record. *Science* 252:1831–33

James MA, Ansell AD, Collins MJ, Curry GB, Peck LS, Rhodes MC. 1992. Recent advances in the study of living brachiopods. *Adv. Mar. Biol.* 28:175–387

Kattner G, Graeve M, Calcagno JA, Lovrich GA, Thatje S, Anger K. 2003. Lipid, fatty acid and protein utilization during lecithotrophic larval development of *Lithodes santolla* (Molina) and *Paralomis granulosa* (Jacquinot). *J. Exp. Mar. Biol. Ecol.* 292:61–74

Kennett JP, Warnke DA, eds. 1992. *The Antarctic Paleoenvironment: A Perspective on Global Change, Part 1.* Washington, DC: Am. Geophys. Union, Antarct. Res. Ser. 56

Klages M, Gutt J, Starmans A, Bruns T. 1995. Stone crabs close to the Antarctic continent: *Lithodes murrayi* Henderson, 1888 (Crustacea; Decapoda; Anomura) off Peter I Island (68°51′ S, 91°51′ W). *Polar Biol.* 15:73–75

Knox GA, Lowry JK. 1977. A comparison between the benthos of the Southern Ocean and the North Polar Ocean with special reference to the Amphipoda and the Polychaeta. In *Polar Oceans*, ed. MJ Dunbar, pp. 423–62. Calgary: Arct. Inst. North Am.

Kowalewski M, Hoffmeister AP, Baumiller TK, Bambach RK. 2005. Secondary evolutionary escalation between brachiopods and enemies of other prey. *Science* 308:1774–77

Lipps JH, Hickman CS. 1982. Origin, age and evolution of Antarctic and deep-sea faunas. In *Environment of the Deep Sea*, ed. WG Ernst, JG Morris, pp. 324–56. Englewood Cliffs: Prentice Hall

Long DJ. 1992. Paleoecology of Eocene antarctic sharks. See Kennett & Warnke 1992, pp. 131–39

Long DJ. 1994. Quaternary colonization or Paleogene persistence? Historical biogeography of skates (Chondrichthyes: Rajidae) in the Antarctic ichthyofauna. *Paleobiology* 20:215–28

Lovrich GA, Vinuesa JH. 1999. Reproductive potential of the lithodids *Lithodes santolla* and *Paralomis granulosa* (Anomura, Decapoda) in the Beagle Channel, Argentina. *Sci. Mar.* 63(Suppl. 1):355–60

Mackensen A, Ehrmann WU. 1992. Middle Eocene through early Oligocene climate history and paleoceanography in the Southern Ocean: stable oxygen and carbon isotopes from ODP sites on Maud Rise and Kerguelen Plateau. *Mar. Geol.* 108:1–27

McClintock JB, Baker BJ. 1997. A review of the chemical ecology of Antarctic marine invertebrates. *Am. Zool.* 37:329–42

McKinney FK, Hageman SJ. 2006. Paleozoic to modern marine ecological shift displayed in the northern Adriatic Sea. *Geology* 34:881–84

Menzies RJ, George RY, Rowe GT. 1973. *Abyssal Environment and Ecology of the World Oceans*. New York: Wiley. 488 pp.

Meredith MP, King JC. 2005. Rapid climate change in the ocean west of the Antarctic Peninsula during the second half of the 20th century. *Geophys. Res. Lett.* 32:L19604

Meyer DL, Macurda DB Jr. 1977. Adaptive radiation of the comatulid crinoids. *Paleobiology* 3:74–82

O'Connor MI, Bruno JF, Gaines SD, Halpern BS, Lester SE, et al. 2007. Temperature control of larval dispersal and the implications for marine ecology, evolution, and conservation. *Proc. Natl. Acad. Sci. USA* 104:1266–71

Oji T. 1996. Is predation intensity reduced with increasing depth? Evidence from the west Atlantic stalked crinoid *Endoxocrinus parrae* (Gervais) and implications for the Mesozoic marine revolution. *Paleobiology* 22:339–51

Olbers D, Borowski D, Völker C, Wölff J-O. 2004. The dynamical balance, transport and circulation of the Antarctic Circumpolar Current. *Antarct. Sci.* 16:439–70

Packard A. 1972. Cephalopods and fish: the limits of convergence. *Biol. Rev. Camb. Philos. Soc.* 47:241–307

Page TJ, Linse K. 2002. More evidence of speciation and dispersal across the Antarctic Polar Front through molecular systematics of Southern Ocean *Limatula* (Bivalvia: Limidae). *Polar Biol.* 25:818–26

Pearse JS, McClintock JB, Bosch I. 1991. Reproduction of Antarctic benthic marine invertebrates: tempos, modes, and timing. *Am. Zool.* 421:37–42

Peck LS. 1993. The tissues of articulate brachiopods and their value to predators. *Philos. Trans. R. Soc. London Ser. B* 339:17–32

Peck LS. 2002. Ecophysiology of Antarctic marine ectotherms: limits to life. *Polar Biol.* 25:31–40

Peck LS. 2005. Prospects for survival in the Southern Ocean: vulnerability of benthic species to temperature change. *Antarct. Sci.* 17:497–507

Peck LS, Convey P, Barnes DKA. 2006. Environmental constraints on life histories in Antarctic ecosystems: tempos, timings and predictability. *Biol. Rev. Camb. Philos. Soc.* 81:75–109

Peck LS, Webb K, Bailey D. 2004. Extreme sensitivity of biological function to temperature in Antarctic marine species. *Funct. Ecol.* 18:625–30

Piatkowski U, Allcock L, Vecchione M. 2003. Cephalopod diversity and ecology. *Ber. Polarforsch.* 470:32–38

Pörtner H-O. 2002. Physiological basis of temperature-dependent biogeography: trade-offs in muscle design and performance in polar ectotherms. *J. Exp. Biol.* 205:2217–30

Pörtner H-O, Peck LS, Hirse T. 2006. Hyperoxia alleviates thermal stress in the Antarctic bivalve *Laternula elliptica*: evidence for oxygen limited thermal tolerance. *Polar Biol.* 29:688–93

Priede IG, Fröse R, Bailey DM, Bergstad OA, Collins MA, et al. 2006. The absence of sharks from abyssal regions of the world's oceans. *Proc. R. Soc. London Ser. B* 273:1435–41

Quilty PG. 1990. Significance of evidence for changes in the Antarctic marine environment over the last 5 million years. In *Antarctic Ecosystems: Ecological Change and Conservation*, ed. KR Kerry, G Hempel, pp. 3–8. Berlin: Springer-Verlag

Raguá-Gil JM, Gutt J, Clarke A, Arntz WE. 2004. Antarctic shallow-water mega-epibenthos: shaped by circumpolar dispersion or local conditions? *Mar. Biol.* 144:29–40

Revelle R, Fairbridge R. 1957. Carbonates and carbon dioxide. *Geol. Soc. Am. Mem.* 67:239–96

Rhodes MC, Thompson RJ. 1993. Comparative physiology of suspension-feeding in living brachiopods and bivalves: evolutionary implications. *Paleobiology* 19:322–34

Samerotte AL, Drazen JC, Brand GL, Seibel BA, Yancy PH. 2007. Correlation of trimethylamine oxide and habitat depth within and among species of teleost fish: an analysis of causation. *Physiol. Biochem. Zool.* 80:197–208

Scher HD, Martin EE. 2006. Timing and climatic consequences of the opening of Drake Passage. *Science* 312:428–30

Seibel BA. 2007. On the depth and scale of metabolic rate variation: scaling of oxygen consumption rates and enzymatic activity in the Class Cephalopoda (Mollusca). *J. Exp. Biol.* 210:1–11

Seibel BA, Drazen JC. 2007. The rate of metabolism in marine animals: environmental constraints, ecological demands and energetic opportunities. *Philos. Trans. R. Soc. London Ser. B.* In press. doi:10.1098/rstb.2007.2101

Sepkoski JJ Jr. 1991. Diversity in the Phanerozoic oceans: a partisan view. In *The Unity of Evolutionary Biology: Proc. 4th Int. Congr. Syst. Evol. Biol.*, Vol. 1, ed. EC Dudley, pp. 210–36. Portland, Oregon: Dioscorides Press

Smetacek V, Nicol S. 2005. Polar ocean ecosystems in a changing world. *Nature* 437:362–68

Snelgrove PVR. 2001. Marine sediments. In *Encyclopedia of Biodiversity*, ed. SA Levin, pp. 71–84. San Diego: Academic

Somero GN, De Vries AL. 1967. Temperature tolerance of some Antarctic fishes. *Science* 156:257–58

Stachowitz JJ, Terwin JR, Whitlatch RB, Osman RW. 2002. Linking climate change and biological invasions: ocean warming facilitates nonindigenous species invasions. *Proc. Natl. Acad. Sci. USA* 99:15497–500

Stanley SM. 1977. Trends, rates and patterns of evolution in the Bivalvia. In *Patterns of Evolution as Illustrated by the Fossil Record*, ed. A Hallam, pp. 209–50. Amsterdam: Elsevier

Stilwell JD, Zinsmei. ter WJ. 1992. *Molluscan Systematics and Biostratigraphy: Lower Tertiary La Meseta Formation, Seymour Island, Antarctic Peninsula*. Washington, DC: Am. Geophys. Union, Antarct. Res. Ser. 55, 192 pp.

Tavares M, De Melo GAS. 2004. Discovery of the first known benthic invasive species in the Southern Ocean: the North Atlantic spider crab *Hyas araneus* found in the Antarctic Peninsula. *Antarct. Sci.* 16:129–31

Thatje S. 2005. The future fate of the Antarctic marine biota? *TREE* 20:418–19

Thatje S, Anger K, Calcagno JA, Lovrich GA, Pörtner H-O, Arntz WE. 2005. Challenging the cold: crabs reconquer the Antarctic. *Ecology* 86:619–25

Thatje S, Arntz WE. 2004. Antarctic reptant decapods: more than a myth? *Polar Biol.* 27:195–201

Thatje S, Fuentes V. 2003. First record of anomuran and brachyuran larvae (Crustacea: Decapoda) from Antarctic waters. *Polar Biol.* 26:279–82

Thatje S, Lörz AN. 2005. First record of lithodid crabs from Antarctic waters off the Balleny Islands. *Polar Biol.* 28:334–37

Thatje S, Schnack-Schiel S, Arntz WE. 2003. Developmental trade-offs in Subantarctic meroplankton communities and the enigma of low decapod diversity at high southern latitudes. *Mar. Ecol. Prog. Ser.* 260:195–207

Thies D, Reif W-E. 1985. Phylogeny and evolutionary ecology of Mesozoic Neoselachii. *N. Jb. Geol. Paläont., Abh.* 169:333–61

Thorson G. 1950. Reproduction and larval ecology of marine bottom invertebrates. *Biol. Rev. Camb. Philos. Soc.* 25:1–45

Tripati A, Backman J, Elderfield H, Ferretti P. 2005. Eocene bipolar glaciation associated with global carbon cycle changes. *Nature* 436:341–46

Vermeij GJ. 1977. The Mesozoic marine revolution: evidence from snails, predators and grazers. *Paleobiology* 3:245–58

Vermeij GJ. 1978. *Biogeography and Adaptation: Patterns of Marine Life*. Cambridge, MA: Harvard Univ. Press. 332 pp.

Vermeij GJ. 1987. *Evolution and Escalation: An Ecological History of Life*. Princeton: Princeton Univ. Press. 527 pp.

Wagner PJ, Kosnik MA, Lidgard S. 2006. Abundance distributions imply elevated complexity of post-Paleozoic marine ecosystems. *Science* 314:1289–92

Wakeling JW, Johnston IA. 1998. Muscle power output limits fast-start performance in fish. *J. Exp. Biol.* 201:1505–26

Walsh JJ. 1988. *On the Nature of Continental Shelves*. San Diego: Academic. 520 pp.

Walther G-R, Post E, Convey P, Menzel A, Parmesan C, et al. 2002. Ecological responses to recent climate change. *Nature* 416:389–95

Watling L, Thurston MH. 1989. Antarctica as an evolutionary incubator: evidence from the cladistic biogeography of the amphipod family Iphimediidae. In *Origins and Evolution of the Antarctica Biota*, ed. JA Crame, pp. 297–309. Bath, United Kingdom: Geol. Soc.

Werner JE, Blake DB, Aronson RB. 2004. Effects of late Eocene cooling on Antarctic marine communities. *Boll. Geofis.* 45(Suppl. 2):262–65 (Abstr.)

Wiedman LA, Feldmann RM. 1988. Brachiopoda from the La Meseta Formation (Eocene), Seymour Island, Antarctica. See Feldman & Woodburne 1988, pp. 449–57

Woll AK, Burmeister AD. 2002. Occurrence of northern stone crab (*Lithodes maja*) at southeast Greenland. In *Crabs in Cold-Water Regions: Biology, Management, and Economics*, ed. AJ Paul, EG Dawe, R Elner, GS Jamieson, GH Kruse, et al. pp. 733–49. Fairbanks: Univ. Alaska Sea Grant Coll. Program, AK-SG-02–01

Zachos J, Pagani M, Sloan L, Thomas E, Billups K. 2001. Trends, rhythms, and aberrations in global climate 65 Ma to present. *Science* 292:686–93

Zinsmeister WJ, Feldmann RM. 1984. Cenozoic high latitude heterochroneity of Southern Hemisphere marine faunas. *Science* 224:281–83

Zwally HJ, Comiso JC, Parkinson CL, Cavalieri DJ, Gloersen P. 2002. Variability of Antarctic sea ice 1979–1998. *J. Geophys. Res.* 107:3041–59

Spatiotemporal Dimensions of Visual Signals in Animal Communication

Gil G. Rosenthal

Department of Biology, Texas A&M University, College Station, Texas 77843-3258;
email: grosenthal@mail.bio.tamu.edu

Annu. Rev. Ecol. Evol. Syst. 2007. 38:155–78

First published online as a Review in Advance on
July 13, 2007

The *Annual Review of Ecology, Evolution, and
Systematics* is online at
http://ecolsys.annualreviews.org

This article's doi:
10.1146/annurev.ecolsys.38.091206.095745

Key Words

form, motion, pattern, texture, vision, visual ecology

Abstract

Much of the information in visual signals is encoded in motion,
form, and texture. Current knowledge about the mechanisms underlying visual communication is spread across diverse disciplines.
Contemporary perspectives on the physics, psychology, and genetics
of visual signal generation and perception can be synthesized into
a conceptually integrative approach. Developmental mechanisms of
pattern formation suggest that small changes in gene regulation or
structure can result in major shifts in signal architecture. Animals in
many species have been shown to attend to variation in higher-order
stimulus properties. Preferences for these properties can be innately
specified or learned, and may also show large shifts or reversals. Perceptual mechanisms, particularly visual attention, associated with
spatiotemporal features are likely to be a major force in shaping the
design of visual signals.

INTRODUCTION

The leopard's spots, the high-contrast bands on the wings of a *Heliconius* butterfly, the intricate filigree on the face of a Picasso triggerfish—so much of what we find compelling about animal diversity has to do with complex visual signals. The information in visual signals is encoded in spatiotemporal patterns: variation in color and intensity across time and two- or three-dimensional space (**Figure 1**). These patterns pose a challenge across disciplines. To developmental biologists, spatial patterns are a compelling model for studying the ontogeny and evolution of form (Brunetti et al. 2001, Parichy 2003). To sensory ecologists, they offer the challenge of understanding how multivariate signals are transmitted in heterogeneous environments (Endler & Mielke 2005, Fleishman et al. 2006). The perception of texture, motion, and shape

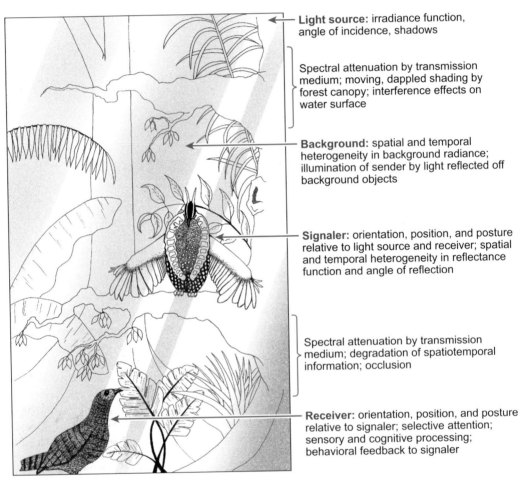

Light source: irradiance function, angle of incidence, shadows

Spectral attenuation by transmission medium; moving, dappled shading by forest canopy; interference effects on water surface

Background: spatial and temporal heterogeneity in background radiance; illumination of sender by light reflected off background objects

Signaler: orientation, position, and posture relative to light source and receiver; spatial and temporal heterogeneity in reflectance function and angle of reflection

Spectral attenuation by transmission medium; degradation of spatiotemporal information; occlusion

Receiver: orientation, position, and posture relative to signaler; selective attention; sensory and cognitive processing; behavioral feedback to signaler

Figure 1

Schematic of the steps involved in production, transmission, and reception of visual signals. Drawing by Nick Ratterman.

encompasses some of the key problems in cognitive psychology (Kovacs & Julesz 1994, Maunsell & Treue 2006), and analysis of visual images is the domain of physicists and applied mathematicians (Billock et al. 2001, Chubb & Yellott 2000). Current knowledge of the biology of complex visual signaling is thus spread across vastly different disciplines. An encyclopedic review of the relevant literature is impossible, and this brief treatment of a massively broad topic is necessarily both superficial and incomplete. The aim of this review, nevertheless, is to encourage the synthesis of these diverse perspectives into a conceptually coherent approach to studying visual communication.

THE MAGNITUDE OF THE PROBLEM

Figure 1 illustrates the scope of understanding complex visual communication with a hypothetical male signaler courting a female with a complex visuomotor display. Photic energy is provided by sunlight, which varies in intensity, color, and angle as a function of season and time of day. Light is filtered by the atmosphere, affecting the characteristics of the photic environment in which animals find themselves communicating; dynamic patches of light and shade are further produced by motion of the forest canopy (Endler 1993) or waves on the water surface (Loew & McFarland 1990). Some of this light is reflected off other objects in the environment before striking the signaler. The male's position, orientation, and posture at any given time determine how this complex visual field illuminates his body: both the path taken by the light (e.g., direct sunlight versus sunlight filtered by leaves and reflected off the ground) and the angle of incidence determine the radiance of light reaching the body surface at any given point in space and time.

The male's body surface then reflects light in a variety of ways, resulting in the richly textured spatial pattern that we see. For example, the pigment in some feathers selectively absorbs high-energy photons, producing a bright red, whereas air spaces in some feathers scatter all wavelengths in all directions, producing a bright, diffuse white light (Finger 1995). Eyes, shiny scales, and some exoskeletons, however, act like mirrors, producing specular highlights (Blake & Buelthoff 1990, Zeil & Hofmann 2001). This spatial variation is complemented by temporal variation in the male's posture and orientation in the course of a courtship display. En route to the female, water or humid air changes the spectral quality of the light reflected off the male (Lythgoe 1979); particles in the transmission medium and occluding objects degrade spatial and temporal information (Kenward et al. 2004). All of these factors combine to affect particular components of the visual signal, and the efficacy of that signal, before it ever reaches the eyes of the female.

How the female perceives the signal depends on how her two eyes are placed relative to the male and the light source—if he is backlit by sunlight, he appears as a black silhouette; if they are facing each other straight on (Dantzker et al. 1999), the reticulate pattern on his chest appears to be regularly spaced and the iridescence on his throat shines a brilliant blue (Rutowski et al. 2007). The movement of the female's head, eye, and body supply her with information about the three-dimensional structure of the scene. Her gaze is drawn to his plumage ornaments, and the patterns these project onto her foveas are parsed by spatial-frequency, temporal-frequency, color, and orientation filters in her retina and brain (Kandel & Wurtz 2000). This information is ultimately integrated into a behavioral response that in turn influences subsequent actions by the male (Patricelli et al. 2002).

Perhaps because of the immensity of the problem, the study of the visual communication process has been somewhat atomized. The production of spatiotemporal patterns is almost exclusively the domain of developmental biologists, while their perception is largely studied by cognitive psychologists. Integrative studies—focusing on signal production, transmission, and perception—have mostly fallen into two categories: evaluating behavioral responses to qualitative variation in stimulus characteristics, and quantitative studies involving color. As **Figure 2** illustrates, however, it is impossible to interpret color data without taking spatial information into account. For example, Barry & Hawryshyn (1999) found that colors reflected by Hawaiian coral reef fish on average matched the background irradiance (**Figure 2d**). Clearly, however, some pattern elements (like the yellow tail of the queen angelfish in **Figure 2a**) present a sharp contrast with both the background and other elements

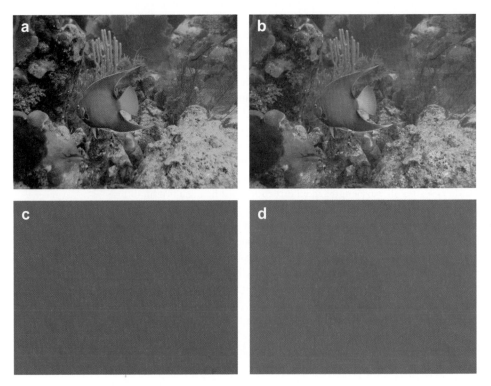

Figure 2

Queen angelfish (*Holacanthus ciliaris*) against a complex coral reef background (Glover's Reef, Belize). (*a*) Unmanipulated digital color image; (*b*) chromatic information removed; (*c*) spatial information removed; (*d*) spatial information removed separately from angelfish and background.

on the animal's skin. Marshall (2000) suggested this allows many reef fish to remain camouflaged from a distance yet conspicuous at close range, via smearing of high-spatial-frequency, high-contrast color patterns like blue and yellow. Microhabitat selection may further modify contrast relationships.

I begin by reviewing mechanisms of signal production and perception, and then discuss how knowledge of perception can help us quantify complex visual signals. I then review the burgeoning literature on the integrative biology of complex visual communication.

TYPES OF SPATIOTEMPORAL SIGNALS

An encyclopedic work by Cott (1940) remains the most exhaustive treatment of the diversity of complex visual signals. Spatiotemporal signals can either be cryptic, whereby they are designed to minimize detection or recognition by receivers, or conspicuous, where they serve to maximize information transfer. The spatial component can involve variation in the pigment or structural composition of the skin, exoskeleton, hair, feather, or scales; as well as modified appendages, horns, and other morphological structures. Many signals have repeating components, known as textures.

The temporal component, which is often absent in crypsis, typically involves muscle-mediated changes in orientation, posture, and position relative to the receiver, and often raising and lowering of hair, feathers, or appendages. In birds and mammals, modulation of blood flow can effect limited color change in the skin (Bradbury & Vehrencamp 1998). In ectothermic vertebrates, neurally-mediated control of chromatophore cells (Bagnara & Hadley 1973) allows for temporal variation in pattern over a range of time scales, both within the course of a signaling display and over longer periods of time (Rosenthal & Lobel 2006). Male haplochromine cichlids *Astatotilapia burtoni*, for example, express a drab, female-like pattern when socially subdominant and a high-contrast sexually dimorphic pattern when dominant and reproductive; although this pattern is maintained for the duration of the male's social status, it is reversible and can change rapidly (Fernald 1990). Similarly, in birds, plumage badges serve as indicators of dominance status, and are modified during seasonal molts (e.g., Qvarnstrom 1997) or hidden and revealed through postural changes (Hansen & Rohwer 1986).

Temporal and spatial components of signals are typically expressed in concert. In courtship displays, conspicuous male ornaments are usually coupled with stereotyped movements that present them to the best advantage (Rosenthal et al. 1996). The reverse is true for cryptic species: a remarkable case is the reef cornetfish *Fistularia commersonii*, which expresses a disruptive banded pattern when stationary, and rapidly transitions to uniform silvery coloration when in motion (Thomson et al. 2000).

SIGNS AND RITUALS: SPATIOTEMPORAL PATTERNS THROUGH THE MID-TWENTIETH CENTURY

Because we need no special instrumentation to detect most aspects of visual signals (with the notable exceptions of polarization patterns and UV cues) it is not surprising that they have generated a long history of scholarship. The role of complex visual signals in animal communication was well established early on (Cott 1940). Numerous experiments in a broad variety of taxa demonstrated that animals attended to nuances in color, motion, texture, and form (Eibl-Eibesfeldt 1970).

Much of mid-twentieth century ethology was concerned with systematizing rules of cause and effect in animal behavior. The finding that complex behavioral responses could be elicited by simple stimuli, called innate releasing mechanisms, had a fundamental effect on contemporary thinking about visual signals. Crude models, which resembled natural signals in only a few, key salient ways, could elicit the same responses as parents, potential mates, or competitors. Even so, spatial relationships among signal elements played a critical role. Tinbergen (1951), for example, found that male sticklebacks produced aggressive behavior only when models were presented in the proper orientation, with red coloration on the underside of the object. Furthermore, Eibl-Eibesfeldt (1970) cited a number of studies showing that though releasers seemed to play a large role in aggressive interactions, a mating response required more nuanced stimuli. Cott (1940) also remarked that mating signals were generally far more complex than aposematic signals, arguing that simpler patterns were more likely to be memorable to predators.

Visual signals were therefore classed into distinct, "simple" and "complex" categories. Lorenz (cited in Eibl-Eibesfeldt 1970) dichotomized visual signals into simple, innately specified "sign stimuli" and complex, learned "gestalten":

> Where an animal can be 'tricked' into responding to simple models, we have a response by an innate releasing mechanism; where it cannot be thus confused, we have an acquired recognition of a gestalt.

Lorenz's assertion, though implicitly accepted, remains to be thoroughly explored. In some cases, behavior toward spatiotemporal patterns has a strong learned component (Engeszer et al. 2004). We know that low-level visual processing can be highly dependent on early visual experience; for example, the color of ambient light during a critical period can influence the expression of visual pigments in retinal photoreceptors (Fuller et al. 2005) and, as shown in classic studies by Hubel & Wiesel (1963), cortical cells that parse spatial frequency and orientation are also sensitive to early visual input. On the other hand, work on auditory imprinting in birds (Marler 1997) showed that individuals have an innate predisposition to learn conspecific song, which differs from that of heterospecifics in multiple acoustic dimensions.

Complex stimuli may thus be innately specified in some cases. Two bird species innately recognize the color, orientation, and shape of patterns characteristic of venomous coral snakes (Smith 1975, 1977), which differ from nonvenomous heterospecifics primarily in the spatial organization of repeated pattern elements. The intricate, species-typical color patterns of coral reef fishes are also likely to have an innate component. Most marine fishes release their gametes into the plankton, leaving no possibility for learning parental phenotypes.

A simple, innately specified set of cues may also serve as a trigger to attend to and imprint on a more complex stimulus. For example, newborn humans attend preferentially to faces, but this appears to be specified by a general bias toward up-down asymmetry in a spatial configuration; abstract patterns with more elements in the top part of the image are more attractive than upside-down faces (Turati et al. 2002). A subset of image components, or even a cue in another modality, may thus direct learned recognition of a complex image.

DEVELOPMENTAL EVOLUTION OF SIGNALS

Despite the profusion of complex visual signals, an animal's body is not a blank canvas. Pattern expression is constrained by evolutionary history in a number of ways. Complex motor patterns are often modified, or ritualized, from nonsignaling antecedents; Hurd et al. (1995) used neural network algorithms to corroborate Darwin's (1872) principle of antithesis, showing that selection favors signal complexes that are maximally different in opposing contexts, e.g., aggressive escalation versus submission. Although signal motor patterns tend to have nonsignaling precursors, many animals have evolved specialized musculature and motor control systems dedicated to communication (e.g., Ma 1995a,b).

Spatial patterning also involves nonsignaling precursors, both in ontogeny and phylogeny. The patterning of spatial signal elements is often developmentally dictated by the location of nonsignaling structures. In salamanders and zebrafish, for example, lateral stripes are specified by the position of lateral-line precursors (Parichy 2003).

The problem of skin patterning has received some attention from developmental biologists. A diversity of repeating patterns, or textures, can be produced by subtle changes to a relatively simple molecular mechanism. Turing (1952) proposed the reaction-diffusion system, whereby spot and stripe patterns arise as a result of instabilities in the diffusion of two or more morphogenetic chemicals (morphogens), in the skin during early development. Differences in boundary and initial conditions, and in the number of morphogens and diffusion properties of each morphogen, can interact to produce an array of stripe and spot patterns (Kondo & Asai 1995, Murray 1981, Painter et al. 1999). Shoji & Iwasa (2003) further suggested that a reaction-diffusion system could act in concert with small differences in diffusion anisotropy of morphogens (caused by structures with directional conformation such as scales) to induce directionality in stripe patterns.

Recent molecular-genetic studies are illuminating mechanisms of pattern formation. Parichy (2003) provides a concise review of the developmental genetics of stripe formation in zebrafish (*Danio rerio*) and their close relatives. Asai et al. (1999) argued that the zebrafish *leopard* gene may be acting as a component of a reaction-diffusion system. Small changes in gene structure or regulation could thus induce qualitative changes in pattern elements (Painter et al. 1999; see **Figure 3a**), producing dramatic differences in signal structure. Such a process could account for the striking differences in pattern sometimes seen among closely related species (**Figure 3b**).

Eyespots on butterflies provide insight into the development of aperiodic signals, and suggest another set of mechanisms for generating signal diversity (Brunetti et al. 2001). The location of a focal morphogen—which will become the center of the eyespot in the adult—is specified during early development. The focus induces rings of regulatory gene expression that determine the size, color, and arrangement of concentric pattern elements. Variation in these properties can arise through several

Figure 3

(*a*) Diversity of patterns generated by reaction-diffusion models. Top two rows show heterochrony in a stripe-splitting sequence; third row, changes in parameter values produce stripe addition instead of splitting; bottom two rows, stripes change into spots as domain size increases. (*b*) Distribution of red body pigmentation of fifteen species of Cameroonian *Aphyosemion* killifish [adapted from Painter et al. 1999 (National Academy of Sciences, USA) and Amiet 1987].

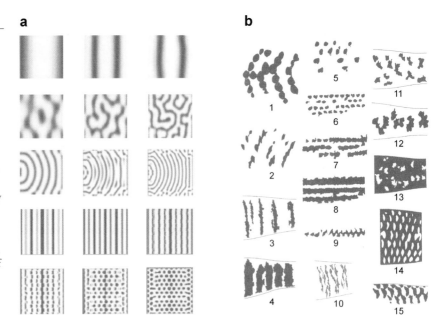

channels: differences in the regulatory domains of transcription factors operating downstream of the focus, different combinations of regulatory genes recruited to specify pattern properties, and differences in the regulation of structural genes involved in scale pigmentation. As with Turing patterns, eyespot formation suggests that small genetic changes could yield substantial differences in pattern structure (Brunetti et al. 2001).

In some cases, patterns are intricate and highly specified. The aptly named Picasso triggerfish, *Rhinecanthus aculeatus*, for example, exhibits a black-and-white polka-dot pattern just anterior to the caudal fin; four white stripes running diagonally from the insertion of the anal fin to the midline of the body; brown and gray bands on the dorsum; a filigree of electric blue lines originating parallel at the eye and curving to meet on the gill cover; and a blue "mustache" above the mouth contrasting with a broad russet band that slopes down to meet the white belly. How are the locations, spatial extents, and chromatic properties of all these elements specified?

In some cases, ontogenetic changes in pattern expression continue through adult life. Plumage badges in birds, for example, intensify with age (Qvarnstrom 1997). A dramatic example is provided by sequential hermaphrodites, where female patterns are replaced by dramatically different male patterns over a period of a few days (Warner & Swearer 1991).

Pattern ontogeny of both motor patterns and spatial patterns is susceptible to environmental influences. For example, in the grasshopper *Schistocerca gregaria*, development of a conspicuous, black-and-yellow aposematic pattern requires both use of a toxic host, and high densities during larval instar stages (Sword 1999). In satin bowerbirds, juvenile males solicit courtship from older males, suggesting that there may be a learned component to these intricate dances (Collis & Borgia 1993).

SIGNAL RECEPTION

The reception of complex visual signals can be broadly divided into receiver decisions, which determine how light reflected from the signaler and background enters the visual system as a function of space and time; and receiver processing, which determines how the visual system analyzes the spatial, temporal, and chromatic distribution of light. Receiver decisions have three components (**Figure 1**). First, a general property of communication, which will not be discussed further, is that receivers provide behavioral feedback that influences the behavior of the signaler. Second, receiver behavior influences the three-dimensional structure of the scene: how is the receiver's field of view oriented and positioned relative to the light source, the background, and the signaler (Heindl & Winkler 2003; see **Figure 1**)? Third, how does the receiver attend to different components of the scene, of the signal, and of the background?

Receiver Decisions

Receiver behavior. Spatiotemporal relationships among signalers, backgrounds, receivers, and illuminant sources are crucial variables in visual communication. Receiver motion provides information about three-dimensional depth and about spatial relationships among scene elements (Wexler & van Boxtel 2005). The geometry of signal illumination, and the position and orientation of the receiver, are early determinants of how the signal will be perceived (Fleishman et al. 2006). The positioning of receivers thus has a profound influence on signal expression. Body patterning in pelagic animals, for example, is often countershaded, with a dark dorsal surface and a light ventral surface that minimize contrast with the substrate and water surface, respectively (Cott 1940). At the other extreme, many courtship displays are expressed in contexts that maximize conspicuousness to receivers: spatial patterns are expressed in concert with motor patterns that showcase them to maximal advantage. In a wide variety of species, male courtship displays involve movements at characteristic viewing distances directly in front of the female (Dantzker et al. 1999). Male orange sulphur butterflies (*Colias eurytheme*) direct their courtship so as to maximize the brightness of iridescent UV coloration (Rutowski et al. 2007).

Receiver attention. Despite a growing literature on attentional constraints and foraging ecology (Dukas 2004), attentional processes have received little explicit consideration in animal communication. Attention plays two roles that are of major importance in signaling interactions. First, some signal features are involved in initially directing attention toward or away from the signaler. Ocelli (false eyespots) on many animals have long been thought to redirect predator attention in a way that facilitates escape (Cott 1940; Stevens 2005).

In contrast, early stages of courtship and aggressive interactions often involve high temporal frequency movements in a characteristic direction, readily detectable in the visual periphery, that direct a viewer to focus on and attend to the signaler (a 'visual grasp reflex', Peters et al. 2002). Signals may also be designed to hold receiver attention. Habituation is a universal property of nervous systems, and large, diverse song repertoires in songbirds maximize attractiveness in part by delaying the habituation of receivers (Ryan 1998). Intricate dances and patterns in courtship may play a similar role.

Second, attentional processes play a primary role in how receivers assign importance to different aspects of a complex stimulus: What are receivers actually looking at? In satin bowerbirds, for example, young females select males based on the static attributes of their bower constructions, whereas older, more experienced females attend mainly to variation in the relatively dynamic temporal features of the male courtship display (Coleman et al. 2004). Attention is not merely a question of concentrating on a spatial dimension of the visual field; subjects can shift their attention between different features of a scene, e.g., shape, color, or velocity (Maunsell & Treue 2006). Such dynamic shifts in attention may explain how receivers weight multiple traits differently across comparisons (Kirkpatrick et al. 2006).

A recent study on humans (Rowe et al. 2007) showed that the breadth of attentional focus in a visual task (correctly identifying a central letter flanked by other letters) depends on affect; positive mood was associated with impaired performance, because individuals were attending to the flanking letters as well as the central target. "Sad" individuals focused their attention on the central target. The multifaceted suites of male traits present in most courtship signals may have evolved in order to satisfy the inconstant attention of happy females. Conversely, because arousal in a negative context is associated with a constriction in attentional focus (Easterbrook 1959), one might expect graded aggressive signals to address progressively narrower attentional demands.

Attentional constraints, which have been amply considered in the context of foraging behavior (Dukas 2004) may therefore play an important role in signal evolution. Assays like the flanker task in Rowe et al. (2007) are straightforward to apply to nonhuman animals. Gaze direction, a widely used indicator of attention in human studies, can now be studied noninvasively in animals thanks to novel tracking methodologies using neural network algorithms (Piratla & Jayasumana 2002). These approaches make it feasible to begin studying the role of attentional mechanisms in animal communication.

Receiver Processing

The sensory periphery influences two key aspects of visual perception: color sensitivity (reviewed in Yokoyama & Yokoyama 1996 [vertebrates] and Briscoe & Chittka 2001 [insects]) and spatial acuity, i.e., the spatial resolving power of the eye. Spatial acuity increases with the size of the eye (by increasing the eye's focal length, analogous to a camera F-stop), with photoreceptor density on the retina, and with the ratio of cones to rods (because of the reduced summation of cones to downstream cells). There is thus a fundamental size constraint on spatial acuity. Moreover, there is a tradeoff between acuity and visual sensitivity, resulting in reduced acuity for animals in low-light environments. As reviewed by Bradbury & Vehrencamp (1998), acuity varies by more than three orders of magnitude across taxa. Temporal frequency resolution, or critical flicker fusion frequency, is also inversely related to sensitivity and shows substantial variation among species (Frank 2000).

Sensory constraints thus place limits on the spatial and temporal frequency of visual signals. From a signaler's point of view, this can impact both the design of spatiotemporal components and the optimal viewing distance. For example, males in some species of swordtails (genus *Xiphophorus*) exhibit conspicuous vertical bars during courtship, coupled with a back-and-forth lateral display (Ryan & Rosenthal 2001). Casual observation suggests that display speed is higher in species lacking bars, with males sometimes executing a 180° turn in 100 ms or less. The signal value of the bars would be lost if they were presented at a rate exceeding the female's critical flicker fusion frequency of about 43 Hz (Rosenthal & Evans 1998). One should also expect environmental demands on sensory systems to constrain the evolution of spatiotemporal patterns, much as has been demonstrated for color (Endler 1992). Specifically, displays in species that are active in low-light environments, and therefore have reduced spatial and temporal resolving power, should be slower and spatially simpler.

Although spatial and temporal resolution impose an upper bound on information transfer, it is unlikely that there is tight coevolution between signal structure and the spatial- and temporal-frequency tuning of receivers. The reason is that the angular size of the same signal on the retina (and therefore the angular speed as well) will vary with viewing distance. One of the challenges of visual processing is to use unstable, two-dimensional images to extract stable information from a complex, three-dimensional world. Size constancy, the ability to judge the real size of objects independent of distance, is ubiquitous in vertebrates (Douglas et al. 1988). Size constancy requires integrating at least two streams of information: the visual angle an object subtends on the retina, and spatial or spatiotemporal indicators of distance such as binocular disparity or accommodation (focus) cues. Similarly, most visual systems are likely to exhibit some degree of constancy with respect to rotation, shape, brightness, color, and texture (Dyer 2006, Kandel & Wurtz 2000).

It is nevertheless probable that there are ecological correlates of higher-order visual processing. Simple cells in the macaque primary visual cortex exhibit diverse, apparently haphazard responses to specific combinations of stimulus orientation, spatial frequency, and speed. van Hateren and van der Schaaf (1998) showed that the distribution of simple cell characteristics corresponded to that of natural image properties, leading these researchers to conclude that "the apparent randomness of simple cell properties may not be the sign of a sloppy design, nor of random variability in development, but may in fact be a deliberate attempt to match the requirements of processing natural images." Selection may thus favor distinct emphases in early visual processing. For example, most salient stimuli for large ungulates are likely to be confined to the horizontal plane, whereas small rodents, preyed upon by swooping aerial predators, should additionally attend to the vertical. Párraga et al. (2002) used natural image statistics and psychophysical data to argue that in humans, sensitivity to chromatic modulation at low spatial frequencies has evolved in the context of a specific ecological task, namely detecting ripe fruit against a complex background. Parsing of natural scenes is thus likely to reflect strong selection for performance in particular contexts.

In addition to extracting invariant features of stimuli from the environment, the other principal challenge of visual processing is image segmentation: distinguishing between the salient stimulus (the signaler) and the background. The question of how visual processing of local scene elements is integrated into a unitary percept is one of the key foci of cognitive psychology (Kovacs 1996; Kovacs & Julesz 1994). This is far from a trivial task (**Figure 2**): Both visual backgrounds and signaler properties are heterogeneous and depend heavily on lighting, distance, and perspective, and the problem is further complicated by occluding objects between signaler and receiver. Edge detection begins with center-surround cells in the retina (Rodieck 1998) and provides early information about discontinuities in the visual scene. Balboa & Grzywacz (2000) showed that occluding objects produce sharp disjunctions between image features and can be detected in early retinal processing. As discussed below, edge-detection mechanisms are exploited by cryptic patterns in camouflage. High-contrast edges are, however, a ubiquitous feature of conspicuous displays. Because these are almost always expressed in concert with motor patterns, motion coherence (Sekuler & Blake 1994) of the patterns on the signaler should facilitate separation of the signaler from the background. Many of the high-contrast markings thought to be 'amplifiers' (Hasson 1989), allowing receivers to more reliably assess honest indicators, may simply provide a means for facilitating figure-ground assignment by receivers.

Once an image is segmented into "figure" and "ground," how is complex spatiotemporal information integrated into recognition of a potential mate or a likely rival? Some units in higher-level visual processing are highly specific. For example, some neurons in the macaque inferotemporal cortex are maximally tuned to average, identity-ambiguous faces of other macaques (Leopold et al. 2006). Biederman and colleagues (Hayworth & Biederman 2006) have suggested that object recognition relies on parsing a visual stimulus into geons—simple geometric shapes that are robust to changes in rotation, scale, and noise, like cylinders and cones. Remarkably, Chen et al. (2003) showed that honeybees (*Apis mellifera*) learned to generalize patterns based on topological invariance, i.e., the number of holes contained, independent of local stimulus cues. Topology is often robust to many aspects of rotation, scale, and translation, and therefore provides reliable information about global properties.

There are therefore multiple higher-order attributes of visual stimuli that animals may use in assessing signals. Many of these share the property of categorical attribution, e.g., one hole versus none, face versus nonface, cylinder versus cube. Just as small changes in gene expression can produce discontinuities in signal expression, it may be that similarly small alterations can effect qualitative changes in both the perceptual salience and the valence of stimuli.

CHARACTERIZING SPATIOTEMPORAL SIGNALS

In order to study how complex visual stimuli and receiver properties might be coevolving, we need to be able to describe signals quantitatively, in the currency used by receivers. Fourier analysis has accomplished this, with great success, for acoustic communication. Any complex, periodic signal can be decomposed into an infinite series of pure sine waves, and the vertebrate ear acts as a frequency analyzer itself. There is thus a unified, mathematically elegant way to describe the production, transmission, and perception of sound.

Similarly, Fourier analysis is integral to studies of visual perception (e.g., Párraga et al. 2002, Shapley & Lennie 1985), and models of visual processing assume that cortical neurons act as Fourier filters (Shapley & Lennie 1985). A difficulty arises, however, when attempting to characterize the visual scenes and signals that are being processed. Although an acoustic signal impinging on an eardrum can be approximated as three-dimensional (frequency × amplitude × time), a projection of a moving scene onto a single retina has five dimensions (wavelength × intensity × vertical × horizontal × time). Perhaps largely because of this added complexity, there is no unified approach to analyzing visual signals, and analyses of spatial, temporal, and spectral components of visual scenes have generally been conducted separately.

Temporal Components

Pioneering studies by Fleishman (1988, 1992) used one-dimensional Fourier analysis of the vertical component of the head-bob display in *Anolis* lizards to quantify conspicuousness relative to background motion. High-frequency components of vertical motion (the jerkiness of the head-bob display) enabled the lizard to stand out from the movement of background vegetation. Peters et al. (2002) obtained a similar result for the complex display of the agamid lizard *Amphibolurus muricatus*, using the spatial distribution of velocity vectors, or optic flow fields, in successive images of a movie to calculate ellipses describing the distribution of direction and speed of movement in both lizard displays and background vegetation (**Figure 4**).

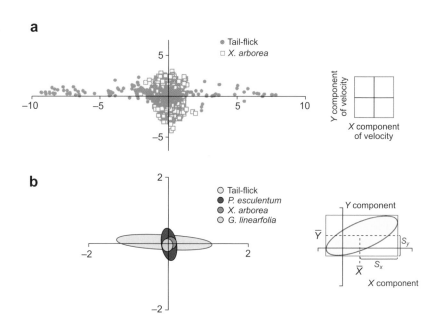

Figure 4

(*a*) Velocity signature of a Jacky dragon (*Amphibolurus muricatus*) tail-flick overlaid on a representative sequence of vegetation. (*b*) The standard ellipse for the tail-flick and each of the vegetation sequences at low wind speed. The center of each ellipse is given by the vector of *x* and *y* means, while the standard deviations and correlation coefficient define the shape and orientation of the ellipse in the X/Y plane (modified from Peters et al. 2002).

Spectral Components

Quantitative studies of vision have overwhelmingly focused on color (Grether et al. 2004). Spectroradiometry is used to measure the spectral characteristics of the visual environment (e.g., downwelling and sidewelling irradiance), and reflectance spectroradiometry to measure reflectance, commonly at point samples along the body surface. Leal & Fleishman (2004) and Fleishman et al. (2006) have pointed out the importance of illuminance geometry and orientation to this process. The principal advantage is that we can estimate the distribution of photons impinging on the retina; we are measuring directly in the currency of the visual system.

Spatial Components

Analysis of two-dimensional static images has proven far more challenging than that of temporal and spectral components of visual signals. Approaches to analyzing spatial pattern variation can be divided into signal-based analyses and scene-based analyses.

Signal-based analysis. Signal-based analysis of spatial components is by far the most common approach used in studies of signal design and evolution. Given a priori reasons to believe that a particular set of traits is salient in communication, we can qualitatively or quantitatively describe phenotypes with respect to these traits. There are two main limitations to this approach. First, we may be ignoring some spatial properties that are important to the receiver; second, our characterization is unlikely to be meaningful from the point of view of visual processing. Given our human visual bias, it is often easiest to simply parse signals into discrete categories by appearance, either in terms of individual components (e.g., tail tips, vertical stripes, horizontal stripes, and spots; Ortolani 1999) or signaling complexes (Crook 1997). Although this approach has yielded powerful insights into the evolution of spatial patterns, there is an inherent subjectivity with respect to category boundaries, and variation within categories may be important; for example, the aspect ratio or relative intensity of spots may be salient to receivers. A similar approach can be used to express spatial components in quantitative terms. Numerous studies have quantitatively measured and manipulated morphological traits thought to be important in visual signaling (e.g., Andersson 1982, Sinervo & Basolo 1996, Wong & Rosenthal 2006).

An intriguing approach to signal-based analysis was taken by Mojsilović et al. (2000), who effectively asked receivers to classify patterns. They presented human subjects with all 190 possible pairwise combinations of a set of 20 color textures taken from an interior design catalog, then asked them to rate each pair as to its similarity. Using hierarchical clustering analysis, they identified a "basic vocabulary of color patterns" whose strongest determinant was pattern equivalence: if the spatial pattern between two textures was identical, they were rated as highly similar independent of color. Conditioned-response and habituation/dishabituation studies may make such an approach feasible for nonhuman animals as well.

Scene-based analysis strives to characterize the properties of images as perceived by receivers. Most scene-based analysis relies on quantitative methods, but Changizi et al. (2006) used categorization by human observers to assess the distribution of topological configurations at contrast discontinuities in natural scenes. Much as one-dimensional Fourier analysis can be used to analyze sounds, two-dimensional Fourier analysis can be used to analyze the global properties of two-dimensional grayscale images. Godfrey et al. (1987) used this approach to evaluate the cryptic effect of zebra and tiger stripes based on analysis of images of these animals in their natural habitats; at lower spatial frequencies, both animals were moderately cryptic, but at higher frequencies only the tiger remained cryptic.

Kiltie & Laine (1992) and Kiltie et al. (1995) argued that wavelet-based methods, which perform localized analysis of frequency and orientation, approximate the response properties of visual neurons by analyzing signals across ranges of spatial scales and orientations, providing a better approximation of local analyzers in the visual system. For example, the orientation of spatial patterns, which Fourier analysis would not detect, had a major effect on the conspicuousness of zebra and tiger coat patterns analyzed by Godfrey (1987). van Hateren & van der Schaaf (1998) used a technique called independent components analysis, which is roughly analogous to wavelets, to evaluate the match between tuning of simple cells in primate visual cortex and the statistics of natural images.

Can Visual Signals Be Comprehensively Characterized?

It would clearly be desirable to simultaneously quantify the salient properties of visual signals: motion, color, and form. This poses a technical challenge and a conceptual one. First, we need to be able to collect data on the wavelength distribution of photons striking a (hypothetical) retina as a function of space and time. Hyperspectral imaging (Chiao et al. 2000a,b), which produces a spectral radiance function for each pixel of a digital image, may provide a solution. Although hyperspectral imaging is in broad use for remote-sensing applications, it has not been widely used in studies of communication (but see Zeil & Hofmann 2001).

Second, if we use expensive technology to collect massive volumes of data from even a brief visual scene (a 10-second display at a crude 100-ms temporal resolution, 300×300-pixel resolution, and 10-nm wavelength resolution from 350–700 nm would produce over 300 million intensity values), we still need a coherent body of theory, not to mention a computational methodology, to make sense of this information. The independent-components and wavelet-based approaches discussed above are extensible to temporal and chromatic dimensions.

The key constraint is that it is impossible to take the receiver out of the picture, because the receiver's position and attention are key determinants of how the scene is going to be processed. One might begin by imaging from the receiver's point of view; this can be approximated from careful observation (Zeil & Hofmann 2001), or directly estimated through "critter cam" views or the use of camera-equipped robot receivers (Patricelli et al. 2002). Joint analysis of orientation, posture, and eye movement can be used to estimate gaze direction over the course of the signaling interaction. This provides a measure of how the receiver attends to different scene components over time, and may provide a heuristic estimate of how the receiver is segmenting the image. Wavelets and similar methods can then be used to identify the features that distinguish the figure from the background.

COMPLEX VISUAL COMMUNICATION IN THE NATURAL ENVIRONMENT

Broadly, patterns have evolved to either maximize or minimize conspicuousness—and sometimes both, whether at different scales or to different receivers (Marshall 2000). A handful of studies have used interspecific correlations to infer function for patterns. Ortolani (1999), for example, used a phylogenetic analysis to identify correlations between ecology and carnivore color patterns, concluding that stripes and spots had evolved in the context of camouflage, whereas high-contrast tail tips might have a communicative function.

Camouflage and Crypsis: A Case Study

By far the most sophisticated studies of complex visual stimuli involve patterns that are not signals in the traditional sense, but rather have evolved to minimize the probability of detection, recognition, or attack by predators. As defined by Endler (1978) crypsis is a function of how much a pattern resembles the visual background from a predator's point of view, which depends on the properties of its visual system (including color and polarization sensitivity, and spatial and temporal resolution) as well as the viewing angle and distance with respect to the putative target. As discussed above, Kiltie et al. (1995) developed a wavelet-based method for quantifying crypsis, providing a method for quantifying pattern characteristics using spatially localized analysis of orientation and frequency components. Selection favors male guppies whose skin patterns match the grain size (i.e., the spatial frequency, and presumably the orientation as well) of background elements (Endler 1980).

Cephalopods, among others, can dynamically adjust skin patterns according to their background. The pattern they adopt depends on the relative size, contrast, and density of background elements; on a fine-grained artificial checkerboard, cuttlefish (*Sepia officinalis*) exhibit a mottled texture on the same scale as the grain. Above a size threshold, however, they cease to background-match and switch to a disruptive 'white square' pattern (Chiao & Hanlon 2001). If edges (by application of a low-pass spatial-frequency filter) and contrast (by application of a high-pass filter) in pattern elements are removed, cuttlefish no longer produce the white square (Chiao et al. 2005).

Such disruptive coloration, where spatial patterns serve to impair visual recognition by breaking up an animal's outline, has been the subject of two recent studies. Merilaita & Lind (2005) presented great tits (*Parus major*) with artificial stimuli that either matched average background characteristics or disrupted the stimulus. Background-matching worked only if the grain and orientation of background elements were similar to those of the stimulus, which was not always the case owing to chance variation in background structure. Disruptive coloration, however, conferred a protective advantage to the stimuli.

How does disruptive coloration actually work? Although disruptive coloration has often been described in terms of breaking up a Gestalt-like "search image" for a stimulus, Stevens & Cuthill (2006) provided a plausible mechanism for disruptive coloration relying purely on low-level processing of edges. Using a computational model of edge detection, based partly on empirical data on avian visual physiology, they evaluated the function of disruptive coloration. By breaking up the outline of the animal, disruptive patterns on moth models were difficult to detect from background at multiple spatial scales. Furthermore, disruptive coloration often interacts with motor patterns to hinder detection and recognition (King 1993).

Conspicuousness and Beauty

By definition, something is minimally conspicuous when it is indistinguishable from a random sample of visual background. It is less clear, however, what makes a complex stimulus maximally conspicuous in multiple dimensions; and even less clear what makes one stimulus more attractive than another. Conspicuous signals minimize the signal-to-noise ratio (Ryan & Keddy-Hector 1992) for perceptual systems that are adapted to a general suite of tasks in the environment [Endler's (1992) sensory drive hypothesis]. Changizi et al. (2006) provided compelling evidence that sensory drive has framed the development of written language. The topology of symbols used in human writing matches that of contour junctions commonly found in natural scenes, even though many of these are more difficult to draw than are many configurations that are more rare in nature.

A host of studies have shown that females prefer more symmetric males in mate choice, although in most cases it is not clear that females attend to natural variation in visual asymmetry (Kirkpatrick & Rosenthal 1994). Artificial neural networks selected to recognize a particular stimulus configuration independent of translation and rotation exhibited strong responses to symmetric signals (Enquist & Arak 1994), presumably because these are more self-similar.

Kenward et al. (2004) used neural networks that mimicked retinas to address selection on repetitive elements in spatial patterns. Ubiquitous scene characteristics like translation, reflection, and occlusion all favored the evolution of repetition, as did the initial presence of a fixed, nonsignaling feature (like an eye), and the presence of edge detectors in the receiver. The evolution of repetition in response to these near-universals of visual communication may help explain the ubiquity of repeated patterns, spatial and temporal, in animal signals.

How might higher-order processing affect attractiveness? In some cases, a major determinant of attractiveness is the number of discrete components; for example, in swordtails, females prefer males expressing a greater number of vertical bars. At one extreme, this could simply be the result of a preference for increased area or perimeter of a particular color; on the other hand, it could reflect a preference based on the number of distinct objects that are perceived.

TOWARD AN INTEGRATIVE BIOLOGY OF COMPLEX VISUAL SIGNALING

"What is it like to be a bat?" Timothy Sprigge's classic question, popularized by Thomas Nagel (1974), provides a compelling metaphor for students of animal behavior. How is the world perceived by an animal whose primary source of information—echolocation—is so alien to our everyday experience? Because we lack an intuitive understanding of the bat's Umwelt, we are forced to approach this question from the bat's perspective, by physically characterizing echolocation signatures and measuring neural and behavioral responses. This approach has served us well, and though we can never share a bat's frame of reference, we have developed a sophisticated understanding of how bats use echolocation to make sense of their environment.

So what is it like to be an eagle? We rarely ask ourselves what the visual world looks like through the eyes of another species. As humans, our nominal sense of reality is so intertwined with visual perception that it is often difficult to separate the two. We never say "You smell? There's nothing there." The previous paragraph was difficult to write without resorting to visual metaphors like "seeing the world" and "points of view." We intuitively assume that the way we process and perceive spatial and temporal relationships among parts of a visual scene gives us a veridical representation of our environment, that the contrasts and shadings and textures we attend to are the best solution to the complicated engineering problem of empirically measuring the universe. Yet differences among species in ecology, perception, and cognitive architecture dictate that there is as much variation in how animals process visual information as there is in other aspects of their biology. Paradoxically, our visual bias has made this harder to appreciate. Although we have to describe an echolocation signal in terms of decibels and Doppler shifts, we tend to describe spatiotemporal variation in terms of our phenomenological experience rather than in the sensory currency of the animals we study.

In the Future Issues, I suggest some key unresolved questions that could be fruitfully addressed by applying a quantitative perspective to the biology of spatiotemporal signals. An important implication that emerges is that there are nonlinearities in both production and perception, such that small genetic changes could result in major changes in signalers and receivers. The zebrafish *leopard* gene, a putative constituent of a reaction-diffusion system, seems a ripe candidate for comparative study.

Second, attentional mechanisms are likely to produce the most important initial filters of information. Although amply studied in human psychology, and implicitly considered in numerous studies of visual preferences, animal attention has received little scrutiny. We would like to know which components of a visual signal are salient to receivers, and how signals are designed to draw, hold, or divert receiver attention.

The intricacies of animal signals are appealing to us in part because of their exuberance. To what extent are receivers actually attending to this complexity, and how is their perception specified; how do individuals acquire preferences for complex visual stimuli?

Our appreciation of the beauty of the natural world is intertwined with the way in which our visual systems make sense of it. Mechanistic studies of communication have focused almost exclusively on detection, recognition, and discrimination; but what proximate and ultimate forces determine the valence of stimuli? Does Darwin's (1871) "taste for the beautiful" exist, a vocabulary of beauty shared across taxa? We are perhaps fortunate to find the natural world so attractive (though not, evidently, attractive enough to preserve). The eminent naturalist Alexander Skutch (1992, p. 10) may have put it best: "Whether in the productions of nature or of art, these four—form, color, pattern, and texture—are the elements of which beauty is compounded."

FUTURE ISSUES

1. How are biases for complex spatial patterns innately specified?

2. What are the major cognitive constraints on pattern learning, and how do they vary by taxon?

3. Do spatiotemporal patterns coevolve with sensitivity?

4. Can small genetic changes lead to qualitative jumps in pattern production and perception?

5. How do ecological constraints on spatial and temporal resolution structure signal design?

6. How does the environment structure the processing of complex visual signals?

7. Is there a relationship between pattern complexity and signal context?

8. How do attentional mechanisms shape the evolution of signals?

9. Is there a broadly shared taste for the beautiful?

10. How much of pattern complexity is meaningful to receivers?

DISCLOSURE STATEMENT

The author is not aware of any biases that might be perceived as affecting the objectivity of this review.

LITERATURE CITED

Amiet J.-L. 1987. *Le genre* Aphyosemion *Meyers* (Pisces, Telostei, Cyprinodontiformes). Compiègne, France: Sciences Nat. 262 pp.

Andersson M. 1982. Female choice selects for extreme tail length in a widowbird. *Nat. London* 299:818–20

Asai R, Taguchi E, Kume Y, Saito M, Kondo S. 1999. Zebrafish leopard gene as a component of the putative reaction-diffusion system. *Mech. Dev.* 89:87–92

Bagnara JT, Hadley ME. 1973. *Chromatophores and Color Change.* Englewood Cliffs, NJ: Prentice-Hall

Balboa RM, Grzywacz NM. 2000. Occlusions and their relationship with the distribution of contrasts in natural images. *Vis. Res.* 40:2661–69

Barry KL, Hawryshyn CW. 1999. Effects of incident light and background conditions on potential conspicuousness of Hawaiian coral reef fish. *J. Mar. Biol. Assoc. UK* 79:495–508

Billock VA, de Guzman GC, Kelso JAS. 2001. Fractal time and 1/*f* spectra in dynamic images and human vision. *Phys. D* 148:136–46

Blake A, Buelthoff H. 1990. Does the brain know the physics of specular reflection? *Nature* 343:165–68

Bradbury JW, Vehrencamp SL. 1998. *Principles of Animal Communication*. Sunderland, MA: Sinauer

Briscoe AD, Chittka L. 2001. The evolution of color vision in insects. *Annu. Rev. Entomol.* 46:471–510

Brunetti CR, Selegue JE, Monteiro A, French V, Brakefield PM, Carroll SB. 2001. The generation and diversification of butterfly eyespot and color patterns. *Curr. Biol.* 11:1578–85

Changizi MA, Zhang Q, Ye H, Shimojo S. 2006. The structures of letters and symbols throughout human history are selected to match those found in objects in natural scenes. *Am. Nat.* 167:E117–39

Chen L, Zhang S, Srinivasan MV. 2003. Global perception in small brains: Topological pattern recognition in honey bees. *Proc. Natl. Acad. Sci. USA* 100:6884–89

Chiao CC, Hanlon RT. 2001. Cuttlefish camouflage: Visual perception of size, contrast and number of white squares on artificial checkerboard substrata initiates disruptive coloration. *J. Exp. Biol.* 204:2119–25

Chiao CC, Kelman EJ, Hanlon RT. 2005. Disruptive body patterning of cuttlefish (Sepia officinalis) requires visual information regarding edges and contrast of objects in natural substrate backgrounds. *Biol. Bull.* 208:7–11

Chiao CC, Osorio D, Vorobyev M, Cronin TW. 2000. Characterization of natural illuminants in forests and the use of digital video data to reconstruct illuminant spectra. *J. Opt. Soc. Am. A* 17:1713–21

Chiao CC, Vorobyev M, Cronin TW, Osorio D. 2000. Spectral tuning of dichromats to natural scenes. *Vis. Res.* 40:3257–71

Chubb C, Yellott JI. 2000. Every discrete, finite image is uniquely determined by its dipole histogram. *Vis. Res.* 40:485–92

Coleman SW, Patricelli GL, Borgia G. 2004. Variable female preferences drive complex male displays. *Nature* 428:742–45

Collis K, Borgia G. 1993. The cost of male display and delayed plumage maturation in the satin bowerbird (*Ptilonorhynchus violaceus*). *Ethology* 94:59–71

Cott HB. 1940. *Adaptive Coloration in Animals*. London: Methuen. 438 pp.

Crook AC. 1997. Colour patterns in a coral reef fish: is background complexity important? *J. Exp. Mar. Biol. Ecol.* 217:237–52

Dantzker MS, Deane GB, Bradbury JW. 1999. Directional acoustic radiation in the strut display of male sage grouse *Centrocercus urophasianus*. *J. Exp. Biol.* 202:2893–909

Darwin C. 1871. *The Descent of Man, and Selection in Relation to Sex*. London: John Murray. 475 pp.

Darwin C. 1872. *The Expression of the Emotions in Man and Animals*. Chicago: Univ. Chicago

Douglas RH, Eva J, Guttridge N. 1988. Size constancy in goldfish (*Carassius auratus*). *Behav. Brain Res.* 30:37–42

Dukas R. 2004. Evolutionary biology of animal cognition. *Annu. Rev. Ecol. Evol. Syst.* 35:347–74

Dyer AG. 2006. Bumblebees directly perceive variations in the spectral quality of illumination. *J. Comp. Physiol. A* 192:333–38

Easterbrook JA. 1959. The effect of emotion on cue utilization and the organization of behavior. *Psychol. Rev.* 66:183–201

Eibl-Eibesfeldt I. 1970. *Ethology: The Biology of Behavior*. New York: Holt, Rinehart & Winston. 530 pp.

Endler JA. 1978. A predator's view of animal color patterns. *Evol. Biol.* 11:319–54

Endler JA. 1980. Natural selection on color patterns in *Poecilia reticulata*. *Evolution* 34:76–91

Endler JA. 1992. Signals, signal conditions, and the direction of evolution. *Am. Nat.* 139:S125–53

Endler JA. 1993. The color of light in forests and its implications. *Ecol. Monogr.* 63:1–27

Endler JA, Mielke PW. 2005. Comparing entire colour patterns as birds see them. *Biol. J. Linn. Soc.* 86:405–31

Engeszer RE, Ryan MJ, Parichy DM. 2004. Learned social preference in zebrafish. *Curr. Biol.* 14:881–84

Enquist M, Arak A. 1994. Symmetry, beauty and evolution. *Nature* 372:169–72

Fernald RD. 1990. *Haplochromis burtoni*: a case study. In *The Visual System of Fish*, ed. RH Douglas, MBA Djamgoz, pp. 443–63. London: Chapman & Hall

Finger E. 1995. Visible and UV coloration in birds: Mie scattering as the basis of color in many bird feathers. *Naturwiss* 82:570–73

Fleishman LJ. 1988. Sensory influences on physical design of a visual display. *Anim. Behav.* 36:1420–24

Fleishman LJ. 1992. The influence of the sensory system and the environment on motion patterns in the visual displays of anoline lizards and other vertebrates. *Am. Nat.* 139:S36–61

Fleishman LJ, Leal M, Sheehan J. 2006. Illumination geometry, detector position and the objective determination of animal signal colours in natural light. *Anim. Behav.* 71:463–74

Frank TM. 2000. Temporal resolution in mesopelagic crustaceans. *Philos. Trans. R. Soc. London Ser. B* 355:1195–98

Fuller RC, Carleton KL, Fadool JM, Spady TC, Travis J. 2005. Genetic and environmental variation in the visual properties of bluefin killifish, *Lucania goodei*. *J. Evol. Biol.* 18:516–23

Godfrey D, Lythgoe JN, Rumball DA. 1987. Zebra stripes and tiger stripes: the spatial frequency distribution of the pattern compared to that of the background is significant in display and crypsis. *Biol. J. Linn. Soc.* 32:427–33

Grether GF, Kolluru GR, Nersissian K. 2004. Individual colour patches as multi-component signals. *Biol. Rev.* 79:583–610

Hansen AJ, Rohwer S. 1986. Coverable badges and resource defence in birds. *Anim. Behav.* 34:69–76

Hasson O. 1989. Amplifiers and the handicap principle in sexual selection: A different emphasis. *Proc. R. Soc. London Ser. B* 235:383–406

Hayworth K, Biederman I. 2006. Neural evidence for intermediate representations in object recognition. *Vis. Res.* 46:4024–31

Heindl M, Winkler H. 2003. Vertical lek placement of forest-dwelling manakin species (Aves, Pipridae) is associated with vertical gradients of ambient light. *Biol. J. Linn. Soc.* 80:647–58

Hubel D, Wiesel T. 1963. Receptive fields of cells in striate cortex of very young, visually inexperienced kittens. *J. Neurophysiol.* 26:994–1002

Hurd PL, Wachtmeister CA, Enquist M. 1995. Darwin's principle of antithesis revisited—a role for perceptual biases in the evolution of intraspecific signals. *Proc. R. Soc. London Ser. B* 259:201–5

Kandel E, Wurtz R. 2000. Constructing the visual image. In *Principles of Neural Science*, ed. E Kandel, J Schwartz, T Jessell, pp. 492–506. New York: McGraw-Hill

Kenward B, Wactmeister CA, Ghirlanda S, Enquist M. 2004. Spots and stripes: the evolution of repetition in visual signal form. *J. Theor. Biol.* 230:407–19

Kiltie RA, Fan J, Laine AF. 1995. A wavelet-based metric for visual texture discrimination with applications in evolutionary ecology. *Math. Biosci.* 126:21–39

Kiltie RA, Laine AF. 1992. Visual textures, machine vision and animal camouflage. *Trends Ecol. Evol.* 7:163–65

King RB. 1993. Color-pattern variation in Lake Erie water snakes: Prediction and measurement of natural selection. *Evolution* 47:1819–33

Kirkpatrick M, Rand AS, Ryan MJ. 2006. Mate choice rules in animals. *Anim. Behav.* 71:1215–25

Kirkpatrick M, Rosenthal GG. 1994. Symmetry without fear. *Nature* 372:134–35

Kondo S, Asai R. 1995. A reaction-diffusion wave on the skin of the marine anglefish *Pomacanthus*. *Nature* 376:765–68

Kovacs I. 1996. Gestalten of today: early processing of visual contours and surfaces. *Behav. Brain Res.* 82:1–11

Kovacs I, Julesz B. 1994. Perceptual sensitivity maps within globally defined visual shapes. *Nature* 370:644–46

Leal M, Fleishman LJ. 2004. Differences in visual signal design and detectability between allopatric populations of *Anolis* lizard. *Am. Nat.* 163:26–39

Leopold DA, Bondar IV, Giese MA. 2006. Norm-based face encoding by single neurons in the monkey inferotemporal cortex. *Nature* 442:572–75

Loew ER, McFarland WN. 1990. The underwater visual environment. In *The Visual System of Fish*, ed. R Douglas, M Djamgoz, pp. 1–43. New York: Chapman & Hall

Lythgoe JN. 1979. *The Ecology of Vision*. Oxford: Clarendon

Ma PM. 1995a. On the agonistic display of the Siamese fighting fish. 1. The frontal display apparatus. *Brain Behav. Evol.* 45:301–13

Ma PM. 1995b. On the agonistic display of the Siamese fighting fish. 2. The distribution, number and morphology of opercular display motoneurons. *Brain Behav. Evol.* 45:314–26

Marler P. 1997. Three models of song learning: evidence from behavior. *J. Neurobiol.* 33:501–16

Marshall NJ. 2000. Communication and camouflage with the same 'bright' colours in reef fishes. *Philos. Trans. R. Soc. Ser. B* 355:1243–48

Maunsell JHR, Treue S. 2006. Feature-based attention in visual cortex. *Trends Neurosci.* 29:317–22

Merilaita S, Lind J. 2005. Background-matching and disruptive coloration, and the evolution of cryptic coloration. *Proc. R. Soc. London Ser. B* 272:665–70

Mojsilović A, Kovačević J, Kall D, Safranek RJ, Ganapathy K. 2000. The vocabulary and grammar of color patterns. *IEEE Trans. Image Process.* 9:417–31

Murray J. 1981. A pre-pattern formation mechanism for animal coat markings. *J. Theor. Biol.* 88:161–99

Nagel T. 1974. What is it like to be a bat? *Philos. Rev.* 83:2–14

Ortolani A. 1999. Spots, stripes, tail tips and dark eyes: Predicting the function of carnivore colour patterns using the comparative method. *Biol. J. Linn. Soc.* 67:433–76

Painter KJ, Maini PK, Othmer HG. 1999. Stripe formation in juvenile *Pomacanthus* explained by a generalized Turing mechanism with chemotaxis. *Proc. Natl. Acad. Sci. USA* 96:5549–54

Parichy DM. 2003. Pigment patterns: Fish in stripes and spots. *Curr. Biol.* 13:947–50

Parraga CA, Troscianko T, Tolhurst DJ. 2002. Spatiochromatic properties of natural images and human vision. *Curr. Biol.* 12:483–87

Patricelli GL, Uy JAC, Walsh G, Borgia G. 2002. Male displays adjusted to female's response. *Nature* 415:279–80

Peters RA, Clifford CWG, Evans CS. 2002. Measuring the structure of dynamic visual signals. *Anim. Behav.* 64:131–46

Piratla NM, Jayasumana AP. 2002. A neural network based real-time gaze tracker. *J. Netw. Comput. Appl.* 25:179–96

Qvarnstrom A. 1997. Experimentally increased badge size increases male competition and reduces male parental care in the collared flycatcher. *Proc. R. Soc. London Ser. B* 264:1225–31

Rodieck RW. 1998. *The First Steps in Seeing.* Sunderland, MA: Sinauer Assoc. 546 pp.

Rosenthal GG, Evans CS. 1998. Female preference for swords in *Xiphophorus helleri* reflects a bias for large apparent size. *Proc. Natl. Acad. Sci. USA* 95:4431–36

Rosenthal GG, Evans CS, Miller WL. 1996. Female preference for a dynamic trait in the green swordtail, *Xiphophorus helleri. Anim. Behav.* 51:811–20

Rosenthal GG, Lobel P. 2006. Communication. In *Behaviour and Physiology of Fish*, ed. K Sloman, S Balshine, R Wilson, pp. 39–78. Oxford: Academic

Rowe G, Hirsh JB, Anderson AK. 2007. Positive affect increases the breadth of attentional selection. *Proc. Natl. Acad. Sci. USA* 104:383–88

Rutowski RL, Macedonia JM, Merry JW, Morehouse NI, Yturralde K, et al. 2007. Iridescent ultraviolet signal in the orange sulphur butterfly (Colias eurytheme): spatial, temporal and spectral properties. *Biol. J. Linn. Soc.* 90:349–64

Ryan MJ. 1998. Sexual selection, receiver biases, and the evolution of sex differences. *Science* 281:1999–2003

Ryan MJ, Keddy-Hector A. 1992. Directional patterns of female mate choice and the role of sensory biases. *Am. Nat.* 139:S4–35

Ryan MJ, Rosenthal GG. 2001. Variation and selection in swordtails. In *Model Systems in Behavioral Ecology*, ed. LA Dugatkin, pp. 133–48. Princeton, NJ: Princeton Univ. Press

Sekuler R, Blake R. 1994. *Perception*. New York: McGraw-Hill. 572 pp.

Shapley R, Lennie P. 1985. Spatial frequency analysis in the visual system. *Annu. Rev. Neurosci.* 8:547–83

Shoji H, Iwasa Y. 2003. Pattern selection and the direction of stripes in two-dimensional Turing systems for skin pattern formation of fishes. *Forma* 18:3–18

Sinervo B, Basolo AL. 1996. Testing adaptation with phenotypic manipulations. In *Evolutionary Biology of Adaptation*, ed. MR Rose, GV Lauder, pp. 149–85. New York: Academic

Skutch AF. 1992. *Origins of Nature's Beauty*. Austin: Univ. Tex. Press. 292 pp.

Smith SM. 1975. Innate recognition of coral snake pattern by a possible avian predator. *Science* 187:759–60

Smith SM. 1977. Coral-snake pattern recognition and stimulus generalisation by naive great kiskadees (Aves: Tyrannidae). *Nature* 265:535–36

Stevens M. 2005. The role of eyespots as anti-predator mechanisms, principally demonstrated in the Lepidoptera. *Biol. Rev.* 80:573–88

Stevens M, Cuthill I. 2006. Disruptive coloration, crypsis and edge detection in early visual processing. *Proc. R. Soc. London Ser. B* 273:2141–47

Sword GA. 1999. Density-dependent warning coloration. *Nature* 397:217

Thomson DA, Findlay LT, Kerstitch AN, Ehrlich PR, Van Dyke CM. 2000. *Reef Fishes of the Sea of Cortez: The Rocky Shore Fishes of the Gulf of California*. Austin: Univ. Tex. Press. 374 pp.

Tinbergen N. 1951. *The Study of Instinct*. Oxford: Clarendon. 228 pp.

Turati C, Simion F, Milani I, Umilta C. 2002. Newborns' preference for faces: what is crucial? *Dev. Psychol.* 38:875–82

Turing A. 1952. The chemical basis of morphogenesis. *Philos. Trans. R. Soc. London Ser. B* 237:37–72

van Hateren JH, van der Schaaf A. 1998. Independent component filters of natural images compared with simple cells in primary visual cortex. *Proc. R. Soc. London Ser. B* 265:359–66

Warner RR, Swearer SE. 1991. Social control of sex change in the bluehead wrasse, *Thalassoma bifasciatum* (Pisces: Labridae). *Biol. Bull.* 181:199–204

Wexler M, van Boxtel JJA. 2005. Depth perception by the active observer. *Trends Cogn. Sci.* 9:431–38

Wong BBM, Rosenthal GG. 2006. Female disdain for swords in a swordtail fish. *Am. Nat.* 167:136–40

Yokoyama S, Yokoyama R. 1996. Adaptive evolution of photoreceptors and visual pigments in vertebrates. *Annu. Rev. Ecol. Syst.* 27:543–67

Zeil J, Hofmann M. 2001. Signals from 'crabworld': cuticular reflections in a fiddler crab colony. *J. Exp. Biol.* 204:2561–69

Gliding and the Functional Origins of Flight: Biomechanical Novelty or Necessity?

Robert Dudley,[1] Greg Byrnes,[1] Stephen P. Yanoviak,[2] Brendan Borrell,[1] Rafe M. Brown,[3] and Jimmy A. McGuire[1,4]

[1] Department of Integrative Biology, University of California, Berkeley, California 94720; email: wings@socrates.berkeley.edu

[2] Department of Pathology, University of Texas Medical Branch, Galveston, Texas 77555, and Florida Medical Entomology Laboratory, Vero Beach, Florida 32962

[3] Natural History Museum, Biodiversity Research Center, and Department of Ecology and Evolutionary Biology, University of Kansas, Lawrence, Kansas 66045

[4] Museum of Vertebrate Zoology, University of California, Berkeley, California 94720

Annu. Rev. Ecol. Evol. Syst. 2007. 38:179–201

The *Annual Review of Ecology, Evolution, and Systematics* is online at
http://ecolsys.annualreviews.org

This article's doi:
10.1146/annurev.ecolsys.37.091305.110014

Key Words

aerodynamics, arboreality, descent, falling, hexapod, trees, vertebrate, wing

Abstract

A biomechanically parsimonious hypothesis for the evolution of flapping flight in terrestrial vertebrates suggests progression within an arboreal context from jumping to directed aerial descent, gliding with control via appendicular motions, and ultimately to powered flight. The more than 30 phylogenetically independent lineages of arboreal vertebrate gliders lend strong indirect support to the ecological feasibility of such a trajectory. Insect flight evolution likely followed a similar sequence, but is unresolved paleontologically. Recently described falling behaviors in arboreal ants provide the first evidence demonstrating the biomechanical capacity for directed aerial descent in the complete absence of wings. Intentional control of body trajectories as animals fall from heights (and usually from vegetation) likely characterizes many more taxa than is currently recognized. Understanding the sensory and biomechanical mechanisms used by extant gliding animals to control and orient their descent is central to deciphering pathways involved in flight evolution.

INTRODUCTION

The evolution of novel locomotor modes plays an important role in the invasion of new habitats, partitioning of resources within those habitats, and ultimately in the generation of organismal diversity. Extreme habitat transitions, such as those between aquatic and terrestrial environments or between land and air, represent major themes in the history of life and involve both the co-option of existing traits as well as genuine "key" innovations (Vermeij 2006, Vermeij & Dudley 2000). The origin of flight represents one such important transition and requires the integration of a suite of morphological, physiological, and behavioral features. Although many biologists view flight as a specialized or even rare form of locomotion characteristic of only one extinct lineage (the pterosaurs) and three extant clades (birds, bats, and the pterygote insects), controlled aerial behaviors are much more widespread among animals. In addition to powered flapping flight, gliders with obvious wings or wing-like structures have evolved at least thirty times among terrestrial vertebrates, including mammals, reptiles, and amphibians (Norberg 1990, Rayner 1988). In addition to more classically described gliding, directed aerial descent (Yanoviak et al. 2005) occurs in the absence of obvious aerodynamic surfaces and is likely characteristic of many more taxa, both vertebrate and invertebrate. Here, we examine the full continuum of such aerial behaviors and place the origin of flight within a specific functional context relating to arboreality and either inadvertent or intentional descent. In particular, we suggest that the numerous evolutionary experiments in gliding and controlled descent may be inevitable consequences of living within vegetational structures elevated above the ground.

Definitionally, it is important to specify what is meant by the word flight. The Oxford English Dictionary (second edition) defines flight as the "action or manner of flying or moving through the air with or as with wings." Biomechanically, we here use the term to indicate any locomotor behavior in the air that involves active control of aerodynamic forces. Parachuting with no regulation of the magnitude or orientation of the ensuing drag force can be truly passive, but all other aerial behaviors involve the generation and often intentional regulation of lift and drag to slow descent, reorient the body, and alter the flight trajectory. A conceptual distinction has been historically made between gliding and parachuting, with the former characterized arbitrarily by a descent angle less than 45° relative to horizontal, and the latter with a descent angle greater than 45° (Oliver 1951). These definitions assume steady-state conditions of a constant speed and orientation of the body in the air, as well as the equilibrium of forces. However, such a discrete characterization of what is a continuous variable, namely the glide angle, is clearly inappropriate. Individual flying lizards in the genus *Draco*, for example, may glide at relatively shallow angles, but can also plummet at angles steeper than 45° according to the behavioral context (McGuire & Dudley 2005). The mechanisms of aerodynamic control during descent are similar in these two cases, differing only in magnitude and not fundamentally in kind. Many features of aerial behavior in gliding animals are also unsteady, involving time-dependent changes in orientation and speed of appendages and of the body itself. Therefore, we use the term gliding interchangeably with the phrase directed aerial descent to mean

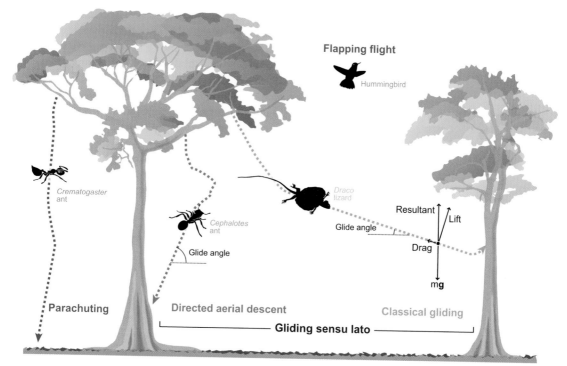

Flapping flight

Hummingbird

Crematogaster ant

Cephalotes ant

Draco lizard

Resultant | Lift

Glide angle

Glide angle

Drag

m**g**

Parachuting

Directed aerial descent

Classical gliding

Gliding sensu lato

Figure 1

Diversity of aerial behaviors and a diagrammatic scenario for the evolution of flight.
Parachuting (*left*), directed aerial descent at steep angles (*center*), and classical gliding at
shallow angles (*right*) represent different stages of aerodynamic control and force production
that characterize a broad diversity of arboreal taxa. Flapping flight (*top*) is hypothesized to
derive from controlled aerial behaviors that phylogenetically precede fully articulated wings.
m**g**, force of gravity.

any controlled descent by an organism that converts gravitational potential energy to
useful aerodynamic work (**Figure 1**). In many if not all cases, such gliding is associated
with volitional horizontal, lateral, and rotational motions independent of particular
values of the instantaneous descent angle relative to horizontal. Following Maynard
Smith (1952), we suggest that the control of aerial trajectory when accelerating under
gravity is essential to the evolution of both gliding and flapping flight. Controlled
aerial behavior may accordingly precede the origin of wings per se. Here, we review a
variety of evidence for both arthropods and terrestrial vertebrates in support of these
possibilities.

MORPHOLOGICAL ADAPTATIONS FOR FLIGHT

The diversity of anatomical structures used aerodynamically by flying animals is impressive. The fluid-dynamic features of such biological airfoils have been discussed
extensively elsewhere (Dudley 2000, Norberg 1990, Vogel 1994). True flapping wings

are easily recognizable as such; in the flying vertebrates they are modified limbs. The wing structures of bats, birds, and pterosaurs all involve the forelimb, and in bats also attach to the hindlimb. Insect wings are homologous neither with limbs nor with any other extant arthropod structure. Instead, these wings comprise thin cuticular membranes supported by venation extending from the thorax and are moved by muscles inserting at the wing base. In addition to the use of flapping wings, or such rudimentary wings as the feathered forelimbs of avian precursors, additional anatomical structures are employed for aerodynamic purpose. In mammals and reptiles, a patagial membrane is stretched laterally from the body, and is suspended between either bones or cartilaginous structures. Patagial membranes are strikingly variable. In mammals, the minimally nine independent origins of gliding flight are accompanied by substantial anatomical differences in the gliding membrane (Jackson 2000, Thorington 1984). Flying lizards of the genus *Draco* suspend the patagial membrane between elongated ribs (Colbert 1967, Russell & Dijkstra 2001); the extinct reptile *Sharovipteryx* sported a patagial membrane held only between the hindlimbs and tail (Dyke et al. 2006, Gans et al. 1987). The Permian reptile *Coelurosaurus* was characterized by a patagial membrane supported by a series of rod-like bones apparently not found in any other gliding animal (Frey et al. 1997). In the absence of a patagial membrane, flattening of the body (Arnold 2002, Losos et al. 1989, Socha 2002), use of lateral skin flaps (as in the gliding lizard *Ptychozoon*; Russell et al. 2001), flattened or relatively long tails (Thorington & Heaney 1981, Thorington et al. 2002), and the spreading of finger and toe webbing (as in hylid and rhacophorid gliding frogs, and in the lizard *Ptychozoon*) all serve to increase effective aerodynamic surface area and to improve lift:drag performance of the body as a whole.

In equilibrium gliding (i.e., moving at constant airspeed and glide angle), airspeed varies in proportion to the square root of the morphological parameter termed wing loading, the ratio of body weight to sustaining aerodynamic area (Norberg 1990). Therefore, species with higher wing loading glide faster when in equilibrium, independent of the lift:drag ratio of the animal. To reach these higher velocities, larger animals must fall vertically under gravity both a greater distance and over a longer time period to attain an equilibrium glide. Consistent with this expectation, gliding *Draco* lizards exhibit a significant correlation between wing loading and height lost over a standardized glide distance, with no evidence of physiological or behavioral compensation for increased body mass (McGuire 2003, McGuire & Dudley 2005). Given this general relationship between wing loading and glide performance, it is important to consider the allometry of aerodynamic surfaces (i.e., change in shape of the force-producing structures relative to change in body size). Relevant morphologies have not been studied in gliding arthropods (see below), but investigations of wing allometries in terrestrial vertebrate gliders have found isometric scaling (i.e., wing area increases in proportion to mass raised to the 2/3rds power) for the two cases of flying squirrels (Thorington & Heaney 1981) and *Draco* lizards (McGuire 2003). By contrast, flying fish exhibit a negative allometry in wing area because of functional constraints on pectoral fin retraction relative to tail beating in water (Davenport 2003). Selection toward smaller body size might thus be expected for terrestrial gliders if only equilibrium flight performance is of concern. Suggestively, large body size only evolves in *Draco* lizards in the context of multispecies sympatry, an outcome that may derive from interspecific competition (McGuire & Dudley 2005). Increased wing loading also limits some aspects of flight maneuverability (Pennycuick 1975), further hindering aerial performance in larger gliders. Conversely, the biomechanical advantages of augmented surface area may be substantial, particularly as even small surfaces can nonetheless generate substantial aerodynamic torque and body rotations if positioned sufficiently far from rotational axes (Dudley 2002).

COMPONENTS OF AERIAL BEHAVIOR

Flight in animals involves a diversity of behaviors, including falls, startle jumps, volitional takeoff, accelerations, moving at constant airspeed, maneuvers, and landing. Many otherwise seemingly nonaerial taxa exhibit behavioral adaptations to allow them to decrease their rate of descent when falling (Dunbar 1988, Oliver 1951, Pellis et al. 1989). In order to initiate a glide, organisms must become airborne by either leaping or falling from structural or habitat heterogeneities. In the cypselurid flying fish, gliding is initiated by breaking through the air-water interface at a shallow angle, unfurling the large lateral fins, and rapidly beating the tail in the water prior to actual liftoff (Davenport 1994). Gliding squid similarly eject from the ocean's surface (Azuma 1992), but no analogous behavior (i.e., horizontal running to effect takeoff) has been identified in terrestrial gliders. All extant gliders are exclusively arboreal and use gravitational potential energy to accelerate downward, albeit initiating the behavior with a jump or fall. However, in many species, an active leap is involved. In the gliding snake, *Chrysopelea paradisi*, the body forms a "J"-shaped loop, hanging beneath the branch to initiate a glide. The anterior body is then accelerated upward and forward to leave the substrate at a horizontal speed of nearly 2 ms^{-1} (Socha 2002, 2006). More conventional gliders, including flying squirrels, typically leap from perches to initiate glides. The flying squirrel *Glaucomys volans*, for example, takes off at a mean speed of 2.5 ms^{-1} (Essner 2002). Furthermore, leaping motions in this flying squirrel closely resemble those of the nongliding tree squirrels and chipmunks. Takeoff of other gliding mammals has been described more qualitatively. For example, dermopterans hang vertically in a head-up position on the boles of trees before launching into a glide (R. Brown & G. Byrnes, personal observation). During this launch, hindlimbs are extended to effect takeoff while the body rolls through 180° (Mendoza & Custodio 2000). Flying lizards of the genus *Draco* employ the same mechanism when initiating a glide from a head-up, vertically perched position, whereas launching from near-horizontal surfaces (e.g., tree limbs) involves an initial leap (R. Brown, J. McGuire, & R. Dudley, personal observation).

Despite large morphological differences among extant gliders, aerodynamic commonalities pertain. Many gliders outstretch their limbs and spread their toes (e.g., aerial mammals and frogs), maximizing the area of gliding and control surfaces (Brown et al. 1997, Heyer & Pongsapipatana 1970). By contrast, nonarboreal taxa never maintain a stable posture when falling (Heyer & Pongsapipatana 1970, Oliver 1951). When their feet are bound, gliding geckos in the genus *Ptychozoon* fall like nongliders (Young et al. 2002). Impairment of body flaps hinders aerodynamic performance in other lizards (Losos et al. 1989, Marcellini & Keefer 1976). Glide trajectories are highly dynamic, but can be broken down into distinct phases: accelerating descent, equilibrium gliding, and landing. Recent research conflates these features by considering only locomotor performance as averaged over the entire trajectory, typically by measuring the vertical height lost and total horizontal distance traveled during a gliding episode (e.g., Ando & Shiraishi 1993, Jackson 2000, Scholey 1986). However, each phase of the glide trajectory has distinctive biomechanical features that influence overall performance. Whereas either time-averaged or instantaneous glide angles are the most commonly reported performance measures, many other characteristics of aerial behavior may be equally relevant according to context, including takeoff speed, glide duration, and capacity to maneuver. Once in the air, gliding animals typically accelerate until the resultant of lift and drag forces acting on the organism equals the weight of the animal, producing a constant airspeed and glide angle (see **Figure 1**). Recent studies, however, have found that an equilibrium phase is uncommon in *Draco* (i.e., only about 50% of studied glides; McGuire & Dudley 2005), in the gliding snake *Chrysopelea* (Socha & LaBarbera 2005, Socha et al. 2005), and in the southern flying squirrel (Bishop 2006), at least over the spatial scales under consideration.

The capacity of gliding animals to maneuver once airborne has often been noted anecdotally (e.g., Colbert 1967, Dolan & Carter 1977, Jackson 2000). The complex structure of the forested habitat characteristic of most gliders sometimes requires alteration in the speed and direction of aerial trajectories. Air turbulence may similarly require dynamic course correction (McCay 2003). Aerial maneuverability and the negotiation of structurally complex environments require components of both axial and torsional agility (Dudley 2002). Axial agility describes the ability to accelerate along any of the three body axes, whereas torsional agility relates to the ability to rotate about these three axes. Patagial and propatagial membranes are under muscular control in many gliders (Colbert 1970, Johnson-Murray 1987, Wilkinson et al. 2006), allowing rapid adjustment of membrane configuration and camber either symmetrically or asymmetrically relative to the animal's longitudinal axis. Furthermore, gliding animals are able to use limb and whole-body movements to effect axial or torsional maneuvers. Adjusting the orientation of the propatagium (i.e., the membrane between the forelimbs and neck in gliding mammals) might also influence aerodynamic force production. Studies of gliding frogs (Emerson & Koehl 1990, Emerson et al. 1990, McCay 2001) illustrate just how complex such maneuvers can be, and furthermore illustrate the importance of the interactions between morphological structures and their displacement to effect whole-body directional change. Only recently, however, has three-dimensional analysis of a free-flying glider undertaking maneuvers been implemented (Socha & LaBarbera 2005, Socha et al. 2005). Many gliding animals exhibit distinct landing maneuvers to reduce flight speed and to increase control while landing. Qualitative descriptions of the landing maneuver exist for gliding marsupials (Jackson 2000, McKay 1983), anomalurids (Kingdon 1974), and flying squirrels (Nowak 1991, Scholey 1986). In all cases, both fore- and hindlimbs are moved forward and downward, increasing patagial billowing. The body pitches nose-upward, orientation of the flight membrane relative to oncoming flow increases, and lift production by the wing/body structure (i.e., aerodynamic stall) decreases rapidly as the animal approaches the landing site (Nachtigall 1979), with a substantial reduction in speed prior to impact (G. Byrnes, unpublished observations). Quantitative descriptions of landing behavior in gliding taxa are otherwise not available.

THE ARBOREAL CONTEXT OF GLIDING FLIGHT

The evolutionary impetus and selective advantages of gliding and powered flight are associated with diverse and sometimes taxon-specific adaptive explanations (e.g., Beard 1990; Dudley 2000; Kingsolver & Koehl 1994; Norberg 1985, 1990; Padian & Chiappe 1998; Scholey 1986; Videler 2005). The frequency with which controlled aerial descent and gliding have evolved suggests that the requisite morphological and behavioral modifications are not necessarily difficult to implement. A variety of advantages may accrue to those animals capable of flight. Aerial behavior may have initially evolved to aid in pursuit of other organisms, or may have served as a means to escape predation, as described below. Powered flight in particular allows organisms to extend dramatically either foraging or breeding territories, to migrate seasonally between divergent habitats, and even to inhabit unpredictable environments or environments with highly dispersed resources (Norberg 1985). For organisms living within a vegetational canopy, gliding or controlled falling may offer an efficient mechanism for travel from branch to branch or from tree to tree. The instantaneous energetic cost of gliding is likely very low relative to active wing flapping given that only minor postural adjustments are required once the animal is airborne. Gliders may thus move more efficiently over longer distances relative to nongliders. For example, Norberg (1983) showed it was more efficient for birds to climb upward and then to glide between trees than to fly directly if the distance between successive trees exceeded about one-half the maximum glide distance. Similar biomechanical and energetic arguments have been advanced for the selective advantages of smaller body size in other gliders (Dial 2003, McGuire 2003, Scheibe & Robins 1998, Thorington & Heaney 1981).

All extant vertebrate gliders are exclusively arboreal and initiate their flights with a jump. Moreover, jumping via a startle response is widespread among animals (Eaton 1984), and one potential commonality among both flying vertebrates and insects may have been the initial acquisition of flight via the pathway of jumping and subsequent aerial trajectories to escape predation. Jump-initiated glides that increased survivorship during predation attempts would then potentially select for greater aerodynamic performance. The increased longevity of flying animals relative to their nonflying counterparts has been adduced as support for the hypothesis that predation avoidance is a major factor underlying the evolution of aerial locomotion (Holmes & Austad 1994, Pomeroy 1990). Arboreality obviously enhances the efficacy of jumping escapes given gravitationally accelerated body motion and access to a three-dimensional physical environment. Greater takeoff heights should facilitate such escapes and promote the evolution of aerial behaviors. For example, the canopy of lowland Indo-Malayan rainforests is typically tens of meters higher than that of African or New World counterparts, and residence at such heights may have structurally facilitated the multiple independent evolution of dedicated vertebrate gliders in this region relative to other tropical areas (de Gouvenain & Silander

2003, Dudley & DeVries 1990). Moreover, even arboreal taxa that are not gliders become inadvertently airborne from time to time. Falling from trees, for example, is a commonplace occurrence in certain frogs and lizards that otherwise exhibit no aerial abilities (Schlesinger et al. 1993, Stewart 1985).

Potential arboreal origins for winged vertebrates have been discussed at length elsewhere (e.g., Padian & Chiappe 1998, Rayner 1988, Zhou 2004). Here, we focus on aerial behavior in arboreal but wingless arthropods that are both common and diverse in forest canopies. Prominent examples include ant and termite workers, isopods, mites, spiders, and thysanurans, as well as the immature stages of many insect orders, especially the Orthoptera, Mantodea, Blattaria, Phasmida, Homoptera, and Hemiptera (Stork et al. 1997). Falling out of or being knocked down from tree crowns is a significant hazard for these and other wingless arboreal arthropods. The branching structure of forest canopies confines walking or running to a limited number of high-traffic routes. As a result, walking or running individuals face an enhanced probability of attracting predators and of becoming dislodged by physical disturbances (e.g., wind, rain), which potentially leads either to escape jumps or to accidental falls.

Falling wingless animals are inevitably displaced to lower vegetational levels or to the understory for all but the smallest forms. Whereas landing on midstory branches, saplings, or shrubs may be inconsequential to small arboreal taxa, landing on the forest floor presents new problems, including structurally complex terrain (e.g., leaf litter) and an unfamiliar suite of predators. One important example concerns the flood plains of large tropical rivers (e.g., varzéa). Landing in the understory of such flooded forests reveals the greatest hazard, namely the abundance of surface-feeding carnivorous fish in this aquatic setting (de Mérona & Rankin-de-Mérona 2004, Saint-Paul et al. 2000). A small fallen animal will be maimed or consumed within seconds of hitting the water. Thus, in both upland and inundated forests, selection should favor traits that either prevent a fall (e.g., modified adhesive appendages; Beutel & Gorb 2001), or that allow the animal to influence how and where it lands (e.g., controlled descent and gliding). Controlled gliding in wingless arthropods has been postulated as now extinct intermediate behavior in the acquisition of flapping wings and powered flight (Dudley 2000, Hasenfuss 2002, Maynard Smith 1952). The recent discovery of directed aerial descent in ants (Yanoviak et al. 2005), an evolutionarily derived lineage, suggests that the behavior may be widespread among extant arboreal arthropods.

Ants compose a substantial fraction of insect abundance and biomass in tropical forests (Davidson et al. 2003, Fittkau & Klinge 1973, Stork et al. 1997, Tobin 1995), and are the most conspicuous arthropods in the crowns of tropical trees. Arboreal ant workers, which are wingless, fall from trees with high frequency in the phenomenon known as "ant rain" (Haemig 1997, Longino & Colwell 1997). Fallen workers that become lost or depredated are costly to the colony, and selection has favored multiple behavioral and morphological traits in arboreal ants to preclude falling. Nonetheless, ants frequently fall from tree crowns (Haemig 1997, Longino & Colwell 1997), and some jump from branch surfaces to escape disturbance (Weber 1957). Whereas many ants simply free-fall once airborne, workers of numerous species can direct their aerial descent to return to their home tree trunk (Yanoviak et al. 2005). Whether initiated by a jump or involuntarily, the subsequent directed descent leading back to the tree trunk is "J"-shaped and can be described as a three-part process: (*a*) an initial uncontrolled vertical drop with appendages extended (i.e., parachuting), (*b*) a rapid turn that results in alignment of the abdomen toward the tree trunk, and (*c*) a steep backward glide to the trunk (Yanoviak & Dudley 2006, Yanoviak et al. 2005; see **Figure 1**). Gliding behavior is now known to occur in at least seven arboreal ant genera (some of which are not closely related), demonstrating ample need for effective aerial performance in diverse taxa. More generally, these studies suggest that the occurrence of controlled gliding in arthropods and possible diversity of underlying aerodynamic mechanisms are substantially underestimated.

EVOLUTIONARY EXPERIMENTS IN VERTEBRATE GLIDING

In contrast to aforementioned findings with gliding ants, the diversity of terrestrial vertebrate gliders has long been appreciated. The 60 extant species of gliding mammals are a remarkably diverse group deriving from minimally nine independent evolutionary origins (Jackson 2000, Mein & Romaggi 1991, Meng et al. 2006, Scheibe & Robins 1998, Storch et al. 1996). Furthermore, this group exhibits tremendous size variation, ranging over two orders of magnitude in body mass (Nowak 1991). The rodent family Sciuridae includes no fewer than 44 species of flying squirrels for which a high degree of maneuverability is typically noted. It has been hypothesized that larger gliding squirrels have relatively longer tails to aid in steering (Thorington & Heaney 1981, Thorington et al. 2002). However, using independent contrasts analysis (as implemented in CAIC; Purvis & Rambaut 1995) and recent sciurid phylogenies (Mercer & Roth 2003, Thorington et al. 2002), we find no support for a positively allometric relationship in tail length within either gliding or nongliding squirrels. All large arboreal squirrels, gliders or otherwise, use their large round tail as a counterweight for balance along narrow branches. By contrast, small gliding squirrels, and the smallest mammalian glider of all, the marsupial feathertail glider, have distichous (i.e., flattened) tails that may serve an important aerodynamic function in the absence of selection pressure for a large counterbalancing tail. Dermopterans, by contrast, are the only large mammalian gliders with a tail that is completely subtended by the patagial membrane; not surprisingly, they are exclusively suspensory but are nonetheless aerially maneuverable.

In contrast to the mammalian gliders, other gliding vertebrates are not well known. Here we review the fascinating taxonomic and morphological diversity of gliding in reptiles and amphibians. Aerial descent has been reported for many anuran taxa in the context of breeding aggregations. Gliding descent at angles less than 45° has been observed in at least two anuran families, the Rhacophoridae and Hylidae, whereas steeper trajectories have been observed in both of these families as well as in the Leptodactylidae. Controlled aerial behaviors are well documented in numerous New World species of the family Hylidae, including *Ecnomiohyla miliaria*, *Phyllomedusa callidryas*, *Pachymedusa dacnicolor*, *Agalychnis spurrelli*, *A. saltator*, and *Scinax ruber* (Duellman 2001, Faivovich et al. 2005, Pauly et al. 2005, Pyburn 1970, Roberts 1994, Scott & Starrett 1974). Some of these species possess minimal aerodynamic surfaces in the form of webbed hands and feet and/or the presence of supplementary integumentary folds bordering the limbs (e.g., *P. callidryas, A. saltator*), but nonetheless are sufficiently aerial as to regularly use gliding locomotion. Intermediate degrees of webbing are exhibited by several documented gliders (*P. dacnicolor, Cruziohyla calcarifer*) and extensive webbing characterizes *Hyla miliaria* and *A. spurrelli*. Additional species with morphologies similar to those of well-documented gliders can be assumed to carry out some form of controlled aerial descent. These include various taxa with intermediate degrees of hand and feet webbing (e.g., *Smilisca sordida, S. sila, Hypsiboas rufitelus, H. salvaje, Ptychohyla dendrophasma, Ecnomiohyla valancifer, E. minera, Agalychnis annae,* and *A. calcarifer*), as well as species with full or nearly complete webbing of hands and feet (e.g., *Hypsiboas boans, Ecnomiohyla thysanota,* and *Agalychnis litodryas*).

Additional predictions may be made from the wide range of morphological variation exhibited by hylid frogs (Cott 1926, Duellman 2001). Nearly all species for which aerial descent has been documented are relatively slender forms with dorsoventrally flattened bodies. It is important to note that just because a species has moderate webbing and/or supplementary dermal flaps does not necessitate that it is a functional glider. Heavy-bodied species that are not capable of aerial descent may possess interdigital webbing and dermal ornamentation as a result of other adaptive contexts (e.g., crypsis). Some species that are endowed with relatively extensive webbing are nevertheless heavy-bodied and are not expected to be gliders (e.g., *Charadrahyla nephila, C. trux,* and members of the genus *Plectrohyla*). Slender-bodied forms such as *H. boans, H. rosenbergi, H. wavrini,* and *Osteopilus vastus* are morphologically likely candidates for gliding behavior, as is *Cruziohyla craspedopus,* a species with extensive cutaneous flaps and fringes on the outer edges of the fore- and hindlimbs. One New World anuran species in the family Leptodactylidae (*Eleutherodactylus coqui*) has been documented to parachute (Stewart 1985), but with little or no horizontal transit, this mode of aerial descent is far from specialized.

The Old World tree frogs (Family Rhacophoridae) have independently converged on gliding morphologies and behavior, with at least five species in the genus *Rhacophorus* documented to be aerially proficient (Boulenger 1912, Emerson & Koehl 1990, Inger 1966, Liem 1970, McCay 2001). Additional species that possess at best limited webbing of hands and feet (e.g., *Polypedates leucomystax, P. macrotis,* and *P. otilophus*) are nonetheless capable of substantial lift generation and controlled descent under experimental conditions (Emerson & Koehl 1990, Emerson et al. 1990). Extensive webbing of the hands and feet together with accessory flaps bordering the limbs suggest that additional rhacophorids may be proficient gliders, including *Rhacophorus harrissoni, R. dulitensis, R. georgii, R. prominanus, R. maximus, R. feae,* and *R. rufipes.* The numerous *Rhacophorus* species with more limited morphological specializations may well be capable of simple parachuting but have not been studied. An additional anuran glider is *Hyperolius castaneus* (Hyperoliidae) in Rwanda and the eastern Congo. Arboreal and semiarboreal habits are typical of these relatively unstudied anuran taxa.

Reptiles exhibit a diversity of aerial locomotor behaviors ranging from simple parachuting in many lizards and snakes to the powered flight of birds and pterosaurs. Here, we focus our attention on the nonpowered fliers, both among extant squamates and early diverging but now extinct reptilian lineages. As discussed above in the context of arthropods, it seems likely that the number of arboreal species capable of either gliding or controlled aerial descent can be dramatically underestimated if we assume that gliding species will exhibit obvious morphological innovations. For example, two lizard species, the polychrotid green anole (*Anolis carolinensis*) and the lacertid *Holaspis guentheri*, can generate glides with less than 45°descent angles although neither species has a patagium, elaborated skin folds, or toe webbing that one might expect in a gliding lizard [Oliver 1951, Schiøtz & Volsøe 1959; note that *Holaspis* exhibits a number of less obvious modifications associated with gliding (Arnold 2002)]. This finding is not surprising given that aerodynamic lift is proportional to the square of speed—if a lizard falls rapidly enough and assumes a proper body orientation, it will be able to produce forward momentum. Thus, the first innovation in an incipiently gliding lineage is likely to be behavioral and to involve appropriate positioning of the body and limbs.

Numerous squamate lineages exhibit morphological and behavioral adaptations for flight. Several species of Southeast Asian geckos in the genera *Ptychozoon*, *Luperosaurus*, and *Cosymbotus* have fully webbed hands and feet, flaps or folds of skin along the lateral body wall, and dorsoventrally flattened tails with or without marginal crenulations that increase surface areas (Brown & Diesmos 2000; Brown et al. 1997, 2000, 2007; Russell 1979; Russell et al. 2001). Although only *Ptychozoon* among these genera has been studied in any detail with respect to gliding performance or body motions (Heyer & Pongsapipatana 1970, Marcellini & Keefer 1976, Young et al. 2002), all are likely capable of highly directed aerial descent. Several species of *Ptychozoon* and two species of *Cosymbotus* have been observed to jump from the trunk of a tree, glide some distance lower, and return to the same tree (Brown et al. 1997, Honda et al. 1997; J. McGuire, personal observations). The agamid flying lizards (genus *Draco*) of Southeast Asia and southern India are clearly the most accomplished squamate gliders. This genus comprises approximately 45 species (McGuire & Kiew 2001), all of which have similar glide membranes composed of a patagium supported by 5–6 elongated thoracic ribs, as well as laterally projectable throat lappets controlled by the hyoid apparatus (Colbert 1967, McGuire 2003, Russell & Dijkstra 2001) that possibly function as canards (i.e., an aerodynamic control surface mounted in front of an aircraft). A number of nonsquamate fossil lineages were putative gliders with morphologies analogous to that of *Draco* in that they had patagia supported by elongated ribs or bony rib-like structures. These lineages include the Late Triassic kuehneosaurids *Icarosaurus seifkeri* and *Kuehneosaurus*, the Late Permian *Coelurosauravus jaekeli*, and the Late Triassic *Sharovipteryx mirabilis*, each of which had its own peculiarities with respect to patagial morphology (Carroll 1978, Dyke et al. 2006, Evans 1982, Frey et al. 1997, Gans et al. 1987, Robinson 1962). The Southeast Asian flying snakes (genus *Chrysopelea*) are thought to lack morphological innovations related to gliding locomotion (Socha 2002, Socha & LaBarbera 2005). However, these snakes have hinged ventral scales (as do a number of other arboreal snake taxa in Southeast Asia such as *Dendrelaphis* and *Ahaetulla*) that could contribute to the concave body form exhibited by gliding *Chrysopelea*.

Among gliding reptiles, the ecological contexts of flight are only well-documented for the lizard genus *Draco*, for which gliding is the primary means of movement within a home range that typically encompasses multiple trees (Alcala 1967; Mori & Hikida 1993, 1994). Flying lizards appear to avoid coming to the ground (except when females come to the ground to deposit eggs in the substrate) and thus movement from one tree to another usually involves gliding. *Draco* also will take to the air to avoid predators, chase conspecific males intruding on their territories, seek mating opportunities with females, and move between trees during foraging. For the lacertid *H. guentheri*, Schiøtz & Volsøe (1959) observed gliding behavior similar to that observed for *Draco*, with lizards initiating glides from one tree to another in several contexts. In one case, the lizard crossed a small river, and in other instances, the lizard immediately began foraging upon landing. Our knowledge of the ecological context of gliding in other reptiles is poorly known. It is clear that the geckos *Ptychozoon*, *Cosymbotus*, and *Luperosaurus*, and the flying snakes (*Chrysopelea*) utilize gliding for close-context escape (Honda et al. 1997; J. McGuire, personal observation). However, it is unclear if any of these taxa utilize gliding as a means of routine displacement (e.g., while foraging) within the forest canopy.

EVOLUTIONARY TRANSITIONS TO FLAPPING FLIGHT

Powered flapping flight has evolved independently four times in birds, bats, pterosaurs, and pterygote insects. The paleobiological context and biomechanical features of transitional forms are controversial in each case and have long been the subject of intense speculation. In general, the functional transition from gliding to flapping and associated production of thrust is biomechanically feasible (Norberg 1985). Recent findings on avian feathering and flight origins have been reviewed elsewhere (Norell & Xu 2005, Prum 2002), as has been pterosaur evolution (Buffetaut & Mazin 2003). No paleontological evidence is available for the morphologies of aerodynamically transitional bats, although their arboreal origins seem clear (Gunnell & Simmons 2005). For the evolution of bird flight, a recent hypothesis (Dial 2003a,b) suggests that forewing-generated aerodynamic forces facilitated hindlimb traction during running ascent on inclined or vertical surfaces (i.e., wing-assisted incline running); whereas the mechanics of this behavior are well documented in some extant birds (Bundle & Dial 2003) and are particularly important for juveniles with reduced wing area relative to adults (Dial et al. 2006), the phylogenetic distribution of this trait has not yet been assessed. The behavior may derive from ancestral use of wings in inadvertent falls, escape from nest predators, or during the ontogenetic acquisition of flapping (i.e., wing-assisted descent). In this vein, a diversity of recent biomechanical and paleontological studies support arboreal origins and gliding intermediates for flight in birds (Chatterjee & Templin 2007, Geist & Feduccia 2000, Long et al. 2003, Longrich 2006, Zhou 2004). Here, we review in detail the relevant literature pertaining to the origins of insect flight.

For the winged (i.e., pterygote) insects, historical origins are indeterminate but probably lie in the Upper Devonian or early Lower Carboniferous. Wingless hexapods are known from 395–390 Mya (Labandeira et al. 1988, Shear et al. 1984), whereas fossils of pterygote hexapods (i.e., winged insects) date from approximately 325 Mya (Nelson & Tidwell 1987). By the Upper Carboniferous, pterygotes are impressively diversified into about fifteen orders (Grimaldi & Engel 2005, Labandeira & Sepkoski 1993). Although pterygote insects are likely monophyletic (Grimaldi & Engel 2005, Regier et al. 2005), the morphological origins of wings remain obscure. Wings have been proposed to derive either from fixed paranotal outgrowths of thoracic and abdominal segments in terrestrial taxa (Bitsch 1994, Rasnitsyn 1981) or from ancestrally mobile gills, gill covers, leg structures, or styli in aquatic forms (Averof & Akam 1995, Kukalová-Peck 1983, Wigglesworth 1973). An intermediate possibility involves a terrestrial origin of wings derived from pre-existing leg, thoracic, or abdominal structures.

Unfortunately, no transitional forms are known between the wingless apterygotes and the winged pterygote insects, and the biology of early winged forms remains speculative and contentious. Of particular interest to the origins of flight is ancestral habitat association of early winged insects—were these animals terrestrial or aquatic? Phylogenetically, the closest sister taxa to the pterygote insects, the apterygote insect orders Zygentoma and Archaeognatha (Thysanura sensu *lato*), are exclusively terrestrial. Deeper within the phylogeny, the sister group to the insectan hexapods is the entognathan hexapods, the Collembola and the Diplura. The few aquatic species of collembolans are clearly derived (D'Haese 2002), and the remainder of the Collembola and all of the Diplura are terrestrial taxa. All hexapods, in turn, derive from a terrestrial crustacean lineage (Regier et al. 2005). An abundance of phylogenetic evidence is now clear on these points: hexapods evolved terrestrially, and the extant lineages closest to the winged insects are exclusively terrestrial. Apterygote insects, and particularly the thysanurans, thus offer the closest similarities of all extant taxa to predecessors of the winged insects.

Additional evidence, particularly that relating to the physiology and origins of the insect tracheal system, indicates that winged insects evolved from terrestrial apterygote ancestors (Dudley 2000, Grimaldi & Engel 2005, Messner 1988, Pritchard et al. 1993, Resh & Solem 1984). Aquatic larvae, particularly those of the extant and phylogenetically basal Odonata and Ephemeroptera, appear to be secondarily derived (Pritchard et al. 1993). Independent of habitat association, however, both larvae and adults of ancestral winged insects probably expressed lateral lobed structures on the abdominal as well as the thoracic segments (Carroll et al. 1995, Kukalová-Peck 1987). If winglets or wings derived initially from fixed paranotal lobes or from modified leg styli, flapping motions might have emerged indirectly through the action of dorsoventral leg muscles that insert on the thorax, as characterizes so-called bifunctional muscles in many extant insects (Fourtner & Randall 1982, Wilson 1962). A general question relating to wing origins concerns the possible evolution of novel wing-like structures, as opposed to modification of pre-existing morphological features. Acquisition of wings from ancestrally mobile structures might seem more parsimonious than the derivation of flapping wings from stationary paranotal lobes, although the neontological and paleontological data available at present are insufficient to prove unequivocally either of these two hypotheses (Dudley 2000, Grimaldi & Engel 2005).

A variety of possible functional roles have been attributed to transitional winglets or early wings, including aerodynamic utility, epigamic display during courtship, and thermoregulation (Douglas 1981; Ellington 1991; Kingsolver & Koehl 1985, 1994). Hydrodynamic use for features that ultimately became aerodynamic structures has been proposed for ancestrally aquatic forms. Hexapods could also have used winglike structures in air either to drift passively, row, or skim actively along water surfaces (i.e., an "airboat" hypothesis; Marden & Kramer 1994, 1995). These behaviors are clearly derived rather than retained ancestral traits of winged insects given their rare occurrence and derived condition in the Paleoptera (Ruffieux et al. 1998, Samways 1996), their multiple independent origins within the Neoptera—including plecopterans, several dipteran taxa, and some trichopterans (Dudley 2000, Will 1995)—and the phylogenetic improbability of ancestrally aquatic pterygotes (Grimaldi & Engel 2005).

Importantly, surface rowing by certain plecopteran taxa, which represents a putatively ancestral biomechanical condition relative to flapping of wings in air, occurs in a highly derived group of stoneflies (Marden & Thomas 2003, Thomas et al. 2000). Biomechanical considerations also suggest that postulated aquatic precursors would have been unlikely to evolve wings that served aerodynamic functions. Water and air differ by almost three orders of magnitude in density, with a corresponding difference in the Reynolds number and in the nature of forces generated by oscillating structures. The functionality of wing designs intermediate to either hydrodynamic or aerodynamic force generation is correspondingly unclear (Dudley 2000). Forces of surface tension would present a formidable physical barrier to partial body emergence as well as to projection and oscillation of flattened structures, particularly for the body sizes (2–4 cm) deduced as characteristic of ancestral pterygotes (Labandeira et al. 1988, Wootton 1976).

Given the assumption of terrestrial pterygote ancestors, a standard explanation for the evolution of wings has been that these structures aerodynamically facilitate jumping escapes from predators on land. Suggestively, a suite of morphological and behavioral protoadaptations for jump-mediated glides is evident among extant apterygote hexapods, the terrestrial sister taxon of the winged insects. Thoracic paranotal lobes as well as styli on the legs and abdominal segments of extant apterygotes could potentially have served in ancestral taxa to generate lift and to facilitate saltatorial escape. Neurobiological studies also support the ancestral presence of dedicated sensorimotor pathways underlying escape behavior in both apterygotes and pterygotes (Edwards & Reddy 1986, Ritzmann 1984). The startle response of ancestral apterygote insects was then apparently co-opted during pterygote evolution to stimulate jumping, wing flapping, and even evasive flight once airborne (Edwards 1997, Hasenfuss 2002, Libersat 1994). The historical context of early pterygote evolution was appropriate for imposition of intense predatory pressure by both invertebrates and vertebrates, with a diversity of insectivorous arthropods (particularly arachnids), amphibians, and reptiles found in Devonian and Carboniferous terrestrial ecosystems (Behrensmeyer et al. 1992, Rolfe 1985, Shear & Kukalová-Peck 1990). Furthermore, the increasing arborescence and geometrical complexity of terrestrial vegetation through the Devonian and into the Carboniferous (Dilcher et al. 2004, Kenrick & Crane 1997) would have provided suitable three-dimensional substrate suitable for aerial escape and maneuvers.

CONCLUSIONS

Locomotor evolution within an arboreal context has yielded diverse vertebrate and arthropod taxa capable of controlled falls, directed aerial descent, and sophisticated gliding. The regulation of aerodynamic forces produced while falling from trees can occur in the complete absence of wings and is enabled by movement of both axial and appendicular structures. Subsequent evolution of more dedicated aerodynamic surfaces such as patagial membranes and true wings likely follows the initial acquisition of aerial maneuverability. Morphological and behavioral intermediates to true flapping flight can accordingly exhibit progressive functionality as they become more elaborate. It may seem unnecessary to emphasize the origins of flight within an arboreal and thus aerial context. However, unlike competing running and even aquatic hypotheses for flight origins in certain groups, this scenario for the evolution of flight is not taxon-specific. Instead, we have emphasized here the relatively undocumented and understudied diversity of controlled aerial behaviors found across a broad range of arboreal taxa. As with other complex evolutionary outcomes (e.g., endothermy), flapping flight represents a specialization attained by only a minority of lineages. Nonetheless, the study of biomechanical and physiological intermediates can reveal both underlying functional demands and alternative strategies to problems imposed by common selective environments. Arboreality may be one such environment for which flight and its control are desirable outcomes.

DISCLOSURE STATEMENT

The authors are not aware of any biases that might be perceived as affecting the objectivity of this review.

ACKNOWLEDGMENTS

We thank Bill Duellman, Ardian Jusufi, Shai Revzen, and Yonatan Munk for comments on the manuscript.

LITERATURE CITED

Alcala AC. 1967. Population biology of the "flying" lizard, *Draco volans*, on Negros Island, Philippines. *Nat. Appl. Sci. Bull.* 20:335–72

Ando M, Shiraishi S. 1993. Gliding flight in the Japanese Giant Flying Squirrel *Petaurista leucogenys*. *J. Mammal. Soc. Jpn.* 18:19–32

Arnold EN. 2002. *Holaspis*, a lizard that glided by accident: mosaics of cooption and adaptation in a tropical forest lacertid (Reptilia, Lacertidae). *Bull. Nat. Hist. Mus. London Zool.* 68:155–63

Averof M, Akam M. 1995. Insect-crustacean relationships: insights from comparative developmental and molecular studies. *Philos. Trans. R. Soc. London Ser. B* 347:293–303

Azuma A. 1992. *The Biokinetics of Flying and Swimming*. Tokyo: Springer-Verlag. 265 pp.

Beard KC. 1990. Gliding behavior and palaeoecology of the alleged primate family Paromomyidae (Mammalia, Dermoptera). *Nature* 345:340–41

Behrensmeyer AK, Damuth JD, DiMichele WA, Sues HD, Wing SL. 1992. *Terrestrial Ecosystems Time: Evolutionary Paleoecology of Terrestrial Plants and Animals.* Chicago: Univ. Chicago Press. 568 pp.

Beutel RG, Gorb SN. 2001. Ultrastructure of attachment specializations of hexapods (Arthropoda): evolutionary patterns inferred from a revised ordinal phylogeny. *J. Zool. Syst. Evol. Res.* 39:177–207

Bishop KL. 2006. The relationship between 3-D kinematics and gliding performance in the southern flying squirrel, *Glaucomys volans. J. Exp. Biol.* 209:689–701

Bitsch J. 1994. The morphological groundplan of Hexapoda: critical review of recent concepts. *Ann. Soc. Entomol. Fr. (NS)* 30:103–29

Boulenger GA. 1912. *A Vertebrate Fauna of the Malay Peninsula from the Isthmus of Kra to Singapore Including the Adjacent Islands. Reptilia and Batrachia.* London: Taylor & Frances

Brown RM, Diesmos AC. 2000. The lizard genus *Luperosaurus*: taxonomy, history, and conservation prospects for some of the world's rarest lizards. *Sylvatrop Tech. J. Philipp. Ecosyst. Nat. Res.* 10:107–24

Brown RM, Diesmos AC, Duya MV. 2007. A new *Luperosaurus* (Squamata: Gekkonidae) from the Sierra Madre of Luzon Island, Philippines. *Raffles Bull. Zool.* 55:167–74

Brown RM, Ferner JW, Diesmos AC. 1997. Definition of the Philippine parachute gecko, *Ptychozoon intermedium* Taylor 1915 (Reptilia: Squamata: Gekkonidae): redescription, designation of a neotype, and comparisons with related species. *Herpetologica* 53:357–73

Brown RM, Supriatna J, Ota H. 2000. Discovery of a new species of *Luperosaurus* (Squamata; Gekkonidae) from Sulawesi, with a phylogenetic analysis of the genus, and comments on the status of *Luperosaurus serraticaudus. Copeia* 2000:191–209

Buffetaut E, Mazin J-M, eds. 2003. *Evolution and Palaeobiology of Pterosaurs.* Geol. Soc. Spec. Publ. 217:1–347

Bundle MW, Dial KP. 2003. Mechanics of wing-assisted incline running (WAIR). *J. Exp. Biol.* 206:4533–64

Carroll RL. 1978. Permo-Triassic "lizards" from the Karoo system. Part II. A gliding reptile from the Upper Permian of Madagascar. *Palaeontol. Afr.* 21:143–59

Carroll SB, Weatherbee SD, Langeland JA. 1995. Homeotic genes and the regulation and evolution of insect wing number. *Nature* 375:58–61

Chatterjee S, Templin RJ. 2007. Biplane wing planform and flight performance of the feathered dinosaur *Microraptor gui. Proc. Natl. Acad. Sci. USA* 104:1576–80

Colbert EH. 1967. Adaptations for gliding in the lizard *Draco. Am. Mus. Novit.* 2283:1–20

Colbert EH. 1970. The Triassic gliding reptile *Icarosaurus. Bull. Am. Mus. Nat. Hist.* 143:121–42

Cott HB. 1926. Observations on the life-habits of some batrachians and reptiles from the lower Amazon: and a note on some mammals from Marajo Island. *Proc. Zool. Soc. London* 2:1159–78

Davenport J. 1994. How and why do flying fish fly? *Rev. Fish Biol. Fisheries* 4:184–214

Davenport J. 2003. Allometric constraints on stability and maximum size in flying fishes: implications for their evolution. *J. Fish Biol.* 62:455–63

Davidson DW, Cook SC, Snelling RR, Chua TH. 2003. Explaining the abundance of ants in lowland tropical rainforest canopies. *Science* 300:969–72

de Gouvenain RC, Silander JA. 2003. Do tropical storm regimes influence the structure of tropical lowland rain forests? *Biotropica* 35:166–80

de Mérona B, Rankin-de-Mérona J. 2004. Food resource partitioning in a fish community of the central Amazon floodplain. *Neotrop. Ichthyol.* 2:75–84

D'Haese CA. 2002. Were the first springtails semiaquatic? A phylogenetic approach by means of 28S rDNA and optimization alignment. *Proc. R. Soc. London Ser. B* 269:1143–51

Dial KP. 2003a. Evolution of avian locomotion: correlates of flight style, locomotor modules, nesting biology, body size, development, and the origin of flapping flight. *The Auk* 120:941–52

Dial KP. 2003b. Wing-assisted incline running and the evolution of flight. *Science* 299:402–4

Dial KP, Randall RJ, Dial TR. 2006. What use is half a wing in the ecology and evolution of birds? *BioScience* 56:437–45

Dilcher DL, Lott TA, Wang X, Wang Q. 2004. A history of tree canopies. In *Forest Canopies*, ed. MD Lowman, HB Rinker, pp. 118–37. Burlington, MA: Elsevier Acad. 2nd ed.

Dolan PG, Carter DC. 1977. *Glaucomys volans. Mammal. Species* 78:1–6

Douglas MM. 1981. Thermoregulatory significance of thoracic lobes in the evolution of insect wings. *Science* 211:84–86

Dudley R. 2000. *The Biomechanics of Insect Flight: Form, Function, Evolution*. Princeton: Princeton Univ. Press. 476 pp.

Dudley R. 2002. Mechanisms and implications of animal flight maneuverability. *Integ. Comp. Biol.* 42:135–40

Dudley R, DeVries PJ. 1990. Tropical rain forest structure and the geographical distribution of gliding vertebrates. *Biotropica* 22:432–34

Duellman WE. 2001. *The Hylid Frogs of Middle America*. Ithaca, NY: Soc. Study Amphib. Reptil.

Dunbar DC. 1988. Aerial maneuvers of leaping lemurs: the physics of whole-body rotations while airborne. *Am. J. Primatol.* 16:291–304

Dyke GJ, Nudds RL, Rayner JMV. 2006. Flight of *Sharovipteryx mirabilis*: the world's first delta-winged glider. *J. Evol. Biol.* 19:1040–43

Eaton RC, ed. 1984. *Neural Mechanisms of Startle Behavior*. New York: Plenum. 377 pp.

Edwards JS. 1997. The evolution of insect flight: implications for the evolution of the nervous system. *Brain Behav. Evol.* 50:8–12

Edwards JS, Reddy GR. 1986. Mechanosensory appendages in the firebrat (*Thermobia domestica*, Thysanura): a prototype system for terrestrial predator evasion. *J. Comp. Neurol.* 243:535–46

Ellington CP. 1991. Aerodynamics and the origin of insect flight. *Adv. Insect Physiol.* 23:171–210

Emerson SB, Koehl MAR. 1990. The interaction of behavioral and morphological change in the evolution of a novel locomotor type: "flying" frogs. *Evolution* 44:1931–46

Emerson SB, Travis J, Koehl MAR. 1990. Functional complexes and additivity in performance: a test case with "flying" frogs. *Evolution* 44:2153–57

Essner RL. 2002. Three-dimensional launch kinematics in leaping, parachuting and gliding squirrels. *J. Exp. Biol.* 205:2469–77

Evans SE. 1982. The gliding reptiles of the upper Permian. *Zool. J. Linn. Soc.* 76:97–123

Faivovich J, Haddad CFB, Garcia PCA, Frost DR, Campbell JA, Wheeler WC. 2005. Systematic review of the frog family Hylidae, with special reference to Hylinae: phylogenetic analysis and taxonomic revision. *Bull. Am. Mus. Nat. Hist.* 294:1–240

Fittkau EJ, Klinge H. 1973. On biomass and trophic structure of the central Amazonian rain forest ecosystem. *Biotropica* 5:2–14

Fourtner CR, Randall JB. 1982. Studies on cockroach flight: the role of continuous neural activation of nonflight muscles. *J. Exp. Zool.* 221:143–54

Frey E, Sues HD, Munk W. 1997. Gliding mechanism in the late Permian reptile *Coelurosauravus*. *Science* 275:1450–52

Gans C, Darevski I, Tatarinov LP. 1987. *Sharovipteryx*, a reptilian glider? *Paleobiology* 13:415–26

Geist NR, Feduccia A. 2000. Gravity-defying behaviors: identifying models for Protoaves. *Am. Zool.* 40:664–75

Grimaldi D, Engel MS. 2005. *Evolution of the Insects*. Cambridge, UK: Cambridge Univ. Press. 755 pp.

Gunnell GF, Simmons NB. 2005. Fossil evidence and the origin of bats. *J. Mammal. Evol.* 12:209–46

Haemig PD. 1997. Effects of birds on the intensity of ant rain: a terrestrial form of invertebrate drift. *Anim. Behav.* 54:89–97

Hasenfuss I. 2002. A possible evolutionary pathway to insect flight starting from lepismatid organization. *J. Zool. Syst. Evol. Res.* 40:65–81

Heyer WR, Pongsapipatana S. 1970. Gliding speeds of *Ptychozoon lionatum* (Reptilia: Gekkonidae) and *Chrysopelea ornata* (Reptilia: Colubridae). *Herpetologica* 26:317–31

Holmes DJ, Austad SN. 1994. Fly now, die later: life-history correlates of gliding and flying in mammals. *J. Mammal.* 75:224–26

Honda M, Hikida T, Araya K, Ota H, Nabjitabhata J, Hoi-Sen Y. 1997. *Cosymbotus craspedotus* (Frilly Gecko) and *C. platyurus* (Flat-tailed Gecko). Gliding behavior. *Herpetol. Rev.* 28:42–43

Inger RF. 1966. The systematics and zoogeography of the Amphibia of Borneo. *Fieldiana Zool.* 52:1–402

Jackson SM. 2000. Glide angle in the genus *Petaurus* and a review of gliding in mammals. *Mamm. Rev.* 30:9–30

Johnson-Murray JL. 1987. The comparative myology of the gliding membranes of *Acrobates*, *Petauroides* and *Petaurus* contrasted with the cutaneous myology of

Hemibelideus and *Pseudocheirus* (Marsupalia: Phalangeridae) and with selected gliding Rodentia (Sciuridae and Anamoluridae). *Aust. J. Zool.* 35:101–13

Kenrick P, Crane PR. 1997. The origin and early evolution of plants on land. *Nature* 389:33–39

Kingdon J. 1974. *East African Mammals—An Atlas of Evolution in Africa.* Vol. 2A. *Hares and Rodents.* New York: Academic

Kingsolver JG, Koehl MAR. 1985. Aerodynamics, thermoregulation, and the evolution of insect wings: differential scaling and evolutionary change. *Evolution* 39:488–504

Kingsolver JG, Koehl MAR. 1994. Selective factors in the evolution of insect wings. *Annu. Rev. Entomol.* 39:425–51

Kukalová-Peck J. 1983. Origin of the insect wing and wing articulation from the arthropodan leg. *Can. J. Zool.* 61:1618–69

Kukalová-Peck J. 1987. New Carboniferous Diplura, Monura, and Thysanura, the hexapod ground plan, and the role of thoracic lobes in the origin of wings (Insecta). *Can. J. Zool.* 65:2327–45

Labandeira CC, Beall BS, Hueber FM. 1988. Early insect diversification: evidence from a Lower Devonian bristletail from Québec. *Science* 242:913–16

Labandeira CC, Sepkoski JJ. 1993. Insect diversity in the fossil record. *Science* 261:310–15

Libersat F. 1994. The dorsal giant interneurons mediate evasive behavior in flying cockroaches. *J. Exp. Biol.* 197:405–11

Liem S. 1970. The morphology, systematics, and evolution of the Old World treefrogs (Rhacophoridae and Hyperolidae). *Fieldiana Zool.* 57:1–145

Long CA, Zhang GP, George TF, Long CF. 2003. Physical theory, origin of flight, and a synthesis proposed for birds. *J. Theor. Biol.* 224:9–26

Longino JT, Colwell RK. 1997. Biodiversity assessment using structured inventory: capturing the ant fauna of a tropical rain forest. *Ecol. Appl.* 7:1263–77

Longrich N. 2006. Structure and function of hindlimb feathers in *Archaeopteryx lithographica.* *Paleobiology* 32:417–31

Losos JB, Papenfuss TJ, Macey JR. 1989. Correlates of sprinting, jumping and parachuting performance in the butterfly lizard, *Leiolepis belliani.* *J. Zool. London* 217:559–68

Marcellini DL, Keefer TE. 1976. Analysis of the gliding behavior of *Ptychozoon lionatum* (Reptilia: Gekkonidae). *Herpetologica* 32:362–66

Marden JH, Kramer MG. 1994. Surface-skimming stoneflies: a possible intermediate stage in insect flight evolution. *Science* 266:427–30

Marden JH, Kramer MG. 1995. Locomotor performance of insects with rudimentary wings. *Nature* 377:332–34

Marden JH, Thomas MA. 2003. Rowing locomotion by a stonefly that possesses the ancestral pterygote condition of co-occurring wings and abdominal gills. *Biol. J. Linn. Soc.* 79:341–49

Maynard Smith J. 1952. The importance of the nervous system in the evolution of animal flight. *Evolution* 6:127–29

McCay MG. 2001. Aerodynamic stability and maneuverability of the gliding frog *Polypedates dennysi.* *J. Exp. Biol.* 204:2817–26

McCay MG. 2003. Winds under the rain forest canopy: the aerodynamic environment of gliding tree frogs. *Biotropica* 35:94–102

McGuire JA. 2003. Allometric prediction of locomotor performance: an example from Southeast Asian flying lizards. *Am. Nat.* 161:337–49

McGuire JA, Dudley R. 2005. The cost of living large: comparative gliding performance in flying lizards (Agamidae: *Draco*). *Am. Nat.* 166:93–106

McGuire JA, Kiew BH. 2001. Phylogenetic systematics of Southeast Asian flying lizards (Iguania: Agamidae: *Draco*) as inferred from mitochondrial DNA sequence data. *Biol. J. Linn. Soc.* 72:203–29

McKay GM. 1983. Greater glider. In *Complete Book of Australian Mammals*, ed. R Strahan, pp. 134–35. Sydney: Angus & Robertson

Mein P, Romaggi JP. 1991. Un gliridé (Mammalia, Rodentia) planeur dans le Miocène supérieur de l'Ardèche: une adaptation non retrouvée dans la nature actuelle. *Geobios* 13(Memo. Spec.):45–50

Mendoza MM, Custodio CC. 2000. Field observations on the Philippine flying lemur (*Cynocephalus volans*). In *Biology of Gliding Mammals*, ed. RL Goldingay, JS Scheibe, pp. 272–80. Fürth: Filander Verlag

Meng J, Hu Y, Wang Y, Wang X, Li C. 2006. A Mesozoic gliding mammal from northeastern China. *Nature* 444:889–93

Mercer JM, Roth VL. 2003. The effects of Cenozoic global change on squirrel phylogeny. *Science* 299:1568–72

Messner B. 1988. Sind die Insekten primäre oder sekundäre Wasserbewohner? *Dtsch. Entomol. Z. NF* 35:355–60

Mori A, Hikida T. 1993. Natural history observations of the flying lizard, *Draco volans sumatranus* (Agamidae, Squamata) from Sarawak, Malaysia. *Raffles Bull. Zool.* 41:83–94

Mori A, Hikida T. 1994. Field observations on the social behavior of the flying lizard, *Draco volans sumatranus*, in Borneo. *Copeia* 1994:124–30

Nachtigall W. 1979. Gleitflug des Flugbeutlers *Petaurus breviceps papuanus* II. Filmanalysen zur Einstellung von Gleitbahn und Rumpf sowie zur Steuerung des Gleitflugs. *J. Comp. Physiol.* 133:89–95

Nelson CR, Tidwell WD. 1987. *Brodioptera stricklani* n. sp. (Megasecoptera: Brodiopteridae), a new fossil insect from the Upper Manning Canyon Shale Formation, Utah (lowermost Namurian B). *Psyche* 94:309–16

Norberg RA. 1983. Optimal locomotion modes of foraging birds in trees. *Ibis* 125:172–80

Norberg UM. 1985. Evolution of vertebrate flight: an aerodynamic model for the transition from gliding to active flight. *Am. Nat.* 126:303–27

Norberg UM. 1985. Flying, gliding and soaring. In *Functional Vertebrate Morphology*, ed. M Hildebrand, pp. 129–58. Cambridge: Harvard Univ. Press

Norberg UM. 1990. *Vertebrate Flight*. Berlin: Springer-Verlag. 291 pp.

Norell MA, Xu X. 2005. Feathered dinosaurs. *Annu. Rev. Earth Planet. Sci.* 33:277–99

Nowak RM. 1991. *Walker's Mammals of the World*. Baltimore: Johns Hopkins Univ. Press. 5th ed.

Oliver JA. 1951. "Gliding" in amphibians and reptiles, with a remark on an arboreal adaptation in the lizard, *Anolis carolinensis carolinensis* Voigt. *Am. Nat.* 85:171–76

Padian K, Chiappe LM. 1998. The origin and early evolution of birds. *Biol. Rev.* 73:1–42

Pauly GB, Bernal X, Taylor RC. 2005. *Scinax ruber* (Red Snouted Treefrog). Arboreality and parachuting. *Herpetol. Rev.* 36:308–9

Pellis SM, Pellis VC, Morrissey TK, Teitelbaum P. 1989. Visual modulation of vestibularly-triggered air-righting in the rat. *Behav. Brain Res.* 35:23–26

Pennycuick CJ. 1975. Mechanics of flight. In *Avian Biology*, ed. DS Farner, JR King, 5:1–75. London: Academic

Pomeroy D. 1990. Why fly? The possible benefits for lower mortality. *Biol. J. Linn. Soc.* 40:53–65

Pritchard G, McKee MH, Pike EM, Scrimgeour GJ, Zloty J. 1993. Did the first insects live in water or in air? *Biol. J. Linn. Soc.* 49:31–44

Prum RO. 2002. Why ornithologists should care about the theropod origin of birds. *The Auk* 119:1–17

Purvis A, Rambaut A. 1995. Comparative analysis by independent contrasts (CAIC): an Apple Macintosh application for analyzing comparative data. *Comput. Appl. Biosci.* 11:247–51

Pyburn WF. 1970. Breeding behavior of the leaf-frogs *Phyllomedusa callidryas* and *Phyllomedusa dacnicolor* in Mexico. *Copeia* 1970:209–18

Rasnitsyn AP. 1981. A modified paranotal theory of insect wing origin. *J. Morphol.* 168:331–38

Rayner JMV. 1988. The evolution of vertebrate flight. *Biol. J. Linn. Soc.* 34:269–87

Regier JC, Shultz JW, Kambic RE. 2005. Pancrustacean phylogeny: hexapods are terrestrial crustaceans and maxillopods are not monophyletic. *Proc. R. Soc. London Ser. B* 272:395–401

Resh VH, Solem JO. 1984. Phylogenetic relationships and evolutionary adaptations of aquatic insects. In *An Introduction to the Aquatic Insects of North America*, ed. RW Merritt, KW Cummins, pp. 66–75. Dubuque: Kendall/Hunt. 2nd ed.

Ritzmann RE. 1984. The cockroach escape response. In *Neural Mechanisms of Startle Behavior*, ed. RC Eaton, pp. 93–131. New York: Plenum

Roberts WE. 1994. Explosive breeding aggregations and parachuting in a neotropical frog. *J. Herpetol.* 28:193–99

Robinson PL. 1962. Gliding lizards from the Upper Keuper of Great Britain. *Proc. Geol. Soc. London* 1601:137–46

Rolfe WD. 1985. Early terrestrial arthropods: a fragmentary record. *Philos. Trans. R. Soc. London Ser. B* 309:207–18

Ruffieux L, Elouard JM, Sartori M. 1998. Flightlessness in mayflies and its relevance to hypotheses on the origin of insect flight. *Proc. R. Soc. London B* 265:2135–40

Russell AP. 1979. The origins of parachuting locomotion in gekkonid lizards (Reptilia: Gekkonidae). *Zool. J. Linn. Soc.* 65:233–49

Russell AP, Dijkstra LD. 2001. Patagial morphology of *Draco volans* (Reptilia, Agamidae) and the origin of glissant locomotion in flying dragons. *J. Zool. London* 253:457–71

Russell AP, Dijkstra LD, Powell GL. 2001. Structural characteristics of the patagium of *Ptychozoon kuhli* (Reptilia: Gekkonidae) in relation to parachuting locomotion. *J. Morphol.* 247:252–63

Saint-Paul U, Zuanon J, Correa MAV, García M, Fabré NN, et al. 2000. Fish communities in central Amazonian white- and blackwater floodplains. *Environ. Biol. Fish.* 57:235–50

Samways MJ. 1996. Skimming and insect evolution. *Trends Ecol. Evol.* 11:471

Scheibe JS, Robins JH. 1998. Morphological and performance attributes of gliding mammals. In *Ecology and Evolutionary Biology of Tree Squirrels*, ed. MA Steele, JF Merritt, DA Zegers, pp. 6:131–44. Va. Mus. Nat. Hist. Spec. Publ.

Schiøtz A, Volsøe H. 1959. The gliding flight of *Holaspis guentheri* Gray, a west-African lacertid. *Copeia* 1959:259–60

Schlesinger WH, Knops JMH, Nash TH. 1993. Arboreal sprint failure: lizardfall in a California oak woodland. *Ecology* 74:2465–67

Scholey K. 1986. The climbing and gliding locomotion of the giant red flying squirrel *Petaurista petaurista* (Sciuridae). In *Bat Flight-Fledermausflug*, ed. W Nachtigall, pp. 187–204. Stuttgart: Gustav-Fischer Verlag

Scott NJ, Starrett A. 1974. An unusual breeding aggregation of frogs, with notes on the ecology of *Agalychnis spurrelli* (Anura: Hylidae). *Bull. South. Calif. Acad. Sci.* 73:86–94

Shear WA, Grearson JD, Wolfe WDI, Smith EL, Norton RA. 1984. Early land animals in North America: evidence from Devonian age arthropods from Gilboa, New York. *Science* 224:492–94

Shear WA, Kukalová-Peck J. 1990. The ecology of Paleozoic terrestrial arthropods: the fossil evidence. *Can. J. Zool.* 68:1807–34

Socha JJ. 2002. Gliding flight in the paradise tree snake. *Nature* 418:603–4

Socha JJ. 2006. Becoming airborne without legs: the kinematics of take-off in a flying snake, *Chrysopelea paradisi*. *J. Exp. Biol.* 209:3358–69

Socha JJ, LaBarbera M. 2005. Effects of size and behavior on the aerial performance of two species of flying snake (*Chrysopelea*). *J. Exp. Biol.* 208:1835–47

Socha JJ, O'Dempsey TO, LaBarbera M. 2005. A 3-D kinematic analysis of gliding in a flying snake, *Chrysopelea paradisi*. *J. Exp. Biol.* 208:1817–33

Stewart MM. 1985. Arboreal habitat use and parachuting by a subtropical forest frog. *J. Herpetol.* 19:391–401

Storch G, Engesser B, Wuttke M. 1996. Oldest fossil record of gliding in rodents. *Nature* 379:439–41

Stork NE, Adis J, Didham RK. 1997. *Canopy Arthropods*. London: Chapman & Hall. 567 pp.

Thomas MA, Walsh KA, Wolf MR, McPheron BA, Marden JH. 2000. Molecular phylogenetic analysis of evolutionary trends in stonefly wing structure and locomotor behavior. *Proc. Natl. Acad. Sci. USA* 97:13178–83

Thorington RW. 1984. Flying squirrels are monophyletic. *Science* 225:1048–50

Thorington RW, Heaney LR. 1981. Body proportions and gliding adaptations of flying squirrels (Petauristinae). *J. Mammal.* 62:101–14

Thorington RW, Pitassy D, Jansa SA. 2002. Phylogenies of flying squirrels (Pteromyinae). *J. Mammal. Evol.* 9:99–135

Tobin JE. 1995. Ecology and diversity of tropical forest canopy ants. In *Forest Canopies*, ed. MD Lowman, NM Nadkarni, pp. 129–47. San Diego, CA: Academic

Vermeij GJ. 2006. Historical contingency and the purported uniqueness of evolutionary innovations. *Proc. Natl. Acad. Sci. USA* 103:1804–9

Vermeij GJ, Dudley R. 2000. Why are there so few evolutionary transitions between aquatic and terrestrial environments? *Biol. J. Linn. Soc.* 70:541–54

Videler JJ. 2005. *Avian Flight*. Oxford: Oxford Univ. Press. 258 pp.

Vogel S. 1994. *Life in Moving Fluids: The Physical Biology of Flow*. Princeton: Princeton Univ. Press. 467 pp.

Weber NA. 1957. The nest of an anomalous colony of the arboreal ant *Cephalotes atratus*. *Psyche* 64:60–69

Wigglesworth VB. 1973. Evolution of insect wings and flight. *Nature* 246:127–29

Wilkinson MT, Unwin DM, Ellington CP. 2006. High lift function of the pteroid bone and forewing of pterosaurs. *Proc. R. Soc. London B* 273:119–26

Will KW. 1995. Plecopteran surface-skimming and insect flight evolution. *Science* 270:1684–85

Wilson DM. 1962. Bifunctional muscles in the thorax of grasshoppers. *J. Exp. Biol.* 39:669–77

Wootton RJ. 1976. The fossil record and insect flight. In *Insect Flight*, ed. RC Rainey, pp. 235–54. Oxford: Blackwell Sci.

Yanoviak SP, Dudley R. 2006. Color preference during directed aerial descent of *Cephalotes atratus* workers (Hymenoptera: Formicidae). *J. Exp. Biol.* 209:1777–83

Yanoviak SP, Dudley R, Kaspari M. 2005. Directed aerial descent in arboreal ants. *Nature* 433:624–26

Young BA, Lee CE, Daley KM. 2002. On a flap and a foot: aerial locomotion in the "flying" gecko, *Ptychozoon kuhli*. *J. Herpetol.* 36:412–18

Zhou Z. 2004. The origin and early evolution of birds: discoveries, disputes, and perspectives from fossil evidence. *Naturwissenschaften* 91:455–71

How Mutational Networks Shape Evolution: Lessons from RNA Models

Matthew C. Cowperthwaite[1] and Lauren Ancel Meyers[2]

[1]Institute for Cellular and Molecular Biology, University of Texas, Austin, Texas 78712; email: mattccowp@mac.com

[2]Section of Integrative Biology and Institute for Cellular and Molecular Biology, University of Texas, Austin, Texas 78712; email: laurenmeyers@mail.utexas.edu

Annu. Rev. Ecol. Evol. Syst. 2007. 38:203–30

First published online as a Review in Advance on July 31, 2007

The *Annual Review of Ecology, Evolution, and Systematics* is online at http://ecolsys.annualreviews.org

This article's doi: 10.1146/annurev.ecolsys.38.091206.095507

Key Words

RNA, robustness, evolutionary dynamics, fitness landscape, genotype-phenotype map

Abstract

Recent advances in molecular biology and computation have enabled evolutionary biologists to develop models that explicitly capture molecular structure. By including complex and realistic maps from genotypes to phenotypes, such models are yielding important new insights into evolutionary processes. In particular, computer simulations of evolving RNA structure have inspired a new conceptual framework for thinking about patterns of mutational connectivity and general theories about the nature of evolutionary transitions, the evolutionary ascent of nonoptimal phenotypes, and the origins of mutational robustness and modular structures. Here, we describe this class of RNA models and review the major conceptual contributions they have made to evolutionary biology.

1. INTRODUCTION

1.1. Overview

Evolutionary biologists have long sought to understand the evolutionary processes that transcend any particular biological system. Models are indispensable tools for gaining such insights. During the twentieth century, evolutionary theoreticians built a powerful conceptual framework on simple mathematical models. Recently, however, thanks to startling advances in molecular biology and computational power, a new generation of higher resolution quantitative models is changing our perspectives on the origins and processes that have led to the current diversity of life on Earth.

1.2. Motivation

Detailed models of RNA structural evolution have enabled advances in evolutionary biology. The success of this model system stems partly from the biological importance of RNA and partly from our ability to rapidly and reliably predict the structures of these molecules.

1.2.1. RNA is central to biology. DNA, RNA, and proteins are the three essential biological macromolecules. Although RNA mediates information transfer from DNA genes to functional proteins and thus lies at the heart of the "central dogma of molecular biology," it has historically been overshadowed by DNA and proteins. Several recent discoveries, however, have brought RNA to center stage. RNA plays a vital regulatory role (for recent reviews see Mattick & Makunin 2006, Niwa & Slack 2007, Winkler & Breaker 2005) in many cellular processes and is the primary genetic material for a large number of viruses, including influenza and HIV. Molecular biologists are thus working hard to characterize the molecular structure of RNA and the relationship between RNA structure and biological function.

1.2.2. RNA makes a great evolutionary model. Evolutionary biologists have harnessed the efforts of RNA molecular biologists. They have built evolutionary models that explicitly consider the relationship between RNA sequence (genotype) and RNA structure (phenotype) which are vastly more biologically realistic than traditional mathematical models. Through computational simulations of evolutionary dynamics, these models yield rapid results, yet incorporate significantly greater biological detail than traditional mathematical models. Simulations have been used to study a wide range of evolutionary patterns and processes, such as the evolution of robustness (Ancel & Fontana 2000), the distribution of fitness effects of beneficial mutations (Cowperthwaite et al. 2005, 2006), the causes and implications of neutral evolution (van Nimwegen et al. 1999), evolutionary transitions (Fontana & Schuster 1998a, Huynen et al. 1996), and the structures of fitness landscapes (Schushter et al. 1994).

2. THE MODELS

2.1. History

This modeling framework originates in the work of Manfred Eigen and, later, Peter Schuster (Eigen 1971, Eigen & Schushter 1979). They sought to address origin-of-life questions, and, in particular, develop a general theory for the emergence of biological information and self-replicating life from "molecular chaos." Based on the assumption that early life must have undergone highly error-prone replication, Eigen sought to understand the evolutionary consequences of high mutation rates (Eigen 1971).

Two influential concepts emerged from this work. Eigen & Schuster used mathematical models to demonstrate that the balance between mutation and selection could result in a quasi-species—a population that stably includes not only the wild type (best type) but also suboptimal mutants of that wild type (Eigen & Schushter 1979). At very high mutation rates, a population may, in fact, include only very few wild-type genotypes and many less-fit variants. The quasi-species concept has often been thought to describe an entirely novel set of evolutionary principles. Recently, however, this concept has been shown to be an extension of classic mutation-selection balance theory (Bull et al. 2005, Wilke 2005). The concept has been embraced by virologists who regularly observe that rapidly mutating viral strains may achieve high levels of diversity, yet there is debate as to whether these viruses truly evolve as quasi-species (Domingo 2002, Eigen 1996, Holmes & Moya 2002).

Eigen's second influential concept is the error catastrophe, the genetic meltdown of a population experiencing excessively high mutation rates. He showed mathematically that, under fairly reasonable assumptions, there would be a critical mutation rate below which populations would stably persist as a quasi-species and above which the wild-type and its close mutants would disappear entirely. Based on these ideas, virologists have sought to cure viral infections by using chemical mutagens to induce error catastrophes.

To test these ideas, Eigen encouraged the development of mathematical and computer models of evolving molecular structures (Eigen 1971). He recognized that such biologically grounded and highly detailed models would elucidate evolutionary dynamics at a higher level of resolution than previously possible. Many researchers have taken his charge and developed models of evolving RNA and protein molecules (see Chan & Bornberg-Bauer 2002 and references therein). Here, we focus on RNA-based evolutionary models. We describe the structures of these models, diverse methods for analyzing them, and the resulting insights into evolutionary processes.

2.2. RNA Folding

RNA molecules are composed of four nucleotides—adenine (A), guanine (G), cytosine (C), and uracil (U). Pairs of nucleotides in an RNA molecule can form stable electrostatic interactions, thus holding two parts of a molecule close together. The strength of an interaction varies with the specific combination of nucleotides, and more stable

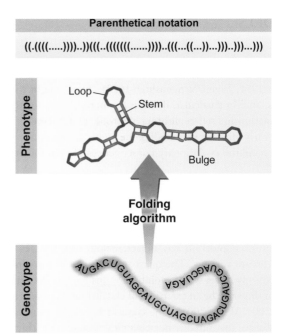

Parenthetical notation

((.((((.....))))..))(((..((((((......))))..(((...((...))...)))..))))...)))

Phenotype

Loop

Stem

Bulge

Folding
algorithm

Genotype

AUGACUGUAGCAUGCUAGCUAGCUAGACUGACUAGCUGUAGCUAGA

Figure 1

Diagram of a genotype-to-phenotype map in the RNA model system. The genotype is the primary nucleotide sequence and the phenotype is the most probable secondary structure (shape). Shape is predicted from sequence using thermodynamic folding algorithms. We label the primary RNA shape motifs. Stems are contiguous stretches of base pairs that include at least two base pairs, loops are unpaired bases that connect the two halves of a stem, and a bulge is an unpaired region in the middle of a stem. Parenthetical notation represents paired bases as matching parentheses and unpaired bases as dots. This notation contains all of the structural information given in the graphical representation.

interactions tend to form at the expense of less stable interactions. Through such pairing, RNA molecules fold into secondary structures (hereafter shapes). The shape of an RNA molecule is composed of combinations of familiar motifs, such as stems (helical base-paired regions) and loops/bulges (unpaired regions) (**Figure 1**).

The shape of an RNA molecule may be vital to its function, particularly for functional RNA molecules (as opposed to protein-coding RNA molecules) like ribosomal RNA, micro-RNA molecules, and ribozymes. In fact, the function of a molecule depends not only on its (two-dimensional) shape, but on its (three-dimensional) tertiary configuration, which includes additional far-reaching pairings between nucleotides already participating in secondary-structure motifs. The formation of tertiary interactions, however, is not particularly well understood. Fortunately, RNA shapes can be predicted with reasonable accuracy, and constitute most of the full structure of a typical molecule (Hofacker et al. 1994; Mathews et al. 2004, 1999; Zuker & Stiegler 1981).

Theoreticians originally developed a set of rules for predicting RNA shape, based on thermodynamic considerations (Waterman 1978). These rules assume that a molecule will fold into the shape that releases the most energy upon formation, and thus is the most stable configuration. This is called the minimum-free-energy (mfe) shape of a molecule. The RNA folding rules are much simpler than the analogous set for proteins, largely because RNA has a smaller set of building blocks (four nucleotides versus twenty amino acids), and generally forms simpler secondary structural motifs. Michael Zuker and colleagues developed the first efficient computer algorithms to predict RNA shape using this approach (Zuker 1989, Zuker & Stiegler 1981). Their software, called mFold, is still actively developed and freely available at **http://www.bioinfo.rpi.edu/applications/mfold/**. More recently, Ivo Hofacker and colleagues have been developing and maintaining the ViennaRNA package, which includes many computational tools for folding and analyzing RNA structures and is freely available from **http://www.tbi.univie.ac.at/~ivo/RNA/** (Hofacker et al. 1994). Researchers are continually improving the accuracy and scope of these folding algorithms. For example, new versions can predict the shapes of RNA molecules during interactions with other molecules (Bernhart et al. 2006, Mathews 2006, Mathews & Turner 2006).

These thermodynamic folding algorithms make several simplifying assumptions. Notably, they cannot predict pseudoknots (a common tertiary motif) or noncanonical base interactions (Hofacker et al. 1994). Researchers have developed comparative-genomics-based approaches that generally yield more accurate predictions of RNA shape, particularly for large RNA molecules (Gutell et al. 2002). The comparative approach, however, is much slower than the thermodynamic approach and requires large sets of homologous sequences to predict the shape of any given sequence. Thus it is not computationally tractable for evolutionary simulations.

2.3. Model Overview

RNA models typically simulate a large population of RNA molecules evolving via mutation and natural selection. The fitness of any given molecule is determined by first predicting its shape(s) and then applying a prespecified fitness function to these predictions (described in detail below). Molecules replicate in proportion to their fitnesses and, upon replication, bases mutate randomly at a prespecified rate. Some versions assume discrete generations (Cowperthwaite et al. 2006), whereas others assume a continuous individual-based birth-death process (Ancel & Fontana 2000, Fontana & Schushter 1998a, Huynen et al. 1996, van Nimwegen et al. 1999).

Analogies can be drawn between the particulars of this model and any other evolutionary system. Each nucleotide is a genetic locus with four possible alleles (A, C, G, or U); interactions among these loci determine the phenotype; and mutations can cause a locus to switch from one allele to another, which, depending on the rest of the molecule, may alter the phenotype. These models do not make many of the assumptions often found in evolutionary models. For instance, fitness stems from a biologically grounded model of molecular folding. Thus the fitness of a given mutant does not come from an assumed probability distribution, but rather is determined organically. The likelihood that a mutation is beneficial or deleterious, and the nature of epistatic (nonadditive) interactions among loci are similarly unconstrained.

2.4. The Genotype-to-Phenotype Map

Phenotypes are produced by manifold interactions between genetic, cellular, organismal, and environmental factors. The term genotype-to-phenotype map refers to this complicated route from genotype to phenotype. The phenotypes of an organism (physiological and behavioral) collectively interact with the environment (including other organisms) to determine fitness. Ultimately, evolutionary biologists aspire to characterize these complex processes and their evolutionary consequences, but these studies have just begun.

The main advantage of RNA folding models is their realistic genotype-to-phenotype map. Unlike many traditional population genetic models, which completely ignore phenotype and assume simple one-to-one maps from genotype to fitness, the phenotypes in the RNA models result from detailed interactions among genes and their microenvironment (Eigen 1971, Fontana & Schushter 1987). In particular, the genotypes are primary nucleotide sequences and the phenotypes are the shapes predicted from these sequences via thermodynamic folding algorithms. Thus the algorithms serve as biologically motivated genotype-to-phenotype maps (Schushter et al. 1994).

The original RNA models consider only the single most-stable (mfe) shape of each molecule (**Figure 2a**). We refer to these as simple models. In reality, however, an RNA molecule may not necessarily fold into its mfe shape, and may even spontaneously switch among several thermodynamically probable shapes. Thus researchers introduced a more complex, but perhaps more biologically realistic, model in which sequences are mapped to the set lowest free energy shapes (**Figure 2b**) (Ancel & Fontana 2000). We refer to these as plastic models because they capture structural plasticity produced by Brownian motion.

2.5. Fitness Functions

As discussed above, fitness is determined in two steps. First the shape(s) of a molecule is predicted using thermodynamic algorithms, and then a fitness value is attained via a function from shapes to real numbers. Here, we use the term fitness function to refer just to this second function from phenotype to fitness and the term fitness landscape to describe the projection of a large set of genotypes (a so-called sequence space) to their ultimate fitness values.

The fitness functions used in RNA models are often based on the similarity of a molecule's shape(s) to a predetermined ideal target shape. Fitness typically decreases monotonically as a function of the distance to the target shape. These models thereby use shape as a proxy for function and do not model function explicitly. This is justified (at least somewhat) by the dominant role typically played by shape in functional tertiary structure and the extreme conservation of shape throughout the evolutionary history of most functional RNA molecules (Doudna 2000).

In simple models that consider only the mfe shape, fitness is determined by the distance between those structures and the target shape. In the plastic models that consider the ensemble of favorable shapes, fitness is determined by the distances between all shapes in the ensemble and the target. In particular, each shape contributes

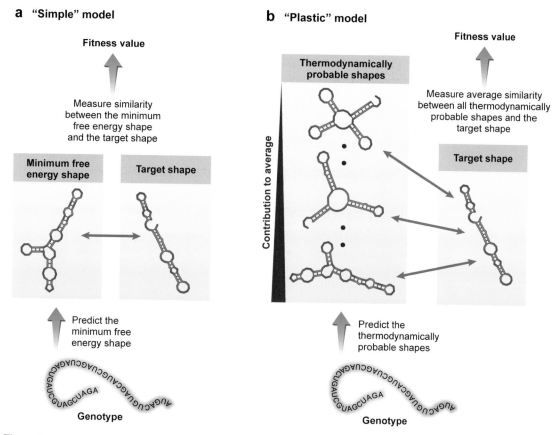

a "Simple" model

Fitness value

Measure similarity
between the minimum
free energy shape
and the target shape

Minimum free
energy shape

Target shape

Predict the
minimum free
energy shape

Genotype

b "Plastic" model

Fitness value

Thermodynamically
probable shapes

Contribution to average

Measure average similarity
between all thermodynamically
probable shapes and the
target shape

Target shape

Predict the
thermodynamically
probable shapes

Genotype

Figure 2

The two fitness models for RNA. (*a*) Under the simple model, the fitness of an RNA genotype
depends only on the similarity of its minimum free energy shape to the target shape. (*b*) Under
the plastic model, the fitness of a molecule is determined by the entire ensemble of probable
(lowest free energy) shapes. The similarity of any given shape to the target contributes to the
final fitness in proportion to its Boltzmann factor, which is an estimate of the thermodynamic
stability of a shape.

to the overall fitness of the molecule in proportion to its thermodynamic likelihood,
which is typically estimated by the Boltzmann coefficient (Ancel & Fontana 2000).
Assuming thermodynamic equilibrium, the Boltzmann coefficient of a shape estimates the fraction of time an RNA molecule spends in that shape and is calculated
using an algorithm developed by McCaskill and coworkers (McCaskill 1990).

There are several methods for quantifying the structural distance between two
shapes (Ancel & Fontana 2000, Stadler et al. 2001). For example, one can represent
each shape in parenthetical notation, where dots stand for unpaired bases and matching parenthesis stand for paired bases (as in **Figure 2**), and then compute a Hamming
distance between two such representations of shapes. Alternatively, the tree-edit
distance measures the differences between the binary-tree representations of two

shapes. Although researchers have used a variety of shape distance metrics, several studies suggest that most observations in RNA models are relatively robust to the specific choice of distance metric (Ancel & Fontana 2000, Fontana & Schushter 1998b).

The form of the fitness function, that is, how exactly fitness declines as distance to the target grows, can profoundly influence the outcome of evolution. One might naively assume that this is linear, such that any unit decrease in similarity to the target shape results in the same loss of fitness. Given that RNA structures are highly evolutionarily conserved, however, it is more likely that fitness declines faster than similarity. That is, even slight deviations from the ideal shape result in substantial loss of function. Many studies have therefore assumed hyperbolic fitness functions (Ancel & Fontana 2000; Cowperthwaite et al. 2005, 2006; Fontana & Schushter 1998a).

3. EVOLUTIONARY INSIGHTS INTO FITNESS LANDSCAPES

Since Sewall Wright introduced fitness landscapes in 1932, the concept has profoundly influenced evolutionary thinking (Wright 1932). Fitness landscapes are maps from large sets of genotypes to their fitnesses. Metaphorically, as populations evolve, they traverse the surfaces of fitness landscapes with mutation and recombination sampling new regions and natural selection pushing uphill. Though fitness landscapes are extremely high dimensional for most real biological systems, they are often illustrated as two-dimensional surfaces in three-dimensional Euclidean space. The structure of a fitness landscape is thought to constrain many micro- and macroevolutionary processes, including the rates of adaptation and speciation (Gavrilets 2004).

With the advent of high-throughput laboratory methodologies and modern computation, researchers are starting to undertake large-scale characterizations of fitness landscapes (Cowperthwaite et al. 2005; Fontana & Schushter 1998b; Gruner et al. 1996a,b; Li et al. 1996; Lunzer et al. 2005; Weinreich et al. 2006). The RNA model system offers the ideal balance of biological complexity and computational tractability for such studies. Some of the earliest and most exciting ideas about fitness landscapes have come out of this body of work (Cowperthwaite et al. 2005; Fontana & Schushter 1998a,b; Gruner et al. 1996a,b; Schushter et al. 1994).

Technically, an RNA fitness landscape is a projection from genotype space—the set of all possible sequences of a given length—to fitness space (often the real numbers). Recall, however, that these models use shape as a proxy for fitness. Consequently, the landscapes that have been characterized are actually maps from sequence space to shape space, where the mapping functions are thermodynamic folding algorithms. All of the RNA landscape studies so far are based on the simple map from sequence to mfe shape (which ignores alternative low free energy structures).

The total number of sequences of a specific length n is 4^n. There is extensive degeneracy in the map from sequences to shapes, with many sequences folding into the same mfe shape, which means the size of the shape space will always be less than the size of the sequence space (Schushter et al. 1994). Waterman first proposed an upper bound for the number of shapes of length n—$S_n = 1.4848 \times n^{-\frac{3}{2}}(1.8488)^n$ based on several assumptions about the nature of the shapes, such as stem length and loop size (Waterman 1978). In the first large-scale computational surveys to estimate the extent of redundancy, Gruner and colleagues folded all 30-nucleotide binary RNA molecules (composed of only A/C or G/U). Approximately one billion unique sequences folded into approximately 220,000 and 1,000 unique shapes in the G/C and A/U landscapes, respectively (Gruner et al. 1996a,b). Evidence for similar degeneracy was found in partial surveys of four-nucleotide RNA landscapes (Fontana & Schushter 1998b, Schushter et al. 1994). Recently, we characterized several complete landscapes for short RNA molecules and found that the number of unique shapes exceeds Waterman's theoretical upper bound, but we did not completely meet the assumptions of Waterman's theory (M.C. Cowperthwaite, E.P. Economo, L.A. Meyers, in review).

A many-to-one relationship between genotypes and phenotypes is not unique to RNA. For instance, there is considerable sequence divergence in 16S rDNA sequences, yet there is extensive functional conservation. As a result, these are key molecules for phylogenetic analysis (Delsuc et al. 2005). Degeneracy has been observed in proteins based on lattice models of protein structure (Chan & Bornberg-Bauer 2002) and is at the heart of the neutral theory of molecular evolution—which asserts that most mutations have negligible phenotypic consequences (Kimura 1968)—and the molecular clock hypothesis (Zuckerkandl & Pauling 1962). As we discuss below, this redundancy profoundly affects the evolutionary dynamics of RNA.

3.1. Mutational Networks

Evolutionary transitions from one phenotype to another are mediated by mutations to their underlying genotypes. Historically, evolutionary biologists have thought of mutations in terms of distributions of fitness effects and have sought to measure the fractions of mutations that are typically beneficial, neutral, and deleterious. Although these distributions are critical determinants of local evolutionary dynamics, they provide little information about larger-scale processes. To this end, it is useful to think in terms of mutational paths connecting distant genotypes and, more generally, in terms of the large-scale patterns of mutational connectivity within genotype spaces.

Specifically, the space of all genotypes can be construed as a mutational network in which each genotype is a node and mutations between genotypes are edges. In other words, any two genotypes that differ by exactly a single point mutation are connected by an edge (**Figure 3**, *bottom*). One can then represent phenotypes (or fitness values) as colors. The coloration in **Figure 3** illustrates the degeneracy in the sequence-shape relationship discussed above. The colored edges represent neutral mutations that preserve the phenotype, and black edges represent non-neutral mutations that may be beneficial or deleterious. RNA mutational networks are regular graphs, that is, each genotype is mutationally connected to exactly $3L$ other genotypes, where L is the sequence length.

3.2. Neutral Networks

Each colored patch in **Figure 3** is a neutral network—a mutationally connected set of genotypes that produces the same phenotype (or fitness value). This concept

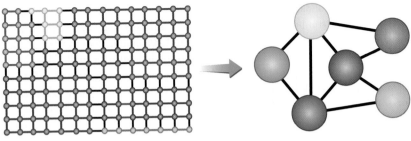

Figure 3

Mutational networks capture patterns of mutational connectivity among genotypes and phenotypes. In the left network, each node is a genotype and each edge is a point mutation. Colors represent phenotypes, and each group of genotypes that share the same color forms a neutral network. The right half shows a phenotype network in which each phenotype is condensed into a single node and two phenotypes are connected by an edge if there is at least one point mutation that converts one phenotype to the other.

originated and has been studied extensively in the RNA model system (Fontana et al. 1993b; Gruner et al. 1996a,b; Huynen et al. 1996; Schushter et al. 1994; van Nimwegen et al. 1999). Following Eigen's quasi-species theory, it is perhaps the most influential idea to emerge from this body of work.

3.2.1. Neutral network structure. Consider a phenotype in a fitness landscape. The structure of its neutral network and its mutational connectivity to the neutral networks of other phenotypes determines the likelihood that it will evolve, and if so, whether it will give rise to other phenotypes. To understand constraints on phenotypic evolution, we must address questions like the following: Are neutral networks confined to small sections of sequence space or do they span the entire space? Do phenotypes have single contiguous neutral networks or several disjoint components? What patterns of adjacency exist between neutral networks for different phenotypes?

The first generality to emerge from neutral network studies is that "not all phenotypes are equal" (Fontana & Schushter 1998b, Schushter et al. 1994). Within an RNA fitness landscape, any given shape may be realized by many or only a few sequences. In other words, the sizes of the neutral networks vary considerably. Henceforth we use the term phenotype abundance to refer to the number of genotypes that map to a particular phenotype. The distributions of phenotype abundances within RNA fitness landscapes have been shown to follow a generalized Zipf's law, a type of semiexponential distribution (Fontana et al. 1993a,b; Schushter et al. 1994; also M.C. Cowperthwaite, E.P. Economo, L.A. Meyers, in review). The critical implication is that most RNA shapes are relatively rare while a few are quite abundant.

The neutral network of a particular phenotype may be composed of a single component or multiple disjoint components (Gruner et al. 1996a,b). A component is a set in which all genotypes are connected by paths of neutral mutations. If a neutral network is comprised of disjoint components, then it contains two or more components that are not connected to each other by neutral mutations. Surprisingly, the number of disjoint components in a phenotype's neutral network does not appear to correlate with its abundance (M.C. Cowperthwaite & L.A. Meyers, unpublished).

The neutral networks of highly abundant phenotypes have been shown typically to span entire fitness landscapes (Fontana et al. 1993b, Schushter et al. 1994). In other words, it is possible to mutate (in succession) every nucleotide in a sequence, all the while preserving its shape. Maynard Smith proposed a similar phenomena in protein fitness landscapes (Maynard Smith 1970). This suggests that neutral networks may facilitate evolution by allowing populations to explore vast expanses of genotype space (via mutation) while maintaining constant fitness (Kirschner & Gerhart 1998, Wagner 2005).

3.2.2. Phenotype networks. As illustrated in **Figure 3**, mutational networks connecting genotypes give rise to mutational networks connecting phenotypes, or phenotype networks. In particular, we aggregate all genotypes that produce a particular phenotype into a single node and connect two phenotypes with an edge if there is at least one point mutation that converts one phenotype to the other. For RNA, we say that two shapes A and B are mutationally adjacent if there exist at least two sequences a and b that differ by exactly one mutation and produce A and B, respectively. Mutationally adjacent shapes are connected by edges in the corresponding phenotype network.

RNA phenotype networks appear to be highly irregular, with few nodes connected to many others and most nodes connected to few others (Schushter et al. 1994, Stadler et al. 2001; also M.C. Cowperthwaite, E.P. Economo, L.A. Meyers, in review). In contrast, classical population genetic models often assume that genotypes map one-to-one onto phenotypes, and that the mutational connectivity among phenotypes is fairly homogeneous. Thus the RNA model system can offer valuable insights into patterns of mutational connectivity and the evolutionary implications of such patterns (Fontana & Schushter 1998a, Huynen et al. 1996; also M.C. Cowperthwaite, E.P. Economo, L.A. Meyers, in review).

One of the first studies to characterize the mutational adjacencies of RNA shapes found that almost any genotype is surrounded by a specific set of highly abundant phenotypes (Schushter et al. 1994). In other words, almost any genotype is within one or a few point mutations of the most common shapes in the landscape; and vice versa, these common shapes are mutationally close to most other phenotypes in the landscape. This hypothesis is called shape-space covering. In phenotype network terms, abundant shapes are connected to almost every other shape. We similarly found a positive correlation between shape abundance and the number of mutationally adjacent shapes for small RNA molecules (M.C. Cowperthwaite, E.P. Economo, L.A. Meyers, in review).

Fontana and colleagues developed a formal theory to describe the genetic accessibility among mutationally adjacent phenotypes and the implications of different mutational structures on evolutionary dynamics (Stadler et al. 2001). Mutationally adjacent shapes are those shapes for which there exists at least one point mutation that can cause a change between those two shapes. These efforts and earlier simulation studies suggest that the degree of mutational connectivity is not simply a binary property (connected or unconnected by point mutations) (Fontana & Schushter 1998a,b; Huynen et al. 1996). Rather some mutationally adjacent phenotypes are nearer to each other than other mutationally adjacent phenotypes, meaning that they are more likely to reach each other via mutation (Fontana & Schushter 1998b). Furthermore, this connectivity is always asymmetrical, resulting from the nonuniform boundaries among adjacent neutral networks (Fontana & Schushter 1998b, Stadler et al. 2001). For example, consider two phenotypes A and B: Asymmetry means that mutating from A frequently produces B, whereas mutating from B does not frequently produce A. In phenotype network terms, this variation in connectivity can be represented as weighted, directed edges between nodes. The weight on an edge pointing from A to B indicates the probability that any given genotype in the neutral network for A will mutate to phenotype B, and, vice versa, the weight on the edge pointing in the opposite direction indicates the fraction of mutations to genotypes in the neutral network for B that produce phenotype A.

3.2.3. Rugged neutral networks: an important caveat.

Most RNA neutral network studies have assumed the simple model in which the fitness of a molecule is determined entirely by its mfe shape. The neutral networks in these studies are simply sets of RNA molecules that fold into the same mfe shape. In reality, however, the fitness of a molecule is determined by other factors, notably the kinetics and energetics of folding. Two molecules that share the same mfe shape may have very different thermodynamic properties and, consequently, different fitnesses. Thus, so-called neutral networks may not truly be neutral.

The plastic model, introduced by Ancel & Fontana (2000), inserts ruggedness into neutral networks. Recall that, in this model, the fitness of an RNA molecule is determined by its entire ensemble of energetically favorable shapes (the specific structures in the ensemble and their relative thermostabilities). Whereas the simple fitness function was discrete (only a finite set of possible values corresponding to a finite set of mfe shapes), the plastic fitness function is continuous (infinite possibilities). In general, any two molecules that share the same mfe shape will have different fitnesses under this model. Ancel & Fontana found that neutral networks have distinct patterns of heterogeneity, with the most thermodynamically stable molecules lying at the dense centers of neutral networks, where most mutations preserve the mfe shape. Thus, if fitness positively correlates with thermodynamic stability, then mfe neutral networks are no longer plateaus but rather mounds that may impede the neutral drift of a population toward alternative phenotypes.

Given that the plastic model is probably more realistic than the simple model, one might be tempted to reject the notion of a neutral network altogether. We argue, however, that the concept remains instructive. The mfe shape is the most likely structure and an important determinant of fitness. Although neutral networks may be more rugged than often assumed, they still contain expansive sets of mutationally connected molecules with roughly similar fitness.

4. EVOLUTIONARY DYNAMICS

4.1. Introduction

Intuitively, the structures of fitness landscapes fundamentally constrain evolution. In this section we review a number of theories linking mutational connectivity to evolutionary dynamics that originated in and/or have been tested using the RNA model system. First we focus specifically on the evolutionary consequences of mutational networks and then turn to more general studies of mutations and their interactions.

4.2. Evolutionary Dynamics on Mutational Networks

Natural selection acts on variation, and thus requires mutations to new phenotypes. The likelihood that a novel mutant phenotype will arise in a population, however, depends on the underlying mutational network and can itself evolve as the population traverses this network.

4.2.1. Evolvability: neutral networks enable evolution.

There is a widely believed claim that neutral networks increase evolvability (Kirschner & Gerhart 1998, Stadler et al. 2001, Wagner 2005). The rationale is that populations evolving on neutral networks may undergo significant genetic change with only negligible phenotypic change, and can thereby explore fitness landscapes. In other words, neutral mutations can accumulate until a genetic background arises that is poised for beneficial change. Under this scenario, neutral mutations will be transient, ultimately facilitating adaptation by subsequent beneficial mutations (Wagner 2005).

Several RNA simulation studies have shown that populations evolving toward a target shape tend to experience long periods of phenotypic stasis, interspersed with short periods of rapid phenotypic change (Ancel & Fontana 2000; Cowperthwaite et al. 2006; Fontana & Schuster 1987, 1998a,b; Huynen et al. 1996). The last of these studies showed that the number of unique sequences in the population increased during periods of phenotypic stasis and used multidimensional scaling to illustrate the genetic dispersal of the population. The population typically subdivides into several genetically different yet phenotypically equivalent subpopulations, each exploring a different region of the fitness landscape via mutation and natural selection.

According to this theory, the more expansive a neutral network, the more likely a population will be able to discover higher fitness phenotypes and thus evolve away from that network. In recent work, however, we systematically asked whether the abundance of a phenotype (the size of its neutral network) increases the likelihood of (*a*) evolving that particular phenotype and/or (*b*) evolving from that phenotype to other new phenotypes (M.C. Cowperthwaite, E.P. Economo, L.A. Meyers, in review). We found that phenotype abundance positively correlates with the number of mutationally adjacent phenotypes, which, on the surface, supports both claims. Yet, in silico simulations suggest that populations evolving on large neutral networks (of abundant phenotypes) did not adapt more quickly than those evolving smaller neutral networks. This stems from the fact that, as the abundance of a phenotype increases, the probability of locating adjacent phenotypes rapidly diminishes (M.C. Cowperthwaite,

E.P. Economo, L.A. Meyers, in review). Populations that evolve abundant phenotypes may therefore face a "needle in the haystack" problem and be unlikely to further adapt even if superior phenotypes are just a single mutation away.

The size of a phenotype's neutral network does, however, affect the probability that the phenotype will arise (M.C. Cowperthwaite, E.P. Economo, L.A. Meyers, in review). Specifically, our simulated populations were more likely to evolve to abundant phenotypes than rare phenotypes. Thus, the structure of RNA mutational networks may bias evolution towards abundant shapes, whether or not those shapes are optimal.

In the same study, we turned to real RNA molecules to test this provocative hypothesis (M.C. Cowperthwaite, E.P. Economo, L.A. Meyers, in review). Unfortunately, we are far from having the computational resources necessary to characterize entire fitness landscapes for large molecules. We thus developed a new statistical shortcut for estimating shape abundance: the "contiguity statistic" measures the cohesiveness of a shape and significantly correlates with abundance (based on an exhaustive folding of all molecules of length 12 through 18). **Figure 4** details the calculation of the contiguity statistic. By calculating the contiguity statistic for thousands of naturally occurring functional RNA molecules in Rfam—a curated database of functional RNA genes (Griffiths-Jones et al. 2005)—it was found that natural phenotypes, indeed, have

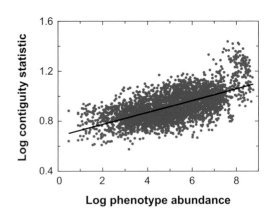

a

$$C = \frac{\text{Total stem-loop length} + \text{number of paired bases}}{\text{Number of contiguous stacks}}$$

Shape	Contiguity statistic	Abundance
	$C = \dfrac{11 + 6}{2} = 8.5$	2,260
	$C = \dfrac{12 + 8}{1} = 20$	117,213

b

Log phenotype abundance

Log contiguity statistic

Figure 4

The contiguity statistic was developed to estimate phenotype abundance for large RNA molecules. This statistic came out of a study in which we folded all molecules of length 12 through 18 and directly measured abundances of all unique shapes (M.C. Cowperthwaite, E.P. Economo, L.A. Meyers, in review). (*b*) The contiguity statistic formula (*C*) captures the cohesiveness of the shape. We calculate this statistic on two simple shapes of length 12 and give the abundance of each shape. (*b*) The contiguity statistic strongly correlates with phenotype abundance for RNA molecules of length 18 ($R \approx 0.80$; $P < 2 \times 10^{-16}$). The graph shows the abundances and contiguity statistics for all 3211 unique shapes realized by molecules of this length. This correlation is equally strong for molecules of length 12 through 18.

significantly higher contiguity values (and thus higher abundance, perhaps) than expected for molecules of similar length and base composition (M.C. Cowperthwaite, E.P. Economo, L.A. Meyers, in review).

RNA molecules may therefore be constrained by both functionality and mutational accessibility, a phenomenon we termed ascent of the abundant (M.C. Cowperthwaite, E.P. Economo, L.A. Meyers, in review). This suggests not only that RNA shapes (and other phenotypes) may be suboptimal, but also that evolution may be more repeatable and predictable than previously thought by virtue of underlying mutational constraints.

4.2.2. Punctuated equilibria: crossing from one neutral network to the next.

One striking feature of the fossil record is the extensive discontinuity in forms (Eldredge et al. 2005), that is, periods of rapid phenotypic change are often separated by longer periods of relative stability. Although this may stem partly from observational biases (Eldredge et al. 2005), punctuated equilibria have also been observed in RNA models (Ancel & Fontana 2000, Cowperthwaite et al. 2006; Fontana & Schushter 1987, 1998a,b; Huynen et al. 1996), protein models (Chan & Bornberg-Bauer 2002), digital organisms (Wilke et al. 2001) and microorganisms (Burch & Chao 1999).

Figure 5 shows a typical simulation of RNA molecules evolving toward a target shape. As described earlier, populations disperse through neutral networks during the

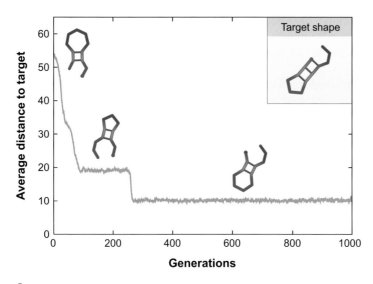

Figure 5

Typical evolutionary dynamics in the RNA model system. Evolving populations experience relatively long epochs of phenotypic stasis interspersed with short periods of rapid phenotypic change. This figure is based on a simulation of a population containing 500 RNA molecules, in which selection favors molecules that resemble the target shape (*upper right*). The Y-axis gives the average phenotypic distance of the population to the target shape, and thus low values correspond to high fitness. Shapes that dominate the population are depicted above the curve.

long periods of stasis. Fontana and colleagues set out to characterize the evolutionary transitions between these epochs (Fontana & Schushter 1998a). They claimed that there were two types of transitions—continuous and discontinuous—and proposed a simple criterion to distinguish them (Fontana & Schushter 1998b, Stadler et al. 2001). Recall that phenotypes differ greatly in their nearness, and a phenotype is said to be near any other phenotype that is likely to be produced by mutation. Continuous transitions are those that involve nearby phenotypes and discontinuous transitions are those that involve phenotypes that are relatively distant (unlikely to be realized by a single mutation). This study reconstructed the steps leading to each major transition. The initial period of rapid adaptation in the simulations occurred primarily through continuous phenotypic transitions; however, the transitions taking place during the subsequent punctuated dynamics were predominantly discontinuous. Thus, major adaptations are hypothesized to occur through fairly improbable jumps between barely adjacent neutral networks.

These jumps are thought to be mediated by extensive neutral drift (Fontana & Schushter 1998a, Huynen et al. 1996). Genotypes that produce one phenotype but are converted to a very different (but better) phenotype by a single mutation must precede these jumps. Such genotypes are likely to be very rare, and may only appear after long periods of evolutionary wandering through neutral networks (Fontana & Schushter 1998a, Schushter & Fontana 1999).

In a related in vitro RNA study, researchers synthesized a single RNA sequence that assumes two entirely different phenotypes, each of which catalyzes a distinct ribozyme reaction (Joyce 2000, Schultes & Bartel 2000). By making relatively few mutational changes to this sequence, these researchers could produce new ribozymes that were highly active for one or the other ribozyme reaction. Thus this single sequence lies at the intersection of the two neutral networks for each function. Schultes & Bartel (2000) suggest that intersection sequences (those that realize both phenotypes) may mediate discontinuous transitions between phenotypes.

4.2.3. Genetic robustness: evolving to the heart of a neutral network. Organisms exist in an ever-changing world. They must evolve to withstand heterogeneous conditions, which include both environmental and genetic perturbations (Meyers & Bull 2002). Evolutionary biologists seek to identify the mechanisms for achieving environmental and genetic robustness as well as the evolutionary origins of those mechanisms.

Genotypes are genetically robust when mutations (or recombination) leave the resulting phenotype unchanged. In mutational network terms, genetically robust genotypes lie in the "dense" regions of neutral networks, where most mutations are likely to create genotypes within the same neutral network. In **Figure 3**, a genotype in the middle of a colored region would be completely robust because all of its mutations are neutral.

Although it is easy to envision natural selection favoring organisms that can cope with environmental variation (Meyers & Bull 2002), the origins of genetic robustness are less intuitive (de Visser et al. 2003). Because a deleterious germ-line mutation does not manifest itself until the next generation, there is no immediate natural selection to prevent it. Under certain circumstances, however, natural selection can act over several generations to reduce the burden of such mutations (de Visser et al. 2003, van Nimwegen et al. 1999). There are several other theories for the origins of genetic robustness, some of which are nonevolutionary (de Visser et al. 2003, Gibson & Wagner 2000).

This discussion goes back to the founders of the modern synthesis—Haldane, Fisher, and Wright—who offered different theories for the evolution of dominance. Dominance is a simple mechanism for robustness by which potentially deleterious mutations at a diploid locus are silenced by the dominant allele. Evolutionary biologists have focused on three scenarios that could give rise to genetic robustness: (*a*) adaptive robustness—robustness evolves by natural selection, (*b*) intrinsic robustness—robustness is a correlated byproduct of character selection, and (*c*) congruent robustness—genetic robustness is a correlated byproduct of selection for environmental robustness (de Visser et al. 2003). These mechanisms are not necessarily mutually exclusive.

Natural RNA molecules and RNA viruses appear to be both environmentally (thermodynamically) and genetically robust (Meyers et al. 2004; Sanjuán et al. 2006a,b; Wagner & Stadler 1999). Studies using the RNA model system have contributed significantly to our understanding of genetic robustness, particularly scenarios *a* and *c* above. For scenario *a*, van Nimwegen and colleagues developed an elegant mathematical model to show that the trans-generational costs of deleterious mutations are enough to drive populations into the hearts of neutral networks, in other words, that adaptive robustness is possible (van Nimwegen et al. 1999). In particular, this model considers a population evolving on an arbitrary neutral network and assumes that all mutations off the network are lethal. They successfully tested the predictions of their model using RNA simulations. Genetic robustness only evolved in these models, however, under relatively high mutation rates.

Turning to scenario *c*, Wagner was the first to hypothesize that genetic robustness may evolve as a by-product of selection for environmental robustness (Wagner et al. 1997). The first semiempirical support for this hypothesis came somewhat accidentally from an RNA study (Ancel & Fontana 2000). Microenvironmental thermal fluctuations can cause an RNA molecule to wiggle between alternative low free energy shapes. An environmentally robust molecule is one that will fold rapidly and reliably into its optimal shape despite these fluctuations.

To study the evolution of environmental robustness, Ancel & Fontana introduced the plastic model, which maps sequences to their ensembles of thermodynamically favorable shapes (described above). Selection for stable folding into a target shape indeed yielded populations of highly stable (environmentally robust) molecules. Surprisingly, the dominant shapes in the evolved populations looked nothing like the target shape. This was in dramatic contrast to natural selection under the simple (mfe shape) model, which almost always led populations to the target shape.

Why did selection for environmental robustness drive populations into apparent evolutionary dead ends? The evolved populations were also highly genetically robust, to the extent that mutations almost never produced phenotypic novelty, thus precluding further adaptation. The researchers eventually connected the dots when they discovered a correlation between the alternate shapes that a molecule produces under thermodynamic noise and the shapes it produces upon mutation. They called this general property of the map from genotype-to-phenotype "plastogenetic congruence" (Ancel & Fontana 2000). As a consequence, molecules that are insensitive to thermal noise are also insensitive to the effects of mutation. A similar correlation has been observed for proteins (Bornberg-Bauer & Chan 1999, Bussmaker et al. 1997, Vendruscolo et al. 1997). Extreme genetic robustness, to the point of an evolutionary standstill, thus evolved simply as a byproduct of environmental robustness.

Ancel & Fontana's study has other evolutionary implications. First, plastogenetic congruence may extend beyond biopolymers and be a general feature of genotype-to-phenotype maps. Phenocopies—epigenetic mimics of genetically based phenotypes—provide anecdotal evidence for plastogenetic congruence in other complex phenotypes (Queitsch et al. 2002; Rutherford & Lindquist 1998; True et al. 2004; Waddington 1950, 1959). This may shed new light on Waddington's theory of developmental canalization from the 1950s (Waddington 1950, 1959). He was among the first to argue that organisms have evolved developmental pathways that are robust to both environmental and genetic perturbations, and thus produce standard phenotypes in the face of variable environments and mutation. He does not, however, claim that these two forms of robustness share a common evolutionary origin. If plastogenetic congruence holds for organismal phenotypes, then Ancel & Fontana's study suggests that genetic canalization may arise as a byproduct of environmental canalization.

Second, the extremely robust molecules found at the end of the evolutionary simulations were also extremely modular (Ancel & Fontana 2000). They can be easily partitioned into structural subunits that withstand thermodynamic perturbations or genetic changes elsewhere in the molecule. Modularity, as it shifts the syntax of genetic variation, opens new avenues for phenotypic innovation. Though this advantage is compelling, it does not explain the origins of modularity in the first place. We have a chicken-and-egg predicament: Until both the modules themselves and recombinational mechanisms are in place, it is not clear that natural selection would favor such organization. The RNA study suggests an origin of modularity that does not rely on the eventual evolutionary benefits modularity might provide. In particular, it arises as a (second) byproduct of selection for environmental robustness. Consider a rough analogy between RNA folding and organismal development. Interactions between nucleotides influence the kinetic pathway of the molecule and its robustness to both the environment and mutations. Similarly, interactions between genes determine the outcome and stability of developmental pathways. Perhaps natural selection for environmental stability similarly sets the stage for modularity in genetic networks.

4.2.4. Survival of the flattest: quasi-species and error thresholds in complex mutational networks. Recall that populations evolving under moderate mutation rates can form quasi-species—mutational clouds around a wild-type (optimal) genotype (Eigen 1971). Quasi-species have been observed in simulated populations of evolving RNA (Ancel & Fontana 2000), proteins (Wilke et al. 2001), and digital organisms (Wilke et al. 2001). Many RNA viruses are believed to exist as quasi-species, though there has been considerable debate over the utility of the term (Holmes & Moya 2002, Moya et al. 2000, Wilke 2005).

Recall further that error catastrophes occur when mutation swamps selection and a population is unable to maintain the wild type or its close relatives. Eigen originally discovered the error threshold (the critical mutation rate above which error catastrophes occur) in a model that assumes there is a single wild-type genotype and all other genotypes have identical significantly lower fitnesses (Eigen & Schushter 1979). What happens when the wild-type phenotype is produced by an entire neutral network of genotypes and not just one? Roughly speaking, an error threshold still exists, but it increases with the breadth of the neutral network, that is, the number of and mutational connectivity among genotypes contained within it. The larger and more connected the neutral network, the more likely a mutation will preserve the wild-type phenotype.

Similar reasoning suggests that neutral network breadth may influence the likelihood that a population will evolve one phenotype versus another. Imagine a population evolving in a complex mutational network where the topologies of neutral networks vary considerably among phenotypes. Under high mutation rates, phenotypes that have high fitness but small neutral networks may be easily displaced by less fit but more robust phenotypes. The extent to which neutral networks influence such competition among phenotypes depends on the mutation rate. Under very low mutation rates, fitness considerations alone dictate dynamics, whereas under high mutation rates, the breadth of neutral networks can be as or more important than fitness. This hypothesis has been called "survival of the flattest" (Wilke et al. 2001) and is a natural extension of Eigen's theory.

Survival of the flattest has been developed and tested in a series of mathematical models and simulations of evolving RNA and digital organisms (Bull et al. 2005, Wilke et al. 2001). In the Wilke et al. study, populations of digital organisms were evolved under two distinct mutation rates (high and low). When subsequently placed in competition under high mutation rates, populations that originally evolved under high mutation rates out-competed those that evolved under low mutation rates even though they had lower fitnesses. More recently, a plant virus competition experiment has suggested that similar tradeoffs may hold for plant viral pathogens (Codõner et al. 2006).

Although virologists have latched onto these ideas and harnessed them to develop effective antiviral strategies (Domingo 2003), Bull and colleagues have suggested that the theory may be widely misinterpreted (Bull et al. 2005). In particular, they distinguish between error catastrophes, in which high mutation rates lead to the complete loss of the wild type in favor of suboptimal genotypes, and extinction catastrophes, in which lethal mutations are so common that no viable genotype can persist. The use of mutation-inducing drugs may not drive viral populations toward error catastrophes as has been claimed (reviewed in Anderson et al. 2004) but rather toward extinction catastrophes.

4.3. The Mutational Spectra of RNA

The phenotypic effects of mutations determine the rate and outcome of evolution. Evolutionary biologists have thus sought to characterize the distributions of fitness effects based on theoretical considerations (Gillespie 1984; Orr 2002, 2003) as well as laboratory mutation accumulation, knockout, and mutagenesis experiments (Estes et al.2004, Sanjuán et al. 2004, Rosen et al. 2002, Imhof & Schlotterer 2001). The RNA model system offers a pseudoexperimental compromise approach to estimating these distributions. It is more biologically grounded than the theoretical models yet yields vastly more information than experimental approaches. Here we review a series of RNA studies that offer new perspectives on local mutational structure, as opposed to global properties of entire mutational networks.

4.3.1. Beneficial fitness effects: many small mutations and few large ones. Beneficial mutations are those that increase the fitness of individuals carrying them, and are the fuel of adaptation. Somewhat counterintuitively, recent theoretical work suggests that distributions of beneficial fitness effects are similar for many fitness landscapes (Gillespie 1984, Orr 2003). This theory is based on Gillespie's mutational landscape model, which considers a high fitness wild type that has just experienced a minor environmental change (Gillespie 1984). The model assumes that the environmental perturbation was small, and thus the wild-type genotype remains reasonably fit, such that fit genotypes are rare in the fitness landscape and that the fitness of any given mutant is chosen at random from the distribution of all fitnesses. Gillespie claimed that the distribution of beneficial mutations could be predicted using extreme value theory (EVT), and Orr subsequently derived the shape of this distribution (Orr 2003). EVT states that, for a large class of common distributions, the differences between the top few values in a large random sample will be exponentially distributed. According to Gillespie's assumptions, the wild type would be among the largest values in a random sample from the distribution of all fitnesses and thus the fitness effects of any beneficial mutations would fall within the purview of EVT (Gillespie 1984). Orr concluded that the fitness effects of beneficial mutations should therefore be exponentially distributed regardless of biological system (Orr 2003).

Several groups have attempted to test this hypothesis experimentally, with most offering mixed support of the Orr-Gillespie theory (Rokyta et al. 2005, Sanjuán et al. 2004, Imhof & Schlotterer 2001). The most comprehensive of these studies used the RNA virus ϕX174 and supported a modified version of the model that incorporated a mutation bias, which could account for the higher frequency of transitions than transversions (Rokyta et al. 2005). Another study in vesicular stomatitis virus (VSV), however, measured beneficial fitness effects that did not appear to be exponentially distributed (Sanjuán et al. 2004).

Recently, the Orr-Gillespie theory was tested in the RNA model system (Cowperthwaite et al. 2005). First, the researchers randomly chose two large sets of sequences and measured the fitness effects of every possible point mutation to each sequence in the set. These sets of genotypes differed in their average fitness— one set had relatively low fitness and the other set had relatively high fitness. The distributions of beneficial fitness effects in both low and high fitness regions of the landscape were decidedly nonexponential. There was a significant overabundance of small-effect mutations; and the distribution appeared exponential only upon truncation of the lower 99% of it.

The discrepancy between the theory and the RNA study rests on a fairly unbiological assumption of the Orr-Gillespie model—that the fitness of any given mutant is essentially a random draw from the distribution of all fitnesses (Cowperthwaite et al. 2005). Intuitively, the fitnesses of mutants are often highly correlated to the fitnesses of their parents, as has been demonstrated in RNA and proteins (Atchley et al. 2000, Fontana et al. 1993b, Parsch et al. 2000). The RNA study suggests that a predictive theory of beneficial fitness effects must consider fitness correlations. Orr recently extended his mathematical analysis to consider fitness correlations, and found that EVT does indeed break down under extreme correlations (Orr 2006).

4.3.2. Epistasis: mutational effects vary with genetic background.
The RNA model system determines fitness from first principles of molecular folding. The shapes of molecules arise out of complex thermodynamic interactions among the nucleotides in the primary sequence. The contribution of any particular nucleotide to the shape (and thus fitness) of the molecule often intricately depends on the nucleotides at several other sites. For example, see **Figure 6**. Epistasis—when the action of one

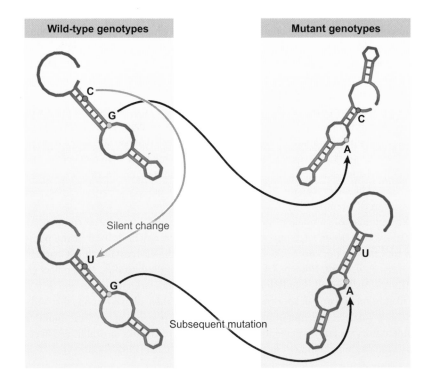

Wild-type genotypes **Mutant genotypes**

Silent change

Subsequent mutation

Figure 6

Epistasis in RNA results when the phenotypic effects of mutations depend on the surrounding nucleotides. The two molecules on the left differ at one position (*red*) but fold into the same shape. Mutations at the same site in each of these molecules (*yellow*) produce very different shapes. Thus, through epistasis, a silent change in background (*gray arrow*) dramatically influences the fitness effect of the subsequent mutation (*black arrows*).

gene is modified by one or more other genes—is thus a ubiquitous property of RNA fitness landscapes.

The presence, magnitude, and direction of epistasis are key inputs to many evolutionary theories, including those that seek to explain the evolution and maintenance of sexual reproduction and the rate of adaptation in asexual organisms (Peters & Otto 2003, Whitlock et al. 1995). Epistatic interactions are often divided into two classes: (*a*) antagonistic epistasis, which occurs when simultaneous mutations at interacting sites yield a smaller fitness effect than the sum (or product) of their individual effects, and (*b*) synergistic epistasis, which occurs when the combined effect of the mutations is greater than the sum (or product) of their individual effects. A third form of epistasis has recently appeared in the literature: sign epistasis, which occurs when the direction of a fitness effect (deleterious or beneficial) is reversed by interactions with other mutations (Weinreich & Chao 2005). One study in the RNA system suggests that most interactions are antagonistic (Wilke et al. 2003). In particular, starting from a high fitness genotype, as deleterious mutations accumulate, the rate of fitness decline decreases, regardless of the order of those mutations.

4.3.3. Compensatory evolution.
Although beneficial mutations are essential for evolution, it is more likely that mutations entering a population will be neutral or deleterious. There is well-developed evolutionary theory that predicts the fates of deleterious mutations in evolving populations (Crow & Kimura 1970, Gillespie 2004). Deleterious mutations are likely to be eliminated by natural selection, but can occasionally reach fixation by chance (drift) alone, particularly in small populations, or by hitchhiking along with beneficial mutations elsewhere in the genome (Johnson & Barton 2002, Kim & Stephan 2000, Peck 1994). A recent RNA study has shown that, under high mutation rates, a third process, compensatory evolution, may lead to the fixation of deleterious mutations much more frequently than either of these other well-studied processes (Cowperthwaite et al. 2006).

Consider a new deleterious mutation. It is possible that, when combined with a subsequent mutation, the original mutation becomes less deleterious, or even beneficial. For example, a mutation to a paired base may break that pairing, to the detriment of the molecule. A subsequent mutation at the matching site may recover that pairing, or perhaps even strengthen (or weaken) the interaction, to the benefit of the molecule. The latter scenario is an example of compensatory evolution through sign epistasis in RNA molecules.

Prior studies of compensatory evolution have focused primarily on compensatory mutations that occur after initially deleterious mutations have fixed in the population, and thus do not contribute the fixation events themselves (Burch & Chao 1999, Escarmis et al. 1999, Poon & Chao 2005). In one of these studies, researchers grew an RNA virus at small population sizes to increase the strength of genetic drift and the likelihood of fixing deleterious mutations. They then allowed strains that had experienced a deleterious mutation to evolve at larger population sizes and found that compensatory mutations generally afforded modest recoveries in viral fitness in comparison to the initial deleterious mutation (Burch & Chao 1999). A later study found that compensatory evolution mediated fitness recoveries in roughly three-quarters of populations in which deleterious mutations fixed (Poon & Chao 2005). There is further evidence for compensatory evolution across many natural and model systems (Poon et al. 2005).

Compensatory evolution may occur prior to fixation of the initial deleterious mutation and, consequently, make fixation of the mutation more likely. As recently illustrated in the RNA model system, this is common under relatively high mutation rates (like those found in RNA viruses) (Cowperthwaite et al. 2006). In evolutionary simulations, initially deleterious mutations fixed far more frequently than was expected by drift alone. Initially harmful mutations interacted with subsequent mutations to increase fitness beyond that of the ancestor and, thus, brought about fitness reversals. Such compensatory events explained as many as 70% of the deleterious fixation events.

Comparative genomic studies have identified possible fixed deleterious mutations in insect and human genomes (Kondrashov et al. 2002, Kulathinal et al. 2004). These observations must be interpreted with caution, however, because the order in which the mutations entered the genome is unknown, and currently deleterious mutations may not have been so when they first appeared. Nonetheless, these studies highlight the complicated nature of mutational interactions and suggest that deleterious mutations may be more than just temporary nuisances. Metaphorically speaking, they may provide stepping stones to distant adaptive peaks.

5. CONCLUSION

In the past two decades, a new generation of computationally intensive and biologically grounded models have changed our perspectives on evolutionary dynamics. We now have a more global understanding of mutational relationships and how they constrain evolution. Here we have reviewed a class of models that have been particularly fruitful. Detailed simulations of evolving RNA structures have inspired general predictive theories about the nature of adaptation, the determinants of evolvability, the origins and mechanisms of robustness, and more. As volumes of biological data accumulate and computational power grows, these models will improve and continue to enrich comprehension of the natural world.

DISCLOSURE STATEMENT

The authors are not aware of any biases that might be perceived as affecting the objectivity of this review.

ACKNOWLEDGMENTS

This work was supported by a grant from the James S. McDonnell Foundation to L.A.M. and an NSF IGERT graduate training grant in Computational Phylogenetics and Applications to Biology to M.C.C.

LITERATURE CITED

Ancel L, Fontana W. 2000. Plasticity, modularity and evolvability in RNA. *J. Exp. Zool.* 288:242–83

Anderson JP, Daifuku R, Loeb LA. 2004. Viral error catastrophe by mutagenic nucleosides. *Annu. Rev. Microbiol.* 58:183–205

Atchley WR, Wollenberg KR, Fitch WM, Terhalle W, Dress AW. 2000. Correlations among amino acid sites in bHLH protein domains: an information theoretic analysis. *Mol. Biol. Evol.* 17:164–78

Bernhart S, Tafer H, Muckstein U, Flamm C, Stadler P, Hofacker I. 2006. Partition function and base pairing probabilities of RNA heterodimers. *Alg. Mol. Biol.* 1:3

Bornberg-Bauer E, Chan HS. 1999. Modeling evolutionary landscapes: mutational stability, topology, and superfunnels in sequence space. *Proc. Natl. Acad. Sci. USA* 96:10689–94

Bull JJ, Meyers LA, Lachmann M. 2005. Quasispecies made simple. *PLoS Comp. Biol.* 1:450–60

Burch CL, Chao L. 1999. Evolution by small steps and rugged landscapes in the RNA virus ø6. *Genetics* 151:921–27

Bussemaker HJ, Thirumalai D, Bhattacharjee JK. 1997. Thermodynamic stability of folded proteins against mutations. *Phys. Rev. Lett.* 79:3530–33

Chan HS, Bornberg-Bauer E. 2002. Perspectives on protein evolution from simple exact models. *Appl. Bioinf.* 1:121–44

Codoñer FM, Darós JA, Solé RV, Elena SF. 2006. The fittest versus the flattest: experimental confirmation of the quasispecies effect with subviral pathogens. *PLoS Pathogens* 2:1187–93

Cowperthwaite MC, Bull JJ, Meyers LA. 2005. Distributions of beneficial fitness effects in RNA. *Genetics* 170:1449–57

Cowperthwaite MC, Bull JJ, Meyers LA. 2006. From bad to good: fitness reversals and the ascent of deleterious mutations. *PLoS Comp. Biol.* 2:1292–300

Crow JF, Kimura M. 1970. *An Introduction to Population Genetics Theory*. New York: Harper & Row

de Visser JAGM, Hermisson J, Wagner GP, Meyers LA, Bagheri-Chaichian H, et al. 2003. Perspective: evolution and detection of genetic robustness. *Evolution* 57:1959–72

Delsuc F, Brinkmann H, Philippe H. 2005. Phylogenomics and the reconstruction of the tree of life. *Nat. Rev. Genet.* 6:361–75

Domingo E. 2002. Quasispecies theory in virology. *J. Vir.* 76:463–65

Domingo E. 2003. Quasispecies and the development of new antiviral strategies. *Prog. Drug. Res.* 60:133–58

Doudna JA. 2000. Structural genomics of RNA. *Nat. Struct. Biol.* 7:954–56

Eigen M. 1971. Self-organization of matter and the evolution of biological macromolecules. *Naturwissenschaften* 58:465–523

Eigen M. 1996. On the nature of virus quasispecies. *Trends Microbiol.* 4:216–18

Eigen M, Schuster P. 1979. *The Hypercycle: A Principle of Natural Self-Organization*. Berlin: Springer-Verlag

Eldredge N, Thompson JN, Brakefield PM, Gavrilets S, Jablonski D, et al. 2005. The dynamics of evolutionary stasis. *Paleobiology* 31:133–45

Escarmis C, Davila M, Domingo E. 1999. Multiple molecular pathways for fitness recovery of an RNA virus debilitated by operation of Muller's ratchet. *J. Mol. Biol.* 285:495–505

Estes S, Phillips PC, Denver DR, Thomas WK, Lynch M. 2004. Mutation accumulation in populations of varying size: the distribution of mutational effects for fitness correlates in *Caenorhabditis elegans*. *Genetics* 166:1269–79

Fontana W, Konings DA, Stadler PF, Schuster P. 1993a. Statistics of RNA secondary structures. *Biopolymers* 33:1389–404

Fontana W, Schuster P. 1987. A computer model of evolutionary optimization. *Biophys. Chem.* 26:123–47

Fontana W, Schuster P. 1998a. Continuity in evolution: on the nature of transitions. *Science* 280:1451–55

Fontana W, Schuster P. 1998b. Shaping space: the possible and the attainable in RNA genotype-phenotype mapping. *J. Theor. Biol.* 194:491–515

Fontana W, Stadler PF, Bornberg-Bauer E, Griesmacher T, Hofacker IL, et al. 1993b. RNA folding and combinatory landscapes. *Phys. Rev. E* 47:2083–99

Gavrilets S. 2004. *Fitness Landscapes and the Origin of Species*, Vol. 41, *Monographs in Population Biology*. Princeton, NJ: Princeton Univ. Press

Gibson G, Wagner GP. 2000. Canalization in evolutionary genetics: a stabilizing theory? *BioEssays* 22:372–80

Gillespie JH. 1984. Molecular evolution over the mutational landscape. *Evolution* 38:1116–29

Gillespie JH. 2004. *Population Genetics: A Concise Guide*. Baltimore, MD: Johns Hopkins Univ. Press

Griffiths-Jones S, Moxon S, Marshall M, Khanna A, Eddy SR, Bateman A. 2005. Rfam: annotating noncoding RNAs in complete genomes. *Nucleic. Acids Res.* 33:D121–24

Gruner W, Giegerich U, Strothmann D, Reidys C, Weber J, et al. 1996a. Analysis of RNA sequence structure maps by exhaustive enumeration. I. Neutral networks. *Monatsh. Chem.* 127:355–74

Gruner W, Giegerich U, Strothmann D, Reidys C, Weber J, et al. 1996b. Analysis of RNA sequence structure maps by exhaustive enumeration. II. Structure of neutral networks and shape space covering. *Monatsh. Chem.* 127:375–89

Gutell RR, Lee JC, Cannone JJ. 2002. The accuracy of ribosomal RNA comparative structure models. *Curr. Opin. Struct. Biol.* 12:301–10

Hofacker IL, Fontana W, Stadler PF, Bonhoeffer LS, Tacker M, Schuster P. 1994. Fast folding and comparison of RNA secondary structures. *Montaschr. Chem.* 125:167–88

Holmes EC, Moya A. 2002. Is the quasispecies concept relevant to RNA viruses? *J. Vir.* 76:460–62

Huynen MA, Stadler PF, Fontana W. 1996. Smoothness within ruggedness: the role of neutrality in adaptation. *Proc. Natl. Acad. Sci. USA* 93:397–401

Imhof M, Schlotterer C. 2001. Fitness effects of advantageous mutations in evolving *Escherichia coli* populations. *Proc. Natl. Acad. Sci. USA* 98:1113–17

Johnson T, Barton NH. 2002. The effect of deleterious alleles on adaptation in asexual populations. *Genetics* 162:395–411

Joyce GF. 2000. RNA structure: ribozyme evolution at the crossroads. *Science* 289:401–2

Kim Y, Stephan W. 2000. Joint effects of genetic hitchhiking and background selection on neutral variation. *Genetics* 155:1415–27

Kimura M. 1968. Evolutionary rate at the molecular level. *Nature* 217:624–26

Kirschner M, Gerhart J. 1998. Evolvability. *Proc. Natl. Acad. Sci. USA* 95:8420–27

Kondrashov AS, Sunyaev S, Kondrashov FA. 2002. Dobzhansky-Muller incompatibilities in protein evolution. *Proc. Natl. Acad. Sci. USA* 99:14878–83

Kulathinal RJ, Bettencourt BR, Hartl DL. 2004. Compensated deleterious mutations in insect genomes. *Science* 306:1553–54

Li H, Helling R, Tang C, Wingreen N. 1996. Emergence of preferred structures in a simple model of protein folding. *Science* 273:666–69

Lunzer M, Miller SP, Felsheim R, Dean AM. 2005. The biochemical architecture of an ancient adaptive landscape. *Science* 310:499–501

Maynard Smith JM. 1970. Natural selection and the concept of a protein space. *Nature* 225:563–64

Mathews DH. 2006. Revolutions in RNA secondary structure prediction. *J. Mol. Biol.* 359:526–32

Mathews DH, Disney MD, Childs JL, Schroeder SJ, Zuker M, Turner DH. 2004. Incorporating chemical modification constraints into a dynamic programming algorithm for prediction of RNA secondary structure. *Proc. Natl. Acad. Sci. USA* 101:7287–92

Mathews DH, Sabina J, Zuker M, Turner DH. 1999. Expanded sequence dependence of thermodynamic parameters improves prediction of RNA secondary structure. *J. Mol. Biol.* 288:911–40

Mathews DH, Turner DH. 2006. Prediction of RNA secondary structure by free energy minimization. *Curr. Opin. Struct. Biol.* 16:270–78

Mattick JS, Makunin IV. 2006. Non-coding RNA. *Hum. Mol. Genet.* 15:R17–29

McCaskill J. 1990. The equilibrium partition function and base pair binding probabilities for RNA secondary structure. *Biopolymers* 29:1105–9

Meyers LA, Bull JJ. 2002. Fighting change with change: adaptive variation in an uncertain world. *Trends Ecol. Evol.* 17:551–57

Meyers LA, Lee JF, Cowperthwaite M, Ellington AD. 2004. The robustness of naturally and artificially selected nucleic acid secondary structures. *J. Mol. Evol.* 58:618–25

Moya A, Elena SF, Bracho A, Miralles R, Barrio E. 2000. The evolution of RNA viruses: a population genetics view. *Proc. Natl. Acad. Sci. USA* 97:6967–73

Niwa R, Slack F. 2007. The evolution of animal microRNA function. *Curr. Opin. Genet. Dev.* 17:145–50

Orr HA. 2002. The population genetics of adaptation: the adaptation of DNA sequences. *Evolution* 56:1317–30

Orr HA. 2003. The distribution of fitness effects among beneficial mutations. *Genetics* 163:1519–26

Orr HA. 2006. The population genetics of adaptation on correlated fitness landscapes: the block model. *Evolution* 60:1113–24

Parsch J, Braverman JM, Stephan W. 2000. Comparative sequence analysis and patterns of covariation in RNA secondary structures. *Genetics* 154:909–21

Peck JR. 1994. A ruby in the rubbish: beneficial mutations, deleterious mutations and the evolution of sex. *Genetics* 137:597–606

Peters AD, Otto SP. 2003. Liberating genetic variance through sex. *BioEssays* 25:533–37

Poon A, Chao L. 2005. The rate of compensatory mutation in the DNA bacteriophage ϕX174. *Genetics* 170:989–99

Poon A, Davis BH, Chao L. 2005. The coupon collector and the suppressor mutation: estimating the number of compensatory mutations by maximum likelihood. *Genetics* 170:1323–32

Queitsch C, Sangster TA, Lindquist S. 2002. Hsp90 as a capacitor of phenotypic variation. *Nature* 417:618–24

Rokyta DR, Joyce P, Caudle BB, Wichman HA. 2005. An empirical test of the mutational landscape model of adaptation using a single-stranded DNA virus. *Nat. Genet.* 37:441–44

Rozen DE, de Visser JAGM, Gerrish PJ. 2002. Fitness effects of fixed beneficial mutations in microbial populations. *Curr. Biol.* 12:1040–45

Rutherford SL, Lindquist S. 1998. Hsp90 as a capacitor for morphological evolution. *Nature* 396:336–42

Sanjuán R, Forment J, Elena SF. 2006a. In silico predicted robustness of viroid RNA secondary structures. I. the effect of single mutations. *Mol. Biol. Evol.* 23:1427–36

Sanjuán R, Forment J, Elena SF. 2006b. In silico predicted robustness of viroid RNA secondary structures. II. Interaction between mutation pairs. *Mol. Biol. Evol.* 23:2123–30

Sanjuán R, Moya A, Elena SF. 2004. The distribution of fitness effects caused by single-nucleotide substitutions in an RNA virus. *Proc. Natl. Acad. Sci. USA* 101:8396–401

Schultes EA, Bartel DP. 2000. One sequence, two ribozymes: implications for the emergence of new ribozyme folds. *Science* 289:448–52

Schuster P, Fontana W. 1999. Chance and necessity in evolution: lessons from RNA. *Physics D* 133:427–52

Schuster P, Fontana W, Stadler PF, Hofacker IL. 1994. From sequences to shapes and back: a case study in RNA secondary structures. *Proc. Roy. Soc. London Ser. B* 255:279–84

Stadler BMR, Stadler P, Wagner GP, Fontana W. 2001. The topology of the possible: formal spaces underlying patterns of evolutionary change. *J. Theor. Biol.* 213:241–74

True HL, Berlin I, Lindquist SL. 2004. Epigenetic regulation of translation reveals hidden genetic variation to produce complex traits. *Nature* 431:184–87

van Nimwegen E, Crutchfield J, Huynen M. 1999. Neutral evolution of mutational robustness. *Proc. Natl. Acad. Sci. USA* 17:9716–20

Vendruscolo M, Maritan A, Banavar JR. 1997. Stability threshold as a selection principle for protein design. *Phys. Rev. Lett.* 78:3967–70

Waddington C. 1950. Genetic assimilation of an acquired character. *Evolution* 7:118–26

Waddington CH. 1959. Canalization of development and genetic assimilation of acquired characters. *Nature* 183:1654–55

Wagner A. 2005. Robustness, evolvability, and neutrality. *FEBS Lett.* 579:1772–78

Wagner A, Stadler PF. 1999. Viral RNA and evolved mutational robustness. *J. Exp. Zool.* 285:119–27

Wagner GP, Booth G, Bagheri-Chaichian H. 1997. A population genetic theory of canalization. *Evolution* 51:329–47

Waterman M. 1978. Secondary structure of single-stranded nucleic acids, pp. 167–212. In *Studies on Foundations and Combinatorics, Advances in Mathematics Supplementary Studies.* Vol. 1. New York: Academic

Weinreich DM, Chao L. 2005. Rapid evolutionary escape by large populations from local fitness peaks is likely in nature. *Evolution* 59:1175–82

Weinreich DM, Delaney NF, DePristo MA, Hartl DL. 2006. Darwinian evolution can follow only very few mutational paths to fitter proteins. *Science* 312:111–14

Whitlock MC, Phillips PC, Moore FB, Tonsor SJ. 1995. Multiple fitness peaks and epistasis. *Annu. Rev. Ecol. Sys.* 26:601–29

Wilke C. 2005. Quasispecies theory in the context of population genetics. *BMC Evol. Biol.* 5:44

Wilke C, Lenski R, Adami C. 2003. Compensatory mutations cause excess of antagonistic epistasis in RNA secondary structure folding. *BMC Evol. Biol.* 3:3

Wilke CO, Wang JL, Ofria C, Lenski RE, Adami C. 2001. Evolution of digital organisms at high mutation rates leads to survival of the flattest. *Nature* 412:331–33

Winkler WC, Breaker RR. 2005. Regulation of bacterial gene expression by riboswitches. *Annu. Rev. Microbiol.* 59:487–517

Wright S. 1932. The roles of mutation, inbreeding, crossbreeding and selection in evolution. *Proc. VI Int. Congr. Genet.* 1:356–66

Zuckerkandl E, Pauling L. 1962. Molecular disease, evolution, and genetic heterogeneity. In *Horizons in Biochemistry*, ed. M Kasha, B Pullman, pp. 189–25. New York: Academic

Zuker M. 1989. On finding all suboptimal foldings of an RNA molecule. *Science* 244:48–52

Zuker M, Stiegler P. 1981. Optimal computer folding of large RNA sequences using thermodynamics and auxiliary information. *Nucleic Acids. Res.* 9:133–48

How Does It Feel to Be Like a Rolling Stone? Ten Questions About Dispersal Evolution

Ophélie Ronce

Institut des Sciences de l'Evolution de Montpellier, UMR-CNRS 5554, Equipe
Génétique et Environnement, Université Montpellier II, 34095 Montpellier cedex 5,
France; email: ronce@isem.univ-montp2.fr

Annu. Rev. Ecol. Evol. Syst. 2007. 38:231–53

First published online as a Review in Advance on
August 3, 2007

The *Annual Review of Ecology, Evolution, and
Systematics* is online at
http://ecolsys.annualreviews.org

This article's doi:
10.1146/annurev.ecolsys.38.091206.095611

Key Words

cooperation, demographic stochasticity, kin competition,
inbreeding depression, local adaptation

Abstract

This review proposes ten tentative answers to frequently asked questions about dispersal evolution. I examine methodological issues,
model assumptions and predictions, and their relation to empirical
data. Study of dispersal evolution points to the many ecological and
genetic feedbacks affecting the evolution of this complex trait, which
has contributed to our better understanding of life-history evolution
in spatially structured populations. Several lines of research are suggested to ameliorate the exchanges between theoretical and empirical
studies of dispersal evolution.

INTRODUCTION

Several good reviews have appeared recently on dispersal (see in particular Bowler & Benton 2005, Clobert et al. 2004, Levin et al. 2003, Olivieri & Gouyon 1997), including whole volumes devoted to the question (Bullock et al. 2002, Clobert et al. 2001, Dingle 1996). The increasing awareness of dispersal's crucial role in the context of global habitat fragmentation, climate change, and biological invasions motivates to a large extent such recent interest [see the special issue in *Science* (volume 313, issue 11) and in particular Kokko & Lopez-Sepulcre 2006]. I have organized the present review around tentative answers to ten frequently asked questions about dispersal evolution. Answers to the first five questions seek to clarify how different methodological constraints, both in theoretical and empirical studies, might affect our understanding of dispersal evolution. In particular, questions two and three set the stage by reviewing briefly the data on dispersal evolution, whereas questions four and five address the general assumptions of models and their relation to data. Answers to questions six through ten build on the previous methodological clarification and attempt to dissipate some confusion about specific selective forces acting on dispersal evolution, using both theory and data. I conclude by addressing the general successes and failures of dispersal evolution studies.

1. WHAT IS DISPERSAL AND WHY IS IT IMPORTANT?

I here define dispersal as any movement of individuals or propagules with potential consequences for gene flow across space. Such definition thus includes both natal dispersal and breeding dispersal. Dispersal movement comprises three stages: (*a*) departure (or emigration), (*b*) a vagrant stage, and (*c*) settling (or immigration). There is no restriction on the ploidy of the dispersing stage, including pollen dispersal. The literature often uses the terms migration and dispersal interchangeably (but see Dingle 1996). The dispersal kernel and dispersal rate are two metrics often used to summarize the consequences of dispersal movements, even though they provide an incomplete description of the dispersal process (Bowler & Benton 2005).

Dispersal holds a central role for both the dynamics and evolution of spatially structured populations, allowing the genetic cohesion of a species across space, its global persistence despite local extinction, and the tracking of favorable environmental conditions in an ever changing world. Dispersal can rescue a small population from local extinction (Brown & Kodric-Brown 1977), but by increasing synchrony in population dynamics, high levels of dispersal can also increase global extinction risk (for experimental evidence, see Molofsky & Ferdy 2005). Dispersal affects the distribution of genetic diversity through space, by increasing the proportion of total diversity contained within rather than between populations (Wright 1969). In particular, dispersal can help mitigate the effect of drift in small populations, decrease mutation load, and thereby reduce the risk of extinction (for theoretical predictions, see Higgins & Lynch 2001; for a review, see Tallmon et al. 2004). Gene flow mediated by dispersal can both impede the evolution of local adaptation (Lenormand 2002) and accelerate it (Gandon et al. 1996). Dispersal affects the evolution of speciation (see review in Barton 2001), inbreeding depression (Roze & Rousset 2003), cooperation

---

Natal dispersal: movement by which an individual leaves its birthplace to engage into mating or reproduction somewhere else

Breeding dispersal: movement between two reproduction events for the same individual

Dispersal kernel: the probability density that an individual initially at coordinates (0,0) is found at coordinates (x,y) after dispersal

Dispersal rate: rate at which individuals leave a patch of habitat, without clear specification about the distance moved once out of this patch

Local adaptation: the higher fitness of resident genotypes in their native environment relative to that of immigrant genotypes in the same environment

and sociality (Le Galliard et al. 2005), and many life-history traits (e.g., Pen 2000). Finally, dispersal plays a key role in community dynamics (see review in Leibold et al. 2004). Understanding dispersal, and also its evolution, is therefore crucial to improve the management of natural populations, as illustrated by the evolutionary suicide of the rare endemic plant *Centaurea corymbosa*. Because of the high risk of dispersal in this cliff species, seed traits enhancing long-distance dispersal have been counter-selected, resulting in the absence of colonization and exchange between populations (Colas et al. 1997). Without human-assisted colonization, the long-term persistence of the species solely relies on that of six small populations, all within 3 km^2, subject to both demographic and environmental stochasticity (Fréville et al. 2004).

2. IS DISPERSAL PLASTIC?

Studying dispersal does not imply simply quantifying a single dispersal kernel, but also assessing how this dispersal kernel varies with individual, social, and ecological conditions. For instance, the heteromorphic plant *Crepis sancta* produces a larger fraction of fruits equipped with a dispersal structure when subjected to experimental nutrient depletion in the soil (Imbert & Ronce 2001). Maternal condition during gestation, such as age (Ronce et al. 1998) or parasitic load (Sorci et al. 1994), affects the dispersal behavior of juveniles in the common lizard *Lacerta vivipara* with complex interactions (Massot et al. 2002). In that same species, manipulation of the social context by either the absence of relatives (Le Galliard et al. 2003) or the presence of frustrated dispersers (Boudjemadi et al. 1999) deeply modifies the rate of dispersal but also the nature, morphology, and colonization success of the dispersers. Density affects emigration and immigration rates in animals (see Clobert et al. 2004) but also seed dispersal in plants, with effects carried over several generations (Donohue 1999). In collared flycatchers, the reproductive success of congeners is assessed by prospecting individuals, and breeding-habitat selection is based on such public information (Doligez et al. 2002).

Reviews of the literature have repeatedly shown that conditional dispersal expression does not reflect only the variation of constraints on the dispersal process, but also the great plasticity in the organism's broad response to various environmental cues during emigration, vagrancy, and immigration (Bowler & Benton 2005, Clobert et al. 2004, Ims & Hjermann 2001, Ronce et al. 2001). There are good theoretical reasons to believe that informed dispersal decisions would confer an evolutionary advantage over a blind process, unless patterns of variation in habitat quality are totally unpredictable or information acquisition is costly (Ronce et al. 2001). I invite the reader to refer to the cited reviews above for further examples and a more general discussion of the benefits and downsides of conditional dispersal (for the latter in particular, see Kokko & Lopez-Sepulcre 2006).

3. HOW FAST CAN DISPERSAL EVOLVE?

Rapid evolution of dispersal requires both the presence of heritable genetic variation for traits affecting dispersal behavior and strong selection acting on these traits.

Macroptery: proportion of individuals carrying functional wings

Evolutionary potential for dispersal is present in many natural populations. In several beetle species, wing polymorphism is under the control of a single gene with two alleles (Roff 1986). Dispersal behavior variation in the Glanville fritillary butterfly, *Melitaea cinxia*, seems to be closely associated with allelic variation at the *pgi* enzymatic locus (Haag et al. 2005). Determinism of variation for dispersal or dispersal-related traits is, however, often polygenic, with heritability typically greater than 0.3 (e.g., for a study on seed heteromorphism in *C. sancta*, see Imbert 2001; for a review in animals, see Roff & Fairbarn 2001). Donohue et al.'s (2005) work on *Arabidopsis thaliana* is unique in that they quantified how the heritability of phenotypic traits translated into the heritability of the dispersal kernel itself. Moreover, they investigated how dispersal kernels were affected by the manipulation of density and found significant genotype by environment interactions (Donohue et al. 2005). Given the predominance of conditional dispersal in nature, similar information about the genetic architecture of dispersal's response to environmental conditions would much improve our understanding of dispersal evolution.

In agreement with measures of heritability, the response of dispersal traits to artificial selection can be fast (see review in Roff & Fairbarn 2001). Genetic variation for dispersal propensity in short-lived organisms, such as *Caenorhabditis elegans*, also allows in vitro evolution experiments, in which characteristics of artificial patchy landscapes are manipulated (Friedenberg 2003). Evidence for the short-term evolution of dispersal in nature comes from specific situations, such as oceanic islands (Denno et al. 2001, Roff 1990), variation in landscape fragmentation (Hanski et al. 2004, Hill et al. 1999, Schtickzelle et al. 2006), ecological succession (Olivieri & Gouyon 1985, Peroni 1994), and biological invasions (Phillips et al. 2006). For instance, Simmons & Thomas (2004) found genetic changes in both the mean macroptery and, more interestingly, its response to population density in expanding edge populations of two bush crickets. Many of the examples above do not elucidate entirely the relative role of genetic changes and plasticity in explaining fast phenotypic changes (e.g., less than ten generations in Cody & Overton 1996). Empirical quantification of the strength of selection acting on dispersal traits in natural populations is almost entirely lacking (but see Donohue 1999).

4. WHICH MODEL ASSUMPTIONS MATTER?

Models of dispersal evolution widely differ in their assumptions, methodology, and ways to describe the dispersal process. The diversity of theoretical approaches sometimes makes the synthesis of their conclusions or the comparison with empirical data quite difficult. Yet different models have also shed light on different evolutionary forces acting on dispersal, unraveling the complex nature of this trait responding to multiple selection pressures (Ronce et al. 2001). Modeling choices, motivated by technical reasons or convenience, can have deep consequences in terms of potential selective forces at stake in the model, which is not always acknowledged with sufficient clarity (Ronce et al. 2001). I discuss in particular four important assumptions about landscape structure, namely the total number of sites between which dispersal occurs, the number of individuals per site, their spatial organization, and the extent of intrinsic spatial heterogeneity in habitat quality in the landscape.

A simplifying assumption frequently made in analytical models for technical reasons is that of an infinite number of sites harboring individuals in the landscape. When the number of sites is finite but large, model results converge quickly toward this limiting case (see, for instance, Rousset 2006). Such an assumption is particularly convenient when modeling stochastic demographic dynamics, as it allows the elimination of stochasticity at the global metapopulation scale and simplifies greatly the analysis (see, e.g., Rousset & Ronce 2004). For that very reason, however, it means that dispersal has no bet-hedging effect in models making this assumption (see discussion in Ronce et al. 2001 and in Section 8). Some analytical models and deterministic simulations also manipulate population densities rather than discrete numbers of individuals in each site, which implies the neglect of stochastic processes related to finite local population size. Consequently, such models ignore spatial genetic structure due to drift and the associated kin-selection phenomena affecting dispersal evolution (see Ronce et al. 2001 and Section 6). They should, however, provide a limiting case toward which stochastic models converge when local population size increases (see Gandon & Michalakis 1999). Convergence, however, is not always checked with accuracy.

An apparent paradox of many dispersal evolution models is their absence of an explicit description of space. Dispersal is often described through an emigration rate out of a given spatial unit harboring a variable number of individuals, and migrants are distributed randomly across the landscape. Such an island model of migration may describe correctly dispersal in some biological systems, but its popularity among modelers results essentially from its analytical tractability. Analytical descriptions of the evolution of space-limited movements have been developed, however (Bolker & Pacala 1999, Comins 1982, Ezoe 1998, Gandon & Rousset 1999, Le Galliard et al. 2005, Rousset & Gandon 2002), including two-patches models (Billiard & Lenormand 2005, Leturque & Rousset 2002, McPeek & Holt 1992), but they often remain mathematically difficult. With the emergence of increasingly powerful computers, individual-based, spatially explicit simulations of dispersal evolution have recently become popular (Heino & Hanski 2001, Hovestadt et al. 2001, Murrell et al. 2002, Travis & Dytham 2002). Fortunately, main qualitative conclusions of spatially implicit models about dispersal evolution, including those assuming global dispersal, still hold when put in a spatial context (see, for instance, the comparison of Gandon & Michalakis 1999 and Heino & Hanski 2001 in Ronce & Olivieri 2004). The main difficulty in such theoretical predictions is evaluating their relevance for empirical measures of dispersal rates, which are necessarily scale specific. A way forward could involve a hierarchical approach of space, distinguishing different types of dispersal movements (see Fontanillas et al. 2004 for an empirical example and Ravigné et al. 2006 for a theoretical example).

Introducing spatial heterogeneity in the landscape, or any other type of temporal autocorrelation in habitat quality, has repeatedly modified predictions of homogeneous models by facilitating either the coexistence of different dispersal strategies (Doebeli & Ruxton 1997, Mathias et al. 2001, Parvinen 2002) or the evolution of conditional dispersal (Doligez et al. 2003, McPeek & Holt 1992). Despite the prevalence

Stochastic demographic dynamics: variation in population numbers either due to environmental or demographic stochasticity

Metapopulation: a set of discrete populations connected by dispersal

Bet-hedging: a strategy that reduces the variance of its gains through time, thereby increasing the geometric mean of its gains

of conditional dispersal in nature, early dispersal evolution models have mainly considered dispersal strategies with a pure genetic determinism. Theoretical studies of conditional dispersal strategies, however, have much increased in frequency in recent years.

5. HOW DO WE MODEL SELECTION ON DISPERSAL?

Because dispersal affects both the spatial distribution of genetic diversity and population dynamics, it alters the selective environment for different genotypes. Selection on genotypes differing in their dispersal propensity is thus generally frequency dependent. In the absence of precise knowledge about genetic variation in dispersal kernels, short-term quantitative genetic predictions about evolution rates are entirely lacking. Instead, models have aimed at predicting which mean dispersal phenotypes or coalition of different phenotypes would dominate in the long term. Game theory, therefore, has been and still is the preferred approach to investigate dispersal evolution patterns. Models have sought to identify convergence stable and evolutionarily stable dispersal strategies (see Geritz et al. 1998). Both stability criteria necessitate computing the fate of a mutant allele conferring a deviant dispersal phenotype when confronted with a resident allele.

Given the complexity of the ecological scenarios envisioned in dispersal evolution models (involving interactions between related individuals, spatial structure, and various sources of stochasticity), researchers have largely discussed the relevant analytical measure of fitness and its most useful approximations. Different measures have been developed to capture the long-term evolutionary consequences of the nonrandom spatial arrangement of population numbers and genetic diversity. They all have shown that selection on deviant dispersal strategies depends on (*a*) the expression of modified dispersal by mutant individuals, but also on (*b*) the statistical association between the phenotype of a mutant and the phenotype of its neighbors, as well as on (*c*) how such neighborhood might modify the local demography (see Rousset & Ronce 2004; Le Galliard et al. 2005). I now briefly review which are those measures, their assumptions, and to which situations they have been most successfully applied.

In patchy populations with either global or limited dispersal, direct fitness methods (Taylor & Frank 1996) have been used to derive inclusive fitness measures (Hamilton 1964), which are functions of relatedness coefficients between different pairs of individuals at various spatial distances (e.g., Frank 1986, Gandon & Rousset 1999, Irwin & Taylor 2000, Taylor 1988). Note that relatedness coefficients used in kin-selection models of dispersal evolution are not fixed parameters but dynamically emerge from localized ecological interactions between individuals and the process of genetic coalescence in a spatially structured population. In particular, they jointly evolve with dispersal. Convergence stability measures derived from inclusive fitness arguments can be used to compute the fixation probability of a deviant mutant classically considered by population genetics (Rousset 2006).

In metapopulations with infinite size and global dispersal, Metz & Gyllenberg (2001) have proposed the *Rm* fitness measure, which has the elegant property of being the spatial equivalent of the lifetime reproductive success *R0* in class-structured

---

Game theory: in the context of population genetics, theory seeking approximations for the long-term evolution of traits, when frequency dependence is expected

Convergence stable strategy: a phenotype toward which the population evolves by successive allelic substitution

Evolutionarily stable strategy: a phenotype characterizing a population such that any rare mutant with deviant phenotype is counterselected

Inclusive fitness: effects of a deviant phenotype on the fitness of individuals (*a*) expressing this phenotype, (*b*) when the deviant phenotype is expressed by others, weighted by a measure of genetic similarity between interacting individuals

Relatedness: a function of probabilities of genetic identity that measures the increased probability of recent coalescence between some pair of genes relative to another

Coalescence: time in the past when two particular gene copies of the present population had their most recent common ancestor

Rm **fitness:** the overall production of successful mutant emigrants from a patch, from initial colonization by a single mutant to the extinction of the mutant lineage in that patch

populations (for applications, see, e.g., Crowley & McLetchie 2002, Gyllenberg et al. 2002, Parvinen 2002, Parvinen et al. 2003). Metz & Gyllenberg (2001) also provided efficient numerical recipes to compute Rm fitness in various scenarios, which explains in part its success. Ajar (2003) has shown how the Rm fitness measure (Metz & Gyllenberg 2001), and its derivatives used to compute convergence and evolutionary stability relate to inclusive fitness concepts (Hamilton 1964).

At the other extreme, in lattice models with very localized dispersal, investigators have used pair approximation methods (van Baalen & Rand 1998) to derive spatial invasion fitness from the dynamics in the frequency of simple spatial configurations, such as neighboring pairs of sites harboring either one or two individuals with the same or different alleles (see Harada 1999, Ferrière & Le Galliard 2001, Le Galliard et al. 2005). The accuracy of approximations used to close the system of equations used in such models, however, has been variable (van Baalen & Rand 1998). The structural form of selection measures for dispersal derived from pair approximation methods bears close resemblance to inclusive fitness measures, including parameters that could be interpreted as relatedness coefficients (see Ferrière & Le Galliard 2001). Although feedbacks between demography and evolution had been at the core of pair approximation developments, they have been incorporated into inclusive fitness approaches only recently, with still few successful applications (Rousset & Ronce 2004).

The rarity of the deviant strategy is often invoked to derive approximations for fitness measures. Yet the above-mentioned approximations rely more exactly on the assumption of small phenotypic differences between the competing genotypes and therefore weak selection (Rousset 2006). More precisely, under weak selection, approximations for convergence stability require computing only the first-order effects of selection, which was shown to be independent of the allelic frequencies for both limited and global dispersal when the number of sites in the metapopulation is large (Rousset 2006). Weak selection also justifies the computation of relatedness coefficients, assuming neutrality at the dispersal modifying locus in this case. Conversely, second-order effects of selection, which determine evolutionary stability, are frequency dependent and require computing the effect of selection on genetic similarity measures (Ajar 2003).

The majority of dispersal evolution models have assumed either clonal reproduction or a haploid life cycle with a single locus determining dispersal ability (however, see Ravigné et al. 2006, Roze & Rousset 2005, Taylor 1988 for diploid models with sexual reproduction and codominance between alleles affecting dispersal). Multilocus models in which a dispersal modifier locus recombines with another locus affecting fitness (Balkau & Feldman 1973, Wiener & Feldman 1993) had interesting developments recently. These recent models (Billiard & Lenormand 2005, Roze & Rousset 2005) take into account both the statistical associations between different loci in the same individual (due to selection, migration, and limited recombination) and their interaction with genetic associations between different individuals (due to genetic coalescence in finite populations). In particular, such models showed how measures of relatedness at dispersal modifier loci were affected by indirect selection at linked loci (see Sections 7 and 9).

6. DOES DISPERSAL ALLOW ESCAPE FROM COMPETITION?

Competition is at the core of many theoretical and empirical studies of dispersal (Lambin et al. 2001). Yet one must distinguish the effects of competition with conspecifics in general from those of competition with relatives. Investigators recognized early on that escaping conspecific competition is a major potential benefit of dispersal [e.g., the first model of dispersal evolution (van Valen 1971)]. In landscapes with variable density through space, dispersal by simple diffusion results in a net flow of individuals from highly populated to less crowded regions. If patterns of spatial variation in density do not match perfectly those in resource availability (Hastings 1983), this confers a selective advantage to genotypes with increased dispersal tendency, as they are more likely to exploit patches of abundant resources with few competitors. In particular, this is the case at the invasion front in expanding populations (see predictions in Travis & Dytham 2002). In extant metapopulation models with both a very large number of sites and very large numbers of individuals per site, escape from overcrowding is the only cause for dispersal evolution (Gyllenberg et al. 2002, Levin et al. 1984, Mathias et al. 2001, Olivieri et al. 1995). Stochastic demographic dynamics generate the conditions for such partially independent temporal variation in density at different sites. Local catastrophic extinctions are an extreme form of such variability. Chaotic population dynamics or asynchronous cycles among sites due to strong density dependence have the same effect (Doebeli & Ruxton 1997, Parvinen 1999). Dispersal strategies conditional on population density in the natal patch were predicted to be more efficient than fixed dispersal strategies at exploiting such heterogeneity (Jánosi & Scheuring 1997, Levin et al. 1984, Metz & Gyllenberg 2001, Poethke & Hovestadt 2002), in agreement with the abundant empirical evidence for density-dependent dispersal (see also Clobert et al. 2004, Ims & Hjermann 2001). It is, however, not always obvious from empirical data that dispersal indeed allows escape from overcrowding. Dispersal agent behavior, directed dispersal, and habitat selection may indeed often result in postdispersal aggregation of high density, as observed, for instance, in *Trilium grandiflorum* (Kalisz et al. 1999).

Even if it did not decrease conspecific aggregation, seed dispersal in *T. grandiflorum* still resulted in decreased relatedness among competing seedlings (Kalisz et al. 1999). Escaping sibling competition could provide ecological benefits for a dispersed individual, such as escaping specialized pests (for a theoretical treatment, see Muller-Landau et al. 2003) or competing with individuals with different ecological niches (e.g., Cheplick & Kane 2004). Yet theory (Hamilton & May 1977) has shown early on that such ecological benefits are not a necessary requirement for dispersal to evolve as a kin-competition avoidance strategy in stable habitats [later generalized by Frank (1986)]. Dispersal evolution can then be understood as an altruistic behavior, providing no direct ecological benefit to the dispersed individual, but alleviating competition for its kin. Spatial genetic aggregation generates spatial variation in postdispersal juvenile density, and highly dispersive genotypes benefit from relaxed competitive conditions, not through the emigrating individuals, but through the progeny that is not dispersed. When the intensity of kin competition varies owing to some heterogeneity in the population, dispersal strategies conditional on cues reflecting such heterogeneity, such as habitat carrying capacity (Leturque & Rousset 2002), maternal age (Ronce et al. 2000a), or family size (Kisdi 2004), have been predicted to evolve. The efficiency of dispersal as a strategy of kin-competition avoidance, however, may be reduced when the dispersal movements of related individuals are strongly correlated [see, e.g., the blue morphs of side-blotched lizards (Sinervo & Clobert 2003)], as when whole sibling families are dispersed in the same fruit.

Distinguishing between the relative effects of kin competition and demographic stochasticity on the evolution of dispersal is often not obvious. This is especially the case in models in which small finite local population sizes generate both strong local genetic relatedness and random variation in population characteristics (Cadet et al. 2003, Heino & Hanski 2001, Le Galliard et al. 2005, Parvinen et al. 2003). Theoretical frameworks allowing the sequential neglect of one or the other type of effects (see, for instance, Ronce et al. 2000a) can help disentangle the respective role of different evolutionary mechanisms. Interactions between kin selection and demographic stochasticity can also lead to counterintuitive emergent properties, such as the evolution of increasing dispersal rates with increasing dispersal cost (Comins et al. 1980, Gandon & Michalakis 1999, Heino & Hanski 2001; see further discussion of such a result in Ronce & Olivieri 2004).

A final reason to escape the maternal environment in species with overlapping generations is to escape competition with parents. Increasing adult life span selects for increasing juvenile mobility in models allowing for empty sites to be colonized (Olivieri et al. 1995). Variation in adult survival rates due to senescence also selects for juvenile dispersal conditional on maternal age, as observed in the common lizard (Ronce et al. 1998). More generally, manipulations of maternal condition or presence suggest that escape from maternal competition is a major determinant of female progeny dispersal in that species (e.g., Le Galliard et al. 2003, Massot et al. 2002).

7. IS DISPERSAL AN INBREEDING AVOIDANCE STRATEGY?

Experimental removal of parents of one sex in high-density populations of white-footed mice causes a delay in the dispersal of the opposite sex progeny (Wolff 1992). More generally, sex-specific dispersal rates in animals and pollen dispersal in plants have often been interpreted as mechanisms for inbreeding avoidance, even though alternative explanations involving kin-competition avoidance (see, e.g., Ravigné et al. 2006 for a model of pollen-dispersal evolution) can also explain the same patterns. Recent theoretical work has helped clarify these issues. Heterosis favoring dispersal is most commonly thought to be the result of the uneven distribution of deleterious recessive alleles among populations and the masking of such alleles in interpopulation crosses. Heterosis thus is expected in metapopulations with strong genetic structure, in which kin competition is also intense (for theory, see Glémin et al. 2003, Whitlock et al. 2000; for an empirical example, see Willi & Fischer 2005). Kin-competition and inbreeding avoidance then cannot be perceived as alternative causes of dispersal evolution (Gandon 1999, Perrin & Goudet 2001). Whereas heterosis favors divergence in sex-specific dispersal rates, kin competition tends to have a stabilizing effect (Gandon

1999, Perrin & Mazalov 2000). Evolution of female preference for immigrant or related males, which further affects the evolution of male dispersal, depends on the balance between heterosis and kin selection (Lehmann & Perrin 2003). Complex interactions affect the evolution of dispersal under the joint influence of heterosis and kin competition, as the former increases the effective migration rate, decreasing relatedness and weakening the incentive effects of kin competition on dispersal (Gandon 1999, Roze & Rousset 2005).

Until recently, however, dispersal evolution models did not account for the full spectrum of interactions between kin competition and heterosis because the latter parameter was considered to be fixed (Motro 1991) or to be a simple mathematical function of the probability of coancestry among individuals in the same patch (Gandon 1999, Perrin & Mazalov 2000). Investigators have shown that heterosis varies with the average effect of mutations, their dominance and rate of occurrence, but also with the intensity of gene flow among local populations and their size (Glémin et al. 2003, Whitlock et al. 2000). Roze & Rousset's (2005) multilocus analytical model allows the quantification of direct selective effects on a dispersal modifier locus owing to kin competition and indirect selection through its association with deleterious alleles at loci contributing to heterosis. Their model predicts that heterosis increases the selected dispersal rate by an order of magnitude in some situations, but also that increasing heterosis can select unexpectedly for decreased dispersal. Simulations have cast doubts about the quantitative importance of heterosis for the evolution of dispersal when the total metapopulation size is small, as heterosis then vanishes rapidly when higher dispersal evolves (Guillaume & Perrin 2006, Ravigné et al. 2006).

8. IS DISPERSAL AN ADAPTATION TO EPHEMERAL HABITATS?

Increased rates of patch destruction in artificial metapopulations of *C. elegans* resulted in the increasing frequency of more dispersive mutants (Friedenberg 2003). Habitat persistence correlates negatively with intraspecific variation in macroptery in several species of insects (Denno et al. 1996). There are two distinct theoretical reasons why dispersal may provide adaptation to ephemeral habitats, which are not always distinguished clearly in the literature. First, dispersal allows tracking patches of favorable habitat and escape from deteriorating local conditions. Local disturbances generate patches with underexploited resources, allowing dispersers to escape from overcrowding (see Section 6). Predictable habitat deterioration through time (due to overexploitation, ecological succession, or any factor generating favorable patches of habitat with a finite life span) creates additional selection pressures favoring dispersal because less dispersive genotypes tend to be more frequent in older patches of habitat of lesser quality (Olivieri et al. 1995). Again, conditional dispersal and habitat selection should provide a strong advantage in variable environments, as long as habitat deterioration can be accurately predicted from some ecological cues [e.g., see the models by Doligez et al. (2003) and Ronce et al. (2005)]. Low conspecific density, for instance, can then convey different types of information about habitat quality, reflecting either low competition for resources or deteriorating ecological conditions such as with root voles for which emigration rates increase with decreasing density (Ims & Andreassen 2000; see a more general discussion in Clobert et al. 2004).

The second argument frequently invoked is that dispersal acts as a bet-hedging strategy in temporally variable environments. By spreading their progeny more evenly among different sites, genotypes with a higher dispersal ability better sample habitat variation within a generation, thus reducing the generation-to-generation variance in their mean performance. Venable & Brown (1988) clearly showed that such an argument does not hold in models in which the number of occupied sites is assumed to be infinite. Indeed, stochastic variance in mean performance between generations then vanishes for all genotypes. In models with a finite number of patches, demographic stochasticity, and density dependence (Doebeli & Ruxton 1997, Kisdi 2002, Parvinen 1999), the respective roles of bet hedging and escape from crowding in shaping the evolution of dispersal have never been quantified clearly.

The relationship between habitat instability and the dispersal propensity of organisms might also not be as straightforward as originally thought. Complex relationships between dispersal and the frequency of local extinction emerge because of feedbacks between population dynamics and evolution (see a review in Ronce & Olivieri 2004). Indeed, more frequent disturbance can result in slower population growth and less incentive to disperse to escape overcrowding (Ronce et al. 2000b), whereas changes in dispersal may also affect the probability of local extinction (Poethke et al. 2003).

9. HOW COSTLY IS DISPERSAL?

Dispersal is a risky behavior. First, mortality may be increased during the vagrancy stage of dispersal owing, for instance, to the transient use of nonoptimal habitat and increased predation risk, such as observed in dispersing root voles (Ims & Andreassen 2000). Estimations of mortality during dispersal from mark-recapture data, such as with the virtual migration model (Hanski et al. 2000), have shown a large amount of variation depending on both species traits and landscape characteristics (Matter 2006, Schtickzelle et al. 2006).

Dispersal is also risky when habitat selection during the settlement stage is constrained or limited, leading to frequent immigration into nonfavorable habitat. A large fraction of wind-dispersed pollen, for instance, never ends up on the stigma of a receptive flower of the same species. Nonephemeral patterns of spatial heterogeneity in habitat quality thus tend to select against passive dispersal (Hastings 1983, McPeek & Holt 1992). In particular, mismatch between postdispersal environment and phenotypes having developed in some other environment could contribute to immigrant inadequacy [but see also examples of dispersal based on phenotype matching (e.g., Cote & Clobert 2007)]. Local adaptation emerges when different alleles have different effects on fitness in different environments. It has been extensively documented in the field (Lenormand 2002) and indirectly selects against dispersal because immigrant individuals are less likely to carry locally favored alleles [see theoretical predictions by Wiener & Feldman (1993)]. The strength of such indirect selection, however, varies with dispersal itself, which attenuates genetic differences among sites and weakens local adaptation (Billiard & Lenormand 2005). Both theory (Gandon et al. 1996) and data (e.g., Morgan et al. 2005), however, have shown that higher dispersal may provide an evolutionary advantage to either host or pathogens engaged in a coevolutionary arms race, increasing the probability of local adaptation. How such an arms race affects the evolution of dispersal has not been explored.

Dispersal may also result in a loss of social status when joining a new group of individuals, involving an exposure to xenophobic behaviors (O'Riain & Jarvis 1997) or a loss of cooperation with related individuals (see a review in Lambin et al. 2001). Perrin & Goudet (2001) showed theoretically that kin cooperation in groups of philopatric females, as observed in many mammals, could counteract kin-competition effects and lead to the evolution of greater sex bias in dispersal. Evolution of dispersal and cooperative behaviors, however, is not always antagonistic, as shown by models of the joint evolution of altruism and mobility (Le Galliard et al. 2005). In the side-blotched lizards, blue morphs actively cooperate, and their nonrandom dispersal movements result in their postdispersal spatial aggregation (Sinervo & Clobert 2003).

Finally, increased mobility may trade off for other life-history traits. Smaller seed size might increase dispersal distance but compromises the survival and competitive ability of seedlings after germination. Flight-fecundity trade-offs, and their physiological basis, have been studied to a great extent in wing dimorphic insects (for a review, see Roff & Fairbarn 2001). In particular, in *Gryllus firmus*, individuals from genetic lineages characterized by a higher macroptery proportion are less fecund, independent from their individual dispersal morph (Roff & Fairbarn 2001). Dispersing individuals, however, may exhibit different sets of life-history traits while achieving the same lifetime reproductive success (see review in Bélichon et al. 1996).

The measurement of dispersal costs is plagued with numerous methodological difficulties and pitfalls in interpretation (for a review, see Bélichon et al. 1996, Clobert et al. 2004). Difficulties in measuring dispersal costs also emerge from the fact that selection may act to reduce such costs (see empirical evidence in Schtickzelle et al. 2006). Dispersing individuals are generally not a random subset of the population and have behavioral, physiological, and morphological attributes to reduce mortality during dispersal and increase settlement success in new patches of habitat (see, e.g., Gundersen et al. 2002). More generally, we lack an efficient framework to combine information about mortality during dispersal, genetic correlations between dispersal propensity and life-history traits, and phenotypic differences between dispersing and philopatric individuals to relate such estimates to parameters describing the cost of dispersal in evolutionary models.

Dispersal cost is indeed a salient feature of dispersal evolution models (see, e.g., Ravigné et al. 2006). In most models, the simplest interpretation of this parameter corresponds to a measure of increased mortality of dispersers during vagrancy or settlement. Its relationship to landscape characteristics is rarely explicit (see, however, Travis & Dytham 1999, Heino & Hanski 2001, Hovestadt et al. 2001), and it is generally not assumed to evolve. Models of joint evolution of local adaptation and dispersal (Billiard & Lenormand 2005, Kisdi 2002) provide an exception to this rule and have led to original predictions. Billiard & Lenormand (2005) found that, for the same set of parameters, evolution could lead to either very high or very low dispersal depending on initial mobility, owing to positive feedbacks in the joint evolution of dispersal and dispersal cost. The cost of dispersal in their model indeed depends tightly on the linkage disequilibrium between loci controlling dispersal propensity and local adaptation. As linkage disequilibrium peaks at intermediate dispersal, this can result in disruptive selection on dispersal.

10. IS LONG-DISTANCE DISPERSAL A CONSEQUENCE OF SELECTION FOR SHORT-DISTANCE DISPERSAL?

We may generally question the fact that different parts of the kernel (for instance, long- and short-distance dispersal) evolve independently. If the same traits enhance the probability of short- and long-distance journeys, occasional long-distance dispersal might be interpreted as a side effect of selection for short-distance movement. Conversely, traits enhancing dispersal may be selected for essentially because they allow long-distance movement, and the large fraction of individuals dispersing at short distance, as in seed or pollen dispersal, may simply be failures to do so. However, there are increasing suggestions that short- and long-distance dispersal events rely on different mechanisms (Higgins et al. 2003) or are accomplished by different types of individuals.

Despite the number of theoretical studies focusing on dispersal evolution, the evolution of dispersal distance rather than dispersal rate has been examined theoretically only recently. Some of these studies (Ezoe 1998, Murrell et al. 2002) have constrained the dispersal kernel to belong to a fixed (Gaussian or exponential) unique distribution. The evolution of short- and long-distance dispersal is completely linked under such assumptions. Two recent models have addressed the question of the evolution of the dispersal kernel's shape in a simple ecological context, without any constraints on the dispersal-distance distribution (Hovestadt et al. 2001, Rousset & Gandon 2002). In both models, dispersal evolves as a strategy for kin-competition avoidance, and dispersal costs vary with distance dispersed. Fat-tailed dispersal kernels deviating from the Gaussian and exponential distribution evolved readily, which suggests that kin competition alone can select for long-distance dispersal. What happens when several selective forces (such as kin-competition avoidance and recolonization of empty space) acting at different spatial scales affect the evolution of dispersal kernels has not been investigated. Clarifying those issues implies a better understanding of the mechanisms and specific traits affecting the dispersal kernel's shape and its different parts (see Donohue et al. 2005), as well as incorporating such constraints into evolutionary models to quantify the intensity of selection acting in different parts of the dispersal kernel.

11. CONCLUSIONS

Both theory and empirical studies have shown that searching for a unique cause for dispersal evolution is misleading (Ronce et al. 2001). Researchers have also claimed that using a single term, dispersal, to describe movements with different spatial scales and different ultimate or proximate motivations is equally misleading (Bowler & Benton 2005). However, I feel that different evolutionary explanations always will be entangled to some extent in promoting dispersal in any realistic situation. A better understanding of their interactions represents the exciting challenge of future theoretical and experimental studies of dispersal.

Theory about dispersal evolution has taught us much more than just the selective pressures acting on a particular life-history trait. It has motivated the study of whole life-history syndromes through the joint evolution of dispersal with dormancy (Venable & Brown 1988; see review in Olivieri 2001), reproductive effort (Crowley & McLetchie 2002, Ronce et al. 2000c), senescence (Dytham & Travis 2006), ecological specialization (Billiard & Lenormand 2005, Kisdi 2002), altruism (Le Galliard et al. 2005), kin recognition (Lehmann & Perrin 2003), sex ratio (Leturque & Rousset 2004), and mating strategies (Ravigné et al. 2006).

From a technical point of view, problems posed by dispersal evolution have stimulated many methodological advances. They have helped clarify theoretical approaches of evolution in the presence of multiple levels of selection and the concepts of inclusive fitness. They have drawn attention to the role of spatial interactions on selective processes. They have stimulated more careful thinking about the effect of genetic drift in life-history evolution. They have forced us to study the many complex feedbacks between population dynamics and evolution. Finally, they have allowed the integration of complex parts of evolutionary theory, such as that of neutral population differentiation, game theory, mutation load, and life-history evolution. Recent multilocus models of dispersal evolution (Billiard & Lenormand 2005, Roze & Rousset 2005), for instance, have contributed to a better conceptual unification in the treatment of genetic associations between and within individuals in a spatial context. Transferring these methodological advances to the study of other phenotypic traits and taking into account the spatial dimension in their evolution are now timely (Ronce & Olivieri 2004).

Such exciting conceptual and methodological challenges probably explain in part the abundant theoretical production on dispersal evolution. There are, however, two downsides to this prolific theoretical development. First, the increasing ecological complexity incorporated in analytical dispersal evolution models has often implied relying on the numerical evaluation of complex mathematical terms (Rousset & Ronce 2004) with diminishing insight into the exact evolutionary mechanisms at stake in the models, making the contribution of such models more similar to that of simulations. This is particularly the case in models with complex demographical dynamics. A pessimistic view would consider that we have reached the limit of what we can extract from these types of models. An optimistic perspective is that we need to put even greater efforts into thinking about the relevant parameters that we should compute to better test ideas about the selective forces explaining evolutionary patterns. The relative ease with which large spatially explicit individual-based simulations can be run currently (Travis & French 2000) should not make us forget that the contribution of such simulations to the understanding of evolutionary mechanisms is always increased by a careful comparison to analytical predictions.

The second downside is that the production of empirical results specifically testing model predictions has comparatively lagged behind (Ronce et al. 2001). This resulted in part because of the difficulty of measuring dispersal in nature, but the lack of dialogue between theory and experiments in this field of research has more complex causes. The very difficulty of the modeling exercises of dispersal evolution, and probably the excitement about methodology, has long distracted theoreticians from empirical evidence, such as the predominance of conditional dispersal, the complex shapes of dispersal kernels, the dynamical nature of dispersal costs, and the simultaneous action of several selective forces. Evolution of the reaction norms of dispersal to various environmental factors has received more theoretical interest recently, as well as the interaction between kin competition, heterosis, stochastic demographic dynamics, or local adaptation. We, however, still lack a general theory for the ontogeny of dispersal that would allow an understanding of the complex interactions between different ultimate causes for dispersal, between different environment effects at different stages of the life history (Ronce et al. 2001), and how such interactions shape the whole dispersal kernel (Ronce et al. 2001; see Section 10). The focus on long-term evolutionary equilibrium has also prevented theory from saying much about transient patterns and the rates of dispersal evolution following ongoing changes, such as climate warming or habitat fragmentation. It is, however, with the latter type of evidence that empiricists are currently confronted. Designing different types of models aiming at describing short-term patterns of evolution, both grounded on empirical estimates of heritability for dispersal characteristics and incorporating the complex demographic and genetic feedbacks revealed by previous theory, might represent a new theoretical challenge.

Refined measurements of dispersal variation in nature (Schtickzelle et al. 2006) and the identification of candidate genes affecting this behavior (Haag et al. 2005) should lead to better empirical insights into dispersal evolution. Given the multiple selective forces and feedbacks affecting dispersal, descriptive approaches, however, may tell us little about the evolutionary mechanisms at stake. Manipulative approaches are unfortunately seldom used in the study of dispersal and have mainly explored its plasticity. Proximate causes for dispersal, however, might not inform us about the ultimate causes having shaped the evolution of its reaction norm. Further development of experiments of the artificial evolution of dispersal with short-lived organisms (Friedenberg 2003), aiming at testing theoretical predictions, is therefore encouraged.

SUMMARY POINTS

1. Expression of dispersal behavior is sensitive to many aspects of the environment in the broad sense (including individual condition, social context, and abiotic factors). Differences among genotypes in their dispersal properties also depend on the environmental context.

2. Some model assumptions preclude the study of particular forces acting on dispersal evolution. This allows the disentanglement of selective pressures but spreads confusion when it is not clearly acknowledged.

3. The consequences of inbreeding depression for the evolution of dispersal cannot be understood without taking into account its complex interactions with kin competition.

4. Distinct consequences of dispersal may be advantageous in ephemeral habitats but feedbacks between population dynamics and the evolution of dispersal make the relationship between dispersal and local extinction frequency highly nonlinear.

5. Potential costs of dispersal are multiple, yet empirical evidence is ambiguous. Both data and theory suggest that these costs can evolve.

6. Theory about dispersal evolution has given us the opportunity to much refine our understanding of evolution in spatially structured systems.

FUTURE ISSUES

1. We need more information about both the heritability of dispersal kernels and the heritability of dispersal reaction norms to environmental cues.

2. We need models to focus on the evolution of the distribution of dispersal distances to better understand how the different evolutionary forces shape such distribution and to better relate theoretical predictions to data.

3. We need to better understand empirically and model theoretically the ontogeny of dispersal to predict the evolution of environmental effects on dispersal.

4. We need to produce short-term predictions about evolutionary changes in dispersal in the context of global change.

DISCLOSURE STATEMENT

The author is not aware of any biases that might be perceived as affecting the objectivity of this review.

ACKNOWLEDGMENTS

I thank I. Olivieri, F. Rousset, E. Imbert, V. Calcagno, J. Robledo-Arnuncio, S. Gandon, T. Lenormand, J. Clobert, and M. Whitlock for many comments, discussions, and useful suggestions, and the French Ministry of Research for funding through the ANR program (contract ANR-05-BDIV-014) and the ACI "Ecologie quantitative" program (contract allocated to I. Olivieri and D. Jolly). This is publication ISEM-2007-006 of the Institut des Sciences de l'Evolution de Montpellier.

LITERATURE CITED

Ajar E. 2003. Analysis of disruptive selection in subdivided populations. *BMC Evol. Biol.* 3:22

Balkau BJ, Feldman MW. 1973. Selection for migration modification. *Genetics* 74:171–74

Barton NH. 2001. The evolutionary consequences of gene flow and local adaptation: future approaches. See Clobert et al. 2001, pp. 329–40

Bélichon S, Clobert J, Massot M. 1996. Are there differences in fitness components between philopatric and dispersing individuals? *Acta Oecol.* 17:503–17

Billiard S, Lenormand T. 2005. Evolution of migration under kin selection and local adaptation. *Evolution* 59:13–23

Bolker BM, Pacala SW. 1999. Spatial moment equations for plant competition: understanding spatial strategies and the advantages of short dispersal. *Am. Nat.* 153:575–602

Boudjemadi K, Lecomte J, Clobert J. 1999. Influence of connectivity on demography and dispersal in two contrasting habitats: an experimental approach. *J. Anim. Ecol.* 68:1207–24

Bowler DE, Benton TG. 2005. Causes and consequences of animal dispersal strategies: relating individual behaviour to spatial dynamics. *Biol. Rev.* 80:205–25

Brown JH, Kodric-Brown A. 1977. Turnover rates in insular biogeography: effect of migration on extinction. *Ecology* 58:445–49

Bullock JM, Kenward RE, Hails RS, eds. 2002. *Dispersal Ecology.* Oxford, UK: Blackwell

Cadet C, Ferrière R, Metz JAJ, van Baalen M. 2003. The evolution of dispersal under demographic stochasticity. *Am. Nat.* 162:427–41

Cheplick GP, Kane KH. 2004. Genetic relatedness and competition in *Triplasis purpurea* (Poaceae): resource partitioning or kin selection? *Int. J. Plant Sci.* 165:623–30

Clobert J, Danchin E, Dhondt AA, Nichols JD, eds. 2001. *Dispersal.* New York: Oxford Univ. Press

Clobert J, Ims RA, Rousset F. 2004. Causes, mechanisms and consequences of dispersal. In *Ecology, Genetics and Evolution of Metapopulations*, ed. I Hanski, O Gaggiotti, pp. 307–35. Amsterdam: Academic

Cody ML, Overton JM. 1996. Short-term evolution of reduced dispersal in island plant populations. *J. Ecol.* 84:53–61

Colas B, Olivieri I, Riba M. 1997. *Centaurea corymbosa*, a cliff dwelling species tottering on the brink of extinction: a demographic and genetic study. *Proc. Natl. Acad. Sci. USA* 94:3471–76

Comins HN. 1982. Evolutionarily stable strategies for localized dispersal in two dimensions. *J. Theor. Biol.* 94:579–606

Comins HN, Hamilton WD, May RM. 1980. Evolutionary stable dispersal strategies. *J. Theor. Biol.* 82:205–30

Cote J, Clobert J. 2007. Social personalities influence natal dispersal in a lizard. *Proc. R. Soc. London Ser. B Biol. Sci.* 274:383–90

Crowley PH, McLetchie DN. 2002. Trade-offs and spatial life-history strategies in classical metapopulations. *Am. Nat.* 159:190–208

Denno RF, Hawthorne DJ, Thorne BL, Gratton C. 2001. Reduced flight capability in British Virgin Island populations of a wing-dimorphic insect: the role of habitat isolation, persistence, and structure. *Ecol. Entomol.* 26:25–36

A complete synthesis of the evolutionary ecology of dispersal, with a large emphasis on dispersal plasticity.

Denno RF, Roderick GK, Peterson MA, Huberty AF, Döbel HG, et al. 1996. Habitat persistence underlies intraspecific variation in the dispersal strategies of plant-hoppers. *Ecol. Monogr.* 66:389–408

Dingle H. 1996. *Migration: The Biology of Life on the Move.* New York: Oxford Univ. Press

Doebeli M, Ruxton GD. 1997. Evolution of dispersal rates in metapopulation models: branching and cyclic dynamics in phenotype space. *Evolution* 51:1730–41

Doligez B, Cadet C, Danchin E, Boulinier T. 2003. When to use public information for breeding habitat selection? The role of environmental predictability and density dependence. *Anim. Behav.* 66:973–88

Doligez B, Danchin E, Clobert J. 2002. Public information and breeding habitat selection in a wild bird population. *Science* 297:1168–70

Donohue K. 1999. Seed dispersal as a maternally influenced character: mechanistic basis of maternal effects and selection on maternal characters in an annual plant. *Am. Nat.* 154:674–89

Donohue K, Polisetty CR, Wender NJ. 2005. Genetic basis and consequences of niche construction: plasticity-induced genetic constraints on the evolution of seed dispersal in *Arabidopsis thaliana*. *Am. Nat.* 165:537–50

Dytham C, Travis JMJ. 2006. Evolving dispersal and age at death. *Oikos* 113:530–38

Ezoe H. 1998. Optimal dispersal range and seed size in a stable environment. *J. Theor. Biol.* 190:287–93

Ferrière R, Le Galliard JF. 2001. Invasion fitness and adaptive dynamics in spatial population models. See Clobert at al. 2001, pp. 57–79

Fontanillas P, Petit E, Perrin N. 2004. Estimating sex-specific dispersal rates with autosomal markers in hierarchically structured populations. *Evolution* 58:886–94

Frank SA. 1986. Dispersal polymorphism in subdivided populations. *J. Theor. Biol.* 122:303–9

Fréville H, Colas B, Riba M, Caswell H, Mignot A, et al. 2004. Spatial and temporal demographic variability in the endemic plant species *Centaurea corymbosa* (Asteraceae). *Ecology* 85:694–703

Friedenberg NA. 2003. Experimental evolution of dispersal in spatiotemporally variable microcosms. *Ecol. Lett.* 6:953–59

Gandon S. 1999. Kin competition, the cost of inbreeding and the evolution of dispersal. *J. Theor. Biol.* 200:345–64

Gandon S, Capowiez Y, Dubois Y, Michalakis Y, Olivieri I. 1996. Local adaptation and gene-for-gene coevolution in a metapopulation model. *Proc. R. Soc. London Ser. B Biol. Sci.* 263:1003–9

Gandon S, Michalakis Y. 1999. Evolutionary stable dispersal rate in a metapopulation with extinctions and kin competition. *J. Theor. Biol.* 3:275–90

Gandon S, Rousset F. 1999. Evolution of stepping-stone dispersal rates. *Proc. R. Soc. London Ser. B Biol. Sci.* 266:2507–13

Geritz SAH, Kisdi E, Meszena G, Metz JAJ. 1998. Evolutionarily singular strategies and the adaptive growth and branching of the evolutionary tree. *Evol. Ecol.* 12:35–57

Glémin S, Ronfort J, Bataillon T. 2003. Patterns of inbreeding depression and architecture of the load in subdivided populations. *Genetics* 165:2193–212

First attempt to measure heritability and genetic correlations of dispersal kernels' characteristics.

Investigates the evolution of dispersal in the lab, using *C. elegans* in artificial landscapes.

Guillaume F, Perrin N. 2006. Joint evolution of dispersal and inbreeding load. *Genetics* 173:497–509

Gundersen G, Andreassen HP, Ims RA. 2002. Individual and population level determinants of immigration success on local habitat patches: an experimental approach. *Ecol. Lett.* 5:294–301

Gyllenberg M, Parvinen K, Dieckmann U. 2002. Evolutionary suicide and evolution of dispersal in structured metapopulations. *J. Math. Biol.* 45:79–105

Haag CR, Saastamoinen M, Marden JH, Hanski I. 2005. A candidate locus for variation in dispersal rate in a butterfly metapopulation. *Proc. R. Soc. London Ser. B Biol. Sci.* 272:2449–56

Hamilton WD. 1964. The genetical evolution of social behaviour, I. *J. Theor. Biol.* 7:1–16

Hamilton WD, May RM. 1977. Dispersal in stable habitats. *Nature* 269:578–81

Hanski I, Alho J, Moilanen A. 2000. Estimating the parameters of survival and migration of individuals in metapopulations. *Ecology* 81:239–51

Hanski I, Eralahti C, Kankare M, Ovaskainen O, Siren H. 2004. Variation in migration propensity among individuals maintained by landscape structure. *Ecol. Lett.* 7:958–66

Harada Y. 1999. Short- vs. long-range disperser: the evolutionarily stable allocation in a lattice-structured habitat. *J. Theor. Biol.* 201:171–87

Hastings A. 1983. Can spatial variation alone lead to selection for dispersal? *Theor. Popul. Biol.* 24:244–51

Heino M, Hanski I. 2001. Evolution of migration rate in a spatially realistic metapopulation model. *Am. Nat.* 157:495–511

Higgins K, Lynch M. 2001. Metapopulation extinction caused by mutation accumulation. *Proc. Natl. Acad. Sci. USA* 98:2928–33

Higgins SI, Nathan R, Cain ML. 2003. Are long-distance dispersal events in plants usually caused by nonstandard means of dispersal? *Ecology* 84:1945–56

Hill JK, Thomas CD, Lewis OT. 1999. Flight morphology in fragmented populations of a rare British butterfly, *Hesperia comma*. *Biol. Conserv.* 87:277–83

Hovestadt T, Messner S, Poethke HJ. 2001. Evolution of reduced dispersal mortality and 'fat-tailed' dispersal kernels in autocorrelated landscapes. *Proc. R. Soc. London Ser. B Biol. Sci.* 268:385–91

Imbert E. 2001. Capitulum characters in a seed heteromorphic plant, *Crepis sancta* (Asteraceae): variance partitioning and inference for the evolution of dispersal rate. *Heredity* 86:78–86

Imbert E, Ronce O. 2001. Phenotypic plasticity for dispersal ability in the seed heteromorphic *Crepis sancta* (Asteraceae). *Oikos* 93:126–34

Ims RA, Andreassen HP. 2000. Spatial synchronization of vole population dynamics by predatory birds. *Nature* 408:194–96

Ims RA, Hjermann D. 2001. Condition dependent dispersal. In *Dispersal*, ed. J Clobert, E Danchin, A Dhondt, JD Nichols, pp. 203–16. New York: Oxford Univ. Press

Irwin AJ, Taylor PD. 2000. Evolution of dispersal in a stepping-stone population with overlapping generations. *Theor. Popul. Biol.* 58:321–28

First theoretical demonstration of the effect of kin selection on dispersal.

Jánosi IM, Scheuring I. 1997. On the evolution of density dependent dispersal in a spatially structured population model. *J. Theor. Biol.* 187:397–408

Kalisz S, Hanzawa FM, Tonsor SJ, Thiede DA, Voigt S. 1999. Ant-mediated seed dispersal alters pattern of relatedness in a population of *Trillium grandiflorum*. *Ecology* 80:2620–34

Kisdi E. 2002. Dispersal: risk spreading versus local adaptation. *Am. Nat.* 159:579–96

Kisdi E. 2004. Conditional dispersal under kin competition: extension of the Hamilton-May model to brood size-dependent dispersal. *Theor. Popul. Biol.* 66:369–80

Kokko H, Lopez-Sepulcre A. 2006. From individual dispersal to species ranges: perspectives for a changing world. *Science* 313:789–91

Lambin X, Aars J, Piertney SB. 2001. Interspecific competition, kin competition and kin facilitation: a review of empirical evidence. See Clobert et al. 2001, pp. 110–22

Le Galliard JF, Ferrière R, Clobert J. 2003. Mother-offspring interactions affect natal dispersal in a lizard. *Proc. R. Soc. Lond. Ser. B Biol. Sci.* 270:1163–69

Le Galliard JF, Ferrière R, Dieckmann U. 2005. Adaptive evolution of social traits: origin, trajectories, and correlations of altruism and mobility. *Am. Nat.* 165:206–24

Lehmann L, Perrin N. 2003. Inbreeding avoidance through kin recognition: Choosy females boost male dispersal. *Am. Nat.* 162:638–52

Leibold MA, Holyoak M, Mouquet N, Amarasekare P, Chase JM, et al. 2004. The metacommunity concept: a framework for multi-scale community ecology. *Ecol. Lett.* 7:601–13

Lenormand T. 2002. Gene flow and the limits to natural selection. *Trends Ecol. Evol.* 17:183–89

Leturque H, Rousset F. 2002. Dispersal, kin competition, and the ideal free distribution in a spatially heterogeneous population. *Theor. Popul. Biol.* 62:169–80

Leturque H, Rousset F. 2004. Intersexual competition as an explanation for sex-ratio and dispersal biases in polygynous species. *Evolution* 58:2398–408

Levin SA, Cohen D, Hastings A. 1984. Dispersal strategies in patchy environments. *Theor. Popul. Biol.* 26:165–91

Levin SA, Muller-Landau HC, Nathan R, Chave J. 2003. The ecology and evolution of seed dispersal: a theoretical perspective. *Annu. Rev. Ecol. Evol. Syst.* 34:575–604

Massot M, Clobert J, Lorenzon P, Rossi J-M. 2002. Condition-dependent dispersal and ontogeny of the dispersal behaviour: an experimental approach. *J. Anim. Ecol.* 71:253–61

Mathias A, Kisdi E, Olivieri I. 2001. Divergent evolution of dispersal in a heterogeneous landscape. *Evolution* 55:246–59

Matter SF. 2006. Changes in landscape structure decrease mortality during migration. *Oecologia* 150:8–16

McPeek MA, Holt RD. 1992. The evolution of dispersal in spatially and temporally varying environments. *Am. Nat.* 140:1010–27

Metz JAJ, Gyllenberg M. 2001. How should we define fitness in structured metapopulation models? Including an application to the calculation of ES dispersal strategies. *Proc. R. Soc. London Ser. B Biol. Sci.* 268:499–508

First theoretical demonstration that conditional dispersal may much attenuate the cost of dispersal.

Molofsky J, Ferdy JB. 2005. Extinction dynamics in experimental metapopulations. *Proc. Natl. Acad. Sci. USA* 102:3726–31

Morgan AD, Gandon S, Buckling A. 2005. The effect of migration on local adaptation in a coevolving host-parasite system. *Nature* 437:253–56

Motro U. 1991. Avoiding inbreeding and sibling competition: the evolution of sexual dimorphism for dispersal. *Am. Nat.* 137:108–15

Muller-Landau HC, Levin SA, Keymer JE. 2003. Theoretical perspectives on evolution of long-distance dispersal and the example of specialized pests. *Ecology* 84:1957–67

Murrell DJ, Travis JMJ, Dytham C. 2002. The evolution of dispersal distance in spatially-structured populations. *Oikos* 97:229–36

O'Riain MJ, Jarvis JUM. 1997. Colony member recognition and xenophobia in the naked mole-rat. *Anim. Behav.* 53:487–98

Olivieri I. 2001. The evolution of seed heteromorphism in a metapopulation: interactions between dispersal and dormancy. In *Integrating Ecology and Evolution in a Spatial Context*, ed. J Silvertown, J Antonovics, pp. 245–68. Oxford, United Kingdom: Br. Ecol. Soc. & Blackwell Sci.

Olivieri I, Gouyon P-H. 1985. Seed dimorphism for dispersal: theory and implications. In *Structure and Functioning of Plant Populations*, ed. J Haeck, JW Woldendrop, pp. 77–90. Amsterdam: North-Holland Pub.

Olivieri I, Gouyon P-H. 1997. Evolution of migration rate and other traits: the metapopulation effect. In *Metapopulation Biology: Ecology, Genetics, and Evolution*, ed. I Hanski, ME Gilpin, pp. 293–323. San Diego: Academic

Olivieri I, Michalakis Y, Gouyon P-H. 1995. Metapopulation genetics and the evolution of dispersal. *Am. Nat.* 146:202–28

Parvinen K. 1999. Evolution of migration in a metapopulation. *Bull. Math. Biol.* 61:531–50

Parvinen K. 2002. Evolutionary branching of dispersal strategies in structured metapopulations. *J. Math. Biol.* 45:106–24

Parvinen K, Dieckmann U, Gyllenberg M, Metz JAJ. 2003. Evolution of dispersal in metapopulations with local density dependence and demographic stochasticity. *J. Evol. Biol.* 16:143–53

Pen I. 2000. Reproductive effort in viscous populations. *Evolution* 54:293–97

Peroni PA. 1994. Seed size and dispersal potential of *Acer rubrum* (Aceraceae) samaras produced by populations in early and late successional environments. *Am. J. Bot.* 81:1428–34

Perrin N, Goudet J. 2001. Inbreeding and dispersal. See Clobert et al. 2001, pp. 123–42

Perrin N, Mazalov V. 2000. Local competition, inbreeding, and the evolution of sex-biased dispersal. *Am. Nat.* 155:116–27

Phillips BL, Brown GP, Webb JK, Shine R. 2006. Invasion and the evolution of speed in toads. *Nature* 439:803

Poethke HJ, Hovestadt T. 2002. Evolution of density-and-patch-size-dependent dispersal rates. *Proc. R. Soc. London Ser. B Biol. Sci.* 269:637–45

Poethke HJ, Hovestadt T, Mitesser O. 2003. Local extinction and the evolution of dispersal rates: causes and correlations. *Am. Nat.* 161:631–40

Ravigné V, Olivieri I, Gonzalez-Martinez SC, Rousset F. 2006. Selective interactions between short distance pollen and seed dispersal in self-compatible species. *Evolution* 60:2257–71

Roff DA. 1986. The evolution of wing dimorphism in insects. *Evolution* 40:1009–20

Roff DA. 1990. The evolution of flightlessness in insects. *Ecol. Monogr.* 60:389–421

Roff DA, Fairbarn DJ. 2001. The genetic basis of dispersal and migration and its consequences for the evolution of correlated traits. See Clobert et al. 2001, pp. 191–202

Ronce O, Brachet S, Olivieri I, Gouyon PH, Clobert J. 2005. Plastic changes in seed dispersal along ecological succession: theoretical predictions from an evolutionary model. *J. Ecol.* 93:431–40

Ronce O, Clobert J, Massot M. 1998. Natal dispersal and senescence. *Proc. Natl. Acad. Sci. USA* 95:600–5

Ronce O, Gandon S, Rousset F. 2000a. Kin selection and natal dispersal in an age-structured population. *Theor. Popul. Biol.* 58:143–59

Ronce O, Olivieri I. 2004. Life history evolution in metapopulations. In *Ecology, Genetics, and Evolution of Metapopulations*, ed. I Hanski, OE Gaggiotti, pp. 227–58. Amsterdam: Academic

Ronce O, Olivieri I, Clobert J, Danchin E. 2001. Perspectives on the study of dispersal evolution. See Clobert et al. 2001, pp. 341–57

Ronce O, Perret F, Olivieri I. 2000b. Evolutionarily stable dispersal rates do not always increase with local extinction rates. *Am. Nat.* 155:485–96

Ronce O, Perret F, Olivieri I. 2000c. Landscape dynamics and evolution of colonizer syndromes: interactions between reproductive effort and dispersal in a metapopulation. *Evol. Ecol.* 14:233–60

Rousset F. 2006. Separation of time scales, fixation probabilities and convergence to evolutionarily stable states under isolation by distance. *Theor. Popul. Biol.* 69:165–79

Rousset F, Gandon S. 2002. Evolution of the distribution of dispersal distance under distance-dependent cost of dispersal. *J. Evol. Biol.* 15:515–23

Rousset F, Ronce O. 2004. Inclusive fitness for traits affecting metapopulation demography. *Theor. Popul. Biol.* 65:127–41

Roze D, Rousset F. 2003. Selection and drift in subdivided populations: a straightforward method for deriving diffusion approximations and applications involving dominance, selfing and local extinctions. *Genetics* 165:2153–66

Roze D, Rousset F. 2005. Inbreeding depression and the evolution of dispersal rates: a multilocus model. *Am. Nat.* 166:708–21

Schtickzelle N, Mennechez G, Baguette M. 2006. Dispersal depression with habitat fragmentation in the bog fritillary butterfly. *Ecology* 87:1057–65

Simmons AD, Thomas CD. 2004. Changes in dispersal during species' range expansions. *Am. Nat.* 164:378–95

Sinervo B, Clobert J. 2003. Morphs, dispersal behavior, genetic similarity, and the evolution of cooperation. *Science* 300:1949–51

Sorci G, Massot M, Clobert J. 1994. Maternal parasite load increase sprint speed and philopatry in female offspring of the common lizard. *Am. Nat.* 144:153–64

Analyzes how kin competition shapes the evolution of the whole dispersal kernel.

Quantifies the interactions between direct and indirect selection on dispersal (see also Billiard & Lenormand 2005), and illustrates how kin selection operates in multilocus models.

Tallmon DA, Luikart G, Waples RS. 2004. The alluring simplicity and complex reality of genetic rescue. *Trends Ecol. Evol.* 19:489–96

Taylor PD. 1988. An inclusive fitness model for dispersal of offspring. *J. Theor. Biol.* 130:363–78

Taylor PD, Frank SA. 1996. How to make a kin selection model. *J. Theor. Biol.* 180:27–37

Travis JMJ, Dytham C. 1999. Habitat persistence, habitat availability and the evolution of dispersal. *Proc. R. Soc. London Ser. B Biol. Sci.* 266:723–28

Travis JMJ, Dytham C. 2002. Dispersal evolution during invasions. *Evol. Ecol. Res.* 4:1119–29

Travis JMJ, French DR. 2000. Dispersal functions and spatial models: expanding our dispersal toolbox. *Ecol. Lett.* 3:163–65

van Baalen M, Rand DA. 1998. The unit of selection in viscous populations and the evolution of altruism. *J. Theor. Biol.* 193:631–48

van Valen L. 1971. Group selection and the evolution of dispersal. *Evolution* 25:591–98

Venable DL, Brown JS. 1988. The selective interactions of dispersal, dormancy, and seed size as adaptations for reducing risk in variable environments. *Am. Nat.* 131:360–84

Whitlock MC, Ingvarsson PK, Hatfield T. 2000. Local drift load and the heterosis of interconnected populations. *Heredity* 84:452–57

Wiener P, Feldman MW. 1993. The effects of the mating system on the evolution of migration in a spatially heterogeneous population. *Evol. Ecol.* 7:251–69

Willi Y, Fischer M. 2005. Genetic rescue in interconnected populations of small and large size of the self-incompatible *Ranunculus reptans*. *Heredity* 95:437–43

Wolff JO. 1992. Parents suppress reproduction and stimulate dispersal in opposite-sex juvenile white-footed mice. *Nature* 359:409–10

Wright S. 1969. *The Theory of Gene Frequencies*. Chicago: Univ. Chicago Press

Historically, the first model of dispersal evolution.

Exploring Cyanobacterial Mutualisms

Kayley M. Usher,[1] Birgitta Bergman,[2]
and John A. Raven[3]

[1]School of Plant Biology, The University of Western Australia, Crawley, Western Australia, 6009 Australia; email: kusher@cyllene.uwa.edu.au

[2]Department of Botany, Stockholm University, SE-106 91 Stockholm, Sweden; email: birgitta.bergman@botan.su.se

[3]Plant Research Unit, University of Dundee at SCRI, Scottish Crop Research Institute, Invergowrie, Dundee DD2 5DA, United Kingdom; email: j.a.raven@dundee.ac.uk

Annu. Rev. Ecol. Evol. Syst. 2007. 38:255–73

First published online as a Review in Advance on August 6, 2007

The *Annual Review of Ecology, Evolution, and Systematics* is online at
http://ecolsys.annualreviews.org

This article's doi:
10.1146/annurev.ecolsys.38.091206.095641

Key Words

evolution of symbiosis, symbiosis, transmission

Abstract

Cyanobacterial symbioses with eukaryotes are ancient associations that are widely distributed in aquatic and terrestrial environments. Cyanobacteria are a significant driving force in the evolution of their hosts, providing a range of services including photosynthesis, nitrogen fixation, UV protection, and defensive toxins. Although widespread, cyanobacteria occur in a limited range of hosts. Terrestrial symbioses are typically restricted to lichens and early evolved plants, and aquatic symbioses to sessile or slow-moving organisms. This review examines the underlying evolutionary processes that may have lead to these patterns. It also examines the facts that the degree of integration between symbiont and host, and the mode of transmission of the symbiont, do not appear to be an indication of how old the symbiosis is or how important it is to host well-being. Biparental transmission of symbionts may prolong the survival of gametes that persist in the environment, increasing chances of fertilization.

SYMBIOTIC CYANOBACTERIA

We use the word symbiosis in a sense similar to that of de Bary (1879), i.e., "the living together of differently named organisms." This definition includes mutualism and parasitism. Cyanobacteria established symbioses with eukaryotes at least 2.1 billion years ago in an association that ultimately led to chloroplasts (Lockhart et al. 1992, Raven 2002a,b; cf. Kopp et al. 2005). It is possible that the movement of cyanobacteria to land was assisted by fungi via the establishment of a symbiosis to form lichens (Taylor et al. 1995). Recent evidence from fossilized lichens suggests that cyanobacteria and/or algae evolved symbioses with fungi between 551 and 635 Mya in shallow subtidal regions (Yuan et al. 2005), more than 150 million years before the first known lichens on land (Taylor et al. 1995).

Today cyanobacterial symbioses are widely distributed in aquatic and terrestrial environments and involve taxonomically diverse hosts. In the ocean some diatoms, dinoflagellates, sponges, cnidarians, ascidians, echiuroid worms, and corals contain cyanobacterial symbionts (Carpenter 2002; Carpenter & Foster 2002; Foster et al. 2006; Janson 2002; Karl et al. 2002; Lesser et al. 2004; Raven 2002a,b). There are also intertidal (Janson et al. 1993, Raven et al. 1990), and even fully submerged, cyanolichens (Carpenter & Foster 2002). Among the embryophytes (land plants) cyanobacterial symbionts are found in liverworts, hornworts, and mosses (bryophytes), the water-fern *Azolla* (pteridophytes), cycads (gymnosperms) and the flowering plant *Gunnera* (angiosperm) (Bergman et al. 2007; DeLuca et al. 2002; Raven 2002a,b). The floating fern *Azolla* is the only well-characterized freshwater plant-cyanobacterial symbiosis (Lechno-Yosef & Nierzwicki-Bauer 2002). Among diatoms with cyanobionts, all species of *Denticula*, some species of *Epithemia*, and a few species of *Rhopalodia* live in freshwater (Krammer & Lange-Bertalot 1988).

The other main symbiosis involving cyanobacteria on land is in lichens. Half (about 13,500) of the ascomycete fungi are lichenized (Kirk et al. 2001), although only a minority have cyanobionts (10% of lichens are bipartite cyanolichens and 3–4% are tripartite cyanolichens: see Rai & Bergman 2002), with the rest containing algal symbionts. In all, there are about 1000 species of cyanolichens, most of which have an ascomycete fungal partner (Rai & Bergman 2002), although there is also the enigmatic glomeromycete *Geosiphon* that is symbiotic with *Nostoc* (Kluge et al. 2002).

Cyanobacteria allow nonphotosynthetic partners to obtain energy from the sun, for example, marine invertebrates and terrestrial and marine lichens (only bipartite cyanolichens). However, in plant-cyanobacterial symbioses, tripartite cyanolichens, diatoms, and possibly in photosynthetic corals, the cyanobiont makes a negligible contribution to the energy and C (carbon) budget of the association, even when the cyanobacteria are exposed to sunlight. Most of these cyanobionts fix N_2 (nitrogen gas), a function of crucial importance that permits hosts to grow in areas that are low in combined N. Nitrogen fixation by cyanobacterial symbionts in a sponge was also reported by Wilkinson & Fay (1979), although others have been unable to find evidence of this (Borowitzka & Hinde 1999, Diaz & Ward 1997).

The cyanobionts may also provide UV protection to their symbiotic partners. Cyanobacteria produce two types of sunscreen compounds, mycosporine-like amino acids (MAAs: Shick & Dunlap 2002) and scytonemin (Proteau et al. 1993), both of which contain N. These are produced by a wide range of cyanobacterial taxa and probably evolved very early in the cyanobacterial lineage (Garcia-Pichel 1998). In addition, cyanobacteria can produce a wide range of powerful toxins that include antibiotics and antifeedants (Borowitzka & Hinde 1999). Hepatotoxins such as microcystin and nodularin are produced by many planktonic cyanobacteria and have also been reported from a lichen cyanobiont (Moffitt & Neilan 2004, Oksanen et al. 2004). β-N-methylamino-L-alanine (BMAA), a potent neurotoxin, is produced by the majority of *Nostoc* strains isolated from lichens and a range of plants, including *Cycas micronesica*, *Azolla filiculoides*, and *Gunnera kauaiensis* (Cox et al. 2005). Free-living members of all five cyanobacterial sections also produce BMAA (Cox et al. 2005). Such a potentially ancient chemical defense may have promoted the development of mutualistic symbioses with cyanobacteria by providing protection to invertebrate and plant hosts. In a similar manner, chemical defense is provided by some endophytic fungi for the roots and seeds of host grasses (Clay & Holah 1999, Saikkonen et al. 1998).

FACTORS AFFECTING THE EVOLUTION OF CYANOBACTERIAL SYMBIOSES

Given the multiple and significant benefits provided by cyanobacterial symbioses, the question arises as to why they are not more common. Terrestrial symbioses are restricted, involving fungi (lichens) and a limited range of plants, although these symbioses are distributed throughout the plant kingdom.

It is apparent that the selection pressure for the acquisition of endosymbiotic cyanobacteria in plants (chloroplasts) was to gain photosynthesis, a process that allows not only tapping of a renewable energy source via capturing light, but in addition provides a mechanism by which to fix C from the atmosphere. The enormous advantages gained in this monophyletic endosymbiotic event are reflected in the success of plants today. However, the reason why more plants have not taken advantage of cyanobionts for diazotrophy (nitrogen fixation) may be related to the fact that many plants have access to enough combined N in their natural environment, the lack of symbiotic competence in hosts, the costs associated with hosting cyanobacteria, or the ability to acquire N via alternative methods, such as parasitism, scavenging, or other types of symbioses (e.g., mycorrhizal). Plant-bacterial symbioses are generally rare; however, cyanobacteria may have been more competent at establishing symbiosis early in the evolution of plants than were heterotrophic bacteria, as indicated by their symbiosis with a range of ancient plants and lichens. *Frankia* and rhizobial symbioses appear to be recently evolved and restricted to angiosperms.

The selection pressure for acquiring fixed N via symbiosis may be far less pronounced for organisms today than when the first cyanobacterial N_2-fixing symbioses arose. This is suggested by the fact that present-day N_2-fixing cyanobacterial symbioses grow in N-poor areas and that the hosts are among the oldest representatives within their higher taxon. For instance, the earliest evidence for terrestrial cyanobacterial symbioses is from a fossilized lichen, with (probably) a glomeromycete (part of the zygomycetes *sensu lato*) as the mycobiont, and a coccoid cyanobiont, dating to 400 Mya (Taylor et al. 1995, 1997, 2004); earlier fossil lichens (551–635 Mya) were intertidal and may have had chlorobionts rather than cyanobionts. With the exception of the flowering plant *Gunnera* (dated, as *Tricolpites* pollen grains, to 90 Mya; Wanntorp et al. 2004), extant plant species with which cyanobacteria form symbioses are ancient lineages: cycads (265–290 Mya: Brenner et al. 2003), bryophytes (470–350 Mya), and ferns (350 Mya, although heterosporous water ferns such as *Azolla* evolved approximately 137 Mya; Schneider et al. 2004).

These organisms evolved in a warmer and generally wetter environment than exists today, and such conditions may have stimulated the development of cyanobacterial symbioses via a close and constant association of partners (**Figure 1**). Also, the rapid plant growth resulting from these conditions may have required larger quantities

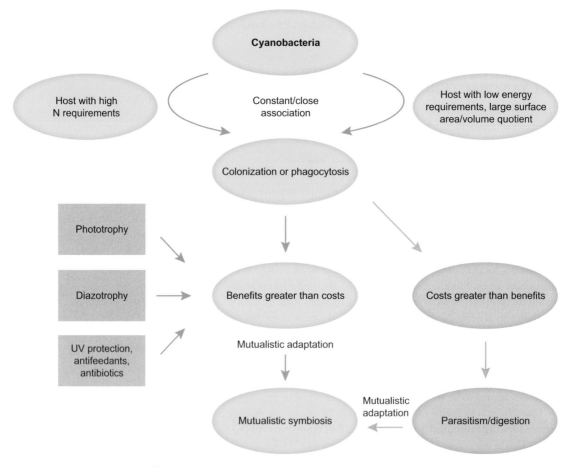

Figure 1.

Inputs and processes that predispose evolution toward cyanobacterial mutualism, parasitism, or digestion.

of N (Vessey et al. 2004). Indeed, most extant plant-cyanobacterial symbioses, and *Geosiphon*, grow in moist or wet habitats (bryophytes, the water-fern *Azolla*, the angiosperm *Gunnera*) (Osborne & Sprent 2002). The cyanobacterial infection units, i.e. the motile hormogonial stage of the life cycle, (Bergman et al. 2007) require at least a film of water in which to move to, and infect, a plant organ. During wet periods the cyanobacteria are metabolically active, and hormogonia may be induced as a response to released plant substances (Bergman et al. 2007). Once the cyanobiont is residing in a vascular plant (leaves, roots, or stem glands) it is protected from desiccation by the homoiohydric mechanisms of the plant (Raven & Sprent 1989, Sprent & Raven 1985). The lack of plant-cyanobacterial symbioses therefore hints at a requirement for wet conditions to facilitate the constant association of the partners while the mutualism is evolving. An on-again–off-again association may not cause the necessary (and initially costly) evolutionary changes to occur in the partners that would lead to a successful partnership. New symbioses should therefore primarily be expected to evolve in lakes and oceans in the present environmental conditions.

Terrestrial and Aquatic Metazoan Symbioses

That terrestrial animals have not formed symbioses with cyanobacteria is possibly due to the large energy costs of locomotion in such multicellular organisms. As a result of these costs, the energy provided by photosynthetic organisms would be small in comparison to the total energy required. In addition, the considerably smaller surface area to volume quotient of an animal, compared to a plant, would result in a much smaller ratio of symbiont cells to host cells, leading to such a small energy contribution that the costs of symbiosis may outweigh the benefits. This is more of a problem for terrestrial than for aquatic metazoans, as flying and walking/running are considerably more metabolically costly than swimming, where body mass is supported (Withers 1992). However, locomotion becomes energetically costly for marine animals that move fast. Slow-moving animals such as sacoglossan sea slugs with photosynthetically functional kleptoplastids (derived from seaweeds; Williams & Walker 1999) generate virtually no resistance from water when crawling, especially within the turbulent boundary layer. These sea slugs and the flatworm *Convoluta* (symbiotic with green microalgae) are relatively small and, as a result, have a large surface area to volume quotient. Some sea slugs increase this quotient further with leaf-like dorsal flaps (Williams & Walker 1999).

In addition, the very large water losses involved in acquiring CO_2 from the atmosphere may also affect the viability of photosynthetic symbioses with terrestrial metazoans. Plants need to have their stomata open during the day to capture enough C to fix because of the very low CO_2 content in air (380 ppmv). The combination of water loss from transpiration and water consumption during photosynthesis uses much more water than metazoan respiration [water vapor lost during the acquisition of O_2 (oxygen) from the atmosphere; Raven 1985] per molecule of metabolized gas consumed. Motile terrestrial animals, unlike almost all terrestrial plants, are not plumbed in to a supply of water, so the addition of photosynthesis by symbiosis may not be desirable. The small gain in energy that could be contributed through partial photosynthetic refixation of respired CO_2 (which could occur without loss of water) may come at too great a cost owing to the need to support the photosynthetic apparatus.

The attributes of slow motion in an aquatic environment and large surface area to volume quotient mean that some marine animals are able to take advantage of photosynthesis where other animals cannot. The least metabolically costly option is to not move at all, and a diverse array of sessile marine invertebrates pursue this strategy, taking advantage of the density of particulate organic matter in the water column. However, even remaining attached to the substratum in the face of currents and waves requires metabolic investment in structural material (Vogel 1974). The hard corals are well known for their dinoflagellate photobionts, whose contribution of energy complements that gained by catching prey. Other sessile invertebrates, including a wide range of sponges, some ascidians, and a cnidarian (Larkum & Kühl 2005, Lesser et al. 2004, Smith & Douglas 1987), contain cyanobacterial symbionts that possibly add the full repertoire of benefits from photosynthesis to chemical defense, to sunscreens, and N metabolism. Nitrogenase activity was detected in two sponges that contained cyanobacterial symbionts, but not in a third that lacked them Wilkinson & Fay 1979). However, marine *Synechococcus*, common in symbioses with marine sponges, lack *nif* genes essential for N_2 fixation but can use ammonium (Glover 1985), a waste product of animals.

The symbioses between cyanobacteria and sessile marine invertebrates that have been investigated occur in shallow waters. However, they are little studied, and examples that occur in deeper water may be discovered, given the ability of some cyanobacteria to photosynthesize in low light (Glover 1985, Larkum & Kühl 2005).

Richelia intracellularis is a heterocystous cyanobacterium that is an intracellular symbiont of some species of the planktonic diatoms *Hemiaulus* and *Rhizosolenia* (Carpenter 2002, Janson 2002, Janson et al. 1999). Some species of *Rhizosolenia* migrate vertically in surface ocean waters by changing the solute composition of the cell vacuole; these movements allow fixation of C (by the diatom) and N (by the cyanobiont) in the nutrient-limited surface waters, and migration from the photic zone allows them to take up phosphate from relatively high-phosphate but poorly illuminated deeper waters.

Although animals, and especially herbivores, can be N-limited in their natural environment, they generally rely (as do primary producers) on effective uptake and use of combined N. Some fruitflies, termites, and cockroaches on land and shipworms in the sea have diazotrophic bacterial symbionts that help to compensate for the low levels of N in some fruits and wood (Behar et al. 2005, Karl et al. 2002, Lo et al. 2003). Why cyanobacteria are not used for this purpose remains unknown, but microbes need to meet at least three criteria before a successful mutualistic (animal or plant) symbiosis can evolve:

1. *The benefits provided by the microbe must be equal to or greater than the costs of housing it*. The cost of synthesizing and maintaining the photosynthetic apparatus (of a cyanobiont) that is of no benefit to the host (e.g., plants) could represent a net cost in some situations. For example, the photosynthetic and

N_2-fixing apparatus requires Fe (iron), and this need can be a limiting resource in many habitats (Raven 2002a,b). Even in complete darkness cyanobacterial symbionts maintain photosystem I, with its high Fe requirement, potentially giving an additional Fe requirement for cyanobacterial, rather than bacterial, diazotrophic symbionts. Where mutualism arises from parasitism the requirement for greater benefits than costs must be met during the evolution of the association.

2. *The cyanobacterium must remain in close contact with the host for a mutualistic association to evolve.* Cyanobacteria are commonly widespread in terrestrial and aquatic environments, but forest soils are usually acidic and therefore contain few cyanobacteria (Rai et al. 2000) (although some forest mosses attract cyanobacteria; DeLuca et al. 2002). Cyanobacteria are not metabolically active in dry conditions, limiting their activities in terrestrial habitats where soils dry out periodically (see argument above for plant-cyanobacterial symbioses).

3. *Where other symbionts are present in a host, the addition of the new symbiont must not upset the operations of the consortium.* Plants and many animals that are symbiotic with cyanobacteria also have other microbial symbionts, and all plants have chloroplasts (originating from cyanobacteria). In these situations the symbionts are either fully compatible with each other or the host separates them physically. For example, highly compartmentalized microenvironments (leaf cavities) in *Azolla* separate cyanobacterial and bacterial symbionts from the host (Lechno-Yosef & Nierzwicki-Bauer 2002). However, a lack of compatibility between symbionts may not be readily overcome for some hosts.

Microbes other than cyanobacteria may be used by a host to provide phototrophy or diazotrophy. Which symbiont ultimately ends up providing the service required by a host may be an accident of evolution or may result from a particular symbiont providing a superior service to alternative microbes. For example, rhizobial/*Frankia* symbioses on land may fix more N_2 per unit of symbiont biomass than do cyanobacterial symbioses or free-living diazotrophs.

EVOLUTION OF CYANOBACTERIAL SYMBIOSES

Cyanobacterial symbionts vary widely in the degree of integration with their host. They may be compartmentalized intercellularly or intracellularly, grouped in specialized host vesicles or in organs such as bladders and cavities in fronds, thalli, root nodules, and stem glands. When compartmentalized intracellularly (as in *Gunnera* spp.), the cyanobacterium still remains outside the plant plasmalemma, like all other bacteria forming symbioses with angiosperms, such as *Frankia* and rhizobia (Bergman 2002, Vessey et al. 2004).

The degree of intimacy between symbiont and host is sometimes considered to reflect the importance of the symbiosis to the host, and obligate symbioses are often thought to be older. Because organelles evolved from intracellular symbionts and are sometimes considered to be the most advanced form of mutualism (Ewald 1987), it is tempting to regard intracellular symbionts as more advanced, and by implication more important, than intercellular ones. Vertical transmission is likewise considered to be important for the maintenance of stable mutualisms (Herre et al. 1999).

Certainly chloroplasts, the most ancient cyanobacterial symbiosis known, fit these criteria well. Chloroplasts evolved from obligate, intracellular cyanobacteria that were vertically transmitted, and most of the genes originally present in the cyanobiont genome are not now present. Of genes necessary for the functioning of the plastid, at least 90% are located in the host nuclear genome, while some other genes (probably including those unique to N_2 fixation) were lost (Martin et al. 2002). Further, about 18% of the nuclear genes of the flowering plant *Arabidopsis* (whose products are not targeted to plastids) were derived from the cyanobiont genome (Martin et al. 2002). The eukaryotic host cell–cyanobiont symbiosis that gave rise to chloroplasts is, arguably, one of the most important symbioses to have evolved. However, among the cyanobacterial symbioses with photosynthetic hosts that have been examined, the only obligate symbioses occur in the aquatic fern *Azolla* and the diatom *Rhopalodia*, which have intercellular cyanobionts (Rai et al. 2000), whereas the intracellular cyanobionts of the glomeromycete fungus *Geosiphon* are facultative. The cyanobacterial symbionts of sponges are usually intercellular, and this is thought to be a very ancient association (Brunton & Dixon 1994, Wilkinson 1992). The dinoflagellate/hard coral symbioses are considered to be obligate, yet in many cases the symbionts are transmitted horizontally (LaJeunesse et al. 2004), as are the N_2-fixing bacteria *Frankia* and Rhizobiaceae in plants. N_2-fixing plant-cyanobacterial symbioses encompass a range of horizontally transmitted symbioses (bryophytes, gymnosperms and the angiosperm *Gunnera*) and one vertically (the *Azolla* cyanobiont) transmitted one. However, irrespective of the mode of transmission, all have evolved into highly successful, long-lived, and persistent symbioses. It therefore appears that there is no pattern relating the mode of transmission of the symbiont, the host organ infected, or the inter- or intracellular location of cyanobionts to the importance of the symbiosis to host well being and reproduction, or the evolutionary age of the symbioses. This fact stresses the view that symbioses have evolved several times and in a variety of ways depending on circumstances and that broad generalizations regarding the importance of a symbiosis based on these characteristics may be misleading.

So how do mutualisms evolve, and how are they initiated and maintained without parasitism arising? Evidence suggests that the ability of each partner to use the by-products of the other is an important factor in establishing and maintaining mutualistic associations (Genkai-Kato & Yamamura 1999, Hay et al. 2004). Mutualisms may begin as commensal or parasitic associations that become beneficial over time. Cyanophages (cyanobacterial viruses) are an example of an infection that has evolved in a way that benefits the host (cyanobacteria) and therefore the symbiont. Many cyanophages of free-living, phototrophic marine cyanobacteria have acquired photosystem II genes that are expressed during infection, thus enhancing the photosynthetic capabilities of cyanobacteria (Sullivan et al. 2006).

Alternatively, organisms may be taken in during the feeding process and phagocytosed, then persist within the host. This second mechanism is the way by which chloroplasts evolved from cyanobacteria, giving rise to the glaucocystophytes, rhodophytes, and chlorophytes. A continuation of this process led to algae being engulfed by a range of organisms and becoming diatoms, cryptophytes and higher plants. In these secondary and tertiary endosymbioses the host obtained plastids and their genomes (and related nuclear genes) (Douglas & Raven 2003). A second example of the evolution of cyanobacterium to chloroplast may be the photosynthetic inclusions in the filose amoeba *Paulinella chromatophora* (Marin et al. 2005). These support the photoautotrophic existence of the amoeba and are a sister group to *Synechococcus/Prochlorococcus* cyanobacteria. It is not yet known if these inclusions are genetically reduced and should be classed as organelles.

Phagocytosis is also likely to be the mechanism by which cyanobacteria evolved mutualistic associations with many marine invertebrates, and photosynthetic saccoglossans also acquire chloroplasts (kleptoplasty) in this manner. Okamoto & Inouye (2005) report what may be an evolving example of phagocytosis leading to secondary endosymbiosis. The flagellate *Hatena* contains the green algae *Nephroselmis* in a strain-specific symbiosis. However, when a symbiotic *Hatena* divides, one cell inherits the algae while the other is asymbiotic. The asymbiotic cell must acquire its symbiont from the water, whereupon it loses its feeding apparatus and becomes autotrophic. Both the symbiont and the host undergo dramatic morphological changes when in symbiosis.

Once inside the host the cyanobacterial symbiont must accommodate to different levels of pH, O_2, and light and avoid or escape the host's immune response. A protective sheath is thought to help some microbes in metazoan hosts evade the immune system, but it is likely that in many cases mutualists are recognized and encouraged by the host. Growth of the host and reproduction of the symbiont must become tightly coupled so that an evolutionarily appropriate ratio is maintained (Raven 2004). The host may be responsible for constraining cell division of symbionts and may restrict nutrient supply (Bergman et al. 2007) or digest excess symbionts. It is also possible that quorum-sensing has a role in restricting growth of cyanobacteria in symbiosis. Quorum-sensing is the use of chemical signals by many gram-negative bacteria to monitor their own population density and modulate gene expression accordingly. There is some evidence to support the idea of interspecies communication using acyl-homoserine lactone (acyl-HSL) (Parsek & Greenberg 2000). This form of quorum-sensing is commonly found in gram-negative bacteria that interact with plant and animal hosts. An example of this phenomenon is the control of light production in *Vibrio fischeri*, which is mutualistic with certain marine animals (Parsek & Greenberg 2000). Another example is the marine alga *Ulva*, the zoospores of which settle on bacterial biofilms that are sensed via their production of quorum-sensing signal molecules (Joint et al. 2002).

Horizontal versus Vertical Transmission—Is Vertical Always Best?

Vertical transmission is widely considered to be a key factor in the evolution and maintenance of mutualisms. It typically occurs via asexual or sexual propagules, and where cyanobacteria are involved, it may increase the survival time of gametes in the environment by providing energy from photosynthesis. It is almost always maternal (via eggs in oogamous organisms and in megasporocarps in *Azolla*), although transmission of cyanobacterial symbionts in sponge sperm has been reported (Usher et al. 2005). Theory predicts that host reproductive systems will evolve toward uniparental inheritance in order to exclude parasites (Law & Hutson 1992), but as host recognition is required for the uptake and incorporation of mutualists into eggs, limiting the uptake of parasites may not be a problem in some associations. Vertical transmission ensures that offspring have immediate access to their symbionts, giving hosts a competitive edge from an early age.

So why don't all mutualisms demonstrate vertical transmission? Wilkinson & Sherratt (2001) identify four factors that are commonly cited as favoring vertical transmission of symbionts in mutualistic associations.

1. The success of the symbiont is tied to the reproduction of the host, so it does not pay for the symbiont to overexploit the host. However, this argument has vertical transmission as a premise and mutualism as a conclusion, rather than vice versa. Nevertheless, vertically transmitted symbionts have a greatly increased likelihood of reaching a suitable host, driving the evolution of the symbiont to favor vertical transmission.

2. Vertical transmission avoids the danger of offspring being unable to locate a suitable symbiont.

3. Vertical transmission leads to symbiont-sorting and genetic uniformity of symbionts within a host, thereby avoiding the evolution of parasitism.

4. Vertical transmission can allow whole communities of symbionts to be inherited, ensuring that all the microbes necessary for success are obtained at an early stage (this hypothesis runs counter to the symbiont-sorting idea).

In the complex world of symbiosis, some of these factors probably apply some of the time, often together with other factors. However, parasitism is unlikely to arise if interacting individuals can respond directly to one another's actions, and a cheating mutualist is unlikely to gain an advantage by withholding its by-products, as these are produced at little cost (Wilkinson & Sherratt 2001). Where local horizontal transmission occurs, hosts and symbionts will have a high chance of encountering one another (Wilkinson & Sherratt 2001), so vertical transmission is not necessary. In addition, evolving a functional process for vertical transmission may not be readily achieved. Horizontal transmission does not require the host to provide costly mechanisms to ensure their offspring have a complement of symbionts and also gives the host a choice as to which symbionts it associates with. Cyanobionts vary in their effectiveness and competitiveness (Nilsson et al. 2005), and horizontal transmission allows the host to obtain the most efficient cyanobiont species from the environment in which they find themselves.

Obtaining symbionts horizontally from the environment (typical in plant symbioses) requires recognition during the establishment phase to avoid the uptake of nonbeneficial symbionts. However, there is a trade-off between having the flexibility to establish symbiosis with a variety of symbionts and having the ability to be highly selective in the uptake of horizontally transmitted symbionts. A host that has the flexibility to use a different symbiont in different environments may take up a less effective symbiont when a better one is available, as in the case of the flatworm

Convoluta roscoffensis and its symbiotic algae. Some flatworms take up the less effective of two algal species and are significantly less successful as a result (Douglas 1998). Plants are often specific to cyanobacteria from the genus *Nostoc*, and recognition involves control by both partners, with the host responding to cell surface markers of the cyanobacterium and the symbiont responding to chemical signals to identify suitable hosts. Some cyanobacterial strains are more efficient at colonizing hosts than others, and competition between closely related *Nostoc* strains exists (Nilsson et al. 2005). Lectins have been identified as being involved in both host and symbiont recognition in plant and lichen symbioses with cyanobacteria (Bergman et al. 1992). In hard corals that use horizontal transmission it appears that many hosts recognize specific dinoflagellate genotypes. Other dinoflagellates were either not able to establish symbiosis or did not establish stable symbioses (Weiss et al. 2001).

Although vertical transmission may restrict host options in the short term, it is becoming clear that it does not completely prevent the uptake of new symbionts. There is evidence that lineage jumping, multiple loss, and acquisition of cyanobacterial symbionts have occurred in hosts, with some symbionts possibly moving in and out of symbiosis a number of times despite being vertically transmitted. In lichens switching of symbiont genotypes occurs repeatedly (Piercey-Normore & DePriest 2001), although the association is transmitted in propagules that contain both the fungal and cyanobacterial partners. *Montipora* corals also have a diversity of dinoflagellate symbionts despite vertical transmission (van Oppen 2004). Host-switching of vertically transmitted cyanobacterial symbionts appears to have occurred in sponges (Ridley et al. 2005, Usher et al. 2004), and multiple acquisition of chloroplasts by dinoflagellates has also occurred (Cavalier-Smith 1993), possibly because their feeding strategy of phagocytosis and selective removal of a prey's protoplasm promotes secondary and tertiary endosymbioses of chloroplasts (Stoebe & Maier 2002).

The oviparous sponge *Chondrilla australiensis* does not incorporate cyanobacterial symbionts in all the eggs, and some rare females do not vertically transmit cyanobacterial symbionts at all (Usher et al. 2005). This trait may leave some sponge settlers the option of acquiring a different species of cyanobacteria from the water column, although this would probably occur only rarely. Multiple acquisition events may perhaps be inevitable in light of the great age of these symbioses, but they provide insights into the evolution and flexibility of intimate associations that appear to have perfect vertical transmission of symbionts. It is becoming apparent that when alterations occur in a reef environment, the members of what appear to be specific and old partnerships may change (Wood 1998).

Mathematical models have been applied to assess the relative benefits of vertical versus horizontal transmission and to provide insights into the evolution of mutualisms. However, models are limited by the degree of complexity that can be computed, and most make broad assumptions and significantly reduce the complexity of situations. For example, vertical transmission and genetic uniformity of symbionts have been considered essential factors in the evolution of obligate symbioses; yet, as discussed above, there are many examples of obligate symbionts that are horizontally transmitted and of mutualisms that involve mixed symbiont species. Small increases in the complexity of model systems produce dramatically different conclusions

(Hoeksema & Bruna 2000), and these limitations need to be considered when interpreting theoretical approaches to the evolution and stability of mutualisms.

Vertical Transmission via Male Gametes

Interesting parallels exist between fertilization in dioecious (or otherwise outbreeding) sessile aquatic animals and wind pollination in land plants. The sperm of some sessile animals such as tunicates and hydroids have evolved to survive for extended periods in the water column until they come into contact with a female and internal fertilization occurs (Bishop 1998). This is similar to wind pollination in land plants, and it is interesting that vertical transmission of symbionts (including mitochondria and plastids) sometimes occurs in the male line in both land plants (Barr et al. 2005, Morgensen & Rusche 2000, Yang et al. 2000) and sessile marine animals. In the bivalve *Mytilus*, doubly uniparental inheritance of mitochondria occurs (Ladoukalis & Zouros 2001, Obata & Komaru 2005, Zouros et al. 1994) whereby female progeny only inherit mitochondria from their mother and male progeny only inherit mitochondria from their father, and a mechanism for biparental transmission of cyanobacterial symbionts in the sponge *Chondrilla australiensis* has been observed (Usher et al. 2005). All of these organisms (*Mytilus*, *Chondrilla*, and various terrestrial seed plants) have nonmotile adults (gamete-producing phases), a factor that may cause selection to occur at the level of the gametes rather than the adults. When gametes are transported by the media in which the adult resides, adult mate choice and sexual selection via attractiveness to various pollinators may not occur.

Paternal transmission of cyanobacteria/plastids and mitochondria must be a polyphyletic trait because oogamy presumably evolved independently in the ancestor of sponges and other metazoa (and oogamous choanoflagellates) and in the charophycean ancestor of the embryophytes. Barr et al. (2005) suggest that low amounts of biparental mitochondrial transmission may allow organisms to facilitate beneficial recombination and mutational clearance and still restrict the spread of selfish genetic elements, and the phenomenon of biparental transmission may occur more frequently than has been recognized.

SUMMARY POINTS

1. Cyanobacteria are a significant driving force in the evolution of their hosts, enabling successful colonization and propagation in habitats that would not otherwise support the host.

2. Constant association and close proximity of the partners may be necessary for the evolution of a mutualism.

3. The net contribution made by cyanobacteria to the host energy budget must be greater than the costs of housing it.

4. The degree of integration between symbiont and host and the mode of transmission of the symbiont are not indications of how old or important the symbiosis is to host well-being.

5. Biparental transmission of symbionts may prolong the survival of gametes that persist in the environment, increasing chances of fertilization.

FUTURE ISSUES

1. The genetic identity and diversity of both hosts and symbionts must be established to understand symbioses properly. The identification and naming of symbionts and hosts promotes understanding of phylogenetic relationships. The naming of cyanobacteria is now covered under the Bacteriological Code, but this convention requires that living, axenic cultures be obtained, which is difficult for some symbiotic organisms. In addition, new species cannot be named until the genus has been described, and only five cyanobacterial genera (and five species) have been validly published under the Bacteriological Code (Oren 2004).

2. A clearer understanding of the benefits that the host and symbiont derive, how this affects their survival and reproduction, and how they have adapted to maximize the advantages of the association must be pursued. The identification of physiological and molecular mechanisms involved in the infection processes, including signaling, recognition, and attachment, is a crucial area to explore if we are to fully understand symbioses and the evolutionary forces behind their existence.

3. We must investigate the evolutionary age of different categories of cyanobacterial symbioses (e.g., facultative and obligate) with a focus on those that may be in the process of developing, or are recently established, to gain insights into the processes and the evolutionary forces involved and the types of environments in which they form.

4. We must screen for additional novel symbioses. Emphasis on aquatic environments is key.

DISCLOSURE STATEMENT

The authors are not aware of any biases that might be perceived as affecting the objectivity of this review.

ACKNOWLEDGMENTS

Many thanks to Andrew Usher for assistance with the manuscript (by KU). Financial support from the Swedish Research Council is gratefully acknowledged (to BB). Discussion with Dr. Pietà Schofield has been very helpful (to JAR).

LITERATURE CITED

Barr CM, Neiman M, Taylor DR. 2005. Inheritance and recombination of mitochondrial genomes in plants, fungi and animals. *New Phytol.* 168:39–50

Behar A, Yuval B, Jurkevitch E. 2005. Enterobacteria-mediated nitrogen fixation in natural populations of the fruit fly *Ceratitis capitata*. *Mol. Ecol.* 14:2637–43

Bergman B. 2002. The *Nostoc-Gunnera* symbiosis. See Rai et al. 2002, pp. 207–32

Bergman B, Rai AN, Johansson C, Söderbäck E. 1992. Cyanobacterial-plant symbioses. *Symbiosis* 14:61–81

Bergman B, Rasmussen U, Rai AN. 2007. Cyanobacterial associations. In *Associative and Endophytic Nitrogen-Fixing Bacteria and Cyanobacterial Associations*, ed. C Elmerich, WE Newton, 5:257–301. Dordrecht: Kluwer Acad.

Bishop JDD. 1998. Fertilization in the sea: are the hazards of broadcast spawning avoided when free-spawned sperm fertilize retaned eggs? *Proc. R. Soc. London Ser. B* 265:725–31

Borowitzka MA, Hinde R. 1999. Sponges symbiotic with algae or cyanobacteria in coral reefs. In *Assessment and Monitoring of Marine System*, ed. S Lokman, NAM Shazili, MS Nasir, MA Borowitzka, pp. 7–17. Teregganu, Kuala Teregganu: Univ. Putra Malaysia

Brenner ED, Stevenson DW, Twigg RW. 2003. Cycads: Evolutionary innovations and the role of plant-derived neurotoxins. *Trends Plant Sci.* 8:446–52

Brunton FR, Dixon OA. 1994. Siliceous sponge-microbe biotic associations and their recurrence through the Phanerozoic reef mound constructors. *Palaios* 9:370–87

Carpenter EJ. 2002. Marine cyanobacterial symbioises. *Biol. Environ. Proc. R. Ir. Acad.* 102B:15–18

Carpenter EJ, Foster RA. 2002. Marine cyanobacterial symbioses. See Rai et al. 2002, pp. 11–17

Cavalier-Smith T. 1993. Kingdom Protozoa and its 18 phyla. *Microbiol. Rev.* 57:953–94

Clay K, Holah J. 1999. Fungal endophyte symbiosis and plant diversity in successional fields. *Science* 285:1742–44

Cox PA, Banack SA, Murch SJ, Rasmussen U, Tien G, et al. 2005. Diverse taxa of cyanobacteria produce β-N-methylamino-L-alanine, a neurotoxic amino acid. *Proc. Natl. Acad. Sci. USA* 102:5074–78

de Bary A. 1879. Die Erscheinung der Symbiose. In *Vortrag auf der Versammlung der Naturforscher und Ärtze zu Cassel*, ed. KJ von Trubner, pp. 1–30. Strassburg: Verlag

DeLuca T, Zackrisson O, Nilsson M-C, Sellstedt A. 2002. Quantifying nitrogen-fixation in feather moss carpets of boreal forests. *Nature* 419:917–20

Diaz MC, Ward BB. 1997. Sponge-mediated nitrification in tropical benthic communities. *Mar. Ecol. Prog. Ser.* 156:97–107

Douglas AE. 1998. Host benefit and the evolution of specialization in symbiosis. *Heredity* 81:599–603

Douglas AE, Raven JA. 2003. Genomes at the interface between bacteria and organelles. *Philos. Trans. R. Soc. London Ser. B* 358:5–18

Ewald PW. 1987. Transmission of modes and evolution of the parasitism-mutualism continuum. *Ann. NY Acad. Sci.* 503:295–306

Foster R, Carpenter EJ, Bergman B. 2006. Unicellular cyanobionts in open ocean dinoflagellates, radiolarians, and tintinnides: ultrastructural characterization and immuno-localization of phycoerythrin and nitrogenase. *J. Phycol.* 42:453–63

Garcia-Pichel F. 1998. Solar UV and the evolutionary history of cyanobacteria. *Orig. Life Evol. Biosph.* 28:321–47

Genkai-Kato M, Yamamura N. 1999. Evolution of mutualistic symbiosis without vertical transmission. *Theor. Popul. Biol.* 55:309–23

Glover HE. 1985. The physiology and ecology of the marine cyanobacterial genus *Synechococcus*. *Adv. Aquat. Microbiol.* 3:49–107

Hay ME, Parker JD, Burkepile DE, Caudill CC, Wilson AE, et al. 2004. Mutualisms and aquatic community structure. The enemy of my enemy is my friend. *Annu. Rev. Ecol. Evol. Syst.* 35:175–97

Herre EA, Knowlton N, Mueller UG, Rehner SA. 1999. The evolution of mutualisms: Exploring the paths between conflict and cooperation. *TREE* 14:49–53

Hoeksema JD, Bruna EM. 2000. Pursuing the big questions about interspecific mutualism: a review of theoretical approaches. *Oecologia* 125:321–30

Janson S. 2002. Cyanobacteria in symbiosis with diatoms. See Rai et al. 2002, pp. 1–10

Janson S, Rai AN, Bergman B. 1993. The marine lichen *Lichina confinis* (O.F. Mull.) C. Ag.: ultrastructure and localization of nitrogenase, glutamine synthetase, phycoerythrin and ribulose 1,5-bisphosphate carboxylase/oxygenase in the cyanobiont. *New Phytol.* 124:149–60

Janson S, Wouters J, Bergman B, Carpenter EJ. 1999. Host specificity in the *Richelia*-diatom symbiosis revealed by *hetR* gene sequence analysis. *Environ. Microbiol.* 1:431–38

Joint I, Tait K, Callow ME, Callow JA, Milton D, et al. 2002. Cell-to-cell communication across the prokaryote-eukaryote boundary. *Science* 298:1207

Karl D, Michaels A, Bergman B, Capone D, Carpenter E, et al. 2002. Dinitrogen fixation in the world's oceans. *Biogeochemistry* 87/88:47–98

Kirk PM, Cannon PF, David JC, Stalpers JA, eds. 2001. *Dictionary of the Fungi*. Wallingford, UK: CABI Biosci. 9th ed.

Kluge M, Mollenhauer D, Wolf E, Schüßler A. 2002. The *Nostoc-Geosiphon* symbiosis. See Rai et al. 2002, pp. 19–30

Kopp RE, Kischvink JL, Hilburn IA, Nash CZ. 2005. The Palaeoproterozoic snowball Earth: A climate disaster triggered by the evolution of oxygenic photosynthesis. *Proc. Natl. Acad. Sci. USA* 102:11131–36

Krammer K, Lange-Bertalot H. 1988. Bacillariophyceae. 2. Teil: Bacillariaceae, Epithemiaceae, Surirelaceae. In *Süsswasserflora von Mitteleuropa*, ed. H Ettl, J Gerloff, H Heynig, D Mollenhauer, 596 pp. New York: Gustaff Fischer Verlag

Ladoukalis ED, Zouros E. 2001. Direct evidence for homologous recombination in mussel (*Mytilus galloprovincialis*) mitochondrial DNA. *Mol. Biol. Evol.* 18:1168–75

LaJeunesse TC, Thornhill DJ, Cox EF, Stanton FG, Fitt WK, Schmidt GW. 2004. High diversity and host specificity observed among symbiotic dinoflagellates in reef coral communities from Hawaii. *Coral Reefs* 23:596–603

Larkum AWD, Kühl M. 2005. Chlorophyll *d*: the puzzle resolved. *Trends Plant Sci.* 10:355–57

Law R, Hutson V. 1992. Intracellular symbionts and the evolution of uniparental cytoplasmic inheritance. *Proc. R. Soc. London Ser. B* 248:69–77

Lechno-Yosef S, Nierzwicki-Bauer SA. 2002. *Azolla-Anabaena* symbiosis. See Rai et al. 2002, pp. 153–78

Lesser MP, Mazel CH, Gorbunov MY, Falkowski PG. 2004. Discovery of symbiotic N_2-fixing cyanobacteria in corals. *Science* 305:997–1000

Lo N, Bandi C, Watanabe H, Nalepa C, Beninati T. 2003. Evidence for cocladogenesis between diverse Dictyopteran lineages and their intracellular endosymbionts. *Mol. Biol. Evol.* 20:907–13

Lockhart PJ, Beanland TJ, Howe CJ, Larkum AWD. 1992. Sequence of *Prochloron didemni atp*BE and the inference of chloroplast origins. *Proc. Natl. Acad. Sci. USA* 89:2742–46

Marin B, Nowack ECM, Melkonian M. 2005. A plastid in the making: Evidence for a second primary endosymbiosis. *Protist* 156:425–32

Martin W, Rujan T, Richly E, Hansen A, Cornelson S, et al. 2002. Evolutionary analysis of *Arabidopsis*. cyanobacterial and chloroplast genomes reveals plastid phylogeny and thousands of cyanobacterial genes in the nucleus. *Proc. Natl. Acad. Sci. USA* 99:12246–51

Moffitt MC, Neilan BA. 2004. Characterization of the nodularin synthetase gene cluster and proposed theory of the evolution of cyanobacterial hepatotoxins. *Appl. Environ. Microbiol.* 70:6353–62

Morgensen HL, Rusche ML. 2000. Occurrence of plastids in rye (Poaceae) sperm cells. *Am. J. Bot.* 87:1189–92

Nilsson M, Rasmussen U, Bergman B. 2005. Competition and competitive fitness among symbiotic cyanobacterial *Nostoc* strains forming artificial association with rice (*Oryza sativa*). *FEMS Microbiol. Lett.* 245:139–44

Obata M, Komaru A. 2005. Specific location of sperm mitochondria in mussel *Mytilus galloprovincialis* zygotes studied by MitTracker. *Dev. Growth Differ.* 47:255–63

Okamoto N, Inouye I. 2005. A secondary symbiosis in progress? *Science* 310:287

Oksanen I, Jokela J, Fewer DP, Wahlsten M, Rikkinen J, Sivonen K. 2004. Discovery of rare and highly toxic microcystins from lichen-associated cyanobacterium *Nostoc* sp. strain IO-102-I. *Appl. Environ. Microbiobiol.* 70:5736–63

Oren A. 2004. A proposal for further integration of the cyanobacteria under the Bacteriological Code. *Int. J. Syst. Evol. Microbiol.* 54:1895–902

Osborne BA, Sprent JI. 2002. Ecology of the *Nostoc-Gunnera* symbiosis. See Rai et al. 2002, pp. 233–51

Parsek MR, Greenberg EP. 2000. Acyl-homoserine lactone quorum sensing in Gram-negative bacteria: A signalling mechanism involved in associations with higher organisms. *Proc. Natl. Acad. Sci. USA* 97:8789–93

Piercey-Normore MD, DePriest PT. 2001. Algal switching among lichen symbioses. *Am. J. Bot.* 88:1490–98

Proteau PJ, Gerwick WH, Garcia-Pichel F, Castenholz R. 1993. The structure of scytonemin, an UV pigment from the sheaths of cyanobacteria. *Experientia* 49:825–29

Rai AN, Bergman B. 2002. Cyanolichens. *Biol. Environ. Proc. R. Ir. Acad.* 102B:19–22

Rai AN, Bergman B, Rasmussen U, eds. 2002. *Cyanobacteria in Symbiosis.* Dordrecht: Kluwer Acad.

Rai AN, Söderbäck E, Bergman B. 2000. Cyanobacterium—plant symbioses. *New Phytol.* 147:449–81

Raven JA. 1985. The comparative physiology of the adaptation of plants and arthropods to life on land. *Philos. Trans. R. Soc. London Ser. B* 309:273–88

Raven JA. 2002a. The evolution of cyanobacterial symbioses. *Biol. Environ. Proc. R. Ir. Acad.* 102B:3–6

Raven JA. 2002b. Evolution of cyanobacterial symbioses. See Rai et al. 2002, pp. 329–46

Raven JA. 2004. Symbiosis, size and celerity. *Symbiosis* 37:281–91

Raven JA, Johnston AM, Handley LL, McInroy SG. 1990. Transport and assimilation of inorganic carbon by *Lichina pygmaea* under emersed and submersed conditions. *New Phytol.* 114:407–17

Raven JA, Sprent JI. 1989. Phototrophy, diazotrophy and palaeoatmospheres: Biological catalysis and the H, C, N and O cycles. *J. Geol. Soc.* 146:161–70

Ridley CP, Bergquist PR, Harper MK, Faulkner DJ, Hooper JNA, Haygood MG. 2005. Speciation and biosynthetic variation in four Dictyoceratid sponges and their cyanobacterial symbiont, *Oscillatoria spongeliae. Chem. Biol.* 12:397–406

Saikkonen K, Faeth SH, Helander M, Sullivan TJ. 1998. Fungal endophytes: a continuum of interactions with host plants. *Annu. Rev. Ecol. Syst.* 29:319–43

Schneider H, Schuettpelz E, Pryer KM, Cranfiull R, Megallon S, Lupia R. 2004. Ferns diversified in the shadow of the angiosperms. *Nature* 428:553–57

Shick JM, Dunlap WC. 2002. Mycosporine-like amino acids and related gadusols: Biosynthesis, accumulation and UV-protective functions in aquatic organisms. *Annu. Rev. Physiol.* 64:223–62

Smith DC, Douglas AE. 1987. *The Biology of Symbiosis.* London: Edward Arnold

Sprent JI, Raven JA. 1985. The evolution of nitrogen-fixing symbioses. *Proc. R. Soc. Edinb. B* 85:215–37

Stoebe B, Maier U-G. 2002. One, two, three: Nature's tool box for building plastids. *Protoplasma* 219:123–30

Sullivan MB, Lindell D, Lee JA, Thompson LR, Bielawski JP, Chisholm SW. 2006. Prevalence and evolution of core Photosystem II genes in marine cyanobacterial viruses and their hosts. *PLoS Biol.* 4:e234

Taylor TN, Hass H, Kerp H. 1997. A cyanolichen from the Lower Devonian Rhynie chert. *Am. J. Bot.* 84:992–1004

Taylor TN, Hass H, Remy W, Kerp H. 1995. The oldest fossil lichen. *Nature* 378:244

Taylor TN, Klavins D, Krings M, Taylor EL, Kerp H, Hass H. 2004. Fungi from the Rhynie chert: a view from the dark side. *Trans. R. Soc. Edinb.: Earth Sci.* 94:457–73

Usher KM, Fromont J, Sutton DC, Toze S. 2004. The biogeography and phylogeny of unicellular cyanobacterial symbionts in selected sponges from Australia and the Mediterranean. *Microb. Ecol.* 48:167–77

Usher KM, Sutton DC, Toze S, Kuo J, Fromont J. 2005. Inter-generational transmission of microbial symbionts in the marine sponge *Chondrilla australiensis* (Desmospongiae). *Mar. Freshw. Res.* 56:125–31

van Oppen MJH. 2004. Mode of zooxanthella transmission does not affect zooxanthellae diversity in acroporid corals. *Mar. Biol.* 144:1–7

Vessey JK, Pawlowski K, Bergman B. 2004. Root-based N_2-fixing symbioses: Legumes, actinorhizal plants, *Parasponia* sp. and cycads. *Plant Soil* 266:205–30

Vogel S. 1974. Current-induced flow through the sponge *Halichondria*. *Biol. Bull.* 147:443–56

Wanntorp L, Dettmann ME, Jarzen DM. 2004. Tracking the Mesozoic distribution of *Gunnera*: Comparison with the fossil pollen species *Tricolpites reticulates* Cookson. *Rev. Palaeobot. Palynol.* 132:163–74

Weiss VM, Reynolds WS, deBoer MD, Krupp DA. 2001. Host-symbiont specificity during onset of symbiosis between the dinoflagellates *Symbiodinium* spp. and planula larvae of the scleractinin coral *Fungia scutaria*. *Coral Reefs* 20:301–8

Wilkinson CR. 1992. Symbiotic interactions between marine sponges and algae. In *Algae and Symbiosis: Plants, Animals, Fungi, Viruses. Interactions Explored*, ed. W Reisser, pp. 112–51. England: Biopress LTD

Wilkinson CR, Fay P. 1979. Nitrogen fixation in coral reef sponges with symbiotic cyanobacteria. *Nature* 279:52

Wilkinson DM, Sherratt TN. 2001. Horizontally acquired mutualisms, an unsolved problem in ecology? *Oikos* 92:377–84

Williams SI, Walker DI. 1999. Mesoherbivore-macroalgal interactions: feeding ecology of sacoglossan sea slugs (Mollusca, Opisthobrachia) and their effects on their food algae. *Ocean. Mar. Biol.* 37:87–128

Withers PC. 1992. *Comparative Animal Physiology*. New York: Saunders Coll.

Wood R. 1998. The ecological evolution of reefs. *Annu. Rev. Ecol. Syst.* 29:179–206

Yang TW, Yang YA, Xiong Z. 2000. Paternal inheritance of chloroplast DNA in interspecific hybrids in the genus *Larrea* (Zygophyllaceae). *Am. J. Bot.* 87:1452–58

Yuan X, Xiao S, Taylor TN. 2005. Lichen-like symbioses 600 mya. *Science* 308:1017–20

Zouros E, Ball AO, Saavedra C, Freeman KR. 1994. An unusual type of mitochondrial DNA inheritance in the blue mussel *Mytilus*. *Proc. Natl. Acad. Sci. USA* 91:7463–67

RELATED REVIEWS

DePriest PT. 2004. Early molecular investigations of lichen-forming symbionts: 1986–2001. *Annu. Rev. Microbiol.* 58:273–301

Honegger R. 1991. Functional aspects of the lichen symbiosis. *Annu Rev. Plant Physiol. Plant Mol. Biol.* 42:553–78

Peters GA, Meeks JC. 1989. The azolla-anabaena symbiosis: Basic biology. *Annu Rev. Plant Physiol. Plant Mol. Biol.* 40:193–210

Shick JM, Dunlap WC. 2002. Mycosporine-like amino acids and related gadusols: Biosynthesis, accumulation, and UV-protective functions in aquatic organisms. *Annu. Rev. Physiol.* 64: 223–62

Stanier RY, Bazine GC. 1977. Phototrophic prokaryotes: the cyanobacteria. *Annu Rev. Microbiol.* 31:225–74

Human Impacts in Pine Forests: Past, Present, and Future*

David M. Richardson,[1] Philip W. Rundel,[2] Stephen T. Jackson,[3] Robert O. Teskey,[4] James Aronson,[5] Andrzej Bytnerowicz,[6] Michael J. Wingfield,[7] and Şerban Proches[1]

[1] Centre of Excellence for Invasion Biology, Department of Botany & Zoology, Stellenbosch University, Matieland 7602, Republic of South Africa; email: rich@sun.ac.za

[2] Department of Ecology and Evolutionary Biology and Center for Embedded Networked Sensing, University of California, Los Angeles, California 90095-1606

[3] Department of Botany, University of Wyoming, Laramie, Wyoming 82071

[4] Warnell School of Forestry and Natural Resources, University of Georgia, Athens, Georgia 30602

[5] Centre d'Ecologie Fonctionnelle et Evolutive, U.P.R. 5175-C.N.R.S., 34293 Montpellier, France and Missouri Botanical Garden, St. Louis, Missouri 63110

[6] USDA Forest Service, Pacific Southwest Research Station, Riverside Fire Laboratory, Riverside, California 92507

[7] Forestry and Agricultural Biotechnology Institute, University of Pretoria, Pretoria 0002, South Africa

Annu. Rev. Ecol. Evol. Syst. 2007. 38:275–97

First published online as a Review in Advance on September 5, 2007

The *Annual Review of Ecology, Evolution, and Systematics* is online at http://ecolsys.annualreviews.org

This article's doi: 10.1146/annurev.ecolsys.38.091206.095650

Key Words

air pollution, biological invasions, conservation, fire, land use

Abstract

Pines (genus *Pinus*) form the dominant tree cover over large parts of the Northern Hemisphere. Human activities have affected the distribution, composition, and structure of pine forests for millennia. Different human-mediated factors have affected different pine species in different ways in different regions. The most important factors affecting pine forests are altered fire regimes, altered grazing/browsing regimes, various harvesting/construction activities, land clearance and abandonment, purposeful planting and other manipulations of natural ecosystems, alteration of biotas through species reshuffling, and pollution. These changes are occurring against a backdrop of natural and anthropogenically driven climate change. We review past and current influence of humans in pine forests, seeking broad generalizations. These insights are combined with perspectives from paleoecology to suggest probable trajectories in the face of escalating human pressure. The immense scale of impacts and the complex synergies between agents of change calls for urgent and multifaceted action.

INTRODUCTION

Pinus is arguably the most important genus of trees in the world. Extensive natural forests and woodlands throughout the Northern Hemisphere are dominated by pines, from subtropical to subarctic latitudes and from coastal plains to mountain ranges and high plateaus.

Humans have interacted with pines since early hominids first encountered these trees in the Mediterranean Basin about a million years ago. Ancient Greeks, Romans, and many other civilizations harvested and managed pines extensively all around the Mediterranean in various ways and for many purposes. The economic importance of Mediterranean pines has continued into the modern era. Although the impacts of human activities on pine forests are most obvious in the Mediterranean region (Barbéro et al. 1998, Le Maitre 1998), pine forests in other parts of Eurasia have also been shaped by humans, most prominently within the past century (e.g., the northern part of northeast China; Xu et al. 2002).

Pines were also important forest constituents in much of North America when humans arrived at least 14,000 years ago, and they have remained important ever since (Williams et al. 2004). These trees were exploited for wood, resins, and seeds by pre-Columbian cultures from Central America to the boreal forests of Canada. Large population centers developed in Mexico, northern Central America, and the southwestern United States during the past 2500 years, with strong local impacts on pine forests. In the five centuries since European conquest and colonization, much of the natural vegetation of the Americas has been radically altered by human activities. Pines, valued for naval timbers, naval stores, and construction materials, were harvested in an east-to-west pattern across North America, concurrent with the spread of Euro-American civilization. Harvesting accelerated in the nineteenth and twentieth centuries with the development of steam-powered sawmills and railroad networks, leaving persistent imprints on the landscape.

This review examines the multitude of human-induced impacts on pine forests, deriving some generalizations from examples at localities throughout the range of pines and involving a broad cross-section of pine species. Examples chosen to show how human-driven impacts have had both positive and negative consequences on the extent and condition of pine forests are compiled in an extensive **Supplemental Appendix** available online. (Follow the Supplemental Material link from the Annual Reviews home page at **http://www.annualreviews.org/**.) Sixty-two of the 111 species from 14 of the 16 subsections of *Pinus* recognized by Price et al. (1998), whose nomenclature we follow, are cited in the text and **Supplemental Appendix**.

Important changes to pine forests can usually be attributed to one or several of the following factors, with interactions between factors evident in most cases: changes in fire regimes, changes in grazing/browsing regimes, harvesting and construction activities, agricultural land clearance and subsequent abandonment, purposeful planting and other manipulations of natural ecosystems, alteration of biotas through species reshuffling, and anthropogenic pollution. We discuss the most important aspects of each of these categories, as well as some interactions, and consider the implications of apparent trends and likely trajectories of alterations in pine forests and pine abundances with escalating global change. We focus on human impacts on pine-dominated forests within the natural range of *Pinus*. However, pines are also widely planted outside their natural range, notably in the Southern Hemisphere. Such plantings have a wide range of impacts on ecosystems and subsequent human-mediated factors affect many aspects of these plantations. In many areas, plantings have served as sources for the widespread invasion of natural vegetation by pines (Richardson 2006). Such impacts are not covered here.

CHANGES TO FIRE REGIMES

Fire has played a pivotal role in the evolution and spread of pines since they evolved in the early Cretaceous. Many pines have evolved morphological and life-history adaptations to fire, conferring resilience to natural fire regimes. Human-altered fire regimes have had a major impact on vegetation in most fire-prone systems, resulting in dramatic effects on pines throughout their range (Vale 2002).

Early human populations in Europe used fire in their landscapes as far back as 400,000 years ago. Intentional ignitions have been an important source of fire in the Mediterranean Basin and elsewhere in Europe for millennia and have massively influenced the extent and composition of pine forests. *Pinus halepensis*, for example, which is well adapted to frequent intense fires, probably expanded its range in the last century into areas formerly covered with forests of less fire-tolerant trees (Barbéro et al. 1998). The extensive use of fire by humans has a much shorter history in North and Central America (likely less than 10,000 years), but human-induced changes to fire regimes, together with many other impacts, have altered forest structure and dynamics over wide areas, with particularly dramatic impacts accumulating since Euro-American settlement (Covington & Moore 1994, Mitchell et al. 2006).

In many parts of the natural range of pines, human activities have increased fire frequencies (Agee 1998). In Central America, Mexico, and Southeast Asia, this has often arisen through the agency of swidden agriculture, which has led to the expansion and increased abundance and range of pines in some areas, e.g., for *P. kesiya* and *P. merkusii* in mainland Southeast Asia (Werner 1993), but in most cases such land use has reduced pine cover, as in Central America (Perry 1991). Euro-American settlement of the Rocky Mountains in the mid-1800s, and associated climatic variability, led to a marked increase in fire frequency, which left a legacy of dense, even-aged stands in montane *P. ponderosa* forests (Veblen et al. 2000).

Heavy grazing of rangelands has reduced fire frequency in many parts of the American West by reducing fuel loads, and this has had a major impact on vegetation dynamics (e.g., Belsky & Blumenthal 1997). However, grazing may also allow flammable grasses to invade open pine forests, thereby increasing fire frequency. Fires have been purposefully excluded from pine forests in several parts of the Northern Hemisphere. This situation and its short- and long-term impacts have been best documented for North America (e.g., Gruell et al. 1982, Parsons & DeBenedetti 1979). *Pinus palustris* forests of the southeastern Coastal Plain, sustained by high-frequency

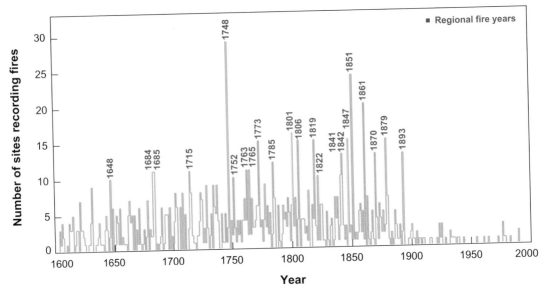

Figure 1

Fire history in southwestern forests and woodlands of the United States from 1600 to the present based on composite fire-scar chronologies from 55 sites in Arizona, New Mexico, and northern Mexico (adapted from figure 5 of Swetnam et al. 1999). Most of the sites were *Pinus ponderosa* forest, or mixed conifer forests with substantial *P. ponderosa*. The y-axis indicates the number of sites with evidence of surface fires in each year. The specific years noted as "Regional fire years" were those in which 10 or more fires occurred among the sites. Fire occurrences before ca. 1900 were controlled primarily by El Niño-Southern Oscillation variation, which governs fine-fuel accumulation (i.e., understory and savanna grasses) and fire-season drought. The dramatic decrease in regional fire frequency in the early 1900s is correlated with intense livestock grazing, which reduced fine fuels, followed by continued grazing and fire-suppression practices.

surface fires, are being replaced by dense slash pine and oak forests in the absence of fire, with consequences for biodiversity, endangered species, and timber production (Kirkman & Mitchell 2006, Mitchell et al. 2006). *Pinus ponderosa* forests in the southwestern United States have been radically altered by Euro-American land uses, including livestock grazing, fire suppression (**Figure 1**), and logging. Dense thickets of young trees now abound, old-growth stands and biodiversity have declined, and human and ecological communities are increasingly vulnerable to destructive crown fires (Allen et al. 2002).

Fire exclusion has allowed pines to spread into some areas where the natural fire regime formerly excluded them (e.g., parks and grasslands of semiarid regions) and has changed forest composition in others (e.g., the southeastern longleaf pine forests). Some impacts of fire suppression in pine forests arose through the disruption of complex relationships between pines, fire, pathogens, and insects (de Groot & Turgeon 1998, Harrington & Wingfield 1998).

GRAZING AND BROWSING

Altered grazing pressure has triggered changes in pine distribution in many regions, but this phenomenon has been best studied in arid and semiarid regions of North America. Changes in stocking rates may initiate, sustain, or halt range expansions, depending on the circumstances. Grazing may interact with climate change, fire, and invasive species, but in all cases the proximal impact is on pine-seedling establishment. Moderate to heavy grazing often facilitates seedling recruitment by reducing grass cover. Local expansion of pine populations has been documented for grasslands, meadows, and steppe. Evidence that pine recruitment decreases when herbivores are excluded lends additional support to the notion that grazing at intermediate levels favors pine expansion (Richardson & Bond 1991). Areas subjected to heavy grazing may remain susceptible to pine colonization long after grazing pressure has been reduced or eliminated.

The expansion of pinyon pines, notably *P. edulis* and *P. monophylla*, in the semi-arid southwest of North America during the twentieth century has been attributed to grazing. The debate on whether pinyons have indeed invaded millions of hectares of former grassland or shrubland, or whether this "invasion" constitutes the re-establishment of woodlands on sites from which trees were eliminated or greatly reduced by human activities, has continued for decades. Richardson & Bond (1991) summarized the invasion scenario, Lanner (1981) provided a thorough criticism of the rationale behind such views, and West (1988) gave a good general account of human-induced changes to pinyon-juniper woodlands. However, recent woodland expansions have occurred against a backdrop of long-term climate variation and range expansion (e.g., Gray et al. 2006, Lyford et al. 2003).

Rabbits have influenced pine regeneration in many areas (e.g., *P. sylvestris* in Britain; see **Supplemental Appendix**). Introduced goats have severe impacts on pine regeneration in many areas. In at least one case (*P. radiata* var. *binata* on Guadalupe Island), such impacts are threatening the continued survival of a pine taxon (see **Supplemental Appendix**). There are, however, still relatively few published accounts of changes in pine distribution or density as a result of browsing, perhaps because marked increases in the density of browsers are relatively recent in many areas. Browsers do most damage to young trees rather than mature ones, and the damage to forest composition is not readily visible. The primary effect of changes such as increased moose densities in *P. sylvestris* forests in Sweden are therefore on rates of recruitment to adult growth stages (Edenius et al. 1995). There are thus substantial time lags between changes in browser density and conspicuous change in forest structure. The literature on the impacts of grazing and browsing on pines is therefore biased in favor of cases that lead to range expansions, which are much more obvious.

HARVESTING

Humans have harvested pines and their products for thousands of years. Four categories of harvesting activities have clearly influenced pine forests. These are: harvesting of nuts; fuel wood gathering; logging of pines; and logging of broad-leaved trees.

The harvesting of pine seeds (nuts) for human consumption has a very long history and has been documented for at least 29 pine species (Kunkel 1984), including *P. cembra*, *P. koraiensis*, *P. pinea*, and *P. sibirica* in Eurasia and *P. coulteri*, *P. lambertiana*, *P. sabiniana*, and all of the pinyons in Central and North America. In some societies, pine seeds from natural forests are a crucial economic resource. For example, during good years, the harvest of *P. gerardiana* seeds provides income for about 13,000 people in the Suleiman Mountains of Pakistan (Martin 1995). In the Indian part of its range, such intense harvesting of *P. gerardiana* seeds and browsing by sheep and goats prevent regeneration (see **Supplemental Appendix**). *Pinus maximartinezii* is in danger of extinction in its tiny Mexican range, where nut collectors often lop off whole branches from trees to collect cones (Styles 1992). Fire, together with browsing by goats, donkeys, and cattle, add to the precarious conservation status of this pine (Martin 1995).

Almost all pine species have been used as fuel wood for centuries (examples in **Supplemental Appendix**). Wood harvesting for fuel has impacted pine populations on local to regional scales. Reduction or extirpation of pinyon populations as a result of firewood harvesting by Native Americans has been well documented, most notably in New Mexico (Betancourt & Van Devender 1981). Recent pinyon expansions in some areas may represent a rebound from harvesting, although climate change and grazing are also implicated. The need for fuel wood in many parts of the natural range of pines still accounts for a large part of the total area of pine forest cleared every year. The situation with respect to fuel wood resources in some developing countries is desperate; see for example Perry's (1991, p. 211) chilling account of the situation in El Salvador. Fuel wood use affects pines as much as any group of trees, despite the fact that pine resin makes them less desirable than hardwoods for use in ovens and fireplaces. The ecological consequences of fuel wood gathering, however, can sometimes have differing short- and long-term effects. Fuel wood gathering in plantations of *P. massoniana* in China reduced the growth rate of mature pines but increased regeneration through enhanced seedling establishment (Kong & Mo 2002).

Le Maitre (1998) discussed the history of logging of pine forests of Europe to supply timber for construction and shipbuilding. Historians have been debating the impact of this logging on the extent of the forests over the whole Mediterranean Basin. There is, however, no doubt that these activities had substantial impacts in many areas (Thirgood 1981). The examples in the **Supplemental Appendix** illustrate a wide range of influences of pine logging in other parts of the world. There is an extensive literature on the history, policies, politics, and practices of logging in different parts of the range of pines. A good example of the complexity of the relationship between logging impacts and vegetation structure is described by McDonald (1976) for a Californian mixed conifer forest. Logging and harvesting of pines for timber and naval stores in southwestern Florida during early European colonization resulted in a dramatic decline in populations of *P. palustris* and, to a lesser extent, *P. elliottii* (Walker 2000).

Special attention has been given in recent decades to assessing the impacts of various logging strategies on animals associated with pine forests. Two bird species especially well studied in North America are the red-cockaded woodpecker (*Picoides borealis*) and Kirtland's warbler (*Dendroica kirtlandii*). The former species is associated with mature forests of *P. echinata*, *P. elliottii*, *P. palustris*, and *P. taeda* in the southern United States, and the latter inhabits young *P. banksiana* forests in Canada and parts of the northern United States. Harvesting of these pine forests and changed fire regimes have altered the pine age structure, jeopardizing bird populations. The many published studies of the habitat requirements of these species, and of the measures required to ensure their survival (Probst & Weinrich 1993; Wilson et al. 1995), provide valuable models for the study and management of other threatened animals associated with pine forests.

In many areas, logging of trees other than pines has had a major influence on pine forests. For example, the clearing of broad-leaved forests in parts of Asia has created suitable conditions for pines, allowing species such as *P. massoniana* and *P. yunnanensis* to expand their ranges.

RURAL EXODUS AND ABANDONMENT OF AGRICULTURAL LAND

Pines have spread into lands following the abandonment of thousand-year-old agriculture in many parts of their natural ranges in the Old World. Pine forests in central Japan, for example, increased with abandonment of agriculture in hilly areas between 1880 and 1980 (Fujihara & Shirai 2001). Many studies have also documented the spread of pines and other trees onto abandoned farmlands in the southeastern United States. Pines, notably *P. taeda*, usually invade such sites within 10 years of abandonment and rapidly form closed stands that are gradually replaced by hardwoods in the absence of major disturbance (e.g., Golley et al. 1994). Similar trends are evident from nearly all parts of the range of pines.

PURPOSEFUL MANIPULATION

Pines have been widely used and planted in the Mediterranean Basin since prehistoric times. Le Maitre (1998) reviewed how this practice has influenced the distribution of *P. brutia*, *P. halepensis*, *P. pinaster*, and *P. pinea*. Other examples (see **Supplemental Appendix**) range from the local planting of pines by Native Americans, which probably had little effect on their general distribution patterns, to the establishment of modern large-scale pine plantations, which has led to a huge artificial increase in the area under pines. Widespread planting of pines has also reshuffled genetic material. Large-scale afforestation started in the second half of the nineteenth century in southern Europe (e.g., Vallauri et al. 2002). Sustained, large-scale forestry, however, was not widespread in Europe until the early twentieth century and expanded to other parts of the world only in the second half of the past century (Le Maitre 1998).

ALTERED BIOTA

The human-orchestrated reshuffling of biotas has affected pine-dominated systems through the increased impacts of pathogens, both native and introduced, and invasive alien species. One of the earliest examples of an introduced pine pathogen is that of the white-pine blister rust (*Cronartium ribicola*), introduced to North America through trade in pine seedlings. Despite a vigorous and longlasting effort to eradicate currants and gooseberries (*Ribes* spp.), alternate hosts of the pathogen, this disease continues to have major impacts on several pines in North America (Kinloch 2003). The disease is expected to become pandemic throughout most of the range of *P. lambertiana* and other species within the next 50–75 years, and in conjunction with heavy logging, the rust is already significantly reducing breeding populations (**Supplemental Appendix**). In Glacier National Park, Kendall & Arno (1990) estimated that 90% of *P. albicaulis* trees have succumbed to the rust. Lanner (1996) gives a good account of the actual and potential ecosystem-level effects of the elimination of this species. The disease has now spread to *P. flexilis*, with dramatic mortality throughout Alberta, Montana, and Wyoming at the time this review was written. A second locus of introduction of *Cronartium*, in New England, resulted in the spread of white pine blister rust through much of the range of *P. strobus*. Although the disease has not been as severe (except locally) as on *P. monticola*, the threat of its presence has limited the dissemination of planting stock of this species (R.D. Westfall, personal communication).

The pine pitch canker fungus, *Fusarium circinatum*, was first recorded on various pine species in the southeastern United States in 1946. The disease was not particularly problematic in native pine stands; most damage has occurred in managed forests and seed orchards. In contrast, when the fungus appeared on *P. radiata* in California in 1986, it spread rapidly from roadside and amenity plantings to native populations on the Monterey Peninsula. It is now common and a serious pathogen throughout coastal regions of California from San Diego to San Francisco. The long-term survival of the species in the wild is a matter of much concern.

One of the most devastating diseases of natural pine forests is pine wilt caused by the pine wood nematode (*Bursaphelenchus xylophilus*), vectored by longhorn beetles (Coleoptera: Cerambycidae). This nematode is responsible for devastating losses to native pines in Southeast Asia (Mamiya 1983). The disease was discovered in the United States in 1979 (Dropkin et al. 1982). It is now thought that the nematode was introduced to Japan from North America on pine logs and it was able to develop an association with the native Japanese beetle *Monochamus alternatus*, which apparently invaded the islands as a hitchhiker on imported nematode-infested timber (Dwinell 1997).

Various human activities have exacerbated the problem of native diseases and parasites in pine forests. These include the replacement of disease-resistant species with more susceptible species and fire suppression, which favors alternate hosts (e.g., oaks; *Quercus* spp.) over disease-resistant and fire-tolerant species (e.g., *Pinus palustris*). This problem is exemplified by fusiform rust (*Cronartium fusiforme*) in the southern United States, where the relatively disease-susceptible *P. elliottii* and *P. taeda* were planted to replace the more resistant *P. palustris* following heavy logging of the longleaf pine in the nineteenth century (Harrington & Wingfield 1998). Insect outbreaks are generally more common in human-altered monocultures, although other factors are also implicated (e.g., poor sites conditions and stress caused by drought and/or pollution). In southern Europe, infestations of pines by mistletoe (*Viscum album* subsp.

austriacum) and scale insects (*Matsucoccus feytaudi*) are clearly on the increase as a result of single-species afforestation projects (Vallauri et al. 2002).

The impact of invasive alien plants on native pines is probably much greater than is currently reflected in the literature. In particular, the negative effects of invasive alien grasses on pine regeneration in many parts of the Northern Hemisphere are understated. *Imperata cylindrica*, native to Southeast Asia, is having a profound influence on the dynamics of the pine forests of the southeastern United States (Jose et al. 2002). Such changes certainly alter the fire regime and must significantly affect pine regeneration. Similar impacts probably apply to many other taxa in other regions, and the magnitude of such problems is increasing rapidly as invasive alien plants become more widespread and abundant.

An additional and contrasting concern is that several pine species need to be controlled themselves, or even locally extirpated, given the rapidly changing ecological conditions worldwide and the fact that some of them have become invasive encroachers and even noxious weeds. Where the species are indisputably alien (i.e., introduced far outside their natural range in recent times), the issues are generally clear (Richardson 2006), but when the species are within their natural range, or close to it, issues are more complex.

AIR POLLUTION

Toxic effects of industrial air pollution on forests have been recorded since the nineteenth century in Central Europe (Godzik & Sienkiewicz 1990) as well as the western United States and British Columbia (Miller 1989). Intensive industrialization caused uncontrolled emissions of noxious gases such as sulfur oxides (SO_x), nitrogen oxides (NO_x), hydrogen chloride (HCl), hydrogen fluoride (HF), and industrial dust toxic to vegetation in the vicinity of smelters, mines, and power plants (Krupa 1997). With increasing industrialization, air pollutants were dispersed over larger areas, causing serious regional air pollution problems including toxic effects on pines in Europe and North America (**Figure 2**).

Recently, however, the most widespread air pollution effects on forests have been caused by ambient ozone (O_3), a highly phytotoxic component of photochemical smog. Fowler et al. (1999) project that about 50% of the Northern Hemisphere forests will be affected by toxic levels of O_3 by the year 2100. Long-range transport can result in dispersion of the pollutant over large areas, resulting in phytotoxic concentrations in remote locations such as Siberia or mountainous areas of the western United States.

Pinus halepensis, widely distributed in the Mediterranean Basin, appears to be sensitive to O_3, since symptoms of this pollutant have been observed in Israel (Naveh et al. 1980), Greece (Gimeno et al. 1992), Spain (Gimeno et al. 1992), and Italy (Ferretti et al. 2003). The health of *P. cembra*, an important conifer in the timberline ecotone of the European mountains, has declined over the past decade in the Maritime Alps of southern France (Bianco & Dalstein 1999, Vollenweider et al. 2003) and in the Carpathian Mountains (Manning & Godzik 2004). In both areas, the observed injury in *P. cembra* can be ascribed to the O_3-induced foliar symptoms on older leaves. In contrast, there is no evidence of O_3 injury in the conifers of the timberline ecotone of

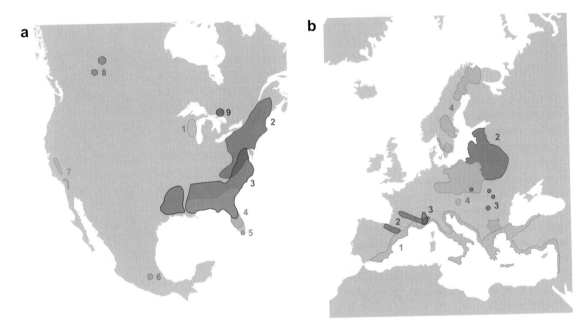

Figure 2

(*a*) Areas in North America where pines are affected by air pollution: (1) Eastern Wisconsin where effects of ozone on *Pinus strobus* have been confirmed, (2) Areas on the East Coast, mostly New England, with suspected effects of ozone on *P. strobus*, (3) Areas in the South where effects on *P. taeda* have been suspected, (4) Parts of the South, mostly in Florida and Alabama, where effects of ozone on *P. elliottii* have been suspected, (5) Area in south Florida with confirmed effects of ozone on *P. elliottii* var. *densa*, (6) Areas near Mexico City with pronounced effects of ozone on *P. hartweggi*, *P. leiophylla*, and *P. montezumae*, (7) San Bernardino and San Gabriel Mountains in Southern California and western slopes of the Sierra Nevada where effects of ozone on *P. jeffreyi* and *P. ponderosa* have been confirmed, (8) Areas in Alberta with effects of sulfurous air pollutants on *P. contorta*, and *P. banksiana* and their hybrids *P. contorta* × *P. banksiana*. (9) Area near metal smelter at Sudbury, Ontario, where SO_2 emissions have affected *P. strobus* and *P. banksiana*. (*b*) Areas in Europe where pines are affected by air pollution: (1) Mediterranean coast areas where ozone affects *Pinus halepensis*, (2) Baltic countries and parts of France and Spain with *P. sylvestris* mildly affected by air pollution, (3) Areas in south-western Alps and the Carpathians with *P. cembra* expressing ozone injury symptoms, (4) Areas in Central Europe and Scandinavia with *P. sylvestris* expressing pronounced air pollution effects.

the central European Alps (Wieser & Havranek 2001), although O_3 regimes are comparable to those in the Maritime Alps (Bianco & Dalstein 1999) and the Carpathians (Bytnerowicz et al. 2002). Other European pine species appear tolerant to ambient levels of ozone.

Pinus ponderosa and *P. jeffreyi* have been affected by O_3 in the western United States, and *P. attenuata*, *P. coulteri*, *P. monticola*, *P. radiata* are potentially sensitive (Miller et al. 1983). Typical effects of O_3 on these species are foliar chlorotic mottle, premature needle senescence leading to crown thinning, and weakening of trees. Severely weakened trees are susceptible to drought and bark beetle attacks, often resulting in the death of the most sensitive individuals (Miller 1989). In the late 1960s, *P. ponderosa* and *P. jeffreyi* were found to be moderately to severely damaged in areas most exposed to O_3 in the San Bernardino and Sierra Nevada Mountains in California (Miller et al. 1983). At present, as the result of several years of drought, long-term effects of air pollution and bark beetle infestations in forests in the San Bernardino Mountains have left millions of dead *P. jeffreyi* and *P. ponderosa* individuals (Keeley et al. 2004). *Pinus strobus* started showing severe symptoms of ozone toxicity over much of the eastern United States in the 1960s (Berry & Ripperton 1963, Rezabek et al. 1989). Sensitive genotypes of *P. strobus* showed reduced photosynthesis, needle necrosis, and decreased height, diameter, and needle length compared to the more tolerant individuals (Irving 1991, Reich et al. 1987). In the South toxic effects of ozone on *P. taeda* have been suspected (Sheffield et al. 1985), and such effects were confirmed for *P. elliotti* var. *densa* (Evans & Fitzgerald, 1993). Forests to the south and southwest of Mexico City have suffered from the effects of urban smog, with some of the highest ambient levels of O_3 in the world. Many *P. hartwegii* and *P. leiophylla* trees showed symptoms of damage similar to those observed on *P. jeffreyi* and *P. ponderosa* in the mountains of Southern California (Miller et al. 2002).

Pines located near metal smelters or power plants can suffer from industrial emissions (**Figure 2**). Known sensitive species in Canada and the northeastern United States include *P. contorta*, *P. banksiana*, and *P. strobus* (Guderian 1977, Legge et al. 1999). *Pinus contorta* and *P. banksiana* hybrids in western Canada have also been affected by SO_2 emissions from extraction of oil from oil sands and gas processing (Mayo et al. 1992). *Pinus sylvestris*, the most widely distributed pine in Eurasia, is well known for its sensitivity to SO_2 (Godzik & Sienkiewicz 1990), HCl, HF (Guderian 1977), and NO_x (Jakubczak & Pieta 1968). The observed crown thinning in this species over large parts of Europe could be partially caused by a synergy between SO_2 and O_3 (Tingey & Reinert 1975). In general, however, the condition of *P. sylvestris* has significantly improved in recent years (Lorenz et al. 2005).

Phototoxic effects of air pollutants have lessened in many European and North American forests because of effective control measures initiated in the 1980s and 1990s against SO_2 and NO_x industrial pollution and the precursors of tropospheric ozone formation (NO_x and volatile organic compounds). With the rapid development of industrial activity in Asia, especially in China, the impacts of industrial air pollution on pine forests will increase (Bytnerowicz et al. 2007).

EFFECTS OF CLIMATE CHANGE ON PINE FORESTS

Pines have been subjected to climatic changes throughout their evolutionary history. The well-documented history of the past 20,000 years, spanning the last glacial-interglacial transition, is especially relevant for understanding the future of pines in the context of ongoing and future global climate change. Some species (*P. banksiana*, *P. sibirica*) have undergone complete geographic displacements since the last glacial maximum 20,000 years ago, whereas others (*P. strobus*, *P. edulis*, *P. ponderosa* var. *scopulorum*) have expanded northward from small, isolated populations to cover vast territories (Jackson et al. 1997, Kremenetski et al. 1998, Lanner & Van Devender 1998). Other species (*P. flexilis*, *P. remota*) have undergone dramatic range contraction and fragmentation in the same period (Betancourt et al. 1990, Lanner & Van Devender 1998). A crucial implication of these patterns is that broad geographic ranges and high abundances provide no guarantee of stability in range or abundance under altered climate regimes of the future.

Seed dispersal was not a major limiting factor to postglacial pine migration; both wind-dispersed (e.g., *P. banksiana*, *P. ponderosa*) and bird-dispersed species (*P. edulis*) expanded their ranges northward rapidly. However, the unprecedented rates of climate change predicted in many future scenarios may outstrip the natural capacity of pines to disperse into newly suitable territory.

Within already established ranges, pine forests have undergone progressive changes in composition and disturbance regime as climate changed. Fire regimes in montane and boreal pine forests have varied dramatically in the past 10,000 years as temperature and moisture regimes changed (Brunelle et al. 2005, Carcaillet et al. 2001). Climate-driven expansions and contractions of pines at upper and lower treeline and at parkland/forest boundaries are well documented (e.g., Lloyd & Graumlich 1997, Lyford et al. 2003, Lynch 1998). Pine populations in forests have increased and decreased in the past few thousand years in response to climate change (Booth & Jackson 2003, Schauffler & Jacobson 2002). Dendroecological studies indicate demographic and fire-regime responses to climate changes over the past few centuries in boreal and montane pine forests and semiarid pine woodlands of North America (Bergeron et al. 2004, Gray et al. 2006, Swetnam et al. 1999).

Ongoing and future climate changes will affect pine-dominated ecosystems in complex ways. Climatic changes will change disturbance regimes, demographic structure, growth rates, and stand composition. Such changes may be under way. For instance, increased radial growth of *P. ponderosa* in the Pacific Northwest of North America observed after 1950, particularly in drought years, has been attributed to increased atmospheric CO_2 concentrations (Soulé & Knapp 2006), and the frequency of large-scale fires has increased in western North American pine forests during the past three decades, corresponding to increasingly early snowmelt (Westerling et al. 2006). Widespread pine mortality events are being induced by the continuing drought in much of the western United States. *Pinus edulis* populations have been decimated over much of the American Southwest as a result of rampant bark beetle infestations in the drought-stressed populations (Breshears et al. 2005). Outbreaks of mountain pine beetle are devastating *P. contorta* populations across wide swaths in western Canada and the United States; these outbreaks are related in part to recent anomalies in winter temperatures (Waring & Pitman 1985). The extensive areas of dead trees increase the risk of wildfire and colonization by invasive species and also alter regional stand age structure.

A longer-term concern is how pines will adjust biogeographically under altered climate regimes. Models of realized distributions of pines species to climate-change scenarios indicate that the ranges of many pine species will be displaced, often dramatically, in a greenhouse world (**Figure 3**). Although these scenarios are probably poorly suited for predicting the future distributions of pines and other species, they provide good indications of the magnitude of biogeographic change that we should

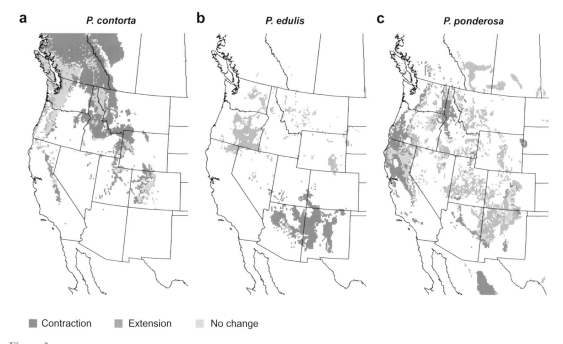

a *P. contorta* **b** *P. edulis* **c** *P. ponderosa*

■ Contraction ▓ Extension ░ No change

Figure 3

Potential changes in abundance and distribution of three *Pinus* species in response to climate change (Thompson et al. 1998). A regional climate model for the western United States, RegCM (Dickinson et al. 1989), was used to simulate the present-day climate and a simulated climate for the region based on a doubling of the current atmospheric CO_2 concentration. The modern climate and species distributions were used to develop empirical models of the climatic responses of the respective species. These models were used in turn to simulate future distributions under the doubled-CO_2 climate scenario. The distribution of *P. contorta* (*a*) is predicted to shift to higher elevations with a reduction in total area to 35% of its present extent. In contrast, *P. edulis* (*b*) would cover approximately the same land area, but would shift dramatically northward, abandoning nearly all of the territory it currently occupies. *Pinus ponderosa* (*c*) is predicted to move to higher elevations and into the Great Basin, with an overall area expansion of 36%. It is important to note that these shifts in distribution are speculative, because of uncertainties in the accuracy of future climate simulations and the adequacy of the models in simulating the environmental factors controlling species distributions. Also, simulations of this type ignore the potential effects of genotypic variation in determining how species will respond to climate change (e.g., Norris et al. 2006). However, the simulations serve to illustrate that *Pinus* forests will likely experience dramatic changes due to climate change and that the magnitude and direction of the effects will differ between species.

expect. The models ignore the underlying genetic diversity that may play an important role in the ability of pines and other species to adjust to climate change (Rehfeldt et al. 2001). Greenhouse gas climate change may cause novel climates to appear across broad regions (Williams et al. 2007). These changes will probably cause the disappearance of some extant communities and the emergence of novel communities, as has occurred before (Jackson & Williams 2004). Predicted climates for the near future are in some cases as different from today's climate as those of the last deglaciation (Jackson & Williams 2004, Williams et al. 2007)—a time when many pines that are abundant now were rare, and other pines that are scarce now were abundant.

CONSERVATION AND RESTORATION EFFORTS IN PINE FORESTS

The growing pressure from logging and land-use changes is rapidly eroding the genetic diversity of pines. At least one-third of *Pinus* species are either threatened in their entirety or have threatened subspecies or varieties (Richardson & Rundel 1998). Some of these taxa were rare before human intervention (e.g., *P. radiata* and *P. torreyana*), but human activities are threatening other widespread taxa as well, such as numerous tropical and subtropical pines in Central America and Mexico (Dvorak 1990). Although many other pine taxa still occupy large ranges, the loss of specific habitats may threaten their genetic diversity, potentially reducing their ability to respond to changing environmental conditions. Low-elevation populations of *P. ponderosa* in California were eliminated a century ago because they were the most accessible timber sources. Ledig (1993) suggests that the alleles that enabled *P. ponderosa* to survive on dry, low-elevation sites may have been lost when those stands were logged. Low- and middle-elevation populations of many other pine species have been reduced or eliminated as forests have been cut for lumber and firewood and land converted to pasture and crops. In Yunnan Province of China, all but the most inaccessible stands of *P. yunnanensis* and other pines have been cleared to make way for agriculture. In the mid-1960s, 60% of the province was forested, but this figure had dropped to less than 30% by 1985 (H. Hutchins, unpublished information). Similarly, Perry (1991, p. 207) describes the plight of Mexico's forests as follows: "They ... continue to disappear, first becoming fragmented then reduced to small patches and parcels and finally to small groups of trees on inaccessible mountain slopes and valleys." The net effect of these "secret extinctions" (Ledig 1993) on the ability of species to deal with rapidly changing environmental conditions will never be known, but must be severe. Recent genecological studies of *P. contorta* highlight the importance of genetic diversity in facilitating responses to future climate change (Rehfeldt et al. 1999, 2001).

Pinus pinaster, which occurs in fragmented populations in the western Mediterranean, has had its distribution reshaped and its genetic structure highly modified in the past two centuries by afforestation, especially in France, Spain, and Portugal, as foresters used lineages of different geographical origin (Gonzaléz-Martinéz et al. 2004). A similar situation occurs at the eastern end of the Mediterranean for *Pinus halepensis* (Barbéro et al. 1998).

Hybridization was probably a major factor in the natural diversification of *Pinus* (see e.g., Wang et al. 2001). As in the eastern Mediterranean case cited above, it is likely and perhaps inevitable that pine plantations bringing together lineages from different regions will sooner or later lead to new hybrids. In the native range of the genus, these could potentially alter the genetic structure of natural pine forests on a large scale.

The loss of genetic diversity in pine forests bears directly on human well-being on a global scale. Although tree breeding has already improved productivity (e.g., Lavery & Mead 1998) and disease resistance (Harrington & Wingfield 1998) of pines in exotic plantations, breeders have barely scratched the surface in their efforts to modify pines for human goods and services (Ledig 1998). Clearly, native gene pools, especially of species with commercial importance, need to be conserved if options are to be kept open for further tree breeding selections. Among the economically important (or potentially so) pines of Mexico and Central America, provenances of at least seven species or varieties are threatened. For some species, ex situ gene conservation banks are the only place where these unique gene pools still exist (Dvorak & Donahue 1992). Among the threatened Mexican and Central American pines, *P. tecunumanii, P. chiapensis, P. greggii*, and *P. maximinoi* have been identified as potential replacements for commercial plantings of temperate pines *P. taeda* and *P. elliottii* in tropical and subtropical regions (Lambeth & McCullough 1997).

Judging from the experience in agriculture, species with less obvious commercial value may nevertheless also be valuable in breeding. An example is *P. washoensis*, a narrow endemic from the mountains on the western rim of the Great Basin in California and northwest Nevada, which persists in a few populations and is on the verge of extinction, largely because of logging in the nineteenth century. Genetic studies suggest that this pine may be a valuable genetic resource for the yellow pines of North America (Niebling & Conkle 1990).

THE FUTURE

We have described some of the many human influences on pine forests. The selected examples show that a wide range of human activities over many centuries have substantially altered pine ecosystems. Many pine species are tenacious, persisting, and even flourishing under severe disturbance regimes. Increasing direct impacts, such as logging and planting, and indirect impacts, such as alteration of disturbance regimes and the composition of biotas, globally increasing ambient ozone and regionally elevated levels of industrial gases and particulate matter, rapidly changing climatic conditions and the further introduction of invasive alien species, including pests and pathogens, are all expected in the future. Such pressures may push many pine taxa beyond their tolerance thresholds.

The immense scale of the impacts and the complex synergies between agents of change call for urgent action on many fronts. For example, (*a*) ecological restoration is increasingly required for almost all pine woodlands and forests to safeguard key ecosystem services. (*b*) Deleterious effects on pine ecosystems of the deliberate or inadvertent reduction of fire frequency must be recognized and effective management strategies devised. (*c*) Relictual fragments of natural pine forests require enhanced protection and reinforcement as well as expansion and reintegration at landscape scales. (*d*) Identification of reference ecosystems as benchmarks will help define appropriate restoration and management goals and facilitate project monitoring and evaluation (Aronson et al. 1995, Egan & Howell 2000, White & Walker 1997). (*e*) Conservation and protection of gene pools is urgently needed for threatened, endangered, and sensitive pine taxa. (*f*) Autecological and synecological information is needed for many narrowly endemic or regionally rare taxa to facilitate conservation and reintroduction efforts. (*g*) Invasive alien pests and pathogens have had huge impacts on native pine ecosystems, and systems to reduce the incidence of new incursions in the future will be essential. (*h*) Interactions among climate change, land use, fire regime, pollution, diseases and pathogens, and invasive species need to be squarely addressed in research and management applications—ecological surprises are inevitable. Instruments for achieving some of these actions are set out in the Conservation Action Plan of the World Conservation Union's Conifer Specialist Group (Farjon & Page 1999).

Efforts are being made to respond to some of these critical needs. For example, Scots pine (*P. sylvestris* var. *scotica*) now occupies only about 1% of its original range in Scotland (Buckley et al. 2001). Forestry Enterprise Scotland has adopted an ambitious approach to active forest management and restoration of both native pine forest and broad-leaved woodland (Peterken & Stevenson 2004). Projects are under way to conserve genetic resources of some of the most threatened tropical *Pinus* species (Dvorak 1990). Examples of the types of field studies needed to provide key autecological and synecological information on rare species are those concerning the Sierra Nevada pine (*Pinus sylvestris* subsp. *nevadensis*) in the mountains of southern Spain (Hódar et al. 2003) and the endemic *P. culminicola* in northeastern Mexico (Jiménez et al. 2005). Kirkman (2005) described landscape-scale attempts at reintroducing frequent low-intensity fires to prevent oaks, evergreen shrubs, and other hardwoods from crowding out the understory flora associated with the *P. palustris* forests in the southeastern United States, where less than 2% of the original forest remains (Means 1996). Mechanisms to limit the global movement of pests and pathogens are continuously being developed, such as those regulating the movement of wood packaging material. There remains a crucial need to improve quarantine and the transport of invasive alien microbes and insects that threaten pine ecosystems.

Rethinking and reorientation of management practices are needed to improve the health and resilience of pine-dominated ecosystems while also addressing human needs and values (e.g., Etienne 2000). Clearly formulated and pragmatic criteria are needed on where and when range expansions and/or densification is desirable or acceptable, and where and when management intervention is warranted. Reintroducing fire is crucial in many contexts, perhaps in conjunction with the removal of hardwoods, management of invasive species, and the reintroduction of rare or framework species of plants, animals, or microorganisms. Sensitive pine species could benefit from effective air pollution control programs aimed at reducing levels of criteria pollutants (ozone, sulfur, and nitrogen oxides or particulate matter) affecting human health. Beyond such purely ecological challenges, however, there is a pressing need for the development of more harmonious integration of forest and wildlife conservation, active, integrated management of ecosystems and landscapes, and ecological restoration where needed, especially in contexts where local people are motivated to participate. The key is to wed ecological and socio-economic values driving the various efforts (Aronson et al. 2007, Clewell & Aronson 2006).

DISCLOSURE STATEMENT

The authors are not aware of any biases that might be perceived as affecting the objectivity of this review.

ACKNOWLEDGMENTS

D.M.R. and S.P. acknowledge financial support from the DST-NRF Center of Excellence for Invasion Biology. S.T.J.'s contribution was supported by the National Science Foundation (Ecology and Earth System History Programs). M.J.W. acknowledges the DST-NRF Center of Excellence in Tree Health Biotechnology. For help with preparation of figures, we thank Tom Swetnam (Laboratory of Tree-Ring Research, University of Arizona; **Figure 1**), Susan Schilling (USDA Forest Service Pacific Southwest Research Station; **Figure 2**), and Mary Anne McGuire (University of Georgia, School of Forestry and Natural Resources; **Figures 1** and **3**).

LITERATURE CITED

Agee JK. 1998. Fire and pine ecosystems. See Richardson 1998, pp. 193–218

Allen CD, Savage M, Falk DA, Suckling KF, Swetnam TW, et al. 2002. Ecological restoration of southwestern ponderosa pine ecosystems: A broad perspective. *Ecol. Appl.* 12:1418–33

Aronson J, Dhillion S, Le Floc'h E. 1995. On the need to select an ecosystem of reference, however imperfect: A reply to Pickett & Parker. *Restor. Ecol.* 3:1–3

Aronson J, Milton SJ, Blignaut J, eds. 2007. *Restoring Natural Capital: Science, Business and Practice*. Washington, DC: Island Press

Barbéro M, Loisel R, Quézel P, Richardson DM, Romane F. 1998. Pines of the Mediterranean Basin. See Richardson 1998, pp. 153–70

Belsky AJ, Blumenthal DM. 1997. Effects of livestock grazing on stand dynamics and soils in upland forest of the interior west. *Conserv. Biol.* 11:315–27

Bergeron Y, Gauthier S, Flannigan M, Kafka V. 2004. Fire regimes at the transition between mixed wood and coniferous boreal forest in northwestern Quebec. *Ecology* 85:1916–32

Berry CR, Ripperton LA. 1963. Ozone, a possible cause of white pine emergence tipburn. *Phytopathology* 53:552–57

Betancourt JL, Van Devender TR. 1981. Holocene vegetation in Chaco Canyon, New Mexico. *Science* 214:656–58

Betancourt JL, Van Devender TR, Martin PS, eds. 1990. *Packrat Middens: The Last 40,000 Years of Biotic Change*. Tucson: Univ. Ariz. Press

Bianco J, Dalstein W. 1999. Abscisic acid in needles of *Pinus cembra* in relation to ozone exposure. *Tree Physiol.* 19:787–91

Booth RK, Jackson ST. 2003. A high-resolution record of Late Holocene moisture variability from a Michigan raised bog. *The Holocene* 13:865–78

Breshears DD, Cobb NS, Rich PM, Price KP, Allen CD, et al. 2005. Regional vegetation die-off in response to global-change-type drought. *Proc. Natl. Acad. Sci. USA* 102:15144–48

Brunelle A, Whitlock C, Bartlein P, Kipfmueller K. 2005. Holocene fire and vegetation along environmental gradients in the Northern Rocky Mountains. *Quat. Sci. Rev.* 24:2281–300

Buckley P, Ito S, McMachlan S. 2001. Temperate woodlands. In *Handbook of Ecological Restoration*, ed. MR Perrow, AJ Davy, 2:503–38. Cambridge, UK: Cambridge Univ. Press

Bytnerowicz A, Godzik B, Frączek W, Grodzińska K, Krywult M, et al. 2002. Distribution of ozone and other air pollutants in forests of the Carpathian Mountains in central Europe. *Environ. Pollut.* 116:3–25

Bytnerowicz A, Omasa K, Paoletti E. 2007. Integrated effects of air pollution and climate change on forests: a Northern Hemisphere perspective. *Environ. Pollut.* 147:438–45

Carcaillet C, Bergeron Y, Richard PJH, Fréchette B, Gauthier S, Prairie YT. 2001. Change of fire frequency in the eastern Canadian boreal forests during the Holocene: Does vegetation composition or climate trigger the fire regime? *J. Ecol.* 89:930–46

Clewell AF, Aronson J. 2006. Motivations for the restoration of ecosystems. *Conserv. Biol.* 20:420–28

Covington WW, Moore MM. 1994. Southwestern ponderosa forest structure— changes since Euro-American settlement. *J. For.* 92:39–47

De Groot P, Turgeon JT. 1998. Insect-pine interactions. See Richardson 1998, pp. 153–70

Dickinson RE, Errico RM, Giorgi F, Bates GT. 1989. A regional climate model for the western United States. *Clim. Change* 15:383–422

Dropkin VH, Foudin A, Kondo E, Linit M, Smith M, Robbins K. 1982. Pinewood nematode: A threat to U.S. Forests? *Plant Dis.* 65:1022–27

Dvorak WS. 1990. CAMCORE: Industry and government's efforts to conserve threatened forest species in Guatemala, Honduras and Mexico. *For. Ecol. Manag.* 35:151–57

Dvorak WS, Donahue JK. 1992. *CAMCORE Cooperative Research Review 1980–1992*. Raleigh: Dep. For., N.C. State Univ.

Dwinell LD. 1997. The pine wood nematode: Regulation and mitigation. *Annu. Rev. Phytopathol.* 35:153–66

Edenius L, Danell K, Nyquist H. 1995. Effects of simulated moose browsing on growth, mortality, and fecundity in Scots pine: Relations to plant productivity. *Can. J. For. Res.* 25:529–35

Egan D, Howell L. 2000. *The Historical Ecology Handbook: A Restorationist's Guide to Reference Ecosystems*. Washington, DC: Island Press

Etienne M. 2000. Pine trees—invaders or forerunners in Mediterranean-type ecosystems: a controversial point of view. *J. Mediterr. Ecol.* 2:221–31

Evans LS, Fitzgerald GA. 1993. Histological effects of ozone on slash pine (*Pinus elliottii* var. *densa*). *Environ. Exp. Bot.* 33:505–13

Farjon A, Page CN, (compilers). 1999. *Conifers. Status Survey and Conservation Action Plan*. Gland, Switz: IUCN/SSC Conifer Spec. Group

Ferretti M, Brusasca G, Buffoni A, Bussotti F, Cozzi A, et al. 2003. Ozone risk in the permanent plots of the Italian intensive monitoring of forest ecosystems—an introduction. *Ann. Sper. Selvic.* 30(Suppl. 1):3–28

Fowler D, Cape JN, Coyle M, Flechard C, Kuylestierna J, et al. 1999. The global exposure of forests to air pollutants. In *Forest Growth Responses to the Pollution Climate of the 21st Century*, ed. LJ Shepard, JN Cape, pp. 5–32. Dordrecht: Kluwer Acad.

Fujihara M, Shirai Y. 2001. Comparison of landscape structure in the 1880s and the 1980s at five areas of the Boso Peninsula, central Japan. *Nat. Hist. Res.* 6:83–96

Gimeno BS, Velissariou D, Barnes JD, Inclán R, Peña JM, Davison AW. 1992. Daños visibles por ozono en acículas de *Pinus halepensis* Mill. en Grecia y España. *Ecología* 6:131–34

Godzik S, Sienkiewicz J. 1990. Air pollution and forest health in Central Europe: Poland, Czechoslovakia, and the German Democratic Republic. In *Ecological Risks—Perspectives from Poland and the United States*, ed. W Grodzinski, EB Cowling, AI Breymeyer, pp. 155–70. Washington, DC: Natl. Acad.

Golley FB, Pinder JE III, Smallidge PJ, Lambert NJ. 1994. Limited invasion and reproduction of loblolly pines in a large South Carolina old field. *Oikos* 69:21–27

Gonzaléz-Martinéz SC, Mariette S, Ribeiro MM, Burban C, Raffin A, et al. 2004. Genetic resource in maritime pine (*Pinus pinaster* Aiton): Molecular and quantitative measures of genetic variation and differentiation among maternal lineages. *For. Ecol. Manag.* 197:103–15

Gray ST, Betancourt JL, Jackson ST, Eddy RG. 2006. Role of multidecadal climatic variability in a range extension of pinyon pine. *Ecology* 87:1124–30

Gruell GE, Schmidt WC, Arno SF, Reich WJ. 1982. Seventy years of vegetative change in a managed Ponderosa Pine Forest in Western Montana—implications for resource management. *USDA For. Serv. Gen. Tech. Rep. INT–130*

Guderian R. 1977. *Air Pollution: Phytotoxicity of Acidic Gases and its Significance in Air Pollution Control. Ecol. Ser. 22.* Berlin: Springer-Verlag

Harrington TC, Wingfield MJ. 1998. Diseases and the ecology of indigenous and exotic pines. See Richardson 1998, pp. 381–404

Hódar JA, Castro J, Zamora R. 2003. Pine processionary caterpillar, *Thaumetopoea pityocampa*, as a new threat for relict Mediterranean Scots pine forests under climatic warming. *Biol. Conserv.* 110:123–29

Irving PM, ed. 1991. *Acidic Deposition: State of Science and Technology.* Summ. Rep. U.S. Natl. Acid Precip. Assess. Program, Washington, DC

Jackson ST, Overpeck JT, Webb T III, Keattch SE, Anderson KH. 1997. Mapped plant macrofossil and pollen records of Late Quaternary vegetation change in eastern North America. *Quat. Sci. Rev.* 16:1–70

Jackson ST, Williams JW. 2004. Modern analogs in Quaternary paleoecology: Here today, gone yesterday, gone tomorrow? *Annu. Rev. Earth Planet. Sci.* 32:495–537

Jakubczak Z, Pieta J. 1968. Some morphological variations of the needles and shoots of Scots pine (*Pinus silvestris* L.) in the polluted surroundings of nitrogen plant in Pulawy. In *Materials from the 6th Int. Conf. "Effects of Air Pollution on Forests,"* Katowice, Poland, Sept. 9–14, ed. J Paluch, S Godzik, pp. 327–35

Jiménez J, Jurado E, Aguirre O, Estrada E. 2005. Effect of grazing on restoration of endemic dwarf pine (*Pinus culminicola* Andresen & Beaman) populations in northeastern Mexico. *Restor. Ecol.* 13:103–7

Jose SJ, Cox J, Miller L, Shilling DG, Merritt S. 2002. Alien plant invasions: the story of cogongrass in southeastern forests. *J. For.* 100:41–44

Keeley JE, Fotheringham CJ, Moritz MA. 2004. Lessons from the October 2003 wildfires in Southern California. *J. For.* 102(7):26–31

Kendall KC, Arno SF. 1990. Whitebark pine: an important but endangered wildlife resource. In *Proc. Symp. Whitebark Pine Ecosystems: Ecology and Management of a High-Mountain Resource*, ed. WC Schmidt, KJ McDonald, pp. 264–73. Bozeman, MT: USDA For. Serv. Gen. Tech. Rep. INT–270

Kinloch BB. 2003. White pine blister rust in North America: past and prognosis. *Phytopathology* 93:1044–47

Kirkman K. 2005. Conserving biodiversity in the longleaf pine savannas of the USA. *Plant Talk* 39:30–32

Kirkman LK, Mitchell RJ. 2006. Conservation management of *Pinus palustris* ecosystems from a landscape perspective. *Appl. Veg. Sci.* 9:67–74

Kong GH, Mo JM. 2002. Plant population dynamics of a human-impacted Masson pine plantation in Dinghushan. *J. Trop. Subtrop. Bot.* 10:193–200

Kremenetski CV, Sulerzhitsky LD, Hantemirov R. 1998. Holocene history of the northern range limits of some trees and shrubs in Russia. *Arct. Alp. Res.* 30:317–33

Krupa SV. 1997. *Air Pollution, People, and Plants*. St. Paul, MN: APS Press

Kunkel G. 1984. *Plants for Human Consumption*. Koenigstein, Ger: Koeltz Sci. Books

Lambeth CC, McCullough RB. 1997. Genetic diversity in managed loblolly pine forests in the southeastern United States: Perspective of the private industrial forest land owner. *Can. J. For. Res.* 27:409–14

Lanner RM. 1981. *The Piñon Pine: A Natural and Cultural History*. Reno: Univ. Nev. Press

Lanner RM. 1996. *Made For Each Other: A Symbiosis of Birds And Pines*. New York: Oxford Univ. Press

Lanner RM, van Devender TR. 1998. The recent history of pinyon pines in the American Southwest. See Richardson 1998, pp. 171–82

Lavery PB, Mead DJ. 1998. *Pinus radiata*: a narrow endemic from North American takes on the world. See Richardson 1998, pp. 432–49

Ledig FT. 1993. Secret extinctions: the loss of genetic diversity in forest ecosystems. In *Our Living Legacy: Proceedings of a Symposium on Biological Diversity*, ed. MA Fenger, EH Miller, JA Johnson, EJR Williams, pp. 127–40. Victoria, Can: R. B.C. Mus.

Ledig FT. 1998. Genetic variation in *Pinus*. See Richardson 1998, pp. 251–80

Legge AH, Jager H-J, Krupa SV. 1999. Sulfur dioxide. In *Recognition of Air Pollution Injury to Vegetation—A Pictorial Atlas*, ed. R Flagler, 3:1–42. Pittsburgh: Air & Waste Manag. Assoc.

Le Maitre DC. 1998. Pines in cultivation: a global view. See Richardson 1998, pp. 409–31

Lloyd AH, Graumlich LJ. 1997. A 3500 year record of changes in the structure and distribution of forests at treeline in the Sierra Nevada, California. *Ecology* 78:1199–2010

Lorenz M, Becher G, Mues V, Becker R, Dise N, et al. 2005. *Forest Condition in Europe, 2005 Tech. Rep. ICP For.* Hamburg, Ger: Fed. Res. Cent. For. For. Products (BFH)

Lyford ME, Jackson ST, Betancourt JL, Gray ST. 2003. Influence of landscape structure and climate variability on a late Holocene plant migration. *Ecol. Monogr.* 73:567–83

Lynch EA. 1998. Origin of a park-forest vegetation mosaic in the Wind River Range, Wyoming. *Ecology* 79:1320–38

Mamiya Y. 1983. Pathology of pine wilt disease caused by *Bursaphelenchus xylophilus*. *Annu. Rev. Phytopathol.* 21:201–20

Manning WJ, Godzik B. 2004. Bioindicator plants for ambient ozone in Central and Eastern Europe. *Environ. Pollut.* 130:33–39

Martin GJ. 1995. *Ethnobotany*. London: Chapman & Hall

Mayo JM, Legge AH, Yeung EC, Krupa SV, Bogner JC. 1992. The effects of sulphur gas and elemental sulphur dust deposition on *Pinus contorta* x *Pinus banksiana*: Cell walls and water relations. *Environ. Pollut.* 76:43–50

McDonald PM. 1976. Forest regeneration and seedling growth from five major cutting methods in North-Central California. *Res. Pap. PSW–115.* Berkeley, CA: U.S. For. Serv. Pac. Southwest Res. Stn.

Means DB. 1996. Longleaf pine forest, going, going . . . In *Eastern Old-Growth Forests: Prospects for Rediscovery and Recovery*, ed. MB Davis, pp. 210–29. Washington, DC: Island Press

Miller PR. 1989. Concept of forest decline in relation to western U.S. forests. In *Air Pollution Toll on Forest and Crops*, ed. JJ MacKenzie, MT El-Ashry, pp. 75–112. New Haven, CT: A World Resourc. Inst., Yale Univ. Press

Miller PR, de Bauer LI, Hernandez-Tejeda T. 2002. Oxidant exposure and effects on pines in forests in the Mexico City and Los Angeles air basins. In *Urban Air Pollution and Forests*, ed. ME Fenn, LI de Bauer, T Hernandez-Tejeda, *Springer Ecol. Ser.* 156:225–42

Miller PR, Longbotham GJ, Longbotham CR. 1983. Sensitivity of selected western conifers to ozone. *Plant Dis.* 67:1113–15

Mitchell RJ, Hiers JK, O'Brien JJ, Jack SB, Engstrom RT. 2006. Silviculture that sustains: the nexus between silviculture, frequent prescribed fire, and conservation of biodiversity in longleaf pine forests of the southeastern United States. *Can. J. For. Res.* 36:2724–36

Naveh Z, Steinberger EH, Chaim S, Rotman A. 1980. Photochemical oxidants—a threat to Mediterranean forest and upland ecosystems. *Environ. Conserv.* 7:301–9

Niebling CR, Conkle MT. 1990. Diversity of Washoe pine and comparisons with allozymes of ponderosa pine races. *Can. J. For. Res.* 20:298–308

Norris J, Jackson ST, Betancourt JL. 2006. Classification tree and minimum-volume ellipsoid analyses of the distribution of ponderosa pine in the western USA. *J. Biogeogr.* 33:342–60

Parsons D, DeBenedetti S. 1979. Impact of fire suppression on a mixed-conifer forest. *For. Ecol. Manag.* 2:21–33

Perry JP. 1991. *The Pines of Mexico and Central America*. Portland, OR: Timber Press

Peterken GF, Stevenson AW. 2004. *A New Dawn for Native Woodland Restoration on the Forestry Commission Estate in Scotland*. Edinburgh: For. Comm. Scotland

Price RA, Liston A, Strauss SH. 1998. Phylogeny and systematics of *Pinus*. See Richardson 1998, pp. 49–68

Probst JR, Weinrich J. 1993. Relating Kirtlands warbler population to changing landscape composition and structure. *Landsc. Ecol.* 8:257–71

Rehfeldt GE, Ying CC, Spittlehouse DL, Hamilton DA. 1999. Genetic responses to climate in *Pinus contorta*: Niche breadth, climate change, and reforestation. *Ecol. Monogr.* 69:375–407

Rehfeldt GE, Wykoff WR, Ying CC. 2001. Physiologic plasticity, evolution, and impacts of a changing climate on *Pinus contorta*. *Clim. Change* 50:355–76

Reich PB, Schoettle AW, Stroo HF, Troiano J, Amundson RG. 1987. Effects of ozone and acid rain on white pine (*Pinus strobus*) seedlings grown in five soils. I. Net photosynthesis and growth. *Can. J. Bot.* 65:977–87

Rezabek CL, Morton JA, Mosher EC. 1989. Regional effects of sulfur dioxide and ozone on eastern white pine (*Pinus strobus*) in eastern Wisconsin. *Plant Dis.* 73:70–73

Richardson DM, ed. 1998. *Ecology and Biogeography of Pinus*. Cambridge: Cambridge Univ. Press

Richardson DM. 2006. *Pinus*: a model group for unlocking the secrets of alien plant invasions? *Preslia* 78:375–88

Richardson DM, Bond WJ. 1991. Determinants of plant distribution: Evidence from pine invasions. *Am. Nat.* 137:639–68

Richardson DM, Rundel PW. 1998. Ecology and biogeography of *Pinus*: an introduction. See Richardson 1998, pp. 3–46

Schauffler M, Jacobson GL Jr. 2002. Persistence of coastal spruce refugia during the Holocene in northern New England, USA, detected by stand-scale pollen stratigraphies. *J. Ecol.* 90:235–50

Sheffield RM, Cost ND, Bechtold WA, McClure JP. 1985. Pine growth reductions in the Southwest. *USDA For. Serv. Resourc. Bull. SE-83*. Asheville, NC: USDA For. Serv. Southwest Res. Exp. Stn.

Soulé PT, Knapp PA. 2006. Radial growth rate increases in naturally occurring ponderosa pine trees: a late-20th century CO_2 fertilization effect? *New Phytol.* 171:379–90

Styles BT. 1992. Genus *Pinus*: A Mexican purview. In *Biological Diversity of Mexico: Origins and Distribution*, ed. TP Ramamoorthy, R Bye, A Lot, J Fa, pp. 397–420. New York: Oxford Univ. Press

Swetnam TW, Allen CD, Betancourt JL. 1999. Applied historical ecology: using the past to manage for the future. *Ecol. Appl.* 9:1189–206

Thirgood JV. 1981. *Man and the Mediterranean Forest. A History of Resource Depletion*. London: Academic

Thompson RS, Hostetler SW, Bartlein PJ, Anderson KH. 1998. A strategy for assessing potential future changes in climate, hydrology, and vegetation in the Western United States. *U.S. Geol. Surv. Circ. 1153.* Washington, DC: U.S. GPO

Tingey DT, Reinert RA. 1975. The effects of ozone and sulphur dioxide singly and in combination on plant growth. *Environ. Pollut.* 9:117–25

Vale TR, ed. 2002. *Fire, Native Peoples, and the Natural Landscape.* Washington, DC: Island Press

Vallauri D, Aronson J, Barbéro M. 2002. An analysis of forest restoration 120 years after reforestation on badlands in the southwestern Alps. *Restor. Ecol.* 10:16–26

Veblen TT, Kitzberger T, Donnegan J. 2000. Climatic and human influences on fire regimes in ponderosa pine forests in the Colorado Front Range. *Ecol. Appl.* 10:1178–95

Vollenweider P, Ottiger M, Günthardt-Goerg MS. 2003. Validation of leaf ozone symptoms in natural vegetation using microscopical methods. *Environ. Pollut.* 124:101–18

Walker KJ. 2000. Historical ecology of the southeastern longleaf and slash pine flatwoods: A Southwest Florida perspective. *J. Ethnobiol.* 20:269–99

Wang X-R, Szmidt AE, Savolainen O. 2001. Genetic composition and diploid hybrid speciation of a high mountain pine, *Pinus densata*, native to the Tibetan plateau. *Genetics* 159:337–46

Waring RH, Pitman GB. 1985. Modifying lodgepole pine stands to change susceptibility to mountain pine beetle attack. *Ecology* 66:889–97

Werner WL. 1993. *Pinus in Thailand. Geoecol. Res.* 7. Stuttgart: Franz Steiner Verlag. 286 pp.

West NE. 1988. Intermountain deserts, shrub steppes, and woodlands. In *North American Terrestrial Vegetation*, ed. MG Barbour, WB Billings, pp. 210–30. Cambridge, NY: Cambridge Univ. Press

Westerling AL, Hidalgo HG, Cayan DR, Swetnam TW. 2006. Warming and earlier spring increase western U.S. forest wildfire activity. *Science* 313:940–43

White PS, Walker JL. 1997. Approximating nature's variation: Selecting and using reference information in restoration ecology. *Restor. Ecol.* 5:338–49

Wieser G, Havranek WM. 2001. Effects of ozone on conifers in the timberline ecotone. In *Trends in European Forest Tree Physiology Research*, ed. S Huttunen, H Heikkilä, J Buchcer, B Sundberg, P Jarvis, R Matyssek, pp. 115–25. Dordrecht: Kluwer Acad.

Williams JW, Jackson ST, Kutzbach JE. 2007. Projected distributions of novel and disappearing climates by 2100 AD. *Proc. Natl. Acad. Sci. USA* 104:5738–42

Williams JW, Shuman BN, Webb T III, Bartlein PJ, Leduc PL. 2004. Late Quaternary vegetation dynamics in North America: Scaling from taxa to biomes. *Ecol. Monogr.* 74:309–34

Wilson CW, Masters RE, Bukenhofer GA. 1995. Breeding bird response to pine grassland community restoration for red-cockaded woodpeckers. *J. Wildl. Manag.* 59:56–67

Xu Q-H, Kong Z-C, Yang X-L, Liang W-D, Sun L-M. 2002. Vegetation changes and human influences on Qian'an Basin since the middle Holocene. *Acta Bot. Sin.* 44:611–16

Chemical Complexity and the Genetics of Aging

Scott D. Pletcher,[1,2] Hadise Kabil,[1]
and Linda Partridge[3]

[1] Huffington Center on Aging, Baylor College of Medicine, Houston, Texas 77030

[2] Department of Molecular and Human Genetics, Baylor College of Medicine, Houston, Texas 77030; email: pletcher@bcm.tmc.edu, kabil@bcm.tmc.edu

[3] Center for Research on Ageing, University College London, Department of Biology, Darwin Building, London WC1E6BT, England; email: l.partridge@ucl.ac.uk

Annu. Rev. Ecol. Evol. Syst. 2007. 38:299–326

First published online as a Review in Advance on August 6, 2007

The *Annual Review of Ecology, Evolution, and Systematics* is online at http://ecolsys.annualreviews.org

This article's doi: 10.1146/annurev.ecolsys.38.091206.095634

Key Words

evolutionary conservation, innate immunity, sensory perception, xenobiotic metabolism

Abstract

We examine how aging is influenced by various chemical challenges that organisms face and by the molecular mechanisms that have evolved to modulate life span in response to them. For example, environmental information, which is detected and processed through sensory systems, can modulate life span by providing information about the presence and quality of food as well as presence and density of conspecifics and predators. In addition, the diverse forms of molecular damage that result from constant exposure to toxic chemicals that are generated from the environment and from metabolism pose an informatic and energetic challenge for detoxification systems, which are important in ensuring longevity. Finally, systems of innate immunity are vital for recognizing and combating pathogens but are also increasingly seen as of importance in causing the aging process. Integrating ideas of molecular mechanism with context derived from evolutionary considerations will lead to exciting new insights into the evolution of aging.

1. INTRODUCTION

1.1. The Evolution of the Biology of Aging

From the pioneering work of Medawar, Williams, and Hamilton in the 1950s until the early 1990s, evolutionary biology dominated the intellectual landscape surrounding analysis of not only why we age but how (Hamilton 1966, Medawar 1952, Williams 1957). Aging, a decline in condition with increasing age apparent as a reduction in survival and reproductive output, is apparently disadvantageous for the individual (Partridge & Barton 1993). Why then, does it exist at all? Is there some hidden advantage to aging? Evolutionary biologists tackled these questions early on and concluded that aging does not have a function and exists only because natural selection is less powerful late in life. Aging can then evolve through the accumulation of mutations with deleterious effects at later ages or as the deleterious late-life consequence of traits that are beneficial in youth (Charlesworth 1994, Partridge & Barton 1993). More recent work has revealed that the force of natural selection does not always decline, and even increases, over some age classes (Baudisch 2005) and that kin selection and intergenerational transfers (Lee 2003) are important. Nonetheless, these two basic tenets, that aging is due to a declining force of natural selection and is not adaptive, still form the conceptual foundation of the biology of aging (Ackerman & Pletcher 2007, Hughes & Reynolds 2005).

A prediction of the traditional evolutionary models was that patterns of aging would be highly polygenic. Accumulation of random mutations with deleterious effects at older ages would mean that single-gene manipulations would, in principle, have minimal effect. A fix for the effects of one mutation would leave the deleterious effects of the vast majority unaltered, and would hence have only a minor effect on the overall rate of aging (Martin 2002, Rose 1991). Even if trade-offs between youth and old age were important, survival and reproductive success would be highly polygenic traits; any gene that did not affect one or the other would undergo mutational death. A second prediction was that the mechanisms that protect against aging would be lineage-specific, the rationale for this being that each species would have a particular lifestyle associated with a particular set of environmental and physiological impacts. There would hence be great diversity in the types of challenges that different kinds of organisms would experience. Aging was a trait that was not expected to show evolutionary conservation of mechanisms (Martin 2002).

Near the end of the twentieth century the predictive foundation of the evolutionary biology of aging was severely shaken by the experiments of molecular geneticists using laboratory model organisms. These established that alterations in the expression or function of single genes were sufficient to retard aging and increase life span, sometimes up to 20-fold (Arantes-Oliveira et al. 2003, Kenyon 2005, Klass 1983). More recently, particular types of intervention have been shown to be effective in species ranging from budding yeast and the nematode worm to the laboratory mouse, establishing the existence of conserved mechanisms of aging across a broad range of taxa (Partridge & Gems 2002).

These findings came as a surprise and not only to evolutionary biologists. During aging, multiple forms of tissue-specific damage and pathology accumulate (Kirkwood et al. 1999). The process can also vary greatly between individuals (Kirkwood et al. 2005). This complexity and variability suggested that aging would be intractable to medical and experimental intervention. The discoveries that single gene mutations can extend life span by keeping organisms healthy and youthful for longer and that their effects show evolutionary conservation has galvanized research into mechanisms of aging.

1.2. Chemical Complexity and the Diversity of Aging

As alluded to earlier, aging is not selected *for*—selection declines in strength with increasing age—but it can be selected *against*, and an important challenge is to discover how slow aging is accomplished. The types of ecological circumstances that select for slow aging are well specified, and include low levels of extrinsic risk of death from disease, predation, or accidents (Charlesworth 1994, Medawar 1952), periods of resource depletion, and overall variability and unpredictability of the environment (Orzack 1989). These factors have contributed to enormous biodiversity in patterns and rates of aging, as a consequence of evolved differences between taxa. Comparative analysis of the ecological and lifestyle determinants of slow aging (Austad 1997, Holmes & Austad 1995, Ricklefs et al. 2003) are insightful but have generally been less successful in identifying mechanisms, because of the presence of multiple, correlated, candidate traits. An additional, important limitation is the availability of genetic information on relevant species. This new field of "evo-gero" (Partridge & Gems 2006), much like "evo-devo," would greatly benefit from sequencing of the genomes of species with extraordinary life spans (de Magalhaes et al. 2005).

Although it is fascinating that a mouse lives three years and a parrot may live up to 100 (a difference of 33-fold), it is arguably more stunning that individuals of identical genotype can vary enormously in their average life expectancy, by at least up to 80-fold (Gardner et al. 2006). This remarkable phenotypic plasticity in life span demonstrates that the same genome can give rise to very different patterns of aging, with the strong supposition that at least some of this naturally occurring plasticity represents adaptive responses to different environmental circumstances. Presumably, the manifestation of plasticity requires mechanisms that initiate and maintain differences in gene expression. Although work has begun, the details of this process are largely unknown (Jaenisch & Bird 2003).

The major theme of this review is that mechanisms underlying variation in patterns of aging among and within species may often revolve around the issue of chemical complexity. By definition, living long increases the time for which the organism is exposed to myriad, potentially damaging conditions. An important aspect of survival, therefore, must invoke mechanisms of somatic endurance, including protection against physical and biological challenges as well as systems to deal with exogenous or endogenous chemical toxicity. Chemicals are not all bad, however, and even potentially harmful molecules may serve as cues that can be used as sources of information. Both ensuring survival and producing appropriate patterns of phenotypic plasticity in response to environmental variation require that organisms gather information about their environment. This information often comes in the form of particular chemicals that are usually associated with particular states of the physical environment, food, conspecifics, and other species, including predators and pathogens. This information can be used to guide mechanisms of phenotypic plasticity, including of life history and aging. Indeed, it is intriguing that many of the mutations that extend life span in model organisms impinge on stress resistance or affect signaling pathways that detect physical stress, nutrients or other organisms.

In addition to discovering how particular organisms resist aging, a second important challenge is to understand at what levels evolutionary conservation occurs as well as why it occurs. Gross pathologies of aging can differ greatly in different organisms: Humans can die from stroke and cancer, whereas nematodes and fruit flies do not. There are at least some differences at the molecular level too: for example, accumulation of extrachromosomal ribosomal DNA circles contributes to aging in budding yeast (*Saccharomyces cerevisiae*) (Sinclair & Guarente 1997), and extrachromosomal mitochondrial DNA circles (senDNAs) contribute to aging in the filamentous fungus *Podospora anserine* (Osiewacz 2002); neither contribute to aging in mammals. Thus, at least some mechanisms of aging are private (lineage-specific) rather than public (evolutionarily conserved) (Martin et al. 1996). It is important to discover whether any general principles are involved here. For instance, there could be little or no conservation of the specific forms of pathology that result from similar forms of molecular damage, or conserved sensing pathways could regulate different types of cellular biochemistry in different evolutionary lineages.

In this review we do not attempt to examine the whole literature on the evolution of aging. Rather, we explore the idea that a primary factor modulating life span is how organisms deal with chemical complexity, both in their environment and internally. The ways in which animals detect, decode, and respond to a range of chemical factors they encounter may be critical for establishing somatic endurance and for specifying decisions that implement plasticity of the aging process. In the first section, we examine chemical sensing and adaptive plasticity of aging. Next, we explore the role of damage control systems, such as detoxification, which ensure somatic endurance in the face of continuous chemical insult. Lastly we touch on the impact of well-controlled pathogen defense mechanisms on aging. It is becoming increasingly clear that the chemicals that confront individual species may be unique and diverse, but many aspects of the underlying response mechanisms may be evolutionarily conserved. Thus, we argue for serious consideration of ecological, environmental, and other chemical cues as potent regulators of potentially evolutionarily conserved molecular processes.

2. CHEMICAL SENSING AND THE PLASTICITY OF AGING

2.1. Chemicals as Information

Ecologists and evolutionary biologists have long recognized the primary importance of chemical stimuli from the environment for influencing the phenotypes of individuals. Sensory perception of particular chemical cues can influence response to predation (Leonard et al. 1999), habitat selection (Morse 1991), mating (Brennan & Kendrick 2006), and the expression of costly defense mechanisms (Darst et al. 2006).

The presence of predators can also induce dramatic physiological or developmental alterations (Voronezhskaya et al. 2004), such as the growth of protective helmets in various species of *Daphnia* in response to waterborne cues from predators (Pijanowska & Kloc 2004). Organisms can also respond to chemical cues that indicate the density or presence of conspecifics or competing species, a process that has been subjected to incisive evolutionary analysis in microorganisms (Keller & Surette 2006, West et al. 2006). The function and potency of specific cues can also depend on the age of the organism (Zimmer et al. 2006).

DR: dietary restriction

Chemical cues from the environment can also influence the plasticity of aging, which can occur on different timescales. Broadly speaking, one form is largely developmental in nature and is often associated with the formation of different morphs within a given life cycle. Social insects, for example, exhibit caste differences in life span (Chapuisat & Keller 2002), with the evolution of eusociality strongly associated with the evolution of increased life span in reproductive females (Keller & Genoud 1997). The nematode *Strongyloides ratti* can adopt a free-living form, which lives roughly 5 days, or a parasitic form, which can live up to 400 days (Gardner et al. 2006). Other forms of plasticity in aging are more malleable. The common laboratory nematode *Caenorhabditis elegans* has a well-known alternate development stage called the dauer (Cassada & Russell 1975), which is reversible (discussed in detail in Section 2.3). In the fruit fly, *Drosophila melanogaster*, nutrient availability affects patterns of aging in a completely reversible manner (Mair et al. 2003).

2.2. Environmental Perception

Chemicals can provide information about environmental variables that can be used to inform life-history decisions. For instance, sensing the nutritional value of food may often involve the use of indirect chemical information. Nutritional content of food is usually considered either in terms of its contribution of critical components to biochemical reactions or of its overall energetic content. Many hypotheses concerning the mechanism of dietary restriction [DR, an environmental manipulation whereby reduced nutrient intake without malnutrition increases life span of a wide range of species (see Sections 2.3, 2.4)] thus focus on the nutritional content of the food. These hypotheses postulate that increased longevity under DR results from changes in metabolic rate (Faulks et al. 2006), alterations in nutritional resource allocation (Shanley & Kirkwood 2000), and/or cellular responses to energy depletion (Lin et al. 2000). However, changes in metabolic rate are often not involved in the response to DR (Houthoofd et al. 2007, Hulbert et al. 2004). Furthermore, many aspects of DR can be achieved by manipulating specific components of the diet (Mair et al. 2005) even with negligible change in overall energy content of the food (Miller et al. 2005, Zimmerman et al. 2003).

The focus on nutrients as energy has drawn attention away from the possibility that specific components of the food or aspects of the environment may provide powerful cues that alter biological processes. In their simplest manifestations, such as nutrient-directed chemotaxis, environmental perception and response have obvious advantages. After detecting the presence of food through olfactory or gustatory sensory systems, for example, most organisms will alter their behavior to isolate and converge on the source of the stimulus. Such basic information-processing capabilities are often considered critical for survival and are essentially present in nearly all living things (Keller & Surette 2006). The impact of relevant chemical cues might be mediated through traditional sensory systems, such as smell, taste, or touch, or they may act at the cellular level through direct molecular interactions with target proteins.

In this section we consider the emerging literature on how specific chemical cues, which may indicate the overall status of the environment or the availability of nutrients, induce inclusive phenotypes such as physiology and aging. Work in laboratory model organisms indicates that activity of the external sensors of the perceptual system is sufficient to influence life span and a range of developmental and physiological processes. Furthermore, there is growing evidence of the importance of internal physiological and cellular processes, a phenomenon that we term molecular sensing, where specific chemical compounds derived from an organism's environment bypass sensory systems and induce biological changes by directly interacting with target cellular proteins. In a few instances, the specific cues that trigger these responses and the downstream molecular pathways that implement them have been identified. Direct experimental evidence for an important role of environmental perception on aging is limited mostly to yeast, flies, and worms, but there is considerable indication that the pathways that implement response to environmental cues, if not the cues themselves, are evolutionarily conserved.

2.3. Canonical Sensory Systems

Genetically, one of the well-understood examples of environment-induced plasticity in development and life span is seen in *C. elegans*. At the first larval stage, the worm makes a developmental decision. Either it proceeds through normal development or it adopts an alternative third larval stage called dauer diapause. Normal larval development is usually complete within 3 days. Adult reproductive development follows and the most common adult form, a self-fertilizing hermaphrodite, lives approximately 14–20 days and produces roughly 250–300 eggs (Cassada & Russell 1975). Dauer animals undergo large-scale tissue remodeling and recycling of cellular components. They are nonfeeding, remain sexually immature, and exhibit social behaviors not generally seen in hermaphrodite adults (Cassada & Russell 1975). Metabolism is shifted to favor fat and carbohydrate storage, and the dauer animal is stress resistant and can live for several months (Burnell et al. 2005, Golden & Riddle 1982).

The decision to enter dauer or proceed with normal development is largely determined by sensory integration of two cues: food availability and a dauer pheromone called the daumone (Golden & Riddle 1982). The daumone is constitutively secreted by the animals and consists of a pyranose sugar conjugated to a fatty acid (Jeong et al. 2005). Specific amphid neurons (ADF, ASI, ASG) likely sense the daumone (Bargmann 2006) and appropriate responses require G protein signaling and the genes *gpa-2* and *gpa-3* (Zwaal et al. 1997). Too high a population density will lead to increased local concentration of daumone and will force dauer development.

Currently unidentified food cues are thought to be sensed by the same neurons and in this way increased food resources antagonize the daumone. Thus, the nematode senses environmental quality and induces adaptive metabolic decisions (Gerisch & Antebi 2004). In addition to their role in developmental decisions, the nematode neurosensory organs can modulate life span in normal adult worms. Mutations that disrupt the structure and/or function of sensory neurons result in increased longevity, as does laser ablation of the two amphid sheath cells, which support the amphid neurons (Apfeld & Kenyon 1999). Laser ablation of specific sensory neurons has variable effects; removal of specific gustatory and olfactory neurons can increase, decrease, or have no effect on life span. Thus, some sensory neurons produce signals that antagonize long life, whereas others promote it (Antebi 2004).

IIS: insulin/Igf1-like signaling

Evidence from similar work in *D. melanogaster* shows that the regulation of aging by canonical sensory input is evolutionarily conserved. When flies are confined to a regime of DR, exposure to food-based odorants reduces life span (Libert et al. 2007). This effect is absent when flies are fully-fed, suggesting that perception of nutrient availability may reverse a subset of the physiological alterations induced by DR and may limit its benefits. Loss of function of the odorant receptor, Or83b, which results in flies with severe olfactory defects, results in a striking longevity phenotype: more than a 50% increase in mean life span is observed compared to wild-type animals. In addition to longevity, other general aspects of adult physiology are altered in flies with defective olfaction. Whereas Or83b mutant flies have normal size and metabolic rate, they have increased triglyceride storage and are resistant to starvation and hyperoxia.

There are a few examples where specific sensory cues capable of stimulating alterations in adult physiology and life span have been conclusively identified. It is currently of some debate whether exposure of the adult *C. elegans* to daumone is sufficient to extend life span. Two reports measuring the effects of crude lipid extracts containing the dauer pheromone are inconsistent: One reported no effect of the pheromone (Alcedo & Kenyon 2004) and the other reported a significant increase in life span (Kawano et al. 2005). Further experiments using the purified daumone (Jeong et al. 2005) would help resolve the issue. Data suggesting that specific compounds reduce life span should be interpreted with caution because toxins or other unpleasant molecules may shorten life span without accelerating the normal aging process. With this in mind, odorants from live yeast paste are sufficient to reduce life span when *D. melanogaster* are diet-restricted (Libert et al. 2007). In the ovoviviparous cockroach, *Nauphoeta cinerea*, specific components of the male pheromone may reduce longevity, although this effect may be confounded with mating activity (Moore et al. 2003).

The downstream mechanisms that are triggered by canonical sensory perception to modulate life span are largely unknown. In *C. elegans*, daumone may suppress daf-7/TGF-β expression, insulin-like peptide (daf-28), and the production of the lipophilic hormone dafachronic acid, which together act to prescribe the dauer phenotype and may influence dauer life span (Motola et al. 2006, Gerisch et al. 2007). Dauer pheromone results in increased fat storage, a phenotype that is common in manipulations that extend life span by down-regulation of insulin/IGF-1 signaling (IIS) (Tatar et al. 2003). Indeed, in adult *C. elegans*, sensory neurons produce insulin-like peptides. Because these peptides may act as receptor agonists or antagonists (Pierce et al. 2001), direct modulation of the insulin pathway is an attractive hypothesis for the longevity effects (Antebi 2004). Complicating a simple interpretation, however, is the observation that olfactory neurons regulate aging somewhat differently than gustatory neurons. Although gustatory neurons appear to affect life span by perturbing IIS, life span extension produced by ablating olfactory neurons is less dependent on IIS and appears to act in a pathway that involves the reproductive system (Alcedo & Kenyon 2004). Reduced olfactory function in *D. melanogaster* is correlated with increased triglyceride storage, suggesting that, as in *C. elegans*, the effect may be regulated at least partly through IIS (Libert et al. 2007). There is, however, currently no genetic evidence linking IIS with modulation of aging by olfaction in flies.

2.4. Molecular Sensors

It has also been suggested that DR exerts its effects in yeast, worms, and flies by activating the protein deacetylase Sir2. Overexpression of Sir2 extends life span in several species (Longo & Kennedy 2006). Because Sir2 is an NAD-dependent histone deacetylase, it has been postulated that it may serve as a molecular sensor of the metabolic state of the cell. This has led to the hypothesis that overexpression of Sir2 mimics a diet-restriction condition (Lin et al. 2000). Consistent with this interpretation, specific strains of yeast cells and fly lines that lack Sir2 protein fail to respond to diet, and the longevity of Sir2 overexpressing animals is not further extended by reduced nutrient availability (Lin et al. 2000). It should be noted, however, that the molecular and physiological basis of the effects of Sir2 and its role in dietary restriction is controversial (Kennedy et al. 2005).

Pharmocological activators of Sir2 have been identified (Howitz et al. 2003) and several of these compounds are plant-derived polyphenols, which show a diversity of structures, ranging from rather simple molecules (monomers and oligomers) to polymers (Cheynier 2005). In some studies these small molecules increased the life span of several species (Howitz et al. 2003, Wood et al. 2004). The most well-known of these compounds, Resveratrol, ameliorated liver pathology and increased life span in mice fed a high-fat diet (Baur et al. 2006). Resveratrol also seems to impact many aspects of mammalian health (Ates et al. 2007, Rahman et al. 2006), although the mechanism is still unclear. Many of the polyphenols that influence aging and physiology are produced by plants in times of stress. Their longevity-promoting effects may therefore be a type of molecular sensory perception and, like dietary restriction, may provide an accurate assessment of environmental condition (Lamming et al. 2004).

Several other types of molecular sensors have been implicated in modulation of longevity. The energy status of the cell is monitored by the AMP-activated protein kinase (AMPK) system, and this pathway promotes longevity in nematode worms (Apfeld et al. 2004). Chemicals that induce oxidative stress are sensed by MAPK signaling modules, and activation of this pathway increases life span in both flies (Wang et al. 2005) and worms (Oh et al. 2005). Health benefits to livestock fed antibiotics may represent a neurohormonal response triggered by alterations in gut flora that are recognized by the immune response (Yun et al. 2006).

As alluded to earlier, manipulation of IIS slows down aging in evolutionarily diverse organisms, as does alteration of signaling in the TOR pathways (Kaeberlein et al. 2005, Kapahi et al. 2004, Vellai et al. 2003). Both are involved in cellular sensing of nutrients, but it is at present not clear whether this role is important in relation to the effect of the pathways on aging, or whether instead these functions are attributable to different mechanisms. Both IIS manipulation and the TOR pathways have been implicated in the response to DR (Bonkowski et al. 2006, Clancy et al. 2002, Kaeberlein et al. 2005, Kapahi et al. 2004), but the upstream signaling mechanisms by which these pathways extend life span and interact with nutrients and other chemicals await elucidation. The deep evolutionary conservation of the effects of these pathways on aging suggests that variation in their expression could explain some of the biodiversity and phenotypic plasticity in life span, a conjecture supported by recent work with honey bees (Corona et al. 2007).

2.5. Evolutionary Perspectives

Pathways that are important in sensory regulation of life span in simple model systems have counterparts conserved in mammals (Partridge & Gems 2002, Tatar et al. 2003). Indeed, we postulate that the regulatory pathways are public mechanisms of aging and are shared by many different organisms. The specific chemical compounds that initiate regulation are probably environment- and species-specific and thus involve private mechanisms of aging. This view predicts that the large number of genes that have been shown to have small effects on aging (Murphy et al. 2003) may be subject to coordinated regulation. Such coordination may underlie plastic responses to the environment and form an evolutionarily ancient aspect of aging in all organisms (Keller & Surette 2006). Experimental manipulations that slow the aging process may, therefore, at least in part, tap into evolved mechanisms of detecting and responding to informative chemicals. If perception mediates an adaptive response to alterations in environmental conditions, then our hypothesis predicts that manipulation of these regulatory systems should, in some cases, have adverse fitness consequences. For example, a DR mimetic, despite increasing life span, might ultimately be maladaptive if resources are abundant. In *C. elegans*, mutations in genes encoding components of the insulin signaling pathway significantly increase worm life span in noncompetitive conditions but result in severely reduced individual fitness in the competitive laboratory environments (Jenkins et al. 2004, Van Voorhies et al. 2005). Life-span extending mutations are often pleiotropic and have effects that are environment-dependent (Partridge et al. 2005). Genetic manipulations and experimental conditions that target sensory cues may actually trigger inappropriate life-history decisions.

Might such sensory systems be a useful target for therapeutic manipulations in humans? There are indications this may be the case. Humans in westernized societies are in general not short of food, voluntarily limit their fecundity and are not subject to many of the challenges that are encountered in nature. One result is the epidemic of obesity, a major risk factor for many aging-related diseases. Different life-history decisions from those that evolved in the past might therefore be associated with benefits to health, and could lead to significant natural selection in present-day populations (Drenos et al. 2006). The cephalic phase response, a group of physiological changes elicited by the sight, smell, or taste of food, involves preabsorptive release of enzymes and hormones elicited by neural activation, as opposed to nutrient-induced stimulation (Teff 2000). The primary aspect of this response is the short-term release of pancreatic polypeptides, insulin, and glucagon, and it is generally viewed as a mechanism to optimize nutrient digestion, absorption, and metabolism (Yun et al. 2006). Such endocrine responses are reminiscent of those induced in mice by DR or by genetic mutations that increase life span (Lindemann 2001, Mattes 1997). Notably, the cephalic phase insulin response is significantly greater in women that are considered "restrained" eaters (Crystal & Teff 2006), suggesting that, similar to *Drosophila*, the physiological impact of nutrient perception depends on current environmental conditions. Olive oil, a well-established but poorly understood cardioprotective agent, has potent anti-inflammatory properties that may arise in part from signaling effects (Beauchamp et al. 2005), and the taste of certain pharmaceutical compounds correlates with their pharmacological activity (Fischer et al. 1965).

Some basic evolutionary questions remain outstanding. For instance, increase in life span in response to DR may represent a case of genuine homology of mechanisms but it could, instead, be a case of evolutionary convergence. We therefore do not yet know what role may be played in humans by mechanisms of DR discovered in animals. Much of the work described above has been performed on laboratory model organisms. These are far from a random sample of biodiversity, because particular life-history traits, such as high fecundity, lead to ease of laboratory culture. Furthermore, these animals are in general kept in relatively pathogen-free environments and scarcely have to move to obtain their food. Although such studies will be challenging, it is important that this work in the laboratory is related to the ecology and evolution of these organisms. The role of DR in natural conditions is not clear, and its status as an evolved response to natural conditions of food shortage has recently been questioned (Miller et al. 2002). To answer the last question, it will be necessary to move from the laboratory into natural conditions.

3. CHEMICAL INSULTS AND AGING

3.1. Aging and Molecular Damage

Aging is characterized by loss of functional capacity and is presumably caused by some constellation of the multiple forms of damage and pathology that accumulate with age. The biochemical routes to the generation of molecular damage are diverse. The need to maintain homeostasis in the face of physical, chemical, and biological challenges has driven the evolution of an array of gene families and pathways that protect against and repair damage. Directly relating specific losses of function to specific forms of damage has proved elusive, and there is currently no real consensus on whether specific types of molecular damage are of predominant importance in causing aging. Nor is it clear if the biochemical processes that generate damage and the types of damage that they produce show evolutionary conservation, on which rests the utility of model organisms and evolved differences in life span for understanding the biochemistry of human aging. We need to understand how deep into the causal chain leading to aging any evolutionary conservation extends. This provides fertile ground for evolutionary biologists to undertake comparative and evolutionary genomic analysis and work toward understanding mechanisms of evolved phenotypic plasticity. The tools are at hand to make progress in these important areas. Recent findings point to the critical importance of diverse toxins, potentially of both exogenous and endogenous origin, in causing the aging process and, hence, to the possible key role of detoxification systems in defense against aging.

ROS: reactive oxygen species

3.2. The Free Radical Theory

The main candidate mechanism for the generation of aging-related damage, predominant for half a century, is the free radical theory (Harman 1956). This points to the key role of damage by reactive oxygen species (ROS) (Martin et al. 1996, Sohal & Weindruch 1996). ROS (e.g., superoxide, hydroxyl radicals and hydrogen peroxide) are produced as a by-product of normal cellular metabolism, mainly from mitochondria. The degree of oxidative stress and, hence, oxidative damage to macromolecules are determined by the balance between ROS production and a finely tuned antioxidant defense system, including enzymatic (e.g., superoxide dismutase, catalase) and nonenzymatic (e.g., ubiquinol, Vitamin E, ascorbate, glutathione) scavengers together with a variety of molecular repair processes. Oxidative damage is a universal accompaniment to aging (Sohal & Weindruch 1996). Furthermore, both generation of ROS from isolated mitochondria and rate of accumulation of oxidative damage during aging are lower in long-lived species and in animals subjected to DR (Ruiz et al. 2005, Sanz et al. 2006). Endogenous antioxidant defenses are, in general, similar or lower in DR animals and long-lived species and also in long-lived ant (Parker et al. 2004) and honey bee queens (Corona et al. 2005), suggesting that long-lived forms may be subject to lower levels of oxidative challenge.

Despite these suggestive correlations, direct experimental evidence for the free radical theory is scant and at times contradictory (Gems & McElwee 2005). For instance, in *C. elegans*, catalytic antioxidants fed to the worm and in *Drosophila* overexpression of the genes for cytoplasmic (Cu/Zn-SOD) and mitochondrial (MnSOD) superoxide dismutases have both been reported to extend life span. However, the work on catalytic antioxidants was not repeatable in another lab (Keaney et al. 2004) and the degree of extension of life span from overexpression of antioxidant defenses in *Drosophila* is greater in short-lived strains, suggesting that it might extend life span by rescuing pathology in strains that have become inbred or adapted to the laboratory environment (Orr & Sohal 2003, Spencer et al. 2003).

There are many reasons why the kinds of negative results just described could occur yet the free radical theory could, in fact, be correct. For instance, manipulations that alter oxidative stress could also interfere with cellular signaling processes that require strict control of the redox state. The almost universal correlative evidence in favor of the importance of free radicals would make it surprising if they are eventually found genuinely to have no role in aging. However, other processes may be as, or more, important, and recent work in comparative genomics has started to identify what they may be.

3.3. Endobiotics, Xenobiotics, and Aging

The genome revolution has opened up a different and potentially much more powerful approach to identifying candidate mechanisms of aging, which does not involve exploring the preconceptions of the experimenter. Genome-wide surveys of changes in gene expression associated with slowing of the aging process can potentially be conducted using unbiased procedures, allowing the data to point to the mechanisms at work. This approach has so far proved particularly powerful for analysis of the mechanisms by which mutations in single genes extend life span, but it also has great potential in the context of evolved differences and phenotypic plasticity in the rate of aging.

Analysis of single gene mutations has produced clear evidence for an evolutionarily conserved role of signaling pathways in the determination of life span. The best-studied case is an endocrine system that resembles the mammalian insulin and insulin-like growth factor 1 systems: the IIS pathway. This pathway evolved early in animal phylogeny and may, indeed, have played a crucial role in the evolution of the metazoa (Skorokhod et al. 1999). It has diverse roles in regulation of growth and body size, blood sugar and metabolism, stress resistance, and fecundity. Its role in aging was originally identified in *C. elegans* (Klass 1983), and it was subsequently found to play a similar role in *Drosophila* and mice (reviewed in Piper et al. 2005).

Several of the genes in *C. elegans* that encode components of the IIS pathway first came to light because mutations in them caused the worm to form dauer larvae in the absence of any crowding or food shortage. Because dauer larvae are very long-lived, the extended life span of the IIS mutant adult worms was suggested to act through the same biochemical processes (Kenyon et al. 1993). A comparison of RNA expression profiles of dauer larvae and of long-lived IIS mutant adults confirmed this possibility; there was a highly significant overlap between the genes that were up- or down-regulated upon entry into the dauer larva and during extension of life span by IIS mutations (McElwee et al. 2004). Particular categories of genes, identified objectively using the functional annotation of the genome, were statistically over- or under-represented in the genes that were differentially regulated in both dauers and long-lived adults. Reassuringly, genes already known to be essential for extension of life span by IIS, such as those encoding small heat shock proteins, were identified by this procedure. In addition, cytochrome P450s, short-chain dehydrogenase/reductases, and UDP-glucuronosyltransferases were significantly up-regulated. In mammals these three enzyme classes act in concert to dispose of toxic endobiotic or xenobiotic compounds. In the long-lived adult worms, glutathione-S-transferases were up-regulated, and these are also linked to detoxification. The toxic compounds that these gene categories target are therefore implicated as a cause of aging (McElwee 2007).

These findings led to the Green Theory of aging (Gems & McElwee 2005), which suggests that cells constantly have to deal with mainly lipophilic metabolic waste products and xenobiotics. These are chemically highly diverse, which poses an informatic challenge for the cell and, hence, potentially explains the size of the gene families involved and the broad spectrum of substrates of each gene product. The detoxification and removal of this toxic waste are energetically expensive, because it must be not only solubilized and detoxified, but the resulting products have to be excreted. This contrasts markedly with the machinery for removing ROS such as superoxide and hydrogen peroxide from the cell, where the detoxification process is rapid, specific, and does not require energy, and where the detoxification products do not require excretion.

3.4. An Evolutionarily Conserved Biochemistry of Aging?

RNA expression profiling of long-lived mutant mice provided some evidence that the role of cellular detoxification processes in slowing aging may be evolutionarily conserved. Ames dwarf and Little mice are both long-lived, as a result of mutations in single genes encoding a transcription factor involved in pituitary development and the growth hormone releasing hormone receptor, respectively. Interestingly, both have reduced circulating levels of Igf-1, suggesting that their extended life span may be related to altered IIS. RNA expression profiling of the livers of these mice showed that, among other changes, there was a marked elevation of expression levels of gene classes involved in endobiotic and xenobiotic metabolism, paralleling the findings with *C. elegans* (Amador-Noguez et al. 2004). Subsequent work has demonstrated that this altered pattern of gene expression has the expected physiological consequence; Little mice are resistant to the effects of a variety of xenobiotic compounds (Amador-Noguez & Darlington 2007).

These results suggest that these cellular detoxification mechanisms could be a public mechanism for resisting the aging process. A recent three-way species comparison of RNA expression in long-lived IIS mutant *C. elegans* and *Drosophila* with that of the Ames and Little mouse has both provided further support for this idea and illuminated the issue of the level at which this system shows evolutionary conservation. Genes involved in cellular detoxification were up-regulated in long-lived mutant IIS representatives of all three species. However, this pattern was apparent only at the level of the gene families involved. There was no tendency toward common expression changes in genes that qualified as orthologs in the different species on the basis of their sequence similarity (McElwee 2007). For this system, longevity assurance mechanisms through which altered IIS extends life span thus appear to be lineage specific at the gene level (private), but conserved at the process level (or semipublic).

The reason for the evolutionary conservation at the level of gene families but not of genes is explicable from the way that the large gene families involved in these detoxification processes evolved (**Figure 1**). They tend to undergo lineage-specific expansion, indicating that their diversification occurred after the divergence of *C. elegans*, *Drosophila*, and the mouse from each other. This is typical of proteins whose function entails recognizing diverse chemical moieties in a changing chemical environment. Such proteins include chemoreceptors and antigen recognition proteins of the innate and acquired immune systems, as well as those involved in cellular detoxification (Lespinet et al. 2002, Thomas et al. 2005). Other processes involved in longevity assurance may show more conventional gene-to-gene orthology. For

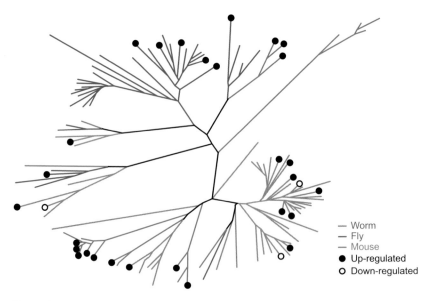

— Worm
— Fly
— Mouse
● Up-regulated
○ Down-regulated

Figure 1

Phylogenetic tree of the GST (glutathione-S-transferase) gene families from worms, flies, and mice. Genes from each species are color-coded and, significantly, differentially expressed genes (q < 0.1) in long-lived IIS mutant *C. elegans*, *D. melanogaster*, and mice are shown by closed (up-regulated) or open (down-regulated) circles. There is little gene-to-gene orthology between species, because the extant genes are largely the product of lineage-specific expansions. The GSTs in each species therefore usually have a more recent common ancestor than the common ancestor of *C. elegans*, *D. melanogaster*, and mice.

instance, the three-species comparison also identified down-regulation of protein synthesis as consistently associated with extension of life span.

It thus appears that the mechanisms of aging against which longevity-assurance mechanisms act are to some extent public and also to some extent lineage-specific (private). The fact that the gene families involved in detoxification undergo lineage-specific expansion suggests in itself that the biochemical challenges faced by the system change during evolution. Detoxification genes are known to be of ecological importance and are, for instance, specifically induced in association with different hosts in cactophilic *Drosophila* (Matzkin et al. 2006) as well as in response to drugs and insecticides (Willoughby et al. 2006). In addition, differences in diet and metabolism could also result in the generation of a different spectrum of endobiotics in different evolutionary lineages.

4. CHEMICALS AS DEFENSES

4.1. Ancient Systems of Pathogen Recognition and Response

Recognizing pathogenic invaders and responding accordingly is the primary function of immunity. The highly complex mammalian immune response has its roots in unicellular organisms that possess enzyme systems to protect against viral infections. More complex mechanisms evolved in different eukaryotic lineages, such as insects, fish, and reptiles. Immunity in these organisms depends primarily on innate (natural) immunity, where receptors recognize patterns of molecular structures on the surface of microorganisms that are absent from eukaryotic cells. Outcomes of this recognition include induction of genes that encode antimicrobial peptides and initiation of an inflammatory response (Renshaw et al. 2002). An additional function of innate immunity is to recruit cells involved in adaptive immunity, which has appeared more recently, in the ancestors of cartilaginous fishes (Hoffmann & Reichhart 2002). Adaptive immune systems use a large repertoire of antigen-recognizing receptors to establish an immunological memory, which provides better protection from fast-evolving pathogens and a more rapid response to subsequent rechallenge from particular infectious agents.

Although most immunological research over the past 20 years has been focused on adaptive immunity, the role of innate immunity in aging has become a subject of considerable attention. There are at least two reasons for this. First, there is an increasing realization that the innate immune response triggers and orients the adaptive response (Fearon & Locksley 1996) and that age-dependent changes in traditional components of innate immunity may lie at the root of increased susceptibility to infections in old individuals (Plowden et al. 2004). Second, genetic mechanisms of innate immunity are highly conserved across species (Hoffmann & Reichhart 2002), and remarkable progress in understanding recognition and signaling pathways involved in the innate response has come from work in model systems, such as *D. melanogaster*, which also serve as important models for the study of aging (Hoffmann & Reichhart 2002).

In insects and many invertebrates, innate immunity is mediated by generic molecules, including peptidoglycan recognition proteins, which have evolved to recognize the chemical structure of the pathogens and antimicrobial peptides, which are effective at eliminating them (Hoffmann & Reichhart 2002). Cellular mechanisms are also present and involve phagocytosis and encapsulation of foreign structures or organisms by hemocytes that circulate in the blood. Recognition of invading pathogens and activation of an immune response in these organisms are homologous to the innate immune defense systems of vertebrates. The most characterized aspect of insect innate immunity surrounds signaling cascades in two pathways, the Imd/Relish pathway and the Toll/Dif pathway (Hultmark 2003), which are evolutionarily conserved and are homologous to mammalian Toll-like receptor and tumor necrosis factor receptor signaling pathways, respectively (Kaneko & Silverman 2005). The two pathways differ in their specificity to pathogen-associated molecular patterns; fungal and gram (+) bacterial infections primarily result in stimulation of the Toll signaling pathway, whereas the Imd pathway is stimulated by infections of gram (–) bacteria (Hoffmann & Reichhart 2002).

PGRP: peptidoglycan recognition protein

4.2. Aging and the Innate Immune Response

The mammalian innate immune system mounts inflammatory responses, and age-related changes in these responses have been postulated to be a causative factor for aging (Franceschi & Bonafe 2003). An inflammatory state has also been reported in old *D. melanogaster* (Pletcher et al. 2005), suggesting that the causes and consequences of age-dependent inflammation may be evolutionarily conserved. Under normal circumstances, inflammation is triggered by the innate system as an initial step of defense (Nguyen et al. 2002). Chronic inflammatory status with aging is believed to be caused by continuous exposure to molecules with a chemical structure that triggers the response (Franceschi & Bonafe 2003). A constant inflammatory state exhausts the innate system, activates various stress responses, and leads to accumulation of cellular and molecular damage. Chronic inflammation is considered to be a driving force for age-related pathologies including neurodegenerative diseases (Wilson et al. 2002), atherosclerosis (Mach 2005), diabetes (Craft 2006), and cancer (Schwartsburd 2004).

Many factors may play a role in aging-related changes in immune function. In the honey bee, *Apis mellifera*, reduced immune function is associated with a change in social behavior (Amdam et al. 2005). At roughly 23 days of age, worker bees transition from colony tasks, such as nursing, to more active duties, including foraging. At this time, there is a severe decline in immunity, including apoptotic loss of hemocytes, which are responsible for producing antimicrobial peptides and other immunity molecules. The loss of hemocytes is likely caused by increases in juvenile hormone titer. When older worker bees are forced to resume colony maintenance, juvenile hormone levels are reduced, vitellogenin levels are increased, and immune function is restored. These data suggest that organisms may exhibit adaptive plasticity of immunoscenescence (Amdam et al. 2005).

Changes in the activity of the innate immune system are associated with aging-related diseases in humans. Expression of interleukin-6 (IL-6), a proinflammatory cytokine, increases with age (Franceschi & Bonafe 2003), and expression level is associated with disability and mortality (Ferrucci et al. 1999). Consistent with the idea that immune activation limits life span, the level of IL-6 is low in centenarians (Franceschi & Bonafe 2003). Constitutive activation of nuclear factor κB (NFκB), a key mediator of inflammation, is associated with a variety of cancers (Haefner 2006). Also, in mammals, activation of NFκB-related transcription factors leads to up-regulation of antiapoptotic pathways (Zamorano et al. 2001). It has been suggested that aging in general and deterioration of the immune system in particular are consequences of a proinflammatory response that was mainly selected for resistance to infections in youth (Franceschi et al. 2000, van den Biggelaar et al. 2004, Vasto et al. 2006). Age-related decline in the efficiency of the immune response and system-wide inflammatory processes are attenuated by DR (Morgan et al. 2007). In rodents, DR decreases the age-associated susceptibility to infectious diseases. It also improves the maintenance and production of circulating immune cells, preserves receptor repertoire diversity, and reduces the production of inflammatory cytokines (Messaoudi et al. 2006).

4.3. Aging and the Costs of Chemical Defenses

Despite the importance of the innate immune system in promoting pathogen resistance, activation of the response can have adverse consequences, particularly for aging and aging-related diseases (Fedorka et al. 2004, Libert et al. 2006). Maintenance and the initiation of the immune response are costly and are associated with trade-offs (Demas et al. 1997). Ubiquitous activation of immune pathways throughout larval stages in *Drosophila* produces developmental abnormalities, including activation of the prophenoloxidase cascade (Takehana et al. 2002) and short-lived adults (DeVeale et al. 2004). Adult-specific, chronic activation of these same pathways results in an inflammation-like state and reduces life span (Libert et al. 2006). Immune activation in bumblebee *Bombus terrestris* results in shorter life span under starvation conditions (Moret & Schmid-Hempel 2000). Several lines of evidence show that testosterone is associated with suppression of the immune response as a cost of the development of secondary sexual traits (Alonso-Alvarez et al. 2007).

One of the emerging links between innate chemical defense mechanisms and aging focuses on the adaptive use of those traditional enemies of longevity, ROS. In addition to antimicrobial proteins produced by the innate immune response, organisms also ward off pathogens by producing microbicidal bursts of ROS. Indeed, cells need to generate massive amounts of ROS to kill the invading pathogens. Although they are effective weapons, the disproportionate production of ROS can oxidize and damage cellular macromolecules and jeopardize host cellular metabolism and integrity. In *Drosophila*, the NADPH oxidase-dependent oxidative burst is critical for surviving natural infection (Ha et al. 2005a), but it can also severely shorten life span if not effectively controlled (Ha et al. 2005b). There are other examples where free radicals serve as important immune signaling molecules. Superoxide anion or low micromolar concentrations of hydrogen peroxide increase the production of the T-cell growth factor IL-2 in activated T cells (Roth & Droge 1987). NFκB is activated by hydrogen peroxide during the launching of an immune response (Schreck et al. 1991).

From flies to humans, signaling networks associated both with aging and with immunity are evolutionarily conserved. Effector molecules of these networks are emerging as correlates of aging-related disease in humans and as key regulators of longevity in invertebrate systems. A critical path for future research leads toward dissecting the mechanistic links between these processes and the extent to which those links are evolutionarily conserved. Using laboratory model systems to dissect the molecular interactions between these pathways in combination with ecological studies on related species provides a critical direction, and an examination of the role of adaptive production of ROS may be a first step.

5. CONCLUSIONS AND FUTURE ISSUES

There is growing evidence that organismal aging and physiology are strongly modulated by mechanisms that detect and decode environmental chemical cues, initiate defenses to endogenous and exogenous chemicals, and mediate physiological decisions in response to chemical insults (**Figure 2**). The particular chemical information and insults encountered are likely to be strongly influenced by environment and lifestyle, and hence to show little evolutionary conservation. However, the systems that decode and respond to them represent mechanisms that are, at least in part, public. If we understood more about the ecology of laboratory model organisms, we

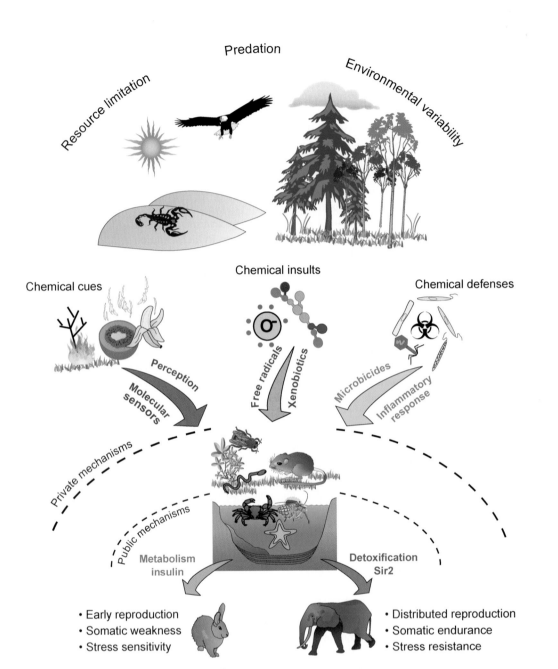

Predation

Resource limitation

Environmental variability

Chemical insults

Chemical cues

Chemical defenses

Perception

Molecular sensors

Free radicals

Xenobiotics

Microbicides

Inflammatory response

Private mechanisms

Public mechanisms

Metabolism insulin

Detoxification Sir2

• Early reproduction
• Somatic weakness
• Stress sensitivity

• Distributed reproduction
• Somatic endurance
• Stress resistance

could learn which aspects of biology of their aging processes are likely to be lineage specific. For instance, recent work suggested that inhibition of protein translation extends life span in *C. elegans*. Could this be a private mechanism for dealing with the presence of antibiotics, which often interact with ribosomes, in the soil environment that *C. elegans* inhabits in nature?

Figure 2

Chemical complexity and aging: Ecological factors such as variability, resource scarcity, and predation provide selection pressures for alternative life histories and survival mechanisms. To cope with environmental challenges, organisms have evolved ways to handle various aspects of chemical complexity, and these systems play a significant role in aging. Through perceptual systems, chemicals may act as cues to trigger physiological decisions and alternative life histories. To resist aging, organisms need to protect themselves from chemical insults, such as molecular damage from free radicals and endobiotic toxins. Mechanisms of natural immunity are also critical for survival in nature, and mechanisms that detect and respond to pathogens can impact life span. Nearly all species are exposed to some form of these challenges, but the exact nature of the threats is largely species specific. The downstream processes that specify and enact life-history decisions (such as Sir2 and insulin signaling), however, may be public mechanisms. Adapted and extended from Ackerman & Pletcher (2007).

We suggest that other, currently unknown, environmental conditions may affect aging through the filter of sensory perception. Recent work in several model systems has begun to characterize mechanisms of thermosensation (Rosenzweig et al. 2005), and there is reason to believe that such mechanisms are conserved in mammals. Is it possible that the near universal relationship between temperature and longevity in poikilotherms, which has long been thought to result from a direct effect of temperature on metabolic reactions, is due, at least in part, to molecular sensing leading to an adaptive temperature response?

Evolutionary biologists were the first to pose rigorous scientific questions concerning how and why organisms age. The simple, yet effective, models that resulted from this early work form the foundation for all of the current research in this area. However, the power of molecular biology has begun to reveal the detailed processes through which organisms respond to environmental conditions and enact systems that slow aging. In so doing, many of the traditional views about aging, that it was species specific and that it would prove resistant to genetic or pharmacological intervention, were overthrown. This has, in our opinion, resulted in a relative stagnation in the advance of evolutionary research into aging, with most work still struggling to distinguish hypotheses of mutational accumulation from those of antagonistic pleiotropy (Ackerman & Pletcher 2007). To maintain its importance in the future, evolutionary biology must join forces with with molecular biology to address the role of conserved mechanisms of aging. In the end, a synergistic approach that combines ecological and evolutionarily defined cues and chemicals with molecular, genetic manipulations may provide unique insight into the evolution of patterns of normal aging and may help in pinpointing manipulations by which healthy life span in humans may be extended.

SUMMARY POINTS

1. Until the 1990s work on the biology of aging was dominated by evolutionary analysis, which led to the predictions that mechanisms of aging were likely to be highly polygenic and lineage specific. This view has been challenged by the discovery of mutations in single genes that extend healthy life span in evolutionarily diverse model organisms.

2. The pathways discovered to modulate aging in the laboratory are often involved in various forms of sensing, particularly of chemicals. Chemical cues can indicate the overall status of the environment, the availability of nutrients, or the presence of other organisms, and they can be detected and decoded through canonical sensory systems such as smell and taste or through molecules that function to sense the cellular milieu. Single-gene mutations that extend life span may therefore tap into evolved mechanisms to ensure longevity and match organismal life-history decisions to the environment.

3. Aging is caused by accumulation of damage, and recent work indicates that detoxification of xenobiotic and endobiotic toxins may play an important role in combating aging. These toxins pose an informatic challenge to detoxification systems, and the gene families involved often undergo lineage-specific expansions.

4. Mechanisms that detect environmental cues and the cues themselves are likely to be private (species-specific) mechanisms of aging, whereas the downstream mechanisms that coordinate adaptive responses to these cues may be public.

5. The innate immune system mounts inflammatory responses, and age-related changes in these responses have been postulated to be a causative factor for aging. Age-related immune activation has been reported in several species, suggesting an intimate link between immune function and the causes and consequences of aging.

6. An emerging area of interest is the link between aging and the adaptive use of reactive oxygen species (ROS) to protect against pathogenic organisms.

7. Environmental variables, such as temperature, that are traditionally considered to act directly on ectothermic organisms by altering their rate of metabolism, could instead be sensed and responded to with evolved phenotypic plasticity.

8. Integrating ideas of molecular mechanism with context derived from evolutionary considerations will lead to exciting new insights into the evolution of aging.

DISCLOSURE STATEMENT

The authors are not aware of any biases that might be perceived as affecting the objectivity of this review.

LITERATURE CITED

Ackerman M, Pletcher SD. 2007. Evolutionary biology as a foundation for studying aging and aging-related disease. In *Evolution in Health and Disease*, ed. SC Stearns, JC Koella, Chapt. 18. Oxford, United Kingdom: Oxford Univ. Press

Alcedo J, Kenyon C. 2004. Regulation of *C. elegans* longevity by specific gustatory and olfactory neurons. *Neuron* 41:45–55

Alonso-Alvarez C, Bertrand S, Faivre B, Chastel O, Sorci G. 2007. Testosterone and oxidative stress: the oxidation handicap hypothesis. *Proc. Biol. Sci.* 274:819–25

Amador-Noguez D, Dean A, Huang W, Setchell K, Moore D, Darlington G. 2007. Alterations in xenobiotic metabolism in the long-lived Little mice. *Aging Cell* 6:453–70

Amador-Noguez D, Yagi K, Venable S, Darlington G. 2004. Gene expression profile of long-lived Ames dwarf mice and Little mice. *Aging Cell* 3:423–41

Amdam GV, Aase AL, Seehuus SC, Kim Fondrk M, Norberg K, et al. 2005. Social reversal of immunosenescence in honey bee workers. *Exp. Gerontol.* 40:939–47

Antebi A. 2004. Long life: a matter of taste (and smell). *Neuron* 41:1–3

Apfeld J, Kenyon C. 1999. Regulation of lifespan by sensory perception in *Caenorhabditis elegans*. *Nature* 402:804–9

Apfeld J, O'Connor G, McDonagh T, DiStefano PS, Curtis R. 2004. The AMP-activated protein kinase AAK-2 links energy levels and insulin-like signals to lifespan in *C. elegans*. *Genes Dev.* 18:3004–9

Arantes-Oliveira N, Berman JR, Kenyon C. 2003. Healthy animals with extreme longevity. *Science* 302:611

Ates O, Cayli S, Altinoz E, Gurses I, Yucel N, et al. 2007. Neuroprotection by resveratrol against traumatic brain injury in rats. *Mol. Cell Biochem.* 294:137–44

Austad SN. 1997. Small nonhuman primates as potential models of human aging. *ILAR J.* 38:142–47

Bargmann CI. 2006. *Chemosensation in C. elegans* (October 25, 2006). In *WormBook*, ed. The *C. elegans* Res. Community. doi:10.1895/wormbook.1.123.1. **http://www.wormbook.org**

Baudisch A. 2005. Hamilton's indicators of the force of selection. *Proc. Natl. Acad. Sci. USA* 102:8263–68

Baur JA, Pearson KJ, Price NL, Jamieson HA, Lerin C, et al. 2006. Resveratrol improves health and survival of mice on a high-calorie diet. *Nature* 444:337–42

Beauchamp GK, Keast RS, Morel D, Lin J, Pika J, et al. 2005. Phytochemistry: Ibuprofen-like activity in extravirgin olive oil. *Nature* 437:45–46

Bonkowski MS, Rocha JS, Masternak MM, Al Regaiey KA, Bartke A. 2006. Targeted disruption of growth hormone receptor interferes with the beneficial actions of calorie restriction. *Proc. Natl. Acad. Sci. USA* 103:7901–5

Brennan PA, Kendrick KM. 2006. Mammalian social odours: attraction and individual recognition. *Philos. Trans. R. Soc. London Ser. B* 361:2061–78

Burnell AM, Houthoofd K, O'Hanlon K, Vanfleteren JR. 2005. Alternate metabolism during the dauer stage of the nematode *Caenorhabditis elegans*. *Exp. Gerontol.* 40:850–56

Cassada RC, Russell RL. 1975. The dauerlarva, a postembryonic developmental variant of the nematode *Caenorhabditis elegans*. *Dev. Biol.* 46:326–42

Chapuisat M, Keller L. 2002. Division of labor influences the rate of ageing in weaver ant workers. *Proc. Biol. Sci.* 269:909–13

Charlesworth B. 1994. *Evolution in Age-Structured Populations*. Cambridge, UK: Cambridge Univ. Press

Cheynier V. 2005. Polyphenols in foods are more complex than often thought. *Am. J. Clin. Nutr.* 81:S223–29

Clancy DJ, Gems D, Hafen E, Leevers SJ, Partridge L. 2002. Dietary restriction in long-lived dwarf flies. *Science* 296:319

Corona M, Hughes KA, Weaver DB, Robinson GE. 2005. Gene expression patterns associated with queen honey bee longevity. *Mech. Ageing Dev.* 126:1230–38

Corona M, Velarde RA, Remolina S, Moran-Lauter A, Wanget Y, et al. 2007. Vitellogenin, juvenile hormone, insulin signaling, and queen honey bee longevity. *Proc. Natl. Acad. Sci. USA* 104:7128–33

Craft S. 2006. Insulin resistance syndrome and Alzheimer disease: Pathophysiologic mechanisms and therapeutic implications. *Alzheimer Dis. Assoc. Disord.* 20:298–301

Crystal SR, Teff KL. 2006. Tasting fat: Cephalic phase hormonal responses and food intake in restrained and unrestrained eaters. *Physiol. Behav.* 89:213–20

Darst CR, Cummings ME, Cannatella DC. 2006. A mechanism for diversity in warning signals: conspicuousness versus toxicity in poison frogs. *Proc. Natl. Acad. Sci. USA* 103:5852–57

de Magalhaes JP, Costa J, Toussaint O. 2005. HAGR: The Human Ageing Genomic Resources. *Nucleic Acids Res.* 33:D537–43

Demas GE, Chefer V, Talan MI, Nelson RJ. 1997. Metabolic costs of mounting an antigen-stimulated immune response in adult and aged C57BL/6J mice. *Am. J. Physiol.* 273:R1631–37

DeVeale B, Brummel T, Seroude L. 2004. Immunity and aging: the enemy within? *Aging Cell* 3:195–208

Drenos F, Westendorp RG, Kirkwood TB. 2006. Trade-off mediated effects on the genetics of human survival caused by increasingly benign living conditions. *Biogerontology* 7:287–295

Faulks SC, Turner N, Else PL, Hulbert AJ. 2006. Calorie restriction in mice: effects on body composition, daily activity, metabolic rate, mitochondrial reactive oxygen species production, and membrane fatty acid composition. *J. Gerontol. A* 61:781–94

Fearon DT, Locksley RM. 1996. The instructive role of innate immunity in the acquired immune response. *Science* 272:50–53

Fedorka KM, Zuk M, Mousseau TA. 2004. Immune suppression and the cost of reproduction in the ground cricket, Allonemobius socius. *Evol. Int. J. Org. Evol.* 58:2478–85

Ferrucci L, Harris TB, Guralnik JM, Tracy RP, Corti MC, et al. 1999. Serum IL-6 level and the development of disability in older persons. *J. Am. Geriatr. Soc.* 47:639–46

Fischer R, Griffin F, Archer RC, Zinsmeister SC, Jastram PS. 1965. Weber ratio in gustatory chemoreception; an indicator of systemic (drug) reactivity. *Nature* 207:1049–53

Franceschi C, Bonafe M. 2003. Centenarians as a model for healthy aging. *Biochem. Soc. Trans.* 31:457–61

Franceschi C, Bonafe M, Valensin S, Olivieri F, De Luca M, et al. 2000. Inflammaging. An evolutionary perspective on immunosenescence. *Ann. NY Acad. Sci.* 908:244–54

Gardner MP, Gems D, Viney ME. 2006. Extraordinary plasticity in aging in Strongyloides ratti implies a gene-regulatory mechanism of lifespan evolution. *Aging Cell* 5:315–23

Gems D, McElwee JJ. 2005. Broad spectrum detoxification: the major longevity assurance process regulated by insulin/IGF-1 signaling? *Mech. Ageing Dev.* 126:381–87

Gerisch B, Antebi A. 2004. Hormonal signals produced by DAF-9/cytochrome P450 regulate *C. elegans* dauer diapause in response to environmental cues. *Development* 131:1765–76

Gerisch B, Rottiers V, Li D, Motola DL, Cummins CL, et al. 2007. A bile acid-like steroid modulates Caenorhabditis elegans lifespan through nuclear receptor signaling. *Proc. Natl. Acad. Sci. USA* 104:5014–9

Golden JW, Riddle DL. 1982. A pheromone influences larval development in the nematode *Caenorhabditis elegans. Science* 218:578–80

Ha EM, Oh CT, Bae YS, Lee WJ. 2005a. A direct role for dual oxidase in *Drosophila* gut immunity. *Science* 310:847–50

Ha EM, Oh CT, Ryu JH, Bae YS, Kang SW, et al. 2005b. An antioxidant system required for host protection against gut infection in *Drosophila. Dev. Cell* 8:125–32

Haefner B. 2006. Targeting NF-κB in anticancer adjunctive chemotherapy. *Cancer Treat. Res.* 130:219–45

Hamilton WD. 1966. The moulding of senesence by natural selection. *J. Theor. Biol.* 12:12–45

Harman D. 1956. Aging: A theory based on free radical and radiation chemistry. *J. Gerontol.* 11:298–300

Hoffmann JA, Reichhart JM. 2002. *Drosophila* innate immunity: an evolutionary perspective. *Nat. Immunol.* 3:121–26

Holmes DJ, Austad SN. 1995. Birds as animal models for the comparative biology of aging: a prospectus. *J. Gerontol. A* 50:B59–66

Houthoofd K, Gems D, Johnson TE, Vanfleteren JR. 2007. Dietary restriction in the nematode *Caenorhabditis elegans. Interdiscip. Top. Gerontol.* 35:98–114

Howitz KT, Bitterman KJ, Cohen HY, Lamming DW, Lavu S, et al. 2003. Small molecule activators of sirtuins extend Saccharomyces cerevisiae lifespan. *Nature* 425:191–96

Hughes KA, Reynolds RM. 2005. Evolutionary and mechanistic theories of aging. *Annu. Rev. Entomol.* 50:421–45

Hulbert AJ, Clancy DJ, Mair W, Braeckman BP, Gems D, et al. 2004. Metabolic rate is not reduced by dietary-restriction or by lowered insulin/IGF-1 signaling and is not correlated with individual lifespan in *Drosophila melanogaster. Exp. Gerontol.* 39:1137–43

Hultmark D. 2003. *Drosophila* immunity: Paths and patterns. *Curr. Opin. Immunol.* 15:12–19

Jaenisch R, Bird A. 2003. Epigenetic regulation of gene expression: how the genome integrates intrinsic and environmental signals. *Nat. Genet.* 33(Suppl.):245–54

Jenkins NL, McColl G, Lithgow GJ. 2004. Fitness cost of extended lifespan in *Caenorhabditis elegans*. *Proc. Biol. Sci.* 271:2523–26

Jeong PY, Jung M, Yim YH, Kim H, Park M, et al. 2005. Chemical structure and biological activity of the *Caenorhabditis elegans* dauer-inducing pheromone. *Nature* 433:541–45

Kaeberlein M, Powers RW 3rd, Steffen KK, Westman EA, Hu D, et al. 2005. Regulation of yeast replicative life span by TOR and Sch9 in response to nutrients. *Science* 310:1193–96

Kaneko T, Silverman N. 2005. Bacterial recognition and signaling by the *Drosophila* IMD pathway. *Cell Microbiol.* 7:461–69

Kapahi P, Zid BM, Harper T, Koslover D, Sapin V, et al. 2004. Regulation of lifespan in *Drosophila* by modulation of genes in the TOR signaling pathway. *Curr. Biol.* 14:885–90

Kawano T, Kataoka N, Abe S, Ohtani M, Honda Y, et al. 2005. Lifespan extending activity of substances secreted by the nematode *Caenorhabditis elegans* that include the dauer-inducing pheromone. *Biosci. Biotechnol. Biochem.* 69:2479–81

Keaney M, Matthijssens F, Sharpe M, Vanfleteren J, Gems D. 2004. Superoxide dismutase mimetics elevate superoxide dismutase activity in vivo but do not retard aging in the nematode *Caenorhabditis elegans*. *Free Radic. Biol. Med.* 37:239–50

Keller L, Genoud M. 1997. Extraordinary life spans in ants: a test of evolutionary theories of ageing. *Nature* 389:958–60

Keller L, Surette MG. 2006. Communication in bacteria: an ecological and evolutionary perspective. *Nat. Rev. Microbiol.* 4:249–58

Kennedy BK, Smith ED, Kaeberlein M. 2005. The enigmatic role of Sir2 in aging. *Cell* 123:548–50

Kenyon C. 2005. The plasticity of aging: Insights from long-lived mutants. *Cell* 120:449–60

Kenyon C, Chang J, Gensch E, Rudner A, Tabtiang R. 1993. A *C. elegans* mutant that lives twice as long as wild type. *Nature* 366:461–64

Kirkwood TB, Feder M, Finch CE, Franceschi C, Globerson A, et al. 2005. What accounts for the wide variation in life span of genetically identical organisms reared in a constant environment? *Mech. Ageing Dev.* 126:439–43

Kirkwood TBL, Martin GM, Partridge L. 1999. Evolution, senescence and health in old age. In *Evolution in Health and Disease*, ed. SC Stearns, pp. 219–30. Oxford, UK: Oxford Univ. Press

Klass MR. 1983. A method for the isolation of longevity mutants in the nematode *Caenorhabditis elegans* and initial results. *Mech. Ageing Dev.* 22:279–86

Lamming DW, Wood JG, Sinclair DA. 2004. Small molecules that regulate lifespan: Evidence for xenohormesis. *Mol. Microbiol.* 53:1003–9

Lee RD. 2003. Rethinking the evolutionary theory of aging: transfers, not births, shape senescence in social species. *Proc. Natl. Acad. Sci. USA* 100:9637–42

Leonard GH, Bertness MD, Yund PO. 1999. Crab predation, waterborne cues, and inducible defenses in the blue mussel, *Mytilus Edulis*. *Ecology* 80:1–14

Lespinet O, Wolf YI, Koonin EV, Aravind L. 2002. The role of lineage-specific gene family expansion in the evolution of eukaryotes. *Genome Res.* 12:1048–59

Libert S, Chao Y, Chu X, Pletcher SD. 2006. Trade-offs between longevity and pathogen resistance in *Drosophila melanogaster* are mediated by NFκB signaling. *Aging Cell* 5:533–43

Libert S, Zwiener J, Chu X, VanVoorhies WA, Roman G, Pletcher SD. 2007. Regulation of *Drosophila* life span by olfaction and food-derived odors. *Science* 315:1133–37

Lin K, Dorman JB, Rodan A, Kenyon C. 1997. *daf-16*: An HNF-3/forkhead family member that can function to double the life-span of *Caenorhabditis elegans*. *Science* 278:1319–22

Lin SJ, Defossez PA, Guarente L. 2000. Requirement of NAD and SIR2 for life-span extension by calorie restriction in Saccharomyces cerevisiae. *Science* 289:2126–28

Lindemann B. 2001. Receptors and transduction in taste. *Nature* 413:219–25

Longo VD, Kennedy BK. 2006. Sirtuins in aging and age-related disease. *Cell* 126:257–68

Mach F. 2005. Inflammation is a crucial feature of atherosclerosis and a potential target to reduce cardiovascular events. *Handb. Exp. Pharmacol.* 2005:697–722

Mair W, Goymer P, Pletcher SD, Partridge L. 2003. Demography of dietary restriction and death in *Drosophila*. *Science* 301:1731–33

Mair W, Piper MDW, Partridge L. 2005. Calories do not explain extension of lifespan by dietary restriction in *Drosophila*. *PLoS Biol.* 3:e223

Martin GM. 2002. Keynote: Mechanisms of senescence—complicationists versus simplificationists. *Mech. Ageing Dev.* 123:65–73

Martin GM, Austad SN, Johnson TE. 1996. Genetic analysis of ageing: Role of oxidative damage and environmental stresses. *Nat. Genet.* 13:25–34

Mattes RD. 1997. Physiologic responses to sensory stimulation by food: Nutritional implications. *J. Am. Diet. Assoc.* 97:406–13

Matzkin LM, Watts TD, Bitler BG, Machado CA, Markow TA. 2006. Functional genomics of cactus host shifts in *Drosophila* mojavensis. *Mol. Ecol.* 15:4635–43

McElwee JJ, Schuster E, Balnc E, Piper MD, Thomas JH, et al. 2007. Evolutionary conservation of regulated longevity assurance mechanisms. *Genome Biol.* 8:R132

McElwee JJ, Schuster E, Blanc E, Thomas JH, Gems D. 2004. Shared transcriptional signature in *Caenorhabditis elegans* Dauer larvae and long-lived daf-2 mutants implicates detoxification system in longevity assurance. *J. Biol. Chem.* 279:44533–43

Medawar PB. 1952. *An Unsolved Problem in Biology*. London: Lewis

Messaoudi I, Warner J, Fischer M, Park B, Hill B, et al. 2006. Delay of T cell senescence by caloric restriction in aged long-lived nonhuman primates. *Proc. Natl. Acad. Sci. USA* 103:19448–53

Miller RA, Buehner G, Chang Y, Harper JM, Sigler R, et al. 2005. Methionine-deficient diet extends mouse lifespan, slows immune and lens aging, alters glucose, T4, IGF-I and insulin levels, and increases hepatocyte MIF levels and stress resistance. *Aging Cell* 4:119–25

Miller RA, Harper JM, Dysko RC, Durkee SJ, Austad SN. 2002. Longer life spans and delayed maturation in wild-derived mice. *Exp. Biol. Med. (Maywood)* 227:500–8

Moore AJ, Gowaty PA, Moore PJ. 2003. Females avoid manipulative males and live longer. *J. Evol. Biol.* 16:523–30

Moret Y, Schmid-Hempel P. 2000. Survival for immunity: the price of immune system activation for bumblebee workers. *Science* 290:1166–68

Morgan TE, Wong AM, Finch CE. 2007. Anti-inflammatory mechanisms of dietary restriction in slowing aging processes. *Interdiscip. Top. Gerontol.* 35:83–97

Morse ANC. 1991. How do planktonic larvae know where to settle? *Am. Sci.* 79:154–67

Motola DL, Cummins CL, Rottiers V, Sharma KK, Li T, et al. 2006. Identification of ligands for DAF-12 that govern dauer formation and reproduction in *C. elegans*. *Cell* 124:1209–23

Murphy CT, McCarroll SA, Bargmann CI, Fraser A, Kamath RS, et al. 2003. Genes that act downstream of DAF-16 to influence the lifespan of *Caenorhabditis elegans*. *Nature* 424:277–83

Nguyen MD, Julien JP, Rivest S. 2002. Innate immunity: the missing link in neuro-protection and neurodegeneration? *Nat. Rev. Neurosci.* 3:216–27

Oh SW, Mukhopadhyay A, Svrzikapa N, Jiang F, Davis RJ, et al. 2005. JNK regulates lifespan in *Caenorhabditis elegans* by modulating nuclear translocation of forkhead transcription factor/DAF-16. *Proc. Natl. Acad. Sci. USA* 102:4494–99

Orr WC, Sohal RS. 2003. Does overexpression of Cu,Zn-SOD extend life span in *Drosophila melanogaster*? *Exp. Gerontol.* 38:227–30

Orzack SH, Tuljapurkar S. 1989. Population dynamics in variable environments. VII. The demography and evolution of iteroparity. *Am. Nat.* 133:901–23

Osiewacz HD. 2002. Aging in fungi: role of mitochondria in Podospora anserina. *Mech. Ageing Dev.* 123:755–64

Parker JD, Parker KM, Sohal BH, Sohal RS, Keller L. 2004. Decreased expression of Cu-Zn superoxide dismutase 1 in ants with extreme lifespan. *Proc. Natl. Acad. Sci. USA* 101:3486–89

Partridge L, Barton NH. 1993. Optimality, mutation and the evolution of ageing. *Nature* 362:305–11

Partridge L, Gems D. 2002. Mechanisms of ageing: Public or private? *Nat. Rev. Genet.* 3:165–75

Partridge L, Gems D. 2006. Beyond the evolutionary theory of ageing, from functional genomics to evo-gero. *Trends Ecol. Evol.* 21:334–40

Partridge L, Gems D, Withers DJ. 2005. Sex and death: What is the connection? *Cell* 120:461–72

Pierce SB, Costa M, Wisotzkey R, Devadhar S, Homburger SA, et al. 2001. Regulation of DAF-2 receptor signaling by human insulin and ins-1, a member of the unusually large and diverse *C. elegans* insulin gene family. *Genes Dev.* 15:672–86

Pijanowska J, Kloc M. 2004. Daphnia response to predation threat involves heat-shock proteins and the actin and tubulin cytoskeleton. *Genesis* 38:81–86

Piper MDW, Selman C, McElwee JJ, Partridge L. 2005. Models of insulin signaling and longevity. *Drug Discov. Today: Dis. Models* 2:249–56

Pletcher SD, Libert S, Skorupa D. 2005. Flies and their golden apples: the effect of dietary restriction on *Drosophila* aging and age-dependent gene expression. *Ageing Res. Rev.* 4:451–80

Plowden J, Renshaw-Hoelscher M, Engleman C, Katz J, Sambhara S. 2004. Innate immunity in aging: Impact on macrophage function. *Aging Cell* 3:161–67

Rahman I, Biswas SK, Kirkham PA. 2006. Regulation of inflammation and redox signaling by dietary polyphenols. *Biochem. Pharmacol.* 72:1439–52

Renshaw M, Rockwell J, Engleman C, Gewirtz A, Katz J, Sambhara S. 2002. Cutting edge: Impaired Toll-like receptor expression and function in aging. *J. Immunol.* 169:4697–701

Ricklefs RE, Scheuerlein A, Cohen A. 2003. Age-related patterns of fertility in captive populations of birds and mammals. *Exp. Gerontol.* 38:741–45

Rose MR. 1991. *Evolutionary Biology of Aging*. Oxford, NY: Oxford Univ. Press

Rosenzweig M, Brennan KM, Tayler TD, Phelps PO, Patapoutian A, et al. 2005. The *Drosophila* ortholog of vertebrate TRPA1 regulates thermotaxis. *Genes Dev.* 19:419–24

Roth S, Droge W. 1987. Regulation of T-cell activation and T-cell growth factor (TCGF) production by hydrogen peroxide. *Cell Immunol.* 108:417–24

Rottiers V, Antebi A. 2007. Regulation of *C. elegans* aging by a bile acid. *Proc. Natl. Acad. Sci. USA* 104:5014–19

Ruiz MC, Ayala V, Portero-Otin M, Requena JR, Barja G, et al. 2005. Protein methionine content and MDA-lysine adducts are inversely related to maximum life span in the heart of mammals. *Mech. Ageing Dev.* 126:1106–14

Sanz A, Caro P, Ayala V, Portero-Otin M, Pamplona R, et al. 2006. Methionine restriction decreases mitochondrial oxygen radical generation and leak as well as oxidative damage to mitochondrial DNA and proteins. *FASEB J.* 20:1064–73

Schreck R, Rieber P, Baeuerle PA. 1991. Reactive oxygen intermediates as apparently widely used messengers in the activation of the NF-κB transcription factor and HIV-1. *EMBO J.* 10:2247–58

Schwartsburd PM. 2004. Age-promoted creation of a procancer microenvironment by inflammation: Pathogenesis of dyscoordinated feedback control. *Mech. Ageing Dev.* 125:581–90

Shanley DP, Kirkwood TB. 2000. Calorie restriction and aging: a life-history analysis. *Evol. Int. J. Org. Evol.* 54:740–50

Sinclair D, Guarente L. 1997. Extrachromosomal rDNA circles—a cause of aging in yeast. *Cell* 91:1033–42

Skorokhod A, Gamulin V, Gundacker D, Kavsan V, Muller IM, et al. 1999. Origin of insulin receptor-like tyrosine kinases in marine sponges. *Biol. Bull.* 197:198–206

Sohal RS, Weindruch R. 1996. Oxidative stress, caloric restriction, and aging. *Science* 273:59–63

Spencer CC, Howell CE, Wright AR, Promislow DE. 2003. Testing an 'aging gene' in long-lived *Drosophila* strains: Increased longevity depends on sex and genetic background. *Aging Cell* 2:123–30

Takehana A, Katsuyama T, Yano T, Oshima Y, Takada H, et al. 2002. Overexpression of a pattern-recognition receptor, peptidoglycan-recognition protein-LE, activates imd/relish-mediated antibacterial defense and the prophenoloxidase cascade in *Drosophila* larvae. *Proc. Natl. Acad. Sci. USA* 99:13705–10

Tatar M, Bartke A, Antebi A. 2003. The endocrine regulation of aging by insulin-like signals. *Science* 299:1346–51

Teff K. 2000. Nutritional implications of the cephalic-phase reflexes: Endocrine responses. *Appetite* 34:206–13

Thomas JH, Kelley JL, Robertson HM, Ly K, Swanson WJ. 2005. Adaptive evolution in the SRZ chemoreceptor families of *Caenorhabditis elegans* and *Caenorhabditis briggsae*. *Proc. Natl. Acad. Sci. USA* 102:4476–81

van den Biggelaar AH, Huizinga TW, de Craen AJ, Gussekloo J, Heijmans BT, et al. 2004. Impaired innate immunity predicts frailty in old age. The Leiden 85-plus study. *Exp. Gerontol.* 39:1407–14

Van Voorhies WA, Fuchs J, Thomas S. 2005. The longevity of *Caenorhabditis elegans* in soil. *Biol. Lett.* 1:247–49

Vasto S, Malavolta M, Pawelec G. 2006. Age and immunity. *Immun. Ageing* 3:2

Vellai T, Takacs-Vellai K, Zhang Y, Kovacs AL, Orosz L, Müller F. 2003. Genetics: Influence of TOR kinase on lifespan in *C. elegans*. *Nature* 426:620

Voronezhskaya EE, Khabarova MY, Nezlin LP. 2004. Apical sensory neurons mediate developmental retardation induced by conspecific environmental stimuli in freshwater pulmonate snails. *Development* 131:3671–80

Wang MC, Bohmann D, Jasper H. 2005. JNK extends life span and limits growth by antagonizing cellular and organism-wide responses to insulin signaling. *Cell* 121:115–25

West SA, Griffin AS, Gardner A, Diggle SP. 2006. Social evolution theory for microorganisms. *Nat. Rev. Microbiol.* 4:597–607

Williams GC. 1957. Pleiotropy, natural selection, and the evolution of senescence. *Evolution* 11:398–411

Willoughby L, Chung H, Lumb C, Robin C, Batterham P, et al. 2006. A comparison of *Drosophila melanogaster* detoxification gene induction responses for six insecticides, caffeine and phenobarbital. *Insect. Biochem. Mol. Biol.* 36:934–42

Wilson CJ, Finch CE, Cohen HJ. 2002. Cytokines and cognition—the case for a head-to-toe inflammatory paradigm. *J. Am. Geriatr. Soc.* 50:2041–56

Wood JG, Rogina B, Lavu S, Howitz K, Helfand SL, et al. 2004. Sirtuin activators mimic caloric restriction and delay ageing in metazoans. *Nature* 430:686–89

Yun AJ, Lee PY, Doux JD. 2006. Are we eating more than we think? Illegitimate signaling and xenohormesis as participants in the pathogenesis of obesity. *Med. Hypotheses* 67:36–40

Zamorano J, Mora AL, Boothby M, Keegan AD. 2001. NF-κB activation plays an important role in the IL-4-induced protection from apoptosis. *Int. Immunol.* 13:1479–87

Zimmer RK, Schar DW, Ferrer RP, Krug PJ, Kats LB, et al. 2006. The scent of danger: Tetrodotoxin (Ttx) as an olfactory cue of predation risk. *Ecol. Monogr.* 76:585–600

Zimmerman JA, Malloy V, Krajcik R, Orentreich N. 2003. Nutritional control of aging. *Exp. Gerontol.* 38:47–52

Zwaal RR, Mendel JE, Sternberg PW, Plasterk RH. 1997. Two neuronal G proteins are involved in chemosensation of the *Caenorhabditis elegans* Dauer-inducing pheromone. *Genetics* 145:715–27

A Global Review of the Distribution, Taxonomy, and Impacts of Introduced Seaweeds

Susan L. Williams[1] and Jennifer E. Smith[2]

[1] Section of Evolution and Ecology and Bodega Marine Laboratory, University of California, Davis, California 94923; email: slwilliams@ucdavis.edu

[2] National Center for Ecological Analysis and Synthesis, University of California, Santa Barbara, California 93106; email: jsmith@nceas.ucsb.edu

Annu. Rev. Ecol. Evol. Syst. 2007. 38:327–59

First published online as a Review in Advance on August 8, 2007

The *Annual Review of Ecology, Evolution, and Systematics* is online at http://ecolsys.annualreviews.org

This article's doi:
10.1146/annurev.ecolsys.38.091206.095543

Key Words

impact, invasive, macroalgae, native range, nonindigenous, vector

Abstract

We reviewed over 407 global seaweed introduction events and have increased the total number of introduced seaweed species to 277. Using binomial tests we show that several algal families contain more successful invaders than would be expected by chance, highlighting groups that should be targeted for management. Hull-fouling and aquaculture are the most significant sources of seaweed invaders and should be carefully regulated. The ecological effects of introduced seaweeds have been studied in only 6% of the species, but these studies show mostly negative effects or changes to the native biota. Herbivores generally prefer native to introduced seaweeds, and are unlikely to control spread, as they can do in other habitats. Undisturbed marine communities can be at least initially resistant to most introduced seaweeds aside from the siphonous green species; however, disturbances and eutrophication can facilitate invasion. Major research gaps include community-level ecological studies and economic assessments.

"Considering our present knowledge it will be many years before we will be able to predict with any degree of certainty the effects an introduced species may have on an existing ecosystem."

Phycologist L. Druehl in a letter to *Science* (Druehl 1973) predicting the establishment of *Sargassum muticum* in the eastern Atlantic after Japanese oysters were introduced from British Columbia to France.

INTRODUCTION AND SCOPE OF REVIEW

Seaweeds make a substantial contribution to marine primary production and provide habitat for nearshore benthic communities (Mann 1973). Over 200 seaweed species support an international economy in primarily phycocolloid (algins, agars, and carrageenans) and food products valued at over U.S. $6.2 billion (Zemke-White & Ohno 1999). Seaweed production has more than doubled over the past two decades. Through human activities including aquaculture, seaweeds have been introduced to non-native locations around the world. Excellent reviews on seaweed introductions have been recently published (Inderjit et al. 2006, Ribera & Boudouresque 1995, Ribera Siguan 2003, Schaffelke et al. 2006, Trowbridge 2006) as have case histories on specific species (Chapman 1998, Mathieson et al. 2003, Meinesz 1999). Regional reviews and checklists help document the arrival of new invaders into specific locations (Boudouresque & Verlaque 2002a, Castilla et al. 2005, Maggs & Stegenga 1999, Orensanz et al. 2002, Ribera Siguan 2002). The Global Invasive Species Database (**http://www.issg.org/database**) of the International Union for the Conservation of Nature (IUCN) provides extensive data (native/introduced ranges, references) on introduced seaweeds.

Our objective was to update and integrate the known information on global seaweed introductions and their ecological effects. Specifically we quantitatively assessed (*a*) the taxonomic affinities of introduced seaweeds, (*b*) their morphological characteristics (functional groups), (*c*) the native and introduced ranges of seaweed invaders, (*d*) the vectors of introduction, and (*e*) the ecological effects of introduced seaweeds on native biota and vice versa. Details of the physiology, molecular evidence of biogeographic affinities, economic impacts, and management of introduced seaweeds were beyond the scope of this review.

CLASSIFICATION AND ANALYSIS

We define an introduced seaweed (or invader) as a species belonging to the Phyla Charophyta, Chlorophyta, Ochrophyta (formerly Phaeophyta) or Rhodophyta that has been introduced beyond its native range through human activities and has become successfully established in the new locale. Information on each unique introduction was compiled from published scientific literature, books, and Algaebase (Guiry & Guiry 2007) and is shown in **Supplemental Table 1** (follow the Supplemental Material link from the Annual Reviews home page at **http://www.annualreviews.org/**). Taxonomic information included basic classification using the nomenclature accepted by Algaebase (Guiry & Guiry 2007) on February 2, 2007. Functional group categories were assigned to each species after Steneck & Dethier (1994) and included filamentous, crustose, corticated foliose, corticated macrophyte, leathery macrophyte, articulated calcareous, and sipohonous (a unique category for the Bryopsidales, see Vroom & Smith 2001).

Vector of Introduction

The vector for each unique introduction was recorded as reported in the literature using the following designations (Ruiz et al. 2000): unreported (no clear indication of mechanism), aquarium introductions (release of aquarium organisms into the wild), aquaculture (the intentional introduction of algae for cultivation), shellfish farming (the introduction of algae growing on or associated with cultured shellfish), ballast (propagules transported within ballast water or attached to ballast rocks), hull fouling (species attached to oceanic vessels), fishing gear, research, and "lessespian" immigrants (species that migrated through the Suez Canal since 1869).

Native and Introduced Ranges

The native distributional range for each introduced seaweed was recorded as reported in Algaebase (Guiry & Guiry 2007). To assess global patterns, broad geographic regions were defined as follows: NE, NW, SE, SW, and central Atlantic and Pacific oceans; Caribbean; Australia and New Zealand; Mediterranean; Indian; Antarctic; Arctic; and the Black and Caspian Seas.

Each introduction to a new region was entered as a separate entry in **Supplemental Table 1** [e.g., *Caulerpa taxifolia* has been introduced to three regions: the Mediterranean, California (NE Pacific), and Australia]. However, species that have secondarily spread to states or countries within a region were only recorded once.

Analyses and Data Summary

We used the total number of unique seaweed introductions (where species may be counted more than once) for most of our analyses. However, for all questions related to the taxonomic affiliation of invaders, the total number of species was used.

Because taxonomy has been shown to be a useful predictor for identifying potential invaders (flowering plants, Daehler 1998; and birds, Lockwood 1999), we tested whether certain algal families contained more introduced species than would be expected by chance. We used the binomial distribution to generate an expected number of invaders per family for all groups containing invaders. The number of species (n) in each family was taken from Algaebase (Guiry & Guiry 2007). The expected proportion (p) of invaders was calculated by summing the total number of invaders and dividing by the total number of known marine algal species. The probability of obtaining a value equal to the observed (X) or more extreme than the observed (both higher and lower) was calculated using a two-tailed binomial test (Zar 1999) using

Crustose: an alga exhibiting a crust-like morphology that tightly adheres to the substratum and can be fleshy or calcified

Corticated: algal morphology consisting of multiple cell layers, most regularly an outer pigmented cortical layer and an inner unpigmented medualary layer

Macrophyte: a macroscopic alga (seaweed)

the following equation:

$$R = (n!/X!(n - X)!)p^{x}q^{n-x},$$

where n is family size, X is the number of observed invaders per group, p is the proportion of species expected to be invasive and $q = 1 - p$. Within each family the cumulative two-tailed R value was then considered significant if it was <0.05 and marginally significant if it was <0.1 (arbitrary adjustments for multiple tests were not performed).

Another database for ecological effects was compiled using data from studies published in primarily international scientific journals, excluding proceedings and reports owing to their uncertain peer review. We separated studies into observational (lacking statistical analysis, limited comparisons, models, calculations), mensurative, or experimental categories. Mensurative and experimental studies included a replicated statistical design; experimental studies involved manipulations of native organisms and/or the introduced seaweed. Response variables were sorted into abundance, diversity, community structure (relative abundances), community function (primary productivity, nitrogen fixation), individual performance (survival, growth, reproduction, size), and feeding response (preference of introduced seaweed relative to native as food source or foraging habitat). We accepted the peer-reviewed statistical results if the effect could be categorized as changed (community structure), negative (lower compared to preintroduction or to native biota), no effect, or enhanced (higher relative to preintroduction or native biota) at alpha $= 0.05$. "Case" refers to a single-response variable in a single study.

RESULTS

We estimate that the global number of introduced seaweed species is 277, increasing the previous estimates from recent reviews by at least 27 species (Ribera Siguan 2003, Schaffelke et al. 2006, Trowbridge 2006), with a total of 408 unique introductions (some species have been introduced to multiple regions) (see **Supplemental Table 1**).

The taxonomic distribution of introduced seaweeds includes 165 Rhodophytes (red algae), 66 Ochrophytes (brown algae), 45 Chlorophytes (green algae) and 1 Charophyte. Some groups, specifically larger families, have many more invaders than others (**Figure 1a–1c**), and in fact family size was positively correlated with invader number (**Figure 1d**). These results suggest that there may simply be proportional

Figure 1

The total number of species and the number of introduced species in a given family for (a) Chlorophytes, (b) Ochrophytes and (c) Rhodophytes. Bold + or – symbols designate families with significantly more ($p < 0.05$) or fewer introduced species than would be expected by chance, respectively, based on Binomial probability, whereas nonbold symbols show marginal significance at $p < 0.1$, respectively. (d) Pearson correlation between family size and invader number.

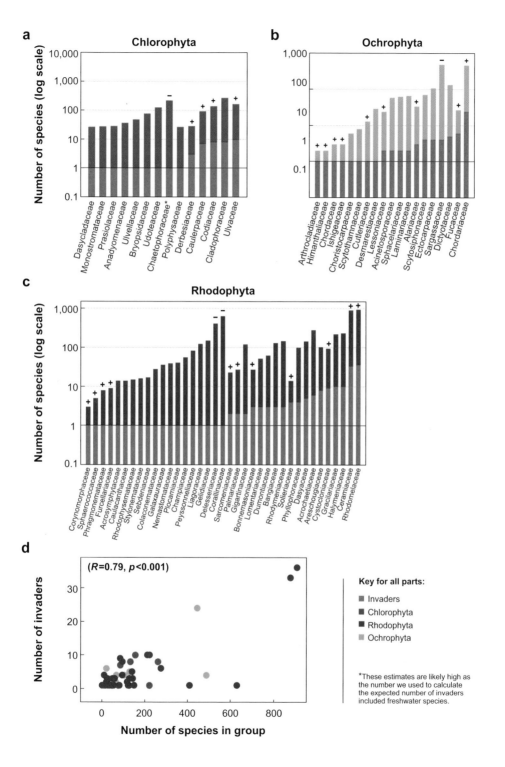

increases in invaders with family size, or perhaps greater evolutionary diversity and physiological strategies in these larger groups enhance invasion success.

Do Some Families Contain More Invaders Than Expected by Chance?

We formally tested whether the number of invaders in a given family was significantly higher or lower than expected by chance. The two green algal families Caulerpaceae and Codiaceae, which contain some of the most well-known introduced seaweed genera (*Caulerpa* and *Codium*), contain significantly more invaders than expected by chance (**Figure 1a**). The Derbesiaceae and Ulvaceae also contain more introduced species than expected by chance. Many species in the Ulvaceae are weedy and are known for their capacity to form nuisance blooms in response to nutrient pollution (Valiela et al. 1997). The green algal family Chaetophoraceae contains significantly fewer species than expected by chance because it is a large family with few invaders (and many freshwater species).

Of the brown algal families the Chordariaceae, Fucaceae, and Alariaceae (marginally significant) all contain more introduced species than expected by chance (**Figure 1b**). The Chordariaceae is a very diverse group containing over 444 species and 24 invaders, including the filamentous and often epiphytic genera *Punctaria*, *Sphaerotrichia*, and *Asperococcus*, and larger more fleshy genera such as *Leathesia* and *Hydroclathrus*. Among the Fucaceaean algae, 6 of the 16 species in the genus *Fucus* have been successfully introduced, suggesting that this genus is highly invasive. The Alariaceae contains 31 species and 3 introduced species including the widely introduced and highly successful *Undaria pinnatifida*. Four other families also contain significantly more invaders than expected because they have fewer than 20 total species and at least one invader. Only the Sargassaceae, one of the largest brown algal families with 487 species, contains significantly fewer introduced species than would be expected by chance.

Among the Rhodophytes, many of the largest families contain significantly more introduced species than expected by chance (**Figure 1c**). Specifically, the Rhodomelaceae and the Ceramiaceae, with 906 and 876 species, respectively, each contain more than 30 introduced species. Most species in these groups are known for their ability to fragment, are uniaxial, and have relatively simple morphologies. The Gracilariaceae is marginally significant with 10 invaders (mainly aquaculture species) and 212 species. The Solieriaceae has a total of 10 species with 4 invaders. Among the Cystocloniaceae, 9 of 85 species have been successful invaders, but all belong to the genus *Hypnea*, suggesting that *Hypnea* may be an exceptionally invasive genus. Within the Areschougiaceae, 8 of 96 total species have been successful invaders, including members of the genera *Kappaphycus*, *Eucheuma*, and *Sarconema*, many of which are intentionally introduced around the globe for commercial carrageenan production (Zemke-White & Smith 2006). Some other families have more invaders than would be expected by chance because they have a small number of known species and at least one invader (**Figure 1c**). Lastly, two red algal families had fewer introduced species than would be expected by chance. Interestingly, these include the rather diverse Corallinaceae (630 spp.) and the Delessariaceae (409 spp.), each with only one invader.

These taxonomic analyses highlight algal families with species that are more or less likely to be successful invaders based on current taxonomy and provide information useful for predicting future invasions. The groups that seem to have a great proclivity for invasion (green families Derbesiaceae, Codiaceae, Ulvaceae, and Caulerpaceae; the brown families Chordariaceae and Fucaceae; and the red families Rhodomelaceae, Ceramiaceae, Cystocloniaceae, and Areschougiaceae) should be carefully monitored.

Morphology and Functional Groups

Broad morphological and anatomical classifications may be useful to understanding if certain groups of algae tend to be more or less successful at invading new regions than others. Our analyses show that the majority of introduced seaweeds are either corticated macrophytes or are filamentous followed by corticated foliose, siphonous, and leathery macrophytes, and finally crustose algae (**Figure 2a**). The distribution and success of these different functional groups are likely to be the result of both the strength of the vector and the characteristics of the invaded habitat. We report the functional groups for only the large macroscopic phases of a given alga's life cycle, but many species of algae have a heteromorphic alternation of generations with spore- and gamete-producing generations having two distinct morphological entities for the same organism. Thus, for many brown algae in the Laminariales (such as *Undaria*), the sporophyte (a leathery macrophyte) is usually easy to detect, but the filamentous gametophyte is microscopic. Heteromorphic species may easily go undetected until reproduction has occurred and the larger macroscopic phase grows up.

Habitats with low physical disturbance and high productivity potential are expected to be dominated by corticated or leathery macrophytes (Steneck & Dethier

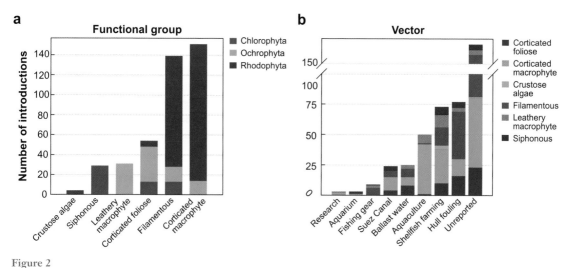

Figure 2

The number of seaweed introductions that (*a*) fall into different functional groups (Steneck & Dethier 1994) and Phyla and (*b*) are accounted for by different vectors or modes of introduction.

1994). This group includes genera such as *Gracilaria*, *Hypnea*, *Eucheuma*, *Fucus*, and *Undaria* that have been introduced to shallow near-shore habitats around the world in association with aquaculture. Many of the genera in this group can also regenerate from vegetative propagules. Filamentous species are expected to be more successful in areas with higher disturbance potential, and here include many Rhodophytes (such as *Acanthophora*, *Polysiphonia*, and *Womersleyella*), Ochrophytes (such as *Ectocarpus* and *Stictyosiphon*), and Chlorophytes (including species of *Cladophora* and *Chaetomorpha*). Many of these filamentous species can readily fragment, are early colonizers, and can be found in a number of disturbed environments including harbors. Corticated foliose species are common in areas with moderate disturbance and productivity potential and include brown algae such as *Padina*, *Dictyota*, and *Colpomenia*, green algae such as *Ulva*, and fewer reds but include species of *Porphyra*. The large leathery macrophytes are expected to be most successful in areas with high productivity potential, as these species support proportionally more nonphotosynthetic tissue than the other functional groups (Littler et al. 1991). Thus, genera such as *Undaria*, *Sargassum*, and *Fucus* have largely been most successful in temperate, high-nutrient regions. The majority of the successfully introduced green algae are siphonous, can occupy a broad range of habitats, and belong to the order Bryopsidales. Some of the common invaders include species from the genera *Codium*, *Caulerpa*, and *Bryopsis*. These siphonous algae are all unicellular and are composed of either simple multinucleate tubes (e.g., *Bryopsis*) or form elaborate morphological configurations from a complex network of interconnected tubes with millions of nuclei (e.g., *Caulerpa*). The siphonous construction of these algae can allow for rapid growth, wound healing and fragment generation, and propagation (Wright & Davis 2006)—characteristics that have likely influenced the invasion success of many of these species, most notably *Caulerpa* spp. (Smith & Walters 1999). Lastly, crustose algae are expected to dominate in areas with high physical disturbance and low productivity potential, that is, environments that are not likely to receive many anthropogenic introductions. Further, only a single crustose alga, *Lithophyllum yessoense*, has been reportedly introduced to the Mediterranean and the NE Atlantic. Interestingly, not a single species in the diverse articulated calcareous functional group (including the genera *Halimeda* and *Corallina*) has been introduced. Perhaps these calcified species have very restrictive physiological tolerances, or they may simply lack appropriate propagules to disperse via anthropogenic activities. In summary, a combination of both the characteristics of the recipient environment and the type of vector can be useful in predicting the type of algal functional group that will likely be successful.

Vectors and Modes of Introduction

Information regarding the mechanism of introduction was lacking for 40% of the known algal introductions, yet there seemed to be strong evidence that these species were indeed introduced (e.g., discontinuous distribution in relation to the native range). Interestingly, almost half of these cases were filamentous algae, which are difficult to identify without microscopy and thus can go undetected until thorough surveys are completed. Of the reported modes of introduction, hull fouling

(77 introductions) and shellfish farming (73 introductions) are nearly equivalent in terms of the number of successful introductions (**Figure 2b**). More than 50% of the species introduced via hull fouling are filamentous, suggesting that this functional group along with some of the weedy corticated foliose genera such as *Ulva* can withstand transport in fouling communities.

Red corticated macrophytes are commonly introduced in association with shellfish farming. Many of the other species introduced via shellfish aquaculture come from a diverse array of functional groups, as any species that can colonize the shells of the commercial species can "hitchhike" into new environments (Naylor et al. 2001). Additionally, species used as packing material for fish or shellfish can easily be transported to new regions, as has occurred with species of *Fucus* in Europe and *Ascophyllum* on the west coast of North America.

The seaweed species that are most commonly cultivated around the world, either for colloid production (carrageenan and agar) or for food, are corticated or leathery macrophytes. The success of cultivated species in the new region is not surprising because many have been selected specifically for their robustness, tolerance to a wide range of physical conditions, propensity for rapid growth, and ability to propagate vegetatively (Naylor et al. 2001).

Although ballast introductions are generally common for marine species (Carlton & Geller 1993), especially invertebrates, they account for only 10% of reported seaweed introductions. Of these most are split among red and brown corticated or foliose macrophytes, leathery macrophytes, and filamentous species. Surprisingly very few species of green algae account for reported ballast introductions, despite their motile microscopic spores and gametes. Ballast is a very diverse means of transportation for many different types of species because propagules can travel variously within the water as vegetative or sexual propagules or, more historically, settled on ballast rock. The absence of light for photosynthesis during transport might diminish their viability and help explain the low proportion of successful seaweed invasions that have occurred through ballast release.

The opening of the Suez Canal in 1869 has been responsible for less than 10% of documented seaweed introductions. These Lessespian immigrants include 24 algal taxa distributed across several functional groups.

Fishing gear has accounted for just over 3% of reported seaweed introductions and includes mainly filamentous species as well as some corticated macrophytes and siphonous species. This vector is likely to be the more important for secondary introduction or spread once a species has become established in a given region as many fishing vessels do not cross broad oceanographic regions; however, derelict nets may.

The rapidly expanding aquarium industry only accounts for 1% of the total number of known introductions but is responsible for the most well-publicized and well-documented seaweed invasion, that of *Caulerpa taxifolia* in the Mediterranean (Meinesz 1999). Although public or commercial aquaria can easily prevent species introductions by taking the proper precautions, individuals with home aquaria may dump non-native organisms into the environment. However, proper education and outreach along with new legislation banning the possession of highly threatening species (such as *Caulerpa*) can help to prevent this type of introduction (Padilla & Williams 2004, Stam et al. 2006, Walters et al. 2006).

The more general categories of shipping (hull fouling plus ballast) and aquaculture (direct cultivation of seaweeds and the indirect introduction via shellfish aquaculture) compose over 85% of known vectors for documented introductions (with 102 and 121 introductions, respectively). These results closely parallel those found by Ribera Siguan (2003) despite the near doubling of species included in the present analysis.

Native and Introduced Species Distributions

The Mediterranean is the most heavily invaded region in the world for introduced seaweeds, with over 132 invasion events accounting for more than 33% of the total number of invasions (**Figure 3*a***). This region has been extensively studied, having many phycologists and many studies to catalogue introduced species (Boudouresque et al. 1994, Boudouresque & Verlaque 2002b, Occhipinti Ambrogi 2000, Ribera

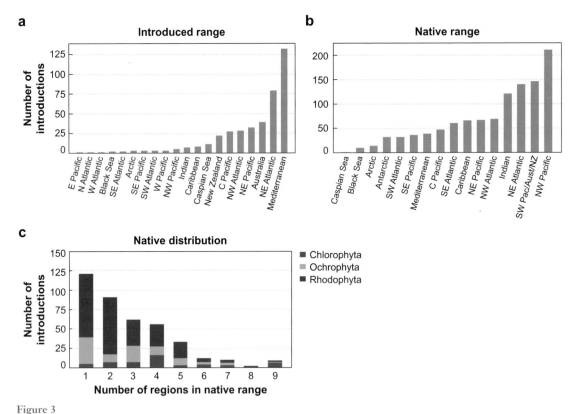

Figure 3

The (*a*) introduced range and (*b*) native range of all seaweed introductions, and (*c*) the number of regions in which each seaweed invader is native.

Siguan 2002). The NE Atlantic, including the North Sea, the Baltic Sea, and the eastern Atlantic Islands, has had 79 seaweed invasions accounting for just under 20% of the total. This region is also rich with historical data and species inventories, making detection of introduced species probable. Australia has documented 39 introductions accounting for approximately 10% of the global total. New Zealand, the NE Pacific, the NW Atlantic, and the central Pacific have all had somewhere between 20–32 invasion events making up between 5% and 8% of the total number of invasions. Many other regions have documented between 1–11 invasion events each, accounting for less than 5% of the total. Interestingly there is very little information available on introduced seaweeds from the tropics (Coles & Eldredge 2002). The above patterns are likely related to a number of factors including the history of research and phycological expertise in a given region and, hence, the ability to detect an invader and the frequency and intensity of inoculation events.

Epiphyte: a plant or animal species that lives on (attached to) a typically larger plant

We were unable to conduct an assessment of the specific source regions for all introduced species and were instead interested in determining if there was overlap in the native ranges of invaders. These results suggest that the majority of successfully introduced seaweed species are native to the NW and Indo-Pacific, an area containing the highest levels of biodiversity for a number of different taxa, including algae (Kerswell 2006); this is followed closely by Australia and New Zealand (**Figure 3b**). Species native to the European Atlantic coast are also successful invaders, as are species from the Indian Ocean, many of which were Lessespian immigrants.

Finally, we hypothesized that the more broadly distributed or cosmopolitan species would be more likely to succeed in new regions than more narrowly distributed species. Contrary to this expectation, species native to only a single region make up a larger proportion of invaders than species with larger native ranges (**Figure 3c**). Clearly, species having narrow distributional ranges will be more easily detected in a new locale owing to discontinuous distributions. Furthermore, many species currently known to have cosmopolitan distributions may have in fact been early introductions prior to exhaustive taxonomic assessments. However, based on these analyses it seems that species that are native to only one oceanic region (and based on the previous section, the NW/Indo-Pacific) are more likely to be identified as successful invaders than species with more wide-ranging distributions.

Evidence for Ecological Effects

In the first systematic summary of the evidence for ecological effects of introduced seaweeds on native communities, we located 68 scientific journal articles with relevant information (11 observational studies, 27 mensurative, 28 experimental, and 2 combined from over 900 sources reviewed). Effects on native seaweeds and epiphytes were most common (**Figure 4**). Several different response variable categories were measured but changes in the abundance of native biota were most commonly reported (40% of 173 unique cases) (**Figure 4**; see also **Table 1**). The majority of the mensurative and experimental studies of introduced seaweeds showed negative effects (48%, 76 cases), particularly on native seaweeds. An additional 8% (13 cases) represented significant changes in community structure, whereas there was no detectable effect

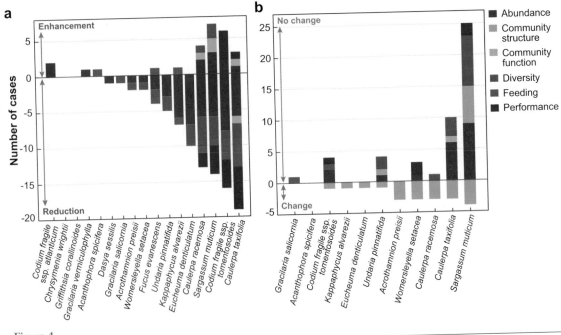

Figure 4

The number of cases (single-response variable in single study) that have documented (*a*) positive or negative effects and (*b*) change or no significant change to native species or communities at alpha = 5%.

reported for 30% (48 cases, mostly epiphytes), and an enhancement was found for 15% (23 cases).

The ecological effects of only 17 out of 277 introduced seaweeds have been studied, including *Caulerpa taxifolia* and *Undaria pinnatifida*, the only seaweeds listed among the world's 100 most invasive species (Lowe et al. 2004). *Sargassum muticum* and *C. taxifolia* were studied the most, whereas *U. pinnatifida* and red seaweeds were studied the least. No exclusively negative results were reported for any species; effects varied across studies, sites (Buschbaum et al. 2006, Wikström et al. 2006), and response variables. Only two species were reported as having no detectable negative effect [but see economic impact of *Gracilaria vermiculophylla* (Freshwater et al. 2006)]. Surprisingly, the ecological effects of numerous introductions of *Grateloupia turuturu* have not been studied.

Herbivores and introduced seaweeds. Compared to studies of effects on seaweeds and epiphytes, far fewer studies were devoted to effects on animals, primarily herbivores. Interactions between introduced seaweeds and native herbivores could alter trophic dynamics and seaweed spread if herbivores avoid eating the invaders ["enemy release" hypothesis (Keane & Crawley 2002)]. In the majority of feeding experiments, although introduced seaweeds were not preferred by generalist herbivores (littorines,

Table 1 Ecological effects of introduced seaweeds reported in observational, mensurative, and experimental studies

Introduced seaweed(s)	Effect[a]	Response variable[b]	Location (citation)
Observational			
Caulerpa racemosa	−	Introduced seaweed cover	Mediterranean (Piazzi et al. 2003a)
Caulerpa taxifolia	−	Seagrass biomass	NW Pacific (Williams & Grosholz 2002)
	−	Sea urchin feeding, behavior, spines, gonads	Mediterranean (Boudouresque et al. 1996)
	+	Nitrogen fixation	Mediterranean (Chisholm & Moulin 2003)
	−	Seaweed photosynthesis	Mediterranean (Ferrer et al. 1997)
Codium fragile ssp. *tomentosoides* and ssp. *atlanticum*	0	Seaweed density	NE Atlantic (Trowbridge 2001)
Codium fragile ssp. *tomentosoides*	−	Kelp canopy cover	NW Atlantic (Mathieson et al. 2003)
	−	Seagrass survival	NW Atlantic (Garbary et al. 2004)
Sargassum muticum	+	Primary production, decomposition	NE Atlantic (Pedersen et al. 2005)
Acanthophora spicifera, Gracilaria salicornia, Kappaphycus alvarezii	−, +, 0	Herbivorous fishes feeding choice	Central Pacific (Stimson et al. 2001)
Dasya sessilis, Chrysymenia wrightii, Griffithsia corallinoides	Δ	Seaweed distribution	Mediterranean (Vincent et al. 2006)
Mensurative			
Caulerpa racemosa	+	Polychaete species richness, abundance; macrofaunal species richness, abundance	Mediterranean (Argyrou et al. 1999)
	−	Seaweed richness, diversity, cover	Mediterranean (Piazzi et al. 2001)
	0	Seagrass leaf longevity	Mediterranean (Dumay et al. 2002a)
	+	Epiphyte biomass; seagrass primary productivity[c]	
	−	Seaweed species richness, cover	Mediterranean (Piazzi et al. 2003a)
	−	Encrusting and erect algae cover; total algal cover	Mediterranean (Balata et al. 2004)
Caulerpa taxifolia	−	Seaweed species richness, cover	Mediterranean (Piazzi et al. 2003a)

(Continued)

Table 1 (*Continued*)

Introduced seaweed(s)	Effect[a]	Response variable[b]	Location (citation)
	−	Encrusting and erect algae cover; total algal cover	Mediterranean (Balata et al. 2004)
	+ −	Epiphyte biomass; seagrass primary production[c] Seagrass leaf longevity	Mediterranean (Dumay et al. 2002a)
	0	Seagrass areal extent	Mediterranean (Jaubert et al. 1999, 2003)
	0 −	Fish group size, group number, individual size Fish foraging	Mediterranean (Levi & Francour 2004)
	− 0	Fish foraging, large size class Fish group size	Mediterranean (Longepierre et al. 2005)
	−, 0	Herbivore abundance	SW Pacific (Gollan & Wright 2006)
	−	Bivalve reproduction	SW Pacific (Gribben & Wright 2006b)
	− 0 Δ	Fish species richness Fish abundance Fish community structure	SW Pacific (York et al. 2006)
Codium fragile ssp. *tomentosoides*	− +	Epifauna diversity, density Epiflora density	NW Atlantic (Schmidt & Scheibling 2006)
Fucus evanescens	−	Epiflora biomass	NE Atlantic (Schueller & Peters 1994)
	−	Epiphyte biomass, species richness	NE Atlantic (Wikström & Kautsky 2004)
Sargassum muticum	0 Δ −	Epifauna abundance; gastropod abundance Epifauna community structure Isopod abundance, size; amphipod abundance	NE Atlantic (Viejo 1999)
	0 −	Seaweed species richness, diversity Leathery and coarse seaweed cover	North Sea (Stæhr et al. 2000)
	0	Epiflora diversity	North Sea (Bjærke & Fredriksen 2003)
	0 +	Canopy seaweed biomass Epibiota biomass	North Sea (Wernberg et al. 2004)
	− + Δ	Seaweed biomass Epiflora biomass Seaweed community structure	NE Atlantic (Sánchez & Fernández 2005)

Table 1   (*Continued*)

Introduced seaweed(s)	Effect[a]	Response variable[b]	Location (citation)
	−, +	Epibiota species richness	North Sea (Buschbaum et al. 2006)
	0, Δ	Epibiota community structure	
	Δ	Infauna community structure	NE Atlantic (Strong et al. 2006)
	−, 0	Infauna diversity, evenness, body length	
	+, 0	Infauna dominance, abundance	
	+	Seaweed species richness, ephemeral seaweed cover	North Sea (Thomsen et al. 2006b)
	0	Leathery, branched, and encrusting algal functional groups cover; herbivore abundance	
	−, 0	Herbivore richness, abundance	North Sea (Wikström et al. 2006)
Undaria pinnatifida	0	Canopy seaweed cover; seaweed species richness; faunal species richness; community structure	SW Pacific (Forrest & Taylor 2002)
Acrothamnion preisii, *Womersleyella setacea*, turf	Δ	Seaweed community structure	Mediterranean (Piazzi & Cinelli 2003)
Acrothamnion preisii, *Womersleyella setacea*, *Caulerpa racemosa*	Δ	Seaweed community structure	Mediterranean (Piazzi & Cinelli 2003)
Acrothamnion preisii, *Womersleyella setacea*, *Caulerpa racemosa*, *Caulerpa taxifolia*	Δ	Seaweed community structure	Mediterranean (Piazzi & Cinelli 2003)
Eucheuma denticulatum and *Kappaphycus alvarezii* farm	−	Seaweed cover; seagrass density, biomass, cover; macrofaunal abundance, biomass	Indian Ocean (Eklöf et al. 2005)
Gracilaria vermiculophylla	+	Filamentous algal species richness, biomass	NW Atlantic (Thomsen et al. 2006a)
15 species	0	Seaweed species richness	Mediterranean (Klein et al. 2005)
Experimental			
Caulerpa racemosa	−, +	Seagrass density	Mediterranean (Ceccherelli & Campo 2002)
	+	Seagrass sexual reproduction[c]	
	−	Sea slug food choice	Mediterranean (Gianguzza et al. 2002)
	−	Introduced seaweed (*Caulerpa taxifolia*) size, growth	Mediterranean (Piazzi & Ceccherelli 2002)
	−	Seaweed species richness, cover	Mediterranean (Piazzi & Ceccherelli 2006)

(Continued)

Table 1 (*Continued*)

Introduced seaweed(s)	Effect[a]	Response variable[b]	Location (citation)
Caulerpa taxifolia	−	Seagrass density	Mediterranean (Ceccherelli & Cinelli 1997)
	0	Seagrass density	Mediterranean (Ceccherelli & Sechi 2002)
	−	Sea slug food choice	Mediterranean (Gianguzza et al. 2002)
	0	Snail, sea hare, and fish food choice	SW Pacific (Davis et al. 2005)
	−	Herbivore food choice, habitat choice	SW Pacific (Gollan & Wright 2006)
	−, 0	Herbivore survivorship	
	+	Bivalve recruitment	SW Pacific (Gribben & Wright 2006a)
Codium fragile ssp. *tomentosoides*	−, 0	Sea urchin grazing	NW Atlantic (Prince & Leblanc 1992)
	−	Sea urchin grazing, reproduction	NW Atlantic (Scheibling & Anthony 2001)
	0	Sea urchin growth	
	+	Sea slug metamorphosis, postlarval growth, development	NE Atlantic (Trowbridge & Todd 2001)
	0	Kelp growth; crab and lobster abundance	NW Atlantic (Levin et al. 2002)
	−	Kelp recruitment; seaweed cover; sea urchin and littorine snail grazing; predatory fish abundance	
	−	Littorine snail density, feeding, growth	NW Atlantic (Chavanich & Harris 2004)
	−	Sea urchin grazing; kelp survival	NW Atlantic (Sumi & Scheibling 2005)
	+	Mussel recruitment, survival, density, size	Adriatic (Bulleri et al. 2006)
	−	Seaweed density, biomass	NW Atlantic (Scheibling & Gagnon 2006)
Fucus evanescens	−	Isopod food choice	Baltic Sea (Schaffelke et al. 1995)
	−, +	Littorine snails food choice; isopod food choice	NE Atlantic, North Sea (Wikström et al. 2006)
Sargassum muticum	−	Kelp recruitment	NE Pacific (Ambrose & Nelson 1982)
	−	Seaweed cover	NE Pacific (DeWreede 1983)
	−	Seaweed abundance, growth; sea urchin feeding	NE Pacific (Britton-Simmons 2004)
	0	Seaweed cover, species richness, diversity, community structure, succession	NE Atlantic (Sánchez & Fernández 2005)

Table 1 (*Continued*)

Introduced seaweed(s)	Effect[a]	Response variable[b]	Location (citation)
Undaria pinnatifida	−	Seaweed richness, diversity	SW Atlantic (Casas et al. 2004)
	0	Seaweed cover	Tasman Sea (Valentine & Johnson 2005)
	Δ	Seaweed community structure	Tasman Sea (Valentine & Johnson 2003)
	−	Kelps; tunicate	NE Atlantic (Farrell & Fletcher 2006)
Gracilaria salicornia	−	Fish food choice	central Pacific (Smith et al. 2004)
Womersleyella setacea	−	Native seaweed cover	Mediterranean (Airoldi 1998, 2000)
	0	Algal crust cover, mortality, fertility	

[a]Δ, change; 0, no detectable change or effect; −, negative effect on, or not preferred to, native biota; +, enhanced effect on, or preferred to, native biota. Multiple entries indicate different results at different sites or times or with different native species within a single study. All mensurative and experimental effects were tested statistically in the cited study and reported here at alpha = 5%.

[b]Response of native biota unless indicated as "introduced."

[c]Enhancements considered evidence of stress, not facilitation.

isopods, polychaetes, sea urchins, fishes), they were eaten, including *Caulerpa taxifolia* with its unique deterrents (Dumay et al. 2002b). In a few cases introduced seaweeds were preferred over at least one, often unpalatable, native species [e.g., the kelp *Agarum* (Britton-Simmons 2004, Prince & Leblanc 1992)]. Thus, introduced seaweeds do not escape completely from novel herbivores.

Despite eating introduced seaweeds, native herbivores have not been documented to control invader spread (Britton-Simmons 2004, Chavanich & Harris 2004, Conklin & Smith 2005, Davis et al. 2005, Gollan & Wright 2006, Levin et al. 2002, Sumi & Scheibling 2005, Trowbridge 1995). Strikingly, even *Undaria pinnatifida* escapes herbivore control despite being highly edible (farmed for human consumption) and rapidly consumed by herbivores (Thornber et al. 2004). In the Tasman Sea, sea urchins cannot keep up with annual growth of *Undaria* and they actually facilitate its spread by consuming native perennial seaweeds and opening space for settlement (Edgar et al. 2004; Valentine & Johnson 2003, 2005). Similarly, herbivores also facilitate highly unpalatable *Asparagopsis armata* in the Mediterranean (Sala & Boudouresque 1997). These general patterns where herbivores cannot control the spread of introduced seaweeds contrast with a recent meta-analysis showing that native generalist herbivores, particularly large vertebrates, provide biotic resistance to plant invasions on land and in freshwater and saltwater marshes (Parker et al. 2006). These differences may be due to the fact that large vertebrate herbivores are not as common in seaweed-dominated habitats.

Strict specialist herbivores are often sought for biocontrol of invasive plants with minimal effects on native species, although with great caution (Secord 2003). However, ascoglossan mollusks (sea slugs) specializing on siphonous green seaweeds were rejected as biocontrols for introduced *Caulerpa* and *Codium* because they can shift their host preference over a short period (Thibaut et al. 2001, Trowbridge & Todd 2001) and can actually enhance the spread of introduced seaweeds through fragmentation (Harris & Mathieson 1999, Zůljevic et al. 2001).

The majority of the studies concerning introduced seaweeds and herbivores have been laboratory feeding preference tests or field inclosures/exclosures of herbivores. Although food choices are informative, they can change over time (Trowbridge 1995) as can seaweed allocation to chemical deterrents (Wikström et al. 2006). How herbivores interact with introduced seaweeds also depends on their relative distributions and abundances, their encounter rate, and on whether predators exert top-down control.

Effects on native marine communities. Community-level ecological interactions involving introduced seaweeds constitute a major research gap. Indirect effects between trophic levels, the mobility of consumers, and restrictions on replication present research challenges. Recent community-level field experiments revealed important indirect effects between trophic levels. Such effects are not evident in single-response variables but can be elucidated through techniques such as structural equation modeling (Britton-Simmons 2004). For example, *Sargassum muticum* had an indirect negative effect on sea urchins through shading native kelp, their preferred food. An indirect effect was also evident in the replacement of a native kelp forest with *Codium fragile* ssp. *tomentosoides* (Levin et al. 2002, Scheibling & Gagnon 2006). This transition was facilitated by another introduced species ["invasional meltdown" (Simberloff & Von Holle 1999); when introduced species facilitate each other's abundance or negative effects, their impacts on negative communities can compound], the bryozoan *Membranipora membranacea*, which severely fouls kelp to the point of decline, opening space for *Codium*. Additional effects of *Codium* on the community varied across consumer functional groups; however, trophic support for the dominant predatory fish was predicted to change because its prey are associated with understory kelps. Other studies of mobile animals and foraging behavior in introduced seaweed communities are too few to generalize but most have reported at least a qualitative change (**Table 1**). Major shifts in community structure can occur even if species richness and biodiversity remain unchanged (Sax et al. 2005), as has occurred where *Sargassum muticum* has invaded (Sánchez et al. 2005, Stæhr et al. 2000). In addition, resilience of native communities may be reduced after invasion by seaweeds (Piazzi & Ceccherelli 2006, Valentine & Johnson 2003).

Studies on the long-term effects of introduced seaweeds on ecosystem processes are sorely needed (primary and secondary production, nutrient cycling; **Table 1**). The hypothesis that introduced seaweeds increase primary productivity, which could lead to higher consumer abundance (Pedersen et al. 2005, Thomsen et al. 2006a, Viejo 1999, Wernberg et al. 2004), requires investigation in natural communities. Generally unpalatable themselves, introduced seaweeds can support high abundances of palatable epiphytes, but epiphyte populations can be notoriously ephemeral food sources for consumers. We found no studies that assessed introduced seaweeds as trophic support for detritivores. Finally, an understanding of how introduced seaweeds alter the flow of matter and energy through ecosystems must be considered along with any effects on biodiversity.

The invasibility of marine communities and the role of disturbance. We searched for studies on introduced species that addressed two core questions in invasion biology: (*a*) what conditions or properties (e.g., biodiversity) confer native communities with resistance to introduced seaweeds, and (*b*) does disturbance increase invasion potential? Factors that influence the invasibility of marine communities, including spatial scale, demographic and functional attributes of resident species, and positive interactions, are just beginning to be revealed in marine studies, which often only involve native species (Arenas et al. 2006, Dunstan & Johnson 2004, France & Duffy 2006, Sax et al. 2005). Furthermore, most research indicates that disturbance tends to increase invasibility of marine communities (Byers 2002, Ruiz et al. 1999). We found only 18 studies on introduced seaweeds that were relevant to the above questions (**Table 2**).

Invasibility: the susceptibility of a native community to the establishment of an introduced species

Undisturbed algal communities (turfs, foliose species, large canopy-forming kelps) can resist seaweed invasions, but only rarely enough to limit their impacts and spread (Andrew & Viejo 1998). The preemptive competitors *Codium fragile* ssp. *tomentosoides*, *Sargassum muticum*, *Undaria pinnatifida*, and *Wormersleyella setacea* are able to invade readily upon disturbance and then persist (**Table 2, Figure 5**). Initial resistance of native marine communities can be overcome by high invader growth rates (Airoldi 2000), during different life history stages (Britton-Simmons 2006, Sánchez & Fernández 2006, Scheibling & Gagnon 2006), and under varying ecological conditions (Bulleri & Airoldi 2005). Long-term studies are thus important to document any weakening in the biotic resistance of native communities (e.g., Harris & Jones 2005).

The siphonous green seaweeds (*Codium fragile* ssp. *tomentosoides*, *Caulerpa taxifolia*, *C. racemosa* var. *cylindracea*) are strong interference (direct) competitors of seaweeds (Piazzi et al. 2001) and seagrasses (Ceccherelli & Cinelli 1997, Ceccherelli et al. 2002, Garbary et al. 2004). Native seagrass resistance to introduced seaweeds is important because seagrasses are declining in many areas around the world (Orth et al. 2006). Of the seagrasses studied in the Mediterranean, only continuous or dense patches of *Posidonia oceanica*, one of the largest and longest-lived seagrass species, can effectively resist introduced *Caulerpa* spp. (Ceccherelli & Campo 2002, Ceccherelli & Cinelli 1999a, Ceccherelli et al. 2000).

Does nutrient enrichment facilitate introduced seaweeds? Nutrient enrichment is among the most significant threats to coastal marine ecosystems and often leads to deleterious algal blooms (Howarth et al. 2002). Introduced seaweeds can be numerous and abundant in areas subjected to nutrient pollution (Boudouresque & Verlaque 2002b, Chisholm et al. 1997, Occhipinti Ambrogi 2000, Schueller & Peters 1994), suggesting that nutrient enrichment may enhance invasion success (Chisholm & Moulin 2003, Chisholm et al. 1997, Fernex et al. 2001, Jaubert et al. 2003, Lapointe et al. 2005, but see Klein et al. 2005). However, vectors and other factors that covary with pollution (**Figure 3**; Sant et al. 1996) can confound causality.

Table 2 Response of introduced seaweeds to experimental removals of native biota and to nutrient enrichment

Introduced seaweed	Response variable	Effect[a]	Type of experiment	Location (citation)
Caulerpa racemosa	Growth	+	Seagrass (*Posidonia oceanica*) removal	Mediterranean (Ceccherelli et al. 2000)
Caulerpa racemosa	Recolonization	–	Algal turf removal	Mediterranean (Piazzi et al. 2003b)
Caulerpa taxifolia	Density	0	Seagrass (*Cymodocea nodosa*) removal	Mediterranean (Ceccherelli & Cinelli 1997)
Caulerpa taxifolia	Density	+	Nutrient enrichment	Mediterranean (Ceccherelli & Cinelli 1997)
Caulerpa taxifolia	Density	0	Seagrass (*Cymodocea nodosa*) removal	Mediterranean (Ceccherelli & Sechi 2002)
Caulerpa taxifolia	Density	0	Nutrient enrichment	Mediterranean (Ceccherelli & Sechi 2002)
	Density, biomass, survival	0		
Codium fragile ssp. tomentosoides	Cover	+	Mussel removal, spring	Adriatic (Bulleri & Airoldi 2005)
		–	Mussel removal exposed, summer	
		+	Mussel removal sheltered, summer	
Codium fragile ssp. tomentosoides	Cover, growth	+	Seaweed removal	NE Atlantic (Scheibling & Gagnon 2006)
Fucus evanescens	Germling growth, survival	0	Nutrient enrichment (laboratory)	North Sea (Steen & Rueness 2004)
Sargassum muticum	Recruitment	+	Seaweed canopy removal	NE Pacific (Deysher & Norton 1981)
Sargassum muticum	Recruitment	+	Seaweed removal	NE Atlantic (Andrew & Viejo 1998)
Sargassum muticum	Germling growth, survival	0	Nutrient enrichment (laboratory)	North Sea (Steen & Rueness 2004)
Sargassum muticum	Recruitment	0	Seaweed removal	NE Pacific (Britton-Simmons 2006)
		0	Understory seaweed removal	
		+	Crustose + turf seaweed removal	
	Survival	+	Canopy + understory seaweed removal	
Sargassum muticum	Cover, length, density, recruitment	+	Nutrient enrichment	NE Atlantic (Sánchez & Fernández 2006)

Table 2 (*Continued*)

Introduced seaweed	Response variable	Effect[a]	Type of experiment	Location (citation)
	Cover, length, density	0	Seaweed canopy removal	
	Recruitment	−	Seaweed canopy removal	
Undaria pinnatifida	Recruitment	+	Seaweed canopy removal	Tasman Sea (Valentine & Johnson 2003)
Undaria pinnatifida	Recruitment	+	Seaweed canopy removal	Tasman Sea (Edgar et al. 2004)
Undaria pinnatifida	Recruitment	+	Seaweed canopy mortality	Tasman Sea (Valentine & Johnson 2004)
Undaria pinnatifida	Recruitment	0	Seaweed canopy removal	NE Atlantic (Farrell & Fletcher 2006)
Asparagopsis armata	Biomass	−	Herbivore exclusion	Mediterranean (Sala & Boudouresque 1997)
Womerseleyella setacea	Cover	+	Seaweed removal	Mediterranean (Airoldi 1998)
Womerseleyella setacea	Cover	+	Algal crust removal	(Airoldi 2000)

[a]Effect indicates how the introduced seaweed responded to the experimental manipulation (disturbance) of the native biota or to nutrient enrichment. +, suggests resistance of native biota to introduced seaweed or enhanced response to nutrient enrichment; 0, indicates no detectable effect; −, suggests facilitation by native biota. All effects were statistically tested in the cited study and reported here at alpha = 5%.

Direct testing of the nutrient enhancement hypothesis has been limited, and experimental results are mixed (**Table 2**). Sánchez & Fernández (2006) found that nutrient enrichment enhances *Sargassum muticum*. Because the native seaweed canopy offers little resistance to introduced *Sargassum*, eutrophication could promote its further spread. The most recent experimental studies have not supported the hypothesis that nutrient or organic enrichment enhances *Caulerpa taxifolia* (Ceccherelli & Sechi 2002, Terrados & Marbà 2006; see also Delgado et al. 1996), contrasting partially with earlier studies (Ceccherelli & Cinelli 1997, 1999b) and indirect evidence cited above.

WHEN INTRODUCED SEAWEEDS BECOME INVASIVE

Introduced species are considered to be invasive when they incur or are likely to incur negative ecological or economic impacts. Quantitative evidence from mensurative and experimental studies on just 6% of the seaweeds introduced to date shows that 13 of them are invasive by this definition and that native marine communities have little lasting resistance to these invaders, particularly if disturbed. This review provides a foundation for further analyses, risk assessment, and targeted management of introduced seaweeds. Of particular concern are the siphonous green seaweeds such as *Caulerpa* and *Codium*, which are well-known invaders, but also species of *Fucus* and *Hypnea*, which are less well known. In addition, the commercially cultivated red seaweeds (*Eucheuma denticulatum, Kappaphycus alvarezii, Gracilaria salicornia, Hypnea musciformis*) are also a concern because they are farmed next to coral reefs and over seagrasses in regions where labor is inexpensive. Reports of the economic and

Figure 5

The number of cases
documenting that native
communities (*a*) resisted,
(*b*) facilitated, or (*c*) had no
effect on an introduced
seaweed.

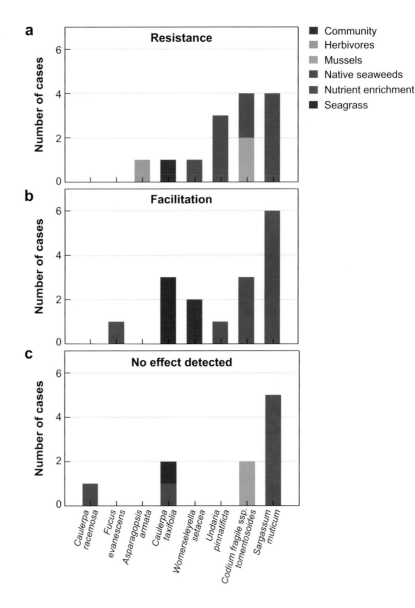

social costs of these seaweeds are just coming to light (see Related Resources), but we
predict they will have a major ecological impact in the future unless preventive steps
are taken now, such as engaging phycologists and the industry to make the culture
environmentally sustainable or to cultivate native species.

Existing recommendations for managing invasive species, including many seaweeds, in many affected countries and regions are too numerous to review here,
but all share common elements: prevention as the most effective means to reduce
future costs, early detection and rapid response when prevention fails, eradication

if possible, control as a last resort, public education and stakeholder engagement, and research in support of management (Lodge et al. 2006). As both shellfish and seaweed aquaculture and the aquarium trade are expanding rapidly, seaweed introductions will undoubtedly increase (Naylor et al. 2001, Padilla & Williams 2004), and it is likely that genetically engineered seaweeds will become available for culture (Walker et al. 2005). Evidence suggests that introduced seaweeds can have negative economic impacts and are incredibly difficult and costly to eradicate. Thus, we reinforce the recommendations that put forth repeatedly to sustain and increase efforts to prevent new introductions and control spread as quickly as possible (Lodge et al. 2006, Schaffelke et al. 2006).

SUMMARY POINTS

1. Some algal families contain significantly more introduced species than would be expected by chance, suggesting that these families are likely to be more invasive.

2. Siphonous green seaweeds, which include *Caulerpa taxifolia*, *C. racemosa*, *Codium fragile* spp. *tomentosoides* and spp. *atlanticum*, and several other *Caulerpa* spp., are highly successful invaders that compete directly with native species.

3. The Mediterranean and the NE Atlantic support the highest number of successful seaweed introductions.

4. Most of the introduced seaweed species in the world are native to the NW and Indo-Pacific. Species with narrower distributional ranges had higher numbers of introductions than the more cosmopolitan species, contrary to what has been predicted.

5. The most important vectors for seaweed introductions are fouling of vessel hulls and aquaculture (direct and indirect). Ballast water introductions are less common for seaweeds than documented for other marine species.

6. The ecological effects of only a limited number (6%) of introduced seaweeds have been tested; from these studies, the impacts tend to be diverse but are generally negative.

7. Native herbivores will consume introduced seaweeds, but they often prefer native species, they cannot control even edible introduced seaweeds, and in some cases they facilitate seaweed spread.

8. Native marine communities have little lasting biotic resistance to introduced seaweeds, particularly if a disturbance opens space for invasion.

FUTURE ISSUES

1. Management to prevent future seaweed introductions should focus on hull fouling, particularly by inconspicuous filamentous species and aquaculture

(both the direct introduction of algae and indirect introduction through shellfish farming) and species in the families with higher than expected probabilities of successful invasions because species in these families are among the most widespread and well-documented seaweed invaders in the world [including *Caulerpa* and *Codium* (the siphonous greens), *Fucus*, *Undaria*, *Asparagopsis* and *Hypnea*].

2. Manipulative community-level field studies in combination with modeling are needed to identify not only the impacts of introduced seaweeds on native communities but also the factors that influence invasibility for a more integrative understanding of invasive seaweed ecology.

3. Understanding the potential role of nutrient enrichment in facilitating seaweed introductions will require experiments that directly test the hypothesis and separate other factors that covary with increasing coastal eutrophication.

4. The economic costs of introduced seaweeds are emerging, and future assessments should include externalities [costs to society or native biota in addition to the identifiable direct costs associated with the specific economy (aquaculture products, eradication programs)].

DISCLOSURE STATEMENT

The authors are not aware of any biases that might be perceived as affecting the objectivity of this review.

ACKNOWLEDGMENTS

We thank James Carlton for valuable insight, Keith Hayes at CSIRO for contributing information on Australian species, the National Center for Ecological Analysis and Synthesis for logistics and support, Michael Guiry for assistance with taxonomy and for maintaining Algaebase, Stuart Sandin for help in programming, Daniel Simberloff for editing, and Molly Engelbrecht and Amanda Newsom for reference research and management. This is contribution 2377 from Bodega Marine Laboratory, University of California at Davis.

LITERATURE CITED

Airoldi L. 1998. Roles of disturbance, sediment stress, and substratum retention on spatial dominance in algal turf. *Ecology* 79:2759–70

Airoldi L. 2000. Effects of disturbance, life histories, and overgrowth on coexistence of algal crusts and turfs. *Ecology* 81:798–814

Ambrose RF, Nelson BV. 1982. Inhibition of giant kelp recruitment by an introduced brown alga. *Bot. Mar.* 25:265–68

Andrew NL, Viejo RM. 1998. Ecological limits to the invasion of *Sargassum muticum* in northern Spain. *Aquat. Bot.* 60:251–63

Arenas F, Sánchez I, Hawkins SJ, Jenkins SR. 2006. The invasibility of marine algal assemblages: Role of functional diversity and identity. *Ecology* 87:2851–61

Argyrou M, Demetropoulos A, Hadjichristophorou M. 1999. Expansion of the macroalga *Caulerpa racemosa* and changes in softbottom macrofaunal assemblages in Moni Bay, Cyprus. *Oceanol. Acta* 22:517–28

Balata D, Piazzi L, Cinelli F. 2004. A comparison among assemblages in areas invaded by *Caulerpa taxifolia* and *C. racemosa* on a subtidal Mediterranean rocky bottom. *Mar. Ecol.* 25:1–13

Bjærke MR, Fredriksen S. 2003. Epiphytic macroalgae on the introduced brown seaweed *Sargassum muticum* (Yendo) Fensholt (Phaeophyceae) in Norway. *Sarsia* 88:353–64

Boudouresque CF, Briand F, Nolan C. 1994. Introduced species in European coastal waters. *Eur. Comm. Ecosyst. Res. Rep.* 8:76–84

Boudouresque CF, Lemee R, Mari X, Meinesz A. 1996. The invasive alga *Caulerpa taxifolia* is not a suitable diet for the sea urchin *Paracentrotus lividus*. *Aquat. Bot.* 53:245–50

Boudouresque CF, Verlaque M. 2002a. Assessing scale and impact of ship-transported alien macrophytes in the Mediterranean Sea. In *CIESM Workshop Monogr. No. 20*, pp. 53–61

Boudouresque CF, Verlaque M. 2002b. Biological pollution in the Mediterranean Sea: Invasive vs introduced macrophytes. *Mar. Pollut. Bull.* 44:32–38

Britton-Simmons KH. 2004. Direct and indirect effects of the introduced alga *Sargassum muticum* on benthic, subtidal communities of Washington State, USA. *Mar. Ecol. Prog. Ser.* 277:61–78

Britton-Simmons KH. 2006. Functional group diversity, resource preemption and the genesis of invasion resistance in a community of marine algae. *Oikos* 113:395–401

Bulleri F, Airoldi L. 2005. Artificial marine structures facilitate the spread of a non-indigenous green alga, *Codium fragile* ssp. *tomentosoides*, in the north Adriatic Sea. *J. Appl. Ecol.* 42:1063–72

Bulleri F, Airoldi L, Branca GM, Abbiati M. 2006. Positive effects of the introduced green alga, *Codium fragile* ssp. *tomentosoides*, on recruitment and survival of mussels. *Mar. Biol.* 148:1213–20

Buschbaum C, Chapman AS, Saier B. 2006. How an introduced seaweed can affect epibiota diversity in different coastal systems. *Mar. Biol.* 148:743–54

Byers JE. 2002. Impact of nonindigenous species on natives enhanced by anthropogenic alteration of selection regimes. *Oikos* 97:449–58

Carlton JT, Geller JB. 1993. Ecological roulette—the global transport of nonindigenous marine organisms. *Science* 261:78–82

Casas G, Scrosati R, Piriz ML. 2004. The invasive kelp *Undaria pinnatifida* (Phaeophyceae, Laminariales) reduces native seaweed diversity in Nuevo Gulf (Patagonia, Argentina). *Biol. Invasions* 6:411–16

Uses the novel approach of structural equation modeling to elucidate indirect interactions between introduced *Sargassum muticum* and the invaded community.

Castilla JC, Uribe M, Bahamonde N, Clarke M, Desqueyroux-Faundez R, et al. 2005. Down under the southeastern Pacific: Marine nonindigenous species in Chile. *Biol. Invasions* 7:213–32

Ceccherelli G, Campo D. 2002. Different effects of *Caulerpa racemosa* on two co-occurring seagrasses in the Mediterranean. *Bot. Mar.* 45:71–76

Ceccherelli G, Cinelli F. 1997. Short-term effects of nutrient enrichment of the sediment and interactions between the seagrass *Cymodocea nodosa* and the introduced green alga *Caulerpa taxifolia* in a Mediterranean bay. *J. Exp. Mar. Biol. Ecol.* 217:165–77

Ceccherelli G, Cinelli F. 1999a. Effects of *Posidonia oceanica* canopy on *Caulerpa taxifolia* size in a north-western Mediterranean bay. *J. Exp. Mar. Biol. Ecol.* 240:19–36

Ceccherelli G, Cinelli F. 1999b. A pilot study of nutrient enriched sediments in a *Cymodocea nodosa* bed invaded by the introduced alga *Caulerpa taxifolia*. *Bot. Mar.* 42:409–17

Ceccherelli G, Piazzi L, Balata D. 2002. Spread of introduced *Caulerpa* species in macroalgal habitats. *J. Exp. Mar. Biol. Ecol.* 280:1–11

Ceccherelli G, Piazzi L, Cinelli F. 2000. Response of the nonindigenous *Caulerpa racemosa* (Forsskål) J. Agardh to the native seagrass *Posidonia oceanica* (L.) Delile: Effect of density of shoots and orientation of edges of meadows. *J. Exp. Mar. Biol. Ecol.* 243:227–40

Ceccherelli G, Sechi N. 2002. Nutrient availability in the sediment and the reciprocal effects between the native seagrass *Cymodocea nodosa* and the introduced rhizophytic alga *Caulerpa taxifolia*. *Hydrobiologia* 474:57–66

Chapman AS. 1998. From introduced species to invader: What determines variation in the success of *Codium fragile* ssp. *tomentosoides* (Chlorophyta) in the North Atlantic Ocean? *Helgol. Mar. Res.* 52:277–89

Chavanich S, Harris LG. 2004. Impact of the non-native macroalga *Codium fragile* (Sur.) Hariot ssp. *tomentosoides* (van Goor) Silva on the native snail *Lacuna vincta* (Montagu, 1803) in the Gulf of Maine. *Veliger* 47:85–90

Chisholm JRM, Fernex FE, Mathieu D, Jaubert JM. 1997. Wastewater discharge, seagrass decline and algal proliferation on the Cote d'Azur. *Mar. Pollut. Bull.* 34:78–84

Chisholm JRM, Moulin P. 2003. Stimulation of nitrogen fixation in refractory organic sediments by *Caulerpa taxifolia*. *Limnol. Oceanogr.* 48:787–94

Coles SL, Eldredge LG. 2002. Nonindigenous species introductions on coral reefs: A need for information. *Pac. Sci.* 56:191–209

Conklin EJ, Smith JE. 2005. Abundance and spread of the invasive red algae, *Kappaphycus* spp., in Kane'ohe Bay, Hawai'i and an experimental assessment of management options. *Biol. Invasions* 7:1029–39

Critchley AT, Ohno M, Largo DB, eds. 2006. *World Seaweed Resources.* Paris, France: UNESCO (DVD-ROM)

Daehler CC. 1998. The taxonomic distribution of invasive angiosperm plants: Ecological insights and comparison to agricultural weeds. *Biol. Conserv.* 84:167–80

Davis AR, Benkendorff K, Ward DW. 2005. Responses of common SE Australian herbivores to three suspected invasive *Caulerpa* spp. *Mar. Biol.* 146:859–68

Highlights the lack of information on introduced species from tropical waters, most notably coral reefs.

Delgado O, Rodriguez-Prieto C, Gacia E, Ballesteros E. 1996. Lack of severe nutrient limitation in *Caulerpa taxifolia* (Vahl) C. Agardh, an introduced seaweed spreading over the oligotrophic northwestern Mediterranean. *Bot. Mar.* 39:61–67

DeWreede RE. 1983. *Sargassum muticum* (Fucales, Phaeophyta)—regrowth and interaction with *Rhodomela larix* (Ceramiales, Rhodophyta). *Phycologia* 22:153–60

Deysher L, Norton TA. 1981. Dispersal and colonization in *Sargassum muticum* (Yendo) Fensholt. *J. Exp. Mar. Biol. Ecol.* 56:179–95

Druehl LD. 1973. Marine transplantations. *Science* 179:12

Dumay O, Fernandez C, Pergent G. 2002a. Primary production and vegetative cycle in *Posidonia oceanica* when in competition with the green algae *Caulerpa taxifolia* and *Caulerpa racemosa*. *J. Mar. Biol. Assoc. UK* 82:379–87

Dumay O, Pergent G, Pergent-Martini C, Amade P. 2002b. Variations in caulerpenyne contents in *Caulerpa taxifolia* and *Caulerpa racemosa*. *J. Chem. Ecol.* 28:343–52

Dunstan PK, Johnson CR. 2004. Invasion rates increase with species richness in a marine epibenthic community by two mechanisms. *Oecologia* 138:285–92

Edgar GJ, Barrett NS, Morton AJ, Samson CR. 2004. Effects of algal canopy clearance on plant, fish and macroinvertebrate communities on eastern Tasmanian reefs. *J. Exp. Mar. Biol. Ecol.* 312:67–87

Eklöf JS, de la Torre Castro M, Adelskold L, Jiddawi NS, Kautsky N. 2005. Differences in macrofaunal and seagrass assemblages in seagrass beds with and without seaweed farms. *Estuar. Coastal Shelf Sci.* 63:385–96

Farrell P, Fletcher RL. 2006. An investigation of dispersal of the introduced brown alga *Undaria pinnatifida* (Harvey) Suringar and its competition with some species on the man-made structures of Torquay Marina (Devon, UK). *J. Exp. Mar. Biol. Ecol.* 334:236–43

Fernex FE, Migon C, Chisholm JRM. 2001. Entrapment of pollutants in Mediterranean sediments and biogeochemical indicators of their impact. *Hydrobiologia* 450:31–46

Ferrer E, Garreta AG, Ribera MA. 1997. Effect of *Caulerpa taxifolia* on the productivity of two Mediterranean macrophytes. *Mar. Ecol. Prog. Ser.* 149:279–87

Forrest BM, Taylor MD. 2002. Assessing invasion impact: survey design considerations and implications for management of an invasive marine plant. *Biol. Invasions* 4:375–86

France KE, Duffy JE. 2006. Consumer diversity mediates invasion dynamics at multiple trophic levels. *Oikos* 113:515–29

Freshwater DW, Montgomery F, Greene JK, Hamner RM, Williams M, Whitfield PE. 2006. Distribution and identification of an invasive *Gracilaria* species that is hampering commercial fishing operations in southeastern North Carolina, USA. *Biol. Invasions* 8:631–37

Garbary DJ, Fraser SJ, Hubbard C, Kim KY. 2004. *Codium fragile*: Rhizomatous growth in the *Zostera* thief of eastern Canada. *Helgol. Mar. Res.* 58:141–46

Gianguzza P, Airoldi L, Chemello R, Todd CD, Riggio S. 2002. Feeding preferences of *Oxynoe olivacea* (Opisthobranchia: Sacoglossa) among three *Caulerpa* species. *J. Molluscan Stud.* 68:289–90

Recommends Before-After-Control-Impact studies for identifying invasive seaweed impacts along with application of a precautionary management approach.

Gollan JR, Wright JT. 2006. Limited grazing pressure by native herbivores on the invasive seaweed *Caulerpa taxifolia* in a temperate Australian estuary. *Mar. Freshw. Res.* 57:685–94

Gribben PE, Wright JT. 2006a. Invasive seaweed enhances recruitment of a native bivalve: roles of refuge from predation and the habitat choice of recruits. *Mar. Ecol. Prog. Ser.* 318:177–85

Gribben PE, Wright JT. 2006b. Sublethal effects on reproduction in native fauna: are females more vulnerable to biological invasion? *Oecologia* 149:352–61

Guiry MD, Guiry GM. 2007. *AlgaeBase version 4.2*. World-wide electronic publication. Galway: National Univ. Ireland. http://www.algaebase.org

Harris LG, Jones AC. 2005. Temperature, herbivory and epibiont acquisition as factors controlling the distribution and ecological role of an invasive seaweed. *Biol. Invasions* 7:913–24

Harris LG, Mathieson AC. 1999. Patterns of range expansion, niche shift and predator acquisition in *Codium fragile ssp. tomentosoides* and *Membranipora membranacea* in the Gulf of Maine. *Rep. NHU-R-99-006*. Durham: NH Sea Grant

Howarth RW, Sharpley A, Walker D. 2002. Sources of nutrient pollution to coastal waters in the United States: Implications for achieving coastal water quality goals. *Estuaries* 25:656–76

Inderjit, Chapman DJ, Ranelletti M, Kaushik S. 2006. Invasive marine algae: An ecological perspective. *Bot. Rev.* 72:153–78

Jaubert JM, Chisholm JRM, Ducrot D, Ripley HT, Roy L, Passeron-Seitre G. 1999. No deleterious alterations in *Posidonia* beds in the Bay of Menton (France) eight years after *Caulerpa taxifolia* colonization. *J. Phycol.* 35:1113–19

Jaubert JM, Chisholm JRM, Minghelli-Roman A, Marchioretti M, Morrow JH, Ripley HT. 2003. Re-evaluation of the extent of *Caulerpa taxifolia* development in the northern Mediterranean using airborne spectrographic sensing. *Mar. Ecol. Prog. Ser.* 263:75–82

Keane RM, Crawley MJ. 2002. Exotic plant invasions and the enemy release hypothesis. *Trends Ecol. Evol.* 17:164–70

Kerswell AP. 2006. Global biodiversity patterns of benthic marine algae. *Ecology* 87:2479–88

Klein J, Ruitton S, Verlaque M, Boudouresque CF. 2005. Species introductions, diversity and disturbances in marine macrophyte assemblages of the northwestern Mediterranean Sea. *Mar. Ecol. Prog. Ser.* 290:79–88

Lapointe BE, Barile PJ, Wynne MJ, Yentsch CS. 2005. Reciprocal invasion: Mediterranean native *Caulerpa ollivieri* in the Bahamas supported by human nitrogen enrichment. *Aquat. Invaders* 16:3–5

Levi F, Francour P. 2004. Behavioural response of *Mullus surmuletus* to habitat modification by the invasive macroalga *Caulerpa taxifolia*. *J. Fish Biol.* 64:55–64

Levin PS, Coyer JA, Petrik R, Good TP. 2002. Community-wide effects of nonindigenous species on temperate rocky reefs. *Ecology* 83:3182–93

Littler MM, Littler DS, Titlyanov EA. 1991. Producers of organic matter on tropical reefs and their relative dominance. *Mar. Biol.* 6:3–14

Lockwood JL. 1999. Using taxonomy to predict success among introduced avifauna: Relative importance of transport and establishment. *Conserv. Biol.* 13:560–67

The single-source database for algal taxonomy and species' distributions.

Asserts that further advancements in remote sensing technology are needed to detect introduced seaweeds over large areas.

Lodge DM, Williams S, MacIsaac HJ, Hayes KR, Leung B, et al. 2006. Biological invasions: Recommendations for US policy and management. *Ecol. Appl.* 16:2035–54

Longepierre S, Robert A, Levi F, Francour P. 2005. How an invasive alga species (*Caulerpa taxifolia*) induces changes in foraging strategies of the benthivorous fish *Mullus surmuletus* in coastal Mediterranean ecosystems. *Biodivers. Conserv.* 14:365–76

Lowe S, Browne M, Boudjelas S, De Poorter M. 2004. *100 of the World's Worst Invasive Alien Species*. Auckland, NZ: Univ. Auckland

Maggs CA, Stegenga H. 1999. Red algal exotics on North Sea coasts. *Helgol. Mar. Res.* 52:243–58

Mann K. 1973. Seaweeds: Their productivity and strategy for growth. *Science* 182:975–81

Mathieson AC, Dawes CJ, Harris LG, Hehre EJ. 2003. Expansion of the Asiatic green alga *Codium fragile* ssp. *tomentosoides* in the Gulf of Maine. *Rhodora* 105:1–53

Meinesz A. 1999. *Killer Algae*, ed. D Simberloff. Chicago, IL: Univ. Chicago Press

Naylor RL, Williams SL, Strong DR. 2001. Aquaculture—A gateway for exotic species. *Science* 294:1655–56

Occhipinti Ambrogi A. 2000. Biotic invasions in a Mediterranean lagoon. *Biol. Invasions* 2:165–76

Occhipinti Ambrogi A, Savini D. 2003. Biological invasions as a component of global change in stressed marine ecosystems. *Mar. Pollut. Bull.* 46:542–51

Orensanz JM, Schwindt E, Pastorino G, Bortolus A, Casas G, et al. 2002. No longer the pristine confines of the world ocean: A survey of exotic marine species in the southwestern Atlantic. *Biol. Invasions* 4:115–43

Orth RJ, Carruthers TJB, Dennison WC, Duarte CM, Fourqurean JW, et al. 2006. A global crisis for seagrass ecosystems. *BioScience* 56:987–96

Padilla DK, Williams SL. 2004. Beyond ballast water: Aquarium and ornamental trades as sources of invasive species in aquatic ecosystems. *Front. Ecol. Environ.* 2:131–38

Parker JD, Burkepile DE, Hay ME. 2006. Opposing effects of native and exotic herbivores on plant invasions. *Science* 311:1459–61

Pedersen MF, Stæhr PA, Wernberg T, Thomsen MS. 2005. Biomass dynamics of exotic *Sargassum muticum* and native *Halidrys siliquosa* in Limfjorden, Denmark— Implications of species replacements on turnover rates. *Aquat. Bot.* 83:31–47

Piazzi L, Balata D, Cecchi E, Cinelli F. 2003a. Co-occurrence of *Caulerpa taxifolia* and *C. racemosa* in the Mediterranean Sea: Interspecific interactions and influence on native macroalgal assemblages. *Cryptogam. Algol.* 24:233–43

Piazzi L, Ceccherelli G. 2002. Effects of competition between two introduced *Caulerpa*. *Mar. Ecol. Prog. Ser.* 225:189–95

Piazzi L, Ceccherelli G. 2006. Persistence of biological invasion effects: Recovery of macroalgal assemblages after removal of *Caulerpa racemosa* var. *cylindracea*. *Estuar. Coastal Shelf Sci.* 68:455–61

A classical paper documenting the global importance of seaweeds as marine primary producers.

Raises global consciousness regarding *Caulerpa taxifolia*, providing the impetus for its eradication in California and management in Australia.

Piazzi L, Ceccherelli G, Balata D, Cinelli F. 2003b. Early patterns of *Caulerpa racemosa* recovery in the Mediterranean Sea: The influence of algal turfs. *J. Mar. Biol. Assoc. UK* 83:27–29

Piazzi L, Ceccherelli G, Cinelli F. 2001. Threat to macroalgal diversity: effects of the introduced green alga *Caulerpa racemosa* in the Mediterranean. *Mar. Ecol. Prog. Ser.* 210:149–59

Piazzi L, Cinelli F. 2003. Evaluation of benthic macroalgal invasion in a harbour area of the western Mediterranean Sea. *Eur. J. Phycol.* 38:223–31

Prince JS, Leblanc WG. 1992. Comparative feeding preference of *Strongylocentrotus droebachiensis* (Echinoidea) for the invasive seaweed *Codium fragile* ssp. *tomentosoides* (Chlorophyceae) and 4 other seaweeds. *Mar. Biol.* 113:159–63

Ribera MA, Boudouresque CF. 1995. Introduced marine plants, with special reference to macroalgae: mechanisms and impact. In *Progress in Phycological Research*, ed. FE Round, DJ Chapman, pp. 187–268. Bristol, UK: Biopress

Ribera Siguan MA. 2002. Review of non-native marine plants in the Mediterranean Sea. In *Invasive Aquatic Species of Europe—Distribution, Impacts and Management*, ed. E Leppäkoski, S Gollasch, S Olenin, pp. 291–310. Dordrecht: Kluwer Acad.

Ribera Siguan MA. 2003. Pathways of biological invasions of marine plants. In *Invasive Species: Vectors and Management Strategies*, ed. GM Ruiz, JT Carlton, pp. 183–226. Washington, DC: Island Press

Ruiz GM, Fofonoff P, Hines AH, Grosholz ED. 1999. Non-indigenous species as stressors in estuarine and marine communities: Assessing invasion impacts and interactions. *Limnol. Oceanogr.* 44:950–72

Ruiz GM, Fofonoff PW, Carlton JT, Wonham MJ, Hines AH. 2000. Invasion of coastal marine communities in North America: Apparent patterns, processes, and biases. *Annu. Rev. Ecol. Syst.* 31:481–531

Sala E, Boudouresque CF. 1997. The role of fishes in the organization of a Mediterranean sublittoral community. 1. Algal communities. *J. Exp. Mar. Biol. Ecol.* 212:25–44

Sánchez I, Fernández C. 2005. Impact of the invasive seaweed *Sargassum muticum* (Phaeophyta) on an intertidal macroalgal assemblage. *J. Phycol.* 41:923–30

Sánchez I, Fernández C. 2006. Resource availability and invasibility in an intertidal macroalgal assemblage. *Mar. Ecol. Prog. Ser.* 313:85–94

Sánchez I, Fernández C, Arrontes J. 2005. Long-term changes in the structure of intertidal assemblages after invasion by *Sargassum muticum* (Phaeophyta). *J. Phycol.* 41:942–49

Sant N, Delgado O, Rodriguez-Prieto C, Ballesteros E. 1996. The spreading of the introduced seaweed *Caulerpa taxifolia* (Vahl) C. Agardh in the Mediterranean sea: Testing the boat transportation hypothesis. *Bot. Mar.* 39:427–30

Sax DF, Kinlan BP, Smith KF. 2005. A conceptual framework for comparing species assemblages in native and exotic habitats. *Oikos* 108:457–64

Schaffelke B, Evers D, Walhorn A. 1995. Selective grazing of the isopod *Idotea baltica* between *Fucus evanescens* and *F. vesiculosus* from Kiel Fjord (western Baltic). *Mar. Biol.* 124:215–18

Schaffelke B, Smith JE, Hewitt CL. 2006. Introduced macroalgae—a growing concern. *J. Appl. Phycol.* 18:529–41

Scheibling RE, Anthony SX. 2001. Feeding, growth and reproduction of sea urchins (*Strongylocentrotus droebachiensis*) on single and mixed diets of kelp (*Laminaria* spp.) and the invasive alga *Codium fragile* ssp. *tomentosoides*. *Mar. Biol.* 139:139–46

Scheibling RE, Gagnon P. 2006. Competitive interactions between the invasive green alga *Codium fragile* ssp. *tomentosoides* and native canopy-forming seaweed. *Mar. Ecol. Prog. Ser.* 325:1–14

Schmidt AL, Scheibling RE. 2006. A comparison of epifauna and epiphytes on native kelps (*Laminaria* species) and an invasive alga (*Codium fragile ssp tomentosoides*) in Nova Scotia, Canada. *Bot. Mar.* 49:315–30

Schueller GH, Peters AF. 1994. Arrival of *Fucus evanescens* (Phaeophyceae) in Kiel Bight (western Baltic). *Bot. Mar.* 37:471–77

Secord D. 2003. Biological control of marine invasive species: cautionary tales and land-based lessons. *Biol. Invasions* 5:117–31

Simberloff D, Von Holle B. 1999. Positive interactions of nonindigenous species: Invasional meltdown? *Biol. Invasions* 1:21–32

Smith CM, Walters LJ. 1999. Fragmentation as a strategy for *Caulerpa* species: Fates of fragments and implications for management of an invasive weed. *Mar. Ecol.* 20:307–19

Smith JE, Most R, Sauvage T, Hunter C, Squair C, Conklin E. 2004. Ecology of the invasive red alga *Gracilaria salicornia* in Waikiki and possible mitigation strategies. *Pac. Sci.* 58:325–43

Stæhr PA, Pedersen MF, Thomsen MS, Wernberg T, Krause-Jensen D. 2000. Invasion of *Sargassum muticum* in Limfjorden (Denmark) and its possible impact on the indigenous macroalgal community. *Mar. Ecol. Prog. Ser.* 207:79–88

Stam WT, Olsen JL, Zaleski SF, Murray SN, Brown KR, Walters LJ. 2006. A forensic and phylogenetic survey of *Caulerpa* species (Caulerpales, Chlorophyta) from the Florida coast, local aquarium shops, and e-commerce: Establishing a proactive baseline for early detection. *J. Phycol.* 42:1113–24

Steen H, Rueness J. 2004. Comparison of survival and growth in germlings of six fucoid species (Fucales, Phaeophyceae) at two different temperature and nutrient levels. *Sarsia* 89:175–83

Steneck RS, Dethier MN. 1994. A functional-group approach to the structure of agal-dominated communities. *Oikos* 69:476–98

Stimson J, Larned ST, Conklin E. 2001. Effects of herbivory, nutrient levels, and introduced algae on the distribution and abundance of the invasive macroalga *Dictyosphaeria cavernosa* in Kaneohe Bay, Hawaii. *Coral Reefs* 19:343–57

Strong JA, Dring MJ, Maggs CA. 2006. Colonisation and modification of soft substratum habitats by the invasive macroalga *Sargassum muticum*. *Mar. Ecol. Prog. Ser.* 321:87–97

Sumi CBT, Scheibling RE. 2005. Role of grazing by sea urchins *Strongylocentrotus droebachiensis* in regulating the invasive alga *Codium fragile* ssp. *tomentosoides* in Nova Scotia. *Mar. Ecol. Prog. Ser.* 292:203–12

Terrados J, Marbà N. 2006. Is the vegetative development of the invasive chlorophycean, *Caulerpa taxifolia*, favored in sediments with a high content of organic matter? *Bot. Mar.* 49:331–38

Exemplifies the growing use of molecular techniques to determine the identity, origin, and vector for an introduced seaweed.

Thibaut T, Meinesz A, Amade P, Charrier S, De Angelis K, et al. 2001. *Elysia subornata* (Mollusca) a potential control agent of the alga *Caulerpa taxifolia* (Chlorophyta) in the Mediterranean Sea. *J. Mar. Biol. Assoc. UK* 81:497–504

Thomsen MS, McGlathery KJ, Tyler AC. 2006a. Macroalgal distribution patterns in a shallow, soft-bottom lagoon, with emphasis on the nonnative *Gracilaria vermiculophylla* and *Codium fragile*. *Estuar. Coasts* 29:465–73

Thomsen MS, Wernberg T, Stæhr PA, Pedersen MF. 2006b. Spatio-temporal distribution patterns of the invasive macroalga *Sargassum muticum* within a Danish *Sargassum* bed. *Helgol. Mar. Res.* 60:50–58

Thornber CS, Kinlan BP, Graham MH, Stachowicz JJ. 2004. Population ecology of the invasive kelp *Undaria pinnatifida* in California: Environmental and biological controls on demography. *Mar. Ecol. Prog. Ser.* 268:69–80

Trowbridge C. 2006. A global proliferation of non-native marine and brackish macroalgae. See Critchley et al. 2006

Trowbridge CD. 1995. Establishment of green alga *Codium fragile* ssp. *tomentosoides* on New Zealand rocky shores: Current distribution and invertebrate grazers. *J. Ecol.* 83:949–65

Trowbridge CD. 2001. Coexistence of introduced and native congeneric algae: *Codium fragile* and *C. tomentosum* on Irish rocky intertidal shores. *J. Mar. Biol. Assoc. UK* 81:931–37

Trowbridge CD, Todd CD. 2001. Host-plant change in marine specialist herbivores: Ascoglossan sea slugs on introduced macroalgae. *Ecol. Monogr.* 71:219–43

Valentine JP, Johnson CR. 2003. Establishment of the introduced kelp *Undaria pinnatifida* in Tasmania depends on disturbance to native algal assemblages. *J. Exp. Mar. Biol. Ecol.* 295:63–90

Valentine JP, Johnson CR. 2004. Establishment of the introduced kelp *Undaria pinnatifida* following dieback of the native macroalga *Phyllospora comosa* in Tasmania, Australia. *Mar. Freshw. Res.* 55:223–30

Valentine JP, Johnson CR. 2005. Persistence of the exotic kelp *Undaria pinnatifida* does not depend on sea urchin grazing. *Mar. Ecol. Prog. Ser.* 285:43–55

Valiela I, McClelland J, Hauxwell J, Behr PJ, Hersh D, Foreman K. 1997. Macroalgal blooms in shallow estuaries: Controls and ecophysiological and ecosystem consequences. *Limnol. Oceanogr.* 42:1105–18

Viejo RM. 1999. Mobile epifauna inhabiting the invasive *Sargassum muticum* and two local seaweeds in northern Spain. *Aquat. Bot.* 64:131–49

Vincent C, Mouillot D, Lauret M, Do Chi T, Troussellier M, Aliaume C. 2006. Contribution of exotic species, environmental factors and spatial components to the macrophyte assemblages in a Mediterranean lagoon (Thau lagoon, Southern France). *Ecol. Model.* 193:119–31

Vroom PS, Smith CM. 2001. The challenge of siphonous green algae. *Am. Sci.* 89:524–31

Walker TL, Collet C, Purton S. 2005. Algal transgenics in the genomic era. *J. Phycol.* 41:1077–93

Walters LJ, Brown KR, Stam WT, Olsen JL. 2006. E-commerce and *Caulerpa*: Unregulated dispersal of invasive species. *Front. Ecol. Environ.* 4:75–79

Wernberg T, Thomsen MS, Stæhr PA, Pedersen MF. 2004. Epibiota communities of the introduced and indigenous macroalgal relatives *Sargassum muticum* and *Halidrys siliquosa* in Limfjorden (Denmark). *Helgol. Mar. Res.* 58:154–61

Wikström SA, Kautsky L. 2004. Invasion of a habitat-forming seaweed: effects on associated biota. *Biol. Invasions* 6:141–50

Wikström SA, Steinarsdottir MB, Kautsky L, Pavia H. 2006. Increased chemical resistance explains low herbivore colonization of introduced seaweed. *Oecologia* 148:593–601

Williams SL, Grosholz ED. 2002. Preliminary reports from the *Caulerpa taxifolia* invasion in southern California. *Mar. Ecol. Prog. Ser.* 233:307–10

Wright JT, Davis AR. 2006. Demographic feedback between clonal growth and fragmentation in an invasive seaweed. *Ecology* 87:1744–54

York PH, Booth DJ, Glasby TM, Pease BC. 2006. Fish assemblages in habitats dominated by *Caulerpa taxifolia* and native seagrasses in south-eastern Australia. *Mar. Ecol. Prog. Ser.* 312:223–34

Zar JH. 1999. *Biostatistical Analysis.* Upper Saddle River, NJ: Prentice Hall

Zemke-White WL, Ohno M. 1999. World seaweed utilization: An end of the century summary. *J. Appl. Phycol.* 11:369–76

Zemke-White WL, Smith JE. 2006. Environmental impacts of seaweed farming in the tropics. See Critchley et al. 2006

Zŭljevic A, Thibaut T, Elloukal H, Meinesz A. 2001. Sea slug disperses the invasive *Caulerpa taxifolia. J. Mar. Biol. Assoc. UK* 81:343–44

RELATED RESOURCES

Colautti RI, Bailey SA, van Overdijk CDA, Amundsen K, MacIsaac HJ. 2006. Characterised and projected costs of nonindigenous species in Canada. *Biol. Invasions* 8:45–59

Int. Counc. Explor. Sea. 1995. *Code of Practice on the Introductions and Transfers of Marine Organisms,* 1994. Copenhagen: ICES

Neill PE, Alcalde O, Faugeron S, Navarrete SA, Correa JA. 2006. Invasion of *Codium fragile* ssp. *tomentosoides* in northern Chile: a new threat for *Gracilaria* farming. *Aquaculture* 259:202–10

Van Beukering PJH, Cesar H. 2004. Ecological economic modeling of coral reefs: evaluating tourist overuse at Hanauma Bay and algae blooms at the Kihei coast, Hawai'i. *Pac. Sci.* 58:243–51

The Very Early Stages of Biological Evolution and the Nature of the Last Common Ancestor of the Three Major Cell Domains

Arturo Becerra, Luis Delaye, Sara Islas, and Antonio Lazcano

Facultad de Ciencias, Universidad Nacional Autónoma de México, 04510 México, D.F., Mexico; email: alar@correo.unam.mx

Annu. Rev. Ecol. Evol. Syst. 2007. 38:361–79

The *Annual Review of Ecology, Evolution, and Systematics* is online at
http://ecolsys.annualreviews.org

This article's doi:
10.1146/annurev.ecolsys.38.091206.095825

Key Words

cenancestor, early cell evolution, LCA, LCA gene complement, LUCA, RNA/protein world

Abstract

Quantitative estimates of the gene complement of the last common ancestor of all extant organisms, that is, the cenancestor, may be hindered by ancient horizontal gene transfer events and polyphyletic gene losses, as well as by biases in genome databases and methodological artifacts. Nevertheless, most reports agree that the last common ancestor resembled extant prokaryotes. A significant number of the highly conserved genes are sequences involved in the synthesis, degradation, and binding of RNA, including transcription and translation. Although the gene complement of the cenancestor includes sequences that may have originated in different epochs, the extraordinary conservation of RNA-related sequences supports the hypothesis that the last common ancestor was an evolutionary outcome of the so-called RNA/protein world. The available evidence suggests that the cenancestor was not a hyperthermophile, but it is currently not possible to assess its ecological niche or its mode of energy acquisition and carbon sources.

INTRODUCTION

LCA: last common ancestor

RNA world: an evolutionary stage prior to proteins and DNA genomes during which life was based on catalytic and replicative RNAs

"All the organic beings which have ever lived on this Earth," wrote Charles Darwin in the *Origin of Species*, "may be descended from some one primordial form." Everything in contemporary biology, including molecular cladistics and comparative genomics, has confirmed Darwin's extraordinary insight. As shown by the construction of a trifurcated, unrooted tree based on comparisons of 16S/18S rRNA sequences, all organisms can be grouped into one of three major monophyletic cell lineages, that is, the domains Bacteria, Archaea, and Eucarya (Woese et al. 1990). These are all derived from a common ancestral form, the last common ancestor (LCA). All organisms share the same genetic code, the same essential features of genome replication and gene expression, basic anabolic reactions, and membrane-associated ATPase-mediated energy production. Minor variations can easily be explained as the outcome of divergent processes from an ancestral life form that predated the separation of the three major biological domains.

It is unlikely that such traits were already present in the first living systems. The discovery of catalytically active RNA molecules has given considerable credibility to suggestions of an evolutionary stage prior to the development of proteins and DNA genomes during which early life forms based largely on ribozymes may have existed (**Figure 1**). The difficulty involved with the prebiotic synthesis and accumulation of ribonucleotides and RNA molecules has led to the suggestion that the RNA world itself was the evolutionary outcome of some earlier primordial living systems, referred to as pre-RNA worlds (Joyce 2002). However, the chemical nature of the first genetic polymers and the catalytic agents that may have formed the hypothetical pre-RNA worlds can only be surmised and cannot be deduced from comparative genomics or deep phylogenetics.

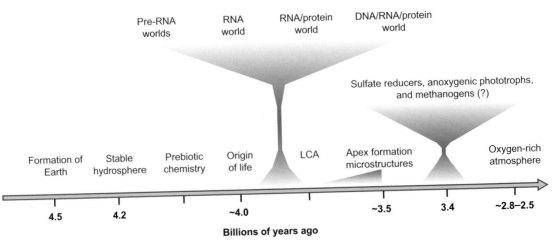

Figure 1

Timeline of the events leading to the origin and early evolution of life. LCA, last common ancestor. Figure adapted with permission from Joyce 2002.

There is little or no geological evidence for the environmental conditions on early Earth at the time of the origin and early evolution of life. It is not possible to assign a precise chronology to the origin and earliest evolution of cells (**Figure 1**), and identification of the oldest paleontological traces of life remains a contentious issue. The early Archean geological record is scarce and controversial, and most of the sediments preserved from such times have been metamorphosed to a considerable extent. Although the biological origin of the microstructures present in the 3.5×10^9 years Apex cherts of the Australian Warrawoona formation (Schopf 1993) has been disputed, currently the weight of evidence favors the idea that life existed 3.5 billion years ago (Altermann & Kazmierczak 2003).

Although no paleontological remains exist for the LCA, insight into the nature of the Archean biosphere is possible. Isotopic fractionation data and other biomarkers support the possibility of a metabolically diverse Archean microbial biosphere, which may have included members of the archaeal kingdom. The proposed timing of the onset of microbial methanogenesis, based on the low ^{13}C values in methane inclusions found in hydrothermally precipitated quartz in the 3.5-billion-year-old Dresser Formation in Australia (Ueno et al. 2006), has been challenged (Lollar & McCollom 2006). However, sulfur isotope investigations of the same site indicate biological sulfate-reducing activity (Shen et al. 2001), and analyses of 3.4×10^9-million-year-old South African cherts suggest that they were inhabited by anaerobic photosynthetic prokaryotes in a marine environment (Tice & Lowe 2004). These results support the idea that the early Archean Earth was teeming with prokaryotes, which included anoxygenic phototrophs, sulfate reducers, and, perhaps, methanogenic archaea (Canfield 2006).

Progenote: a hypothetical biological entity in which phenotype and genotype had an imprecise, rudimentary linkage relationship

Cenancestor: the organism ancestral to domains Bacteria, Archaea, and Eucarya, i.e., the last common ancestor

THE SEARCH FOR THE LAST COMMON ANCESTOR: WHAT'S IN A NAME?

Recognition of the significant differences that exist between the transcriptional and translational machineries of the Bacteria, Archaea, and Eucarya were assumed to be the result of independent evolutionary refinements. This led to the conclusion that the primary branches were the descendants of a progenote, a hypothetical biological entity in which phenotype and genotype still had an imprecise, rudimentary linkage relationship (Woese & Fox 1977). However, the analysis of homologous traits found among some of its descendants suggested that it was not a direct, immediate descendant of the RNA world, a protocell, or any other prelife progenitor system, but that it was a complex organism, much like extant bacteria (Lazcano et al. 1992).

This conclusion has been disputed. The availability of an increasingly large number of completely sequenced cellular genomes has sparked new debates, rekindling the discussion on the nature of the ancestral entity. This is shown, for instance, in the diversity of names that have been coined to describe the LCA: progenote (Woese & Fox 1977), cenancestor (Fitch & Upper 1987), last universal common ancestor (LUCA) (Kyrpides et al. 1999), or, later on, last universal cellular ancestor (Philippe & Forterre 1999), universal ancestor (Doolittle 2000), last common community, (Line 2002), and most recent common ancestor (Zhaxybayeva & Gogarten 2004), among

HGT: horizontal gene
transfer

others. These terms are not truly synonymous, and they reflect the current contro-
versies on the nature of the universal ancestor and the evolutionary processes that
shaped it.

From a cladistic viewpoint, the LCA is merely an inferred inventory of features
shared among extant organisms, all of which are located at the tip of the branches of
molecular phylogenies. Accordingly, it could be argued that the most parsimonious
characterization of the cenancestor could be achieved by summarizing the features of
the oldest recognizable nodes of universal cladograms. However, large-scale studies
based on the availability of genomic data have revealed major discrepancies with the
topology of rRNA trees (Doolittle 2000). Very often these differences have been
interpreted as evidence of horizontal gene transfer (HGT) events between different
species, questioning the feasibility of the reconstruction and proper understanding
of early biological history.

Reticulate phylogenies greatly complicate the inference of ancestral traits and can
lead to overestimates of the LCA gene content. Recognition of HGT has also led
to proposals suggesting that populations of precellular ancestral entities occupied, as
a whole, the node located at the bottom of universal trees (Kandler 1994, Koonin
& Martin 2005), including the suggestion that the LCA was not a single organismic
entity but rather a highly diverse population of metabolically complementary, cellular progenotes endowed with multiple small, linear chromosome-like genomes that
benefited from massive multidirectional horizontal transfer events (Woese 1998).

Such entities may have existed, but universally distributed features such as the
genetic code and the gene expression machinery are indications of their ultimate
monophyletic origin. Even if such communal progenote ancestors diverged sharply
into the three domains soon after the appearance of the code and the establishment of
translation, the origin of the sequences ancestral to those found in extant organisms
and the separation of the Bacteria, Archaea, and Eucarya are different events. In other
words, the divergence of the primary domains took place later, perhaps even much
later, than the appearance of the genetic components of their LCA.

Universal gene-based phylogenies ultimately reach a single ancestor, but the LCA
must have been part of a population of entities similar to it that existed throughout
the same period. Its siblings may have not survived, but some of their genes did if
they became integrated via lateral transfer into the LCA genome (Zhaxybayeva &
Gogarten 2004). Accordingly, the cenancestor should be considered one of the last
evolutionary outcomes of a tree trunk of unknown length, during which the history
of a long, but not necessarily slow, series of ancestral events including lateral gene
transfer, gene losses, and paralogous duplications probably played a significant role
in the accretion of complex genomes (Castresana 2001, Glansdorff 2000, Lazcano
et al. 1992).

AN EARLY EXPANSION OF GENE FAMILIES

The extraordinary similarities at very basic biochemical and genetic levels among all
known life forms can be interpreted as propinquity of descent, that is, all organisms

are of monophyletic origin. The molecular details of these universal processes not only provide direct evidence of the monophyletic origin of all extant forms of life, but also imply that the sets of genes encoding the components of these complex traits were frozen a long time ago; that is, major changes in them are strongly selected against. Although these complex, multigenic traits must have evolved through a series of simpler states, no evolutionary intermediate stages or ancient simplified versions of ATP production, DNA replication, or ribosome-mediated protein synthesis have been discovered in extant organisms.

Protein sequence comparisons have confirmed the role that many ancient gene duplications have played in the evolution of genomes (Becerra & Lazcano 1998). Clues to the genetic organization and biochemical complexity of primitive entities from which the cenancestor evolved may be derived from the analysis of paralogous gene families. The number of sequences that have undergone such duplications prior to the divergence of the three lineages includes genes encoding for a variety of enzymes that participate in widely different processes such as translation, DNA replication, CO_2 fixation, nitrogen metabolism, and biosynthetic pathways. Whole-genome analysis has revealed the impressive expansion of sequences involved in membrane transport phenomena, such as ABC transporters, P-type ATPases, and ion-coupled permeases (Clayton et al. 1997). Structural studies of proteins provide evidence of another group of paralogous duplications. A number of enzymes—including protein disulfide oxidoreductase (PDO) (Ren et al. 1998), the large subunit of carbomoyl phosphate synthetase (Alcántara et al. 2000), and HisA, a histidine biosynthetic isomerase (Alifano et al. 1996)—are formed by two tandem homologous modules. This indicates that the size and structure of a number of proteins are the evolutionary outcome of paralogous duplications followed by gene fusion events that took place prior to the divergence of the three primary kingdoms.

A third group of paralogous genes can be recognized. All cells are endowed with different sets formed by a relatively small number of paralogous sequences. The list includes, among others, the pair of homologous genes encoding the EF-Tu and EF-G elongation factors (Iwabe et al. 1989). Another is formed by the duplicated sequences encoding the F-type ATPase hydrophilic alpha and beta subunits (Gogarten et al. 1989). No cell is known that is endowed with only one EF factor or only one type of F-type ATPase hydrophilic subunit. However, the extraordinary conservation of these duplicates implies that the LCA was preceded by a simpler cell with a smaller genome in which only one copy of each of these genes existed, that is, by cells in which protein synthesis involved only one elongation factor and with ATPases with limited regulatory abilities.

Paralogous families of metabolic genes support the proposal that anabolic pathways were assembled by the recruitment of primitive enzymes that could react with a wide range of chemically related substrates, that is, the so-called patchwork assembly of biosynthetic routes (Jensen 1976). Such relatively slow, unspecific enzymes may have represented a mechanism by which primitive cells with small genomes could have overcome their limited coding abilities.

GENOMICS AND THE TRACES OF THE RNA/PROTEIN WORLD

Studies of deep phylogenies (Brown et al. 2001, Daubin et al. 2001, Doolittle 2000, Moreira & López-García 2006) and comparative genomics (**Table 1**) can provide important insights into the gene complement of the LCA. However, if the term universal distribution is restricted to its most obvious sense, that is, that of traits found in all completely sequenced genomes, then quite surprisingly the resulting repertoire

Table 1   Estimates of the LCA gene complement based on quantitative genomic analysis

Cenancestral traits	Methodology	Number of sequences and functional categories
LCA		**80 universally conserved COGs (50 of them exhibiting three domain phylogenies)**
Relatively efficient transcription and ribosome structure; functions linked to membranes; ability to synthesize long DNA molecules (Harris et al. 2003)	Identification of universally conserved COGs showing three domain phylogenies	Translation and transcription: (63/80) DNA replication and repair: (5/80) Membrane-associated proteins: (1/80) Nucleotide and sugar metabolism: (4/80) Amino acid metabolism: (1/80) Protein management: (2/80) Others: (2/80)
LUCA		**~600 genes assigned to LUCA (COGs)**
Nearly sufficient genes to sustain a functioning organism (Mirkin et al. 2003)	Construction of parsimonious scenarios for individual sets of COGs based on species trees	Translation and transcription: (112/600) DNA replication and repair: (30/600) Membrane-associated proteins and metabolism: (287/600) Protein management: (25/600) Others: (94/600)
LUCA		**~63 ubiquitous genes (proteins)**
Simple, with a small number of genes; lack of a modern-type DNA genome and replication system (Koonin 2003)	Sequence comparison of ~100 genomes	Translation and transcription: (56/63) DNA replication and repair: (3/63) Membrane-associated proteins: (3/63) Protein management: (1/63)
LCA		**49 universally distributed protein folds (SCOP superfamilies)**
With a very sophisticated genetic inventory of structural equipment (Yang et al. 2005)	Distribution of SCOP protein superfamilies in 174 complete genomes	Translation and transcription: (32/49) DNA replication and repair: (5/49) Metabolism: (5/49) Protein management: (1/49) Others: (5/49)
LCA		**~115 protein domains (Pfam domains)**
Similar to extant cells in genetic complexity (Delaye et al. 2005)	BLAST sequence comparison of 20 genomes and orthologous identification based on the Pfam database	Translation and transcription: (56/115) DNA replication and repair: (6/115) Membrane-associated proteins: (7/115) Nucleotide and sugar metabolism: (33/115) Amino acid metabolism: (12/115) Protein management: (1/115)

Table 1   *(Continued)*

Cenancestral traits	Methodology	Number of sequences and functional categories
LUCA		**20 described motifs (octapeptides in proteins with known 3D structure)**
Total amount of ancestral octamers may be in the order of several hundreds (Sobolevsky & Trifonov 2006)	Identification of omnipresent (or nearly so) octamer protein motifs	Translation and transcription; DNA replication and repair; protein management; membrane-associated proteins
LUCA		**~1000 genes, with a minimum of 561 to 669 sequence/function categories (proteins)**
Fairly complex genome similar to those of free-living prokaryotes (Ouzounis et al. 2006)	Identification of homologous sequences among 184 genomes, using a method that corrects for gene losses	Translation and transcription: (34/659) DNA replication and repair: (35/659) Membrane-associated proteins: (120/659) Metabolism: (309/659) Others: (161/659)
LUCA		**140 ancestral protein domains (CATH superfamiles)**
Genetically complex entity, with practically all essential traits present in extant organisms (Ranea et al. 2006)	Distribution of CATH protein superfamilies in 114 complete genomes	Translation and transcription: (52/140) DNA replication and repair: (12/140) Membrane associated proteins: (2/140) Metabolism (46/140) Others: (28/140)

Acronyms: BLAST, basic local alignment search tool; CATH, protein structure classification; COGs, clusters of orthologous genes; LCA, last common ancestor; LUCA, last universal common ancestor; SCOP, structural classification of proteins.
For an extended version of this table, follow the Supplemental Material link from the Annual Reviews home page at **http://www.annualreviews.org**.

is formed by relatively few features and by incompletely represented biochemical processes (**Table 1**). Reconstructions of gene complements of distant ancestors are mere statistical approximations of the biological past because their accuracy depends upon manifold factors, including the possible biases in the construction of genome databases, the levels of HGT, the significant variations in substitution rates of different proteins, and the degree of secondary losses, as well as methodological caveats (Becerra et al. 1997, Mirkin et al. 2003).

As shown in **Table 1**, the results of different attempts to characterize the LCA include gene sequences from incompletely represented basic biological processes, such as transcription, translation, energy metabolism, biosynthesis of nucleotides and amino acids, and folding of proteins, as well as some sequences related to replication, repair, and cellular transport. Despite the different methodological approaches, these inventories provide significant insights into (*a*) the biological complexity of the LCA, (*b*) evidence pertaining to the chemical nature of the cenancestral genome, and (*c*) the existence of an ancient RNA/protein world.

Although estimates may vary, many of the conserved domains correspond to proteins that interact directly with RNA (such as ribosomal proteins, DEAD-type helicases, aminoacyl tRNA synthetases, and elongation factors, among others) or take

RNA/protein world: an early stage in cell evolution during which ribosome-mediated protein synthesis had already evolved but DNA genomes had not emerged

part in RNA and nucleotide biosyntheses, including the DNA-dependent RNA polymerase β and β′ subunits, dimethyladenosine transferase, adenyl-succinate lyases, dihydroorotate oxidase, and ribose-phosphate pyrophosphokinase, among many others (Anantharam et al. 2002, Delaye & Lazcano 2000, Delaye et al. 2005). Together with the high conservation of ATP-dependent RNA helicases (see below), the presence of these sequences is consistent with the hypothesis that the cenancestor was an evolutionary outcome of the so-called RNA/protein world.

ATP-dependent RNA helicases are highly conserved proteins that participate in a variety of cellular functions involving the unwinding and rearrangement of RNA molecules, including translation initiation, RNA splicing, ribosome assembly, and degradosome-mediated mRNA decay (Schmid & Linder 1992). The degradosome is a multienzymatic complex involved in mRNA processing and breakdown that is formed by polynucleotide phosphorylase, polyphosphate kinase, ATP-dependent DEAD/H-type RNA helicase, and enolase, a glycolytic enzyme that catalyzes the conversion of 2-phosphoglycerate to phosphoenolpyruvate and water (Blum et al. 1997).

Although RNA hydrolysis is an exergonic process, degradosome-mediated mRNA turnover plays a key role as a regulatory mechanism for gene expression in both prokaryotes and eukaryotes (Blum et al. 1997). A possible explanation for the conservation of DEAD-type RNA helicases may lie in their role in protein biosynthesis and in mRNA degradation. If this interpretation is correct, then it could be argued that degradosome-mediated mRNA turnover is an ancient control mechanism at RNA level that was established prior to the divergence of the three primary kingdoms. Together with other lines of evidence, including the observation that the most highly conserved gene clusters in several (eu)bacterial genomes are regulated at RNA level (Siefert et al. 1997), the results reported here are fully consistent with the hypothesis that during early stages of cell evolution, RNA molecules played a more conspicuous role in cellular processes, that is, the so-called RNA/protein world (**Figure 1**).

THE NATURE OF THE CENANCESTRAL GENOME: DNA OR RNA?

Because all extant cells are endowed with DNA genomes, the most parsimonious conclusion is that this genetic polymer was already present in the cenancestral population. Although it is possible to recognize the evolutionary relatedness of various orthologous proteins involved with DNA replication and repair (ATP-dependent clamp loader proteins, topoisomerases, gyrases, and 5′-3′ exonucleases) across the entire phylogenetic spectrum (Edgell & Doolittle 1997, Harris et al. 2003, Leipe et al. 1999, Olsen & Woese 1997, Penny & Poole 1999), comparative proteome analysis has shown that (eu)bacterial replicative polymerases and primases lack homologs in the two other primary kingdoms.

The peculiar distribution of the DNA replication machinery has led to suggestions not only of an LCA endowed with an RNA genome, but also of the polyphyletic origins of DNA and of many enzymes associated with its replication (Koonin & Martin 2005, Leipe et al. 1999) in which viruses may have played a central role (Forterre 2006). Koonin & Martin (2005) have argued that the LCA was an acellular entity endowed with high numbers of RNA viral-like molecules that had originated abiotically within the cavities of a hydrothermal mound. This idea, which has little, if any, empirical support, does not take into account the problems involved with the abiotic synthesis and accumulation of ribonucleotides and polyribonucleotides, nor does it explain the emergence of functional RNA molecules.

Forterre (2006) also argued that the ultimate origins of cellular DNA genomes lie in viral systems, which gave rise to polyphyletic deoxyribonucleotide biosyntheses. According to a rather complex hypothetical scheme, gene transfers mediated by viral takeovers took place three times, giving origin to the DNA genomes of the three primary kingdoms. The invasion of the ancestor of the bacterial domain by a DNA virus eventually led to a replacement of its cellular RNA genes by DNA sequences, whereas the archaeal and eucaryal DNA replication enzymes resulted from an invasion by closely related DNA viruses (Forterre 2006).

We find it difficult to accept these different schemes. There are indeed manifold indications that RNA genomes existed during early stages of cellular evolution (Lazcano et al. 1988i), but, as argued below, it is likely that double-stranded DNA genomes had become firmly established prior to the divergence of the three primary domains. The major arguments supporting this possibility follow:

1. In sharp contrast to other energetically favorable biochemical reactions (such as phosphodiester backbone hydrolysis or the transfer of amino groups), the direct removal of the oxygen from the 2'-C ribonucleotide pentose ring to form the corresponding deoxy-equivalents is a thermodynamically much-less-favored reaction that considerably reduces the likelihood of multiple, independent origins of biological ribonucleotide reduction.

2. Demonstration of the monophyletic origin of ribonucleotide reductases is greatly complicated by their highly divergent primary sequences and the different mechanisms by which they generate the substrate 3'-radical species required for the removal of the 2'-OH group. However, sequence analysis and biochemical characterization of archaeal ribonucleotide reductases have shown their similarities with their bacterial and eucaryal counterparts, confirming their ultimate monophyletic origin (Stubbe et al. 2001).

3. Sequence similarities shared by many ancient, large proteins found in all three domains suggest that considerable fidelity existed in the operative genetic system of their common ancestor. Despite claims to the contrary (Poole & Logan 2005), such fidelity is unlikely to be found in RNA-based genetic systems (Lazcano et al. 1992, Reanney 1987) that do not replicate using the multiunit cellular DNA-dependent RNA polymerase.

THE DNA POLYMERASE I PALM DOMAIN: A MOLECULAR PENTIMENTO

Enzyme evolution often involves the acquisition of new catalytic or binding properties by an existing protein scaffold. The structural homology of functional domains of DNA and RNA helicases (Caruthers et al. 2000, Theis et al. 1999) suggests that DNA helicases evolved from a nonspecific helicase inherited from the RNA/protein world.

However, this model of enzyme evolution cannot account for the diversity of extant polymerases. Although it has been argued that all polymerases have a monophyletic origin (Lazcano et al. 1988a), the available evidence suggests that this is not the case, as shown by the identification of several nonhomologous classes of polymerases: primases, DNA polymerases, DNA-dependent RNA polymerases, replicases, and poly(A) polymerase, among others (Steitz 1999).

Detailed analysis of the three-dimensional structures of the DNA pol I, DNA pol II, pol Y, reverse transcriptases, and several viral replicases has shown that they all share homologous palm subdomains, which catalyze the formation of the phosphodiester bond (Steitz 1999). Homologous palm subdomains have also been identified in the viral T7 DNA and RNA polymerases (Jeruzalmi & Steitz 1998), indicating that they can catalyze the template-dependent polymerization of either ribo- or deoxyribonucleotides.

These observations are consistent with the hypothesis that the conserved palm subdomain, which is formed by approximately 150 amino acids, is one of the oldest recognizable components of an ancestral cellular polymerase that may have been both a replicase and a transcriptase during the RNA/protein world stage (Delaye et al. 2001). This hypothesis, which Koonin (2006) incorporated into his proposal on the evolution of viral and cellular polymerases, is supported by the presence of homologs of this domain in adenylate cyclase, the eukaryotic RNA recognition protein, pseudouridine synthetase, and several ribosomal proteins (Aravind et al. 2002).

If our hypothesis is correct, the lack of absolute template and substrate specificity of polymerases (Chaput et al. 2003, Lazcano et al. 1988a) suggests that relatively few mutations would have been required for the evolution of this RNA replicase into a DNA polymerase prior to the divergence of the three domains. If the palm domain was part of the ancestral replicase during the RNA/protein world stage, then the presence of the *Escherichia coli* replicative DNA pol III (DNA pol C) and its homologs can be explained by a nonorthologous displacement (Delaye et al. 2001). By analogy with the yeast and animal mitochondrial RNA polymerases, which play a dual role in transcription and in the initial priming required for DNA replication (Schinkel & Tabak 1989), we propose that the original RNA polymerase endowed with the palm domain described above catalyzed the formation of the RNA primer required for DNA replication. This ancestral polymerase may have also acted as a transcriptase during the RNA/protein stage, but the distribution of the highly conserved sequences of the oligomeric DNA-dependent RNA polymerase (**Table 1**) indicates that prior to the evolutionary divergence of the three primary kingdoms, a modern type of transcription had evolved. How this complex oligomeric transcriptase came into being can only be surmised for the time being.

WAS THE LAST COMMON ANCESTOR AN EXTREMOPHILE?

The discovery of a number of archaeal and bacterial species that thrive under extreme environmental conditions including high temperature and low pH values has led to suggestions that extremophiles may be considered models of primordial organisms (Di Giulio 2001, 2003). Additionally, the discovery has led to speculation that their lifestyles may provide insights into extraterrestrial habitats where life could develop (Cleaves & Chalmers 2004). With the exception of heat-loving prokaryotes, however, the phylogenetic distribution of other extremophiles in molecular cladograms does not provide clues of possible antiquity.

Within the Bacteria, the earliest branching organisms are represented by Aquificales and Thermotogales, whereas among the Archaea, the deepest and shortest branches, that is, the slowest-evolving clades, correspond to the Nanoarchaeota, Pyrodictiacea, and Methanopyraceae (Stetter 2006). The position and length of the branches of these thermophiles and hyperthermophiles in rRNA trees support the idea that the LCA was a thermophile (Di Giulio 2001, Stetter 2006).

However, the recognition that the deepest branches in rooted rRNA universal phylogenies are occupied by hyperthermophiles does not provide by itself conclusive proof of a heat-loving LCA. The hypothesis for a thermophilic LCA is further weakened by evidence that calls into question the early branching of *Thermotoga* and *Aquifex*, two bacterial thermophilic species (Daubin et al. 2001, Forterre et al. 1993, Gogarten & Townsend 2005, Klenk et al. 1994). A nonhyperthermophilic LCA is also supported by the analysis of optimal growth temperatures in prokaryotes correlated with the G+C nucleotide content of 40 rRNA sequences. This has led Galtier et al. (1999) to conclude that the universal ancestor was a mesophile, although alternative opinions exist (Di Giulio 2001).

The possibility that the LCA was not a heat-loving entity is further supported by recent phylogenetic studies of the above-mentioned PDO sequences. These enzymes are involved in dithiol-disulfide exchange reactions (Pedone et al. 2004, Ren et al. 1998), and computational analysis of genomic data suggests that they play a major role in the formation and/or stabilization of intracellular protein disulfide bonds in disulfide-rich thermophiles (Ladenstein & Ren 2006). The available evidence suggests that the paralogous duplication that led to PDOs, which are formed by two homologous domains, first took place in the crenarcheaota and spread from them into the Bacteria by HGT via the euryarchaeota (Becerra et al. 2007), confirming previous suggestions that the bacterial PDOs have an archaeal origin (Pedone et al. 2004). This is consistent with the hypothesis that significant gene exchanges took place between archaea and thermophilic bacteria (Forterre et al. 2000, Makarova & Koonin 2003).

CENANCESTRAL METABOLISM AND MEMBRANES: TWO OPEN ISSUES

The number of highly conserved metabolic genes is surprisingly small (**Table 1**). The inventory includes many sugar-metabolism-related sequences, such as enolase, thioredoxin (*trxB*), phosphoribosyl-pyrophosphate synthase (*prs*), and UDP-galactose 4-epimerase (*galE*). Very likely, the evolutionary conservation of the *trxB* and *prsA* genes is best understood in terms of the key roles they play in nucleotide biosynthesis. The role of UDP-galactose 4-epimerase in complex carbohydrate synthesis via the interconversion of the galactosyl and glucosyl groups is well known. Although the uniqueness of the enzyme mechanism has been acknowledged, it is possible that the conservation of UDP-galactose 4-epimerase is due to an unknown participation in other basic processes, as is the case for enolase.

The conclusion that the LCA was a prokaryote-like organism similar to extant bacteria does not say much about its mode of energy acquisition and carbon sources. The patchy distribution of metabolic genes hinders our understanding of the sources of carbon, nitrogen, and energy for the LCA and its immediate predecessors (Moreira & López-García 2006), and for now it is difficult to assess the cenancestral metabolism. However, if multiple copies of every major gene family are assumed to have been present in the LCA genome (Glansdorff 2000), then the observed complex distribution patterns of bioenergetic and biosynthetic genes can be explained as the outcome of polyphyletic gene losses as the cenancestor descendants adapted to a wide variety of environments under different selection pressures (Castresana 2001).

It has been argued that the LCA was an acellular entity (Koga et al. 1998, Koonin & Martin 2005). However, the high conservation and wide phylogenetic distribution of membrane-bound proteins and multiunit enzymes such as the ATPase hydrophilic subunits (Gogarten et al. 1989), signal recognition particles (Gribaldo & Cammarano 1998), and ABC transporters (Delaye et al. 2005) are consistent with the idea that the cenancestor was a membrane-bounded cell, which may have been endowed with heterochiral lipids composed of a mixture of glycerol-1-phosphate and glycerol-3-phosphate (Peretó et al. 2004, Wächstershäuser 2003).

The conservation of membrane-bound proton pump ATPase subunits (**Table 1**) suggests that the cenancestor produced a chemically driven protein gradient across its cell membrane using a variety of oxidized inorganic molecules as molecular acceptors (Castresana & Moreira 1999). The conservation of ABC transporters, P-type ATPases, and ion-coupled permeases is an indication of the high conservation of membrane transport phenomena throughout evolution (Clayton et al. 1997). The conservation of ABC transporters (**Table 1**) involved in the import of metabolic substrates is also consistent with the possibility of a heterotrophic LCA that depended upon external sources of organic compounds.

CONCLUSIONS AND OUTLOOK

Regardless of the qualitative and quantitative differences in the methodological approaches used to identify the gene complement of the cenancestor, the inventories shown in **Table 1** indicate an overlap that reflects an impressive level of conservation for a significant number of sequences involved in basic biological processes. Current descriptions of the LCA are limited by the scant information available: It is hard to understand in full the evolutionary forces that acted upon our distant ancestors, whose environments and detailed biological characteristics are forever beyond our ken. By definition, the node located at the bottom of a cladogram is the root of a phylogenetic tree and corresponds to the common ancestor of the group under study. But names may be misleading. What we have been calling the root of the universal tree is in fact the tip of its trunk: Inventories of LCA genes include sequences that originated in different precenancestral epochs. For instance, a number of highly conserved ribosomal proteins may have originated during the RNA/protein world

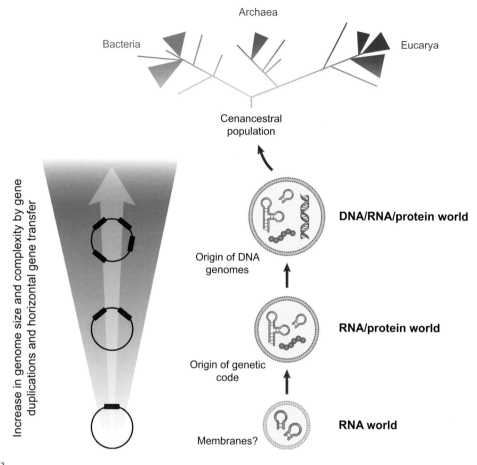

Figure 2

Evolutionary events that may have preceded the last common ancestor and the divergence of the three major cell lineages.

stage, whereas thymidine kinase and thioredoxin reductase, which are involved in deoxyribonucleotide biosynthesis, evolved at a later stage.

Although we favor a bacterial-like cenancestor, for now it is difficult to assess the sources of energy and carbon used by the LCA or the environmental conditions in which it thrived. Biological evolution prior to the divergence of the three domains was not a continuous, unbroken chain of progressive transformation steadily proceeding toward the LCA (**Figure 2**). RNA-binding domains (Delaye & Lazcano 2000) and invariant sequences that exhibit a surprising degree of conservation, such as GHVDHGKT, DTPGHVDF, and GAGKSTL (Goto et al. 2002), are among the oldest recognizable motifs found in extant databases and may provide insights onto the nature of the evolutionary processes that shaped ribosome-mediated protein biosynthesis. Older stages are not yet amenable to phylogenetic analysis, and it is difficult to see how the applicability of molecular cladistics and comparative genomics can be extended beyond a threshold that corresponds to a period of cellular evolution in which protein biosynthesis was already in operation. Many details of the processes that led to the beginnings of life are shrouded in mystery and may remain so, but current developments in comparative genomics suggest that important insights on very early stages of biological evolution can be achieved, even if the possible intermediates that may have once existed have long since vanished.

SUMMARY POINTS

1. Theoretical estimates of the gene content of the LCA's genome suggest that it was not a progenote or a protocell, but an entity similar to extant prokaryotes.

2. The presence of a core of highly conserved RNA-related sequences supports the hypothesis that the LCA was preceded by earlier entities in which RNA molecules played a more conspicuous role in cellular processes and in which ribosome-mediated protein synthesis had already evolved.

3. Whole-genome analysis has revealed high levels of sequence redundancy, which demonstrates the significance of paralogous duplications in shaping the size and complexity of cell genomes. This redundancy suggests that during early stages of biological evolution, anabolic pathways and other biological processes were catalyzed by less-specific enzymes that could react with a wide range of chemically related substrates.

4. The chemistry of ribonucleotide reduction, combined with sequence analysis, supports the hypothesis of a monophyletic origin of DNA that took place prior to the evolutionary divergence of the three main cell domains. The available evidence suggests that the bacterial DNA polymerase I palm domain, and its homologs, is a descendant of a component of an ancestral RNA-dependent RNA polymerase that may have played dual roles as a replicase and a transcriptase during the RNA/protein stage.

5. The availability of genomic data has revealed major discrepancies with the topology of rRNA trees. This has led researchers to question the early branching of *Thermotoga* and *Aquifex*, two bacterial thermophilic species. These conclusions suggest not only that heat-loving bacteria may have been recipients of archaeal hyperthermophilic traits, but also that the LCA was not an extremophile.

FUTURE ISSUES

1. The identification and proper annotation of highly conserved open reading frames found in all cell genomes remain important open issues.

2. Characterization of the cenancestor requires proper assessments of the frequency of ancestral events, including lateral gene transfer, gene losses, and paralogous duplications.

3. Understanding of the evolution of central metabolic pathways is hampered by the unexplained absence of one or more biosynthetic genes in the genomes of manifold free-living prokaryotes that have been sequenced.

4. The development of models of the carbon sources and energy acquisition mechanisms of the LCA and its immediate predecessors should be addressed.

5. It is important to assess the oldest paleontological evidence for the domain Archaea.

DISCLOSURE STATEMENT

The authors are not aware of any biases that might be perceived as affecting the objectivity of this review.

ACKNOWLEDGMENTS

We are indebted to Dr. Janet Siefert for her careful editing of the manuscript and many useful suggestions. Support from CONACYT-México (Project 50520-Q) to A.L. is gratefully acknowledged.

LITERATURE CITED

Alcántara C, Cervera J, Rubio V. 2000. Carbamate kinase can replace in vivo carbamoyl phosphate synthetase. Implications for the evolution of carbamoyl phosphate biosynthesis. *FEBS Lett.* 484:261–64

Alifano P, Fani R, Liò P, Lazcano A, Bazzicalupo M, et al. 1996. Histidine biosynthetic pathway and genes: structure, regulation, and evolution. *Microbiol. Rev.* 60:44–69

Altermann W, Kazmierczak J. 2003. Archean microfossils: a reappraisal of early life on Earth. *Res. Microbiol.* 154:611–17

Anantharaman V, Koonin EV, Aravind L. 2002. Comparative genomics and evolution of proteins involved in RNA metabolism. *Nucleic Acids Res.* 30:1427–64

Aravind L, Mazumderer R, Vasudevan S, Koonin EV. 2002. Trends in protein evolution, inferred from sequence and structure analysis. *Curr. Opin. Struct. Biol.* 12:392–99

Becerra A, Delaye L, Lazcano A, Orgel L. 2007. Protein disulfide oxidoreductases and the evolution of thermophily: Was the last common ancestor a heat-loving microbe? *J. Mol. Evol.* In press

Becerra A, Islas S, Leguina JI, Silva E, Lazcano A. 1997. Polyphyletic gene losses can bias backtrack characterizations of the cenancestor. *J. Mol. Evol.* 45:115–18

Becerra A, Lazcano A. 1998. The role of gene duplication in the evolution of purine nucleotide salvage pathways. *Orig. Life Evol. Biosph.* 28:539–53

Blum E, Py B, Carpousis AJ, Higgins CF. 1997. Polyphosphate kinase is a component of the *Escherichia coli* RNA degradosome. *Mol. Microbiol.* 26:387–98

Brown JR, Douady CJ, Italia MJ, Marshall WE, Stanhope MJ. 2001. Universal trees based on large combined protein sequence datasets. *Nat. Genet.* 28:281–85

Canfield DE. 2006. Biochemistry: gas with an ancient history. *Nature* 440:426–27

Caruthers JM, Johnson ER, McKay DB. 2000. Crystal structure of yeast initiation factor 4A, a DEAD-box RNA helicase. *Proc. Natl. Acad. Sci. USA* 97:13080–85

Castresana J. 2001. Comparative genomics and bioenergetics. *Biochem. Biophys. Acta* 1506:147–62

Castresana J, Moreira D. 1999. Respiratory chains in the last common ancestor of living organisms. *J. Mol. Evol.* 49:453–60

Chaput JC, Ichida JK, Szostak JW. 2003. DNA polymerase-mediated DNA synthesis on TNA template. *J. Am. Chem. Soc.* 125:856–57

Clayton RA, White O, Ketchum KA, Venter CJ. 1997. The genome from the third domain of life. *Nature* 387:459–62

Cleaves HJ, Chalmers JH. 2004. Extremophiles may be irrelevant to the origin of life. *Astrobiology* 4:1–9

Daubin V, Gouy M, Perriere G. 2001. A phylogenomic approach to bacterial phylogeny: evidence for a core of genes sharing a common history. *Genome Res.* 12:1080–90

Delaye L, Becerra A, Lazcano A. 2005. The last common ancestor: What's in a name? *Orig. Life Evol. Biosph.* 35:537–54

Delaye L, Lazcano A. 2000. RNA-binding peptides as molecular fossils. In *Origins from the Big-Bang to Biology: Proceedings of the First Ibero-American School of Astrobiology*, ed. J Chela-Flores, G Lemerchand, J Oró, pp. 285–88. Dordrecht: Klüwer Acad.

Delaye L, Vázquez H, Lazcano A. 2001. The cenancestor and its contemporary biological relics: the case of nucleic acid polymerases. In *First Steps in the Origin of Life in the Universe*, ed. J Chela-Flores, T Owen, F Raulin, pp. 223–30. Dordrecht: Kluwer Acad.

Di Giulio M. 2001. The universal ancestor was a thermophile or a hyperthermophile. *Gene* 281:11–17

Di Giulio M. 2003. The universal ancestor and the ancestor of bacteria were hyperthermophiles. *J. Mol. Evol.* 57:721–30

Doolittle WF. 2000. The nature of the universal ancestor and the evolution of the proteome. *Curr. Opin. Struct. Biol.* 10:355–58

Edgell RD, Doolittle WF. 1997. Archaea and the origins of DNA replication proteins. *Cell* 89:995–98

Fitch WM, Upper K. 1987. The phylogeny of tRNA sequences provides evidence of ambiguity reduction in the origin of the genetic code. *Cold Spring Harbor Symp. Quant. Biol.* 52:759–67

Forterre P. 2006. Three RNA cells for ribosomal lineages and three DNA viruses to replicate their genomes: a hypothesis for the origin of cellular domain. *Proc. Natl. Acad. Sci. USA* 103:3669–74

Forterre P, Benachenhou-Lahfa N, Confalonieri F, Duguet M, Elie C, Labedan B. 1993. The nature of the last universal ancestor and the root of the tree of life, still open questions. *BioSystems* 28:15–32

Forterre P, Bouthier de la Tour C, Philippe H, Duguet M. 2000. Reverse gyrase from hyperthermophiles: probable transfer of a thermoadaptation trait from Archaea to bacteria. *Trends Genet.* 16:152–54

Galtier N, Tourasse N, Gouy M. 1999. A nonhyperthermophilic common ancestor to extant life forms. *Science* 283:220–21

Glansdorff N. 2000. About the last common ancestor, the universal life-tree and lateral gene transfer: a reappraisal. *Mol. Microbiol.* 38:177–85

Gogarten JP, Kibak H, Dittrich P, Taiz L, Bowman EJ, et al. 1989. Evolution of the vacuolar H^+-ATPase, implications for the origin of eukaryotes. *Proc. Natl. Acad. Sci. USA* 86:6661–65

Gogarten JP, Townsend JP. 2005. Horizontal gene transfer, genome innovation and evolution. *Nat. Rev. Microbiol.* 3:679–87

Goto N, Kurokawa K, Yasunaga T. 2002. Finding conserved amino acid sequences among prokaryotic proteomes. *Genome Inform.* 13:443–44

Gribaldo S, Cammarano P. 1998. The root of the universal tree of life inferred from anciently duplicated genes encoding components of the protein-targeting machinery. *J. Mol. Evol.* 47:508–16

Harris JK, Kelley ST, Spiegelman GB, Pace NR. 2003. The genetic core of the universal ancestor. *Genome Res.* 13:407–12

Iwabe N, Kuma K, Hasegawa M, Osawa S, Miyata T. 1989. Evolutionary relationship of archaebacteria, eubacteria, and eukaryotes inferred from phylogenetic trees of duplicated genes. *Proc. Natl. Acad. Sci. USA* 86:9355–59

Jensen RA. 1976. Enzyme recruitment in evolution of new function. *Annu. Rev. Microbiol.* 30:409–25

Jeruzalmi D, Steitz TA. 1998. Structure of T7 RNA polymerase complexed to the transcriptional inhibitor T7 lysozyme. *EMBO J.* 17:4101–13

Joyce GF. 2002. The antiquity of RNA-based evolution. *Nature* 418:214–21

Kandler O. 1994. The early diversification of life. In *Early Life on Earth, Nobel Symp. 84*, ed. S Bengtson, pp. 152–60. New York: Columbia Univ. Press

Klenk HP, Palm P, Zillig W. 1994. DNA-dependent RNA polymerases as phylogenetic marker molecules. *Syst. Appl. Microbiol.* 16:638–47

Koga Y, Kyuragi T, Nishihara M, Sone N. 1998. Did archaeal and bacterial cells arise independently from noncellular precursors? A hypothesis stating that the advent of membrane phospholipids with enantiomeric glycerophosphate backbones caused the separation of the two lines of descent. *J. Mol. Evol.* 46:54–63

Koonin EV. 2003. Comparative genomics, minimal gene-sets and the last universal common ancestor. *Nat. Rev.* 1:127–36

Koonin EV. 2006. Temporal order of evolution of DNA replication systems inferred by comparison of cellular and viral DNA polymerases. *Biol. Direct.* 1:39

Koonin EV, Martin W. 2005. On the origin of genomes and cells within inorganic compartments. *Trends Genet.* 21:647–54

Kyrpides N, Overbeek R, Ouzounis C. 1999. Universal protein families and the functional content of the last universal common ancestor. *J. Mol. Evol.* 49:413–23

Ladenstein R, Ren B. 2006. Protein disulfides and protein disulfide oxidoreductases in hyperthermophiles. *FEBS J.* 273:4170–85

Lazcano A, Fastag J, Gariglio P, Ramírez C, Oró J. 1988a. On the early evolution of RNA polymerase. *J. Mol. Evol.* 27:365–76

Lazcano A, Fox GE, Oró J. 1992. Life before DNA: the origin and early evolution of early Archean cells. In *The Evolution of Metabolic Function*, ed. RP Mortlock, pp. 237–95. Boca Raton, FL: CRC

Lazcano A, Guerrero R, Margulis L, Oró J. 1988b. The evolutionary transition from RNA to DNA in early cells. *J. Mol. Evol.* 27:283–90

Leipe DD, Aravind L, Koonin EV. 1999. Did DNA replication evolve twice independently? *Nucleic Acids Res.* 27:3389–401

Line MA. 2002. The enigma of the origin of life and its timing. *Microbiology* 148:21–27

Lollar BS, McCollom TM. 2006. Biosignatures and abiotic constrainst on early life. *Nature* 444:E18

Makarova KS, Koonin EV. 2003. Comparative genomics of archaea: How much have we learned in six years, and what's next? *Genome Biol.* 4:115–45

Mirkin BG, Fenner TI, Galperin MY, Koonin EV. 2003. Algorithms for computing parsimonious evolutionary scenarios for genome evolution, the last common ancestor, and dominance of horizontal gene transfer in the evolution of prokaryotes. *BMC Evol. Biol.* 3:2

Moreira D, López-García P. 2006. The last common ancestor. *Earth Moon Planets* 98:187–93

Olsen G, Woese CR. 1997. Archaeal genomics: an overview. *Cell* 89:991–94

Ouzounis AC, Kunin V, Darzentas N, Goldovsky L. 2006. A minimal estimate for gene content of the last universal common ancestor—exobiology from a terrestrial perspective. *Res. Microbiol.* 157:57–68

Pedone E, Ren B, Ladenstein R, Rossi M, Bartolucci S. 2004. Functional properties of the protein disulfide oxidoreductase from the archaeon *Pyrococcus furiosus*. *FEBS Eur. J. Biochem.* 271:3437–48

Penny D, Poole A. 1999. The nature of the last common ancestor. *Curr. Opin. Gen. Dev.* 9:672–77

Peretó J, López-García P, Moreira D. 2004. Ancestral lipid biosynthesis and early membrane evolution. *Trends Biochem. Sci.* 29:469–77

Philippe H, Forterre P. 1999. The rooting of the universal tree of life is not reliable. *J. Mol. Evol.* 49:509–23

Poole A, Logan DT. 2005. Modern mRNA proofreading and repair: clues that the last universal common ancestor possessed an RNA genome? *Mol. Biol. Evol.* 22:1444–55

Ranea AG, Sillero A, Thorton MJ, Orengo AC. 2006. Protein superfamily evolution and the last universal common ancestor (LUCA). *J. Mol. Evol.* 63:513–25

Reanney DC. 1987. Genetic error and genome design. *Cold Spring Harbor Symp. Quant. Biol.* 52:751–57

Ren B, Tibbelin G, de Pascale D, Rossi M, Bartolucci S, Ladenstein R. 1998. A protein disulfide oxidoreductase from the archaeo *Pyrococcus furiosus* contains two thioredoxin fold units. *Nat. Struct. Biol.* 7:602–11

Schinkel AH, Tabak HF. 1989. Mitochondrial RNA polymerase: dual role in transcription and replication. *Trends Genet.* 5:149–54

Schmid SR, Linder P. 1992. D-E-A-D protein family of putative RNA helicases. *Mol. Microbiol.* 6:283–92

Schopf JW. 1993. Microfossils of the early Archaean Apex chert: new evidence for the antiquity of life. *Science* 260:640–46

Shen Y, Buick R, Canfield DE. 2001. Isotopic evidence for microbial sulphate reduction in the early Archaean era. *Nature* 410:77–81

Siefert JL, Martin KA, Abdi F, Wagner WR, Fox GE. 1997. Conserved gene clusters in bacterial genomes provide further support for the primacy of RNA. *J. Mol. Evol.* 45:467–72

Sobolevsky Y, Trifonov EN. 2006. Protein modules conserved since LUCA. *J. Mol. Evol.* 63:622–34

Steitz TA. 1999. DNA polymerases: structural diversity and common mechanisms. *J. Biol. Chem.* 274:17395–98

Stetter KO. 2006. Hyperthermophiles in the history of life. *Philos. Trans. R. Soc. London Ser. B* 361:1837–43

Stubbe J, Ge J, Yee CS. 2001. The evolution of ribonucleotide reduction revisited. *Trends Biochem. Sci.* 26:93–99

Theis K, Chen PJ, Skorvaga M, Van Houten B, Kisker C. 1999. Crystal structure of UvrB, a DNA helicase adapted for nucleotide excision repair. *EMBO J.* 18:6899–907

Tice MM, Lowe DR. 2004. Photosynthetic microbial mats in the 3416-Myr-old ocean. *Nature* 431:549–52

Ueno Y, Yamada K, Yoshida N, Maruyama S, Isozaki Y. 2006. Evidence from fluid inclusions for microbial methanogenesis in the early Archaean era. *Nature* 440:516–19

Wächstershäuser G. 2003. From precells to Eukarya—a tale of two lipids. *Mol. Microbiol.* 47:13–22

Woese CR. 1998. The universal ancestor. *Proc. Natl. Acad. Sci. USA* 95:6854–59

Woese CR, Fox GE. 1977. The concept of cellular evolution. *J. Mol. Evol.* 10:1–6

Woese CR, Kandler O, Wheelis ML. 1990. Towards a natural system of organisms, proposal for the domains Archaea, Bacteria, and Eucarya. *Proc. Natl. Acad. Sci. USA* 87:4576–79

Yang S, Doolittle RF, Bourne PE. 2005. Phylogeny determined by protein domain content. *Proc. Natl. Acad. Sci. USA* 102:373–78

Zhaxybayeva O, Gogarten PJ. 2004. Cladogenesis, coalescence and the evolution of the three domains of life. *Trends Genet.* 20:182–87

Functional Versus Morphological Diversity in Macroevolution

Peter C. Wainwright

Section of Evolution & Ecology, University of California, Davis, California 95616;
email: pcwainwright@ucdavis.edu

Annu. Rev. Ecol. Evol. Syst. 2007. 38:381–401

First published online as a Review in Advance on
August 8, 2007

The *Annual Review of Ecology, Evolution, and
Systematics* is online at
http://ecolsys.annualreviews.org

This article's doi:
10.1146/annurev.ecolsys.38.091206.095706

Key Words

disparity, innovation, many-to-one mapping

Abstract

Studies of the evolution of phenotypic diversity have gained momentum among neontologists interested in the uneven distribution of diversity across the tree of life. Potential morphological diversity in a lineage is a function of the number of independent parameters required to describe the form, and innovations such as structural duplication and functional decoupling can enhance the potential for diversity in a given clade. The functional properties of organisms are determined by underlying parts, but any property that is determined by three or more parts expresses many-to-one mapping of form to function, in which many morphologies will have the same functional property. This ubiquitous feature of organismal design results in surfaces of morphological variation that are neutral with respect to the functional property, and enhances the potential for simultaneously optimizing two or more functions of the system.

INTRODUCTION

This review examines general principles and repeating themes in the evolution of complex functional systems, with a particular focus on intrinsic features of organismal design that influence diversity. There is an increasing interest in the patterns and causes of morphological diversity in the tree of life (Carroll 2001, Collar et al. 2005, Erwin 2007, Foote 1997, Gavrilets & Vose 2005, Harmon et al. 2003, Losos & Miles 2002, Losos et al. 1998, Lovette et al. 2002, Niklas 1986, 2004, Ricklefs 2006, Schaefer & Lauder 1996, Stadler et al. 2001, Vermeij 1973, Wagner et al. 2006, Warheit et al. 1999). This area has had a strong tradition in paleontology, but the rapid emergence of new phylogenies and some methodological advances have ushered in an era of phylogenetically informed comparative analyses of morphological diversity in living groups. Empirical studies of functional diversity are much less common (Collar & Wainwright 2006), in part because of the intense effort required to characterize the functional properties of the large number of taxa required for a comparative analysis, but simulations and theoretical analyses have augmented the harder-to-come-by empirical data on functional diversity (Alfaro et al. 2004, 2005; Gavrilets 1999; McGhee & McKinney 2000).

Of special concern is how innovation in design and the nature of the form-function relationship impact the evolution of diversity in functional systems. Not all evolutionary novelties are equally potent. Some, such as the feathery crest on the head of some birds, seem to have little effect on the ecological potential and success of the lineage in possession, whereas others, such as powered flight or endothermy, are major breakthroughs in functional design that drastically changed the ecology and evolution of the lineages in which they evolved. The history of life is characterized by the periodic introductions of novelties that have had significant effects on subsequent ecological and evolutionary diversity: multicellularity, genome duplications, body segmentation, flowers, jaws, etc. To the extent that innovations in design are an important causative agent in spurring bouts of morphological, functional, and ecological diversification, the study of innovation takes on great significance as we try to understand the uneven distribution of diversity across the tree of life. But, how exactly do innovations influence diversity and how do we go about testing hypotheses about their effects on macroevolution? These are issues I hope to get at in the following pages.

I begin by reviewing some general features of the relationship between performance traits and their underlying basis that are fundamental to understanding how this relationship evolves. I then consider several categories of innovation that can influence functional diversity. This leads to a discussion of intrinsic properties of the relationship between form and function that influence patterns of diversity, and I attempt to draw some general conclusions about how the relationship between performance and its underlying basis is likely to shape patterns of diversity. In the final section, I review a recently developed method for conducting comparisons of phenotypic diversity between phylogenetic groups while accounting for differences in time and shared history within the lineages.

PERFORMANCE, FUNCTION, AND MORPHOLOGY

It is generally accepted that organismal performance traits are a major target of natural selection (performance being the ability of individuals to do the tasks that fill their lives). Performance traits typically have a complex underlying basis in the size, shape, and various properties of component parts, and the interplay between performance and its underlying basis has revealed a number of interesting evolutionary dynamics. Throughout this review I refer to functional properties, by which I mean the emergent physiological and mechanical properties of specific organ systems. The mechanical advantage of a lever system is an example of a functional property: It is a predictable property of the system equal to the ratio of the input lever divided by the output lever. Organisms are full of functional properties. The study of these properties and their underlying mechanisms traditionally falls under the purview of disciplines such as physiology, biomechanics, and biochemistry. When I refer to functional diversity, I mean diversity in functional properties, as opposed to diversity in the underlying parts. The connection to performance is that functional properties underlie performance capacity. The maximum bite force that an individual can exert is partly a function of the mechanical advantage of the jaw muscle acting across the jaw joint. Thus, there is a basic hierarchy to organismal design that is a theme in this review: The details of the phenotype come together to determine functional properties and functional properties determine the performance capacity of the individual. Understanding exactly how functional properties are related to underlying design is necessary before we can contrast the evolutionary dynamics of form, function, and performance.

The dependence of performance on organismal design is most easily grasped when the performance trait has a simple mechanical basis in design (Wainwright 1988), and it becomes less clear when behavior mediates the impact of the underlying mechanical system on performance. Thus, though musculoskeletal design may determine the capacity for jump height, the animal must be motivated to ever make use of that capacity, so it is possible for the underlying basis of jump performance to be obscured by the complicating influence of motivation. It is also true that organisms may infrequently, or even never, exhibit peak performance capacity during their lives (Husak 2006, Husak & Fox 2006, Irschick et al. 2005). Further, the nature of the map of phenotype to performance may involve regions of parameter space where phenotypic variation does not influence the expectations for performance capacity (Alfaro et al. 2004, Koehl 1996). These and other factors can make it difficult or very complex to fully determine the basis of performance, and many performance traits are situation-dependent (Helmuth et al. 2005). However, performance capacity and functional properties are emergent properties of organismal construction, and cause and effect are fundamentally discoverable.

Hierarchical Nature of Performance

Performance has a natural hierarchical structure, with the most proximal measures being the most directly linked to particular systems and the highest-order traits involving the integration of multiple functional systems. Consider locomotion as an example. Force output from a single leg muscle can be estimated with remarkable accuracy from morphological measurements of the muscle and a little knowledge about its biochemistry (Powell et al. 1984). Sprint speed involves the coordination of mechanical output from many muscles and their gearing as they work across skeletal linkage systems (Vanhooydonck et al. 2006). The ability to escape from a predator may only partly depend on sprint speed, because it also depends on the ability to change direction quickly and to respond appropriately to the predator's movements (Irschick et al. 2005). The point is that muscle contraction force, sprint speed, and escape ability range continuously from functional properties of the muscle to whole-organism performance traits, and each is based on the design of underlying systems, but that the more proximal attributes, such as force and sprint speed, have a simpler basis because they involve fewer components of the integrated organism. Fitness itself is the ultimate performance trait, integrating across all performance properties and all levels of design. The hierarchical nature of performance has important implications for evolution.

Trade-Offs Are Fundamental to Adaptive Diversity

Organismal design involves trade-offs. One of the keys to understanding diversity in functional systems is to identify the trade-offs that are associated with the construction of individual systems. Mechanical and physiological mechanisms involve inherent trade-offs where modifications that improve performance in one aspect come at the cost of another property (Ghalambor et al. 2004, Toro et al. 2004, Vogel 1994, Wainwright & Richard 1995). An example of this can be seen in the mechanical advantage of a simple lever system. As the ratio of input lever to output lever increases, the force transmission of the system increases, but the amount of movement that is transmitted through the system decreases. Trade-offs like this one are constraints on evolution. One significant category of innovation is the type of qualitative design change that decouples two performance traits that are primitively linked to a trade-off.

INNOVATION AND DIVERSITY

Two categories of innovations can be recognized in terms of how they impact the evolutionary dynamics and diversity of functional systems: those that directly influence the potential for phenotypic variation and those that allow the lineage to move into new regions of the adaptive landscape where new variants are favored. Innovations in the first category change the potential morphospace of the body plan possessed by the lineage. Innovations in the second category represent breakthroughs in organismal performance that allow the lineage to move into a novel region of the adaptive landscape where a variety of new adaptive peaks can be reached. These two classes of innovations mirror the basic distinction in morphospace biology between the morphospace of the theoretically possible (Hickman 1993, Raup 1966) and the adaptive landscape that is produced by mapping fitness into the morphospace (Arnold 2003, McGhee 1999, Wright 1932).

I want to emphasize the distinction between species richness and other forms of diversity, such as morphological diversity, functional diversity, or ecological diversity. Here I am concerned with these latter forms of diversity. In addition to morphospace expansion, I illustrate that the nature of how morphology maps onto function can influence patterns of diversification in evolving lineages. Advancements in organismal design that affect this mapping have emerged as a new class of innovations that can impact phenotypic diversity. The science of doing comparative analyses of morphological and ecological diversity has lagged behind similar studies of lineage diversification rate, or species richness, and I review some recent methodological and conceptual progress on this front. Morphological and functional diversity are frequently measured as variance among the members of the group under consideration (Foote 1997) and, unless otherwise stated, this is what I mean when referring to diversity.

Expanding the Theoretical Morphospace

The potential morphospace occupied by a body plan is determined by the number of independent parameters that are required to define the morphospace (Niklas 2004, Raup 1966, Sanchez 2004, Vermeij 1973). In his classic work on the mollusk shell, Raup identified three parameters that generated a morphospace of all mollusk shells (Raup 1966). A novelty that increases the number of parameters required to describe the form increases the size of the potential morphospace and provides an opportunity for greater diversity. One conceptually simple novelty that results in this sort of increase in morphospace is a structural duplication or subdivision event that results in increases in the number of elements that make up the form, and thus increases the dimensionality of potential morphospace (Friel & Wainwright 1997, Schaefer & Lauder 1986).

A common anatomical form of duplication among metazoans is body segmentation, a phenomenon that typically involves replicated body regions each with the same basic plan. Segmentation illustrates the general ways in which duplication enhances diversity. Repeated body segments allow retention of a role in a primitive function in one or more segments, while other segments can become modified for novel functions. Arthropods offer a classic example of this pattern, in which the body is segmented into units that each possess axial structures and limb elements. Anterior body segments are modified for performance in sensory systems and jaws, and in more posterior segments for locomotor specialization. In decapod shrimps, some body segments are modified for walking, whereas other segments are specialized for burst locomotion, which is used during escape from predators. In arthropods, the duplication of body segments with the same suite of anatomical elements permits an expansion of the potential morphospace because each element in each segment becomes a new axis in morphospace. Some researchers have thought of the diversity of form among repeated elements as a form of complexity (McShea 1996). In practice, only a few case studies have explored the consequences of duplication or subdivision events for morphospace expansion and phenotypic diversity (Friel & Wainwright 1997; Schaefer & Lauder 1986, 1996; Wainwright & Turingan 1993). The idea here would be to test whether duplication events result in higher phenotypic diversity among lineages within that clade, as compared to a comparison clade—ideally the sister group—that lacks the duplication.

Redundancy creates the potential for different body segments to perform different functions, such that the specialization of a segment for, say, burst locomotion need not be constrained by the necessity for the same segment to maintain performance in a second function, such as walking behavior. This consequence of duplication events, known as functional decoupling, can enhance diversity because it more readily leads to body designs that exhibit higher overall performance capacities (Liem 1973, Wainwright & Turingan 1993). Thus, for example, when walking and burst swimming are performed by different body regions in shrimp, adaptation for higher performance is not constrained by the need for a single region to maintain both functions. Functional decoupling through redundancy in design appears to be a widespread and powerful way in which novelties become innovations that lead to increased diversity. This mechanism is well known in molecular evolution (Burmester et al. 2006, Chung et al. 2006, Lynch 2003, Ohno 1970, Spady et al. 2005), developmental biology (Carroll 2001, Hughes & Friedman 2005) and organismal functional design (Emerson 1988, Friel & Wainwright 1997, Lauder 1990, Schaefer & Lauder 1996).

The idea that a single anatomical system may be required to perform multiple functions has been viewed as a major constraint on evolutionary diversification, and conversely, the decoupling of such a constraint is seen as a major avenue to increasing diversity. In fact, it is a basic feature of organisms that body regions are involved in multiple functions, and the need to maintain functionality across the range of functions is likely to be an important factor in shaping adaptive evolution. Although structural duplication is one way to decouple functions, there are other routes. Gatesy & Middleton (1997) argued that the introduction of winged locomotion in birds released a constraint on the hindlimbs of theropods. Their proposal was that the theropod hindlimb had been morphologically constrained because this body region was the sole system used for locomotion. With the origin of forelimb-powered locomotion (flight), they argued that the bird hindlimb was free to become modified for a greater diversity of locomotor specializations. They showed evidence for increased diversity of the hindlimb in birds as compared to theropods (Gatesy & Middleton 1997, Middleton & Gatesy 2000). Similar arguments have been made by other researchers. The origin of a specialized condition of the pharyngeal jaw apparatus in cichlid fishes, by providing a second set of jaws that could specialize on prey processing functions, was hypothesized to result in increased diversity of the oral jaws, which are used in prey capture (Liem 1973, Liem & Osse 1975), and there has been some support for this idea (Hulsey et al. 2006).

Nature of the Form-Function Map

In recent years there has been mounting evidence of intrinsic design features that can relieve functional constraints on anatomical systems in the absence of duplication events. One general mechanism involves many-to-one mapping of morphology to function (Alfaro et al. 2004, 2005; Hulsey & Wainwright 2002; Wainwright et al.

2005). One inherent feature of a complex functional system is that multiple morphologies can have the same functional property. This occurs when the functional property depends on three or more underlying parameters and is contrasted with simple systems that are determined by one or two parameters. Many functions determined by only two parameters show many-to-one mapping (e.g., y = a + b). Consider the example of the mechanical advantage of a muscle acting across a joint. Mechanical advantage is the ratio of two distances: an input lever length and an output lever length. Once scale is removed, there is only a single combination of input lever length and output lever length that results in a particular value of mechanical advantage. This system exhibits one-to-one mapping of form to mechanical property. Now, consider instead the force exerted at the end of the outlever. This is a function of the mechanical advantage of the lever and a third parameter, the input force exerted by the muscle acting on the lever. Output force is equal to input force times the ratio of input lever to output lever. There are numerous combinations of the three parameters that all give the same value of output force. For example, mechanical advantage of 0.5 (input lever length/output lever length = 0.5/1) and input force of 8 give the same output force as a mechanical advantage of 1.0 (input lever/output lever = 1/1) and an input force of 4. In this case, there is a many-to-one mapping of musculo-skeletal design-to-force output of the system.

Many-to-one mapping of phenotype to functional property has been shown in many systems (Blob et al. 2006, Collar & Wainwright 2006, Guderley et al. 2006, Hulsey & Wainwright 2002, Lappin & Husak 2005, Marks & Lechowicz 2006, Stayton 2006, Toro et al. 2004). One well-explored example is suction index (SI) (**Figure 1**), the capacity of an individual fish to generate suction pressure during suction feeding. Suction index (SI) is based on the strength of a muscle and the morphology of the linkage system that transmits this force to the expanding mouth cavity during suction feeding. Because most biologically relevant functional properties have a basis in three or more elements, it can be expected that many-to-one mapping is a general feature of organismal design. Therefore, any macroevolutionary consequences of many-to-one mapping have the potential to be quite widespread.

The implications of many-to-one mapping of form to function for diversity are at least twofold. First, because many forms can have the same functional property, there are surfaces in morphospace that define mechanically neutral morphological variation (**Figure 2**). These regions of neutral change in morphology (or, more generally, in the parameters that determine the functional property) offer opportunities for lineages to explore regions in morphospace without altering the functional property of the mechanism. The simplest consequence of this is that there may be phenotypic variation among individuals in a population, or among species, that does not translate into diversity in the functional property. It is possible to see conservation of function in the presence of substantial morphological changes.

A second consequence involves the observation that most body parts serve multiple functions. Anatomical regions of the body almost always have multiple functional properties that are important to the organism and are potentially exposed to natural selection. Many-to-one mapping and the resulting regions of neutral morphological variation with respect to one functional property allow flexibility in design so that

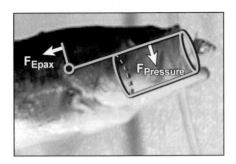

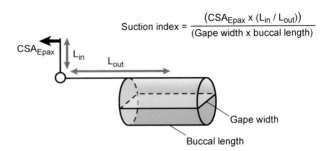

$$\text{Suction index} = \frac{(CSA_{Epax} \times (L_{in} / L_{out}))}{(\text{Gape width} \times \text{buccal length})}$$

Figure 1

Model of suction feeding capacity in fishes that allows calculation of the capacity of an individual fish to generate suction as a function of several morphological features. The model treats the feeding mechanism as a lever system that transmits force and movement from the contracting epaxial muscles to the expanding buccal cavity (Carroll et al. 2004). It is assumed that the ability to generate a buccal pressure gradient is limited by the forces that the expansion muscles can generate and the ability of the skeletal elements to resist these forces. The expanding buccal cavity of centrarchids can be modeled as an expanding cylinder with pressure being distributed across its surface. The magnitude of the expansion force is equal to the magnitude of the buccal pressure multiplied by the projected area of the buccal cavity. This force exerts a torque on the neurocranium, directed ventrally at the buccal cavity. The force generated by the epaxial muscles (and matched by the ventral, sternohyoideus muscle) must be greater or equal to the resolved force of buccal pressure as it is transmitted through the lever system of the neurocranium. The magnitude of the pressure gradient that a fish can generate is therefore a function of the amount of force that the epaxial muscles can generate (proportional to physiological cross sectional area, PCSA), the moment arm of the epaxialis (L_{in}), the moment arm of the buccal cavity (L_{out}), and the projected area of the buccal cavity (buccal length $\times$ buccal width). Force generation of the epaxial muscles is based on force per unit of the cross-sectional area and PCSA and, by omitting the former from the equation, allows one to generate a suction index (SI) that involves the morphological parameters of the relationship, but does not make assumptions about how force per unit area of the muscle may vary among feeding events and across taxa. This model was tested by making measurements of peak suction pressure in 45 individual centrarchid fishes, ranging across five species, each ranging about 2.5-fold in body length. Morphological measurements were made from each specimen to parameterize the model, and the predictions were compared against realized performance. Suction index (SI) shows many-to-one mapping because many different combinations of the morphological parameters will have the same value of SI. That is, in terms of SI, many different-shaped fish skulls are functionally similar.

the structure can become modified for other functions while maintaining the original function.

Many-to-one mapping leads to several macroevolutionary expectations that have been explored in simulations and empirically in a few natural systems. First, many-to-one mapping can partially decouple the accumulation of morphological and mechanical diversity within clades. This pattern was found in simulations of the evolution of a four-parameter lever mechanism found in fish jaws, the 4-bar linkage, even when some constraints were placed on the evolution of the four skeletal elements from observed variance-covariance relationships (Alfaro et al. 2004, 2005). A weak positive relationship ($r = 0.5$) was found between the disparity of morphological traits

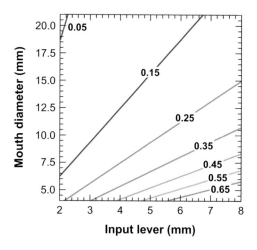

Figure 2

Plot of suction index (SI) in a two-dimensional region of morphospace defined by the input lever of the epaxial muscle and the diameter of the mouth. All other parameters in the SI model are held constant for this figure. The lines on the plot represent isoclines of constant value of SI, and indicate that many combinations of in-lever and mouth diameter result in the same value of SI. This many-to-one mapping is a general feature of complex systems: Any functional property determined by three or more parameters will show many-to-one mapping of morphology to the functional property.

(variance across tips) and the mechanical property of the 4-bar linkage after 1000 simulations on clades with 500 terminals (Alfaro et al. 2004). An empirical study of SI in centrarchid fishes found no relationship between variance among species in SI and variance in the parameters that make up SI across the three major clades in this group (Collar & Wainwright 2006). These studies consistently reveal the sobering pattern that morphological diversity may not predict functional diversity, even when the morphological parameters are being used to calculate the functional property. Indeed, if a weak or no correlation is the generally expected relationship between the diversity of functional properties and the diversity of the parameters used to calculate those properties, this is a major blow to research programs that try to interpret morphological diversity as being indicative of functional or ecological diversity based on the general assumption that different morphologies have different functional properties and result in different ecological patterns (Briggs et al. 1992, Valentine 1980). Studies of the evolution of functional diversity should measure functional properties and not rely upon the assumption that morphology can be used as a proxy for functional diversity.

A second link to macroevolution concerns expectations for convergent evolution. Using simulations of natural selection acting on a mechanical property of a complex lever system in the jaws of some fishes (the 4-bar linkage), Alfaro et al. (2004) showed that the morphology arrived at in response to selection depends on the starting morphology in the simulation, even when all cases result in the population reaching the same value of the mechanical property. Mechanical convergence only created morphological convergence when the starting forms were similar. Many-to-one mapping tends to amplify the contingency of evolution (Foote 1998) and results in different lineages taking different routes through morphospace, even if they are exposed to identical histories of selection.

One important property of organismal design is that body parts are almost always involved in determining multiple functional properties. This means that though selection is working to optimize many performance traits at once, the underlying basis for those traits often involves shared parts, and in some cases involves 100% overlap in parts. Although this connectedness is often seen as a major constraint on the evolution of design, one of the features of many-to-one mapping of form to function is that the surfaces of neutral variation with respect to one functional trait permit changes that alter other functional traits. In other words, the complexity of most organismal functional properties increases the capacity for optimizing multiple functions. Because of these neutral surfaces of change, there are pathways of change that need not necessarily involve trade-offs in function. Specific examples of this phenomenon are developed in recent papers that involve a combination of simulation and empirical data (Alfaro et al. 2004, 2005). The evolutionary flexibility that is provided by many-to-one mapping of form to function appears to be a major factor in facilitating phenotypic diversification.

Breakthroughs that Change the Adaptive Landscape

Although changes in body plan can alter the range of potential morphologies, it is also clear that some novelties have sufficiently radical consequences for the performance capacity of the organism such that they effectively open up whole new possible ways of life and can lead to subsequent diversification. Sometimes referred to as major innovations or key innovations, such an innovation is a breakthrough in design that moves the lineage into a new region of the adaptive landscape (Erwin 1992, Hunter 1998, Liem 1973, Vermeij 1995). There are many compelling examples (Bateman & DiMichelle 1994, Jernvall 1995, Jernvall et al. 1996, Norris 1996). With the origin of powered flight in birds, new feeding strategies, life history patterns, and opportunities for novel habitat use became possible, and the resulting radiation appears to be anchored on this design breakthrough. Similarly, the origin of jaws in vertebrates was followed by a successful radiation of highly predatory organisms. Note that in these examples, the innovations involve a significant enhancement in some aspect of the behavioral performance capacity of the organisms that opened up the exploitation of novel resources, such as new food types, new habitats, or new life-history patterns.

In recent years researchers have tended to focus on the implications of putative key innovations on lineage diversification rate (Bond & Opell 1998, Gianoli 2004, Hodges 1996, Sanderson & Donoghue 1996), and there has not been as much work done testing the impact of innovations on morphological, functional, or ecological diversity (Gatesy & Middleton 1997). This is in part owing to the difficulty and work involved in assembling data for a large number of taxa, and it can be anticipated that such studies will become more frequent in the future.

Both increases in the potential morphospace and breakthroughs in design only open up the potential for subsequent radiations; they do not make such radiations inevitable. There is an important role for stochastic processes and the appropriate ecological conditions of disruptive selection to realize the potential change in diversity. In this sense the innovations only set the stage for changes in diversity; they do not, by themselves, cause the change (Labandeira 1997, Wagner 2000). Changes in morphological diversification also need not be tied to changes in speciation rate, extinction rate, or net diversification rate (Ricklefs 2004, 2006).

COMPARATIVE ANALYSIS OF DIVERSITY

The discussions above suggest a research program in the history and biology of innovations. There are many questions that one might like to ask about a putative innovation and its consequences for diversity. Was the innovation followed by an increase in the diversity of the functional systems affected by the breakthrough or of other functional systems? Does the innovation result in a qualitative shift in some aspect of performance capacity or resource use? Given that the innovation is associated with an increase in diversity, what is the tempo of that change? Is there a period of relative stasis before the diversity is accelerated, or does the change happen coincident with the innovation? These questions are inherently historical, and all of them can be addressed with the use of a time-calibrated phylogeny of the group and its relatives as a basis for comparisons. Methods and concepts for conducting phylogenetically correct comparisons of morphological, functional, and ecological diversity (Foote 1996, Garland 1992, O'Meara et al. 2006) between lineages have lagged behind the development of methods for doing comparative studies of lineage diversification rate (Magallon & Sanderson 2001, Sanderson & Donoghue 1994, Slowinski & Guyer 1994). Important insights about the role of phylogenetic history in species diversity have resulted in the emergence in recent years of a focus on lineage diversification rate. But, just as diversification rate should be recognized as the phylogenetically corrected metric of species richness in a clade, similarly, rate of morphological evolution provides a phylogenetically corrected metric of trait diversity (Collar et al. 2005, Garland 1992, Martins 1994, O'Meara et al. 2006). Below I review an approach for comparing rates of morphological evolution (that is, morphological diversity) between two lineages.

Functional and Morphological Diversity

There are several widely used metrics of morphological diversity, but variance and range are the most widely used (reviewed by Ciampaglio et al. 2001, Erwin 2007, Foote 1997). Range is of interest because it reflects information about the farthest regions of morphospace that have been reached by members of the group in question (Pie & Weitz 2005). Range may be useful in addressing questions about which regions of morphospace have been occupied by a group and which have not (Stebbins 1951, Van Valkenburgh 1988). The multivariate measure of range is usually some version of an N-dimensional minimum polygon that encloses all individuals in the group. Although range is of particular interest in some case studies, the statistical properties of a range make doing careful quantitative comparisons between groups awkward. Also, in a Gausian distribution, range scales with sample size, which further complicates comparisons.

Variance of traits is the most widely used metric of morphological diversity (Foote 1997, McClain 2005, Roy & Foote 1997). Variance captures the dispersion of members of the group in morphospace, is not so susceptible to the effects of a few outliers, and does not scale with sample size so the metric is versatile and amenable to statistical tests. Variance also relates directly to the most commonly used model of character evolution, Brownian motion, and its derivatives so that the connection between this model of evolution and variance among evolving lineages is strong (Martins & Hansen 1997, O'Meara et al. 2006, Pagel 1999, Purvis 2004).

It is intuitive that morphological diversity among species is affected by their phylogenetic history. After all, species usually most resemble their closest relatives. But exactly how do we expect phylogeny to relate to morphological diversity, and how can we use this knowledge in framing comparative tests of morphological diversity? To get at these issues we first need a model of trait evolution. Perhaps the most straightforward model that is used is Brownian motion (Martins & Hansen 1997), the model used in calculating independent contrasts (Felsenstein 1985) and the estimation of ancestral states (Schluter et al. 1997). Under Brownian motion the potential for trait change occurs at some designated time interval, with the magnitude of the change being drawn from a normal distribution with a mean of zero and some variance. The variance of this distribution of potential trait change is referred to as the Brownian rate parameter, and the expected variance of the trait among lineages in a phylogeny is equal to the number of opportunities for trait change (proportional to time in the Brownian model) times this variance in the distribution of potential trait changes, or the Brownian rate parameter. The larger the rate parameter, the greater the expected variance among similar-aged lineages descending from a common ancestor. Thus, diversity among the lineages within a clade develops as a function of the time in the phylogeny, the amount of shared history between lineages, and the rate of evolution of the trait (Ackerly & Nyffeler 2004, Garland 1992, Gittleman et al. 1996, Mooers et al. 1999, O'Meara et al. 2006, Thomas et al. 2006). A key insight that emerges from this is that a phylogenetically corrected comparison of morphological diversity between two clades involves removing the confounding effects of time and shared history, and comparing the rate of evolution of the traits of interest (Garland 1992, O'Meara et al. 2006). To compare diversity in two clades after removing the effects of time and phylogeny, one can compare estimates of the Brownian rate parameter. The Brownian rate parameter can be estimated with average values for the phenotypic traits in each tip taxon, and a phylogeny can be estimated with branch lengths proportional to time.

As a simple illustration of this effect of time, consider two monophyletic groups of birds, one that shows considerable variation among species in bill morphology and the other that shows minimal differences among species (**Figure 3**). In each group, bill morphology has been diversifying since the time of the most recent common ancestor (MRCA). If the age of the MRCA of the diverse group is considerably older than the age of the MRCA in the low-diversity group (**Figure 3a**), then time may be a

a

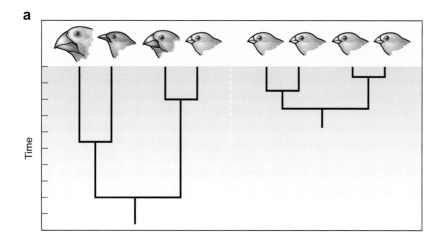

b

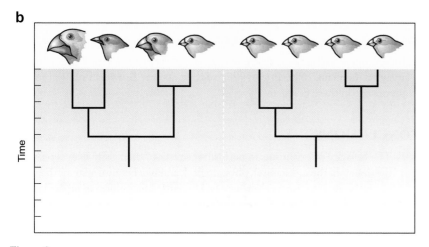

Figure 3

Diagrams illustrating the importance of time in the interpretation of differences in morphological diversity between two monophyletic groups of birds. (*a*) The high-diversity group is much older than the low-diversity group. In this case the difference in diversity between groups may be due to differences in the amount of time the two groups have had to diversify. (*b*) Here, the age of the two groups and the total time in the phylogenies are the same, suggesting that the rate of bill evolution in the high-diversity group would have been higher than in the low-diversity group. The approach described in this chapter is designed to separate the effects of time and rate of evolution on the observed diversity in a group of terminal taxa.

trivial explanation for the difference in diversity between groups. But, if the MRCAs are of similar age (**Figure 3*b***), or the age of the low-diversity group is actually older, then we would infer that the rate of evolution of bill morphology has been higher in the diverse group.

This framework is formalized in recently developed software that accepts as input a phylogeny with branch lengths in time, or relative time, and trait values for the tips of the phylogeny (Collar et al. 2005, O'Meara et al. 2006). The program, "Brownie," then estimates the Brownian rate parameter and allows one to compare it between two clades, or between a clade and its paraphyletic outgroup. This program is well designed for testing hypotheses of the effects of specific innovations, or synapomorphies, on morphological diversity. In these tests, it is actually the Brownian rate parameter, or the rate of trait evolution, that is compared between groups, thus removing the confounding affects of time and shared history. Preliminary tests should be run to verify that the exisiting trait distribution fits expectations of Brownian motion and that the trait value is correlated with phylogeny (Blomberg et al. 2003, Pagel 1997).

An important discussion concerns whether Brownian motion is the most appropriate model of character evolution (Butler & King 2004, Freckleton & Harvey 2006, Martins & Hansen 1997), and whether it is the best for comparative analyses of diversity (Diaz-Uriarte & Garland 1996, O'Meara et al. 2006). Brownian motion is well suited to this particular hypothesis-testing framework because it makes the fewest assumptions about constraints on trait evolution. However, it may not always capture the important ecological forces and processes that occur during radiations. Other models strive to capture the dynamics under expectations of multiple adaptive peaks superimposed upon Brownian motion (Butler & King 2004, Hansen 1997, Pagel 1997). But niche-filling models may yield quite different patterns of diversification (Harvey & Rambaut 2000, Price 1997), suggesting that Brownian motion may be inappropriate in these cases.

CONCLUSIONS

1. The capacity of organisms to perform the tasks of their daily lives is rooted in the design of the mechanical, physiological, and biochemical systems that make up the body. Performance capacity is inherently hierarchical in organization.

2. The theoretical morphospace can be expanded by increasing the number of parameters required to define the morphospace. This means that innovations that increase the number of parameters required to define the form can have the effect of increasing the potential diversity of lineages. Duplication and decoupling events are major mechanisms of morphospace expansion and can lead to increased diversity. They are documented to work at virtually all levels of organismal design.

3. Many-to-one mapping, in which many different morphological combinations have the same functional property, is an inherent property of all functions that are determined by three or more parameters. This ubiquitous feature of organismal design can encourage diversity by allowing morphological variation that is neutral with respect to the functional property. This flexibility in form allows parts of the body to accommodate multiple functions by adapting to one function through changes in form, while maintaining a second function by moving through regions of neutral morphological change.

4. Many-to-one mapping can lead to a disconnect between morphological diversity and functional diversity, a pattern that has been confirmed in several natural systems. This decoupling of diversity at adjacent levels of design poses a major

obstacle to attempts to infer ecological diversity from data sets on morphology, a program that has an especially rich history in paleontology.

5. In order to conduct phylogenetically correct comparisons of morphological or functional diversity one must separate the confounding affects of time, shared phylogenetic history, and rate of trait evolution, which combine to yield standing clade diversity. Methods for comparing rate of character evolution have recently been developed that require as input phylogenies with branch lengths proportional to time and trait values for taxa in the tree. These methods can be used to compare diversity in two clades, test for the effects of putative innovations on diversity, or study the tempo of evolution.

DISCLOSURE STATEMENT

The author is not aware of any biases that might be perceived as affecting the objectivity of this review.

ACKNOWLEDGMENTS

I thank Michael Alfaro, Dan Bolnick, Rose Carlson, David Collar, Ted Garland, Darrin Hulsey, Brian O'Meara, and Michael Sanderson for many valuable discussions on innovation and diversity. Thanks to M. Alfaro for preparing **Figure 2**.

LITERATURE CITED

Ackerly DD, Nyffeler R. 2004. Evolutionary diversification of continuous traits: phylogenetic tests and application to seed size in the California flora. *Evol. Ecol. Res.* 28:249–72

Alfaro ME, Bolnick DI, Wainwright PC. 2004. Evolutionary dynamics of complex biomechanical systems: an example using the four-bar mechanism. *Evolution* 58:495–503

Alfaro ME, Bolnick DI, Wainwright PC. 2005. Evolutionary consequences of many-to-one mapping of jaw morphology to mechanics in labrid fishes. *Am. Nat.* 165:E140–54

Arnold SJ. 2003. Performance surfaces and adaptive landscapes. *Integr. Comp. Biol.* 43:367–75

Bateman RM, DiMichelle WA. 1994. Heterospory: the most iterative key innovation in the evolutionary history of the plant kingdom. *Biol. Rev.* 69:345–417

Blob RW, Rai R, Julius ML, Schoenfuss HL. 2006. Functional diversity in extreme environments: effects of locomotor style and substrate texture on the waterfall-climbing performance of Hawaiian gobiid fishes. *J. Zool.* 268:315–24

Blomberg SP, Garland T, Ives AR. 2003. Testing for phylogenetic signal in comparative data: Behavioral traits are more labile. *Evolution* 57:717–45

Bond JE, Opell BD. 1998. Testing adaptive radiation and key innovation hypotheses in spiders. *Evolution* 52:403–14

Briggs DEG, Fortey RA, Wills MA. 1992. Morphological disparity in the Cambrian. *Science* 256:1670–73

Burmester T, Storf J, Hasenjäger A, Klawitter S, Hankeln T. 2006. The hemoglobin genes of *Drosophila*. *FEBS J.* 273:468–80

Butler MA, King AA. 2004. Phylogenetic comparative analysis: a modeling approach for adaptive evolution. *Am. Nat.* 164:683–95

Carroll AM, Wainwright PC, Huskey SH, Collar DC, Turingan RG. 2004. Morphology predicts suction feeding performance in centrarchid fishes. *J. Exp. Biol.* 207:3873–81

Carroll SB. 2001. Chance and necessity: The evolution of morphological complexity and diversity. *Nature* 409:1102–9

Chung WY, Albert R, Albert I, Nekrutenko A, Makova KD. 2006. Rapid and asymmetric divergence of duplicate genes in the human gene coexpression network. *BMC Bionformatics* 7:46

Ciampaglio CN, Kemp M, McShea DW. 2001. Detecting changes in morphospace occupation patterns in the fossil record: Characterization and analysis of measures of disparity. *Paleobiology* 27:695–715

Collar DC, Near TJ, Wainwright PC. 2005. Comparative analysis of morphological diversity: trophic evolution in centrarchid fishes. *Evolution* 59:1783–94

Collar DC, Wainwright PC. 2006. Incongruent morphological and mechanical diversity in the feeding mechanisms of centrarchid fishes. *Evolution* 60:2575–84

Diaz-Uriarte R, Garland T. 1996. Testing hypotheses of correlated evolution using phylogenetically independent contrasts: Sensitivity to deviations from Brownian motion. *Syst. Biol.* 45:27–47

Emerson S. 1988. Testing for historical patterns of change: a case study with frog pectoral girdles. *Paleobiology* 14:174–86

Erwin DH. 1992. A preliminary classification of evolutionary radiations. *Hist. Biol.* 6:133–47

Erwin DH. 2007. Disparity: Morphological pattern and developmental context. *Palaeontology* 50:57–73

Felsenstein J. 1985. Phylogenies and the comparative method. *Am. Nat.* 125:1–15

Foote M. 1996. Models of morphological diversification. In *Evolutionary Paleobiology*, ed. D Jablonski, DH Erwin, JH Lipps, pp. 62–86. Chicago, IL: Univ. Chicago Press

Foote M. 1997. The evolution of morphological diversity. *Annu. Rev. Ecol. Syst.* 28:129–52

Foote M. 1998. Contingency and convergence. *Science* 280:2068–69

Freckleton RP, Harvey PH. 2006. Detecting non-Brownian trait evolution in adaptive radiations. *PLoS Biol.* 4:2104–11

Friel JP, Wainwright PC. 1997. A model system of structural duplication: homologies of the adductor mandibulae muscles of tetraodontiform fishes. *Syst. Biol.* 46:441–63

Garland TJ, Harvey PH, Ives AR. 1992. Procedures for the analysis of comparative data using phylogenetically idependent contrasts. *Syst. Biol.* 41:18–32

Gatesy SM, Middleton KM. 1997. Bipedalism, flight, and the evolution of theropod locomotor diversity. *J. Vertebr. Paelontol.* 17:308–29

Gavrilets S. 1999. Dynamics of clade diversification on the morphological hypercube. *Proc. R. Soc. London Ser. B* 266:817–24

Gavrilets S, Vose A. 2005. Dynamic patterns of adaptive radiation. *Proc. Natl. Acad. Sci. USA* 102:18040–45

Ghalambor CK, Reznick DN, Walker JA. 2004. Constraints on adaptive evolution: the functional trade-off between reproduction and fast start escape performance in the guppy (*Poecilia reticulata*). *Am. Nat.* 164:38–50

Gianoli E. 2004. Evolution of a climbing habit promotes diversification in flowering plants. *Proc. R. Soc. London Ser. B* 271:2011–15

Gittleman J, Anderson C, Kot M, Luh H-K. 1996. Comparative tests of evolutionary lability and rates using molecular phylogenies. In *New Uses for New Phylogenies*, ed. PH Harvey, AJ Leigh Brown, J Maynard Smith, S Nee, pp. 289–307. Oxford: Oxford Univ. Press

Guderley H, Houle-Leroy P, Diffee GM, Camp DM, Garland T. 2006. Morphometry, ultrastructure, myosin isoforms, and metabolic capacities of the "mini muscles" favoured by selection for high activity in house mice. *Comp. Biochem. Physiol. B* 144:271–82

Hansen TF. 1997. Stabilizing selection and the comparative analysis of adaptation. *Evolution* 51:1341–51

Harmon LJ, Schulte JA, Larson A, Losos JB. 2003. Tempo and mode of evolutionary radiation in iguanian lizards. *Science* 301:961–64

Harvey PH, Rambaut A. 2000. Comparative analyses for adaptive radiations. *Philos. Trans. R. Soc. London Ser. B* 355:1599–606

Helmuth B, Kingsolver JG, Carrington E. 2005. Biophysics, physiologicalecology, and climate change: Does mechanism matter? *Annu. Rev. Physiol.* 67:177–201

Hickman CS. 1993. Theoretical design space: a new program for the analysis of structural diversity. In *Progress in Contructional Morphology. Neues Jahrb. Geol. Paleontol. Abh.*, ed. A Seilacher, K Chinzei, 190:169–82

Hodges SA. 1996. Key evolutionary innovations: Floral nectar spurs and diversification in Aquilegia. *Am. J. Bot.* 83:31

Hughes AL, Friedman R. 2005. Gene duplication and the properties of biological networks. *J. Mol. Evol.* 61:758–64

Hulsey CD, De Leon FJG, Rodiles-Hernandez R. 2006. Micro- and macroevolutionary decoupling of cichlid jaws: A test of Liem's key innovation hypothesis. *Evolution* 60:2096–109

Hulsey CD, Wainright PC. 2002. Projecting mechanics into morphospace: Disparity in the feeding system of labrid fishes. *Proc. R. Soc. London Ser. B* 269:317–26

Hunter JP. 1998. Key innovations and the ecology of macroevolution. *Trends Ecol. Evol.* 13:31–36

Husak JF. 2006. Does survival depend on how fast you can run or how fast you do run? *Funct. Ecol.* 20:1080–86

Husak JF, Fox SF. 2006. Field use of maximal sprint speed by collared lizards (Crotaphytus collaris): Compensation and sexual selection. *Evolution* 60:1888–95

Irschick DJ, Herrel AV, Vanhooydonck B, Huyghe K, Van Damme R. 2005. Locomotor compensation creates a mismatch between laboratory and field estimates of escape speed in lizards: A cautionary tale for performance-to-fitness studies. *Evolution* 59:1579–87

Jernvall J. 1995. Mammalian molar cusp patterns: Developmental mechanisms of diversity. *Acta Zool.* 198:1–61

Jernvall J, Hunter JP, Fortelius M. 1996. Molar tooth diversity, disparity, and ecology in Cenozoic ungulate radiations. *Science* 274:1489–92

Koehl MAR. 1996. When does morphology matter? *Annu. Rev. Ecol. Syst.* 27:501–42

Labandeira CC. 1997. Insect mouthparts: ascertaining the paleobiology of insect feeding strategies. *Annu. Rev. Ecol. Syst.* 28:153–93

Lappin AK, Husak JF. 2005. Weapon performance, not size, determines mating success and potential reproductive output in the collared lizard (*Crotaphytus collaris*). *Am. Nat.* 166:426–36

Lauder GV. 1990. Functional morphology and systematics: studying functional patterns in an historical context. *Annu. Rev. Ecol. Syst.* 21:317–40

Liem KF. 1973. Evolutionary strategies and morphological innovations: cichlid pharyngeal jaws. *Syst. Zool.* 22:425–41

Liem KF, Osse JWM. 1975. Biological versatility, evolution, and food resource exploitation in African cichlid fishes. *Am. Zool.* 15:427–54

Losos JB, Jackman TR, Larson A, de Queiroz K, Rodriguez-Schettino L. 1998. Contingency and determinism in replicated adaptive radiations of island lizards. *Science* 279:2115–18

Losos JB, Miles DB. 2002. Testing the hypothesis that a clade has adaptively radiated: Iguanid lizard clades as a case study. *Am. Nat.* 160:147–57

Lovette IJ, Bermingham E, Ricklefs RE. 2002. Clade-specific morphological diversification and adaptive radiation in Hawaiian songbirds. *Proc. R. Soc. London Ser. B* 269:37–42

Lynch M. 2003. The origins of genome complexity. *Science* 302:1401

Magallon S, Sanderson MJ. 2001. Absolute diversification rates in angiosperm clades. *Evolution* 55:1762–80

Marks CO, Lechowicz MJ. 2006. Alternative designs and the evolution of functional diversity. *Am. Nat.* 167:55–66

Martins EP. 1994. Estimating the rate of phenotypic evolution from comparative data. *Am. Nat.* 144:193–209

Martins EP, Hansen TF. 1997. Phylogenies and the comparative method: a general approach to incorporating phylogenetic information into the analysis of interspecific data. *Am. Nat.* 149:646–67

McClain CR. 2005. Bathymetric patterns of morphological disparity in deep-sea gastropods from the western North Atlantic Basin. *Evolution* 59:1492–99

McGhee GR. 1999. *Theoretical Morphology: The Concept and its Application*. New York: Columbia Univ. Press

McGhee GR, McKinney FK. 2000. A theoretical morphological analysis of convergently evolved erect helical form in the Bryozoa. *Paleobiology* 26:556–77

McShea DW. 1996. Metazoan complexity and evolution: Is there a trend? *Evolution* 50:477–92

Middleton KM, Gatesy SM. 2000. Theropod forelimb design and evolution. *Zool. J. Linn. Soc.* 128:149–87

Mooers AØ, Vamosi SM, Schluter D. 1999. Using phylogenies to test macroevolutionary hypotheses of trait evolution in Cranes (Gruinae). *Am. Nat.* 154:249–59

Niklas KJ. 1986. Computer simulations of branching patterns and their implications for the evolution of plants. *Lect. Math. Life Sci.* 18:1–50

Niklas KJ. 2004. Computer models of early land plant evolution. *Annu. Rev. Earth Planet. Sci.* 32:45–65

Norris RD. 1996. Symbiosis as an evolutionary innovation in the radiation of Paleocene planktonic foraminifera. *Paleobiology* 22:461–80

Ohno S. 1970. *Evolution by Gene Duplication.* New York: Springer-Verlag

O'Meara BC, Ané CM, Sanderson MJ, Wainwright PC. 2006. Testing for different rates of continuous trait evolution in different groups using likelihood. *Evolution* 60:922–33

Pagel M. 1997. Inferring evolutionary processes from phylogenies. *Zool. Scr.* 26:331–48

Pagel M. 1999. Inferring the historical patterns of biological evolution. *Nature* 401:877–84

Pie MR, Weitz JS. 2005. A null model of morphospace occupation. *Am. Nat.* 166:E1–13

Powell P, Roy RR, Kanim P, Bello MA, Edgerton V. 1984. Predictability of skeletal muscle tension from architectural determinations in guinea pig hindlimbs. *J. Appl. Physiol.* 57:1715–21

Price T. 1997. Correlated evolution and independent contrasts. *Trans. R. Soc. London Ser. B* 352:519–29

Purvis A. 2004. Evolution: How do characters evolve? *Nature* 430(6997):338–41; doi:10.1038/nature03092

Raup DM. 1966. Geometric analysis of shell coiling: general problems. *J. Paleontol.* 40:1178–90

Ricklefs RE. 2004. Cladogenesis and morphological diversification in passerine birds. *Nature* 430:338–41

Ricklefs RE. 2006. Time, species, and the generation of trait variance in clades. *Syst. Biol.* 55:151–59

Roy K, Foote M. 1997. Morphological approaches to measuring biodiversity. *Trends Ecol. Evol.* 12:277–81

Sanchez JA. 2004. Evolution and dynamics of branching colonial form in marine modular cnidarians: gorgonian octocorals. *Hydrobiologia* 530:283–90

Sanderson MJ, Donoghue MJ. 1994. Shifts in diversification rate with the origin of angiosperms. *Science* 264:1590–93

Sanderson MJ, Donoghue MJ. 1996. Reconstructing shifts in diversification rates on phylogenetic trees. *Trends Ecol. Evol.* 11:15–20

Schaefer SA, Lauder GV. 1986. Historical transformation of functional design: Evolutionary morphology of the feeding mechanisms of loricariod catfishes. *Syst. Zool.* 35:489–508

Schaefer SA, Lauder GV. 1996. Testing historical hypotheses of morphological change: Biomechanical decoupling in loricarioid catfishes. *Evolution* 50:1661–75

Schluter D, Price T, Mooers AØ, Ludwig D. 1997. Likelihood of ancestor states in adaptive radiation. *Evolution* 51:1699–711

Slowinski JB, Guyer C. 1994. Testing whether certain traits have caused amplified diversification: an improved method based on a model of random speciation and extinction. *Am. Nat.* 142:1019–24

Spady TC, Seehausen O, Loew ER, Jordan RC, Kocher TD, Carleton KL. 2005. Adaptive molecular evolution in the opsin genes of rapidly speciating cichlid species. *Mol. Biol. Evol.* 22:1412–22

Stadler BMR, Stadler PF, Wagner GP, Fontana W. 2001. The topology of the possible: Formal spaces underlying patterns of evolutionary change. *J. Theor. Biol.* 213:241–74

Stayton CT. 2006. Testing hypotheses of convergence with multivariate data: Morphological and functional convergence among herbivorous lizards. *Evolution* 60:824–41

Stebbins GL. 1951. Natural selection and the differentiation of angiosperm families. *Evolution* 5:299–24

Thomas GH, Freckleton RP, Szekely T. 2006. Comparative analyses of the influence of developmental mode on phenotypic diversification rates in shorebirds. *Proc. R. Soc. London Ser. B* 273:1619–24

Toro E, Herrel AV, Irschick DL. 2004. The evolution of jumping performance in Caribbean Anolis lizards: Solutions to biomechanical trade-offs. *Am. Nat.* 163:844–56

Valentine JW. 1980. Determinants of diversity in higher taxonomic catagories. *Paleobiology* 6:444–50

Vanhooydonck B, Herrel A, Van Damme R, Irschick DJ. 2006. The quick and the fast: The evolution of acceleration capacity in Anolis lizards. *Evolution* 60:2137–47

Van Valkenburgh B. 1988. Trophic diveristy in past and present guilds of marge predatory mammals. *Paleobiology* 14:156–73

Vermeij GJ. 1973. Biological versatility and earth history. *Proc. Natl. Acad. Sci. USA* 70:1936–38

Vermeij GJ. 1995. Economics, volcanoes, and Phanerozoic revolutions. *Paleobiology* 21:125–52

Vogel S. 1994. *Life in Moving Fluids*. Princeton, NJ: Princeton Univ. Press

Wagner PJ. 2000. Exhaustion of morphologic character states among fossil taxa. *Evolution* 54:365–86

Wagner PJ, Ruta M, Coates MI. 2006. Evolutionary patterns in early tetrapods. II. Differing constraints on available character space among clades. *Proc. R. Soc. London Ser. B* 273:2107–11

Wainwright PC. 1988. Morphology and ecology: the functional basis of feeding constraints in Caribbean labrid fishes. *Ecology* 69:635–45

Wainwright PC, Alfaro ME, Bolnick DI, Hulsey CD. 2005. Many-to-one mapping of form to function: a general principle in organismal design? *Integr. Comp. Biol.* 45:256–62

Wainwright PC, Richard BA. 1995. Predicting patterns of prey use from morphology with fishes. *Environ. Biol. Fishes* 44:97–113

Wainwright PC, Turingan RG. 1993. Coupled vs uncoupled functional systems—motor plasticity in the Queen Triggerfish, *Balistes vetula*. *J. Exp. Biol.* 180:209–27

Warheit KI, Forman JD, Losos JB, Miles DB. 1999. Morphological diversification and adaptive radiation: A comparison of two diverse lizard clades. *Evolution* 53:1226–34

Wright S. 1932. The roles of mutation, inbreeding, crossbreeding and selection in evolution. *Proc. 6th Int. Congr. Genet.* 1:356–66

Evolutionary Game Theory and Adaptive Dynamics of Continuous Traits

Brian J. McGill[1] and Joel S. Brown[2]

[1]Department of Biology, McGill University, Stewart Biology Building, Montreal, QC H3A 1B1 Canada; email: mail@brianmcgill.org

[2]Department of Biological Sciences, University of Illinois Chicago, Chicago, Illinois 60607; email: squirrel@uic.edu

Annu. Rev. Ecol. Evol. Syst. 2007. 38:403–35

First published online as a Review in Advance on August 8, 2007

The *Annual Review of Ecology, Evolution, and Systematics* is online at
http://ecolsys.annualreviews.org

This article's doi:
10.1146/annurev.ecolsys.36.091704.175517

Keywords

branching point, evolutionarily stable strategy (ESS)

Abstract

Continuous-trait game theory fills the niche of enabling analytically solvable models of the evolution of biologically realistically complex traits. Game theory provides a mathematical language for understanding evolution by natural selection. Continuous-trait game theory starts with the notion of an evolutionarily stable strategy (ESS) and adds the concept of convergence stability (that the ESS is an evolutionary attractor). With these basic tools in hand, continuous-trait game theory can be easily extended to model evolution under conditions of disruptive selection and speciation, nonequilibrium population dynamics, stochastic environments, coevolution, and more. Many models applying these tools to evolutionary ecology and coevolution have been developed in the past two decades. Going forward we emphasize the communication of the conceptual simplicity and underlying unity of ideas inherent in continuous-trait game theory and the development of new applications to biological questions.

INTRODUCTION

Life is a game. Games have players, strategies, and payoffs. In the evolutionary game, individual organisms are the players, their heritable phenotypes are their strategies, and their per capita growth rates (fitness) are their payoffs. The game happens because the fitness of an individual is simultaneously influenced by its own strategy, the strategies of others, and other features of the abiotic and biotic environment. Evolutionary game theory provides a key mathematical language to understanding natural selection better. Here we review this approach and describe its current applicability within the context of adaptive dynamics, evolutionary stability, and continuous (quantitative) traits.

Evolutionary game theory has advanced greatly from Maynard Smith & Price's (1973) pioneering concept of the evolutionarily stable strategy (ESS). Game theory has become the lingua franca of most concepts in animal behavior (Reeve & Dugatkin 1998) relating to mate choice, breeding strategies, animal contests, social groups, and the evolution of cooperation via forms of reciprocal altruism. Such matrix games as "Hawk-Dove" and "Prisoner's Dilemma" are used to teach many aspects of animal behavior. The 1970s also saw evolutionary game theory applied to continuous traits such as body size and other characteristics that might influence population dynamics within species and between species (Lawlor & Maynard-Smith 1976). Just as in animal social behaviors, evolutionary game theory provides an obvious mathematical language for evolutionary ecology. Evolutionary ecology (broadly defined) explicitly requires the use of conceptual and modeling tools that can make predictions about evolutionary trajectories and outcomes of selection on heritable quantitative traits such as flowering times, age at first reproduction, optimal habitat choice, etc.

Despite advances in the past 20 years, continuous-trait evolutionary game theory remains outside the mainstream of evolutionists, evolutionary ecologists, and their textbooks and courses. Why the disjunction between rapid advances in evolutionary game theory and the slower integration of these advances into the mainstream? First, these new applications fall into what has traditionally been the purview of population and quantitative genetics and must therefore displace a resident approach. Second, the literature on continuous-trait evolutionary game theory can be a confusing Babel of terms, concepts, definitions, and notations. Third, it has been a while (Vincent & Brown 1988) since evolutionary game theory has been reviewed in a succinct and accessible manner. This review aims to address these issues.

HISTORY OF GAME THEORY

Most people (Luce & Raiffa 1957) consider von Neumann the father of game theory (von Neumann & Morgenstern 1944). Primarily interested in economic applications, von Neumann and other pioneers dealt primarily with two-player zero-sum games. In a zero-sum game, one player's gain matches another's loss. Although accurate for some games (for example, casino games), most evolutionary games are likely nonzero.

John Nash (1950) developed the Nash equilibrium in a two-page paper that brought him a Nobel prize. In a Nash solution all players possess a no-regret strategy.

No single player can increase personal payoffs by unilaterally changing strategies. The Nash equilibrium revolutionized game theory. It applied equally well to many-player games, asymmetric games, and non-zero-sum games.

Price, a graduate student of Maynard Smith, introduced two seminal ideas in evolution: evolutionary game theory and the Price equation (Frank 1997) before abandoning the field for other interests, studying Christianity, helping the homeless, and eventually dying homeless himself (Frank 1997). Maynard Smith took it upon himself to publish Price's idea (Maynard Smith & Price 1973) and followed with a seminal book (Maynard-Smith 1982). As Maynard Smith notes, the idea of an ESS had been presaged by the study of sex ratios as a frequency-dependent problem (Fisher 1930, Hamilton 1967). An ESS was defined as a strategy (or set of strategies) that, when common in the population, cannot be invaded by rare alternative strategies—a unilateral change in strategy by one mutant individual will not increase its payoff. Hence, the Nash equilibrium is embedded within the ESS concept (see Bulmer 1994 for a discussion of the similarities and differences), although Maynard Smith developed the idea independently.

Advances continued apace. Auslander et al. (1978) applied the Nash equilibrium to evolutionary games with continuous traits. Several researchers, recognizing the utility of continuous-trait game theory, merged Maynard Smith's evolutionary game theory with more conventional game theory (Brown & Vincent 1987a,b; Roughgarden 1976; Vincent & Brown 1984). In addition, models emerged viewing an ESS as the endpoint of a dynamical evolutionary process (Christiansen 1991; Eshel 1983; Maynard-Smith 1981, 1982; Vincent & Brown 1984, 1987; Vincent et al. 1993). Economists too have realized the power of this evolutionary approach to game theory and incorporated it into their work (Fudenberg & Harris 1992, Gintis 2000, Hofbauer 1996).

Within biology, continuous-trait evolutionary game theory moved in two directions. First, some evolutionary ecologists adapted the tools to solve specific problems of biological interest (**Figure 1**). We examine some of this work at the end of this review. Second, a group of scientists, primarily in continental Europe, advanced the dynamic aspect of evolutionary game theory, renaming the subject adaptive dynamics (Diekmann et al. 1996, Dieckmann & Law 1996, Hofbauer 1996). But with the new insights has come a confusing proliferation of similar and/or identical definitions and terms. As we shall see, the new millennium has continued to refine, advance, and apply the mathematical tools of adaptive dynamics and continuous-trait game theory.

CONTINUOUS-TRAIT GAMES

Continuous-trait games consider strategies that form a mathematical continuum (that is, a section of the real number line), such as [0,1] or $(-\infty, +\infty)$. This is in contrast to discrete (also called matrix) games where the set of strategies is a finite, unordered list (for example, fight, run, wait). Most traits studied in evolutionary ecology are continuous (or nearly so), such as date of first flowering, time to maturity, an animal's body size, the bill dimensions on a bird, or the allocation of resources to roots, stems, and reproductive tissues on a plant. The developments of matrix and continuous

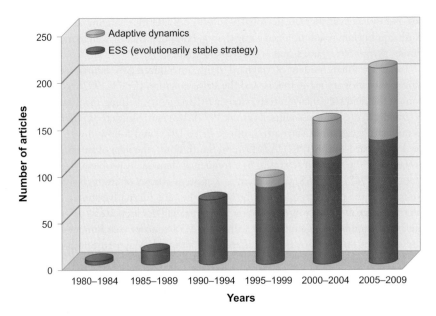

Figure 1

Growth of articles using the terms "ESS" or "Adaptive dynamics" over time. The ISI Web of Science was queried for the number of articles containing either "ESS OR EVOLUTIONARILY STABLE STRATEG*" or "ADAPTIVE DYNAMICS" in their titles, abstracts, or keywords for five-year periods. Note that this includes matrix games as well as continuous-trait games owing to the difficulty in separating them out. The years 2005–2009 are estimated based on multiplying the totals for 2005–2006 by 5/2 (probably an underestimate if rates are in fact increasing). The search was limited to 13 mainstream ecology, evolution, and evolutionary ecology journals to avoid medical and engineering terms that also abbreviate to "ESS": *American Naturalist*; *Annual Review of Ecology, Evolution, and Systematics*; *Ecology*; *Ecology Letters*; *Evolution*; *Evolutionary Ecology*; *Evolutionary Ecology Research*; *Journal of Animal Ecology*; *Journal of Ecology*; *Trends in Ecology Evolution*; *Oikos*; *Oecologia*; and *Theoretical Population Biology*. In keeping with our focus on evolutionary ecology, journals primarily focused on behavior were intentionally excluded although they contain a great many additional articles using game theory.

games have followed somewhat independent paths, with various researchers relating the two fields (Day & Taylor 2003, Vincent & Cressman 2000). Although our focus is on continuous traits, the **Supplemental Appendix** (follow the Supplemental Material link from the Annual Reviews home page at **http://www.annualreviews.org/**) shows how matrix games are a special case of continuous-trait games.

Basic Tool

The recipe for an evolutionary game begins with an ecological model of population dynamics where we define fitness as per capita growth rate (Crow & Kimura 1970):

$$W(u, U, N) = \frac{1}{N}\frac{dN}{dt},$$ 1.

where

- W denotes fitness. This has a long tradition in population genetics, but many alternative notations have been used in the literature on continuous-trait game theory, including F, G, ρ, σ, etc.

- u denotes the strategy played by the player of interest. Here a strategy is synonymous with some heritable trait such as body size or flowering date.

- U denotes the strategy/phenotype played by the opponent, the resident population.

- N denotes the population size of the resident population.

The **Supplemental Appendix** provides an example of deriving and developing a continous game: $W(u, U, N)$.

Making fitness a function of various subsets of the $u/U/N$ trio produces a variety of established approaches as special cases. $W(N)$ gives population dynamics. $W(u)$ gives classical optimization theory. $W(u, N)$ gives density-dependent selection, which is known to evolve to the value of u that maximizes the equilibrium value of N (Roughgarden 1979). In contrast, $W(u, U)$ gives frequency dependence without density dependence, a common simplification within game theory. Thus, the u/U pair (frequency dependence) is the *sine qua non* of continuous-trait game theory, giving a twist not found in any earlier approaches. At the cost of losing the strongly mechanistic Mendelian foundations of population genetics, we gain the ability to realistically solve evolutionary models with complex ecologies.

A key innovation of evolutionary game theory involves extending the classical notion of facing a single opponent playing strategy U to facing a population playing strategy U. In Maynard Smith's (1982, p. 23) idea of "playing the field," the individual does not interact in a pair-wise fashion with other individuals; rather the individual faces an opponent that is the population at large. For example, the consequences of flowering date to an individual may be influenced by the interaction between its flowering date (u) and the flowering dates of all of the neighboring plants. In this case we can think of the entire population playing the single strategy U. Under this interpretation we see u as the strategy of a mutant individual or a focal individual, and U is the resident strategy or the strategy found among the N individuals of the population (or $N - 1$ with 1 for the mutant). Even if individuals in the resident population show variation, the playing-the-field approach can still work if the fitness of the target individual is well approximated by considering the average strategy of the resident population, $\bar{U}$, that is, $W(u, \bar{U}, N) \approx W(u, \{U_1, U_2, \dots\}, N)$. Later we explore extensions where the population has variation in strategy values among individuals that cannot be reduced to the average strategy value.

Not surprisingly, an evolutionary game specified by $W(u, U, N)$ invites three distinct dynamics: the fate of the mutant playing u, changes in the population-wide strategy U, and changes in population size N. We discuss the first dynamic in the next section and the second and third dynamics in the following section.

Evolutionary Stability: Resistance to Invasion

Maynard Smith & Price (1973) considered the fate of a rare mutant or invader (playing strategy u) playing against some resident population (playing U). For evolutionary stability, they suggest "a strategy such that, if most of the members of a population adopt it, there is no 'mutant' strategy that would give higher reproductive fitness" (p. 15). Mathematically, the condition is

$$W(u, U^*) < W(U^*, U^*). \qquad\qquad 2.$$

This means that U^* is resistant to invasion by any rare alternative strategy u or in population dynamic terms the population size of the mutant or invader population will decrease owing to lower relative fitness. Mathematically, Equation 2 is equivalent to saying fitness with respect to u takes on a maximum at $u = U^*$ when everyone else is also playing U^* (see **Figure 2**)—U^* is the best response to itself.

Maynard Smith & Price call the strategy U^* an ESS. As recognition grew that the equilibrium identified in Equation 2 is the outcome of only one of three dynamics involved in a game (specifically the fate of a mutant playing u), various researchers gave it more specialized names such as evolutionarily unbeatable strategy (Eshel 1983), δ-stability (Taylor 1989), internal stability (Lessard 1990), evolutionary stability (Christiansen 1991), ESS maximum (Abrams et al. 1993b), and the ESS maximum principle (Vincent & Brown 1988, 2005). Regardless of terminology, it recognizes the same property, namely that an individual cannot increase its fitness by unilaterally changing its strategy.

We need to be careful about two points. First, if W is also a function of N, then we must (*a*) determine the equilibrium population size N^* corresponding to U^*, (*b*) replace Equation 2 with $W(u, U^*, N^*) < W(U^*, U^*, N^*)$, and (*c*) require that $W(U^*, U^*, N^*) = 0$ (that is, no population growth if the resident population is of size N^* and playing U^*). Second, the ESS usually allows $W(u, U^*) = W(U^*, U^*)$ so long as the further condition $W(u, u) < W(U^*, U^*)$ holds, because this will still prevent the rare mutant population from growing (Maynard Smith 1982). This second condition can be relevant for matrix games, but it rarely applies to continuous games.

Not only does Equation 2 give a test for an ESS, but, if we remember calculus, it gives a means to find ESS strategies. Specifically an ESS occurs at maxima of the function W for the variable u, which can be found by requiring

$$\frac{\partial}{\partial u} W(u, U^*) = 0 \qquad\qquad 3.$$

and

$$\frac{\partial^2}{\partial u^2} W(u, U^*) < 0 \qquad\qquad 4.$$

when evaluated at $u = U = U^*$ (and $N = N^*$ if population dynamics are included). The **Supplemental Appendix** applies Equations 2–4 to an example of a population choosing between two food patches.

Resistance to invasion is a static concept. It guarantees that a population at (U^*, N^*) can maintain its position against a rare invader, but it says nothing about what would happen if the population starts at (or was perturbed to) a nearby point $U^* + \delta$.

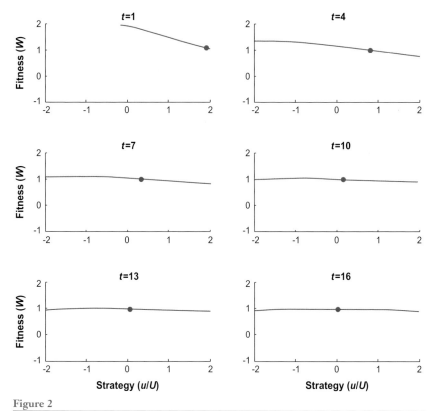

Figure 2

Population moving across adaptive landscape. This figure shows a sequence of snapshots (from $t = 1$ to $t = 16$ with 4 stops in between) of a population evolving in a continuous-trait game model of the Levene model of competiton (Geritz 1998, Levene 1953) (also used in **Figures 3** and **4**). The vertical axis is for fitness, W. The horizontal axis is for the strategy, u/U. The solid line gives the adaptive landscape with the fitness for a mutant of strategy u at each time t when the population is at U_t given by the location of the asterisk. These figures were calculated using the discrete equation for the dynamics of U, that is, Equation 5. Note the following points: (a) the population moves up hill, (b) the speed of movement is faster up steep slopes and much slower on nearly flat slopes, (c) the shape of the adaptive landscape changes as the population moves (U changes), (d) this dynamic results in stopping at the top of the hill, (e) this stopping point is an equilibrium point which meets the requirements of Equations 3 and 4 [the slope is zero at the evolutionarily stable strategy (ESS), and the second derivative is negative implying a maximum] and is therefore an ESS.

In fact, it is possible that $W(U^*, U^* + \delta) < W(U^* + \delta, U^* + \delta)$ (that is, a point U^* that is resistant to invasion is itself incapable of invading a population with a resident strategy close to but different than U^*; Geritz et al. 2002). Therefore invasion resistance by itself is a poor predictor of the evolutionary dynamics of a whole population. An additional concept is needed known as convergence stability, which we discuss in the next section.

Evolutionary Stability: Adaptive Dynamics and Convergence Stability

To model the evolution of the resident population phenotype U, continuous-trait game theory extends Fisher's (1930) Fundamental Theorem of Natural Selection, which states that single locus genetic models evolve (change allele frequencies) according to $\Delta p = k \, dw/dp$ where p is allele frequency. Visually, the change in gene frequency is in the direction of and proportional to the slope of the adaptive landscape. By extension of the Fundamental Theorem of Natural Selection to evolutionary strategies, the rate of change in the value of a population's strategy, U, is proportional to the slope of the adaptive landscape ($\partial W/\partial u$) and in the direction of this fitness gradient (Roughgarden 1983). This leads to a dynamical equation for the change in the resident strategy, U:

$$\Delta U = k \frac{\partial}{\partial u} W(u, U, N) \quad \text{or} \quad U_{t+1} = U_t + k \frac{\partial}{\partial u} W(u_t, U_t, N_t). \qquad 5.$$

The constant of proportionality k is sometimes broken into components involving heritability h^2 and/or additive genetic variance σ^2. If this evolutionary "speed" term is too large, then evolutionary dynamics may become nonequilibrial (particularly for the difference equation dynamic). Equation 5 is sometimes called the canonical equation of adaptive dynamics (Dieckmann & Law 1996). A continuous version is given by

$$\frac{dU}{dt} = k \frac{\partial}{\partial u} W(u, U, N). \qquad 6.$$

Equations 5 and 6 give the second dynamic inherent in $W(u, U, N)$ (changes in U).

The third dynamic concerns population size, N. By definition of W (Equation 1),

$$\frac{dN}{dt} = NW(u, U, N) \quad \text{or in discrete terms} \quad N_{t+1} = N_t + N_t W(u_t, U_t, N_t). \qquad 7.$$

Together Equations 5, 6, and 7 describe the dynamics of the state variables U and N. For continuous-trait evolutionary game theory, these equations were first suggested without proof by Brown & Vincent (1987a). The 1990s saw rigorous and formal derivations of these equations (Abrams et al. 1993b, Geritz 1998, Metz et al. 1996, Vincent et al. 1993) with perhaps the most detailed being given by Dieckmann & Law (1996). Many different flavors of Equations 5 and 6 exist with different notations, definitions of k, and initial assumptions. Fortunately, these evolutionary dynamics are remarkably robust to the slight differences in interpretation.

At least three other approaches to phenotypic evolution have been taken. First, computer simulations of mutating asexual populations can be used when analytical solutions are unavailable. Second, several researchers have shown that learning behavior can lead to strategy dynamics similar to Equation 5. Finally, researchers have explored evolution toward an ESS using population or quantitative genetics models. Although focusing on the purely phenotypic approach to the dynamics, we discuss these three alternative dynamics in the **Supplemental Appendix**.

It is straightforward to use the dynamical Equations 5 and 6 to identify the convergent stable endpoints of the dynamic (Abrams et al. 1993b, Bulmer 1994, Geritz et al. 1998, Vincent 1990, Vincent et al. 1993). Specifically U^* is an evolutionary endpoint if Equation 3 holds and if

$$\frac{\partial^2}{\partial u^2}W(u, U) + \frac{\partial^2}{\partial u \partial U}W(u, U) < 0. \qquad 8.$$

Or more generally if the dependence of fitness, W, on N is modeled, then a third term is added to the left-hand side: $+(\partial^2 W/\partial u \partial N)(\partial N^*/\partial U)$ with $W(u, U^*, N^*) = 0$ (Vincent & Brown 2005). The proliferation of terms referring to the stability of the adaptive dynamic include continuous evolutionary stability, m-stability, ESS (coopting the earlier meaning) and convergence stability. We will refer to strategies U^* that meet Equation 8 as convergent stable. When U^* is convergent stable it is an evolutionary attractor; populations with U different from U^* will evolve toward U^*. If the opposite of Equation 8 holds (LHS > 0) for a point U^* where Equation 3 holds, then it is an evolutionary repellor; populations just slightly off this point will move further away from it. The **Supplemental Appendix** provides a worked example of these equations.

The conditions for convergence stability and for resistance to invasion have some important similarities. The first-order conditions of $\partial W/\partial u = 0$ are identical. The second-order condition for convergence stability contains one or two additional terms beyond $\partial^2 W/\partial u^2$. The evaluation of resistance to invasion only involves unilateral changes in an individual's strategy with no changes in the population's strategy or population size. Convergence stability additionally involves the consequences of collective changes in strategy ($\partial^2 W/\partial u \partial U$) and the effect of the populations strategy on population size ($\partial^2 W/\partial u \partial N)(\partial N^*/\partial U$). The term $\partial^2 W/\partial u \partial U = \partial(\partial W/\partial u)/\partial U$ is critical. It captures the idea that the shape of the adaptive landscape ($\partial W/\partial u$) changes as U changes (see **Figure 2**). This is the famous idea of the adaptive landscape as a rubber sheet that gets stretched as a population moves across it. Wright (1930) was well aware of this effect of frequency dependence and generally avoided it. Continuous-trait game theory, on the contrary, embraces this addition. But it can have profound implications, making evolution short-sighted (Roughgarden 1979)—that is, meaning it evolves in the direction that currently causes the greatest increase of fitness, not necessarily in the direction that will bring it to the highest collective fitness.

Because of the possibility of $\partial^2 W/\partial u \partial U$ being positive or negative, $\partial^2 W/\partial u^2$ can be negative yet Equation 8 can fail, or $\partial^2 W/\partial u^2$ can be positive yet Equation 8 can be true. Thus the two types of evolutionary stability (invasion resistance and convergence stability) do not imply each other.

BESTIARY OF EVOLUTIONARY GAME OUTCOMES

As a consequences of the independence of the two types of stability there are four possible outcomes when evaluating a strategy U^* that satisfies $\partial W/\partial u = 0$. The strategy U^* may exhibit any one of the following outcomes: (*a*) resistant to invasion and convergent stable, (*b*) resistant to invasion and not convergent stable, (*c*) invadable and convergent stable, and (*d*) invadable and not convergent stable (Cohen et al. 1999, Geritz et al. 1998). This creates an exciting bestiary of outcomes.

Understanding the independence of invasion resistance and convergence stability and the resulting four combinations has been the most important advance since the initial formulation of evolutionary games. However, as one can imagine given the half-dozen terms for each of two types of stability, there has also been a proliferation of terms for the four outcomes.

We have already given proposed unifying terms for the two types of stability (invasion resistant = meets Equation 2, and convergence stable = meets Equation 8). The four permutations of these two terms suffice to describe all evolutionary outcomes in their full generality. However, we also propose that the four outcomes can be meaningfully further classified into three groups, only two of which are biologically interesting:

ESS: We suggest defining outcome *a* as the ESS. This means the ESS is an uninvadable fitness maximum and convergent stable. Although Maynard Smith's original definition lacked an appreciation for convergence stability, he fully intended the concept to describe the likely outcome of evolution by natural selection acting on games. We favor updating Maynard Smith's concept for two reasons. First, the term ESS has the greatest cachet in the dictionary of evolutionary ecology—why lose that? Second, fitness maxima that are evolutionary repellors (not convergent stable) will not evolve. Applying the ESS to outcome *a* follows Maynard Smith's original definition with the necessary addition of convergence stability (which was generally true at a fitness maximum in the discrete games he studied; see a detailed discussion in the **Supplemental Appendix** for matrix games).

Branching point: The term branching point describes outcome *c* where a strategy is both a fitness minimum and convergent stable. As noted by Brown & Pavlovic (1992) and then extensively developed by Abrams et al. (1993b), Dieckmann & Doebeli (1999), Doebeli & Dieckmann (2000), and Geritz et al. (1997), populations may evolve to these branching points (**Figures 3** and **4**) and then under the right conditions diverge into two separate populations or species with distinct strategies—our topic for the next section.

Repelling points: Outcomes *b* and *d* are not convergent stable. Evolution will not move strategies to these repelling points. These points should not be observed in nature. It remains an open question on how these repellors may affect evolution toward ESS and branching points (Doebeli et al. 2004). For example, repelling points may serve to divide basins of attraction between alternate ESSs. Or, like branching points, the existence of repelling points may presage an ESS that contains more than one strategy (see the section entitled "Coalitions and Evolutionary Stable Strategies with Multiple Strategies Played").

An ESS (convergent stable maximum) can be global or local in each of the two types of evolutionary stability. For invasion stability, an ESS can resist only nearby invaders (a local ESS) or all invaders (a global ESS) (Vincent & Brown 2005 prefer to further restrict the usage of ESS to global maxima). Even a global ESS may not be the sole ESS on the landscape—there may be other strategies that are ESSs but they are masked as soon as one of the ESSs is achieved. For convergence stability,

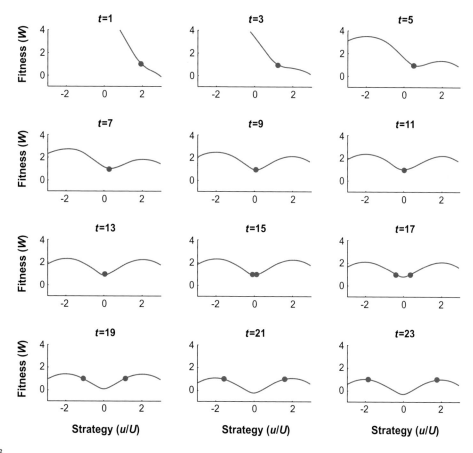

Figure 3

Evolution toward minimum fitness. This is the same model of competition used in **Figure 2**, but the parameter for niche width has been decreased to allow two distinct fitness peaks to emerge. This example shows how evolutionarily stable strategy (ESS) dynamics can in fact cause a population to evolve toward a fitness minimum. This depends critically on the fact that the landscape changes shape as the population evolves. The basic effect is that the landscape changes shape faster than the population moves. Just before time $t = 15$ the population was split into two subpopulations starting just slightly to either side of the fitness minimum. With two populations, the populations are now able to evolve apart toward the two fitness maxima (ESSs). This is an example of a branching point. This is also an example where a coalition of two is able to invade an equilibrium that is uninvadable by any coalition of one.

all possible starting strategies can converge to the ESS (globally convergent) or only some subset of initial strategies (called the basin of attraction) will evolve to the ESS (locally convergent). A globally convergent landscape can have only one ESS whereas systems with multiple ESSs (necessarily only locally convergent) will have a distinct basin of attraction associated with each ESS.

　　With these local and global properties applied to each outcome of invadability and convergence stability, the bestiary of possibilities grows! For instance, the ESS may

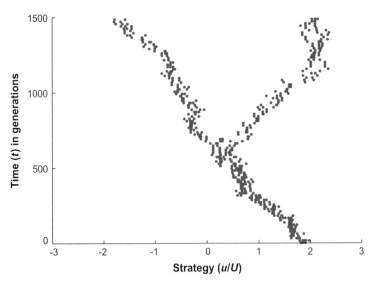

Figure 4

Individual-based models and branching points. This figure is an example of using a computer simulation of an individual-based model (IBM) of a population reproducing asexually with occasional mutation (see **Supplemental Appendix**). In this simulation the same game as in **Figure 3** is modeled. The result is very similar—the population evolves to the branching point at $U = 0$, then splits into two populations (a coalition) moving to the two peaks near +2 and –2. The main difference is that the population already had a spread of phenotypes and did not require the specific intervention of adding a second population at the branching point. There is considerable debate about whether this branching will occur when the organism reproduces sexually.

be globally convergent stable but not a global maximum. Once this ESS has evolved, a distant strategy can invade. But once this distant strategy starts to become common it either (*a*) moves toward U^* and merges back into U^* or (*b*) goes extinct. This is known as the resident strikes back scenario (Dieckmann et al. 1999). This outcome requires weird mathematical conditions and it may or may not be biologically likely (Geritz et al. 2002).

This brings us to the current frontier of evolutionary game theory where strategy dynamics lead to a dizzying array of outcomes. When population dynamics are overlaid on the strategy dynamics, the possibilities become Byzantine. Geritz et al. (1999) point out that this diversity of outcomes occurs because of interactions between the two stability properties of resistance to invasion and convergence stability. A single model (Geritz et al. 1999) exhibited at least five different types of bifurcations, as follows:

1. Global ESS becomes local ESS with no other equilibria appearing,

2. Repellor and branching point appear simultaneously (with branching point adding a strategy or causing extinction),

3. ESS and repellor collide and annihilate each other,

4. Global ESS becomes local ESS simultaneously with appearance of branching point and repellor, and
5. Branching point leads to a coalition of two strategies, one of whose equilibria is a continuation of the local ESS.

The exploration of bifurcations in adaptive dynamic systems remains perhaps the most important theoretical topic left to pursue in continuous-trait game theory.

ADAPTIVE SPECIATION AND BRANCHING POINTS

Branching points are evolutionary attractors that are fitness minima (that is, Equation 2 is false but Equation 8 is true). In some models, the evolution of the resident population strategy U can evolve uphill along the adaptive landscape yet come to rest at a minimum. This occurs because the adaptive landscape changes even as the strategy moves along it (see **Figure 3** for an example of this counterintuitive dynamic). When the population is at strategy $U^* - \delta$ ($\delta > 0$ here), the landscape takes on a positive slope. At $U^* + \delta$ the landscape takes on a negative slope; but when the population plays exactly U^*, the signs reverse leaving U^* at a minimum. Even if the system starts with two evolving species with strategies that are close together but distinct, the process of branching occurs. At first, the two evolving populations may climb the same slope toward what would be the branch point, but as the strategies approach the minimum, the position of the minimum itself moves to a point between the species. The species do not need to cross the valley of the landscape, rather the valley crosses under them.

Multiple fates of these two separately evolving populations can be imagined:

1. Both populations continue to evolve and move toward something like an ESS; they remain two populations with distinct strategies—the ESS may contain more than one distinct strategy.
2. One or both populations diverge to strategies that are themselves branching points, permitting the process of strategy diversification to continue as an adaptive radiation.
3. One population evolves toward its own extinction whereas the other moves away from the branching point toward an ESS.
4. Both populations become extinct.

Outcomes 1 and 2 raise two questions: Can we describe these branching, separately evolving populations as species? And how do we define an ESS with two resident strategies instead of one?

Outcome 3 might seem counterintuitive: How can a population evolve in the direction of increasing fitness until it goes extinct? In fact, several biologically realistic examples of this have been identified (Geritz et al. 1999, 2002). Runaway selection provides one example where some trait of the population becomes ever more extreme owing to evolutionary forces, but the equilibrium population size for the more extreme trait becomes progressively smaller until extinction occurs. Thus we need an adjective to describe different types of branching points. We propose calling outcomes 1 and 2 (two or more permanently evolving populations) coalition branching points for reasons that are made clear two sections below. We propose calling outcome 3 extinction branching points. Ferrière and coworkers called outcome 4 evolutionary suicide (Ferrière 2000), although this may require mathematically precise and biologically unrealistic conditions (Geritz et al. 2002).

Speciation: Do Branching Points Really Branch?

The identification of branching points may lead to major advances in the studies of sympatric speciation and adaptive radiations. That these convergent stable minima are a real phenomenon of game theory is demonstrable fact. What they mean in nature is an open and exciting empirical question. Are there species in nature with strategies that reside at one of these minima? There has been debate for (Cohen et al. 1999, 2001) and against (Abrams 2001, Waxman & Gavrilets 2005) the relevance of branching points to natural systems.

Early models and tests of sympatric speciation (Maynard Smith 1966, Thoday & Gibson 1962) used disruptive selection as the driving mechanism of speciation. As shown in most evolution textbooks, this disruptive selection explicitly or implicitly assumes that the resident phenotype of the population is at a minimum of the adaptive landscape. But how did the population's phenotype get to this point of disruptive selection in the first place?

Evolutionary game theory and the phenomenon of branching points (convergent stable minima) resolve these problems with nongame theoretic models of sympatric speciation. Natural selection evolving up the fitness gradient can itself arrive at the minimum of frequency-dependent adaptive landscapes. When the number of resident strategies is below that of the ESS (see the next section below on Coalitions and Evolutionary Stable Strategies with Multiple Strategies Played), natural selection may drive the strategy to the branching point, exert disruptive selection, and permit speciation. Rosenzweig (1978) anticipated some of these properties of frequency- and density-dependent adaptive landscapes in what he called competitive speciation.

Current modeling and conceptual research addresses the question of how branching points can permit speciation. With asexually reproducing organisms branching occurs easily (**Figure 4** and Geritz & Kisdi 2000). But sexual reproduction creates a blending or interbreeding between separating populations that may preclude splitting the population into subpopulations with distinct resident strategies (Tregenza & Butlin 1999). This fact is captured in the discontinuous jump that was required between time steps 13 and 15 in **Figure 3**—the researcher who conducted the computer simulations had to arbitrarily introduce two populations as the progenitor population approached the branching point.

Several mechanisms may allow speciation even under sexual reproduction. First, species may assortatively mate on the trait that is subject to disruptive selection. This is conceivable, for example, for body size. In fact, if the species persists at a branching point, selection strongly favors assortative mating! Second, a marker trait or sexually selected trait such as throat color may become linked to the trait under disruptive selection and allow assortative mating on the marker trait. The assortative mating becomes adaptive because of the lesser fitness of intermediate types (Dieckmann & Doebeli 1999, Doebeli & Dieckmann 2000, Geritz & Kisdi 2000, Kisdi & Geritz 1999). Third, and perhaps most likely, these traits may cause organisms to be more commonly found at sites (host plants, elevations, microhabitat) to which they are most adapted (either through active choice by the organism or by Darwinian selection), and the spatial separation correlated with the trait leads to assortative mating as a byproduct of proximity (Doebeli & Dieckmann 2003). It is important to note that these four modes of speciation (asexual and the three assortative mating mechanisms) not only are not unique to adaptive dynamics but were developed first and most extensively elsewhere (Kondrashov 1986, Kondrashov & Kondrashov 1999, Rosenzweig 1978).

The novel and wonderful contribution of adaptive dynamics to sympatric speciation is in showing how the disruptive selection comes about. Prior to game theory, disruptive selection was often considered somewhat pathological or a chance event. But with the inclusion of density and frequency dependence into standard models of population dynamics, we see disruptive selection occurring easily in the form of branching points. These branching points emerge as a very common and widespread property of continuous-trait games.

Coalitions and Evolutionary Stable Strategies with Multiple Strategies Played

In matrix game theory it was recognized that a single strategy may not be an ESS. For instance, in a Hawk-Dove game, the strategy Dove may have a higher payoff than Hawk in an all-Hawk world and vice versa for Hawk having a higher payoff in an all-Dove world. Consequently, the two strategies coexist. Just like matrix games, the ESS for continuous-trait evolutionary games may possess a coalition of coexisting strategies. The existence of branching points in models suggests that the model may have an ESS coalition with two or more strategies.

The ESS concept extends nicely to cover ESS coalitions of n populations, playing strategies $U_1, U_2, \ldots U_n$. The extension of notation is straightforward:

$$(1/N_i)(dN_i/dt) = W(u, U_1, U_2, \ldots, U_n, N_1, N_2, \ldots N_3) = W(u, \mathbf{U}, \mathbf{N}) \quad \text{for} \quad u = U_i,$$
9.

where the boldfaced $\mathbf{U}$ and $\mathbf{N}$ represent vectors of the resident strategies and their population sizes for $i = 1, \ldots, n$. Brown & Vincent first developed the idea of an ESS coalition of more than one strategy, referring to Equation 9 as the fitness generating function (using G instead of W). Although not explicit, they considered the ESS (given in Equations 10 and 11 below) to be both resistant to invasion and convergent stable (Brown & Vincent 1987b; Vincent & Brown 1984, 1988; Vincent et al. 1996). With the discovery of branching points, the ESS coalition has become an important and necessary extension, and others have readily adopted it (Geritz et al. 1998, 1999). By setting $u = U_i$ in the function, one generates the fitness function for the population using strategy i.

The necessary conditions for $\mathbf{U}^*$ to be resistant to invasion represent a straightforward extension of Equations 3 and 4 (Brown & Vincent 1987b; Vincent & Brown

1984, 1988; Vincent et al. 1996):

$$\frac{\partial}{\partial u} W(u, U_1, U_2, \ldots, U_n, N_1, \ldots, N_n) = 0 \quad \text{for} \quad u = U_i \quad \text{for each} \quad i \qquad 10.$$

and

$$\frac{\partial^2}{\partial u^2} W(u, U_1, U_2, \ldots, U_n, N_1, \ldots, N_n) < 0 \quad \text{for} \quad u = U_i \quad \text{for each} \quad i, \qquad 11.$$

where each population of the coalition must exist at positive population size: $N_i^* > 0$. Resistance to invasion requires that each strategy of the ESS resides on a separate peak of the adaptive landscape and that each of these peaks has the same fitness, which is zero at $\mathbf{N}^*$: $W(u, \mathbf{U}^*, \mathbf{N}^*) \leq W(U_i, \mathbf{U}, \mathbf{N}) = 0$ for all $i = 1, \ldots n$. Changes in N_is balance the adaptive landscape so that all strategies of the coalition reside on peaks of equal fitness.

The concept of convergence stability also remains the same for a coalition with more than one strategy. The necessary first order condition remains the same as Equation 3, but the second order conditions analogous to Equation 8 become onerous and quite intractable above a coalition of three or more strategies (Cohen et al. 1999, Geritz et al. 1998). But the idea continues to be that $\mathbf{U}^*$ is convergent stable if strategy dynamics will return the populations' strategies to $\mathbf{U}^*$ following a perturbation of δ_i to one or all of the strategies of the coalition (where $U_i = U_i^* + \delta_i$ refers to a perturbation of strategy i). Graphical methods using invasion cones for analyzing 2-coalitions often work better (Geritz et al. 1998, Matessi & Di Pasquale 1996), and other times simply visually verifying the uninvadability of an n-coalition is adequate (for example, Brown 1990a). Generally the convergence stability of a candidate $\mathbf{U}^*$ is evaluated using adaptive dynamics to see what happens to coalitions with starting conditions close to $\mathbf{U}^*$. Additional work is continuing in this area. There are currently no general results indicating when a model or circumstance will yield an ESS with specifically n-strategies.

Recently, Cohen (2003) has introduced a new approach that is analytically more difficult but biologically more realistic than the idea of a coalition. Rather than a point or set of points along the strategy continuum, Cohen models the evolution of a probability distribution on the strategy continuum. Thus, the resident population could play strategies that are, for example, normally distributed along the continuum. When the mutant "plays the field," the field is then this probability distribution. This has much in common with the use of quantitative genetics to model phenotypic evolution (Abrams et al. 1993a). Although an ongoing and exciting development, the approach remains rather intractable.

EXPANDING THE DOMAIN OF EVOLUTIONARY GAME THEORY

Thus far, we have developed the machinery for predicting phenotypic evolution of a single trait for deterministic systems with stable and fast population dynamics. Can this machinery be expanded to cover more complicated situations? We explore some of these extensions.

Density Dependence and Nonequilibrium Population Dynamics

So far we have largely ignored the dynamics of population size, N. The justification for doing this is a fast/slow argument—evolution proceeds slowly relative to population dynamics. Hence, one can assume that strategy dynamics occur mostly when the populations are at or near N^*.

Of course, this may not always hold. First, strategy dynamics may be fast (for example, selection strong enough) and with a comparable time scale to population dynamics (Yoshida et al. 2003). Second, population dynamics may exhibit oscillations or even chaos. Both may happen for traits associated with predator-prey interactions. Predator/prey models frequently show cyclical dynamics, and the evolution of predation traits may also cycle as the evolutionary responses oscillate with the population dynamics (Abrams & Matsuda 1997, Dieckmann et al. 1995).

When evolution is almost as fast as population dynamics, or if the population dynamics do not have a stable point equilibrium, then we have two modeling choices. We can simulate the system using Equations 5 and 7, or we can use an analytical solution. The analytical solution looks at long-term fitness using the idea of a Lyapunov exponent (Caswell 2001, p. 542 and 561; Ferrière & Gatto 1995, Rand et al. 1994). A Lyapunov exponent can be thought of as the log geometric mean of fitness:

$$
\begin{aligned}
W_{longterm}(u, U, N) &= [\log |W(u, U_1, N_1)W(u, U_2, N_2)\ldots W(u, U_T, N_T)|]^{1/T} \\
&= 1/T \, \Sigma_i \, \log |W(u, U_i, N_i)|. \qquad\qquad 12.
\end{aligned}
$$

So if the population dynamics follow a five-cycle, then we geometrically average the fitness of the invading phenotype u, across the five different population sizes N_i in the cycle and the corresponding population strategies U_i. The geometric average is used because population growth is multiplicative ($N_T = W_T \ldots W_2 W_1 N_0$), whereas the log is mathematically convenient. If chaotic rather than cyclic dynamics occur, then we must take the limit as $T \to \infty$. When using Lyapunov exponents, modelers generally favor using a discrete population dynamic based on $N_{t+1} = N_t W(u, U_t, N_t)$, where equilibrium occurs when $W = 1$.

With Lyapunov exponents we say that if $W_{longterm}(U, U) > W_{longterm}(u, U)$ for all (nearby) u (that is, U is a global or local maximum in $W_{longterm}$), then U is an ESS under nonequilibrium dynamics. Although a logical extension of the idea that an ESS is a fitness maximum, it is now only true in the long-term. It is possible for an invading mutant to have negative fitness in one half of a population cycle, positive fitness in the other half, but overall a positive Lyapunov exponent. If the mutant begins its invasion in the half of the cycle where fitness is negative, then it will fail, but in the long-term such a mutant should invade if repeated attempts occur. Thus, we can achieve a strong analytic result (using Lyapunov exponents), but it holds true only in a long-term, averaged-across-many-replicates sense. Finally, the property of convergence stability remains poorly understood for nonequilibrium populations and/or fast evolutionary dynamics.

Stochastic Environments

We have developed the ESS concept as a deterministic process, but the world experienced by organisms is inherently stochastic. Temperature, precipitation, and other factors fluctuate randomly. Many interesting questions in evolutionary ecology derive from such stochasticities (Cohen 1966, Roff 2002).

Stochasticity has the potential to change the evolutionary outcome when the model has multiple ESSs by stochastically bouncing from one basin of evolutionary attraction to another. We can define an uninvadable strategy in a stochastic environment by using mathematics similar to that for nonequilibrium population dynamics. We denote environmental conditions (for example, temperature) here by a little e and subscript it by t for time to suggest that this takes on varying values over time according to the probability distribution $\mathscr{E}$. We take the same concept of a Lyapunov exponent, except here we average not over the sequence of population sizes, N_t, but over the environmental states, giving

$$W_{stochastic}(u, U, \mathscr{E}) = E_{\mathscr{E}}\{\log(u, U, \mathscr{E})\} = 1/T \Sigma_t \log W(u, U, e_t), \qquad 13.$$

where $E_{\mathscr{E}}$ denotes taking the expectation with respect to $\mathscr{E}$. In practice this can be done by taking a simple average over a number of samples from $\mathscr{E}$. In more advanced models it is important to know not just $\mathscr{E}$, but the autocorrelation structure across time (for example, do good years tend to follow good years) (Tuljapurkar 1990, Tuljapurkar & Orzack 1980).

Equation 13 suggests that $W_{stochastic}$ should parallel $W_{longterm}$, and indeed similar issues arise. We can make a weak statement using $W_{stochastic}$ as an invasion exponent (Metz et al. 1992, Rand et al. 1994) and determine which phenotypes can invade against all other phenotypes. This identifies a coalition of phenotypes likely to be present and is sometimes called an evolutionarily stable combination (ESC) (Cohen & Levin 1991, Ellner & Hairston 1994, Ludwig & Levin 1991). This avoids making a strong statement about the one phenotype to win. A stronger statement about a local or global maximum of $W_{stochastic}$ (a stochastic ESS) emerges by looking for a fitness maximum in $W_{stochastic}$, but declaring a precise ESS strategy can be misleading when the biological reality will likely involve a cloud of indistinguishable strategies around this ESS. Furthermore, as before, the convergence stability of systems using Lyapunov exponents for either fluctuating populations or environments remains an unsolved problem.

Structured Populations

Thus far we have ignored age or stage structure. In reality populations are subdivided by age, size, and spatial structure, and there may be evolutionary strategies associated with different stages or ages, or a given strategy may influence different stages differently. Matrix models can examine the population dynamics of structured populations (Caswell 2001). The ESS machinery applies to matrix models of population dynamics. The dominant eigenvalue of the matrix determines the long-term behavior of the population dynamics. As an evolutionary game, we simply equate fitness

$W(u, U, N)$ to the dominant eigenvalue (Caswell 2001; Doebeli & Ruxton 1997; Vincent & Brown 2001, 2005) $W(u, U, N) = \lambda(u, U, N)$, and apply the standard conditions for seeking resistance to invasion (maxima of W is an ESS, etc). The conditions for convergence stability also remain the same with the caveat that one assumes that a stable age or stage distribution is established quickly. Like the assumption of fast population dynamics to a stable equilibrium, we now assume fast dynamics on maintaining and adjusting the stable stage distribution.

Structured (matrix) populations in a stochastic or nonequilibrium environment can also be studied through a combination of Lyapunov values and dominant eigenvalues. Although conceptually straightforward, the modeling begins to invoke complicated mathematical machinery (Caswell 2001; Doebeli & Ruxton 1997; Tuljapurkar 1990, 1997; Tuljapurkar & Orzack 1980). Furthermore, the actual time distribution of population sizes may result from an interaction of resident strategies and stochasticity. For such a case, one must use a simulation to produce a distribution of states and then examine the best response to this distribution (Schmidt et al. 2000).

Multiple Traits

One often studies the evolution of two traits simultaneously. Equations 2–4 and 5–8 have exact analogues for the evolution of multiple traits, but the terms are now vectors and matrices instead of scalars. For instance, a strategy, u or U, is now a vector where each element represents one trait. Details are in Leimar (2005). Most importantly, the scalar k from Equation 5 turns into a matrix. If the two traits are independent, then k is diagonal, and it can be shown that the traits can be modeled separately. However it is common for traits to covary owing to linkage, epistasis, or other covariances. For example, life history traits such as age at maturity, body size, fecundity, and lifespan all have well-known positive and negative correlations (Roff 2002).

When traits do covary, k is no longer diagonal. Strategy dynamics now occur on a two-dimensional landscape with each trait forming one axis, and maxima look like hills. In a two-dimensional landscape, the strategy vector has many possible directions for moving uphill. The exact path up the landscape is determined by the sequence of mutations, which are in turn constrained by the covariance of the traits (Dieckmann & Law 1996, Leimar 2005, Matessi & Jayakar 1976). This outcome is already known in quantitative genetics (Lande 1979). In unusual circumstances, such as when there are two close peaks with a saddle (pass) between them, the indirect route of adaptive dynamics can cause evolution to leave the basin of attraction of one peak and evolve to the other peak (Leimar 2005, Matessi & Jayakar 1976). Leimar (2001) refers to this change in evolutionary outcome resulting from the sequence of mutations as a Darwinian demon, but in practice the conditions in which this can happen may be quite unlikely (Leimar 2005). A promising new approach to the evolution of multiple traits extends Levins' visual fitness set diagrams (Levins 1968) to the frequency-dependent scenario (de Mazancourt & Dieckmann 2004).

Regardless of whether the two traits covary, evolving in two-space offers a new feature. Instead of just having attractors and repellors, equilibrium points can now be saddle points, attracting in some directions in trait space and repelling in others.

Although saddle points are unstable as an evolutionary endpoint, evolution can spend large amounts of time flying by the saddle points, giving them a strong structuring role in the dynamics (Cushing et al. 2002).

Under vector-valued strategies the ESS concept likely remains unchanged. Evolutionary branching points can still occur and behave as sources of speciation. But, now we have the opportunity for the vector of traits to become coadapted within the organism. In this way, many of the covariances seen across species in certain traits (bill length and bill depth in many birds) may be the result of coadaptation and not genetic constraints.

Multiple Species and Coevolution

Continuous-trait game theory provides an excellent tool to model the coevolution of species. In fact, coevolution is really just a special case of the multitrait case described above where the traits are independent of each other (that is, the covariance matrix, k, is diagonal) because they are different species.

For example, for a two-species system with one trait coevolving in each species we have

$$W_1(u, U, V, N, M) \quad \text{and} \quad W_2(v, U, V, N, M), \qquad 14.$$

where the fitness function is subscripted by species and may take quite different forms in an asymmetric coevolution model such as predator/prey. As always, we explore the fitness of a mutant of species 1 playing the strategy u against a population of species 1 playing the strategy U and the population of species 2 playing the strategy V. Population sizes of N and M refer to species 1 and species 2, respectively. (Note that any rare mutants in species 2 playing strategy v are assumed to have no effect on a rare mutant of species 1 playing strategy u.) Likewise we explore the fitness of a mutant in species 2 playing strategy v against the same context.

The model can easily be extended to more than two species. For example, $W_1(u_i, \mathbf{U}, \mathbf{V}, \mathbf{N}, \mathbf{M})$, where $\mathbf{U}$ and $\mathbf{N}$ are vectors, gives the strategies and population sizes for different prey species, while $W_2(v, \mathbf{U}, \mathbf{V}, \mathbf{N}, \mathbf{M})$ may be the predator's fitness function, where elements of $\mathbf{V}$ and $\mathbf{M}$ represent different predator species. Note that all predators share the same basic fitness function, W_2, but there is one equation for each u_i and v_i (an invading mutant for each species). This is sometimes referred to as a fitness generating function (Brown & Vincent 1992).

Evolution is now occurring along several adaptive landscapes, one for each species. The strategies and population sizes of all species may influence the shape of each species' adaptive landscape. The conditions for resistance to invasion remain unchanged; $\mathbf{U}^*$ and $\mathbf{V}^*$ must simultaneously take on maxima of their respective landscapes and, at equilibrium populations sizes, these maxima will yield 0 fitness. The concept of convergence stability remains the same but the conditions are quite unwieldy and have yet to be fully described with more than two species. The ESSs can still be defined as coalitions of strategies that are convergent stable and resistant to invasion but can now involve multiple species of prey and predators. The prey and/or predator's adaptive dynamics may lead to evolutionary branching, nonequilibrial evolutionary dynamics, and local maxima (Dieckmann et al. 1995). Such a model allows for coevolution both within and between the predator and prey species.

APPLICATIONS OF CONTINUOUS-TRAIT GAME THEORY

It is impossible to summarize in a single review article all biological problems that have been studied using continuous-trait games (see **Figure 1**). Here is a very brief summary. Broadly speaking, biological applications of continuous-trait game theory fall into two categories, models of evolution of a single species (evolutionary ecology) and models of evolution between two species (coevolution).

Evolutionary Ecology

In 1983 three nearly simultaneous papers (Bulmer 1983, Iwasa et al. 1983, Parker & Courtney 1983) used ESS models to explain the well-known phenomenon of protandry in insects (males emerge before the females). These models suggested not only that early emergence should occur but that there should be an abrupt drop-off in male emergence before the end of female emergence. Additional predictions include: environmental stochasticity should lead to increased variation in emergence time under all conditions and tracking of the female emergence times if this is predictable (Iwasa & Haccou 1994); a variety of modifications occur in divoltine populations (Wiklund et al. 1992); arrival order of migratory birds should depend on individual condition (Kokko 1999); and if disturbances (for example, late freezes) influence emergence times, then optimal emergence strategies depend on the spatial scales of disturbance and population regulation (Iwasa & Levin 1995).

Cohen (1966) studied dormancy in annual plants (seed banks) without density dependence and showed that dormancy (partial germination) was optimal in a stochastically varying environment. Seger & Brockman (1987) pointed out that this same line of thinking applies to diapausing insects and named this argument bet-hedging. In the mid-1980s several researchers (Bulmer 1984; Ellner 1985a,b; Goodman 1984) nearly simultaneously applied game theory to identify ESS solutions showing that the inclusion of density and frequency dependence caused an increase in the optimal dormancy fraction. Tuljapurkar & Istock (1993) studied the effect of environmental harshness and temporal weather autocorrelations on diapausing. Simultaneous temporal stochasticity and deterministic spatial heterogeneity (with limited dispersal) can lead to branching points with the coexistence of multiple germination strategies (Mathias & Kisdi 2002).

At this writing (February 2007), over 95 papers have analyzed the evolution of dispersal using a game theoretic approach. Game theory can explicitly allow for density and frequency dependence and/or environmental variation. Several researchers (Comins et al. 1980, Motro 1982, 1983) suggested that dispersal would be an ESS to avoid kin competition. Kin selection in ESS models is discussed generally, with an example based on dispersal, by Taylor & Frank (1996). Levin et al. (1984) showed that stochastic spatial heterogeneity could also make dispersal an ESS, a conclusion expanded in several directions (Lemel et al. 1997, Mathias et al. 2001). Several researchers have explored how nonequilibrium population dynamics (for example, chaotic) favor dispersal (Doebeli & Ruxton 1997, Gyllenberg & Metz 2001, Holt & McPeek 1996, Parvinen 1999). A number of researchers also explore local (seed) dispersal and possible trade-offs on seed size and/or competition (Ezoe 1998, Lavorel et al. 1994, Levin & Muller-Landau 2000, Winkler & Fischer 1999). Dispersal in a metapopulation (Gyllenberg et al. 2002) can evolve to the point where the species drives itself extinct. Evolution of dispersal can lead to branching points and dimorphisms in dispersal strategies (Doebeli & Ruxton 1997, Mathias et al. 2001) or to phenotypically plastic reaction norms (Ezoe & Iwasa 1997). The link between dispersal and various other life history strategies such as age of reproduction (Ronce et al. 2000) and brood size (Kisdi 2004) has been explored. Other analyses have identified the ESS strategy for migration versus overwintering (Kaitala et al. 1993) and joining or leaving social groups (Kokko & Johnstone 1999, Stephens et al. 2005).

The evolution of seed size versus seed number (or any offspring number versus offspring quality trade-off) is also amenable to ESS modeling. As with dormancy and dispersal, optimization models make one prediction, but ESS models that include density dependence and/or environmental variation make different predictions. Specifically, the Smith-Fretwell model (1974) predicts an intermediate seed size, but empirical observations show considerable variation in seed size within a single parent. Using game theory one can introduce a variety of factors that predict variation in seed size within a single parent, including the introduction of a stochastic environment (Yoshimura & Clark 1991), the assumption of asymmetric competition favoring larger seeds (Geritz 1995, Geritz et al. 1998), and the existence of size-selective predators (Geritz 1998). As already discussed, seed size is often correlated with dispersal ability, resulting in several ESS analyses of the trade-off (Ezoe 1998, Levin & Muller-Landau 2000). The evolution of optimal body size at maturity in animals is similarly enhanced by the use of game theory to include density dependence and/or stochastic environments (Lytle 2001). The existence of asymmetric competition also explains allocation to inefficient support structures required to achieve large body sizes (for example, woody growth in trees) and may explain the mixture of body sizes observed within a community (Falster & Westoby 2003).

Venable & Brown (1988, 1993) presented a synthesis that ties the last three paragraphs together. Seed size, dormancy, and dispersal are all strategies that reduce risk from temporal variability in three factors: environmental stochasticity, crowding, and sibling competition. They showed that a change in one trait leads to a change in the optimum of the others, ultimately leading to the existence of trade-offs resulting from bet-hedging.

Habitat choice and the evolution of specialists and generalists are also frequently analyzed using game theory because choice of habitat clearly depends on what habitats other individuals are choosing. Although Fretwell never mentioned game theory, his concept of an ideal free distribution or IFD (Fretwell 1972, Fretwell & Lucas 1969) is a game, as shown by the worked example in the **Supplemental Appendix**. Rosenzweig and Abramsky have done extensive theoretical and field work on habitat choice (Abramsky et al. 1991, 1997; Rosenzweig 1981, 1987). Joel Brown has extended this work in an explicitly game theoretic context (1988, 1990b, 1992, 1999).

Brew (1982) used game theory to extend this work from discrete patches to a spatial continuum. It has also become clear that the game of habitat selection can lead to the evolution of both generalists and specialists, and to the coexistence of multiple species (Brown 1998, Kisdi 2002, Schmidt et al. 2000).

Coevolution

Game theory becomes a tool for modeling coevolution and the evolution of niches once different strategies of an ESS coalition represent different species, and branching points provide opportunities for speciation. For modeling the coevolution of competitors, Lotka-Volterra competition equations, consumer-resource dynamics, and Levene hard selection and plant growth models have been used as the ecological starting point for W (Apaloo et al. 2005, Brown & Vincent 1987a, Cohen et al. 1999, Cressman & Garay 2003, Flaxman & Reeve 2006, Geritz et al. 1998, Rees & Westoby 1997, Vincent et al. 1996). As we have already seen (**Figures 3** and **4**), competition can lead to branching points and coalitions with multiple species, in some cases through repeated cycles of additional branching (that is, coexistence of many species). But although branching points can be sufficient to achieve the ESS community of n-species, the process in Lotka-Volterra models often proceeds nicely to $n - 1$ species, but then an insurmountable valley emerges separating an occupied peak from an unoccupied peak of even higher fitness (Vincent & Brown 2005). Even when the niche space is continuous, game theory models of competition generally result in a finite number of distinct species with strategies spread out along the niche axis (Mitchell 2000). This result parallels the idea of limiting similarity (MacArthur 1967), but the ESS generally supports fewer species with wider niches than possible under strict nonevolutionary niche packing.

Models of predator-prey coevolution require at least two independent fitness functions built around an ecological model for the prey and another for the predator (Abrams & Matsuda 1997, Bowers et al. 2003, Brown & Vincent 1992, Dieckmann et al. 1995). Any standard model of predator-prey population dynamics can be fashioned into an evolutionary game where the prey have a strategy or vector of strategies influencing resource acquisition, competition, and susceptibility to predation. The predators possess a strategy that influences their ability to catch prey, which is simultaneously influenced by the prey's strategy. These models reveal several important outcomes. First, predator-prey coevolution frequently produces branching points that enhance the number of prey and/or predator species in the ESS (Brown & Vincent 1992, Kisdi 2006). Second, a predator species may be evolutionarily keystone and necessary for promoting the presence of additional prey species within the ESS (but once the prey species has evolved, the predator species may not still be necessary for prey coexistence). Third, sometimes when the number of predator species and prey species is well below the numbers of the coalition ESS, the strategy dynamics will result in perpetual cycling where the prey strategy evolves away from the predator and the predator's strategy chases after it. Or the system may self-annihilate as the predator strategy chases the prey strategy to such extreme values that the prey evolves to extinction.

Coevolution of mutualisms offers a new frontier for evolutionary game theory (de Mazancourt et al. 2005). Like other forms of species interaction, frequency dependence can create ESS coalitions with many coexisting species (Bever 1999). But a number of issues unique to mutualism are emerging. Mutualisms face a Prisoner's Dilemma–like game in which taking from the partner but not giving is the optimal strategy. A variety of mechanisms have been identified that allow a mutualism to coevolve (McGill 2005). The coevolution of mutualisms can in turn affect the evolution of other life history traits (like root:shoot ratio) (Geritz et al. 2006).

Despite the literally hundreds of papers already published (**Figure 1**), continuous-trait game theory is only just beginning to tap into its potential impact on ecology, evolution, coevolution, community assemblages, and the evolution of niches.

SUMMARY POINTS

1. Continuous-trait game theory is a modeling tool whose use is rapidly growing owing to its unique ability to address real-world questions in evolutionary ecology that involve complex traits in a density- and frequency-dependent context.

2. The key innovation of game theory relative to optimization theory is examining the fitness $W(u, U)$ of a rare mutant, u, playing against the field, U. An evolutionarily stable strategy (ESS) is one that is its own best response; at an ESS an individual maximizes its fitness by playing the same strategy as the population, which makes an ESS resistant to invasion by rare alternative strategies.

3. Adaptive dynamics extends the static nature of game theory by exploring the dynamic processes that lead a population to evolve toward an ESS. Surprisingly, evolution can lead to fitness minima as well as maxima being evolutionary repellors. Adaptive dynamics highlights the importance of convergence stability (being an evolutionary attractor).

4. We propose clarification of the convoluted terminology of game theory. We suggest defining an ESS as a strategy or coalition of strategies that is a fitness maximum (invasion-resistant) and convergent stable (evolutionarily attracting). Branching points are fitness minima yet still evolutionary attractors (convergent stable). Finally, evolutionary repellors are strategies that are not convergent stable regardless of whether they are maxima or minima.

5. The notion of an ESS extends easily to (*a*) coalitions of coexisting populations, (*b*) nonequilibrium population dynamics, (*c*) stochastic environments (*d*) stage-structured populations, (*e*) multiple traits, and (*f*) coevolution of multiple species.

6. Continuous-trait game theory has been applied to a wide variety of questions in evolutionary ecology, including evolution of phenology, germination, nutrient foraging in plants, predator-prey foraging, offspring size-number, dispersal, and coevolution.

FUTURE ISSUES

1. Continuous-trait game theory needs to rise to the challenge of its critics (Waxman & Gavrilets 2005) and produce strong predictions that can be tested. We believe the best way to do this is through the development of applied models rather than more theoretical development. This goal leads immediately to the next three items below.

2. The mathematical development of game theory is important but must stop reveling in small differences between theories and instead emphasize the underlying unity of game theory as a language for natural selection. This will make game theory more attractive and accessible to those simply interested in evolution and not game theory in particular. Similarity of new work to preexisting concepts should be highlighted. The proliferation of terminology is excessive and has been detrimental to the field.

3. Collaborations between mathematical modelers and field ecologists are needed to develop applications of continuous-trait game theory to biologically important questions. This requires models to be less heuristic and more amenable to empirical test.

4. Models need to make stronger predictions. Many existing models produce weak predictions. For example, a prediction that an intermediate germination fraction (that is, >0 and <1) is an ESS counts as a weak prediction. Stronger predictions would include predictions that the germination fraction will increase or decrease with increasing environmental noise, environmental autocorrelation, etc.

5. Although developing new and better applications remains the most important future direction, additional mathematical development is also needed on specific topics such as the evolution of multiple traits, evolution in a stochastic environment, tractable conditions for convergence stability in Lyapunov and multitrait/multispecies models, and bifurcation theory. Further exploration of unusual outcomes that occur only in atypical models and of evolution in cyclic or chaotic populations is probably not a high priority.

DISCLOSURE STATEMENT

The authors are not aware of any biases that might be perceived as affecting the objectivity of this review.

LITERATURE CITED

Abrams PA. 2001. Adaptive dynamics: neither F nor G. *Evol. Ecol. Res.* 3:369–73

Abrams PA, Harada Y, Matsuda H. 1993a. On the relationship between quantitative genetics and ESS models. *Evolution* 47:982–85

Abrams PA, Matsuda H. 1997a. Fitness minimization and dynamic instability as a consequence of predator-prey coevolution. *Evol. Ecol.* 11:1–20

Abrams PA, Matsuda H. 1997b. Prey adaptation as a cause of predator-prey cycles. *Evolution* 51:1742–50

Abrams PA, Matsuda H, Harada Y. 1993b. Evolutionarily unstable fitness maxima and stable fitness minima of continuous traits. *Evol. Ecol.* 7:465–87

Abramsky Z, Rosenzweig ML, Pinshow B. 1991. The shape of a gerbil isocline measured using principles of optimal habitat selection. *Ecology* 72:329–40

Abramsky Z, Rosenzweig ML, Subach A. 1997. Gerbils under threat of owl predation: isoclines and isodars. *Oikos* 78:81–90

Apaloo J, Muir PW, Hearne JW. 2005. Multi-species evolutionary dynamics. *Evol. Ecol.* 19:55–77

Auslander D, Guckenheimer J, Oster G. 1978. Random evolutionarily stable strategies. *Theor. Popul. Biol.* 13:276–93

Bever JD. 1999. Dynamics within mutualism and the maintenance of diversity: inference from a model of interguild frequency dependence. *Ecol. Lett.* 2:52–61

Bowers RG, White A, Boots M, Geritz SAH, Kisdi E. 2003. Evolutionary branching/speciation: contrasting results from systems with explicit or emergent carrying capacities. *Evol. Ecol. Res.* 5:883–91

Brew JS. 1982. Niche shift and the minimization of competition. *Theor. Popul. Biol.* 22:367–81

Brown JS. 1988. Patch use as an indicator of habitat preference, predation risk, and competition. *Behav. Ecol. Soc.* 22:37–47

Brown JS. 1990a. Community organization under predator-prey coevolution. See Vincent et al. 1990, pp. 263–88

Brown JS. 1990b. Habitat selection as an evolutionary game. *Evolution* 44:732–46

Brown JS. 1992. Evolution in heterogeneous environments: effects of migration on habitat specialization. *Evol. Ecol.* 6:360–82

Brown JS. 1998. Game theory and habitat selection. In *Game Theory and Animal Behavior*, ed. LA Dugatkin, HK Reeve, pp. 188–220. New York: Oxford Univ. Press

Brown JS. 1999. Vigilance, patch use and habitat selection: foraging under predation risk. *Evol. Ecol. Res.* 1:49–71

Brown JS, Pavlovic NB. 1992. Evolution in heterogeneous environments: effects of migration on habitat specialization. *Evol. Ecol.* 6:360–82

Brown JS, Vincent TL. 1987a. Coevolution as an evolutionary game. *Evolution* 41:66–79

Brown JS, Vincent TL. 1987b. A theory for the evolutionary game. *Theor. Popul. Biol.* 31:140–66

Brown JS, Vincent TL. 1992. Organization of predator-prey communities as an evolutionary game. *Evolution* 46:1269–83

Bulmer M. 1983. Models for the evolution of protandry in insects. *Theor. Popul. Biol.* 23:314–22

Bulmer M. 1984. Delayed germination of seeds: Cohen's model revisited. *Theor. Popul. Biol.* 26:367–77

Bulmer M. 1994. *Theoretical Evolutionary Ecology.* Sunderland, MA: Sinauer

Caswell H. 2001. *Matrix Population Models: Construction, Analysis, and Interpretation.* Sunderland, MA: Sinauer

Christiansen FB. 1991. On conditions for evolutionary stability for a continuously varying character. *Am. Nat.* 138:37–50

Cohen D. 1966. Optimizing reproduction in a randomly varying environment. *J. Theor. Biol.* 12:119–29

Cohen D, Levin SA. 1991. Dispersal in patchy environments: the effects of temporal and spatial structure. *Theor. Popul. Biol.* 39:63–99

Cohen Y. 2003. Distributed evolutionary games. *Evol. Ecol. Res.* 5:383–96

Cohen Y, Vincent TL, Brown JS. 1999. A G-function approach to fitness minima, fitness maxima, evolutionarily stable strategies and adaptive landscapes. *Evol. Ecol. Res.* 1:923–42

Cohen Y, Vincent TL, Brown JS. 2001. Does the G-function deserve an F? *Evol. Ecol. Res.* 3:375–77

Comins HN, Hamilton WD, May RM. 1980. Evolutionarily stable dispersal strategies. *J. Theor. Biol.* 82:205–30

Cressman R, Garay J. 2003. Stability in n-species coevolutionary systems. *Theor. Popul. Biol.* 64:519–33

Crow JF, Kimura M. 1970. *An Introduction to Population Genetics Theory.* New York: Harper & Row

Cushing JM, Costantino RF, Dennis B, Desharnis R, Henson SM. 2002. *Chaos in Ecology: Experimental Nonlinear Dynamics.* San Diego, CA: Academic

Day T, Taylor PD. 2003. Evolutionary dynamics and stability in discrete and continuous games. *Evol. Ecol. Res.* 5:605–13

de Mazancourt C, Dieckmann U. 2004. Trade-off geometries and frequency-dependent selection. *Am. Nat.* 164:765–78

de Mazancourt C, Loreau M, Dieckmann U. 2005. Understanding mutualism when there is adaptation to the partner. *J. Ecol.* 93:305–14

Dieckmann U, Doebeli M. 1999. On the origin of species by sympatric speciation. *Nature* 400:354–57

Dieckmann U, Law R. 1996. The dynamical theory of coevolution: a derivation from stochastic ecological processes. *J. Math. Biol.* 34:569–612

Dieckmann U, Marrow P, Law R. 1995. Evolutionary cycling in predator-prey interactions: population dynamics and the red queen. *J. Theor. Biol.* 176:91–102

Diekmann O, Christiansen F, Law R. 1996. Evolutionary dynamics. Editorial. *J. Math. Biol.* 34:483

Diekmann O, Mylius SD, ten Donkelaar JR. 1999. Saumon a la Kaitala et Getz, sauce hollandaise. *Evol. Ecol. Res.* 1:261–75

Doebeli M, Dieckmann U. 2000. Evolutionary branching and sympatric speciation caused by different types of ecological interactions. *Am. Nat.* 156:S77–101

Doebeli M, Dieckmann U. 2003. Speciation along environmental gradients. *Nature* 421:259–64

Doebeli M, Hauart C, Killingback T. 2004. The evolutionary origin of cooperators and defectors. *Science* 306:859–62

Doebeli M, Ruxton GD. 1997. Evolution of dispersal rates in metapopulation models: branching and cyclic dynamics in phenotype space. *Evolution* 51:1730–41

Ellner S. 1985a. ESS germination strategies in randomly varying environments. I. Logistic-type models. *Theor. Popul. Biol.* 28:50–79

Ellner S. 1985b. ESS germination strategies in randomly varying environments. II Reciprocal yield-law models. *Theor. Popul. Biol.* 28:80–116

Ellner S, Hairston NG Jr. 1994. Role of overlapping generations in maintaining genetic variation in a fluctuating environment. *Am. Nat.* 143:403–17

Eshel I. 1983. Evolutionary and continuous stability. *J. Theor. Biol.* 103:99–111

Ezoe H. 1998. Optimal dispersal range and seed size in a stable environment. *J. Theor. Biol.* 190:287–93

Ezoe H, Iwasa Y. 1997. Evolution of condition-dependent dispersals: a genetic algorithm search for the ESS reaction norm. *Res. Popul. Ecol.* 39:127–37

Falster DS, Westoby M. 2003. Plant height and evolutionary games. *Trends Ecol. Evol.* 18:337–43

Ferrière R. 2000. Adaptive responses to environmental threats: evolutionary suicide, insurance and rescue. *Options 2000*, pp. 12–16. Laxenburg, Austria: Int. Inst. Appl. Syst. Anal.

Ferrière R, Gatto M. 1995. Lyapunov exponents and the mathematics of invasion in oscillatory or chaotic populations. *Theor. Popul. Biol.* 48:126–71

Fisher RA. 1930. *The Genetical Theory of Natural Selection.* Oxford: Clarendon Press

Flaxman SM, Reeve HK. 2006. Putting competition strategies into ideal free distribution models: habitat selection as a tug of war. *J. Theor. Biol.* 243:587–93

Frank SA. 1997. The Price equation, Fisher's fundamental theorem, kin selection, and causal analysis. *Evolution* 51:1712–29

Fretwell SD. 1972. *Populations in a Seasonal Environment.* Princeton, NJ: Princeton Univ. Press

Fretwell SD, Lucas HLJ. 1969. On territorial behavior and other factors influencing habitat distribution in birds. *Acta Biotheor.* 19:16–36

Fudenberg D, Harris C. 1992. Evolutionary dynamics with aggregate shocks. *J. Econ. Theory* 57:420–41

Geritz SAH. 1995. Evolutionarily stable seed polymorphism and small-scale spatial variation in seedling density. *Am. Nat.* 146:685–707

Geritz SAH. 1998. Co-evolution of seed size and seed predation. *Evol. Ecol.* 12:891–911

Geritz SAH, Gyllenberg M, Jacobs FJA, Parvinen K. 2002. Invasion dynamics and attractor inheritance. *Math. Biol.* 44:548–60

Geritz SAH, Gyllenberg M, Yan P. 2006. Plant growth and the optimal sharing of photsynthetic products with a mycorrhizal symbiont. *Evol. Ecol. Res.* 8:577–90

Geritz SAH, Kisdi E. 2000. Adaptive dynamics in diploid, sexual populations and the evolution of reproductive isolation. *Proc. R. Soc. London Ser. B* 267:1671–78

Geritz SAH, Kisdi E, Meszena G, Metz JAJ. 1998. Evolutionarily singular strategies and the adaptive growth and branching of the evolutionary tree. *Evol. Ecol.* 12:35–57

Geritz SAH, Metz JAJ, Kisdi E, Meszena G. 1997. Dynamics of adaptation and evolutionary branching. *Phys. Rev. Lett.* 78:2024–27

Geritz SAH, Van Der Meijden E, Metz JAJ. 1999. Evolutionary dynamics of seed size and seedling competitive ability. *Theor. Popul. Biol.* 55:324–43

Gintis H. 2000. *Game Theory Evolving: A Problem-Centered Introduction to Modeling Strategic Interaction*. Princeton, NJ: Princeton Univ. Press

Goodman D. 1984. Risk spreading as an adaptive strategy in iteroparous life histories. *Theor. Popul. Biol.* 25:1–20

Gyllenberg M, Metz JAJ. 2001. On fitness in structured metapopulations. *J. Math. Biol.* 43:545–60

Gyllenberg M, Parvinen K, Dieckmann U. 2002. Evolutionary suicide and evolution of dispersal in structured metapopulations. *J. Math. Biol.* 45:79–105

Hamilton WD. 1967. Extraordinary sex ratios. *Science* 156:477–88

Hofbauer J. 1996. Evolutionary dynamics for bimatrix games: a Hamiltonian system? *J. Math. Biol.* 34:675–88

Holt RD, McPeek MA. 1996. Chaotic population dynamics favors the evolution of dispersal. *Am. Nat.* 148:709–18

Iwasa Y, Haccou P. 1994. ESS emergence pattern of male butterflies in stochastic environments. *Evol. Ecol.* 8:503–23

Iwasa Y, Levin SA. 1995. The timing of life-history events. *J. Theor. Biol.* 172:33–42

Iwasa Y, Odendaal FJ, Murphy DD, Ehrlich PR, Launer AE. 1983. Emergence patterns in male butterflies: a hypothesis and a test. *Theor. Popul. Biol.* 23:363–79

Kaitala A, Kaitala V, Lundberg P. 1993. A theory of partial migration. *Am. Nat.* 142:59–81

Kisdi E. 2002. Dispersal: risk spreading vs local adaptation. *Am. Nat.* 159:579–96

Kisdi E. 2004. Conditional dispersal under kin competition: extension of the Hamilton-May model to brood size-dependent dispersal. *Theor. Popul. Biol.* 66:369–80

Kisdi E. 2006. Trade-off geometries and the adaptive dynamics of two coevolving species. *Evol. Ecol. Res.* 8:959–73

Kisdi E, Geritz SAH. 1999. Adaptive dynamics in allele space: evolution of genetic polymorphism by small mutations in a heterogeneous environment. *Evolution* 53:993–1008

Kokko H. 1999. Competition for early arrival in migratory birds. *J. Anim. Ecol.* 68:940–50

Kokko H, Johnstone RA. 1999. Social queuing in animal societies: a dynamic model of reproductive skew. *Proc. R. Soc. London Ser. B* 266:571–78

Kondrashov AS. 1986. Sympatric speciation: when is it possible. *Biol. J. Linn. Soc.* 27:201–23

Kondrashov AS, Kondrashov FA. 1999. Interactions among quantitative traits in the course of sympatric speciation. *Nature* 400:351–54

Lande R. 1979. Quantitative genetic analysis of multivariate evolution applied to brain-body size allometry. *Evolution* 33:402–16

Lavorel S, ONeill RV, Gardner RH. 1994. Spatiotemporal dispersal strategies and annual plant-species coexistence in a structured landscape. *Oikos* 71:75–88

Lawlor LR, Maynard-Smith J. 1976. The coevolution and stability of competing species. *Am. Nat.* 110:79–99

Leimar O. 2001. Evolutionary change and Darwinian demons. *Selection* 2:65–72

Leimar O. 2007. Multidimensional convergence stability and the canonical adaptive dynamics. In *Elements of Adaptive Dynamics*, ed. U Dieckmann, JAJ Metz. Cambridge, UK: Cambridge Univ. Press. In press

Lemel JY, Belichon S, Clobert J, Hochberg ME. 1997. The evolution of dispersal in a two-patch system: some consequences of diferences between migrants and residents. *Evol. Ecol.* 11:613–29

Lessard S. 1990. Evolutionary stability: One concept, several meanings. *Theor. Popul. Biol.* 37:159–70

Levene H. 1953. Genetic equilibrium when more than one ecological niche is available. *Am. Nat.* 87:331–32

Levin SA, Cohen D, Hastings A. 1984. Dispersal strategies in patchy environments. *Theor. Popul. Biol.* 26:165–91

Levin SA, Muller-Landau HC. 2000. The evolution of dispersal and seed size in plant communities. *Evol. Ecol. Res.* 2:409–35

Levins R. 1968. *Evolution in Changing Environments*. Princeton, NJ: Princeton Univ. Press. ix+120 pp.

Luce RD, Raiffa H. 1957. *Games and Decisions: Introduction and Critical Survey*. New York: Wiley

Ludwig D, Levin SA. 1991. Evolutionary stability of plant communities and the maintenance of multiple dispersal types. *Theor. Popul. Biol.* 40:285–307

Lytle DA. 2001. Disturbance regimes and life-history evolution. *Am. Nat.* 157:525–36

MacArthur RH. 1967. The limiting similarity, convergence and divergence of coexisting species. *Am. Nat.* 101:377–85

Matessi C, Di Pasquale C. 1996. Long-term evolution of multilocus traits. *J. Math. Biol.* 34:613–53

Matessi C, Jayakar SD. 1976. Conditions for the evolution of altruism under Darwinian selection. *Theor. Popul. Biol.* 9:360–87

Mathias A, Kisdi E. 2002. Adaptive diversification of germination strategies. *Proc. R. Soc. London Ser. B* 269:151–55

Mathias A, Kisdi E, Olivieri I. 2001. Divergent evolution of dispersal in a heterogeneous landscape. *Evolution* 55:246–59

Maynard-Smith J. 1981. Will a sexual population evolve to an ESS? *Am. Nat.* 117:1015–18

Maynard-Smith J. 1982. *Evolution and the Theory of Games*. Cambridge, UK: Cambridge Univ. Press. 224 pp.

Maynard Smith J. 1966. Sympatric speciation. *Am. Nat.* 100:637–50

Maynard Smith J, Price GR. 1973. The logic of animal conflict. *Nature* 246:15–18

McGill BJ. 2005. A mechanistic model of a mutualism and its ecological and evolutionary dynamics. *Ecol. Model.* 187:413–25

Metz JAJ, Geritz SAH, Meszena G, Jacobs FJA, van Heerwaarden J. 1996. Adaptive dynamics: a geometrical study of the consequences of nearly faithful reproduction. In *Stochastic and Spatial Structures of Dynamical Systems*, ed. SJ van Strien, SM Verduyn-Lunel, pp. 183–231. Amsterdam: KNAW Verhandelingen

Metz JAJ, Nisbet RM, Geritz SAH. 1992. How should we define 'fitness' for general ecological scenarios? *Trends Ecol. Evol.* 7:198–202

Mitchell WA. 2000. Limits to species richness in a continuum of habitat heterogeneity: an ESS approach. *Evol. Ecol. Res.* 2:293–316

Motro U. 1982. Optimal rates of dispersal I. Haploid populations. *Theor. Popul. Biol.* 21:394–411

Motro U. 1983. Optimal rates of dispersal. III. Parent-offspring conflict. *Theor. Popul. Biol.* 23:159–68

Nash J. 1950. Equilibrium points in n-person games. *Proc. Natl. Acad Sci. USA* 36:48–49

Parker GA, Courtney P. 1983. Seasonal incidence: adaptive variation in the timing of life history stages. *J. Theor. Biol.* 105:147–55

Parvinen K. 1999. Evolution of migration in a metapopultaion. *Bull. Math. Biol.* 61:531–50

Rand DA, Wilson HB, McGlade JM. 1994. Dynamics and evolution: evolutionarily stable attractors, invasion exponents and phenotype dynamics. *Philos. Trans. R. Soc. London Ser. B* 343:261–83

Rees M, Westoby M. 1997. Game-theoretical evolution of seed mass in multi-species ecological models. *Oikos* 78:116–26

Reeve HK, Dugatkin LA, eds. 1998. *Game Theory and Animal Behavior*. New York: Oxford Univ. Press

Roff D. 2002. *Life History Evolution*. Sunderland, MA: Sinauer

Ronce O, Gandon S, Rousset F. 2000. Kin selection and natal dispersal in an age-structured population. *Theor. Popul. Biol.* 58:143–59

Rosenzweig ML. 1978. Competitive speciation. *Biol. J. Linn. Soc.* 10:275–89

Rosenzweig ML. 1981. A theory of habitat selection. *Ecology* 62:327–35

Rosenzweig ML. 1987. Habitat selection as a source of biological diversity. *Evol. Ecol.* 1:315–30

Roughgarden J. 1976. Resource partitioning among competing species—a coevolutionary approach. *Theor. Popul. Biol.* 9:338–424

Roughgarden J. 1979. *Theory of Population Genetics and Evolutionary Ecology: An Introduction*. New York: Macmillan. 612 pp.

Roughgarden J. 1983. The theory of coevolution. In *Coevolution*, ed. DJ Futuyma, M Slatkin, pp. 33–64. Sunderland, MA: Sinauer

Schmidt KA, Earnhardt JM, Brown JS, Holt RD. 2000. Habitat selection under temporal heterogeneity: exorcising the ghost of competition past. *Ecology* 81:2622–30

Seger J, Brockmann HJ. 1987. What is bet-hedging? In *Oxford Surveys in Evolutionary Biology*, ed. PH Harvey, L Partridge, pp. 183–211. Oxford: Oxford Univ. Press

Smith CC, Fretwell SD. 1974. The optimal balance between size and number of offspring. *Am. Nat.* 108:499–506

Stephens PA, Russell AF, Young AJ, Sutherland WJ, Clutton-Brock TH. 2005. Dispersal, eviction and conflict in meerkats (Suricata suricata): an evolutionarily stable strategy model. *Am. Nat.* 165:120–35

Taylor PD. 1989. Local stability in seed provisioning model. *J. Theor. Biol.* 136:1–12

Taylor PD, Frank SA. 1996. How to make a kin selection model. *J. Theor. Biol.* 180:27–37

Thoday JM, Gibson JB. 1962. Isolation by disruptive selection. *Nature* 193:1164–66

Tregenza T, Butlin RK. 1999. Speciation without isolation. *Nature* 400:311–12

Tuljapurkar SD. 1990. *Population Dynamics in Variable Environments*. New York: Springer-Verlag

Tuljapurkar SD. 1997. Stochastic matrix models. In *Structured-Population Models in Marine, Terrestrial and Freshwater Systems*, ed. SD Tuljapurkar, H Caswell, pp. 59–87. New York: Chapman & Hall

Tuljapurkar SD, Istock C. 1993. Environmental uncertainty and variable diapuse. *Theor. Popul. Biol.* 43:251–80

Tuljapurkar SD, Orzack SH. 1980. Population dynamics in variable environments. I. Long-run growth rates and extinction. *Theor. Popul. Biol.* 18:314–42

Venable DL, Brown JS. 1988. The selective interactions of dispersal, dormancy and seed size as adaptations for reducing risk in variable environments. *Am. Nat.* 131:360–84

Venable DL, Brown JS. 1993. The seed population-dynamic functions of seed dispersal. In *Frugivory and Seed Dispersal: Ecological and Evolutionary Aspects. Vegetatio* 107/108, ed. TH Fleming, A Estrada, pp. 31–55. Dordrecht: Kluwer

Vincent TL. 1990. Strategy dynamics and the ESS. See Vincent et al. 1990, pp. 236–62

Vincent TL, Brown JS. 1984. Stability in an evolutionary game. *Theor. Popul. Biol.* 26:408–27

Vincent TL, Brown JS. 1987. Evolution under nonequilibrium dynamics. *Math. Model.* 8:766–71

Vincent TL, Brown JS. 1988. The evolution of ESS theory. *Annu. Rev. Ecol. Syst.* 19:423–43

Vincent TL, Brown JS. 2001. Evolutionarily stable strategies in multistage biological systems. *Selection* 2:85–102

Vincent TL, Brown JS. 2005. *Evolutionary Game Theory, Natural Selection, and Darwinian Dynamics*. Cambridge, UK: Cambridge Univ. Press. 400 pp.

Vincent TL, Cohen Y, Brown JS. 1993. Evolution via strategy dynamics. *Theor. Popul. Biol.* 44:149–76

Vincent TL, Cressman R. 2000. An ESS maximum principle for matrix games. *Theor. Popul. Biol.* 58:173–86

Vincent TL, Mees AI, Jennings LS, eds. 1990. *Dynamics of Complex Interconnected Biological Systems*. Boston: Birkhauser

Vincent TL, Van MV, Goh BS. 1996. Ecological stability, evolutionary stability and the ESS maximum principle. *Evol. Ecol.* 10:567–91

Vincent TLS, Scheel D, Brown JS, Vincent TL. 1996. Trade-offs and coexistence in consumer-resource models: It all depends what and where you eat. *Am. Nat.* 148:1038–58

von Neumann J, Morgenstern O. 1944. *Theory of Games and Economic Behaviour*. Princeton, NJ: Princeton University Press

Waxman D, Gavrilets S. 2005. 20 questions on adaptive dynamics. *J. Evol. Biol.* 18:1139–54

Wiklund C, Wickman PO, Nylin S. 1992. A sex difference in the propensity to enter direct/diapause development—a result of selection for protandry. *Evolution* 46:519–28

Winkler E, Fischer M. 1999. Two fitness measures for clonal plants and the importance of spatial aspects. *Plant Ecol.* 141:191–99

Wright S. 1930. Evolution in Mendellian populations. *Genetics* 16:97–159

Yoshida T, Jones LE, Ellner SP, Fussmann GF, Hairston NGJ. 2003. Rapid evolution drives ecological dynamics in a predator-prey system. *Nature* 424:303–6

Yoshimura J, Clark CW. 1991. Individual adaptation in stochastic environments. *Evol. Ecol.* 5:173–92

The Maintenance of Outcrossing in Predominantly Selfing Species: Ideas and Evidence from Cleistogamous Species

Christopher G. Oakley, Ken S. Moriuchi, and Alice A. Winn

Department of Biological Science, Florida State University, Tallahassee, Florida 32306-1100; email: coakley@bio.fsu.edu; moriuchi@bio.fsu.edu; winn@bio.fsu.edu

Annu. Rev. Ecol. Evol. Syst. 2007. 38:437–57

First published online as a Review in Advance on August 8, 2007

The *Annual Review of Ecology, Evolution, and Systematics* is online at http://ecolsys.annualreviews.org

This article's doi: 10.1146/annurev.ecolsys.38.091206.095654

Key Words

autogamy, inbreeding depression, mixed mating, reproductive strategy, self-fertilization

Abstract

Cleistogamous species present strong evidence for the stability of mixed mating, but are generally not considered in this context. Individuals of cleistogamous species produce both obligately selfing cleistogamous flowers (CL) and potentially outcrossed chasmogamous flowers (CH) with distinct morphologies. Greater energetic economy and reliability of CL relative to CH suggest that forces that maintain selection for outcrossing may be stronger in these species than in mixed maters with monomorphic flowers. We reviewed data from 60 studies of cleistogamous species to evaluate proposed explanations for the evolutionary stability of mixed cleistogamous and chasmogamous reproduction and to quantify the magnitude of selection necessary to account for the maintenance of CH. We found circumstantial support for existing hypotheses for the stability of cleistogamy, and that forces that maintain CH must account for a 15–342% advantage of reproduction via CL. We suggest that heterosis and the effects of mass action pollination should be considered.

INTRODUCTION

Mixed mating: A combination of selfing and outcrossing within an individual

Cleistogamous species: Species in which individuals are capable of producing both chasmogamous and cleistogamous flowers

Cleistogamous flowers (CL): Small inconspicuous permanently closed flowers that self-fertilize in the bud

Chasmogamous flowers (CH): Open flowers that are potentially outcrossed

The pattern of mating among individuals of sexually reproducing species has substantial influence on the process of evolution. The average degree of relatedness among mates determines how genetic variation is distributed among lineages and therefore influences the potential of a species to respond to natural selection. Much of the theory and empirical work concerning the evolution of mating systems has focused on hermaphroditic plants because they exhibit a wide range of variation and a high degree of evolutionary lability in their mating systems. A recent review of this body of work (Goodwillie et al. 2005) highlights the persistent puzzle of what evolutionary forces maintain stable mixtures of self-fertilization and outcrossing. The most general models of mating system evolution conclude that only complete selfing and complete outcrossing are evolutionarily stable, but Goodwillie et al. (2005) report that 42% of 345 species for which data on mating systems is currently available exhibit mixed mating, defined as rates of selfing between 0.2 and 0.8. Although the estimated frequency of mixed mating may be biased by unequal sampling of taxa (e.g., undersampling of obligately outcrossing taxa; Igic & Kohn 2006), it is clear that a substantial number of species do engage in mixed mating.

The appreciable frequency of mixed mating species has stimulated the development of models that incorporate additional forces that could influence mating system evolution. In particular, models have added ecological forces related to pollen transfer with the result that many have identified conditions that allow for stable mixed mating. Goodwillie et al. (2005) compile a list of more than twenty models that have been proposed to explain the adaptive significance and stability of mixed mating, and note that a lack of empirical data rather than theoretical explanations limits our current understanding of mating system evolution.

Cleistogamous species, in which individuals produce a mixture of cleistogamous flowers (CL) and chasmogamous flowers (CH), present a particularly striking challenge to the prediction that mixed mating is not evolutionarily stable. Cleistogamous species may also offer particular insight into selective forces that maintain outcrossing once selfing evolves. CL are highly reduced structures that obligately self-fertilize in the bud. Because CL never reach anthesis, they can neither donate nor receive outcrossed pollen. Most species that produce CL also produce open CH that can participate in outcrossing. The production of both CH and CL by individuals inherently facilitates a mixture of selfing and outcrossing (Darwin 1896). The production of CL appears to have evolved independently in as many as 50 plant families (Culley & Klooster 2007, Lord 1981, Plitmann 1995, Uphof 1938), and there are few reported evolutionary losses of cleistogamy (Campbell et al. 1983, Culley & Klooster 2007, Lord 1981), suggesting that the conditions that favor the evolution and maintenance of cleistogamy are common. Understanding the conditions that favor the stable production of both CH and CL could provide insight into the conditions that favor stable mixed mating in general. Although cleistogamy is well known among botanists, it has rarely been exploited as a tool for testing mating system theory because it entails unusual features that are not accounted for in general models.

The most general models of mating system evolution emphasize the balance between two genetic forces. The major force favoring the evolution of self-fertilization is the automatic fitness advantage to mixed selfing and outcrossing relative to outcrossing alone (Fisher 1941, Jain 1976). This advantage arises because an individual that both selfs and outcrosses gains fitness through siring its own ovules and those of other individuals, whereas a pure outcrosser can only sire ovules produced by other individuals, resulting in a 50% fitness advantage, on average, to mixed selfing and outcrossing. The primary force believed to counter the automatic fitness advantage is inbreeding depression, the reduction in fitness of selfed relative to outcrossed progeny (Charlesworth & Charlesworth 1987, Lande & Schemske 1985, Lloyd 1979). Assuming that outcrossed mates are unrelated, if inbreeding depression is greater than 50%, complete outcrossing is predicted to evolve, and if inbreeding depression is less than 50%, complete selfing should evolve. The automatic fitness advantage is reduced when selfing reduces the amount of pollen a selfer contributes to outcrossing (Feldman & Christiansen 1984, Holsinger et al. 1984, Nagylaki 1976), an effect described as pollen discounting. Pollen discounting lowers the threshold inbreeding depression required for complete outcrossing to evolve (Holsinger et al. 1984), and recent theory indicates that pollen discounting, along with other aspects of the dynamics of pollen transfer may be essential to the evolutionary stability of mixed mating (Johnston 1998, Morgan & Wilson 2005, Porcher & Lande 2005).

The mating system of cleistogamous species (used here to designate species in which individuals can produce both CH and CL) is not well described by models that emphasize the balance between inbreeding depression and the automatic fitness advantage. Because CL never reach anthesis, they cannot contribute any pollen to outcrossing and therefore do not reap the automatic fitness advantage (i.e., they suffer 100% pollen discounting). On the other hand, CL are typically considerably smaller and therefore cheaper to produce than CH (Lord 1981, Schemske 1978, Waller 1979) and can produce seed even when pollen vectors fail. The greater economy and reliability of CL provide advantages to selfing that do not occur in mixed mating species with monomorphic flowers. The other main genetic force in models of mating system evolution, inbreeding depression, is expected to be low in regularly selfing species (Husband & Schemske 1996), including cleistogamous taxa. The economy and reliability of CL combined with the expectation of low inbreeding depression suggest that once cleistogamy arises, complete selfing via CL should be favored. Because selection favoring selfing will be particularly strong in cleistogamous species, the forces that maintain outcrossing via the production of CH must be exceptionally pronounced. These features make cleistogamous species particularly favorable empirical systems for identifying factors that maintain the potential for outcrossing once selfing evolves. The relative ease of measuring costs and benefits of reproduction via CH and CL also facilitates quantitative estimation of the magnitude of selection necessary to maintain mixed mating in cleistogamous species, which could be used to evaluate the likelihood that specific mechanisms are sufficient to explain the maintenance of outcrossing in cleistogamous species.

Although theory for mating system evolution has typically treated cleistogamy as a special case, if at all (but see Lu 2002, Masuda et al. 2001, Schoen & Lloyd

Inbreeding depression: The reduction of fitness in selfed (relative to outcrossed) progeny

Pollen discounting: Reduction in the amount of pollen available for outcrossing as a function of self-fertilization

1984), the pronounced dimorphism associated with mating system in cleistogamous species has made them attractive subjects for empirical studies of the patterns and consequences of selfing and outcrossing (e.g., Clay & Antonovics 1985, Lu 2000, Schemske 1978, Schmitt & Gamble 1990, Waller 1979). Consequently, the literature contains quantitative data on relative costs of reproduction by the two flower morphs, the relative fitness of progeny produced by each flower type, CH outcrossing rates, and environmental factors associated with increased or decreased proportion of CL produced. The possible insight that cleistogamy could provide into mating system evolution appears to have been overlooked.

Here we summarize the theoretical and empirical literature on cleistogamous species that is relevant to mating system evolution and consider how cleistogamy might be further exploited as a tool in the study of the evolution of plant mating systems. We begin by summarizing the formal models and verbal arguments that have been advanced to explain the evolutionary stability of cleistogamy. We then review data from 60 empirical studies of cleistogamous species that are relevant to the predictions and assumptions of these models. Finally, we describe the evolutionary scenarios that appear most likely to account for the maintenance of outcrossing once selfing arises in cleistogamous species.

Variations on the Theme of Cleistogamy

True cleistogamy was defined by Lord (1981) as the presence of two developmentally distinct flower types. CL are consistently characterized by reduction in the size of the androecium and by precocious maturation of anthers relative to CH of the same species (Lord 1981). Beyond these commonalities, there are many variations on the theme of cleistogamy.

The degree of morphological divergence of CH and CL can range from primarily a matter of differences in organ size as in *Lamium amplexicaule*, to substantial differences in organ shape as in *Viola odorata* (Lord & Hill 1987). In a number of species, seeds produced by the two flower types differ in size as well as in germination behavior (see e.g., McNamara & Quinn 1977, Zeide 1978). Some cleistogamous species are also amphicarpic, producing both underground seeds with limited dispersal potential and more readily dispersed above-ground seeds (Schnee & Waller 1986, Weiss 1980). Floral phenology of cleistogamous species ranges from simultaneous to sequential, nonoverlapping production of the two floral forms.

The degree of cleistogamy, measured as the proportion of flowers produced by an individual that are cleistogamous, varies both among and within species in relation to plant size and growing conditions (see e.g., Clay 1983, Le Corff 1993, Weiss 1980). Little is known about the genetic basis for variation in degree of cleistogamy in nonagricultural species, although Clay (1982) reported a broad sense heritability of 52.6% in the field and 71.6% in the greenhouse for percent CL for a population of the grass *Danthonia spicata*. Crosses between lines of the cultivated annual plant *Salpiglossis sinuata* that differ in degree of cleistogamy suggest that a dominant allele at a single major locus confers cleistogamy and that modifiers influence the degree of cleistogamy (Lee et al. 1976). Recent work with cultivated barley, *Hordeum vulgare* spp. *vulgare*, suggests that two major genes control cleistogamy and that the direction of dominance is opposite for the two loci (Turuspekov et al. 2004).

Cleistogamy is particularly common among members of the grass family, Poaceae (Campbell et al. 1983), and in dicot families characterized by CH with pronounced adaptation for zoophily such as the Balsaminaceae, Fabaceae, and Violaceae (Lord 1981, Plitmann 1995). Herbaceous annuals and perennials are approximately equally represented among cleistogamous taxa, but few woody species are cleistogamous (Plitmann 1995).

THEORIES FOR THE EVOLUTIONARY STABILITY OF CLEISTOGAMY

Hypotheses for the Maintenance of Chasmogamous Flowers

Three mathematical models (Lu 2002, Masuda et al. 2001, Schoen & Lloyd 1984) have formally addressed the conditions that can favor maintenance of production of both CH and CL by individuals (**Table 1**). Schoen & Lloyd (1984) added the unique features of cleistogamy to the general modeling framework for the evolution of plant mating systems based on the balance between automatic selection and inbreeding depression. Schoen & Lloyd's model incorporates the relative costs of reproduction by the two flower types, the complete inability of CL to participate in outcrossing

Table 1 Mathematical and verbal models proposed to maintain the production of both chasmogamous flowers (CH) and cleistogamous flowers (CL)

Model citation	Critical assumptions	Force proposed to maintain both CH and CL	Predictions for empirical data
Schoen & Lloyd 1984	Two environments, with parental fitness being maximized by a different flower type in each environment	Adaptive phenotypic plasticity	Each flower type produced under conditions in which it makes a greater contribution to parental fitness
Masuda et al. 2001	Pure chasmogamous flowering favored in the absence of geitonogamy, which increases with number of CH	Geitonogamy	Simultaneous production of CH and CL
Lu 2002	Selfing lineages purge lethal mutations, but different lines fix different, mildly deleterious alleles	Heterosis	Heterosis in crosses between lines that have undergone uninterrupted selfing
Waller 1980, Zeide 1978	Production of two flower morphs reduces variance in fitness over time	Variance discounting	Inverse relationship between plant size and fraction of flowers that are cleistogamous
Waller 1980, 1984; Schmitt & Ehrhardt 1987	Temporal variation in sibling density and less competition among sibling progeny of CH than among progeny of CL	Variance discounting	Greater fitness of progeny of CH than progeny of CL when grown in competition with siblings

(i.e., 100% pollen discounting), and potential differences in the relative fitnesses of progeny derived from the two flower types, including both the effects of flower type per se (e.g., differences in seed size) and inbreeding depression. Initial analysis of the model suggests that a mutation resulting in the production of CL would be favored when costs of male function are high and/or the structure of CH limits reproductive assurance via autogamy. The correspondence of these conditions with the high frequency of cleistogamy in the grasses and in families with pronounced adaptation for zoophily suggest that the model does capture some of the essential features of cleistogamy.

Schoen & Lloyd (1984) used their model to compare the fitnesses of alternate phenotypes that differ only in their degree of cleistogamy (proportion of total flowers that are cleistogamous). Adding the details of cleistogamy alone produced the same conclusions as have general models for mating system evolution (e.g., Charlesworth & Charlesworth 1987, Lande & Schemske 1985, Lloyd 1979). Either complete cleistogamous selfing, or complete chasmogamous outcrossing was favored depending on the relative costs of chasmogamous and cleistogamous reproduction, the selfing rate of CH, and the relative viabilities of offspring produced via CL and CH.

The predictions were strikingly different when temporal variation in the environment of reproducing parental plants was added to the model (Schoen & Lloyd 1984). This version of the model assumes that during a portion of the flowering season, reproduction via CH contributes more to parental fitness than reproduction via CL and that the reverse is true under other conditions. The model predicts that a phenotype that is able to respond perfectly to an environmental cue by producing the appropriate flower type will have greater fitness than one that responds to a lesser extent, making temporally separated production of both CH and CL stable. This is implicitly a model of cleistogamy as adaptive phenotypic plasticity, which requires that selection favors different phenotypes (in this case flower types) in different environments. If there is only one selective environment, or if plants cannot detect and respond to a cue indicating which flower type is favored, production of both CH and CL would not be stable.

The model of cleistogamy as a form of adaptive plasticity is plausible, but relies on the presumed existence of alternate environmental states in which each floral morph is favored. Schoen & Lloyd (1984) suggested temporal variation in the availability of pollinators as one scenario that might impose such a pattern of selection. The production of CH would be favored when pollinators were active, and reproduction via CL would contribute more to parental fitness when they were scarce or absent. Schoen & Lloyd (1984) also explored a variant of their model that includes differential dispersal of seeds from CH and CL, but because this model has limited generality, we will not consider it further here.

Masuda et al. (2001) later added the effect of increased geitonogamous chasmogamous selfing (selfing between CH of the same individual) as a function of increased CH production to Schoen & Lloyd's framework. They implicitly imposed conditions that favor chasmogamous reproduction, and asked if an increase in geitonogamy with increased production of CH can stabilize the production of both CH and CL by limiting the advantages of producing only CH. The model predicts that the production

Autogamy:
Self-fertilization within a flower

Geitonogamy:
Self-fertilization between flowers on the same plant

of both CH and CL can be stable, although it does not specify the conditions that favor the production of CH in the absence of geitonogamy. This model has several parallels among the models for mixed mating in species with monomorphic flowers that suggest that selfing persists in primarily outcrossing taxa as a byproduct of adaptations to increase pollinator attraction (e.g., Johnston 1998, Porcher & Lande 2005, Scofield & Schultz 2006). This scenario seems much less likely for cleistogamous species because CL would not contribute to pollinator attraction but would be expected to exact a cost owing to inbreeding depression, especially in a primarily outcrossing species. In addition, regular cleistogamous selfing would be expected to reduce inbreeding depression via purging and eventually shift selection toward complete selfing.

A distinctly different idea for the maintenance of both CH and CL has been suggested by Lu (2002) who proposed that the two floral morphs are maintained as a balance between the benefits of the economy of purging in CL and those of heterosis, which can only be realized via chasmogamous outcrossing. Lu reasons that CL provide an economical means of purging inbreeding depression caused by recessive lethals, but that uninterrupted selfing that would result from complete CL production could permit the fixation of more mildly deleterious mutations within selfing lineages. This load could be reduced by heterosis in crosses between serially selfed lines that have fixed different mutations, which requires the production of CH. Lu modified the fitness equations of the Schoen & Lloyd model to include effects of purging in CL and of heterosis for crosses between two parents, each of which was derived from cleistogamous seeds from a different selfing lineage. Her model suggests that if CL provide cheap purging of lethals and outcrossing confers heterosis, then as the selfing rate in a population increases, the benefit of crossing between lines also increases and that this force favors the maintenance of chasmogamous outcrossing. This model seems plausible, although the greater economy of CL alone might be sufficient to favor their production, even without their role in purging, leaving the potential for heterosis as the novel explanation for the maintenance of CH.

Several verbal hypotheses have advanced the idea that unpredictable variation in the environment periodically provides an advantage to reproduction via chasmogamous outcrossing that balances selection favoring the economy and reliability of reproduction via cleistogamous selfing (Waller 1980, Zeide 1978). In contrast to the suggestion of adaptive plastic response to predictable variation invoked by Schoen & Lloyd (1984), these explanations portray the production of both CH and CL as a variance discounting strategy (Real 1980, Seger & Brockmann 1987) that confers greater geometric mean fitness than either pure CH or CL when the environment varies unpredictably through time. The mechanisms that have been proposed to underlie the advantage of producing genetically variable progeny parallel mechanisms proposed to explain the maintenance of sexual reproduction.

Waller (1980, 1984) and Zeide (1978) suggest that although reproduction by CL is cheaper and more reliable, if there are periods of time during which CH are favored, then producing both flower types could reduce the long-term variance in fitness and therefore maximize geometric mean fitness. Because outcrossing in CH allows greater recombination than does self-fertilization, one possible mechanism favoring chasmogamous reproduction is that the greater genetic variation generated by chasmogamous outcrossing could facilitate adaptation when the environment changes (Waller 1984). This scenario parallels the suggestion that the production of genetic variation via sexual reproduction allows adaptation to environmental change and favors the maintenance of sex (Kondrashov 1993, Maynard Smith 1978, Williams 1975).

Avoidance of sibling competition has also been invoked both to explain the advantage of chasmogamous reproduction (Waller 1980) and to favor the maintenance of sexual reproduction (Maynard Smith 1978, Price & Waser 1982, Williams 1975, Young 1981). The argument for cleistogamy is that half-sibs produced by chasmogamous outcrossing are genetically more dissimilar than full-sibs that result from cleistogamous selfing and are therefore expected to experience a smaller reduction in fitness when grown in competition with each other than would full-sib selfed offspring produced via CL. The production of both CH and CL would be favored by unpredictable temporal variation in the degree to which siblings experience competition. Periods of intense competition would favor chasmogamous reproduction, and times of reduced competition would favor cleistogamous reproduction for the sake of energetic economy.

Data Relevant to Evaluating Hypotheses for the Stability of Cleistogamy

None of the models for the maintenance of production of both CH and CL has been tested directly, and the complete data necessary for such tests is not available in the literature. However, published empirical data on cleistogamous species can be used to evaluate the likelihood of the scenarios proposed by different models, to suggest what would constitute appropriate tests of these models and to determine the magnitude of the advantage that must be conferred by any mechanism proposed to maintain investment in CH despite their greater energetic cost and lower reliability.

The frequency and phenology of the production of CH and CL could illuminate the roles of adaptive plasticity and geitonogamy in the maintenance of cleistogamy. If mixed production of CH and CL constitutes adaptive plasticity (Schoen & Lloyd 1984), then each flower type should be produced in the environment in which it contributes most to parental fitness. If, for example, variation in pollinator availability selects for the production of CH and CL during different seasons, then CH and CL should be produced at different times and each during the season in which it is favored. Data on changes in the proportion of flowers that are CH in response to the environment could suggest what specific environmental factors favor CH reproduction. A decrease in the number of CH with total flower number would be consistent with avoidance of geitonogamy (Masuda et al. 2001) because the opportunity for geitonogamous pollination would be expected to increase with display size (see e.g., Barrett et al. 1994, Klinkhamer & de Jong 1993, Routley & Husband 2003). In costrast to the scenario for adaptive plasticity in response to seasonal variation in pollinator availablity, avoidance of geitonogamy should favor simultaneous rather than sequential production of the two flower morphs.

The hypothesis that the production of both CH and CL reflects adaptation to an unpredictably varying environment is difficult to test empirically because demonstrating variance discounting requires a long temporal sequence of fitness estimates, but data on the degree of cleistogamy and on sibling competition can support or refute the mechanisms purported to underlie selection for mixed production of CH and CL in these models. Zeide (1978) suggested that individuals with limited resources should favor the production of the lower risk phenotype. Thus we might expect smaller plants and those in poorer condition to produce CL in preference to CH. The feasibility of the sibling competition mechanism would be supported by evidence of stronger competition among more related cleistogamous siblings than among less related outcrossed chasmogamous siblings.

Data on the frequency of fruit set of each flower type and on the energetic cost and relative performance of seeds produced by CH and CL can be combined to estimate the magnitude of advantage to CH necessary to justify their continued production. This estimate could serve as a benchmark to evaluate the likelihood that any particular mechanism is sufficient to account for the continued production of CH despite their greater energetic expense and reliance on pollen vectors. Dissection of the cost of CH into its component parts can also identify specific forces that are or are not likely to account for their maintenance.

REVIEW OF THE EMPIRICAL DATA

We reviewed 60 published studies of cleistogamous species and summarized the data on the phenology of flower production, the identity of environmental factors associated with differences in the percent of flowers produced that are cleistogamous, and the relative fitness costs and benefits of reproduction via CH and CL, including the rate of selfing in CH and the magnitude of inbreeding depression.

Phenology and Plasticity

We found 37 studies containing information on the timing of production of CH and CL and 18 describing plastic changes in degree of cleistogamy. We report these data by genus to correct for the disproportionately large representation of species within certain genera (e.g., *Impatiens*, *Oxalis*, and *Viola*). Unexpectedly, we found a difference in the phenology of flower dimorphism between annual and perennial cleistogamous taxa (**Supplemental Table 1**; follow the Supplemental Material link from the Annual Reviews home page at **http://www.annualreviews.org**). For every annual species (N = 13 species in 11 genera) for which we found data, there was overlap in the production of the two flower types, and these species always produced CL first (**Supplemental Table 1**), followed by simultaneous production of CH and CL, and occasionally a second period of only CL. Among perennials, about half of the species (total N = 17 species in 10 genera) produced the two flower types sequentially in distinct seasons. Also unlike the annuals, most perennials initiated production of CL either simultaneously with or after the production of CH within a flowering season (**Supplemental Table 1**). In at least some perennial cleistogamous species [e.g., some species of *Ruellia* (Long & Uttal 1962) and *Viola septemloba* (A.A. Winn, unpublished data)], smaller or younger individuals produce only CL, which suggests that the production of CL has priority over the production of CH in perennials as well.

All 18 studies that examined variation in the degree of cleistogamy reported evidence for phenotypic plasticity (**Supplemental Table 2**). Sixteen of these studies reported a decrease in the proportion of CL in response to putatively more favorable environmental conditions (e.g., increased light, water or nutrients, or unspecified conditions that increase plant size), or an increase in the proportion of CL in response to less favorable conditions (e.g., drought, herbivory, disturbance). Both flower phenology and responses to environmental conditions suggest that the production of CL has priority over CH.

Relative Costs of Reproduction via Chasmogamous and Cleistogamous Flowers

Quantifying the advantage of chasmogamous reproduction necessary to prevent evolutionary replacement by CL requires measures of the contribution of each of the two flower types to parental fitness (**Figure 1**). These contributions depend on the relative energetic cost per seed of reproduction by CH and CL, and the fitness of offspring produced by each type. Energetic cost per seed is determined by the fertility (probability of successful fruit set), seed number per flower, and the cost of producing reproductive structures (flower, fruit, and seeds) for each flower type. Differences in the fitnesses of chasmogamous and cleistogamous progeny may be due to both flower type per se, as in cases where seed size differs between CH and CL (e.g., McNamara & Quinn 1977, Zeide 1978), and genetic differences between offspring produced by selfing and outcrossing. CH are self-fertile, and if selfing occurs in CH, it will reduce the benefits of reproduction by CH that are based on genetic differences between CH and CL progeny.

Eleven studies of 15 species provided data relevant to the relative energetic costs of reproduction via CH and CL. The data suggest a consistent trend for CL to be less costly to produce (as estimated by dry biomass) and to have greater probability of fruit set, resulting in a lower cost per seed for CL than for chasmogamous reproduction

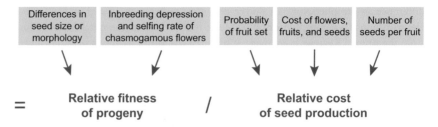

Figure 1

Components of the relative contribution of cleistogamous and chasmogamous reproduction to parental fitness.

Table 2 Estimates of the cost per seed for reproduction by cleistogamous flowers (CL) relative to that by chasmogamous flowers (CH); relative fertility (probability of fruit set) and relative flower cost represent two components of relative cost per seed[a]

Species (Family)	Relative fertility	Relative flower cost	Relative cost per seed	Reference
Amphicarpum purshii (Poaceae)	4.89			McNamara & Quinn 1977
Commelina benghalensis (Commelinaceae)	1.11			Kaul et al. 2002
Collomia grandiflora (Polemoniaceae)		0.9		Wilken 1982
Centaurea melitensis (Asteraceae)	1.02			Porras & Alvarez 1999
Impatiens capensis (Balsaminaceae)	1.4	0.01	0.57	Schemske 1978, Waller 1979
Impatiens pallida (Balsaminaceae)	1.77	0.01	0.38	Schemske 1978
Microlaena polynoda (Poaceae)	4.26	0.78		Schoen 1984
Oxalis acetosella (Oxalidaceae)	1.25			Berg 2003
Scutellaria indica (Lamiaceae)	19.2	0.15		Sun 1999
Viola hirta (Violaceae)	1.01			Berg 2003
Viola mirabilis (Violaceae)	1.08			Berg 2003
Viola pubescens (Violaceae)	0.43			Culley 2002
Viola riviniana (Violaceae)	0.99			Berg 2003
Viola septemloba (Violaceae)	1.43	0.44	0.67	A.A. Winn, unpublished data
Mean	3.06	0.38	0.54	
SD	5.02	0.39	0.15	

[a]All values are the ratio of CL to CH.

(Table 2). Complete data necessary to estimate the average cost of seed production via each flower type were available for three species and indicate that the cost of seed production via CL is slightly more than half the cost via CH (Table 2). Estimates of relative fertility (probability that a flower will successfully set fruit) of CH and CL were reported for 13 species, and on average, fertility of CL was 3.06 times that of CH (Table 2). Estimates of relative fertility varied widely, from 0.43 for *Viola pubescens* (Culley 2002), to 19.2 in *Scutellaria indica* (Sun 1999), but 11 of the 13 estimates were greater than or equal to 1, indicating that greater fertility of CH is probably rare. Although CH were as much as 100 times more energetically costly to produce, the cost of a flower is typically not a large proportion of the total cost per seed, less than 1% for CL and approximately 10% for CH in *Impatiens capensis*, for example (Schemske 1978). Consequently, flower cost alone contributed very little compared with fertility cost to differences in the energetic cost of seed production by CH and CL.

The fitness of progeny produced via CL is often equivalent to if not greater than the fitness of progeny produced via CH (Supplemental Table 3). The average fitness of CL relative to CH progeny for 19 studies of 14 cleistogamous species was 1.12 (standard deviation 0.43) (Supplemental Table 3). This is a rough approximation because many estimates are based on a limited portion of the life cycle (e.g., seed germination or early growth only) and/or were obtained from individuals grown in artificial conditions (e.g., greenhouse), which may not accurately reflect fitness differences that accrue under natural conditions (cf. Armbruster & Reed 2005).

Both differences in nongenetic phenotypic characteristics of seeds produced by CH and CL (e.g., maternally controlled differences in size, dispersal potential, and germination behavior) and inbreeding depression can contribute to differences in the fitness of chasmogamous and cleistogamous progeny. Studies comparing open-pollinated chasmogamous progeny with progeny from CL are appropriate for determining the contributions of reproduction by each flower type to parental fitness, but estimates of inbreeding depression could be used to partition this difference into effects owing to nongenetic differences in seed traits produced by CH and CL and those owing to differences in the genetic make-up of seeds from CH and CL. We found only three estimates of the relative fitness of progeny from hand outcrossed and selfed CH, which are not confounded by nongenetic differences in seed phenotype or by selfing in CH. Culley (2000) reported inbreeding depression [$1-(W_{CH\ selfed}/W_{CH\ outcrossed})$] of 10% in *Viola pubescens* (this estimate excludes differences in percent germination, which was unusually low and considerably greater for seeds produced by selfing than those produced by outcrossing). Studies of *Impatiens capensis* (Lu 2002) and *Viola septemloba* (C.G. Oakley & A.A. Winn, submitted) estimated inbreeding depression of 22% and 43%, respectively.

Self-fertilization of CH could reduce the genetic advantage that CH progeny are expected to enjoy. Selfing rates of CH have been reported in five studies, three of which present estimates for *Impatiens capensis*. The overall average is 50% (standard deviation 28%), with estimates ranging from 25% to 48% in *Impatiens capensis* (Lu 2000, Mitchell-Olds & Waller 1985, Waller & Knight 1989) to 36% in *Viola* (Culley 2002), to greater than 96% in *Lespedeza capitata* (Cole & Biesboer 1992). Because estimates of selfing we report are for CH only, and all CL are selfing, total selfing rates of cleistogamous species are likely to be high. For example, the selfing rate in CH of *Viola pubescens* was estimated to be 36% (Culley 2002), but 55% of the flowers produced are cleistogamous, so the total selfing rate would approach 71%.

How Strong Must Selection for Outcrossing be?

We can estimate the magnitude of the fitness advantage to cleistogamous reproduction relative to chasmogamous reproduction by dividing the estimates of the relative fitness of cleistogamous progeny by the estimates of the relative cost per seed of cleistogamous reproduction (see **Figure 1**). Complete data are available for two species and indicate a fitness advantage to cleistogamous reproduction of 15% for *Impatiens capensis* and of 231% for *Viola septemloba*. In other words, after accounting for differences in energetic cost of production and for differences in progeny fitness including inbreeding depression, some evolutionary force must provide an advantage to reproduction by CH of some magnitude to favor their continued production.

We also estimated a composite average relative advantage of cleistogamous reproduction using an average of the costs and benefits of reproduction via each flower type over all species for which we found data on fertility of CH and CL (**Table 2**) or the fitness of progeny from CH and CL (**Supplementary Table 3**). This composite estimate assumes that the combined weight of flowers, fruits, and seeds is similar for reproduction via CH and CL, and that seed number per fruit is likewise similar. Given these assumptions, some evolutionary force must provide an advantage to CH greater than 342% that of CL reproduction to justify their continued production. This value is 193% if the extreme differential fertility reported for *Scutellaria indica* (**Table 2**) is ignored.

Although the accuracy of the estimated fitness advantage of reproduction via cleistogamy might be improved by more data on the relative energetic costs of CH and CL (which would probably raise the estimate) and more complete estimates of inbreeding depression (which might increase or decrease it), much of the estimated cost of chasmogamous reproduction derives from differences in the economy and reliability of seed production, for which data were reasonably abundant and accurate.

EVALUATION OF THE HYPOTHESES

Collectively, the empirical data from studies of cleistogamous species show that on average, CL are cheaper to produce, have greater fruit set (**Table 2**), and produce progeny that suffer little, if any, reduction in fitness relative to progeny of CH (**Supplemental Table 3**). High overall rates of selfing and low estimates of inbreeding depression suggest that cleistogamous species can effectively purge inbreeding depression via selfing in CL. After accounting for differences in energetic costs of production and for differential progeny fitness, chasmogamous reproduction must repay a fitness cost of between 15% and 342% of the benefit of cleistogamous reproduction to parental fitness to prevent evolutionary replacement by pure CL. The majority of the advantage to CL appears to derive from their greater likelihood of setting fruit rather than the lower energetic cost of individual CL.

The empirical data provide some circumstantial support for each of the proposed mechanisms for the evolutionary stability of producing both flower forms. However, most of these mechanisms appear to be either unlikely to explain the persistence of CH and/or unable to generate the magnitude of selection necessary to counter the advantage of CL flowers that we detected.

Sequential, temporally separated production of CH and CL, reported for some perennials (**Supplemental Table 1**), is consistent with adaptive plastic response to seasonal variation in factors that influence the relative benefits of reproduction via the two flower types. However, this interpretation rests on the assumption that there are environmental states in which each flower morph contributes more to parental fitness than the alternative, which has not been demonstrated for CH. Schoen & Lloyd (1984) suggested that chasmogamous flower production might be favored when pollinators were abundant and cleistogamous flower production favored when they were scarce or absent but, given that CL do not require pollinators to set seed, are energetically cheaper to produce, and suffer little if at all from the effects of inbreeding depression, it is difficult to envision why CL should not be produced in all environments. Adaptive phenotypic plasticity might in principle produce strong selection favoring the ability to produce alternate phenotypes (Via & Lande 1985), but unless conditions that favor reproduction via CH over CL can be identified, the Schoen & Lloyd model cannot explain the stable persistence of the two flower forms.

We suggested at the outset that the hypothesis of avoiding geitonogamy (Masuda et al. 2001) was an unlikely candidate for explaining the stability of cleistogamy. Although simultaneous production of CH and CL in many species (**Supplemental Table 1**) is consistent with the hypothesis that avoiding geitonogamy contributes to the maintenance of cleistogamy, two other patterns do not fit the geitonogamy scenario. The plastic decrease in proportion of CL with increasing plant size (**Supplemental Table 2**) runs counter to the expectation that avoiding geitonogamy would become more important as plant size, and presumably total number of flowers, increases (Barrett et al. 1994, Klinkhamer & de Jong 1993, Routley & Husband 2003). The apparent priority of the production of CL observed in all annuals and at least some perennials also conflicts with the geitonogamy hypothesis.

We were not able to address the hypothesis that heterosis could maintain selection for CH as suggested by Lu (2002), because we found no data on the fitness of crosses between inbred lines relative to crosses between outbred lines. Nearly all the data on the costs and benefits of the production of the two flower types support selection favoring CL, supporting our suggestion that purging per se is not required for Lu's hypothesis. The available data on inbreeding depression in cleistogamous species is not informative with respect to heterosis because we do not know the inbreeding histories of the parents of the crosses.

As expected, we found no estimates of geometric mean fitness as a function of degree of cleistogamy, which would be necessary for a direct test of the variance discounting hypothesis for the maintenance of cleistogamy. The common pattern of a plastic decrease in the degree of cleistogamy with increasing plant size and resource availability (**Supplementary Table 2**) suggests that chasmogamous flower production is relatively more advantageous under conditions that are favorable for growth. Although this pattern could be interpreted as support for an adaptive plastic response to spatial environmental variation in conditions affecting the fitness benefits of reproduction via CH and CL, such a scenario seems likely to select for a response resulting in either all CL or all CH within each environment rather than a mixture. We concur with Waller (1980) who noted that a plastic increase in the proportion of CH with plant size or environmental favorability is consistent with risk averse variance discounting, although it is only circumstantial support.

Several studies provided data necessary to test the assumption of greater effects of competition among siblings derived from CL than from CH, which is required for avoidance of sibling competition to explain the stability of mixed chasmogamous and cleistogamous reproduction. Chasmogamously derived half-siblings grown in competition did not outperform CL derived full-siblings grown in competition in any of five tests of four different species (Berg & Redbo-Torstensson 1999, 2000; McCall et al. 1989; Schmitt & Ehrhardt 1987; Schmitt et al. 1985). Additional data from noncleistogamous species also provide little support for this hypothesis. In a general review of 22 studies of plants and insects that compared competition between similar and dissimilar genotypes, Price & Waser (1982) reported an average 3% advantage of dissimilar genotypes compared to similar genotypes. Additional subsequent studies have reported either that siblings grown in competition outperformed nonsiblings grown in competition (Donohue 2003, Tonsor 1989), or that there was no detectable difference (Willson et al. 1987; but see Kelley 1989, Schmitt & Antonovics 1986). An average advantage of 3% is much too weak relative to the chasmogamous advantage required (1.15- to 3.42-fold) to support sibling competition as a general explanation for the maintenance of CH.

SYNTHESIS AND PROSPECTUS

Evaluation of the assumptions and predictions of hypothesis for the stability of cleistogamy with published data suggests that adaptive plasticity, avoidance of geitonogamy, and avoidance of sibling competition are unlikely to explain the evolutionary stability of mixed mating in cleistogamous species. Although other mechanisms of variance discounting and/or heterosis might contribute to the maintenance of outcrossing in cleistogamous species, additional forces that have not been included in models of cleistogamy might also be relevant. In particular, more recent general models for the evolution of mixed mating find that the dynamics of pollen transfer may be critical to understanding the stability of mixed mating (Holsinger 1991, Johnston 1998, Porcher & Lande 2005), and they may also play a role in maintaining cleistogamy.

Candidates for the Maintenance of Cleistogamy

Chasmogamous flowers can perform two functions that CL cannot; they can produce outcrossed offspring, which experience a greater degree of genetic recombination than selfed offspring, and they can contribute to male fitness by siring seeds produced by different individuals. The question raised by our review of the data for cleistogamous species is, What forces related to these two functions could be of large enough magnitude (either alone or in concert) to repay the 15–342% fitness cost of the more expensive and less reliable seed production of CH? Two candidates are pollen dynamics and heterosis in outcrossing between inbred lines.

A recent model of mating system evolution suggests that genetic forces acting on mating system and many aspects of pollination ecology combine to prevent the evolution of complete self-fertilizaton (Porcher & Lande 2005; cf. Holsinger 1991, Johnston 1998). The model predicts that when the selfing rate is very high, the joint effects of pollen limitation, pollen discounting, and inbreeding depression combine to generate selection preventing further increase in the selfing rate. The magnitude of selection that can be generated by this scenario is not known, but it might be exaggerated in cleistogamous species because it would need to counter exceptionally strong selection favoring the economy and reliability of selfing via CL. The model could be modified to account for the unusual features of cleistogamy and parameterized with data from a particular species to determine the likelihood that this force could maintain the observed investment in CH.

The role of heterosis in favoring the maintenance of CH in Lu's model invokes a scenario similar to that portrayed by models of metapopulation genetics. Several models of metapopulation genetics support the potential for rare gene flow between populations to lead to substantial heterosis (Roze & Rousset 2004, Theodorou & Couvet 2002, Whitlock et al. 2000), especially for small, predominantly selfing populations. Crosses between populations may result in heterosis because a portion of the genetic load may become fixed within populations owing to drift. Drift is expected to fix alternate alleles in different populations, such that crosses between populations can mask previously expressed deleterious mutations (Roze & Rousset 2004, Theodorou & Couvet 2002, Whitlock et al. 2000). Empirical evidence of substantial heterosis in interpopulation crosses (e.g., Fenster 1991, Fenster & Galloway 2000, Paland & Schmid 2003, van Treuren et al. 1993, Weller et al. 2005) is consistent with this theory and supports a potentially large advantage to maintaining the potential for periodic outcrossing.

In some respects, Lu's model depicts an extreme metapopulation in which high rates of selfing via CL generate selfing lineages, each with an effective population size of one, within a larger population. These selfing lineages can build up a genetic load that can only be relieved by outcrossing. Although the heterosis resulting from crosses between inbred lines may need to be large to justify the expense of CH in cleistogamous species, this mechanism could also operate in species with monomorphic flowers whenever selfing rates are high enough to create the appropriate genetic structure. A first step in determining whether this scenario is plausible would be to determine how much load accumulates in selfing lineages. Studies in noncleistogamous species have shown that uninterrupted selfing over 5 to 6 generations can build up substantial load as revealed by crosses between selfing lines (Barrett & Charlesworth 1991, Dudash et al. 1997; cf. Johansen-Morris & Latta 2006).

Additional Approaches to the Study of Cleistogamy

The mechanisms that favor the maintenance of CH in cleistogamous species are also likely to operate in mixed mating species with monomorphic flowers. Because the proposed effects of these mechanisms are likely to be more pronounced in cleistogamous taxa, these species may provide particularly suitable systems in which to pursue the empirical work necessary to test their feasibility and likely relative importance. Considerable existing data on the fitness costs and benefits of reproduction via CL and CH available for species of *Impatiens* and *Viola* make them particularly suitable for obtaining the additional data needed to determine which of the proposed mechanisms could be of sufficient magnitude to justify the continued production of outcrossing flowers. The additional data needed include estimates of fitness accrued by siring outcrossed progeny and the rates of evolution of inbreeding depression and fixed mutational load in natural populations.

A potentially valuable empirical approach that has not yet been pursued in cleistogamous taxa is the comparison of closely related species that differ in their degree of cleistogamy. A comparative approach could permit more focused, testable predictions about what factors are important for the evolution or maintenance of cleistogamy in a given taxonomic group. Culley & Klooster (2007) recently mapped the incidence of cleistogamy onto a composite phylogeny for the angiosperms. They identified several possible instances of evolutionary loss of CL. Scrutiny of the factors associated with such losses may be valuable for identifying the conditions that favor the maintenance of CH.

Cleistogamous species present an interesting evolutionary enigma in their own right, but can also contribute to resolving the longstanding question of what maintains mixed mating in general and especially what maintains outcrossing once effective selfing evolves. General features of cleistogamous species are consistent with the current view that the forces of inbreeding depression and automatic selection alone are not sufficient to explain the stability of mixed mating. The relative ease with which the fitness consequences of reproduction via selfing and outcrossing can be quantified provides a yardstick for evaluating which of the many additional forces that have been proposed are most likely to play a role in the maintenance of mixed mating. Rather than making them arcane exceptions to mating system theory, many of the unusual features of cleistogamous species make them valuable tools for elucidating the adaptive significance of mixed mating.

DISCLOSURE STATEMENT

The authors are not aware of any biases that might be perceived as affecting the objectivity of this review.

ACKNOWLEDGMENTS

We thank G.P. Cheplick and T.M. Culley for comments on the manuscript and for sharing their unpublished manuscripts with us. D.W. Schemske and D.R. Levitan also provided valuable input for improving the manuscript.

LITERATURE CITED

Armbruster P, Reed DH. 2005. Inbreeding depression in benign and stressful environments. *Heredity* 95:235–42

Barrett SCH, Charlesworth D. 1991. Effects of a change in the level of inbreeding on the genetic load. *Nature* 352:522–24

Barrett SCH, Harder LD, Cole WW. 1994. Effects of flower number and position on self-fertilization in experimental populations of *Eichhornia paniculata* (Pontederiaceae). *Funct. Ecol.* 8:526–35

Berg H. 2003. Factors influencing seed:ovule ratios and reproductive success in four cleistogamous species: a comparison between two flower types. *Plant Biol.* 5:194–202

Berg H, Redbo-Torstensson P. 1999. Offspring performance in three cleistogamous Viola species. *Plant Ecol.* 145:49–58

Berg H, Redbo-Torstensson P. 2000. Offspring performance in *Oxalis acetosella*, a cleistogamous perennial herb. *Plant Biol.* 2:638–45

Campbell CS, Quinn JA, Cheplick GP, Bell TJ. 1983. Cleistogamy in grasses. *Annu. Rev. Ecol. Syst.* 14:411–41

Charlesworth D, Charlesworth B. 1987. Inbreeding depression and its evolutionary consequences. *Annu. Rev. Ecol. Syst.* 18:237–68

Clay K. 1982. Environmental and genetic determinants of cleistogamy in a natural population of the grass *Danthonia spicata*. *Evolution* 36:734–41

Clay K. 1983. Variation in the degree of cleistogamy within and among species of the grass *Danthonia*. *Am. J. Bot.* 70:835–43

Clay K, Antonovics J. 1985. Demographic genetics of the grass *Danthonia spicata*: success of progeny from chasmogamous and cleistogamous flowers. *Evolution* 39:205–10

Cole CT, Biesboer DD. 1992. Monomorphism, reduced gene flow, and cleistogamy in rare and common species of *Lespedeza* (Fabaceae). *Am. J. Bot.* 79:567–75

Culley TM. 2000. Inbreeding depression and floral type fitness differences in *Viola canadensis* (Violaceae), a species with chasmogamous and cleistogamous flowers. *Can. J. Bot.* 78:1420–29

Culley TM. 2002. Reproductive biology and delayed selfing in *Viola pubescens* (Violaceae), an understory herb with chasmogamous and cleistogamous flowers. *Int. J. Plant Sci.* 163:113–22

Culley TM, Klooster MR. 2007. The cleistogamous breeding system: a review of its frequency, evolution, and ecology in angiosperms. *Bot. Rev.* 73:1–30

Darwin C. 1896. *The Different Forms of Flowers on Plants of the Same Species.* New York: Appleton. 352 pp.

Donohue K. 2003. The influence of neighbor relatedness on multilevel selection in the Great Lakes sea rocket. *Am. Nat.* 162:77–92

Dudash MR, Carr DE, Fenster CB. 1997. Five generations of enforced selfing and outcrossing in *Mimulus guttatus*: inbreeding depression variation at the population and family level. *Evolution* 51:54–65

Feldman MW, Christiansen FB. 1984. Population genetic theory of the cost of inbreeding. *Am. Nat.* 123:642–53

Fenster CB. 1991. Gene flow in *Chamaecrista fasciculata* (Leguminosae). 2. Gene establishment. *Evolution* 45:410–22

Fenster CB, Galloway LF. 2000. Population differentiation in an annual legume: genetic architecture. *Evolution* 54:1157–72

Fisher RA. 1941. Average excess and average effect of a gene substitution. *Ann. Eugen.* 11:53–63

Goodwillie C, Kalisz S, Eckert CG. 2005. The evolutionary enigma of mixed mating systems in plants: occurrence, theoretical explanations, and empirical evidence. *Annu. Rev. Ecol. Syst.* 36:47–79

Holsinger KE. 1991. Mass-action models of plant mating systems: the evolutionary stability of mixed mating systems. *Am. Nat.* 138:606–22

Holsinger KE, Feldman MW, Christiansen FB. 1984. The evolution of self-fertilization in plants: a population genetic model. *Am. Nat.* 124:446–53

Husband BC, Schemske DW. 1996. Evolution of the magnitude and timing of inbreeding depression in plants. *Evolution* 50:54–70

Igic B, Kohn JR. 2006. The distribution of plant mating systems: study bias against obligately outcrossing species. *Evolution* 60:1098–103

Jain SK. 1976. The evolution of inbreeding in plants. *Annu. Rev. Ecol. Syst.* 7:469–95

Johansen-Morris AD, Latta RG. 2006. Fitness consequences of hybridization between ecotypes of *Avena barbata*: hybrid breakdown, hybrid vigor, and transgressive segregation. *Evolution* 60:1585–95

Johnston MO. 1998. Evolution of intermediate selfing rates in plants: pollination ecology versus deleterious mutations. *Genetica* 102:267–78

Kaul V, Sharma N, Koul AK. 2002. Reproductive effort and sex allocation strategy in *Commelina benghalensis* L., a common monsoon weed. *Bot. J. Linn. Soc.* 140:403–13

Kelley SE. 1989. Experimental studies of the evolutionary significance of sexual reproduction. 5. A field-test of the sib-competition lottery hypothesis. *Evolution* 43:1054–65

Klinkhamer PGL, de Jong TJ. 1993. Attractiveness to pollinators: a plant's dilemma. *Oikos* 66:180–84

Kondrashov AS. 1993. Classification of hypotheses on the advantage of amphimixis. *J. Hered.* 84:372–87

Lande R, Schemske DW. 1985. The evolution of self-fertilization and inbreeding depression in plants. 1. Genetic models. *Evolution* 39:24–40

Le Corff J. 1993. Effects of light and nutrient availability on chasmogamy and cleistogamy in an understory tropical herb, *Calathea micans* (Marantaceae). *Am. J. Bot.* 80:1392–99

Lee CW, Erickson HT, Janick J. 1976. Inheritance of cleistogamy in *Salpiglossis sinuata. J. Hered.* 67:267–70

Lloyd DG. 1979. Some reproductive factors affecting the selection of self-fertilization in plants. *Am. Nat.* 113:67–79

Long RW, Uttal LJ. 1962. Some observations on flowering in *Ruellia* (Acanthaceae). *Rhodora* 64:200–6

Lord EM. 1981. Cleistogamy: a tool for the study of floral morphogenesis, function and evolution. *Bot. Rev.* 47:421–49

Lord EM, Hill JP. 1987. Evidence for heterochrony in the evolution of plant form. In *Development as an Evolutionary Process: Proc. Meet. Mar. Biol. Lab., Woods Hole, Mass., Aug. 23, 24, 1985*, ed. RA Raff, CA Raff, pp. 47–70. New York: Liss

Lu Y. 2000. Effects of density on mixed mating systems and reproduction in natural populations of *Impatiens capensis. Int. J. Plant Sci.* 161:671–81

Lu Y. 2002. Why is cleistogamy a selected reproductive strategy in *Impatiens capensis* (Balsaminaceae)? *Biol. J. Linn. Soc.* 75:543–53. Erratum. 2002. *Biol. J. Linn. Soc.* 76:463

Masuda M, Yahara T, Maki M. 2001. An ESS model for the mixed production of cleistogamous and chasmogamous flowers in a facultative cleistogamous plant. *Evol. Ecol. Res.* 3:429–39

Maynard Smith J. 1978. *The Evolution of Sex*. Cambridge, UK: Cambridge Univ. Press. 22 pp.

McCall C, Mitchell-Olds T, Waller DM. 1989. Fitness consequences of outcrossing in *Impatiens capensis*: tests of the frequency-dependent and sib-competition models. *Evolution* 43:1075–84

McNamara J, Quinn JA. 1977. Resource allocation and reproduction in populations of *Amphicarpum purshii* (Gramineae). *Am. J. Bot.* 64:17–23

Mitchell-Olds T, Waller DM. 1985. Relative performance of selfed and outcrossed progeny in *Impatiens capensis*. *Evolution* 39:533–44

Morgan MT, Wilson WG. 2005. Self-fertilization and the escape from pollen limitation in variable pollination environments. *Evolution* 59:1143–48

Nagylaki T. 1976. A model for the evolution of self-fertilization and vegetative reproduction. *J. Theor. Biol.* 58:55–58

Paland S, Schmid B. 2003. Population size and the nature of genetic load in *Gentianella germanica*. *Evolution* 57:2242–51

Plitmann U. 1995. Distributions of dimorphic flowers as related to other elements of the reproductive strategy. *Plant Species Biol.* 10:53–80

Porcher E, Lande R. 2005. The evolution of self-fertilization and inbreeding depression under pollen discounting and pollen limitation. *J. Evol. Biol.* 18:497–508

Porras R, Alvarez JMM. 1999. Breeding system in the cleistogamous species *Centaurea melitensis* (Asteraceae). *Can. J. Bot.* 77:1632–40

Price MV, Waser NM. 1982. Population structure, frequency-dependent selection, and the maintenance of sexual reproduction. *Evolution* 36:35–43

Real LA. 1980. Fitness, uncertainty, and the role of diversification in evolution and behavior. *Am. Nat.* 115:623–38

Routley MB, Husband BC. 2003. The effect of protandry on siring success in *Chamerion angustifolium* (Onagraceae) with different inflorescence sizes. *Evolution* 57:240–48

Roze D, Rousset F. 2004. Joint effects of self-fertilization and population structure on mutation load, inbreeding depression and heterosis. *Genetics* 167:1001–15

Schemske DW. 1978. Evolution of reproductive characteristics in *Impatiens* (Balsaminaceae): significance of cleistogamy and chasmogamy. *Ecology* 59:596–613

Schmitt J, Antonovics J. 1986. Experimental studies of the evolutionary significance of sexual reproduction. 4. Effect of neighbor relatedness and aphid infestation on seedling performance. *Evolution* 40:830–36

Schmitt J, Ehrhardt D, Swartz D. 1985. Differential dispersal of self-fertilized and outcrossed progeny in jewelweed (*Impatiens capensis*). *Am. Nat.* 126:570–75

Schmitt J, Ehrhardt DW. 1987. A test of the sib-competition hypothesis for outcrossing advantage in *Impatiens capensis*. *Evolution* 41:579–90

Schmitt J, Gamble SE. 1990. The effect of distance from the parental site on offspring performance and inbreeding depression in *Impatiens capensis*: a test of the local adaptation hypothesis. *Evolution* 44:2022–30

Schnee BK, Waller DM. 1986. Reproductive-behavior of *Amphicarpaea bracteata* (Leguminosae), an amphicarpic annual. *Am. J. Bot.* 73:376–86

Schoen DJ. 1984. Cleistogamy in *Microlaena polynoda* (Gramineae): an examination of some model predictions. *Am. J. Bot.* 71:711–19

Schoen DJ, Lloyd DG. 1984. The selection of cleistogamy and heteromorphic diaspores. *Biol. J. Linn. Soc.* 23:303–22

Scofield DG, Schultz ST. 2006. Mitosis, stature and evolution of plant mating systems: low-Phi and high-Phi plants. *Proc. R. Soc. B* 273:275–82

Seger J, Brockmann HJ. 1987. What is bet-hedging? *Oxford Surv. Evol. Biol.* 4:182–211

Sun M. 1999. Cleistogamy in *Scutellaria indica* (Labiatae): effective mating system and population genetic structure. *Mol. Ecol.* 8:1285–95

Theodorou K, Couvet D. 2002. Inbreeding depression and heterosis in a subdivided population: influence of the mating system. *Genet. Res.* 80:107–16

Tonsor SJ. 1989. Relatedness and intraspecific competition in *Plantago lanceolata*. *Am. Nat.* 134:897–906

Turuspekov Y, Mano Y, Honda I, Kawada N, Watanabe Y, Komatsuda T. 2004. Identification and mapping of cleistogamy genes in barley. *Theor. Appl. Genet.* 109:480–87

Uphof JCT. 1938. Cleistogamic flowers. *Bot. Rev.* 4:21–49

van Treuren R, Bijlsma R, Ouborg NJ, van Delden W. 1993. The significance of genetic erosion in the process of extinction. 4. Inbreeding depression and heterosis effects caused by selfing and outcrossing in *Scabiosa columbaria*. *Evolution* 47:1669–80

Via S, Lande R. 1985. Genotype-environment interaction and the evolution of phenotypic plasticity. *Evolution* 39:505–22

Waller DM. 1979. The relative costs of self-fertilized and cross-fertilized seeds in *Impatiens capensis* (Balsaminaceae). *Am. J. Bot.* 66:313–20

Waller DM. 1980. Environmental determinants of outcrossing in *Impatiens capensis* (Balsaminaceae). *Evolution* 34:747–61

Waller DM. 1984. Differences in fitness between seedlings derived from cleistogamous and chasmogamous flowers in *Impatiens capensis*. *Evolution* 38:427–40

Waller DM, Knight SE. 1989. Genetic consequences of outcrossing in the cleistogamous annual, *Impatiens capensis*. 2. Outcrossing rates and genotypic correlations. *Evolution* 43:860–69

Weiss PW. 1980. Germination, reproduction and interference in the amphicarpic annual *Emex spinosa* (L.) Campd. *Oecologia* 45:244–51

Weller SG, Sakai AK, Thai DA, Tom J, Rankin AE. 2005. Inbreeding depression and heterosis in populations of *Schiedea viscosa*, a highly selfing species. *J. Evol. Biol.* 18:1434–44

Whitlock MC, Ingvarsson PK, Hatfield T. 2000. Local drift load and the heterosis of interconnected populations. *Heredity* 84:452–57

Wilken DH. 1982. The balance between chasmogamy and cleistogamy in *Collomia grandiflora* (Polemoniaceae). *Am. J. Bot.* 69:1326–33

Williams GC. 1975. *Sex and Evolution*. Princeton, NJ: Princeton Univ. Press. 200 pp.

Willson MF, Hoppes WG, Goldman DA, Thomas PA, Katusic-Malmborg PL, Bothwell JL. 1987. Sibling competition in plants: an experimental study. *Am. Nat.* 129: 304–11

Young JPW. 1981. Sib competition can favor sex in two ways. *J. Theor. Biol.* 88:755–56

Zeide B. 1978. Reproductive behavior of plants in time. *Am. Nat.* 112:636–39

Sympatric Speciation: Models and Empirical Evidence

Daniel I. Bolnick[1] and Benjamin M. Fitzpatrick[2]

[1] Section of Integrative Biology, University of Texas, Austin, Texas 78712; email: danbolnick@mail.utexas.edu

[2] Department of Ecology and Evolutionary Biology, University of Tennessee, Knoxville, Tennessee 37996; email: benfitz@utk.edu

Annu. Rev. Ecol. Evol. Syst. 2007. 38:459–87

First published online as a Review in Advance on August 8, 2007

The *Annual Review of Ecology, Evolution, and Systematics* is online at http://ecolsys.annualreviews.org

This article's doi: 10.1146/annurev.ecolsys.38.091206.095804

Key Words

assortative mating, disruptive selection, reinforcement reproductive isolation

Abstract

Sympatric speciation, the evolution of reproductive isolation without geographic barriers, remains highly contentious. As a result of new empirical examples and theory, it is now generally accepted that sympatric speciation has occurred in at least a few instances, and is theoretically plausible. Instead, debate has shifted to whether sympatric speciation is common, and whether models' assumptions are generally met in nature. The relative frequency of sympatric speciation will be difficult to resolve, because biogeographic changes have obscured geographical patterns underlying many past speciation events. In contrast, progress is being made on evaluating the empirical validity of key theoretical conditions for sympatric speciation. Disruptive selection and direct selection on mating traits, which should facilitate sympatric speciation, are biologically well supported. Conversely, costs to assortative mating are also widely documented, but inhibit speciation. Evaluating the joint incidence of these key factors may illuminate why sympatric speciation appears to be relatively uncommon.

INTRODUCTION

Speciation: The evolution of genetically distinct populations (clusters), maintained by reproductive isolation in the case of sexual taxa

Sympatric speciation is among the most persistently contested topics in evolution, dating back to correspondence between Darwin and Wagner. Early geneticists claimed that new species arise instantaneously via mutation from within their ancestral range (de Vries 1901–1903), whereas naturalists countered that related species in nature were always separated by geographic barriers (Jordan 1905). Debate waned after Mayr (1963) outlined a compelling case against sympatric divergence, arguing that it was theoretically unlikely. The key problem is that mating and recombination rapidly break down linkage disequilibrium, preventing formation of genetically distinct subgroups. Mayr therefore argued that overlapping ranges are better explained by secondary contact between allopatrically derived species. However, Mayr presciently predicted that "the issue will be raised again at regular intervals. Sympatric speciation is like the Lernaean Hydra which grew two new heads whenever one of its old heads was cut off" (p. 451).

Since 1990, the number of papers on sympatric speciation has increased exponentially. This revival of the Hydra can be attributed in part to molecular phylogenetics, which provides a new source of data to evaluate Mayr's alternative hypothesis of secondary contact. There are now a few widely accepted examples of sympatric speciation, which in turn inspired a proliferation of theoretical models. Many skeptics now concede that sympatric speciation is theoretically possible and has probably occurred in nature (Coyne & Orr 2004). Debate has shifted to the still more difficult questions of how frequent sympatric speciation may be, and what mechanisms drive speciation. In this review, we describe recent empirical results and the theory underlying this shift, and discuss prospects for future progress.

WHEN IS SPECIATION SYMPATRIC?

Mayr (1963) defined sympatric speciation as evolution of reproductive isolation "without geographic isolation" (p. 449). Geographic isolation occurs when the distance between populations exceeds individuals' ability to disperse between them (cruising range) or when intervening environments are inhospitable, precluding dispersal. However, it is not always easy to determine when isolation is geographic, as opposed to biological isolation arising from intrinsic differences in habitat preference among groups. The key distinction is that geographic barriers are extrinsic features of the environment that affect all individuals, independent of genotype. Ambiguity arises because biological and geographic barriers may be difficult to distinguish empirically and may interact if extrinsic barriers are conditional upon individuals' biological traits. Another difficulty is that isolation is often a matter of degree and may depend on spatial scale. For instance, phytophagous insects using two or more host plants may be considered geographically overlapping at a large scale, and isolated at a small scale.

Some definitions of sympatric speciation add a population genetic stipulation that the initial population must be panmictic. Sympatric speciation is the evolution of reproductive isolation "within a single interbreeding unit" (Mayr 1942, p. 189) or between populations that initially use two distinct habitats but disperse freely between them (Futuyma & Mayer 1980). Gavrilets (2003) provided the most precise statement of the population genetic view: "sympatric speciation is the emergence of new species from a population where mating is random with respect to the birthplace of the mating partners." (p. 2198) As with isolation, panmixia is a matter of degree: Are we to conclude that two populations are not sympatric if only 40% of individuals move between populations each generation rather than the 50% expected with perfect panmixia? Although population genetic criteria focus on the initial conditions of a speciating lineage, they also implicitly adopt the biogeographic criterion that no extrinsic barriers arise subsequently. In practice, theoreticians tend to adopt this more precise population genetic definition as an initial condition in models. Empiricists rarely have information about the population structure of ancestral species and so tend to adopt the biogeographic definition, using current distribution patterns and phylogenetics to infer past geographic isolation. This semantic and methodological divide may contribute to disagreements over what constitutes sufficient evidence for sympatric speciation.

What Evidence Distinguishes Sympatric from Nonsympatric Speciation?

Coyne & Orr (2004) proposed four criteria for identifying cases of sympatric speciation, which we modify slightly:

1. Species thought to have arisen via sympatric speciation should have largely or completely overlapping geographic ranges. The spatial scale used to determine overlap should be commensurate with the dispersal ability of the organisms. In principle, sympatrically derived species could become allopatric over time (Baack 2004, Stuessy et al. 2004), but it is not clear how to unambiguously demonstrate secondary allopatry.

2. Speciation must be complete. Ongoing divergence is an important subject of study for speciation biology. However, we cannot declare a case of sympatric speciation if speciation has not occurred, because partial divergence may be a stable outcome (Matessi et al. 2001). Evaluating this criterion is subjective, because one must impose a binary decision on a continuous process of divergence. For instance, one must decide how much hybridization (biological species concept) or how much phenotypic or genetic overlap (cluster species concept) is allowed between distinct species.

3. Clades thought to arise via sympatric speciation must be sister species or monophyletic endemic species flocks. Evidence for monophyly should not be based on a single locus, owing to the risk of introgression. This criterion is conservative and not strictly necessary. Consider a sympatric speciation event that produces sister species A and B on an island. Later, A colonizes another island and speciates allopatrically into A and A'. Strictly speaking, A and B are no longer sister species, but the initial speciation event remains sympatric. Nonetheless, we retain this criterion because a biologist given modern distributions of A, A', and

B could not readily distinguish between sympatric speciation and secondary contact of A and B.

4. Coyne & Orr's fourth criterion stated that "the biogeographic and evolutionary history of the groups must make the existence of an allopatric phase *very unlikely*" (p. 142, their italics). We agree that to establish sympatric speciation, one must reject alternative hypotheses (allopatry and parapatry; see sidebar, Allopatry as a Null Model). However, this criterion is redundant, simply restating Mayr's biogeographic definition of sympatric speciation. The critical question is, what evidence is required to reject any past period of geographic isolation and conclude that current sympatry is representative of biogeographic patterns during speciation? The most common approach is to focus on sister species inhabiting a uniform and isolated geographic area that makes secondary contact unlikely (e.g., Savolainen et al. 2006, Schliewen et al. 1994). One problem with this approach is the need to explain one and only one colonization event. In continental settings, biogeographic data are less informative because range shifts are more likely. However, sympatric speciation may still be reasonable if speciation is very recent relative to climatological or geological events that might alter distributions, and if neither population exhibits phylogeographic signatures of range expansion. A final line of evidence comes from speciation events such as hybrid speciation, autopolyploidy, and transitions to selfing, which are likely to occur within the range of the parental species. Host shifts, though easiest in sympatry, may occur in parapatry or during colonization events and so require additional biogeographic evidence that the shift occurred between sympatric host taxa.

Sympatry: Describes populations with broadly overlapping geographic ranges

EMPIRICAL SUPPORT FOR SYMPATRIC SPECIATION IN NATURE

Few studies have managed to satisfy all four criteria (Barluenga et al. 2006b, Gislason et al. 1999, Savolainen et al. 2006, Schliewen et al. 1994, Sorenson et al. 2003), and all have their critics. Lacking room to review all case studies, we focus on three main types of evidence marshaled in support of sympatric speciation. (See Supplemental Material for a more complete list of putative examples. Follow the Supplemental Material link from the Annual Reviews home page at **http://www.annualreviews.org/**.)

Evidence from Isolated Environments

Some of the most compelling cases of sympatric speciation are found in small isolated environments like oceanic islands (Savolainen et al. 2006), postglacial lakes (Gislason et al. 1999), and crater lakes (Schliewen et al. 1994). However, even in isolated sites, double invasion and introgression can lead to sympatric species that may be misinterpreted as cases of sympatric speciation (Taylor & McPhail 2000). The crater Lake Apoyo (23,000 years old) is one of the more recently documented cases of putative sympatric speciation, between the endemic arrow cichlid (*Amphilophus zaliosus*) and the widely distributed midas cichlid (*A. citrinellus*). The lake's small size and uniform

ALLOPATRY AS A NULL MODEL

Coyne & Orr (2004) claimed that allopatric speciation is a null model for speciation, being theoretically uncontroversial and empirically well documented. We disagree for several reasons. First, allopatry is difficult to falsify, but failure to reject allopatry does not mean that it is supported. Cases with insufficient data are simply inconclusive. Because alternative hypotheses cannot be rigorously evaluated for many taxa, adopting an alternative as a default risks an unknown Type II error rate that will bias comparative estimates of the frequency of different modes of speciation. Second, in practice, studies rarely conduct statistical tests of null predictions. Rather, the relative credibility of past events is evaluated based on subjective judgments. If allopatry is treated as a null, ad hoc hypotheses favoring allopatry will be given more credibility than equally ad hoc stories of sympatry. Finally, there is not a simple dichotomous choice between allopatric and sympatric speciation, as both parapatry and geographic changes are well documented. Mayr's biogeographic definition of sympatric speciation excludes any period of geographic isolation, however brief. If we apply an equally stringent definition to allopatric speciation requiring zero gene flow during speciation, we might find that mixed-geography speciation, in which both allopatry and gene flow contribute to divergence, is relatively common.

topography make intralacustrine geographic barriers implausible. Genetic data support monophyly of Lake Apoyo cichlids with respect to other *A. citrinellus* populations (Barluenga et al. 2006b). All mitochondrial haplotypes are unique to Lake Apoyo and share a distinctive substitution. Microsatellite and amplified fragment length polymorphism (AFLP) data exhibit separate but closely related clusters within Apoyo. Barluenga and colleagues have been criticized for omitting one of the potential outgroups in nearby Lake Nicaragua (Schliewen et al. 2006, but see the Barluenga et al. 2006a rebuttal). Also, although nuclear data indicate that the Apoyo fish are closely related, Apoyo *citrinellus* are genetically intermediate between the co-occurring *zaliosus* and the more widespread *citrinellus*. Critics say this is consistent with secondary contact and a period of introgression (Schliewen et al. 2006), though it may also be explained by *zaliosus* arising from within Apoyo *citrinellus*, consistent with the lower diversity of *zaliosus* and difficulty of colonizing the lake (Barluenga et al. 2006b). If, as seems likely, the crater lake actually harbors three to four endemic species, secondary contact and introgression would be still less parsimonious (A. Meyer, personal communication). We describe this case not because the case is closed, but because it epitomizes the ambiguities that plague even the best examples of sympatric speciation. Genetic data may not always be able to distinguish between introgression and incomplete lineage sorting, particularly in very recently diverged taxa.

Cameroon crater lake cichlids (Schliewen & Klee 2004, Schliewen et al. 1994) and Lord Howe island palms (Savolainen et al. 2006) are other leading cases of sympatric speciation in isolated environments, with multilocus monophyly and thoroughly sampled outgroups. However, there are a number of putative cases of sympatric speciation that inhabit similar geographic settings, but have not yet been thoroughly studied (Berrebi & Valiushok 1998, Dimmick & Edds 2002, Klemetsen et al. 2002, Wilson et al. 2004). For instance, 25–50% of Arctic char populations (*Salvelinus alpinus*) exhibit genetically distinct ecomorphs (Wilson et al. 2004). Although these char morphs are thought to arise sympatrically, detailed phylogeographic data are available for only one lake (Gislason et al. 1999).

Evidence from Host Shifts

It is difficult to rule out secondary contact in continental or oceanic habitats where range expansions are easier. In such settings, case studies may rest on ecological rather than biogeographic evidence against geographic isolation. Specifically, host shifts are generally expected to occur within a population's native range, though this is not strictly necessary. Because host shifts could directly confer reproductive isolation if mating occurs on the host, there are fewer theoretical objections to such sympatric divergence. Consequently, sympatric sister species using different hosts provide an alternative line of evidence for sympatric speciation. The most famous example is *Rhagoletis pomonella*, which specialized on hawthorn fruits prior to the colonial-era introduction of apples into North America. The shift to apples occurred within the native range of hawthorn flies and led to reproductive isolation by divergent mate timing and habitat choice (Feder & Filchak 1999, Linn et al. 2003). However, the host races are not classified as distinct species owing to low levels of hybridization. Consequently, this case is usually portrayed as demonstrating plausibility of sympatric speciation via host shift rather than an actual case of sympatric speciation. Recent genetic evidence suggests that the host shift involved a chromosomal inversion that arose in allopatry (in Mexico), arrived in the northeast via gene flow, and was later co-opted for speciation (Feder et al. 2003). Although definitions of sympatric speciation do not stipulate that genetic variation must arise in sympatry, speciation may or may not have been possible without this allopatrically derived preadaptation.

Two recent reviews of insect host shifts (Berlocher & Feder 2002, Drés & Mallet 2002) highlight several examples of more distinct sympatric sister species that use different hosts [*Enchenopa binotata* complex (Wood & Keese 1990); *Spodoptera frugiperda* on corn and rice (Prowell et al. 2004); *Nilaparvata lugens* on weed grass and rice (Claridge et al. 1997, Sezer & Butlin 1998); and two other species pairs in the *Rhagoletis pomonella* group (Berlocher 1999)]. Instances of sympatric host-shift speciation have also been inferred in vertebrates. Genetically distinct sympatric species of brood parasitic indigobirds (*Vidua*) in Africa rely on different host species to raise their young. Because male indigobirds learn (and females imprint on) their host's courtship songs, reproductive isolation can be virtually instantaneous if eggs are laid in a new species' nest (Sorenson et al. 2003). Indopacific gobies feed and mate exclusively on *Acropora* corals: a young species of goby was recently described that uses a novel coral host and is completely contained within the range of its sister species (Munday et al. 2004). An equivalent process can occur in plants via adaptation to different edaphic conditions (Savolainen et al. 2006).

Evidence from Instantaneous Speciation

Instantaneous speciation necessarily occurs in sympatry with parental populations, but is often dismissed as rare. For instance, meiotic errors can produce polyploid offspring reproductively isolated from their parents (*autopolyploid speciation*; Ramsey & Schemske 1998). Less than 7% of plant speciation events are estimated to have involved changes in ploidy (Otto & Whitton 2000). Not all of these are autopolyploids, and not all polyploidy generates instantaneous speciation (Ramsey & Schemske 1998). This still may represent a substantial number of speciation events. Interspecific hybridization can also produce populations that are reproductively isolated from their sympatric parents by ploidy, karyotype, ecology, or mating behavior (Gompert et al. 2006, Mavarez et al. 2006, Rieseberg et al. 1995, Schwarz et al. 2005). Although hybridization must occur where both parental species are present, it is ambiguous whether this constitutes sympatric speciation because it does not arise from a single panmictic population and may occur in hybrid zones at the periphery of adjoining parental distributions. However, hybrid parapatry may also reflect niche partitioning between sympatrically derived taxa (Gompert et al. 2006, Rieseberg et al. 1995). Finally, sexual populations can produce asexual or selfing lineages via mutation, such as the selfing annual plant *Stephanomeria malheurensis* and its outcrossing sympatric sister species *S. exigua* (Gottlieb 1979). Such sympatric speciation is uncontroversial because recombination does not oppose divergence.

HOW GENERAL IS SYMPATRIC SPECIATION?

The preceding case studies establish that sympatric speciation has probably occurred, something now acknowledged even by many skeptics (Coyne & Orr 2004). Debate has shifted to whether sympatric speciation is common. This requires estimates of relative frequencies of sym-, para-, and allopatric speciation in nature. Tallying individual well-supported cases of each geographic mode is out of the question owing to several sources of ascertainment bias. Whereas extensive work is required to demonstrate sympatric speciation, allopatric pairs are accepted with little consideration (see sidebar, Allopatry as a Null Model). Rates of allopatric speciation may be over- or underestimated as a result of taxonomic practices (Agapow et al. 2004): decisions to lump or split allopatric variants into one or more species can raise or lower the number of allopatric speciation events. Keeping such biases in mind, biologists have tried two coarse approaches to estimate the frequency of sympatric speciation.

Counting Examples

The simplest approach is to count the number of putative sister species that exhibit a given amount of range overlap. Collectively, studies taking this approach

represent 309 speciation events, of which 9.4% resulted in sister species with over 90%
contemporary range overlap, compared with 72.2% with zero range overlap
(Berlocher 1998, Coyne & Price 2000, Fitzpatrick & Turelli 2006, Lynch 1989).
These data support the claim that sympatric speciation is rare, because fewer than 1
in 10 speciation events are even candidates. It seems reasonable to expect that many
more sister species would be sympatric if sympatric speciation were very common.
However, these studies do not include lacustrine fish species flocks or phytophagous
insects, which represent the majority of putative cases of sympatric speciation (**Supplemental Table 1**). This counting exercise is also flawed in two ways. First, it
is sensitive to how allopatric variants are treated by taxonomists. For instance, the
freshwater sunfish *Lepomis punctatus* was widely sympatric with its sister species *L. microlophus* until *punctatus* was split into two allopatric taxa, ending the sister relationship
with *microlophus* (Near et al. 2005). Second, it assumes that present geographic overlap
accurately reflects geography during speciation. These counts include cases of sympatric species known to have come into contact after allopatric speciation (Chesser &
Zink 1994, Coyne & Orr 2004, Coyne & Price 2000), or after a period of allopatry
followed by reinforcement in sympatry as in benthic and limnetic stickleback species
pairs (Taylor & McPhail 2000). Conversely, some unknown fraction of allopatric or
parapatric species may have originated in sympatry (Stuessy et al. 2004). One study to
effectively avoid these biases is Coyne & Price's (2000) survey of oceanic island birds,
which found little support for sympatric speciation in remote settings that remove
many of the biogeographic ambiguities plaguing other studies. It would be valuable
to carry out such studies on a wider range of taxonomic groups in isolated sites such
as oceanic islands, lakes, and caves.

Age-Range Correlation

Several researchers suggested a method of assessing the prevalence of sympatric versus allopatric speciation that explicitly accounts for potential postspeciation range
shifts (Barraclough & Vogler 2000, Lynch 1989). If allopatric speciation were the
rule, young species pairs would be predominantly allopatric, with increasing sympatry for older pairs due to range shifts. Conversely, sympatric speciation would lead
to decreasing sympatry with age. This age-range correlation (ARC) has been widely
criticized (Fitzpatrick & Turelli 2006, Losos & Glor 2003). First, the null distribution of range overlaps is poorly defined and must be estimated by randomizing
the data in each study. Second, results may depend on how range overlap between
clades is measured. Third, biogeographic changes associated with Pleistocene climate fluctuations are very recent compared with even the youngest speciation events
in many groups (Avise et al. 1998, Fitzpatrick & Turelli 2006), so geographic evidence may have been compromised for all species pairs. Finally, ARC will yield
conclusive results only if a single geographic mode of speciation has been considerably more common than others; a mixture of sympatric, parapatric, and allopatric
speciation may result in an uninformative distribution of geographic range overlaps,
resembling a null distribution of randomized range overlaps. Consequently, ARC
studies are generally inconclusive, failing to demonstrate any consistent relationship

between geographic range overlap and phylogenetic relationships (Fitzpatrick & Turelli 2006).

Empirical Summary: The Frequency of Sympatric Speciation

Disruptive selection:
Selection against phenotypically intermediate members of a population, favoring increased variance

Available data continue to support the orthodox position that sympatric speciation is less common than parapatric or allopatric speciation. However, there are enough well-supported examples to confidently say that sympatric speciation can occur in nature, and other putative cases of sympatric speciation remain understudied (**Supplemental Table 1**). This raises an important question: Are we likely to ever obtain an unbiased estimate of the frequency of sympatric speciation? We suggest not, because (*a*) ascertainment biases may favor recognition of nonsympatric speciation events to an unknown extent, (*b*) relationships between current and past biogeography can be rapidly disrupted by global climate change or range expansion, and (*c*) potential incipient species such as sympatric host races may not be reliable indicators of the likelihood of speciation.

What, then, are the prospects for future research on sympatric speciation? Clearly, much insight can be gained from more detailed dissection of many individual case studies. As for assessing the general incidence of sympatric speciation, we recommend an approach that complements counting and ARC studies. Rather than ask what fraction of past speciation events were sympatric, we suggest a pair of questions: Theoretically, what assumptions are conducive to sympatric speciation? Empirically, how often are these assumptions met in natural systems? Below, we summarize some major models of sympatric speciation, highlighting crucial assumptions (see Coyne & Orr 2004, Dieckmann et al. 2004, Gavrilets 2004, Kirkpatrick & Ravigne 2002 for more extensive reviews). We devote the final section of this review to evaluating empirical data on three key assumptions.

THEORIES OF SYMPATRIC SPECIATION

Nearly all models of sympatric speciation share a common framework. In an initially panmictic population, disruptive selection drives an evolutionary (rather than biogeographic) change in mating patterns. The result is reproductive isolation between subsets of the population's descendents, maintaining strong linkage or Hardy-Weinberg disequilibrium. The problem is that even low rates of mating between diverging subpopulations leads to recombination that increases frequencies of intermediate phenotypes. Models of sympatric speciation must therefore explain how divergent selection is able to overcome recombination to establish Hardy-Weinberg disequilibrium at mating loci, and perhaps linkage disequilibrium between mating and ecological genes. To date, there are over 70 models of sympatric speciation (**Supplemental Table 2**). Following Kirkpatrick & Ravigne (2002), these can be organized by their assumptions regarding four key issues: (*a*) the cause of disruptive selection, (*b*) how individuals select mates, (*c*) whether selection acts directly or indirectly on mating characters, and (*d*) the genetic basis of changes in mating patterns (1- versus 2-allele models). We discuss each of these assumptions, highlighting selected models as examples.

Cause of Disruptive Selection

Runaway sexual selection. The burst of interest in sympatric speciation in the 1990s coincided with a sharp increase in research on species flocks of East African cichlids. These rapidly speciating fish are often characterized by striking divergence in sexually dimorphic color patterns between ecologically very similar species (Seehausen 1997), inspiring speculation that sympatric speciation could arise through disruptive sexual selection alone. Higashi et al. (1999) modeled a population in which, to attract females, males express a quantitative trait ranging between two extremes (e.g., red to blue). Females vary in a second quantitative trait ranging from strongly preferring red, through random choice, to strongly preferring blue. Given an initial population composed of intermediate but polymorphic males (purple) and females (random mating), the population rapidly diverges into two species characterized by blue or red males and females that prefer blue or red. This is because any initially extreme-colored males are favored by the few choosy females, while intermediate males have no corresponding benefit because intermediate females mate randomly. Matings between extreme males and choosy females will produce offspring with both traits, resulting in disruptive two-tailed runaway sexual selection that builds up linkage disequilibrium until the population splits into two distinct groups.

Although this model is deeply flawed, its failings are highly informative. First, costs to assortative mating impose selection against stringent mate preferences, eliminating genetic variation for choosiness and preventing speciation (Arnegard & Kondrashov 2004). Second, Higashi's model assumed an initially polymorphic population without explaining how this polymorphism could arise. Even minor perturbations from Higashi's initially symmetric polymorphism can prevent speciation because runaway sexual selection becomes predominantly directional rather than disruptive (Arnegard & Kondrashov 2004). The current consensus is that sexual selection alone is unlikely to drive sympatric speciation (Arnegard & Kondrashov 2004, Coyne & Orr 2004, Gavrilets 2003, van Doorn et al. 2004). Additional sources of disruptive selection are required to stabilize mating polymorphisms that are later sorted into distinct species and/or compensate for costs of female choice. This disruptive selection can arise from a large variety of negative frequency-dependent interactions. These include sexual conflict (Gavrilets & Waxman 2002), male-male competition (van Doorn et al. 2004), opposing sexual and natural selection (Turner & Burrows 1995), and meiotic drive genes affecting sex ratio and secondary sexual traits (Lande et al. 2001). However, the vast majority of models invoke disruptive selection arising from ecological interactions such as resource competition.

Ecological disruptive selection. Stable disruptive selection from ecological interactions is generally modeled in one of two ways, which we will refer to as Levene and unimodal models. In Levene models, a population inhabits two distinct habitats such as host plants, which impose divergent adaptive demands. Two different alleles can stably coexist in this setting if each is favored in a different habitat with separate density regulation (Levene 1953). Stable disruptive selection ensues when heterozygotes have a lower mean fitness (averaging across habitats) than either homozygote

(Gavrilets 2006, Wilson & Turelli 1986). Despite its implicit spatial structure, this model satisfies the population genetic definition of sympatry if initial migration rates between habitats are high ($m \sim 0.5$) or if low migration follows a host shift from an initially panmictic population specializing on a single habitat.

In contrast, unimodal models assume a single environment containing resources continuously distributed (usually normal) along a trait axis such as prey size. Consumers are characterized by a continuous trait (e.g., gape width) governing resource use. Competition is less intense among phenotypically divergent individuals (Bürger 2005, Slatkin 1979). The consumer population's mean phenotype will evolve to match the most abundant prey trait. This will represent an evolutionary equilibrium for the mean phenotype, but not necessarily the variance (Roughgarden 1972). The equilibrium phenotype variance depends on both the diversity of resources available, and the width of individuals' niches. If each individual uses the full range of resources, then no among-individual variance will be maintained; the population will be subject to stabilizing selection (Bürger 2005). If, however, individuals use small subsets of the available resources, then the equilibrium trait variance may be large. Disruptive selection occurs when the existing trait variance is less than its equilibrium. Phenotypically extreme individuals then experience relatively weak resource competition, even given their less abundant resources. This disruptive selection will persist until increased genetic variation, plasticity, sexual dimorphism, or speciation raises the trait variance to its equilibrium level. In the final section of this review, we evaluate empirical support for frequency-dependent competition and disruptive selection.

How Individuals Select Mates

In sexual taxa, disruptive selection alone is insufficient to cause speciation. Random mating quickly breaks down any Hardy-Weinberg or linkage disequilibrium that arises from selection. Only extraordinarily strong frequency-dependent disruptive selection, in which few if any intermediates survive, can maintain distinct and reproductively isolated groups. Under more realistic levels of selection, ecological divergence must be accompanied by reproductive divergence. We therefore need to specify both how mating occurs and how it responds to selection.

Two general mating schemes are commonly used: assortment and trait-preference models. Assortment occurs when individuals (usually assumed to be females) prefer mates that are similar to themselves at some phenotypic trait expressed in both sexes. This matching leads to a correlation between trait values of mated pairs. Traits used for matching can include ecological traits under disruptive selection (Dieckmann & Doebeli 1999, Maynard Smith 1966), location or timing of mating (Fry 2003), or signal traits such as color or pheromones (Maynard Smith 1966, Udovic 1980). In contrast, trait-preference models assume that each female expresses a preference for males with a particular trait, genetically independent of her own genotype for that trait. In principle, male traits could be ecological or secondary sexual traits, though nearly all trait-preference models assume the latter (Doebeli 2005, Maynard Smith 1966). Assortment models are often said to be less general, but are more conducive to speciation (Gavrilets 2004, Kirkpatrick & Ravigne 2002).

Whether Selection Acts Directly or Indirectly on Mating Characters

Disruptive selection can directly affect mating patterns if traits used in mating are also directly relevant to ecological performance or genetically correlated with ecological traits. Ecological differences may reduce mating rates by affecting mate timing or location (Fry 2003) or if assortment is based on ecologically functional traits like morphology (Bürger et al. 2006). If ecologically divergent individuals are unlikely to mate from the outset, then when disruptive natural selection reduces the frequency of intermediate ecological genotypes, the remaining phenotypic extremes mate largely within themselves. Consequently, the population bifurcates into two noninterbreeding groups. This scenario greatly facilitates sympatric speciation, so we review existing empirical evidence for mating/ecology pleiotropy in the final section of this review.

There are two main objections to models of mating/ecology pleiotropy (sometimes called magic trait models; Gavrilets 2004). First, some researchers question whether traits controlling both ecology and mating are biologically realistic (Felsenstein 1981). Second, these models assume, rather than explain, strong assortative mating within the original population (Gavrilets 2005). Other models have tried to explain how assortative mating might initially evolve from random-mating populations (Dieckmann & Doebeli 1999, Fry 2003, Seger 1985). These models focus on indirect selection on mating traits in a process analogous to reinforcement (Servedio & Noor 2003), but without a period of allopatric divergence. Disruptive selection means that individuals who mate randomly risk producing intermediate offspring with lower fitness, indirectly favoring individuals that mate with their own ecotype. Selection is indirect because increased choosiness does not alter an individual's own fitness. Instead, choosy females avoid producing ecologically intermediate and less-fit offspring.

How Mating Patterns Evolve

There are two general ways in which disruptive selection can indirectly favor increased assortative mating through a reinforcement-like process. First, mating within ecotypes can arise by an evolutionary increase in assortment if the stringency of mate preferences (choosiness) can evolve. Alternatively, linkage disequilibrium can build up between previously independent mating and ecological traits. These correspond to the one- and two-allele models of Felsenstein (1981).

One-allele models. Consider a Levene-style population using two habitats favoring genotypes AA and aa respectively, resulting in disruptive selection. Next, assume that a new allele M arises at a second independent locus, which causes AA individuals to mate with AA, and aa individuals to mate with aa. By reducing production of less-fit heterozygous offspring, this allele is favored in both AA and aa individuals over the original m allele that conferred random mating. Because this model involves substitution of a single allele in both emerging daughter species, it is called a one-allele model (Felsenstein 1981). Recombination does not oppose the spread of the M allele, so speciation is relatively straightforward, albeit less robust than when selection acts directly on mating characters (Felsenstein 1981, Gavrilets 2006). More generally, one-allele models occur whenever emerging daughter species fix the same set of mating alleles. One-allele models include the evolution of reduced migration (Balkau & Feldman 1973), adaptive habitat selection (Fry 2003), or imprinting (Verzijden et al. 2005). They can also include the evolution of stronger assortative mating based on ecological traits (Dieckmann & Doebeli 1999). This requires that females are able to evaluate the match between their own trait and that of a prospective mate. Although such magic trait assortative mating is frequently used in models, there are questions about its biological generality (Gavrilets 2004).

Trait-preference mating:
When females prefer to mate with males exhibiting specific mating characters, independent of the females' expression of that character

Two-allele models. Sympatric speciation is more difficult when reproductive isolation requires trait-preference mating or assortment for nonecological traits, because divergent species must fix different mating alleles. Continuing the example used above, females might be polymorphic for a mating locus with alleles m for random mating, M_A preferring AA ecotypes, and M_a preferring aa. Randomly mating individuals (m) risk producing low-fitness Aa heterozygotes, so selection should favor M_A/AA and M_a/aa females. Emerging daughter species must fix different mating alleles, so this is called a two-allele model (Felsenstein 1981). The problem is that unless selection against heterozygotes is extraordinarily strong, recombination breaks down linkage disequilibrium between M_A and A, and between M_a and a, so females mate randomly with respect to their own ecological trait (Udovic 1980). Consequently, recombination-selection antagonism greatly reduces the ease of sympatric speciation.

When female preferences are for neutral third traits such as male color or song, emerging daughter species must fix both different preference alleles and different marker trait alleles, both of which may recombine with the ecological trait under disruptive selection. For instance, if $M_B M_B$ and $M_b M_b$ each mate exclusively with their own genotype, recombination between M and A loci still opposes speciation. Models suggest that sympatric speciation is very difficult but still possible in this relatively complex scenario (Dieckmann & Doebeli 1999, Doebeli 2005, Udovic 1980). To see how, consider a population under disruptive selection for an ecological trait such as jaw size, which is also polymorphic for independent loci affecting male color (blue or red) and female preferences for blue or red. Because color and preference are independent of jaw size, red-preferring females gain no fitness benefit from mating with red males. However, if linkage disequilibrium arises stochastically between the three traits, females can use male color as a weak proxy for ecology. Females who happen to be choosier will be more likely to mate with their own ecotype, reducing their risk of producing less-fit intermediate offspring. If disruptive selection is sufficiently strong, this reinforcement-like process can outpace the breakdown of linkage disequilibrium by recombination, leading to speciation (Dieckmann & Doebeli 1999, Doebeli 2005). Much like pure sexual selection models, these models often assume initially polymorphic mating traits despite the purifying effect of assortative mating (Kirkpatrick & Nuismer 2004).

Criticisms of Sympatric Speciation Models

In conclusion, sympatric speciation is easiest with strong disruptive ecological selection, strong mate preferences, low costs of being choosy, low recombination between mating and ecological loci, and large initial trait variances or high mutation rates (Gavrilets 2004). Culturally transmitted mate preferences and spatial structure can further facilitate sympatric divergence. Although simulations and analytical theory confirm that sympatric speciation is theoretically possible, they do not guarantee that it actually occurs in natural populations. The models outlined above adopt a variety of assumptions that have been widely criticized as being biologically unrealistic (Barton & Polechova 2005; Bolnick 2004a; Gavrilets 2005; Kirkpatrick & Nuismer 2004; Matessi et al. 2001; Polechova & Barton 2005; Waxman & Gavrilets 2005a,b). Sympatric speciation can be greatly delayed or prevented if certain key assumptions are relaxed (Bolnick 2004a, Bürger et al. 2006, Gavrilets 2005, Matessi et al. 2001, Waxman & Gavrilets 2005a).

Three assumptions have been singled out as particularly problematic. First, ecological assumptions required for disruptive selection may rarely be satisfied. Natural populations may not exhibit the narrow individual niche width required for frequency-dependent competition (Ackermann & Doebeli 2004, Roughgarden 1972) or phenotypic variances may not be constrained to be less than their equilibrium level (Bolnick & Doebeli 2003, Polechova & Barton 2005). Second, trait-preference mating, using traits unrelated to ecology, may be more general than magic trait assortative mating based on ecological characters (Gavrilets 2005). Third, females with stringent mate preferences may experience fitness costs that oppose the evolution of increased assortative mating, slowing or preventing sympatric speciation (Bolnick 2004a, Bürger et al. 2006, Kirkpatrick & Nuismer 2004, Schneider & Bürger 2006). Other criticisms highlight the use of constant ecological parameters, highly symmetrical and polymorphic initial conditions, very stringent assortative mating (Bolnick 2004a), a stable environment and resource distribution (Johansson & Ripa 2006), and soft selection (Demeeus et al. 1993). In conclusion, theoretical plausibility of sympatric speciation may be irrelevant if theoretical assumptions are not met in natural populations.

ARE THEORETICAL ASSUMPTIONS EMPIRICALLY JUSTIFIED?

Unfortunately, judgments about the biological realism of model assumptions are usually made on the basis of intuition rather than data. For instance, after showing that two-allele speciation is quite difficult, Felsenstein (1981) stated, "I find it easier to imagine genetic variation of the two-allele sort than of the one-allele sort" (p. 135). The lack of empirical grounding is unfortunate, because it may be easier to evaluate the frequency with which key assumptions are met than to estimate the frequency of sympatric speciation itself. One could then use the frequency with which key assumptions are jointly satisfied as a rough guide to the potential for sympatric speciation in nature. Conversely, finding that assumptions are rarely satisfied, alone or in combination, may help explain why sympatric speciation is uncommon. In this final section, we review empirical support for three leading conditions widely thought to facilitate sympatric speciation (though no one condition is entirely necessary or sufficient): (*a*) natural populations experience frequency-dependent disruptive selection, (*b*) natural selection operates directly on assortative mating patterns, and (*c*) assortative mating imposes weak or no costs.

Condition 1: Natural Populations Experience Frequency-Dependent Disruptive Selection

Most sympatric speciation models invoke ecologically-driven disruptive selection to (*a*) maintain polymorphisms that are later sorted by assortative mating, (*b*) drive divergence in traits that pleiotropically cause reproductive isolation, or (*c*) indirectly favor reinforcement of reproductive isolation. Disruptive selection is often assumed to arise from frequency-dependent competition among ecologically heterogeneous individuals within a population. Such ecological heterogeneity is widely documented, ranging from host races (Drés & Mallet 2002) and discrete polymorphisms (Smith & Skulason 1996), to more subtle individual-level niche variation (Bolnick et al. 2003). Further, competition has been shown to be stronger among phenotypically more similar individuals (Benkman 1996, Schluter 1994, Smith 1990, Swanson et al. 2003). This can give rare phenotypes an advantage or suppress the fitness of intermediate individuals during periods of intense competition, as confirmed by experiments in the laboratory (Bolnick 2001, Rainey & Travisano 1998) and field (Bolnick 2004b). Thus, frequency-dependent competition does drive disruptive selection on trophic traits in some natural populations (Benkman 1996, Bolnick 2004b, Hori 1993, McLaughlin et al. 1999, Pfennig et al. 2007, Robinson & Wilson 1996, Smith 1993).

How common is this disruptive selection? A recent meta-analysis concluded that 8% of selection estimates yielded significant positive quadratic curvatures (Kingsolver et al. 2001). Because positive quadratic coefficients can occur without a true fitness minimum (e.g., L-shaped fitness functions), we tentatively conclude that ≤8% of natural populations are subject to the kind of disruptive selection invoked in speciation models. However, this study may underestimate the frequency of disruptive selection because it omitted Levene-style metapopulations such as insect host races where disruptive selection is particularly likely. In conclusion, sympatric speciation models invoking disruptive selection are well justified, though they probably apply to only a minority of natural populations. Why, then, do we not see evidence of sympatric speciation in oceanic island birds, which are likely candidates for character release and disruptive selection (Coyne & Price 2000, Werner & Sherry 1987)? The answer is that disruptive selection is insufficient for sympatric speciation—if island birds primarily use preference-trait mating rules or experience significant costs to mate choice, speciation may be prevented.

Condition 2: Ecological Selection Operates Directly on Mating Characters (Mating/Ecology Pleiotropy)

Perhaps the clearest consensus to emerge from theory is that disruptive selection can lead to speciation when selected traits directly confer reproductive isolation (Gavrilets 2004). Where proponents and opponents of sympatric speciation disagree is whether such pleiotropic assortative mating is biologically realistic. We have no estimate of the frequency of pleiotropy between ecology and mating. Nonetheless, it is clear that such pleiotropy does exist and can arise from a variety of mechanisms.

The simplest form of mating/ecology pleiotropy arises when mating occurs on resources such as host plants. In a classic laboratory experiment, Rice & Salt (1990) confirmed that assortative mating can arise via evolution of divergent habitat preferences in sympatry. This is one of the few successful laboratory tests of sympatric speciation (Rice & Hostert 1993), perhaps because it applied selection directly on mating patterns in addition to several trade-offs between niches. *Drosophila melanogaster* populations were maintained in cages containing spatially separated divergent food sources. Each generation, newly hatched flies were mixed in the center of the cage and allowed to select an environment, where they mated and laid eggs. Disruptive selection was imposed by retaining only eggs laid on the two most divergent habitats. This clearly corresponds to sympatry because spatial segregation arose from genetic differences in habitat preference, rather than physical barriers to movement, from an initially panmictic population. By the 25th generation over 98% of flies quickly traveled to their parents' habitat, resulting in strong assortative mating. In contrast, laboratory selection experiments looking for reproductive isolation under indirect selection have failed (Rice & Hostert 1993), with one exception that has not been successfully replicated (Thoday & Gibson 1962).

Does habitat-specific mating occur in nature? The answer is a clear yes. Many phytophagous insects mate on the same host on which they feed (Berlocher & Feder 2002, Drés & Mallet 2002, Katakura et al. 1989, Wood & Keese 1990), as do some vertebrates (Munday et al. 2004), so evolutionary shifts to new hosts may automatically confer some reproductive isolation from conspecifics on the ancestral host. Recent hybridization between two *Rhagoletis* species produced a lineage that is unable to recognize either ancestral host plant and instead mates and oviposits on an introduced honeysuckle that is avoided by both parental species, resulting in reproductive isolation (Schwarz et al. 2005). It would be very useful to know what fraction of phytophagous insect species exhibit such host-specific mating.

Ecological divergence may also result in temporal reproductive isolation, because many species time their breeding to coincide with peak resource availability. Insect host races frequently diverge to match alternate host plant phenologies (Feder & Filchak 1999, Pratt 1994, Smith 1988, Wood & Keese 1990) or exploit different stages of fruit development within a single host (Weiblen & Bush 2002). Sympatric populations of lacustrine fish exhibit divergent spawning times, possibly coinciding with the availability of alternate prey (Palstra et al. 2004, Skulason et al. 1989, Taylor & Bentzen 1993, Wood & Foote 1996). In plants, the use of different edaphic conditions can lead to flowering time divergence (Savolainen et al. 2006).

The preceding mechanisms prevent proximity during mating, but ecological divergence can also directly modify mate-attraction cues. Host plant chemistry, for example, influences cuticular hydrocarbons and pheromones of phytophagous insects (Landolt & Phillips 1997, Stennett & Etges 1997). Changes in bill morphology in Darwin's finches influences vocal performance during courtship song (Podos 2001). However, direct selection on male signal traits will not facilitate sympatric speciation unless matched by parallel pleiotropic effects on female preferences. Correlated male signals and female preferences might evolve in three ways. First, female preferences may result from imprinting on parental phenotypes, as in several bird species (Grant & Grant 1997, Irwin & Price 1999, Sorenson et al. 2003). In contrast, insects do not appear to exhibit imprinting on their larval host (Barron 2001, vanEmden et al. 1996). Second, a population may already exhibit positive assortative mating based on ecologically important traits, which disruptive selection simply makes more effective by increasing trait variance. There is extensive evidence for assortative mating within natural populations, revealed by phenotypic correlations between mated male/female pairs for body size (Johannesson et al. 1995, McKaye 1986, Schliewen et al. 2001), color (Reynolds & Fitzpatrick 2007), major histocompatibility complex (MHC) genotype (Aeschlimann et al. 2003), or diet (Ward et al. 2004). Finally, female preferences and male signals may be controlled by the same gene or closely linked genes. This scenario has generally been disregarded based on the logic that genetic control of signal production is likely to be very different from that of signal perception (Boake 1991). However, a number of examples occur where the same gene(s) influence both production and discrimination of mating signals [e.g., *desat1* in *Drosophila melanogaster* (Marcillac et al. 2005); bindin in *Echinometra* sea urchins (Palumbi 1999); MHC loci in sticklebacks (Aeschlimann et al. 2003)]. In *Heliconius* butterflies, one quantitative trait locus (QTL) influences male preference, female wing-color, and predator avoidance (Kronforst et al. 2006). This QTL may be an inversion that suppresses recombination between signal and preference loci, rather than a single pleiotropic gene, but the distinction is unimportant for speciation models.

In conclusion, selection clearly does operate directly on traits involved in assortative mating owing to mating/ecology pleiotropy. When disruptive selection coincides with such pleiotropy, sympatric speciation may be relatively easy. However, there are also many examples in which mating traits are independent of ecological characters (Hager & Teale 1996, Nosil et al. 2006). In addition, it may be difficult, in retrospect, to distinguish cases in which ecology/mating pleiotropy preceded ecological divergence from the less robust scenario of one-allele reinforcement, in which such pleiotropy evolves in response to disruptive selection. To date, only one study has demonstrated one-allele reinforcement, during secondary contact between two allopatrically diverged *Drosophila* species (Ortiz-Barrientos & Noor 2005).

Condition 3: Assortative Mating Confers Weak or No Costs

If highly choosy females incur fitness costs, sympatric speciation may be drastically slowed or prevented (Bolnick 2004a, Bürger et al. 2006, Kirkpatrick & Nuismer 2004,

Schneider & Bürger 2006). This is because direct costs of searching for a suitable mate may exceed the indirect fitness benefits of assortative mating. Many models impose costs by setting a limit on C, the number of males (or sperm) that a female (or egg) is able to reject before entirely losing her ability to reproduce (Bolnick 2004a). Models suggested that speciation only failed when such costs were strong [$C < 30$ (Bolnick 2004a); or $C < 10$ (Bürger et al. 2006)]. However, judgments about what constitutes strong costs should be based on empirical data. Although we have no direct measures of C, it is sometimes possible to determine the number of males that females encounter during mate choice. Existing data (mostly from birds) indicate that females evaluate anywhere from 1 to 100 males prior to breeding, but generally fewer than 10 (see **Supplemental Table 3**). This suggests that costs of mate choice may frequently override benefits of selecting rare male phenotypes.

Fitness costs may often arise via incremental risks or expenses, rather than an abrupt cut-off (Alatalo et al. 1988, Slagsvold & Dale 1991). Experiments manipulating actual or perceived predation risk have shown that females mitigate higher risks by becoming less choosy (Forsgren 1992, Godin & Briggs 1996, Hedrick & Dill 1993, Jennions & Petrie 1997, but see Reid & Stamps 1997). For instance, guppies were less choosy between more- and less-colorful males when predaceous fish were visible (Godin & Briggs 1996). Mate assessment also imposes energetic costs owing to movement (Milinski & Bakker 1992, Slagsvold & Dale 1994), courtship displays (Wikelski et al. 2001), and lost foraging time. Consequently, females may become less choosy when resources are limited (Palokangas et al. 1992, Reid & Stamps 1997). Over 90% of pronghorn antelope females visit multiple male territories before selecting a mate, but in a drought year 81% took the first mate they encountered (Byers et al. 2006). This poses an interesting catch-22 for sympatric speciation models: They often invoke food limitation to drive disruptive selection, yet this may also undermine females' ability to exercise stringent mate choice, preventing speciation.

Despite the costs, many natural populations do exhibit assortative mating and genetic variation for mating preferences (Jennions & Petrie 1997). Laboratory selection experiments have repeatedly led to the evolution of increased assortative mating (Rice & Hostert 1993), proving that there is genetic variation for mate preferences within populations. Far fewer studies have evaluated whether there is genetic variation for the stringency of these preferences (choosiness). In the clearest example, isofemale lines of brown planthoppers had identical mean preferences for the frequency of male courtship vibrations, but differed in the range of vibrations that they would accept (Butlin 1993). Such results confirm that natural populations can harbor standing genetic variation for degree of assortative mating, as assumed in several models (Dieckmann & Doebeli 1999) but questioned by some critics (Gavrilets 2005, Waxman & Gavrilets 2005a). The presence of genetic variation for assortative mating in natural populations suggests that the costs of mate choice are not always overwhelmingly strong. This is supported by some studies finding only weak costs to mate choice: daily expenditure in female sage grouse only rises by 1% when visiting a lek (Gibson & Bachman 1992). Our understanding of sympatric speciation would be greatly enhanced by more empirical data on costs of and genetic variation for preference, signal traits, and choosiness in natural populations, and integrating such data into theoretical models.

SUMMARY

In 1963, Mayr concluded an extensive critique by saying that the burden of proof was on proponents of sympatric speciation. This remains true, notwithstanding the handful of compelling empirical cases and the pile of supportive theoretical models: We should continue to demand rigorous evidence before accepting a case study as a good example of sympatric speciation. However, we should also expect sound empirical evidence for claims of allopatric or parapatric speciation. Because past range expansions or secondary colonization events can be difficult to conclusively reject or demonstrate, many pairs of sympatric sister taxa may remain unresolved. This ambiguity is preferable to accepting a null model (which would bias estimates of the frequency of different modes of speciation). However, we may therefore be unlikely to achieve any confidence when estimating the frequency of different geographic modes of speciation. Based on current evidence, it seems reasonable to conclude that sympatric speciation occurs in nature, but is relatively rare, though more taxonomic groups should be investigated.

An alternative approach is to evaluate the frequency with which the prerequisites for straightforward sympatric speciation occur in nature. This may help us understand both when sympatric speciation is possible and why it is not more common. The empirical evidence available to date is by no means sufficient to judge the generality of any one factor, let alone their joint occurrence. Nonetheless, it is clear that niche variation, frequency-dependent competition, and resulting disruptive selection occur in a variety of systems. Mating/ecology pleiotropy, once dismissed by theoreticians as unlikely (Maynard Smith 1966), clearly does occur in nature. Costs to assortative mating are more difficult to assess, because available data cannot be easily compared with model parameters. Costs appear to be strong in some systems, whereas in others females are able to evaluate many males, leaving the door open for strong assortative mating. These empirical observations should guide choices of theoretical models and parameter ranges. Conversely, the rapidly growing theoretical literature on sympatric speciation is helping to identify key biological phenomena that facilitate or constrain sympatric speciation, pointing out profitable directions for future empirical efforts.

Given the difficulty of clearly distinguishing geographic scenarios, has the geography of speciation outlived its utility? Are we better off focusing on the mechanisms driving speciation, such as natural or sexual selection, reinforcement, drift, and hybridization? While such questions are doubtless useful and perhaps more tractable, such a research program still requires understanding the geography of speciation. Geographic structure influences what mechanisms can operate and how strong they must be to cause reproductive isolation. Furthermore, all speciation events occur in some geographic context, and the spatial distribution of biodiversity may depend on what speciation mechanisms are possible. Hence, we believe that geography will continue to play a central role in speciation research. This does not mean that the traditional sympatric/parapatric/allopatric distinction should be accepted uncritically. In particular, it is clear that diverging populations can shift from allopatry to sympatry (Jordal et al. 2006, Taylor & McPhail 2000) or vice versa (Baack 2004, Stuessy et al. 2004) during the process of divergence. These geographic changes may play a fundamental role in facilitating reproductive isolation, through reinforcement (Servedio & Noor 2003) or hybrid speciation (Rieseberg et al. 1995). More subtle still, genetic variation at some loci may arise in allopatry but be sorted into reproductively isolated groups within a sympatric population (Feder et al. 2003). Such temporal changes offer some profound new directions in studying the geography of speciation, which will further blur what it means for speciation to be sympatric.

SUMMARY POINTS

1. To demonstrate that sympatric speciation has occurred, one must rule out alternative hypotheses (parapatry, allopatry). This is not equivalent to using allopatry as a null hypothesis or default explanation.

2. Two major types of evidence can support sympatric speciation: (*a*) sister species in an isolated environment that makes secondary contact unlikely, and (*b*) speciation mechanisms that occur most easily in sympatry (host shifts or instantaneous speciation).

3. Comparative approaches to estimate the relative frequency of sympatric and allopatric speciation are severely undermined by past biogeographic changes.

4. Although theoretical models indicate that sympatric speciation is possible in principle, the models are laden with numerous assumptions whose empirical validity is not well known.

5. Disruptive ecological selection, which many models invoke to maintain polymorphism and drive speciation, occurs at a low to moderate frequency in nature.

6. Magic traits, which influence both mating patterns and ecological fitness, are widely used in theoretical models and are known to occur in natural populations.

7. Theoretical models suggest that costs to mate choice may slow or prevent sympatric speciation. Empirical studies support the idea that mate choice costs can be significant.

FUTURE ISSUES

1. Phylogeographic studies within species can detect range expansions, and so might be useful in distinguishing cases of secondary contact from cases of true long-term sympatry.

2. Biogeographic comparative studies of range overlaps have not been conducted for the two groups most widely thought to exhibit sympatric speciation, phytophagous insects and lacustrine fishes.

3. More studies such as that by Coyne & Price (2000) are needed to evaluate the frequency of sympatric sister taxa in isolated environments such as islands, caves, and lakes.

4. The distinction between allopatry, parapatry, and sympatry is blurred by cases where speciation involves temporal shifts between biogeographic modes. This added temporal dimension needs to be better integrated into biogeography of speciation.

5. How often are natural populations subject to frequency-dependent disruptive selection invoked by theoretical models? How strong and how persistent is this selection?

6. How strong is assortative mating within populations?

7. Are ecologically functional traits commonly used as the basis for assortative mating?

8. Are the costs of mate choice in natural populations sufficient to oppose the evolution of assortative mating?

DISCLOSURE STATEMENT

The authors are not aware of any biases that might be perceived as affecting the objectivity of this review.

ACKNOWLEDGMENTS

We would like to thank D. Agashe, R. Bürger, J. Coyne, M. Forister, S. Gavrilets, M. Kirkpatrick, H. López-Fernández, C. McBride, A. Meyer, P. Nosil, W.E. Stutz, R. Svanbäck, M. Turelli, and G.S. van Doorn for comments on drafts of this manuscript. The authors were supported by NSF grant DEB-0412802 (D.I.B.) the University of Texas at Austin (D.I.B.), and the University of Tennessee (B.M.F.).

LITERATURE CITED

Ackermann M, Doebeli M. 2004. Evolution of niche width and adaptive diversification. *Evolution* 58:2599–612

Aeschlimann PB, Haberli MA, Reusch TBH, Boehm T, Milinski M. 2003. Female sticklebacks *Gasterosteus aculeatus* use self-reference to optimize MHC allele number during mate selection. *Behav. Ecol. Sociobiol.* 54:19–26

Agapow PM, Bininda-Emonds ORP, Crandall KA, Gittleman JL, Mace GM, et al. 2004. The impact of species concept on biodiversity studies. *Q. Rev. Biol.* 79:161–79

Alatalo RV, Carlson A, Lundberg A. 1988. The search cost in mate choice of the pied flycatcher. *Anim. Behav.* 36:289–91

Arnegard ME, Kondrashov AS. 2004. Sympatric speciation by sexual selection alone is unlikely. *Evolution* 58:222–37

Avise JC, Walker D, Johns GC. 1998. Speciation durations and Pleistocene effects on vertebrate phylogeography. *Proc. R. Soc. London Ser. B* 265:1707–12

Baack EJ. 2004. Cytotype segregation on regional and microgeographic scales in snow buttercups (*Ranunculus adoneus*: Ranunculaceae). *Am. J. Bot.* 91:1783–88

Balkau B, Feldman MW. 1973. Selection for migration modification. *Genetics* 74:171–74

Barluenga M, Stoelting KN, Salzburger W, Muschick M, Meyer A. 2006a. Evolutionary biology—Evidence for sympatric speciation? Reply. *Nature* 444:E13

Barluenga M, Stoelting KN, Salzburger W, Muschick M, Meyer A. 2006b. Sympatric speciation in Nicaraguan crater lake cichlid fish. *Nature* 439:719–23

Barraclough TG, Vogler AP. 2000. Detecting the geographical pattern of speciation from species-level phylogenies. *Am. Nat.* 155:419–34

Barron AB. 2001. The life and death of Hopkins' host-selection principle. *J. Insect Behav.* 14:725–37

Barton NH, Polechova J. 2005. The limitations of adaptive dynamics as a model of evolution. *J. Evol. Biol.* 18:1186–90

Benkman CW. 1996. Are the ratios of bill crossing morphs in crossbills the result of frequency-dependent selection? *Evol. Biol.* 10:119–26

Berlocher SH. 1998. Can sympatric speciation be proven from biogeographic and phylogenetic evidence? In *Endless Forms: Species and Speciation*, ed. DJ Howard, SH Berlocher, pp. 99–113. New York: Oxford Univ. Press

Berlocher SH. 1999. Host race or species? Allozyme characterization of the 'flowering dogwood fly', a member of the *Rhagoletis pomonella* complex. *Heredity* 83:652–62

Berlocher SH, Feder JL. 2002. Sympatric speciation in phytophagous insects: moving beyond controversy? *Annu. Rev. Entomol.* 47:773–815

Berrebi P, Valiushok D. 1998. Genetic divergence among morphotypes of Lake Tana (Ethiopia) barbs. *Biol. J. Linn. Soc.* 64:369–84

Boake CRB. 1991. Coevolution of senders and receivers of sexual signals-genetic coupling and genetic correlations. *Trends Ecol. Evol.* 6:225–27

Bolnick DI. 2001. Intraspecific competition favours niche width expansion in *Drosophila melanogaster*. *Nature* 410:463–66

Bolnick DI. 2004a. Waiting for sympatric speciation. *Evolution* 87:895–99

Bolnick DI. 2004b. Can intraspecific competition drive disruptive selection? An experimental test in natural populations of sticklebacks. *Evolution* 87:608–18

Bolnick DI, Doebeli M. 2003. Sexual dimorphism and adaptive speciation: two sides of the same ecological coin. *Evolution* 57:2433–49

Bolnick DI, Svanbäck R, Fordyce JA, Yang LH, Davis JM, et al. 2003. The ecology of individuals: incidence and implications of individual specialization. *Am. Nat.* 161:1–28

Bürger R. 2005. A multilocus analysis of intraspecific competition and stabilizing selection on a quantitative trait. *J. Math. Biol.* 50:355–96

Bürger R, Schneider KA, Willensdorfer M. 2006. The conditions for speciation through intraspecific competition. *Evolution* 60:2185–206

Butlin R. 1993. The variability of mating signals and preferences in the brown plan-thopper, *Nilaparvata lugens* (Homoptera, Delphacidae). *J. Insect Behav.* 6:125–40

Byers JA, Byers AA, Dunn SJ. 2006. A dry summer diminishes mate search effort by pronghorn females: Evidence for a significant cost of mate search. *Ethology* 112:74–80

Chesser RT, Zink RM. 1994. Modes of speciation in birds: a test of Lynch's method. *Evolution* 48:490–97

Claridge MF, Dawah HA, Wilson MR. 1997. *Species: the Units of Biodiversity*. London: Chapman & Hall. 460 pp.

Coyne JA, Orr HA. 2004. *Speciation*. Sunderland, MA: Sinauer. 545 pp.

Coyne JA, Price T. 2000. Little evidence for sympatric speciation in island birds. *Evolution* 54:2166–71

Demeeus T, Michalakis Y, Renaud F, Olivieri I. 1993. Polymorphism in heteroge-neous environments, evolution of habitat selection and sympatric speciation—hard and soft selection models. *Evol. Ecol.* 7:175–98

de Vries H. 1901–1903. *De Mutationstheorie*, Vols. I, II. Leipzig: Verlag von Veit

Dieckmann U, Doebeli M. 1999. On the origin of species by sympatric speciation. *Nature* 400:354–57

Dieckmann U, Doebeli M, Metz JAJ, Tautz D, eds. 2004. *Adaptive Speciation*. New York: Cambridge Univ. Press. 476 pp.

Dimmick WW, Edds DR. 2002. Evolutionary genetics of the endemic *Schizoathicine* (Cypriniformes: Cyprinidae) fishes of Lake Rara, Nepal. *Biochem. Syst. Ecol.* 30:919–29

Doebeli M. 2005. Adaptive speciation when assortative mating is based on female preference for male marker traits. *J. Evol. Biol.* 18:1587–600

Drés M, Mallet J. 2002. Host races in plant-feeding insects and their importance in sympatric speciation. *Philos. Trans. R. Soc. London Ser. B* 357:471–92

Feder JL, Berlocher SH, Roethele JB, Dambroski H, Smith JJ, et al. 2003. Allopatric genetic origins for sympatric host-plant shifts and race formation in *Rhagoletis*. *Proc. Natl. Acad. Sci. USA* 100:10314–19

Feder JL, Filchak KE. 1999. It's about time: the evidence for host plant-mediated selection in the apple maggot fly, *Rhagoletis pomonella*, and its implications for fitness trade-offs in phytophagous insects. *Entomol. Exp. Appl.* 91:211–25

Felsenstein J. 1981. Skepticism towards Santa Rosalia, or why are there so few kinds of animals? *Evolution* 35:124–38

Fitzpatrick BM, Turelli M. 2006. The geography of mammalian speciation: Mixed signals from phylogenies and range maps. *Evolution* 60:601–15

Forsgren E. 1992. Predation risk affects mate choice in a Gobiid fish. *Am. Nat.* 140:1041–49

Fry JD. 2003. Multilocus models of sympatric speciation: Bush vs Rice vs Felsenstein. *Evolution* 57:1735–46

Futuyma DJ, Mayer GC. 1980. Non-allopatric speciation in animals. *Syst. Zool.* 29:254–71

Gavrilets S. 2003. Models of speciation: what have we learned in 40 years? *Evolution* 57:2197–215

Gavrilets S. 2004. *Fitness Landscapes and the Origin of Species*. Princeton, NJ: Princeton Univ. Press. 476 pp.

Gavrilets S. 2005. "Adaptive speciation"—it is not that easy: a reply to Doebeli et al. *Evolution* 59:696–99

Gavrilets S. 2006. The Maynard Smith model of sympatric speciation. *J. Theor. Biol.* 239:172–82

Gavrilets S, Waxman D. 2002. Sympatric speciation by sexual conflict. *Proc. Natl. Acad. Sci. USA* 99:10533–38

Gibson RM, Bachman GC. 1992. The costs of female choice in a lekking bird. *Behav. Ecol.* 3:300–9

Gislason D, Ferguson MM, Skulason S, Snorrason SS. 1999. Rapid and coupled phenotypic and genetic divergence in Icelandic arctic char (*Salvelinus alpinus*). *Can. J. Fish. Aquat. Sci.* 56:2229–34

Godin JGJ, Briggs SE. 1996. Female mate choice under predation risk in the guppy. *Anim. Behav.* 51:117–30

Gompert Z, Fordyce JA, Forister ML, Shapiro AM, Nice CC. 2006. Homoploid hybrid speciation in an extreme environment. *Science* 314:1923–25

Gottlieb LD. 1979. The origin of phenotype in a recently evolved species. In *Topics in Plant Population Biology*, ed. OT Solbrig, S Jain, GD Johnson, PH Raven, pp. 264–86. New York: Columbia Univ. Press

Grant PR, Grant BR. 1997. Hybridization, sexual imprinting, and mate choice. *Am. Nat.* 149:1–28

Hager BJ, Teale SA. 1996. The genetic control of pheromone production and response in the pine engraver beetle *Ips pini*. *Heredity* 77:100–7

Hedrick AV, Dill LM. 1993. Mate choice by female crickes is influenced by predation risk. *Anim. Behav.* 46:193–96

Higashi M, Takimoto G, Yamamura N. 1999. Sympatric speciation by sexual selection. *Nature* 402:523–26

Hori M. 1993. Frequency-dependent natural selection in the handedness of scale-eating cichlid fish. *Science* 270:216–19

Irwin DE, Price T. 1999. Sexual imprinting, learning and speciation. *Heredity* 82:347–54

Jennions MD, Petrie M. 1997. Variation in mate choice and mating preferences: A review of causes and consequences. *Biol. Rev. Cambridge Philos. Soc.* 72:283–327

Johannesson K, Rolan-Alvarez E, Ekendahl A. 1995. Incipient reproductive isolation between two sympatric morphs of the intertidal snail *Littorina saxatilis*. *Evolution* 49:1180–90

Johansson J, Ripa J. 2006. Will sympatric speciation fail due to stochastic competitive exclusion? *Am. Nat.* 168:572–78

Jordal BH, Emerson BC, Hewitt GM. 2006. Apparent 'sympatric' speciation in ecologically similar herbivorous beetles facilitated by multiple colonizations of an island. *Mol. Ecol.* 15:2935–47

Jordan DS. 1905. The origin of species through isolation. *Science* 22:545–62

Katakura H, Shioi M, Kira Y. 1989. Reproductive isolation by host specificity in a pair of phytophagous ladybird beetles. *Evolution* 43:1045–53

Kingsolver JG, Hoekstra HE, Hoekstra JM, Berrigan D, Vignieri SN, et al. 2001. The strength of phenotypic selection in natural populations. *Am. Nat.* 157:245–61

Kirkpatrick M, Nuismer SL. 2004. Sexual selection can constrain sympatric speciation. *Proc. R. Soc. London Ser. B* 271:687–93

Kirkpatrick M, Ravigne V. 2002. Speciation by natural and sexual selection: Models and experiments. *Am. Nat.* 159:S22–35

Klemetsen A, Elliott JM, Knudsen R, Sorensen P. 2002. Evidence for genetic differences in the offspring of two sympatric morphs of Arctic charr. *J. Fish Biol.* 60:933–50

Kronforst MR, Young LG, Kapan DD, McNeely C, O'Neill RJ, Gilbert LE. 2006. Linkage of butterfly mate preference and wing color preference cue at the genomic location of *wingless*. *Proc. Natl. Acad. Sci. USA* 103:6575–80

Lande R, Seehausen O, van Alphen JJM. 2001. Mechanisms of rapid sympatric speciation by sex reversal and sexual selection in cichlid fish. *Genetica* 112:435–43

Landolt PJ, Phillips TW. 1997. Host plant influences on sex pheromone behavior of phytophagous insects. *Annu. Rev. Entomol.* 42:371–91

Levene H. 1953. Genetic equilibrium when more than one ecological niche is available. *Am. Nat.* 87:331–33

Linn C, Feder JL, Nojima S, Dambroski HR, Berlocher SH, Roelofs W. 2003. Fruit odor discrimination and sympatric host race formation in *Rhagoletis*. *Proc. Natl. Acad. Sci. USA* 100:11490–93

Losos JB, Glor RE. 2003. Phylogenetic comparative methods and the geography of speciation. *Trends Ecol. Evol.* 18:220–27

Lynch JD. 1989. The gauge of speciation: on the frequencies of modes of speciation. In *Speciation and its Consequences*, ed. D Otte, JA Endler, pp. 527–53. Sunderland, MA: Sinauer

Marcillac F, Grosjean Y, Ferveur JF. 2005. A single mutation alters production and discrimination of *Drosophila* sex pheromones. *Proc. R. Soc. London Ser. B* 272:303–9

Matessi C, Gimelfarb A, Gavrilets S. 2001. Long-term buildup of reproductive isolation promoted by disruptive selection: how far does it go? *Selection* 2:41–64

Mavarez J, Salazar CA, Bermingham E, Salcedo C, Jiggins CD, Linares M. 2006. Speciation by hybridization in *Heliconius* butterflies. *Nature* 441:868–71

Maynard Smith J. 1966. Sympatric speciation. *Am. Nat.* 100:637–50

Mayr E. 1942. *Systematics and the Origin of Species*. New York: Columbia Univ. Press. 382 pp.

Mayr E. 1963. *Animal Species and Evolution*. Cambridge, MA: Belknap. 811 pp.

McKaye KR. 1986. Mate choice and size assortative pairing by the cichlid fishes of Lake Jiloa, Nicaragua. *J. Fish Biol.* 29:135–50

McLaughlin RL, Ferguson MM, Noakes DLG. 1999. Adaptive peaks and alternative foraging tactics in brook charr: evidence of short-term divergent selection for sitting-and-waiting and actively searching. *Behav. Ecol. Sociobiol.* 45:386–95

Milinski M, Bakker TCM. 1992. Costs influence sequential mate choice in sticklebacks, *Gasterosteus aculeatus*. *Proc. R. Soc. London Ser. B* 250:229–33

Munday PL, van Herwerden L, Dudgeon CL. 2004. Evidence for sympatric speciation by host shift in the sea. *Curr. Biol.* 14:1498–504

Near TJ, Bolnick DI, Wainwright PC. 2005. Fossil calibrations and molecular divergence time estimates in centrarchid fishes (Teleostei: Centrarchidae). *Evolution*
59:1768–82

Nosil P, Crespi BJ, Sandoval CP, Kirkpatrick M. 2006. Migration and the genetic
covariance between habitat preference and performance. *Am. Nat.* 167:E66–78

Ortiz-Barrientos D, Noor MAF. 2005. Evidence for a one-allele assortative mating
locus. *Science* 310:1467

Otto SP, Whitton J. 2000. Polyploid incidence and evolution. *Annu. Rev. Genet.*
34:401–37

Palokangas P, Alatalo RV, Korpimaki E. 1992. Female choice in the kestrel under
different availability of mating options. *Anim. Behav.* 43:659–65

Palstra AP, de Graaf M, Sibbing FA. 2004. Riverine spawning and reproductive segregation in a lacustrine cyprinid species flock, facilitated by homing? *Anim. Biol.*
54:393–415

Palumbi SR. 1999. All males are not created equal: Fertility differences depend on
gamete recognition polymorphisms in sea urchins. *Proc. Natl. Acad. Sci. USA*
96:12632–37

Pfennig DW, Rice AM, Martin RA. 2007. Field and experimental evidence for competition's role in phenotypic divergence. *Evolution* 61:257–71

Podos J. 2001. Correlated evolution of morphology and vocal signal structure in
Darwin's finches. *Nature* 409:185–88

Polechova J, Barton NH. 2005. Speciation through competition: a critical review.
Evolution 59:1194–210

Pratt GF. 1994. Evolution of *Euphilotes* (Lepidoptera: Lycaenidae) by seasonal and
host shifts. *Biol. J. Linn. Soc.* 51:387–416

Prowell DP, McMichael M, Silvain JF. 2004. Multilocus genetic analysis of host
use, introgression, and speciation in host strains of fall armyworm (Lepidoptera:
Noctuidae). *Ann. Entomol. Soc. Am.* 97:1034–44

Rainey PB, Travisano M. 1998. Adaptive radiation in a heterogeneous environment.
Nature 394:69–72

Ramsey J, Schemske DW. 1998. Pathways, mechanisms, and rates of polyploid formation in flowering plants. *Annu. Rev. Ecol. Syst.* 29:467–501

Reid ML, Stamps JA. 1997. Female mate choice tactics in a resource-based mating
system: Field tests of alternative models. *Am. Nat.* 150:98–121

Reynolds RG, Fitzpatrick BM. 2007. Assortative mating in poison-dart frogs based
on an ecologically important trait. *Evolution.* 61(9):2253–59

Rice WR, Hostert EE. 1993. Laboratory experiments on speciation: What have we
learned in 40 years? *Evolution* 47:1637–53

Rice WR, Salt GW. 1990. The evolution of reproductive isolation as a correlated
character under sympatric conditions: experimental evidence. *Evolution* 44:1140–
52

Rieseberg LH, Van Fossen C, Desrochers AM. 1995. Hybrid speciation accompanied
by genomic reorganization in wild sunflowers. *Nature* 375:313–16

Robinson BW, Wilson DS. 1996. Trade-offs of ecological specialization: An intraspecific comparison of pumpkinseed sunfish phenotypes. *Ecology* 77:170–78

Roughgarden J. 1972. Evolution of niche width. *Am. Nat.* 106:683–718

Savolainen V, Anstett M-C, Lexer C, Hutton I, Clarkson JJ, et al. 2006. Sympatric speciation in palms on an oceanic island. *Nature* 441:210–13

Schliewen U, Rassmann K, Markmann M, Markert J, Kocher T, Tautz D. 2001. Genetic and ecological divergence of a monophyletic cichlid species pair under fully sympatric conditions in Lake Ejagham, Cameroon. *Mol. Ecol.* 10:1471–88

Schliewen UK, Klee B. 2004. Reticulate sympatric speciation in Cameroonian crater lake cichlids. *Front. Zool.* 1:1–12

Schliewen UK, Kocher TD, McKaye KR, Seehausen O, Tautz D. 2006. Evolutionary biology—Evidence for sympatric speciation? *Nature* 444:E12–13

Schliewen UK, Tautz D, Paabo S. 1994. Sympatric speciation suggested by monophyly of crater lake cichlids. *Nature* 368:629–32

Schluter D. 1994. Experimental evidence that competition promotes divergence in adaptive radiation. *Science* 266:798–801

Schneider KA, Bürger R. 2006. Does competitive divergence occur if assortative mating is costly? *J. Evol. Biol.* 19:570–88

Schwarz D, Matta BM, Shakir-Botteri NL, McPheron BA. 2005. Host shift to an invasive plant triggers rapid animal hybrid speciation. *Nature* 436:546–49

Seehausen O. 1997. Distribution of and reproductive isolation among color morphs of a rock-dwelling Lake Victoria cichlid (*Haplochromis nyererei*). *Ecol. Freshw. Fish* 6:59–66

Seger J. 1985. Intraspecific resource competition as a cause of sympatric speciation. In *Evolution: Essays in Honor of John Maynard Smith*, ed. PJ Greenwood, PH Harvey, M Slatkin, pp. 43–53. Cambridge, UK: Cambridge Univ. Press

Servedio MR, Noor MAF. 2003. The role of reinforcement in speciation: Theory and data. *Annu. Rev. Ecol. Evol. Syst.* 34:339–64

Sezer M, Butlin RK. 1998. The genetic basis of host plant adaptation in the brown planthopper (*Nilaparvata lugens*). *Heredity* 80:499–508

Skulason S, Snorrason SS, Noakes DLG, Ferguson MM, Malmquist HJ. 1989. Segregation in spawning and early life-history among polymorphic arctic charr, *Salvelinus alpinus* in Thingvallavatn, Iceland. *J. Fish Biol.* 35:225–32

Slagsvold T, Dale S. 1991. Mate choice models—can cost of searching and cost of courtship explain mating patterns of female pied flycatchers. *Ornis Scan.* 22:319–26

Slagsvold T, Dale S. 1994. Why do female pied flycatchers mate with already mated males—deception or restricted mate sampling? *Behav. Ecol. Sociobiol.* 34:239–50

Slatkin M. 1979. Frequency- and density-dependent selection on a quantitative character. *Genetics* 93:755–71

Smith DC. 1988. Heritable divergence of *Rhagoletis pomonella* host races by seasonal asynchrony. *Nature* 336:66–67

Smith TB. 1990. Resource use by bill morphs of an African finch: evidence for intraspecific competition. *Ecology* 71:1246–57

Smith TB. 1993. Disruptive selection and the genetic basis of bill size polymorphism in the African finch *Pyrenestes*. *Nature* 363:618–20

Smith TB, Skulason S. 1996. Evolutionary significance of resource polymorphisms in fishes, amphibians, and birds. *Annu. Rev. Ecol. Syst.* 27:111–33

Sorenson MD, Sefc KM, Payne RB. 2003. Speciation by host switch in brood parasitic indigobirds. *Nature* 424:928–31

Stennett MD, Etges WJ. 1997. Premating isolation is determined by larval rearing substrates in cactophilic *Drosophila mojavensis*. III. Epicuticular hydrocarbon variation is determined by use of different host plants in *Drosophila mojavensis* and *Drosophila arizonae*. *J. Chem. Ecol.* 23:2803–24

Stuessy TF, Weiss-Schneeweiss H, Keil DJ. 2004. Diploid and polyploid cytotype distribution in *Melampodium cinereum* and *M-leucanthum* (Asteraceae, Heliantheae). *Am. J. Bot.* 91:889–98

Swanson BO, Gibb AC, Marks JC, Hendrickson DA. 2003. Trophic polymorphism and behavioral differences decrease intraspecific competition in a cichlid, *Herichthys minckleyi*. *Ecology* 84:1441–46

Taylor EB, Bentzen P. 1993. Evidence for multiple origins and sympatric divergence of trophic ecotypes of smelt (Osmerus) in northeastern North America. *Evolution* 47:813–32

Taylor EB, McPhail JD. 2000. Historical contingency and ecological determinism interact to prime speciation in sticklebacks, *Gasterosteus*. *Proc. R. Soc. London Ser. B* 267:2375–84

Thoday JM, Gibson JB. 1962. Isolation by disruptive selection. *Nature* 193:1164–66

Turner GF, Burrows MT. 1995. A model of sympatric speciation by sexual selection. *Proc. R. Soc. London Ser. B* 260:287–92

Udovic D. 1980. Frequency-dependent selection, disruptive selection, and the evolution of reproductive isolation. *Am. Nat.* 116:621–41

van Doorn GS, Dieckmann U, Weissing FJ. 2004. Sympatric speciation by sexual selection: a critical reevaluation. *Am. Nat.* 163:709–25

vanEmden HF, Sponagl B, Baker T, Ganguly S, Douloumpaka S. 1996. Hopkins 'host selection principle', another nail in its coffin. *Physiol. Entomol.* 21:325–28

Verzijden MN, Lachlan RF, Servedio MR. 2005. Female mate-choice behavior and sympatric speciation. *Evolution* 59:2097–108

Ward AJW, Hart PJB, Krause J. 2004. The effects of habitat- and diet-based cues on association preferences in three-spined sticklebacks. *Behav. Ecol.* 15:925–29

Waxman D, Gavrilets S. 2005a. 20 questions on adaptive dynamics: a target review. *J. Evol. Biol.* 18:1139–54

Waxman D, Gavrilets S. 2005b. Issues of terminology, gradient dynamics and the ease of sympatric speciation in Adaptive Dynamics. *J. Evol. Biol.* 18:1214–19

Weiblen GD, Bush GL. 2002. Speciation in fig pollinators and parasites. *Mol. Ecol.* 11:1573–78

Werner TK, Sherry TW. 1987. Behavioral feeding specialization in *Pinaroloxias inornata*, the "Darwin's Finch" of Cocos Island, Costa Rica. *Proc. Natl. Acad. Sci. USA* 84:5506–10

Wikelski M, Carbone C, Bednekoff PA, Choudhury S, Tebbich S. 2001. Why is female choice not unanimous? Insights from costly mate sampling in marine iguanas. *Ethology* 107:623–38

Wilson AJ, Gislason D, Skulason S, Snorrason SS, Adams CE, et al. 2004. Population genetic structure of Arctic Charr, *Salvelinus alpinus* from northwest Europe on large and small spatial scales. *Mol. Ecol.* 13:1129–42

Wilson DS, Turelli M. 1986. Stable underdominance and the evolutionary invasion of empty niches. *Am. Nat.* 127:835–50

Wood CC, Foote CJ. 1996. Evidence for sympatric genetic divergence of anadromous and nonanadromous morphs of sockeye salmon (*Oncorhynchus nerka*). *Evolution* 50:1265–79

Wood TK, Keese M. 1990. Host-plant-induced assortative mating in *Echenopa* treehoppers. *Evolution* 44:619–28

The Evolution of Color Polymorphism: Crypticity, Searching Images, and Apostatic Selection

Alan B. Bond

School of Biological Sciences, University of Nebraska, Lincoln, Nebraska 68588;
email: abond1@unl.edu

Annu. Rev. Ecol. Evol. Syst. 2007. 38:489–514

First published online as a Review in Advance on
August 15, 2007

The *Annual Review of Ecology, Evolution, and Systematics* is online at
http://ecolsys.annualreviews.org

This article's doi:
10.1146/annurev.ecolsys.38.091206.095728

Key Words

background complexity, detectability, perceptual switching,
selective attention

Abstract

The development and maintenance of color polymorphism in cryptic prey species is a source of enduring fascination, in part because it appears to result from selective processes operating across multiple levels of analysis, ranging from cognitive psychology to population ecology. Since the 1960s, prey species with diverse phenotypes have been viewed as the evolved reflection of the perceptual and cognitive characteristics of their predators. Because it is harder to search simultaneously for two or more cryptic prey types than to search for only one, visual predators should tend to focus on the most abundant forms and effectively overlook the others. The result should be frequency-dependent, apostatic selection, which will tend to stabilize the prey polymorphism. Validating this elegant hypothesis has been difficult, and many details have been established only relatively recently. This review clarifies the argument for a perceptual selective mechanism and examines the relevant experimental evidence.

INTRODUCTION

Edward Poulton's classic work, *The Colours of Animals* (1890), provided the first evolutionary account of cryptic coloration, focusing particularly on woodland moths that adopt the color patterns of leaves, twigs, and bark. Poulton noted that these insects were often polymorphic, occurring in multiple distinctive pattern variants, and he suggested that this served to reduce the efficiency of predatory search. He wrote that polymorphism provides "a wider range of objects for which [predators] may mistake the larvae, and the search must occupy more time, for equivalent results, than in the case of other species which are not dimorphic" (Poulton 1890, p. 47). Polymorphism was thus viewed from the outset as an adaptive response to the behavior of visual predators, but confirming the selective mechanism has proved a subtle and complex task. Although other aspects of the evolution of polymorphism have recently been reviewed (Hedrick 2006, Sinervo & Calsbeek 2006), the selective role of predator cognition has not been fully developed and critically discussed in many years. The argument has three main premises: (*a*) Visual search is subject to processing limitations, in that it is more difficult to search for two or more dissimilar, cryptic prey types simultaneously than to search for only one. (*b*) Predators consequently tend to focus on the most rewarding prey type and effectively overlook the others. (*c*) The result is frequency-dependent, apostatic selection, a higher relative mortality among more abundant prey types that serves to stabilize prey polymorphism. We briefly discuss cryptic coloration and color polymorphism, and then consider each premise of the argument in turn.

COMPONENTS OF CRYPTICITY

Cryptic or concealing coloration allows prey animals to blend into the background, reducing their vulnerability to visually searching predators (Cott 1957, Edmunds 1974). Crypticity is an intuitively accessible concept, but it resists easy quantification. It is usually used to describe whether a prey animal looks like its surroundings or whether human observers can easily detect it. This usage might be better termed detectability than crypticity, as it varies with the observation conditions, the characteristics of the predator's visual system, and short-term changes in the predator's focus of attention and searching speed. To reduce ambiguity, Endler (1991) defined a cryptic color pattern as one that "resembles a random sample of the visual background," conditioned by the biophysics of predator vision and the characteristics of the ambient illumination. This was a significant conceptual advance, but the notion of random sampling is open to competing methodological interpretations. Merilaita & Lind (2005) extracted prey-shaped pieces from arbitrary locations in a textured background. They showed that the resulting stimuli were highly variable in appearance, and many were very easy to detect.

Using a cookie-cutter on the background may, however, be less appropriate as a model of crypticity than it is of pattern mimicry. In this phenomenon, which has also been termed special resemblance (Cott 1957) or masquerade (Endler 1981), the animal evolves a detailed likeness to some specific, local feature of its environment

(Edmunds 1990, Robinson 1969, Vane-Wright 1980). Pattern mimicry has produced some of the most spectacular examples of protective coloration, such as stick caterpillars (de Ruiter 1952), leaf-mimicking katydids (Castner & Nickle 1995), and leafy seadragons (Connolly et al. 2002). However, most cryptically colored species display a more global, statistical resemblance to the background, in which the wavelength, intensity, and size of prey color patches converge on the mean of a multivariate distribution of background features. Endler (1978, 1984) proposed to extract such statistical distributions for both the background and the prey and to measure crypticity in terms of the correspondence between the distributions. In studies of the ability of blue jays to detect artificial moths on randomly textured backgrounds, a similar statistical measure accounted for 30%–40% of the variance in detection accuracy (Bond & Kamil 1999, Kamil & Bond 2001).

In addition to detectability and crypticity, a third component in the effectiveness of concealing coloration is the configuration of the background. An intricate and complex background containing many contrasting elements over a range of spatial frequencies can require a prolonged and careful search, even if the prey are not otherwise particularly cryptic (Bond & Kamil 2006, Merilaita 2003, Robinson 1990). This is the basis for the visual puzzles in many children's books (e.g., Handford 1987), in which the task is to locate a particular item in a highly detailed and cluttered image. There may be only a limited resemblance between the item and the distracting elements of the background, but because there is so much information to be processed, the search can be very laborious. The impact of complexity is illustrated in **Figure 1**, which displays a set of six artificial moths on four differently textured backgrounds. The intensity histogram of the background in **Figure 1b** corresponds closely to that of the moths, but the average spatial frequency is somewhat higher. The distinctive patterns of local spatial correlations in the moths draw the viewer's attention, forming a contrast to the uncorrelated features of the background. In **Figure 1c**, the background pixels were drawn from independent light and dark intensity distributions that minimally overlapped those of the moths, but the average spatial frequency was reduced, increasing the visual complexity of the texture. These moths are less cryptic, in terms of statistical resemblance, than those in **Figure 1b**, but both humans and blue jays find them far more difficult to detect.

CATEGORIES OF COLOR POLYMORPHISM

Color polymorphism is widely distributed among cryptic prey species. It has been studied extensively in pulmonate snails (Cain & Sheppard 1954, Clarke et al. 1978, Cook 1998), but it is common among other mollusks as well (e.g., Atkinson & Warwick 1983, Whiteley et al. 1997). There are cryptic polymorphic spiders (Bonte & Maelfait 2004, Oxford & Gillespie 1998), crustaceans (Devin et al. 2004, Hargeby et al. 2004, Merilaita et al. 2001), and even vertebrates (Olendorf et al. 2006, Wente & Phillips 2003). Examples occur throughout the Acridid and Tetrigid grasshoppers (**Figure 2**) (Dearn 1990, Nabours et al. 1933), where many species are so variable that color is wholly inadequate as a basis for classification (Rowell 1971). Many mantids (Edmunds 1976) and katydids (Owen 1980) are polymorphic, as are water boatmen

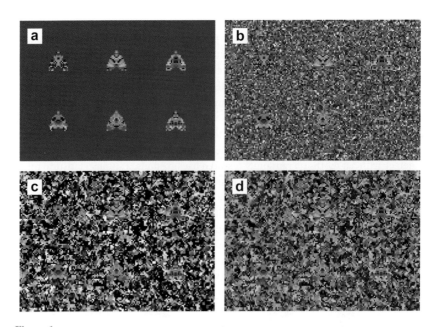

Figure 1

Artificial moths on backgrounds of varying crypticity and complexity. (*a*) Uniform dark gray background. (*b*) Intensity histogram of background matches that of the moths; mean spatial frequency is somewhat higher. (*c*) Background constructed from a bimodal intensity histogram that minimally overlaps the moth distribution; mean spatial frequency is comparable to that of moths. (*d*) Background matches moths in both pixel intensity and spatial frequency.

(Popham 1941), walking sticks (Sandoval 1994), and spittlebugs (Halkka & Halkka 1990). Perhaps the best-researched examples of color polymorphism are found among moths, in which both the larvae and the adults are usually cryptically colored and frequently polymorphic (Greene 1996, Janzen 1984, Kettlewell 1973, Poulton 1890, Sargent 1976). In some moth taxa, polymorphism is pervasive: Roughly 40% of the species in Barnes & McDunnough's (1918) survey of the underwing moths of North America occur in multiple distinctive forms.

Polymorphic species can be separated into two broad categories that appear to reflect distinguishable evolutionary strategies. One grouping consists of species that occur in many disparate forms, all of which bear a general resemblance to the same background. The number of morphs can range from perhaps three or four to dozens, as in grouse locusts (*Acrydium arenosum*) (**Figure 2**) (Nabours et al. 1933), land snails (*Cepaea nemoralis*) (Cain & Sheppard 1954), or underwing moths (*Catocala micronympha*) (Barnes & McDunnough 1918). Such generalist polymorphisms are mainly associated with relatively homogeneous but visually complex environments, such as temperate grasslands, exposed rock and soil, leaf litter, or beach gravel (Clarke et al. 1978, Dearn 1990, Rowell 1971, Whiteley et al. 1997). These habitats tend to be fine grained (Levins 1968), in that there are no large patches of distinctive substrate, but the visual complexity enables a range of diverse forms to be equally difficult

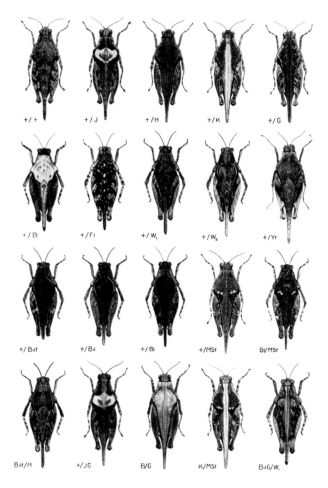

Figure 2

A selection of morphs of *Acrydium arenosum*, a highly polymorphic grouse locust from central Kansas. Reprinted from Nabours et al. (1933) by permission of the Genetics Society of America. Copyright is retained by the Genetics Society of America.

to detect. In the most extreme cases, generalist species can be massively polymorphic, occurring in an essentially unlimited number of forms. The original example is *Donacilla cornea*, a European beach clam (Whiteley et al. 1997), but there are other organisms that may be similarly unrestricted in their morph diversity [e.g., bogong moths: *Agrotis infusa* (Common 1954) or brittlestars: *Ophiopholis aculeata* (Moment 1962)].

A second group of polymorphic species are associated with heterogeneous environments that are divided into large, disparate substrate patches (Bond & Kamil 2006, Merilaita et al. 2001). In such coarse-grained habitats (Levins 1968), an individual can occupy only one substrate type at a time and, because of the disparity in patch appearance, cannot match all substrates equally well. The result is selection for maximum crypticity on only one of the available substrates. The scale of heterogeneity in these cases ranges from regional contrasts in soil reflectance, which produce matching polymorphisms in desert reptiles (Norris & Lowe 1964), to variation among species of food plants in a single area of chaparral, where each of the forms of a polymorphic walking stick is maximally abundant on the plant species that it most closely resembles (Sandoval 1994). We refer to this category as specialist polymorphisms. Because diversity is limited by the number of substrate types, specialist polymorphisms generally involve fewer, more distinctive forms, often only green and brown (Dearn 1990, Edmunds 1976, Owen 1980, Poulton 1890, Wente & Phillips 2003) or dark and light (Kettlewell 1973). In many cases, the dimorphism is genetically determined; in others, it is a polyphenism, a developmental difference in color pattern cued by physical or chemical signals (Edmunds 1976, Greene 1996, Rowell 1971). Because each morph specializes on a particular substrate type, these species may be highly selective in choosing a resting location, actively avoiding nonmatching backgrounds (Edmunds 1976, Owen 1980, Sargent 1981).

Although the structure of the environment helps determine whether a given species will show a specialized or generalized polymorphism, the ecological relationships can be complicated. *Catocala relicta* is an underwing moth that occurs in at least five distinctive forms (Barnes & McDunnough 1918). Their favored host plants are white birches and poplars, and when given a choice between oaks and birches, all five forms are differentially attracted to birch trees (Sargent 1981). Three of the morphs are white with dark markings that make them maximally cryptic on younger birch trees and branches. The other two are more uniformly dark, perhaps resembling the texture of the trunks of more mature trees. *C. relicta* is thus a specialist on a particular substrate type, but occurs in multiple, disparate forms within its specialized niche.

COGNITIVE DEMANDS OF VISUAL SEARCH

The rate of information processing in visual systems is subject to inherent limitations. Bottlenecks are encountered at many different levels, but perhaps the clearest illustration of the problem is that the number of sensory receptors in the vertebrate retina is several orders of magnitude higher than the number of axons in the optic nerve. Ganglion cells redundantly encode the essential features of the visual world to minimize the loss of information (Barlow 2001), but even under optimal conditions, the retina transmits data at only about the rate of an Ethernet connection (Koch et al. 2006). Very demanding perceptual tasks, such as detecting camouflaged prey or dealing with a cluttered visual field, take up a lot of bandwidth, so attentional filters are an essential feature of any complex visual system. They are analogous to packet priority schemes in networking software, which allocate bandwidth to specific categories of information in accordance with dynamically changing criteria of their relative importance. Attentional filtering occurs at a variety of different levels in the nervous system and has been a principal focus of research in cognitive psychology for half a century (Parasuraman & Davies 1984, Pashler 1998).

Dukas (2002, 2004) has argued persuasively for the central importance of attention in behavioral ecology. Competing informational demands require that animals allocate their attentional resources so as to optimize their chances of finding food and avoiding predators (Clark & Dukas 2003). And because the environment is constantly changing, their attentional allocation must be flexible. Predators must decide where to look for prey, what visual features to focus on, and how long to persist in searching, and they must be able to revise their decisions rapidly on the basis of recent experiences of success and failure. Their transitory expectations are expressed in the form of stimulus filters that determine which prey items will be detected and which will be overlooked. The role of expectancy in the perception of cryptic stimuli is familiar to any naturalist. Croze (1970) cites a particularly evocative description from one of Ian Fleming's novels: "When you are looking for one particular species under water...you have to keep your brain and your eyes focussed for that one individual pattern. The riot of color and movement and the endless variety of light and shadows fight your concentration all the time" (Fleming 1965, p. 312). Note that Fleming emphasized the complexity of the coral reef background, rather than the cryptic coloration of the fish James Bond was seeking, as the primary justification for attentional search.

Von Uexküll (1934) provided the original description of attentional filtering in visual search. In his best-known narrative, he tells of his being accustomed to having water served at mealtime in an earthenware pitcher. One day, the pitcher had been broken, so the servant replaced it with a glass carafe. When von Uexküll came to the table, he looked in vain for the pitcher. He asked for water and was told that it was in its usual place. Only then did the scattered reflections from the carafe assemble themselves into a coherent image. Von Uexküll claimed that the expected stimulus configuration of the pitcher, which he termed a searching image, overrode the perceptual input and prevented him from recognizing the carafe. A striking aspect of all of von Uexküll's searching image anecdotes is that the overlooked object is invariably relatively cryptic, consistent with the view that expectancy is most influential when there are substantial demands on processing capacity. A searching image for an earthenware pitcher might, for example, have little effect on perception of a red ceramic teapot. When the visual system is heavily engaged, an increase in the detectability of one type of stimulus occurs only at the cost of overlooking other, alternative stimuli. Interference with the detection of unexpected stimuli is a characteristic feature of selective attention in humans (Kahneman 1973, Posner & Snyder 1975) and has consequently been adopted as the operationally defining feature of an attentional process (Mackintosh 1975, Plaisted & Mackintosh 1995).

For many years, this interference effect proved to be exceedingly difficult to confirm in animal subjects. It has, however, recently been clearly demonstrated by Dukas & Kamil (2000, 2001) with blue jays in an operant apparatus, using a design that compelled the birds to divide their attention either between spatial locations or between stimulus types. The results showed a striking reduction in performance under divided attention conditions, confirming the limitation on information-processing capacity. Unequivocal evidence of interference in selective, as opposed to divided, attention has been harder to obtain because of the large number of potential confounds that must be excluded. In the best controlled studies, however, birds induced to search for particular cryptic stimuli have shown a significantly reduced ability to detect alternatives, consistent with an attentional explanation (Blough 1989, 1992; Bond & Kamil

1999; Bond & Riley 1991; Langley 1996; Reid & Shettleworth 1992). Zentall (2005) has reviewed the evidence from a range of different psychological experiments and has concluded that animals are clearly capable of attending to selective aspects of a stimulus display.

SEARCHING IMAGES AND PREDATOR ECOLOGY

The diet of generalist predators is seldom a simple linear function of prey availability. Infrequently encountered prey types tend to be underrepresented in the diet, whereas more abundant ones are consumed in excess. The phenomenon is referred to as switching (Murdoch 1969, Murdoch & Oaten 1974) or displaying a sigmoid functional response (Holling 1965). Frequency-dependent switching patterns can occur for many reasons, ranging from spatial heterogeneity to the availability of prey refuges to nutritional constraints (Greenwood 1984, Sherratt & Harvey 1993, Sinervo & Calsbeek 2006, Staddon 1983). When Tinbergen (1960) studied the predation patterns of insectivorous birds in pine woods, however, he attributed the predators' frequency-dependent prey choices to the effects of searching images. Tinbergen tracked the abundance of various cryptic, arboreal caterpillars and recorded the frequency with which the birds brought them to their nest boxes. There was generally a delay of two to three days between the initial emergence of a prey species and its first appearance in the diet, and Tinbergen inferred that the birds overlooked new prey types because they were not expecting to find them. Only after a set of chance encounters with a newly abundant prey, he hypothesized, did the birds acquire a searching image for them, resulting in a rapid increase in their representation in the diet.

Tinbergen's hypothesis was concerned not with whether searching images existed—whether animals actually used selective attention in visual search—but with how searching images were evoked. To emphasize this distinction, we refer to Tinbergen's hypothesis as perceptual switching. The mechanism is presumably a simple positive feedback. If adoption of a searching image improves the predator's ability to detect cryptic prey items, then prey that are detected more often or more recently will be found more readily than those with which the predator has had less experience. This should produce a disproportionately greater frequency of predation on relatively abundant prey types. And if a searching image for one prey type reduces the ability to detect other, disparate ones, the bias in favor of abundant prey types should be matched by a tendency to overlook rarer forms. The result will be a sigmoid switching function (**Figure 3**), derived entirely from perceptual dynamics. Clarke (1962) subsequently extended Tinbergen's model, arguing that perceptual switching could account for the maintenance of color polymorphism in cryptic prey species.

Clarke noted that frequency-dependent predation on polymorphic prey should select in favor of rare, deviant phenotypes. With a sigmoid switching function, abundant morphs will be taken disproportionately often, driving down their numbers, while rare ones will be ignored and allowed to increase. As the rarer morphs become more abundant, they will, in turn, become the focus of the predators' interest and will be taken in disproportionate numbers. If mortality is mainly a consequence of visual

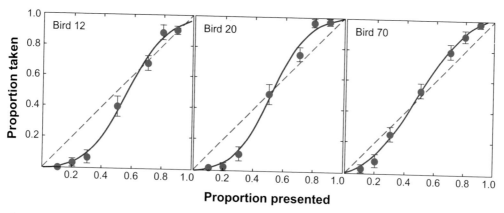

Figure 3

Perceptual switching curves for three pigeons searching for cryptically colored black beans and red wheat on a complex gravel background. The proportion of beans in the sample is shown on the abscissa; the proportion taken by the bird is on the ordinate. Mean values for each treatment condition are plotted with filled circles; hash marks indicate two standard errors. The solid line is a least-squares fit of the raw data to a perceptual switching model; the dashed line indicates the null hypothesis of indifference or lack of selection. Figure adapted with permission from Bond (1983).

predation, this predation pattern could serve to stabilize the numbers of an array of alternative morphs. Clarke's theoretical mechanism, which he called apostatic selection, was attractive, but it was based on very limited empirical results: Tinbergen's invocation of searching images went well beyond what could legitimately be inferred from his uncontrolled field observations (Croze 1970, Royama 1970). This prompted a contentious literature on whether searching images exist, how they can be experimentally demonstrated, and whether they provide a satisfactory account of the evolution of color polymorphism (Curio 1976, Endler 1991; reviewed in Krebs 1973, Punzalan et al. 2005, Shettleworth et al. 1993).

Validation of the perceptual switching hypothesis seems straightforward, in principle. It only requires, first, that prey types that have been detected more frequently should subsequently be detected more reliably, and second, that when the discriminability of one prey type is maximized, the others should become less detectable. Researchers were slow to realize, however, the number of controls that were required to eliminate other possible interpretations. Many early studies, for example, focused on the initial acquisition of a feeding response to novel prey types (Fullick & Greenwood 1979, Willis et al. 1980), or on subsequent feeding preferences for familiar prey items (Allen 1974, 1984), or on learning to discriminate a familiar prey in novel, cryptic circumstances (Dawkins 1971a, Lawrence 1985). However, disparities in training or familiarity only induce transitory response biases. They cannot sustain a continuing pattern of perceptual switching (Bond 1983, Langley et al. 1995). A hypothetical predator influenced only by long-term learning would eventually become familiar with all prey in its environment and would consume them in direct proportion to their abundance. For an appropriate test of perceptual switching, predators have to be thoroughly experienced with all prey types prior to beginning experimental trials.

Other possible sources of frequency-dependent switching also have to be controlled. To avoid diet selection biases, only prey of roughly comparable size and nutritional value should be used (Punzalan et al. 2005). When multiple prey items can be viewed simultaneously, local visual contrasts between items become influential, and response biases reduce the effect of perceptual switching (Allen & Weale 2005, Bond 1983, Church et al. 1997). The prey thus need to be randomly intermixed and overdispersed, so as to insure that they are encountered sequentially (Bond 1983, Gendron 1982, Getty & Pulliam 1993). To produce a polymorphism of cryptically colored stimuli, many studies have used heterogeneous backgrounds with corresponding cryptic and conspicuous food items (e.g., Dawkins 1971a, Getty & Pulliam 1993, Reid & Shettleworth 1992). Interpretation of the results of these experiments can be complicated, however, as the difference in backgrounds provides an independent cue that can bias the predator's search and increase the variance in prey detection (Bond & Kamil 2006, Kono et al. 1998, Royama 1970). Finally, the prey items should all be roughly equivalently cryptic on the same substrate (a generalist polymorphism) to maximize analytical sensitivity. Given these constraints, there have been three primary experimental approaches to studying searching images, each of which has a different balance of control and realism: free-choice experiments, sequence analyses, and operant detection tasks.

Much of the literature on searching images and color polymorphism consists of free-choice experiments that provide animals with two or more distinguishable prey types at different relative abundances. If varying the abundances results in a sigmoid switching function, and if other factors have been carefully controlled, this design should in principle provide robust evidence of the dynamics of perceptual switching. For example, Bond (1983) studied captive pigeons that fed from trays containing varying proportions of familiar, natural food grains randomly dispersed on a complex gravel background. To control for the possibility of prior preferences and response biases, the same grain types were displayed on both cryptic and conspicuous backgrounds. The birds all showed significant perceptual switching effects when the grains were cryptic (**Figure 3**), effects that disappeared when the grains were conspicuous. The response rate was highest on samples containing a single grain type and decreased as the relative proportions approached equality, suggesting a reduction in searching efficiency as a consequence of switching back and forth between food types (Dukas & Kamil 2000). The occurrence of perceptual switching has been demonstrated in several other free-choice experiments as well (Cooper 1984, Cooper & Allen 1994, Langley et al. 1995, Reid & Shettleworth 1992, Tucker 1991). The method produces robust and consistent results, but prey crypticity is difficult to manipulate in a systematic fashion. And because the sequence of prey encounters cannot be controlled, and has seldom even been directly recorded, the results cannot confirm a reduction in the detectability of alternative prey types (Bond 1983, Kamil & Bond 2006).

If recently experienced prey types become easier to detect, while alternative ones become harder, predators should tend to take prey in homogeneous runs (McNair

1980). Dawkins (1971b) identified this runs effect in studies of chicks feeding on colored rice grains and related it to attentional searching. Her design can be carried further, however. Given the underlying dynamics, the time between successive items within a run should be significantly shorter than the time between runs of different prey types. More cryptic prey should also be taken in longer runs and should show longer delays between runs of different types. These predictions have been confirmed in human subjects sorting colored wooden beads (Bond 1982), but the analytical power of sequential analysis has been relatively neglected in free-choice experiments on animals. Dawkins' results did, however, suggest that adoption of a searching image might be induced by presenting cryptic visual stimuli one at a time and manipulating the imposed sequence to produce a run of a particular stimulus type. This rationale transforms visual search for multiple prey types into an operant serial detection task, a standard psychological procedure in which an animal is presented with a succession of visual displays.

Using this design, Pietrewicz & Kamil (1977, 1979) trained blue jays to detect cryptic moths in slide images and found that runs in the imposed stimulus sequence significantly improved the probability both of detecting moths when they were present (positive trials) and of correctly rejecting images without moths (negative trials). No such changes were observed when moths of different types were presented in random sequences, providing strong evidence for improvement in discriminability over successive encounters with the same prey type. Similar results have since been obtained in studies with pigeons, using images of cryptic grains (Bond & Riley 1991, Langley 1996) or alphanumeric characters (Blough 1989, 1991, 1992), and have been repeatedly confirmed in experiments with blue jays, using artificial stimuli with predetermined levels of crypticity (Bond & Kamil 1999; Dukas & Kamil 2000, 2001). Several operant studies have also confirmed the predicted reduction in the detectability of alternative stimuli, under conditions that effectively excluded nonattentional explanations (Blough 1989, 1992; Bond & Kamil 1999; Bond & Riley 1991; Langley 1996).

Under controlled experimental conditions, therefore, animals exhibit attentional biases consistent with Tinbergen's hypothesis of perceptual switching. Despite this apparent triumph of the operant approach, however, it is unclear how the results relate to the evolution of color polymorphism in the field (Punzalan et al. 2005). In particular, the ecological and evolutionary effects of perceptual switching depend crucially upon how searching images are deployed. How should predators choose which prey type to attend to at any given moment, how rapidly should they scan their environment, and how long should they persist in searching for a prey type that is no longer abundant? These are questions not of the perceptual mechanism itself, but of the allocation of cognitive resources to make optimal use of the animal's perceptual capabilities (Staddon 1980). Decision processes have been heavily investigated in other areas of foraging behavior, particularly in patch and diet selection (Shettleworth et al. 1993, Stephens & Krebs 1986). The dynamics of perceptual switching have, as yet, received far less consideration, although a rough outline of the relevant issues can be gathered from the existing literature.

DYNAMICS OF PERCEPTUAL SWITCHING

The improvement in discriminability resulting from attention is, in part, a function of the demands of the detection task, including the crypticity of the prey, the complexity of the background, and the magnitude of competing attentional requirements. Several authors have assumed that predators alternate between a selective search mode in which they focus exclusively on one prey type and a broader, general mode that is less efficient but receptive to multiple stimulus alternatives (Bond 1983, McNair 1980, 1981). This model is almost certainly an oversimplification. Attentional search is not exclusive, but actually somewhat porous, providing little interference to the detection of more salient, nonattended items (Blough 1992, Bond & Kamil 1999). Allocation of attentional resources appears to be graded, with longer runs of the same prey type eliciting larger attentional allotments (Dukas & Ellner 1993, Dukas & Kamil 2001, Gendron 1986). And there is no evidence of a generally receptive search mode: Animals can locate multiple, disparate, conspicuous items using pre-attentive parallel search mechanisms (Bond & Kamil 2006, Dukas 2002), but cryptic prey cannot be successfully detected with a nonattentive search (Wolfe 1994, Wolfe et al. 2002). Estimating the economic benefits of particular patterns of attentional allocation will require more extensive research and modeling.

A predator that persists in searching for prey that have become locally depleted might be said to show functional response hysteresis, in that he shifts only gradually between states in response to imposed changes in resource abundance (Getty & Krebs 1985, Murdoch 1970). How readily apostatic selection can stabilize a prey polymorphism will depend upon the level of hysteresis. A predator that is slow to switch will produce wide fluctuations in relative abundance, increasing the chance of local morph extinctions. The processes that initiate and terminate attentional search are, thus, of major importance in determining the ecological effects of perceptual switching. Searching images are evoked by a contiguous series of detections of the same stimulus item (Blough 1991, 1992; Bond & Kamil 1999; Kamil & Bond 2006). The number of required detections seems invariably more than one, but generally fewer than five or six (Bond & Kamil 1999, Bond & Riley 1991, Croze 1970, Pietrewicz & Kamil 1979). Depending upon the species, time intervals between detections can also have a significant impact. Attentional search can be elicited in blue jays even with negative trials interspersed among the positives in a run (Bond & Kamil 1999, Pietrewicz & Kamil 1979), but pigeons require contiguous blocks of positive trials (Bond & Riley 1991). This may account for the strong effects of intertrial intervals in visual search in pigeons (Plaisted 1997), effects that could not be replicated using blue jays (Bond & Kamil 1999).

The reduction in attention to a prey type that has declined in numbers and become unprofitable has been considered as an analog to the patch-departure problem in optimal foraging theory. Several authors have modeled this decision with a giving-up time that determines a predator's persistence in searching for a specific prey type (Bond 1983, Croze 1970, Dukas & Ellner 1993). Using this model, Bond (1983) suggested that pigeons searching for two disparate cryptic grains could be switching their attentional focus in response to as little as 100 ms of unrewarded searching. In a later pigeon study, however, Langley et al. (1995) evoked a searching image with one grain type and subsequently tested it using equal numbers of both grains. A delay of 30 s between evocation and testing had no effect, but when the delay was increased to 3 min, the birds showed a decreased responsiveness to the attended grain type. Searching images can thus be discarded if a sufficient delay is imposed, but a simple temporal threshold does not appear to capture the dynamics of the switching process. It is possible that searching images are seldom actually relinquished, but, rather, are gradually attenuated when prey detections become infrequent, allowing alternative stimuli to be more readily noticed. By this reasoning, perceptual switching could be modeled as a graded fluctuation among alternative attentional states (Falmagne 1965, Maljkovic & Martini 2005, Schweickert 1993). This model would help to explain why strong searching image effects can be observed even when prey appearance is continuously variable (Bond & Kamil 2002, 2006).

In addition to perceptual factors, the dynamics of perceptual switching are strongly influenced by the distribution of searching effort in space and time, measured either as the rate of movement through the habitat (Gendron 1986; Gendron & Staddon 1983, 1984) or as the duration of scanning at successive discrete locations (Bond & Riley 1991, Endler 1991, Getty & Pulliam 1991). This is an essential aspect of visual search, as it regulates and focuses the flow of novel visual information. Limiting the search rate produces a spatial allocation of attentional resources that complements the discriminability (Dukas 2004, Getty et al. 1987) or the probability (Wolfe et al. 2005, 2007) of particular visual stimuli. When Gendron & Staddon (1983, 1984) modeled the effects of varying the search rate, they found an inverse relationship between search rate and prey detectability. Conspicuous stimuli are reliably detected at greater distances, so they can be searched for at greater speeds. Cryptic stimuli, in contrast, require the predator to slow down and scan its surroundings more thoroughly. Subsequent studies have substantially confirmed this speed/accuracy trade-off in prey detection (Gendron 1986, Getty et al. 1987) and have clarified the complementary roles of perceptual and tactical constraints in the allocation of search effort (Getty & Pulliam 1991, 1993).

The complementarity of visual and spatial attention makes disentangling their relative influences a significant challenge, particularly in free-choice experiments. Guilford & Dawkins (1987) noted that predators that optimized their search rates would have asymmetrical effects on mixtures of prey types. Relatively cryptic prey would be detected less frequently than expected, whereas more conspicuous items would be found more readily. In some circumstances, these effects could be confused with perceptual switching. Several studies have since confirmed the occurrence of perceptual switching when search rates are either controlled or irrelevant (reviewed in Endler 1991, Kamil & Bond 2006, Shettleworth et al. 1993) or have demonstrated simultaneous, independent effects of both search rate and searching images (Bond & Riley 1991, Getty et al. 1987). Asymmetrical results have been pervasive, however, in studies that used small numbers of cryptic prey types. Attention to relatively conspicuous stimuli invariably reduces the detection of more cryptic alternatives, but attention to relatively cryptic stimuli generally has less of an effect (Blough 1989, 1992; Bond & Kamil 1999; Bond & Riley 1991; Reid & Shettleworth 1992).

Simulation results suggest that this asymmetry may be an inherent consequence of the optimal allocation of attention, a manifestation of the compromise between speed and accuracy (Bond & Riley 1991).

APOSTATIC SELECTION AND BALANCED POLYMORPHISM

The development and maintenance of color polymorphism is one of the great persisting issues in ecology and evolution, in part because it is often unclear whether natural polymorphisms constitute a stable array of forms maintained by frequency-dependent predation (Oxford & Gillespie 1998), or just a transient mix of alleles resulting from genetic drift and dispersal (Fisher 1930, Ford 1975). Many natural color polymorphisms, when closely examined, can be attributed to nonselective processes, such as regional selection combined with migration and dispersal (Dearn 1984, King & Lawson 1995, Reillo & Wise 1988), or genetic drift and local population bottlenecks (Brakefield 1990). Even the world's most intensively studied polymorphism, that of the *Cepaea* land snails, is sustained in part by migration, drift, and founder effects (Bellido et al. 2002, Cook 1998).

Observational studies of color polymorphism that have successfully excluded nonselective explanations commonly invoke apostatic selection by visual predators as the likely agency (e.g., Atkinson & Warwick 1983, Jormalainen et al. 1995, Oxford & Gillespie 1998). These inferences may well be correct, but they often only reflect the absence of evidence to the contrary. Allen and his colleagues have conducted numerous experiments in which colored baits are broadcast in the field and fed on by wild birds (reviewed in Allen 1988, Allen & Weale 2005, Cooper & Allen 1994). Their results are consistent with apostatic selection, but because they generally did not control which birds had access to the baits and did not record the sequence and timing of items taken by individual birds, the nature of the selective mechanism is open to alternative interpretations. Translocation experiments on natural polymorphisms (e.g., Cox & Cox 1974, Halkka et al. 1975, Sandoval 1994) have provided compelling field evidence of stabilizing selection in specialist polymorphisms. Field studies that have gone further and manipulated relative morph abundances directly, as in Olendorf et al.'s (2006) work on guppies or Reid's (1987) study of mangrove snails, have demonstrated apostatic selection in favor of rare color morphs. Even in these cases, however, the selective mechanism has been inferred from the pattern of prey survival rather than from observations of predatory behavior. Whether selection has resulted from perceptual switching in individual predators has not been established.

The only direct evaluation of the selective effects of visual predation on prey polymorphism has been conducted by Bond & Kamil (1998, 2002, 2006) in a laboratory system that allowed for complex interactions between the abundance, the appearance, and the detection of prey. Their methodology, based on operant studies of searching images, involved presenting captive blue jays with digital moths imbedded in complex textured backgrounds. An initial study with this preparation (Bond & Kamil 1999) showed clear evidence of perceptual switching: During runs of a single moth type, discriminability improved for the focal moth and declined for alternative, disparate forms. The experimental rationale was thus the inverse of a typical field study. First, they developed a system in which perceptual switching was known to be displayed, and then they tested it for indications of apostatic selection.

To determine whether perceptual switching could maintain a balanced polymorphism, Bond & Kamil (1998) created a digital moth population with equal numbers of each of three distinctive morphs and exposed them to predation by a squad of blue jays. At the end of each day's session, moths that had been detected were considered killed and were removed from the population. Those that were overlooked were cloned in proportion to their relative numbers, bringing the population up to its previous level for the following day. Over the course of 50 generations, morph abundances rapidly achieved a characteristic oscillatory equilibrium that included all three morphs and that was independent of their initial population levels. Additional analyses demonstrated that the equilibrium was a result of apostatic selection, a positive correlation between prey abundance and detectability. The experiment confirmed that perceptual switching can produce apostatic selection and can thereby maintain a balanced polymorphism.

In the second phase of their research, Bond & Kamil (2002) eliminated the constraint of fixed asexual morphs. Moth phenotypes were specified by haploid genomes that evolved over successive generations in accordance with a genetic algorithm. Moths that the blue jays overlooked or were slow to detect were more likely to be chosen to breed. Under this regime, initially monomorphic parental populations evolved through 100 progeny generations without further experimental intervention. To test whether crypticity and phenotypic variance were selected for in the experimental lines, the results were contrasted with those from two sets of control lineages, one resulting only from genetic drift and the other using frequency-independent directional selection for increased crypticity. In each experimental lineage, the moths became significantly harder to detect, but the experimental lines also evolved significantly greater phenotypic variance than the frequency-independent controls (**Figure 4**). This indicated that apostatic selection could actively promote the evolution of multiple disparate forms in a previously monomorphic population.

The backgrounds used in this study were fine grained and relatively homogenous, resembling coarse gravel. Selection on this substrate resulted in massive, generalist polymorphisms, similar to those in beach clams (Whiteley et al. 1997). Only in a later study using heterogeneous backgrounds did Bond & Kamil observe any signs of specialist polymorphisms (Bond & Kamil 2006). This was an attractive finding, as it was consistent both with ecological theory of the effects of coarse-grained habitats (Kisdi 2001, Levins 1968, Merilaita et al. 1999) and with the known association between patchy habitats and genetic polymorphism (Hedrick 2006). Even in their most heterogeneous treatments, however, Bond & Kamil still did not obtain a classical, discrete polymorphism of a handful of disparate forms. Some additional causal factors appear to be operating in natural environments to limit morph diversity, perhaps involving more extreme contrasts between patch types, biases among prey in their choice of resting locations, or disruptive selection against hybrid individuals (Hedrick 2006, Maynard Smith 1962).

Analysis of the pattern of detection as a function of trial sequence in these experiments showed that blue jays consistently overlooked atypical cryptic moths,

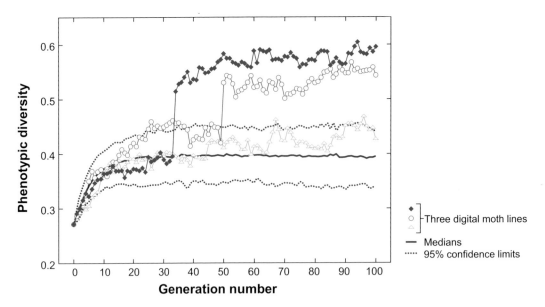

Figure 4

Changes in phenotypic variance of a population of digital moths in three experimental lines
(*plotted in blue with triangles, circles, and diamonds, respectively*) under selection by captive blue
jays. Experimental results are contrasted to the phenotypic variance in 200 replicate control
lines that underwent only frequency-independent selection for crypticity. Medians and 95%
confidence limits for the control lines are shown in red. Phenotypic variance increased to some
degree in all treatments, but the increase was significantly greater in the experimental lines
than in the controls. Two of the experimental lines exhibited an abrupt shift to a higher level of
phenotypic variance at some point in the course of selection trials, apparently produced by the
explosive spread of mutant regulatory genes. Figure adapted with permission from Bond &
Kamil (2002).

confirming perceptual switching as the primary agency of selection. This is a somewhat puzzling finding in that the digital moth populations were extraordinarily variable, far more so than the stimuli used in any previous study of searching images.
This diversity might be expected to reduce the birds' tendency to attend to particular stimulus features, but in fact it had the opposite effect. Perceptual switching was
shown at some crypticity levels in all experimental treatments, but it was spectacularly
evident among moderately cryptic moths on very complex backgrounds (**Figure 5**,
plot d2). Searching images did not show as large an effect with highly cryptic moths
on these backgrounds (**Figure 5, plot d3**). Presumably, the limitation on the number
of features that can reliably distinguish these stimuli from the background restricts
the benefits of visual attention (Bond 1983). However, the birds were clearly still
conducting a serial attentional search, even in these circumstances, because highly
cryptic, atypical moths were almost completely overlooked.

The results of these virtual ecology studies thus bear an encouraging resemblance
to the ecological associations of natural polymorphisms. Combined with the findings of other experiments on attentional search, perceptual switching, and apostatic

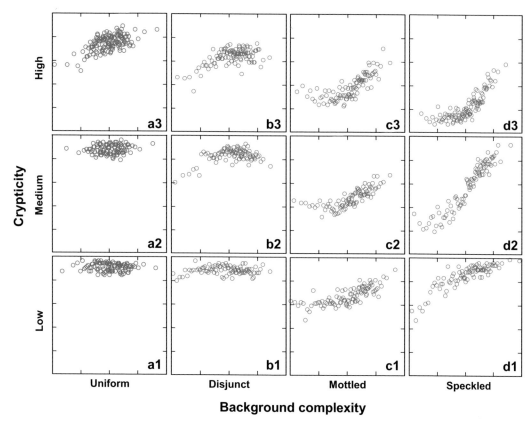

Cryptricity (vertical axis label, with High, Medium, Low)

High — a3, b3, c3, d3
Medium — a2, b2, c2, d2
Low — a1, b1, c1, d1

Uniform   Disjunct   Mottled   Speckled

Background complexity

Figure 5

Each panel displays the detection accuracy in blocks of 100 trials (on the ordinate, ranging from 0% to 100%) as a function of the morphological similarity between the current digital moth and the previous correctly detected one (on the abscissa). In these plots, apostatic selection appears as a positive regression slope, indicating higher accuracies for moths that were more similar to previously detected targets. Moth crypticity increases from the bottom to the top panel, dividing the range of crypticity indices within each treatment into percentile groupings: low (0 to 33rd), medium (34th to 66th), and high (67th to 100th). Visual complexity increases from the left panels (a1–a3) to the right ones (d1–d3), ranging from the uniform texture used in Bond & Kamil (2002) to the three heterogeneity treatments used in Bond & Kamil (2006). Significant levels of apostatic selection were displayed in high crypticity moths on the Uniform and Disjunct backgrounds (a3 and b3) and in all crypticity groupings on the Mottled and Speckled backgrounds (c1–c3 and d1–d3).

selection, the literature now provides a persuasive and well-supported causal narrative. Punzalan et al. (2005) remain unconvinced, asserting that there is no unequivocal evidence of a role for searching images in sustaining any natural polymorphism. Their insistence on ecological realism seems overly stringent, however. Given the requisite level of experimental control, it is difficult to envision any similarly complex behavioral mechanism that could be definitively demonstrated under field conditions. The classical field studies of visual search, such as Croze's (1970) experiments on carrion crows and Murton's (1971) investigations of wood pigeons, make clear that foraging behavior in the wild is a rich and complex phenomenon, one that resists our desire for unequivocal answers. Using established methods and analytical techniques, however, it should be possible to design a well-controlled free-choice experiment that would definitively demonstrate the occurrence of apostatic selection among the forms of a naturally polymorphic prey species.

To regulate and record interactions between predators and prey, one would need to conduct an aviary study, similar to Getty & Pulliam's (1993) elegant work on the foraging behavior of captive sparrows. Getty & Pulliam placed cryptic seeds at predetermined random locations on natural soil backgrounds, released single white-throated sparrows to search for them, and videotaped all movements and detection successes of the birds. The authors then analyzed the tapes using random search models developed for naval reconnaissance, which allowed them to determine not just which seeds were found by the birds, but which ones were overlooked. Although Getty & Pulliam (1993) were not concerned with searching image effects, it would be possible to use their design with, perhaps, a random array of *Cepaea* land snails on a complex natural background at a range of relative and absolute abundances. Other species would certainly work as well, but predation on *Cepaea* has been well researched in the field (Bantock & Bayley 1973, Bantock et al. 1975) and in aviaries (Tucker 1991), and the snail shells can readily be converted to motionless stimuli (Allen & Weale 2005, Harvey et al. 1975). To control for response bias effects, tests would also be run on a similarly structured alternative substrate on which the shells were relatively conspicuous. Such research would constitute a valuable step in linking established laboratory results to the evolutionary dynamics of natural color polymorphisms.

GOALS FOR FUTURE WORK

Experiments have confirmed the role of selective attention in the search for cryptic prey, have shown that predators switch among attentional states in response to short-term changes in prey detection, and have demonstrated that perceptual switching can promote and maintain cryptic color polymorphism. A great deal still remains to be done, however. Whether perceptual switching plays a role in maintaining natural polymorphisms depends upon the predator's allocation of cognitive resources in space and time, and this crucial aspect of foraging dynamics has been little explored. The structure and pattern of the background has direct effects on visual search and prey detectability that probably constrain the evolution of prey phenotypes in ways that we do not yet understand. And finally, apostatic selection has been shown to produce increases in phenotypic variance, but has yet to generate a characteristic polymorphism of a limited number of distinctive forms. There are clearly major components in the biology of polymorphism that have yet to be included in an explanatory account. We have only begun to explore the richness and complexity of visual search, predator cognition, and their evolutionary effects on prey coloration.

DISCLOSURE STATEMENT

The author is not aware of any biases that might be perceived as affecting the objectivity of this review.

ACKNOWLEDGMENTS

Production of this manuscript was supported in part by grants from the National Science Foundation (IOB-0234441) and the National Institutes for Mental Health (R01-MH068426).

LITERATURE CITED

Allen JA. 1974. Further evidence for apostatic selection by wild passerine birds: training experiments. *Heredity* 33:361–72

Allen JA. 1984. Wild birds prefer to eat the more familiar of artificial morphs that are similar in colour. *Heredity* 53:705–15

Allen JA. 1988. Frequency-dependent selection by predators. *Philos. Trans. R. Soc. London Ser. B* 319:485–503

Allen JA, Weale ME. 2005. Anti-apostatic selection by wild birds on quasi-natural morphs of the land snail *Cepaea hortensis*: a generalised linear mixed models approach. *Oikos* 108:335–43

Atkinson WD, Warwick T. 1983. The role of selection in the colour polymorphism of *Littorina rudis* Maton and *Littorina arcana* Hannaford-Ellis (Prosobranchia: Littorinidae). *Biol. J. Linn. Soc.* 20:137–51

Bantock CR, Bayley JA. 1973. Visual selection for shell size in *Cepaea* (Held.). *J. Anim. Ecol.* 42:247–61

Bantock CR, Bayley JA, Harvey PH. 1975. Simultaneous selective predation on two features of a mixed sibling species population. *Evolution* 29:636–49

Barlow H. 2001. Redundancy reduction revisited. *Network* 12:241–53

Barnes W, McDunnough J. 1918. Illustrations of the North American species of the genus *Catocala*. *Mem. Am. Mus. Nat. Hist.*, New Ser. Ver. 3, Pt. 1. New York: Am. Mus. Nat. Hist.

Bellido A, Madec L, Arnaud J-F, Guiller A. 2002. Spatial structure of shell polychromatism in populations of *Cepaea nemoralis*: new techniques for an old debate. *Heredity* 88:75–82

Blough PM. 1989. Attentional priming and visual search in pigeons. *J. Exp. Psychol. Anim. B* 15:358–65

Blough PM. 1991. Selective attention and search images in pigeons. *J. Exp. Psychol. Anim. B* 17:292–98

Blough PM. 1992. Detectability and choice during visual search: joint effects of sequential priming and discriminability. *Anim. Learn. Behav.* 20:293–300

Bond AB. 1982. The bead game: response strategies in free assortment. *Hum. Factors* 24:101–10

Bond AB. 1983. Visual search and selection of natural stimuli in the pigeon: the attention threshold hypothesis. *J. Exp. Psychol. Anim. B* 9:292–306

Bond AB, Kamil AC. 1998. Apostatic selection by blue jays produces balanced polymorphism in virtual prey. *Nature* 395:594–96

Bond AB, Kamil AC. 1999. Searching image in blue jays: facilitation and interference in sequential priming. *Anim. Learn. Behav.* 27:461–71

Bond AB, Kamil AC. 2002. Visual predators select for crypticity and polymorphism in virtual prey. *Nature* 415:609–14

Bond AB, Kamil AC. 2006. Spatial heterogeneity, predator cognition, and the evolution of color polymorphism in virtual prey. *Proc. Natl. Acad. Sci. USA* 103:3214–19

Bond AB, Riley DA. 1991. Searching image in the pigeon: a test of three hypothetical mechanisms. *Ethology* 87:203–24

Bonte D, Maelfait JP. 2004. Colour variation and crypsis in relation to habitat selection in the males of the crab spider *Xysticus sabulosus* (Hahn, 1832) (Araneae: Thomisidae). *Belg. J. Zool.* 134:3–7

Brakefield PM. 1990. Genetic drift and patterns of diversity among colour-polymorphic populations of the homopteran *Philaenus spumarius* in an island archipelago. *Biol. J. Linn. Soc.* 39:219–37

Cain AJ, Sheppard PM. 1954. Natural selection in *Cepaea*. *Genetics* 39:89–116

Castner JL, Nickle DA. 1995. Intraspecific color polymorphism in leaf-mimicking katydids (Orthoptera: Terrigoniidae: Pseudophyllinae: Pterochrozini). *J. Orthop. Res.* 4:99–103

Church SC, Jowers M, Allen JA. 1997. Does prey dispersion affect frequency-dependent selection by wild birds? *Oecologia* 111:292–96

Clark CW, Dukas R. 2003. The behavioral ecology of a cognitive constraint: limited attention. *Behav. Ecol.* 14:151–56

Clarke BC. 1962. Balanced polymorphism and the diversity of sympatric species. In *Taxonomy and Geography*, ed. D Nichols, pp. 47–70. Oxford: Syst. Assoc.

Clarke BC, Arthur W, Horseley DT, Parkin DT. 1978. Genetic variation and natural selection in pulmonate snails. In *The Pulmonates. Vol. 2A: Systematics, Evolution, and Ecology*, ed. V Fretter, J Peake, pp. 220–70. New York: Academic

Common IFB. 1954. A study of the ecology of the adult bogong moth, *Agrotis infusa* (Boisd.) (Lepidoptera: Noctuidae), with special reference to its behaviour during migration and aestivation. *Aust. J. Zool.* 2:223–63

Connolly RM, Melville AJ, Keesing JK. 2002. Abundance, movement and individual identification of leafy seadragons. *Mar. Freshw. Res.* 53:777–80

Cook LM. 1998. A two-stage model for *Cepaea* polymorphism. *Philos. Trans. R. Soc. London Ser. B* 353:1577–93

Cooper JM. 1984. Apostatic selection on prey that match the background. *Biol. J. Linn. Soc.* 23:221–28

Cooper JM, Allen JA. 1994. Selection by wild birds on artificial dimorphic prey on varied backgrounds. *Biol. J. Linn. Soc.* 51:433–46

Cott HB. 1957. *Adaptive Coloration in Animals*. London: Methuen. 508 pp.

Cox GW, Cox DG. 1974. Substrate color matching in the grasshopper *Circotettix rabula*. *Great Basin Nat.* 34:60–70

Croze H. 1970. *Searching Image in Carrion Crows*. Berlin: Paul Parey. 85 pp.

Curio E. 1976. *The Ethology of Predation*. Berlin: Springer-Verlag. 250 pp.

Dawkins M. 1971a. Perceptual changes in chicks: another look at the "search image" concept. *Anim. Behav.* 19:566–74

Dawkins M. 1971b. Shifts of "attention" in chicks during feeding. *Anim. Behav.* 19:575–82

de Ruiter L. 1952. Some experiments on the camouflage of stick caterpillars. *Behaviour* 4:222–32

Dearn JM. 1984. Colour pattern polymorphism in the grasshopper *Phaulacridium vittatum* I. Geographic variation in Victoria and evidence of habitat association. *Aust. J. Zool.* 32:239–49

Dearn JM. 1990. Color pattern polymorphism. In *Biology of Grasshoppers*, ed. RF Chapman, A Joern, pp. 517–49. New York: Wiley

Devin S, Bollache L, Beisle J-N, Moreteau J-C, Perrot-Minnot M-J. 2004. Pigmentation polymorphism in the invasive amphipod *Dikerogammarus villosus*: some insights into its maintenance. *J. Zool.* 264:391–97

Dukas R. 2002. Behavioural and ecological consequences of limited attention. *Philos. Trans. R. Soc. London Ser. B* 357:1539–47

Dukas R. 2004. Causes and consequences of limited attention. *Brain Behav. Evol.* 63:197–210

Dukas R, Ellner S. 1993. Information processing and prey detection. *Ecology* 74:1337–46

Dukas R, Kamil AC. 2000. The cost of limited attention in blue jays. *Behav. Ecol.* 11:502–6

Dukas R, Kamil AC. 2001. Limited attention: the constraint underlying search image. *Behav. Ecol.* 12:192–99

Edmunds M. 1974. *Defense in Animals*. Burnt Mill, UK: Longman. 357 pp.

Edmunds M. 1976. The defensive behaviour of Ghanaian praying mantids with a discussion of territoriality. *Zool. J. Linn. Soc.* 58:1–37

Edmunds M. 1990. The evolution of cryptic coloration. In *Insect Defenses: Adaptive Mechanisms and Strategies of Prey and Predators*, ed. DL Evans, JO Schmidt, pp. 3–21. Albany, NY: SUNY Press

Endler JA. 1978. A predator's view of animal color patterns. *Evol. Biol.* 11:319–64

Endler JA. 1981. An overview of the relationships between mimicry and crypsis. *Biol. J. Linn. Soc.* 16:25–31

Endler JA. 1984. Progressive background matching in moths, and a quantitative measure of crypsis. *Biol. J. Linn. Soc.* 22:187–231

Endler JA. 1991. Interactions between predators and prey. In *Behavior Ecology*, ed. JR Krebs, NB Davies, pp. 169–96. London: Blackwell

Falmagne JC. 1965. Stochastic models for choice reaction time with applications to experimental results. *J. Math. Psychol.* 2:77–124

Fisher RA. 1930. *The Genetical Theory of Natural Seleciton*. Oxford: Clarendon. 272 pp.

Fleming I. 1965. *Bonded Fleming: A James Bond Omnibus*. New York: Viking. 439 pp.

Ford EB. 1975. *Ecological Genetics*. London: Chapman & Hall. 442 pp.

Fullick TG, Greenwood JJD. 1979. Frequency-dependent food selection in relation to two models. *Am. Nat.* 113:762–65

Gendron RP. 1982. *The foraging behavior of bobwhite quail searching for cryptic prey*. PhD thesis. Duke Univ. 118 pp.

Gendron RP. 1986. Searching for cryptic prey: evidence for optimal search rates and the formation of search images in quail. *Anim. Behav.* 34:898–912

Gendron RP, Staddon JER. 1983. Searching for cryptic prey: the effect of search rate. *Am. Nat.* 121:172–86

Gendron RP, Staddon JER. 1984. A laboratory simulation of foraging behavior: the effect of search rate on the probability of detecting prey. *Am. Nat.* 124:407–15

Getty T, Kamil AC, Real PG. 1987. Signal detection theory and foraging for cryptic or mimetic prey. In *Foraging Behavior*, ed. AC Kamil, JR Krebs, HR Pulliam, pp. 525–49. New York: Plenum

Getty T, Krebs JR. 1985. Lagging partial preferences for cryptic prey: a signal detection analysis of great tit foraging. *Am. Nat.* 125:39–60

Getty T, Pulliam HR. 1991. Random prey detection with pause-travel search. *Am. Nat.* 138:1459–77

Getty T, Pulliam HR. 1993. Search and prey detection by foraging sparrows. *Ecology* 74:734–42

Greene E. 1996. Effect of light quality and larval diet on morph induction in the polymorphic caterpillar *Nemoria arizonaria* (Lepidoptera: Geometridae). *Biol. J. Linn. Soc.* 58:277–85

Greenwood JJD. 1984. The functional basis of frequency-dependent food selection. *Biol. J. Linn. Soc.* 23:177–99

Guilford T, Dawkins MS. 1987. Search images not proven: a reappraisal of recent evidence. *Anim. Behav.* 35:1838–45

Halkka O, Halkka L. 1990. Population genetics of the polymorphic meadow spittlebug, *Philaenus spumarius* (L.). *Evol. Biol.* 24:149–91

Halkka O, Halkka L, Raatikainen M. 1975. Transfer of individuals as a means of investigating natural selection in operation. *Heredity* 51:581–606

Handford M. 1987. *Where's Waldo?* Boston: Little, Brown. 27 pp.

Hargeby A, Johansson J, Ahnesjo J. 2004. Habitat-specific pigmentation in a freshwater isopod: adaptive evolution over a small spatiotemporal scale. *Evolution* 58:81–94

Harvey PH, Birley N, Blackstock TH. 1975. The effect of experience on the selective behaviour of song thrushes feeding on artificial populations of *Cepaea* (Held). *Genetica* 45:211–16

Hedrick PW. 2006. Genetic polymorphism in heterogeneous environments: the age of genomics. *Annu. Rev. Ecol. Evol. Syst.* 37:67–93

Holling CS. 1965. The functional response of predators to prey density and its role in mimicry and population regulation. *Mem. Entomol. Soc. Can.* 45:1–60

Janzen DH. 1984. Two ways to be a tropical big moth: Santa Rosa saturniids and sphingids. *Oxf. Surv. Evol. Biol.* 1:85–140

Jormalainen V, Merilaita S, Tuomi J. 1995. Differential predation on sexes affects colour polymorphism of the isopod *Idotea baltica* (Pallas). *Biol. J. Linn. Soc.* 55:45–68

Kahneman D. 1973. *Attention and Effort*. Englewood Cliffs, NJ: Prentice-Hall. 246 pp.

Kamil AC, Bond AB. 2001. The evolution of virtual ecology. In *Model Systems in Behavioral Ecology*, ed. LA Dugatkin, pp. 288–310. Princeton, NJ: Princeton Univ. Press

Kamil AC, Bond AB. 2006. Selective attention, priming, and foraging behavior. In *Comparative Cognition: Experimental Explorations of Animal Intelligence*, ed. EA Wasserman, TR Zentall. Oxford: Oxford Univ. Press

Kettlewell HBD. 1973. *The Evolution of Melanism*. Oxford: Clarendon. 423 pp.

King RB, Lawson R. 1995. Color pattern variation in Lake Erie water snakes: the role of gene flow. *Evolution* 47:1819–33

Kisdi E. 2001. Long-term adaptive diversity in Levene-type models. *Evol. Ecol. Res.* 3:721–27

Koch K, McLean J, Segev R, Freed MA, Berry MJII, et al. 2006. How *much* the eye tells the brain. *Curr. Biol.* 16:1428–34

Kono H, Reid PJ, Kamil AC. 1998. The effect of background cuing on prey detection. *Anim. Behav.* 56:963–72

Krebs JR. 1973. Behavioral aspects of predation. In *Perspectives in Ethology*, ed. PPG Bateson, PH Klopfer. New York: Plenum

Langley CM. 1996. Search images: selective attention to specific visual features. *J. Exp. Psychol. Anim. B.* 22:152–63

Langley CM, Riley DA, Bond AB, Goel N. 1995. Visual search for natural grains in pigeons: search images and selective attention. *J. Exp. Psychol. Anim. B.* 22:139–51

Lawrence ES. 1985. Evidence for search image in blackbirds (*Turdus merula* L.): Short-term learning. *Anim. Behav.* 33:929–37

Levins R. 1968. *Evolution in Changing Environments*. Princeton, NJ: Princeton Univ. Press. 120 pp.

Mackintosh NJ. 1975. A theory of attention: variations in the associability of stimuli with reinforcement. *Psychol. Rev.* 82:276–98

Maljkovic V, Martini P. 2005. Implicit short-term memory and event frequency effects in visual search. *Vis. Res.* 45:2831–46

Maynard Smith J. 1962. Disruptive selection, polymorphism and sympatric speciation. *Nature* 195:60–62

McNair JN. 1980. A stochastic foraging model with predator training effects: I. Functional response, switching, and run lengths. *Theor. Popul. Biol.* 17:141–66

McNair JN. 1981. A stochastic foraging model with predator training effects: II. Optimal diets. *Theor. Popul. Biol.* 19:147–62

Merilaita S. 2003. Visual background complexity facilitates the evolution of camouflage. *Evolution* 57:1248–54

Merilaita S, Lind J. 2005. Background-matching and disruptive coloration, and the evolution of cryptic coloration. *Proc. R. Soc. London Ser. B* 272:665–70

Merilaita S, Lyytinen A, Mappes J. 2001. Selection for cryptic coloration in a visually heterogeneous habitat. *Proc. R. Soc. London Ser. B* 268:1925–29

Merilaita S, Tuomi J, Jormalainen V. 1999. Optimization of cryptic coloration in heterogeneous habitats. *Biol. J. Linn. Soc.* 67:151–61

Moment GB. 1962. Reflexive selection: a possible answer to an old puzzle. *Science* 136:262–63

Murdoch WW. 1969. Switching in general predators: experiments on predator specificity and stability of prey populations. *Ecol. Monogr.* 39:335–54

Murdoch WW. 1970. Population regulation and population inertia. *Ecology* 51:497–502

Murdoch WW, Oaten A. 1974. Predation and population stability. *Adv. Ecol. Res.* 9:1–131

Murton RK. 1971. The significance of a specific search image in the feeding behaviour of the wood-pigeon. *Behaviour* 40:10–42

Nabours RK, Larson I, Hartwig N. 1933. Inheritance of color patterns in the grouse locust *Acrydium arenosum* Burmeister (Tettigidae). *Genetics* 18:159–71

Norris KS, Lowe CH. 1964. An analysis of background color-matching in amphibians and reptiles. *Ecology* 45:565–80

Olendorf R, Rodd FH, Punzalan D, Houde AE, Hurt C, et al. 2006. Frequency-dependent survival in natural guppy populations. *Nature* 441:633–36

Owen DF. 1980. *Camouflage and Mimicry.* Chicago: Univ. Chicago Press. 158 pp.

Oxford GS, Gillespie RG. 1998. Evolution and ecology of spider coloration. *Annu. Rev. Entomol.* 43:619–43

Parasuraman R, Davies DR. 1984. *Varieties of Attention.* New York: Academic. 554 pp.

Pashler HE. 1998. *The Psychology of Attention.* Mahwah, NJ: Lawrence Erlbaum. 494 pp.

Pietrewicz AT, Kamil AC. 1977. Visual detection of cryptic prey by blue jays (*Cyanocitta cristata*). *Science* 195:580–82

Pietrewicz AT, Kamil AC. 1979. Search image formation in the blue jay (*Cyanocitta cristata*). *Science* 204:1332–33

Plaisted K. 1997. The effect of interstimulus interval on the discrimination of cryptic targets. *J. Exp. Psychol. Anim. B.* 23:248–59

Plaisted KC, Mackintosh NJ. 1995. Visual search for cryptic stimuli in pigeons: implications for the search image and search rate hypotheses. *Anim. Behav.* 50:1219–32

Popham EJ. 1941. The variation in the colour of certain species of *Arctorcorisa* (Hemiptera, Corixidae) and its significance. *Proc. Zool. Soc. London Ser. A* 111:135–72

Posner MI, Snyder CRR. 1975. Facilitation and inhibition in the processing of signals. In *Attention and Performance*, ed. PM Rabbitt, S Dornic, pp. 669–82. San Diego, CA: Academic

Poulton EB. 1890. *The Colours of Animals: Their Meaning and Use, Especially Considered in the Case of Insects.* New York: Appleton. 360 pp.

Punzalan D, Rodd FH, Hughes KA. 2005. Perceptual processes and the maintenance of polymorphism through frequency-dependent predation. *Evol. Ecol.* 19:303–20

Reid DG. 1987. Natural selection for apostasy and crypsis acting on the shell colour polymorphism of a mangrove snail, *Littoraria filosa* (Sowerby) (Gastropoda: Littorinidae). *Biol. J. Linn. Soc.* 30:1–24

Reid PJ, Shettleworth SJ. 1992. Detection of cryptic prey: search image or search rate? *J. Exp. Psychol. Anim. B* 18:273–86

Reillo PR, Wise DH. 1988. An experimental evaluation of selection on color morphs of the polymorphic spider *Enoplognatha ovata* (Araneae: Theridiidae). *Evolution* 42:1172–89

Robinson MH. 1969. Defenses against visually hunting predators. *Evol. Biol.* 3:225–59

Robinson MH. 1990. Predator-prey interactions, informational complexity, and the origins of intelligence. In *Insect Defenses: Adaptive Mechanisms and Strategies of Prey and Predators*, ed. DL Evans, JO Schmidt, pp. 129–49. Albany, NY: SUNY Press

Rowell CHF. 1971. The variable coloration of the acridoid grasshoppers. *Adv. Insect Physiol.* 8:145–98

Royama T. 1970. Factors governing the hunting behaviour and selection of food by the great tit (*Parus major* L.). *J. Anim. Ecol.* 39:619–68

Sandoval CP. 1994. Differential visual predation on morphs of *Timema cristinae* (Phasmatodeae: Timemidae) and its consequences for host range. *Biol. J. Linn. Soc.* 52:341–56

Sargent TD. 1976. *Legion of Night: The Underwing Moths*. Amherst, MA: Univ. Mass. Press. 222 pp.

Sargent TD. 1981. Antipredator adaptations of underwing moths. In *Foraging Behavior: Ecological, Ethological and Psychological Approaches*, ed. AC Kamil, TD Sargent, pp. 259–87. New York: Garland STPM

Schweickert R. 1993. Information, time and the structure of mental events: a 25-year review. In *Attention and Performance*, ed. DE Meyer, S Kornblum, pp. 535–66. Cambridge, MA: MIT Press

Sherratt TN, Harvey IF. 1993. Frequency-dependent food selection by arthropods: a review. *Biol. J. Linn. Soc.* 48:167–86

Shettleworth SJ, Reid PJ, Plowright CMS. 1993. The psychology of diet selection. In *Diet Selection: An Interdisciplinary Approach to Foraging Behaviour*, ed. RN Hughes, pp. 56–77. Oxford: Blackwell Sci.

Sinervo B, Calsbeek R. 2006. The developmental, physiological, neural and genetical causes and consequences of frequency-dependent selection in the wild. *Annu. Rev. Ecol. Evol. Syst.* 37:581–610

Staddon JER, ed. 1980. *Limits to Action: The Allocation of Individual Behavior*. New York: Academic. 308 pp.

Staddon JER. 1983. *Adaptive Behaviour and Learning*. London: Cambridge Univ. Press. 555 pp.

Stephens DW, Krebs JR. 1986. *Foraging Theory*. Princeton, NJ: Princeton Univ. Press. 247 pp.

Tinbergen L. 1960. The natural control of insects in pine woods I. Factors influencing the intensity of predation by songbirds. *Arch. Néerl. Zool.* 13:265–343

Tucker GM. 1991. Apostatic selection by song thrushes (*Turdus philomelos*) feeding on the snail *Cepaea hortensis*. *Biol. J. Linn. Soc.* 43:149–56

Vane-Wright RI. 1980. On the definition of mimicry. *Biol. J. Linn. Soc.* 13:1–6

von Uexküll J. 1934. *Streifzüge durch die Umwelten von Tieren und Menschen*. Berlin: Springer. 101 pp.

Wente WH, Phillips JB. 2003. Fixed green and brown color morphs and a novel color-changing morph of the Pacific tree frog *Hyla regilla*. *Am. Nat.* 162:462–73

Whiteley DAA, Owen DF, Smith DAS. 1997. Massive polymorphism and natural selection in *Donacilla cornea* (Poli, 1791) (Bivalvia: Mesodesmatidae). *Biol. J. Linn. Soc.* 62:475–94

Willis AJ, McEwan JWT, Greenwood JJD, Elton RA. 1980. Food selection by chicks: effects of colour, density, and frequency of food types. *Anim. Behav.* 28:874–79

Wolfe JM. 1994. Visual search in continuous, naturalistic stimuli. *Vis. Res.* 34:1187–95

Wolfe JM, Horowitz TS, Kenner NM. 2005. Rare items often missed in visual searches. *Nature* 435:439–40

Wolfe JM, Horowitz TS, Van Wert MJ, Kenner NM, Place SS, Kibbi N. 2007. Low target prevalence is a stubborn source of errors in visual search tasks. *J. Exp. Psychol. Gen.* 136: In press

Wolfe JM, Oliva A, Horowitz TS, Butcher SJ, Bompas A. 2002. Segmentation of objects from backgrounds in visual search tasks. *Vis. Res.* 42:2985–3004

Zentall TR. 2005. Selective and divided attention in animals. *Behav. Processes* 69:1–15

Point, Counterpoint: The Evolution of Pathogenic Viruses and their Human Hosts

Michael Worobey, Adam Bjork, and Joel O. Wertheim

Ecology and Evolutionary Biology, University of Arizona, Tucson, Arizona 85721;
email: worobey@email.arizona.edu, bjork@email.arizona.edu,
wertheim@email.arizona.edu

Annu. Rev. Ecol. Evol. Syst. 2007. 38:515–40

First published online as a Review in Advance on
August 15, 2007

The *Annual Review of Ecology, Evolution, and
Systematics* is online at
http://ecolsys.annualreviews.org

This article's doi:
10.1146/annurev.ecolsys.38.091206.095722

1543-592X/07/1201-0515$20.00

Key Words

human-virus coevolution, immune response, intrinsic immunity,
HIV evolution, major histocompatibility complex

Abstract

Viral pathogens play a prominent role in human health owing to their
ability to rapidly evolve creative new ways to exploit their hosts. As
elegant and deceptive as many viral adaptations are, humans and
their ancestors have repeatedly answered their call with equally impressive adaptations. Here we argue that the coevolutionary arms
race between humans and their viral pathogens is one of the most
important forces in human molecular evolution, past and present.
With a focus on HIV-1 and other RNA viruses, we highlight recent developments in our understanding of the human innate and
adaptive immune systems and how the selective pressures exerted
by viruses have shaped the human genome. We also discuss how the
antiviral function of cellular machinery like RNAi and APOBEC3G
blur the lines between innate and adaptive immunity. The remarkable power of natural selection is revealed in each host-pathogen
arms race examined.

INTRODUCTION

The battles played out between humans and their pathogenic viruses are some of the most dramatic of all the conflicts studied by evolutionary biologists and ecologists. Viral pathogens play a dominant role in human health in terms of both morbidity and mortality and, as such, they represent one of the most potent selective forces acting on human populations (Vallender & Lahn 2004, World Health Org. 2003). Their habit of hiding within host cells, their large population sizes, rapid reproductive rates, and, above all, high rates of mutation and recombination (Domingo & Holland 1997, Worobey & Holmes 2001) make them especially formidable adversaries. Viruses are masterpieces of nature: compact, streamlined, highly flexible bundles of protein-coated genetic material with the audacious ability to hijack the sophisticated biochemical machinery within their hosts' cells.

It is becoming increasingly clear that humans (and their predecessors and relatives) have risen to the evolutionary challenge posed by viral pathogens with some elegant defenses and remarkably rapid immunological, genetic, and genomic changes of their own, ones that reveal a long history of host-virus coevolution that holds valuable lessons for biologists and clinicians. How can giant, long-lived targets like human beings defend against the asymmetric threat posed by viral pathogens? The answer lies, in large part, in another evolutionary masterpiece: the vertebrate immune system. This gives some of the most complex, slowly evolving organisms the ability, at both the individual and population level, to keep up with pathogens that represent the other evolutionary extreme. Given that 23 of 94 vertebrate-specific gene families and (conservatively) 1 out of every 30 human genes code for defense and immunity proteins (Int. Hum. Genome Consort. 2001), it is clear that much of the last 400 million years of our own evolutionary story has involved the battle with pathogens.

Here, to bring into sharper relief what we believe might be the most crucial part of that story, we attempt to answer the following questions: How have viruses shaped the human immune system, our genes, and our genomes? Have they done so differently than other pathogens, and, if so, in what ways? What evolutionary pressures have humans and other vertebrates imposed onto viruses, and what are the consequences of this feedback for them and us? To answer these questions we dissect a variety of host-pathogen arms races and conflicts in light of recent advances in the understanding of the human genome and virus-relevant innate and adaptive immunity, with an emphasis on HIV-1. We also consider newly discovered antivirus defenses that do not fit neatly within the conventional definitions of innate or adaptive immunity. These include RNA interference, as well as intrinsic immunity, whereby constitutively expressed host gene products can poison viral genomes or otherwise disrupt viral life cycles, even in the absence of virus-triggered signaling (Bieniasz 2004). Throughout, we highlight how lessons learned at the intersection of evolution, ecology, genetics, virology, and immunology can contribute both to a better understanding of each of these fields and to improving therapeutic and preventative control measures against viral pathogens.

INNATE IMMUNITY: THE FIRST LINE OF DEFENSE

The front-line component of the immune system, the innate immune response, has extremely deep roots. Every multicellular organism has a complex innate immune system that allows it to discriminate between self and nonself on the basis of a limited set of more or less generic cues, molecular patterns normally present in the invader but not the host (Beutler 2004). Receptors recognizing different classes of pathogens, if activated, rapidly unleash anti-invader responses, such as the interferon response that temporarily makes the body much less hospitable to viruses (Akira et al. 2006). It is the interferon response that is behind the sore muscles, fever, and other flu-like symptoms associated with so many viral infections.

The defining characteristic of the innate immune system is its ability to mount rapid and effective responses to a wide variety of pathogens without the requirement of previous exposure. However, recent findings demonstrate that the innate immune system can operate with a previously underappreciated level of specificity. The Toll-like receptors (TLRs) provide a good example.

The Toll receptor pathway was first identified in *Drosophila* for its role in establishing dorsal-ventral polarity in the developing embryo (Hashimoto et al. 1988), but it was later hypothesized (Belvin & Anderson 1996) and confirmed (Lemaitre et al. 1996) to play an important role in innate immunity as well. A large, multigene family of homologous proteins, known as Toll-like receptors, has since been identified in vertebrates, including mammals (reviewed in Takeda et al. 2003).

TLRs are transmembrane proteins that distinguish self from nonself by detecting pathogen-associated molecular patterns (PAMPs). Recognition of these ligands on invading pathogens, by TLRs located on the surface of immune cells, triggers the induction of genes responsible for the inflammatory (innate) immune response and also contributes to stimulating the adaptive immune response (reviewed in Akira & Hemmi 2003). Most mammalian genomes code for between 10 and 15 TLRs; 10 have been identified in humans (TLR1–TLR10).

Some TLRs can recognize a variety of viral and/or bacterial pathogens. On the viral side, for example, TLR2 can identify PAMPs found on measles virus, cytomegalovirus, and hepatitis C virus (HCV) (Bieback et al. 2002, Compton et al. 2003, Dolganiuc et al. 2004). Others are more restricted: Respiratory syncytial virus and herpes simplex virus are identified by TLR4 and TLR9, respectively (Kurt-Jones et al. 2000, Lund et al. 2003). TLR3 is something of an all-purpose virus alarm, as it identifies double-stranded RNA (dsRNA) (Alexopoulou et al. 2001), found in most viruses at some point in their life cycle; it is the trigger for the interferon response.

A key point here is that, unlike the main players in the adaptive immune system—for example, major histocompatibility complex (MHC) genes—although adept at what they do, TLRs are not very evolutionarily flexible. Viral antigen recognition mediated by MHC genes is characterized by an impressive amplification in the gene family, unrivalled levels of polymorphism at each locus (**Figure 1**), and striking levels of positive selection (MHC Seq. Consort. 1999). TLRs, however, represent the opposite: a set of good, generic signal receptors, held in place by strong purifying selection. The function and nucleic acid sequences of TLRs are highly conserved

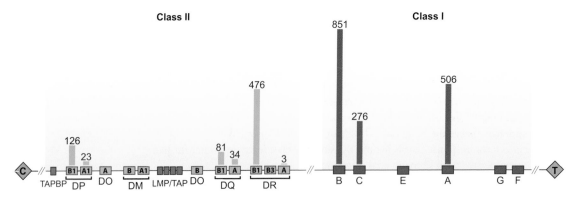

Figure 1

Human major histocompatibility complex (MHC) class I and II loci. This is a schematic diagram (not to scale) of some of the key loci in the MHC class I and II regions. Included are the nine classical human leukocyte antigen (HLA) loci, each of which is labeled with the number of alleles found so far in the human population. Several other HLA loci exist, some of which are shown here. Many non-*HLA* genes with immune function are also found in the MHC. *TAPBP*, *LMP*, and *TAP*, which are involved in antigen processing by the proteasome and transport of peptides to MHC molecules, are shown here as examples. C, centromere; T, telomere.

across vertebrates, and the TLR arsenal of extant vertebrates is remarkably similar to that of the ancestral vertebrate, even down to the level of gene copy number, which has typically remained at just one for each TLR (Roach et al. 2005). The PAMPs targeted by TLRs are presumably not at liberty to evolve to avoid recognition: hence the evolutionary stability (and the reliability) of TLRs (Smith et al. 2003).

INTRINSIC IMMUNITY: MUTATIONAL MELTDOWN AND OTHER WAYS TO KILL A VIRUS WITHOUT EVER KNOWING IT IS THERE

Our cells, it turns out, constitutively produce potent antiviral drugs. One of the most dramatic examples of the evolutionary arms race between pathogens and their hosts—one that no one knew about until just a few years ago—is the case of retroviruses (and retroelements) versus the *APOBEC* gene family of Old World primates. One player in this arms race, the HIV protein Vif (viral infectivity factor), has long been known to be important; viruses with defective or absent *vif* genes (Δvif) are unable to replicate in most human cell lines (Gabuzda et al. 1992). However, the reason why HIV needs Vif in order to infect human cells remained largely obscure for over a decade. During this same time period, HIV researchers also began coming across viruses that had a bizarre mutational pattern. Their genomes had been hypermutated—fatally riddled with G-to-A (guanine-to-adenine) mutations that resulted in defective progeny (Borman et al. 1995). The solution to this small mystery would reveal a surprisingly broad new line of defense against viruses.

The breakthrough came when Sheehy et al. (2002) compared gene expression in human cell lines and identified a protein, APOBEC3G, present only in cell lines in which *vif*-defective viruses were unable to replicate. Zhang et al. (2003) and Mangeat et al. (2003) independently drew the connection that it was APOBEC3G's function as a cytidine deaminase—a DNA-editing enzyme—that could explain both the necessity of a functional Vif protein and the hypermutation of HIV genomes. During the viral life cycle, APOBEC3G is incorporated into the viral capsid. Typically, it is subsequently degraded by the virus's Vif protein, which is effectively the antidote to APOBEC3G. However, in the absence of a functional Vif, APOBEC3G is not degraded; upon infection of a new host cell, it hypermutates the viral genome, via deamination, while the virus is reverse-transcribing its RNA genome to DNA.

It is now known that APOBEC3G is actually one of many APOBEC proteins that comprise a recently expanded part of a large gene family that includes AID (activation-induced deaminase) (Conticello et al. 2005). AID is a protein responsible for the somatic hypermutation that takes place in the immunoglobulin (Ig) genes of cells that produce antibodies, allowing them to selectively fine-tune their antigen recognition (Muramatsu et al. 2000). The common ancestry, and mechanism, of APOBEC3G and AID is a shining example of how natural selection is a great tinkerer—in this case co-opting a mechanism crucial to adaptive immunity and shaping it into a totally distinct weapon.

The antiretroviral properties of APOBEC3G have been observed in other viruses too, including HTLV-I (human T cell lymphotrophic virus), SIV (simian immunodeficiency virus), and hepatitis B virus (HBV), as well as in retroelements such as retrotransposons present in the human genome (Bogerd et al. 2006a, 2006b; Muckenfuss et al. 2006; Turelli et al. 2004). Retroelements compose approximately 42% of the human genome (Int. Hum. Genome Consort. 2001), and it has been suggested that their activity may sometimes provide an adaptive advantage to individual hosts by helping to create new genes, including some with roles in adaptive immunity (Agrawal et al. 1998, van de Lagemaat et al. 2003). However, retroelement activity has demonstrably negative fitness effects as well, such as disrupting the expression of crucial genes and causing malignancies and autoimmune disorders (reviewed in Bannert & Kurth 2004). In addition, gene products encoded by some retroelements may be toxic for the host organism (Boissinot et al. 2006). These elements may have overall deleterious host fitness effects and could have selected for the expansion and diversification of the *APOBEC/AID* gene family (Sawyer et al. 2004).

Several other members of the *ABOBEC* gene family have demonstrable antiretroviral activity; the radiation of this gene family has been accompanied by remarkably strong selective pressure, with some of the highest dN/dS ratios ever measured (OhAinle et al. 2006, Sawyer et al. 2004, Turelli & Trono 2005, Zheng et al. 2004). Additional unrelated genes such as *TRIM5α* have also been highly effective at restricting retrovirus propagation and, accordingly, also exhibit evidence of extremely intense positive selection in both humans and other primate species (Sawyer et al. 2005). TRIM5α appears to disrupt the uncoating of retroviruses, which leads to their death (Stremlau et al. 2004). This mechanism of action implies that endogenous and exogenous retroviruses, rather than retroelements, likely provided the selection for TRIM5α antiretroviral activity. Much of the most exciting research on intrinsic immunity proteins is focused on their potential use as drug treatments for HIV and other retroviral infections (Mangeat & Trono 2005).

RNA INTERFERENCE AND GENE SILENCING: AN ANCIENT VIRAL DEFENSE MECHANISM?

While the antiretroviral effects of the *APOBEC* genes likely resulted from the modification of previously existing proteins functioning in the adaptive immune system, an ancient, putatively antiviral mechanism called RNAi (RNA-mediated interference) has been co-opted and modified to perform unrelated cellular functions: transcriptional and posttranscriptional gene regulation. Surprisingly, it was these derived functions of RNAi that led to its discovery. One of the first documented instances of RNAi activity occurred not when researchers attempted to suppress viral infection but when they introduced a second copy of a gene for purple pigment into a petunia plant with the intention of producing more pigment. The result of this experiment was not the production of more purple pigment, but less (Jorgensen 1995).

Gene silencing phenomena, whereby a specific gene could be suppressed by the introduction of the homologous sequence, was subsequently observed in a variety other organisms, but it was Fire et al.'s (1998) pioneering work on *Caenorhabditis elegans* that first identified the role of dsRNA in RNAi and gene silencing. In addition, they noted that owing to the small amount of dsRNA required to initiate gene suppression, a catalytic protein cascade must be involved in this RNA-mediated gene suppression.

A crucial part of RNAi activity is its specificity, which is achieved by using the original dsRNA template as a guide for identifying all other copies of the gene of interest. This template is created by the enzyme Dicer (Bernstein et al. 2001), which cleaves the dsRNA into short (approximately 22-bp) RNA sequences known as small interfering RNAs (siRNAs). Once the dsRNA is cleaved, the siRNA is incorporated into the RNA-induced silencing complex. This protein complex then binds to and cleaves other copies of the target RNA, which essentially suppresses all protein products that would have been produced from the original RNA sequence (reviewed in Stram & Kuzntzova 2006).

Despite its recent discovery, RNAi is now understood to be an extremely important regulator of gene expression; however, the most parsimonious proposal for the ancestral function of this RNAi machinery is defense against RNA viruses and transposable elements, given the conservation of this function across the eukaryotic spectrum (Cerutti & Casas-Mollano 2006). The antiviral effects of RNAi, owing to its ability to identify and degrade homologous nucleotide sequences, are thus a good example of tremendous specificity in organisms traditionally thought to have only nonspecific innate immune responses.

Like the interferon response, RNAi is triggered specifically by dsRNA, albeit by shorter nucleotide strands. Unlike the interferon system, which can result in a more general response that upregulates a variety of antiviral functions such as proteasome function and MHC class I production, the RNAi machinery specifically targets the gene of interest, allowing the rest of the cell to continue to function normally. More-over, RNAi is a more general antiviral mechanism than interferon, as the latter is expressed only in a limited range of cell types.

Note that although RNAi is still used as an antiviral defense in plants and inverte-brates, it is possible that in mammals its antiviral function has been largely supplanted by the interferon and adaptive immune responses (Cerutti & Casas-Mollano 2006). Regardless of whether it survives only as a derived, gene-regulatory mechanism in mammals, it can still induce potent antiviral responses. One of the most promising avenues of biomedical research today involves the artificial induction of RNAi in humans by introducing synthetic siRNAs designed to target specific genes, including essential viral genes, which can lead to the suppression of viral replication within the host cell. Encouraging results have been seen in a large group of viruses ranging from single-stranded RNA viruses (influenza virus and SARS) to retroviruses (HIV-1), hepadnaviruses (HBV), and even DNA viruses (herpesviruses). These studies are re-viewed in Stram & Kuzntzova (2006), who note that RNAi may have an impact on human health in the coming century comparable to that of antibiotics in the previous century. Like antibiotic use, therapeutic RNAi represents the successful harnessing, for medical purposes, of evolutionary adaptations devised by natural selection during ancient host-pathogen arms races.

ADAPTIVE IMMUNITY: EDUCATION BY PATHOGEN

What is Adaptive Immunity?

The vertebrate adaptive immune response is highly specific and extremely dynamic, and it can recognize and respond to virtually any pathogen, including viruses within infected cells. It mimics many of the features that make viruses dangerous, including the ability to rapidly generate extensive genetic diversity through recombination and mutation and select for successful variants.

Unlike in viruses, this selection in B and T cells takes place across somatic gen-erations; successful variants possess a cell-surface receptor that physically binds with high affinity to some foreign molecule (i.e., an antigen), allowing the immune system to see an invading pathogen, to remember it, and to respond rapidly in the event of subsequent encounters. B cell receptors are known as immunoglobulins. They search for, and bind to, macromolecular antigens on foreign invaders. Daughter B cells then secrete increasingly specific Ig (i.e., antibodies) that circulate throughout the inter-cellular space. T cell receptors, however, bind specifically to smaller peptide regions presented by special molecules encoded by class I and class II genes of the MHC (**Figure 2**).

During development, human fetuses generate an extremely large population of B cells and T cells, each one bearing a slightly different cell-surface receptor. Cells that recognize antigen are culled at this point because only self antigens are present (Burnet 1959). An almost unlimited combinatorial explosion of receptor diversity is generated through somatic recombination, whereby sets of gene segments drawn from a diverse but finite pool are fused into unique, irreversible mosaics (Janeway et al. 2005).

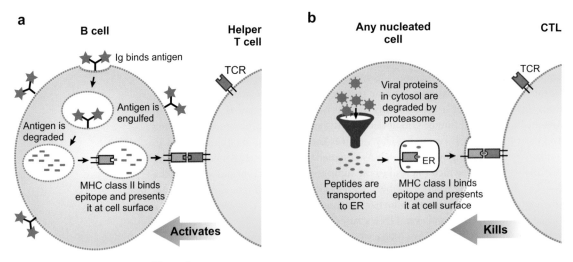

a B cell / Helper T cell

Ig binds antigen

TCR

Antigen is engulfed

Antigen is degraded

MHC class II binds epitope and presents it at cell surface

Activates

b Any nucleated cell / CTL

TCR

Viral proteins in cytosol are degraded by proteasome

Peptides are transported to ER

ER

MHC class I binds epitope and presents it at cell surface

Kills

Figure 2

Antigen presentation by major histocompatibility complex (MHC) molecules. (*a*) B cells bind viral antigen with their immunoglobulin receptors and engulf it. The antigen is then degraded within vesicles into peptides 13 amino acids or greater in length. Peptides that bind MHC class II molecules are then presented at the cell surface where the epitope/MHC complex can be recognized by helper T cells with matching T cell receptors. If a match (binding) occurs, the helper T cell activates the B cell (signals it to proliferate). (*b*) Cells infected with virus degrade viral proteins found in the cytosol, via the proteasome. The resulting 8–10 amino acid peptides are then transported to the endoplasmic reticulum (ER). Peptides that bind MHC class I molecules are presented at the cell surface. Cytotoxic T lymphocytes (CTLs), whose receptors bind the epitope/MHC complex, induce apoptosis in the virus-infected cell.

B cells further the refinement of Ig specificity via somatic hypermutation, driven by AID (see above discussion of ABOBEC3G). It is the remaining pool of hundreds of millions of circulating lymphocytes that is called upon when the host is challenged by pathogens. One or more lymphocytes, by chance, might bear a receptor that is a good fit for some antigenic region; the tighter the fit, the stronger the signal for the lymphocyte to proliferate. The upshot, after several rounds of lymphocyte division, is an adapted population of lymphocytes. The successful variant in the original pool is amplified into a massive army of clones with the ability to home in on the pathogen with great accuracy and precision, then kill or neutralize it. These cells remain ready for quick action should the pathogen ever return.

Major Histocompatibilty Complex Molecules versus Viral Epitopes

Any attempt to understand how selection imposed by viruses has shaped the human genome rests on an appreciation of the MHC (**Figure 1**). The MHC is an extremely gene-dense region of chromosome 6, housing approximately 130 expressed genes, almost half of which have immune system function (Meyer & Thomson 2001, MHC Seq. Consort. 1999). These include the diverse and rapidly evolving MHC class I and

II genes that broker the antigen recognition performed by T cells and B cells, as well as many other genes with immune function. Without MHC molecules—better known as human leukocyte antigen (HLA) in humans—there can be no effective neutralizing antibody (NAb) response from B cells, of the sort that helps us eliminate influenza virus (**Figure 2a**), and no cytotoxic T lymphocyte (CTL) response to destroy cells infected with virus, of the sort that helps us fight HIV-1 (**Figure 2b**).

MHC class I molecules have a direct and special connection to viruses: They play the central role in the task of alerting CTLs to cells that have been breached by virus and are an unambiguous antivirus adaptation (**Figure 2**). Expressed by all nucleated cells, class I molecules bind to and transport intracellular, cytosolic viral peptides. These are derived from larger proteins that have been minced into short fragments in the proteasome, a sort of meat grinder that cleaves any proteins found in the cytosol into bite-size pieces for MHC molecules. They present these peptides on the infected cell's surface (physically bound in pockets or grooves) for inspection by roving T cells; because of the specificity discussed above, a given T cell will recognize only specific epitope/MHC complexes (Doherty & Zinkernagel 1975, Zinkernagel & Doherty 1974). If a CTL binds to the peptide/MHC complex, it induces apoptosis in the cell that presented the antigen. MHC delivers a constant efflux of cytosolic peptides to the cell surface, allowing the immune system to effectively see what is going on deep within almost every cell of the body and to respond accordingly.

MHC class II molecules, however, are involved in transporting and displaying antigens derived from vesicles within the cell (**Figure 2**). These can be of two main sorts. First, B cell–mediated responses, including the NAb response, require a go-ahead signal from CD4+ helper T cells. This is where MHC class II genes come into play. A successful antibody response to a viral pathogen begins with a helper T cell recognizing a viral T cell epitope presented by the B cell: The B cell must capture viral antigens with its Ig receptor, engulf them, degrade them into small peptides, then spit out these minced up fragments onto its cell surface with the aid of MHC class II molecules (**Figure 2**).

Second, there are vesicles in which pathogens such as *Mycobacterium tuberculosis* seek intracellular refuge from immune attack. Class II molecules deliver antigen from such invaders to the cell surface where it can be recognized by CD4+ T cells, which command the infected cell to flood the vesicle with bactericidal poison (not shown). This sort of antigen presentation, although not directly connected to viruses, is still related: One of the reasons so many AIDS patients die from tuberculosis is that the virus destroys the T cells that would normally recognize the *M. tuberculosis*–derived epitopes bound to MHC class II molecules on the surface of an infected cell.

There is a large degree of specificity between the binding of MHC and viral peptides, and this is a crucial point for understanding the evolution of both viruses and humans. A given MHC/HLA variant will only bind one or a few short peptide fragments, and changing just one amino acid can often obliterate binding (reviewed in Frank 2002). This sets the stage for CTL epitope escape, on the part of viruses (discussed below). But it also evidently puts tremendous selective pressure on humans to evolve and maintain a diverse array of MHC molecules. Heterozygosity at *HLA* loci can lead to a powerfully diverse immune response (Carrington et al. 1999,

Doherty & Zinkernagel 1975). Accordingly, *HLA* loci are by far the most polymorphic of any in our genome (MHC Seq. Consort. 1999). *HLA-B*, an MHC class I gene, leads the way with 851 alleles, followed by *HLA-A*, another class I locus, with 506, and *HLA-DRB1*, a class II locus, with 476 (**Figure 1**) (Robinson et al. 2003; **http://www.ebi.ac.uk/imgt/hla/**).

Most *HLA* polymorphism is also evidently generated de novo in each host species (Shiina et al. 2006), contrary to earlier assertions that it represents ancient variation predating the origin of humans. Shiina et al. traced single-nucleotide variation generated before and after human-chimpanzee speciation and showed conclusively that trans-species diversity could account for only a small fraction of MHC class I diversity in either humans or chimpanzees. This is important not only because it illustrates the strong incentive for constant generation and maintenance of a diverse HLA repertoire in response to a constantly evolving microbial antigenic repertoire, but also because it finally puts the MHC in line with other evolutionary genetics data that conclude that a narrow bottleneck occurred at the origin of our species (Hammer 1995).

With respect to nonself antigens, MHC class I molecules are virtually totally devoted to viruses, the only pathogens that normally replicate within the cytosol (Janeway et al. 2005). It appears, once again, that it is this group of pathogens that has driven our genes to the extremes of diversification. Viruses might even be one of the ultimate causes behind some of our more pleasurable, proximate activities: MHC gene products are aromatic, and we might unwittingly choose potential mates in part by sniffing out good, complementary MHC alleles—ones that might tilt the odds in favor of producing offspring with more robust immune systems (reviewed in Milinski 2006).

Natural experiments—cases where patients are born with defective MHC class II antigen presentation—suggest that viruses might be the main selective agents at class II loci as well: In one cohort of 10 children with MHC class II deficiency, 8 died during the study; 6 deaths were attributed directly to viral infection, and an additional death was likely due to complications arising from a viral infection (Saleem et al. 2007). This is an unsettling reminder of what lies just behind the veil of our immune system.

The MHC is associated with more genetic diseases than any other region of the genome (MHC Seq. Consort. 1999). This includes most autoimmune diseases, such as rheumatoid arthritis and diabetes. It seems that a good deal of human genetic disease has arisen via hitchhiking of deleterious mutations (MHC Seq. Consort. 1999), and it is probably not a coincidence that hitchhiking diversity reaches its peaks in the vicinity of the antigen-presenting *HLA* loci (Shiina et al. 2006). In other words, the long-standing and continuing battle to generate HLA diversity, in response to the Red Queen vicissitudes presented by pathogens, has been marked by a lot of costly collateral genetic damage—yet another indication of the utmost importance of our defenses against pathogens. The magnitude of debris dragged in the wake of selection on MHC loci says a lot about the strength of the selection and, in turn, the agents behind it.

HUMAN-VIRUS BATTLEGROUNDS WITHIN, AND BETWEEN, HOSTS

The Human Leukocyte Antigen Environment, Cytotoxic T Lymphocyte Escape, and Persistent Viruses

MHC class I–mediated antigen recognition, which places viral epitopes in the crosshairs of CTLs, puts strong selective pressure on viruses to alter those epitopes. Typically, the number of viral peptides detectable by a host's CTL population is, at any given time, quite small. Often only two or three immunodominant epitopes are detectable (Yewdell & Bennink 1999). This is why genetic variation (resulting from high viral mutation and replication rates, and large population sizes) is so crucial to viral persistence; a mutant virus with just the right amino acid substitution becomes invisible by moving outside the highly focused spotlight of the prevailing CTL response. Such CTL epitope escape mutations allow a few lucky variants to reproduce without the risk of their host cells being killed off through CTL-mediated damage control. This can allow the viral population to persist within the host for long periods—years or decades in some cases—with successive rounds of selective sweeps as viral epitopes are seen by CTLs, then altered by mutations that lead to CTL escape, then seen again by a new set of CTLs, and so on.

In HIV-1, within-host population dynamics leads to distinctive ladder-like phylogenetic trees. When samples are collected across many time points, it can be seen that the fate of most viral lineages within the host is to go extinct and that viral genetic diversity is repeatedly pruned down to one or a few variants that, for a short time, escape from immune responses. This is a nice illustration of the connectedness of viral ecology (predator-prey population dynamics) and evolution (genetic changes) and how aspects of both can be inferred from phylogenetic patterns. Grenfell et al. (2004) have termed this unified view of ecological and evolutionary dynamics phylodynamics. Interestingly, the phylodynamic patterns of HIV-1 at the population level are very different, with little evidence for selective sweeps that would be indicative of differences in host-to-host transmission fitness: The population-level phylogeny is very bushy, not ladder-like. Influenza virus phylogenies at the population level, however, look like within-host HIV-1 phylogenies. This pattern might reflect antigenic drift, whereby viruses that have mutated to escape previously infected hosts' antibody responses enjoy a selective advantage because they can reinfect previously exposed hosts (Bush et al. 1999). Recent findings, however, support a more nuanced perspective on influenza phylodynamics, one that includes a bigger role for stochastic effects than previously appreciated. Local introductions and extinctions of different viral variants often appear to be due to chance rather than fitness differences (Nelson et al. 2006), a finding that suggests that the ladder-like phylogenies of influenza virus might not be completely due to the host's population-level immune selection.

A similar ladder-like phylogeny has been inferred for Ebola virus isolates sampled over the past 30 years across central Africa. Walsh et al. (2005) sought to determine if the Ebola virus was endemic to the outbreak regions or if each outbreak represented a new introduction of the virus to a geographical region. The inferred phylogeny indicated that the virus responsible for each new Ebola outbreak during this time period was a descendant of a preceding outbreak elsewhere and therefore was a recent introduction. Furthermore, these outbreaks followed a pattern of migration that was reflected in the viral phylogeny. This example shows that spatial and temporal dynamics (wave-like spread)—in the absence of strong immune selection—can clearly generate ladder-like phylogenies reminiscent of those shaped by immunity-driven selective sweeps. It also demonstrates that phylodynamics can be used to predict the future spread of emerging viruses.

The MHC class I alleles inherited from one's parents determine the immune system's ability to present pathogen antigens to T cells. An individual's *HLA* genotype can have far-reaching effects on the outcome of a variety of viral infections and is predictive, for example, of whether HIV is likely to kill you quickly or slowly, should you become infected. At a general level, Carrington et al. (1999) found that heterozygosity for *HLA-A*, *-B*, and *-C* translated into significantly delayed onset to AIDS among patients infected with HIV-1. More specifically, they also identified two deleterious alleles—one *HLA-B* and one *HLA-C*—associated with rapid progression to AIDS symptoms. Interestingly, they also noticed that the relative hazard of progression to AIDS or death was two- to threefold higher for homozygosity at the *HLA-B* locus versus *HLA-A*, a pattern that mirrors the higher polymorphism observed in *HLA-B* in the population at large.

The class I locus *HLA-B* turns out to be particularly important in defining the HLA environment that HIV-1 must adapt to in order to maintain a chronic infection (Bihl et al. 2006, Kiepiela et al. 2004). Kiepiela et al.'s study was a particularly elegant demonstration that *HLA-B* plays the dominant role in influencing HIV disease outcome. Among hundreds of HIV-positive patients in a South African cohort, a significantly greater number of CTL responses were HLA-B-restricted compared with HLA-A, and expression of particular *HLA-B* (but not *HLA-A*) variants was associated with extreme disease outcomes (fast or slow progression to AIDS). This is directly relevant to vaccine design because it suggests that artificial stimulation of CTLs directed against epitopes presented by HLA-A might be much less protective than ones designed for HLA-B, and that such CTL vaccines could be expected to perform very differently in patients with distinct complements of HLA alleles.

This observation also suggests an answer to the riddle of why *HLA-B* is more polymorphic in the human population than *HLA-A*, and why *HLA-B* has evolved more rapidly at the molecular level as well (Kiepiela et al. 2004). With HIV, and perhaps many other viruses, *HLA-B* is where the main action is in the battle between the pathogen and the host's CTLs. Moreover, it has recently become clear that the immune system's most effective CTL responses against HIV-1 target Gag polyprotein epitopes (Kiepiela et al. 2007, Zuniga et al. 2006). The *gag* gene, which codes for important structural components of the virus, is one of the most conserved in the entire HIV-1 genome. It is probably not a coincidence that the best place for the CTL response to lock onto viral epitopes is in a region where the virus is obviously under strong evolutionary constraints. Presumably, many potential CTLs escape mutations in *gag* come with too high a cost to be selected. At any rate, the battle within an HIV-1-infected host seems to be fought primarily between the Gag protein of the virus and the one or two *HLA-B* alleles available to carry Gag epitopes to CTLs.

Evasion of adaptive immune responses is of course a crucial feature of other viruses that can establish persistent infections. Specific HLA alleles with either negative or positive associations with disease have also been found for HBV and HCV (Gaudieri et al. 2006, Heeney et al. 2006, Thio et al. 2003). With respect to HCV, which most often establishes a persistent infection, the HLA class II alleles *DQB1*0301* and *DRB1*11* are associated with self-limiting infection, predominantly in Caucasian populations (Yee 2004). Individuals with these alleles can likely eliminate the virus via an effective antibody response, triggered by a helper T cell population that binds particularly well to HCV peptides presented by these alleles. Polymorphisms at nonimmunity-related genes, such as *CCR5*, which codes for a cell-surface receptor that mediates the entry of HIV-1 into cells, can also have dramatic consequences for disease progression, including making some hosts effectively impervious to HIV-1 (see sidebar).

Kiepiela et al. (2004) found that HIV-1-infected infants in their Zulu/Xhosa study population had a much higher frequency of deleterious *HLA-B* alleles (B*18 and B*5802) and a much lower frequency of protective alleles (B*57 and B*5801) than the population at large. This suggests that HIV-1 can be expected to cause rapid changes in allele frequencies at HLA loci, especially *HLA-B*. It follows logically from the mechanics of MHC-based antigen presentation, and the diverse array of viral pathogens like HIV-1 and HCV-pulling antigen-recognition genes in many different

CCR5-Δ32: A VIRAL-RESISTANCE ALLELE

CCR5-Δ32 is a truncated *CCR5* allele found at high frequencies in Europeans. It leads to a defective cell-surface coreceptor, preventing HIV infection in homozygotes and decreasing the risk of infection in heterozygotes (Liu et al. 1996, Samson et al. 1996). *CCR5-Δ32* carriers also have a decreased likelihood of contracting hepatitis B virus (Thio et al. 2007) and improved outcomes during hepatitis C virus infection (Goulding et al. 2005).

The HIV/AIDS pandemic is too recent to account for current *CCR5-Δ32* frequencies. Bubonic plague and smallpox have been proposed as selective agents (Stephens et al. 1998, Galvani & Slatkin 2003), although recently it has been argued that neutral genetic drift may better explain the current frequency of *CCR5-Δ32* (Sabeti et al. 2005).

Biomedical researchers are optimistic that development of *CCR5*-antagonists (e.g., using RNAi) could lead to important therapeutic advancements for the fight against HIV (Qin et al. 2003), although there could be trade-offs. *CCR5-Δ32* carriers are now known to be at an increased risk for symptomatic West Nile virus infection (Glass et al. 2006) and are more susceptible to noninfectious diseases such as rheumatoid arthritis (Garred et al. 1998) and inflammatory bowel disease (Martin et al. 2001).

directions, that the evolutionary outcome was both an amplification in the number of *HLA* genes, and an accumulation of a diverse pool of allelic variants at these loci.

Viral Interference with Antigen Presentation, and the Host Response: Natural Killer Cells

Viruses do not limit themselves merely to fleeing the CTL response by quick evolutionary change. They also go on the offensive. They have evolved a suite of adaptations that target the heart of the antigen-recognition system: the presentation of viral epitopes by MHC molecules. In the arms race between viruses and humans (**Table 1**), it perhaps in this arena that host-virus coevolution has reached the most extraordinary level of escalation.

In HIV-1, to take one well-studied example, the accessory protein Vpu actively degrades the class I molecules, abrogating CTL responses (Kerkau et al. 1997). Another protein, Nef, downregulates MHC expression by the infected cell (Piguet & Trono 1999). Patients with a rare, *nef*-deleted virus tend to have extremely slow progression to AIDS (Deacon et al. 1995, Kirchhoff et al. 1995). In other words, downregulation of MHC seems to make the difference between a killer virus and a near-commensal virus. Other viruses have evolved similar techniques for interfering with virtually every step necessary for antigen presentation by both MHC class I and class II molecules (reviewed in Brodsky et al. 1999 and Ploegh 1998). MHC class I interference by viruses is not unexpected, but the fact that so many viruses go to the trouble of interfering with MHC II antigen presentation (Brodsky et al. 1999) is further evidence that much of the evolution in this MHC class, too, is virus induced.

The arms race does not end there, however. Our immune system includes a special set of cells, known as natural killer (NK) cells, which deal with MHC interference. These cells do not attempt to recognize viral antigens themselves. Rather, they express cell-surface receptors, called killer Ig-like receptors (KIR), that bind to MHC class I. Unless an interrogated cell can satisfy a curious NK cell that is free from interference with antigen presentation, by binding its KIR receptors with MHC, it is commanded to undergo apoptosis, just to be on the safe side. Incidentally, because NK-cell receptors need to keep up with MHC gene evolution, they evolve extremely rapidly (Khakoo et al. 2000) and have undergone amplification and exist as a large cluster of related loci on chromosome 19 (Nolan et al. 2006). The innate immune system, therefore, is not restricted to static solutions like TLRs, but can also display rapid molecular and genomic change when called upon by viral selective pressure.

Not to be outdone, viruses have evolved mechanisms for avoiding the NK response. The HIV-1 Nef protein, for instance, while downregulating the expression of *HLA-B* and *HLA-A*, the class I loci most likely to be associated with a vigorous CTL response, leaves *HLA-C*, which is much less likely to stimulate a strong CTL response, to be expressed at high levels. In doing so, it hobbles the host's CTL response but leaves enough impotent MHC molecules on the cell's surface to inhibit NK-cell activity (Collins & Baltimore 1999). It is this sort of adaptation, whereby a virus with a genome only 10 kb in length so subtly thwarts all of the intrinsic, innate, and adaptive immunity weapons deployed against it, that justifies the use of the term masterpiece.

Table 1 Key host-virus interactions and their consequences

Host-virus conflict/ interaction	Consequences for virus (within and between hosts)	Consequences for host (individual and population)
Toll-like receptors v. PAMPs (innate immunity)	Potentially unavoidable recognition	Rapid antiviral response
	↓ replication	Clearance or modulation of infection
	↓ transmission	Purifying selection on TLRs
	selection to disrupt signaling downstream of TLR binding	↓ susceptibility to cross-species transmission
RNAi v. viral dsRNA	Potentially unavoidable recognition	Specific antiviral response
	↓ replication	Clearance or modulation of infection
	↓ transmission	Tool for controlling gene expression
	Selection to usurp host RNA silencing to virus's advantage	↓ susceptibility to cross-species transmission
APOBEC3G v. Vif (intrinsic immunity)	Hypermutation of viral genome	Automatic antiviral response without need for recognition
	↓ replication	Prevention, clearance, or modulation of infection
	↓ transmission	Positive selection on *APOBEC3G*
	Selection on *vif* to block APOBEC3G	↓ susceptibility to cross-species transmission
Antibodies v. viral epitopes (B cell–mediated adaptive immunity)	↓ replication	Highly specific antiviral response
	↓ transmission	Prevention, clearance, or modulation of infection
	↓ reinfection	Selection for ↑ diversity of MHC class II molecules (gene amplification, positive/balancing selection)
	Positive selection for epitope escape mutations within hosts (persistence) and between hosts (reinfection)	
	Selection for glycan shield and other antibody-avoidance tactics	
CTLs v. viral epitopes (T cell–mediated adaptive immunity)	↓ replication	Highly specific antiviral response
	↓ transmission	Prevention, clearance, or modulation of infection
	↓ reinfection	Selection for ↑ diversity of MHC class I molecules
	Positive selection for epitope escape mutations within and between hosts	Selection to respond to viral interference with antigen recognition (NK cells)
	Selection for interference with MHC-based antigen recognition	
Natural killer cells v. viral interference with antigen recognition (innate immunity)	↓ replication	Clearance or modulation of infection by eliminating cryptic infected cells
	↓ transmission	Selection for ↑ diversity of *KIR* genes
	Selection to deceptively upregulate expression of ineffective MHC genes	

Within-Host Adaptations Can Be Costly: Cytotoxic T Lymphocyte Escape Mutant Transmission, HIV-1 Attenuation, and the Glycan Shield

In principle it is possible that selective pressures acting on viruses within hosts could conflict with those acting at, or after, transmission from one host to the next. In the case of HIV-1's resistance to NAbs, such conflicting pressures might help explain why the virus persists within humans and might expose a chink in the armor of HIV-1 through which vaccination strategies may be directed.

Thus far, attempts to develop an effective HIV-1 vaccine have been spectacularly unsuccessful. Essentially, researchers are hoping to use a vaccine to induce an immune response more effective than the body's own natural response. Although the human immune system can raise NAb against the HIV Env glycoprotein (the only viral protein exposed to the external environment), the virus has proven extremely successful at avoiding it (Albert et al. 1990). It is not just the rapid rate of mutation that helps HIV escape NAb; many of these mutations lead to the development and/or rearrangement of N-glycosylated sites, special amino acids where host sugar molecules can attach to the protein. Under NAb-driven selection, N-glycosylated sites accumulate, leading to a glycan shield that obscures the viral Env protein behind a layer of host-derived sugars (Wei et al. 2003).

This glycan shield seems like an impenetrable barrier behind which HIV should be free to replicate and transmit without worry of an NAb response. A ray of hope comes from a study of discordantly infected heterosexual couples (where one member of a presently monogamous couple is infected with HIV). Derdeyn et al. (2004) noticed that the HIV variants transmitted between these partners were significantly more likely to be ones with diminutive sugar shields. They hypothesized that while the glycan shield may be advantageous for HIV within a chronically infected host, it is a hindrance during transmission to a naïve host with no HIV-1 antibodies and likely must be lost and then re-evolved within each newly infected host. A vaccine directed at raising NAb may be more effective during this transmission event, when the glycan shield is absent, and might therefore be able to block HIV transmission.

As promising as this may be, a second study looking at homosexual male discordant couples did not find the same pattern of loss of the glycan shield upon transmission (Frost et al. 2005). Note that these patients were infected with a strain of HIV-1 that is distantly related to the strain studied in the heterosexual couples. It is not clear if this, or the different mode of transmission, explains the different glycan shield dynamics observed in the two studies. Clearly, more research is needed on the selective forces that govern the glycan shield formation/loss and how this relates to HIV transmissibility and future vaccine design.

While antibody-neutralization-sensitive variants might be selectively transmitted between hosts, there is no evidence for such selective transmission of HIV-1 CTL escape mutants (Frater et al. 2006). However, even if they are transmitted at background frequency, escape mutants clearly do sometimes move from one host to the next. What is the consequence of this for the host and virus? That depends upon the HLA environment within that host. Escape mutants must sometimes arrive in new hosts with similar HLA to the host in which they evolved, putting the new host at an immediate disadvantage. But because of the standing HLA diversity in the human population, most often escape mutants simply go extinct because the new host has a different MHC environment (Goulder et al. 2001, Kiepiela et al. 2004, Phillips et al. 1991). There is little evidence that CTL responses within hosts are rapidly leading to a situation where most strains of HIV-1 are CTL escape mutants.

Ariën et al. (2005) found evidence that HIV might be attenuating over time: When competed against each other in head-to-head assays, strains from the 1980s exhibited significantly higher replicative fitness than strains circulating 20 years later. In other words, the later strains appeared to have lost some of their ability to infect new host cells, at least when competing directly against early strains in these tightly controlled experiments. Ariën et al. (2005, 2007) speculate that this drop in replicative fitness might be the result of the series of population bottlenecks experienced by HIV-1 each time it enters a new host, and might result in the evolution of milder strains of virus that take longer to progress to AIDS. The idea is that, despite the observation that replicative fitness actually increases throughout chronic infection within a host (Troyer et al. 2005), the genetic bottleneck that occurs when the virus is transmitted from one host to the next could obliterate and even reverse any fitness gains achieved within the first host. Moreover, if the genetic environment of the newly infected host is different, in particular at *HLA* loci, that will further penalize the newly transmitted virus: Its hard-won CTL escape mutations might no longer be beneficial because mutations that provide a cloak of invisibility in one host might have the opposite effect in the next host. With each such transmission between genetically mismatched hosts, the virus must evolve to escape the T cell responses of the new host; such evolution appears to come at the cost of replicative fitness (Martinez-Picado et al. 2006). It remains to be seen whether a decline in replicative fitness is a general feature of the unfolding HIV-1 pandemic and, if so, whether it has a real connection to intrinsic virulence, a property about which it is extremely difficult to make reliable predictions.

What Factors Prevent or Permit Viral Emergence?

Successful between-host transmission—involving hosts from different species—is one of the landmarks that every new emerging infectious disease agent must pass, and it pays to consider how it might, or might not, occur. The existence of intrinsic immunity genes that restrict viral host range may help explain why humans have acquired so few retroviruses from the many other infected primate species (Worobey 2007). Experimental evidence suggests that human APOBEC3G and its orthologs in Old World monkeys can restrict the ability of viruses to jump into new species. For example, human APOBEC3G blocks SIV from the African green monkey from infecting humans; the monkey homolog of APOBEC3G, in turn, blocks HIV. Surprisingly, the ability of these proteins to restrict viruses from other species is determined by a single amino acid substitution, which presumably prevents mismatched Vif protein from degrading APOBEC3G (Bogerd et al. 2004, Mangeat et al. 2004, Schrofelbauer et al. 2004). But for that single amino acid, there might be no human AIDS pandemic, at least as we know it.

Single amino acids can also make the difference on the viral side. Anishchenko et al. (2006) found that Venezuelan equine encephalitis virus, a mosquito-borne RNA virus that sometimes emerges out of its rodent reservoir and establishes outbreaks in horses and humans, requires just a single change in its envelope glycoprotein to do so. In this case, the main barrier to cross-species transmission might thus be ecological (rare opportunities for viruses to encounter abnormal host species) rather than evolutionary. More often, though, moving from one fitness peak to another (in a new host) might require multiple mutations, and such jumps are presumably difficult even for rapidly evolving viruses (Holmes 2006).

What is remarkable, and reassuring, is how effective our defenses must be: For every new pathogen that has become established in human populations, how many more must have died out after having infected just one or a handful of individuals? And for every hopeful cross-species transmission event where at least one human individual became productively infected, how many other exposures must have failed to even cause an initial infection, with host-restriction elements or innate immunity responses eliminating the pathogen before it gained a foothold in the host?

CONCLUSION

Clearly, a large proportion of the human genome is given over to genes that help shield us from pathogens. At some of the loci known to be directly involved in viral defense, we see allelic diversity that is unrivalled by other types of genes. We also see some of the most rapid molecular evolution and strong balancing and directional positive selection. Moreover, the genome itself may have experienced its most dynamic changes in response to viruses. Segmental duplications across the genome generally tend to be enriched for immunity genes (Gonzalez et al. 2005), and virus-relevant gene families such as MHC class I and II, *KIR*, and *APOBEC* are well represented in the top ranks of amplified loci. In the landscape of the human genome, the MHC is the richest biodiversity hot spot, and viruses are likely the main reason for this.

MHC/HLA genomic and allelic diversity represents one of the human population's most valuable genetic endowments, and the diversity of *HLA* alleles in the human population has no doubt buffered us from extinction. Nonrandom distributions of *HLA* alleles in different populations most likely echo local battles with pathogens that have come and gone. The overall diversity observed is all the more remarkable in light of the fact that periodic selective sweeps have probably been a feature of this history. The resulting MHC and other immunity-related genetic diversity presents both challenges and opportunities for therapies and immunization. We cannot, for example, expect vaccines designed to elicit beneficial T cell responses to work uniformly across patients because different individuals bring to the table different complements of antigen-recognition alleles. In the short run, this complicates matters; in the long run, we can look forward not only to drugs tailored to individual genomes, but perhaps also vaccines.

We believe the observations brought together in this review argue strongly in favor of viruses being the dominant agents shaping vertebrate and human defense and immunity, both present and past. Other ideas, such as that adaptive immunity might have evolved in vertebrates because of a need to recognize and manage beneficial gut microbial communities (McFall-Ngai 2007), seem at odds both with the virus-driven genetic and genomic patterns inherent in the MHC and other immunity genes. The unparalleled levels of polymorphism and deleterious genetic diversity associated with the virus-specialized class I *HLA* loci, plus the fact that individuals with defective antigen-presentation genes die rapidly and almost exclusively of overwhelming viral infection, suggest that viruses have been, and remain, more important selective agents than bacterial and eukaryotic pathogens.

Viruses have shaped our battle-worn genotypes and phenotypes and may be an impetus behind many characteristics that seem far removed from pathogens, from complex gene regulation, to diabetes and arthritis, to mate choice and sexual reproduction. They are among the most important forces in human evolution, and future insights into human-virus interactions will likely play a key role both in controlling their emergence and spread and in understanding our own genetic heritage.

DISCLOSURE STATEMENT

The authors are not aware of any biases that might be perceived as affecting the objectivity of this review.

ACKNOWLEDGMENTS

We thank Bernie Crespi for insightful editorial advice and David Robertson and Tom Gilbert for helpful comments on the manuscript.

LITERATURE CITED

Agrawal A, Eastman QM, Schatz DG. 1998. Transposition mediated by RAG1 and RAG2 and its implications for the evolution of the immune system. *Nature* 394:744–51

Akira S, Hemmi H. 2003. Recognition of pathogen-associated molecular patterns by TLR family. *Immunol. Lett.* 85:85–95

Akira S, Uematsu S, Takeuchi O. 2006. Pathogen recognition and innate immunity. *Cell* 124:783–801

Albert J, Abrahamsson B, Nagy K, Aurelius E, Gaines H, et al. 1990. Rapid development of isolate-specific neutralizing antibodies after primary HIV-1 infection and consequent emergence of virus variants which resist neutralization by autologous sera. *Aids* 4:107–12

Alexopoulou L, Holt AC, Medzhitov R, Flavell RA. 2001. Recognition of double-stranded RNA and activation of NF-κB by Toll-like receptor 3. *Nature* 413:732–38

Anishchenko M, Bowen RA, Paessler S, Austgen L, Greene IP, Weaver SC. 2006. Venezuelan encephalitis emergence mediated by a phylogenetically predicted viral mutation. *Proc. Natl. Acad. Sci. USA* 103:4994–99

Ariën KK, Troyer RM, Gali Y, Colebunders RL, Arts EJ, Vanham G. 2005. Replicative fitness of historical and recent HIV-1 isolates suggests HIV-1 attenuation over time. *Aids* 19:1555–64

Ariën KK, Vanham G, Arts EJ. 2007. Is HIV-1 evolving to a less virulent form in humans? *Nat. Rev. Microbiol.* 5:141–51

Bannert N, Kurth R. 2004. Retroelements and the human genome: new perspectives on an old relation. *Proc. Natl. Acad. Sci. USA* 101(Suppl. 2):14572–79

Belvin MP, Anderson KV. 1996. A conserved signaling pathway: the *Drosophila* toll-dorsal pathway. *Annu. Rev. Cell Dev. Biol.* 12:393–416

Bernstein E, Caudy AA, Hammond SM, Hannon GJ. 2001. Role for a bidentate ribonuclease in the initiation step of RNA interference. *Nature* 409:363–66

Beutler B. 2004. Innate immunity: an overview. *Mol. Immunol.* 40:845–59

Bieback K, Lien E, Klagge IM, Avota E, Schneider-Schaulies J, et al. 2002. Hemagglutinin protein of wild-type measles virus activates toll-like receptor 2 signaling. *J. Virol.* 76:8729–36

Bieniasz PD. 2004. Intrinsic immunity: a front-line defense against viral attack. *Nat. Immunol.* 5:1109–15

Bihl F, Frahm N, Di Giammarino L, Sidney J, John M, et al. 2006. Impact of HLA-B alleles, epitope binding affinity, functional avidity, and viral coinfection on the immunodominance of virus-specific CTL responses. *J. Immunol.* 176:4094–101

Bogerd HP, Doehle BP, Wiegand HL, Cullen BR. 2004. A single amino acid difference in the host APOBEC3G protein controls the primate species specificity of HIV type 1 virion infectivity factor. *Proc. Natl. Acad. Sci. USA* 101:3770–74

Bogerd HP, Wiegand HL, Doehle BP, Lueders KK, Cullen BR. 2006a. APOBEC3A and APOBEC3B are potent inhibitors of LTR-retrotransposon function in human cells. *Nucleic Acids Res.* 34:89–95

Bogerd HP, Wiegand HL, Hulme AE, Garcia-Perez JL, O'Shea KS, et al. 2006b. Cellular inhibitors of long interspersed element 1 and Alu retrotransposition. *Proc. Natl. Acad. Sci. USA* 103:8780–85

Boissinot S, Davis J, Entezam A, Petrov D, Furano AV. 2006. Fitness cost of LINE-1 (L1) activity in humans. *Proc. Natl. Acad. Sci. USA* 103:9590–94

Borman AM, Quillent C, Charneau P, Kean KM, Clavel F. 1995. A highly defective HIV-1 group O provirus: evidence for the role of local sequence determinants in G→A hypermutation during negative-strand viral DNA synthesis. *Virology* 208:601–9

Brodsky FM, Lem L, Solache A, Bennett EM. 1999. Human pathogen subversion of antigen presentation. *Immunol. Rev.* 168:199–215

Burnet FM. 1959. *The Clonal Selection Theory of Acquired Immunity*. London: Cambridge Univ. Press

Bush RM, Fitch WM, Bender CA, Cox NJ. 1999. Positive selection on the H3 hemagglutinin gene of human influenza virus A. *Mol. Biol. Evol.* 16:1457–65

Carrington M, Nelson GW, Martin MP, Kissner T, Vlahov D, et al. 1999. HLA and HIV-1: heterozygote advantage and B*35-Cw*04 disadvantage. *Science* 283:1748–52

Cerutti H, Casas-Mollano JA. 2006. On the origin and functions of RNA-mediated silencing: from protists to man. *Curr. Genet.* 50:81–99

Collins KL, Baltimore D. 1999. HIV's evasion of the cellular immune response. *Immunol. Rev.* 168:65–74

Compton T, Kurt-Jones EA, Boehme KW, Belko J, Latz E, et al. 2003. Human cytomegalovirus activates inflammatory cytokine responses via CD14 and Toll-like receptor 2. *J. Virol.* 77:4588–96

Conticello SG, Thomas CJ, Petersen-Mahrt SK, Neuberger MS. 2005. Evolution of the AID/APOBEC family of polynucleotide (deoxy)cytidine deaminases. *Mol. Biol. Evol.* 22:367–77

Deacon NJ, Tsykin A, Solomon A, Smith K, Ludford-Menting M, et al. 1995. Genomic structure of an attenuated quasi species of HIV-1 from a blood transfusion donor and recipients. *Science* 270:988–91

Derdeyn CA, Decker JM, Bibollet-Ruche F, Mokili JL, Muldoon M, et al. 2004. Envelope-constrained neutralization-sensitive HIV-1 after heterosexual transmission. *Science* 303:2019–22

Doherty PC, Zinkernagel RM. 1975. A biological role for the major histocompatibility antigens. *Lancet* 1:1406–9

Dolganiuc A, Oak S, Kodys K, Golenbock DT, Finberg RW, et al. 2004. Hepatitis C core and nonstructural 3 proteins trigger Toll-like receptor 2-mediated pathways and inflammatory activation. *Gastroenterology* 127:1513–24

Domingo E, Holland JJ. 1997. RNA virus mutations and fitness for survival. *Annu. Rev. Microbiol.* 51:151–78

Fire A, Xu S, Montgomery MK, Kostas SA, Driver SE, Mello CC. 1998. Potent and specific genetic interference by double-stranded RNA in *Caenorhabditis elegans*. *Nature* 391:806–11

Frank SA. 2002. *Immunology and Evolution of Infectious Disease.* Princeton: Princeton Univ. Press

Frater AJ, Edwards CT, McCarthy N, Fox J, Brown H, et al. 2006. Passive sexual transmission of human immunodeficiency virus type 1 variants and adaptation in new hosts. *J. Virol.* 80:7226–34

Frost SD, Liu Y, Pond SL, Chappey C, Wrin T, et al. 2005. Characterization of human immunodeficiency virus type 1 (HIV-1) envelope variation and neutralizing antibody responses during transmission of HIV-1 subtype B. *J. Virol.* 79:6523–27

Gabuzda DH, Lawrence K, Langhoff E, Terwilliger E, Dorfman T, et al. 1992. Role of vif in replication of human immunodeficiency virus type 1 in CD4+ T lymphocytes. *J. Virol.* 66:6489–95

Galvani AP, Slatkin M. 2003. Evaluating plague and smallpox as historical selective pressures for the *CCR5-Δ32* HIV-resistance allele. *Proc. Natl. Acad. Sci. USA* 100:15276–79

Garred P, Madsen HO, Petersen J, Marquart H, Hansen TM, et al. 1998. CC chemokine receptor 5 polymorphism in rheumatoid arthritis. *J. Rheumatol.* 25:1462–65

Gaudieri S, Rauch A, Park LP, Freitas E, Herrmann S, et al. 2006. Evidence of viral adaptation to HLA class I-restricted immune pressure in chronic hepatitis C virus infection. *J. Virol.* 80:11094–104

Glass WG, McDermott DH, Lim JK, Lekhong S, Yu SF, et al. 2006. CCR5 deficiency increases risk of symptomatic West Nile virus infection. *J. Exp. Med.* 203:35–40

Gonzalez E, Kulkarni H, Bolivar H, Mangano A, Sanchez R, et al. 2005. The influence of *CCL3L1* gene-containing segmental duplications on HIV-1/AIDS susceptibility. *Science* 307:1434–40

Goulder PJ, Brander C, Tang Y, Tremblay C, Colbert RA, et al. 2001. Evolution and transmission of stable CTL escape mutations in HIV infection. *Nature* 412:334–38

Goulding C, McManus R, Murphy A, MacDonald G, Barrett S, et al. 2005. The *CCR5-Δ32* mutation: impact on disease outcome in individuals with hepatitis C infection from a single source. *Gut* 54:1157–61

Grenfell BT, Pybus OG, Gog JR, Wood JL, Daly JM, et al. 2004. Unifying the epidemiological and evolutionary dynamics of pathogens. *Science* 303:327–32

Hammer MF. 1995. A recent common ancestry for human Y chromosomes. *Nature* 378:376–78

Hashimoto C, Hudson KL, Anderson KV. 1988. The Toll gene of *Drosophila*, required for dorsal-ventral embryonic polarity, appears to encode a transmembrane protein. *Cell* 52:269–79

Heeney JL, Dalgleish AG, Weiss RA. 2006. Origins of HIV and the evolution of resistance to AIDS. *Science* 313:462–66

Holmes EC. 2006. The evolution of viral emergence. *Proc. Natl. Acad. Sci. USA* 103:4803–4

Int. Hum. Genome Consort. 2001. Initial sequencing and analysis of the human genome. *Nature* 409:860–921

Janeway CA, Travers P, Walport M, Schlomchik MJ. 2005. *Immunobiology: The Immune System in Health and Disease.* New York: Garland Sci.

Jorgensen RA. 1995. Cosuppression, flower color patterns, and metastable gene expression states. *Science* 268:686–91

Kerkau T, Bacik I, Bennink JR, Yewdell JW, Hunig T, et al. 1997. The human immunodeficiency virus type 1 (HIV-1) Vpu protein interferes with an early step in the biosynthesis of major histocompatibility complex (MHC) class I molecules. *J. Exp. Med.* 185:1295–305

Khakoo SI, Rajalingam R, Shum BP, Weidenbach K, Flodin L, et al. 2000. Rapid evolution of NK cell receptor systems demonstrated by comparison of chimpanzees and humans. *Immunity* 12:687–98

Kiepiela P, Leslie AJ, Honeyborne I, Ramduth D, Thobakgale C, et al. 2004. Dominant influence of HLA-B in mediating the potential coevolution of HIV and HLA. *Nature* 432:769–75

Kiepiela P, Ngumbela K, Thobakgale C, Ramduth D, Honeyborne I, et al. 2007. CD8+ T cell responses to different HIV proteins have discordant associations with viral load. *Nat. Med.* 13:46–53

Kirchhoff F, Greenough TC, Brettler DB, Sullivan JL, Desrosiers RC. 1995. Brief report: absence of intact nef sequences in a long-term survivor with nonprogressive HIV-1 infection. *N. Engl. J. Med.* 332:228–32

Kurt-Jones EA, Popova L, Kwinn L, Haynes LM, Jones LP, et al. 2000. Pattern recognition receptors TLR4 and CD14 mediate response to respiratory syncytial virus. *Nat. Immunol.* 1:398–401

Lemaitre B, Nicolas E, Michaut L, Reichhart JM, Hoffmann JA. 1996. The dorsoventral regulatory gene cassette spatzle/Toll/cactus controls the potent antifungal response in *Drosophila* adults. *Cell* 86:973–83

Liu R, Paxton WA, Choe S, Ceradini D, Martin SR, et al. 1996. Homozygous defect in HIV-1 coreceptor accounts for resistance of some multiply-exposed individuals to HIV-1 infection. *Cell* 86:367–77

Lund J, Sato A, Akira S, Medzhitov R, Iwasaki A. 2003. Toll-like receptor 9-mediated recognition of Herpes simplex virus-2 by plasmacytoid dendritic cells. *J. Exp. Med.* 198:513–20

Mangeat B, Trono D. 2005. Lentiviral vectors and antiretroviral intrinsic immunity. *Hum. Gene Ther.* 16:913–20

Mangeat B, Turelli P, Caron G, Friedli M, Perrin L, Trono D. 2003. Broad antiretroviral defence by human APOBEC3G through lethal editing of nascent reverse transcripts. *Nature* 424:99–103

Mangeat B, Turelli P, Liao S, Trono D. 2004. A single amino acid determinant governs the species-specific sensitivity of APOBEC3G to Vif action. *J. Biol. Chem.* 279:14481–83

Martin K, Heinzlmann M, Borchers R, Mack M, Loeschke K, Folwaczny C. 2001. Δ32 mutation of the chemokine-receptor 5 gene in inflammatory bowel disease. *Clin. Immunol.* 98:18–22

Martinez-Picado J, Prado JG, Fry EE, Pfafferott K, Leslie A, et al. 2006. Fitness cost of escape mutations in p24 Gag in association with control of human immunodeficiency virus type 1. *J. Virol.* 80:3617–23

McFall-Ngai M. 2007. Adaptive immunity: care for the community. *Nature* 445:153

Meyer D, Thomson G. 2001. How selection shapes variation of the human major histocompatibility complex: a review. *Ann. Hum. Genet.* 65:1–26

MHC Seq. Consort. 1999. Complete sequence and gene map of a human major histocompatibility complex. The MHC sequencing consortium. *Nature* 401:921–23

Milinski M. 2006. The major histocompatibility complex, sexual selection, and mate choice. *Annu. Rev. Ecol. Evol. Syst.* 37:159–86

Muckenfuss H, Hamdorf M, Held U, Perkovic M, Lower J, et al. 2006. APOBEC3 proteins inhibit human LINE-1 retrotransposition. *J. Biol. Chem.* 281:22161–72

Muramatsu M, Kinoshita K, Fagarasan S, Yamada S, Shinkai Y, Honjo T. 2000. Class switch recombination and hypermutation require activation-induced cytidine deaminase (AID), a potential RNA editing enzyme. *Cell* 102:553–63

Nelson MI, Simonsen L, Viboud C, Miller MA, Taylor J, et al. 2006. Stochastic processes are key determinants of short-term evolution in influenza A virus. *PLoS Pathog.* 2:e125

Nolan D, Gaudieri S, Mallal S. 2006. Host genetics and viral infections: immunology taught by viruses, virology taught by the immune system. *Curr. Opin. Immunol.* 18:413–21

OhAinle M, Kerns JA, Malik HS, Emerman M. 2006. Adaptive evolution and antiviral activity of the conserved mammalian cytidine deaminase APOBEC3H. *J. Virol.* 80:3853–62

Phillips RE, Rowland-Jones S, Nixon DF, Gotch FM, Edwards JP, et al. 1991. Human immunodeficiency virus genetic variation that can escape cytotoxic T cell recognition. *Nature* 354:453–59

Piguet V, Trono D. 1999. The Nef protein of primate lentiviruses. *Rev. Med. Virol.* 9:111–20

Ploegh HL. 1998. Viral strategies of immune evasion. *Science* 280:248–53

Qin XF, An DS, Chen IS, Baltimore D. 2003. Inhibiting HIV-1 infection in human T cells by lentiviral-mediated delivery of small interfering RNA against CCR5. *Proc. Natl. Acad. Sci. USA* 100:183–88

Roach JC, Glusman G, Rowen L, Kaur A, Purcell MK, et al. 2005. The evolution of vertebrate Toll-like receptors. *Proc. Natl. Acad. Sci. USA* 102:9577–82

Robinson J, Waller MJ, Parham P, de Groot N, Bontrop R, et al. 2003. IMGT/HLA and IMGT/MHC: sequence databases for the study of the major histocompatibility complex. *Nucleic Acids Res.* 31:311–14

Sabeti PC, Walsh E, Schaffner SF, Varilly P, Fry B, et al. 2005. The case for selection at *CCR5-Δ32*. *PLoS Biol.* 3:e378

Saleem MA, Arkwright PD, Davies EG, Cant AJ, Veys PA. 2007. Clinical course of patients with major histocompatibility complex class II deficiency. *Arch. Dis. Child 2000* 83:356–59

Samson M, Libert F, Doranz BJ, Rucker J, Liesnard C, et al. 1996. Resistance to HIV-1 infection in Caucasian individuals bearing mutant alleles of the CCR-5 chemokine receptor gene. *Nature* 382:722–25

Sawyer SL, Emerman M, Malik HS. 2004. Ancient adaptive evolution of the primate antiviral DNA-editing enzyme APOBEC3G. *PLoS Biol.* 2:e275

Sawyer SL, Wu LI, Emerman M, Malik HS. 2005. Positive selection of primate TRIM5α identifies a critical species-specific retroviral restriction domain. *Proc. Natl. Acad Sci. USA* 102:2832–37

Schrofelbauer B, Chen D, Landau NR. 2004. A single amino acid of APOBEC3G controls its species-specific interaction with virion infectivity factor (Vif). *Proc. Natl. Acad. Sci. USA* 101:3927–32

Sheehy AM, Gaddis NC, Choi JD, Malim MH. 2002. Isolation of a human gene that inhibits HIV-1 infection and is suppressed by the viral Vif protein. *Nature* 418:646–50

Shiina T, Ota M, Shimizu S, Katsuyama Y, Hashimoto N, et al. 2006. Rapid evolution of major histocompatibility complex class I genes in primates generates new disease alleles in humans via hitchhiking diversity. *Genetics* 173:1555–70

Smith KD, Andersen-Nissen E, Hayashi F, Strobe K, Bergman MA, et al. 2003. Toll-like receptor 5 recognizes a conserved site on flagellin required for protofilament formation and bacterial motility. *Nat. Immunol.* 4:1247–53

Stephens JC, Reich DE, Goldstein DB, Shin HD, Smith MW, et al. 1998. Dating the origin of the *CCR5-Δ32* AIDS-resistance allele by the coalescence of haplotypes. *Am. J. Hum. Genet.* 62:1507–15

Stram Y, Kuzntzova L. 2006. Inhibition of viruses by RNA interference. *Virus Genes.* 32:299–306

Stremlau M, Owens CM, Perron MJ, Kiessling M, Autissier P, Sodroski J. 2004. The cytoplasmic body component TRIM5α restricts HIV-1 infection in Old World monkeys. *Nature* 427:848–53

Takeda K, Kaisho T, Akira S. 2003. Toll-like receptors. *Annu. Rev. Immunol.* 21:335–76

Thio CL, Astemborski J, Bashirova A, Mosbruger T, Greer S, et al. 2007. Genetic protection against hepatitis B virus conferred by *CCR5Δ32*: evidence that CCR5 contributes to viral persistence. *J. Virol.* 81:441–45

Thio CL, Thomas DL, Karacki P, Gao X, Marti D, et al. 2003. Comprehensive analysis of class I and class II HLA antigens and chronic hepatitis B virus infection. *J. Virol.* 77:12083–87

Troyer RM, Collins KR, Abraha A, Fraundorf E, Moore DM, et al. 2005. Changes in human immunodeficiency virus type 1 fitness and genetic diversity during disease progression. *J. Virol.* 79:9006–18

Turelli P, Mangeat B, Jost S, Vianin S, Trono D. 2004. Inhibition of hepatitis B virus replication by APOBEC3G. *Science* 303:1829

Turelli P, Trono D. 2005. Editing at the crossroad of innate and adaptive immunity. *Science* 307:1061–65

Vallender EJ, Lahn BT. 2004. Positive selection on the human genome. *Hum. Mol. Genet.* 13(Spec. No. 2):R245–54

van de Lagemaat LN, Landry JR, Mager DL, Medstrand P. 2003. Transposable elements in mammals promote regulatory variation and diversification of genes with specialized functions. *Trends Genet.* 19:530–36

Walsh PD, Biek R, Real LA. 2005. Wave-like spread of Ebola Zaire. *PLoS Biol.* 3:e371

Wei X, Decker JM, Wang S, Hui H, Kappes JC, et al. 2003. Antibody neutralization and escape by HIV-1. *Nature* 422:307–12

World Health Org. 2003. *World health report 2003: shaping the future.* Geneva: WHO

Worobey M, Holmes EC. 2001. Homologous recombination in GB virus C/hepatitis G virus. *Mol. Biol. Evol.* 18:254–61

Worobey M. 2007. The origins and diversification of human immunodeficiency virus. In *Global HIV/AIDS Medicine*, ed. MA Sande, PA Volberding, P Lange. In press. Philadelphia: Elsevier

Yee LJ. 2004. Host genetic determinants in hepatitis C virus infection. *Genes Immun.* 5:237–45

Yewdell JW, Bennink JR. 1999. Immunodominance in major histocompatibility complex class I-restricted T lymphocyte responses. *Annu. Rev. Immunol.* 17:51–88

Zhang H, Yang B, Pomerantz RJ, Zhang C, Arunachalam SC, Gao L. 2003. The cytidine deaminase CEM15 induces hypermutation in newly synthesized HIV-1 DNA. *Nature* 424:94–98

Zheng YH, Irwin D, Kurosu T, Tokunaga K, Sata T, Peterlin BM. 2004. Human APOBEC3F is another host factor that blocks human immunodeficiency virus type 1 replication. *J. Virol.* 78:6073–76

Zinkernagel RM, Doherty PC. 1974. Restriction of in vitro T cell-mediated cyto-
toxicity in lymphocytic choriomeningitis within a syngeneic or semiallogeneic
system. *Nature* 248:701–2

Zuniga R, Lucchetti A, Galvan P, Sanchez S, Sanchez C, et al. 2006. Relative dominance of Gag p24-specific cytotoxic T lymphocytes is associated with human
immunodeficiency virus control. *J. Virol.* 80:3122–25

The Evolution of Resistance and Tolerance to Herbivores

Juan Núñez-Farfán,[1] Juan Fornoni,[1] and Pedro Luis Valverde[2]

[1]Laboratorio de Genética Ecológica y Evolución, Departamento de Ecología Evolutiva, Instituto de Ecología, Universidad Nacional Autónoma de México, A.P. 70-275 Distrito Federal 04510, México; email: farfan@servidor.unam.mx

[2]Departamento de Biología, Universidad Autónoma Metropolitana-Iztapalapa, A.P. 55-535 Distrito Federal 09340, México

Annu. Rev. Ecol. Evol. Syst. 2007. 38:541–66

First published online as a Review in Advance on August 16, 2007

The *Annual Review of Ecology, Evolution, and Systematics* is online at http://ecolsys.annualreviews.org

This article's doi: 10.1146/annurev.ecolsys.38.091206.095822

Key Words

herbivory, adaptive evolution, EES, plant defenses, mixed defense strategies, costs and benefits of defense

Abstract

Tolerance and resistance are two different plant defense strategies against herbivores. Empirical evidence in natural populations reveals that individual plants allocate resources simultaneously to both strategies, thus plants exhibit a mixed pattern of defense. In this review we examine the conditions that promote the evolutionary stability of mixed defense strategies in the light of available empirical and theoretical evidence. Given that plant tolerance and resistance are heritable and subject to environmentally dependent selection and genetic constraints, the joint evolution of tolerance and resistance is analyzed, with consideration of multiple species interactions and the plant mating system. The existence of mixed defense strategies in plants makes it necessary to re-explore the coevolutionary process between plants and herbivores, which centered historically on resistance as the only defensive mechanism. In addition, we recognize briefly the potential use of plant tolerance for pest management. Finally, we highlight unresolved issues for future development in this field of evolutionary ecology.

INTRODUCTION

Resistance: constitutive or induced response of plants against herbivory to avoid or reduce the amount of damage

Tolerance: response of plants induced after consumption to buffer the negative fitness effect of damage

Alternative redundant strategies of defense: two or more strategies that confer similar fitness benefits against the same selective pressure

The evolution of plant defense against natural enemies, such as herbivores and pathogens, has been the motif of study of evolutionary biologists during the past three decades, and has retained a central role because the theory of coevolution emerged directly from the study of interactions between plants and herbivores (Ehrlich & Raven 1964). Specifically, studies aimed to understand the origin and maintenance of plant and animal adaptations that serve as mediators of the interactions (Rausher 1996). These studies encompass an ample range of foci, from cell and molecular responses to herbivore damage, to phylogenetic patterns of relationships (Agrawal & Fishbein 2006, Becerra 1997, de Meaux & Mitchell-Olds 2003).

After the recognition arose that many plant attributes function as defenses against herbivores and pathogens (Ehrlich & Raven 1964, Painter 1958), the understanding of trait evolution, and coevolutionary relationships, received considerable attention (Bergelson et al. 2001; Rausher 1996, 2001). Today, resistance and tolerance are well recognized as two components of plant defense against natural enemies, which involve different plant traits and genetic backgrounds, and with different effects on the fitness of both the plant and the enemy (Rausher 1996, 2001; Rosenthal & Kotanen 1994; Stowe et al. 2000; Strauss & Agrawal 1999). Both strategies are considered adaptive, and both are assumed to imply fitness costs to a plant genotype. Thus, all else being equal, the evolution and maintenance of each defensive strategy in plant populations should be affected by the relative fitness costs and benefits that each strategy involves.

Seven years ago, Stowe et al. (2000) reviewed the topic of the evolution of plant tolerance to herbivores and raised many relevant questions about the joint evolution of both tolerance and resistance. Despite the many remaining unanswered questions, there have been advances derived from empirical and theoretical work. We here highlight these advances in the study of the evolution of plant defense against herbivores (and other natural enemies). Particularly, we review the models for the joint evolution of tolerance and resistance, the assumptions and predictions of these models, and empirical work related to this issue. We propose lines of research in this field of evolutionary ecology.

A cornerstone of the theoretical and empirical development of this field was the proposal that both tolerance and resistance can function as alternative redundant strategies of defense (Simms & Triplett 1994). This hypothesis establishes that natural selection would not act to increase resistance if no fitness reduction is observed in the presence of herbivores because of tolerance. Thus, highly tolerant genotypes would not experience selection on resistance, and vice versa. The hypothesis predicts a negative genetic correlation between tolerance and resistance if they are redundant in terms of fitness and both imply similar costs. Fitness redundancy can occur when a similar reproductive output can be attained through either tolerance or resistance in the presence of herbivores (Valverde et al. 2003). Both the absence of redundancy and/or differences in the magnitude of the cost that tolerance and resistance involve may explain the absence of a negative genetic correlation between the two strategies of defense. Although two studies have demonstrated that costs of tolerance and resistance differ within populations (Fornoni et al. 2004b, Pilson 2000), we do not know much about the environmental conditions that determine the extent of redundancy between tolerance and resistance (Mauricio 2000, Valverde et al. 2003).

In contrast, an increasing accumulation of empirical evidence shows that plants allocate resources simultaneously to both tolerance and resistance. This pattern was gathered from the observation that in a few instances, individuals (genotypes) with relatively high levels of resistance have low levels of tolerance, and vice versa (Leimu & Koricheva 2006a). Thus, the theoretical expectation of a negative genetic correlation between tolerance and resistance seems to have rather low empirical support (Fineblum & Rausher 1995, Fornoni et al. 2003a, Pilson 2000, Stowe 1998) and, at the same time, suggests that a pattern of simultaneous allocation to both tolerance and resistance should be common in natural populations.

In this review we define a mixed pattern of defense allocation of an individual plant (genotype) as the simultaneous allocation of resources to component traits of resistance and tolerance to herbivory.

THE SIMULTANEOUS EVOLUTION OF RESISTANCE AND TOLERANCE

Theoretical studies predicted that a mixed pattern of defense allocation may or may not constitute an evolutionarily stable phenotype within populations (Abrahamson & Weis 1997, Fineblum & Rausher 1995, Fornoni et al. 2004a, Mauricio et al. 1997, Restiff & Koella 2004, Roy & Kirchner 2000, Tiffin 2000a). The hypotheses provided by these models can be grouped according to whether they propose that a mixed pattern of allocation to tolerance and resistance does or does not constitute an evolutionary stable equilibrium. Below, each hypothesis is explained and discussed in the context of the available evidence.

Mixed Patterns of Defense Allocation as Evolutionary Stable Strategy

Intermediate levels of tolerance and resistance are favored by natural selection.
A mixed pattern of defense allocation could be selected if maximum fitness is attained at intermediate levels of both strategies, or if a combination of both mechanisms of defense pays higher fitness benefits than either of the strategies alone. This hypothesis predicts that stabilizing selection or positive correlational selection acts on tolerance and resistance to herbivory. Theoretical analyses indicate stabilizing selection may be expected when the costs or benefits of tolerance and resistance are nonlinear functions of allocation to defense (Fornoni et al. 2004a, Tiffin & Rausher 1999) or the relative magnitude of the costs of tolerance and resistance differ within populations (Fornoni et al. 2004a, Tiffin 2000a). In addition, correlational selection is expected when a combination of both tolerance and resistance pays higher fitness benefits than either strategy alone. That is, correlational selection is expected when the benefits of allocating resources simultaneously to tolerance and resistance are more than additive (Fornoni et al. 2004a).

Several reviews demonstrated the generalized existence of costs of resistance characters (Bergelson & Purrington 1996, Bergelson et al. 2001, Purrington 2000, Strauss

Table 1 Review of papers that estimated direct allocational costs of tolerance to herbivory[a]

Plant species	Source of damage	Magnitude of cost	Fitness measure	F/G	References
Brassica rapa	Phyllotrera cruciferae (insect folivory)	−0.87	Total seeds	F	Pilson 2000
	Ceutorhynchus assimilis (insect folivory)	−0.34			
Asclepias syriaca	Artificial defoliation (50%)		Total biomass	G	Hochwender et al. 2000
	Low soil nutrient	−0.69			
	High soil nutrient	−0.43			
Ipomoea purpurea	Insect folivory	−0.325	Total seeds	F	Tiffin & Rausher 1999
	Insect apical meristem damage	−0.189			
Ipomoea hederacea	Deer herbivory		Total seeds	F	Stinchcombe 2002a
	Control	−0.457			
	Fungicide	−0.743			
	Insecticide	−0.156			
	Dual-spray	−0.086			
Datura stramonium	Artificial defoliation (>30%)	0.77	Total seeds	G	Fornoni & Núñez-Farfán 2000
	Tolerance to insect folivory (Pop 1)	−0.49	Total seeds	F	Fornoni et al. 2004b
	Tolerance to insect folivory (Pop 2)	−0.11			
Arabidopsis thaliana	Tolerance to insect folivory	−0.46	Total fruits	F	Mauricio et al. 1997
	Tolerance to rabbit apical meristem damage	−0.129	Total fruits	F	Weinig et al. 2003a
Arabis perennans	Insect folivory (Plutella xylostella)		Growth rate	F	Siemens et al. 2003
	Low competition	−0.447			
	High competition	0.132			

[a]The magnitude of costs of tolerance corresponds to the product-moment correlation coefficient between fitness in the absence of damage [or an estimation following Mauricio et al. (1997)] and tolerance. Estimates significantly different from zero are indicated in bold. F: field experiment; G: greenhouse experiment.

et al. 2002), and the major role of the environment on the expression and magnitude of costs (Koricheva 2002, Marak et al. 2003, Osier & Lindroth 2006). In addition, although fewer studies examined the presence of fitness costs of tolerance (see references in **Table 1**), available data support the expectation that tolerance to herbivory is costly (Mauricio et al. 1997, Pilson 2000, Tiffin & Rausher 1999). Almost all plant species surveyed so far expressed significant fitness costs of tolerance to herbivory (**Table 1**). However, few studies have demonstrated that the environment affects the magnitude of the costs of tolerance and resistance (Fornoni et al. 2004b, Hochwender et al. 2000, Pilson 2000, Siemens et al. 2003).

Available evidence indicates variable results regarding the shape of the cost and benefit functions for tolerance and resistance. Some studies detected a nonlinear cost or benefit function for tolerance or resistance (e.g., Bergelson et al. 2001, Fornoni et al. 2004b, Mauricio et al. 1997, Pilson 2000), whereas others failed to detect a nonlinear component (e.g., Stinchcombe 2002a, Tiffin & Rausher 1999). Theoretical studies indicate that a nonlinear cost function may arise when the balance relationship between limiting resources for two traits (tolerance and resistance) changes across the environment (see Yoshida 2006 for a theoretical analysis of the shape of the trade-off functions). A nonlinear benefit function is expected when the detrimental effects of herbivory are eliminated completely (Simms & Rausher 1987). Hence a unit of increase in allocation to resistance or tolerance will not increase the net benefits. To date we are aware of no study that has manipulated the environment explicitly to explore its consequences on the shape of the cost or benefit function of a defensive trait and the resulting selection pattern. Provided the shape of the cost and benefit function can have strong effects on the expected pattern of selection and the evolution of plant defense (Fornoni et al. 2004a, Restiff & Koella 2004), future studies should explore how the environment conditions the occurrence of a nonlinear cost or benefit function (Skogsmyr & Fagerström 1992).

ESS: evolutionary stable strategy

Studies that estimated natural selection acting on tolerance and resistance to herbivory simultaneously did not show evidence of stabilizing selection (e.g., Fornoni et al. 2004b, Mauricio et al. 1997, Pilson 2000, Tiffin & Rausher 1999, Weinig et al. 2003a). Instead, studies showed directional selection to increase or to reduce both or either strategy (Fornoni et al. 2004b, Mauricio et al. 1997, Pilson 2000, Tiffin & Rausher 1999). In particular, two studies found correlational selection indicating that a combination of resistance and tolerance was favored (Pilson 2000, Tiffin & Rausher 1999). Although previous studies support the theoretical assumption that fitness benefits of both strategies can interact (Fornoni et al. 2004a), the causes of the epistatic interaction in terms of fitness between tolerance and resistance remains unexplained. Finally, although the assumptions and predictions of theory were confirmed partially in some instances, more empirical work is needed to reject the hypothesis that a mixed defensive strategy constitutes an evolutionary stable strategy (ESS) maintained by natural selection.

Frequency-dependent selection maintains a mixed pattern of defense allocation. Only a few studies have examined the role of frequency-dependent selection on plant defenses against herbivory (i.e., Berenbaum & Zangerl 1998, Roy 1998, Siemens & Roy 2005). Traditionally, more attention is given to the analysis of frequency-dependent selection on polymorphic traits (Krebs & Davis 1981, but see Castillo et al. 2002). The results of theoretical analyses suggest that the dynamics of stable cycling commonly found in simple Mendelian traits are less common in polygenic traits (Seger 1983). Most attempts to detect frequency-dependent selection were conducted on discrete host resistance polymorphisms, usually found among resistance components of plant-pathogen interactions (Dybdahl & Lively 1998, Parker 1989; however, also see Dirzo & Harper 1982).

No study has manipulated the frequency of resistant/tolerant host genotypes explicitly to examine if the evolution of plant defensive strategies is subject to frequency-dependent selection. However, some studies tested some of the predictions of the frequency-dependence selection hypothesis (Berenbaum & Zangerl 1998, Roy 1998).

QTL: quantitative trait loci

The study by Berenbaum and Zangerl (1998) found that in *Pastinaca sativa*, the range of variation in the level of a resistance component (coumarin) against its consumer *Depressaria pastinacella* did not change from 1873 to present, suggesting a cyclical dynamic promoted by frequency-dependent selection. In a reciprocal transplant experiment Roy (1998) did not find evidence that common hosts are eaten more and have lower fitness than rare hosts.

In a recent paper, Siemens and Roy (2005) evaluated the following: (*a*) if the amount of damage experienced by the most common host genotype was higher than that expected by its frequency, and the amount of damage experienced by the most rare host genotypes was lower than that expected by its frequency (assortative damage hypothesis); (*b*) if the most common host genotype had higher levels of damage and lower average fitness compared with the more rare host genotypes; and (*c*) if the amount of damage correlated positively with the frequency of the most common host genotype. This study implemented molecular marker techniques to determine the frequency of genotypes for quantitative traits within populations (Siemens & Roy 2005). Unfortunately, their results provide little support for the hypothesis of frequency-dependent selection acting on plant defenses against herbivory. However, the increasing use of quantitative trait loci (QTL) in the past few years offers new possibilities to examine if frequency-dependent selection acts on quantitative traits. QTLs of major effects could be used to explore the relationship between the frequency of genotypes and fitness. Although two studies have explored the potential for frequency-dependent selection on plant resistance, no study has yet examined this possibility on plant tolerance. Before frequency-dependent selection can be put aside as an explanation of the maintenance of variation on tolerance and resistance, explicit experimental tests of the hypothesis are required. If tolerance, resistance, or both are subject to frequency-dependent selection and both are redundant in terms of fitness, a mixed pattern of allocation to tolerance and resistance by a given plant genotype may be evolutionarily stable.

Mixed Patterns of Defense Allocation as Non-Evolutionary Stable Strategy

To date there is substantial evidence to propose that plant species have the potential to express simultaneously the machinery to resist and tolerate damage by natural enemies (Rausher 2001). Whether one or both alternatives are activated by the same plant or population has been the subject of recent theoretical debate (Fornoni et al. 2004a, Mauricio et al. 1997, Restiff & Koella 2004, Tiffin 2000a). In a previous model Mauricio et al. (1997) proposed that if tolerance and resistance function as redundant alternative strategies, the initial condition (founder effect) will determine which one evolves. Also, population differences in the cost/benefit ratio for tolerance and resistance can vary, promoting the evolution of either tolerance or resistance but not both (Fornoni et al. 2004a). Hence an evolutionarily unstable pattern of simultaneous allocation to tolerance and resistance may result in a selection mosaic across populations derived from the interplay among migration, natural selection, and history (Thompson 2005). However, there is little evidence for this hypothesis

(Fornoni et al. 2004b, Juenger et al. 2000) and future efforts are needed to search for geographic variation of tolerance and resistance to herbivory.

However, a mixed pattern of defense allocation may be evolutionarily unstable if the selection pattern is not consistent with the observed response to selection. In other words, a mixed pattern of defense allocation may be evolutionarily unstable when genetic or environmental constraints reduce the possibility of predicting the evolutionary trajectory of plant defenses within populations. This unpredictability in the response to selection may be caused if (*a*) tolerance and resistance are also selected by factors other than herbivory, (*b*) genetic correlations between tolerance/resistance and other traits constrain their evolution, (*c*) fluctuating selection on tolerance and resistance maintains their variation, or (*d*) the adaptive value of tolerance and resistance fluctuates temporally. In particular, when variation in the biotic environment promotes fluctuating responses to selection on plant defenses, diffuse evolution is expected (Strauss et al. 2005). Below we describe four scenarios that account for the presence of a non-ESS mixed pattern of defense allocation.

Tolerance and resistance are also selected by factors other than herbivory. Although component traits of both resistance and tolerance are hypothesized to be the target of selection by agents other than herbivory (Aarssen 1995, Fornoni et al. 2003a, Stowe et al. 2000), to our knowledge there have been few attempts to address simultaneously how selection on a defensive trait changes in the presence/absence of herbivory and other putative selection pressures (however, see Løe et al. 2007). Because putative component traits of tolerance are associated directly or indirectly with plant growth (see Rosenthal & Kotanen 1994, Stowe et al. 2000, Strauss & Agrawal 1999, Tiffin 2000b), it is reasonable to suspect that traits expressed by plants to tolerate herbivory damage may also serve other functions and hence be selected by agents other than herbivory. Thus, any selective agent acting on growth could select potentially to increase or decrease tolerance. Moreover, if characters that confer tolerance also serve other functions, their associated costs may be reduced. Although simulating leaf damage can help to understand the response of plants after damage (Fornoni & Núñez-Farfán 2000, Pilson & Decker 2002), there are only two studies that found a positive genetic correlation between a putative tolerance trait and the operational estimation of tolerance (slope of the relationship between fitness and damage for each individual plant genotype) (Juenger & Bergelson 2000, Weinig et al. 2003a). In both cases higher tolerance was related to an increase in branch production after damage and early flowering, respectively. This evidence suggests that component traits of tolerance to herbivory may have other important ecological functions. Hence, whether the variation on tolerance is subject to selection imposed by agents other than herbivory remains an untested hypothesis. Partial support for this possibility can be found among studies that manipulate the presence of damage and examine changes in the pattern and intensity of selection on component traits of defense (Berenbaum et al. 1986, Juenger & Bergelson 2000, Mauricio 2000, Mauricio & Rausher 1997, Shonle & Bergelson 2000).

Following this experimental approach it is interesting to consider three scenarios, among others, as follows: (1) Selection gradient on a defensive trait is favored in the presence and absence of damage but the intensity of selection changes when the abundance of herbivores is manipulated. Under this condition other factors besides herbivory may be acting as selection agents on plant defense (Juenger & Bergelson 2000). (2) Selection gradient on a defensive trait is favored equally in the presence and absence of damage. This condition strongly suggests that although the putative defensive trait is equally adaptive, it is unlikely that herbivory is the main selection agent (Juenger & Bergelson 2000, Juenger et al. 2005, Mauricio 2000). (3) Selection gradient on a defensive trait is favored in the presence of damage, but not selected for or against in the absence of damage. This condition satisfies the expectation that herbivores are responsible for the evolution of the defensive trait (Berenbaum et al. 1986, Juenger & Bergelson 2000, Mauricio & Rausher 1997). (4) Selection against a defensive trait is observed in the presence and absence of herbivores. This condition suggests that other factors obscure the possible effect of herbivory, or levels of damage between treatments were insufficient to express the benefits of defense, thus only the costs of defense are expressed (Mauricio & Rausher 1997, Shonle & Bergelson 2000). Accordingly, conditions 1 and 2 support the hypothesis that other factors besides herbivory also impose selection on tolerance and resistance to herbivory. Empirical evidence indicates that selection on resistance traits expresses the pattern outlined under conditions 2, 3, and 4, whereas selection on tolerance fits conditions 2 and 3. It is important that all these studies usually are set up to detect an effect of herbivory and may not render completely the selection environment experienced by plants under natural conditions.

It is particularly noteworthy that the same operational definition of tolerance may be used to estimate tolerance to different stressful factors simultaneously (Fornoni et al. 2003b, Simms 2000). In turn, this experimental approach would allow analysis of the correlation among tolerances to different environmental conditions. A positive correlation on tolerances to different environmental stresses would suggest tolerance to herbivory reflects the activation of similar genes and the expression of a general mechanism of plants under stressful conditions. Evidence favoring this hypothesis would indicate that correlated selection on tolerance traits may constitute a general mechanism to maintain tolerance to herbivory at intermediate levels (Weinig et al. 2003a). However, the only study that addressed this issue found no evidence that tolerance to herbivory could also be selected in a competitive environment (Tiffin 2002, but see Jones et al. 2006).

A possible profitable approach to test the possible dual function of plant tolerance and resistance is the simultaneous manipulation of putative selective agents other than herbivory that may also act directly on defensive traits. If the expression of additive variance and/or the selection differential (i.e., the response to selection) of a defensive trait changes after the manipulation of selective agents other than herbivory, the evolution of defensive traits may also be determined by other factors. If other selective forces constrain the evolution of maximal resistance or tolerance, a mixed pattern of allocation to both strategies would be evolutionarily unstable.

Genetic correlations constrain the evolution of tolerance and resistance. Because plants are usually eaten by several species of natural enemies, trade-offs between resistance to different species of enemies may maintain overall resistance at intermediate levels. This hypothesis assumes that deterring one species of herbivore would reduce the amount of resources to defend against other species. A recent meta-analysis indicated that resistance to multiple natural enemies usually tends to be positively correlated (Leimu & Koricheva 2006b). The studies that estimated tolerance to several natural enemies simultaneously found that tolerance to different natural enemies was independent or positively correlated (Pilson 2000, Tiffin & Rausher 1999). Taken together, this evidence suggests that general mechanisms of defense rather than species-specific responses are likely to be favored because they lower the cost of defense. Thus, the simultaneous evolution of defense to multiple enemies may not account for the presence of intermediate levels of tolerance and resistance as a result of genetic constraints.

Several studies found that defensive traits (particularly resistance traits) are correlated with other fitness-enhancing traits, which represents a constraint to the evolution of increasing levels of resistance (Mutikainen et al. 2002, Purrington 2000, Rausher 1996, Strauss et al. 2002). This cost of allocating resources to defense may be expressed as a direct reduction in fitness in an herbivore-free environment (Rausher 1996, Strauss et al. 2002), or as an indirect effect of altering the phenotypic expression of other traits that also increase fitness (Adler 2000, Domínguez & Dirzo 1994, Strauss et al. 1999). Thus, a negative correlation between defense and other ecologically important traits may result in a potential cost of defense (ecological cost). The prevalence of fitness costs of defensive traits resulting from a pleiotropic relationship with other plant characters indicates that a correlated response to selection could constrain the evolution of plant defense. Although several studies reported the presence of ecological costs (Mutikainen et al. 2002, Orians et al. 2003, Pilson 2000, Puustinen et al. 2004, Rausher 1996, Stowe 1998, Stowe et al. 2000, Strauss et al. 1999), relatively few studies complemented these results with estimations of natural selection on correlated traits (Nuñez-Farfán & Dirzo 1994, Shonle & Bergelson 2000, Weinig et al. 2003a). Even though the occurrence of genetic correlations among traits is highly supported by theory and data, relatively few studies demonstrated that the pattern of selection on traits correlated with defense can constrain the evolution of tolerance or resistance effectively. To date the paucity of evidence of genetic correlations between tolerance and resistance to herbivory and other ecologically important traits (see below) suggests that it is not possible that ecological costs account for the presence of intermediate levels of tolerance (however, see Juenger & Bergelson 2000).

Fluctuating selection constrains the evolution of tolerance and resistance. Several studies demonstrated that changes in the physical and biotic environment can change the phenotypic (genotypic) expression of plant defensive traits (Cronin & Abrahamson 1999, Fritz & Simms 1992, Herms & Mattson 1992, Hutha et al. 2000, Rosenthal & Kotanen 1994, Strauss & Agrawal 1999, Stowe et al. 2000, Strauss & Irwin 2004, Strauss et al. 2005, Valverde et al. 2001). However, comparatively little empirical data exist about the evolutionary consequences of this type of genotype × environment interaction (Fritz & Simms 1992, Strauss et al. 2005, Thompson 2005). In particular, in the context of plant-herbivore interaction, the effort has been to examine theoretically and empirically the effect of the biotic environment on the expression of the genetic variance-covariances and/or the selective value of tolerance or resistance (Hougen-Eitzman & Rausher 1994; Iwao & Rausher 1997; Juenger & Bergelson 1998; Pilson 1996; Stinchcombe & Rausher 2001, 2002; Tiffin 2002). Among five studies that manipulated the presence/absence of herbivore species and estimated natural selection in a factorial design, four of them detected significant evidence of diffuse selection acting on plant defenses (Juenger & Bergelson 1998; Pilson 1996; Stinchcombe & Rausher 2001, 2002). Most studies found that the intensity of selection on plant resistance or tolerance to one of the natural herbivores changed in the presence/absence of the other herbivore species (Juenger & Bergelson 1998; Stinchcombe & Rausher 2001, 2002). Hence, fluctuation in the presence/absence of herbivore species would not change the direction of the response to selection. However, when fluctuating selection results in contrasting selection, the response to selection could not be predicted. The study by Pilson (1996) found contrasting selection (positive versus negative directional selection) on flea beetle damage when manipulating the presence/absence of a diamondback moth. Thus, although the presence of diffuse selection is indicative of an ecological interaction between natural enemies affecting plant fitness, only when the pattern of selection expresses an opposite trend would changes in the biotic environment lead to diffuse evolution. A similar outcome can be expected if the presence/absence of a particular species of herbivore alters the expression of the additive genetic variance or covariance in plant defenses (Strauss et al. 2005). However, none of the previous studies detected a significant effect on the genetic variance-covariance of defensive traits when manipulating the presence/absence of particular natural enemies.

The simultaneous expression of tolerance and resistance throughout ontogeny was recently examined (Boege et al. 2007). Plant genotypes of *Brassica rapa* that were highly resistant at the juvenile stage had low levels of tolerance at the adult stage, and vice versa. This finding was complemented with a theoretical analysis suggesting that the observed trade-off may result from changes in the magnitude of costs and benefits of plant defenses during plant development (Boege et al. 2007). To our knowledge, this study is the first attempt to examine possible ontogenetic constraints on the expression of tolerance and resistance against herbivory. If the environment affects the ontogenetic trajectories of tolerance and resistance through variation in the costs of each strategy, variation in the phenotypic expression of each strategy would be expected.

In general, studies that address the occurrence of diffuse selection (evolution) focus only on the presence/absence of different species of natural enemies from a pairwise perspective (see references in Strauss et al. 2005). However, other aspects of the biotic environment may also alter the host plant's response to selection by a particular herbivore species. In a recent paper Bennett et al. (2006) synthesized the existing evidence and discussed how the mutualistic interactions between plants and mycorrhizal fungi can alter the expression of plant tolerance and resistance. Also, the presence/absence of invasive plant and herbivore species can affect the selective value of tolerance and resistance of native species against natural enemies (J. Lau, personal communication).

The interaction between plants and the natural enemies of herbivores (parasites or predators) can also affect the adaptive value of plant defenses. For instance, if plant resistance against herbivory negatively affects herbivore enemies, the presence of the latter would reduce the fitness benefits of resistance for the host plant (Abrahamson & Weis 1997, Gassman & Hare 2005). In turn, the presence of the natural enemies of insect herbivores may also reduce the amount of tissue removed from the host plants, lowering the selective pressure imposed by herbivores on their hosts. In this sense, if plant tolerance does not impose selection on herbivores, allowing an increase of their population size (Garrido-Espinosa & Fornoni 2006), the presence of parasites or predators may help the host plant by reducing the intensity of herbivory. Although the evolutionary ecology of tritrophic interactions is a novel area development (Hare 2002), we are still far from understanding their implications for the maintenance of intermediate levels of tolerance and resistance. It is expected that spatio-temporal fluctuations in the presence and abundance of the third trophic level would alter the expression and adaptive value of different strategies of plant defense.

Similarly, the interaction between plants and pollinators (through the extent of selfing suffered by plants) may also alter the adaptive value of plant defenses when mating systems covary with plant defenses (see below). To our knowledge, we are just starting to explore the role of multispecific interactions in the simultaneous evolution of tolerance and resistance. It is important that although diffuse selection suggests that fluctuating selection may occur, it does not constitute a definite test of temporal fluctuation in selection. Thus, temporal (across generation) exploration of the selective value of tolerance and resistance is desirable to conclude if fluctuating selection accounts for the presence of mixed patterns of plant defense allocation.

Local adaptation of natural enemies and the evolution of tolerance and resistance. Coevolutionary theory predicts that both plants and herbivores will be involved in an arms race resulting from a continuous reciprocal adaptive dynamic (Berenbaum & Zangerl 1998, Futuyma 1983, Rausher 1988). However, the study of the simultaneous evolution of tolerance and resistance has not yet considered the possibility that herbivores can coevolve with their host plants (however, see Restiff & Koella 2003). One of the possible outcomes of the coevolutionary interaction between plants and herbivores is that host resistance can be bypassed by the evolution of counter-resistance traits by consumers (Futuyma 1998, Thompson 2005). The selection pressure imposed by host plant resistance on their herbivores can promote local adaptation within populations and differentiation among populations of phytophagous insects (Mopper & Strauss 1998, Van Zandt & Mopper 1998). Hence, the adaptive value of plant resistance would depend not only on the costs but also on the possible reduction in their benefits driven by insect adaptation (Jokela et al. 2000, Simms & Rausher 1987).

As herbivores adapt to their host plant, the effectiveness of resistance declines whereas the extent of damage on plants is expected to increase through an increment in the per capita rate of consumption or in the population size of the herbivore (Jokela et al. 2000). The response of plants against rapidly evolving natural enemies is not well documented and little evidence supports the expectation of reciprocal local adaptation between plants and herbivores (Berenbaum & Zangerl 1998). The evidence indicates that insect herbivores more often adapt to their host plant than vice versa (Mopper & Strauss 1998). This pattern is also supported by several studies that address how fast insect herbivores adapt to insecticides and plant resistance (Gould 1998). If insect herbivores become locally adapted to their host plants more rapidly than their hosts can adapt, the adaptive value of host resistance will be strongly conditioned by the extent of local adaptation of the natural enemies. A theoretical study suggests that when the effectiveness of host resistance is reduced, host tolerance represents the only profitable strategy to cope with an increasing amount of damage (Jokela et al. 2000). As herbivores become locally adapted to the level of resistance of their host population, the adaptive value of resistance would be expected to decrease, even though the importance of tolerance may increase. If tolerance does not impose selection on the herbivores, promoting an increase in the amount of damage (Garrido-Espinosa & Fornoni 2006), and resistance can eventually become ineffective when herbivores become locally adapted, neither tolerance nor resistance alone would be evolutionarily stable.

The hypothesis presented above leads to two predictions that can be examined among populations: (*a*) A positive correlation across populations is expected between the adaptive value of tolerance and the extent of local adaptation of the herbivore populations to their corresponding host population; (*b*) a negative correlation is expected between the effectiveness of host resistance and the extent of local adaptation of the herbivore population. To date, no study has examined the possible relationship across populations between the extent of local adaptation of insect herbivores and the adaptive value of tolerance and resistance. In turn, the presence of intermediate levels of tolerance and resistance would represent evolutionarily unstable states in the temporal fluctuating dynamic of both strategies.

Interaction between the Mating System and Plant Defenses

Plants can produce offspring by mating with themselves and relatives or interchanging gametes with other, genetically unrelated individuals. The mating system adopted by individuals and populations has important evolutionary consequences for individual selection, genetic diversity, and population structure (Hamrick & Godt 1996). Ample evidence demonstrates that plant progenies produced through outcrossing differ in morphological, physiological, and fitness-related traits from plant progenies produced through selfing owing to inbreeding depression (Charlesworth & Charlesworth 1987, Husband & Schemske 1996). The genetic causes behind inbreeding depression are the loss of the advantage of overdominace in fitness loci, the expression of deleterious alleles in the homozygous condition, or both (Falconer & Mackay 1996). In turn, this negative effect of inbreeding on fitness can also be the result of a reduction in the defensive response of the plant against natural enemies (Núñez-Farfán et al. 1996).

At the same time, theoretical and empirical evidence shows that herbivores and other natural enemies can affect the mating system of plants (Agrawal & Lively 2001, Elle & Hare 2002, Lively & Howard 1994, Steets et al. 2006). The consumption of plant tissue by herbivores can impose a cost by reducing the amount of resources for reproduction, which may alter the mating system through changes in the proportion of self and cross-progeny produced (Strauss et al. 1996). The mating system could be altered by a reduction or changes in vigor, reproductive allocation to flowers available for outcrossing (Steets & Ashman 2004, Steets et al. 2006), floral display size (Elle & Hare 2002, Strauss et al. 1996), attractiveness to pollinators (i.e., rewards, flower size; Ivey & Carr 2005, Steets & Ashman 2004, Strauss et al. 1996), pollen production and exportation (e.g., Knight et al. 2005, Strauss et al. 1996), and alterations in the pollinator composition, abundance, and behavior (e.g., Steets & Ashman 2004, Steets et al. 2006, Strauss 1997). Finally, the consumption of flowers by herbivores (florivory) affects both pollen exportation and seed formation, and may have consequences for the mating system (see McCall & Irwin 2006 for a review). Thus, a consensus exists regarding the reciprocal interaction between the plant mating system and defensive mechanism against herbivory. A detailed examination of the effect of the mating system on plant tolerance and resistance is presented below.

The effect of the mating system on defense in plants. Recently, a growing interest has been given to the consequences of the mating system adopted by plants on defense against herbivores (Armbruster & Reed 2005, Carr & Eubanks 2002, Núñez-Farfán et al. 1996, Strauss & Karban 1994) and other natural enemies (Carr et al. 2003, Levri & Real 1998, Ouborg et al. 2000, Stephenson et al. 2004). Given that the mating system affects genetic diversity throughout the genome, it is expected that inbreeding would reduce variability at the loci controlling the expression of defenses against natural enemies.

To what extent does inbreeding affect tolerance and resistance to herbivores differentially? Available evidence from QTL mapping indicates that one to nine loci are potentially related to resistance against insect herbivory (**Figure 1**). We found 52 studies in our search, and obtained data for resistance to 25 herbivore species. Half of these studies found fewer than four QTL for resistance against herbivores (**Figure 1**). Eleven out of 12 studied species correspond to resistance QTL analysis on cultivated hosts. In contrast, to our knowledge, only 4 studies searched for the presence of QTL for tolerance to herbivory (Agrama et al. 2002, Alam & Cohen 1998, Soundararajan et al. 2004, Weinig et al. 2003b). From this evidence, we can conclude one to several loci are involved in the expression of tolerance. For rice, the number of QTL for resistance and tolerance to a brown planthopper varies from one to four (Alam & Cohen 1998, Soundararajan et al. 2004). In sorghum, there are nine QTL for resistance and tolerance (Agrama et al. 2002).

Conversely, in *Arabidopsis thaliana* the number of loci controlling tolerance to rabbit herbivory is thought to be higher than the number of loci controlling resistance to the same natural enemy (Weinig et al. 2003b). To the extent that the number of loci (and potential interactions between them) related to the expression of resistance and tolerance differ, the negative effect of inbreeding will vary between both strategies. Also, differences in the magnitude of inbreeding depression on tolerance and resistance may depend on the history of the population, past selection, the number and magnitude of deleterious alleles for each defensive strategy, linkage, and intralocus (dominance) and among loci (epistasis) interactions (see Fritz et al. 2003, Kliebentein

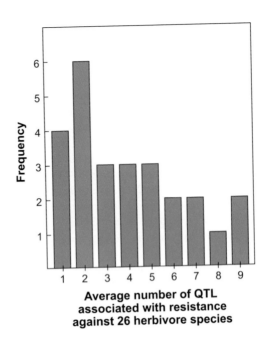

Figure 1

Frequency distribution of the number of QTL detected for host plant resistance against
herbivore insect species. Studies were found by performing keyword searches in the electronic
bibliographic databases Web of Science and Cambridge Scientific Abstracts. The results
contain studies from the following journals: *Crop Science, Heredity, Theoretical and Applied
Genetics, Plant Breeding, Genetics, Maydica, Genetica, Molecular Ecology, Entomologia
Experimentalis et Applicata, Proceedings of the National Academy of Sciences, Journal of Economic
Entomology*, and *Advances in Agronomy*. The final data set contains information about resistance
against 26 herbivore species obtained from 53 studies published between 1970 and 2007. The
information describes 11 cultivated (*Solanum, Oriza, Zea, Glycine, Phaseolus, Brassica, Hordeum,
Citrus, Salix, Sorghum*) and one wild (*Arabidopsis thaliana*) species. The search is complemented
with a previous review on the topic (Yencho et al. 2000). One study examined the presence of
QTL against rabbit resistance (Weinig et al. 2003b).

et al. 2002). Thus, given that tolerance and resistance are complex traits, lack of
knowledge about the genetic architecture of defenses against herbivores makes it difficult to predict precisely the consequences of the mating system on each defensive
strategy. Furthermore, variation in ecological conditions can generate genotype ×
environment interactions that affect the amount of phenotypic variation of traits and
the opportunity for selection on defensive loci (Rönnberg-Wästljung et al. 2006,
Weinig et al. 2003b).

Table 2 summarizes the effects of the mating system on the defenses of plants
against herbivores. From eight studies corresponding to five herbaceous plant species
preyed upon by insect herbivores, the results are mixed but a negative effect of inbreeding is more common. In the case of resistance, three out of five assessments
indicate that outcrossing improves defense. However, in one of these three cases, evidence indicates that outcrossing may not be advantageous by itself but in relation to

Table 2   Effects on resistance (R) and/or tolerance (T) of outcrossed and selfed progenies of different plant species[a]

Plant species	Insect herbivores	Genetic material	Defensive strategy	Mating system effect	Reference
Datura stramonium	*Epitrix parvula* *Sphenarium purpurascens*	S/O	Resistance	Means of damage produced by each herbivore did not differ between selfed and outcrossed progenies (R⇔)	Núñez-Farfán et al. 1996
Mimulus guttatus	*Philaenus spumarius*	S/O 2P	Tolerance	Plant biomass and flower production were more affected by herbivores in inbred than in outcrossed plants in one of the two studied populations (T↓)	Carr & Eubanks 2002
Solidago glaucus	*Apterothrips apteris*	S/O	Resistance	Crosses of high × low infestation clones produced progenies with intermediate level of resistance (R↑)	Strauss & Karban 1994
Mimulus guttatus	*Philaenus spumarius*	S/O 5P	Tolerance	At the population level, inbred progenies had lower plant aboveground biomass than outcrossed progenies (T↓) No differences among progenies for other fitness-related characters (probability of bolting and producing flowers, survival, and reproductive effort) (T⇔)	Ivey et al. 2004
Ipomoea heredacea	*Deloyala guttata* *Charidotella bicolor* *Spodoptera exigua*	S/O	1) Tolerance to (a) and (b) 2) Tolerance to (c)	1) Damaged outcrossed plants showed a higher biomass reduction relative to its control than inbreed plants (T↑) 2) Damaged inbred plants showed a higher reproductive effort reduction relative to its control than outcrossed plants (T↓)	Hull-Sanders & Eubanks 2005
Gilia achilleifolia	Unidentified	$s_0 s_1 s_2$	1) Resistance 2) Tolerance	1) No effect on the percentage of grazed plants in each group (R ⇔) 2) Higher percentage of plants attaining reproduction after grazing in the outcrossed progenies (T↓)	Schoen 1983
Cucurbita pepo	*Diabrotica undecimpunctata*	s_0 $s_1 s_2 s_3 s_4$	Resistance	Higher leaf damage by herbivores in inbred than in outcrossed plants (R↓)	Hayes et al. 2004
Cucurbita pepo	*Diabrotica undecimopunctata*	S/O	Resistance	Inbred plants had more damage than outcrossed plants (R↓)	Stephenson et al. 2004

[a]S/O = selfed and outcrossed progenies; s_0, s_1, s_2, s_3, and s_4 stand for the number of generations of selfing (i.e., s_0 equals f = 0 up to s_4 equals f = 0.875); ⇔, ↓, and ↑ = nil, negative, or positive, respectively, effects on the defensive strategy in selfed progenies; 2P and 5P = number of populations studied.

the alleles contributed to progeny; outcrossing may disrupt local adaptation of natural enemies to certain host genotypes, thus maintaining the advantages of recombination (Strauss & Karban 1994). In the two other studies, resistance was unaltered by inbreeding (Núñez-Farfán et al. 1996, Schoen 1983). Regarding tolerance, three studies measured tolerance in different fitness components and/or populations. These studies offer evidence that the effects of inbreeding on tolerance vary (*a*) among populations (Carr & Eubanks 2002, Ivey et al. 2004), (*b*) among components of tolerance (Carr & Eubanks 2002, Hull-Sanders & Eubanks 2005, Ivey et al. 2004), and (*c*) between specialist and generalist herbivores (Hull-Sanders & Eubanks 2005). Overall, negative effects of inbreeding on tolerance are more prevalent (**Table 2**).

The available evidence partially supports the expectation of negative effects of inbreeding on tolerance and resistance to herbivores. However, it is not possible to ascertain which strategy would be more affected for several reasons: (*a*) Resistance and tolerance have not been estimated simultaneously for inbred and outbred progenies of the same populations in natural conditions; (*b*) it is necessary to identify the components of tolerance related to reproductive fitness; (*c*) it is necessary to take into account the abundance and kind of herbivores and the amount of damage exerted on plants, given that phenotypic variances of tolerance and resistance vary among populations; and (*d*) knowledge of the history of populations is needed (i.e., inbreeding coefficient of the base population; Schoen 1983) because, for instance, differences between selfed and outcrossed progenies would be negligible in highly inbred populations. To the extent that the plant mating system differentially affects tolerance and resistance to herbivory, fluctuations in the levels of selfing within and among populations may explain the presence of intermediate levels of tolerance and resistance.

CAVEATS ON THE SIMULTANEOUS ESTIMATION OF TOLERANCE AND RESISTANCE

An important aspect of the simultaneous estimation of tolerance and resistance is the use of independent predictors of both traits to avoid covariation between them (e.g., Mauricio et al. 1997). For a genetic correlation between tolerance and resistance, it is preferable to obtain independent estimates of both strategies, particularly when natural rather than artificial damage is used. Statistically, it may be feasible to use different plant replicates of the same genotype to estimate tolerance and resistance, although increasing the sample size of the experiments may impose practical limitations.

The estimation of natural selection on tolerance can be affected by a statistical artifact when mean fitness across damage levels is correlated with fitness variance. For instance, the estimation can be affected when genotypes with low vigor (low mean fitness across damage levels) have low variance in fitness and a flat slope of tolerance. To solve this problem Agrawal et al. (2004) standardized the fitness of each family to a mean of zero and a variance of one before estimating tolerance, thus accounting for any differences in plant vigor and fitness variance among genotypes.

There is an additional pitfall in the simultaneous estimation of tolerance and resistance. The use of a unique estimate of resistance or tolerance to the whole community of natural enemies (Fornoni et al. 2004b, Mauricio et al. 1997) may lead to misleading conclusions about the selective value of a combination of tolerance and resistance. If tolerance and resistance are selected by different species of natural enemies, in the presence of both consumers the joint selection pattern may indicate that both strategies coexist. However, this conclusion would depend on the presence/absence of species of natural enemies. Thus, it would be preferable to estimate tolerance and resistance to each species of consumer separately (Pilson 2000, Tiffin & Rausher 1999). Unless tolerance/resistance to one species of consumer is highly correlated with tolerance/resistance to a second species, independent estimates for each strategy of defense should be obtained. This potential bias in the estimation of tolerance/resistance may be solved by manipulating the presence/absence of different species of natural enemies and estimating resistance and tolerance to each species (Juenger & Bergelson 1998, Stinchcombe & Rausher 2002).

MANIPULATING TOLERANCE AND RESISTANCE FOR PEST MANAGEMENT AND THE EVOLUTION OF PEST RESISTANCE

One of the challenges of modern agriculture is the use of genetically modified plant species to increase global food production (Gould 1998). To this goal, plants engineered to express toxic compounds to reduce pest damage have been increasingly used during the past decades (Gould 1998). The practice promoted to farmers by environmental agencies and industry is the high-dose/refuge approach, which involves planting susceptible crop lines (refuge) in combination with resistant lines that possess a high dose of toxic compounds. In theory, this practice can reduce the evolution of pest resistance, in time expanding the effectiveness of genetically modified crops (Gould 2003).

One of the problems with this approach is that farmers are reluctant to assign even a small proportion of their lands for planting susceptible crops because of yield losses (Vacher et al. 2006). Historically, the theoretical and empirical development of pest management practices has ignored the fact that plants can respond against natural enemies by mechanisms other than resistance. Hence the recognition that wild plants are likely to express mixed defensive strategies calls into question their potential use for pest management in agriculture. In those crop species were the reproductive structures have economic value, the use of lines tolerant to foliar damage instead of susceptible lines as refuges may reduce yield losses without promoting pest resistance evolution (Garrido-Espinosa & Fornoni 2006).

Rausher (2001) proposed the possible use of tolerance to herbivory as a complement to pest management. The lack of knowledge about the genetic basis of tolerance may preclude its use in modern crop improvement programs. However, the presence of QTL for tolerance against herbivory was detected recently (see above). This evidence offers the possibility to manipulate plant tolerance to reduce yield decrement by pest damage. If a mixed pattern of allocation to tolerance and resistance allowed plants to track the evolutionary response of herbivores under natural conditions, the incorporation of recent knowledge about the simultaneous evolution of tolerance and resistance may help to improve current practices to control pest resistance and evolution.

CONCLUDING REMARKS

The evidence suggests that the simultaneous allocation of resources to tolerance and resistance against natural enemies is pervasive among host plants. Although several factors have been detected that may explain the presence of a mixed pattern of defense allocation, the causes that would originate such a pattern remain unclear. In particular, it is still unclear how the environment (biotic and physic) determines the pattern of defense allocation to tolerance and resistance. To this goal, manipulating both herbivore species and other putative selective agents on plant defensive traits may clarify which environmental conditions promote the evolution of a mixed pattern of defense allocation. In addition, if the response of plants to herbivory damage through tolerance constitutes a general mechanism against stressful conditions, plant tolerance is likely to be evolutionarily older than resistance (Fornoni et al. 2003b). Tolerance to damage imposed by both frost and desiccation were crucial steps during evolution in terrestrial environments (Agrawal et al. 2004, Oliver et al. 2000), before insect herbivores started imposing selection on plants (Fornoni et al. 2003b). It is interesting to note the lack of studies with a phylogenetic approach to determine how tolerance and resistance characters map onto phylogeny. Increased knowledge about the mechanism of plant tolerance would open new possibilities for analyzing the simultaneous evolution of tolerance and resistance, because both could be estimated independently and links with other traits, interactions, and environmental conditions could be explored. The increasing interest in the reciprocal interaction between plant defenses and mating systems and costs and benefits of tritrophic interactions opens an attractive line of research to examine the indirect costs of defense and the constraints to its evolution.

One of the questions that we probably should be aware of in the near future is whether a pattern of mixed allocation to tolerance and resistance affects the paradigm of coevolution between plants and herbivores. How would the classical arms race or cyclical dynamics between plants and herbivores operate when both tolerance and resistance are expressed simultaneously? The fact that tolerance may slow down the evolution of herbivore counter-resistant responses (Garrido-Espinosa & Fornoni 2006, Stinchcombe 2002b) poses new challenges for the understanding of the co-evolutionary process. In this sense a less phytocentric approach should be followed to understand the simultaneous evolution of plants and herbivore traits involved in the interaction. The consequences of a mixed defensive strategy to the process of coevolution between plants and herbivores have not been discussed yet (however, see Restiff & Koella 2003).

FUTURE ISSUES

Although theoretical work on the evolution of tolerance and resistance has guided empirical studies, to improve existing models several unanswered questions remain. We believe that to understand why plants express a mixed defense strategy more experimentation is needed. From our perspective, the following questions deserve consideration:

1. Is a mixed defense strategy evolutionarily stable?

2. How does the environment affect the proportion of resource allocation to tolerance and resistance for a given plant genotype? How do the costs and benefits of tolerance and resistance, as well as their selective value, change along resource availability gradients?

3. How do ontogenetic patterns of allocation to tolerance and resistance determine the presence of mixed patterns of plant defense? Is there any variation during ontogeny in the expression of tolerance and resistance and their corresponding selective value?

4. What is the role of other interactions (e.g., mating systems, tritrophic interactions) in the maintenance of mixed defense strategies in plants against herbivory?

5. Is herbivory the primary selective agent of plant tolerance?

6. What is the role of tolerance in the coevolutionary process between plants and natural enemies? To what extent can tolerance constrain the coevolution between plants and their natural enemies?

7. What is the number of quantitative trait loci controlling tolerance and resistance in wild species? How can a pattern of mixed defensive strategies of tolerance and resistance help to control pests and diseases in domesticated plant species?

8. How are tolerance and resistance distributed among different phylogenetically related plant lineages?

DISCLOSURE STATEMENT

The authors are not aware of any biases that might be perceived as affecting the objectivity of this review.

ACKNOWLEDGMENTS

We thank John R. Stinchcombe for sharing unpublished data about the cost of tolerance. Financial support was provided by grants CONACYT 42031 and PAPIIT IN226305-3 to JNF, and PAPIIT IN200807-2 to JF.

LITERATURE CITED

Aarssen LW. 1995. Hypothesis for the evolution of apical dominance in plants: implications for the interpretation of overcompensation. *Oikos* 74:149–56

Abrahamson WG, Weis AE. 1997. *Evolutionary Ecology Across Three Trophic Levels: Goldenrods, Gall-makers and Natural Enemies*. Princeton: Princeton Univ. Press

Adler LS. 2000. The ecological significance of toxic nectar. *Oikos* 91:409–20

Agrama HA, Widle GE, Reese JC, Campbell LR, Tuinstra MR. 2002. Genetic mapping of QTLs associated with greenbug resistance and tolerance in *Sorghum bicolor*. *Theor. Appl. Genet.* 104:1373–78

Agrawal AA, Conner JK, Stinchcombe JR. 2004. Evolution of plant resistance and tolerance to frost damage. *Ecol. Lett.* 7:1199–208

Agrawal AA, Fishbein M. 2006. Plant defense syndromes. *Ecology* 87:S132–S49

Agrawal AF, Lively CM. 2001. Parasites and the evolution of self-fertilization. *Evolution* 55:869–79

Alam SN, Cohen MB. 1998. Detection and analysis of QTLs for resistance to the brown planthopper, *Nilaparvata lugens*, in a double-haploid rice population. *Theor. Appl. Genet.* 97:1370–79

Armbruster P, Reed DH. 2005. Inbreeding depression in benign and stressful environments. *Heredity* 95:235–42

Becerra JX. 1997. Insects on plants: Macroevolutionary chemical trends in host use. *Science* 276:253–56

Bennett AE, Alers-Garcia J, Bever JD. 2006. Three-way interactions among mutualistic mycorrhizal fungi, plants, and plant enemies: hypothesis and synthesis. *Am. Nat.* 167:141–52

Berenbaum MR, Zangerl AR, Nitao JK. 1986. Constraints on chemical coevolution: wild parsnips and the parsnip webworm. *Evolution* 40:1215–28

Berenbaum MR, Zangerl AR. 1998. Chemical phenotype matching between a plant and its insect herbivore. *Proc. Natl. Acad. Sci. USA* 95:13743–48

Bergelson J, Dwyer G, Emerson JJ. 2001. Models and data on plant-enemy coevolution. *Annu. Rev. Genet.* 35:469–99

Bergelson J, Purrington CB. 1996. Surveying patterns in the cost of resistance in plants. *Am. Nat.* 148:536–58

Boege K, Dirzo R, Siemens D, Brown P. 2007. Ontogenetic switches from plant resistance to tolerance: minimizing costs with age? *Ecol. Lett.* 10:177–87

Carr DE, Eubanks MD. 2002. Inbreeding alters resistance to insect herbivory and host plant quality in *Mimulus guttatus* (Scrophulariaceae). *Evolution* 56:22–30

Carr DE, Murphy JF, Eubanks MD. 2003. The susceptibility and response of inbred and outbred *Mimulus guttatus* to infection by Cucumber mosaic virus. *Evol. Ecol.* 17:85–103

Castillo RA, Cordero C, Domínguez CA. 2002. Are reward polymorphisms subject to frequency- and density-dependent selection? Evidence from a monoecious species pollinated by deceit. *J. Evol. Biol.* 15:544–52

Charlesworth D, Charlesworth B. 1987. Inbreeding depression and its evolutionary consequences. *Annu. Rev. Ecol. Syst.* 18:237–68

Cronin JT, Abrahamson WG. 1999. Host-plant genotype and other herbivores influence goldenrod stem galler preference and performance. *Oecologia* 121:392–404

de Meaux J, Mitchell-Olds T. 2003. Evolution of plant resistance at the molecular level: ecological context of species interactions. *Heredity* 91:345–52

Dirzo R, Harper JL. 1982. Experimental studies on slug-plant interactions. IV. The performance of cyanogenic and acyanogenic morphs of *Trifolium repens* in the field. *J. Ecol.* 70:119–38

A key study that used the definition of tolerance and resistance to herbivores to other stresses (frost damage).

First demonstration of an ontogenetic trade-off between tolerance and resistance.

Domínguez CA, Dirzo R. 1994. Effect of defoliation on *Erythroxylum havanense*, a tropical proleptic species. *Ecology* 75:1896–902

Dybdahl MF, Lively CM. 1998. Host-parasite coevolution: Evidence for rare advantage and time-lagged selection in a natural population. *Evolution* 52:1057–66

Ehrlich P, Raven P. 1964. Butterflies and plants: A study in coevolution. *Evolution* 18:586–608

Elle E, Hare JD. 2002. Environmentally induced variation in floral traits affects the mating system in *Datura wrightii*. *Funct. Ecol.* 16:79–88

Falconer DS, Mackay TFC. 1996. *Introduction to Quantitative Genetics*. Edinburgh: Addison Wesley Longman

Fineblum WL, Rausher MD. 1995. Trade-off between resistance and tolerance to herbivore damage in a morning glory. *Nature* 377:517–20

Fornoni J, Núñez-Farfán J. 2000. Evolutionary ecology of *Datura stramonium*: genetic variation and costs for tolerance to defoliation. *Evolution* 54:789–97

Fornoni J, Valverde PL, Núñez-Farfán J. 2003a. Quantitative genetics of plant tolerance and resistance against natural enemies of two natural populations of *Datura stramonium*. *Evol. Ecol. Res.* 5:1049–65

Fornoni J, Núñez-Farfán J, Valverde PL. 2003b. Evolutionary ecology of tolerance to herbivory: Advances and perspectives. *Comm. Theor. Biol.* 8:643–66

Fornoni J, Núñez-Farfán J, Valverde PL, Rausher MD. 2004a. Evolution of mixed strategies of plant defense allocation against natural enemies. *Evolution* 58:1685–95

Fornoni J, Valverde PL, Núñez-Farfán J. 2004b. Population variation in the cost and benefit of tolerance and resistance against herbivory in *Datura stramonium*. *Evolution* 58:1696–704

Fritz RS, Simms EL. 1992. *Plant Resistance to Herbivores and Pathogens: Ecology, Evolution and Genetics*. Chicago: Univ. Chicago Press

Fritz RS, Hochwender CG, Brunsfeld SJ, Roche BM. 2003. Genetic architecture of susceptibility to herbivores in hybrid willows. *J. Evol. Biol.* 16:1115–26

Futuyma DJ. 1983. Evolutionary interactions among herbivorous insects and plants. In *Coevolution*, ed. DJ Futuyma, M Slatkin, pp. 207–31. Sunderland: Sinauer

Futuyma DJ. 1998. *Evolutionary Biology*. Sunderland: Sinauer

Garrido-Espinosa E, Fornoni J. 2006. Host tolerance does not impose selection on natural enemies. *New Phytol.* 170:609–14

Gassmann AJ, Hare JD. 2005. Indirect cost of a defensive trait: variation in trichome type affects the natural enemies of herbivorous insects on *Datura wrightii*. *Oecologia* 144:62–71

Gould F. 1998. Sustainability of transgenic insecticidal cultivars: Integrating pest genetics and ecology. *Annu. Rev. Entomol.* 43:701–26

Gould F. 2003. *Bt*-resistance management-theory meets data. *Nat. Biotech.* 21:1450–51

Hamrick JL, Godt MJW. 1996. Effects of life history traits on genetic diversity in plant species. *Philos. Trans. R. Soc. Lond. Ser. B* 351:1291–98

Hare D. 2002. Plant genetic variation in tritrophic interactions. In *Multitrophic Level Interactions*, ed. T Tscharntke, BA Hawkins, pp. 8–43. Cambridge, UK: Cambridge Univ. Press

Experimental evidence that tolerance would not promote a coevolutionary response on the herbivores.

Hayes CN, Winsor JA, Stephenson AG. 2004. Inbreeding influences herbivory in *Cucurbita pepo* ssp. *texana* (Cicirbitaceae). *Oecologia* 140:601–8.

Herms D, Matson W. 1992. The dilemma of plants: To grow or to defend. *Q. Rev. Biol.* 67:283–335

Hochwender C, Marquis RJ, Stowe K. 2000. The potential for and constraints on the evolution of compensatory ability in *Asclepias syriaca*. *Oecologia* 122:361–70

Hougen-Eitzman D, Rausher MD. 1994. Interactions between herbivorous insects and plant-insect coevolution. *Am. Nat.* 143:677–97

Huhta A, Lennartsson T, Tuomi J, Rautio P, Laine K. 2000. Tolerance of *Gentianella campestris* in relation to damage intensity: an interplay between apical dominance and herbivory. *Evol. Ecol.* 14:373–92

Hull-Sanders HM, Eubanks MD. 2005. Plant defense theory provides insight into interactions involving inbred plants and insect herbivores. *Ecology* 86:897–904

Husband BC, Schemske DW. 1996. Evolution of the magnitude and timing of inbreeding depression in plants. *Evolution* 50:54–70

Ivey CT, Carr DE, Eubanks MD. 2004. Effects of inbreeding in *Mimulus guttatus* on tolerance to herbivory in natural environments. *Ecology* 85:567–74

Ivey CT, Carr DE. 2005. Effects of herbivory and inbreeding on the pollinators and mating system of *Mimulus guttatus* (Phrymaceae). *Am. J. Bot.* 92:1641–49

Iwao K, Rausher MD. 1997. Evolution of plant resistance to multiple herbivores: quantifying diffuse coevolution. *Am. Nat.* 149:316–35

Jokela J, Schmid-Hempel P, Rigby MC. 2000. Dr. Pangloss restrained by the Red Queen – steps towards a unified defence theory. *Oikos* 89:267–74

Jones T, Kulseth S, Mechtenberg K, Jorgenson C, Zehfus M, et al. 2006. Simultaneous evolution of competitiveness and defense: induced switching in *Arabis drummondii*. *Plant Ecol.* 184:245–57

Juenger T, Bergelson J. 1998. Pairwise vs diffuse natural selection and the multiple herbivores of scarlet gilia, *Ipomopsis aggregata*. *Evolution* 52:1583–92

Juenger T, Bergelson J. 2000. The evolution of compensatory herbivory in scarlet gilia, *Ipomopsis aggregata*: herbivore-imposed natural selection and the quantitative genetics of tolerance. *Evolution* 54:79–92

Juenger T, Lennartsson T, Tuomi J. 2000. The evolution of tolerance in *Gentianella campestris*: natural selection and the quantitative genetics of tolerance. *Evol. Ecol.* 14:393–419

Juenger T, Morton TC, Miller RE, Bergelson J. 2005. Scarlet gilia resistance to insect herbivory: the effects of early season browsing, plant apparency, and phytochemistry on patterns of seed fly attack. *Evol. Ecol.* 19:79–101

Kliebenstein D, Pedersen D, Barker B, Mitchell-Olds T. 2002. Comparative analysis of quantitative trait loci controlling glucosinolates, myrosinase and insect resistance in *Arabidapsis thaliana*. *Genetics* 161:325–32

Knight TM, Steets JA, Vamosi JC, Mazer SJ, Burd M, et al. 2005. Pollen limitation of plant reproduction: Pattern and process. *Annu. Rev. Ecol. Evol. Syst.* 36:467–97

Koricheva J. 2002. Meta-analysis of sources of variation in fitness costs of plant anti-herbivore defenses. *Ecology* 83:176–90

Krebs JR, Davis NB. 1981. *An Introduction to Behavioural Ecology*. Oxford: Blackwell Scientific Publications

Leimu R, Koricheva J. 2006a. A meta-analysis of tradeoffs between plant tolerance and resistance to herbivores: combining the evidence from ecological and agricultural studies. *Oikos* 112:1–9

Leimu R, Koricheva J. 2006b. A meta-analysis of genetic correlations between plant resistances to multiple enemies. *Am. Nat.* 168:E15–E37

Levri MA, Real LA. 1998. The role of resources and pathogens in mediating the mating system of *Kalmia latifolia*. *Ecology* 79:1602–9

Lively CM, Howard RS. 1994. Selection by parasites for clonal diversity and mixed mating. *Philos. Trans. R. Soc. Lond. Ser. B* 346:271–81

Løe G, Toräng P, Gaudeul M, Ågren J. 2007. Trichome production and spatiotemporal variation in herbivory in the perennial herb *Arabidopsis lyrata*. *Oikos* 116:134–42

Marak HB, Biere A, Van Damme JMM. 2003. Fitness costs of chemical defense in *Plantago lanceolata* L.: effects of nutrient and competition stress. *Evolution* 57:2519–30

Mauricio R, Rausher MD. 1997. Experimental manipulation of putative selective agents provides evidence for the role of natural enemies in the evolution of plant defense. *Evolution* 51:1435–44

Mauricio R, Rausher MD, Burdick DS. 1997. Variation in defense strategies in plants: Are resistance and tolerance mutually exclusive? *Ecology* 78:1301–11

Mauricio R. 2000. Natural selection and the joint evolution of tolerance and resistance as plant defenses. *Evol. Ecol.* 14:491–507

McCall AC, Irwin RE. 2006. Florivory: the intersection of pollination and herbivory. *Ecol. Lett.* 9:1351–65

Mopper S, Strauss SY. 1998. *Genetic Structure and Local Adaptation in Natural Insect Populations*. New York: Chapman & Hall

Mutikainen P, Walls M, Ovaska J, Keinänen M, Julkunen-Tiitto R, Vapaavouri E. 2002. Costs of herbivore resistance in clonal saplings of *Betula pendula*. *Oecologia* 133:364–71

Núñez-Farfán J, Dirzo R. 1994. Evolutionary ecology of *Datura stramonium* L. in Central Mexico: natural selection for resistance to herbivorous insects. *Evolution* 48:423–36

Núñez-Farfán J, Cabrales-Vargas RA, Dirzo R. 1996. Mating system consequences on resistance to herbivory and life history traits in *Datura stramonium*. *Am. J. Bot.* 83:1041–49

Oliver MJ, Tuba Z, Mishler BD. 2000. The evolution of vegetative desiccation tolerance in land plants. *Plant Ecol.* 151:85–100

Orians CM, Lower S, Fritz RS, Roche BM. 2003. The effects of plant genetic variation and soil nutrients on secondary chemistry and growth in shrubby willow, *Salix sericea*: patterns and constraints on the evolution of resistance traits. *Bioch. Syst. Ecol.* 31:233–47

Osier TL, Lindroth RL. 2006. Genotype and environment determine allocation to and costs of resistance in quaking aspen. *Oecologia* 148:293–303

Ouborg NJ, Biere A, Mudde CL. 2000. Inbreeding effects on resistance and transmission-related traits in the *Silene-Microbotryum* pathosystem. *Ecology* 81:520–31

Observational study showing that leaf trichomes may be selected by factors other than herbivory.

The first study that examined the evolutionary consequences of the interaction between mating system and plant resistance.

Painter RH. 1958. Resistance of plants to insects. *Annu. Rev. Entomol.* 3:267–90

Parker M. 1989. Disease impact and local genetic diversity in the clonal plant *Podophyllum peltatum*. *Evolution* 43:540–47

Pilson D. 1996. Two herbivores and constraint on selection for resistance in *Brassica rapa*. *Evolution* 50:1492–1500

Pilson D. 2000. The evolution of plant response to herbivory: simultaneously considering resistance and tolerance in *Brassica rapa*. *Evol. Ecol.* 14:457–89

Pilson D, Decker KL. 2002. Compensation for herbivory in wild sunflower: Response to simulated damage by the head-clipping weevil. *Ecology* 83:3097–107

Purrington CB. 2000. Costs of resistance. *Curr. Opin. Plant Biol.* 3:305–8

Puustinen S, Koskela T, Mutikainen P. 2004. Direct and ecological costs of resistance and tolerance in the stinging nettle. *Oecologia* 139:76–82

Rausher MD. 1988. Is coevolution dead? *Ecology* 69:898–901

Rausher MD. 1996. Genetic analyses of coevolution between plants and their natural enemies. *Trends Genet.* 12:212–17

Rausher MD. 2001. Co-evolution and plant resistance to natural enemies. *Nature* 411:857–64

Restiff O, Koella JC. 2003. Shared control of epidemiological traits in a coevolutionary model of host-parasite interaction. *Am. Nat.* 161:827–36

Restif O, Koella JC. 2004. Concurrent evolution of resistance and tolerance to pathogens. *Am. Nat.* 164:E90–E102

Rönnberg-Wästljung AC, Åhman I, Glynn C, Widenfalk O. 2006. Quantitative trait loci for resistance to herbivores in willow: field experiments with varying soils and climates. *Entomol. Exp. Appl.* 118:163–74

Rosenthal JP, Kotanen PM. 1994. Terrestrial plant tolerance to herbivory. *Trends Ecol. Evol.* 9:145–48

Roy BA. 1998. Differentiating the effects of origin and frequency in reciprocal transplant experiments used to test negative frequency-dependent selection hypotheses. *Oecologia* 115:73–83

Roy B, Kirchner J. 2000. Evolutionary dynamics of pathogen resistance and tolerance. *Evolution* 54:51–63

Schoen DJ. 1983. Relative fitness of selfed and outcrossed progeny in *Gilia achilleifolia* (Polemoniaceae). *Evolution* 37:292–301

Seger J. 1983. Evolution of exploiter-victim relationships. In *Natural Enemies: The population biology of predators, parasites and diseases*, ed. MJ Crawley, pp. 3–25. Oxford: Blackwell Scientific Publications

Shonle I, Bergelson J. 2000. Evolutionary ecology of the tropane alkaloids of *Datura stramonium* L. (Solanaceae). *Evolution* 54:778–88

Siemens DH, Lischke H, Maggiulli N, Schurch S, Roy BA. 2003. Cost of resistance and tolerance under competition: the defense-stress benefit hypothesis. *Evol. Ecol.* 17:247–63

Siemens DH, Roy BA. 2005. Tests for parasite-mediated frequency-dependent selection in natural populations of an asexual plant species. *Evol. Ecol.* 19:321–38

Simms EL, Rausher MD. 1987. Costs and benefits of plant resistance to herbivory. *Am. Nat.* 130:570–81

Insightful review paper that highlights the potential complement between the studies of coevolution in natural and agricultural systems.

The first coevolutionary model that incorporated tolerance and resistance explicitly and simultaneously.

Simms EL, Triplett J. 1994. Costs and benefits of plant responses to disease: Resistance and tolerance. *Evolution* 48:1973–85

Simms EL. 2000. Defining tolerance as a norm of reaction. *Evol. Ecol.* 14:563–70

Skogsmyr I, Fagerström T. 1992. The cost of antiherbivory defence: An evaluation of some ecological and physiological factors. *Oikos* 64:451–57

Soundararajan RP, Kadirvel P, Gunathilagaraj K, Maheswaran M. 2004. Mapping of quantitative trait loci associated with resistance to brown planthopper in rice by means of a doubled haploid population. *Crop Sci.* 44:2214–20

Steets JA, Ashman TL. 2004. Herbivory alters the expression of a mixed-mating system. *Am. J. Bot.* 91:1046–51

Steets JA, Hamrick JL, Ashman TL. 2006. Consequences of vegetative herbivory for maintenance of intermediate outcrossing in an annual plant. *Ecology* 87:2717–27

Stephenson AG, Leyshon B, Travers SE, Hayes CN, Winsor JA. 2004. Interrelationships among inbreeding, herbivory, and disease on reproduction in a wild gourd. *Ecology* 85:3023–34

Stinchcombe JR, Rausher MD. 2001. Diffuse selection on resistance to deer herbivory in the ivyleaf morning glory, Ipomoea hederacea. *Am. Nat.* 158:376–88

Stinchcombe JR, Rausher MD. 2002. The evolution of tolerance to deer herbivory: modifications caused by the abundance of insect herbivores. *Proc. R. Soc. Lond. Ser. B* 269:1241–46

Stinchcombe JR. 2002a. Environmental dependency in the expression of costs of tolerance to deer herbivory. *Evolution* 56:1063–67

Stinchcombe JR. 2002b. Can tolerance traits impose selection on herbivores? *Evol. Ecol.* 15:595–602

Stowe K. 1998. Experimental evolution of resistance in *Brassica rapa*: Correlated response of tolerance in lines selected for glucosinolate content. *Evolution* 52:703–12

Stowe K, Marquis RJ, Hochwender CG, Simms EL. 2000. The evolutionary ecology of tolerance to consumer damage. *Annu. Rev. Ecol. Syst.* 31:565–95

Strauss SY, Karban R. 1994. The significance of outcrossing in an intimate plant-herbivore relationship. I. Does outcrossing provide an escape from herbivores adapted to the parent plant? *Evolution* 48:454–64

Strauss SY, Conner JK, Rush SL. 1996. Foliar herbivory affects floral characters and plant attractiveness to pollinators: Implications for male and female plant fitness. *Am. Nat.* 147:1098–107

Strauss SY. 1997. Floral characters link herbivores, pollinators, and plant fitness. *Ecology* 78:1640–45

Strauss SY, Agrawal A. 1999. The ecology and evolution of tolerance to herbivory. *Trends Ecol. Evol.* 14:179–85

Strauss SY, Siemens DF, Decher MB, Mitchell-Olds T. 1999. Ecological costs of plant resistance to herbivores in the currency of pollination. *Evolution* 53:1105–13

Strauss SY, Rudgers JA, Lau JA, Irwin RE. 2002. Direct and ecological costs of resistance to herbivory. *Trends Ecol. Evol.* 17:278–85

Strauss SY, Irwin RE. 2004. Ecological and evolutionary consequences of multispecies plant-animal interactions. *Annu. Rev. Ecol. Evol. Syst.* 35:435–66

A key report about the main assumption of the trade-off between tolerance and resistance.

A recent extensive review about diffuse selection and its implication for evolution of interactions.

Strauss SY, Sahli H, Conner JK. 2005. Toward a more trait-centered approach to diffuse (co)evolution. *New Phytol.* **165:81–89**

Thompson JN. 2005. *The geographic mosaic of coevolution.* Chicago: Univ. Chicago Press

Tiffin P, Rausher MD. 1999. Genetic constraints and selection acting on tolerance to herbivory in the common morning glory, *Ipomoea purpurea*. *Am. Nat.* 154:700–16

Tiffin P. 2000a. Are tolerance, avoidance, and antibiosis evolutionarily and ecologically equivalent responses of plants to herbivores? *Am. Nat.* 155:128–38

Tiffin P. 2000b. Mechanisms of tolerance to herbivore damage: what do we know? *Evol. Ecol.* 14:523–36

Tiffin P. 2002. Competition and the time of damage affect the pattern of selection acting on plant defense against herbivores. *Ecology* 83:1981–90

Vacher C, Bourguet D, Desquilbet M, Lemarié S, Ambec S, Hochberg ME. 2006. Fees or refuges: which is better for sustainable management of insect resistance to transgenic *Bt* corn? *Biol. Lett.* 2:198–202

Valverde PL, Fornoni J, Núñez-Farfán J. 2001. Defensive role of leaf trichomes in resistance to herbivorous insects in *Datura stramonium*. *J. Evol. Biol.* 14:424–32

The first experimental attempt that demonstrated that resistance and growth (tolerance) may function as redundant defenses against herbivory.

Valverde PL, Fornoni J, Núñez-Farfán J. 2003. Evolutionary ecology of *Datura stramonium*: equal plant fitness benefits of growth and resistance against herbivory. *J. Evol. Biol.* **16:127–37**

Van Zandt PA, Mopper S. 1998. A meta-analysis of adaptive dem formation in phytophagous insect populations. *Am. Nat.* 152:595–604

Weinig C, Stinchcombe JR, Schmitt J. 2003a. Evolutionary genetics of resistance and tolerance to natural herbivory in *Arabidopsis thaliana*. *Evolution* 57:1270–80

Weinig C, Stinchcombe JR, Schmitt J. 2003b. QTL architecture of resistance and tolerance traits in *Arabidopsis thaliana* in natural environments. *Mol. Ecol.* 12:1153–63

Yencho GC, Cohen MB, Byrne PF. 2000. Applications of tagging and mapping insect resistance loci in plants. *Annu. Rev. Entomol.* 45:393–422

Yoshida T. 2006. Ecological stoichiometry and the shape of resource-based trade-offs. *Oikos* 112:406–11

Plant-Animal Mutualistic Networks: The Architecture of Biodiversity

Jordi Bascompte and Pedro Jordano

Integrative Ecology Group, Estación Biológica de Doñana, CSIC, E-41080 Sevilla, Spain; email: bascompte@ebd.csic.es, jordano@ebd.csic.es

Annu. Rev. Ecol. Evol. Syst. 2007. 38:567–93

First published online as a Review in Advance on August 16, 2007

The *Annual Review of Ecology, Evolution, and Systematics* is online at
http://ecolsys.annualreviews.org

This article's doi:
10.1146/annurev.ecolsys.38.091206.095818

Key Words

coevolution, complex networks, mutualism, pollination, seed dispersal

Abstract

The mutually beneficial interactions between plants and their animal pollinators and seed dispersers have been paramount in the generation of Earth's biodiversity. These mutualistic interactions often involve dozens or even hundreds of species that form complex networks of interdependences. Understanding how coevolution proceeds in these highly diversified mutualisms among free-living species presents a conceptual challenge. Recent work has led to the unambiguous conclusion that mutualistic networks are very heterogeneous (the bulk of the species have a few interactions, but a few species are much more connected than expected by chance), nested (specialists interact with subsets of the species with which generalists interact), and built on weak and asymmetric links among species. Both ecological variables (e.g., phenology, local abundance, and geographic range) and past evolutionary history may explain such network patterns. Network structure has important implications for the coexistence and stability of species as well as for the coevolutionary process. Mutualistic networks can thus be regarded as the architecture of biodiversity.

1. INTRODUCTION

Network: a set of nodes
(e.g., species) connected
through links (e.g., trophic
or mutualistic interactions)

Charles Darwin was fascinated by the almost perfect match between the morphology of some orchids and that of the insects that pollinate them (Darwin 1862). Darwin realized that the reproduction of these plants was intimately linked to their interaction with the insects, and even predicted that the extinction of one of the partners would lead to the extinction of the other [p. 202: "If such great moths were to become extinct in Madagascar, assuredly the *Angraecum* would become extinct" (Darwin 1862)]. Since then, a myriad of scientific papers have described the mutually beneficial (mutualistic) interactions between plants and their animal pollinators or seed dispersers.

The classic paper by Ehrlich & Raven (1964) advocated that plant-animal interactions have played a very important role in the generation of Earth's biodiversity. As a matter of fact, flowering plants and insects are two of the major groups of living beings. The origin of flowering plants opened new niches for insect diversification, which in turn may have driven plant speciation. Alternatively, one group may have tracked the previous diversification of the other group without affecting it (Ehrlich & Raven 1964, Pellmyr 1992). In any case, animal-pollinated angiosperm families are more diverse than their abiotically pollinated sister clades (Dodd et al. 1999). These plant-animal interactions are found widely from the mid-Cretaceous period, more than 100 million years ago (Mya), but some preliminary adaptations to mutually beneficial pollination can be tracked as early as approximately the mid-Mesozoic era, almost 200 Mya (Labandeira 2002). Similarly, interactions with animal frugivores also played a central role in the diversification of plant fruit structures and dispersal devices. The early evolution of animal-dispersed fruits in the Pennsylvanian period, together with the diversification of small mammals and birds in the Tertiary period, allowed the widespread occurrence of biotic dispersal in higher plants (Tiffney 2004).

The current importance of mutualisms for biodiversity maintenance is supported by the fact that more than 90% of tropical plant species rely on animals for the dispersal of their seeds (Jordano 2000). Similar figures can be adduced for pollination (Bawa 1990). If these animals disappear, their plant partners may follow. The cascading consequences of the disappearance of large seed dispersers due to hunting or habitat loss, experienced through a reduction in seed dispersal or pollination, is an important threat to biodiversity (Dirzo & Miranda 1990, Kearns et al. 1998, Wright 2003). Information from the fossil record shows clearly that major extinctions of flowering plants resulted from episodes of insect diversity decline, for example during the Middle to Late Pennsylvanian extinction, during the Permian event, and at the Cretaceous/Tertiary boundary (Labandeira 2002, Labandeira et al. 2002). To assess the likelihood and magnitude of these coextinction cascades, we need a network approach to plant-animal mutualistic interactions.

Early studies on mutualistic interactions dealt mainly with species-specific patterns of interactions or reduced subsets of the whole community (Johnson & Steiner 1997, Nilsson 1988). Although examples of these highly specific pairwise interactions exist, such as Darwin's moth and the *Angraecum* (Darwin 1862, Johnson & Steiner 1997, Nilsson 1988), fig wasps and figs (Cook & Rasplus 2003), and yucca moths and yuccas (Pellmyr 2003), their strong emphasis in evolutionary studies probably reflects the aesthetics of such almost perfect matching, more than their frequency in nature (Schemske 1983, Waser et al. 1996). As a consequence, several researchers advocated a community context to address mutualistic interactions (Bronstein 1995, Feinsinger 1978, Fox 1988, Herrera 1982, Inouye & Stinchcombe 2001, Iwao & Rausher 1997, Janzen 1980, Jordano 1987, Petanidou & Ellis 1993, Waser et al. 1996). This opened the path for significant progress in the past two decades in the analysis of how pairwise interactions are shaped within small groups of species across time and space (Parchman & Benkman 2002, Thompson 1994, Thompson & Pellmyr 1992). The next frontier is the extension of these multispecific systems to embrace whole networks, to address the organization of these large assemblages of species by ecological and evolutionary processes, and to infer the consequences of network architecture for the persistence of biodiversity. This is the goal of this review. The rationale for this endeavor is that some questions can be only addressed using a network approach. For example, What are the community-wide consequences of a mutualism disruption or a species invasion? How does coevolution proceed in species-rich communities? Before trying to address these questions we need some tools and concepts that come from the study of other types of networks.

Erdös-Rényi random network: a set of nodes with a probability p that any two nodes chosen at random are connected by a link

2. COMPLEX NETWORKS

2.1. A Network Approach to Complex Systems

The field of complex networks has grown extraordinarily in the last few years (Albert & Barabasi 2002, Amaral et al. 2000, Dorogovtsev & Mendes 2002, Newman 2003, 2004, Newman et al. 2006, Solé & Bascompte 2006, Strogatz 2001). Several systems that range from genetic networks to societies and the Internet have been described with a common framework in which elements (genes, proteins, or ecological species) are nodes connected by links. These links can take the form of gene activation, protein interaction, or species interactions such as predator-prey or mutualism (Albert & Barabasi 2002, Amaral et al. 2000, Dorogovtsev & Mendes 2002, May 2006, Montoya et al. 2006, Newman 2003, Proulx et al. 2005, Strogatz 2001, Watts 2003) (**Figure 1a**). Historically, these networks have been described and analyzed by graph theory, an important field in mathematics.

The great mathematician Paul Erdös, together with Alfred Rényi, built graph theory by studying the simplest network: a random graph (Erdös & Rényi 1959). This random graph is defined by a set of nodes and a probability p that two such nodes chosen at random are connected by a link. Many mathematical theorems on these random graphs have been produced since then. For example, one can derive analytically the probability of finding a node with a specific number of links. The Erdös-Rényi model established a theoretical approach to complex networks. However, as we discuss below, it has limited applications to the real world because the majority of complex networks are much more heterogeneous, i.e., they show a huge variability in the number of links per node.

One of the most interesting contributions from recent research on complex networks is the recognition that several networks, despite differences in the nature of

www.annualreviews.org • Plant-Animal Mutualistic Networks   569

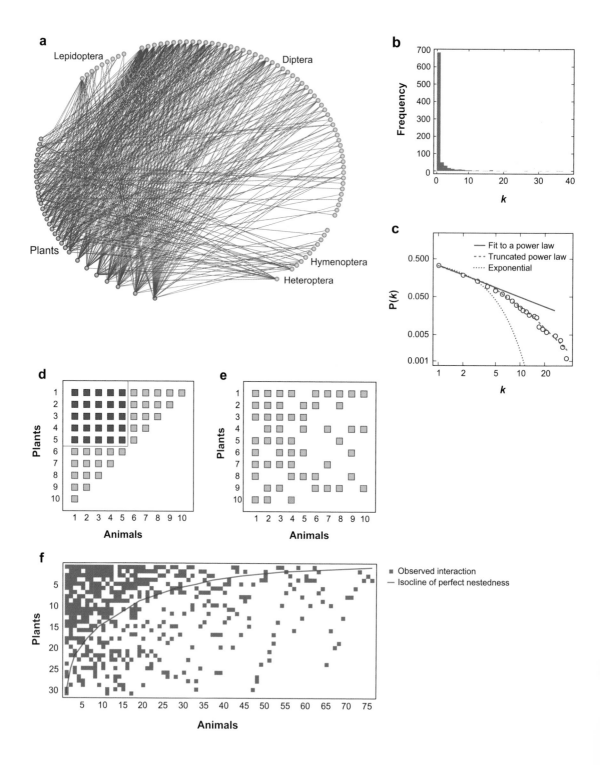

a

Lepidoptera

Diptera

Plants

Hymenoptera

Heteroptera

b

c

Fit to a power law
Truncated power law
Exponential

d

Plants

Animals

e

Plants

Animals

f

Plants

Animals

■ Observed interaction
— Isocline of perfect nestedness

their nodes, exhibit similar statistical properties. This is important for at least two reasons: First, a common architecture may be related to common patterns of network formation. Second, structure greatly influences network robustness, which is measured as the fraction of the species that must become extinct for the resulting network to fragment into several disjointed pieces (Albert et al. 2000). Network theory has certainly led to a new approach in many different fields. For example, molecular biologists used to be limited to the study of a few genes at a time, but they can now develop extensive maps that describe how many such genes depend on each other for genetic control such as gene activation (Luscombe et al. 2004).

Ecology has a long tradition in the study of networks such as food webs (Cohen 1978, Cohen et al. 1990, May 1973, Pimm 1982) and other types of network-related processes (May 2006, Proulx et al. 2005). These studies mainly focused on global descriptors such as connectance (i.e., the fraction of realized links), compartments, fraction of top predators, and similar variables. Researchers recently revisited this area with new tools that illuminate community organization using a new generation of larger, more resolved food webs (Bascompte & Melián 2005, Belgrano et al. 2005, Montoya et al. 2006, Pascual & Dunne 2006).

Species degree: the number of different species a certain species interacts with; also known as generalization level

Degree distribution: the frequency distribution of the number of interactions per species

Scale-free network: a heterogeneous network in which the bulk of the nodes have a few links, but a few nodes are much more connected than expected by chance

2.2. Scale-Free Networks

A node is characterized by its degree, which is defined as the number of links to other nodes. A first measure of network structure is based on the concept of degree distribution, i.e., the frequency distribution of the number of links per node (**Figure 1b**). Erdős-Rényi's random graphs are characterized by a degree distribution that follows a Poisson distribution (or an exponential one if the number of nodes keeps growing). The tail of this distribution is narrow: All nodes have a similar number of links, and the probability of a node having a number of links larger than the average drops very fast. In sum, random graphs are very homogeneous.

Conversely, complex networks such as the Internet are much more heterogeneous; the bulk of the nodes have a few interactions, but a few nodes are much more connected than would be expected by chance. These highly connected nodes are hubs that act as the glue to bring the network together. The average degree is not a good

←————————————————————————————

Figure 1

The architecture of plant-animal interaction networks. (*a*) A plant-pollinator network in Zackenberg Arctic tundra, Greenland that illustrates the pattern of interactions among insects (*orange dots*) and plants (*green dots*) (H. Elberling and J.M. Olesen, unpublished data). (*b*) Frequency distribution of the number of interactions per species, *k*, for pollinators in a temperate forest community in Kibune forest, Kyoto, Japan (Inoue et al. 1990). (*c*) Degree distribution is defined as the probability of one species interacting with *k* species. The lines illustrate the fit to a power-law (*gray line*), truncated power-law (*dotted blue line*), and exponential (*dashed purple line*) distributions. (*d–f*) Interaction matrices illustrate the situation of perfect nestedness (*d*), with a core of interactions among generalist species (*dark blue*), a random interaction pattern (*e*), and a real interaction network (*f*) [*f* corresponds to the graph in *a*]. A filled square indicates an observed interaction between plant species *i* in a row and animal species *j* in a column, and the line represents the isocline of perfect nestedness.

descriptor of such networks, because the variance is much higher. Mathematically, this heterogeneous distribution of degree is described by a power-law function:

$$p(k) \propto k^{-\gamma}, \qquad\qquad 1.$$

where $p(k)$ is the probability of a node having k links and γ is a critical exponent. This distribution is called scale-free because the relationship between k and $p(k)$ is not defined on a particular scale (Schroeder 1991). In a log-log plot, this relationship is given by a straight line of slope $-\gamma$ for the entire range of k values (**Figure 1c**). This is not true for a distribution with an exponential tail that has a specific scale, the average number of links per node. In that case, the relationship between *log k* and *log p(k)* changes as one moves along the x-axis (Schroeder 1991).

2.3. Weighted Networks

Up to this point we have considered binary links. A next step is to characterize the intensity or weight of these interactions. A quantitative extension of degree is that of a node's strength, i.e., the sum of the weights of all its interactions (Barrat et al. 2004). In the case of world-wide air traffic, for example, we would consider the total number of passengers flying from one airport to another airport. The strength of an airport would be defined by the sum of such values, and would give us the quantitative importance of each airport (Barrat et al. 2004). Analogous to the degree distribution, one could also plot the strength distribution, that is, how many airports serve a certain range of passengers (Barrat et al. 2004).

2.4. Mechanisms of Network Buildup

We showed that Erdös-Rényi's random graphs do not reproduce some properties of the architecture of real networks. This finding called for the generation of new models of network formation. Specifically, the discovery that complex networks, such as the Internet and protein networks, have skewed connectivity distributions, and so do not fit the exponential decay expected for random networks, led to research on the simplest mechanism of network formation compatible with such a pattern. The answer was first provided by Simon (1955) and Price (1965), and later by Barabasi and colleagues (Barabasi & Albert 1999). Their model is called preferential attachment. Imagine a core of randomly connected initial nodes. Then, at each time step a new node is introduced and the new node tends to interact with an existing node with a probability proportional to its degree. That is, new nodes tend to interact preferentially with the most-connected nodes, leading to a kind of rich-gets-richer process. This simple self-organizing model generates power-law connectivity distributions as observed in real-world complex networks (Barabasi & Albert 1999).

2.5. One-Mode and Two-Mode Networks

Two main types of networks exist: one-mode and two-mode webs. In one-mode networks nodes belong to a single category, such as airports or genes. In principle, any node may be connected to another node. In two-mode networks there are two well-defined types of nodes, and interactions occur between but not within node types. Examples of two-node networks include social networks in which people are linked to a set of social events (Borgatti & Everett 1997). Two-mode networks are represented by bipartite graphs. Plant-animal mutualistic networks are by definition two-mode networks: Plants (green nodes in **Figure 1a**) are pollinated or dispersed by animals (orange nodes in **Figure 1a**). The bipartite representation of mutualistic networks illustrates explicitly the reciprocity involved in the interaction and helps in the understanding of the complex patterns that arise in highly diversified mutualisms (Bascompte et al. 2003, 2006b; Jordano 1987; Jordano et al. 2003, 2006).

A bipartite graph is defined by an adjacency matrix whose elements a_{ij} will be 1 if plant i and animal j interact, and will be zero otherwise (**Figure 1d,e**). In weighted networks, there are two such adjacency matrices: one for plants (P) and the other for animals (A). $d_{ij}{}^{P}$ represents the mutualism strength or dependence of the plant species i on the animal species j, and $d_{ji}{}^{A}$ depicts the dependence of animal species j on plant species i (**Figure 2b**).

Bipartite graph: the graphical representation of a two-mode network, consists of two sets of nodes with interactions between (but not within) sets

3. THE STRUCTURE OF PLANT-ANIMAL MUTUALISTIC NETWORKS

The concepts from network theory described above allow the visualization of interactions in highly diverse communities, and provide ways to quantify and compare network patterns across communities statistically. The first comparative study of mutualistic interactions from a network perspective is arguably the study by Jordano (1987). The past five years have seen a tremendous explosion of studies on mutualistic networks (Bascompte et al. 2003, 2006b; Jordano et al. 2003; Memmott 1999; Memmott & Waser 2002; Thompson 2006; Vázquez & Aizen 2004). These first papers described the structure of mutualistic networks. We describe briefly the results on network structure in this section. Below, we discuss the potential mechanisms that led to the observed network structure and its implications.

3.1. Degree Distribution

Motivated by the discovery of scale-free networks in the Internet (Albert et al. 2000), and simultaneous to their search in food webs (Camacho et al. 2002, Dunne et al. 2002a, Solé & Montoya 2001), Jordano and colleagues (2003) explored the degree distribution of 29 plant-pollinator networks and 24 plant-frugivore networks in natural communities. The bulk of the cases for both plants and animals (65.6%) showed degree distributions with a power-law regime but decaying as a marked cut-off (i.e., truncated power-laws or broad-scale networks); a few cases (22.2%) showed scale-invariance. The remaining networks either best fit an exponential distribution or showed no fit at all. The truncation of the degree distribution is described by the following equation:

$$p(k) \propto k^{-\gamma} e^{-k/k_c}, \qquad \qquad 2.$$

Figure 2

Weighted networks of plant-animal interactions. In an interaction web (*a*), we can represent the strength of mutual dependences between interacting species by the variable thickness of the links, shown in (*b*). For each pairwise interaction, two values of mutual dependence are obtained: d_{ji}^{A} for the dependence of the animal species *j* on plant species *i* (*orange arrow*), and d_{ij}^{P} for the dependence of plant species *i* on animal species *j* (*green arrow*). (*c*) Examples of the frequency distributions of dependence values for animals and plants in several communities, which illustrates the marked skew in interaction strength. Histograms in green represent dependences of plants on pollinators; orange represents dependences of animal frugivores on plants. Modified from Bascompte et al. (2006b).

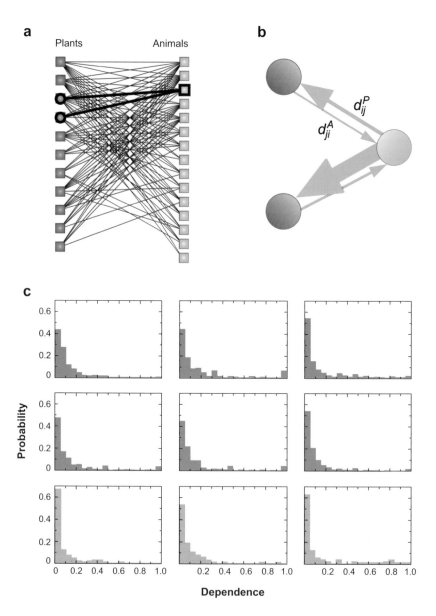

where the new term in relation to Equation 1 is e^{-k/k_c}, which defines the exponential cut-off. Mutualistic networks are still much more heterogeneous than expected by chance (i.e., the bulk of the species have a few interactions and a few species have a very large number of interactions), although not as heterogeneous as scale-free networks. The exponential truncation in Equation 2 signifies that as the number of interactions reaches the critical k_c value, the probability of finding more connected species drops faster than expected for a power-law (**Figure 1*c***). Both plant-pollinator and plant–seed disperser networks show the same degree distribution: The data for different communities collapse to a simple scaling function when the scaled cumulative distributions of links per species, $k^{-\gamma} P(k)$ are plotted versus the scaled links per species (k/k_c) (p. 74, Jordano et al. 2003). This result reveals a shared pattern of internal topology that is independent of scaling considerations (Bersier et al. 1999, Sugihara et al. 1989), which is evident only upon examination of the whole network of interactions, not just isolated species (Jordano et al. 2003). Regardless of the differences in latitude, ecosystem properties, and species composition, mutualistic networks display a common and well-defined connectivity distribution (Jordano et al. 2003).

Nestedness: a pattern of interaction in which specialists interact with species that form perfect subsets of the species with which generalists interact

3.2. Nestedness

This concept of network structure originated in the field of island biogeography and describes the pattern of species presences across islands (Atmar & Patterson 1993, Patterson & Atmar 1986). In the previous section we looked at the number of interactions per species without noticing the identity of the partners. Nestedness relates the set of animals that interact with one plant species, for example, with the set of animals that interact with another plant species (**Figure 1d,e**). Bascompte and colleagues (Bascompte et al. 2003) studied 27 plant-frugivore networks and 25 plant-pollinator networks and concluded that these networks are neither randomly assembled nor organized in compartments arising from tight, reciprocal specialization. Plant-animal mutualistic networks are highly nested (**Figure 1f**). That is, specialists interact with species that form well-defined subsets of the species with which generalists interact. In other words, if we rank plants from the most specialized to the least specialized, we find that the set of animals a plant interacts with are contained in a larger set, which in turn is contained in a larger set, and so on, as in nested Chinese boxes (Bascompte et al. 2003).

Two properties arise from a nested matrix (**Figure 1d**). First, there is a core of generalist plants and animals that interact among themselves. Therefore, a few species may be involved in a large number of interactions, which introduces functional redundancy and the possibility for alternative routes for system persistence if some of these interactions disappear. Thus, all species are very close to each other, which is also noted by the small path lengths or the average minimum number of intermediate species that separate two given species (Olesen et al. 2006). A few species in the core may drive the selective forces experienced by species that are subsequently attached to this network (Thompson 1994). Second, asymmetries exist in the level of specialization. That is, specialists tend to interact with the most generalist species (**Figure 1d**), as noted independently by Vázquez & Aizen (2004). Generalists tend to be more abundant, less-fluctuating species compared with specialists because generalists rely on so many other species. Thus, other things being equal, this asymmetrical structure provides pathways for the persistence of specialists (Bascompte et al. 2003). Essentially, nestedness means that mutualistic networks are very cohesive. As for the connectivity distribution, regardless of the type of mutualism and the ecological details, all communities are organized similarly (**Figure 1f**).

Researchers recently looked for nestedness in other types of mutualisms and different interactions (Guimarães et al. 2007b, Lafferty et al. 2006, Ollerton et al. 2003, Selva & Fortuna 2007). For example, Guimarães et al. (2006) analyzed ants and extrafloral nectary-bearing plants and found that nestedness values are very similar to the values reported previously for pollination and seed dispersal (Bascompte et al. 2003). Nestedness is found in fish parasites (Poulin & Valtonen 2001), marine cleaning mutualisms (Guimarães et al. 2007b), and scavenger communities (Selva & Fortuna 2007). In relation to food webs, analogous two-mode networks such as plants and herbivores or herbivores and carnivores are significantly less nested than mutualistic networks (Bascompte et al. 2003). This result is in agreement with the finding of a larger propensity for compartments in plant-herbivore networks (Lewinsohn et al. 2006). Finally, the consideration of parasites increases the level of nestedness of the food web, which may increase the cohesion and robustness of the whole network (Lafferty et al. 2006). Other properties of food webs, such as their organization in subwebs (Melián & Bascompte 2004) and the correlation between the degree of a species and the average degree of the species with which it interacts (Melián & Bascompte 2002), also show a cohesive core of generalists.

3.3. Dependences and Asymmetries

The network properties discussed above are based on qualitative data. Mutualism strength or the dependence of a plant species on an animal species has been estimated as the relative frequency of floral visits or the relative frequency of fruits consumed by that particular animal species (Bascompte et al. 2006b, Jordano 1987) (**Figure 2a,b**). Similarly, the dependence of an animal species on a plant species has been estimated as the relative frequency of fruits consumed that come from that particular plant species. Pairwise dependence seems to be a good surrogate for the total effects of a pairwise interaction in most networks (Vázquez et al. 2005).

As noted in the examples plotted in **Figure 2c**, the frequency distribution of dependence values is highly skewed, with many weak values and a few strong dependences (Bascompte et al. 2006b, Jordano 1987). This abundance of weak interactions has been reported also in food webs (Bascompte et al. 2005, Fagan & Hurd 1994, Paine 1980, 1992; Raffaelli & Hall 1995, Ulanowicz & Wolff 1991, Wootton 1997) and non-biological networks (Barrat et al. 2004). Interestingly enough, mounting evidence suggests that the dominance of weak interactions in food webs promotes community persistence and stability by buffering the transmission of perturbation through the whole community (Bascompte et al. 2005, Kokkoris et al. 1999, May 1973, McCann et al. 1998). Researchers compiled weighted interaction networks for host-parasitoid interactions (Müller et al. 1999, van Veen et al. 2006), which led to a community-wide quantification of parasitism rates and the role of indirect interactions (Müller et al. 1999).

The combination of the two dependence values within a plant-animal pair (**Figure2b**) is highly asymmetric (Bascompte et al. 2006b, Jordano 1987) but not more asymmetric than expected on the basis of the skewed distribution of dependence values (Bascompte et al. 2006b). However, in the few cases in which a plant species, for example, is highly dependent on an animal species, then that animal tends to rely significantly less on the plant (Bascompte et al. 2006b). Brazil's manduvi tree, for example, depends almost entirely on the Toco toucan to disperse its seeds, but this is not an exclusive relationship: The toucan also depends on a large, diverse group of other fruiting species. This asymmetry may help interdependent groups of species coexist, because if both plant and animal depend strongly on each other, a decrease in plant abundance will be followed by a similar decrease in the animal abundance, which in turn will feed back on its partner. This kind of downward loop is less common in uneven relationships because the plant could recover by relying on a generalist partner that depends on many other species. This verbal argument is shown mathematically as reviewed in section 5.2. Similar constraints in the combination of interaction strength values in food chains (e.g., avoiding strong interactions in long loops or in successive levels of tri-trophic food chains) also enhances food-web stability (Bascompte et al. 2005, Neutel et al. 2002).

Species strength: the sum of dependences or interaction strengths of the animals on a specific plant species, or the sum of dependences of the plants on a specific animal species

3.4. Species Strength

To explore how the weak, asymmetric dependences described in the previous section shape the whole network, we now consider the frequency distribution of species strength. A quantitative extension of species degree, species strength can be defined as the sum of dependences of the animals on a specific plant, or the sum of dependences of the plants on a specific animal. Species strength represents a measure of the quantitative importance of a species for the other set. Species strength increases faster than species degree (Bascompte et al. 2006b). Thus, mutualistic networks are even more heterogeneous when quantitative information is used.

Nestedness can partially explain this higher-than-expected strength of generalists in plant-animal assemblages. Owing to the nested structure, species with high degree interact with specialists. Because specialists by definition interact exclusively with these generalists, specialists contribute largely to increase the strength of generalists. When we described nestedness, we focused on asymmetry at the level of species. In this section we focus on asymmetry at the level of links, which certainly builds on the previous asymmetry. Further work should quantify what component of asymmetry at the link level is explained by asymmetry at the species level.

3.5. Network Structure and Sampling Effort

To conclude this section on network structure, we now briefly consider how robust these patterns are in terms of resolution and sampling effort. Some of the network patterns described here, such as degree distribution or nestedness, are defined using binary data, i.e., assuming that all links are the same. To what degree do these results stand up when quantitative information is used? As noted above, weighted extensions of both species degree (strength) and asymmetry of generalization (link asymmetry), confirm and expand on results obtained by the analysis of binary data. Once more, these networks seem to be very heterogeneous and asymmetric. Thus, the previous results are not an artifact of using binary data. In relation to sampling effort, the only study to our knowledge that addressed the issue of sampling effort on network structure explicitly concluded that nestedness is quite robust. Even when the number of species and, especially, the number of interactions grow with sampling intensity (both in time and space), the value of nestedness converges when a minimum sampling effort is reached (Nielsen & Bascompte 2007).

Forbidden links: pairwise interactions that are impossible to occur, for example, owing to phenological or size mismatch

4. ECOLOGICAL AND EVOLUTIONARY PROCESSES

Once network patterns are described, we may investigate the suite of ecological and evolutionary mechanisms that are responsible for generating such patterns. We saw above that physicists studied preferential attachment to generate some network patterns such as power-law degree distributions. One avenue is to explore to what degree several modifications of these basic mechanisms produce most network patterns. For example, which mechanisms lead to truncated power-law connectivity distributions? The most basic explanation is small size effects, i.e., the truncation of a power-law owing to the fact that the network is not large enough to accommodate extremely connected species (Guimarães et al. 2005, Jordano et al. 2003, Keitt & Stanley 1998, Mossa et al. 2002). However, Jordano and colleagues (2003) found that the frequency of truncated power-laws was not larger among the smallest communities, which suggests that other explanations are at work. Knowledge of the natural history of the mutualisms gives rise to a related explanation, the concept of forbidden links (Jordano et al. 2003, 2006). This refers to the fact that, in opposition to other nonbiological networks, some connections are not currently possible (at least over ecological time scales, e.g., one assumes no adaptation occurs) because of phenological or size constraints. A plant, for example, would not interact with a pollinator that is a late season migrant arriving at the community after the flowering period. A small bird, for example, would not be able to disperse the seeds of a species producing very large fruits. In one of the communities studied by Jordano and coworkers (2003), 51% of the nonobserved interactions were due to phenological uncoupling, and 24% were due to size restrictions.

Conversely, Vázquez (2005) proposed a neutral explanation in which network patterns can be explained on the basis of species abundance and random interactions. Researchers have claimed that abundance is a major factor in the explanation of network patterns, although most recent papers tend to shift the explanation back to the role of forbidden links (Santamaría & Rodríguez-Gironés 2007) or to a combination of abundance and forbidden links (Blüthgen et al. 2006, Stang et al. 2007). Santamaría and Rodríguez-Gironés (2007) discuss three reasons why neutrality should be rejected as the most parsimonious explanation of network topology in favor of forbidden links. First, it is not clear whether generalist species are generalists because they are more abundant or vice versa (Stang et al. 2007). Second, recent work by Blüthgen and coworkers (2006) has challenged the tenet that species abundance determines the frequency of interaction. Third, neutral theory assumes that the phenotypic characteristics of interacting species are irrelevant for network patterns, which contradicts strong confirmations for forbidden links (Jordano et al. 2003).

Forbidden links operate as constraints on the preferential attachment mechanism. Other mechanisms can explain truncations without such constraints. For example, when the initial core of species over which the preferential attachment operates is large enough, power-laws become truncated (Guimarães et al. 2005) without the need to adduce any further explanations. Also, the two-mode nature of these networks imposes truncations that would not be observed in one-mode networks if, for example, one of the sets (plants or animals) is much larger than the other (Guimarães et al. 2007a).

Phylogenetic signal: tendency of phylogenetically similar species to have similar phenotipic attributes

How are these basic mechanisms of network build-up mediated in ecological networks? For example, preferential attachment may be at work, but it may not necessarily act through species degree. If degree is correlated with any other ecological property, such as local abundance or geographic distribution, a new species may tend to become attached to the most abundant or more widely distributed species. Also, species are not independent entities but related to a common evolutionary history. Thus, we must first determine the magnitude of the phylogenetic signal in a species position (e.g., number of interactions, or with whom it interacts) in the network. In sum, we must look at the details of the species forming the network.

4.1. Phylogenetic and Ecological Correlates

Understanding interaction patterns from a biological perspective requires a combination of phylogenetic information and information on species' ecological traits to estimate effects on associations (Ives & Godfray 2006). Phylogenetic signal is the tendency of species closer in the phylogeny to have similar network properties (Blomberg et al. 2003, Freckleton et al. 2002, Garland et al. 2005, Ives & Godfray 2006, Lewinsohn et al. 2005). For example, **Figure 3a** illustrates a case where phylogenetically related species tend to have the same number of interactions, whereas **Figure 3b** shows a scenario with no relationship between phylogenetic proximity and the number of interactions. Finding a phylogenetic signal informs us about the extent to which past evolutionary history determines both the position of species in the network (e.g., their degree and with whom they interact) and the global network architecture. The role of past evolutionary history in explaining network patterns highlights the limitations of explanations based exclusively on ultimate ecological factors (Herrera 1992; Ives & Godfray 2006; P. Jordano & J. Bascompte, submitted; Rezende et al. 2007a,b).

4.1.1. Phylogenies and species positions. In a study on phylogenetic signal on a species position in the network, Rezende and colleagues (2007b) built a large data set with 36 plant pollinator and 23 plant-frugivore mutualistic networks and compiled the phylogenies for each plant and animal community (**Figure 3c**). These researchers found a significant phylogenetic signal in species degree (i.e., number of interactions per species) in approximately half of the largest phylogenies. Therefore, there is a tendency of species close in the phylogeny to have a similar number of interactions. However, species strength, the quantitative importance of a species for the other set, is only significant in 1 of the 38 phylogenies that correspond to weighted networks. The reason for this difference between degree and strength may lie in the fact that

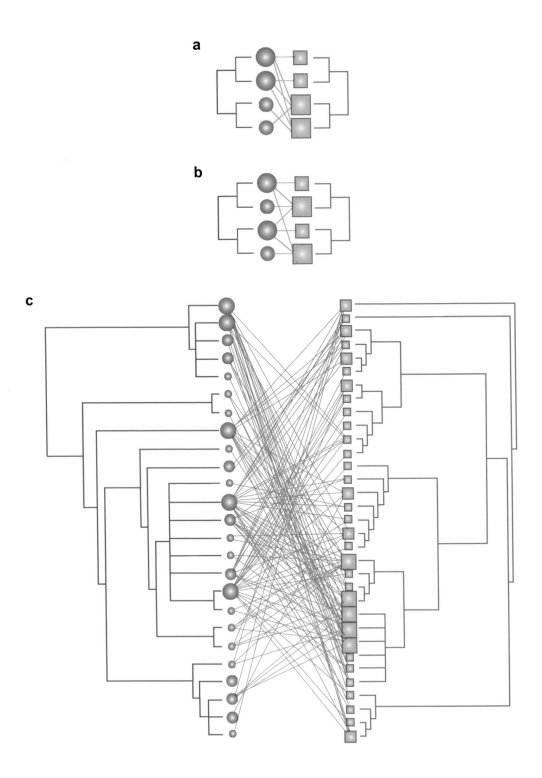

abundance affects species strength strongly and abundance may respond to more proximate, local factors.

The second property of a species position in the network analyzed by Rezende and coworkers (2007b) is with whom each species interacts. One can create a matrix of ecological dissimilarity between species in which two plants, for example, are very distant if they are visited by different pollinator species. Similarly, one can construct a matrix of phylogenetic distance. Rezende and coworkers analyzed the correlation between these two distance matrices and found a significant correlation between ecological and phylogenetic distance in approximately one-half of the 103 available phylogenies. This means that two phylogenetically similar species tend to interact with the same subset of species. Also, ecological and phylogenetic distances correlate better among animals than among plants (more than half the correlations were significant for animals whereas one-third were significant for plants). One potential explanation involves evolutionary differences linked to mobility: By playing a more active role, animals can search and select with whom they interact, which promotes selection for specific floral or fruit phenotypes (Rezende et al. 2007b).

In summary, there is significant phylogenetic signal in both the number of interactions per species and with whom they interact.

4.1.2. Ecological correlates of species positions. P. Jordano and J. Bascompte (submitted) performed phylogenetically independent contrasts between the ecological traits of a species and two measures of its position in the network: degree and eigenvector centrality. Eigenvector centrality is defined as the connectance of a node measured as the number of interactions of the node in question and the number of interactions of the nodes to which it is connected (Borgatti & Everett 1997, Jordano et al. 2006). Their goal was to see to what extent variation in degree and centrality across species correlates with local abundance, body or fruit size, geographic range, and phenological spread after accounting for phylogenetic effects. Jordano and Bascompte used network, phylogenetic, and ecological information from two Mediterranean communities in Southern Spain, and found the following results: Geographic range is significantly correlated with species degree both for plants and animals in the two communities; phenological spread is significantly correlated with species degree in both plants and animals in one community, and only with plants in the second community; and abundance is significantly correlated with species degree only for animals in both communities.

Figure 3

A phylogenetic approach to mutualistic networks. (*a*) A scenario with a strong phylogenetic signal in species degree, so that phylogenetically related species tend to have a similar number of interactions. Green circles represent plants and orange squares represent animals. The size of the node is proportional to its degree. (*b*) A similar case without phylogenetic signal. The phylogenetic information can be incorporated in the analysis of complex webs of interaction as shown in (*c*); this example corresponds to a plant-frugivore community in southeastern Spain. Plant phylogenies in this community are shown in green and animal phylogenies are shown in orange. Modified from Rezende et al. (2007b).

Ecological factors such as abundance are certainly involved in shaping mutualistic networks (Jordano 1987, Jordano et al. 2003, Olesen et al. 2002, Vázquez & Aizen 2004). However, the magnitude and even direction of the correlation between species degree and abundance may change across species and communities (Blüthgen et al. 2006). Also, more explanatory power may exist in the interaction between two such factors, such as abundance and morphological constraints (Jordano et al. 2003, Stang et al. 2007, Santamaría & Rodríguez-Gironés 2007).

4.1.3. Phylogenies and global network patterns. We turn now to phylogenetic effects on global network patterns. P. Jordano and J. Bascompte (submitted) tested the effect of plant and animal phylogenies in the explanation of the global structure of interactions via the use of the statistical methods recently developed by Ives & Godfray (2006). Phylogenetic covariation patterns explain a significant fraction of the total variance of the interaction pattern at the whole-network level (P. Jordano & J. Bascompte, submitted). In this case, as opposed to the findings for a species position in the network, there is a more marked effect of phylogeny for plants. This suggests that the overall pattern of interaction is influenced markedly by the evolutionary history of the plants and is more labile when mapped against the phylogeny of frugivores.

In summary, the results of this explicit use of phylogenetic data in the study of mutualistic networks provide insights into the ongoing assembly process. Both ultimate ecological factors and the evolutionary history conveyed in the phylogenies explain network patterns. The phylogenetic patterns of shared ancestry play a key role in the explanation of both species positions in the network and the overall pattern of mutualistic associations between the two sets of species.

5. IMPLICATIONS OF NETWORK STRUCTURE

5.1. Coevolutionary Implications

Plant-animal mutualistic networks form the physical template on which coevolution may proceed. Heterogeneous, nested networks built on weak and asymmetric interactions confer a predictable pattern of links among species that can both be generated by and affect coevolution. Two coevolutionary forces in combination can potentially generate a nested network: coevolutionary complementarity and coevolutionary convergence (Thompson 2005, 2006). Pairwise interactions build up on traits that are complementary between a plant and an animal, such as the length of the pollinator's tongue and the length of the corolla. This complementarity is key for the success of the pairwise interaction, and it is based on phenotypic traits that play a role in the fitness outcome of the interaction for the two partners. Once this pairwise interaction is defined, other species can become attached to the network through convergence of traits. One example is the syndromes or convergence in fruit shape and color among species that are dispersed by mammals as opposed to birds. Support for the role of coevolutionary complementarity comes from simulations that indicate that phenotypic complementarity, particularly when several traits are involved, produces highly nested networks (Rezende et al. 2007a, Santamaría & Rodríguez-Gironés 2007).

The identity of the species in the core of a nested network, with the potential to drive the coevolution of the whole network, can change geographically. Local communities vary in species composition relative to regional pools of species, which results in different local assemblages of mutualists. This phenomenon provides a link between the two major theories that bring tractability to multispecific coevolutionary studies: network theory and the geographic mosaic theory (Bascompte & Jordano 2006, Thompson 1994, 2005). For example, to what extent is local network structure explained by properties at the landscape level? As we discussed in the previous section, a species degree is correlated with its geographic distribution. The most generalized species, those that form the core of the matrix, will probably be present across communities, whereas specialists may be more variable across communities (Bascompte & Jordano 2006).

5.2. Implications for Network Robustness

The architecture of mutualistic networks may have profound implications for robustness, which is defined as network resistance to species loss (**Figure 4**). Albert and coworkers (2000) illustrated that random networks with exponential degree distributions are very fragile: The network suddenly fragments after the removal of a small fraction of nodes. Conversely, a network with a scale-free degree distribution is very robust to the random loss of nodes, but very fragile to the extinction of the most generalist species (Albert et al. 2000). Similar species-deletion experiments in food webs provided information on the fragility of ecological networks (Dunne et al. 2002b, Pimm 1979, Solé & Montoya 2001) (**Figure 4**).

More recently, Memmott and coworkers (2004) simulated the progressive extinction of pollinators and explored the cumulative secondary extinction of the plants that depend on them (Jordano et al. 2006, Memmott et al. 2004, Morris 2003). Memmott and coworkers (2004) concluded that mutualistic networks are very robust and referred both to the truncated power-law degree distribution and to the nested structure as combined explanations for such robustness. On theoretical grounds, networks with truncated power-law distributions of the species degree (broad-scale, as opposed to the scale-free distribution) are less fragile to the loss of the most-connected nodes (Albert et al. 2000).

A second approach to network robustness is a dynamic approach. This approach describes whether small fluctuations around a steady state will amplify or die out (Bronstein et al. 2004, May 1973, Solé & Bascompte 2006). The analysis of a simple model of multispecies facultative mutualisms revealed that as community size increases, the average product of pair-wise mutualistic effects must decrease for the community to remain stable (Bascompte et al. 2006b). This is in agreement with two network patterns reported above, namely, the high frequency of weak dependence values, and their asymmetry when one dependence is large (Bascompte et al. 2006b). However, as with any analytic results, a number of strong assumptions are required to generate such a clear, straightforward expression (Bascompte et al. 2006a, Holland et al. 2006). One such assumption is that all animals interact with all plants, i.e., the model does not incorporate network structure. Fortuna & Bascompte (2006) took a

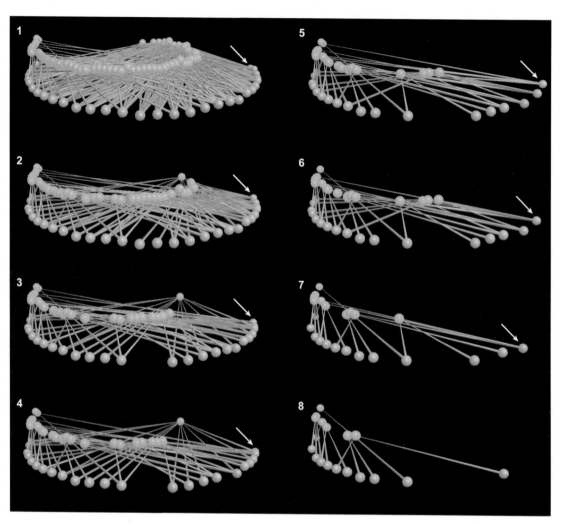

Figure 4

Consequences of species extinction on mutualistic networks. A single plant species is deleted at each step, from (*1*) to (*8*), starting from the most generalist species and proceeding toward the most specialist species. All species that become isolated undergo coextinction. In this case, the network is very fragile, as shown by the large number of secondary extinctions. However, the same network would be very robust (almost all species would persist) if the first species to go extinct were the specialists, or the loss of species was at random. The structure of the network, as described by its degree distribution (**Figure 1*b***) and nestedness (**Figure 1*f***), highly conditions network robustness to species extinctions. Image produced with FoodWeb3D, written by R.J. Williams and provided by the Pacific Ecoinformatics and Computational Ecology Lab (**http://www.foodwebs.org**).

first step toward introducing network structure in dynamical models by studying a metacommunity model in which species interact exactly as in two real mutualistic networks. Real networks, as compared with randomizations, start losing species sooner, but the community as a whole persists for higher values of habitat loss (Fortuna & Bascompte 2006). This robustness is again explained by both the heterogeneous degree distribution and the cohesive organization in nested systems.

5.3. Implications for Conservation Biology

The fact that mutualistic networks form well-defined and predictable patterns of interdependences provides a community-wide perspective for species conservation. For example, because of the asymmetry in specialization, both specialist and generalist plant species exhibit similar reproductive susceptibility to habitat loss: Although specialists depend on a single resource, they tend to interact with the most generalist animal species. Other things being equal, generalist animals tend to be more abundant (Ashworth et al. 2004).

The invasion of exotic species is a leading factor in mutualism disruptions (Bronstein et al. 2004, Traveset & Richardson 2006). At least three papers explored how network structure affects the likelihood of invasions (Memmott & Waser 2002, Morales & Aizen 2006, Olesen et al. 2002). All three studies concluded that invasive species become well integrated into the existing pollination network. Memmott & Waser (2002), for example, conclude that a lower number of pollinator species visited flowers of alien plants compared with native plants, but these insects were extremely generalist (Memmott & Waser 2002). This asymmetry in specialization is in agreement with the predictions of a nested community as noted above. Thus, network architecture provides alien species with more abundant and reliable resources. However, disagreement exists regarding the likelihood of invader complexes, groups of plant and animal invaders that rely more on each other than on native species (Olesen et al. 2002). These invader complexes are important because they can increase invasion speed and establishment success greatly. Whereas Olesen and colleagues (2002) found that introduced plants and pollinators do not interact as much as expected by chance, and so there is no evidence of invader complexes (Olesen et al. 2002), Morales & Aizen (2006) found that alien flower visitors were more closely associated with alien than with native species.

Another issue in conservation biology that requires a network approach is defaunation, which is an increasing problem in tropical ecosystems and has far-reaching consequences for biotic interactions (Dirzo & Miranda 1990). Hunting preferentially targets large species of mammals and birds that play a paramount role in seed dispersal because they are highly mobile and contribute disproportionately to connectivity in fragmented landscapes. The community-wide effects of the extinction of such large species depends on the structure of mutualistic networks and their ecological correlates (see above). For example, are large-bodied frugivores randomly scattered through the matrix of interactions, or are they more likely the generalists forming the core? In the latter case, the nested structure of mutualistic networks implies that losing these few species may induce a collapse of the whole network, which is defined as the fragmentation of a previous single-connected cluster into a set of disconnected subsets.

SUMMARY POINTS

1. Mutually beneficial interactions such as pollination and seed dispersal form heterogeneous, nested networks built on weak and asymmetric links among animal and plant species.

2. Researchers find a common, well-defined network architecture regardless of the type of mutualism, species composition, latitude, and other variables.

3. Mutualistic networks can be approached neither as collections of pair-wise, highly specific interactions nor as diffuse, random assemblages.

4. The above network patterns may facilitate species persistence; mutualistic networks can thus be regarded as the architecture of biodiversity.

5. Several ecological factors and evolutionary history contribute to create the observed network patterns.

FUTURE ISSUES

1. The field requires an exploration of how mutualistic networks change in time and space.

2. Also needed is the development of a mathematical theory for mutualistic networks that is aimed at linking network structure and dynamics.

DISCLOSURE STATEMENT

The authors are not aware of any biases that might be perceived as affecting the objectivity of this review.

ACKNOWLEDGMENTS

We thank J. Bronstein, P.R. Guimarães Jr., C. Müller, J.M. Olesen, D. Schemske, J.N. Thompson, A. Valido, and D. Vázquez for their comments on a previous draft. J.M. Olesen, T. Lewinsohn, C.J. Melián, M.A. Fortuna, E. Rezende, and J.N. Thompson contributed to shape our views on the subjects reviewed here. Our work was funded by the European Heads of Research Councils, the European Science Foundation, and the EC Sixth Framework Program through a EURYI (European Young Investigator) Award (to J.B.), by the Spanish Ministry of Education and Science (Grants REN2003-04774 to J.B. and CGL2006-00373 to P.J.), and by a Junta de Andalucía Excellence Grant (P.J. and J.B.).

LITERATURE CITED

Albert R, Barabasi AL. 2002. Statistical mechanics of complex networks. *Rev. Mod. Phys.* 74:47–97

Albert R, Jeong H, Barabasi AL. 2000. Error and attack tolerance of complex networks. *Nature* 406:378–82

Amaral LA, Scala A, Barthelemy M, Stanley HE. 2000. Classes of small-world networks. *Proc. Natl. Acad. Sci. USA* 97:11149–52

Ashworth L, Aguilar R, Galetto L, Aizen MA. 2004. Why do pollination generalist and specialist plant species show similar reproductive susceptibility to habitat fragmentation? *J. Ecol.* 92:717–19

Atmar W, Patterson BD. 1993. The measure of order and disorder in the distribution of species in fragmented habitat. *Oecologia* 96:373–82

Barabasi AL, Albert R. 1999. Emergence of scaling in random networks. *Science* 286:509–12

Barrat A, Barthelemy M, Pastor-Satorras R, Vespignani A. 2004. The architecture of complex weighted networks. *Proc. Natl. Acad. Sci. USA* 101:3747–52

Bascompte J, Jordano P. 2006. The structure of plant-animal mutualistic networks. See Pascual & Dunne 2006, pp. 143–59

Bascompte J, Jordano P, Melián CJ, Olesen JM. 2003. The nested assembly of plant-animal mutualistic networks. *Proc. Natl. Acad. Sci. USA* 100:9383–87

Bascompte J, Jordano P, Olesen JM. 2006a. Response to comment on "Asymmetric coevolutionary networks facilitate biodiversity maintenance." *Science* 313:1887

Bascompte J, Jordano P, Olesen JM. 2006b. Asymmetric coevolutionary networks facilitate biodiversity maintenance. *Science* 312:431–33

Bascompte J, Melián CJ. 2005. Simple trophic modules for complex food webs. *Ecology* 86:2868–73

Bascompte J, Melián CJ, Sala E. 2005. Interaction strength combinations and the overfishing of a marine food web. *Proc. Natl. Acad. Sci. USA* 102:5443–47

Bawa K. 1990. Plant-pollinator interactions in tropical rain forests. *Annu. Rev. Ecol. Syst.* 21:399–422

Belgrano A, Scharler U, Dunne JA, Ulanowicz RE, eds. 2005. *Aquatic Food Webs: An Ecosystem Approach*. Oxford, UK: Oxford Univ. Press

Bersier L-F, Dixon P, Sugihara G. 1999. Scale-invariant or scale-dependent behavior of the link density property in food webs: a matter of sampling effort? *Am. Nat.* 153:676–82

Blomberg SP, Garland T Jr, Ives AR. 2003. Testing for phylogenetic signal in comparative data: Behavioral traits are more labile. *Evolution* 57:717–45

Blüthgen N, Menzel F, Blüthgen N. 2006. Measuring specialization in species interaction networks. *BMC Ecol.* 6:9

Borgatti SP, Everett MG. 1997. Network analysis of 2-mode data. *Soc. Netw.* 19:243–69

Bronstein J. 1995. The plant-pollinator landscape. In *Mosaic Landscapes and Ecological Processes*, ed. L Hansson, L Fahrig, G Merriam, pp. 257–88. London: Chapman & Hall

Bronstein J, Dieckmann U, Ferrière R. 2004. Coevolutionary dynamics and the conservation of mutualisms. In *Evolutionary Conservation Biology*, ed. R Ferrière, U Dieckmann, D Couvet, pp. 305–26. Cambridge, UK: Cambridge Univ. Press

Camacho J, Guimerà R, Nunes Amaral LA. 2002. Robust patterns in food web structure. *Phys. Rev. Lett.* 88:228102

Cohen JE. 1978. *Food Webs and Niche Space*. Princeton, NJ: Princeton Univ. Press

Cohen JE, Briand F, Newman CM. 1990. *Community Food Webs: Data and Theory*. Berlin: Springer-Verlag

Cook JM, Rasplus JY. 2003. Mutualisms with attitude: Coevolving fig wasps and figs. *Trends Ecol. Evol.* 18:241–48

Darwin C. 1862. *On the Various Contrivances by Which British and Foreign Orchids are Fertilised by Insects, and on the Good Effects of Intercrossing*. London: Murray

Dirzo R, Miranda A. 1990. Contemporary neotropical defaunation and forest structure, function, and diversity. *Conserv. Biol.* 4:444–47

Dodd M, Silvertown J, Chase M. 1999. Phylogenetic analysis of trait evolution and species diversity variation among angiosperm families. *Evolution* 53:732–44

Dorogovtsev S, Mendes J. 2002. Evolution of networks. *Adv. Phys.* 51:1079–187

Dunne JA, Williams RJ, Martinez ND. 2002a. Food-web structure and network theory: The role of connectance and size. *Proc. Natl. Acad. Sci. USA* 99:12917–22

Dunne JA, Williams RJ, Martinez ND. 2002b. Network structure and biodiversity loss in food webs: robustness increases with connectance. *Ecol. Lett.* 5:558–67

Ehrlich P, Raven P. 1964. Butterflies and plants: a study in coevolution. *Evolution* 18:586–608

Erdös P, Rényi A. 1959. On random graphs. *Publ. Math.* 6:290–97

Fagan W, Hurd L. 1994. Hatch density variation of a generalist arthropod predator: population consequences and community impact. *Ecology* 75:2022–32

Feinsinger P. 1978. Ecological interactions between plants and hummingbirds in a successional tropical community. *Ecol. Monogr.* 48:269–87

Fortuna MA, Bascompte J. 2006. Habitat loss and the structure of plant-animal mutualistic networks. *Ecol. Lett.* 9:281–86

Fox LR. 1988. Diffuse coevolution within complex communities. *Ecology* 69:906–7

Freckleton R, Harvey P, Pagel M. 2002. Phylogenetic analysis and comparative data: a test and review of evidence. *Am. Nat.* 160:712–26

Garland T Jr, Bennett A, Rezende E. 2005. Phylogenetic approaches in comparative physiology. *J. Exp. Biol.* 208:3015–35

Guimarães PR Jr, de Aguiar MAM, Bascompte J, Jordano P, Furtado dos Reis S. 2005. Random initial condition in small Barabasi-Albert networks and deviations from the scale-free behavior. *Phys. Rev. E* 71:037101

Guimarães PR Jr, Machado G, de Aguiar MAM, Jordano P, Bascompte J, et al. 2007a. Build-up mechanisms determining the topology of mutualistic networks. *J. Theor. Biol.* In press

Guimarães PR Jr, Rico-Gray V, dos Reis SF, Thompson JN. 2006. Asymmetries in specialization in ant-plant mutualistic networks. *Proc. R. Soc. London Ser. B* 273:2041–47

Guimarães PR Jr, Sazima C, Furtado dos Reis S, Sazima I. 2007b. The nested structure of marine cleaning symbiosis: is it like flowers and bees? *Biol. Lett.* 3:51–54

Herrera C. 1982. Seasonal variation in the quality of fruits and diffuse coevolution between plants and avian dispersers. *Ecology* 63:773–85

Herrera C. 1992. Historical effects and sorting processes as explanations for contemporary ecological patterns: character syndromes in Mediterranean woody plants. *Am. Nat.* 140:421–46

Holland JN, Okuyama T, DeAngelis DL. 2006. Comment on "Asymmetric coevolutionary networks facilitate biodiversity maintenance." *Science* 313:1887

Inoue T, Kato M, Kakutani T, Suka T, Itino T. 1990. Insect-flower relationship in the temperate deciduous forest of Kibune, Kyoto: an overview of the flowering phenology and the seasonal pattern of insect visits. *Contrib. Biol. Lab., Kyoto Univ.* 27:377–463

Inouye B, Stinchcombe JR. 2001. Relationships between ecological interaction modifications and diffuse coevolution: similarities, differences, and causal links. *Oikos* 95:353–60

Ives A, Godfray H. 2006. Phylogenetic analysis of trophic associations. *Am. Nat.* 168:E1–14

Iwao K, Rausher MD. 1997. Evolution of plant resistance to multiple herbivores: Quantifying diffuse coevolution. *Am. Nat.* 149:316–35

Janzen D. 1980. When is it coevolution? *Evolution* 34:611–12

Johnson SD, Steiner KE. 1997. Long-tongued fly pollination and evolution of floral spur length in the Disa draconis complex (Orchidaceae). *Evolution* 51:45–53

Jordano P. 1987. Patterns of mutualistic interactions in pollination and seed dispersal: connectance, dependence asymmetries, and coevolution. *Am. Nat.* 129:657–77

Jordano P. 2000. Fruits and frugivory. In *Seeds: The Ecology of Regeneration in Natural Plant Communities*, ed. M Fenner, pp. 125–66. Wallingford, UK: Commonw. Agric. Bur. Int.

Jordano P, Bascompte J, Olesen JM. 2006. The ecological consequences of complex topology and nested structure in pollination webs. In *Specialization and Generalization in Plant-Pollinator Interactions*, ed. NM Waser, J Ollerton, pp. 173–99. Chicago: Univ. Chicago Press

Jordano P, Bascompte J, Olesen JM. 2003. Invariant properties in coevolutionary networks of plant-animal interactions. *Ecol. Lett.* 6:69–81

Kearns C, Inouye D, Waser N. 1998. Endangered mutualisms: the conservation of plant-pollinator interactions. *Annu. Rev. Ecol. Evol. Syst.* 29:83–112

Keitt T, Stanley H. 1998. Dynamics of North American breeding bird populations. *Nature* 393:257–60

Kokkoris G, Troumbis A, Lawton J. 1999. Patterns of species interaction strength in assembled theoretical competition communities. *Ecol. Lett.* 2:70–74

Labandeira C. 2002. The history of associations between plants and animals. In *Plant-animal Interactions. An Evolutionary Approach*, ed. C Herrera, O Pellmyr, pp. 26–74. Oxford, UK: Blackwell

Labandeira C, Johnson K, Wilf P. 2002. Impact of the terminal cretaceous event on plant-insect associations. *Proc. Natl. Acad. Sci. USA* 99:2061–66

Lafferty K, Dobson A, Kurtis A. 2006. Parasites dominate food web links. *Proc. Natl. Acad. Sci. USA* 103:11211–16

Lewinsohn TM, Novotny V, Basset Y. 2005. Insects on plants: Diversity of Herbivore assemblages revisited. *Annu. Rev. Ecol. Evol. Syst.* 36:597–620

Lewinsohn TM, Prado PI, Jordano P, Bascompte J, Olesen JM. 2006. Structure in plant-animal interaction assemblages. *Oikos* 113:174–84

Luscombe NM, Babu MM, Yu H, Snyder M, Teichmann SA, Gerstein M. 2004. Genomic analysis of regulatory network dynamics reveals large topological changes. *Nature* 431:308–12

May RM. 1973. *Stability and Complexity in Model Ecosystems.* Princeton, NJ: Princeton Univ. Press

May RM. 2006. Network structure and the biology of populations. *Trends Ecol. Evol.* 21:394–99

McCann AK, Hastings A, Huxel G. 1998. Weak trophic interactions and the balance of nature. *Nature* 395:794–98

Melián CJ, Bascompte J. 2002. Complex networks: two ways to be robust? *Ecol. Lett.* 5:705–8

Melián CJ, Bascompte J. 2004. Food web cohesion. *Ecology* 85:352–58

Memmott J. 1999. The structure of a plant-pollination food web. *Ecol. Lett.* 2:276–80

Memmott J, Waser NM. 2002. Integration of alien plants into a native flower-pollinator visitation web. *Proc. Biol. Sci.* 269:2395–99

Memmott J, Waser NM, Price MV. 2004. Tolerance of pollination networks to species extinctions. *Proc. Biol. Sci.* 271:2605–11

Montoya JM, Pimm SL, Solé RV. 2006. Ecological networks and their fragility. *Nature* 442:259–64

Morales CL, Aizen MA. 2006. Invasive mutualisms and the structure of plant-pollinator interactions in the temperate forests of north-west Patagonia, Argentina. *J. Ecol.* 94:171–80

Morris WF. 2003. Which mutualisms are most essential? Buffering of plant reproduction against the extinction of different kinds of pollinator. In *The Importance of Species*, ed. P Kareiva, S Levin, pp. 260–80. Princeton, NJ: Princeton Univ. Press

Mossa S, Barthélémy M, Stanley H, Amaral LAN. 2002. Truncation of power law behavior in 'scale-free' network models due to information filtering. *Phys. Rev. Lett.* 88:138701

Müller CB, Adriaanse ICT, Belshaw R, Godfray HCJ. 1999. The structure of an aphid-parasitoid community. *J. Anim. Ecol.* 68:346–70

Neutel AM, Heesterbeek JA, de Ruiter PC. 2002. Stability in real food webs: Weak links in long loops. *Science* 296:1120–23

Newman MEJ. 2003. The structure and function of complex networks. *SIAM Rev.* 45:167–256

Newman MEJ. 2004. Analysis of weighted networks. *Phys. Rev. E* 70:056131

Newman MEJ, Barabasi AL, Watts DJ. 2006. *The Structure and Dynamics of Networks.* Princeton, NJ: Princeton Univ. Press

Nielsen A, Bascompte J. 2007. Ecological networks, nestedness, and sampling effort. *J. Ecol.* 95:1134–41

Nilsson L. 1988. The evolution of flowers with deep corolla tubes. *Nature* 334:147–49

Olesen J, Bascompte J, Dupont Y, Jordano P. 2006. The smallest of all worlds: pollination networks. *J. Theor. Biol.* 240:270–76

Olesen J, Eskildsen L, Venkatasamy S. 2002. Invasion of pollination networks on oceanic islands: importance of invader complexes and epidemic super generalists. *Divers. Distrib.* 8:181–92

Ollerton J, Johnson SD, Cranmer L, Kellie S. 2003. The pollination ecology of an assemblage of grassland asclepiads in South Africa. *Ann. Bot.* 92:807–34

Paine RT. 1980. Food webs: Linkage, interaction strength and community infrastructure. The third Tansley lecture. *J. Anim. Ecol.* 49:667–85

Paine RT. 1992. Food-web analysis through field measurement of per capita interaction strength. *Nature* 355:73–75

Parchman T, Benkman C. 2002. Diversifying coevolution between crossbills and black spruce on Newfoundland. *Evolution* 56:1663–72

Pascual M, Dunne JA, eds. 2006. *Ecological Networks. Linking Structure to Dynamics in Food Webs.* Oxford, UK: Oxford Univ. Press

Patterson BD, Atmar W. 1986. Nested subsets and the structure of insular mammalian faunas and archpielagos. *Biol. J. Linn. Soc.* 28:65–82

Pellmyr O. 1992. Evolution of insect pollination and angiosperm diversification. *Trends Ecol. Evol.* 7:46–49

Pellmyr O. 2003. Yuccas, yucca moths, and coevolution: a review. *Ann. Mo. Bot. Gard.* 90:35–55

Petanidou T, Ellis W. 1993. Pollinating fauna of a phryganic ecosystem: composition and diversity. *Biodivers. Lett.* 1:22

Pimm SL. 1979. The structure of food webs. *Theor. Popul. Biol.* 16:144–58

Pimm SL. 1982. *Food Webs.* London: Chapman & Hall

Poulin R, Valtonen ET. 2001. Nested assemblages resulting from host size variation: the case of endoparasite communities in fish hosts. *Int. J. Parasitol.* 31:1194–204

Price DJdeS. 1965. Networks of scientific papers. *Science* 149:510–15

Proulx SR, Promislow DE, Phillips PC. 2005. Network thinking in ecology and evolution. *Trends Ecol. Evol.* 20:345–53

Raffaelli DG, Hall SJ. 1995. Assessing the importance of trophic links in food webs. In *Food Webs: Integration of Pattern and Dynamics*, ed. GA Polis, KO Winemiller, pp. 185–91. New York: Chapman & Hall

Rezende E, Jordano P, Bascompte J. 2007a. Effects of phenotypic complementarity and phylogeny on the nested structure of mutualistic networks. *Oikos.* In press

Rezende E, Lavabre J, Guimarães PR Jr, Jordano P, Bascompte J. 2007b. Nonrandom coextinctions in phylogenetically structured mutualistic networks. *Nature.* In press

Santamaría L, Rodríguez-Gironés M. 2007. Linkage rules for plant-pollinator networks: Trait complementarity or exploitation barriers? *PLoS Biol.* 5(2):e31

Schemske DW. 1983. Limits to specialization and coevolution in plant-animal mutualisms. In *Coevolution*, ed. MH Nitecki, pp. 67–110. Chicago: Univ. Chicago Press

Schroeder M. 1991. *Fractals, Chaos, and Power Laws*. New York: Freeman

Selva N, Fortuna M. 2007. The nested structure of a scavenger community. *Proc. R. Soc.* 274:1101–8

Simon H. 1955. On a class of skewed distribution functions. *Biometrika* 42:425–40

Solé RV, Bascompte J. 2006. *Self-Organization in Complex Ecosystems*. Princeton, NJ: Princeton Univ. Press

Solé RV, Montoya JM. 2001. Complexity and fragility in ecological networks. *Proc. R. Soc. B* 268:2039–45

Stang M, Klinkhamer PGL, van der Meijden E. 2007. Asymmetric specialization and extinction risk in plant-flower visitor webs: a matter of morphology or abundance? *Oecologia* 151:442–53

Strogatz SH. 2001. Exploring complex networks. *Nature* 410:268–76

Sugihara G, Schoenly K, Trombla A. 1989. Scale invariance in food web properties. *Science* 245:48–52

Thompson JN. 1994. *The Coevolutionary Process*. Chicago: Univ. Chicago Press

Thompson JN. 2005. *The Geographic Mosaic of Coevolution*. Chicago: Univ. Chicago Press

Thompson JN. 2006. Mutualistic webs of species. *Science* 312:372–73

Thompson JN, Pellmyr O. 1992. Mutualism with pollinating seed parasites amid copollinators: constraints on specialization. *Ecology* 73:1780–91

Tiffney BH. 2004. Vertebrate dispersal of seed plants through time. *Annu. Rev. Ecol. Evol. Syst.* 35:1–29

Traveset A, Richardson DM. 2006. Biological invasions as disruptors of plant reproductive mutualisms. *Trends Ecol. Evol.* 21:208–16

Ulanowicz RE, Wolff WF. 1991. Ecosystem flow networks: loaded dice? *Math. Biosci.* 103:45–68

van Veen FJ, Morris RJ, Godfray HC. 2006. Apparent competition, quantitative food webs, and the structure of phytophagous insect communities. *Annu. Rev. Entomol.* 51:187–208

Vázquez DP. 2005. Degree distribution in plant-animal mutualistic networks: forbidden links or random interactions? *Oikos* 108:421–26

Vázquez DP, Aizen MA. 2004. Asymmetric specialization: a pervasive feature of plant-pollinator interactions. *Ecology* 85:1251–57

Vázquez DP, Morris WF, Jordano P. 2005. Interaction frequency as a surrogate for the total effect of animal mutualists on plants. *Ecol. Lett.* 8:1088–94

Waser NM, Chittka L, Price MV, Williams N, Ollerton J. 1996. Generalization in pollination systems, and why it matters. *Ecology* 77:1043–60

Watts JD. 2003. *Small Worlds: The Dynamics of Networks between Order and Randomness*. Princeton, NJ: Princeton Univ. Press

Wootton JT. 1997. Estimates and tests of per capita interaction strength: diet, abundance, and impact of intertidally foraging birds. *Ecol. Monogr.* 67:45–64

Wright SJ. 2003. The myriad consequences of hunting for vertebrates and plants in tropical forests. *Perspect. Plant Ecol. Evol. Syst.* 6:73–86

RELATED RESOURCES

Software packages for the analysis of complex networks:

Aninhado **http://www.guimaraes.bio.br/sof.html**

> A software package for estimation of nestedness using NTC algorithm but including new null models.
>
> Guimarães PR Jr, Guimarães PR. 2006. Improving the analyses of nestedness for large sets of matrices. *Environ. Model. Softw.* 21:1512–13

Binmatnest **http://www.eeza.csic.es/eeza/personales/rgirones.aspx**

> Another package calculating the nestedness temperature of binary presence-absence matrices in a different way than the NTC.
>
> Rodríguez-Gironés MA, Santamaría L. 2006. A new algorithm to calculate the nestedness temperature of presence-absence matrices. *J. Biogeogr.* 33:924–35

FoodWeb3D **http://www.foodwebs.org/index_page/wow2.html**

> Written by R.J. Williams and provided by the Pacific Ecoinformatics and Computational Ecology Lab (**http://www.foodwebs.org**).

Libraries sna, network, nettheory in R package **http://www.r-project.org/**

NTC **http://www.aics-research.com/nestedness/tempcalc.html**

> The Nestedness Calculator measures the extent of the order present in nested presence-absence matrices.
>
> Atmar W, Patterson BD. 1993. The measure of order and disorder in the distribution of species in fragmented habitat. *Oecologia* 96:373–82

Pajek **http://vlado.fmf.uni-lj.si/pub/networks/pajek/**

> Batagelj V, Mrvar A. 2003. Pajek—Analysis and visualization of large networks. In *Graph Drawing Software*, ed. M. Juenger, P Mutzel, pp. 77–103. Berlin: Springer-Verlag.

Gene Flow and Local Adaptation in Trees

Outi Savolainen, Tanja Pyhäjärvi, and Timo Knürr

Department of Biology, University of Oulu, FIN-90014 Finland; email:
Outi.Savolainen@oulu.fi; Tanja.Pyhajarvi@oulu.fi; Timo.Knurr@oulu.fi

Annu. Rev. Ecol. Evol. Syst. 2007. 38:595–619

First published online as a Review in Advance on
August 17, 2007

The *Annual Review of Ecology, Evolution, and
Systematics* is online at
http://ecolsys.annualreviews.org

This article's doi:
10.1146/annurev.ecolsys.38.091206.095646

Key Words

cline, natural selection, provenance trial, polygenic variation, migration

Abstract

Populations are locally adapted when populations have the highest relative fitness at their home sites, and lower fitness in other parts of the range. Results from the extensive experimental plantations of populations of forest trees from different parts of the range show that populations can survive and grow in broad areas outside the home site. However, intra- and interspecific competition limit the distribution of genotypes. For populations from large parts of the range, relative fitness, compared with the local population, is often highest at the home site. At the edges of the range, this local adaptation may break down. The extent of local adaptation is determined by the balance between gene flow and selection. Genetic differentiation and strong natural selection occur over a range of tens or hundreds of kilometers, but reliable measurements of gene flow are available only for much shorter distances. Current models of spatially varying selection could be made more realistic by the incorporation of strong selection and isolation-by-distance characteristic of tree populations. Many studies suggest that most variation in adaptive traits is based on loci with small effects. Association genetics methods and improved genomic resources are useful for the identification of the loci responsible for this variation. The potential for adaptation to current climate change depends on genetic variation and dispersal and establishment rates.

INTRODUCTION

The evolution of adaptation has received increasing attention in recent years. Theoretical work has generated new predictions on the genetic dynamics and architecture of adaptation (Orr 1998). Specifically, local adaptation (Kawecki & Ebert 2004) and adaptation to clinally varying environments have been addressed (Barton 1999). Genetic tools, such as quantitative trait locus (QTL) mapping (Tanksley 1993) and association studies, are now available for the study of the genetic basis of adaptation. The study of sequence variation also provides means to detect loci responsible for local adaptation (Wright & Gaut 2005). Ongoing climate change has increased interest in the ability of species and populations to adapt to new environmental conditions (Houghton et al. 2001). These developments led to new approaches in the study of adaptation of trees.

Tree species occupy shifting geographic ranges, as documented in pollen and other fossil records (Huntley & Birks 1983, Willis & van Andel 2004). Range expansions and contractions have left their marks both in the plastid DNA and the nuclear genes of current populations (Heuertz et al. 2006, Petit et al. 2003). Comparisons of molecular and quantitative data suggest that recent selection, which occurred after postglacial colonization, is the predominant factor that shapes present quantitative trait variation (Collignon et al. 2002, Kremer et al. 2002).

Local adaptation can occur with respect to many selective factors, such as climate, edaphic factors, and parasites (Hedrick 2006, Kawecki & Ebert 2004, Linhart & Grant 1996), and different traits can respond to selection. Endler (1977) provides examples of small scale clinal variation in trees.

Foresters study adaptation with the establishment of common garden experiments at multiple sites; each contains many provenances (populations from well defined geographic areas) of the species. The goal of these studies is to find the most suitable provenances for planting in different environments. These studies provide extensive information on the effects of provenance transfers and genetic differentiation of tree populations (Langlet 1971, Morgenstern 1996).

In this review, we combine results from these classical studies on trees with evolutionary theory, new findings on migration, and recent extensive genetic work. Specifically, we ask the following questions: First, which genetic models of spatially varying selection are most relevant for forest tree populations? Second, how much gene flow can be detected with recently developed molecular tools? Third, do old provenance trials provide evidence of local adaptation? Fourth, what is the evidence for local adaptation from genetic differentiation studies in common garden experiments? Fifth, what is the genetic architecture of adaptive traits: many loci with small effects or fewer loci with larger effects? Last, is local adaptation possible at range margins, and can trees respond genetically to climate change caused by humans?

BACKGROUND FOR LOCAL ADAPTATION

Spatial variation in the pattern of natural selection can lead to local adaptation and genetic differentiation between populations. Ecologists have studied local adaptation

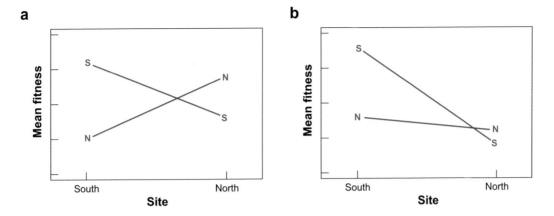

Figure 1

The definition of local adaptation (Kawecki & Ebert 2004). The fitnesses of the northern and southern populations show a genotype by environment interaction. In (a) each has highest absolute and relative fitness at its local site, in (b) both have highest absolute fitness in the south, but each has highest relative fitness at its local site.

in reciprocal transplant experiments. In a classical study, Clausen et al. (1948) transplanted *Achillea* populations between coastal, mid-altitude, and high elevation sites and measured the fitnesses. In such experiments, if the local population always has higher fitness at its home site than nonlocals transplanted to that site, the populations are locally adapted (**Figure 1**) (see Kawecki & Ebert 2004). A relative fitness estimate of a population (when away from its local site) is obtained from a comparison with the local population. For the two examples in **Figure 1**, each population has a relative fitness of 1 at home, and lower fitness elsewhere. This same approach can be used in a continuously varying environment. Notice that in **Figure 1b**, both populations have higher absolute fitness in the south than in the north, but each has the highest relative fitness at home.

Population geneticists examine the conditions in which spatially varying selection can give rise to genetic differentiation and local adaptation. Hedrick (2006) recently reviewed selection in heterogeneous environments at single loci. Many traits related to adaptation have a polygenic basis. The conditions for the maintenance of polygenic variation in an environment that consists of of distinct patches depend on the variation of allelic effects over environments (Barton & Turelli 1989).

For forest trees, the most relevant models deal with polygenic traits in a continuously changing environment (Barton 1999, Endler 1977, Slatkin 1978). Genetic differentiation depends on the scale of environmental heterogeneity and the balance of selection and gene flow. In these models, gene action is additive, there are no environmental effects on the phenotype, and selection is weak relative to genetic variation within populations. A one-dimensional axis describes spatial location. For forest trees, the axis may be from south to north and its scale may be in meters or kilometers, to coincide with units of gene flow. Along this axis, the optimum phenotype for some trait varies. For example, the timing of cessation of growth is an important trait; trees that grow in the north will have an earlier optimum for cessation of growth than those that grow in the south. At each location along the axis, there is a fitness function. The individuals with the optimum phenotype have the highest fitness. As the phenotype deviates from the optimum, the fitness declines. This is described by a Gaussian fitness function

$$W(z) \propto e^{-\frac{1}{2}s(z-z_0)^2},$$

where z is the phenotypic measure of the individual, $W(z)$ is the fitness of an individual with phenotype z, z_0 is the optimum at that location (different between locations), and s describes the strength of selection. A small value of s results in a wide fitness function and weak selection, and a larger s results in stronger selection. Gene flow by gametes follows a symmetric dispersal function, centered on the location of the offspring, with variance σ^2.

Slatkin (1978) showed that a population's ability to track the phenotypic optimum, i.e., to adapt to the different environmental conditions, depends on $\sigma/\sqrt{(V_A s)}$, where V_A is the additive genetic variance of the trait at each location, and σ and s are as described above. Thus, if dispersal distances are short and selection is considerable, adaptation to a fine-scaled environment is possible. If dispersal distances are long and selection is weak, genetic differentiation is not possible.

The results of this kind of model depend on assumptions about weak diversifying selection. In this case, the mean can change owing to sequential narrow allelic frequency clines at individual loci, where each locus has a high level of polymorphism over a short geographical range (Barton 1999).

With a linearly changing optimum and additional assumptions, the mean phenotype of the population at each site will be at the optimum (Felsenstein 1977). However, if densities are heterogeneous, dispersal is asymmetric, the environmental gradient is not linear, or genes do not contribute additively, the population will not match the optimum exactly (Barton 1999, Lenormand 2002).

Instead of a purely analytical approach as described in the models above, Latta (1998) and LeCorre & Kremer (2003) used simulations with varying optima to examine the effects of differential phenotypic selection on the underlying loci. The simulations were based on a different model of migration, the island model, where each subpopulation is of equal size and receives a constant proportion of migrants from the other populations. Their parameter space includes a wide range of intensities of stabilizing selection within populations, partly outside the range of the model used by Slatkin (1978). With strong diversifying selection and high gene flow, phenotypic differentiation can be due to small allelic frequency changes between populations. Under a rather weak intensity of stabilizing selection within populations, there will be a great deal of linkage disequilibrium (correlations between loci) at the between-population level (Latta 1998, LeCorre & Kremer 2003). Pronounced phenotypic differentiation requires a stronger intensity of selection within populations than is considered by Slatkin (1978). In this case, there is more differentiation at the individual loci and weaker linkage disequilibrium between populations.

Quantitative genetic models assume large population sizes and deterministic spatially varying selection. In small populations, factors such as genetic drift may prevent genetic differentiation and local adaptation. Extensive gene flow from adjacent areas may prevent a population from tracking the optimum (García-Ramos & Kirkpatrick 1997). Temporally varying selection may prevent spatial genetic differentiation, and one phenotypically plastic genotype might evolve (see Hedrick 2006). Further, populations may not yet be at equilibrium.

In summary, the analytical models are realistic in their assumptions concerning isolation-by-distance and spatially varying selection pressure, but allow only weak selection and symmetric normally distributed migration. The island model simulations allow stronger stabilizing selection, but are based on an unrealistic migration model. Clearly, more realistic assumptions should be incorporated into analytical and simulation models. Further, many parameters of the models need empirical data.

THE QUANTIFICATON OF GENE FLOW
THROUGH SEEDS AND POLLEN

Before we address selection, it is important to consider the level of gene flow. Compared with herbaceous and annual plants, trees have more extensive gene flow (Hamrick et al. 1992). Gene flow occurs through pollen dispersal, seed dispersal, and establishment of fertile adult trees. Thus, in addition to dispersal, realized gene flow requires successful fertilization, germination, and survival from competition. Early studies used marking or labeling methods for migration estimation (Koski 1970), but we discuss mainly genetic methods, because they come closer to a description of realized gene flow.

On the basis of the differentiation of maternally (mitochondrial DNA or mtDNA), paternally (chloroplast DNA or cpDNA in some gymnosperms), and biparentally inherited (nuclear DNA) genetic markers, gene flow through pollen is more extensive (20 to nearly 200 times higher) than gene flow through seeds, at least in wind-pollinated trees (Ennos 1994). Highly variable genetic markers allow the determination of paternity and thus the estimation of pollinator-offspring distances. Paternity assignment or exclusion methods measure only current dispersal and ignore establishment. Differentiation among the pollen pools of two or several mother trees can be used to estimate the mean pollination distance and the effective number of pollinators (TwoGener method) (Smouse et al. 2001). These two approaches measure different aspects of pollen flow. For example, the mean distances of pollen dispersal, δ, are similar in *Pinus sylvestris* compared with *Quercus alba* and *Quercus lobata* (Sork et al. 2002), but the effective number of pollinators (pollen donors) is much higher for *P. sylvestris* (**Table 1**).

Long-distance pollen flow can be extensive (see table 2 in Petit & Hampe 2006). For example, in an isolated *Pinus sylvestris* population in Spain, 4.3% of fertilizing pollen came from a distance of at least 30 km (Robledo-Arnuncio & Gil 2005). Estimates of δ vary between species (**Table 1**), but are generally hundreds of meters, for both animal (*Dinizia*, *Sorbus*) and wind-pollinated species. To relate these estimates to the polygenic model of spatially varying selection, if dispersal is normally distributed, then $\delta = \sqrt{(2/\pi)}\sigma$. The estimated pollen dispersal kernel functions and shapes vary, but researchers often use an exponential power distribution (Oddou-Muratorio et al.

Table 1 Pollen dispersal estimates of trees, and the methods used for the estimation

Species	N_a[a]	N_o[b]	Dispersal distribution[c]	Method	δ[d]	N_{ep}[e]	Reference
Dinizia excelsa	24	596	Exponential power, b = 0.821	TwoGener, modified	225		Austerlitz et al. (2004)
Fraxinus mandshurica	150	492	Two component model	Paternity	197		Goto et al. (2006)
Pinus sylvestris	34	813	Exponential power, b = 0.67	Paternity	136		Robledo-Arnuncio & Gil (2005)
Pinus sylvestris	60	720	Normal	TwoGener	17–29	71–125	Robledo-Arnuncio et al. (2004)
Quercus alba	54	1586	Negative exponential	TwoGener	17	8.2	Smouse et al. (2001)
Quercus lobata	21	211	Normal	TwoGener	65	3.7	Sork et al. (2002)
Quercus lobata	33	288	Exponential power, b = 0.847	TwoGener, modified	121		Austerlitz et al. (2004)
Quercus petraea, Quercus robur	296	984	Negative exponential for short and uniform for long distances	Paternity	287, 333		Streiff et al. (1999)
Sorbus torminalis	60	1728	Exponential power, b = 0.21–0.33	Paternity	743–1077		Oddou-Muratorio et al. (2005)
Sorbus torminalis	60	1728	Exponential power, b = 0.285–0.565	TwoGener, modified	209–482	7–12	Austerlitz et al. (2004)

[a]Number of sampled adults.
[b]Number of sampled offspring.
[c]b = shape parameter.
[d]Average distance of pollen flow (m).
[e]Effective number of pollinators.

2005, Robledo-Arnuncio et al. 2004, Streiff et al. 1999) (**Table 1**). Distributions are typically fat tailed: More extreme values have a higher probability than in the normal distribution (**Table 1**), indicated by the estimates of the exponential shape parameter *b*, which are typically between 0 and 1 (**Table 1**).

Pollen fossil and macrofossil evidence have provided estimates of seed migration rates over long time spans. In Europe, the ice-free areas were quickly colonized as continental ice retreated over the past 18,000 years. Palynological data for most tree species show migration rates of 50–500 m/year, and 1500 m/year for pines (Hewitt 1999). Marker gene–based estimates of average seed flow are not as extensive as pollen flow (however, see Bacles et al. 2006). In *Pinus pinaster*, researchers estimated the average seed dispersal distance to be ~12 m, 40–60 m for *Jacaranda copaia* (Jones et al. 2005), 14 m for *Quercus pyrenaica*, 42 m for *Quercus petrea* (Valbuena-Carabana et al. 2005), and ~540 m for *Fraxinus mandschurica* var. *japonica* (Goto et al. 2006). Most genetic and trapping-based direct estimates of seed dispersal are lower than those from fossil data. Genetic and trapping methods may miss some long-distance migration, and establishment was probably more successful in the postglacial open landscapes (Clark 1998).

Methods that measure potential seed flow may underestimate the relative amount of realized long-distance gene flow. In the immediate surroundings of the mother tree, establishment success increases with distance from the mother tree (Nathan & Casagrandi 2004). For example, in *Pinus pinaster*, estimated seed dispersal distance is higher for saplings than for seeds (González-Martínez et al. 2006). However, the direct measurement of realized gene flow is difficult in long-lived forest trees.

In contrast, indirect methods based on a population genetic model measure realized gene flow over long time periods. In an island model, the differentiation of populations is described by Wright's F_{ST} statistics (see Whitlock & McCauley 1999). Briefly, Wright's F_{ST} statistics assess the proportion of genetic variation found between subpopulations. At equilibrium, migration (m) and differentiation are related to each other as follows: $Nm = 1 - (F_{ST})/4F_{ST}$. Most forest tree species have estimated F_{ST} smaller than 0.1, which suggests extensive gene flow (Slavov et al. 2004). F_{ST} can be used to estimate the relationship between seed and pollen flow (Ennos 1994). However, forest tree populations deviate in several ways from the assumptions of Wright's island model, and gene flow estimates based on F_{ST} should be interpreted with caution (Whitlock & McCauley 1999). First, long-lived trees may only rarely reach the equilibrium assumed and the distribution of genetic diversity may be mostly influenced by population history and demography, not by current gene flow (Austerlitz et al. 2000). Second, in continuous populations the isolation-by-distance model is more appropriate because genetic differentiation and geographical distance are positively correlated (Wright 1943).

With the isolation-by-distance model, pairwise differentiation between subpopulations is used to estimate gene dispersal relative to effective population density by examination of the relationship of $F_{ST}/(1 - F_{ST})$ with geographical distance (Rousset 1997). For instance, Hardy et al. (2006) performed a small spatial scale study and found the gene dispersal distances in 10 tropical tree species ranged from 150–1200 m.

Many studies indicate the existence of long-distance dispersal of pollen and seed (see Petit & Hampe 2006 for examples). These rare dispersal events are difficult to observe, but can have a strong impact on population structure and adaptation (Nathan 2006). A two-component model that separates short- and long-distance dispersal yields the best fit for many data, both for seed and pollen dispersal (Goto et al. 2006, Jones et al. 2005). Estimation of the long tail of the distribution is difficult and heavily model dependent (Austerlitz et al. 2004, Jones et al. 2005). In summary, gene flow extends at least hundreds of meters, but long-range dispersal is poorly quantified. Widely distributed northern conifers and other wind-pollinated trees may spread a part of their pollen even further.

EVIDENCE OF LOCAL ADAPTATION FROM PROVENANCE TRIAL EXPERIMENTS

The study of local adaptation, as described in **Figure 1**, requires that experiments fulfill the following conditions: First, the experimental sites must include the home sites of the populations. Second, the performance of transferred populations should be compared with the performance of local populations to obtain measures of relative fitness. Third, the traits compared must be reasonable surrogates for fitness.

For close to 200 years, foresters have been transferring seeds and seedlings from different provenances to common garden experiments outside the original location, often in multiple sites (de Vilmorin 1862, Langlet 1971). These experiments provide extensive information on the transfer responses of different populations and species. The main interest of foresters in these studies is to find the most productive (best growing) seed sources for each area, and the experiments serve this purpose well. We summarize some general patterns, and examine whether these studies fulfill the above requirements for the study of local adaptation.

Many forest trees have large ranges over a broad span of environments. For instance, the distribution of *Pinus contorta* spans 33° of latitude and 3900 m of altitude. *P. contorta* grows in areas where the average annual temperature ranges from –7°C to 11°C (Rehfeldt et al. 1999). Trees of individual populations of most species can survive and grow in a broad range of environments, even far from their original growing site, when planted in well tended trials without competition (Eriksson et al. 1980, Heikinheimo 1949). *P. contorta* can maintain good height growth (80% of maximum) in an area where the mean annual temperature spans 6°C (Rehfeldt et al. 1999). In *Fraxinus americanus* plantations in the eastern US, 80% of maximum growth is found 3–5° of latitude north or south from the latitude of origin (Roberds et al. 1990). In *Pinus sylvestris*, for populations between 50°N and 60°N, optimum survival is at a mean annual temperature of 2.7°C, but close to optimum survival is found in areas with mean annual temperatures between 1°C and 5°C (range 4°C) (Rehfeldt et al. 2003). The same study found that several *Larix* species have even broader niches for survival, from 4.6°C to close to 10°C.

Survival and growth may improve away from the home location. For northern parts of the range of *Pinus* and *Larix*, growth and survival can increase by more than 30% when transferred to a location 5° of latitude south, for example (Rehfeldt et al. 1999, 2002, Shutyaev & Giertych 1997, 2000). In contrast, Carter (1996) showed that eight of ten studied tree species in the eastern US would grow better in a cooler climate, with the optimum located north of their current sites. Thus, species and populations from different areas show variable responses.

Even if trees of an individual population can grow and survive in a broad set of climate conditions, the distribution of genotypes changes over much shorter distances. Differences in quantitative traits can be detected between populations that are separated by one degree of latitude (see next section). The distribution of genotypes is governed by relative performance and many biotic interactions.

Analyses of provenance trials rarely aim to address directly the issue of local adaptation. If results are standardized, the comparisons are made against the mean of all provenances at a site, or perhaps the best provenance. In one rare exception, Prescher (1986) analyzed volume production with a comparison of transferred provenances of *Pinus sylvestris* to the local populations to obtain a relative measure of productivity that is closely related to local adaptation. Wu & Ying (2004) compared the height growth of the local population and a predicted best population to study local optimality. Across part of the range, local populations did have the best growth, but in the west and at high elevations this was not the case.

We used data from a transfer trial series of *Pinus sylvestris* to assess local adaptation with a comparison of the fitness of transferred populations to local populations (T. Knürr, K. Kärkkäinen, and O. Savolainen, manuscript in preparation). The trial was established between 1952 and 1954 by Eiche (1966), and involved a total of 69 provenances transferred to 21 experimental sites in central and northern Sweden that encompass different latitudes (60–68°N) and altitudes (5–765 m). The number of provenances tested per site was either 7 or 14. Provenances were represented by 260 trees for each transfer. Full fitness data are unavailable, but because mortality up to age 20 describes the climate-related mortality (Persson 1994), and height is a good fitness surrogate (Wu & Ying 2004), we use provenance survival rates and mean height 9–12 years after planting for fitness components (Eiche 1966). Tall trees may have good competitive ability, they are healthy, and they will probably have good reproductive success. The two components at least partly comprise different aspects of fitness.

Transfer response functions in relation to latitudinal transfers were first obtained separately for the two fitness components, following earlier studies (Eriksson et al. 1980, Prescher 1986). These curves show the predicted height or survival for the population as it is transferred over latitudes. The response curve for survival of two populations, both from an altitude of 200 m, that originated from latitude 60°N and 66°N, respectively, as well as the survival of the corresponding local populations from the same altitude, are displayed in **Figure 2a**. Relative measures for survival are obtained by dividing the survival of the transferred population by the survival of the local population at each site. Relative measures for the height component were calculated analogously (**Figure 2b**). Multiplication of the two components yields the estimate of total fitness of the population, as a function of the latitudinal transfer distance (**Figure 2c**).

These two examples show several results. The survival and height of local populations are lower in the north than in the south, as was found earlier by Eriksson and coworkers (1980). For the population from the central part of the range (60°N), the highest fitness is at its home site. Close to the northern margin (66°N), the local population is not at the fitness optimum. The local population would have higher relative fitness slightly to the south of its location (**Figure 2c**, *right*), although its relative height increases when transferred slightly to the north (**Figure 2b**, *right*). The analysis of relative fitness consisting of two components suggests that populations may be locally adapted in central parts of the range, but local adaptation in *Pinus sylvestris* (in northern Europe) may break down close to the margin of the species distribution.

It is clearly important to consider both components of fitness, because they give somewhat opposing results. The change of environment seems to be more rapid in the north; the fitness of a population (at 66°N) transferred north falls off more rapidly than if it is transferred to the south, as shown by the asymmetric fitness curves. The importance of the two fitness components is different: In the north,

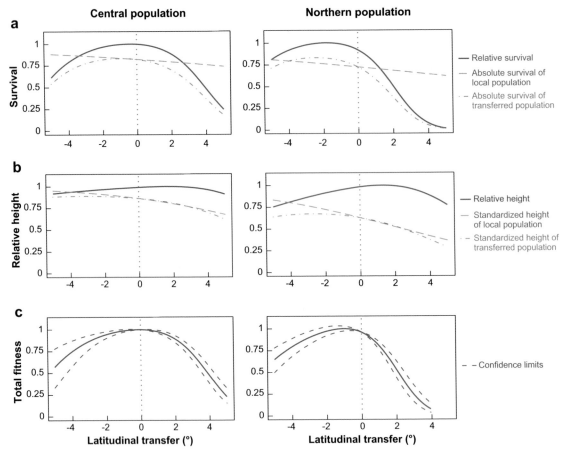

a

Central population

Northern population

— Relative survival
— Absolute survival of local population
-·- Absolute survival of transferred population

b

— Relative height
— Standardized height of local population
-·- Standardized height of transferred population

c

Latitudinal transfer (°)

Latitudinal transfer (°)

- - Confidence limits

Figure 2

Transfer response functions for fitness and its components in *Pinus sylvestris* for a central population from latitude 60°N and a northern population from latitude 66°N. The bold lines show response functions that relate transferred populations to local populations. Positive numbers indicate latitudinal transfer to the north. Latitudes of local populations are the latitudes to which the central and northern populations were transferred. (a) Absolute survival of the transferred population is indicated by the dashed-dotted line; the dashed line corresponds to absolute survival of the local populations. (b) The dashed-dotted line and the dotted line show standardized heights of the transferred population and the corresponding local populations, respectively; i.e., the heights are calculated as proportions of the maximal height observed across all experimental sites and all provenances. (c) The dashed lines show upper (97.5%) and lower (2.5%) confidence limits of the fitness functions.

the differences between populations are larger in survival than in height growth and competitive ability. The importance of different fitness components differs in transplant experiments that involve herbaceous plants as well (Angert & Schemske 2005).

QUANTITATIVE TRAIT GENETIC DIFFERENTIATION IN ADAPTIVE TRAITS

Here we examine clinal variation in trees in relation to the polygenic models. In the simplest models, the population mean tracks the optimum if the environment changes at a scale less than $\sigma/\sqrt{(V_{AS})}$ (Slatkin 1978). The models deal with the distribution of genetic variation after fertilization and before selection. Greenhouses, gardens, or very young trials provide the most appropriate measurements.

The growing season in northern (or very southern) areas is shorter than the growing season close to the equator. The decrease in growing season follows a regular latitudinal pattern in some parts of northern Europe, where longitudinal and altitudinal effects are weak. Many phenotypic traits related to climatic adaptation in northern and central European trees show clines over latitude (**Figure 3**). *Quercus petraea* populations from different latitudes differ with respect to spring bud flush date in a common garden (Ducousso et al. 1996) (**Figure 3a**). *Betula pendula* populations stop growth at different times and day lengths (Viherä-Aarnio et al. 2005) (**Figure 3b**). *Pinus sylvestris* populations set their first year terminal buds in the greenhouse at different times (**Figure 3c**) (García-Gil et al. 2003, Mikola 1982, Notivol et al. 2007), and also display latitudinal clines in cold tolerance (Aho 1994, Hurme et al. 1997). *Picea abies* populations from different latitudes (at altitude 200 m) show different critical night lengths for growth cessation in controlled conditions (Dormling 1979) (**Figure 3d**) or shoot elongation in field trials (Skrøppa & Magnussen 1993).

The examples in Figure 3 were chosen from areas where latitudinal clines are steep. Longitudinal clines also exist, e.g., in cold tolerance for *Pinus sylvestris* (Andersson & Fedorkov 2004). The cline in bud phenology of *Quercus petraea* has an altitudinal component in addition to a latitudinal one (Ducousso et al. 1996). Forest trees in North American mountainous areas, such as *Pseudotsuga menziesii* var. *glauca*, show highly complex patterns of variation that can be accounted for by latitude, longitude, altitude, and slope (Rehfeldt 1989).

In some species, only weak quantitative trait differentiation is found in relation to climate. The clines of *Larix occidentalis* have much gentler slopes than *Pseudotsuga menziesii* or *Pinus contorta* (Rehfeldt 1995), such that genetic differentiation in *Larix occidentalis* is found only with altitudinal differences of 500 m or more. *Prosopis chilensis* provenances over ten degrees of latitude do not vary in frost tolerance (Verzino et al. 2003).

Overall, steep clines may be common where natural selection acts on large populations. Many *Drosophila* species exhibit steep clines across latitudes in traits related to the timing of reproduction (Lankinen 1986). In isolated populations, genetic drift and founder effects may have more influence. For instance, the latitudinal component accounts for much less of the between-population adaptive variation in the weedy annual *Arabidopsis thaliana* than in populations of trees (Stenøien et al. 2002, Stinchcombe et al. 2004).

Spatially varying selection is probably responsible for the patterns seen in **Figure 3**. In principle, genetic drift and isolation-by-distance can also produce these kinds of patterns in neutral traits. However, the tree species represented in **Figure 3**

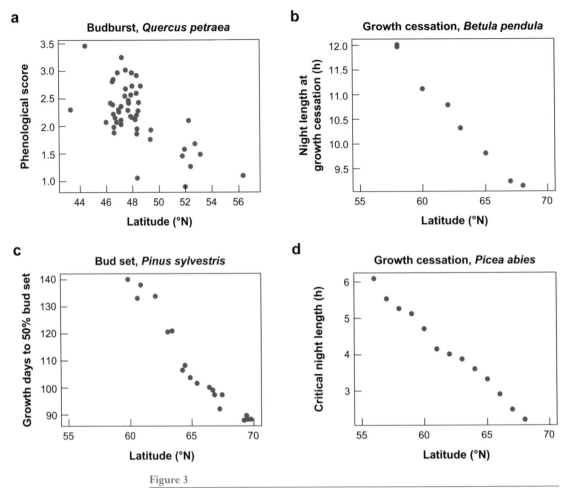

Figure 3

Clinal variation in traits related to timing of growth is shown, redrawn from data from
(*a*) Ducousso et al. (1996), (*b*) Viherä-Aarnio et al. (2005), (*c*) Mikola (1982), (*d*) Dormling
(1979). Note that the traits are different, and the latitude scale is not the same in all figures.

[and many other species (Hamrick et al. 1992)] have low differentiation among populations at marker loci (Karhu et al. 1996, Kremer et al. 1997, Lagercrantz & Ryman 1990), which supports the role of selection.

The most commonly used method to assess the role of selection is based on the island model. The measure F_{ST} (proportion of variation between populations at marker loci) is compared with Q_{ST}, an analogous measure for quantitative traits (Prout & Barker 1993). The latter is defined as $Q_{ST} = V_B/(V_B + 2V_A)$, where V_B is the between-population variance (assumed to be all-additive) and V_A the within-population additive genetic variance. If morphological or other polygenic traits show higher differentiation than neutral markers ($Q_{ST} > F_{ST}$), researchers conclude that divergent selection is responsible (McKay & Latta 2002, Yang et al. 1996).

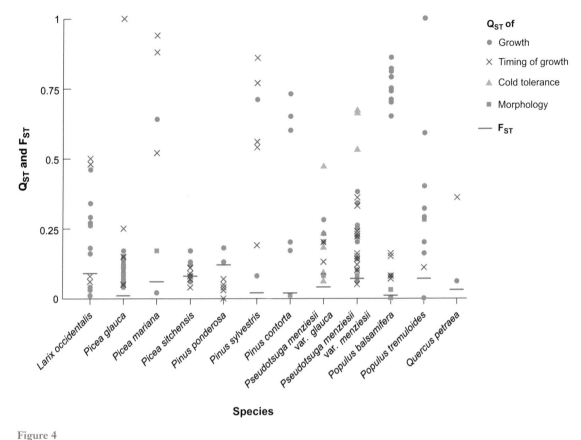

Figure 4

F_{ST} and Q_{ST} values of twelve tree species. For each species, Q_{ST}s of various traits related to morphology (*squares*), growth (*circles*), timing of growth (*crosses*), and cold tolerance (*triangles*) are presented. F_{ST}s are indicated by horizontal bars. Data are taken from the studies listed in an extensive table available online. See the Supplemental Material link in the online version of this article or at **http://www.annualreviews.org/**.

In **Figure 4**, we summarize the F_{ST} and Q_{ST} information for many forest trees [table 1 of Howe et al. (2003) and additional studies; for the full list of references for **Figure 4**, see the Supplemental Material link in the online version of this article or at **http://www.annualreviews.org/**]. The traits are related to morphology, growth, timing of growth, and cold tolerance. The F_{ST} estimates (horizontal bars) are low for all species. Most of the Q_{ST} estimates are higher, which suggests that the quantitative traits are subject to diversifying selection. The comparison of Q_{ST} estimates between traits, species, and studies is more complicated. The estimates of Q_{ST} are highly dependent on the geographical range of sampling and the number of populations within the range. For instance, the estimates of Q_{ST} for Scots pine bud set differ between studies, depending on the range and number of populations. The very low Q_{ST} estimates for Sitka spruce are based on a large number of populations from a single island, which constitutes only a very restricted part of the range. In a new study over the whole range from Alaska to California, Q_{ST} estimates were much higher (values not shown in **Figure 4**) for the same traits related to timing of growth (Mimura & Aitken 2007). Regressions of population trait means on distance in this study are more informative than the island model–based Q_{ST} statistic.

The assessment of the role of different factors that govern the clines is difficult in light of the theory on spatial variation in polygenic traits, because we do not have information on the critical parameters (dispersal distance, strength of selection, and level of additive genetic variation). Long-range dispersal is poorly known. Direct measures of the strength of selection are not available for these large-scale clines [as opposed to many smaller scale studies, see Endler (1977)]. However, additive genetic variances along the cline can be estimated (Notivol et al. 2007). A further complication is that the population means may not track the optima (Barton 1999, Felsenstein 1977). Nonadditive modes of gene action and genetic drift can have a role in the clines, but they are not included in the models.

The cases where there is a difference in clinal variation between species or areas can be informative. In a common garden comparison of populations from different altitudes (600–1800 m), *Abies lasiocarpa* showed a steep cline in height at growth cessation, whereas *Pinus contorta* and *Picea glauca* × *P. engelmannii* were not differentiated (Green 2005). The difference between species may be due to different dispersal distances, but the rate of the change in optima may also differ between species. If clinal variation is examined within a species, for the same trait, in different environmental situations, the balance of selection and gene flow can be more easily evaluated. Frost tolerance in *Pinus sylvestris* varied over latitudes in Sweden (Sundblad & Andersson 1995), but in the same experiment, there was no differentiation between populations covering a similar range of climatic conditions across altitudes, from about the same latitude. The dispersal distances, additive genetic variances, and strength of selection may all be similar in the lowlands and the mountains, but the rate of environmental change per km is much faster in the mountains.

GENETIC ARCHITECTURE AND MOLECULAR NATURE OF GENETICS OF ADAPTATION

Quantitative traits related to adaptation are moderately heritable (Howe et al. 2003). The nature of the underlying genes is still poorly known for most quantitative characters. This information is needed for detailed theoretical predictions on the maintenance of clinal variation. QTL mapping methods estimate the number and location of the loci that govern phenotypic differences between parents of crosses (Mauricio 2001, Tanksley 1993). In trees, mapping is based on existing pedigrees because of the long generation times (Howe et al. 2003).

How many loci are involved in governing variation in a trait and what are the effect sizes? Crosses between *Pseudotsuga menziesii* revealed many (>10) QTLs with relatively small effect (less than 5% of phenotypic variation explained, PVE), related to timing of budset, budflush, and cold tolerance (Jermstad et al. 2001a,b). In *Eucalyptus* and *Salix*, crosses have revealed few (2–5) QTLs for frost tolerance with relatively large effect (5–15% PVE) (Byrne et al. 1997, Tsarouhas et al. 2004). In *Quercus robur* the variation in budburst was governed by many small QTLs, whereas in the same cross, height variation was due to fewer loci (Scotti-Saintagne et al. 2004). In a cross between a northern and southern Finnish population of *Pinus sylvestris*, some cold tolerance QTLs were found with large effects relative to phenotypic variation within and between populations (Hurme et al. 2000). Likewise, other studies detected QTLs with large effects on cold tolerance in *P. sylvestris* (Yazdani et al. 2003). Recent theory on adaptation suggests that directional selection is initially based on the alleles with largest effects, at least within single populations (Orr 1998). Crosses between closely related species *Populus trichocarpa* and *P. deltoides* measured large QTL effects in cessation of growth (5–20% PVE) (Frewen et al. 2000), not directly comparable to within-species effects.

Variation in different species in similar traits may be controlled by the same underlying loci. A comparative mapping study found QTLs for phenological traits in the same areas of homologous chromosomes in *Quercus petraea* and *Castanea sativa* (Casasoli et al. 2006).

Identification of the actual loci that underlie the variation is needed for the examination of the molecular nature of the variation and the effects of individual DNA variants. The improvement in the genomic resources of trees, such as the genome sequence of *Populus trichocarpa*, facilitates this work (Tuskan et al. 2006). The resolution of QTL studies is typically too low to identify individual genes, but studies of sequence variation of candidate loci can provide evidence for the role of a locus (Wright & Gaut 2005). Because forest trees display steep clines of quantitative traits, similar patterns of nucleotide variation may indicate selection. In *Populus tremula*, clinal variation occurs in growth cessation and in single nucleotide polymorphism (SNP) frequency in the phytochrome locus *PhyB2* (Ingvarsson et al. 2006). García-Gil et al. (2003) found no frequency differences in the orthologous locus *PhyP* in *Pinus sylvestris* along a cline in bud set timing.

If the clines in phenotypic traits are due to small frequency differences at many loci, it can be difficult to detect the individual loci (Barton 1999, Latta 1998, LeCorre & Kremer 2003), because linkage disequilibria between populations would be responsible for the change in the mean. So far, few cases of polymorphisms maintained by spatially varying selection have been detected in any species, or any plants in particular (Hedrick 2006, Wright & Gaut 2005). The association of variation of phenotypes with candidate locus genotypes within populations or across clines will provide a powerful method to detect functionally important SNPs in large random mating populations (González-Martínez et al. 2007, Neale & Savolainen 2004). If the effects are small, detection may require very large studies.

SPECIFIC ISSUES AT RANGE MARGINS

When there are no obvious geographical barriers, what limits the spread of a species in constant conditions? Several alternative explanations exist. The populations may lack appropriate genetic variation (Bradshaw & McNeilly 1991), or local adaptation may not have evolved yet after a recent colonization. Finally, gene flow from the more central parts of distribution may prevent adaptation (García-Ramos & Kirkpatrick 1997).

As an example, we consider *Pinus sylvestris* at its margin in northern Scandinavia, because detailed data on its biology are available. *P. sylvestris* has the widest distribution of all pines, but in northern Europe its distribution ends at approximately latitude 69°N without a geographical barrier. During the ice age, until approximately 10,000 years ago, this area was covered by ice. *P. sylvestris* pollen records show that the species arrived in northern Finland approximately 7000–9000 years BP (Hyvärinen 1975). What limits the distribution of *P. sylvestris* now? Are the northern populations poorly adapted?

As discussed above, a population at latitude 66°N would have had higher relative fitness in a more southern area (T. Knürr, K. Kärkkäinen, and O. Savolainen, manuscript in preparation) (**Figure 2c**, *right*). The absolute fitness also declines at the margin. Survival of the local populations is high in warm conditions of 1200 day degrees (d.d., temperature summed for days with average temperature >5°C, which corresponds to approximately latitude 60°N, altitude 200 m), but starts to decrease at about 1000 d.d., and is very low at 600 d.d. (corresponds to approximately latitude 68°N, altitude 400 m) (Persson 1994) (**Figure 5a**). The northern populations also produce less pollen (and seeds) than more southern populations (Koski & Tallqvist 1978) (**Figure 5b**). Trees in the north also are less likely to mature their seeds (Harju et al. 1996). The northern populations show ample genetic variation at marker loci (Karhu et al. 1996) and additive genetic variation in quantitative traits, such as timing of bud set or frost tolerance (Savolainen et al. 2004). High mortality and variation between trees in seed production show that directional selection should be possible,

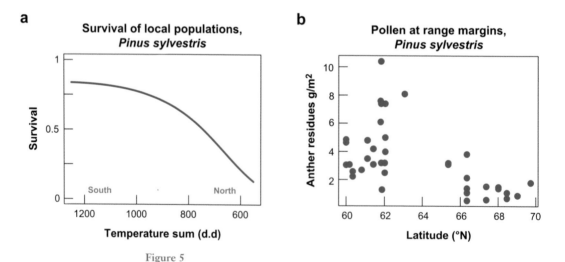

Figure 5

(a) Survival of *Pinus sylvestris* populations to height 2.5 m, as a function of temperature sum in day degrees (d.d), which corresponds to between 60°N at 200 m and 68°N at 400 m, based on data from Persson (1994). (b) Pollen production of *Pinus sylvestris* at different latitudes, based on results presented by Koski & Tallqvist (1978).

and lack of variation does not seem to prevent adaptation. However, the critical traits are often unknown (Hänninen 2006) and may not have been included in these studies.

The hypothesis that asymmetric gene flow from the center prevents adaptation (García-Ramos & Kirkpatrick 1997) can be examined. Pollen production (estimated by counting anther residues) is low at the range margin (Koski & Tallqvist 1978, Sarvas 1962) (**Figure 5b**). The winds from the south that prevail at pollination time bring abundant southern pollen to the northern forests. This nonlocal pollen produces genetic variation, but the progeny are not adapted to these conditions. In experimental crosses, the progeny of north (female) × south (male) crosses have worse frost tolerance than progeny of north × north crosses (Aho 1994). Thus, in *Pinus sylvestris* gene flow may indeed have a role in hampering adaptation. Similar effects may take place at other, less intensively studied range margins of tree species.

ADAPTATION TO FUTURE CLIMATE CHANGE

In northernmost Europe, temperatures are expected to rise by about 4°C in the next 100 years (Houghton et al. 2001). Existing trees will show an immediate phenotypic response that will depend on the species and the population. Migration of species and populations to new areas is one possible response (Davis & Shaw 2001), and the ranges of many species are changing already (Parmesan 2006). The potential for local populations to respond by evolving new genotypic composition has also received some attention (Bradshaw & McNeilly 1991, Davis & Shaw 2001, Parmesan 2006). The general theory on polygenic variation suggests that even under strong selection, the response to selection may be slow (Lynch 1996).

Provenance trials are used to predict the immediate responses to a warmer climate. For many northern species (*Pinus sylvestris*, *P. contorta*, and three species of *Larix*), growth is expected to increase in northern parts of the range, but decrease in the southern range margin (Mátyás 1996, Rehfeldt et al. 1999, 2002, 2003). Will trees be able to evolve the new genotypic compositions required for future local adaptation? Rehfeldt and coworkers (2002) made predictions based on experience of selection efficacy in breeding populations. The evolutionary response depends on genetic change by selection in the current population, but the response is also influenced by dispersal. The evolutionary response requires that there are possibilities of establishment for the new seedlings. The high genetic variation of forest trees can facilitate adaptation to new conditions (Hamrick et al. 1992). The evolutionary response can be slowed by limited seed dispersal and reduced possibilities for establishment in the current filled landscapes, in which existing trees survive better than in the earlier, colder climate (Clark et al. 1999, Savolainen et al. 2004). However, the eventual results depend not only on the characteristics of individual species, but also on responses of other competing species (Kellomäki et al. 2001).

CONCLUSIONS

Gene flow in forest trees is extensive, and has given rise in many cases to uniform neutral allele frequencies. The extent of gene flow through pollen is difficult to estimate because of the long tail of the distribution. In the face of this gene flow, most trees studied show signals of adaptive differentiation in reciprocal transplant experiments or common garden experiments. The strength of selection on the different adaptive traits remains to be quantified. Howe and coworkers (2003) described the quantitative genetics of adaptation at the biometrical level, but studies on the molecular nature of the loci that underly adaptive variation are just beginning. The increasing number of genomic resources for trees will significantly advance these studies, and allow for improved estimates of distributions of allelic effects. Predictions of climate change effects should take evolutionary change into account.

SUMMARY POINTS

1. Foresters have studied adaptation, growth, and survival in forest trees from a forestry perspective extensively, but less often study these issues specifically with respect to the questions of evolutionary biology. For tropical trees, there are few results relating to local adaptation.

2. Gene dispersal estimates of 100–200 m are common for both tropical and temperate trees. This near dispersal can be estimated via the use of genetic markers. The long-distance dispersal may be very important, but it is difficult to estimate.

3. Indirect evidence and the reanalysis of data in an evolutionary perspective show that forest tree populations may be locally adapted over large parts of the range, but at range margins local adaptation may break down.

4. Many forest trees show latitudinal and altitudinal differentiation in adaptive traits, at levels much higher than is observed at neutral genetic markers. This points to strong diversifying selection, but direct estimates of the strength of selection are not available.

5. Current migration of trees is slow relative to climate change. Even if populations have extensive genetic variation, limited dispersal and establishment possibilities may hamper an evolutionary response to climate change.

FUTURE ISSUES

1. Studies of the genetic basis of quantitative variation at the nucleotide level, with sequence analysis and association studies, will help identify the loci that underly quantitative variation. Improved genomic resources will facilitate this work.

2. Soon it will be possible to estimate the distribution of size allelic effects at loci that underly adaptive quantitative variation, and to examine the assumptions of the quantitative genetic models against empirical data.

3. Estimates of selection on quantitative traits in trees and estimates of overall gene flow in trees are needed at the same geographical scale.

4. Much more research is required on adaptation to climate change. Researchers should analyze the potential for evolutionary change in the current ecological context.

DISCLOSURE STATEMENT

The authors are not aware of any biases that might be perceived as affecting the objectivity of this review.

ACKNOWLEDGMENTS

Discussions over a long time on forest tree adaptation with Katri Kärkkäinen and Veikko Koski have been very helpful. Bengt Andersson, Rosario García-Gil, Antoine Kremer, Helmi Kuittinen, Sonja Kujala, Katri Kärkkäinen, Jaakko Lumme, and Witold Wachowiak provided useful comments on the manuscript. We are especially grateful for the comments of Doug Schemske. We thank Anneli Viherä-Aarnio for providing the data for **Figure 3b**. Financial support from the Academy of Finland, by the University of Oulu, from the Finnish Graduate School for Population Genetics, and EU-project TREESNIPS (QLRT-CT-2002-01973) is gratefully acknowledged.

LITERATURE CITED

Aho M-L. 1994. Autumn frost hardening of one-year-old *Pinus sylvestris* (L.) seedlings: effect of origin and parent trees. *Scand. J. For. Res.* 9:17–24

Andersson B, Fedorkov A. 2004. Longitudinal differences in Scots pine frost hardiness. *Silvae Genet.* 53:76–80

Angert AL, Schemske DW. 2005. The evolution of species' distributions: reciprocal transplants across elevational ranges of *Mimulus cardinalis* and *M. lewisii*. *Evolution* 59:1671–84

Austerlitz F, Dick CW, Dutech C, Klein EK, Oddou-Muratorio S, et al. 2004. Using genetic markers to estimate the pollen dispersal curve. *Mol. Ecol.* 13:937–54

Austerlitz F, Mariette S, Machon N, Gouyon P-H, Godelle B. 2000. Effects of colonization processes on genetic diversity: differences between annual plants and tree species. *Genetics* 154:1309–21

Bacles CF, Lowe AJ, Ennos RA. 2006. Effective seed dispersal across a fragmented landscape. *Science* 311:628

Barton NH. 1999. Clines in polygenic traits. *Genet. Res.* 74:223–36

Barton NH, Turelli M. 1989. Evolutionary quantitative genetics: how little do we know? *Annu. Rev. Genet.* 23:337–70

Bradshaw AD, McNeilly T. 1991. Evolutionary response to global climate change. *Ann. Bot.* 67(Suppl. 1):5–14

Careful study that compares the fit of various pollen dispersal kernels to three observed datasets and describes the effect on the estimate of the mean.

Thorough theoretical study on genetic variation in clines.

Byrne M, Murrell JC, Owen JV, Williams ER, Moran GF. 1997. Mapping quantitative trait loci influencing frost tolerance in *Eucalyptus nitens*. *Theor. Appl. Genet.* 95:975–79

Carter KK. 1996. Provenance tests as indicators of growth response to climate change in 10 north temperate tree species. *Can. J. For. Res.* 26:1089–95

Casasoli M, Derory J, Morera-Dutrey C, Brendel O, Porth I, et al. 2006. Comparison of quantitative trait loci for adaptive traits between oak and chestnut based on an expressed sequence tag consensus map. *Genetics* 172:533–46

Clark JS. 1998. Why trees migrate so fast: confronting theory with dispersal biology and the paleorecord. *Am. Nat.* 152:204–24

Clark JS, Silman M, Kern R, Macklin E, HilleRisLambers J. 1999. Seed dispersal near and far: patterns across temperate and tropical forests. *Ecology* 80:1475–94

Clausen J, Keck DD, Hiesey WM. 1948. Experimental studies on the nature of species. III. Environmental responses of climatic races of *Achillea*. *Carnegie Inst. Wash. Publ.* 581:1–189

Collignon A-M, Van de Sype H, Favre J-M. 2002. Geographical variation in random amplified polymorphic DNA and quantitative traits in Norway spruce. *Can. J. For. Res.* 32:266–82

Davis MB, Shaw RG. 2001. Range shifts and adaptive responses to quaternary climate changes. *Science* 292:673–79

de Vilmorin PPA. 1862. Exposé historique et descriptif de l'Ecole forestière des Barres près de Nogent-sur-Vernisson (Loiret). *Mémoires Soc. Imp. Cent. Agric. Fr.*, pp. 297–353

Dormling I. 1979. Influence of light intensity and temperature on photoperiodic response of Norway spruce provenances. In *Proc. IUFRO Jt. Meet. Work. Parties Norway spruce provenances Norway spruce breed.*, pp. 398–407. Bucharest: IUFRO

Ducousso A, Guyon JP, Kremer A. 1996. Latitudinal and altitudinal variation of bud burst in western populations of sessile oak (*Quercus petraea* (Matt) Liebl). *Ann. Sci. For.* 53:775–82

Eiche V. 1966. Cold damage and plant mortality in experimental provenance plantations with Scots pine in northern Sweden. *Stud. For. Suec.* 36:1–218

Endler JA. 1977. *Geographic Variation, Speciation and Clines*. Princeton, NJ: Princeton Univ. Press

Ennos RA. 1994. Estimating relative rates of pollen and seed migration among plant populations. *Heredity* 72:250–59

Eriksson G, Andersson S, Eiche V, Ifver J, Persson A. 1980. Severity index and transfer effects on survival and volume production of *Pinus sylvestris* in northern Sweden. *Stud. For. Suecica* 156:1–32

Felsenstein J. 1977. Multivariate normal genetic models with a finite number of loci. In *Proc. Int. Conf. Quantitative Genetics*, ed. E Pollak, O Kempthorne, TB Bailey Jr., pp. 227–45. Ames: Iowa State Univ. Press

Frewen BE, Chen THH, Howe GT, Davis J, Rohde A, et al. 2000. Quantitative trait loci and candidate gene mapping of bud set and bud flush in *Populus*. *Genetics* 154:837–45

Comparative QTL mapping study that shows the same chromosomal areas govern phenological traits in *Quercus* and *Castanea*.

An old, extensive, well-documented provenance trial, which has been repeatedly analyzed for further questions.

García-Gil MR, Mikkonen M, Savolainen O. 2003. Nucleotide diversity at two phytochrome loci along a latitudinal cline in *Pinus sylvestris*. *Mol. Ecol.* 12:1195–206

García-Ramos G, Kirkpatrick M. 1997. Genetic models of adaptation and gene flow in peripheral populations. *Evolution* 51:21–28

González-Martínez SC, Burczyk J, Nathan R, Nanos N, Gil L, Alía R. 2006. Effective gene dispersal and female reproductive success in Mediterranean maritime pine (*Pinus pinaster* Aiton). *Mol. Ecol.* 15:4577–88

González-Martínez SC, Wheeler NC, Ersoz E, Nelson CD, Neale DB. 2007. Association genetics in *Pinus taeda* L. I. Wood property traits. *Genetics* 175:399–409

Goto S, Shimatani K, Yoshimaru H, Takahashi Y. 2006. Fat-tailed gene flow in the dioecious canopy tree species *Fraxinus mandshurica* var. *japonica* revealed by microsatellites. *Mol. Ecol.* 15:2985–96

Green DS. 2005. Adaptive strategies in seedlings of three co-occurring, ecologically distinct northern coniferous tree species across an elevational gradient. *Can. J. For. Res.* 35:910–17

Hamrick JL, Godt MJW, Sherman-Broyles SL. 1992. Factors influencing levels of genetic diversity in woody plant species. *New For.* 6:95–124

Hardy OL, Maggia L, Bandou E, Breyne P, Caron H, et al. 2006. Fine-scale genetic structure and gene dispersal inferences in 10 Neotropical tree species. *Mol. Ecol.* 15:559–71

Extensive study on gene dispersal in 10 tropical tree species that utilizes a local isolation-by-distance pattern.

Harju AM, Kärkkäinen K, Ruotsalainen S. 1996. Phenotypic and genetic variation in the seed maturity of Scots pine. *Silvae Genet.* 45:205–11

Hedrick PW. 2006. Genetic polymorphism in heterogeneous environments: the age of genomics. *Annu. Rev. Ecol. Evol. Syst.* 37:67–93

Heikinheimo O. 1949. Results of experiments on the geographical races of spruce and pine. *Commun. Inst. For. Fenn.* 37:1–44

Heuertz M, De Paoli E, Källman T, Larsson H, Jurman I, et al. 2006. Multilocus patterns of nucleotide diversity, linkage disequilibrium and demographic history of Norway spruce (*Picea abies* (L.) Karst). *Genetics* 174:2095–105

Hewitt GM. 1999. Post-glacial recolonization of European biota. *Biol. J. Linn. Soc.* 68:87–112

Howe GT, Aitken SN, Neale DB, Jermstad KD, Wheeler NC, Chen THH. 2003. From genotype to phenotype: unraveling the complexities of cold adaptation in forest trees. *Can. J. Bot.* 81:1247–66

A thorough review of adaptive quantitative genetics of trees.

Huntley B, Birks HJB. 1983. *An Atlas of Past and Present Pollen Maps of Europe: 0–13000 Years Ago*. Cambridge, UK: Cambridge Univ. Press

Hurme P, Repo T, Savolainen O, Pääkkönen T. 1997. Climatic adaptation of bud set and frost hardiness in Scots pine (*Pinus sylvestris* L.). *Can. J. For. Res.* 27:716–23

Hurme P, Sillanpää M, Arjas E, Repo T, Savolainen O. 2000. Genetic basis of climatic adaptation in Scots pine by Bayesian quantitative locus analysis. *Genetics* 156:1309–22

Hyvärinen H. 1975. Absolute and relative pollen diagrams from Northernmost Fennoscandia. *Fennia* 142:1–23

Hänninen H. 2006. Climate warming and the risk of frost damage to boreal forest trees: identification of critical ecophysiological traits. *Tree Physiol.* 26:889–98

Houghton JT, Ding Y, Griggs DJ, Noguer M, van der Linden PJ, et al. eds. 2001. *IPCC Climate Change 2001: The Scientific Basis.* Cambridge, UK: Cambridge Univ. Press. 881 pp.

Ingvarsson PK, García MV, Hall D, Luquez V, Jansson S. 2006. Clinal variation in phyB2, a candidate gene for day-length-induced growth cessation and bud set, across a latitudinal gradient in European Aspen (*Populus tremula*). *Genetics* 172:1845–53

Jermstad KD, Bassoni DL, Jech KS, Wheeler NC, Neale DB. 2001a. Mapping of quantitative trait loci controlling adaptive traits in coastal Douglas-fir. I. Timing of vegetative bud flush. *Theor. Appl. Genet.* 102:1142–151

Jermstad KD, Bassoni DL, Wheeler NC, Anekonda TS, Aitken SN, et al. 2001b. Mapping of quantitative trait loci controlling adaptive traits in coastal Douglas-fir. II. Spring and fall cold-hardiness. *Theor. Appl. Genet.* 102:1152–58

Jones FA, Chen J, Weng GJ, Hubbell SP. 2005. A genetic evaluation of seed dispersal in the neotropical tree *Jacaranda copaia* (Bignoniaceae). *Am. Nat.* 166:543–55

Karhu A, Hurme P, Karjalainen M, Karvonen P, Kärkkäinen K, et al. 1996. Do molecular markers reflect patterns of differentiation in adaptive traits of conifers? *Theor. Appl. Genet.* 93:215–21

Kawecki TJ, Ebert D. 2004. Conceptual issues in local adaptation. *Ecol. Lett.* 7:1225–41

Kellomäki S, Rouvinen I, Peltola H, Strandman H, Steinbrecher R. 2001. Impact of global warming on the tree species composition of boreal forests in Finland and effects on emission of isoprenoids. *Glob. Change Biol.* 7:531–44

Koski V. 1970. A study of pollen dispersal as a mechanism of gene flow in conifers. *Commun. Inst. For. Fenn.* 70:1–78

Koski V, Tallqvist R. 1978. Results on long time measurements of the quantity of flowering and seed crop of forest trees. *Folia For.* 364:1–60

Kremer A, Kleinschmit J, Cotrell J, Cundall EP, Deans JD, et al. 2002. Is there a correlation between chloroplastic and nuclear divergence, or what are the roles of history and selection on genetic diversity in European oaks? *For. Ecol. Manag.* 156:75

Kremer A, Petit RJ, Ducousso A, LeCorre V. 1997. General trends of variation of genetic diversity in *Quercus petraea* (Matt.) Liebl. In *Proc. Diversity and Adaptation in Oak Species*, ed. KC Steiner, pp. 81–89. University Park, PA: Penn State Univ. Press

Lagercrantz U, Ryman N. 1990. Genetic structure of Norway spruce (*Picea abies*): concordance of morphological and allozyme variation. *Evolution* 44:38–53

Langlet O. 1971. Two hundred years of genecology. *Taxon* 20:653–721

Lankinen P. 1986. Geographical variation in circadian eclosion rhythm and photoperiodic adult diapause in *Drosophila littoralis*. *J. Comp. Physiol. A* 159:123–42

Latta RG. 1998. Differentiation of allelic frequencies at quantitative trait loci affecting locally adaptive traits. *Am. Nat.* 151:283–92

LeCorre V, Kremer A. 2003. Genetic variability at neutral markers, quantitative trait loci and trait in a subdivided population under selection. *Genetics* 164:1205–19

Lenormand T. 2002. Gene flow and the limits to natural selection. *Trends Ecol. Evol.* 17:183–89

Careful evaluation of basic concepts of local adaptation.

Linhart YB, Grant MC. 1996. Evolutionary significance of local genetic differentiation in plants. *Annu. Rev. Ecol. Syst.* 27:237–77

Lynch M. 1996. A quantitative-genetic perspective on conservation issues. In *Conservation Genetics. Case Histories from Nature*, ed. JC Avise, JL Hamrick, pp. 471–501. New York: Chapman & Hall

Mátyás C. 1996. Climatic adaptation of trees: Rediscovering provenance tests. *Euphytica* 92:45–54

Mauricio R. 2001. Mapping quantitative trait loci in plants: Uses and caveats for evolutionary biology. *Nat. Rev. Genet.* 2:370–81

McKay HK, Latta RG. 2002. Adaptive population divergence: markers, QTL and traits. *Trends Ecol. Evol.* 17:285–91

Mikola J. 1982. Bud-set phenology as an indicator of climatic adaptation of Scots pine in Finland. *Silva Fennica* 16:178–84

Mimura M, Aitken SN. 2007. Adaptive gradients and isolation-by-distance with postglacial migration in *Picea sitchensis*. *Heredity* 99:224–32

Morgenstern EK. 1996. *Geographic Variation in Forest Trees: Genetic Basis and Application of Knowledge in Silviculture*. Vancouver: Univ. B.C. Press

Nathan R. 2006. Long-distance dispersal of plants. *Science* 313:786–88

Nathan R, Casagrandi R. 2004. A simple mechanistic model of seed dispersal, predation and plant establishment: Janzen-Connell and beyond. *J. Ecol.* 92:733–46

Neale DB, Savolainen O. 2004. Association genetics of complex traits in conifers. *Trends Plant Sci.* 9:325–30

Notivol E, García-Gil MR, Alía R, Savolainen O. 2007. Genetic variation of growth rhythm traits in the limits of a latitudinal cline of Scots pine. *Can. J. For. Res.* 37:540–51

Oddou-Muratorio S, Klein EK, Austerlitz F. 2005. Pollen flow in the wildservice tree, *Sorbus torminalis* (L.) Crantz. II. Pollen dispersal and heterogeneity in mating success inferred from parent-offspring analysis. *Mol. Ecol.* 14:4441–52

Orr HA. 1998. The population genetics of adaptation: the distribution of factors fixed during adaptive evolution. *Evolution* 52:935–49

Parmesan C. 2006. Ecological and evolutionary responses to recent climate change. *Annu. Rev. Ecol. Evol. Syst.* 37:637–69

Persson B. 1994. Effects of provenance transfer on survival in nine experimental series with *Pinus sylvestris* (L.) in Northern Sweden. *Scand. J. For. Res.* 9:275–87

Petit RJ, Aguinagalde I, de Beaulieu J-L, Bittkau C, Brewer S, et al. 2003. Glacial refugia: hotspots but not melting pots of genetic diversity. *Science* 300:1563–65

Petit RJ, Hampe A. 2006. Some evolutionary consequences of being a tree. *Annu. Rev. Ecol. Evol. Syst.* 37:187–214

Prescher F. 1986. Transfer effects on volume production of *Pinus sylvestris* L.: a response surface model. *Scand. J. For. Res.* 1:285–92

Prout T, Barker JSF. 1993. F statistics in *Drosophila buzzattii*: Selection, population size and inbreeding. *Genetics* 134:369–75

Rehfeldt GE. 1989. Ecological adaptations in Douglas fir (*Pseudotsuga menziesii* var. *glauca*): a synthesis. *For. Ecol. Manag.* 28:203–15

Rehfeldt GE. 1995. Genetic variation, climate models and the ecological genetics of *Larix occidentalis*. *For. Ecol. Manag.* 78:21–37

Rehfeldt GE, Tschebakova NM, Milyutin LI, Parfenova EI, Wykoff WR, Kouzmina NA. 2003. Assessing population response to climate in *Pinus sylvestris* and *Larix* spp. of Eurasia with climate-transfer models. *Eurasian J. For. Res.* 6:83–98

Rehfeldt GE, Tschebakova NM, Parfenova YI, Wykoff WR, Kuzmina NA, Milyutin LI. 2002. Intraspecific responses to climate in *Pinus sylvestris*. *Glob. Change Biol.* 8:912–29

Rehfeldt GE, Ying CC, Spittlehouse DL, Hamilton DA. 1999. Genetic responses to climate change in *Pinus contorta*: niche breadth, climate change, and reforestation. *Ecol. Monogr.* 69:375–407

Roberds JH, Hyun JO, Namkoong G, Rink G. 1990. Height response functions for white ash provenances grown at different latitudes. *Silvae Genet.* 39:121–29

Robledo-Arnuncio JJ, Gil L. 2005. Patterns of pollen dispersal in a small population of *Pinus sylvestris* L. revealed by total-exclusion paternity analysis. *Heredity* 94:13–22

Robledo-Arnuncio JJ, Smouse PE, Gil L, Alía R. 2004. Pollen movement under alternative silvicultural practices in native populations of Scots pine (*Pinus sylvestris* L.) in central Spain. *For. Ecol. Manag.* 197:245–55

Rousset F. 1997. Genetic differentiation and estimation of gene flow from F-statistics under isolation by distance. *Genetics* 145:1219–28

Sarvas R. 1962. Investigations on the flowering and seed crop of *Pinus sylvestris*. *Commun. Inst. For. Fenn.* 53(4):1–198

Savolainen O, Bokma F, García-Gil MR, Komulainen P, Repo T. 2004. Genetic variation in cessation of growth and frost hardiness and consequences for adaptations of *Pinus sylvestris* to climatic changes. *For. Ecol. Manag.* 197:79–89

Scotti-Saintagne C, Bodenes C, Barreneche T, Bertocchi E, Plomion C, Kremer A. 2004. Detection of quantitative trait loci controlling budburst and height growth in *Quercus robur*. *Theor. Appl. Genet.* 109:1648–59

Shutyaev AM, Giertych M. 1997. Height growth variation in a comprehensive Eurasian provenance experiment of *Pinus sylvestris* L. *Silvae Genet.* 46:332–49

Shutyaev AM, Giertych M. 2000. Genetic subdivisions of the range of Scots pine (*Pinus sylvestris* L.) based on a transcontinental provenance experiment. *Silvae Genet.* 49:137–51

Skrøppa T, Magnussen S. 1993. Provenance variation in shoot growth components of Norway spruce. *Silvae Genet.* 42:111–20

Slatkin M. 1978. Spatial patterns in the distributions of polygenic traits. *J. Theor. Biol.* 70:213–28

Slavov GT, DiFazio SP, Strauss SH. 2004. Gene flow in forest trees: gene migration patterns and landscape modeling of transgene dispersal in hybrid poplar. In *Introgression from Genetically Modified Plants into Wild Relatives*, ed. HCM den Nijs, D Bartsch, J Sweet, pp. 89–106. Wallingford, UK: CAB Int.

Smouse PE, Dyer RJ, Westfall RD, Sork VL. 2001. Two-generation analysis of pollen flow across a landscape. I. Male gamete heterogeneity among females. *Evolution* 55:260–71

Sork VL, Davis FW, Smouse PE, Apsit VJ, Dyer RJ, et al. 2002. Pollen movement in declining populations of California Valley oak, *Quercus lobata*: where have all the fathers gone? *Mol. Ecol.* 11:1657–68

Large-scale thorough analysis in *Pinus sylvestris* of transfer experiments, with respect to climate change effects.

Fundamental results on polygenic variation with spatially varying selection in continuous environments.

Stenøien H, Fenster CB, Kuittinen H, Savolainen O. 2002. Quantifying latitudinal clines to light responses in natural populations of *Arabidopsis thaliana* (Brassicaceae). *Am. J. Bot.* 89:1604–8

Stinchcombe JR, Weinig C, Ungerer M, Olsen KM, Mays C, et al. 2004. A latitudinal cline in flowering time in *Arabidopsis thaliana* modulated by the flowering time gene FRIGIDA. *Proc. Natl. Acad. Sci. USA* 101:4712–17

Streiff R, Ducousso A, Lexer C, Steinkellner H, Gloessl J, Kremer A. 1999. Pollen dispersal inferred from paternity analysis in a mixed oak stand of *Quercus robur* L. and *Q. petraea* (Matt.) Liebl. *Mol. Ecol.* 8:831–41

Sundblad L-G, Andersson B. 1995. No difference in frost hardiness between high and low altitude *Pinus sylvestris* (L.) offspring. *Scand. J. For. Res.* 10:22–26

Tanksley SD. 1993. Mapping polygenes. *Annu. Rev. Genet.* 27:205–33

Tsarouhas V, Gullberg U, Lagercrantz U. 2004. Mapping of quantitative trait loci (QTLs) affecting autumn freezing resistance and phenology in *Salix. Theor. Appl. Genet.* 108:1335–42

Tuskan GA, DiFazio S, Jansson S, Bohlmann J, Grigoriev I, et al. 2006. The genome of black cottonwood, *Populus trichocarpa* (Torr. & Gray). *Science* 313:1596–604

Valbuena-Carabaña M, González-Martínez SC, Sork VL, Collada C, Soto A, et al. 2005. Gene flow and hybridisation in a mixed oak forest (*Quercus pyrenaica* Willd. and *Quercus petraea* (Matts.) Liebl.) in central Spain. *Heredity* 95:457–65

Verzino G, Carrranza C, Ledesma M, Joseau M, DiRienzo J. 2003. Adaptive genetic variation of *Prosopis chilensis* (Mol) Stuntz: Preliminary results from one test-site. *For. Ecol. Manag.* 175:119–29

Viherä-Aarnio A, Häkkinen R, Partanen J, Luomajoki A, Koski V. 2005. Effects of seed origin and sowing time on timing of height growth cessation of *Betula pendula* seedlings. *Tree Physiol.* 25:101–8

Whitlock MC, McCauley DE. 1999. Indirect measures of gene flow and migration: $F_{ST} \neq 1/(4 Nm + 1)$. *Heredity* 82:117–25

Willis KJ, van Andel TH. 2004. Trees or no trees? The environments of central and eastern Europe during the last glaciation. *Quat. Sci. Rev.* 23:2369–87

Wright S. 1943. Isolation-by-distance. *Genetics* 28:114–38

Wright SI, Gaut BS. 2005. Molecular population genetics and the search for adaptive evolution in plants. *Mol. Biol. Evol.* 22:506–19

Wu HX, Ying CC. 2004. Geographic pattern of local optimality in natural populations of lodgepole pine. *For. Ecol. Manag.* 194:177–98

Yang R-C, Yeh FC, Yanchuk AD. 1996. A comparison of isozyme and quantitative genetic variation in *Pinus contorta* ssp. *latifolia* by F_{st}. *Genetics* 142:1045–52

Yazdani R, Nilsson JE, Plomion C, Mathur G. 2003. Marker trait association for autumn cold acclimation and growth rhythm in *Pinus sylvestris. Scand. J. For. Res.* 18:29–38

First genome sequence of a tree, which paves the way for genomic approaches in local adaptation research.

The Evolution of Multicellularity: A Minor Major Transition?

Richard K. Grosberg[1] and Richard R. Strathmann[2]

[1]Center for Population Biology, College of Biological Sciences, University of California, Davis, California 95616; email: rkgrosberg@ucdavis.edu

[2]Friday Harbor Laboratories, University of Washington, Friday Harbor, Washington 98250; email: rrstrath@u.washington.edu

Annu. Rev. Ecol. Evol. Syst. 2007. 38:621–54

First published online as a Review in Advance on August 17, 2007

The *Annual Review of Ecology, Evolution, and Systematics* is online at
http://ecolsys.annualreviews.org

This article's doi:
10.1146/annurev.ecolsys.36.102403.114735

Key Words

multicellular organism, unicellular bottleneck, germ-soma, chimera, life cycle evolution, mutation load, genetic conflict

Abstract

Benefits of increased size and functional specialization of cells have repeatedly promoted the evolution of multicellular organisms from unicellular ancestors. Many requirements for multicellular organization (cell adhesion, cell-cell communication and coordination, programmed cell death) likely evolved in ancestral unicellular organisms. However, the evolution of multicellular organisms from unicellular ancestors may be opposed by genetic conflicts that arise when mutant cell lineages promote their own increase at the expense of the integrity of the multicellular organism. Numerous defenses limit such genetic conflicts, perhaps the most important being development from a unicell, which minimizes conflicts from selection among cell lineages, and redistributes genetic variation arising within multicellular individuals between individuals. With a unicellular bottleneck, defecting cell lineages rarely succeed beyond the life span of the multicellular individual. When multicellularity arises through aggregation of scattered cells or when multicellular organisms fuse to form genetic chimeras, there are more opportunities for propagation of defector cell lineages. Intraorganismal competition may partly explain why multicellular organisms that develop by aggregation generally exhibit less differentiation than organisms that develop clonally.

INTRODUCTION

Beneath the outward harmony of living organisms lies an often contentious history of transitions to ever more inclusive, hierarchically nested levels of biological organization (Bonner 1988, Buss 1987, Carroll 2001, Leigh 1977, Maynard Smith 1988, Maynard Smith & Szathmáry 1995, McShea 2001, Michod 1999). Although views differ on what defines a major evolutionary transition, almost everyone agrees that the following transitions qualify as major (Buss 1987; Maynard Smith & Szathmáry 1995; Queller 1997, 2000): (*a*) the compartmentalization of replicating molecules, yielding the first cells; (*b*) the coalescence of replicating molecules to form chromosomes; (*c*) the use of DNA and proteins as the fundamental elements of the genetic code and replication; (*d*) the consolidation of symbiotic cells to generate the first eukaryotic cells containing chloroplasts and mitochondria; (*e*) sexual reproduction involving the production (by meiosis) and fusion of haploid gametes; (*f*) the evolution of multicellular organisms from unicellular ancestors; and (*g*) the establishment of social groups composed of discrete multicellular individuals.

In every major transition, selection favoring increased levels of biological complexity is opposed by genetic conflicts acting within and across levels of biological organization. As new levels of organization of replicators emerge from previously independently replicating units, how do the fitness interests of lower and higher levels of biological replication become aligned? In other words, what keeps selection acting on the ancestral level of biological organization (within-group selection) from disrupting the integration of the derived level (Maynard Smith & Szathmáry 1995; also see Buss 1987; Frank 2003, Michod 1999, 2003)? In particular, how are defectors or cheaters that selfishly improve their own fitness kept at bay so that the transition becomes established (Buss 1987; Frank 1998, 2003; Griesemer 2000; Leigh 1977, 1991; Maynard Smith 1988; Maynard Smith & Szathmáry 1995; Michod & Nedelcu 2003; Queller 1997, 2000)?

Here, we focus on the evolutionary transition from unicellular to multicellular organization. The first evidence of this transition comes from fossils of prokaryotic filamentous and mat-forming Cyanobacteria-like organisms, dating back 3 to 3.5 billion years (Knoll 2003, Schopf 1993), with signs of cell differentiation more than 2 billion years ago (Tomitani et al. 2006). Multicellular eukaryotes may have existed 1 billion years ago (Knoll et al. 2006), but a major burst of metazoan diversification occurred about 600–700 Mya, at a time of dramatic increases in atmospheric and oceanic oxygen (Carroll 2001, King 2004, Knoll 2003, Maynard Smith & Szathmáry 1995, Pfeiffer et al. 2001).

History has often repeated itself: Multicellular organisms independently originated at least 25 times from unicellular ancestors (**Figure 1**) (Bonner 1998, 2000; Buss 1987; Carroll 2001; Cavalier-Smith 1991; Kaiser 2001; Maynard Smith & Szathmáry 1995; Medina et al. 2003). Multicellularity appears to have originated once for the Metazoa (King 2004), but multiple times (with secondary losses) in plants, fungi, and the Eubacteria (Bonner 2000, Kaiser 2001, Kirk 1998, Medina et al. 2003, Shapiro 1998). Indeed, multicellular organisms continue to evolve from unicellular ancestors (Boraas et al. 1998), and sometimes continue to revert to a

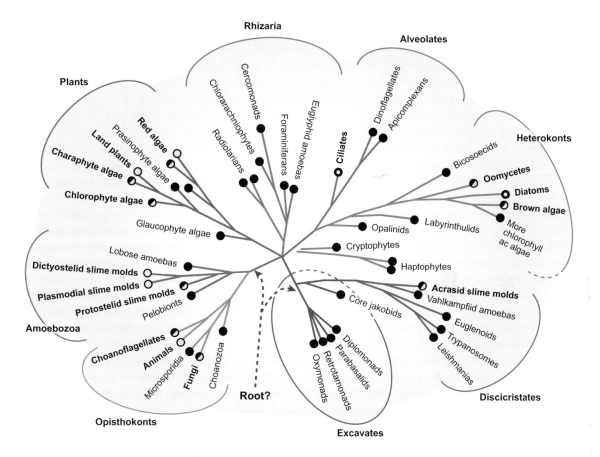

○ All members are multicellular
◐ Clade contains unicellular and colonial/multicellular species
◉ Unicellular with rare multicellular forms

Figure 1

The phylogenetic distribution of multicellularity among eukaryotes. Multicellularity also arose multiply in prokaryotes. Taxa in boldface include at least some multicellular representatives. (After King 2004, from Baldauf 2003.) Figure adapted with permission from Baldauf 2003.

unicellular state [e.g., bacteria (Velicer et al. 1998), mammals (Murgia et al. 2006, Strathmann 1991)].

At least in developmental terms, the transition from uni- to multicellular organization may be easy. In some bacteria (Branda et al. 2001), algae (Lürling & Van Donk 2000), and numerous myxobacteria, myxomycetes, and cellular slime molds (Bonner 1998, Kaiser 2001), the transition to multicellular organization is an inducible response to environmental stimuli. Many of the developmental requirements for multicellular organization, including cell adhesion, cell-cell communication and coordination, and programmed cell death (PCD), likely existed in ancestral unicellular organisms (reviewed in Bonner 1974, 2000; Kaiser 2001; Keller & Surette 2006; King 2004; Lachmann et al. 2003; Miller & Bassler 2001; Shapiro 1998). Moreover, epigenetic modification of patterns of gene expression, a hallmark of cellular differentiation in multicellular organisms, also characterizes many unicellular organisms (e.g., Ausmees & Jacobs-Wagner 2003).

In contrast to rare (or even singular) evolutionary transitions, the developmental and evolutionary lability of the transition to multicellularity permits analysis of both the selective forces favoring the evolution of a major transition and the adaptive mechanisms that control defectors and stabilize the transition. We review (a) the sources and nature of genetic conflicts of interest that could challenge a shift in the unit of selection from the cell to the multicellular organism; (b) the circumstances under which each source of conflict would have the greatest impact on the transition; and (c) the mechanisms that control defecting cells and align the fitness interests of the cells that cooperate to form multicellular organisms. We argue that the most common ways that multicellular organisms develop and propagate align the fitness interests of the replicators that constitute multicellular organisms. Thus, the transition to multicellularity is relatively easy—a minor major transition. Nevertheless, some modes of formation and propagation of multicellular organisms increase the scope for conflicts among replicators and the evolutionary consequences of deleterious mutations.

Finally, we turn to the question posed by Szathmáry & Wolpert (2003, p. 301):

> Is then the evolutionary transition to multicellularity a difficult one or not? The blunt answer is: not at all, since multicellularity has arisen more than twenty times in evolution. . . . However, there are only three lineages that produced complex organisms: plants animals, and fungi. Three hits in 3.5 billion years are not that many.

Three hits is almost certainly an underestimate, because plants include independently derived, complex multicellular organisms, like red and brown algae (Niklas 2000). Nevertheless, although the transition to simple multicellularity may be relatively easy, there appear to be less well-understood obstacles to the evolution of multicellular complexity.

DEVELOPMENT OF MULTICELLULAR ORGANISMS

If there is genetic variation among the cells that constitute a multicellular organism, then selection within its life span can favor the increase of cells of one genotype at the expense of others (Buss 1983a,b, 1987; Grosberg 1988; Hughes 1989; Klekowski 1988; Michod 1997, 1999, 2003; Orive 2001; Otto & Hastings 1998; Otto & Orive 1995). Some modes of development and propagation of multicellular organisms limit the scope for selection to occur within and across generations, whereas others may promote it (**Figure 2**) (Bell & Koufopanou 1991; Crespi 2001; Grosberg & Strathmann 1998; Hamilton 1964b, 1987a,b; Keller & Surette 2006; Kondrashov 1994; Michod 2003; Michod & Roze 1999; Otto & Orive 1995; Queller 2000; Roze & Michod 2001; Seger 1988).

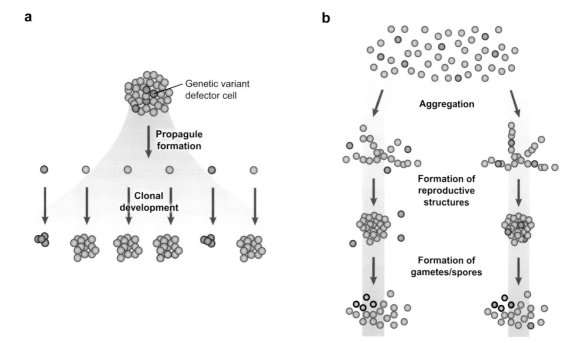

a   **b**

Genetic variant
defector cell

Propagule
formation

Clonal
development

Aggregation

Formation of
reproductive
structures

Formation of
gametes/spores

Figure 2

Generalized life cycles of multicellular organisms and their genetic consequences. (*a*) Clonal
development from unicellular spore or zygote. Genetic variant defectors (red cells) that arise
from somatic mutation within an organism are redistributed among organisms each
generation. (*b*) Aggregative development. Defector cells are denoted in red. In the left
pathway, a recognition system excludes cheaters joining an aggregation; in the right pathway,
cheaters gain access to the spore population. Cells that become spores have black outlines.
(Adapted from Keller & Surette 2006.)

Unicellar/Clonal Development

Clonal development from a unicellular spore or zygote (unitary development, sensu
Queller 2000) characterizes virtually all multicellular aquatic organisms, as well as
the majority of terrestrial forms (**Figure 2*a***) (Bonner 1998, 2000). Multicellularity
arises when cells stay connected to (or even encased within) one another following
division.

Under strictly clonal development, genetic variation among cell lineages within
an organism is low and can arise only through somatic mutation, intergenotypic fusion, or pathogen infection (Dawkins 1976, Grosberg & Strathmann 1998, Hamilton
1964b, Maynard Smith 1988, Seger 1988). Consequently, epigenetic changes in patterns of gene expression (including methylation and chromatin diminution) or posttranscriptional modification, rather than genetic variation among cells within the
organism, are the main sources of specialization of individual cells or groups of
cells (King 2004). Furthermore, when clonal development regularly features passage through a unicellular stage, any genetic variation that has arisen within the

parent becomes redistributed among offspring in the next generation. This regular redistribution of genetic variation from the within- to the among-organism level will further reduce opportunities for within-organism selection. It should also deter defector mutant lineages that enhance their own propagation at the expense of the multicellular organism because advantages based on defection will be curtailed when defectors constitute whole organisms in the next generation (**Figure 2a**).

Vegetative, Multicellular Development

Some organisms that develop clonally also produce multicellular vegetative propagules, with important consequences for the accumulation and distribution of genetic variation within and among individuals (Kondrashov 1994, Michod & Roze 1999, Roze & Michod 2001). (To our knowledge, no multicellular organisms that develop by aggregation also produce multicellular propagules.) The recency with which vegetative propagules passed through a unicellular developmental stage is a fundamentally important determinant of the distribution of within- and among-organism genetic variation (Grosberg & Strathmann 1998, Kondrashov 1994, Roze & Michod 2001). Kondrashov (1994) modeled the effects of propagule size, relatedness of cells, and spatial patterns of cell division on mutation load in vegetative propagules at the equilibrium between mutation and selection against deleterious mutants. These analyses, like those of Roze & Michod (2001), show that at one extreme, if all cells in a vegetative propagule are recently descended from a single ancestral cell, then within-organism genetic variation (in terms of mutation load) does not appreciably increase with the number of cells that constitute the propagule. At the other extreme, if distantly related cells initiate a propagule, mutation load and within-organism genetic variation can dramatically increase. Additionally, if selection acts on the cell lineages that potentially contribute to a multicellular propagule before the propagule is formed, mutation load should decline (Avise 1993, Crow 1988, Otto & Orive 1995).

In many clades of multicellular organisms, including land plants, most groups of algae, Eubacteria, and Archaebacteria, the cells have rigid walls (Niklas 2004). Consequently, mutant cell lineages cannot migrate from where they originate (Buss 1983b, Pineda-Krch & Lehtilä 2004, Sussex 1973), and opportunities for cheaters to invade propagules are reduced. Vegetative propagation can genetically approximate development from a unicell, without reduction to a single-cell propagule, as in a unicellular meristem (Kondrashov 1994). Similarly, in mitotically generated chains of immobile cells, as in diatoms and Cyanobacteria, cells of vegetative fragments will often share a recent common ancestor. Animal cells are far more mobile; consequently, variants can more readily spread through the individual (Buss 1983a,b). If mobile cells join a stolon, bud, fragment, gemmule, or other multicellular propagule, then mutations that have accumulated over repeated mitotic cycles throughout the organism could be included in the propagule. Analogous processes can occur among syncytial nuclei encased within the rigid walls of fungal hyphae; however, a variety of mechanisms, including clamp connections in Basidiomycota, may limit the spread of mutant nuclei (Buss 1987). Thus, control of mutational load and defector cell lineages in animals may require more frequent unicellular bottlenecks than in most multicellular bacteria and plants.

Aggregative Development

Aggregative development of multicellularity occurs in a few groups of terrestrial or semiterrestrial microorganisms (Bonner 2000), including myxobacteria (Dworkin 1972, Shimkets 1990), some ciliates (e.g., Blanton & Olive 1983), myxomycetes (Olive 1975), and cellular slime molds (**Figure 2b**). In the aggregative mode, cells live independently for most of their life cycles, but episodically associate to become multicellular (reviewed in Bonner 1998, 2000; Crespi 2001; Keller & Surette 2006).

In contrast to the clonal mode of development, genetic variation in aggregating multicellular forms can arise through the association of genetically distinct lineages (**Figure 2b**). Depending upon the mating system and population viscosity, these lineages may or may not be closely related. Thus, the primary source of genetic heterogeneity within aggregating forms arises from the association of different genotypes, rather than mutation.

Aggregating life cycles greatly increase the opportunity for cheating cell lineages to exploit cooperating lineages, and to spread and persist in populations (Armstrong 1984; Buss 1982; Crespi 2001; De Angelo et al. 1990; Gadagkar & Bonner 1994; Hamilton 1964b, 1987a,b; Keller & Surette 2006, Matsuda & Harada 1990; Maynard Smith & Szathmáry 1995; Michod & Roze 1999). Furthermore, the aggregative mode will result in more within-organism genetic variance than in organisms that develop in the clonal mode, increasing the scope for within-organism selection relative to among-organism selection (Michod 2003). This may be the reason why differentiation in most obligately aggregating multicellular organisms is limited to the production of multicellular fruiting bodies or foraging slug-like structures (Bonner 2000), the main exceptions being organisms such as colonial marine animals and fungi that facultatively (and perhaps accidentally) fuse following extensive multicellular development (Grosberg 1988). It may also explain why multicellular vegetative propagules are rarely, if ever, formed by aggregation.

ADVANTAGES OF MULTICELLULARITY

The frequent origination and spread of multicellularity suggest that selection favoring the transition is pervasive and that the genetic and developmental obstacles to this transition are relatively easy to overcome. Multiple processes could favor the evolution and persistence of multicellularity and differentiation.

Size-Related Advantages

Morphological, paleontological, and experimental evidence all suggest that the advantages of increased size initially favored, and can still promote, the origination and persistence of multicellularity (Bonner 1965, 1998, 2000, 2004; King 2004). The first multicellular organisms were likely filaments or clusters of undifferentiated cells (Knoll 2003), and the initial size-related benefits of forming such structures were probably ecological (Bonner 2000). The most likely selective agents were phagotrophic organisms that consumed unicellular prey (Stanley 1973). Trade-offs between susceptibility and other aspects of performance (say, buoyancy, physiological exchange, or growth rate) limit size increase for unicellular prey. An apparently easy response was to increase size by increasing the number of cells by cloning and adhesion. Once multicellular predators and suspension feeders evolved, the size race was on between consumers and their victims (Bell 1985).

Experiments using both unicellular and multicellular predators have led either to the facultative induction of multicellularity or to the evolution of multicellular descendants from unicellular ancestors. For example, as the freshwater green alga *Scenedesmus acutus* divides, it retains its daughter cells within its cell walls; these daughter cells can remain attached and form colonies, or they can separate and live as unicells (Van den Hoek et al. 1995). *S. acutus* is normally colonial in field populations, but unicellular in lab cultures (Lürling & Van Donk 2000). Exposure of unicellular *S. acutus* cultures to water from cultures of *Daphnia*, a cladoceran predator on *Scenedesmus*, significantly increased colony formation over controls. The colonial morphs grew and photosynthesized at the same rate as unicells, but sank more rapidly. In a similar experiment, cultures of the unicellular alga *Chlorella vulgaris* repeatedly evolved to form stable, self-replicating colonies within 100 generations in the presence of a phagotrophic predatory flagellate, with the colonies nearly invulnerable to predation (Boraas et al. 1998). The colonial state heritably persisted after removal of the predator.

Even in the absence of phagotrophs (the first fossils of which date back to 750 Mya; see Porter et al. 2003), increased size may have been favored to (*a*) give a competitive advantage in benthic forms, perhaps via overgrowth (Buss 1990); (*b*) provide storage reserves when nutrients are limiting (Koufopanou & Bell 1993, Szathmáry & Wolpert 2003); (*c*) expand feeding opportunities [e.g., group-feeding myxomycetes (Dworkin 1972) and myxobacteria (Shimkets 1990) that produce extracellular digestive enzymes]; (*d*) generate an internal environment protected by an external layer of cells (Gerhart & Kirschner 1997); (*e*) allow novel metabolic opportunities (e.g., Pfeiffer & Bonhoeffer 2003); or (*f*) enhance motility for dispersal or foraging (Foster et al. 2002).

Functional Specialization and Division of Labor

Unicells are capable of differentiation, producing diverse phenotypes in response to environmental cues. However, they can divide labor only in time. Multicellular organisms can simultaneously partition complementary tasks among different cells. By providing a larger pool of laborers, selection for increased size of multicellular organisms can supply both the opportunity for increasing division of labor and the incentive for increasing specialization through economies of scale (e.g., Bell 1985, Bonner 2004, Queller 2000). Moreover, when development occurs clonally, the benefits of division of labor are shared primarily among close relatives, fueling the evolution of division of labor (Hamilton 1964b, 1987b).

The proteobacterium *Caulobacter crescentus* exhibits one of the most basic forms of division of labor: a flagellated swarmer cell attaches to a substrate, loses its flagellum, and becomes a functional stalk, then repeatedly divides to produce more swarmers that detach (England & Gober 2001, Martin & Brun 2000). To the extent that the *Caulobacter* life cycle approximates the earliest phases of multicellular differentiation, it implies that constraints at the cellular level promoted the initial phases of division of labor in multicellular organisms. Here we examine two proposed cell-level constraints: metabolic incompatibilities and trade-offs between motility and cell division.

Metabolic cooperation. Some key metabolic processes cannot concurrently take place within a cell. For example, photosynthesis interferes with nitrogen fixation because nitrogenase does not effectively catalyze fixation in the presence of oxygen (reviewed in Kaiser 2001). Unicellular Cyanobacteria must either forgo nitrogen fixation, or fix nitrogen at night and photosynthesize by day. However, the nitrogen-fixing heterocysts of some filamentous multicellular Cyanobacteria differentiate in response to a shortage of fixed nitrogen. Heterocysts have less permeable cell walls than do photosynthetic cells and so can effectively remain anaerobic and fix N_2, even when adjacent to photosynthesizing cells in a filament. Heterocyst-like structures appeared in the fossil record perhaps 2 billion years ago (Giovannoni et al. 1988). This antiquity, along with the regular spacing of heterocysts along filaments, suggests that genetic machinery for cell-cell interactions and developmental coordination—requirements for differentiation and division of labor—existed in some of the very first multicellular prokaryotes (reviewed in Kaiser 2001, Wolk 2000, Zhang et al. 2006).

Motility-mitosis trade-offs. Comparative and experimental studies of the evolution of germ and soma in volvocacean green algae indicate advantages from division of labor, with the benefits increasing with size (Bell 1985, Bell & Koufopanou 1991, Koufopanou & Bell 1993, reviewed in Kirk 2003, Michod 2003). The clade encompasses unicellular forms (e.g., *Chlamydomonas*), as well as small (<64 cells), undifferentiated colonies (e.g., *Gonium*, *Eudorina*), and larger (up to 50,000 cells) colonies (e.g., *Volvox*) with cellular differentiation. The larger spherical colonial forms consist of two types of cells: (*a*) internal, unflagellated germ cells that can divide and give rise to new colonies, and (*b*) flagellated, external somatic cells that keep the colony suspended (Kirk 1998) and promote nutrient exchange (Solari et al. 2006). The somatic cells appear to be terminally differentiated and have a limited capacity to divide once they form. This pattern of germ-soma differentiation evolved multiple times (and likely was lost several times, too) in the Volvocales (Kirk 1997, 1998, 1999; Coleman 1999). A small number of genetic changes appears to control this transition (Kirk 1997, 2003).

The loss of mitotic activity in the somatic cells of large volvocaceans is associated with a peculiar trade-off between cell division and locomotion (Johnson & Porter 1968, Koufopanou 1994). Some green algae lack rigid cell walls and can divide and swim at the same time (Kirk 1997, 1998), but in volvocaceans, the flagella are fixed to the cell wall so that their basal bodies cannot move to act as centrioles during mitosis while still attached to the flagella. The constraint is not absolute: Flagella can continue beating without their basal bodies for up to five cycles of cell division. Beyond that, mitosis or locomotion must stop. However, by differentiating a nonmotile, internal set of cells (the germ line), a colony can continue swimming while its germ cells divide to form new colonies. Additionally, this arrangement of germ and soma improves the efficiency of nutrient uptake through a source-sink effect (Bell 1985, Koufopanou & Bell 1993, Solari et al. 2006). The beating of the external flagella also enhances nutrient flux to the colony and waste removal from the colony (Solari et al. 2006).

A similar trade-off between motility and cell division appears in metazoans, although with a different structural cause. Metazoan cells, like their protozoan ancestors, have a single microtubule organizing center that serves either in the flagellar basal body or in the mitotic spindle. Consequently, the cells cannot divide while they have a functional flagellum or cilium (Buss 1987, King 2004, Margulis 1981). Buss (1987) argued that this limitation had profoundly affected the evolution of metazoan development and differentiation:

> The blastula can hardly continue to develop while moving, for its surface is covered with ciliated cells, each of which is unable to divide. In this respect the ciliated embryo has inherited a severe liability from its protist ancestors. . . . A mechanism must exist by which the embryo may simultaneously continue to move, yet escape its protistan past. How then can an embryo, covered with cells incapable of dividing, continue to develop while moving?
>
> Buss 1987, pp. 42–44

According to Buss (1987), this ancestral metazoan constraint favored differentiation of a population of flagellated somatic cells (altruistically sacrificing their own reproduction) to propel a group of internal, unflagellated, proliferating cells. In turn, these internal cells were the forerunners of the germ line (Buss 1983a, 1987; King 2004).

Many groups of metazoans, notably taxa that vegetatively propagate, do not sequester a germ line (Blackstone & Jasker 2003; Buss 1983a,b, 1987), but still produce motile embryos and larvae. Do developing embryos or larvae pay a price in terms of motility because of this constraint? Perhaps, but the price should be low because flagellated or ciliated cells need not divide synchronously or shed their cilia or flagella for a long period. For example, the cells of sea urchin blastulae lose cilia briefly (≤ 30 min) and asynchronously (Masuda & Sato 1984); the blastulae do not become bald and they continuously swim while developing. Thus, the cell-level trade-off embodied in the flagellation constraint does not necessarily force the evolution of internal, mitotically active cells (germ) and external, ciliated or flagellated cells (soma).

GENETIC CONFLICTS IN THE TRANSITION

The advantages of multicellularity depend upon cooperation among cells, but cooperation invites cheating. The most widely cited threats to intercellular cooperation during the transition to multicellularity arise from conflicts between genetically distinct cell lineages, some of which may be devoted to the cooperative production of an efficient multicellular organism and others of which are committed to selfish proliferation at the expense of the integrity and performance of the organism (reviewed in Buss 1987, Michod 2003). Mutation, aggregation or fusion with other genotypes, and infection by intracellular symbionts or pathogens are the key sources of defector cell lineages in multicellular organisms.

Kin selection theory predicts that cooperation among cell lineages will be favored when $rb > c$, where r is the genetic relatedness of recipient to actor, b is the fitness benefit to the recipient, and c the cost to the actor (Hamilton 1964a,b). Recent descent of all cells by clonal development from an ancestral unicell provides the greatest assurance of high relatedness ($r \approx 1$). Selection can then favor genes that result in some cells of the organism ceasing division (or dying) to enhance survival and reproduction of the remaining cells that constitute the organism. In Hamilton's (1964b p. 25) words, "...our theory predicts for clones a complete absence of any form of competition which is not to the overall advantage and also the highest degree of mutual altruism. This is borne out well enough by the behavior of clones which make up the bodies of multicellular organisms." However, in multicellular groups that develop by aggregation, close kinship of cells is far from assured.

Genetic conflicts that arose during lower-level transitions remain unsettled in the transitions to multicellularity (Burt & Trivers 2006, Dawkins 1976, Maynard Smith & Szathmáry 1995). Some of these conflicts involve fungible (or interchangeable) units; others occur between nonfungible units (Lachmann et al. 2003, Queller 2000). Fungible units share the same gene pool and directly compete for transmission, and kinship is a key factor influencing cooperation. Organelles within cells, cells within organisms, and conspecific multicellular individuals in societies all represent fungible units. Nonfungible units do not share the same gene pool and compete indirectly; kinship is not a factor in the evolution of cooperation between nonfungible units. Examples of potential conflicts between nonfungible units in the transition to multicellularity include interactions between genes within chromosomes, and—for eukaryotes—interactions between nuclei and organelles. In addition, many multicellular organisms depend upon higher-level, nonfungible cooperative interactions with different species.

Conflicts between Fungible Units

Cell-lineage interactions. When costs of cooperation reduce the growth rate of a cell lineage, mutant noncooperative cells could gain a competitive advantage over cooperating cells, at least within the organism's life span (Michod 1997, Michod & Roze 1999, Orive 2001, Otto & Orive 1995). If these cells gain access to reproductive tissues, then their influence could extend across generations. Buss (1982, 1983a,b, 1987) proposed that cell lineages that evolve behaviors giving them a disproportionately large share of a host's reproductive output represent a major long-term threat to the integrity of multicellular individuals. Germ-line parasites gain their advantage by directing differentiation toward reproductive cells, using the host for somatic support. Somatic cell parasites multiply more rapidly than their host's cells, biasing access to germ lines by a numerical advantage. In the long run, a transmission advantage for either form of defector requires access to the germ line. Thus, the defectors must invade organisms that do not sequester a germ line (and which have mobile cells), strike before a host sequesters a germ line (Buss 1982, Michod 1999), or arise in a germ line or meristem.

Defectors in clonal developers. Genetic variants, presumably owing to in situ mutation, arise with detectable frequency in many multicellular organisms with clonal development (reviewed in Pineda-Krch & Lehtilä 2004). Cancers, primarily in vertebrates, constitute the clearest evidence for defecting cell lineages in clonal developers, but very few are transmitted across generations and manage to re-infect other hosts or succeed on their own. Consequently, cancers rarely have any evolutionary potential beyond the life span of the individual they threaten (Frank 2003, Nunney 1999). The exceptions are interesting and may be more numerous than currently thought. Canine transmissible venereal tumor is the best documented example. Strong genetic evidence indicates that the infectious agent is itself a tumor cell that originated between 200 and 2500 years ago from a wolf or East Asian dog breed (Murgia et al. 2006). The tumor cells are apparently transmitted venereally, as well as by allogrooming and bites (Das & Das 2000). Although the genetic data are less comprehensive, devil facial tumor disease of Tasmanian devils has a similar natural history, and the "infective agent is a rogue cell line that initially evolved in a tumor of unknown origin" (Pearse & Swift 2006). Transmission of sarcoma cells among hamsters by mosquitoes may also occur (Banfield et al. 1965). HeLa cells, originally isolated from a human cervical carcinoma, now thrive by providing social benefits to those who maintain them as unicells in tissue culture (Strathmann 1991).

The general lack of transgenerational success partly reflects the challenges for a defector gaining access to a host's germ line (Buss 1983a, 1987). For example, defector mutants in organisms with rigid cell walls, unless they appear in meristematic tissue, will only succeed in damaging their host (Klekowski 1988). The same is true of defectors in metazoans that sequester their germ line: Mutants that appear in somatic cells have little hope for a future of their own (Buss 1983a,b, 1987, 1990).

Just as importantly, any obligate defector (such as a cancer) that rapidly divides and fails to cooperate with the other cells of the organism is unlikely to perform well on its own. Yet, a unicellular bottleneck would force it to do just that (**Figure 2a**) (Bell 1989, Grosberg & Strathmann 1998, Raff 1988, Van Valen 1988, Wolpert 1990). A multicellular organism composed entirely of defector cells would, at best, have no advantage relative to cooperating genotypes in the next generation and would more likely be at a functional disadvantage (Strathmann 1991). For example, in the social myxobacterium *Myxococcus xanthus*, most defector strains can persist only in the company of normal strains; in isolation, defectors produce abnormal multicellular structures, and perish (Fiegna & Velicer 2003). The same situation should apply to isolated defector mutants in the cellular slime mold *Dictyostelium* (Buss 1982), unless they facultatively produce normal stalks when they develop by themselves (Strassmann et al. 2000).

The more differentiated and integrated a multicellular organism is, the greater the potential that a defector will compromise organismal function (Roberts 2005) and—in the case of clonal development from a unicell—its own future, were it to produce spores or zygotes. As multicellular organisms become more differentiated, a defector cell lineage is more likely to be a defective cell lineage once it leaves the company of its host organism. Facultative defection represents one escape from this quandary, with defectors cooperating with cells carrying the same mutation (Queller 2000). Mutations for such facultative defection seem less probable than a purely defector mutant; moreover, if there is a cost associated with being a facultative defector, then this strategy would lose its advantage once it had produced an organism composed entirely of cells with that mutation.

Defectors in aggregating developers. Multicellular organisms that develop by aggregation offer more opportunities for cheaters to prosper, especially when kinship among the genotypes that co-aggregate is low (**Figure 2b**). It should therefore not be surprising that the best evidence for successful, persistent defecting cell lineages comes from organisms that become multicellular by aggregation.

Several recent reviews have examined evidence for cheating in cellular slime molds, myxobacteria, and pseudomonad Eubacteria (Ackerman & Chao 2004, Crespi 2001, Keller & Surette 2006, Pál & Papp 2000, Strassmann & Queller 2004, Travisano & Velicer 2004, Velicer 2003). Cellular slime molds aggregate when starved, forming a multicellular slug (the pseudoplasmodium) that eventually differentiates into a fruiting body, or sorocarp. About 20% of the sorocarp consists of sterile stalk cells that distally support a mass of spores (the sorus). These spores germinate to form the next generation of free-living amoebae.

Filosa (1962) first clearly documented mutants in natural populations of *Dictyostelium mucoroides* that could act as defectors. In cultures initiated from individual spores isolated from a single sorus, he regularly found a large range of phenotypes in the next generation of sorocarps. At one extreme were cultures with normal stalks and sori; at the other were cultures that lacked stalks and produced only sori. Stalkless mutants would presumably perform poorly in nature because of the stalk's presumed role in dispersal (Bonner 1967, 1982; Huss 1989). However, when aggregating with stalk-producing forms, these stalkless mutants, in principle, could persist.

Buss (1982) isolated a naturally occurring stalkless mutant that could co-aggregate with normal stalked forms. By experimentally varying the ratio of stalkless to stalked spores that initiated a culture, Buss showed that stalkless forms, as expected, consistently produced a greater-than-expected fraction of the spores from fruiting bodies. These mutants, some of which result from a simple deficiency in an F-box protein, appear to be guided into prespore developmental pathways and may also direct normal cells to become part of the sterile stalk (Ennis et al. 2000). All else being equal, the success of such obligately stalkless mutants should exhibit negative frequency dependence in natural populations because, as the frequency of stalkless mutants increased (given their advantage in producing spores), they would increasingly co-aggregate with other stalkless mutants and be ineffective dispersers (Armstrong 1984, Buss 1982, Gadagkar & Bonner 1994).

Recent studies of *Dictyostelium discoideum* have added novel insights to these findings. Strassmann et al. (2000) used microsatellite markers to confirm that clones isolated from field populations combined to form chimeric sorocarps. They identified a high frequency of cheaters from natural isolates, some of which facultatively adjusted their allocation to stalk and spore, depending upon whether they grew in isolation or as a chimera.

In many myxobacteria, there are two social phases of the life cycle in which cheating can occur (Dworkin 1996). During the group-foraging phase, cells secrete an array of chemicals that coordinate movement and digest prey (reviewed in Dworkin 1996, Shimkets 1990). Social parasites could respond to these chemicals, and consume partially digested food, without paying the price to make these substances. In addition, swarming requires motility and coordination, behaviors mediated by extracellular pili and fibrils (Shi & Zusman 1993) that are costly to produce and can potentially be parasitized (Velicer & Yu 2003).

In the second social phase, starved myxobacterial cells aggregate to form fruiting bodies that, like those of many cellular slime molds, consist of supportive somatic cells, destined to die, and cells that will become spores. Nonspore cells lyse, potentially feeding cells that will become spores. The studies of *M. xanthus* by Velicer and colleagues have identified mutants that in mixed-cell cultures obtained a disproportionately high representation of spores, or failed to make spores but outgrew spore-making competitors (e.g., Fiegna & Velicer 2003, Velicer et al. 1998, 2000). Some of these sporeless mutants retained the capacity to make spores, but only when mixed with spore formers (Velicer et al. 2000). These sporeless mutants exploited their spore-making hosts, and some eventually lost the ability to make spores themselves, going extinct (Fiegna & Velicer 2003, Velicer et al. 2000). Remarkably, at least one of these sporeless mutants eventually recovered the ability to make spores, and the recovery required a single point mutation at the locus encoding an acetyltransferase (Fiegna et al. 2006).

Finally, in selection experiments performed on the eubacterial plant pathogen *Pseudomonas fluorescens*, wild-type unicells, when grown in unshaken medium, eventually die from oxygen starvation, but novel cells often evolve that colonize the surface of the culture medium and form a multicellular biofilm that can both respire and access nutrients (Rainey & Rainey 2003). The production of the extracellular adhesive polymer that supports the film is costly to individual cells. Defectors do not produce the matrix of the biofilm and at high frequencies can sink the entire structure.

Selection on the evolution of cooperation in organisms that develop by aggregation depends upon whether the cells that form multicellular structures are clone mates or close kin (Crespi 2001). As Ackerman & Chao (2004) note, the production of adhesives is the foundation of multicellularity in *Myxococcus* and *Pseudomonas*. These environmental adhesives could also increase population viscosity, uniting clone mates or kin as cells divide before differentiating into a fruiting body or biofilm, and providing an incentive for cooperation (Rainey & Rainey 2003, Velicer et al. 2000). Nevertheless, several laboratory studies on the genetic composition of fruiting bodies in *D. mucoroides* and *D. discoideum* suggest that multiple genotypes can co-aggregate and that defectors sometimes join these aggregations (Buss 1982, Filosa 1962, Strassmann et al. 2000).

The opportunities for co-aggregation of multiple clones of microorganisms in the rhizosphere may be substantial. Buss (1982) found at least two strains of *D. mucoroides* in closely spaced soil samples. Francis & Eisenberg (1993) identified multiple clones of *D. discoideum* in spatially restricted samples. A microsatellite-based study by Fortunato et al. (2003) revealed multiple genotypes of *D. discoideum* within samples separated by as little as 6 mm. (Nevertheless, average r among isolates from a single soil sample was approximately 0.5, partly owing to multiple isolates being from the same clone but also because $r \approx 0.15$ for different clones in a sample.) Vos & Velicer (2006) isolated 78 clones of *M. xanthus* containing 21 unique genotypes in a 225-cm² soil sample, suggesting that members of different clones co-occur on spatial scales that would allow them to join the same aggregation. Whether different clones naturally co-aggregate and cooperate in feeding swarms or to form fruiting bodies is not yet settled. *M. xanthus* clones isolated from remote locations are antagonistic, with reduced spore production in forced mixtures (Fiegna & Velicer 2005). It is not yet known whether naturally co-occurring clones behave comparably poorly (Vos & Velicer 2006).

In sum, a great deal remains to be discovered about spatial scales of interaction in natural habitats and the importance of kinship in the evolution of social behavior in aggregating microorganisms (reviewed in Crespi 2001). *D. discoideum* apparently do not distinguish close from distant kin when they aggregate (Strassmann et al. 2000). However, some clones of *D. mucoroides* do not mix (e.g., Buss 1982), and Mehdiabadi et al. (2006) recently showed a strong effect of kinship on the formation and composition of sorocarps in *Dictyostelium purpureum*. Kaushik et al. (2006) reported substantial levels of interclonal incompatibility in *Dictyostelium giganteum*. Even with low levels of relatedness, if the fitness of defector mutants on their own is sufficiently low, kin selection can maintain altruistic behavior (Hudson et al. 2002). Thus, in some species, kinship alone may control conflict in multicellular structures produced by aggregation, whereas in other species, alternative defenses, including greenbeard recognition (Queller et al. 2003) and policing (Frank 1995, 2003; Sachs et al. 2004), may also be important. Moreover, despite the costs from defectors, there may be synergistic benefits to co-aggregation and cooperation that outweigh the costs (Foster et al. 2002), especially when there are diminishing returns to escalating investment in either competition or cooperation (Foster 2004).

Defectors in fusion chimeras. Many sponges, as well as colonial cnidarians, bryozoans, and ascidians, occasionally fuse with genetically distinct conspecifics, forming genetically chimeric individuals (reviewed in Buss 1987, 1990; Grosberg 1988; Hughes 1989, 2002). Intergenotypic fusion also occurs among the hyphae of some fungi (Wu & Glass 2001), the sporelings of red and green algae (Gonzáles & Correa 1996, Santelices 2004), and the roots of some plants (reviewed in Pineda-Krch & Lehtilä 2004). As with aggregative development, intergenotypic fusion opens the door to horizontal transfer of defectors, especially when cells are mobile. Moreover, because virtually all metazoans capable of intergenotypic fusion do not sequester a germ line (Blackstone & Jasker 2003; Buss 1982, 1983a,b, 1987; Nieuwkoop & Sutasurya 1981), and because their cells are mobile, they are especially vulnerable to defectors that invade germ lines and therefore can be vertically transmitted across generations.

Several studies of invertebrate chimeras provide circumstantial evidence for the existence of defector genotypes in natural populations (Pancer et al. 1995, Sabbadin & Zaniolo 1979; reviewed in Buss 1987, 1990; Grosberg 1988). The clearest demonstration of intraspecific cheating in fusion chimeras comes from the colonial ascidian *Botryllus schlosseri*, where some genotypes are predictably overrepresented in the gametic output of chimeric colonies (e.g., Stoner & Weissman 1996, Stoner et al. 1999). Although chimeric organisms may have some synergistic performance advantages over single genotypes that offset the individual cost of being parasitized (Buss 1982, 1990; Foster 2004; Rinkevich & Weissman 1992), they may also often pay substantial functional costs (e.g., Barki et al. 2002, Frank et al. 1997, Maldonado 1998, Rinkevich & Weissman 1992, Santelices et al. 1999). Nevertheless, with some notable exceptions (e.g., Bishop & Sommerfeldt 1999), almost all metazoans that fuse with conspecifics (reviewed in Grosberg 1988), including *Botryllus* (Oka 1970, Scofield et al. 1982), possess highly polymorphic, genetically based self/nonself recognition systems that limit fusion to clone mates and close kin (reviewed in Grosberg 1988, Grosberg et al. 1996). Consequently, the fitness costs of such parasitism should be correspondingly reduced.

Conflicts between Nonfungible Units

Multicellular organisms include diverse intracellular and extracellular symbionts that are nonfungible evolutionary units. As Hamilton (1987b) suggested for insect societies composed of close kin, development from unicells could be costly in that multicellular organisms composed of a clone of cells could be more vulnerable to pathogens. Multicellular organisms have, however, evolved a variety of cellular and humoral defenses against horizontally transmitted pathogens. In addition, host organisms also select among mutualistic partners: Squids choose specific light-emitting *Vibrio* spp. (Fidopiastis et al. 1998), anthozoan cnidarians accept or reject specific algal strains as photosynthetic partners (Baker 2003, Muller-Parker & Secord 2005), and leguminous plants punish or exclude less-productive nitrogen-fixing rhizobia (Denison 2000). Finally, a unicellular bottleneck aids control of uncooperative extracellular symbionts because unicellular offspring cannot harbor large extracellular parasites (Grosberg & Strathmann 1998).

Eukaryotic cells contain vertically transmitted intracellular symbionts, notably mitochondria and chloroplasts, but often also less helpful ones. Selection among replicators within cells can differ from selection between cells, generating conflicts that can increase the fitness of symbionts at the expense of host function (Burt & Trivers 2006). In addition, asexually multiplying endosymbionts are subject to Muller's ratchet, accumulating deleterious mutations. This misalignment of selection on host cells and endosymbionts originated in unicellular organisms and continues, incompletely resolved, in multicellular organisms. Moreover, for a multicellular organism, genetic differences among endosymbionts could be the basis for genetic differences among its cells, leading to competition among its cell lineages.

Uniparental inheritance of mitochondria (or other internal symbionts) partially limits opportunities for conflicts within the cells of a multicellular organism (Birky 1995). There also appears to be a within-cell germ-line bottleneck that further restricts the number of transmitted endosymbionts, limiting the scope for diversification within host cells and thus maintaining performance at the cellular and multicellular levels (Bergstrom & Pritchard 1998, Krakauer & Mira 1999). Although the within-cell bottleneck is not as complete as that for nuclei, it can serve to redistribute variation within a cell to variation between cells, and thus between multicellular offspring (Bergstrom & Pritchard 1998, Sekiguchi et al. 2003).

DEFENSES AGAINST DEFECTORS

Michod (2003) distinguished between two kinds of defense against defectors. The first limits the prospects for transmission or survival of defectors from generation to generation, restricting the opportunity for defectors to spread through a population. The second controls within-organism variation and threats that arise within the life cycle of a multicellular organism. There is a sharp divide between those who argue that the primary conflict mediator in the evolution of multicellularity was and remains a unicellular bottleneck (e.g., Dawkins 1982; Keller & Surette 2006; Maynard Smith 1988; Maynard Smith & Szathmáry 1995; Queller 1997, 2000; Seger 1988; Van Valen 1988; reviewed in Grosberg & Strathmann 1998), and those who either omit the bottleneck entirely or who acknowledge its significance, albeit as a secondary one. For example, Buss (1987) emphasized the roles of maternal control, germ-line sequestration, and self/nonself recognition systems as conflict mediators during transitions to multicellularity, but did not explicitly consider the impacts of a unicellular bottleneck, a point noted in several reviews of his landmark book (e.g., Bell 1989, Seger 1988, Van Valen 1988). These contrasting views persist. Michod & Nedelcu (2003, p. 66) state, "In the case of multicellular groups, conflict mediation may involve the spread of conflict modifiers producing self-policing, maternal control of cell fate, decreased propagule size, determinate growth of the organism, apoptotic responses, or germ line sequestration . . . " In contrast, Queller (2000, p. 1648) stated, "Relatedness has such a simple role in the transition to multicellularity, that for a long time it received little attention. When the multicellular form develops from a single cell undergoing mitotic divisions . . . it is a clone of genetically identical cells . . . "

Unicellular Bottlenecks/Propagule Size

The life cycles of the vast majority of multicellular organisms regularly include a unicellular stage. There may be some ecological advantages to being a unicell, especially with respect to dispersal, and most of the proposed advantages of sexual reproduction require a unicellular stage; however, the developmental and ecological risks must be considerable. Indeed, many organisms expend substantial resources to protect their offspring while they are small (reviewed in Grosberg & Strathmann 1998,

Strathmann 1998). Protection, in turn, often intimately brings together cells of parents and their offspring, as well as those of siblings, heightening the scope and intensity of parent-offspring and sibling conflict (Clutton-Brock 1991, Haig 1993, Grosberg & Strathmann 1998, Parker et al. 2002).

The effects of a unicellular bottleneck on genetic conflicts could offset these risks in several ways. First, a unicellular bottleneck ensures that each generation begins with a group of cells that shares all of their genes by descent (Dawkins 1976, 1982; Grosberg & Strathmann 1998; Hamilton 1964b; Maynard Smith 1988; Seger 1988). When Buss (1987, p. 53) asked, "Why, then, should any cell in a dividing embryo become ciliated, or otherwise differentiated, in a fashion which limits its own capacity for increase?" he answered, "It should not" (Buss 1987, p. 53). However, with clonal development, $r = 1$ in Hamilton's inequality ($rb > c$), and it seems a cell should become differentiated if $b > c$.

Second, in each generation, the variation among replicators that arises within the parental life cycle is partitioned among offspring, rather than continuing within offspring. Some offspring are more fit and some less, rather than all being compromised (Crow 1988; Dawkins 1982; Griesemer et al. 2005; Grosberg & Strathmann 1998; Keller & Surette 2006; Kondrashov 1994; Maynard Smith 1988; Michod & Roze 2001; Queller 1997, 2000). This process not only reduces mutational load, it also limits the prospects for successful survival and transmission of defector cell lineages that fail to produce a multicellular organism that is as fit as organisms produced by cooperator genotypes (Lachmann et al. 2003). Finally, passage through a unicellular bottleneck, in addition to uniparental inheritance of organelles, reduces (but does not eliminate) the potential for conflict within and among host cells carrying genetically heterogeneous populations of organelles (Bergstrom & Pritchard 1998).

Whether a unicellular bottleneck is sufficient to protect multicellular organisms against defector cell lineages depends upon the frequency and types of variation that arise among cell lineages. In particular, if the frequency of mutations that (*a*) increase their own replication rate, (*b*) damage their multicellular host, and (*c*) can be transmitted across generations is high enough, then other mechanisms of control—including germ-line sequestration and policing—may be important (Roze & Michod 2001). However, there is little evidence that defector cell lineages arise by somatic mutation at a frequency that renders a unicellular bottleneck ineffective (Queller 2000). Indeed, Michod (2003, pp. 298–99) concluded, "So long as mutations are selfish, smaller propagule size may be selected, including single cell reproduction."

There are several cases in which multicellular organisms appear to lack a unicellular bottleneck for multiple generations, or in which single cells contain multiple nuclei. In either situation, genetic uniformity of the cell lineages that constitute a clonally developing multicellular organism could be cumulatively eroded.

1. Attine ants farm a variety of basidiomycete fungi as asexual, multicellular hyphae in their nests (reviewed in Mueller 2002, Mueller et al. 2005). Ant foundresses transport their fungal symbionts to new nests as a multicellular bolus held in an infrabuccal pouch (Chapela et al. 1994, Mueller 2002). Although the fungal symbionts of attines appear to be able to form normal basidiocarps and

produce spores, they may rarely, if ever, actually develop from a unicellular spore (Mueller 2002). However, phylogenetic and population genetic data suggest that the fungal symbiont population could be regularly feralized, passing through a standard bottleneck by germinating through spores and then being redomesticated by ant hosts (Mueller 2002).

2. Lichens, some of which produce multicellular propagules that contain both fungal and algal cells (Kondrashov 1994), may include other examples of long-term vegetative propagation without a bottleneck and may also be stabilized by partner fidelity feedback. However, many fungi in the lichen symbiosis periodically reacquire new algal partners (Honegger 1993, Yahr et al. 2004).

3. Laboratory studies of some members of the delta subgroup of magnetotactic proteobacteria suggest that adult colonies produce multicellular daughter colonies, without any intervening unicellular stage (Keim et al. 2004). Nevertheless, the fact that these normally anaerobic organisms were aerobically cultured cautions against accepting the absence of a bottleneck without further study.

4. Arbuscular mycorrhizal fungi are symbionts of terrestrial plants, facilitating the uptake of nutrients for their hosts and receiving carbon in return (Smith & Read 1997). They are thought to be exclusively asexual and to date from the Ordovician. Their unicellular spores contain hundreds of nuclei, which, in the absence of a uninucleate bottleneck, could lead to the accumulation of genetic differences within and among the cells of the developing fungus. However, recent genetic analyses indicate that all nuclei within a spore are genetically uniform (Pawlowska & Taylor 2004), although the mechanism that produces such homokaryosis is not yet understood.

We know of no other documented instances where multicellular or multinucleate vegetative propagation continues for more than a few generations without a unicellular stage.

Self/Nonself Recognition and Policing

Virtually all multicellular organisms—including most metazoans, plants, and fungi, and many algae and bacteria—use genetic, environmental, or physiological cues to distinguish conspecific self from nonself cells or tissues (reviewed in Buss 1983a,b, 1987; Grosberg 1988; Mydlarz et al. 2006). By directly recognizing defector cells or the carriers of defector cells through their phenotypes and genotypes (greenbeard recognition) or by kinship, multicellular organisms can limit the opportunity for invasion by a selfish genotype.

An individual can most efficiently protect itself by detecting and excluding defectors before they can gain the benefits of cooperative multicellularity. The allorecognition systems of many clonal marine invertebrates represent this form of defense, but do not recognize defectors per se, instead limiting cooperative interactions to close relatives with whom they are likely to share genes throughout their genome. The sequencing and annotation of the *D. discoideum* genome have revealed several additional ways that recognition and policing may control defectors in aggregating multicellular organisms. First, in studies of *dimA*, a gene that controls reception of a molecule (DIF-1) necessary for differentiation into sterile prestalk cells, knock-out mutants produce no *dimA* product and cells differentiate only into prespore cells. Such mutants should be very effective cheaters in chimeras. However, aggregations of wild-type cells exclude *dimA* mutants from spores (Foster et al. 2004). This pleiotropic effect of *dimA* means that although mutants should have an advantage in chimeras, they rarely get to exercise it. Second, cells with knockouts of the *csA* gene (which encodes an adhesion protein, gp80) join aggregates with wild-type cells on agar media (Queller et al. 2003), but their weaker intercellular binding properties leave them lagging behind normal amoebae and congregating toward the posterior end of pseudoplasmodia. Cells in this location are very likely to become spores, and so *csA*-deficient mutants should be successful defectors. However, on natural substrates like soil, *csA*⁻ cells have such weak intercellular binding strength that they are generally not entrained into developing pseudoplasmodia or sorocarps. The *csA* locus may qualify as a greenbeard gene (sensu Dawkins 1982) and represents a way that *csA*⁺ (wild-type) amoebae can exclude potential defectors from joining sorocarps. Similar greenbeard effects may also influence interactions between fire ants (Keller & Ross 1998, Krieger & Ross 2002) and bacteria (Haig 1997).

Alternatively, with policing, a recognition system can detect cheaters after they appear within an organism, priming an effector mechanism that sequesters, kills, punishes, or encourages the suicide or PCD of the defectors (Frank 1998, 2003; Michod 2003; Sachs et al. 2004). The vertebrate immune system epitomizes this form of policing, although social hymenoptera exhibit similar behaviors (reviewed in Frank 2003, Sachs et al. 2004).

Programmed Cell Death and Apoptosis

PCD encompasses several mechanisms of cellular suicide in both prokaryotes and eukaryotes (Koonin & Aravind 2002). The occurrence of caspase orthologs and other proteins related to PCD in diverse unicellular organisms (Bidle & Falkowski 2004, Gordeeva et al. 2004) implies that some of the functions of PCD in multicellular organisms existed in unicellular ancestors. Apoptosis is a specialized form of PCD, where a cell commits suicide in response to external or internal signals and dies in a way that avoids release of materials damaging to other cells (reviewed in Lodish et al. 2000). In eukaryotes, mitochondria play an important role in apoptosis, suggesting that apoptosis originated with the manipulation of host cells by ancestral protomitochondria. These protomitochondria may have released caspases or cytochrome *c*, perhaps either to maintain their symbiosis by killing host cells upon disturbance (Burt & Trivers 2006, Kobayashi 1998) or to trigger genetic recombination of the host cell in response to stress (Blackstone & Green 1999).

Michod (1999, p. 122) argued that cell death "should only increase within-organism conflict" because it "increases the number of cell divisions necessary for an organism to reach a given adult size," which consequently increases "the opportunity for within-organism change and variation." However, PCD plays critical roles in normal development and the repair or removal of damaged or infected cells or tissues

(Meier et al. 2000). Indeed, cancers often result from mutations that disrupt or prevent apoptosis (Kaufmann & Gores 2000). In destroying mutant defector or defective cells that arise during development, apoptosis is a form of policing that represents an extreme form of altruism—an unselfish suicidal role that is expected among clone mates (Hamilton 1964a,b). Apoptosis and other forms of PCD do not necessarily involve conflict; rather, PCD illustrates the ease of policing in a clone of cells.

Maternal Control in Early Development

Buss (1987) proposed that selection to limit conflicts among cell lineages favored the evolution of maternal control of early embryonic development of metazoans. It is true that in the earliest stages of embryogenesis of many metazoans, the maternal genome is often well represented as messenger RNA and maternal gene products dominate developmental processes. However, the cells of an embryo are recent descendents from the zygote, with the least potential for mutation to have influenced genetic divergence among cells. Thus, conflicts among cell lineages should be negligible at early stages, and selection for control of conflicts should be minimal (Queller 2000).

There are alternative hypotheses. Multicellular organisms, including metazoans, differ greatly in the extent to which the maternal genome controls events in early development. The importance of maternal transcripts appears to vary with development time and degree of parental care or protection in metazoans. Reliance on maternal mRNA rather than embryonic transcription may therefore reflect selection for speedy development (Strathmann et al. 2002).

Germ-Line Sequestration

Buss (1983a,b, 1987) also argued that sequestration of a germ line is a defense against the origination and proliferation of defector cell lineages and their transmission across generations. Once a cell, or group of cells, is set aside as the progenitor(s) of all reproductive cells, mutations that arise in the much larger group of nonreproductive (somatic) cell lineages are excluded from being incorporated into gametes and transmitted to the next generation. Germ-line sequestration then becomes one of the key breakthroughs allowing the evolution of differentiated, integrated multicellular organisms. Michod (2003) has modeled the influence of the timing of sequestration, the number of cells sequestered, and the number of mitotic cycles on the distribution of mutational load and fitness variance at the level of the cell lineage and multicellular organism. In these models, the main cost of sequestering a germ line is that sequestered cells and their descendants can no longer provide somatic support. The earlier a germ line is sequestered, and the fewer progenitor cells that initiate it, the less likely it is that defector mutants can arise and come to dominate the gamete pool before the germ line is sequestered. Germ lines are typically set aside as groups of cells, often after multiple cycles of cell division. However, at the extreme, when a single cell initiates the germ line, only mutations arising in that cell and its descendants can appear in gametes. Also, germ line cells typically go through fewer division cycles than most somatic cells, the difference being magnified as organisms grow somatically (Michod & Roze 2001). Thus, even if all cells in an organism have the same per-division mutation rate to defectors, gamete cells, during their ontogeny, are less likely to be hit by a defector mutant than are the remaining somatic cells. Moreover, germ lines may experience lower mutation rates than somatic cell lineages (Drake et al. 1998, Maynard Smith & Szathmáry 1995), perhaps because they have lower metabolic rates than most somatic cells (Michod 2003).

The assumptions in these models raise a fundamental question: How often do defector mutants arise during the ontogeny of a multicellular organism? Such mutants would need to have a remarkable combination of features if they were to survive beyond the generation in which they arose. They could not so thoroughly damage the soma of the organism in which they arose that the host no longer functioned well enough to reproduce competitively. The mutation would have to occur in a cell or cell lineage that produced gametes. In organisms that develop clonally from a unicell, defectors would inevitably have to produce a competitive soma on their own. Such mutants are conceivable, but seem most likely in organisms with minimal differentiation and integration and that develop by aggregation rather than clonally.

Although most models for the evolution of germ-line sequestration (e.g., Michod 1996, 1997; Michod & Roze 1999) consider only mutants that are obligate defectors (unconditional sensu Queller 2000), facultative defectors could arise that produce a normal soma when developing as a clone but defect when they arise in a population of cooperating cells (Hudson et al. 2002, Strassmann et al. 2000). In this case, if there were no costs to being a facultative defector, then the mutant should spread to fixation in a population (Matsuda & Harada 1990). If there were a cost, then a substantial increase in the frequency of the mutant should require a low cost or a high mutation rate. Thus, defector mutants that both significantly disrupt the integrity of clonally developing multicellular organisms and that can also spread through subsequent generations should be rare, reaching an equilibrium determined by the rate at which they arise by mutation, the benefits they gain in a chimeric state, and the costs of developing in isolation (Queller 2000).

All told, germ-line sequestration can reduce the threat of defectors in multicellular organisms being incorporated into gametes and vertically transmitted. Although there are no decisive studies, organisms that sequester germ lines might produce gametes with lower deleterious mutational loads than organisms in lineages that do not sequester their germ lines. Nevertheless, many lineages of multicellular organisms, even those with motile cells, do not sequester germ lines but still manage to produce a diversity of organisms with highly differentiated body plans (Blackstone & Jasker 2003, Extavour & Jakam 2003). Moreover, many of the multicellular organisms that do not sequester a germ line early in ontogeny have scattered pluripotent cells (e.g., plant meristems, stem cells in cnidarians, circulating cells in colonial ascidians), some of which may later become epigenetically modified and develop into germ cells. Thus, germ-line sequestration is not an essential defense for the evolution of multicellular differentiation, even in lineages with motile cells.

The question of what selective factors promote the evolution of a sequestered germ line remains unresolved and may have little to do with the control of defectors. The germ-soma distinction may variously be a response to constraints on motility as aquatic multicellular organisms increased in size (Solari et al. 2006), a means of avoiding the challenges of dedifferentiating highly modified cells into totipotent gametes by setting aside undifferentiated cells (Jablonka & Lamb 1995), or simply employing as the germ line those cells left after other cell lineages have begun to differentiate. As Queller (2000, p. 1653) suggested, "the germ line might have originated as a consequence of other cell lineages altruistically removing themselves from the reproductive line, to perform some somatic benefit to the organism." From this perspective, populations of germ and stem cells might be kept small, not so much because of the risk of a mutation occurring, but because minimizing the number of germ cells and stem cells limits the number of cells that are in a less differentiated and less useful state.

EVOLUTION OF COMPLEXITY IN MULTICELLULAR ORGANISMS

To conclude, we return to the question of what limits the complexity, specifically the number of different kinds of cells, of multicellular organisms once the transition has occurred. We ignore the arrangements of cells, although much of the diversity of multicellular organisms lies in how cells are arranged. We also ignore the complexity of individual cells, although cells of multicellular organisms appear to have fewer kinds of visible cell parts than do unicellular organisms (McShea 2002).

One limit on the number of kinds of cells is suggested by lower levels of cellular differentiation in organisms developing by aggregation than in many comparably sized clonal developers (Wolpert & Szathmáry 2002). Conflict control was the fundamental role of a unicellular bottleneck; cooperation results whether cells differentiate or not. Nonetheless, the bottleneck subsequently facilitated the evolution of extensive differentiation and integration, perhaps because the scope for coordinated development among cells is greater when cooperation is assured by genetic identity among cells. (Differentiation of cells is often influenced by symbionts, but the differentiating cells within each partner species are usually clone mates.)

What are the limits on number of cell types in organisms with clonal development from a unicell? Carroll (2001) noted that because "very few regulatory proteins can orchestrate markedly different cell physiologies, it is curious that more multicellular forms have not evolved." The answer may simply be a lack of selective pressure or trade-offs. For example, the vertebrate immune system generates different cells in indefinite variety because variety is favored by selection. We expect that functional requirements limit the number of cell types in organisms that develop from a single cell. Evolving a change in a regulatory pathway that produces additional variation among cells should require few mutational steps, and few mutational steps could then remove the regulatory change. Once the major functional specializations are satisfied, there is a presumably diminishing return in added capabilities from each additional cell type.

This argument implicitly concerns the role of small variants (within cell types) in the generation of cellular diversity, but trends for more distinctively different cell types support this line of reasoning. For example, the number of cell types increases with body length (Bonner 1965) and with log cell number (estimated from body volume), although in the latter case the slope is only 0.056 (Bell & Mooers 1997).

Bell & Mooers (1997) explain this pattern by the small effect of a single cell on the performance of a large organism, a tissue of many similar cells being the smallest effective unit for a task.

Different kinds of organisms appear to require different numbers of cell types. For a given size, number of cell types is greater for animals than for green plants and greater for green plants than for phaeophytes (Bell & Mooers 1997). The difference in number of cell types between animals and plants suggests a relation between complexity (in terms of variety of cells) and motility of either the constituent cells or the whole organism. This difference is also consistent with the apparent ease of evolving multicellularity for phototrophs versus the difficulties for phagotrophs (Cavalier-Smith 1991). Nevertheless, the barriers to the evolution of multicellular phagotrophy are not obvious. For example, several groups of multicellular phagotrophs (sponges, placozoans, and vorticellid colonies) lack a multicellular mouth, suggesting that the need for a mouth was not a serious obstacle. Perhaps importantly, at least for metazoans, both the number of cell types (Valentine et al. 1994) and body lengths (Bonner 1965) appear to have increased from approximately 600 Mya to the present.

Taken together, the available data suggest that just as the evolution of multicellularity itself is a minor major transition, the same may be true for complexity of multicellular organisms, at least in terms of cellular diversity (also see Vermeij 2006). Although there may be genetic, developmental, and phylogenetic constraints on the evolution of cellular diversity, the evidence so far suggests that cellular diversity can evolve easily when functionally called for by selection.

DISCLOSURE STATEMENT

The authors are not aware of any biases that might be perceived as affecting the objectivity of this review.

ACKNOWLEDGMENTS

The University of Washington's Helen R. Whiteley Center and Friday Harbor Laboratories, National Science Foundation grants OCE-0217304 and OCE-0623102 to R.S., and IBN-0416713 and OCE-0623699 to R.K.G. supported this research. The Mellon Foundation also funded R.K.G. The authors benefited from discussions with G. von Dassow, Egbert Leigh, Geerat Vermeij, Rick Michod, Steve Frank, Nicole King, Mike Loeb, Michael Turelli, David Queller, Joan Strassmann, Jim Griesemer, and Steve Dudgeon.

LITERATURE CITED

Ackerman M, Chao L. 2004. Evolution of cooperation: two for one? *Curr. Biol.* 14:R73–74

Armstrong DP. 1984. Why don't cellular slime molds cheat? *J. Theor. Biol.* 109:271–83

Ausmees N, Jacobs-Wagner C. 2003. Spatial and temporal control of differentiation and cell cycle progression in *Caulobacter crescentus*. *Annu. Rev. Microbiol.* 57:225–47

Avise J. 1993. The evolutionary biology of aging, sexual reproduction, and DNA repair. *Evolution* 47:1293–301

Baker AC. 2003. Flexibility and specificity in coral-algal symbiosis: diversity, ecology, and biogeography of *Symbiodinium*. *Annu. Rev. Ecol. Evol. Syst.* 34:661–89

Baldauf SL. 2003. The deep roots of eukaryotes. *Science* 300:1703–1706.

Banfield WG, Woke PA, MacKay CM, Cooper HL. 1965. Mosquito transmission of a reticulum cell sarcoma of hamsters. *Science* 148:1239–40

Barki Y, Gateño D, Graur D, Rinkevich B. 2002. Soft-coral natural chimerism: A window in ontogeny allows the creation of entities comprised of incongruous parts. *Mar. Ecol. Prog. Ser.* 231:91–99

Bell G. 1985. The origin and early evolution of germ cells as illustrated in the Volvocales. In *The Origin and Evolution of Sex*, ed. H Halvorson, A Monroy, pp. 221–56. New York: Liss

Bell G. 1989. Darwin and biology. *J. Heredity* 80:417–21

Bell G, Koufopanou V. 1991. The architecture of the life cycle in small organisms. *Philos. Trans. R. Soc. London Ser. B* 332:81–89

Bell G, Mooers AO. 1997. Size and complexity among multicellular organisms. *Biol. J. Linn. Soc. London* 60:345–63

Bergstrom CT, Pritchard J. 1998. Germline bottlenecks and the evolutionary maintenance of mitochondrial genomes. *Genetics* 149:2135–46

Bidle KD, Falkowski P. 2004. Cell death in planktonic, photosynthetic microorganisms. *Nat. Rev. Microbiol.* 2:643–55

Birky CW. 1995. Uniparental inheritance of mitochondrial and chloroplast genes: mechanisms and evolution. *Proc. Natl. Acad. Sci. USA* 92:11331–38

Bishop JDD, Sommerfeldt AD. 1999. Not like *Botryllus*: indiscriminate postmetamorphic fusion in a compound ascidian. *Proc. R. Soc. London Ser. B* 266:241–48

Blackstone NW, Green DR. 1999. The evolution of a mechanism of cell suicide. *BioEssays* 21:84–88

Blackstone NW, Jasker BD. 2003. Phylogenetic considerations of clonality, coloniality, and mode of germline development in animals. *J. Exp. Zool. B* 297:35–47

Blanton RS, Olive LS. 1983. Ultrastructure of aerial stalk formation by the ciliated protozoan *Sorogena stoianovitchae*. *Protoplasma* 116:125–35

Bonner JT. 1965. *Size and Cycle*. Princeton, NJ: Princeton Univ. Press. 219 pp.

Bonner JT. 1967. *The Cellular Slime Molds*. Princeton, NJ: Princeton Univ. Press. 205 pp. 2nd ed.

Bonner JT. 1974. *On Development: The Biology of Form*. Cambridge, MA: Harvard Univ. Press. 282 pp.

Bonner JT. 1982. Evolutionary strategies and developmental constraints in the cellular slime molds. *Am. Nat.* 119:530–52

Bonner JT. 1988. *The Evolution of Complexity by Means of Natural Selection*. Princeton, NJ: Princeton Univ. Press. 272 pp.

Bonner JT. 1998. The origins of multicellularity. *Integr. Biol.* 1:28–36

Bonner JT. 2000. *First Signals: The Evolution of Multicellular Development*. Princeton, NJ: Princeton Univ. Press. 146 pp.

Bonner JT. 2004. Perspective: the size-complexity rule. *Evolution* 58:1883–90

Boraas ME, Seale DB, Boxhorn JE. 1998. Phagotrophy by a flagellate selects for colonial prey: a possible origin of multicellularity. *Evol. Ecol.* 12:153–64

Branda SS, González-Pastor JE, Ben-Yehuda S, Losick R, Kolter R. 2001. Fruiting body formation by *Bacillus subtilis*. *Proc. Natl. Acad. Sci. USA* 98:11621–26

Burt A, Trivers R. 2006. *Genes in Conflict: The Biology of Selfish Genetic Elements.* Cambridge, MA: Harvard Univ. Press. 602 pp.

Buss LW. 1982. Somatic cell parasitism and the evolution of somatic tissue compatibility. *Proc. Natl. Acad. Sci. USA* 79:5337–41

Buss LW. 1983a. Evolution, development, and the units of selection. *Proc. Natl. Acad. Sci. USA* 80:1387–91

Buss LW. 1983b. Somatic variation and evolution. *Paleobiology* 9:12–16

Buss LW. 1987. *The Evolution of Individuality.* Princeton, NJ: Princeton Univ. Press. 201 pp.

Buss LW. 1990. Competition within and between encrusting colonial marine invertebrates. *Trends Ecol. Evol.* 5:352–56

Carroll SB. 2001. Chance and necessity: the evolution of morphological complexity and diversity. *Nature* 409:1102–9

Cavalier-Smith T. 1991. Cell diversification in heterotrophic flagellates. In *The Biology of Free-living Heterotrophic Flagellates*, ed. DJ Patterson, J Larsen, pp. 113–31. Oxford, UK: Clarendon

Chapela IH, Rehner SA, Schultz TR, Mueller UG. 1994. Evolutionary history of the symbiosis between fungus-growing ants and their fungi. *Science* 266:1691–94

Clutton-Brock TH. 1991. *The Evolution of Parental Care.* Princeton, NJ: Princeton Univ. Press. 352 pp.

Coleman AW. 1999. Phylogenetic analysis of "Volvocaceae" for comparative genetic studies. *Proc. Natl. Acad. Sci. USA* 96:13892–97

Crespi BJ. 2001. The evolution of social behavior in microorganism. *Trends Ecol. Evol.* 16:178–83

Crow JF. 1988. The importance of recombination. In *The Evolution of Sex: An Examination of Current Ideas*, ed. RE Michod, BR Levin, pp. 56–73. Sunderland, MA: Sinauer

Das U, Das AK. 2000. Review of canine transmissible venereal sarcoma. *Vet. Res. Commun.* 24:545–56

Dawkins R. 1976. *The Selfish Gene.* Oxford, UK: Oxford Univ. Press. 224 pp.

Dawkins R. 1982. *The Extended Phenotype.* San Francisco: Freeman. 307 pp.

DeAngelo MJ, Kish VM, Kolmes SA. 1990. Altruism, selfishness, and heterocytosis in slime molds. *Ethol. Ecol. Evol.* 2:439–43

Denison RF. 2000. Legume sanctions and the evolution of symbiotic cooperation by rhizobia. *Am. Nat.* 156:567–76

Drake JW, Charlesworth B, Charlesworth D, Crow JF. 1998. Rates of spontaneous mutation. *Genetics* 148:1667–86

Dworkin M. 1972. The myxobacteria: new directions in studies of prokaryotic development. *Crit. Rev. Microbiol.* 2:435–52

Dworkin M. 1996. Recent advances in the social and developmental biology of the myxobacteria. *Microbiol. Rev.* 60:70–102

England JC, Gober JW. 2001. Cell cycle control of morphogenesis in *Caulobacter*. *Curr. Opin. Microbiol.* 4:674–80

Ennis HL, Dao DN, Pukatzki SU, Kessin RH. 2000. *Dictyostelium* amoebae lacking an F-box protein form spores rather than stalk in chermas with wild type. *Proc. Natl. Acad. Sci. USA* 97:3292–97

Extavour CG, Akam M. 2003. Mechanisms of germ cell specification across the metazoans: epigenesis and preformation. *Development* 130:5869–84

Fidopiastis PM, von Boletzky S, Ruby EG. 1998. A new niche for *Vibrio logei*, the predominant light organ symbiont of squids in the genus *Sepiola*. *J. Bacteriol.* 180:59–64

Fiegna F, Velicer GJ. 2003. Competitive fates of bacterial social parasites: persistence and self-induced extinction of *Myxococcus xanthus* cheaters. *Proc. R. Soc. London Ser. B* 266:493–98

Fiegna F, Velicer GJ. 2005. Exploitative and hierarchical antagonism in a cooperative bacterium. *PLoS Biol.* 3:1980–87

Fiegna F, Yu Y-TN, Kadam SV, Velicer GJ. 2006. Evolution of an obligate social cheater to a superior cooperator. *Nature* 441:310–14

Filosa MF. 1962. Heterocytosis in cellular slime molds. *Am. Nat.* 96:79–91

Fortunato A, Strassmann JE, Santorelli L, Queller DC. 2003. Co-occurrence in nature of different clones of the social amoeba, *Dictyostelium discoideum*. *Mol. Ecol.* 12:1031–38

Foster KR. 2004. Diminishing returns in social evolution: the not so tragic commons. *J. Evol. Biol.* 17:1058–72

Foster KR, Fortunato A, Strassmann JE, Queller DC. 2002. The costs and benefits of being a chimera. *Proc. R. Soc. London Ser. B* 269:2357–62

Foster KR, Shaulsky G, Strassmann JE, Queller DC, Thompson CRL. 2004. Pleiotropy as a mechanism to stabilise cooperation. *Nature* 431:693–96

Francis D, Eisenberg R. 1993. Genetic structure of a natural population of *Dictyostelium discoideum*, a cellular slime mold. *Mol. Ecol.* 2:385–91

Frank SA. 1995. Mutual policing and repression of competition in the evolution of cooperative groups. *Nature* 377:520–22

Frank SA. 1998. *Foundations of Social Evolution*. Princeton, NJ: Princeton Univ. Press. 268 pp.

Frank SA. 2003. Repression of competition and the evolution of cooperation. *Evolution* 57:693–705

Frank U, Reun U, Loya Y, Rinkevich B. 1997. Alloimmune maturation in the coral *Stylophora pistillata* is achieved through three distinctive stages, 4 months post-metamoprhosis. *Proc. R. Soc. London Ser. B* 264:99–104

Gadagkar R, Bonner JT. 1994. Social insects and social amoebae. *J. Biosci.* 19:219–45

Gerhart J, Kirschner M. 1997. *Cells, Embryos, and Evolution: Toward a Cellular and Developmental Understanding of Phenotypic Variation and Evolutionary Adaptability*. Oxford, UK: Blackwell Sci. 642 pp.

Giovannoni SJ, Turner S, Olsen GJ, Barns S, Lane DJ, Pace NR. 1988. Evolutionary relationships among Cyanobacteria and green chloroplast. *J. Bacteriol.* 170:3584–92

González P, Correa JA. 1996. Fusion and histocompatibility in Rhodophyta. *Hydrobiologia* 326/327:387–92

Gordeeva AV, Labas YA, Zvyagilskaya RA. 2004. Apoptosis in unicellular organisms: mechanisms and evolution. *Biochemistry* 69:1055–66

Griesemer J. 2000. The units of evolutionary transition. *Selection* 1:67–80

Griesemer J, Haber MH, Yamashita G, Gannett L. 2005. Critical notice: cycles of contingency—developmental systems and evolution. *Biol. Philos.* 20:517–44

Grosberg RK. 1988. The evolution of allorecognition specificity in clonal invertebrates. *Q. Rev. Biol.* 63:377–412

Grosberg RK, Levitan DR, Cameron BB. 1996. The evolutionary genetics of allorecognition in the colonial hydrozoan *Hydractinia symbiolongicarpus*. *Evolution* 50:2221–40

Grosberg RK, Strathmann RR. 1998. One cell, two cell, red cell, blue cell: the persistence of a unicellular stage in multicellular life histories. *Trends Ecol. Evol.* 13:112–16

Haig D. 1993. Genetic conflicts in human pregnancy. *Q. Rev. Biol.* 68:495–532

Haig D. 1997. The social gene. In *Behavioural Ecology: An Evolutionary Approach*, ed. JR Krebs, NB Davies, pp. 284–304. Oxford, UK: Blackwell Sci. 4th ed.

Hamilton WD. 1964a. The genetical evolution of social behaviour. Part I. *J. Theor. Biol.* 7:1–16

Hamilton WD. 1964b. The genetical evolution of social behaviour. Part II. *J. Theor. Biol.* 7:17–52

Hamilton WD. 1987a. Discriminating nepotism: expectable, common, overlooked. In *Kin Recognition in Animals*, ed. DJC Fletcher, CD Michener, pp. 417–37. New York: Wiley

Hamilton WD. 1987b. Kinship, recognition, disease, and intelligence: constraints of social evolution. In *Animal Societies: Theories and Facts*, ed. J Itô, JL Brown, J Kikkawa, pp. 81–102. Tokyo: Jpn. Sci. Soc.

Hammerstein P, ed. 2003. *Genetic and Cultural Evolution of Cooperation*. Cambridge, MA: MIT Press

Honegger R. 1993. Tansley review No. 60. Developmental biology of lichens. *New Phytol.* 125:695–77

Hudson RE, Aukema JE, Rispe C, Roze D. 2002. Altruism, cheating, and anticheater adaptations in cellular slime molds. *Am. Nat.* 160:31–43

Hughes RN. 1989. *A Functional Biology of Clonal Animals*. New York: Chapman & Hall. 331 pp.

Hughes RN. 2002. Genetic mosaics and chimeras. In *Reproductive Biology of Invertebrates*, Vol. XI. *Progress in Asexual Reproduction*, ed. RN Hughes, pp. 159–73. New York: Wiley

Huss MJ. 1989. Dispersal of cellular slime molds by two soil invertebrates. *Mycologia* 81:677–82

Jablonka E, Lamb MJ. 1995. *Epigenetic Inheritance and Evolution*. Oxford, UK: Oxford Univ. Press

Johnson UG, Porter KR. 1968. Fine structure of cell division in *Chlamydomonas reinhardi*. *J. Cell Biol.* 38:403–25

Kaiser D. 2001. Building a multicellular organism. *Annu. Rev. Genet.* 35:103–23

Kaufmann SH, Gores GJ. 2000. Apoptosis in cancer: cause and cure. *BioEssays* 22:1007–17

Kaushik S, Katoch B, Nanjundiah V. 2006. Social behaviour in genetically heterogeneous groups of *Dictyostelium giganteum*. *Behav. Ecol. Sociobiol.* 59:521–30

Keim CN, Martins JL, Abreu F, Rosado AS, de Barros HL, et al. 2004. Multicellular life cycle of magnetotactic prokaryotes. *FEMS Microbiol. Lett.* 240:203–8

Keller L, Ross KG. 1998. Selfish genes: a green beard in the red fire ant. *Nature* 394:573–75

Keller L, Surette MG. 2006. Communication in bacteria: an ecological and evolutionary perspective. *Nat. Rev. Microbiol.* 4:249–58

King N. 2004. The unicellular ancestry of animal development. *Dev. Cell* 7:313–25

Kirk D. 1997. The genetic program for germ-soma differentiation in *Volvox*. *Annu. Rev. Genet.* 31:359–80

Kirk D. 1998. *Volvox*. Cambridge, UK: Cambridge Univ. Press. 381 pp.

Kirk D. 1999. Evolution of multicellularity in the volvocine algae. *Curr. Opin. Plant Biol.* 2:296–501

Kirk D. 2003. Seeking the ultimate and proximate causes of *Volvox* multicellularity and cellular differentiation. *Integr. Comp. Biol.* 43:247–53

Klekowski EJ Jr. 1988. *Mutation, Developmental Selection, and Plant Evolution*. New York: Columbia Univ. Press. 373 pp.

Knoll AH. 2003. *Life on a Young Planet*. Princeton, NJ: Princeton Univ. Press. 277 pp.

Knoll AH, Javaux EJ, Hewitt D, Cohen P. 2006. Eukaryotic organisms in Proterozoic oceans. *Philos. Trans. R. Soc. B* 361:1023–38

Kobayashi I. 1998. Selfishness and death: raison d'être of restriction, recombination and mitochondria. *Trends Genet.* 14:368–74

Kondrashov AS. 1994. Mutation load under vegetative reproduction and cytoplasmic inheritance. *Genetics* 137:311–18

Koonin EV, Aravind L. 2002. Origin and evolution of eukaryotic apoptosis. *Cell Death Differ.* 9:394–404

Koufopanou V. 1994. The evolution of soma in the Volvocales. *Am. Nat.* 143:907–31

Koufopanou V, Bell G. 1993. Soma and germ: an experimental approach using *Volvox*. *Proc. R. Soc. London Ser. B* 254:107–13

Krakauer DC, Mira A. 1999. Mitochondria and germ-cell death. *Nature* 400:125–26

Krieger MJB, Ross KG. 2002. Identification of a major gene regulating complex social behavior. *Science* 295:328–32

Lachmann M, Blackstone NW, Haig D, Kowald A, Michod RE, et al. 2003. Group report: cooperation and conflict in the evolution of genomes, cells, and multicellular organisms. See Hammerstein 2003, pp. 327–56

Leigh EG. 1977. How does selection reconcile individual advantage with the good of the group? *Proc. Natl. Acad. Sci. USA* 10:4542–46

Leigh EG. 1991. Genes, bees, and ecosystems: the evolution of a common interest among individuals. *Trends Ecol. Evol.* 6:257–62

Lodish H, Berk A, Zipursky SL, Matsudaira P, Baltimore D, Darnell J. 2000. *Molecular Cell Biology*. New York: Freeman. 4th ed.

Lürling M, Van Donk E. 2000. Grazer-induced colony formation in *Scenedesmus*: Are there costs to being colonial? *Oikos* 88:111–18

Maldonado M. 1998. Do chimeric sponges have improved chances of survival? *Mar. Ecol. Prog. Ser.* 164:301–6

Margulis L. 1981. *Symbiosis in Cell Evolution*. San Francisco: Freeman. 419 pp.

Martin ME, Brun YV. 2000. Coordinating development with the cell cycle in *Caulobacter*. *Curr. Opin. Microbiol.* 2:496–501

Masuda M, Sato H. 1984. Reversible resorption of cilia and the centriole cycle in dividing cells of sea urchin blastulae. *Zool. Sci.* 1:445–62

Matsuda H, Harada Y. 1990. Evolutionarily stable stalk to spore ratio in cellular slime molds and the law of equalization of net incomes. *J. Theor. Biol.* 147:329–44

Maynard Smith J. 1988. Evolutionary progress and levels of selection. In *Evolutionary Progress*, ed. MH Nitecki, pp. 219–30. Chicago: Univ. Chicago Press

Maynard Smith J, Szathmáry E. 1995. *The Major Transitions in Evolution*. Oxford, UK: Freeman. 346 pp.

McShea DW. 2001. The minor transitions in evolution and the question of a directional bias. *J. Evol. Biol.* 14:502–18

McShea DW. 2002. A complexity drain on cells in the evolution of multicellularity. *Evolution* 56:441–52

Medina M, Collins A, Taylor J, Valentine JW, Lipps J, et al. 2003. Phylogeny of Opisthokonta and the evolution of multicellularity and complexity in Fungi and Metazoa. *Int. J. Astrobiol.* 2:203–11

Mehdiabadi NJ, Jack CN, Farnham TT, Platt TG, Kalla SE, et al. 2006. Kin preference in a social microbe. *Nature* 442:881–82

Meier P, Finch A, Evan G. 2000. Apoptossis in development. *Nature* 407:796–801

Michod RE. 1996. Evolution of the individual. *Am. Nat.* 150:S5–21

Michod RE. 1997. Cooperation and conflict in the evolution of multicellularity. I. Multilevel selection of the organism. *Am. Nat.* 149:607–45

Michod RE. 1999. *Darwinian Dynamics. Evolutionary Transitions in Fitness and Individuality*. Princeton, NJ: Princeton Univ. Press. 262 pp.

Michod RE. 2003. Cooperation and conflict during the origin of multicellularity. See Hammerstein 2003, pp. 291–307

Michod RE, Nedelcu AM. 2003. On the reorganization of fitness during evolutionary transition in individuality. *Integr. Comp. Biol.* 43:64–73

Michod RE, Roze D. 1999. Cooperation and conflict in the evolution of individuality. III. Transitions in the unit of fitness. In *Mathematical and Computational Biology: Computational Morphogenesis, Hierarchical Complexity, and Digital Evolution*, ed. CL Nehaniv, pp. 47–92. Providence, RI: Am. Math. Soc.

Michod RE, Roze D. 2001. Cooperation and conflict in the evolution of multicellularity. *Heredity* 86:1–7

Miller MB, Bassler BL. 2001. Quorum sensing in bacteria. *Annu. Rev. Microbiol.* 55:165–99

Mueller UG. 2002. Ant vs fungus vs mutualism: ant-cultivar conflict and the deconstruction of the attine ant-fungus symbiosis. *Am. Nat.* 160:S67–98

Mueller UG, Gerardo NM, Aanen DK, Six DL, Schultz TR. 2005. The evolution of agriculture in insects. *Annu. Rev. Ecol. Evol. Syst.* 36:563–95

Muller-Parker G, Secord D. 2005. Symbiont distribution along a light gradient within an intertidal cave. *Limnol. Oceanogr.* 50:272–78

Murgia C, Pritchard JK, Kim SY, Fassati A, Weiss RA. 2006. Clonal origin and evolution of a transmissible cancer. *Cell* 126:477–87

Mydlarz LD, Jones LE, Harvell CD. 2006. Innate immunity, environmental drivers, and disease ecology of marine and freshwater invertebrates. *Annu. Rev. Ecol. Evol. Syst.* 37:251–88

Nieuwkoop PD, Sutasurya LA. 1981. *Primordial Germ Cells in the Invertebrates: From Epigenesis to Preformation.* Cambridge, UK: Cambridge Univ. Press. 258 pp.

Niklas KJ. 2000. The evolution of plant body plans. *Ann. Bot.* 85:411–38

Niklas KJ. 2004. The cell walls that bind the tree of life. *BioScience* 54:831–41

Nunney L. 1999. Lineage selection and the evolution of multistage carcinogenesis. *Proc. R. Sci. London Ser. B* 266:493–98

Oka H. 1970. Colony specificity in compound ascidians. In *Profiles of Japanese Science and Scientists*, ed. M Yukawa, pp. 195–206. Tokyo: Kodansha

Olive LS. 1975. *The Mycetozoans.* New York: Academic. 293 pp.

Orive ME. 2001. Somatic mutations in organisms with complex life histories. *Theor. Popul. Biol.* 59:235–49

Otto SP, Hastings IM. 1998. Mutation and selection within the individual. *Genetica* 102/103:507–24

Otto SP, Orive ME. 1995. Evolutionary consequences of mutation and selection within an individual. *Genetics* 141:1173–87

Pál C, Papp B. 2000. Selfish cells threaten multicellular life. *Trends Ecol. Evol.* 15:351–52

Pancer Z, Gershon H, Rinkevich B. 1995. Coexistence and possible parasitism of somatic and germ-cell lines in chimeras of the colonial urochordate *Botryllus schlosseri. Biol. Bull.* 189:106–12

Parker GA, Royle NJ, Hartley IR. 2002. Intrafamilial conflict and parental investment: a synthesis. *Philos. Trans. R. Soc. London Ser. B* 357:295–307

Pawlowska TE, Taylor JW. 2004. Organization of genetic variation in individuals of arbuscular mycorrhizal fungi. *Nature* 427:733–37

Pearse A-M, Swift K. 2006. Transmission of devil facial-tumour disease. *Nature* 439:549

Pfeiffer T, Bonhoeffer S. 2003. An evolutionary scenario for the transition to undifferentiated multicellularity. *Proc. Natl. Acad. Sci. USA* 100:1095–98

Pfeiffer T, Schuster S, Bonhoeffer S. 2001. Cooperation and competition in the evolution of ATP-producing pathways. *Science* 292:504–7

Pineda-Krch M, Lehtilä K. 2004. Costs and benefits of genetic heterogeneity within organisms. *J. Evol. Biol.* 17:1167–77

Porter SM, Meisterfeld R, Knoll AH. 2003. Vase-shaped microfossils from the Neoproterozoic Chuar Group, Grand Canyon: a classification guided by modern testate amoebae. *J. Paleontol.* 77:409–29

Queller DC. 1997. Cooperators since life began. A review of *The Major Transitions in Evolution. Q. Rev. Biol.* 72:184–88

Queller DC. 2000. Relatedness and the fraternal major transitions. *Philos. Trans. R. Soc. London Ser. B* 355:1647–55

Queller DC, Ponte E, Bozzaro S, Strassmann JE. 2003. Single gene greenbeard effects in the social amoeba *Dictyostelium discoideum*. *Science* 299:105–6

Raff RA. 1988. The selfish cell lineage. *Cell* 54:445–46

Rainey PB, Rainey K. 2003. Evolution of cooperation and conflict in experimental bacterial populations. *Nature* 425:72–74

Rinkevich B, Weissman IL. 1992. Chimeras vs genetically homogeneous individuals: potential fitness costs and benefits. *Oikos* 63:119–24

Roberts G. 2005. Cooperation through interdependence. *Anim. Behav.* 70:901–8

Roze D, Michod RE. 2001. Mutation, multilevel selection, and the evolution of propagule size during the origin of multicellularity. *Am. Nat.* 6:638–54

Sabbadin A, Zaniolo G. 1979. Sexual differentiation and germ cell transfer in the colonial ascidian *Botryllus schlosseri*. *J. Exp. Zool.* 207:289–304

Sachs J, Mueller IG, Wilcox TP, Bull JJ. 2004. The evolution of cooperation. *Q. Rev. Biol.* 79:135–60

Santelices B. 2004. Mosaicism and chimerism as components of intraorganismal genetic heterogeneity. *J. Evol. Biol.* 17:1187–88

Santelices B, Correa JA, Aedo D, Flores V, Hormazábal M, Sanchez P. 1999. Convergent biological processes in coalescing Rhodophtya. *J. Phycol.* 35:1127–49

Schopf JW. 1993. Microfossils of the Archean apex chert: new evidence of the antiquity of life. *Science* 60:640–46

Scofield VL, Schlumpberger JM, West LA, Weissman IL. 1982. Protochordate allorecognition is controlled by a MHC-like gene system. *Nature* 295:499–502

Seger J. 1988. Review of *The Evolution of Individuality* (LW Buss). *Q. Rev. Biol.* 63:336–37

Sekiguchi K, Kasai K, Levin BC. 2003. Inter- and intragenerational transmission of a human mitochondrial DNA heteroplasmy among 13 maternally-related individuals and differences between and within tissues in two family members. *Mitochondrion* 2:401–14

Shapiro JA. 1998. Thinking about bacterial populations as multicellular organisms. *Annu. Rev. Microbiol.* 52:81–104

Shi W, Zusman DR. 1993. The two motility systems of *Myxococcus xanthus* show different selective advantages on various surfaces. *Proc. Natl. Acad. Sci. USA* 90:3378–82

Shimkets LJ. 1990. Social and developmental biology of the myxobacteria. *Microbiol. Rev.* 54:473–501

Smith SE, Read DJ. 1997. *Mycorrhizal Symbiosis*. San Diego: Academic

Solari CA, Ganguly S, Kessler JO, Michod RE, Goldstein RE. 2006. Multicellularity and the functional interdependence of motility and molecular transport. *Proc. Natl. Acad. Sci. USA* 103:1353–58

Stanley SM. 1973. An ecological theory for the sudden origin of multicellular life in the late Precambrian. *Proc. Natl. Acad. Sci. USA* 70:1486–89

Stoner DS, Rinkevich B, Weissman IL. 1999. Heritable germ and somatic cell lineage competitions in chimeric colonial protochordates. *Proc. Natl. Acad. Sci. USA* 96:9148–53

Stoner DS, Weissman IL. 1996. Somatic and germ cell parasitism in a colonial ascidian: a possible role for a polymorphic allorecognition system. *Proc. Natl. Acad. Sci. USA* 93:15,254–59

Strassmann JE, Queller DC. 2004. Sociobiology goes micro. *ASM News* 70:526–32

Strassmann JE, Zhu Y, Queller DC. 2000. Altruism and social cheating in the social amoeba *Dictyostelium discoideum*. *Nature* 408:965–67

Strathmann RR. 1991. From metazoan to protist via competition among cell lineages. *Evol. Theory* 10:67–70

Strathmann RR. 1998. Peculiar constraints on life histories imposed by protective or nutritive devices for embryos. *Am. Zool.* 35:426–33

Strathmann RR, Staver JM, Hoffman JR. 2002. Risk and the evolution of cell cycle durations of embryos. *Evolution* 56:708–20

Sussex IM. 1973. Do concepts of animal development apply to plant systems. *Brookhaven Symp. Biol.* 25:145–51

Szathmáry E, Wolpert L. 2003. The transition from single cells to multicellularity. See Hammerstein 2003, pp. 285–304

Tomitani A, Knoll AH, Cavanaugh CM, Ohno T. 2006. The evolutionary diversification of the Cyanobacteria: molecular-phylogenetic and paleontological perspectives. *Proc. Natl. Acad. Sci. USA* 103:5442–47

Travisano M, Velicer GJ. 2004. Strategies of microbial cheater control. *Trends Microbiol.* 12:72–78

Valentine JW, Collins AG, Meyer CP. 1994. Morphological complexity increase in metazoans. *Paleobiology* 20:131–42

Van den Hoek E, Mann DG, Jahns HM. 1995. *Algae: An Introduction to Phycology*. Cambridge, UK: Cambridge Univ. Press. 623 pp.

Van Valen LM. 1988. Is somatic selection an evolutionary force? *Evol. Theory* 8:163–67

Velicer GJ. 2003. Social strife in the microbial world. *Trends Microbiol.* 11:330–37

Velicer GJ, Kroos L, Lenski RE. 1998. Loss of social behaviors by *Myxococcus xanthus* during evolution in an unstructured habitat. *Proc. Natl. Acad. Sci. USA* 95:12376–80

Velicer GJ, Kroos L, Lenski RE. 2000. Developmental cheating in the social bacterium *Myxococcus xanthus*. *Nature* 404:598–601

Velicer GJ, Yu Y-TN. 2003. Evolution of novel cooperative swarming in the bacterium *Myxococcus xanthus*. *Nature* 425:75–78

Vermeij GJ. 2006. Historical contingency and the purported uniqueness of evolutionary innovations. *Proc. Natl. Acad. Sci. USA* 103:1804–9

Vos M, Velicer GJ. 2006. Genetic population structure of the soil bacterium *Myxococcus xanthus* at the centimeter scale. *Appl. Environ. Microbiol.* 72:3615–25

Wolk CP. 2000. Heterocyst formation in *Anabaena*. In *Prokaryotic Development*, ed. YV Brun, LJ Shimkets, pp. 83–104. Washington, DC: ASM Press

Wolpert L. 1990. The evolution of development. *Biol. J. Linn. Soc. London* 39:109–24

Wolpert L, Szathmáry E. 2002. Evolution and the egg. *Nature* 420:745

Wu J, Glass NL. 2001. Identification of specificity determinants and generation of alleles with novel specificity in the *het-c* heterokaryon incompatibility locus of *Neurospora crassa*. *Mol. Cell. Biol.* 21:1045–57

Yahr R, Vilgalys R, Depriest PT. 2004. Strong fungal specificity and selectivity for algal symbionts in Florida scrub *Cladonia* lichens. *Mol. Ecol.* 13:3367–78

Zhang C-C, Laurent S, Sakr S, Peng L, Bédu S. 2006. Heterocyst differentiation and pattern formation in Cyanobacteria: a chorus of signals. *Mol. Microbiol.* 59:367–75

Developmental Genetics
of Adaptation in Fishes:
The Case for Novelty

J.T. Streelman,[1] C.L. Peichel,[2] and D.M. Parichy[3]

[1] School of Biology, Institute for Bioengineering and Bioscience, Georgia Institute of Technology, Atlanta, Georgia 30332-0230; email: todd.streelman@biology.gatech.edu

[2] Division of Human Biology, Fred Hutchinson Cancer Research Center, Seattle, Washington 98109-1024; email: cpeichel@fhcrc.org

[3] Department of Biology, Institute for Stem Cell and Regenerative Medicine, University of Washington, Seattle, Washington 98195-1800; email: dparichy@u.washington.edu

Annu. Rev. Ecol. Evol. Syst. 2007. 38:655–81

First published online as a Review in Advance on August 17, 2007

The *Annual Review of Ecology, Evolution, and Systematics* is online at http://ecolsys.annualreviews.org

This article's doi: 10.1146/annurev.ecolsys.38.091206.095537

Key Words

color pattern, development, evolution, gender, jaws, teeth

Abstract

During the past decade of study in evolutionary developmental biology, we have seen the focus shift away from the stunning conservation of form and function between distantly related taxa toward the causal explanation of differences between closely related species. A number of fish models have emerged at the forefront of this effort to dissect the developmental genetic and molecular basis of evolutionary novelty and adaptation. We review the highlights of this research, concentrating our attention on skeletal morphology (cranial and postcranial), pigmentation patterning, and sex determination. Thus far, the genes involved in adaptation among fishes belong to well-characterized molecular pathways. We synthesize the current state of knowledge to evaluate theories about the interplay between development and evolution. General rules of evolutionary change have not materialized; however, the field is wide open, and fishes will likely continue to contribute insights to this central biological question.

INTRODUCTION

Fishes, Novelty, and How Development Works

The publication in December 1996 of an entire issue of *Development* dedicated to the zebrafish embryo and its embryogenesis changed the way evolutionary biologists think about fishes. The description of mutants in pathways affecting most aspects of vertebrate morphology (brains, eyes, jaws, fins, pigment) provided resounding evidence of the interplay between genes and development on a comprehensive scale. The simple figures used to document phenotypes (e.g., cleared and stained embryos lacking jaw bones or with duplicated cartilages, fishes without melanophores) provided visual compendia of developmental diversity. Students with favorite traits now had favorite mutants. The landmark issue of *Development* was particularly inspirational to those interested in evolution. The zebrafish mutants, first the domain of biomedicine, contributed to an undercurrent of discovery that adaptation (when development works) was just the flip side of disease (when development fails). Comparative biologists recognized that understanding the key to complex phenotypes and evolutionary novelty, encoded in the genome and unveiled through the developing embryo, was a tractable research objective.

This mindset was accompanied by major challenges. Conceptually, mutant screens are an imperfect metaphor for the identification of genotype-phenotype associations in nature. First, the classical experimental paradigm of forward genetics has sought to minimize complexity by isolating the effects of single mutations. Second, most zebrafish mutants were embryonic lethals; they never developed to function as adults. Subsequently, biologists have inferred how development works by studying how development fails. This approach has advanced our knowledge of gene function, but has also underscored the notion that genes do not operate in a vacuum, that environmental and genomic context matters. As such, a major and complementary objective of current research is to understand the molecular basis of natural diversity. Notably, understanding the origin of biological diversity was named one of the "25 Hard Questions" by *Science* magazine in July 2005, and "Evolution in Action" was *Science*'s 2005 Breakthrough of the Year.

Teleost fishes represent a unique assemblage in which to study the genetics of adaptation and evolutionary novelty, or how development works. First, the group contains bona fide model organisms (*Danio*, *Takifugu*, *Tetraodon*, *Oryzias*), with research programs in forward and reverse genetics, molecular biology, and genomics providing information, hypotheses, and technical insight. Second, the species richness and diversity of fishes are unrivaled among vertebrates. Closely related species differ in a wide range of traits, many of which we explore below. Numerous natural lineages are amenable to genetic and developmental analysis because barriers to hybridization are minimal or absent and embryos are easy to manipulate (e.g., danios, sticklebacks, cichlids). Understanding the genetics of development in natural lineages would provide theoretically novel insights into gene function because (*a*) new genes, not identified in mutant screens, might be involved and (*b*) new mutations, compatible with adult viability, would likely play a role.

Here, we review recent advances in the developmental genetics of adaptation in teleost fishes. We focus on three types of traits: skeletons (including craniofacial and postcranial elements), pigmentation, and sex (gender) determination. These traits have received considerable attention from researchers and fit together conceptually. Skeletal elements and pigment patterns have their cellular origin in the vertebrate cell type called the neural crest (Gans & Northcutt 1983, Hall 1999). Pigment patterns and skeletal variants are sometimes linked to sex chromosomes, and theoretical population genetic models of adaptive speciation predict linkage among these trait types (reviewed in Bolnick & Fitzpatrick 2007). Some of the evolutionary lineages and the traits we highlight have been reviewed elsewhere in the past few years (Cresko et al. 2007, Kazianis 2006). Our goal is to describe and summarize this vast primary literature to examine if diverse adaptations in different fish lineages share common developmental pathways or common gene regulatory logic. We integrate these data to address hypotheses that codify the rules of evolutionary development among closely related organisms.

Neural crest: a pluripotent population of embryonic precursor cells that contributes to numerous vertebrate traits

SKELETONS

Traveling Light: Adaptation via Loss

Recent work has yielded considerable insight into the developmental genetics of trait loss in fishes. Assorted lineages have lost features of the craniofacial (i.e., teeth) and postcranial skeletons (i.e., ribs and fins), as well as body armor, scales, eyes, and pigmentation (see below) (**Table 1**; **Figure 1**). Research to date suggests that trait loss is controlled by a small number of genes of large effect and high penetrance; further study is required to determine if this is a general rule.

Table 1   Summary of genes involved in adaptation among different fish lineages

Trait	Lineage	Gene	Data[a]	Gene type[b]
Pelvic fin loss	Pufferfishes	*hoxd9a*	T	I/O
Pelvic fin loss	Stickleback	*pitx1*	Both	I/O
Eye loss	Cavefish	*shh, twhh*	T	Plug-in
Pigment loss	Cavefish	*oca2*	Both	DGB
Armor loss	Stickleback	*eda*	G	Plug-in
Tooth loss	Cypriniforms	FGF, *dlx2*	T	Plug-in
Jaw function	Cichlid	*bmp4*	Both	Plug-in
Sex determination	Medaka	*dmy*	G	I/O

[a]Data column specifies the type of data [genetic (G), transcriptional (T), or both] used to demonstrate the relationship between genotype and phenotype. Genetic data is an association between genotype and phenotype found by genetic linkage or genetic association analysis. Transcriptional data is an association between genotype and phenotype found by showing a correlation between a phenotypic difference and a difference in a gene's expression pattern.
[b]Gene type column assigns genes according to Davidson & Erwin's (2006) terminology. DGB, differentiation gene battery; I/O, input/output.

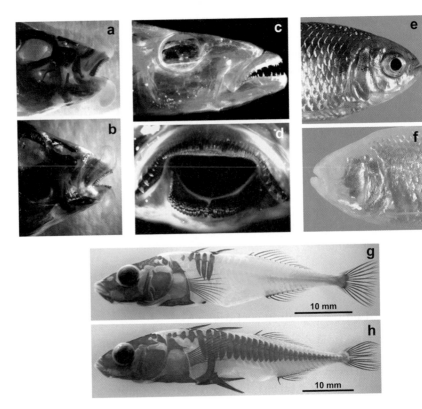

Figure 1

Variation in skeletal morphology and anatomy among model teleosts. (*a*) *Placidochromis milomo* (Lake Malawi) and (*b*) *Lobochilotes labiatus* (Lake Tanganyika), demonstrating parallel evolution of cartilaginous fleshy lips, function unknown. (*c*) *Rhamphochromis esox*, a piscivore from Lake Malawi with a highly kinematic jaw and unicuspid teeth. (*d*) An oral view of *Pseudotropheus elongatus*, an algae eater from Lake Malawi with multiple rows of multicuspid teeth. (*e*) Eyed and pigmented versus (*f*) eyeless and albino tetras, *Astyanax*. (*g, h*) Variation in body armor and pelvic spines among Alaskan sticklebacks. Photos of tetras and sticklebacks are courtesy of Yoshiyuki Yamamoto and William Cresko, respectively.

Understanding the developmental genetic basis of adaptation builds on decades of natural history, field ecology, and evolutionary biology. For instance, Northern Hemisphere stickleback fish have independently colonized freshwater habitats from marine ancestors soon after the last glacial maximum ($\sim$10,000 years ago). Riverine, lacustrine, and stream populations have evolved numerous adaptations, including changes in body size, habitat use, gill raker number, and the reduction of body armor [i.e., scales that are modified to form bony plates, as well as pelvic and dorsal spines (Bell & Foster 1994)] (**Figure 1**). Peichel et al. (2001) mapped the genetic basis of pelvic and armor reduction in backcross progeny of lacustrine benthic versus limnetic threespine sticklebacks from Priest Lake, British Columbia. A single quantitative trait locus (QTL) for pelvic spine length was located on chromosome 8, and QTL

Quantitative trait locus (QTL): a genomic region that has been shown by linkage mapping studies to harbor genetic variation that contributes to segregating phenotypic variation

for body armor (plates) were located on chromosomes 13 and 26. Each of these genomic regions explained a substantial portion of phenotypic variation in the focal trait [percent variance explained (PVE) = ~25%].

Subsequent to this study, numerous reports have refined the story for each trait. Colosimo et al. (2004) used F_2 fishes from an intercross of marine versus Paxton Lake, British Columbia, parents to document a QTL of major effect (PVE > 75%) for body armor on chromosome 4, with four additional minor effect loci on separate chromosomes. The major locus for armored plates on chromosome 4 also segregated in a California stream population. This locus was later identified as *ectodysplasin* (*eda*) by positional cloning, linkage disequilibrium mapping, and transgenesis (Colosimo et al. 2005). Notably, *eda* low-plate alleles segregate at low frequency in marine high-plated ancestral populations, explaining the parallel loss of armor in most freshwater lineages (Colosimo et al. 2005).

Shapiro et al. (2004) used a similar cross-design to identify a major QTL for pelvic reduction on stickleback chromosome 7, with four additional minor effect loci on different chromosomes. Mapping of candidate genes and in situ hybridization strongly suggest that regulatory mutations in *pitx1* (paired-like homeodomain transcription factor 1) are responsible for this phenotype. Similarly, genetic complementation analysis implicated *pitx1* in the pelvic reduction of other freshwater threespine stickleback populations (Shapiro et al. 2004) and distantly related (common ancestor at least 10 mya) ninespine stickleback populations (Shapiro et al. 2006). Cresko et al. (2004) studied the genetics of bony armor loss among Alaskan freshwater threespine stickleback populations and demonstrated parallel Mendelian control of both pelvic and armor phenotypes. Alaskan sticklebacks segregated for a pelvic reduction gene on chromosome 7 (likely *pitx1*), and armor phenotypes mapped to the *eda* locus on chromosome 4 (Miller et al. 2007).

Other fish lineages show analogous loss of scale or pelvic structures; strikingly, these phenotypes result from alterations in the same developmental pathways identified in stickleback. Kondo et al. (2001) reported that the spontaneous medaka mutant *rs-3*, which lacks scales, is encoded by the receptor for *ectodysplasin* (*edar*). Pelvic fin loss in pufferfishes is accompanied by altered expression of the limb-positioning marker *hoxd9a*, which is upstream of *pitx1* (Tanaka et al. 2005). Finally, additional fish groups are characterized by the loss of morphological features, from eyes to oral jaw teeth. Blind cavefishes (*Astyanax*) possess eyes that degenerate during development (**Figure 1**). Cave populations are characterized by expanded sonic hedgehog (*shh*) and tiggy-winkle hedgehog (*twhh*) expression at the embryonic midline when compared to their surface-dwelling eyed ancestors (Yamamoto et al. 2004). Zebrafish and other cypriniform fishes lack teeth on their oral jaws. This may result from altered fibroblast growth factor (FGF) signaling through *dlx2* in oral epithelium (Stock et al. 2006).

Fish Jaws and Dentitions: Elaboration and Complexity

Detailed study of trait loss in fishes provided some of the first evidence that genetic mapping and assays of gene expression could be used to understand the molecular

Percent variance explained (PVE): the amount of segregating phenotypic variation explained by a particular QTL

control of natural adaptations. Of course, trait loss may be a special case of adaptation: what about the more complex morphologies in which individuals differ in the subtler aspects of shape, size, and function? The natural history of fish feeding ecology, functional morphology, and diversity provided a place to begin. Notable features of the fish craniofacial skeleton include (*a*) two sets of toothed jaws (oral and pharyngeal) elaborated to (sometimes) bizarre extremes (**Figure 1**), (*b*) dentitions on jaws and numerous other bony elements replaced continuously through development, and (*c*) a long and perhaps dubious history of these traits as markers of evolutionary relationships.

Cichlid fishes have figured prominently in studies attempting to identify the developmental genetic basis of craniofacial adaptation, largely because they represent closely related species with a wide range of trophic and dental morphologies (Albertson & Kocher 2006). Albertson et al. (2003) mapped QTL for craniofacial morphology in the F_2 of a cross between two Lake Malawi cichlids with divergent feeding strategies. Genes of large effect (10%–25% PVE) for multiple craniofacial phenotypes mapped to common intervals of chromosomes 1, 2, and 16 [reassigned to chromosomes 7, 15, and 19 after comparison to the more extensive tilapia cichlid map (Lee et al. 2005, Streelman & Albertson 2006)], leading to speculation that trait linkage on chromosomes might facilitate the rapid and replicative evolution of jaw design among rift lake cichlids (**Figure 1**). Using a test that compares the direction of QTL effects to a neutral expectation, the authors documented strong directional selection on the oral jaw apparatus and the dentition (Albertson et al. 2003). In 2005, Albertson and colleagues focused on the functional aspects of lower jaw shape that represent a trade-off between the speed and force of jaw opening and closing (Albertson et al. 2005, Hulsey et al. 2005). Importantly, they showed that opening and closing lever systems were genetically decoupled with QTL localized to different chromosomes. They observed that the gene *bmp4* mapped to the closing lever system QTL interval (on chromosome 19) and subsequently demonstrated greater *bmp4* expression in the parental species with more robust jaws [similar to results in Darwin's finches (Abzhanov et al. 2004)]. Finally, they showed that *bmp4* injection into zebrafish embryos was sufficient to recapitulate the lower jaw–shape phenotype observed in cichlids. This study provided a possible explanation for the observation that *bmp4* evolves rapidly and non-neutrally among East African cichlids (Terai et al. 2002b). Given the avid interest in modeling fish jaws as simple versus complex biomechanical systems (Alfaro et al. 2004, Hulsey et al. 2005, Wainwright 2007), the cichlid system is ideal for further exploration in this context.

Recent work in fishes has demonstrated the complexity of dental patterning in vertebrates. Fraser et al. (2004) showed that first-generation teeth on the oral jaw of rainbow trout express *pitx2*, *shh*, and *bmp4* in similar spatiotemporal patterns to the mouse, suggesting the conservation of these molecules in the initiation of odontogenesis since the common ancestor of fish and mammals (~450 mya). However, not all is conserved between mammals and fishes, or even between the oral and pharyngeal jaws of fishes. Notably, Fraser et al. (2004) described differences in *pitx2* expression during continued morphogenesis of trout teeth, with *pitx2* expression present in oral jaw teeth but absent from pharyngeal teeth. Working with zebrafish, Laurenti et al.

(2004) similarly demonstrated differences between pharyngeal first-generation teeth and the oral teeth of mammals (zebrafish lack teeth on the oral jaw so no direct comparison is possible). Specifically, the gene *eve1*, a member of the homeobox-containing evx gene family, not expressed during tooth development in mammals, is expressed during tooth initiation and morphogenesis of the first pharyngeal tooth. Jackman et al. (2004) used chemical knockdown of FGF signaling to show that FGFs are required for zebrafish first-generation tooth development. Furthermore, *fgf8* and *pax9* were not expressed under normal conditions in zebrafish tooth germs (unlike in mouse), and both *Dlx* and *Lhx* genes were expressed in dental mesenchyme (as in mouse molars).

In 2003, Streelman and colleagues demonstrated that tooth number was correlated with tooth cusp number in natural populations of cichlid fish from Lake Malawi, East Africa (Streelman et al. 2003b). Given simple genetic control of tooth shape in this system (Albertson et al. 2003) and the iterative role of certain genes in the stages of tooth development (Peters & Balling 1999), these authors suggested that variation in the expression of a single activating or inhibitory molecule might integrate tooth and cusp number (Streelman et al. 2003b; also Plikus et al. 2005). Streelman & Albertson (2006) subsequently identified a QTL of major effect for tooth shape on cichlid chromosome 5, near genes for orange blotch (OB) color and sex (Streelman et al. 2003a) (see below). Furthermore, they demonstrated, using *bmp4* as a marker of tooth initiation, that tooth number and spacing are specified earlier than tooth shape.

Much is left to learn about fish dentitions. For instance, first-generation teeth are morphologically unlike replacement teeth (Sire et al. 2002), do not show species-specific adult shapes, and exhibit unique gene-expression programs (Fraser et al. 2006). There is great interest in tooth replacement and its molecular mechanisms because subsequent tooth generations may arise from stemlike cells (Huysseune & Thesleff 2004), yet only one study to date has examined gene-expression programs in replacement dentitions (Fraser et al. 2006). No study has investigated the molecular choreography of tooth replacement in species with adult teeth shaped differently than first-generation teeth, and no study has examined how lingual rows of teeth are initiated and patterned (e.g., cichlid species can have more than 15 rows of teeth on the oral jaws). Understanding the molecules involved in the complexity of fish odontogenesis will shed light on the general mechanisms of periodic patterning applicable not only to dentitions (Salazar-Ciudad & Jernvall 2002), but also to other organs such as hair and feathers (Houghton et al. 2005).

PIGMENTATION

Pigment patterns represent one of the most extraordinary illustrations of teleost adaptation (**Figure 2**). Famous examples include coral reef fishes, cichlids of East Africa, and aquarium favorites such as guppies and loaches. The myriad pigment patterns of teleosts serve in a variety of roles, including warning coloration, camouflage, schooling, mate recognition, and mate choice (Couldridge & Alexander 2002, Endler 1988, Engeszer et al. 2004, Jordan et al. 2003, McMillan et al. 1999, Millar et al. 2006, Rosenthal & Ryan 2005).

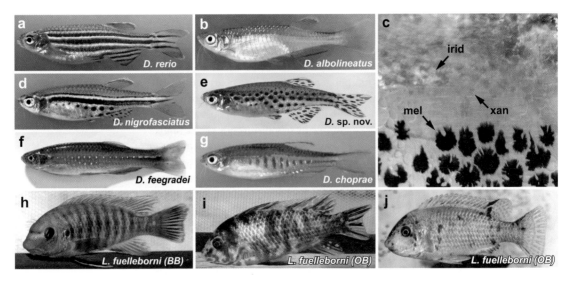

Figure 2

Pigment pattern variation and pigment cells of teleosts. Shown are several species within *Danio* (*a–g*), as well as the cichlid *Labeotropheus fuelleborni* (*h–j*), illustrating differing color patterns associated with the absence and presence of the orange blotch polymorphism [blue-black (BB) and orange blotch (OB), respectively]. Panel *c* shows melanophores (mel), xanthophores (xan), and iridophores (irid) in the *D. rerio* adult pigment pattern. Iridescent iridophores are present throughout but can be seen here only where they catch the light.

Pigment Patterns Through Development

Vertebrate skin pigment cells are derived embryologically from neural crest cells, which also contribute to craniofacial bone, cartilage, and teeth and produce most of the peripheral nervous system (Hall 1999, Le Douarin 1999). Neural crest cells have long been recognized as a key vertebrate innovation (Gans & Northcutt 1983), and pigment patterns, in addition to skeletons (see above), have provided a valuable opportunity to study the developmental and genetic factors responsible for evolutionary changes in the patterning of neural crest–derived traits. In contrast to studies of skeletal diversification, which have focused largely on particular genes and tissues, studies of pigmentation have emphasized the cellular mechanisms of pigment pattern development. The different emphasis reflects the notion that evolutionary changes in gene activity are only interpretable in a cellular context (e.g., Parichy 2005) and this cellular context has thus far been less explored for pigment patterning as compared to skeletogenesis.

Pigment patterns reflect the numbers and arrangements of several classes of pigment cells, or chromatophores. These include black melanophores, yellow or orange xanthophores, red erythrophores, blue cyanophores, white leucophores, and iridescent iridophores (Bagnara & Matsumoto 2006, Parichy et al. 2006). The color of each class of cell results from the particular pigments contained within specialized organelles. By combining different classes of cells, different spatial arrangements of cells, and different pigment concentrations within individual cells, a seemingly infinite range of patterns and colors can be produced.

Most fishes exhibit different pigment patterns during different life-cycle phases. The first pattern to develop arises as embryonic neural crest cells disperse from above the neural tube, differentiating chromatophores during or even prior to their migration and subsequently colonizing specific locations to generate an embryonic/early larval pigment pattern (Kelsh 2004, Raible & Eisen 1994). Commonly this consists of stripes of melanophores dorsally, laterally, and ventrally, with xanthophores broadly scattered over the flank (Lamoreux et al. 2005, Quigley et al. 2004), although a variety of other patterns also occur. The functional significance of these pigment patterns remains unexplored.

The diversity of teleost pigmentation consists mostly of patterns expressed in the adult. In some species, the adult pigment patterns develop during metamorphosis, when the larval form is transformed into a juvenile by remodeling or the initial appearance of a variety of traits [e.g., fins, skin, scales, skeleton, gut, kidney, and sensory systems (Webb 1999)]. Pigment pattern metamorphosis has been most studied in the zebrafish, *Danio rerio* (**Figure 2**). In this species, metamorphic melanophores differentiate scattered over the flank, and then melanophores coalesce at sites of adult stripe formation, with additional metamorphic melanophores differentiating already within the stripes; most embryonic/early larval melanophores die (Parichy & Turner 2003b).

Developmental changes in pigment pattern also can occur during later development, particularly with the onset of sexual maturation, and these may be either permanent or transient, as is the case for nuptial coloration (Beeching et al. 2002, Dickman et al. 1988, Maan et al. 2006, Mabee 1995). To date, virtually nothing is known about the molecular and cellular bases of pigment pattern changes within the adult phases of the life cycle.

Genes Underlying Changes in Pigmentation

One way that teleost pigment patterns evolve is by modifying the quantity or quality of the pigments carried by chromatophores. Two recent studies provide nice examples of how genetic approaches can provide insights into the evolution of pigmentation in fishes and beyond.

In Mexican tetras, *Astyanax*, several cave-dwelling populations exhibit a suite of derived traits including albinism, reduced eyes, and enhancements of other sensory systems (Jeffery 2001, Yamamoto et al. 2004) (**Figure 1**). The phylogeography of these populations is complex, although cave forms have clearly evolved repeatedly (Strecker et al. 2004). Despite their albinism, cavefish retain melanophores (McCauley et al. 2004), and genetic mapping identified a major effect QTL for melanin loss (Protas et al. 2006). By mapping candidate genes associated with mammalian albinism, researchers found a correspondence between the cavefish QTL and *oculocutaneous albinism-2* (*oca2*). Complementation tests showed that albinism in a second cavefish population is associated with the same locus, and molecular analyses revealed that each population harbors different small genomic deletions within *oca2*. The deletions are functionally significant as *oca2* complementary DNA from melanized, surface-dwelling *Astyanax* allows the melanization of murine *oca2*-deficient melanocytes, whereas the two cavefish deletion complementary DNAs do not. This study nicely shows how pigmentation loss can result independently from changes at the same locus and suggests that such parallelism may reflect both an absence of pleiotropic effects and the large size of *oca2*, making it a high-frequency target for selection. These results are reminiscent of recent studies of *MC1R* in mammalian pigmentation (Hoekstra et al. 2006). The cavefish example also illustrates how knowledge of pigment cell genes and development in mammals can be applied to understanding pigment evolution in teleosts.

Knowledge of pigment development in teleosts also can inform us about the evolution of pigment in mammals, including humans. A striking example is the *D. rerio golden* mutant, which has reduced melanin but otherwise normal melanophores. Positional cloning identified *golden* as *slc24a5*, which encodes a sodium/calcium transporter localized to pigment granules within melanophores (Lamason et al. 2005). Mutations in *aim1*, also a transporter involved in melanin synthesis, explain a similar orange-red medaka variant called b (Fukamachi et al. 2001). Remarkably, a polymorphism within human *SLC24A5* is associated with different pigmentation between European and African populations, and significantly reduced heterozygosity indicates past selection at this locus. Whether variation at *slc24a5* or *aim1* has contributed to pigment evolution in teleosts and other taxa remains to be determined.

Mechanistic Bases for Cellular Pattern Diversification

Beyond changes in pigment content, a major factor in teleost pigment pattern diversification has been changes to the numbers and arrangements of chromatophore classes. Such variation has received extensive theoretical attention (Asai et al. 1999, Miguez & Munuzuri 2006, Painter et al. 1999), and recent studies have started to elucidate the underlying mechanisms, primarily using *D. rerio* and its relatives.

One recent insight concerns the origins of chromatophores responsible for pattern diversification. Unlike embryonic/early larval melanophores that differentiate directly from neural crest cells, metamorphic melanophores in *D. rerio* differentiate from latent precursors of presumptive neural crest origin (Johnson et al. 1995, Parichy & Turner 2003b, Parichy et al. 2003). Mounting evidence suggests these precursors are stem cells, able to generate differentiated progeny while themselves remaining undifferentiated (Parichy & Turner 2003a, Yang & Johnson 2006). A sister species, *D. nigrofasciatus*, exhibits superficially similar adult stripes to *D. rerio*, yet cell lineage analyses reveal these stripes are formed largely by reorganizing embryonic/early larval neural crest–derived melanophores rather than by differentiating stem cell–derived metamorphic melanophores (Quigley et al. 2004). Thus, danios exhibit at least two different modes of pigment pattern metamorphosis.

Analyses of danios show that cryptic but genetically distinct populations of metamorphic melanophores differentially contribute to pigment pattern evolution (Johnson et al. 1995; Parichy et al. 1999, 2000a,b). In *D. rerio*, early metamorphic melanophores that are initially dispersed and then migrate into stripes depend on the kit receptor tyrosine kinase, as they are ablated in *kit* mutants. By contrast, late metamorphic melanophores that develop already within stripes do so independently of *kit*; i.e., they persist—in stripes—in *kit* mutants. As distinct populations of *kit*-dependent and *kit*-independent melanocytes have not been found in mammals, these cell populations might be unique to *D. rerio*. To test this idea, a recent study isolated a *kit* mutant in *D. albolineatus*, which normally exhibits uniformly dispersed melanophores. The mutant retained a population of *kit*-independent melanophores, showing conservation of these cellular populations in at least one other danio. Strikingly, and in contrast to the uniform wild-type *D. albolineatus* pattern (**Figure 2**), the *kit*-independent melanophores occurred in stripes. These and other data showed that *D. albolineatus* has latent stripe-forming potential, and that stripe loss in this species occurred in part by a failure of *kit*-dependent melanophores to migrate into stripes, thereby obscuring the stripes formed by *kit*-independent melanophores (Mills et al. 2007, Quigley et al. 2005). These studies show how a manipulative, genetic approach can be used to deconstruct the evolution of an adult phenotype.

Studies of danios also suggest that an important factor in pigment pattern diversification depends on chromatophore interactions. In *D. rerio*, stripes arise through interactions between melanophores and xanthophores, and between cells within each of these classes (Maderspacher & Nusslein-Volhard 2003, Parichy & Turner 2003a, Watanabe et al. 2006). Genetic analyses indicate that variation in danio pigment patterns likely reflect evolutionary modifications to the strength and timing of these interactions, which appear to serve as a pattern-generating mechanism that can be deployed at different times and in different places (Parichy & Turner 2003a, Quigley et al. 2005). Interspecific complementation testing of candidate genes identified as *D. rerio* mutants further revealed that such interactions are likely to be perturbed in *D. albolineatus*—contributing to the uniform pigment pattern—owing to changes in *colony stimulating factor 1 receptor* (*csf1r, fms*), which encodes a receptor tyrosine kinase expressed by cells of the xanthophore lineage (Parichy & Johnson 2001, Quigley et al. 2005).

Although danios are an especially tractable system for analyzing pigment pattern development and evolution, these species represent only a small fraction of teleost pigment pattern diversity. In this regard two additional groups are especially interesting—guppies and cichlids—both because of color pattern variation and because of the deep foundation of ecological and behavioral observations regarding these patterns (Genner & Turner 2005, Lindholm et al. 2004, Seehausen et al. 1999). For cichlids, a particularly exciting recent advance is the ability to map factors genetically using closely related species. For instance, a QTL associated with alternative barred and OB postmetamorphic color patterns in *Metraclima zebra* maps to the vicinity of c-*ski1* on chromosome 5 (Streelman et al. 2003a) (**Figure 2**). As representative cichlid genome sequences become available (**Table 2**), identification of this locus and other inferred genetic factors (Barson et al. 2007, Maan et al. 2006) will provide new and important insights into pigment pattern diversification. Moreover, mechanistic studies of danios and other model organisms should provide inroads to understanding the cellular bases for pattern diversification in these other species.

Table 2 Genomic resources for model teleosts

Resource	Species	Web site
Cichlid Genome Consortium	Cichlids	http://www.cichlidgenome.org
Ensembl	Zebrafish, stickleback pufferfish, medaka	http://www.ensembl.org
JGI	Pufferfish (*Takifugu*)	http://genome.jgi-psf.org/Takru4/Takru4.home.html
Medaka home page	Medaka	http://biol1.bio.nagoya-u.ac.jp:8000
Genoscope	Pufferfish (*Tetraodon*)	http://www.genoscope.cns.fr/externe/tetranew/
Sanger Institute	Zebrafish	http://www.sanger.ac.uk/Projects/D_rerio/
Stanford Genome Evolution Center	Zebrafish, stickleback	http://cegs.stanford.edu/index.jsp
Xiphophorus home page	*Xiphophorus*	http://xiphophorus.org
Zebrafish Information Network	Zebrafish	http://zfin.org

Pigmentation Genes Evolve Rapidly in Teleosts

A problem complementary to the evolution of pigment patterns is the evolution of pigment pattern genes, and several recent studies have assessed naturally occurring variation at such loci. For example, surveys of several cichlid species with diverse color patterns found differential rates of evolution among loci and between recently duplicated paralogous copies, including *csf1r*, which is mentioned above (Braasch et al. 2006, Sugie et al. 2004). An especially intriguing example is *hagoromo*, which encodes an F-box/WD-40 repeat protein that is required for metamorphic melanophore development in *D. rerio* (Kawakami et al. 2000). Analyses of more than a dozen cichlid species reveal accelerated rates of amino acid evolution in specific domains and an extraordinary increase in the complexity of alternatively spliced *hagoromo* transcripts (Terai et al. 2002a, 2003). It will be fascinating to learn how *hagoromo* functions in pigment pattern development and to test its causal involvement in generating species-specific pigment patterns.

SEX (GENDER) DETERMINATION

Sex Determination Mechanisms in Fish Are Diverse

Most developmental pathways, such as those discussed above, are well conserved across disparate taxa. By contrast, the developmental pathways that determine sex are strikingly variable and can even differ between closely related species. Teleost fishes present attractive models to understand the evolution of sex determination pathways, as the entire range of environmental and genetic sex-determining mechanisms is represented across lineages (Devlin & Nagahama 2002). For example, many fishes have environmentally determined sex, which can depend on factors such as temperature or social interactions. Genetic mechanisms of sex determination in fishes may be polygenic or simple and associated either with no cytogenetically visible sex chromosomes or with heteromorphic sex chromosomes in either males (XY systems) or females (ZW systems). This wide diversity of sex determination mechanisms can be found even in closely related fish species (Devlin & Nagahama 2002, Mank et al. 2006).

Cytogenetically visible sex chromosome: in this context heteromorphic chromosomes belonging to one sex that can be observed by examining chromosome squashes under a light microscope

Particularly apposite examples of this diversity are found within poeciliid fishes [guppies, mollies, swordtails, and platyfish (Volff & Schartl 2001)], salmonid fishes (Phillips et al. 2001, Woram et al. 2003), the stickleback family Gastesteidae (Chen & Reisman 1970), and the tilapia genus *Oreochromis* (Lee et al. 2003, 2004). Diversity of sex determination mechanisms in closely related fish species supports the hypothesis that this developmental pathway is evolutionarily plastic and that sex determination mechanisms and sex chromosomes can evolve rapidly.

The plasticity of sex determination mechanisms in fish is highlighted by recent work in medaka (*Oryzias latipes*). With the identification of a duplicated copy of the *dmrt1* gene called *dmrt1bY* or *DMY* as the medaka master sex determination locus (Matsuda et al. 2002, Nanda et al. 2002), there was speculation that this gene would serve a similar role in all fish, just as *Sry* is the master sex determination switch in nearly all mammals (Marshall Graves 2002). Although the *Dmrt* gene family is widely present in fish (Volff et al. 2003b), the *dmrt1bY/DMY* gene is absent from other fish species (Kondo et al. 2003, Veith et al. 2003). In fact, although *dmrt1bY/DMY* is present in a second species, *Oryzias curvinotus* (Kondo et al. 2004, Matsuda et al. 2003), other species within the *Oryzias* genus do not have this gene (Kondo et al. 2003, 2004), suggesting that *dmrt1bY/DMY* has arisen within the *Oryzias* lineage in the past 10 million years (Kondo et al. 2004).

The enormous variation in sex determination pathways in fish presents an opportunity to understand the mechanisms by which sex determination genes arise and sex determination pathways evolve. Remarkably, the mechanisms of sex determination remain unknown for *D. rerio*, although multiple loci and environmental influences are likely to be involved. Currently, efforts are underway to identify the master sex determination genes in platyfish, tilapia, salmonids, and stickleback. This work should identify whether there are common themes that connect the types of genes used as master sex determination loci, as well as provide insights into the evolution of sex determination pathways.

Sex Chromosome Evolution in Teleosts

In addition to the diversity of sex determination mechanisms in fish, there is also great diversity in the presence of sex chromosomes. Approximately 10% of fish species have cytogenetically visible sex chromosomes (Devlin & Nagahama 2002). However, this is likely an underestimate of the number of fish species that have sex chromosome systems because young sex chromosome systems that are in early stages of differentiation are unlikely to be observed by traditional cytogenetic analysis. Many closely related species of fish differ in sex chromosome complement, suggesting that sex chromosomes can arise rapidly in fish. Many fish sex chromosomes are therefore likely to be younger than the stable XX-XY sex chromosome system in mammals, which is over 300 million years old (Graves 2006). Therefore, studying sex chromosomes in fish provides a unique opportunity to investigate the genetic and molecular events that accompany the earliest stages of sex chromosome evolution.

After the acquisition of a sex determination locus, one of the first steps in the evolution of a sex chromosome is the suppression of recombination around a sex determination locus, which has been hypothesized to occur to reduce recombination between the sex determination locus and linked genes with sex-specific fitness effects (Bull 1983, Fisher 1931, Rice 1987a). This suppression of recombination leaves the heterogametic sex with one chromosome in a consistently heterozygous state, which ultimately results in the degeneration of sex-linked loci in the heterogametic sex (Bull 1983, Charlesworth 1991, Rice 1987b). Based on these models, it is predicted that a sex chromosome would show reduction of recombination near the sex determination region, resulting in the loss of homology between the X and the Y chromosome, particularly owing to the accumulation of deleterious mutations, including an increase in transposable elements on the Y chromosome. Chromosome rearrangements may or may not accompany these early stages of sex chromosome evolution. Recent studies of the sex chromosomes of a number of different fish species have begun to illuminate these processes on a molecular level and have also begun to provide insight into the timing of events in sex chromosome evolution.

In particular, recent work in medaka fish (Kondo et al. 2006) has provided a detailed molecular view of the events that accompany the early stages of sex chromosome evolution, just after the evolution of a new sex determination gene. As described above, the sex determination gene in *O. latipes* was recently identified as the *dmrt1bY/DMY* gene, a duplicate copy of the *dmrt1* gene (Matsuda et al. 2002, Nanda et al. 2002). Kondo et al. (2006) cloned and sequenced the regions flanking *dmrt1bY* on both the X and the Y chromosome, as well as the *dmrt1* region. They found a completely Y-specific region that resulted from a duplication of a 43-kb region of chromosome 9 that includes the *dmrt1* gene. A number of repetitive elements have accumulated within the Y-specific region, accounting for an increase in its size to 258 kb. Thus, in this relatively young (less than 10 million years old) sex chromosome system (Kondo et al. 2004), there is evidence for both degeneration of Y-linked sequences and accumulation of repetitive DNA (Kondo et al. 2006).

It may be that the *dmrt1bY/DMY* locus in medaka represents a unique mechanism of sex chromosome evolution. To gain insights into the general mechanisms that underlie the evolution of sex chromosomes, it is important to analyze other sex chromosome systems of differing ages. In fishes, there are a number of other sex chromosome systems in species with the requisite genetic and genomic tools for this analysis. To date, the most well-studied systems have been poeciliid fishes (guppies and platyfish), salmonid species, threespine stickleback (*Gasterosteus aculeatus*), and tilapiine cichlids (*Oreochromis* spp.). In most of these systems, genetic analysis has revealed a genetic basis for sex determination even in the absence of cytogenetically visible sex chromosomes.

There must be some differentiation between sex chromosomes in most of these sex chromosome systems, as researchers have observed reduction in recombination between the X and the Y chromosomes near the sex determination region in threespine stickleback (Peichel et al. 2004), blue tilapia (Lee et al. 2004), and platyfish (Gutbrod & Schartl 1999, Morizot et al. 1991). Given the loss of recombination near sex determination regions of these fish, it is not surprising that there is also evidence that many of these systems have accumulated repetitive DNA. In tilapia, there are subtle differences in the amount of heterochromatin, which consists of repetitive DNA elements that have accumulated on the Y chromosome relative to the X (Griffin et al. 2002, Harvey et al. 2002). Similarly, the sex determination region of lake trout, brown trout, and Atlantic salmon is next to a large heterochromatic block (Artieri et al. 2006, Phillips & Ihssen 1985, Phillips et al. 2002). Sequencing of X- and Y-specific bacterial artificial chromosome clones in threespine stickleback (*G. aculeatus*) and platyfish (*X. maculatus*) revealed that the Y chromosomes in both species had significantly more repetitive and transposable elements than the X chromosomes (Froschauer et al. 2002, Peichel et al. 2004, Schultheis et al. 2006).

Beyond examining the accumulation of transposable elements, investigators have done relatively little to explore the effects of loss of recombination at the sequence level. That viable and fertile YY salmonid (Chevassus 1988), tilapia (Penman & McAndrew 2000), and platyfish (Kallman 1984) males can be generated suggests that genes required for viability and fertility on the Y chromosome have not been rendered nonfunctional. Some sex-linked genes in platyfish appear to be pseudogenes; however, there are a number of duplicate copies of these genes, such that at least one functional copy might remain (Volff et al. 2003a). There are a number of sequence differences between the X and the Y chromosome in the threespine stickleback (Peichel et al. 2004); however, it is not known whether genes on the stickleback Y have become nonfunctional or whether YY individuals can be generated in stickleback. In the future, it will be important to compare the levels of cytogenetic differentiation with levels of sequence divergence and to explore in more detail the molecular changes that have occurred in the regions around a sex determination locus.

Sexually antagonistic gene: a gene with a differential fitness effect in the sexes, so expression in one sex is beneficial but expression in the other sex is detrimental

Pigmentation and Skeletal Traits Are Linked on Sex Chromosomes

Reduction of recombination around a sex determination locus appears to be a general phenomenon in sex chromosome evolution. Theoretical work suggests that this may result from linkage of a sexually antagonistic gene to the sex determination locus, which would select for the loss of recombination to prevent detrimental alleles from being expressed in the wrong sex (Bull 1983, Fisher 1931, Rice 1987a). Thus, we might expect that there would be an excess of sexually antagonistic genes linked to the sex chromosomes. In particular, male display traits, such as color, can be considered sexually antagonistic traits because expression in males is beneficial, but expression in females would be deleterious, as it might expose females to predation and incur production costs (Bull 1983, Endler 1980, Fisher 1931). This model does not exclude species with female display traits; in this case we might simply expect to see linkage of female display traits to a female determining locus. In support of this model, there is good evidence for linkage of (fe)male display traits to sex chromosomes in a number of fish species (Lindholm & Breden 2002).

The poeciliid fish provide some of the most spectacular examples of sex linkage of male display traits (Lindholm & Breden 2002). In guppies (which have an XY sex determination system), pigmentation, fin size and shape, courtship behavior, and male attractiveness are linked to the Y chromosome (Brooks 2000, Brooks & Endler 2001). The Y-linked color patterns are extremely polymorphic in natural populations and differ in their attractiveness to females (Lindholm et al. 2004). Different Y-linked color alleles are associated with increased predation (Endler 1983) and mortality (Brooks 2000), suggesting that a balance between natural and sexual selection contributes to the maintenance of color polymorphisms in guppy populations (Endler 1980).

In another poeciliid fish genus, *Xiphophorus*, a number of traits involved in male attractiveness are closely linked to the sex determination locus on the Y chromosome (Basolo 2006, Cummings et al. 2006, Rosenthal & García de León 2006). As in guppies, pigmentation loci are tightly linked to the sex determination locus and are highly polymorphic within and between *Xiphophorus* populations (Kallman 1975). In addition, the puberty or pituitary locus is tightly linked to the sex determination locus and determines both the onset of sexual maturity (Kallman & Borkoski 1978, Kallman et al. 1973) and reproductive tactics (Zimmerer & Kallman 1989). This locus is also highly polymorphic, leading to alternative mating strategies within populations. Males that mature later are robust and ornamented and have elaborate courtship behaviors, whereas the males that mature early are small, have little ornamentation, and perform sneaker copulations. As for color, this polymorphism is likely to be maintained within populations owing to a balance of natural and sexual selection (Ryan et al. 1992). Although large males are favored by sexual selection and are preferred by females (Ryan et al. 1990), they are not favored by natural selection and are more heavily preyed upon (Rosenthal et al. 2001), providing an advantage for smaller and less conspicuous males.

Traits important for adaptation have been found linked to sex chromosomes in several other fishes. Among Malawi cichlids of the genus *Metriaclima*, sex is determined by a locus on chromosome 7, unless the OB trait is segregating in the family, in which case sex is under the control of a dominant female determiner linked to OB on chromosome 5 (Streelman et al. 2003a; T.D. Kocher, personal communication). Notably, genes for jaw shape and function map to cichlid chromosome 7 (Albertson et al. 2003, 2005; see above), and a QTL of major effect for tooth shape maps to chromosome 5 and is linked to the sex determination locus, as well as to genes for coloration (OB locus) and putative color preference (opsin gene cluster) (Carleton & Kocher 2001, Streelman & Albertson 2006). In tilapiine cichlids, a red color mutant maps close to the sex-determining locus of female heterogametic species on chromosome 3 (Lee et al. 2005). Finally, at least one skeletal trait, the size of the opercle bone, has been mapped to the stickleback sex chromosome (Kimmel et al. 2005). These latter data provide empirical evidence for quantitative genetic models of adaptive speciation that predict gametic association between ecological, marker, and preference traits (Dieckmann & Doebeli 1999, Kondrashov & Kondrashov 1999) on incipient sex chromosomes with reduced recombination.

SYNTHESIS AND PERSPECTIVE

The studies reviewed above have engendered novel insights into the developmental genetic basis of adaptation. Conceptually, this has shifted focus toward studying how development works in diverse and highly complex natural systems. Much has been learned about how genes with manifold pleiotropic functions (e.g., *pitx1*, *bmp4*, *shh*) can be employed specifically in an organ- or tissue-specific manner (Albertson et al.

2005, Shapiro et al. 2004, Yamamoto et al. 2004). Less satisfying, however, is that new genes or new gene functions have not been discovered; the genes involved in the traits we highlight might have been predicted in the context of traditional developmental genetics research. This is either because forward and reverse genetic screens are so thorough that they are redundant or because investigators have thus far studied a biased set of natural mutations [i.e., genes of large effect (Orr 1998)]. The next 5–10 years of research will address this question as new techniques (e.g., Miller et al. 2007) and improved genomic resources (**Table 2**) are used to investigate new traits in more teleost lineages. In summation, we consider a major question in evolutionary biology addressed by the studies reviewed here.

How Does Evolution Happen?

Many authors have discussed whether there are general rules governing evolutionary developmental biology (Carroll 2005, Gerhart & Kirschner 1997, Wilkins 2001). Davidson & Erwin (2006) have codified such rules in terms of gene regulatory networks (GRNs) and the evolutionary scale of change among the components of such networks. At one extreme are kernels, or sets of genes at the core of GRNs, that may be conserved over long periods of evolutionary time. At the other extreme are differentiation gene batteries (DGBs), genes involved in terminal differentiation of tissues or structures; DGBs reside at the periphery of GRNs and might be employed to distinguish among closely related species. In fact, Davidson & Erwin (2006) propose a "relation between the network-component class in which changes might occur and the taxonomic level of morphogenetic effects." According to Davidson & Erwin's (2006; their figure 3) hierarchical scheme, all the genes responsible for adaptive differences among closely related fish species [*pitx1*, *shh*, *twhh*, *oca2*, *eda*, *bmp4*, *dmy* (**Table 1**)] should belong in DGBs. However, six of the seven are better characterized as input/output (I/O) switches or plug-ins, both of which are classes of evolutionarily conserved components of multiple developmental networks. Davidson & Erwin hypothesize that changes in I/O switches and plug-ins explain differences at the taxonomic level of class, order, or family. Only *oca2* fits the definition of a DGB. So why do the data from fish adaptations not fit Davidson & Erwin's schema? The answer seems to lie in the degree of modular function for these I/O switch and plug-in genes. I/O switches and plug-ins can elicit major morphological change (because they regulate other genes through morphogenesis, unlike DGB genes), but the modularity of their regulation allows other pleiotropic functions of the encoded protein to remain unchanged [e.g., fin versus jaw function of *pitx1* (Shapiro et al. 2004)]. Perhaps a better prediction is that the genes involved in adaptation among closely related species will be those genes central to key morphogenetic processes (e.g., cell proliferation, differentiation, death, and migration) whose regulation across tissue- and cell-type is highly modular. In the language of GRNs, these are well-connected hubs, but the genes to which they are connected may vary across tissues and from species to species. The developmental and evolutionary flexibility of GRNs has not been examined among closely related species, but the approach is tractable in vertebrates (Tsaparas et al. 2006).

Gene regulatory networks (GRNs): the sum total of genes and their connections that influence a biological output, often depicted or modeled as wiring diagrams or logic circuits

Modularity: generally, the evolutionary or developmental decoupling of components involved in form and/or function

In summary, the next decade of research, highlighting these and other fish models, will surely contribute important data regarding the developmental genetic basis of adaptation. Further study fusing the power of molecular biology and genomics in fish groups of tremendous morphological, functional, physiological, and behavioral diversity will shape our understanding of how development works.

FUTURE ISSUES

1. Adaptation to new environments involves a wide range of morphological, physiological, and behavioral changes. In particular, the genetic basis of physiological and behavioral diversity has been relatively unexplored in any system. Because the fish models highlighted in this review display enormous morphological, physiological, and behavioral diversity, it should be possible to use the genetic and genomic tools developed for these systems to identify the genetic and molecular basis of any trait of interest. It will be particularly interesting to determine whether the types of mutations, genes, and pathways that are important for morphological adaptation are more generally involved in physiological and behavioral novelty.

2. As technical costs decrease, more fish lineages will become appropriate models to answer key biological questions. The richness and diversity found among teleost fishes are nearly limitless in this regard.

3. Many future research efforts will focus on traits expressed after embryogenesis or in adult life stages. New techniques and the application of standard techniques to new situations (explant culture, tissue- or stage-specific gene knockdown) will be required to rigorously evaluate functional associations between genotype and phenotype.

DISCLOSURE STATEMENT

The authors are not aware of any biases that might be perceived as affecting the objectivity of this review.

ACKNOWLEDGMENTS

We thank members of our laboratories (especially Darrin Hulsey and Gareth Fraser) and our colleagues (Craig Albertson, William Cresko, Yoshiyuki Yamamoto, Tom Kocher, and Felix Breden) for discussion and comments on the manuscript. Our research is funded by the NIH (J.T.S., C.L.P., D.M.P.), the NSF (J.T.S., D.M.P.), the Burroughs Welcome Fund (C.L.P.), and the Alfred P. Sloan Foundation (J.T.S.).

LITERATURE CITED

Abzhanov A, Protas M, Grant BR, Grant PR, Tabin CJ. 2004. *Bmp4* and morphological variation of beaks in Darwin's finches. *Science* 305:1462–65

Albertson RC, Kocher TD. 2006. Genetic and developmental basis of cichlid trophic diversity. *Heredity* 97:211–21

Albertson RC, Streelman JT, Kocher TD. 2003. Directional selection has shaped the oral jaws of Lake Malawi cichlid fishes. *Proc. Natl. Acad. Sci. USA* 100:5252–57

Albertson RC, Streelman JT, Kocher TD, Yelick PC. 2005. Integration and evolution of the cichlid mandible: the molecular basis of alternative feeding strategies. *Proc. Natl. Acad. Sci. USA* 102:16287–92

Alfaro ME, Bolnick DI, Wainwright PC. 2004. Evolutionary dynamics of complex biomechanical systems: an example using the four-bar mechanism. *Evolution* 58:495–503

Artieri CG, Mitchell LA, Ng SHS, Parisotto SE, Danzmann RG, et al. 2006. Identification of the sex-determining locus of Atlantic salmon (*Salmo salar*) on chromosome 2. *Cytogenet. Genome Res.* 112:152–59

Asai R, Taguchi E, Kume Y, Saito M, Kondo S. 1999. Zebrafish *leopard* gene as a component of the putative reaction-diffusion system. *Mech. Dev.* 89:87–92

Bagnara JT, Matsumoto J. 2006. Comparative anatomy and physiology of pigment cells in nonmammalian tissues. See Nordland et al. 2006, pp. 11–60

Barson NJ, Knight ME, Turner GF. 2007. The genetic architecture of male colour differences between a sympatric Lake Malawi cichlid species pair. *J. Evol. Biol.* 20:45–53

Basolo AL. 2006. Genetic linkage and color polymorphism in the southern platyfish (*Xiphophorus maculatus*): a model system for studies of color pattern evolution. *Zebrafish* 3:65–83

Beeching SC, Holt BA, Neiderer MP. 2002. Ontogeny of melanistic color pattern elements in the convict cichlid, *Cichlasoma nigrofasciatum*. *Copeia* 2002:199–203

Bell MA, Foster SA. 1994. *The Evolutionary Biology of the Threespine Stickleback*. Oxford, UK: Oxford Univ. Press

Bolnick D, Fitzpatrick BM. 2007. Sympatric speciation: models and empirical evidence. *Annu. Rev. Ecol. Evol. Syst.* 38:459–87

Braasch I, Salzburger W, Meyer A. 2006. Asymmetric evolution in two fish-specifically duplicated receptor tyrosine kinase paralogons involved in teleost coloration. *Mol. Biol. Evol.* 23:1192–202

Brooks R. 2000. Negative genetic correlation between male sexual attractiveness and survival. *Nature* 406:67–70

Brooks R, Endler JA. 2001. Direct and indirect sexual selection and quantitative genetics of male traits in guppies (*Poecilia reticulata*). *Evolution* 55:1002–15

Bull JJ. 1983. *Evolution of Sex Determining Mechanisms*. Menlo Park, CA: Benjamin Cummings

Carleton KL, Kocher TD. 2001. Cone opsin genes of African cichlid fishes: tuning spectral sensitivity by differential gene expression. *Mol. Biol. Evol.* 18:1540–50

Carroll SB. 2005. *Endless Forms Most Beautiful: The New Science of Evo Devo and the Making of the Animal Kingdom*. New York: Norton

Charlesworth B. 1991. The evolution of sex chromosomes. *Science* 251:1030–33

Chevassus B. 1988. *Modification du phénotype sexuel et du mode de reproduction chez les poissons salmonidés: inversion sexuelle, gynogénèse, hybridation interspécifique et polyploïdisation*. PhD diss. Univ. Paris XI, Orsay

Chen T-R, Reisman HM. 1970. A comparative chromosome study of the North American species of sticklebacks (Teleostei: Gasterosteidae). *Cytogenetics* 9:321–32

Colosimo PF, Hosemann KE, Balabhadra S, Villarreal G, Dickson M, et al. 2005. Widespread parallel evolution in sticklebacks by repeated fixation of *ectodysplasin* alleles. *Science* 307:1928–33

Colosimo PF, Peichel CL, Nereng K, Blackman BK, Shapiro MD, et al. 2004. The genetic architecture of parallel armor plate reduction in threespine sticklebacks. *PLoS Biol.* 2:635–41

Couldridge VCK, Alexander GJ. 2002. Color patterns and species recognition in four closely related species of Lake Malawi cichlid. *Behav. Ecol.* 13:59–64

Cummings ME, García de León FJ, Mollaghan DM, Ryan MJ. 2006. Is UV ornamentation an amplifier in swordtails? *Zebrafish* 3:91–100

Cresko WA, Amores A, Wilson C, Murphy J, Currey M, et al. 2004. Parallel genetic basis for repeated evolution of armor loss in Alaskan threespine stickleback populations. *Proc. Natl. Acad. Sci. USA* 101:6050–55

Cresko WA, McGuigan KL, Phillips PC, Postlethwait JH. 2007. Studies of threespine stickleback developmental evolution: progress and promise. *Genetica* 129:105–26

Davidson EH, Erwin DH. 2006. Gene regulatory networks and the evolution of animal body plans. *Science* 311:796–800

Devlin RH, Nagahama Y. 2002. Sex determination and sex differentiation in fish: an overview of genetic, physiological, and environmental influences. *Aquaculture* 208:191–364

Dickman MC, Schliwa M, Barlow GW. 1988. Melanophore death and disappearance produces color metamorphosis in the polychromatic Midas cichlid (*Cichlasoma citrinellum*). *Cell Tissue Res.* 253:9–14

Dieckmann U, Doebeli M. 1999. On the origin of species by sympatric speciation. *Nature* 400:354–57

Endler JA. 1980. Natural selection on color patterns in *Poecilia reticulata*. *Evolution* 34:76–91

Endler JA. 1988. Sexual selection and predation risk in guppies. *Nature* 332:593–94

Engeszer RE, Ryan MJ, Parichy DM. 2004. Learned social preference in zebrafish. *Curr. Biol.* 14:881–84

Fisher RA. 1931. The evolution of dominance. *Biol. Rev.* 6:345–68

Fraser GJ, Berkovitz, Graham A, Smith MM. 2006. Gene deployment for tooth replacement in the rainbow trout (*Oncorhynchus mykiss*): a developmental model for evolution of the osteichthyan dentition. *Evol. Dev.* 8:446–57

Fraser GJ, Graham A, Smith MM. 2004. Conserved deployment of genes during odontogenesis across osteichthyans. *Proc. R. Soc. London Ser. B* 271:2311–17

Froschauer A, Körting C, Katagiri T, Aoki T, Asakawa S, et al. 2002. Construction and initial analysis of bacterial artificial chromosome (BAC) contigs from the sex-determining region of the platyfish *Xiphophorus maculatus*. *Gene* 295:247–54

Fukamachi S, Shimada A, Shima A. 2001. Mutations in the gene encoding B, a novel transporter protein, reduce melanin content in madaka. *Nat. Genet.* 28:381–85

Gans C, Northcutt RG. 1983. Neural crest and the origin of vertebrates: a new head. *Science* 220:268–74

Genner MJ, Turner GF. 2005. The mbuna cichlids of Lake Malawi: a model for rapid speciation and adaptive radiation. *Fish Fish.* 6:1–34

Gerhart J, Kirschner M. 1997. *Cell, Embryos, and Evolution: Toward a Cellular and Developmental Understanding of Phenotypic Variation and Evolutionary Adaptability.* Malden, MA: Blackwell Sci.

Graves JA. 2006. Sex chromosome specialization and degeneration in mammals. *Cell* 124:901–14

Griffin DK, Harvey SC, Campos-Ramos R, Ayling L-J, Bromage NR, et al. 2002. Early origins of the X and Y chromosomes: lessons from tilapia. *Cytogenet. Genome Res.* 99:157–63

Gutbrod H, Schartl M. 1999. Intragenic sex-chromosomal crossovers of *Xmrk* oncogene alleles affect pigment pattern formation and the severity of melanoma in *Xiphophorus*. *Genetics* 151:773–83

Hall BK. 1999. *The Neural Crest in Development and Evolution*. New York: Springer-Verlag

Harvey SC, Masabanda J, Carrasco LAP, Bromage NR, Penman DJ, Griffin DK. 2002. Molecular-cytogenetic analysis reveals sequence differences between the sex chromosomes of *Oreochromis niloticus*: evidence for an early stage of sex-chromosome differentiation. *Cytogenet. Genome Res.* 79:76–80

Hoekstra HE, Hirschmann RJ, Bundey RA, Insel PA, Crossland JP. 2006. A single amino acid mutation contributes to adaptive beach mouse color pattern. *Science* 313:101–4

Houghton L, Lindon C, Morgan BA. 2005. The ectodysplasin pathway in feather tract development. *Development* 132:863–72

Hulsey CD, Fraser GJ, Streelman JT. 2005. Evolution and development of complex biomechanical systems: 300 million years of fish jaws. *Zebrafish* 2:243–57

Huysseune A, Thesleff I. 2004. Continuous tooth replacement: the possible involvement of epithelial stem cells. *Bioessays* 26:665–71

Jackman WR, Draper BW, Stock DW. 2004. Fgf signaling is required for zebrafish tooth development. *Dev. Biol.* 274:139–57

Jeffery WR. 2001. Cavefish as a model system in evolutionary biology. *Dev. Biol.* 231:1–12

Johnson SL, Africa D, Walker C, Weston JA. 1995. Genetic control of adult pigment stripe development in zebrafish. *Dev. Biol.* 167:27–33

Jordan R, Kellogg K, Juanes F, Stauffer J. 2003. Evaluation of female mate choice cues in a group of Lake Malawi mbuna (Cichlidae). *Copeia* 2003:181–86

Kallman KD. 1975. The platyfish *Xiphophorus maculatus*. In *Handbook of Genetics*, Vol. 4, ed. RC King, pp. 81–132. New York: Plenum

Kallman KD. 1984. A new look at sex determination in poeciliid fishes. In *Evolutionary Genetics of Fishes*, ed. BJ Turner, pp. 95–171. New York: Plenum

Kallman KD, Borkoski V. 1978. A sex-linked gene controlling the onset of sexual maturity in female and male platyfish (*Xiphophorus maculatus*), fecundity in females and adult size in males. *Genetics* 89:79–119

Kallman KD, Schreibman MP, Borkoski V. 1973. Genetic control of gonadotrop differentiation in the platyfish, *Xiphophorus maculatus* (Poeciliidae). *Science* 181:678–80

Kawakami K, Amsterdam A, Shimoda N, Becker T, Mugg J, et al. 2000. Proviral insertions in the zebrafish *hagoromo* gene, encoding an F-box/WD40-repeat protein, cause stripe pattern anomalies. *Curr. Biol.* 10:463–66

Kazianis S. 2006. Historical, present and future use of *Xiphophorus* fishes for research. *Zebrafish* 3:9–10

Kelsh RN. 2004. Genetics and evolution of pigment patterns in fish. *Pigment Cell Res.* 17:326–36

Kimmel CB, Ullman B, Walker C, Wilson C, Currey M, et al. 2005. Evolution and development of facial bone morphology in threespine sticklebacks. *Proc. Natl. Acad. Sci. USA* 102:5791–96

Kocher TD. 2004. Adaptive evolution and explosive speciation: the cichlid fish model. *Nat. Rev. Genet.* 5:288–98

Kondo M, Hornung U, Nanda I, Imai S, Sasaki T, et al. 2006. Genomic organization of the sex-determining and adjacent regions of the sex chromosomes of medaka. *Genome Res.* 16:815–26

Kondo S, Kuwahara Y, Kondo M, Naruse K, Mitani H, et al. 2001. The medaka *rs-3* locus required for scale development encodes ectodysplasin-A receptor. *Curr. Biol.* 11:1202–6

Kondo M, Nanda I, Hornung U, Asakawa S, Shimizu N, et al. 2003. Absence of the candidate male sex-determining gene *dmrt1b(Y)* of medaka from other fish species. *Curr. Biol.* 13:416–20

Kondo M, Nanda I, Hornung U, Schmid M, Schartl M. 2004. Evolutionary origin of the medaka Y chromosome. *Curr. Biol.* 14:1664–69

Kondrashov AS, Kondrashov FA. 1999. Interactions among quantitative traits in the course of sympatric speciation. *Nature* 400:351–54

Lamason RL, Mohideen MA, Mest JR, Wong AC, Norton HL, et al. 2005. SLC24A5, a putative cation exchanger, affects pigmentation in zebrafish and humans. *Science* 310:1782–86

Lamoreux ML, Kelsh RN, Wakamatsu Y, Ozato K. 2005. Pigment pattern formation in the medaka embryo. *Pigment Cell Res.* 18:64–73

Laurenti P, Thaëron C, Allizard F, Huysseune A, Sire J-Y. 2004. Cellular expression of *eve1* suggests its requirement for the differentiation of the ameloblasts and for the initiation and morphogenesis of the first tooth in the zebrafish (*Danio rerio*). *Dev. Dyn.* 230:727–33

Le Douarin NM. 1999. *The Neural Crest*. Cambridge, UK: Cambride Univ. Press

Lee B-Y, Hulata G, Kocher TD. 2004. Two unlinked loci controlling the sex of blue tilapia (*Oreochromis aureus*). *Heredity* 92:543–49

Lee B-Y, Lee W-J, Streelman JT, Carleton KL, Howe AE, et al. 2005. A second-generation genetic linkage map of Tilapia (*Oreochromis* spp.). *Genetics* 170:237–44

Lee B-Y, Penman DJ, Kocher TD. 2003. Identification of a sex-determining region in Nile tilapia (*Oreochromis niloticus*) using bulked segregant analysis. *Anim. Genet.* 34:379–83

Lindholm AK, Breden F. 2002. Sex chromosomes and sexual selection in poeciliid fishes. *Am. Nat.* 160:S214–24

Lindholm AK, Brooks R, Breden F. 2004. Extreme polymorphism in a Y-linked sexually selected trait. *Heredity* 92:156–62

Maan ME, Haesler MP, Seehausen O, van Alphen JJ. 2006. Heritability and hete-rochrony of polychromatism in a Lake Victoria cichlid fish: stepping stones for speciation? *J. Exp. Zool. B Mol. Dev. Evol.* 306:168–76

Mabee PM. 1995. Evolution of pigment pattern development in centrarchid fishes. *Copeia* 1995:586–607

Maderspacher F, Nusslein-Volhard C. 2003. Formation of the adult pigment pattern in zebrafish requires *leopard* and *obelix* dependent cell interactions. *Development* 130:3447–57

Mank JE, Promislow DE, Avise JC. 2006. Evolution of alternative sex-determining mechanisms in teleost fishes. *Biol. J. Linn. Soc.* 87:83–93

Marshall Graves JA. 2002. The rise and fall of *SRY. Trends Genet.* 18:259–64

Matsuda M, Nagahama Y, Shinomiya A, Sato T, Matsuda C, et al. 2002. *DMY* is a Y-specific DM-domain gene required for male development in the medaka fish. *Nature* 417:559–63

Matsuda M, Sato T, Toyazaki Y, Nagahama Y, Hamaguchi S, Sakaizumi M. 2003. *Oyzias curvinotus* has *DMY*, a gene that is required for male development in the medaka, *O. latipes. Zool. Sci.* 20:159–61

McCauley DW, Hixon E, Jeffery WR. 2004. Evolution of pigment cell regression in the cavefish *Astyanax*: a late step in melanogenesis. *Evol. Dev.* 6:209–18

McMillan WO, Weigt LA, Palumbi SR. 1999. Color pattern evolution, assor-tative mating, and genetic differentiation in brightly colored butterflyfishes (Chaetodontidae). *Evolution* 53:247–60

Miguez DG, Munuzuri AP. 2006. On the orientation of stripes in fish skin patterning. *Biophys. Chem.* 124:161–67

Millar NP, Reznick DN, Kinnison MT, Hendry AP. 2006. Disentangling the selec-tive factors that act on male colour in wild guppies. *Oikos* 113:1–12

Miller MR, Dunham JP, Amores A, Cresko WA, Johnson EA. 2007. Rapid and cost-effective polymorphism identification and genotyping using restriction site as-sociated DNA (RAD) markers. *Genome Res.* 17:240–48

Mills MG, Nuckels RJ, Parichy DM. 2007. Deconstructing evolution of adult pheno-types: Genetic analyses of *kit* reveal homology and evolutionary novelty during adult pigment pattern development of *Danio* fishes. *Development* 134:1081–90

Morizot DC, Slaugenhaupt SA, Kallman KD, Chakravarti A. 1991. Genetic linkage map of fish of the genus *Xiphophorus* (Teleostei: Poeciliidae). *Genetics* 127:399–410

Nanda I, Kondo M, Hornung U, Asakawa S, Winkler C, et al. 2002. A duplicated copy of *DMRT1* in the sex-determining region of the Y-chromosome of the medaka, *Oryzias latipes. Proc. Natl. Acad. Sci. USA* 99:11778–83

Nordland JJ, Boissy RE, Hearing VJ, King RA, Oetting WS, Ortonne JP, eds. 2006. *The Pigmentary System: Physiology and Pathophysiology.* New York: Oxford Univ. Press

Orr HA. 1998. The population genetics of adaptation: the distribution of factors fixed during adaptive evolution. *Evolution* 52:935–49

Painter KJ, Maini PK, Othmer HG. 1999. Stripe formation in juvenile *Pomacanthus* explained by a generalized turing mechanism with chemotaxis. *Proc. Natl. Acad. Sci. USA* 96:5549–54

Parichy DM. 2005. Variation and developmental biology: prospects for the future. In *Variation: A Hierarchical Examination of a Central Concept in Biology*, ed. B Hallgrimsson, BK Hall, pp. 475–98. New York: Academic

Parichy DM, Johnson SL. 2001. Zebrafish hybrids suggest genetic mechanisms for pigment pattern diversification in *Danio*. *Dev. Genes Evol.* 211:319–28

Parichy DM, Mellgren EM, Rawls JF, Lopes SS, Kelsh RN, Johnson SL. 2000a. Mutational analysis of *endothelin receptor b1* (*rose*) during neural crest and pigment pattern development in the zebrafish *Danio rerio*. *Dev. Biol.* 227:294–306

Parichy DM, Ransom DG, Paw B, Zon LI, Johnson SL. 2000b. An orthologue of the kit-related gene *fms* is required for development of neural crest-derived xanthophores and a subpopulation of adult melanocytes in the zebrafish, *Danio rerio*. *Development* 127:3031–44

Parichy DM, Rawls JF, Pratt SJ, Whitfield TT, Johnson SL. 1999. Zebrafish *sparse* corresponds to an orthologue of *c-kit* and is required for the morphogenesis of a subpopulation of melanocytes, but is not essential for hematopoiesis or primordial germ cell development. *Development* 126:3425–36

Parichy DM, Reedy MV, Erickson CA. 2006. Regulation of melanoblast migration and differentiation. See Nordland et al. 2006, pp. 108–39

Parichy DM, Turner JM. 2003a. Temporal and cellular requirements for *Fms* signaling during zebrafish adult pigment pattern development. *Development* 130:817–33

Parichy DM, Turner JM. 2003b. Zebrafish *puma* mutant decouples pigment pattern and somatic metamorphosis. *Dev. Biol.* 256:242–57

Parichy DM, Turner JM, Parker NB. 2003. Essential role for *puma* in development of postembryonic neural crest-derived cell lineages in zebrafish. *Dev. Biol.* 256:221–41

Peichel CL, Nereng KS, Ohgi KA, Cole BLE, Colosimo PF, et al. 2001. The genetic architecture of divergence between threespine stickleback species. *Nature* 414:901–5

Peichel CL, Ross JA, Matson CK, Dickson M, Grimwood J, et al. 2004. The master sex-determination locus in threespine sticklebacks is on a nascent Y chromosome. *Curr. Biol.* 14:1416–24

Penman DJ, McAndrew BJ. 2000. Genetics for the management and improvement of cultured tilapias. In *Tilapias: Biology and Exploitation*, ed. BCM Beveridge, BJ McAndrew, pp. 227–66. Dordrecht: Kluwer Acad.

Peters H, Balling R. 1999. Teeth: where and how to make them. *Trends Genet.* 15:59–65

Phillips RB, Ihssen P. 1985. Identification of sex chromosomes in lake trout (*Salvelinus namaycush*). *Cytogenet. Cell Genet.* 39:14–18

Phillips RB, Konkol NR, Reed KM, Stein JD. 2001. Chromosome painting supports lack of homology among sex chromosomes in *Oncorhyncus*, *Salmo*, and *Salvelinus* (Salmonidae). *Genetica* 111:119–23

Phillips RB, Matsuoka MP, Reed KM. 2002. Characterization of charr chromosomes using fluorescence in situ hybridization. *Environ. Biol. Fishes* 64:223–28

Plikus MV, Zeichner-David M, Mayer J-A, Reyna J, Bringas P, et al. 2005. Morphoregulation of teeth: modulating the number, size, shape and differentiation by tuning Bmp activity. *Evol. Dev.* 7:440–57

Protas ME, Hersey C, Kochanek D, Zhou Y, Wilkens H, et al. 2006. Genetic analysis of cavefish reveals molecular convergence in the evolution of albinism. *Nat. Genet.* 38:107–11

Quigley IK, Manuel JL, Roberts RA, Nuckels RJ, Herrington ER, et al. 2005. Evolutionary diversification of pigment pattern in *Danio* fishes: differential *fms* dependence and stripe loss in *D. albolineatus*. *Development* 132:89–104

Quigley IK, Turner JM, Nuckels RJ, Manuel JL, Budi EH, et al. 2004. Pigment pattern evolution by differential deployment of neural crest and postembryonic melanophore lineages in *Danio* fishes. *Development* 131:6053–69

Raible DW, Eisen JS. 1994. Restriction of neural crest cell fate in the trunk of the embryonic zebrafish. *Development* 120:495–503

Rice WR. 1987a. The accumulation of sexually antagonistic genes as a selective agent promoting the evolution of reduced recombination between primitive sex chromosomes. *Evolution* 41:911–14

Rice WR. 1987b. Genetic hitchhiking and the evolution of reduced genetic activity of the Y sex chromosome. *Genetics* 116:161–67

Rosenthal GG, Flores Martinez TY, García de León FJ, Ryan MJ. 2001. Shared preferences by predators and females for male ornaments in swordtails. *Am. Nat.* 158:146–54

Rosenthal GG, García de León FJ. 2006. Sexual behavior, genes and evolution in *Xiphophorus*. *Zebrafish* 3:85–90

Rosenthal GG, Ryan MJ. 2005. Assortative preferences for stripes in danios. *Anim. Behav.* 70:1063–66

Ryan MJ, Hews DK, Wagner WEJ. 1990. Sexual selection on alleles that determine body size in the swordtail *Xiphophorus nigrensis*. *Behav. Ecol. Sociobiol.* 26:231–37

Ryan MJ, Pease CM, Morris MR. 1992. A genetic polymorphism in the swordtail *Xiphophorus nigrensis*: testing the prediction of equal fitnesses. *Am. Nat.* 139:21–31

Salazar-Ciudad I, Jernvall J. 2002. A gene network model accounting for development and evolution of mammalian teeth. *Proc. Natl. Acad. Sci. USA* 99:8116–20

Schultheis C, Zhou Q, Froschauer A, Nanda I, Selz Y, et al. 2006. Molecular analysis of the sex-determining region of the platyfish *Xiphophorus maculatus*. *Zebrafish* 3:299–309

Seehausen O, Mayhew PJ, van Alphen JJM. 1999. Evolution of colour patterns in East African cichlid fish. *J. Evol. Biol.* 12:514–34

Shapiro MD, Bell MA, Kingsley DM. 2006. Parallel genetic origins of pelvic reduction in vertebrates. *Proc. Natl. Acad. Sci. USA* 103:13753–58

Shapiro MD, Marks ME, Peichel CL, Blackman BK, Nereng KS, et al. 2004. Genetic and developmental basis of evolutionary pelvic reduction in threespine sticklebacks. *Nature* 428:717–23

Sire J-Y, Davit-Beal T, Delgato S, van der Heyden C, Huysseune A. 2002. First-generation teeth in nonmammalian lineages: evidence for a conserved ancestral character? *Microsc. Res. Tech.* 59:408–34

Stock DW, Jackman WR, Trapani J. 2006. Developmental genetic mechanisms of evolutionary tooth loss in cypriniform fishes. *Development* 133:3127–37

Strecker U, Faundez VH, Wilkens H. 2004. Phylogeography of surface and cave *Astyanax* (Teleostei) from Central and North America based on cytochrome *b* sequence data. *Mol. Phylogenet. Evol.* 33:469–81

Streelman JT, Albertson RC. 2006. Evolution of novelty in the cichlid dentition. *J. Exp. Zool. B Mol. Dev. Evol.* 306:216–26

Streelman JT, Albertson RC, Kocher TD. 2003a. Genome mapping of the orange blotch colour pattern in cichlid fishes. *Mol. Ecol.* 12:2465–71

Streelman JT, Webb JF, Albertson RC, Kocher TD. 2003b. The cusp of evolution and development: a model of cichlid tooth shape diversity. *Evol. Dev.* 5:600–8

Sugie A, Terai Y, Ota R, Okada N. 2004. The evolution of genes for pigmentation in African cichlid fishes. *Gene* 343:337–46

Tanaka M, Hale LA, Amores A, Yan YL, Cresko WA, et al. 2005. Developmental genetic basis for the evolution of pelvic fin loss in the pufferfish *Takifugu rubripes*. *Dev. Biol.* 281:227–39

Terai Y, Morikawa N, Kawakami K, Okada N. 2002a. Accelerated evolution of the surface amino acids in the WD-repeat domain encoded by the *hagoromo* gene in an explosively speciated lineage of East African cichlid fishes. *Mol. Biol. Evol.* 19:574–78

Terai Y, Morikawa N, Kawakami K, Okada N. 2003. The complexity of alternative splicing of *hagoromo* mRNAs is increased in an explosively speciated lineage in East African cichlids. *Proc. Natl. Acad. Sci. USA* 100:12798–803

Terai Y, Morikawa N, Okada N. 2002b. The evolution of the prodomain of bone morphogenetic protein 4 (*Bmp4*) in an explosively speciated lineage of East African cichlid fishes. *Mol. Biol. Evol.* 19:1628–32

Tsaparas P, Mariño-Ramírez L, Bodenreider O, Koonin EV, Jordan IK. 2006. Global similarity and local divergence in human and mouse gene coexpression networks. *BMC Evol. Biol.* 6:70

Veith A-M, Froschauer A, Korting C, Nandi I, Hanel R, et al. 2003. Cloning of the *dmrt1* gene of *Xiphophorus maculatus*: *dmY/dmrt1Y* is not the master sex-determining gene in the platyfish. *Gene* 317:59–66

Volff J-N, Korting C, Froschauer A, Zhou Q, Wilde B, et al. 2003a. The *Xmrk* oncogene can escape nonfunctionalization in a highly unstable subtelomeric region of the fish *Xiphophorus*. *Genomics* 82:470–79

Volff J-N, Schartl M. 2001. Variability of genetic sex determination in poeciliid fishes. *Genetica* 111:101–10

Volff J-N, Zarkower D, Bardwell VJ, Schartl M. 2003b. Evolutionary dynamics of the DM domain gene family in metazoans. *J. Mol. Evol.* 57:S241–49

Wainwright PC. 2007. Functional versus morphological diversity in macroevolution. *Annu. Rev. Ecol. Evol. Syst.* 38:381–401

Watanabe M, Iwashita M, Ishii M, Kurachi Y, Kawakami A, et al. 2006. Spot pattern of leopard *Danio* is caused by mutation in the zebrafish connexin41.8 gene. *EMBO Rep.* 7:893–97

Webb JF. 1999. Larvae in fish development and evolution. In *The Origin and Evolution of Larval Forms*, ed. BK Hall, MH Wake, pp. 109–58. New York: Academic

Wilkins AS. 2001. *The Evolution of Developmental Pathways*. Sunderland, MA: Sinauer Assoc.

Woram RA, Gharbi K, Sakamoto T, Hoyheim B, Holm L-E, et al. 2003. Comparative genome analysis of the primary sex-determining locus in salmonid fishes. *Genome Res.* 13:272–80

Yamamoto Y, Stock DW, Jeffery WR. 2004. Hedgehog signalling controls eye degeneration in blind cavefish. *Nature* 431:844–47

Yang CT, Johnson SL. 2006. Small molecule-induced ablation and subsequent regeneration of larval zebrafish melanocytes. *Development* 133:3563–73

Zimmerer EJ, Kallman KD. 1989. Genetic basis for alternative reproductive tactics in the pygmy swordtail, *Xiphophorus nigrensis*. *Evolution* 43:1298–307

Terrestrial Carbon–Cycle Feedback to Climate Warming

Yiqi Luo

Department of Botany and Microbiology, University of Oklahoma,
Norman, Oklahoma 73072; email: yluo@ou.edu

Annu. Rev. Ecol. Evol. Syst. 2007. 38:683–712

First published online as a Review in Advance on August 17, 2007

The *Annual Review of Ecology, Evolution, and Systematics* is online at
http://ecolsys.annualreviews.org

This article's doi:
10.1146/annurev.ecolsys.38.091206.095808

Key Words

global change, net ecosystem production, photosynthesis, respiration, soil carbon pools, temperature sensitivity

Abstract

The coupled carbon-climate models reported in the literature all demonstrate a positive feedback between terrestrial carbon cycles and climate warming. A primary mechanism underlying the modeled positive feedback is the kinetic sensitivity of photosynthesis and respiration to temperature. Field experiments, however, suggest much richer mechanisms driving ecosystem responses to climate warming, including extended growing seasons, enhanced nutrient availability, shifted species composition, and altered ecosystem-water dynamics. The diverse mechanisms likely define more possibilities of carbon-climate feedbacks than projected by the kinetics-based models. Nonetheless, experimental results are so variable that we have not generated the necessary insights on ecosystem responses to effectively improve global models. To constrain model projections of carbon-climate feedbacks, we need more empirical data from whole-ecosystem warming experiments across a wide range of biomes, particularly in tropic regions, and closer interactions between models and experiments.

1. INTRODUCTION

Human activities, such as fossil-fuel burning and deforestation, have resulted in a gradual increase in the atmospheric CO_2 concentration from 280 ppm volumetrically in pre-industrial time to ~380 ppm at present and potentially to 700 ppm toward the end of the twenty-first century (Intergov. Panel Clim. Change 2007). As a consequence of the buildup of CO_2 and other greenhouse gases in the atmosphere, Earth's surface temperature has increased by 0.74°C since 1850 and is expected to increase by another 1.1°C ~ 6.4°C by the end of this century (Intergov. Panel Clim. Change 2007). Because temperature affects almost all aspects of terrestrial carbon (C) processes, increasing Earth's surface temperature likely enhances ecosystem C fluxes, potentially feeding back to a buildup of atmospheric CO_2 concentration and climate dynamics. Will climate warming trigger terrestrial carbon–cycle feedback that leads to warmer climate? This is a central question in global-change research that urgently needs to be addressed in the coming years.

Terrestrial feedbacks to climate change involve several greenhouse gases (e.g., CO_2, CH_4, N_2O, and O_3) and are modulated by changes in precipitation, land uses, and nitrogen (N) deposition and invasive species (Field et al. 2007). This review does not cover all those aspects; rather it focuses on ecosystem C uptake and release processes in response to changes in Earth's surface temperature. This review will include neither wetlands and/or peatlands, nor the issues involved in destabilization of peat deposits. First, I critically examine mechanisms that have been incorporated to capture temperature feedback within global models. The models are effective tools to evaluate carbon-climate feedbacks. The accuracy of their projections, however, depends on how closely the models represent the real-world processes of C uptake and release. Second, I review experimental evidence to show that the responses of C uptake and release processes to temperature changes are extremely variable. Third, this article illustrates that several mechanisms underlie the variable responses of major C processes to climate warming. Those mechanisms include changes in phenology and the length of growing seasons, species composition, nutrient dynamics, and ecohydrological processes. The last section of this review briefly discusses various approaches to improve the model representation of terrestrial C processes. I conclude that there is an urgent need for more empirical knowledge from experiments and observations that will permit the quantification of temperature sensitivity at the ecosystem scale and fundamentally improve our ability to predict feedbacks of the terrestrial carbon cycle to climate warming.

2. MODELED POSITIVE FEEDBACK

All global models that have sought to couple climate dynamics and carbon cycles have predicted a positive feedback between carbon cycling and climate warming (Friedlingstein et al. 2006). Cox et al. (2000) evaluated this feedback issue first with three simulations. The first simulation examined the effects of rising atmospheric CO_2 concentration on the land C sink. The model prescribed atmospheric CO_2 concentrations according to the Intergovernment Panel for Climate Change 1992 "Business-as-Usual" scenario (IS92a) (Alcamo et al. 1995) and projected that land ecosystems would sequester nearly 400 Gt (10^{15} g) C owing to CO_2 fertilization in the twenty-first century. The second simulation explored the effects of climate warming on the carbon cycle. Rising atmospheric CO_2 concentration induced climate warming by $5.5°C$ over Earth's land surface. This warming stimulated C loss, resulting in a net source of 60 Gt C from land ecosystems to the atmosphere over the twenty-first century. The third simulation coupled the climate model with the carbon-cycle model, causing the projected atmospheric CO_2 to be 980 ppm in 2100, 40% higher than the 700 ppm predicted by IS92a. The land ecosystems became a net source of 170 Gt C in the coupled carbon-climate simulation. The coupled carbon-climate model projected the temperature to increase by $8.0°C$ over land, $2.5°C$ greater than the climate-model simulation alone.

Friedlingstein et al. (2006) recently examined the climate-carbon feedback using 11 coupled climate change–carbon cycle models with a common protocol. All 11 models unanimously displayed a positive climate warming–carbon cycle feedback (**Figure 1**). By the end of the twenty-first century, the predicted feedback caused additional CO_2 buildup in the atmosphere, from a low of 20 ppm in the Lawrence Livermore National Laboratory climate-carbon model (Thompson et al. 2004) to a high of 200 ppm in the Hadley Center Climate Model coupled with a land-surface model (HadCM3LS). The majority of models projected additional CO_2 buildup between 50 and 100 ppm (**Figure 1a,b**). The additional CO_2 buildup in the atmosphere triggered by the climate-carbon feedback led to an additional climate warming of $0.1°C–1.5°C$. All the 11 models except the University of Maryland model (Zeng et al. 2004) exhibited stronger sensitivities to climate warming for land C storage compared with ocean C storage. For example, the HadCM3LS model projected a loss of 177 Gt C per degree Celsius of warming from land ecosystems, whereas other models projected losses of 20–112 Gt C per degree Celsius of warming (**Figure 1c,d**).

Similarly, a variety of terrestrial ecosystem models that are not coupled with climate dynamics have ubiquitously projected the loss of C from land ecosystems in response to climate warming (Berthelot et al. 2005, Cao & Woodward 1998, Cramer et al. 2001, Ito 2005) regardless of the model structure or climate-change scenario. The high degree of uniformity among model projections stems from the similar representation of carbon-climate relationships among models. The primary mechanism incorporated into these uncoupled and coupled carbon-climate models is the kinetic sensitivity of photosynthesis and respiration to temperature. The temperature sensitivity of photosynthesis is primarily described by either empirical equations (Cox 2001) or biochemical processes in the Farquhar model (Cao & Woodward 1998). The temperature-respiration relationship is usually described by exponential or Arrhenius functions (Luo & Zhou 2006). The sensitivity of photosynthesis and respiration influences net primary productivity (NPP) and ecosystem respiration, respectively, and ultimately determines net changes in land ecosystem C storage in response to climate warming.

In an intercomparison study with the 11 coupled carbon-climate models, two models simulated minor increases in NPP, five models showed little change, and four models simulated large decreases in NPP with global-scale climate warming (Friedlingstein et al. 2006) (**Figure 1e**). Variations in modeled responses of NPP can

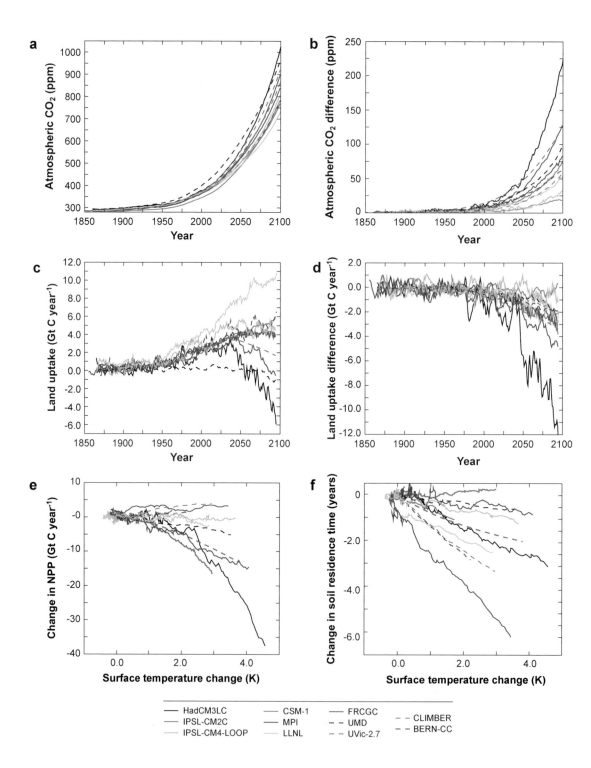

have substantial effects on the strength of the carbon-climate feedback (Matthews et al. 2005). At regional scales, most models simulated a climate-induced increase of NPP at high latitudes at which warming considerably decreased the duration of snow cover and increased the length of the growing season. In tropical regions, the majority of the models simulated a decrease of NPP, although the degree of decrease varied greatly among models. In a perhaps extreme simulation by HadCM3LS, the excess heating under climate warming induced marked soil drying and dieback of the rainforest in the Amazon basin (Cox et al. 2004), causing dramatically decreased atmospheric CO_2 uptake by vegetation and increased C loss from soil. In fact, HadCM3LS simulated the largest NPP sensitivity to climate warming—a global decrease of 8 Gt C in NPP per degree Celsius of warming.

The other major contribution to the modeled C loss in response to climate warming is the decomposition of soil organic matter. In fact, practically all models assume that the decomposition rate of organic matter increases with temperature. When decomposition rates increase under climate warming, the residence time of C pools decreases because the residence time is an inverse of a specific decomposition rate (Luo et al. 2003). In Friedlingstein et al.'s (2006) intercomparison, all models except one simulated a decrease in soil C residence time (**Figure 1f**), indicating that most models assume that specific respiration rates increase with climate warming. The Frontier Research Center for Global Change model (Ito & Oikawa 2002), for example, simulated a decrease in soil C residence time by approximately 2 years per degree Celsius of warming (**Figure 1f**). Most of the models used a temperature-sensitivity index of Q_{10}—a quotient of change in respiration caused by a change in temperature of $10°C$—equaling 2. In contrast, the University of Maryland model has Q_{10} ranging from 1.1 for the slow soil C pool to 2.2 for the fast turnover soil C pool. Carbon loss under climate change is also regulated by the residence time itself. HadCM3LC, which has one single pool with a residence time of 25 years (Jones et al. 2004), projected a strong ecosystem response to climate warming. The National Center for Atmosphere Research–Climate System Model 1 coupled carbon-climate model, which has nine pools, projected a decrease in C sink at low latitudes that nearly canceled an increase at high latitude (Fung et al. 2005).

Overall, the modeled positive feedback between climate warming and global carbon cycling is attributable primarily to stimulated net C release from land ecosystems in response to climate warming. The net land C release results from decreased NPP in most models and increased respiratory C release by all the models under climate warming. Major uncertainties remain about both the direction and degree of the

---

Figure 1

(*a*) Atmospheric CO_2 concentration (ppm) as simulated by 11 coupled carbon-climate models (see Friedlingstein et al. 2006 for detailed description of the 11 models). (*b*) Atmospheric CO_2 difference between the coupled and uncoupled simulations (ppm). (*c*) Land C fluxes for the coupled runs (Gt C year^{-1}). (*d*) Differences between coupled and uncoupled land C fluxes (Gt C year^{-1}). (*e*) Simulated net primary productivity (NPP) sensitivity to climate (coupled run – uncoupled run). (*f*) Simulated soil C turnover time sensitivity to climate (coupled run – uncoupled run). Figure adopted from Friedlingstein et al. 2006.

response of NPP and soil respiration, as well as the possibility of vegetation dieback and soil drying, especially in tropical forests.

3. EXPERIMENTAL EVIDENCE

Models that couple the carbon cycle and climate change are essential for examining biosphere-atmosphere feedbacks at the global scale. Field experiments cannot be used to quantify the global-scale sensitivity of terrestrial ecosystems to climate warming over time spans of decades or centuries. However, models are necessary abstractions of reality, and the accuracy of their projections depends on how well the models represent the mechanisms responsible for the real-world feedback. As Moorcroft (2006) argues, model-assisted "understanding of biosphere-atmosphere feedbacks is a collection of interesting, but largely untested, hypotheses for the future state of terrestrial ecosystem and climate." It is therefore imperative to critically examine experimental evidence about key C uptake and release processes that determine the terrestrial carbon feedback to climate warming.

3.1. Carbon Uptake Under Warming

Most C uptake processes, such as photosynthesis, plant growth, and primary production, are sensitive to changes in temperature. Their responses to climate warming are regulated by other factors and processes, leading to diverse changes observed in warming experiments.

Photosynthesis. Temperature influences the rate of photosynthetic CO_2 uptake through changes in the ratio of $[CO_2]:[O_2]$ dissolved in solution, the specificity of Rubisco for CO_2 and O_2, and rates of carboxylation and oxygenation (Brooks & Farquhar 1985, Long 1991). For C_3 plants, net photosynthesis increases with temperature at its low range, reaches a maximum at optimal temperature, and then declines (**Figure 2a**). The optimal temperature usually varies broadly depending on the local adaptation of different species to their habitats and thermal acclimation over seasons (Pearcy & Ehleringer 1984).

Experiments have shown the diverse effects of warming on photosynthesis, including increases (Bergh & Linder 1999, Loik et al. 2004), decreases (Callaway et al. 1994, Gunderson et al. 2000, He & Dong 2003, Roden & Ball 1996), and no apparent change (Llorens et al. 2004, Loik et al. 2000, Nijs et al. 1996, Starr et al. 2000). Warming air temperature by 3°C–5°C, for example, increased photosynthesis in four vascular species in arctic tundra (Chapin & Shaver 1996) and two dominant tree species and a shrub species in a boreal forest (Beerling 1999). In contrast, a 3.5°C increase in air temperature did not significantly impact the photosynthesis of *Polygonum viviparum* in arctic polar semidesert (Wookey et al. 1994). Leaf photosynthesis increased in spring, decreased in early fall, and did not change in summer and late fall for four species exposed to an air warming of 0.5°C–2.0°C in the southern Great Plains of the United States (Zhou et al. 2007a) (**Figure 3**). The variable responses may result from different methods and/or levels of warming and may reflect diverse

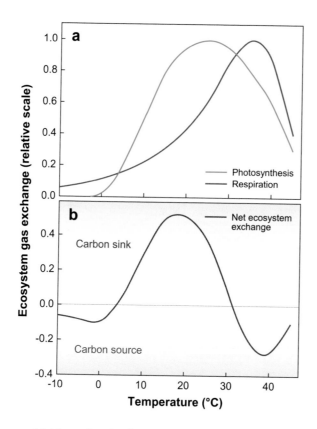

Figure 2

Idealized response functions
of (*a*) plant photosynthesis
and ecosystem respiration
and (*b*) net ecosystem
exchange to temperature,
illustrating that climate
warming can result in either
net C release or net C
uptake by terrestrial
ecosystems purely on the
basis of kinetic sensitivity. In
the very low or very high
temperature ranges in which
respiration is higher than
photosynthesis, ecosystems
have net C release. In an
intermediate temperature
range, photosynthesis is
higher than respiration,
leading to net C uptake in
terrestrial ecosystems.

temperature sensitivities and optimal temperatures of photosynthesis among species
and ecotypes (Chapin et al. 1995, Llorens et al. 2004, Shaw et al. 2000). In addition, other factors may influence the results, such as drought, leaf age, and nutrient
availability (Gunderson et al. 2000).

Plant growth. The effects of warming on plant growth are highly variable. Experimental warming increased leaf production by 50% and shoot production by
26% for *Colobanthus quitensis* but decreased leaf production by 17% for *Deschampsia
antarctic* in Antarctica (Day et al. 1999). Warming stimulated growth of C_4 plants
in a tallgrass prairie over a 6-year experiment, whereas the growth of C_3 plants increased in the first 2 years and then decreased in the last 2 years (Luo et al. 2007).
Field soil-warming experiments showed that herbs and grass were more responsive to elevated temperature than shrubs, whereas tree species were less sensitive in
a temperate forest (Farnsworth et al. 1995). A meta-analysis of 13 tundra experiments similarly showed that the vegetative growth of herbaceous species was more
responsive to warming than woody species (Arft et al. 1999). However, Chapin &
Shaver (1985) observed that evergreen species generally responded more strongly
to warming than deciduous species (except for *Betalu nana*), whereas the growth
of graminoid species did not change or decreased under greenhouse warming in

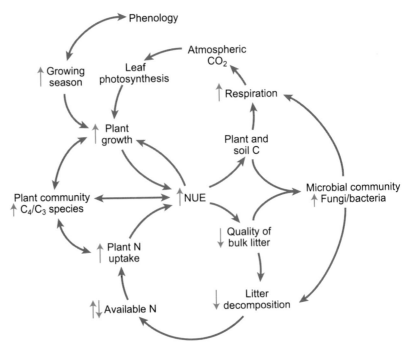

Figure 3

Illustration of regulatory mechanisms of ecosystem temperature sensitivity using results from
an Oklahoma warming experiment. Major carbon (C) and nitrogen (N) processes are affected
by warming in complex ways. Warming extended growing seasons, shifted species composition
toward C_4 plants, and increased plant biomass growth. The increased growth was associated
with increased plant N uptake and N use efficiency (NUE). Warming also increased soil
respiration, which was roughly balanced by increased C uptake via plant growth, leading to
little change in soil C storage, although warming accelerated almost all the rate processes.
Owing to the increased dominance of C_4 plants in the grassland, the quality of bulk litter
decreased under warming, which likely leads to diminished or even decreased soil N
availability over time, although warming may initially increase N availability.

tundra. The individualistic responses to warming reflect differences in optimum
growth temperatures across species, as well as the limitations on growth by other
factors than temperature.

The effects of warming on primary production are also diverse. Soil warming increased the yields of crops by 19%–50% and vegetables by 19%–100% (Rykbost et al.
1975), primarily owing to enhanced growth in early spring. Experimental warming
increased NPP by up to 25% in a tallgrass prairie (Luo et al. 2007) (**Figure 3**). Soil
warming in a Norway spruce forest at Flakaliden in northern Sweden increased the
stem-wood growth of trees in heated plots by 50% relative to controls after 5 years
(Bergh et al. 1999, Jarvis & Linder 2000). A synthesis of data from 20 field warming
experiments indicates that warming, on average, stimulated aboveground plant productivity by 19% (Rustad et al. 2001). In contrast, the aboveground biomass of sugar
maples decreased in response to warming in open top chambers (Norby et al. 1995).

Total aboveground biomass was largely unresponsive to temperature manipulation in tundra (Hobbie & Chapin 1998; Shaver et al. 1986, 1998).

Along a gradient of increasing infrared heating, shrub production increased, whereas graminoid production decreased in a bog. In a fen, graminoids were most productive at high infrared heating, and forbs were most productive at medium infrared heating (Weltzin et al. 2000). In both the bog and fen communities, ratios of belowground to aboveground NPP increased with warming, indicating shifts in C allocation.

3.2. Carbon Release Under Warming

Terrestrial ecosystems release C to the atmosphere through autotrophic (i.e., plant) and heterotrophic (primarily by microbes) respiration. Both autotrophic and heterotrophic respiration is very sensitive to changes in temperature. Since most measurements were made on plant tissues (e.g., leaf or root) to quantify plant respiration and at soil surface to quantify soil respiration, I discuss plant respiration and soil respiration separately although soil respiration includes both root and microbial respiration.

Plant respiration. Actively growing plants respire approximately 50% of the C available from photosynthesis (after photorespiration), with the remainder available for growth and reproduction (Law et al. 1999, Ryan 1991). Respiration increases with temperature in its low range when the respiration rate is mainly limited by biochemical reactions (Atkin et al. 2000). At high temperatures, the transport of substrates and products of the metabolism (e.g., sugar, oxygen, CO_2) mainly via diffusion processes becomes a limiting factor. At very high temperatures, the protoplasm system may start to break down. As a result, respiration usually follows a general temperature-response curve, increasing exponentially with temperature in its low range, reaching a maximum at an optimal temperature, and then declining (**Figure 2a**). During the exponential increase phase, respiration often doubles in response to a 10°C temperature increase ($Q_{10} = 2$) (Amthor 1989, Ryan et al. 1995).

Although leaf respiration is usually stimulated by experimental warming (Zhou et al. 2007a), the acclimation of plant respiration to temperature (Atkin et al. 2006) reduces C loss over extended periods. Based on short-term studies indicating that warming stimulates plant respiration more than photosynthesis, many plant-growth models predict an increase in respiration:photosynthesis ratio at elevated temperatures. Long-term experiments (Gifford 1994, 1995) suggest that respiration:photosynthesis ratio is often remarkably insensitive to growth temperature (Arnone & Körner 1997, Gunn & Farrar 1999, Lambers 1985) because plants acclimate to a new temperature environment over a few days. This acclimation may be controlled by carbohydrate status, the demand for ATP, and/or the reduced production of reactive oxygen species (Atkin et al. 2005). A simple substrate-based model of plant acclimation to temperature shows that respiration is effectively limited by carbohydrate supply from photosynthesis (Dewar et al. 1999). The short-term, positive temperature response of respiration:photosynthesis ratio therefore reflects the transient dynamics of nonstructural carbohydrate and protein pools, whereas the insensitivity of respiration/photosynthesis ratio to temperature on a longer time scale reflects the state behavior of the pools.

Soil respiration. Soil respiration accounts for approximately two-thirds of C loss from terrestrial ecosystems and is generally responsive to temperature changes (Luo & Zhou 2006). Its sensitivity to climate warming has been identified as one of the major sources of uncertainty in model projections of future climate change (Cox et al. 2000, Friedlingstein et al. 2006). As a consequence, scientists have conducted extensive research on the sensitivity of soil and/or ecosystem respiration to climate warming (Davidson & Janssens 2006). When natural ecosystems have been exposed to experimental warming, soil CO_2 efflux generally increases (Melillo et al. 2002, Mertens et al. 2001, Zhou et al. 2007b) (**Figure 3**). A meta-analysis of data collected at 17 sites from tundra, grassland, and forest shows that soil respiration under experimental warming increased at 11 sites, decreased at 1 site, and did not change at 5 sites (Rustad et al. 2001). An increase in soil temperature by 5°C above ambient temperature using buried heating cables, for example, caused additional C release of 538 g m^{-2} year^{-1} from soil in Harvard Forest (Peterjohn et al. 1994). In contrast, infrared heating slightly decreased soil respiration in a Rocky Mountain meadow in Colorado (Saleska et al. 1999) and a grassland in Oklahoma in the first year of the experiment (Luo et al. 2001).

Warming-induced increases in soil respiration likely result from changes in multiple processes (Shaver et al. 2000). Global warming extends the length of growing seasons (Lucht et al. 2002, Norby et al. 2003), alters plant phenology (Dunne et al. 2003, Sherry et al. 2007), stimulates plant growth (Wan et al. 2005), increases mineralization and soil N availability (Melillo et al. 2002, Rustad et al. 2001), reduces soil water content (Harte et al. 1995, Wan et al. 2002), and shifts species composition and community structure (Harte & Shaw 1995, Saleska et al. 2002, Weltzin et al. 2003). Responses also differ across locations. The magnitude of response in soil respiration to soil warming is greater in cold, high-latitude ecosystems than in warm, temperate areas (Kirschbaum 1995). Recent warming has likely caused a great loss of C in tundra and boreal soils (Goulden et al. 1998).

It is commonly observed that the magnitude of response in soil respiration to warming decreases over time (Rustad et al. 2001). The yearly flux of CO_2 from heated plots at the Harvard Forest was ~40% higher than control plots in the first year but gradually declined to the level in the control plots after the 6-year warming treatment (Melillo et al. 2002). This decline can be attributable to acclimatization (Luo et al. 2001) and/or depletion of substrates (Eliasson et al. 2005, Gu et al. 2004, Niinistö et al. 2004, Pajari 1995). In addition, warming caused a shift in the soil microbial community toward more fungi (Zhang et al. 2005), which are more tolerant to high soil temperature and dry environments than bacteria owing to their filamentous nature (Holland & Coleman 1987). Shifted microbial community structure may partially explain observed decreases in the temperature sensitivity of soil CO_2 efflux.

3.3. Net Ecosystem Production

Changes in plant production and ecosystem respiration together determine the long-term effects of warming on ecosystem C balances. If C uptake and release are primarily determined by the kinetic properties of photosynthesis and respiration, respectively, net ecosystem production (NEP) should be negative at high and low temperature ranges and positive (i.e., sink) at an intermediate range along a temperature variation over a season or latitude (**Figure 2b**). In addition, NEP is also regulated by many processes other than photosynthesis and respiration kinetics, leading to complex responses to climate warming. Buried heating cables only warm soil and have generally caused net C loss, such as in experiments at Harvard Forest (Melillo et al. 2002) and the arctic tundra (Billings et al. 1982). Ineson et al. (1998) also showed a net C reduction of approximately 10% after 3 years of heating an upland grassland ecosystem at Great Dun Fell in the United Kingdom.

Whole-ecosystem warming using infrared heaters or greenhouse chambers may decrease, increase, or cause no changes in net ecosystem exchange. Using infrared heating, Marchand et al. (2004) found a 24% increase in canopy C uptake and a nearly 50% increase in net C sink under warming in comparison with that under control in high-arctic tundra. After 8 years of experiment, an average increase of 5.6°C in air temperature with field greenhouse warming did not cause much change in canopy photosynthesis, ecosystem respiration, and net ecosystem C exchange in arctic tundra (Johnson et al. 2000), although warming stimulated early canopy development and extended the length of growing seasons by 3 weeks. The warming experiment at the southern Great Plains did not cause significant changes in soil C stocks or NEP (Luo et al. 2007) (**Figure 3**). Saleska et al. (2002), however, observed a decrease of soil organic carbon by ~ 200 g C m^{-2} ($\sim 8.5\%$ reduction) in warmed plots relative to control plots in a Rocky Mountain meadow.

If warming primarily stimulates the decomposition of litter and the oxidation of soil organic matter, soil C pools will decline over time. If increased temperature strongly stimulates NPP, climate warming may lead to increased terrestrial C storage (Smith & Shugart 1993). There are several mechanisms (e.g., increased N mineralization, extended growing seasons, and shifted species composition) that may enhance NPP.

In summary, warming experiments have not produced many clear and consistent patterns across ecosystems. Warming caused increases, decreases, or no change in photosynthesis, plant growth, primary production, soil respiration, and NEP. There are several reasons for these highly variable responses. First, the levels of temperature increases are not consistent among warming experiments. Second, warming methods vary among experiments, ranging from the infrared heating of whole ecosystems to heating cables for soil warming and passive heating systems using open top chambers and/or nighttime cover. Third, measured variables and measurement methods differ among investigators. Fourth, plants and ecosystems are inherently diverse in their responses to warming. Overall, the highly variable results from field experiments may not have generated the necessary insights on ecosystem responses to climate warming to the detriment of model improvement. Nevertheless, experiments have suggested that the kinetic sensitivities of photosynthesis and respiration, although fundamental to models, are usually overridden by other processes. The latter processes strongly regulate ecosystem responses to climate warming.

4. REGULATORY MECHANISMS OF ECOSYSTEM TEMPERATURE SENSITIVITY

Both carbon release and uptake processes in intact ecosystems are affected by complex mechanisms such as phenology and the length of growing seasons, nutrient dynamics, species composition, and water availability in addition to the kinetic sensitivities of photosynthesis and respiration (**Figures 3** and **4**). It is essential to understand such regulatory mechanisms to develop terrestrial ecosystem models capable of predicting changes in C uptake and release in response to climate warming.

4.1. Changes in Phenology and Length of Growing Seasons

Plant phenology is responsive to environmental cues such as temperature, photoperiod, and moisture (Rathcke & Lacey 1985) and has been used as a sensitive indicator of climate change in Earth's system (Peñuelas & Filella 2001, Walther et al. 2002). Long-term ground-based and remote-sensing measurements indicate that plant phenology has been advanced by 2–3 days in spring and delayed by 0.3–1.6 days in autumn

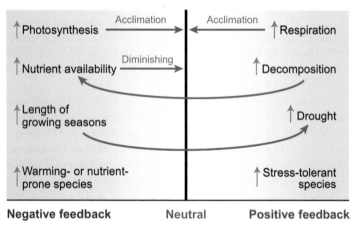

Figure 4

Schematic summary of major regulatory mechanisms that lead to either positive or negative feedbacks of terrestrial C cycles to climate warming. Climate warming instantaneously stimulates photosynthetic C uptake and respiratory release. Acclimation can neutralize their kinetic responses. Warming-stimulated decomposition of soil organic matter is associated with respiratory C release and increases nutrient availability that stimulates plant growth and ecosystem C uptake. The warming-induced increases in nutrient availability may be diminishing over time. Warming extends growing seasons and lengthens C uptake periods. Extended growing seasons and warming in combination can exacerbate drought stress and limit net ecosystem C uptake in some regions. Warming-induced changes in species composition can result in either positive or negative feedbacks of C cycles to climate warming, depending on which species adapt to the new environment.

per decade (Myneni et al. 1997, Parmesan & Yohe 2003) over the past 30–80 years, probably owing to recent climatic warming (Parmesan & Yohe 2003).

Researchers have consistently observed across experiments that plant phenology is responsive to warming (e.g., Cleland et al. 2006). Field experiments via heating and snow removal in alpine ecosystems reveal that a 3°C increase in temperature advances flowering time by 4.5 days (Dunne et al. 2003, Price & Waser 1998). *Maianthemum canadense* and *Uvularia sessilifolia*, the herbaceous dominants, emerged 7–10 days earlier in the growing season under soil-warming conditions (Peterjohn et al. 1993). Leaf bud burst and flowering phenology occurred earlier in warmed plots compared with control plots at 13 different tundra sites (Arft et al. 1999). Sherry et al. (2007) observed that early flowering species advanced phenology by 3–17 days in spring, whereas late flowering species delayed phenology up to 12 days in the Oklahoma warming experiment (**Figure 3**).

Shifts in phenology result in a growing-season extension, with earlier greenness in spring and later senescence in autumn. The divergence of phenology toward early spring and late autumn under experimental warming (Sherry et al. 2007) extended the growing seasons by more than 2 weeks (Wan et al. 2005). Intact field greenhouse warming in arctic tundra extended growing seasons for 3 weeks. Analysis of remote-sensing data has also shown that climate warming has extended the growing season in the past decades (Myneni et al. 1997, Nemani et al. 2003, White et al. 1999).

Those changes in phenology and growing-season length directly affect ecosystem C processes (**Figure 3**). In a deciduous forest, for example, the timing of leaf expansion and senescence influenced interannual shifts in photosynthesis (Goulden et al. 1996). Analysis of satellite data suggested that extending the growing season earlier in the spring and/or later in the autumn increased primary production in terrestrial ecosystems (Nemani et al. 2003). Analysis of data from eddy-flux networks showed that NEP and gross C assimilation increase with decreasing latitude (Falge et al. 2002). Measured NEP from eddy-flux sites is strongly correlated with the length of the C uptake period for temperate broad-leaved forests (Baldocchi & Wilson 2001) and other vegetation types (Churkina et al. 2005) (**Figure 4**).

Changes in phenology and growing-season length also affect C cycling indirectly via changes in species composition, water balance, and nutrient processes (see discussion below). For example, experimental warming in a tallgrass prairie in Oklahoma stimulated the availability and plant uptake of soil N in spring, causing higher leaf N concentration in *Schizachyrium scoparium* compared with the control. The effect was reversed in summer owing to increased soil drying, decreased soil N availability, and increased plant growth (Y. An & Y. Luo, unpublished data).

4.2. Changes in Species Composition

Species composition in ecological communities reflects interactions among organisms under a set of environmental conditions. Climate warming alters essential environmental conditions (such as temperature and soil nutrient and water availabilities) and results in changes in species composition (Peñuelas & Boada 2003). Under experimental warming, for example, species composition changed to favor shrubs over graminoid species in a bog and graminoids over forbs in a fen in northern Wisconsin peatland (Weltzin et al. 2000). Experimental warming differentially affected the growth of C_3 and C_4 species and shifted species composition in favor of C_4 plants in the southern Great Plains (Luo et al. 2007, Wan et al. 2005). In a Rocky Mountain meadow, warming stimulated the relative abundance of shrub species, but depressed forb species (Harte & Shaw 1995). Experimental warming at 11 locations across the tundra biome rapidly altered a whole plant community by increasing the height and cover of deciduous shrubs and graminoids and decreasing the cover of mosses and lichens (Walker et al. 2006). As a consequence, species diversity and evenness decreased in the warming plots in comparison with that in the control. Climate warming also altered the geographical distribution of plants (Parmesan & Yohe 2003, Peñuelas & Boada 2003).

Shifts in species composition result from changes in the competitive balance among species. A shift of species composition occurred in a montane meadow toward shrubs (De Valpine & Harte 2001, Harte & Shaw 1995) because warming lowered leaf water potential (Loik & Harte 1997) and increased photosynthetic rates and water-use efficiency for *Artemisia tridentata* (Shaw et al. 2000). In contrast, warming stimulated mineralization, increased N availability, and favored fast-growing and N-rich species in a moist tundra ecosystem near Toolik Lake, Alaska (Chapin et al. 1995). Mesic sites in tundra had higher species diversity and were more responsive than xeric sites to warming (Walker et al. 2006). In the tallgrass prairie, experimental warming stimulated C_4 plant growth but depressed C_3 plant growth (Luo et al. 2007) because C_4 plants have competitive advantages in a warm and dry environment.

Changes in plant community composition have long-term effects on ecosystem C balance. A shift from forbs to shrubs resulted in decreased litter inputs and decreased soil organic C in an alpine meadow ecosystem (Saleska et al. 2002). Because the decomposition of woody litter is slower than forb litter, increased C residence time potentially leads to long-term recovery of SOC in the warming plot (Saleska et al. 2002). A modeling study that was calibrated to tussock vegetation at Toolik Lake, Alaska, suggested that warming stimulates soil respiration and N availability, thus favoring high-productivity forbs over the shrubs (McKane et al. 1997) and leading to the partial recovery of the initial SOC loss in the long run (Herbert et al. 1999). In an Oklahoma grassland, warming increased C_4 plant growth, causing increased primary production and litter input to the soil (Luo et al. 2007). Increased C_4 litter production with a low C:N ratio slowed down decomposition and increased C accumulation in the litter layers (**Figure 3**). Decreased litter quality in response to warming resulted in a shift in the soil microbial community composition toward a fungi dominance (Zhang et al. 2005). Fungi dominance, in turn, could favor soil aggregation and C storage in ecosystems (Rillig 2004), thus reinforcing physical and biochemical protection of soil C storage (Jastrow 1996, Six et al. 2002).

4.3. Nutrient-Mediated Feedbacks

Many warming experiments have showed that increased temperature causes faster microbial decomposition of organic matter (Grogan & Chapin 2000), increased N mineralization (Chapin et al. 1995, Johnson et al. 2000, Shaver et al. 1998), and increased N uptake by plants (Ineson et al. 1998, Jarvis & Linder 2000, Rustad et al. 2001, Welker et al. 2004). Soil warming in Harvard Forest stimulated the net N mineralization rate for a decade (Melillo et al. 2002). A meta-analysis of net N mineralization rates from 12 ecosystem warming experiments showed a large stimulation by 46% (Rustad et al. 2001). However, Jonasson et al. (1999) did not find any significant change in total soil N and phosphate contents over 5 years of greenhouse warming at Abisko, Sweden, possibly owing to the microbial immobilization of gross mineralized N or small temperature increases.

Warming-induced changes in soil N transformations can trigger long-term feedbacks on ecosystem C balances because N strongly regulates terrestrial C sequestration. Stimulated N mineralization and plant N uptake under warming resulted in increased biomass production in arctic tundra (Gough & Hobbie 2003, Hobbie et al. 2002, Welker et al. 2004). Over time, accelerated decomposition under warming may lower soil organic pools, leading to declines in mineral N availability and constraints on plant N uptake in the long-term (**Figure 4**). Ultimately, this feedback may reduce the stimulation of biomass growth by warming. Furthermore, warming has been shown to decrease bulk litter quality in a tallgrass ecosystem (An et al. 2005) (**Figure 3**), leading to the reduced release of soil N over time. Thus, the increased N demand due to faster plant growth under warming may not be met by the N supply in the long-term, possibly leading to progressive N limitation (Luo et al. 2004).

4.4. Feedbacks Through Hydrological Cycling

Warming usually happens in concert with drought. The 2003 summer heat wave occurred in Europe with combined drought and high temperature, resulting in net release of C from terrestrial ecosystems (Ciais et al. 2005) (**Figure 4**). In general, the influences of climate warming on large-scale hydrological processes (such as precipitation, runoff, tropospheric water vapor, and evaporation) have been extensively studied using modeling and observation approaches (Huntington 2006). Both theoretical analysis and observational evidence suggest that climate warming likely results in increases in evaporation and precipitation at the global scale. Many regions may experience severe drought or moisture surplus (Dai et al. 1998).

The indirect effects of climate warming on C dynamics via changes in ecosystem-scale hydrological cycling have not been carefully studied. In general, climate warming accelerated evapotranspiration, leading to soil drying (Harte et al. 1995, Llorens et al. 2004, Wan et al. 2002) and decreased soil- and leaf-level water potentials. However, Zavaleta et al. (2003) found that experimental warming increased spring soil moisture by 5%–10% in an annual grassland in California owing to the accelerated decline of canopy greenness by inducing earlier plant senescence. Warming-induced increases in soil surface evaporation and plant transpiration accelerate soil water depletion. With an increased soil water deficit, water replenishment in soil may increase, and water runoff decreases after precipitation events as shown in a modeling study (E.S. Weng and Y. Luo, submitted manuscript). If precipitation occurs at such a frequency that soil water is replenished before the soil water content is lowered to such an extent that plant growth is influenced severely, the portion of precipitation water used for plant growth increases under warming in comparison with that in control. However, in years when precipitation is evenly distributed and largely partitioned for evapotranspiration without much surplus for runoff at ambient conditions, warming substantially decreases soil water content (Zhou et al. 2007b) and may not change precipitation partitioning to runoff.

In response to warming and soil drying, plants may adjust ecophysiological processes so as to influence C balance. Warming and associated drought, for example, may stimulate belowground growth, increase the root/shoot ratio, and result in shifts of the plant community to C_4 species, shrubs, and other drought-tolerant species. Warming-induced extension of growing seasons to early spring and late fall can increase ecosystem-level water-use efficiency and production in regions in which winter precipitation does not contribute to plant growth (Luo et al. 2007). Early spring warming in an alpine forest of Colorado, however, was usually related to shallow late-spring snow pack and resulted in low springtime and annual net CO_2 uptake (Monson et al. 2005). Thus, warming may decrease ecosystem productivity in regions in which either summer plant growth depends on winter soil water storage or there is no winter water surplus (**Figure 4**).

5. QUEST FOR PREDICTIVE UNDERSTANDING

One ultimate goal of global-change research is to project future states of climate and ecosystems. Although research over the past years has established a modeling framework that can be used to evaluate feedbacks between climate change and global C cycles, the model assumption that kinetics of photosynthetic and respiratory biochemistry underlie terrestrial carbon–cycle feedback to climate warming is not fully in accordance with experimental results. Conversely, results from experiments and observation have great uncertainties owing to the nature of perturbation experiments (Luo & Reynolds 1999), different experimentation methods, scales of studies, and other issues. A search for predictive understanding from imperfect models and uncertain experimental evidence therefore represents a great challenge. Here I discuss a few approaches that are not mutually exclusive but may simultaneously contribute to our predictive understanding of terrestrial carbon–cycle feedback to climate warming.

5.1. Fundamental Approach

Potential feedbacks of terrestrial ecosystems to climate warming originate from temperature-sensitive processes at biochemical and physical levels. The primary processes include enzyme kinetics involved in photosynthesis, plant respiration, decomposition of litter, and oxidation of soil organic matter. Temperature regulation of enzymatic activities is usually described by an exponential or Arrhenius equation (Luo & Zhou 2006) in most coupled carbon-climate or stand-alone ecosystem models. However, declining phases of photosynthesis and respiration in the high temperature range (**Figure 2**), which are usually observed in laboratory studies but rarely in field, are usually not represented in models (but see Parton et al. 1997). Moreover, although the fundamental nature of these biochemical processes is independent of the hierarchical level (from cellular to global), scaling up of biochemical kinetics to project carbon-climate feedback at the global scale is challenging because many indirect processes (e.g., those discussed in Section 4) can easily override the lower-level biochemical processes. Nevertheless, it is essential to understand and more accurately model kinetics of biochemical processes (Davidson & Janssens 2006, Davidson et al. 2006) to project carbon-climate feedback.

Temperature also directly affects plant development and growth via cell division and differentiation, the occurrence of fires and insects, and water-temperature relationships in permafrost ecosystems. Cell differentiation determines phenological responses to warming and directly regulates the dynamics of leaf area over growing seasons at local scales, whereas temperature effects on fire and insect infestation alter C balance at landscape and regional scales. Most of the processes are regulated by other factors and cannot be described easily by simple temperature functions. For example, phenology is regulated by photoperiod, temperature, and moisture (Peñuelas & Filella 2001, Rathcke & Lacey 1985, Walther et al. 2002). Fire occurrence is determined by fuel loading, moisture content, and temperature (Weise et al. 2005). Most models simulate temperature effects on phenology and fire occurrence based on empirical relationships.

5.2. Pragmatic Approach

Although there are variable responses of all C uptake and release processes to climate warming, two globally coherent patterns have emerged across warming experiments and observations. Global coherence is a common term in economics and refers to a process or event that has a similar effect across multiple systems at different locations throughout the world, even though the mechanisms underlying the coherent process may be different (Parmesan 2006). One globally coherent pattern observed across warming experiments is a consistent response of phenology to warming, leading to the extension of growing seasons. The other commonly observed ecosystem response to climate warming is a shift in species composition. The development of reliable models that can more accurately simulate the globally coherent patterns is a critical step toward improving model projections of future carbon-cycling feedback to climate warming.

Phenology consistently responds to climate warming (Parmesan 2006), although the degree of the response and the mechanisms underlying it may differ (see Section 4.1). To account for such coherent responses, several phenology models have been developed to predict leafing out, senescence, and reproductive events. Leafing out, for example, can be predicted by growing degree days, a combined chilling and forcing temperature (Chuine et al. 1999), a moisture index in water-limited ecosystems (Kramer et al. 2000, White et al. 1997), the timing of the first heavy precipitation in deserts (Beatley 1974), or the occurrence of soil thawing in arctic ecosystems (van Wijk et al. 2003). Jolly et al. (2005) proposed a growing season index to predict foliar phenology at the global scale based on daily temperature, the vapor pressure deficit, and photoperiod. Leaf senescence has been predicted by a frost index (Kramer et al. 2000), environmental fluctuations (Arora & Boer 2005), and specific thresholds for different plant functional types in a dynamic global vegetation model (Sitch et al. 2003).

Warming causes shifts in the species composition of plant communities across many experimental sites (see Section 4.2). Prediction of species shifts at individual levels, however, has been difficult owing to species-specific responses (Chapin & Shaver 1985), whereas responses at the level of plant functional type may be more predictable. Researchers have developed several models to simulate species response to climate change. Herbert et al. (1999) used a multiple-element limitation model to simulate changes in species composition in response to climate change. Increased temperature stimulated N release and therefore favored fast-growing species with low N use efficiency. Peters (2002) developed a mixed life-form, individual plant-based gap dynamics model to examine the consequences of differences in recruitment, resource acquisition, and mortality to patterns in species dominance and composition under a variety of soils and climate conditions. The model predicted that a grass, *Bouteloua eriopoda*, will dominate the Chihuahuan desert if climate change leads to increased summer water availability. If climate change leads to increased winter precipitation, a C_3 shrub, *Larrea tridentata*, may dominate. The major challenge of the individual-based modeling approach is scaling up the simulation of local-level species dynamics to landscape-level changes. Epstein et al. (2001) used a regional-scale model to examine warming effects on species dynamics in arctic tundra at four levels of aggregation—individual species, functional types, life forms, and vegetation types. The level of aggregation affected simulation results of community composition, total community biomass, and NPP.

As we accumulate more experimental evidence, some other ecosystem properties may emerge to be globally coherent so as to assist our predictive understanding of ecosystem responses to climate warming. Results from numerous experiments have demonstrated that ecosystem respiration is tightly coupled with plant photosynthesis. It will be useful to examine how warming affects such coupling between soil respiration and NPP and if the coupling between photosynthesis and respiration leads to any predictive relationships (e.g., relative constant ratio) under different warming regimes. At the leaf level, the ratio of instantaneous respiration to photosynthesis has been found to be relatively constant, even in plants exposed to contrasting growth temperature (Gifford 1995, Ziska & Bunce 1998). When individual plants are exposed to growth temperature within a range plants experienced in the natural habitat, plants adjust biomass allocation to maintain a homeostatic respiration to photosynthesis ratio (Atkin et al. 2007). This constancy, if held for the majority of plant species, offers a potential to account for photosynthetic and respiratory acclimation in coupled carbon-climate models (Gifford 2003).

5.3. Probabilistic Approach

Carbon processes and regulatory mechanisms are inherently variable. The variations stem from many sources, including genetic differences between organisms, environmental variability over time and space (e.g., climate dynamics and heterogeneity in the physical and chemical processes in soil), and diversity in plant and microbial responses to environmental change. Observations on the responses of ecosystem C uptake and release processes to climate warming are also subject to errors in measurement and disparities in experimental methods and treatment levels. Synthesis of experimental data across sites, experiments, and studies all shows great variations in C processes (e.g., Arft et al. 1999, Luo et al. 2006, Rustad et al. 2001). Although it is exceedingly desirable to discover invariant functions that cut across scales or globally coherent patterns to fundamentally improve models, high variability in ecosystem processes within and between studies poses a significant challenge for the development of our predictive understanding of carbon-climate feedbacks. To realistically reflect variability in ecosystem processes, probabilistic approaches such as stochastic modeling and ensemble analysis may be effective tools to assess uncertainty. Stochastic modeling has been widely used in other disciplines (Zhang 2002) and applied to global-change research (e.g., Moorcroft et al. 2001). Ensemble results from the intercomparison of multiple models (e.g., Friedlingstein et al. 2006) can account for variations among model structure and parameters but may not identify systematic bias if some of the key processes have not been integrated into any of the models.

Most simulation models assume that there are some intrinsic constants of parameters for each vegetation type, which are modified by environmental scalars such as temperature and moisture functions to model spatial and temporal variations in ecosystem processes. However, analyses by Hui et al. (2003) and Richardson et al. (2007) indicate that parameters need to vary with years to account for interannual variability in net ecosystem exchange of C. It is also likely that intrinsic parameter values may vary with space. Spatial and temporal variations in parameters propagate to variability in projections of future states of terrestrial ecosystems as quantified by stochastic approaches to data-model assimilation (Xu et al. 2006). We need extensive studies of spatial and temporal variability in parameter values and their propagation to model projections of future states of ecosystems and climate.

6. CONCLUDING REMARKS

The global-modeling community has established a quantitative framework over the past years to evaluate feedbacks between climate change and global C cycles. The coupled carbon-climate models reported in the literature all simulate a positive feedback between terrestrial C cycle and climate warming. The high degree of uniformity among projections by various models results from a similar mechanism underlying the modeled changes in C fluxes. That mechanism is the kinetic sensitivity of photosynthesis and respiration to temperature. Experimental results suggest much richer mechanisms than kinetic sensitivity that drive ecosystem responses to climate warming. Climate warming, for example, consistently affects phenology, leading to extended growing seasons and enhanced biomass growth and C sequestration from the atmosphere. Experiments often show that species composition changes in response to climate warming. Altered species composition can lead to either the net source or net sink of C, depending on the ecophysiology of altered species. Experimental warming also consistently stimulates mineralization and nutrient availability, favoring plant growth. The increase in nutrient availability may be transient, and its impact on plant growth and ecosystem C storage may diminish over time. Climate warming modifies ecosystem-water balance as well, via changes in precipitation, evapotranspiration, and other ecohydrological processes. Diverse mechanisms likely delineate more possibilities of carbon-climate feedbacks than projected by the current, kinetics-based global models.

Experimental evidence on the temperature sensitivity of ecosystem C uptake and release processes is extremely variable. Climate warming causes increases, decreases, and no change in photosynthesis, plant growth, primary production, soil respiration, and NEP although trends of warming-induced changes exist for some variables. The highly variable experimental results have not provided the necessary insights on model improvement to realistically simulate ecosystem responses to climate warming. To improve our understanding of ecosystem temperature sensitivities, we need to improve experimental studies in several aspects. First, we need whole-ecosystem warming experiments to examine the integrated responses of entire ecosystems to climate warming. Second, we need to establish common research protocols among experiments (e.g., levels of temperature increases) to facilitate direct comparison and data synthesis. Third, we need experiments with multiple levels of temperature increases to investigate nonlinear responses of ecosystems to climate warming. Fourth, we need experiments in underrepresented and/or critical biomes (e.g., ecosystems in tropic regions) to develop global views of the temperature sensitivities of ecosystem C uptake and release processes. Fifth, we need long-term experiments to identify ecosystem responses at different timescales. Finally, we need multifactor global-change experiments to investigate the interactive effects of temperature, elevated CO_2, precipitation, N deposition, and invasive species on carbon-climate feedbacks.

To effectively constrain the model projections of future states of climate and ecosystems, we have to not only continue the model representation of fundamental processes (e.g., kinetics of photosynthesis and respiration) but also use pragmatic and probabilistic approaches to model improvement. The pragmatic approach is to develop empirical modules to simulate globally coherent patterns that have consistently emerged from experiments and observations such as phenology and species shifts. The probabilistic approach is to account for variations in ecosystem processes, spatial and temporal variability in model parameters, and the propagation of variations in parameter values and observations to model projections.

DISCLOSURE STATEMENT

The author is not aware of any biases that might be perceived as affecting the objectivity of this review.

ACKNOWLEDGMENTS

Pierre Friedlinstein provided constructive comments on an earlier version of the manuscript. The author was financially supported by U.S. National Science Foundation under DEB0078325 and DEB 0444518, and by the Office of Science

(BER), Department of Energy, Grants No. DE-FG03-99ER62800 and DE-FG02-06ER64319.

LITERATURE CITED

Alcamo JA, Bouwman J, Edmonds A, Grübler T, Morita, Sugandhy A. 1995. An evaluation of the IPCC IS92 emission scenarios. In *Climate Change 1994: Radiative Forcing of Climate Change and an Evaluation of the IPCC IS92 Emission Scenarios*, ed. JT Houghton, LG Meira Filho, J Bruce, H Lee, BA Callander, E Haites, N Harris, K Maskell, pp. 233–304. Cambridge, UK: Cambridge Univ. Press

Amthor JS. 1989. *Respiration and Crop Productivity*. New York: Springer-Verlag

An Y, Wan S, Zhuo X, Subedar AA, Wallace LL, Luo Y. 2005. Plant nitrogen concentration, use efficiency, and contents in a tallgrass prairie ecosystem under experimental warming. *Glob. Change Biol.* 11:1733–44

Arft AM, Walker MD, Gurevitch J, Alatalo JM, Bret-Harte MS, et al. 1999. Responses of tundra plants to experimental warming: meta-analysis of the international tundra experiment. *Ecol. Monogr.* 69:491–511

Arnone JA III, Körner C. 1997. Temperature adaptation and acclimation potential of leaf dark respiration in two species of *Ranunculus* from warm and cold habitats. *Arct. Alp. Res.* 29:122–25

Arora VK, Boer GJ. 2005. A parameterization of leaf phenology for the terrestrial ecosystem component of climate models. *Glob. Change Biol.* 11:39–59

Atkin OK, Bruhn D, Hurry VM, Tjoelker MG. 2005. The hot and the cold: unraveling the variable response of plant respiration to temperature. *Funct. Plant Biol.* 32:87–105

Atkin OK, Edwards EJ, Loveys BR. 2000. Response of root respiration to changes in temperature and its relevance to global warming. *New Phytol.* 147:141–54

Atkin OK, Scheurwater I, Pons TL. 2006. High thermal acclimation potential of both photosynthesis and respiration in two lowland *Plantago* species in contrast to an alpine congeneric. *Glob. Change Biol.* 12:500–15

Atkin OK, Scheurwater I, Pons TL. 2007. Respiration as a percentage of daily photosynthesis in whole plants is homeostatic at moderate, but not high, growth temperatures. *New Phytol.* 174:367–80

Baldocchi DD, Wilson KB. 2001. Modeling CO_2 and water vapor exchange of a temperate broadleaved forest across hourly to decadal time scales. *Ecol. Model.* 142:155–84

Beatley JC. 1974. Phenological events and their environmental triggers in Mojave desert ecosystems. *Ecology* 55:856–63

Beerling DJ. 1999. Long-term responses of boreal vegetation to global change: an experimental and modeling investigation. *Glob. Change Biol.* 5:55–74

Bergh J, Linder S. 1999. Effects of soil warming during spring on photosynthetic recovery in boreal Norway spruce stands. *Glob. Change Biol.* 5: 245–53

Bergh J, Linder S, Lundmark T, Elfving B. 1999. The effect of water and nutrient availability on the productivity of Norway spruce in northern and southern Sweden. *For. Ecol. Manag.* 119:51–62

Berthelot M, Friedlingstein P, Ciais P, Dufresne JL, Monfray P. 2005. How uncertainties in future climate change predictions translate into future terrestrial carbon fluxes. *Glob. Change Biol.* 11:959–70

Billings WD, Luken JO, Mortensen DA, Peterson KM. 1982. Arctic tundra: a source or sink for atmospheric carbon dioxide in a changing environment? *Oecologia* 53:7–11

Brooks A, Farquhar GD. 1985. Effect of temperature on the CO_2/O_2 specificity of ribulose-1, 5-bisphosphate carboxylase/oxygenase and the rate of respiration in the light. *Planta* 165:397–406

Callaway RM, DeLucia EH, Thomas EM, Schlesinger WH. 1994. Compensatory responses of CO_2 exchange and biomass allocation and their effects on the relative growth rate of ponderosa pine in different CO_2 and temperature regimes. *Oecologia* 98:159–66

Cao MK, Woodward FI. 1998. Dynamic responses of terrestrial ecosystem carbon cycling to global climate change. *Nature* 393:249–52

Chapin FS III, Shaver GR. 1985. Individualistic growth response of tundra plant species to environmental manipulations in the field. *Ecology* 66:564–76

Chapin FS III, Shaver GR. 1996. Physiological and growth responses of arctic plants to a field experiment simulating climatic change. *Ecology* 77:822–40

Chapin FS III, Shaver GR, Giblin AE, Nadelhoffer K, Laundre JA. 1995. Responses of arctic tundra to experimental and observed changes in climate. *Ecology* 76:694–711

Chuine I, Cour P, Rousseau DD. 1999. Selecting models to predict the timing of flowering of temperate trees: implications for tree phenology modeling. *Plant Cell Environ.* 22:1–13

Churkina G, Schimel D, Braswell BH, Xiao X. 2005. Spatial analysis of growing season length control over net ecosystem exchange. *Glob. Change Biol.* 11:1777–87

Ciais P, Reichstein M, Viovy N, Granier A, Ogee J, et al. 2005. Europe-wide reduction in primary productivity caused by the heat and drought in 2003. *Nature* 437:529–33

Cleland EE, Chiariello NR, Loarie SR, Mooney HA, Field CB. 2006. Diverse responses of phenology to global changes in a grassland ecosystem. *Proc. Natl. Acad. Sci. USA* 103:13740–44

Cox PM. 2001. Description of the TRIFFID dynamic global vegetation model. *Tech. Note 24*, Hadley Centre, Met Office, London. 16 pp.

Cox PM, Betts RA, Collins M, Harris PP, Huntingford C, Jones CD. 2004. Amazonian forest dieback under climate-carbon cycle projections for the 21st century. *Theor. Appl. Climatol.* 78:137–56

Cox PM, Betts RA, Jones CD, Spall SA, Totterdell IJ. 2000. Acceleration of global warming due to carbon-cycle feedbacks in a coupled climate model. *Nature* 408:184–87

Cramer W, Bondeau A, Woodward FI, Prentice IC, Betts RA, et al. 2001. Global response of terrestrial ecosystem structure and function to CO_2 and climate change: results from six dynamic global vegetation models. *Glob. Change Biol.* 7:357–73

Dai A, Trenberth KE, Karl TR. 1998. Global variations in droughts and wet spells: 1900–1995. *Geophys. Res. Lett.* 25:3367–70

Davidson EA, Janssens IA. 2006. Temperature sensitivity of soil carbon decomposition and feedbacks to climate change. *Nature* 440:165–73

Davidson EA, Janssens I, Luo Y. 2006. On the variability of respiration in terrestrial ecosystems: moving beyond Q_{10}. *Glob. Change Biol.* 12:154–64

Day TA, Ruhland CT, Grobe CW, Xiong F. 1999. Growth and reproduction of Antarctic vascular plants in response to warming and UV radiation reduction in the field. *Oecologia* 119:24–35

De Valpine P, Harte J. 2001. Plant responses to experimental warming in a montane meadow. *Ecology* 82:637–48

Dewar RC, Medlyn BE, McMurtrie RE. 1999. Acclimation of the respiration/photosynthesis ratio to temperature: insights from a model. *Glob. Change Biol.* 5:615–22

Dunne JA, Harte J, Taylor KJ. 2003. Subalpine meadow flowering phenology responses to climate change: integrating experimental and gradient methods. *Ecol. Monogr.* 73:69–86

Eliasson PE, McMurtrie RE, Pepper DA, Strömgren M, Linder S, Ågren GI. 2005. The response of heterotrophic CO_2 flux to soil warming. *Glob. Change Biol.* 11:167–81

Epstein HE, Chapin FS III, Walker MD, Starfield AM. 2001. Analyzing the functional type concept in arctic plants using a dynamic vegetation model. *Oikos* 95:239–52

Falge E, Tenhunen J, Baldocchi D, Aubinet M, Bakwin P, et al. 2002. Phase and amplitude of ecosystem carbon release and uptake potential as derived from FLUXNET measurements. *Agric. For. Meteorol.* 113:75–95

Farnsworth EJ, Nuñez-Farfán J, Careaga SA, Bazzaz FA. 1995. Phenology and growth of three temperate forest life forms in response to artificial soil warming. *J. Ecol.* 83:967–77

Field CB, Lobell DB, Peters HA, Chiariello NR. 2007. Feedbacks of terrestrial ecosystems to climate change. *Annu. Rev. Environ. Resour.* 32:1–29

Friedlingstein P, Cox P, Betts R, Bopp L, von Bloh W, et al. 2006. Climate-carbon cycle feedback analysis: results from the (CMIP)-M-4 model intercomparison. *J. Clim.* 19:3337–53

Fung I, Doney SC, Lindsay K, John J. 2005. Evolution of carbon sinks in a changing climate. *Proc. Natl. Acad. Sci. USA* 102:11201–6

Gifford RM. 1994. The global carbon cycle: a viewpoint on the missing sink. *Aust. J. Plant Physiol.* 21:1–5

Gifford RM. 1995. Whole plant respiration and photosynthesis of wheat under increased CO_2 concentration and temperature: long-term vs short-term distinctions for modeling. *Glob. Change Biol.* 1:249–63

Gifford RM. 2003. Plant respiration in productivity models: conceptualisation, representation and issues for global terrestrial carbon-cycle research. *Funct. Plant Biol.* 30:171–86

Gough L, Hobbie SE. 2003. Responses of moist nonacidic arctic tundra to altered environment: productivity, biomass, and species richness. *Oikos* 103:204–16

Goulden ML, Munger JW, Fan SM, Daube BC, Wofsy SC. 1996. Exchange of carbon dioxide by a deciduous forest: response to interannual climate variability. *Science* 271:1576–78

Goulden ML, Wofsy SC, Harden JW, Trumbore SE, Crill PM, et al. 1998. Sensitivity of boreal forest carbon balance to soil thaw. *Science* 279:214–17

Grogan P, Chapin FS III. 2000. Initial effects of experimental warming on above- and belowground components of net ecosystem CO_2 exchange in arctic tundra. *Oecologia* 125:512–20

Gu L, Post WM, King AW. 2004. Fast labile carbon turnover obscures sensitivity of heterotrophic respiration from soil to temperature, a model analysis. *Glob. Biogeochem. Cycles* 18:1–11

Gunderson CA, Norby RJ, Wullschleger SD. 2000. Acclimation of photosynthesis and respiration to stimulated climatic warming in northern and southern populations of *Acer saccharum*: laboratory and field evidence. *Tree Physiol.* 20:87–96

Gunn S, Farrar JF. 1999. Effects of a 4°C increase in temperature on partitioning of leaf area and dry mass, root respiration and carbohydrates. *Funct. Ecol.* 13 (Suppl. 1):12–20

Harte J, Shaw R. 1995. Shifting dominance within a montane vegetation community: results of a climate-warming experiment. *Science* 267:876–80

Harte J, Torn MS, Chang FR, Feifarek B, Kinzig AP, et al. 1995. Global warming and soil microclimate: results from a meadow-warming experiment. *Ecol. Appl.* 5:132–50

He WM, Dong M. 2003. Plasticity in physiology and growth of *Salix matsudana* in response to simulated atmospheric temperature rise in the Mu Us Sandland. *Photosynthetica* 41:297–300

Herbert DA, Rastetter EB, Shaver GR, Agren GI. 1999. Effects of plant growth characteristics on biogeochemistry and community composition in a changing climate. *Ecosystems* 2:367–82

Hobbie SE, Chapin FS III. 1998. The response of tundra plant biomass, aboveground production, nitrogen, and CO_2 flux to experimental warming. *Ecology* 79:1526–44

Hobbie SE, Nadelhoffer KJ, Hogberg P. 2002. A synthesis: The role of nutrients as constraints on carbon balances in boreal and arctic regions. *Plant and Soil* 242: 163–170

Holland EA, Coleman DC. 1987. Litter placement effects on microbial and organic-matter dynamics in an agroecosystem. *Ecology* 68:425–33

Hui D, Luo Y, Katul G. 2003. Partitioning interannual variability in net ecosystem exchange between climatic variability and function changes. *Tree Physiol.* 23:433–42

Huntington TG. 2006. Evidence for intensification of the global water cycle: review and synthesis. *J. Hydrol.* 319:83–95

Ineson P, Benham DG, Poskitt J, Harrison AF, Taylor K, Woods C. 1998. Effects of climate change on nitrogen dynamics in upland soils. II. A soil warming study. *Glob. Change Biol.* 4:153–62

Intergov. Panel Clim. Change. 2007. *Working group 1: the physical science basis. Summary for policymakers.* **http://ipcc-wg1.ucar.edu/wg1/wg1-report.html**

Ito A. 2005. Climate-related uncertainties in projections of the 21st century terrestrial carbon budget: Off-line model experiments using IPCC greenhouse gas scenarios and AOGCM climate projections. *Climate Dyn.* 24:435–448

Ito A, Oikawa T. 2002. A simulation model of the carbon cycle in land ecosystems (Sim-CYCLE): a description based on dry-matter production theory and plot-scale validation. *Ecol. Model.* 151:147–79

Jarvis PG, Linder S. 2000. Constraints to growth of boreal forests. *Nature* 405:904–5

Jastrow JD. 1996. Soil aggregate formation and the accrual of particulate and mineral-associated organic matter. *Soil Biol. Biochem.* 28:665–76

Johnson LC, Shaver GR, Cades DH, Rastetter E, Nadelhoffer K, et al. 2000. Plant carbon–nutrient interactions control CO_2 exchange in Alaskan wet sedge tundra ecosystems. *Ecology* 81:453–69

Jolly WM, Nemani R, Running SW. 2005. A generalized, bioclimatic index to predict foliar phenology in response to climate. *Glob. Change Biol.* 11:619–32

Jonasson S, Michelson A, Schmidt IK, Nielsen EV. 1999. Responses in microbes and plants to changed temperature, nutrient and light regimes in the arctic. *Ecology* 80:1828–43

Jones CD, McConnell C, Coleman K, Cox P, Falloon P, et al. 2004. Global climate change and soil carbon stocks: predictions from two contrasting models for the turnover of organic carbon in soil. *Glob. Change Biol.* 11:154–66

Kirschbaum MUF. 1995. The temperature dependence of soil organic matter decomposition, and the effect of global warming on soil organic C storage. *Soil Biol. Biochem.* 27:753–60

Kramer K, Leinonen I, Loustau D. 2000. The importance of phenology for the evaluation of impact of climate change on growth of boreal, temperate and Mediterranean forests ecosystems: an overview. *Int. J. Biometeorol.* 44:67–75

Lambers H. 1985. Respiration in intact plants and tissues: its regulation and dependence on environmental factors, metabolism and invaded organisms. In *Encyclopedia of Plant Physiology*, Vol. 18, ed. R Douce, DA Day, pp. 418–73. Berlin: Springer-Verlag

Law BE, Ryan MG, Anthoni PM. 1999. Seasonal and annual respiration of a ponderosa pine ecosystem. *Glob. Change Biol.* 5:169–82

Llorens L, Peñuelas J, Beier C, Emmett B, Estiarte M, Tietema A. 2004. Effects of an experimental increase of temperature and drought on the photosynthetic performance of two ericaceous shrub species along a north-south European gradient. *Ecosystems* 7:613–24

Loik ME, Harte J. 1997. Changes in water relations for leaves exposed to a climate-warming manipulation in the Rocky Mountains of Colorado. *Environ. Exp. Bot.* 37:115–23

Loik ME, Redar SP, Harte J. 2000. Photosynthetic response to acclimation-warming manipulation for contrasting meadow species in the Rocky Mountains, Colorado, USA. *Funct. Ecol.* 14:166–75

Loik ME, Still CJ, Huxman TE, Harte J. 2004. In situ photosynthetic freezing tolerance for plants exposed to a global warming manipulation in the Rocky Mountains, Colorado, USA. *New Phytol.* 162:331–41

Long SP. 1991. Modification of the response of photosynthetic productivity to rising temperature by atmospheric CO_2 concentrations: Has its importance been underestimated? *Plant Cell Environ.* 14:729–39

Lucht W, Prentice IC, Myneni RB, Sitch S, Friedlingstein P, et al. 2002. Climatic control of the high-latitude vegetation greening trend and Pinatubo effect. *Science* 296:1687–89

Luo Y, Hui D, Zhang D. 2006. Elevated carbon dioxide stimulates net accumulations of carbon and nitrogen in terrestrial ecosystems: a meta-analysis. *Ecology* 87:53–63

Luo Y, Reynolds JF. 1999. Validity of extrapolating field CO_2 experiments to predict carbon sequestration in natural ecosystems. *Ecology* 80:1568–83

Luo Y, Sherry R, Zhou X, Wan S. 2007. Plant ecophysiological regulation of terrestrial carbon-cycle feedback to climate warming. *Science* Submitted

Luo Y, Su B, Currie WS, Dukes J, Finzi A, et al. 2004. Progressive nitrogen limitation of ecosystem responses to rising atmospheric CO_2 concentration. *BioScience* 54:731–39

Luo Y, Wan S, Hui F, Wallace LL. 2001. Acclimatization of soil respiration to warming in a tall grass prairie. *Nature* 413:622–25

Luo Y, White L, Canadell J, DeLucia E, Ellsworth D, et al. 2003. Sustainability of terrestrial carbon sequestration: a case study in Duke Forest with inversion approach. *Glob. Biogeochem. Cycles* 17:1021

Luo Y, Zhou X. 2006. *Soil Respiration and the Environment.* San Diego: Academic/Elsevier. 328 pp.

Marchand FL, Nijs I, de Boeck HJ, Mertens S, Beyens L. 2004. Increased turnover but little change in the carbon balance of high-arctic tundra exposed to whole growing season warming. *Arct. Antarct. Alp. Res.* 36:298–307

Matthews HD, Eby M, Weaver AJ, Hawkins BJ. 2005. Primary productivity control of simulated carbon cycle-climate feedbacks. *Geophys. Res. Lett.* 32:L14708

McKane RB, Rastetter EB, Shaver GR, Nadelhoffer KJ, Giblin AE, et al. 1997. Climatic effects on tundra carbon storage inferred from experimental data and a model. *Ecology* 78:1170–87

Melillo JM, Steudler PA, Aber JD, Newkirk K, Lux H, et al. 2002. Soil warming and carbon-cycle feedbacks to the climate system. *Science* 298:2173–76

Mertens S, Nijs I, Heuer M, Kockelbergh F, Beyens L, et al. 2001. Influence of high temperature on end-of-season tundra CO_2 exchange. *Ecosystems* 4:226–36

Monson RK, Sparks JP, Rosenstiel TN, Scott-Denton LE, Huxman TE, et al. 2005. Climatic influences on net ecosystem CO_2 exchange during the transition from wintertime carbon source to springtime carbon sink in a high-elevation, subalpine forest. *Oecologia* 146:130–47

Moorcroft PR. 2006. How close are we to a predictive science of the biosphere? *Trends Ecol. Evol.* 21:400–7

Moorcroft PR, Hurtt GC, Pacala SW. 2001. A method for scaling vegetation dynamics: the ecosystem demography model (ED). *Ecol. Monogr.* 71:557–85

Myneni RB, Keeling CD, Tucker CJ, Asrar G, Nemani RR. 1997. Increased plant growth in the northern high latitudes from 1981 to 1991. *Nature* 386:698–70

Nemani RR, Keeling CD, Hashimoto H, Jolly WM, Piper SC, et al. 2003. Climate-driven increases in global terrestrial net primary production from 1982 to 1999. *Science* 300:1560–63

Niinistö SM, Silvola J, Kellomäki S. 2004. Soil CO_2 efflux in a boreal pine forest under atmospheric CO_2 enrichment and air warming. *Glob. Change Biol.* 10:1–14

Nijs I, Teughels H, Blum H, Hendrey G, Impens I. 1996. Simulation of climate change with infrared heaters reduces the productivity of *Lolium perenne* L. in summer. *Environ. Exp. Bot.* 36:271–80

Norby RJ, Gunderson CA, Edwards NT, Wullschleger SD, O'Neill EG. 1995. TACIT: temperature and CO_2 interactions in trees. Photosynthesis and growth. *Ecol. Soc. Am. Bull.* 76(Suppl.):197

Norby RJ, Hartz-Rubin JS, Verbrugge MJ. 2003. Phenological responses in maple to experimental atmospheric warming and CO_2 enrichment. *Glob. Change Biol.* 9:1792–801

Pajari B. 1995. Soil respiration in a poor upland site of Scots pine stand subjected to elevated temperatures and atmospheric carbon concentration. *Plant Soil* 169:563–70

Parmesan C. 2006. Ecological and evolutionary responses to recent climate change. *Annu. Rev. Ecol. Evol. Syst.* 37:637–69

Parmesan C, Yohe G. 2003. A globally coherent fingerprint of climate change impacts across natural systems. *Nature* 421:37–42

Parton WJ, Schimel DS, Cole CV, Ojima DS. 1987. Analysis of factors controlling soil organic-matter levels in Great-Plains grasslands. *Soil Science Society of America Journal* 51: 1173–1179

Pearcy RW, Ehleringer J. 1984. Comparative ecophysiology of C_3 and C_4 plants. *Plant Cell Environ.* 7:1–13

Peñuelas J, Boada M. 2003. A global change-induced biome shift in the Montseny mountains (NE Spain). *Glob. Change Biol.* 9:131–40

Peñuelas J, Filella I. 2001. Responses to a warming world. *Science* 294:793–95

Peterjohn WT, Melillo JM, Bowles FP, Steudler PA. 1993. Soil warming and trace gas fluxes: experimental design and preliminary flux results. *Oecologia* 93:18–24

Peterjohn WT, Melillo JM, Steudler PA, Newkirk KM, Bowles FP, Aber JD. 1994. Responses of trace gas fluxes and N availability to experimentally elevated soil temperatures. *Ecol. Appl.* 4:617–25

Peters DPC. 2002. Plant species dominance at a grassland-shrubland ecotone: an individual-based gap dynamics model of herbaceous and woody species. *Ecol. Model.* 152:5–32

Price MV, Waser NM. 1998. Effects of experimental warming on plant reproductive phenology in a subalpine meadow. *Ecology* 79:1261–71

Rathcke B, Lacey EP. 1985. Phenological patterns of terrestrial plants. *Annu. Rev. Ecol. Syst.* 16:179–214

Richardson D, Hollinger DY, Aber JD, Ollinger SV, Braswell BH. 2007. Environmental variation is directly responsible for short but not long-term variation in forest-atmosphere carbon exchange. *Glob. Change Biol.* 13:1–16

Rillig MC. 2004. Arbuscular mycorrhizae and terrestrial ecosystem processes. *Ecol. Lett.* 7:740–54

Roden JS, Ball MC. 1996. The effect of elevated [CO_2] on growth and photosynthesis of two eucalyptus species exposed to high temperatures and water deficits. *Physiol. Plant.* 111:909–19

Rustad LE, Campbell JL, Marion GM, Norby RJ, Mitchell MJ, et al. 2001. A meta-analysis of the response of soil respiration, net nitrogen mineralization, and aboveground plant growth to experimental ecosystem warming. *Oecologia* 126:543–62

Ryan MG. 1991. Effects of climate change on plant respiration. *Ecol. Appl.* 1:157–67

Ryan MG, Gower ST, Hubbard RM, Waring RH, Gholz HL, et al. 1995. Woody tissue maintenance respiration of four conifers in contrasting climates. *Oecologia* 101:133–40

Rykbost KA, Boersma L, Mack HJ, Schmisseur WE. 1975. Yield response of soil warming: agronomic crops. *Agronomy J.* 67:733–38

Saleska SR, Jarte J, Torn MS. 1999. The effect of experimental ecosystem warming on CO_2 fluxes in a montane meadow. *Glob. Change Biol.* 5:125–41

Saleska SR, Shaw MR, Fisher ML, Dunne JA, Still CJ, et al. 2002. Plant community composition mediates both large transient decline and predicted long-term recovery of soil carbon under climate warming. *Glob. Biogeochem. Cycles* 16:1055

Shaver GR, Canadell J, Chapin FS III, Gurevitch J, Henry G, et al. 2000. Global warming and terrestrial ecosystems, a conceptual framework for analysis. *BioScience* 50:871–82

Shaver GR, Chapin FS III, Gartner BL. 1986. Factors limiting seasonal growth and peak biomass accumulation in *Eriophorum vaginatum* in Alaskan tussock tundra. *J. Ecol.* 74:257–78

Shaver GR, Johnson LC, Cades DH, Murray G, Laundre JA, et al. 1998. Biomass and CO_2 flux in wet sedge tundras: response to nutrients, temperature, and light. *Ecol. Monogr.* 68:75–99

Shaw MR, Loik ME, Harte J. 2000. Gas exchange and water relations of two Rocky Mountain shrub species exposed to a climate change manipulation. *Plant Ecol.* 146:197–206

Sherry RA, Zhou X, Hui D, Gu S, Arnone JA III, et al. 2007. Divergence of reproductive phenology under climate warming. *Proc. Natl. Acad. Sci. USA* 104:198–202

Sitch S, Smith B, Prentice IC, Arneth A, Bondeau A, et al. 2003. Evaluation of ecosystem dynamics, plant geography and terrestrial carbon in the LPJ dynamic global vegetation model. *Glob. Change Biol.* 9:161–85

Six J, Conant RT, Paul EA, Paustian K. 2002. Stabilization mechanisms of soil organic matter: implications for C-saturation of soils. *Plant Soil* 241:155–76

Smith TM, Shugart HH. 1993. The transient response of terrestrial carbon storage to a perturbed climate. *Nature* 361:523–26

Starr G, Oberbauer SF, Pop EW. 2000. Effects of lengthened growing season and soil warming on the phenology and physiology of *Polygonum bistorta*. *Glob. Change Biol.* 6:357–69

Thompson SL, Govindasamy B, Mirin A, Caldeira K, Delire C, et al. 2004. Quantifying the effects of CO_2-fertilized vegetation on future global climate. *Geophys. Res. Lett.* 31:L23211

van Wijk MT, Williams M, Laundre JA, Shaver GR. 2003. Interannual variability of plant phenology in tussock tundra: modeling interactions of plant productivity, plant phenology, snowmelt and soil thaw. *Glob. Change Biol.* 9:743–58

Walker MD, Wahren CH, Hollister RD, Henry GHR, Ahlquist LE, et al. 2006. Plant community responses to experimental warming across the tundra biome. *Proc. Natl. Acad. Sci. USA* 103:1342–46

Walther GR, Post E, Convey P, Menzel A, Parmesan C, et al. 2002. Ecological responses to recent climate change. *Nature* 416:389–95

Wan S, Hui D, Wallace LL, Luo Y. 2005. Direct and indirect effects of experimental warming on ecosystem carbon processes in a tallgrass prairie. *Glob. Biogeochem. Cycles* 19:GB2014

Wan S, Luo Y, Wallace LL. 2002. Changes in microclimate induced by experimental warming and clipping in tallgrass prairie. *Glob. Change Biol.* 8:754–68

Weise DR, Fujioka FM, Nelson RM. 2005. A comparison of three models of 1-h time lag fuel moisture in Hawaii. *Agric. For. Meteorol.* 133:28–39

Welker JM, Fahnestock JT, Henry GHR, O'Dea KW, Chimner RA. 2004. CO_2 exchange in three Canadian high arctic ecosystems: response to long-term experimental warming. *Glob. Change Biol.* 10:1981–95

Weltzin JF, Bridgham SD, Pastor J, Chen J, Harth C. 2003. Potential effects of warming and drying on peatland plant community composition. *Glob. Change Biol.* 9:141–51

Weltzin JF, Pastor J, Harth C, Bridgham SD, Updegraff K, Chapin CT. 2000. Response of bog and fen plant communities to warming and water-table manipulations. *Ecology* 81:3464–78

White MA, Running SW, Thornton PE. 1999. The impact of growing-season length variability on carbon assimilation and evapotranspiration over 88 years in the eastern US deciduous forest. *Int. J. Biometeorol.* 42:139–45

White MA, Thornton PE, Running SW. 1997. A continental phenology model for monitoring vegetation responses to interannual climatic variability. *Glob. Biogeochem. Cycles* 11:217–34

Wookey PA, Welker JM, Parsons AN, Press MC, Callaghan TV, Lee JA. 1994. Differential growth, allocation and photosynthetic responses of *Polygonum viviparum* to simulated environmental change at a high arctic polar semidesert. *Oikos* 70:131–39

Xu T, White L, Hui D, Luo Y. 2006. Probabilistic inversion of a terrestrial ecosystem model: analysis of uncertainty in parameter estimation and model prediction. *Glob. Biogeochem. Cycles* 20:GB2007

Zavaleta ES, Thomas BD, Chiariello NR, Asner GP, Shaw MR, Field CB. 2003. Plants reverse warming effect on ecosystem water balance. *Proc. Natl. Acad. Sci. USA* 100:9892–93

Zeng N, Qian H, Munoz E, Iacono R. 2004. How strong is carbon cycle-climate feedback under global warming? *Geophys. Res. Lett.* 31:L20203

Zhang D. 2002. *Stochastic Methods for Flow in Porous Media: Coping with Uncertainties.* San Diego: Academic

Zhang W, Parker K, Luo Y, Wan S, Wallace LL, Hu S. 2005. Soil microbial responses to experimental warming and clipping in a tallgrass prairie. *Glob. Change Biol.* 11:266–77

Zhou X, Liu X, Wallace LL, Luo Y. 2007a. Photosynthetic and respiratory acclimation to experimental warming for four species in a tallgrass prairie ecosystem. *J. Integr. Plant Biol.* 49:270–81

Zhou X, Wan S, Luo Y. 2007b. Source components and interannual variability of soil CO_2 efflux under experimental warming and clipping in a grassland ecosystem. *Glob. Change Biol.* 13:761–75

Ziska LH, Bunce JA. 1998. The influence of increasing growth temperature and CO_2 concentration on the ratio of respiration to photosynthesis in whole plants of soybean. *Glob. Change Biol.* 4:637–43

Shortcuts for Biodiversity Conservation Planning: The Effectiveness of Surrogates

Ana S.L. Rodrigues[1] and Thomas M. Brooks[2,3,4]

[1] Department of Zoology, University of Cambridge, Cambridge CB2 3EJ, United Kingdom; email: aslr2@cam.ac.uk

[2] Center for Applied Biodiversity Science, Conservation International, Arlington, Virginia 22202; email: t.brooks@conservation.org

[3] World Agroforestry Center (ICRAF), University of the Philippines Los Baños, Laguna 4031, Philippines

[4] School of Geography and Environmental Studies, University of Tasmania, Hobart TAS 7001, Australia

Annu. Rev. Ecol. Evol. Syst. 2007. 38:713–37

The *Annual Review of Ecology, Evolution, and Systematics* is online at
http://ecolsys.annualreviews.org

This article's doi:
10.1146/annurev.ecolsys.38.091206.095737

Key Words

complementarity, indicators, representation, Species Accumulation Index, surrogacy

Abstract

Biodiversity is not completely known anywhere, so conservation planning is always based on surrogates for which data are available and, hence, assumed effective for the conservation of unknown biodiversity. We review the literature on the effectiveness of surrogates for conservation planning based on complementary representation. We apply a standardized approach based on a Species Accumulation Index of surrogate effectiveness to compare results from 575 tests in 27 studies. Overall, we find positive, but relatively weak, surrogacy power. Cross-taxon surrogates are substantially more effective than surrogates based on environmental data. Within cross-taxon tests, surrogacy was higher for tests within the same realm (terrestrial, marine, freshwater). Surrogacy was higher when extrapolated (rather than field) data were used. Our results suggest that practical conservation planning based on data for well-known taxonomic groups can cautiously proceed under the assumption that it captures species in less well-known taxa, at least within the same realm.

INTRODUCTION

Surrogacy: extent to which conservation planning based on a particular set of biodiversity features (surrogates) effectively represents another set (targets)

The Need for Surrogates in Conservation Planning

Rates of biodiversity loss are accelerating (Pimm et al. 1995) as human dominance of Earth's natural systems increases (Vitousek et al. 1997). The most effective and least expensive way of preserving biodiversity is by maintaining native species in their habitats where there is the greatest chance of success in ensuring long-term conservation. Accordingly, protected areas are recognized as key components of national and international conservation strategies (SCBD 2005). Resources (e.g., land, money) available for conservation are limited, so their allocation must be planned strategically. This recognition prompted the development of complementarity-based methods of systematic conservation planning aimed at ensuring efficient use of scarce conservation resources (Margules & Pressey 2000, Pressey et al. 1993).

Although conservation planning methods have developed quickly, a factor limiting their application is the availability of adequate data (Prendergast et al. 1999). Conservation planning relies fundamentally on spatial information about the distribution of biodiversity (Margules & Pressey 2000), which is still very limited. Existing data are heavily biased against the regions of the world that have the most biodiversity, the least developed protected area networks, and the highest levels of biodiversity loss—precisely those regions most in need of conservation planning (Pimm 2000). Moreover, though some regions (e.g., the British Isles; Thomas et al. 2004) and some taxa (e.g., birds; del Hoyo et al. 1992–2006) benefit from very detailed data, we do not have perfect knowledge of biodiversity and its distribution for any part of the world or any taxonomic group.

Conservation planning is therefore necessarily based on those surrogates for biodiversity for which data can be obtained. Surrogates usually fall into two categories: better-known taxa, typically birds or mammals; and features such as vegetation types or other land classes that can be easily mapped, for example, using remote sensing. Planners assume that conservation efforts made to protect the biodiversity features explicitly incorporated in the conservation planning exercise are also effective for the conservation of unmapped biodiversity (e.g., Faith et al. 2001). That is, they hope that the known biodiversity is a good surrogate for the unknown.

The Need for a Review of Surrogacy Effectiveness

Different surrogates are unlikely to be equally powerful in representing broader biodiversity. It is not surprising that different studies have found strikingly different results in surrogacy effectiveness (e.g., Araújo et al. 2001, Warman et al. 2004a). Understanding which surrogates are more effective for conservation planning would allow for better use of the available information and provide precious guidance to future data acquisition efforts.

Testing the effectiveness of a particular surrogate in representing overall biodiversity is, strictly speaking, impossible because it would require complete knowledge of biodiversity in the corresponding region. Hence, all real-life tests of surrogacy are, in fact, tests of whether one subset of biodiversity (hereinafter, the surrogate; e.g., mammals) is a good surrogate for another (hereinafter, the target; e.g., plants).

Testing the surrogate efficiency of mammals in representing plants in a given region requires data on both taxa—but if such plant data are available, then they can and should be used directly to plan for plant conservation, irrespective of whether or not mammals turn out to be good surrogates. Far from rendering surrogacy tests useless, however, this means that many surrogacy tests are needed. Indeed, in order to identify surrogates that can be used when they are most critically needed (i.e., in situations of poor data), we must extrapolate from the results of a diversity of tests in regions for which better data are available. A single positive or negative surrogacy result is not a useful basis for generalization because results are likely to vary geographically and taxonomically (e.g., Ferrier & Watson 1997).

Dozens of studies have been published in the past few years with hundreds of tests of biodiversity surrogacy in conservation planning. Some are often quoted as evidence that surrogates work (e.g., Howard et al. 1998), while others are claimed as evidence that they do not (e.g., van Jaarsveld et al. 1998); but the full set of studies has never been analyzed collectively (although see Favreau et al. 2006).

Scope of This Review

We reviewed the literature on the effectiveness of biodiversity surrogates looking for patterns in the results to advance the understanding of whether surrogates are useful and which types of surrogates are more likely to be more useful than others. We focused specifically on studies using complementary reserve selection, that is, studies selecting complementary sets of sites based on data for surrogates, and then evaluating their effectiveness in representing targets. Unlike Favreau et al. (2006), we excluded studies that followed other approaches, for example, those testing correlations between species richness across different taxa.

What we term here as surrogates are frequently referred to as indicators in the literature (e.g., Bonn et al. 2002, Gladstone 2002, Howard et al. 1998, Ryti 1992). Indicators, however, commonly refer to measures of the state of the environment or of environmental change (Lawton & Gaston 2001), which we do not address. The term surrogates as used here is not restricted to a small number of species that are frequently the focus of conservation action (flagship, focal, indicator, keystone, and umbrella species; Favreau et al. 2006), but refers to any set of biodiversity features used to guide conservation planning with the expectation of conserving broader biodiversity.

Conservation planning methods based on complementary representation have been developed over the past two decades (Kirkpatrick 1983, Margules & Pressey 2000, Pressey et al. 1993) and are more efficient in identifying networks of areas representing a diversity of biological features (e.g., networks that represent all species in a region) than methods that prioritize areas by ranking them based on variables such as species richness (e.g., Margules et al. 1988, Williams et al. 1996). We therefore focus on complementary representation, excluding studies that followed other approaches for prioritizing areas for conservation. Accordingly, we exclude the plethora of studies

Complementary representation: approach to systematic conservation planning for selecting sets of sites that between them represent all targeted biodiversity features testing predictors of species richness (e.g., Flather et al. 1997, Gaston & Blackburn 1995, Hess et al. 2006, Negi & Gadgil 2002, Pearson & Carroll 1998, Pharo et al. 2000, Vessby et al. 2002). We also exclude studies investigating whether land units defined according to similarities/dissimilarities in certain types of attributes (e.g., environmental characteristics, or species composition of surrogate taxa) show matching patterns of similarity/dissimilarity in terms of species composition of target taxa (e.g., Mac Nally et al. 2002, Oliver et al. 2004, Pharo & Beattie 2001, Su et al. 2004; for example, Su et al. investigated whether pairs of sites with similar bird communities also had similar butterfly communities). Finally, given the need to standardize comparisons, we exclude analyses that assessed the surrogacy value of sets of regional (e.g., Burgess et al. 2002, Grenyer et al. 2006) or site (e.g., Brooks et al. 2001) biodiversity conservation priorities that were selected using systematic conservation planning techniques other than complementary representation.

METHODS FOR EVALUATING AND COMPARING SURROGACY

Different Measures of Surrogacy Value

Published studies of surrogacy value follow three main methods: (*a*) comparing the spatial overlap between complementary sets of sites selected using data for surrogates and for targets (e.g., Grenyer et al. 2006, Reyers et al. 2000, Saetersdal et al. 1993, van Jaarsveld et al. 1998); (*b*) measuring the correspondence between the sequence of complementary site selection for surrogates and targets (e.g., Gaston et al. 1995, Polasky et al. 2001); and (*c*) investigating the effectiveness of sets of sites based on surrogate data in representing the target taxa (e.g., Grenyer et al. 2006, Lawler et al. 2003, Tognelli et al. 2005), also called incidental representation (Warman et al. 2004a). Two other methods have been recently proposed for measuring surrogacy: spatial patterns of site irreplaceability (Ferrier et al. 2000) and complementarity between targets and surrogates (Warman et al. 2004b; Williams et al. 2006), but these are too recent to have been applied elsewhere. These different measures often produce quite different conclusions for any given dataset (e.g., Grenyer et al. 2006, Hopkinson et al. 2001, Reyers & van Jaarsveld 2000).

The first two approaches are based on comparisons between the spatial location of the sites selected using surrogates and target features. However, distinctiveness in spatial pattern is not proof of low surrogacy value. For example, complementary representation problems typically have multiple solutions (e.g., Arthur et al. 1997, Hopkinson et al. 2001), identical in terms of representation of the biodiversity features being analyzed but often very different from each other spatially (e.g., Rodrigues & Gaston 2002). If surrogacy value was based on measures of spatial similarity in such cases, one could reach the nonsensical conclusion that a taxon is a bad surrogate for itself. The relevant question in a surrogacy test is, therefore, what is the extent to which areas selected for surrogates capture the target features (Balmford 1998, Faith et al. 2001)? We therefore consider in this review only those studies that applied this third method.

Even within studies that measured surrogacy as the extent to which areas selected for surrogates capture the target features, different ways of presenting the results pose challenges to interpretation and subsequent comparability across studies (Favreau et al. 2006). To understand these challenges, we describe what we consider to be the ideal approach for presenting and measuring results of a surrogacy test: Ferrier & Watson's Species Accumulation Index (*SAI*) of surrogate efficiency (1997; see also Ferrier 2002).

Species Accumulation Index (*SAI*): quantitative measure of surrogacy value, relating observed surrogacy to the randomly expected and the maximum possible surrogacy

Species Accumulation Index of Surrogate Efficiency

Considering again a test of whether mammals are good surrogates for plants, surrogacy value can be measured by evaluating how well a set of sites selected to maximize the representation of mammals performs in representing plants. When calculated across a diversity of sets of sites measuring different total areas, this produces a surrogate curve (**Figure 1**). The position of this curve on its own is not informative; rather, it must be compared with two reference curves: the optimal curve, indicating the maximum possible representation of plants in sets of sites of particular areas, and a random curve, indicating what the expected representation of plants would be were sites selected randomly. The optimal curve is equivalent to the best possible surrogacy value, in this case a surrogate as effective as using plant data directly. The random curve indicates the representation expected on average in the absence of biological data; in fact, it is a measure of the effectiveness of using area as a surrogate, because the larger the area selected the more target species are predicted to be captured as a result of the species-area relationship (Rosenzweig 1995). The three curves necessarily coincide at 0% (no target species represented) and 100% of total area (all target species represented).

We can assess surrogacy value qualitatively through visual comparison of these three curves (**Figure 1**). The closer the surrogacy curve is to the optimal, the higher the surrogacy value; maximum surrogacy is obtained when they coincide (by definition, no values can be found above the optimal curve). Null surrogacy is obtained when the surrogate and the random curves coincide on average; and negative surrogacy is obtained when the surrogate curve falls below the random curve.

A quantitative measure of surrogacy value can be obtained as $SAI = (S - R)/(O - R)$, where S is the area under the surrogate curve, R is the area under the random curve, and O is the area under the optimal curve (Ferrier 2002). *SAI* equals one when the optimal and surrogate curves coincide (perfect surrogacy), is between one and zero when the surrogate curve is mainly above the random curves (positive surrogacy value), is zero when the surrogate and random curves coincide on average (zero surrogacy), and is negative when the surrogate curve is mainly below the random selection (negative surrogacy value).

In fact, there is no single random curve, but one for each combination of random sites selected. The random curve is therefore best represented as a "random band"—with the mean random curve in the middle—limited by a measure of dispersal such as the limits of the 95% confidence interval (**Figure 1**). Surrogacy is not significantly different from random (hereinafter, null surrogacy) if, on average, the surrogate curve falls within this random band.

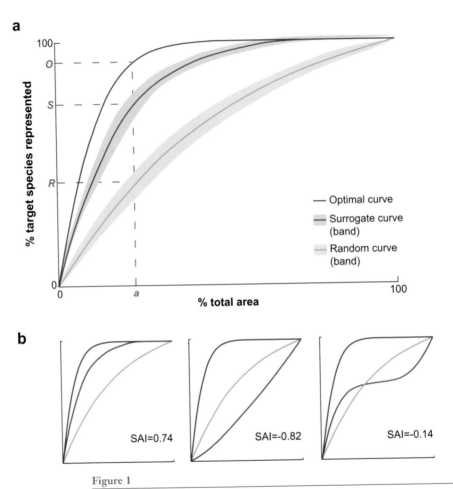

Figure 1

Measuring surrogacy value. (*a*) Surrogacy can be assessed by comparing: a surrogate curve (or band), indicating the average percentage of target species represented when selecting conservation areas using surrogate data; an optimal curve, maximizing the number of target species represented; and a random curve (or band), indicating the average percentage of target species represented when sites are selected randomly. A quantitative comparison can be made by calculating a Species Accumulation Index (*SAI*) of surrogate efficiency (an approach for presenting and measuring results of a surrogacy test). A point estimate of *SAI* can be calculated for a particular area *a* by comparing: a surrogate point, *S*, the average percentage of target species represented in a set of sites of total area *a* selected to maximize surrogate representation; a random point, *R*, the average percentage of target species represented in randomly selected sets of total area *a*; and, an optimal point, *O*, the maximum percentage of target species that can be represented in area *a*. (*b*) Some of the possible relative positions between the optimal, surrogate, and random curves and respective values of *SAI*.

Strictly speaking, there is no single surrogate curve either. As discussed above, there are typically many solutions to the same complementary representation problem, and thus often many equivalent sets of sites maximizing the representation of surrogates in a given area. These sets of sites likely vary in their representation of targets (i.e., in their surrogacy value). For this reason, the surrogacy curve is best defined as the average across a number of optimal solutions—ideally all of them—rather than from one single solution (Rodrigues et al. 2005). As with the random curve, this can also be represented as a band with confidence intervals or a band between the extreme values of surrogacy encountered for each area. To present only the best (e.g., Hopkinson et al. 2001) or worst surrogacy value is not particularly useful.

It is important to stress that a random curve is not the same as a worst-case scenario: a random selection of sites is likely to sample the environmental and biological variation of a region in a reasonably unbiased way, as predicted from basic sampling theory (Sutherland 2006). Results of null surrogacy should be interpreted accordingly—not as bad, but neutral. Current strategies for creating protected-area networks are often less representative than random, as they are typically biased toward particular habitats or regions (e.g., unfertile areas; Pressey et al. 1996).

Review of Published Surrogacy Analyses

We reviewed the literature for studies that evaluated surrogacy value in conservation planning, as defined above. Very few studies included all the information described above as ideal, but we tried to extract as much data as possible from the published analyses by including all studies that presented at least information on the position of a surrogacy curve (or point) and corresponding random curve (or point). For these, at least a qualitative evaluation of surrogacy value was possible; if data on optimal representation were also available, then a quantitative evaluation (calculation of *SAI*) was possible as well. The **Supplemental Appendix** (follow the Supplemental Material link from the Annual Reviews home page at **http://www.annualreviews.org/**) details how we dealt with studies for which less-than-ideal information was provided.

We found 27 studies that met our criteria for inclusion, which together reported 575 surrogacy tests. Of these, 464 tests from 16 studies allowed for a calculation of *SAI*. These are summarized in **Table 1** and comprise two main types of tests: tests of complementary cross-taxon surrogacy, which evaluate the effectiveness of a complementary set of sites selected to represent a particular set of species as a surrogate for representing another set of species; and tests of environmental surrogacy, which evaluate the effectiveness of environmental data (species assemblages or abiotic data) as a surrogate for representing species.

Within cross-taxon tests, we distinguished among tests in which surrogates and targets are nonoverlapping groups, surrogates are subsets of targets, and targets and surrogates are partially overlapping groups. We can predict that the second and third types of tests perform better than the first, as they correspond to situations where surrogates include some information regarding the spatial distribution of targets.

Complementary cross-taxon surrogacy: extent to which conservation planning based on complementary representation of species surrogates effectively represents target species

Environmental surrogacy: extent to which conservation planning based on environmental data effectively represents species as targets

Table 1   Overview of studies that analyzed the performance of surrogates for biodiversity conservation planning based on complementary representation

Reference	Region; selection units	Surrogates	Targets	Surrogacy[1]		
				+	0	−
Cross-taxon surrogacy: Studies in which surrogate and target species are nonoverlapping groups						
Dobson et al. 1997	USA; counties	Plants; molluscs; arthropods; fishes; herptiles; birds; mammals	Set of all species (924) except the surrogate group being tested	6	1	
Virolainen et al. 2000	Finland & Sweden; forest plots	Heteroptera; polypores; vascular plants; beetles	Heteroptera; polypores; vascular plants; beetles	6	1	5
Garson et al. 2002	Canada; 0.2° × 0.2° cells	Breeding birds	Plants at risk + animals at risk	1		
Lund & Rahbek 2002	Denmark; 100 km² cells	Bats; butterflies; large moths; birds; herptiles; click beetles	Set of all species except the surrogate group being tested	3	3	
Beger et al. 2003	Papua New Guinea; reef platforms	Fishes; corals	Fishes; corals	2		
Lawler et al. 2003	USA; ~650 km² hexagonal cells	Freshwater fishes; birds; freshwater mussels; amphibians; mammals; reptiles	Freshwater fishes; birds; freshwater mussels; amphibians; mammals; reptiles	42	15	3
Moore et al. 2003	Africa; degree cells	Mammals; birds; snakes; amphibians; threatened mammals; threatened birds	Threatened birds; threatened mammals	8		
Araújo et al. 2004	Europe; 50 × 50 km cells	Plants; mammals; birds; amphibians + reptiles	Plants; mammals; birds; amphibians + reptiles	8		4
Sauberer et al. 2004	Austria; 600 m × 600 m plots	Bryophytes; vascular plants; gastropods; spiders; orthopterans; carabid beetles; ants; birds	Set of all species except the surrogate group being tested	8		
Warman et al. 2004a	Canada; 10,000 km² hexagons	Mammals; birds; amphibians; reptiles	Mammals; birds; amphibians; reptiles	12		
Chiarucci et al. 2005	Italy; forest plots	Vascular plants; woody plants	Fungi		2	
Tognelli et al. 2005	Chile; 0.5° coastal latitudinal bands	Coastal marine mammals; coastal marine birds; coastal marine reptiles; coastal marine teleost fishes	Coastal marine mammals; coastal marine birds; coastal marine reptiles; coastal marine teleost fishes	9	1	2
McKenna 2006	Indonesia; reef sites	Scleractinian corals; fishes; molluscs	Scleractinian corals; fishes; molluscs	2	3	1

Table 1 *(Continued)*

Reference	Region; selection units	Surrogates	Targets	Surrogacy[1]		
				+	0	–
Cross-taxon surrogacy: Studies in which surrogates are a subgroup of targets						
Game & Peterken 1984	UK; woods	Rare woodland herbs	All woodland herbs	1		
Williams et al. 2000a	Africa; degree cells	Large mammal species	Mammals	1		
Williams et al. 2000b	Africa; degree cells	"Flagship" species; "Big Five" species; large mammals	Mammals + breeding birds	1		2
Bonn et al. 2002	South Africa; 0.25° cells	Threatened birds; endemic birds	Bird species	2	1	
Gladstone 2002	Australia; rocky shore locations	Macroalgae; molluscs	Rocky shore macroscopic species	1	1	
Lund 2002	Denmark; 100 km^2 grid cells	Listed species (EC Habitats Directive): Annex II and IV	Mosses + orchids + water beetles + click beetles + moths + butterflies + bats + hoverflies + amphibians + birds + reptiles	2		
Beger et al. 2003	Papua New Guinea; islands	Corals; fishes	Fishes + corals	2		
Moore et al. 2003	Africa; degree cells	Birds; mammals; snakes; amphibians; threatened mammals; threatened birds	Birds + mammals + snakes + amphibians	6		
Saetersdal et al. 2003	Norway; 0.25 ha sample plots	Vascular plants	Vascular plants + bryophytes + macrolichens + polyphores fungi + carabid beetles + staphylinid beetles + spiders + snails	1		
Kati et al. 2004	Greece; sites	Woody plants; orchids; orthopterans; amphibians; freshwater turtles; lizards; terrestrial tortoises; small terrestrial birds	Woody plants + orchids + orthopterans + amphibians + freshwater turtles + lizards + terrestrial tortoises + small terrestrial birds	6		
Tognelli et al. 2005	Chile; 0.5° coastal latitudinal bands	Coastal marine species: endemics; threatened; mammals; birds; reptiles; teleost fishes	Coastal marine vertebrates	5		1

(Continued)

Table 1 (*Continued*)

Reference	Region; selection units	Surrogates	Targets	Surrogacy[1] +	0	–
Tognelli 2005	South America; 100 × 100 km cells	Mammals: threatened; rare; flagship; large	Terrestrial, nonvolant mammals	2	2	
Grenyer et al. 2006	Global; ~9274 km² cells	Birds; mammals; amphibians	Terrestrial vertebrates	3		
Cross-taxon surrogacy—Studies in which surrogates and targets are partially overlapping groups						
Lund 2002	Denmark; 100 km² cells	Listed species (EC Habitats Directive) Annex II; Annexes II+IV	Red list species of: mosses + orchids + water beetles + click beetles + moths + butterflies + bats + hoverflies + amphibians + birds + reptiles	2		
Lawler et al. 2003	USA; ~650 km² hexagonal cells	At-risk species of: freshwater fishes + birds + freshwater mussels + amphibians + mammals + reptiles	Freshwater fishes; birds; freshwater mussels; amphibians; mammals; reptiles	6		
Warman et al. 2004a	Canada; 10,000 km² hexagons	Mammals; birds; amphibians; reptiles; listed species (mammals + birds + amphibians + reptiles + fishes + plants + molluscs + lepidoptera)	Mammals; birds; amphibians; reptiles; listed species (mammals + birds + amphibians + reptiles + fishes + plants + molluscs + lepidoptera)	8		
Tognelli et al. 2005	Chile; 0.5° coastal latitudinal bands	Coastal marine species: endemics; threatened; mammals; birds; reptiles; teleost fishes	Coastal marine species: endemics; threatened; mammals; birds; reptiles; teleost fishes	16	1	1
Grenyer et al. 2006	Global; ~9274 km² cells	Birds; mammals; amphibians	Rare terrestrial vertebrates; threatened terrestrial vertebrates	6		
Environmental surrogacy—Surrogates based on species assemblages						
Ferrier & Watson 1997	Australia; survey sites	Forest types; vegetation systems	Ants; beetles; spiders; birds; reptiles; bats; rainforest canopy trees; plants	69	11	0
Williams et al. 2000b	Africa; degree cells	Ecoregions	Mammals + breeding birds	1		
Araújo et al. 2004	Europe; 50 × 50 km cells	Assemblage diversity of: plants; breeding birds; mammals; amphibians + reptiles	Plants; breeding birds; mammals; amphibians + reptiles	6	14	52
Trakhtenbrot & Kadmon 2005	Israel; 5 × 5 km cells	Floristic regions	Plants; rare plants	2		

Table 1   (*Continued*)

Reference	Region; selection units	Surrogates	Targets	Surrogacy[1]		
				+	0	–
Environmental surrogacy: Surrogates based on a mix of abiotic and species data						
Ferrier & Watson 1997	Australia; survey sites	Species modelling; canonical ordination	Ants; beetles; spiders; birds; reptiles; bats; plants	21	7	1
Environmental surrogacy: Surrogates based on purely abiotic data						
Ferrier & Watson 1997	Australia; survey sites	Environmental classification; environmental ordination	Ants; beetles; spiders; birds; reptiles; bats; plants	60	75	20
Araújo et al. 2001	Europe; 50 × 50 km cells	Environmental diversity	Plants; breeding birds; mammals; amphibians; reptiles	3	1	1
Bonn & Gaston 2005	South Africa; 0.25° cells	Environmental diversity	Birds			1
Trakhtenbrot & Kadmon 2005	Israel; 5 × 5 km cells	Environmental clusters	Plants; rare plants	2		

[1] Tests were classified into those indicating positive (+), null (0), or negative (–) surrogacy value. See the **Supplemental Appendix** for more details.

We excluded tests in which targets are a subset of surrogates [e.g., Moore et al.'s (2003) test of whether "all birds" are a good surrogate for "threatened birds"] or where data on one taxon is used to create assemblages that are then tested as surrogates for the same taxon (e.g., plant assemblages as surrogates of plants; Araújo et al. 2004). These are irrelevant in practical terms because in these cases data are available for all target species, thus making it possible to optimize site selection for them directly without the need for surrogates.

ANALYSIS OF SURROGACY TESTS

Although the number of surrogacy tests that we found (575) was large, they come from only 27 studies and 19 countries or regions. Moreover, most of the tests come from just three studies (Araújo et al. 2004, Ferrier & Watson 1997, Lawler et al. 2003). The same datasetes are often used for a diversity of tests, sometimes in different studies (e.g., Bonn & Gaston 2005, Bonn et al. 2002). This means that the tests compiled here are not independent. Furthermore, not all the studies provide sufficient information for a quantitative evaluation of surrogacy value; several do not allow assessment of statistical significance, most use heuristic approximations rather than optimization, and the great majority are from species-poor parts of the world (**Table 1**; see also **Supplemental Table 1**).

Given these limitations, we feel that there is not yet enough information to yield detailed patterns regarding which specific surrogates are more effective under which conditions. We therefore limited our analysis to try to answer the broader question of whether the available evidence supports the value of surrogates in conservation planning and in contrasting the effectiveness of cross-taxon and environmental surrogates.

Overall Effectiveness of Biodiversity Surrogates

Of the 575 tests reviewed, we found positive surrogacy in 59%, whereas across the 446 tests for which surrogacy was quantified, the median *SAI* value was 0.12 (**Table 2**). Overall, this indicates a tendency for a positive, but weak, surrogacy value.

Complementary Cross-Taxon Surrogacy

Complementary cross-taxon surrogacy examines the extent to which conservation planning that maximizes representation for a set of surrogate species is effective in representing a set of target species. We found high levels of surrogacy in tests of complementary cross-taxon surrogacy, both in terms of frequency of positive results and in terms of *SAI* values (**Table 2**).

As predicted, we found that the levels of surrogacy were significantly lower [$\chi^2 = 9.1$, d.f. $= 1$; $p \leq 0.01$] in tests in which surrogates and targets are nonoverlapping groups, and higher in tests in which surrogates are either a subgroup of the targets or when there is partial overlap between the two groups (**Table 2**). This effect is also noticeable in the frequency distribution of *SAI* values (for the tests from which this can be derived) (**Figure 2*a***).

Most of the complementary cross-taxon surrogacy tests reviewed were restricted to terrestrial taxa (**Table 2**, **Figure 2*b***). Only four studies referred to surrogacy analysis within the marine realm (Beger et al. 2003, Gladstone 2002, McKenna 2006, Tognelli et al. 2005), and most of the 48 tests conducted among them found positive surrogacy. Only one study (Lawler et al. 2003), including 4 tests, considered cross-taxon surrogacy within freshwater; all 4 tests showed positive surrogacy. High levels of surrogacy were also found within the terrestrial realm (**Table 2**, **Figure 2*b***). Throughout, we considered amphibians to be terrestrial, but similar results are obtained if they are classified as freshwater species (see the **Supplemental Appendix**). Our results tentatively suggest that cross-taxon surrogacy may be stronger in freshwater than in terrestrial systems and weaker in the marine realm (**Figure 2*b***). It would, however, be premature to draw strong conclusions from such small sample sizes.

One study included tests of freshwater taxa as surrogates and/or as targets relative to terrestrial species (Lawler et al. 2003). The frequency of tests that yielded positive surrogacy was significantly higher within realms than across realms ($\chi^2 = 5.98$, d.f. $= 1$; $p \leq 0.05$). Median values of *SAI* were also higher within realms (**Table 2**, **Figure 2*c***), but not significantly so (K-S test, $D = 0.19$). There is therefore some evidence that surrogacy is better within than across realms, which is as expected (Reid 1998).

Although these results provide reasons for optimism in the use of complementary cross-taxon surrogates in conservation planning, they raise more questions than they

Table 2 Comparison of surrogacy levels for different types of studies

| Type of test | Qualitative assessment[1] | | | | | Quantitative assessment[2] | | |
| | No. of studies | No. of tests | Surrogacy (%) | | | No. of studies | No. of tests | Median *SAI* (Q1, Q3) |
			+	0	–			
All tests	27	575	59	24	17	16	464	0.12 (0.03, 0.28)
Complementary cross-taxon surrogacy								
All cross-taxon	23	228	78	14	8	13	121	0.41 (0.10, 0.56)
Nonoverlapping groups	13	148	72	18	10	5	90	0.35 (0.09, 0.53)
Surrogates as subgroups of targets	13	40	83	10	8	8	23	0.52 (0.18, 0.72)
Partial overlap between surrogates and targets	5	40	95	3	3	2	8	0.49 (0.38, 0.59)
Within the marine realm	4	48	75	15	10	3	12	0.19 (0.04, 0.52)
Within the freshwater realm	1	4	100	0	0	1	2	0.72 (0.69, 0.75)
Within the terrestrial realm	16	127	80	10	9	9	74	0.44 (0.10, 0.58)
Freshwater surrogates, terrestrial targets	1	16	50	44	6	1	16	0.17 (0.07, 0.29)
Terrestrial surrogates, freshwater targets	1	16	69	25	6	1	8	0.38 (0.17, 0.52)
Vertebrate surrogates, nonvertebrate targets	5	17	71	24	6	4	11	0.26 (0.07, 0.43)
Vertebrate surrogates and targets	10	128	80	11	9	7	67	0.46 (0.17, 0.58)
Bird surrogates, nonbird targets	11	33	88	6	6	3	14	0.33 (0.15, 0.45)
Threatened/listed species surrogates	6	20	95	5	0	4	10	0.57 (0.41, 0.71)
Threatened/listed species targets	8	40	77	15	8	3	30	0.52 (0.22, 0.64)
Environmental surrogates								
All environmental	6	347	47	31	22	5	343	0.08 (–0.07, 0.20)
Assemblages	4	155	50	16	34	3	153	0.08 (–0.27, 0.22)
Abiotic + species data	1	29	69	28	3	1	29	0.19 (0.09, 0.36)
Abiotic surrogates	4	163	40	45	15	3	161	0.06 (–0.03, 0.15)
Type of data (surrogates-targets)								
Field-field	15	147	40	18	42	13	115	–0.18 (–0.38, 0.09)
Extrapolated-extrapolated	8	154	81	13	6	5	271	0.44 (0.19, 0.58)
Extrapolated-field	4	274	57	34	9	5	78	0.13 (0.02, 0.23)

[1] Surrogacy was qualitatively assessed by classifying tests into those indicating positive (+), null (0), or negative (–) surrogacy value.
[2] Surrogacy was quantitatively evaluated by measuring the median, first quartile (Q1), and third quartile (Q3) of the Species Accumulation Index (*SAI*) of surrogacy value.

answer. The samples are far too small to provide clear insights into some of the obvious questions, such as whether vertebrates are good surrogates for nonvertebrates. Indeed, we only found 17 such tests, indicating a frequent but not particularly strong surrogacy. In contrast, 128 tests were found in which vertebrates are both surrogates and targets and, as expected, levels of surrogacy are relatively high in this case. As to whether birds are adequate surrogates for nonbirds, the results are encouraging, but again derived from a rather small sample (**Table 2**).

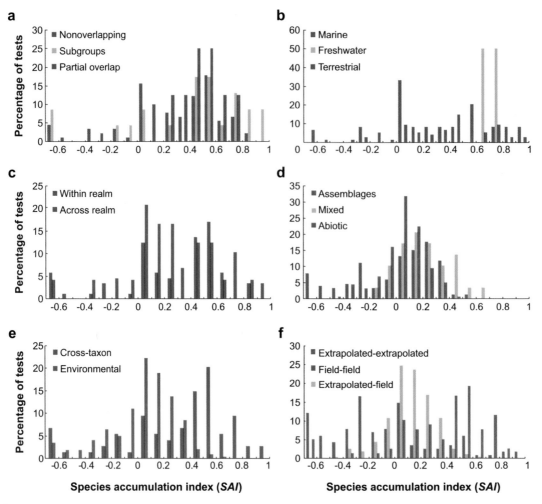

Figure 2

Frequency distribution of *SAI* values for (*a*) tests of cross-taxon surrogacy when surrogates and targets have variable degrees of overlap, (*b*) tests of cross-taxon surrogacy within the marine, freshwater, and terrestrial realms (amphibians considered terrestrial taxa), (*c*) tests of cross-taxon surrogacy within and across realms, (*d*) tests of different types of environmental surrogates, (*e*) comparison between tests on terrestrial cross-taxon and on environmental surrogates, and (*f*) studies with different data sources. The first bar includes all studies with *SAI* values lower than –0.6.

Another relevant question is whether conservation planning focused on species identified as priorities for conservation (listed as threatened or at-risk) is an effective surrogate for a broader set of species. Conversely, it is pertinent to ask if those priority species themselves are well represented by other surrogates. The few tests found that addressed these questions provide some evidence for positive surrogacy in both circumstances (**Table 2**).

Environmental Surrogacy

The surrogacy tests analyzed here examined the extent to which conservation planning based on contextual data such as species assemblages or abiotic information (hereinafter broadly called environmental data) is effective in representing a set of target species. We found 347 such tests, in six studies: nearly half of the tests indicated positive surrogacy, while the median *SAI* value was positive but close to zero (**Table 2**). However, all but 11 of these tests come from two studies (Araújo et al. 2004, Ferrier & Watson 1997), which should be kept in mind in interpreting the results of this analysis.

There are many possible approaches for using environmental information in conservation planning; Ferrier & Watson (1997) alone tested nearly 30 different variations. These variations can be loosely grouped based on the extent to which they are more or less based on abiotic information or integrate some level of biological information. We considered three general types of environmental surrogates:

1) Those based on biological assemblages or communities, including vegetation systems and forest types (Ferrier & Watson 1997), ecoregions (Williams et al. 2000b), floristic provinces (Trakhtenbrot & Kadmon 2005), and assemblage diversity (Araújo et al. 2004).

2) Those in which abiotic data are combined with species' raw distributional data, including canonical ordination, modeling of species' distributions, and modeling of biological distance (Ferrier & Watson 1997).

3) Those based on abiotic environmental data (such as geology, climate, and altitude), including environmental classification; environmental ordination; raw environmental distance; methods based on geology, land systems, land units, and landforms (Ferrier & Watson 1997); environmental diversity (Araújo et al. 2001, Bonn & Gaston 2005); and environmental cluster analysis (Trakhtenbrot & Kadmon 2005).

When considering these three types of methods separately, those tests that are based on extrapolations of species locality data using abiotic data information perform somewhat better than those based purely on abiotic variables or on biological assemblages (**Table 2**). This distinction is significant in terms of percentage of positive results ($\chi^2 = 6.14$, d.f. $= 1$, $p \leq 0.05$) as well as in terms of *SAI* values (K-S test, $D = 0.34$, $p \leq 0.01$), and is also visible on a graph of the frequency distribution of *SAI* in each case (**Figure 2d**).

Comparing Complementary Cross-Taxon Surrogacy and Environmental Surrogacy

Given that all environmental tests are within the terrestrial realm, we restricted comparison of them to terrestrial cross-taxon tests. Tests of complementary cross-taxon surrogacy had a significantly higher frequency of positive results ($\chi^2 = 41.9$, d.f. $= 1$; $p \leq 0.001$; **Table 2**) as well as significantly higher values of *SAI* (K-S test, $D = 0.56$, $p \leq 0.001$; **Table 2**). A noticeably better performance of cross-taxon surrogates in relation to environmental surrogates is also clear from the frequency distribution of values of *SAI* (**Figure 2e**); the former peaks between 0.5 and 0.6, whereas the latter peaks between 0 and 0.1. This suggests that environmental surrogates have lower surrogacy power than cross-taxon surrogates for biodiversity conservation.

Effect of Data Source on Surrogacy Levels

Another way in which the studies reviewed can be classified is according to the source of data for surrogates and targets. Although a wide diversity of data collection methods has been employed, the methods fall into two broad classes:

1. Field data: when all records in the dataset were obtained through field work, including data directly collected by the authors themselves (e.g., Beger et al. 2003) as well as collections of existing species' presence records (including from museums and herbaria; e.g., Trakhtenbrot & Kadmon 2005); often compiled into databases (e.g., Dobson et al. 1997) or atlases (e.g., Araújo et al. 2001).

2. Extrapolated data: these include data predicted or extrapolated through some form of remote mapping and include generalized species range maps (e.g., Moore et al. 2003), modeled species distributions (Ferrier & Watson 1997), land classes (e.g., ecoregions; Williams et al. 2000b), and maps of abiotic variables (e.g., Araújo et al. 2001).

For any given region and scale, collection of field data is generally more expensive and time-consuming than generating extrapolated data from already available information. Extrapolated data are, therefore, more desirable as surrogates, as they are more likely to be available in the data-poor situations when surrogates are most needed.

Field data tend to be prone to omission errors (false absences) by failing to detect species where they truly occur. Extrapolated data, on the other hand, are prone to commission errors (false presences), by presuming species to be present where they are absent. Commission errors are more serious in conservation planning, as they may lead to considering a species conserved when, in fact, it is not (Loiselle et al. 2003). Field data are thus more desirable as targets, as they provide a more stringent test of surrogates.

Studies can be classified into three types according to data source:

1. Field-field: when data for both surrogates and targets are field data (e.g., Virolainen et al. 2000);

2. Extrapolated-extrapolated: when data for both surrogates and targets are extrapolated data (e.g., Grenyer et al. 2006);

3. Extrapolated-field: when surrogates are based on extrapolated data while targets are field data (e.g., Bonn & Gaston 2005).

For the reasons presented above, studies of the third type provide the most desirable tests of surrogacy value. Extrapolated-field studies had typically positive, but relatively low, surrogacy (**Table 2**). Both when considering the percentage of positive studies (**Table 2**) and when analyzing the distribution of *SAI* values (**Table 2**,

Figure 2*f*), these studies had intermediate performance in relation to field-field studies (that performed the worst) and extrapolated-extrapolated studies (that performed the best).

The four extrapolated-field studies analyzed all used environmental surrogates. It is thus not clear if the low surrogacy values found are related to the use of environmental data or to the data sources. Cross-taxon surrogacy studies using extrapolated-field data are therefore a research priority.

Priorities for Future Research

A substantial body of research already exists investigating the effectiveness of different types of surrogates for conservation planning, but the majority of studies are not directly comparable owing to limitations in their analyses or presentation of results (see the **Supplemental Appendix**). As a result, some studies that utilized very impressive datasets were rejected from this review (e.g., Hopkinson et al. 2001, Howard et al. 1998). A much more comprehensive assessment of patterns of surrogacy value could be obtained from already existing data, and we hope that our analysis provides an incentive for such data to be revisited. Over the longer term, we hope that this review provides a methodological framework for future tests, creating the standardization required for comparison of results across types of surrogates and regions.

Throughout, we considered surrogacy value in terms of complementary representation of biodiversity. However, real-world conservation planning requires much more complex approaches, including considerations of scheduling and persistence. Scheduling conservation action is fundamental because it is typically impossible to protect all of the biodiversity features of interest in a region at the same time (Pressey & Taffs 2001). In order to minimize biodiversity loss, some features should be given protection first (Margules & Pressey 2000, Pressey & Taffs 2001): those of high vulnerability (with high likelihood of being lost in the short term; e.g., threatened species) and of high irreplaceability (for which few spatial options for conservation are available; e.g., narrow endemics). Encouragingly, Moritz et al. (2001) found high levels of cross-taxon surrogacy when the surrogate and optimal curves are derived by iterative selection of sites according to irreplaceability of the surrogate and of the target group, respectively. A similar approach can be employed in future studies to investigate levels of surrogacy when sites are selected to account simultaneously for irreplaceability and vulnerability. Prioritizing conservation action based on vulnerability and irreplaceability has been attempted at a diversity of scales, ranging from global (e.g., Rodrigues et al. 2004) to local (Eken et al. 2004, Ricketts et al. 2005). A few studies have investigated the surrogacy value of these approaches in representing particular taxa at broad (Burgess et al. 2002, Grenyer et al. 2006) and fine (Brooks et al. 2001) scales, but none have tested surrogates for identifying sites of high irreplaceability and high threat of target taxa. For example, it would be important to understand if Important Bird Areas, selected for their high irreplaceability and high vulnerability for birds (BirdLife Int. 2004) are adequate surrogates for the identification of Key Biodiversity Areas of high irreplaceability and high vulnerability of other taxa (Eken et al. 2004).

Although ensuring representation is a clear first step in systematic conservation planning (we cannot protect what is not represented in the first place; Margules et al. 1988, Pressey et al. 1993), it does not necessarily ensure persistence in the long term (Margules et al. 1994, Rodrigues et al. 2000b, Virolainen et al. 1999). If spatial data on biodiversity representation are difficult to obtain, data on spatial patterns of expected persistence are much more so, rendering surrogates even more indispensable. A fundamental research front in conservation planning is thus developing biodiversity surrogates for improving the probability of persistence of poorly known biodiversity. This includes testing surrogates for persistence within particular taxa as well as testing cross-taxon persistence surrogacy. Possible surrogates of long-term persistence within a particular taxon include past patterns of species persistence (Rodrigues et al. 2000b), species abundance (Rodrigues et al. 2000a), habitat quality, and connectivity (Araújo & Williams 2000). These surrogates can be tested through the approach developed above, calculating *SAI* for a test in which a taxon is used as a surrogate for itself but with a time lag. The test would therefore evaluate the extent to which sets of sites selected using surrogates of persistence (e.g., targeting peaks of abundance) retain species over time (e.g., Araújo & Williams 2000, Rodrigues et al. 2000a). These tests therefore require time series of good distributional data.

Once good surrogates of persistence have been developed for particular taxa, cross-taxon surrogates of persistence are possible. For example, if peaks of abundance have been found to predict species persistence effectively (Rodrigues et al. 2000a), a relevant test is whether targeting peaks of abundance for a surrogate taxon effectively captures peaks of abundance for a target taxon (e.g., Bonn et al. 2002). Surrogates of persistence are particularly needed for threatened species, which have lower probabilities of long-term persistence in the absence of conservation effort. A promising research line is therefore to understand if conservation planning for the persistence of threatened species in a particular taxonomic group is effective for the persistence of threatened species in a different group. That is, perhaps, expected if there is a coincidence in the spatial patterns of threat incidence for both groups.

CONCLUSIONS

Our analysis provides grounds for optimism on the value of using cross-taxon biodiversity surrogates in conservation planning. Less support was found for using surrogates from one realm—terrestrial, freshwater, or marine—to target biodiversity in another, which implies that a great expansion of aquatic biodiversity data collection and compilation is urgently necessary. Similarly, we found little support for the use of surrogates based on environmental data, especially if they are mainly derived from abiotic data.

This analysis also found low levels of surrogacy in the most relevant type of surrogacy tests in terms of data sources, those in which the surrogate data are obtained through extrapolation while the target data has been collected in fieldwork. This may, however, be confounded by the fact that all of those tests use environmental surrogates, leaving unanswered the question of what their performance would have been if they had used cross-taxon surrogacy.

All of these results need to be interpreted cautiously given the limited number of studies and tests analyzed. A more comprehensive analysis will be required as new studies are published, hopefully presenting their results in a standardized, comparable, format. Future research is also required in developing surrogates for effective scheduling of conservation action and for biodiversity persistence. In the meantime, we judge that the practical selection of conservation areas based on data for well-known taxonomic groups can cautiously proceed under the assumption that these areas capture species in less well-known taxa within the same realm.

SUMMARY POINTS

1. Biodiversity is not completely known anywhere, and so conservation planning is always based on surrogates for which data are available, with the assumption that such planning is effective for the conservation of unknown biodiversity.

2. Obtaining a coherent picture of the effectiveness of different types of surrogates requires standardized comparison of results across many tests. The Species Accumulation Index, *SAI*, is proposed as a standardized measure of surrogate effectiveness.

3. We reviewed 27 studies evaluating surrogacy value in conservation planning based on complementary representation, which together reported on 575 surrogacy tests. Of these, 464 tests allowed for a quantitative evaluation of surrogacy (i.e., a calculation of *SAI*). Most tests indicate positive surrogacy, but generally low *SAI* values.

4. Surrogacy value was substantially higher in complementary cross-taxon tests than in those based on environmental surrogates.

5. Within cross-taxon surrogates, surrogacy value was higher when there was some overlap between the surrogate group and the target group, and for tests within the same realm (terrestrial, marine, or freshwater).

6. Within environmental surrogates, those that used abiotic information to extrapolate from species data performed better than those purely based on abiotic information or those based on species assemblages.

7. Higher surrogacy was found when both surrogates and targets were based on extrapolated data, lower when both were based on field data, and intermediate when surrogates were extrapolated and targets were obtained from field data.

8. Many more tests are required before definitive conclusions can be extracted. In the meantime, we judge that the practical selection of conservation areas based on data for well-known taxonomic groups can cautiously proceed under the assumption that these areas capture species in less well-known taxa within the same realm.

FUTURE ISSUES

1. More surrogacy tests are needed to investigate whether particular (better-known) taxa such as vertebrates are good surrogates for other (poorly known) taxa such as plants and insects.

2. Cross-taxon surrogacy tests are needed in which surrogates are based on extrapolated data while targets are based on field data.

3. Research is urgently needed in developing effective surrogates for scheduling conservation action, that is, capturing areas of high irreplaceability and high vulnerability of the targets.

4. Research is urgently needed on how to use biodiversity surrogates to improve persistence probabilities (over and above the simple representation) of poorly known biodiversity.

DISCLOSURE STATEMENT

The authors are not aware of any biases that might be perceived as affecting the objectivity of this review.

ACKNOWLEDGMENTS

S. Ferrier, B. Phalan, and W. Turner have provided valuable comments that improved this manuscript. We are particularly grateful to S. Ferrier for the suggestion of analyzing the studies according to data source. A.S.L.R is a Marie Curie Intra-European Fellow under the European Community Sixth Framework Program. We are also grateful to the Gordon and Betty Moore Foundation for funding.

LITERATURE CITED

Araújo MB, Densham PJ, Williams PH. 2004. Representing species in reserves from patterns of assemblage diversity. *J. Biogeogr.* 31:1037–50

Araújo MB, Humphries CJ, Densham PJ, Lampinen R, Hagemeijer WJM, et al. 2001. Would environmental diversity be a good surrogate for species diversity? *Ecography* 24:103–10

Araújo MB, Williams PH. 2000. Selecting areas for species persistence using occurrence data. *Biol. Conserv.* 96:331–45

Arthur JL, Hachey M, Sahr K, Huso M, Kiester AR. 1997. Finding all optimal solutions to the reserve site selection problem: Formulation and computational analysis. *Environ. Ecol. Stat.* 4:153–65

Balmford A. 1998. On hotspots and the use of indicators for reserve selection. *Trends Ecol. Evol.* 13:409

Beger M, Jones GP, Munday PL. 2003. Conservation of coral reef biodiversity: a comparison of reserve selection procedures for corals and fishes. *Biol. Conserv.* 111:53–62

BirdLife Int. 2004. *State of the World's Birds 2004—Indicators for Our Changing World*. Cambridge, UK: BirdLife Int. 73 pp.

Bonn A, Gaston KJ. 2005. Capturing biodiversity: Selecting priority areas for conservation using different criteria. *Biodivers. Conserv.* 14:1083–100

Bonn A, Rodrigues ASL, Gaston KJ. 2002. Threatened and endemic species: are they good indicators of patterns of biodiversity? *Ecol. Lett.* 5:733–41

Brooks T, Balmford A, Burgess N, Hansen LA, Moore J, et al. 2001. Conservation priorities for birds and biodiversity: Do East African Important Bird Areas represent species diversity in other terrestrial vertebrate groups? *Ostrich* 15:3–12

Burgess ND, Rahbek C, Larsen FW, Williams P, Balmford A. 2002. How much of the vertebrate diversity of sub-Saharan Africa is catered for by recent conservation proposals? *Biol. Conserv.* 107:327–39

Chiarucci A, D'Auria F, De Dominicis V, Lagana A, Perini C, Salerni E. 2005. Using vascular plants as a surrogate taxon to maximize fungal species richness in reserve design. *Conserv. Biol.* 19:1644–52

del Hoyo J, Elliot A, Christie D, Sartagal J, eds. 1992–2006. *Handbook of the Birds of the World*, Vols. 1–11. Barcelona: Lynx Ed.

Dobson AP, Rodriguez JP, Roberts WM, Wilcove DS. 1997. Geographic distribution of endangered species in the United States. *Science* 275:550–53

Eken G, Bennun L, Brooks TM, Darwall W, Fishpool LDC, et al. 2004. Key biodiversity areas as site conservation targets. *BioScience* 54:1110–18

Faith DP, Margules CR, Walker PA, Stein J, Natera G. 2001. Practical application of biodiversity surrogates and percentage targets for conservation in Papua New Guinea. *Pac. Conserv. Biol.* 6:289–303

Favreau J, Drew C, Hess G, Rubino M, Koch F, Eschelbach KA. 2006. Recommendations for assessing the effectiveness of surrogate species approaches. *Biodivers. Conserv.* 15:3949–69

Ferrier S. 2002. Mapping spatial pattern in biodiversity for regional conservation planning: Where to from here? *Syst. Biol.* 51:331–63

Ferrier S, Watson G. 1997. *An evaluation of the effectiveness of environmental surrogates and modelling techniques in predicting the distribution of biological diversity*. Environ. Aust., Canberra **http://www.deh.gov.au/biodiversity/publications/technical/surrogates/**

Ferrier S, Pressey RL, Barrett TW. 2000. A new predictor of the irreplaceability of areas for achieving a conservation goal, its application to real-world planning, and a research agenda for further refinement. *Biol. Conserv.* 93:303–325

Flather CH, Wilson KR, Dean DJ, McComb WC. 1997. Identifying gaps in conservation networks: Of indicators and uncertainty in geographic-based analyses. *Ecol. Appl.* 7:531–42

Game M, Peterken GF. 1984. Nature reserve selection-strategies in the woodlands of Central Lincolnshire, England. *Biol. Conserv.* 29:157–81

Garson J, Aggarwal A, Sarkar S. 2002. Birds as surrogates for biodiversity: an analysis of a data set from southern Quebec. *J. Biosci.* 27:347–60

Gaston KJ, Blackburn TM. 1995. Mapping biodiversity using surrogates for species richness: Macro-scales and New World birds. *Proc. R. Soc. London Ser. B* 262:335–41

Gaston KJ, Williams PH, Eggleton P, Humphries CJ. 1995. Large-scale patterns of biodiversity—Spatial variation in family richness. *Proc. R. Soc. London Ser. B* 260:149–54

Gladstone W. 2002. The potential value of indicator groups in the selection of marine reserves. *Biol. Conserv.* 104:211–20

Grenyer R, Orme CDL, Jackson SF, Thomas GH, Davies RG, et al. 2006. Global distribution and conservation of rare and threatened vertebrates. *Nature* 444:93–96

Hess GR, Bartel RA, Leidner AK, Rosenfeld KM, Rubino MJ, et al. 2006. Effectiveness of biodiversity indicators varies with extent, grain, and region. *Biol. Conserv.* 132:448–57

Hopkinson P, Travis JMJ, Evans J, Gregory RD, Telfer MG, Williams PH. 2001. Flexibility and the use of indicator taxa in the selection of sites for nature reserves. *Biodivers. Conserv.* 10:271–85

Howard PC, Viskanic P, Davenport TRB, Kigenyi FW, Baltzer M, et al. 1998. Complementarity and the use of indicator groups for reserve selection in Uganda. *Nature* 394:472–75

Kati V, Devillers P, Dufrene M, Legakis A, Vokou D, Lebrun P. 2004. Testing the value of six taxonomic groups as biodiversity indicators at a local scale. *Conserv. Biol.* 18:667–75

Kirkpatrick JB. 1983. An iterative method for establishing priorities for the selection of nature reserves—an example from Tasmania. *Biol. Conserv.* 25:127–34

Lawler JJ, White D, Sifneos JC, Master LL. 2003. Rare species and the use of indicator groups for conservation planning. *Conserv. Biol.* 17:875–82

Lawton JH, Gaston KJ. 2001. Indicator species. In *Encyclopedia of Biodiversity*, ed. SA Levin, 3:437–50. New York: Academic

Loiselle BA, Howell CA, Graham CH, Goerck JM, Brooks T, et al. 2003. Avoiding pitfalls of using species distribution models in conservation planning. *Conserv. Biol.* 17:1591–600

Lund MP. 2002. Performance of the species listed in the European Community 'Habitats' Directive as indicators of species richness in Denmark. *Environ. Sci. Policy* 5:105–12

Lund MP, Rahbek C. 2002. Cross-taxon congruence in complementarity and conservation of temperate biodiversity. *Anim. Conserv.* 5:163–71

Mac Nally R, Bennett AF, Brown GW, Lumsden LF, Yen A, et al. 2002. How well do ecosystem-based planning units represent different components of biodiversity? *Ecol. Appl.* 12:900–12

Margules CR, Nicholls AO, Pressey RL. 1988. Selecting networks of reserves to maximise biological diversity. *Biol. Conserv.* 43:63–76

Margules CR, Nicholls AO, Usher MB. 1994. Apparent species turnover, probability of extinction and the selection of nature-reserves—a case-study of the Ingleborough limestone pavements. *Conserv. Biol.* 8:398–409

Margules CR, Pressey RL. 2000. Systematic conservation planning. *Nature* 405:243–53

McKenna SA. 2006. Use of surrogate taxa in coral reef surveys for marine reserve design and conservation. *Proc. 10th Int. Coral Reef Symp., Okinawa*, pp. 1498–503

Moore JL, Balmford A, Brooks T, Burgess ND, Hansen LA, et al. 2003. Performance of subsaharan vertebrates as indicator groups for identifying priority areas for conservation. *Conserv. Biol.* 17:207–18

Moritz C, Richardson KS, Ferrier S, Monteith GB, Stanisic J, et al. 2001. Biogeographical concordance and efficiency of taxon indicators for establishing conservation priority in a tropical rainforest biota. *Proc. R. Soc. London Ser. B* 268:1875–81

Negi HR, Gadgil M. 2002. Cross-taxon surrogacy of biodiversity in the Indian Garhwal Himalaya. *Biol. Conserv.* 105:143–55

Oliver I, Holmes A, Dangerfield JM, Gillings M, Pik AJ, et al. 2004. Land systems as surrogates for biodiversity in conservation planning. *Ecol. Appl.* 14:485–503

Pearson DL, Carroll SS. 1998. Global patterns of species richness: Spatial models for conservation planning using bioindicator and precipitation data. *Conserv. Biol.* 12:809–21

Pharo EJ, Beattie AJ. 2001. Management forest types as a surrogate for vascular plant, bryophyte and lichen diversity. *Aust. J. Bot.* 49:23–30

Pharo EJ, Beattie AJ, Pressey RL. 2000. Effectiveness of using vascular plants to select reserves for bryophytes and lichens. *Biol. Conserv.* 96:371–78

Pimm SL. 2000. Conservation connections. *Trends Ecol. Evol.* 15:262–63

Pimm SL, Russell GJ, Gittleman JL, Brooks TM. 1995. The future of biodiversity. *Science* 269:347–50

Polasky S, Csuti B, Vossler CA, Meyers SM. 2001. A comparison of taxonomic distinctness vs richness as criteria for setting conservation priorities for North American birds. *Biol. Conserv.* 97:99–105

Prendergast JR, Quinn RM, Lawton JH. 1999. The gaps between theory and practice in selecting nature reserves. *Conserv. Biol.* 13:484–492

Pressey RL, Ferrier S, Hager TC, Woods CA, Tully SL, Weinman KM. 1996. How well protected are the forests of north-eastern New South Wales?—Analyses of forest environments in relation to formal protection measures, land tenure, and vulnerability to clearing. *For. Ecol. Manag.* 85:311–33

Pressey RL, Humphries CJ, Margules CR, Vane-Wright RI, Williams PH. 1993. Beyond opportunism—key principles for systematic reserve selection. *Trends Ecol. Evol.* 8:124–28

Pressey RL, Taffs KH. 2001. Scheduling conservation action in production landscapes: priority areas in western New South Wales defined by irreplaceability and vulnerability to vegetation loss. *Biol. Conserv.* 100:355–76

Reid WV. 1998. Biodiversity hotspots. *Trends Ecol. Evol.* 13:275–80

Reyers B, van Jaarsveld AS. 2000. Assessment techniques for biodiversity surrogates. *S. Afr. J. Sci.* 96:406–8

Reyers B, van Jaarsveld AS, Krüger M. 2000. Complementarity as a biodiversity indicator strategy. *Proc. R. Soc. London Ser. B* 267:505–13

Ricketts TH, Dinerstein E, Boucher T, Brooks TM, Butchart SHM, et al. 2005. Pinpointing and preventing imminent extinctions. *Proc. Natl. Acad. Sci. USA* 102:18497–501

Rodrigues ASL, Andelman SJ, Bakarr MI, Boitani L, Brooks TM, et al. 2004. Effectiveness of the global protected area network in representing species diversity. *Nature* 428:640–43

Rodrigues ASL, Brooks TM, Gaston KJ. 2005. Integrating phylogenetic diversity in the selection of priority areas for conservation: does it make a difference? In *Phylogeny and Conservation*, ed. A Purvis, JL Gittleman, TM Brooks, pp. 101–19. Cambridge, UK: Cambridge Univ. Press

Rodrigues ASL, Gaston KJ. 2002. Maximising phylogenetic diversity in the selection of networks of conservation areas. *Biol. Conserv.* 105:103–11

Rodrigues ASL, Gaston KJ, Gregory RD. 2000a. Using presence-absence data to establish reserve selection procedures that are robust to temporal species turnover. *Proc. R. Soc. London Ser. B* 267:897–902

Rodrigues ASL, Gregory RD, Gaston KJ. 2000b. Robustness of reserve selection procedures under temporal species turnover. *Proc. R. Soc. London Ser. B* 267:49–55

Rosenzweig ML. 1995. *Species Diversity in Space and Time*. Cambridge, NY: Cambridge Univ. Press. 436 pp.

Ryti RT. 1992. Effect of the focal taxon on the selection of nature-reserves. *Ecol. Appl.* 2:404–10

Saetersdal M, Gjerde I, Blom HH, Ihlen PG, Myrseth EW, et al. 2003. Vascular plants as a surrogate species group in complementary site selection for bryophytes, macrolichens, spiders, carabids, staphylinids, snails, and wood living polypore fungi in a northern forest. *Biol. Conserv.* 115:21–31

Saetersdal M, Line JM, Birks HJB. 1993. How to maximize biological diversity in nature-reserve selection—vascular plants and breeding birds in deciduous woodlands, Western Norway. *Biol. Conserv.* 66:131–38

Sauberer N, Zulka KP, Abensperg-Traun M, Berg H-M, Bieringer G, et al. 2004. Surrogate taxa for biodiversity in agricultural landscapes of eastern Austria. *Biol. Conserv.* 117:181–90

Secr. Conv. Biol. Divers. (SCBD) 2005. *Protected areas programme of work.* **http://www.biodiv.org/programmes/cross-cutting/protected/wopo.asp**

Su JC, Debinski DM, Jakubauskas ME, Kindscher K. 2004. Beyond species richness: Community similarity as a measure of cross-taxon congruence for coarse-filter conservation. *Conserv. Biol.* 18:167–73

Sutherland WJ. 2006. *Ecological Census Techniques: A Handbook*. Cambridge, UK: Cambridge Univ. Press. 2nd rev. ed.

Thomas JA, Telfer MG, Roy DB, Preston CD, Greenwood JJD, et al. 2004. Comparative losses of British butterflies, birds, and plants and the global extinction crisis. *Science* 303:1879–81

Tognelli MF. 2005. Assessing the utility of indicator groups for the conservation of South American terrestrial mammals. *Biol. Conserv.* 121:409–17

Tognelli MF, Silva-Garcia C, Labra FA, Marquet PA. 2005. Priority areas for the conservation of coastal marine vertebrates in Chile. *Biol. Conserv.* 126:420–28

Trakhtenbrot A, Kadmon R. 2005. Environmental cluster analysis as a tool for selecting complementary networks of conservation sites. *Ecol. Appl.* 15:335–45 van Jaarsveld AS, Freitag S, Chown SL, Muller C, Koch S, et al. 1998. Biodiversity assessment and conservation strategies. *Science* 279:2106–8

Vessby K, Soderstrom B, Glimskar A, Svensson B. 2002. Species-richness correlations of six different taxa in Swedish seminatural grasslands. *Conserv. Biol.* 16:430–39

Virolainen KM, Ahlroth P, Hyvarinen E, Korkeamaki E, Mattila J, et al. 2000. Hot spots, indicator taxa, complementarity and optimal networks of taiga. *Proc. R. Soc. London Ser. B* 267:1143–47

Virolainen KM, Virola T, Suhonen J, Kuitunen M, Lammi A, Siikamäki P. 1999. Selecting networks of nature reserves: methods do affect the long-term outcome. *Proc. R. Soc. London Ser. B* 266:1141–46

Vitousek PM, Mooney HA, Lubchenco J, Melillo JM. 1997. Human domination of Earth's ecosystems. *Science* 277:494–99

Warman LD, Forsyth DM, Sinclair ARE, Freemark K, Moore HD, et al. 2004a. Species distributions, surrogacy, and important conservation regions in Canada. *Ecol. Lett.* 7:374–79

Warman LD, Sinclair ARE, Scudder GGE, Klinkenberg B, Pressey RL. 2004b. Sensitivity of systematic reserve selection to decisions about scale, biological data, and targets: case study from southern British Columbia. *Conserv. Biol.* 18:655–66

Williams P, Faith D, Manne L, Sechrest W, Preston C. 2006. Complementarity analysis: Mapping the performance of surrogates for biodiversity. *Biol. Conserv.* 128:253–64

Williams P, Gibbons D, Margules C, Rebelo A, Humphries C, Pressey R. 1996. A comparison of richness hotspots, rarity hotspots, and complementary areas for conserving diversity of British birds. *Conserv. Biol.* 10:155–74

Williams PH, Burgess N, Rahbek C. 2000a. Assessing large 'flagship species' for representing the diversity of sub-Saharan mammals. In *Priorities for the Conservation of Mammalian Diversity—Has the Panda Had its Day?* ed. A Entwistle, N Dunstone, pp. 85–99. Cambridge, UK: Cambridge Univ. Press

Williams PH, Burgess ND, Rahbek C. 2000b. Flagship species, ecological complementarity and conserving the diversity of mammals and birds in sub-Saharan Africa. *Anim. Conserv.* 3:249–60

Understanding the Effects of Marine Biodiversity on Communities and Ecosystems

John J. Stachowicz,[1] John F. Bruno,[2] and J. Emmett Duffy[3]

[1] Section of Evolution and Ecology, University of California, Davis, California 95616; email: jjstachowicz@ucdavis.edu

[2] Department of Marine Sciences, University of North Carolina, Chapel Hill, North Carolina 27599-3300

[3] Virginia Institute of Marine Sciences, College of William and Mary, Gloucester Point, Virginia 23062

Annu. Rev. Ecol. Evol. Syst. 2007. 38:739–66

First published online as a Review in Advance on August 20, 2007

The *Annual Review of Ecology, Evolution, and Systematics* is online at http://ecolsys.annualreviews.org

This article's doi: 10.1146/annurev.ecolsys.38.091206.095659

Key Words

diversity, ecosystem function, food webs, productivity, stability, trophic cascade, trophic skew

Abstract

There is growing interest in the effects of changing marine biodiversity on a variety of community properties and ecosystem processes such as nutrient use and cycling, productivity, stability, and trophic transfer. We review published marine experiments that manipulated the number of species, genotypes, or functional groups. This research reveals several emerging generalities. In studies of primary producers and sessile animals, diversity often has a weak effect on production or biomass, especially relative to the strong effect exerted by individual species. However, sessile taxon richness did consistently decrease variability in community properties, and increased resistance to, or recovery from disturbance or invasion. Multitrophic-level studies indicate that, relative to depauperate assemblages of prey species, diverse ones (*a*) are more resistant to top-down control, (*b*) use their own resources more completely, and (*c*) increase consumer fitness. In contrast, predator diversity can either increase or decrease the strength of top-down control because of omnivory and because interactions among predators can have positive and negative effects on herbivores. Recognizing that marine and terrestrial approaches to understanding diversity-function relationships are converging, we close with suggestions for future research that apply across habitats.

INTRODUCTION

Ecologists have long pondered the relationships between species *diversity* (italicized terms are defined in more detail in the **Supplemental Glossary**, follow the Supplemental Material link from the Annual Reviews home page at **http://www.annualreviews.org/**) or complexity and various measures of the *stability* or performance of an ecosystem. Rapid changes in the biological composition and richness of most of Earth's ecosystems as a result of human activities have breathed new urgency into these questions. Stimulated in part by these transformations, theoretical and empirical research in ecology has turned to the relationship between *biodiversity* and *ecosystem functioning*. Ecosystem functioning, as we consider it, includes aggregate, community or ecosystem-level processes and properties such as production, standing biomass, invasion resistance, food web dynamics, element cycling, resource use, and trophic transfer (Chapin et al. 1998, Loreau et al. 2001, Tilman 1999). An influential series of field experiments, conducted primarily in terrestrial grasslands, has demonstrated that the identity and number of plant species in a system can strongly influence ecosystem functioning (Hooper et al. 2005; Loreau et al. 2001, 2002; Tilman 1999; Tilman et al. 2006). Similarly, experiments in laboratory microcosms show that changing biodiversity in multilevel food webs can also have pervasive ecosystem impacts (Naeem & Li 1997, Naeem et al. 1994, Petchey et al. 2002).

Although early reviews lamented that comparable studies were rare in marine systems (e.g., Emmerson & Huxham 2002), this is no longer the case. Experimental manipulations of biodiversity in marine systems have both provided independent tests of generality of results from terrestrial systems, and exploited advantages of marine systems to develop new frontiers in our understanding of the ecological consequences of biodiversity. For example, the stronger top-down control in the sea relative to terrestrial habitats (Shurin et al. 2002) suggests that traditional measures of ecosystem function such as production or biomass may be influenced more by herbivores or predators than by plant diversity in marine systems (Duffy 2003, Paine 2002). Correspondingly, marine studies have greatly influenced the developing theory and empirical understanding of the role of predator and prey biodiversity in regulating the top-down control of populations and communities. Of course, consumers are also important determinants of plant biomass and species composition in terrestrial habitats (Schmitz et al. 2000), so lessons derived from marine studies of consumer diversity may guide future work in terrestrial systems. The focus on the effects of marine diversity at the consumer level has the potential to more directly address conservation concerns (Srivastava & Vellend 2005) because of the bias in extinctions toward higher trophic levels (Byrnes et al. 2007, Duffy 2003, Lotze et al. 2006). Additionally, in marine communities, the importance of diversity change relative to other stressors may be high because of the widespread harvest of wild plants and animals that still occurs in marine systems. Finally, the long history of detailed observational data collection on multiple trophic levels in oceanography has been exploited to assess the effects of natural diversity gradients at large scales, facilitating a better connection between changes in biodiversity and disruptions of *ecosystem services* than is currently possible in terrestrial systems (e.g., Frank et al. 2006, Worm et al. 2006).

Diversity: strictly incorporates information about both the number of entities and their relative abundance (evenness); in practice it is used interchangeably with richness

Stability: measure of community variability, sometimes equated to resistance, but more generally measured as the coefficient of variation over time

Richness: number of entities at a particular hierarchical level; often used interchangeably with diversity

Biodiversity: variety of life at any hierarchical level, including genes, species, functional groups, or ecosystems

Ecosystem functioning: aggregate or emergent aspects of ecosystems (e.g., production, nutrient cycling), carrying no inherent judgment of value

Ecosystem services: functions that are judged to have some clear value to humanity, (e.g., water filtration, production of harvestable fish)

We recognize two principal motivations for understanding the effects of variation in diversity on ecosystem function. One, ultimately motivated by practical and conservation concerns, is understanding and predicting consequences of ongoing diversity loss in nature. We discuss this in more detail below. The second, more basic, rationale involves general understanding of how ecosystems work. For example, biodiversity-function research has spurred experimentalists to compare effects of multiple species together and independently. In the past, such approaches have led to major advances in our understanding of the effects of predators on prey population and community structure (Ives et al. 2005, Sih et al. 1998), and in the maintenance of diversity via intransitive or context-dependent competitive networks (Buss & Jackson 1979). In effect, biodiversity manipulations address the flip side of the coexistence question fundamental to ecology: How do so many species coexist (or not)? Many studies motivated initially by concerns of predicting consequences of declining biodiversity may ultimately prove to have a more enduring value in elucidating the outcome of simultaneous interactions among multiple species and the contingency of the outcome of pairwise interactions on the presence of other species. Additionally, there are well documented natural gradients in marine biodiversity with respect to latitude, longitude, and depth (e.g., Rex et al. 1993, Roberts et al. 2002, Roy et al. 1998, Worm et al. 2003) that are independent of human activity and could affect the stability, consistency, or performance of particular communities. For example, latitudinal gradients in invasion (Sax 2001) or predation pressure and prey defenses (Bertness et al. 1981, Bolser & Hay 1996) that correspond with gradients in diversity are well known, but causal links have rarely been rigorously investigated.

Because the relationship between diversity and ecosystem functioning has been a contentious field, with confusion in terminology often contributing to the contention, we first briefly review key concepts and mechanisms underpinning diversity-function relationships. We then discuss current patterns of biodiversity change in marine systems and how these relate to the types of diversity that have been manipulated in experiments. Next, we review available studies that provide data to assess the effects of marine biodiversity on ecosystem functioning. Finally, we close with a discussion of the generalities that have emerged thus far, and what we perceive to be the most pressing issues for further research.

Richness effect: occurs when diverse communities differ in ecosystem function from the average monoculture. Can be caused by many mechanisms

Identity or composition effects: describe variation among species or particular combinations of species in their influence on an ecosystem function

MECHANISMS AND CONCEPTS

The theoretical basis for a positive relationship between the richness or diversity of plants or sessile invertebrates and production, biomass or resource use is well developed and relatively straightforward (Loreau 2000). Following previous work, we define a *richness effect* as occurring when a mixture (of species, genotypes, functional groups, etc; see definition of biodiversity) performs differently than the average performance of its component species in monoculture (also known as nontransgressive overyielding). Apart from richness, individual species can also differ in their effects on ecosystem processes in different, independent ways; such functional differences among individual species or among combinations of species are often referred to as *identity* or *composition effects*. Richness effects result from two main classes of

Complementarity: greater
performance of a species in
mixture than expected from
its performance in
monoculture caused by
interactions such as resource
partitioning or facilitation

Sampling effect: the
greater statistical
probability of including a
species with a dominant
effect in an assemblage as
species richness increases

Multivariate
complementarity:
phenomenon by which a
diverse assemblage
maximizes multiple
ecosystem functions
simultaneously, because
different species control
different functions

Selection effect: a more
general version of sampling
effect that can be positive or
negative

phenomena: (*a*) complementary properties of species including niche partitioning and facilitation (*complementarity*), and (*b*) strong effects of a dominant species on the function of interest (*sampling effect*). These mechanisms are most clearly distinguished in experiments that include diverse mixtures of species as well as each of the component species in monoculture. When the response of the species mixture is greater than the highest performing monoculture (transgressive overyielding), this can be taken as clear evidence that the richness effect is not accounted for purely by the effects of a dominant species, and that some form of complementarity is operating. The sampling effect occurs when the presence of a particular species drives the relationship between richness and ecosystem function, as a result of two conditions: (*a*) a greater statistical probability of including a species with a particular trait (e.g., high productivity) in an assemblage as species richness increases, and (*b*) the species with highest function in monoculture is also the dominant competitor in a mixed species assemblage (Huston 1997, Tilman et al. 1997). When multiple functions are considered, the highest levels of different processes can sometimes be caused by different species. In such cases, when multiple functions are considered simultaneously or combined into a multivariate index of ecosystem functioning, a sort of *multivariate complementarity* results, in which diverse communities simultaneously maximize multiple functions and thus produce an ecosystem state different from any monoculture (e.g., Duffy et al. 2003).

The premise of the sampling effect, that high production in monoculture and competitive dominance are correlated, is not always met, especially given likely trade-offs between growth rate and competitive ability. Thus, the sampling phenomenon has been generalized to *selection effects* (Hector et al. 2002, Loreau & Hector 2001), which can be either positive or negative depending on whether the dominant species in polyculture displays relatively high or low performance, respectively, when grown alone. Importantly, this means that the absence of a significant relationship between species richness and ecosystem function can result from the counteracting mechanisms of positive effects of complementarity driven by resource partitioning and the tendency for productive species to fare poorly in competition in mixed species plots (negative selection; e.g., Bruno et al. 2005, Hector et al. 2002). Additionally, what appears to be a sampling effect in which mixture performance is equivalent to the best performing monoculture may in fact be the result of positive complementarity balanced by negative selection. This can make predicting the consequences of the loss of diversity for function not only complex but dependent on relative extinction risks of different species.

Although these basic mechanisms and experimental approaches originally developed for sessile organisms also apply to mobile consumers, studies of predator richness effects have typically employed one of two distinct experimental design strategies. First, replacement series (or substitutive) designs control the initial abundance or biomass of organisms, but as a result intraspecific density declines with increasing richness. Second, additive designs hold intraspecific density constant with increasing species richness, but as a result total organism density increases with richness. Most within-trophic-level richness experiments begin as replacement designs to avoid explicitly confounding biomass and richness. But the initial design is somewhat irrelevant for these experiments as they often track population-level processes over longer time frames using organisms with sufficiently rapid generation times that their relative and absolute densities adjust during the experiment owing to birth, death, and recruitment. The choice between additive and replacement designs is more important in experiments of short duration or when density of the manipulated taxon is otherwise prevented from changing over the course of the experiment. A full discussion of the costs and benefits of each design is beyond the scope of our review (see, e.g., Sackville-Hamilton 1994), but we note two cautions. First, additive designs confound total density of organisms with species richness, whereas replacement designs confound intraspecific density with richness, so each is limited in the types of mechanisms and outcomes that it can elucidate. Second, additive designs will become intractable when experiments include a large range of species richness because organism density in polycultures will become unnaturally high and can force interactions that might rarely occur in nature. The optimal design choice may be guided by the particular question of interest or by empirical richness-abundance relationships when known.

In addition to measuring the effects of diversity on processes like production using a single measure approach at the end of an experiment, one can also measure the relationship between richness and variability in these same processes over time (review in Hooper et al. 2005, McCann 2000, Tilman et al. 2006). In general, diversity is predicted to increase the stability (or decrease temporal fluctuations) of aggregate community properties like biomass, while slightly destabilizing population abundance of individual species (Lehman & Tilman 2000, May 1974). This can be the result of several, nonmutually exclusive mechanisms, including statistical averaging (the *portfolio effect*), as well as complementary responses of species to changing environmental conditions (often detected as negative covariances in species abundances in diverse communities), and overyielding, which indicates that species are stably coexisting (Tilman 1999, Tilman et al. 2006). A related idea is that diversity contributes to the *resistance* to, or *resilience* (recovery) from a disturbance (see Pimm 1984 for terminology), which can be caused by the mechanisms above as well as by the inclusion of highly resistant species in diverse assemblages (sampling effect). We group these together because they all involve assessing the effect of richness on variability in ecosystem processes rather than on mean states.

Resistance: capacity of a system to resist change in the face of a perturbation

Resilience: capacity of a system to recover from a disturbance or perturbation

REALISTIC SCENARIOS OF DIVERSITY CHANGE

Even though there have been relatively few documented global-scale extinctions of marine species compared to land, many species are locally extinct and even more have been driven ecologically extinct: Their populations are sufficiently small that they can no longer play a significant ecological role in a particular community (Sala & Knowlton 2006, Steneck et al. 2004). The order in which species go extinct is not likely to be random, and the relative extinction risks of different species can alter the expected correlation between diversity and ecosystem functioning (Solan et al. 2004). For example, the depletion of predators relative to prey by selective harvest and habitat degradation has caused a skewing of trophic structure toward dominance at lower levels and the general alteration of aquatic and terrestrial food webs (Duffy 2003, Jones et al. 2004, Pauly et al. 1998, Petchey et al. 2004).

Extinctions

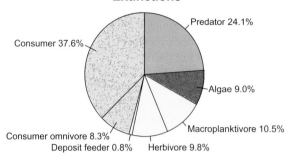

- Predator 24.1%
- Consumer 37.6%
- Algae 9.0%
- Macroplanktivore 10.5%
- Herbivore 9.8%
- Deposit feeder 0.8%
- Consumer omnivore 8.3%

Invasions

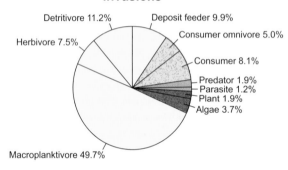

- Detritivore 11.2%
- Deposit feeder 9.9%
- Herbivore 7.5%
- Consumer omnivore 5.0%
- Consumer 8.1%
- Predator 1.9%
- Parasite 1.2%
- Plant 1.9%
- Algae 3.7%
- Macroplanktivore 49.7%

Trophic classifications

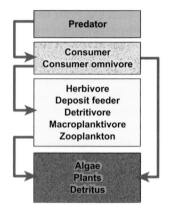

Predator

Consumer
Consumer omnivore

Herbivore
Deposit feeder
Detritivore
Macroplanktivore
Zooplankton

Algae
Plants
Detritus

Figure 1

Changing patterns of trophic skew in coastal/estuarine marine ecosystems as the combined result of species introductions and local extinctions. Data replotted from Byrnes et al. (2007). Species loss is biased toward higher trophic levels, whereas species gain is biased toward lower levels (primary consumers). The functional groups most responsible for this skew were top predators (24.1% of extinctions but 6.1% of invasions on average), secondary consumers (37.6% of extinctions but 2.2% of invasions), and suspension feeding macroplanktivores (10.5% of extinctions but 44.6% of invasions). (Percentages may sum to greater than 100% owing to rounding.)

Trophic skew: altered distribution of species richness among trophic levels because of differential effects of invasion and extinction at each level

However, at the local to regional scale, diversity gains also occur through species introductions, so the net change in species richness is not always clear cut. Because different processes drive extinctions (e.g., overfishing) and invasion (e.g., ballast water transport), the types of species being gained and lost differ (Lotze et al. 2006). Byrnes et al. (2007) classified all documented marine species extinctions from several regions by trophic level and feeding mode and found that 70% of species lost were high-order consumers (trophic level 3 or 4), whereas 70% of invaders were lower order consumers, particularly suspension feeders or deposit feeders (**Figure 1**). Thus the combined effect of both processes has resulted so far in little net change in richness but an enhancement of *trophic skew* by decreasing predator richness while increasing primary consumers and detritivores. Invasions are most numerous in coastal embayments, so their influence on trophic skew may be reduced in open coast or oceanic environments, but the loss of top predators appears to be a global phenomenon. Little is known about how diversity is changing at the local scale at which most experiments are conducted, although local and regional diversity are often strongly correlated.

EMPIRICAL RESEARCH

A summary of results for the most studied ecosystem processes and properties is provided in **Table 1a,b**, and a complete catalog of experimental manipulations is provided in **Supplemental Tables 1–4**. The compilation in **Table 1a,b** includes experimental studies that manipulated the richness of at least three functional groups, species, or genotypes. Though observational studies can be very useful, especially in testing for links between diversity and ecosystem services (see the sidebar, Connecting Diversity to Ecosystem Services), we limit ourselves primarily to experiments in this review. Most experiments manipulate richness in randomly constructed communities, allowing partitioning of effects owing to richness versus identity. We focus on these random assembly experiments, as these more directly address our principal theme of how the number and variety of species per se influence ecosystem properties.

Overall, the majority of experiments and metrics reported in **Table 1a,b** detected a significant effect of richness (85/123). Although publication bias against finding no effect is possible, it is nonetheless clear that richness effects are widespread. Transgressive overyielding, in which the diverse assemblage outperforms the best monoculture was found in far fewer studies (26/105), over half of which were from studies of the effect of animal richness on invasion resistance, resource use, or secondary production.

CONNECTING DIVERSITY TO ECOSYSTEM SERVICES

Worm et al. (2006) analyzed the effects of changes in marine biodiversity on fundamental ecosystem services by combining available data from sources ranging from small scale experiments to global fisheries. At a global scale, they analyzed relationships between species richness and fishery production in 64 large marine ecosystems varying naturally in diversity. Ecosystems with naturally low diversity showed lower fishery productivity, more frequent collapses (strong reductions in fishery yield), and lower resilience than naturally species-rich systems. They suggested that the greater resilience of more diverse ecosystems may result because fishers can switch more readily among target species when there are many species available (high richness), potentially providing overfished taxa with a chance to recover. This mechanism is consistent with theory, small-scale experiments, and with the negative relationship Worm et al. (2006) found between fished taxa richness and variation in catch from year to year. Although the correlative approach employed in their comparison did not allow assignment of causation or mechanisms, it does allow the examination of larger scale processes and the connection between richness and ecosystem services (fish production) with clear value to humans.

Table 1a Summary of experiments manipulating diversity of marine organisms published through early 2007[a]: Within-trophic-level manipulations

Response	Positive	Negative	No effect	Transgressive overyielding
Stability, disturbance, resistance, or resilience[b]	9	1	0	N/A
Plant biomass or production	7	0	6	0/13
Decomposition	0	0	2	0/2
Associated species diversity	0	0	3	0/2
Associated species abundance	2	0	1	0/2
Resource use[b]	6	0	3	4/8[d]
Resource regeneration[c]	4	4	9	1/14
Invader abundance or survival	0	6	1	5/7
Invader settlement	2	0	1	2/3
Secondary production	6	0	1	4/6

[a]Full data for all studies are in **Supplemental Tables 1–4**. Individual studies may be counted multiple times in the table if they either conducted more than one independent experiment or measured more than one potentially independent response variable. Not all studies explicitly tested for transgressive overyielding, and so its existence was inferred in some cases from data available in graphs; in some cases it was impossible to tell because monoculture means were not given, so the total number of possible studies in which trangressive overyielding could be detected is often less than the total number of studies that showed a richness effect (see Supplemental Materials for details).
[b]includes data from plants, sessile and mobile invertebrates.
[c]includes multivariate complementarity.
[d]infauna and epifauna manipulated.

Table 1b Summary of experiments manipulating diversity of marine organisms published through early 2007[a]: Effects of manipulating richness at one level on response by other levels

Taxon manipulated	Response	Positive	Negative	No effect	Transgressive overyielding
Algal prey	Consumer growth	6	0	0	1/6
	Consumer survival	5	0	2	0/7
	Consumer reproduction	5	0	3	1/8
	Integrated production or population growth	6	0	1	5/7
Consumer	Prey biomass	3	8	4	2/15
Predator	Plant biomass (two trophic levels away)	3	2	1	2/5

[a]Full data for all studies are in **Supplemental Tables 1–4**. Individual studies may be counted multiple times in the table if they either conducted more than one independent experiment or measured more than one potentially independent response variable. Not all studies explicitly tested for transgressive overyielding, and so its existence was inferred in some cases from data available in graphs; in some cases it was impossible to tell because monoculture means were not given, so the total number of possible studies in which trangressive overyielding could be detected is often less than the total number of studies that showed a richness effect (see Supplemental Materials for details).

Strikingly, despite a large number of experiments examining plant richness effects on plant community properties or processes, none found evidence of transgressive overyielding (see also Cardinale et al. 2006). The frequency of richness effects also varied among taxa and response metrics, though in most cases the total number of experiments was small. Still, we note that richness effects appear to be less common for within-trophic-level response variables (**Table 1a**, 47/74) than for those that cross trophic levels (**Table 1b**, 38/49), although some experiments were difficult to classify into one of those groups. Nearly all studies (91/99, see data in **Supplemental Tables 1–4**) find significant identity effects, indicating that most experiments find strong effects of particular species, regardless of taxonomic group or metric of response.

Effects of Producer Diversity on Primary Production and Related Processes

The most common ecosystem processes measured in species richness manipulations to date are primary production and biomass accumulation (Hooper et al. 2005). Terrestrial experiments usually measure production as biomass accumulation over a season. Biomass is less reflective of production for algae because the majority of biomass can be removed by a variety of disturbances, transported away from production sites by currents or removed by intense herbivory (Cebrian 1999). For these reasons, experiments examining the effects of marine primary producer (hereafter, plant) diversity on production are often performed in herbivore-free cages or mesocosms that minimize tissue loss due to natural senescence or disturbance. Further, in many marine systems, macrophyte biomass is inversely, or nonlinearly, related to primary productivity due to resource depletion that can limit production when standing stock is high (Carpenter 1986). Production and biomass are thus in some ways separate ecosystem functions in many marine systems, with production measuring energy and material fluxes and biomass measuring habitat characteristics.

Primary production and biomass. A positive relationship between algal richness and biomass was detected in field surveys of highly diverse macroalgal communities in Jamaica (Bruno et al. 2006). This pattern is concordant with the diversity-productivity hypothesis but could clearly be driven by a variety of factors other than algal richness. All the experiments that measured the effects of marine algal or angiosperm richness and identity on primary production detected strong identity effects but only roughly half found evidence for an effect of richness (**Table 1a,b; Supplemental Table 1**). In several studies, the relative strength of these was compared, and all found that the magnitude of the identity or composition effects was roughly 10 times stronger than that of richness, which was generally weak and likely ecologically insignificant (e.g., Bruno et al. 2005, 2006). Thus even where richness effects occurred, identity and richness contributed little to primary production. By comparison, a recent meta-analysis across terrestrial and aquatic habitats (Cardinale et al. 2006) detected more consistent richness effects (67/76 studies for biomass) that the researchers argued were largely explained by the strong effects of particular species present in

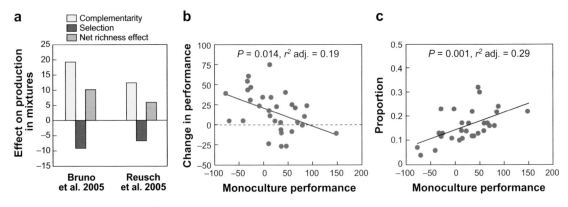

Figure 2

Partitioning of richness effects from two field experiments into complementarity and selection effects. (*a*) Bruno et al. (2005) tested the effects of seaweed richness on production, measured as final wet mass, and Reusch et al. (2005) tested the effects of seagrass genotypic richness on shoot density. Calculations are based on comparing net production in monocultures and mixtures. In both cases the relatively small richness effect is the result of strong positive complementarity offset by negative selection effects. (*b*) Relationship between the performance (% change in wet biomass) of a species in monoculture and its relative performance in polyculture (polyculture% growth-monoculture% growth). (*c*) Relationship between the performance (% change in wet biomass) of a species in monoculture and its proportion of the total final polyculture biomass (that is, dominance). The dashed line in *b* is the 1:1 growth function; points above this line are cases where species grew faster in mixture. (*b* and *c* are redrawn from Bruno et al. 2005.)

polyculture (sampling effects), because species-rich mixtures rarely outperformed the best-performing monoculture.

Simple equations allow the partitioning of diversity effects into components attributable to sampling (or more generally selection) and complementarity (Loreau & Hector 2001). Interestingly, studies that have performed this analysis generally find that complementarity effects are positive and selection effects are negative (**Figure 2**). Positive complementarity occurs when species are more productive on average in mixtures than in monoculture, likely owing to facilitation or resource partitioning (Bruno et al. 2005, Loreau 1998). Negative selection indicates that species that do well in monoculture (e.g., fast growing species) perform relatively poorly when grown with other species. These mechanisms can counteract each other, leading to weak or neutral net richness effects despite strong complementarity among species. This finding contrasts with the sampling effect hypothesis (Huston 1997), which argues that positive effects of richness are driven largely by the random inclusion and ultimate dominance of species with especially high functionality (that is, those with the greatest monoculture performance). In benthic marine communities the most productive genotypes and species often do not dominate polycultures, and species with lower inherent productivity often persist and perform well in diverse communities (**Figure 2**; see also Bracken & Stachowicz 2006, Bruno et al. 2006, Duarte et al. 2000, Hughes

& Stachowicz 2004; Reusch et al. 2005), frequently negating positive effects of biodiversity on algal biomass based on facilitation and complementarity. This could be due in part to trade-offs between fast growth and competitive ability and is not a uniquely marine phenomenon (Hooper & Dukes 2004, Loreau & Hector 2001), although it is not yet known whether positive or negative selection effects are more prevalent in different habitats.

The ability to explicitly partition richness effects between sampling and complementarity is restricted to metrics for which the contribution of each species in polyculture can be unambiguously determined. For some metrics, such as invasion resistance or nutrient uptake or regeneration, partitioning would be impossible because one cannot easily determine the amount of function due to each species in polyculture. However, there are other processes where this partitioning could be profitably applied, including the relative contribution of different predators to prey mortality.

Resistance, resilience, and stability. Virtually all studies that have examined the effects of producer richness on measures of stability have found a positive effect, though the effects on different metrics of stability vary among studies. Allison (2004) manipulated intertidal macroalgal diversity and measured community resistance and resilience in response to thermal stress. Surprisingly, the more diverse plots were less resistant than depauperate ones, losing more biomass to heat stress, but this was largely because they had greater biomass before the stress was imposed. In contrast, there was a positive effect of functional group richness on resilience (recovery), owing to the presence of particularly resilient species and to facilitation promoting recruitment. In contrast, experimentally enhancing eelgrass (*Zostera marina*) genotypic (clonal) richness increased community resistance to grazing by geese (measured as the change in shoot density in response to the disturbances) but had no effect on resilience (rate of recovery) (Hughes & Stachowicz 2004). Two other eelgrass experiments (Reusch et al. 2005, Williams 2001) provide evidence consistent with an effect of intraspecific diversity on processes associated with recovery after disturbance or transplantation.

Kertesz (2006) tested the effects of diversity on stability by crossing macroalgal richness with manipulations of nutrient concentration. The results suggested that richness tended to stabilize biomass production across seasons and in response to variable resource concentrations, as the coefficient of variation in biomass declined with increasing richness. When we compared the results of seven published experiments performed with the same species pool across environmental gradients in time and space (Bruno et al. 2005, 2006), we found that the cross-experiment coefficient of variation in biomass was nearly an order of magnitude higher for algal monocultures, on average, than for the highest-diversity mixtures. Similarly, Worm et al. (2006) reanalyzed the experimental data of Watermann et al. (1999) and found that microalgal biomass accumulation varied less at high than low richness across a factorial combination of three sediment types and three temperatures. Thus, experimental manipulation of marine microalgae, seaweeds, and seagrasses all showed that diversity consistently reduces temporal fluctuations in community biomass (see also Stachowicz et al. 2002) and/or increases stability. This, combined with the relatively weak contribution of richness to average production, suggests that the effects of richness on marine plant production and biomass may be greatest when considering ecosystem processes that involve variability rather than mean responses (**Table 1a,b**).

Decomposition, nutrient availability, and uptake. Bracken & Stachowicz (2006) directly tested the hypothesis that seaweed species richness is positively related to nutrient depletion by manipulating the richness and composition of macroalgae across a gradient of nutrient concentrations in microcosms. They found that species differed in their use of nitrate and ammonium and that nutrient uptake was 22% greater in polycultures than predicted based on a weighted average of species' uptake rates in monoculture. Complementarity among species in total nitrogen uptake only emerged when the use of multiple forms of nitrogen (ammonium and nitrate) were considered simultaneously (multivariate complementarity); diversity had no effect on uptake of either nitrogen form alone. In a separate study, total soil nitrogen accumulation in a restored salt marsh was positively related to plant species richness (Callaway et al. 2003). Despite this, total soil nitrogen availability was also higher in the most diverse plots, probably reflecting the increased organic matter incorporation into soils as a result of higher total aboveground and litter biomass in the species-rich plots. These diversity effects may have resulted from a mix of sampling and complementarity effects, as one species in monoculture (*Salicornia virginica*) did achieve equal biomass to the mixture, whereas no single species had as great an effect on soil nutrient levels as the mixture. Of two experiments examining the effect of seagrass genotypic diversity on sediment porewater ammonium concentration, one found an inverse relationship suggestive of more complete resource use when diversity is high (Hughes & Stachowicz 2004, but not Reusch et al. 2005).

Facilitation of associated species. Macrophyte diversity could affect the structure of communities of epiphytic algae and animals that inhabit them not only by providing enhanced food through greater primary production, but also via creation of larger and more structurally complex or heterogeneous habitats (Bruno & Bertness 2001, Heck & Orth 1980). Results of experimental studies of this phenomenon are mixed. Two studies found that seagrass genotypic diversity had no effect on the diversity of associated invertebrate species but was positively related to epifaunal abundance (Hughes & Stachowicz 2004, Reusch et al. 2005). Higher shoot density in higher diversity treatments did play some role in this (Reusch et al. 2005), but at least one study found an effect of diversity even when controlling for shoot density (Hughes & Stachowicz 2004). Several studies found little effect of manipulating plant or algal species diversity on the animal community despite strong effects of particular plant or algal species (Moore 2006, Parker et al. 2001). In contrast, intertidal seaweed diversity increased the richness and diversity but not the abundance of associated invertebrates, apparently because each algal species harbored a semiunique invertebrate fauna (J. Stachowicz, M. Bracken and M. Graham, in preparation). The generally weak and inconsistent effects of macrophyte richness on associated species richness could be due to the relative rarity of host specialization in the particular systems studied or the generally low host-specificity of marine consumers compared with many insects on land (Hay & Steinberg 1992). Interestingly, effects of grassland plant richness on terrestrial insect abundance and diversity are similarly weak and inconsistent (Haddad et al. 2001, Siemann et al. 1998), probably because insect communities consist of multiple trophic levels, which interact among themselves as well as with plant diversity.

Nutrient Regeneration and Bioturbation

A number of experiments in soft-sediment systems have tested the effects of infaunal species richness on fluxes of nutrients out of the sediments (reviewed in Raffaelli et al. 2003, Waldbusser & Marinelli 2006). Broad-scale correlations between infaunal richness and ammonium flux or biomass provide intriguing evidence for a positive diversity-nutrient efflux correlation (Emmerson & Huxham 2002), although whether this relationship is causal or whether both diversity and ammonium flux are controlled by a third variable is unclear. In experiments, richness effects, when they occur, are usually owing to strong effects of a particular species, typically a bioturbator (Emmerson et al. 2001, Ieno et al. 2006). Further investigation in infaunal systems found that functional richness did enhance ammonium flux, but this effect depended on flow (Biles et al. 2003). Similarly, particular species or combinations of epibenthic grazers have stronger effects than grazer richness on sediment organic matter by influencing the quantity and types of algal biomass accumulating in the seagrass canopy and on the sediment surface and its subsequent processing (Canuel et al. 2007). Manipulation of seagrass species richness had no effect on the rate of seagrass detrital decompostion in litterbags (Moore 2006, Moore & Fairweather 2006). Overall, the conclusion from these experiments is that ecosystem processes can often be predicted from species composition, but not from species richness (**Table 1a,b**; **Supplemental Table 2**).

Although experiments with infaunal invertebrates have often found that a single strong interactor dominates ecosystem function, this is not always the case. Several studies have found that interactions among species can result in underyielding or overyielding of nutrient fluxes relative to expectations based on additivity (e.g., Emmerson et al. 2001, Raffaelli et al. 2003). As one example, Waldbusser et al. (2004) manipulated infaunal polychaete richness and measured effects on phosphate and oxygen flux and on sediment profiles of oxygen and pH. They found strong species-specific effects on particular response variables, but different species controlled different processes, leading to multivariate complementarity. Interactions among species also led to underyielding with respect to both oxygen and phosphate flux in the multispecies communities. Waldbusser and colleagues attributed this to the high oxygen permeability of *Clymenella* tubes leading to greater oxygen content in the deeper sediments; oxic porewaters increased the adsorption of phosphate onto particles, decreasing phosphate flux out of the sediments.

Both of the manipulations of infaunal diversity that measure variability in processes found reduced variability in multispecies assemblages relative to monocultures: either reduced spatial variation in fluxes (Waldbusser et al. 2004) or greater proportion of variance explained in regressions (Emmerson et al. 2001). Although the precise mechanism underlying this effect is unknown, it suggests some sort of complementary effect of species on sediment properties in space (deep versus shallow burrowers) or time, or perhaps owing to context-dependent effects of species (Emmerson et al. 2001).

Resistance to Invasion

Based on the idea that more diverse assemblages more completely use available resources, one might expect diverse communities to be less susceptible than species-poor communities to invasion by new species (Elton 1958). The relationship between species richness and invasibility in terrestrial systems is characterized by apparently contradictory results from experiments, which generally show reduced invasion success with increasing diversity, and observational studies that show the opposite (Fridley et al. 2007). The positive result in surveys is most often explained as a consequence of spatial heterogeneity, which positively affects both native and exotic richness by increasing niche diversity, although alternative explanations exist (Davies et al. 2005; Fridley et al. 2004, 2007).

Marine studies on diversity and invasion do not always follow this pattern, however, and have shed some light on this apparent paradox. Stachowicz et al. (1999, 2002) found that survival and cover of three different sessile invertebrate invaders decreased with increasing resident species richness because resident species were complementary in their temporal patterns of space occupation. Individual species fluctuated in abundance, but these fluctuations were out of phase. Thus, at least one species was always abundant and occupying space in the high-diversity treatments, whereas there were periods of high space availability in the low-diversity treatments. This mechanism appears to operate in the field at larger scales (Stachowicz et al. 2002) driven by complementary temporal niches that arise from seasonal differences in recruitment patterns among species (Stachowicz & Byrnes 2006). Such seasonal or temporal niches may drive diversity effects on invasion resistance in other communities. For example, the biomass of mobile and sessile invertebrate invaders in experimental seagrass mesocosms decreased with increasing species richness of resident mobile invertebrates (France & Duffy 2006a). Grazers in this system do show seasonal abundance patterns (Duffy et al. 2001, Parker et al. 2001), which should produce more complete resource use throughout the season and contribute to this effect.

In contrast to these findings, an experimental study of marine algae found that algal functional group richness did not affect invasion by other native species and that instead functional group identity most strongly affected invasion (Arenas et al. 2006). Although these researchers found that resource availability did control invasion success, algal species identity (and not richness) controlled resource availability. However, in other algal experiments, complementary use of light and space by different functional groups reduced total resource availability and thus invasion success (Britton-Simmons 2006). An overall negative effect of species richness on invader abundance can result even when algal richness enhances initial settlement of invaders through facilitation (White & Shurin 2007). On balance, experimental marine studies generally support an inhibitory effect of increasing diversity on invasion success, mediated in large part by complementary resource use among taxa (**Table 1a,b**; **Supplemental Table 3**).

Observational studies, while they cannot unambiguously assign causation, can illuminate whether the mechanistic effects of richness identified in experiments are sufficiently strong to generate patterns in the context of natural variation in other important factors. Compared with terrestrial systems, there have been surprisingly few observational studies of resident diversity and invasion in the sea. A survey of sessile marine invertebrates in Tasmania found a positive correlation between the number of native species and the species richness (and to a lesser extent the abundance) of both native and non-native settlers (Dunstan & Johnson 2004). They attributed the positive correlation to a combination of interspecific facilitation and the dominance of low-richness communities by a few large colonies, which were difficult to displace. Likewise, a similar study performed across several spatial scales found that the strength and direction of the relationships between native and exotic plant richness and cover in estuarine shoreline plant communities varied among sites and sampling scale, with negative relationships only occurring at smaller spatial scales (Bruno et al. 2004). Several other small-scale studies have found negative correlations between native richness and invader abundance (White & Shurin 2007) or invader richness (Stachowicz et al. 2002). Thus, the effect of native richness can be strong enough to generate field patterns, whereas in other cases it is overwhelmed by other factors.

Using a multiple regression approach, Stachowicz & Byrnes (2006) examined the context dependency of richness effects on invasion. They found that substrate heterogeneity and the availability of primary space markedly influenced the slope of the relationship. Specifically, the substrate heterogeneity and additional settlement space generated by a structurally complex exotic bryozoan (*Watersipora subtorquata*) caused the native-invader richness relationship to shift from negative to positive (**Figure 3**). A negative relationship was only found when facilitators were rare and space was limiting, suggesting that the conditions under which the effect of species richness on invasion is dominant are restricted. Terrestrial studies often agree, finding that the positive effects of heterogeneity or resource levels on both native and exotic richness drive a positive native-exotic richness correlation, particularly at larger spatial scales, whereas smaller scale negative relationships are often (but certainly not always) found, reflective of the more limited heterogeneity at that scale (e.g., Davies et al. 2005, Fridley et al. 2007, Shea & Chesson 2002).

Within-Trophic-Level Effects of Animal Diversity on Secondary Production and Resource use

Given the generally stronger top-down control in marine than in terrestrial ecosystems (Cyr & Pace 1993, Shurin et al. 2002), a key question regarding marine and other aquatic systems is whether diversity of animal consumers has any consistent effect on resource use and production. Mechanisms that might lead to such an effect are similar to those for sessile species discussed previously (see also Duffy 2002). A growing number of studies have addressed potential effects of consumer species richness on

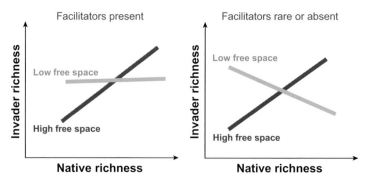

Figure 3

Interactive effects of native richness, resource availability, and facilitation on the richness of exotic species (after Stachowicz & Byrnes 2006). At high levels of open space, there was a strong positive relationship between native and invader richness, likely owing to nonselective disturbance agents that affect native and invader diversity in a similar negative manner, resulting in native and invader richness covarying positively as a reflection of extrinsic factors. At low levels of open space, the slope of the relationship depended on the presence of a foundation species, *Watersipora subtorquata*, which provides secondary space for attachment and thus alleviates space limitation. When *Watersipora* was present there was a consistent positive relationship between native diversity and exotic diversity, regardless of the level of open primary space. When *Watersipora* was absent the native-invader relationship was positive at high levels of open space, but became negative at low levels of open space. One interpretation of these data is that the negative effects of native richness on invasion are only sufficiently strong to be manifest in field patterns when available resources (both primary and secondary space) are in short supply.

ecosystem properties (Duffy et al. 2007), including marine experiments in a variety of estuarine, rocky shore, and subtidal habitats (**Table 1a,b; Supplemental Table 4**). One series of experiments manipulated diversity of crustacean herbivores and followed the effects on development of experimental seagrass ecosystems in mesocosms (Duffy et al. 2001, 2003, 2005; France & Duffy 2006a,b). Most of these studies found that, as predicted by theory (e.g., Holt & Loreau 2002), increasing richness of grazer species resulted in greater grazer biomass and lower standing stocks of their algal prey. Because the experimental seagrass ecosystems allowed natural recruitment of algae and sessile invertebrates, they were able to show that grazer richness affected community succession, not only reducing total resource (algal) biomass but also shifting the composition of the assemblage toward unpalatable cyanobacteria and sessile invertebrates and reducing prey diversity (Duffy et al. 2003, France & Duffy 2006a). These experiments also found strong evidence for multivariate complementarity in which particular species influenced individual response variables such as epiphyte or animal biomass, but the most diverse grazer assemblage maximized each of these response variables simultaneously, producing a community state different than that of any single grazer species.

One potentially important pattern emerging from the seagrass grazer studies is that effects of species richness on ecosystem properties were not detectable with three grazer species (Duffy et al. 2001) but were clear with four (Duffy et al. 2005), six (Duffy et al. 2003), or eight species (France & Duffy 2006b). Results of these experiments suggest that effects of species loss will be less predictable and more idiosyncratic when diversity is initially low, whether naturally or as a result of experimental design (Duffy et al. 2001, O'Connor & Crowe 2005, Schiel 2006).

Within the marine microbial loop, there is also evidence that diversity at a focal trophic level enhances both production and resource use by that level. Increasing richness of herbivorous ciliates strongly decreased total abundance of their algal prey (Gamfeldt et al. 2005), an effect attributable in part to complementarity because the mixture of three ciliate species reduced algal abundance well below the level achieved by any single ciliate species. These researchers also found that total ciliate abundance was strongly enhanced by increasing diversity of either the ciliates themselves or their algal prey. Remarkably, ciliate abundance was more than twice as great than in any other treatment when diversity of both algal prey and herbivorous ciliates was highest.

Diversity in Multitrophic-Level Experiments

Although some of the mechanisms by which diversity acts within trophic levels translate simply to multitrophic-level situations, interactions among mobile heterotrophs can be more varied when considering their effect on adjacent trophic levels (Duffy et al. 2007). A substantial body of experimental work on multiple predator effects (Sih et al. 1998) in a variety of communities including benthic marine systems (e.g., Crowder et al. 1997, Hixon & Carr 1997, Martin et al. 1989) supports many mechanisms by which increasing predator diversity can either decrease (via diet complementarity or predator facilitation) or increase (via intraguild interference or omnivory) herbivory, with correspondingly positive and negative effects on plant biomass. This diversity of potential mechanisms with opposing effects on herbivores can make the prediction of the consequences of changing predator or prey diversity on plant biomass complicated.

Prey diversity and the strength of top-down control. Several hypotheses suggest that prey diversity can affect the strength of top-down control. First, the variance in edibility hypothesis argues that a more diverse prey assemblage is more likely to contain at least one resistant species that can thrive in the presence of consumers (a sampling effect), such that more diverse assemblages will maintain higher biomass under strong consumer pressure (Duffy 2002; Leibold 1989). Thus, in multitrophic systems, the edibility of prey species is expected to be an important mediator of diversity effects because it can foster shifts in species dominance that in turn affect ecosystem functional properties (Thébault & Loreau 2003, 2006). Second, the dilution hypothesis, or its inverse, the resource concentration hypothesis (e.g., Keesing et al. 2006), suggests that a more diverse prey assemblage should reduce the relative and absolute abundances of prey available to specialist consumers, reducing consumer efficiency. When consumers are generalists or have overlapping resource requirements, deletion of a particular species might result in an increase in the abundance of remaining predator species that can compensate for the loss of top-down control. However, O'Connor & Crowe (2005) found that biomass compensation by remaining grazing limpet species after the removal of a dominant could only maintain grazing pressure in the short term.

Few experiments have explicitly tested effects of prey diversity on consumer control in marine systems. Two experiments come from seagrass beds. First, at the plant level, eelgrass (*Zostera marina*) plots planted with higher genotypic richness lost fewer shoots, on average, than low-richness plots when exposed to grazing by geese (Hughes & Stachowicz 2004). Though the mechanism was unclear it was not driven by the dominance of a resistant genotype in mixed genotype plots. Second, at an intermediate trophic level, increasing richness of an assemblage of crustacean herbivores resulted in higher grazer biomass in the presence of crab predators than did the average herbivore monoculture (Duffy et al. 2005), probably because predator-resistant herbivore species came to dominate the assemblage under intense predation pressure. Both of these studies suggest that diversity at the prey level can dampen ecosystem responses to top-down control, stressing the context-dependence of diversity effects at one trophic level on the activities of other levels. In contrast, manipulation of marine microalgal diversity found that algal diversity did not reduce algal susceptibility to herbivore control (Gamfeldt et al. 2005).

The most comprehensive, albeit indirect, evidence that prey diversity can reduce consumer control of aggregate prey standing stock comes from a meta-analysis of 172 aquatic experiments, which found that herbivore control of algal biomass declined with increasing algal diversity (Hillebrand & Cardinale 2004). The underlying cause of this pattern could not be determined, but more diverse algal communities might be more likely to contain unpalatable algal species, have higher rates of recovery owing to more complete resource use, or have a greater incidence of facilitative interactions. The damping effect of prey diversity on top-down control also appears consistent with the general pattern in terrestrial studies of plant-insect interactions (Andow 1991).

Prey diversity and consumer nutrition and production. Mixed algal diets enhanced herbivore growth, biomass accumulation, and/or reproductive output compared to average algal monocultures in nearly all cases examined for taxa as diverse as protozoa, crustaceans, and sea urchins (see **Table 1a,b**; **Supplemental Table 4**; and Worm et al. 2006 for details). In several cases, grazers fed mixed diets performed no differently than those fed the best single food item, perhaps because grazers selectively consumed only the species that led to highest fitness. However, in many studies monospecific diets that produced highest growth led to low survival and vice-versa. Thus, when considering integrated measures of animal performance (e.g., growth × survival × reproductive output) animals fed diverse diets often outperformed even the best single species diet (**Table 1b**). These studies suggest that availability of a diverse prey base may be important to maintaining high production. The mechanisms underlying these effects are unclear, although diverse diets could be better either because of the provision of complementary nutrients, dilution of defensive chemicals, or both (Bernays et al. 1994, DeMott 1998).

Predator diversity and the strength of trophic cascades. Predator diversity could also affect the strength of trophic cascades, either enhancing positive indirect effects on plants where predators have complementary feeding preferences or modes that enhance prey risk, or dampening cascades where predators interfere with or eat one another (Casula et al. 2006, Sih et al. 1998). Only a few experiments have manipulated predator diversity and measured the cascading effects on producers, yet these are clearly critical to understanding the effects of current biodiversity declines on ecosystem functioning (**Figure 1**). These experiments have confirmed that increasing diversity of species that are strict consumers of herbivores (that is, no intraguild predation or omnivory) can indirectly increase plant biomass in salt marsh (Finke & Denno 2005) and subtidal algal ecosystems (Bruno & O'Connor 2005, Byrnes et al. 2006). In most cases, this effect was because predator diversity decreased herbivore activity or per capita feeding rate rather than herbivore density. Positive field correlations between predator diversity and plant biomass (Byrnes et al. 2006) reinforce that these mechanisms likely operate in natural systems.

These manipulations of predator diversity illustrate an important factor influencing the impacts of changing diversity at higher trophic levels that potentially distinguishes them qualitatively from better-studied diversity effects at the plant level, namely the commonness of omnivory and intraguild predation. Experiments in both the subtidal macroalgal system (Bruno & O'Connor 2005) and salt marsh (Finke & Denno 2005) found that when omnivores and intraguild predators were included in the most diverse predator communities, high predator diversity led to lower, not higher, plant biomass. In the algal community this was because some predators also fed on algae, whereas in the salt marsh community predators interfered with or ate other predators, reducing herbivore suppression. Such complex trophic interactions are a hallmark of even very simple natural ecosystems and can, in some cases, reverse expected diversity effects based on niche partitioning and facilitation.

The aspects of food web complexity that lead to predator richness decreasing plant biomass (e.g., omnivory and intraguild predation) and the mechanisms leading to enhanced plant biomass (e.g., complementary prey preferences, predator-predator facilitation), are both predicted to strengthen with increasing species richness. Thus it is perhaps not surprising that a meta-analysis of 114 trophic cascade experiments in a range of systems found no statistical support for an effect of predator diversity on plant biomass (Borer et al. 2005). However, the range of species diversity in the studies analyzed by Borer et al. (2005) was probably insufficient to detect an effect of diversity, even if one existed. Earlier meta-analyses of terrestrial experiments that included a broader range of studies did find greater cascading impacts of predators on plants in systems with low herbivore diversity (e.g., Schmitz et al. 2000). Intriguingly, some oceanographic surveys suggest that at high diversity compensatory population dynamics among predator species contributes to a greater stability of the predator community in the face of intense harvesting, and that this diversity is both a partial cause and a consequence of high primary production and biomass (Frank et al. 2006).

Dispersal and Connectivity

Several recent experiments have explored how connections among habitat patches alter relationships between biodiversity and marine ecosystem properties in meta-communities. In experimental seagrass habitats that were closed to immigration and emigration, more diverse grazer assemblages achieved higher grazer abundance and more effectively cropped algal biomass, compared with less diverse grazer assemblages (France & Duffy 2006a,b), as discussed in detail above. When patches were connected so that grazers could move among them, the dependence of grazer abundance on diversity was erased and patches became more heavily dominated by inedible algae, apparently as grazers were free to move in search of patches with higher quality food (France & Duffy 2006b). Similar results have been found in experiments manipulating grazer diversity and connectivity among habitat patches in Baltic rock pool mesocosms (Matthiessen et al. 2007). Thus, connectivity among habitat patches can strongly modify the biodiversity effects on ecosystem properties found in previous studies of closed systems largely by eliminating differences in realized diversity among patches. Alternatively, variation in connectivity among environments can actually establish local scale variation in diversity, which then affects production and biomass, as shown in a laboratory study of microalgae (Matthiessen & Hillebrand 2006).

SUMMARY POINTS

Recent meta-analyses of biodiversity manipulations [Worm et al. 2006 (see the sidebar, Connecting Diversity to Ecosystem Services), Balvanera et al. 2006, Cardinale et al. 2006], as well as our own compilation (**Table 1a,b**; **Supplemental Tables 1–4**) suggest some generalizations about the effects of marine biodiversity:

1. The richness of species, functional groups, and/or genotypes affects ecosystem functioning in the majority of measurements in the majority of experiments. Diverse assemblages usually perform differently than the average monoculture (85/123 cases, **Table 1a,b**), but are less often better than the best-performing monoculture (26/105 cases). Identity effects were found in nearly all studies (91/99, see **Supplemental Tables 1–4**) and were stronger than richness effects in the few studies for which effect sizes were compared.

2. Positive richness effects in which polycultures outperform the average, but not best, monoculture commonly result from either positive selection (the sampling effect) or a combination of complementarity and negative selection.

3. A significant number of studies provides support for the idea of multivariate complementarity, in which a mixture outperforms all monocultures only when multiple aspects of the total community response are considered simultaneously or integrated into a multivariate index of ecosystem performance. Further research is needed to assess the general frequency and ecological significance of this phenomenon.

4. Although more comparative studies are needed, richness manipulations appear to more strongly and consistently affect metrics related to stability (resistance, resilience, reduced variability) than metrics that reflect mean states.

5. The effect of predator richness on plant biomass is consistently strong, but the direction of the effect is variable (increase versus decrease in plant biomass) and contingent upon the degree of omnivory and intraguild predation.

6. Observational studies suggest that the mechanistic effects of diversity on ecosystem functioning identified in experiments can be strong enough to generate correlations between diversity and function in field surveys. Such studies permit examination of a broader range of richness, larger spatial and temporal scales, more realistic environmental conditions, and provide better potential to test connections with ecosystem services valued by society (see sidebar, Connecting Diversity to Ecosystem Services).

FUTURE ISSUES

Although marine biodiversity-function studies have proliferated in the past 5–10 years and produced some emerging generalizations, these studies also point to gaps in our understanding of biodiversity-ecosystem function relationships in both terrestrial and marine environments. We offer the following suggestions for future work to consider.

1. Incorporate temporal and spatial heterogeneity. Many experiments lack the sort of spatial and temporal heterogeneity within replicates that is often key to coexistence and may enhance the likelihood that complementarity among species will be expressed. Also, experiments over longer timescales will enhance our understanding of whether remaining species can numerically or behaviorally compensate for the loss of superficially similar species, better assessing the degree to which species are redundant.

2. Develop a multitrophic-level perspective. Realistic estimates of diversity change show differential change at different trophic levels at the regional scale. Given that the effect of diversity at a particular trophic level is often contingent on the presence, diversity, and density of organisms at adjacent trophic levels, a better consideration of the food-web context of diversity manipulations is warranted.

3. Assess the relative importance of diversity. Factorial or nested experiments that manipulate richness along with other factors or compare the magnitude of richness and identity effects can help address the importance of diversity relative to other factors. Statistical analysis of survey data using multiple regression, structural equation modeling, data assimilation, or inverse

modeling will also be useful, though care must be taken in inferring causation from correlation.

4. Reconcile reciprocal relationships between diversity and ecosystem processes. Here we reviewed the effect of biodiversity on ecosystem processes, but many of these processes (productivity, stability, nutrient availability, strength of consumer control) are well-known to affect diversity. How these reciprocal relationships are reconciled is an obvious question in need of attention (Worm & Duffy 2003, Hughes et al. 2007).

DISCLOSURE STATEMENT

The authors are not aware of any biases that might be perceived as affecting the objectivity of this review.

ACKNOWLEDGMENTS

The authors thank Drew Harvell, Matt Bracken, Randall Hughes, and Jarrett Byrnes for constructive criticism on the manuscript. The authors also acknowledge the United States National Science Foundation Biological Oceanography program for funding their work on marine biodiversity ecosystem-function relationships, including this manuscript.

LITERATURE CITED

Allison G. 2004. The influence of species diversity and stress intensity on community resistance and resilience. *Ecol. Monogr.* 74:117–34

Andow DA. 1991. Yield loss to arthropods in vegetationally diverse agroecosystems. *Environ. Entomol.* 20:1228–35

Arenas FI, Sanchez I, Hawkins SJ, Jenkins SR. 2006. The invasibility of marine algal assemblages: role of functional diversity and identity. *Ecology* 87:2851–61

Balvanera P, Pfisterer AB, Buchmann N, He J-S, Nakashizuka T, et al. 2006. Quantifying the evidence for biodiversity effects on ecosystem functioning and services. *Ecol. Lett.* 9:1146–56

Bernays EA, Bright KL, Gonzalez N, Angel J. 1994. Dietary mixing in a generalist herbivore: tests of two hypotheses. *Ecology* 75:1997–2006

Bertness MD, Garrity SD, Levings SC. 1981. Predation pressure and gastropod foraging: a tropical-temperate comparison. *Ecology* 78:1976–89

Biles CL, Solan M, Isaksson I, Paterson DM, Emes C, Raffaelli DG. 2003. Flow modifies the effect of biodiversity on ecosystem functioning: An in situ study of estuarine sediments. *J. Exp. Mar. Biol. Ecol.* 285/286:165–77

Bolser RC, Hay ME. 1996. Are tropical plants better defended? Palatability and defenses of temperate vs tropical seaweeds. *Ecology* 77:2269–86

Borer ET, Seabloom EW, Shurin JB, Anderson KE, Blanchette CA, et al. 2005. What determines the strength of a trophic cascade? *Ecology* 86:528–37

Bracken MES, Stachowicz JJ. 2006. Seaweed diversity enhances nitrogen uptake via complementary use of nitrate and ammonium. *Ecology* 87:2397–403

Britton-Simmons KH. 2006. Functional group diversity, resource preemption and the genesis of invasion resistance in a community of marine algae. *Oikos* 113:395–401

Bruno JF, Bertness MD. 2001. Habitat modification and facilitation in benthic marine communities. In *Marine Community Ecology*, ed. MD Bertness, SD Gaines, ME Hay, pp. 201–18. Sunderland, MA: Sinauer

Bruno JF, Boyer KE, Duffy JE, Lee SC, Kertesz JS. 2005. Effects of macroalgal species identity and richness on primary production in benthic marine communities. *Ecol. Lett.* 8:1165–74

Bruno JF, Kennedy CW, Rand TA, Grant MB. 2004. Landscape-scale patterns of biological invasions in shoreline plant communities. *Oikos* 107:531–40

Bruno JF, Lee SC, Kertesz JS, Carpenter RC, Long ZT, Duffy JE. 2006. Partitioning effects of algal species identity and richness on benthic marine primary production. *Oikos* 115:170–78

Bruno JF, O'Connor MI. 2005. Cascading effects of predator diversity and omnivory in a marine food web. *Ecol. Lett.* 8:1048–56

Buss LW, Jackson JB. 1979. Competitive networks: nontransitive competitive relationships in cryptic coral reef environments. *Am. Nat.* 113:223–34

Byrnes JE, Reynolds PL, Stachowicz JJ. 2007. Invasions and extinctions reshape coastal marine food webs. *PloS ONE* 2(3):e295

Byrnes JE, Stachowicz JJ, Hultgren KM, Hughes RA, Olyarnik SV, Thornber CS. 2006. Predator diversity strengthens trophic cascades in kelp forests by modifying herbivore behaviour. *Ecol. Lett.* 9:61–71

Callaway JC, Sullivan G, Zedler JB. 2003. Species-rich plantings increase biomass and nitrogen accumulation in a wetland restoration experiment. *Ecol. Appl.* 13:1626–39

Canuel EA, Spivak AC, Waterson EJ, Duffy JE. 2007. Biodiversity and food web structure influence short-term accumulation of sediment organic matter in an experimental seagrass system. *Limnol. Oceanogr.* 52:590–602

Cardinale BJ, Srivastava DS, Duffy JE, Wright JP, Downing AL, et al. 2006. Effects of biodiversity on the functioning of trophic groups and ecosystems. *Nature* 443:989–92

Carpenter RC. 1986. Partitioning herbivory and its effects on coral-reef algal communities. *Ecol. Monogr.* 56:345–63

Casula P, Wilby A, Thomas MB. 2006. Understanding biodiversity effects on prey in multi-enemy systems. *Ecol. Lett.* 9:995–1004

Cebrian J. 1999. Patterns in the fate of production in plant communities. *Am. Nat.* 154:449–68

Chapin FS III, Sala OE, Burke IC, Grime JP, Hooper DU, et al. 1998. Ecosystem consequences of changing biodiversity. *BioScience* 48:45–52

Crowder LB, Squires DD, Rice JA. 1997. Nonadditive effects of terrestrial and aquatic predators on juvenile estuarine fish. *Ecology* 78:1796–804

Cyr H, Pace ML. 1993. Magnitude and patterns of herbivory in aquatic and terrestrial ecosystems. *Nature* 361:148–50

Davies KF, Chesson P, Harrison S, Inouye BD, Melbourne BA, Rice KJ. 2005. Spatial heterogeneity explains the scale dependence of the native-exotic diversity relationship. *Ecology* 86:1602–10

DeMott WR. 1998. Utilization of a cyanobacterium and a phosphorus-deficient green alga as complementary resources by daphnids. *Ecology* 79:2463–81

Duarte CM, Terrados J, Agawin NSR, Fortes MD. 2000. An experimental test of the occurence of competitive interactions among SE Asian seagrasses. *Mar. Ecol-Prog. Ser.* 197:23

Duffy JE. 2002. Biodiversity and ecosystem function: the consumer connection. *Oikos* 99:201–19

Duffy JE. 2003. Biodiversity loss, trophic skew and ecosystem functioning. *Ecol. Lett.* 6:680–87

Duffy JE, Cardinale BJ, France KE, McIntyre PB, Thébault E, Loreau M. 2007. The functional role of biodiversity in ecosystems: Incorporating trophic complexity. *Ecol. Lett.* 10:522–538

Duffy JE, Macdonald KS, Rhode JM, Parker JD. 2001. Grazer diversity, functional redundancy, and productivity in seagrass beds: an experimental test. *Ecology* 82:2417–34

Duffy JE, Richardson JP, Canuel EA. 2003. Grazer diversity effects on ecosystem functioning in seagrass beds. *Ecol. Lett.* 6:637–45

Duffy JE, Richardson JP, France KE. 2005. Ecosystem consequences of diversity depend on food chain length in estuarine vegetation. *Ecol. Lett.* 8:301–9

Dunstan PK, Johnson CR. 2004. Invasion rates increase with species richness in a marine epibenthic community by two mechanisms. *Oecologia* 138:285–92

Elton CS. 1958. *The Ecology of Invasions by Animals and Plants.* London: Methuen. 181 pp.

Emmerson M, Huxham M. 2002. How can marine ecology contribute to the biodiversity-ecosystem functioning debate? See Loreau et al. 2002, pp. 139–46

Emmerson MC, Solan M, Emes C, Paterson DM, Raffaelli D. 2001. Consistent patterns and the idiosyncratic effects of biodiversity in marine ecosystems. *Nature* 411:73–77

Finke DL, Denno RF. 2005. Predator diversity and the functioning of ecosystems: the role of intraguild predation in dampening trophic cascades. *Ecol. Lett.* 8:1299–306

France KE, Duffy JE. 2006a. Consumer diversity mediates invasion dynamics at multiple trophic levels. *Oikos* 113:515–29

France KE, Duffy JE. 2006b. Diversity and dispersal interactively affect predictability of ecosystem function. *Nature* 441:1139–43

Frank KT, Petrie B, Shackell NL, Choi JS. 2006. Reconciling differences in trophic control in mid-latitude marine ecosystems. *Ecol. Lett.* 9:1096–105

Fridley JD, Brown RL, Bruno JF. 2004. Null models of exotic invasions and scale-dependent patterns of native and exotic richness. *Ecology* 85:3215–22

Fridley JD, Stachowicz JJ, Naeem S, Sax DF, Seabloom E, et al. 2007. The invasion paradox: reconciling pattern and process in species invasions. *Ecology.* 88:3–17

Gamfeldt L, Hillebrand H, Jonsson PR. 2005. Species richness changes across two trophic levels simultaneously affect prey and consumer biomass. *Ecol. Lett.* 8:696–703

Haddad NM, Tilman D, Haarstad J, Ritchie M, Knops JMH. 2001. Contrasting effects of plant richness and composition on insect communities: A field experiment. *Am. Nat.* 158:17–35

Hay ME, Steinberg PD. 1992. The chemical ecology of plant-herbivore interactions in marine vs terrestrial communities. In *Herbivores: Their Interaction with Secondary Metabolites, Evolutionary and Ecological Processes*, Vol. 2, ed. JA Rosenthal, MR Berenbaum, pp. 371–413. San Diego, CA: Academic

Heck KL Jr, Orth RJ. 1980. Seagrass habitats: the roles of habitat complexity, competition and predation in structuring associated fish and motile macroinvertebrate assemblages. In *Estuarine Perspectives*, ed. VS Kennedy, pp. 449–64. New York: Academic

Hector A, Bazeley-White E, Loreau M, Otway S, Schmid B. 2002. Overyielding in grassland communities: testing the sampling effect hypothesis with replicated biodiversity experiments. *Ecol. Lett.* 5:502–11

Hillebrand H, Cardinale BJ. 2004. Consumer effects decline with prey diversity. *Ecol. Lett.* 7:192–201

Hillebrand H, Shurin JB. 2005. Biodiversity and aquatic food webs. In *Aquatic Food Webs: An Ecosystem Approach*, ed. A Belgrano, UM Scharler, J Dunne, RE Ulanowicz, pp. 184–97. New York: Oxford Univ. Press

Hixon MA, Carr MH. 1997. Synergistic predation, density dependence, and population regulation in marine fish. *Science* 277:946–49

Holt RD, Loreau M. 2002. Biodiversity and ecosystem functioning: the role of trophic interactions and the importance of system openness. In *The Functional Consequences of Biodiversity: Empirical Progress and Theoretical Extensions*, ed. AP Kinzig, S Pacala, D Tilman, pp. 246–62. Princeton, NJ: Princeton Univ. Press

Hooper DU, Chapin FSI, Ewel JJ, Hector A, Inchausti P, et al. 2005. Effects of biodiversity on ecosystem functioning: a consensus of current knowledge. *Ecol. Monogr.* 75:3–35

Hooper DU, Dukes JS. 2004. Overyielding among plant functional groups in a long-term experiment. *Ecol. Lett.* 7:95–105

Hughes AR, Stachowicz JJ. 2004. Genetic diversity enhances the resistance of a seagrass ecosystem to disturbance. *Proc. Natl. Acad. Sci. USA* 101:8998–9002

Hughes AR, Byrnes JE, Kimbro DL, Stachowicz JJ. 2007. Reciprocal relationships and potential feedbacks between biodiversity and disturbance. *Ecol. Lett.* 10:849–64

Huston MA. 1997. Hidden treatments in ecological experiments: re-evaluating the ecosystem function of biodiversity. *Oecologia* 110:449–60

Ieno EN, Solan M, Batty P, Pierce GJ. 2006. How biodiversity affects ecosystem functioning: roles of infaunal species richness, identity and density in the marine benthos. *Mar. Ecol-Prog. Ser.* 311:263–71

Ives AR, Cardinale BJ, Snyder WE. 2005. A synthesis of subdisciplines: predator-prey interactions, and biodiversity and ecosystem functioning. *Ecol. Lett.* 8:102–16

Jones GP, McCormick MI, Srinivasan M, Eagle JV. 2004. Coral decline threatens fish biodiversity in marine reserves. *Proc. Natl. Acad. Sci. USA* 101:8251–53

Keesing F, Holt RD, Ostfeld RS. 2006. Effects of species diversity on disease risk. *Ecol. Lett.* 9:485–98

Kertesz JS. 2006. *The role of biodiversity in a fluctuating environment*. MS thesis. San Francisco State Univ. 44 pp.

Lehman CL, Tilman D. 2000. Biodiversity, stability, and productivity in competitive communities. *Am. Nat.* 156:534–52

Leibold MA. 1989. Resource edibility and the effects of predators and productivity on the outcome of trophic interactions. *Am. Nat.* 134:922–49

Loreau M. 1998. Separating sampling and other effects in biodiversity experiments. *Oikos* 82:600–2

Loreau M. 2000. Biodiversity and ecosystem functioning: recent theoretical advances. *Oikos* 91:3–17

Loreau M, Hector A. 2001. Partitioning selection and complementarity in biodiversity experiments. *Nature* 412:72–76

Loreau M, Naeem S, Inchausti P, eds. 2002. *Biodiversity and Ecosystem Functioning: Synthesis and Perspectives.* New York: Oxford Univ. Press

Loreau M, Naeem S, Inchausti P, eds. 2002. Perspectives and challenges. See Loreau et al. 2002, pp. 237–42

Loreau M, Naeem S, Inchausti P, Bengtsson J, Grime JP, et al. 2001. Biodiversity and ecosystem functioning: current knowledge and future challenges. *Science* 294:804–8

Lotze HK, Lenihan HS, Bourque BJ, Bradbury RH, Cooke RG, et al. 2006. Depletion, degradation, and recovery potential of estuaries and coastal seas. *Science* 312:1806–9

Martin TH, Wright RA, Crowder LB. 1989. Non-additive impact of blue crabs and spot on their prey assemblages. *Ecology* 70:1935–42

Matthiessen B, Gamfeldt L, Jonsson PR, Hillebrand H. 2007. Effects of grazer richness and composition on algal biomass in a closed and open marine system. *Ecology.* 88:178–187

Matthiessen B, Hillebrand H. 2006. Dispersal frequency affects local biomass production by controlling local diversity. *Ecol. Lett.* 9:652–62

May RM. 1974. *Stability and Complexity in Model Systems.* Princeton, NJ: Princeton Univ. Press. 2nd ed.

McCann KS. 2000. The diversity–stability debate. *Nature* 405:228–33

Moore T. 2006. *Seagrass meadows: spatial scaling, biodiversity & ecosystem function*. PhD thesis. Flinders Univ., Adelaide, South Australia. 189 pp.

Moore TN, Fairweather PG. 2006. Decay of multiple species of seagrass detritus is dominated by species identity, with an important influence of mixing litters. *Oikos* 114:329–37

Naeem S, Li S. 1997. Biodiversity enhances ecosystem reliability. *Nature* 390:507–9

Naeem S, Thompson LJ, Lawler SP, Lawton JH, Woodfin RM. 1994. Declining biodiversity can alter the performance of ecosystems. *Nature* 368:734–37

O'Connor NE, Crowe TP. 2005. Biodiversity loss and ecosystem functioning: distinguishing between number and identity of species. *Ecology* 86:1783–96

Paine RT. 2002. Trophic control of production in a rocky intertidal community. *Science* 296:736–39

Parker JD, Duffy JE, Orth RJ. 2001. Plant species diversity and composition: experimental effects on marine epifaunal assemblages. *Mar. Ecol. Prog. Ser.* 224:55–67

Pauly D, Christensen V, Dalsgaard J, Froese R, Torres F Jr. 1998. Fishing down marine food webs. *Science* 279:860–63

Petchey OL, Casey T, Jiang L, McPhearson PT, Price J. 2002. Species richness, environmental fluctuations, and temporal change in total community biomass. *Oikos* 99:231–40

Petchey OL, Downing AL, Mittelbach GG, Persson L, Steiner CF, et al. 2004. Species loss and the structure and functioning of multitrophic aquatic systems. *Oikos* 104:467–78

Pimm SL. 1984. The complexity and stability of ecosystems. *Nature* 307:321–26

Raffaelli D, Emmerson M, Solan M, Biles C, Paterson D. 2003. Biodiversity and ecosystem processes in shallow coastal waters: an experimental approach. *J. Sea Res.* 49:133–41

Reusch TBH, Ehlers A, Haemmerli A, Worm B. 2005. Ecosystem recovery after climatic extremes enhanced by genotypic diversity. *Proc. Natl. Acad. Sci. USA* 102:2826–31

Rex MA, Stuart CT, Hessler RR, Allen JA, Sanders HL, Wilson GDF. 1993. Global-scale latitudinal patterns of species diversity in the deep-sea benthos. *Nature* 365:636–39

Roberts CM, McClean CJ, Veron JEN, Hawkins JP, Allen GR, et al. 2002. Marine biodiversity hotspots and conservation priorities for tropical reefs. *Science* 295:1280–84

Roy K, Jablonski D, Valentine JW, Rosenberg G. 1998. Marine latitudinal diversity gradients: tests of causal hypotheses. *Proc. Natl. Acad. Sci. USA* 95:3699–3702

Sackville-Hamilton NR. 1994. Replacement and additive designs for plant competition studies. *J. Appl. Ecol.* 31:599–603

Sala E, Knowlton N. 2006. Global marine biodiversity trends. *Annu. Rev. Environ. Resourc.* 31:93–122

Sax D. 2001. Latitudinal gradients and geographic ranges of exotic species: implications for biogeography. *J. Biogeogr.* 28:139–50

Schiel DR. 2006. Rivets or bolts? When single species count in the function of temperate rocky reef communities. *J. Exp. Mar. Biol. Ecol.* 338:233–52

Schmitz OJ, Hamback PA, Beckerman AP. 2000. Trophic cascades in terrestrial systems: a review of the effects of carnivore removals on plants. *Am. Nat.* 155:141–53

Shea K, Chesson P. 2002. Community ecology theory as a framework for biological invasions. *Trends Ecol. Evol.* 17:170–76

Shurin JB, Borer ET, Seabloom EW, Anderson K, Blanchette CA, et al. 2002. A cross-ecosystem comparison of the strength of trophic cascades. *Ecol. Lett.* 5:785–91

Siemann E, Tilman D, Haartstad J, Ritchie M. 1998. Experimental tests of the dependence of arthropod diversity on plant diversity. *Am. Nat.* 152:738–50

Sih A, Englund G, Wooster D. 1998. Emergent impacts of multiple predators on prey. *Trends Ecol. Evol.* 13:350–55

Solan M, Cardinale BJ, Downing AL, Engelhardt KAM, Ruesink JL, Srivastava DS. 2004. Extinction and ecosystem function in the marine benthos. *Science* 306:1177–80

Srivastava DS, Vellend M. 2005. Biodiversity-ecosystem function research: is it relevant to conservation? *Annu. Rev. Ecol. Evol. Syst.* 36:267–94

Stachowicz JJ, Byrnes JE. 2006. Species diversity, invasion success, and ecosystem functioning: disentangling the influence of resource competition, facilitation, and extrinsic factors. *Mar. Ecol. Prog. Ser.* 311:251–62

Stachowicz JJ, Fried H, Whitlatch RB, Osman RW. 2002. Biodiversity, invasion resistance and marine ecosystem function: reconciling pattern and process. *Ecology* 83:2575–90

Stachowicz JJ, Whitlatch RB, Osman RW. 1999. Species diversity and invasion resistance in a marine ecosystem. *Science* 286:1577–79

Steneck RS, Vavrinec J, Leland AV. 2004. Accelerating trophic level dysfunction in kelp forest ecosystems of the western North Atlantic. *Ecosystems* 7:323–31

Thebault E, Loreau M. 2003. Food-web constraints on biodiversity-ecosystem functioning relationships. *Proc. Natl. Acad. Sci. USA* 100:14949–54

Thebault E, Loreau M. 2006. The relationship between biodiversity and ecosystem functioning in food webs. *Ecol. Res.* 21:17–25

Tilman D. 1999. The ecological consequences of changes in biodiversity: a search for general principles. *Ecology* 80:1455–74

Tilman D, Lehman C, Thompson K. 1997. Plant diversity and ecosystem productivity: theoretical considerations. *Proc. Natl. Acad. Sci. USA* 94:1857–61

Tilman D, Reich PB, Knops JMH. 2006. Biodiversity and ecosystem stability in a decade-long grassland experiment. *Nature* 441:629–32

Waldbusser GG, Marinelli RL. 2006. Macrofaunal modification of porewater advection: Role of species function, species interaction, and kinetics. *Mar. Ecol. Prog. Ser.* 311:217–31

Waldbusser GG, Marinelli RL, Whitlatch RB, Visscher PT. 2004. The effects of infaunal biodiversity on biogeochemistry of coastal marine sediments. *Limnol. Oceanogr.* 49:1482–92

Watermann F, Hillebrand H, Gerdes G, Krumbein WE, Sommer U. 1999. Competition between benthic cyanobacteria and diatoms as influenced by different grain sizes and temperatures. *Mar. Ecol. Prog. Ser.* 187:77–87

White LF, Shurin J. 2007. Diversity effects on invasion vary with life history stage in marine macroalgae. *Oikos.* 116:1193–1203

Williams SL. 2001. Reduced genetic diversity in eelgrass transplantations affects both population growth and individual fitness. *Ecol. Appl.* 11:1472–88

Worm B, Barbier EB, Beaumont N, Duffy JE, Folke C, et al. 2006. Impacts of biodiversity loss on ocean ecosystem services. *Science* 314:787–90

Worm B, Duffy JE. 2003. Biodiversity, productivity and stability in real food webs. *Trends Ecol. Evol.* 18:628–32

Worm B, Lotze HK, Myers RA. 2003. Predator diversity hotspots in the blue ocean. *Proc. Natl. Acad. Sci. USA* 100:9884–88

Stochastic Dynamics of Plant-Water Interactions

Gabriel Katul,[1,2] Amilcare Porporato,[1,2] and Ram Oren[1]

[1] Nicholas School of the Environment and Earth Sciences, Duke University, Durham, North Carolina 27708-0328

[2] Department of Civil and Environmental Engineering, Pratt School of Engineering, Duke University, Durham, North Carolina 27708-0328; email: gaby@duke.edu

Annu. Rev. Ecol. Evol. Syst. 2007. 38:767–91

First published online as a Review in Advance on September 5, 2007

The *Annual Review of Ecology, Evolution, and Systematics* is online at http://ecolsys.annualreviews.org

This article's doi: 10.1146/annurev.ecolsys.38.091206.095748

Key Words

dimension reduction, nonlinearity, scale effects, soil-plant atmosphere system

Abstract

Describing water flow from soil through plants to the atmosphere remains a formidable scientific challenge despite years of research. This challenge is not surprising given the high dimensionality and degree of nonlinearity of the soil-plant system, which evolves in space and time according to complex internal physical, chemical, and biological laws forced by external hydroclimatic variability. Although rigorous microscopic laws for this system still await development, some progress can be made on the formulation of macroscopic laws that upscale known submacroscopic processes and use surrogate stochasticity to preserve the probabilistic and spectral information content of the high dimensional system. The external hydroclimatic forcing is inherently intermittent with variability across all scales, thereby precluding the use of standard approximations employed in analysis of stochastic processes (e.g., small noise perturbations). Examples are provided to show how superposition of stochasticity at multiple space-time scales shapes plant-water interactions.

It is surely one of the triumphs of evolution that Nature discovered how to make highly accurate machines in such a noisy environment.

(Phillips & Quake 2006)

1. INTRODUCTION

We explore the stochasticity in plant-water dynamics with emphasis on its consequences for deriving laws for plant-water interactions at larger scales (referred to as macroscopic laws) starting from what is known at the smaller scales. The theme of this review purposely resembles a centuries-old problem whose answer led to the development of statistical mechanics and linkages to thermodynamics (see Nelson 2004 for a clear introduction to the applications of thermodynamics and statistical mechanics to biological systems; Dewar 2003, Martyushev & Seleznev 2006, Ozawa et al. 2003, Roderick 2001, and Whitfield 2005, among others, advocate thermodynamic principles to explain macroscopic behaviors in complex natural systems). The original problem considered whether knowledge of molecular matter via microscopic laws permitted description of macroscopic behavior. Naturally, the large number of molecules that make up macroscopic systems and their stochastic motion invites statistical treatment. Although plant-water interactions share attributes with molecular systems, including their high dimensionality (because of the large number of interacting processes impacting water movement within the soil-plant system), there are fundamental differences that prevent immediate applications of statistical mechanics to plant-water interactions:

1. The fundamental submacroscopic laws describing water movement within the plant system are not entirely known (e.g., water flow at the root-soil interface, in the xylem, and in the leaf);
2. Averaging the individual properties of these submacroscopic laws may not provide a meaningful description for the next hierarchical level because of nonlinear interactions and lack of scale separation (i.e., presence of significant variability at all scales); and
3. The drivers of many macroscopic laws remain stochastic because the soil-plant system is open to external environmental forcing such as rainfall, temperature, and radiation. In other words, the soil-plant system is open and far from equilibrium.

Our main objective is to present what is known about plant-water interactions within a stochastic framework, building on recent advances in dynamical systems theory, complex systems science, and stochastic processes. Long-term field measurements at various time and spatial scales are becoming increasingly available thereby providing unprecedented details of the spatial and temporal statistics of plant-water interactions (e.g., long-term ecosystem water vapor flux measurements via eddy-covariance methods, gas-exchange measurements, sapflow and soil moisture measurements, tree-ring reconstruction of net primary productivity, and space-borne leaf area index measurements, to name a few). In the absence of a theory for microscopic dynamics, such a stochastic framework may be a first step that inspires the future statistical mechanics theory in ecology and plant-water dynamics.

2. REVIEW OF REVIEWS

Previous reviews of plant-water interactions have been conducted by Noy Meir (1973) for desert ecosystems and Lathwell & Grove (1986) for tropical ecosystems. The former primarily dealt with space-time stochasticity in rainfall, whereas the latter dealt with the physical and chemical properties of soils that impact ecosystem productivity.

At the plant scale, a wealth of review articles has also dealt with physical, chemical, and physiological aspects of plant-water interactions, primarily the effects of water stress on physiological processes in various organs, gas exchange processes, and plant hydrodynamics. We mention in particular the reviews of Hsiao (1973) on plant-water stress, Sack & Holbrook (2006) on leaf hydraulics, and Sperry (2000) and Sperry et al. (2002) on hydraulic constraints to leaf water supply, as well as several books that summarize the state of knowledge up to the past two decades or so (e.g., Jones 1992, Kramer & Boyer 1995, Larcher 1995, Nobel 2005, Steffen & Denmead 1988).

At the watershed scale, a number of reviews presented the consequences of plant-water interactions on the hydrologic cycle with emphasis on potential feedbacks on the climate system (Brutsaert 1982, Eagleson 2002) and biogeochemical cycling (Eagleson 2002, Schlesinger 1997). The interplay between stochasticity and nonlinear dynamics in ecosystems has also received significant attention—evidenced by a recent special issue of *Oecologia* (in 2005; see also Austin et al. 2004, Schwinning & Sala 2004). Finally, the stochastic aspects of hydroclimatic forcing and their propagation within water-limited ecosystems were also the subject of several recent reviews (Kull & Jarvis 1995, Porporato & Rodriguez-Iturbe 2002, Porporato et al. 2002, Rodriguez-Iturbe & Porporato 2004). The scope of this review diverges from these earlier reviews by focusing on stochasticity and how it arises in plant-water dynamics. We also discuss what the consequences are of incorporating such stochasticity in sample contemporary ecological problems.

3. STOCHASTIC PROCESSES: AN ALEATORY ORIENTATION

The soil-plant-atmosphere system can be considered a stochastic dynamical system evolving in space and time according to its internal physical, chemical, and biological laws and subject to external variability. We speak of stochasticity in a broad sense, referring to the practical impossibility of precisely modeling and predicting temporal evolution and spatial configuration of the system at all spatial and temporal scales. We avoid the philosophical discussion on the true nature of this unpredictable and irregular behavior (Ford 1983) and whether it should be attributed to high dimensionality, instabilities, and sensitivities to initial conditions (Boffetta et al. 2002, Cross & Hohenberg 1993, Eckmann & Ruelle 1985), or to simple ignorance of the detailed functioning of the system that by itself requires a statistical approach (Clark & Gelfand 2006, Jaynes 2003, Kass & Raftery 1995, Sivia & Skilling 2006). To some extent, all these are present and somewhat interlinked in each compartment of the soil-plant-atmosphere system.

Stochastic fluctuations are often referred to as random noise, and solutions of equations with random parameters (or random dynamical systems) are called stochastic processes or random functions. The novice may be easily disoriented and daunted by the vast literature in this branch of mathematics, which was inspired by different disciplines such as physics, chemistry, theoretical finance, communication and information theory, hydrology, etc., and which benefited from fundamental contributions by mathematicians and scientists such as Laplace, Poisson, Maxwell, Bolzmann, Einstein, and Kolmogorov, to name a few. The following is an incomplete orientation list of references for applications of random functions (Bendat & Piersol 1971, Papoulis 1991, Vanmarke 1983), stochastic processes (Cox & Miller 1977, Gardiner 2004, Horsthemke & Lefever 1984, Larson & Shubert 1979, van Kampen 1992) and their applications in the natural sciences (Sornette 2004).

The mathematical modeling of noisy dynamics typically requires a considerable degree of abstraction and simplification. In general, when the analysis is limited to the temporal domain, the existing literature is sufficiently developed for some classes of noise. If the governing (or state) variables are or can be approximated as discrete, one can employ the well-established theory of Markov chains (Cox & Miller 1977, Norris 1998); more typically, the state variables are time continuous and one must resort to so-called stochastic differential equations (SDEs). The most typical building blocks of continuous-state stochastic processes are either Gaussian noises (Brownian motion or Wiener process), used when the irregular fluctuations do not appear to be pulsing and intermittent, or jumps compounded with suitable point processes, used when the pulsing components and the space-time intermittency become dominant (e.g., hourly or daily rainfall). The basic models just mentioned are characterized by random independent increments so that they do not add new memory to the system. Because of their flat power spectrum, they also called white noises. Their use always entails some subtle mathematical aberrations at very fine scales such as discontinuities in the state variables, as is the case for jump processes, or in the first derivative, as is the case for Brownian motion and its relatives (Cox & Miller 1977, Gardiner 2004). Two manifestations of these aberrations appear when analyzing the trajectory of a Brownian particle at infinitely small temporal scales using one stochastic differential equation. The first manifestation is its discontinuous velocities, which is interpreted as being produced by an unrealistic infinite force, and the second is that the trajectories cross a given level infinitely many times in an infinitesimal time interval. These pathologies are the necessary price paid for simplicity, and may be acceptable so long as no special physical meaning is attributed to the fluctuations at such short timescales.

Loosely speaking, white noises give rise to Markovian processes or processes in which future states do not directly depend on past states. They can be described mathematically as either SDEs for the state variables, emphasizing a specific realization (e.g., one possible outcome) of the stochastic process, or the ensemble of realizations (e.g., all possible outcomes) in terms of partial differential equations (PDEs) for the probability density functions of the state variables described by the SDEs (Gardiner 2004, Horsthemke & Lefever 1984, van Kampen 1992).

The use of non-Markovian (or colored noises), fractional Brownian motion, and Levy-type processes has gained recent popularity because of their ability to describe fine temporal dynamics, presence of scaling (e.g., power-laws), and long-term memory effects signified by slowly decaying temporal autocorrelations (Horsthemke & Lefever 1984, van Kampen 1992). Unfortunately, this standard mathematical machinery becomes difficult to implement analytically for nonlinear systems with two or more degrees of freedom (Arnold 2003), and even more difficult when considering space-time processes (Bouchaud & Georges 1990, Cross & Hohenberg 1993, Durrett & Levin 1994, Falkovich et al. 2001, Sagues & Sancho 2004).

We conclude this orientation by noting that the presence of noise in the context of plant-water dynamics may not be necessarily a source of disorder within the system but can be a stabilizing factor (e.g., promoting coexistence among plants), a pattern generator, or a mechanism responsible for phase transitions (e.g., abrupt changes such as extinctions and catastrophic shifts), all of which are now gaining attention in a number of fields (D'Odorico et al. 2005, Horsthemke & Lefever 1984, Porporato & D'Odorico 2004, Rietkerk et al. 2004, Sagues & Sancho 2004, Scheffer et al. 2001, Sornette 2004, Vandenbroeck et al. 1994).

4. STOCHASTIC DYNAMICS

The genesis of stochasticity in the dynamics of plant-water uptake may be traced back to both internal and external factors. To illustrate the origins of internal factors, consider the conceptual diagram in **Figure 1** showing on the abscissa the dimensionality of the system and on the ordinate the degree of nonlinearity of the equations describing the interactions among its variables. As we discuss below, water movement in the soil-plant system typically resides in the high-dimensional and nonlinear region of this figure. When the dynamics of these interactions is approximated by a few coupled equations, the high dimensionality is replaced by stochasticity to preserve the probabilistic and spectral information content present in the natural system. As a result the stochasticity appears in model parameters of the simplified system (e.g., hydraulic conductivity of the soil pores and plant conduits, foliage distributions, stomatal properties, etc.) as well as in the external variability of the hydroclimatic forcing (e.g., rainfall, wind, temperature, radiation).

This stochastic representation of plant-water dynamics is logical in the context of propagating projected climatic fluctuations—an example of external forcing—on ecosystem processes. Vitousek's (1992) review "Global Environmental Change" pointed out some of the problems and difficulties in propagating climate projections to ecosystem processes with the current modeling framework. This difficulty is driven by the fact that ecosystems respond to the full range of hydroclimate variability (particularly the extremes). Hence, there are two practical benefits in considering the plant-water system in a stochastic but simplified framework: (*a*) reduced complexity and (*b*) proper accounting of hydroclimatic variability.

4.1. Origin of Internal Stochasticity

Movement of water in the soil-plant-atmosphere system (**Figure 2**) begins with water migrating from wetter to drier soil pores adjacent to the rooting system following potential energy gradients. Once it has reached and entered the rooting system through a patchy and heterogeneous root membrane, water flows through a tortuous and complex network within the xylem. It experiences phase transition within the leaves,

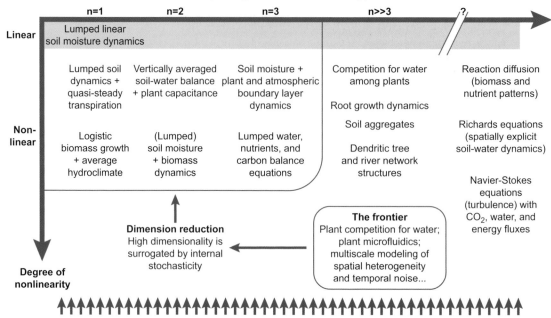

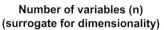

Figure 1

Conceptual diagram showing the complexity of the soil-plant system viewed in the dimensionality/nonlinearity plane. Stochasticity arises when simplified models attempt to capture this complexity (revised from Strogatz 1994). Surrogate stochasticity is added to the simplified models to preserve the probabilistic and spectral information content of the original system. We refer to internal stochasticity as related to the presence of random parameters to distinguish it from the external forcing, which is present at all scales and is stochastic in nature. The "frontier" represents open problems lacking rigorous mathematical formulations.

and exits to the atmosphere in the form of water vapor through leaf stomata. The vapor molecules are then transported by turbulent eddies from within the canopy into the free atmosphere. The transporting energy and sizes of these eddies are partially set by complex interactions among canopy attributes (e.g., leaf area and height), mesoscale forcing (e.g., geostrophic winds and weather patterns), landscape heterogeneity, and the airflow above the canopy.

Resolving all spatial scales needed to describe the trajectory of water in the soil-plant-atmosphere system necessitates a three-dimensional simulation domain spanning 0.1 μm to tens of kilometers, equivalent to requiring $\sim(10^{10})^3$ nodes per time step. The time step must be sufficiently fine to resolve the fastest process, which is the action of viscous dissipation on turbulent fluctuations in the atmosphere (~0.001 s). This high dimensionality in space and time is well beyond the capacity of any

Figure 2

Water moves from the soil to the atmosphere through roots, xylem, and stomata: (*a*) The
root-soil system is heterogeneous at a wide range of scales (smallest ∼0.1 μm). (*b*) The xylem
within the individual branches generate a complex network whose precise details are rarely
known in plant-water dynamics modeling (scale ∼100 μm; after Zimmerman 1983).
(*c*) Plant-atmosphere gas exchange is controlled by stomata (scale ∼10 μm; from Grant &
Vatnick 2004). (*d*) Turbulent eddies transport water vapor from stomata to the free
atmosphere (scale ∼10 m). (*e*) Atmospheric states are modulated by the landscape
heterogeneity (scale ∼10 km). Note that stochasticity is present at all spatial scales.

brute-force computation at present and in the foreseeable future. Furthermore, there
are insurmountable scale issues in attempting to relate water flow in the soil-plant
system with its driving forces. For one, the constitutive laws currently used to describe
water movement in the soil, root, plant, and atmosphere systems do not share the
same representative elementary volume (REV), defined as the minimal spatial scale
of representation.

Consider the constitutive laws in each of the three compartments of the soil-plant-
atmosphere system:

1. Soil. Darcy's law and the water continuity equation are typically combined in
the Richards equation, which describes water movement in unsaturated soils near
the rooting zone at an REV scale containing a sufficiently large number of pore
spaces. This is a nonlinear partial differential equation that provides a space-time

description of water movement but averages out the variability of the soil and root matrix at scales smaller than the REV (Hillel 2004, Jury et al. 1991). It is a major theoretical challenge to up-scale this equation beyond the REV to include the effects of spatial heterogeneities in soil properties, macroporosity, and preferential flows of water and nutrients at discontinuities (e.g., large roots, rocks, etc.). Even the application of Darcy's law within a REV that includes such randomness and heterogeneities is questionable (Steffen & Denmead 1988). In particular, the origin of preferential flows and their impacts on the plant-water relationship remain open areas of research (de Rooij 2000).

2. Plant. Similar problems arise at the plant level. Laminar flow equations (e.g., Hagen-Poiseulle's law for capillary tubes) based on the continuum assumption in fluid mechanics are typically used to describe root and xylem water movement. These assumptions are now being challenged by recent research. For example, predictions of the onset of embolism in the plant xylem require microscale thermodynamic description of air and water microfluid dynamics not captured by Poiseulle's law; the derivation of observed vulnerability curves (Sperry 2000, Sperry et al. 2002) from first principles has not yet been tackled, and the connection between stomatal conductance and the plant-xylem system remains a subject of research (Katul et al. 2003, Oren et al. 1999, Sack & Holbrook 2006, Sperry & Hacke 2002, Sperry et al. 2002).

3. Atmosphere. Mass, momentum, and energy exchanges between the canopy and the lower atmosphere are described by the Navier-Stokes equations and require detailed description of the boundary conditions at the plant-atmosphere interface. Describing the boundary conditions for these equations is complicated by stochasticity at multiple scales. Randomness, beginning with patchiness at the stomatal level and progressing to patchiness in stomatal conductance (Buckley et al. 1997), random leaf distribution, leaf area density, and onward to the atmosphere must be accounted for as dynamic boundary conditions to the Navier-Stokes equations (Albertson et al. 2001). Even for the simple case of a static boundary condition, the Navier-Stokes equations cannot be solved at all the necessary scales (Finnigan 2000).

These constitutive laws (the Richards equation, Poiseuille law, and Navier-Stokes equations) often provide reasonable approximations at a particular scale, typically where microscopic heterogeneities can be averaged out. However, a major challenge is the derivation of effective parameters in simplified models at larger scales (e.g., Bohrer et al. 2005, Chuang et al. 2006, Ewers et al. 2007, Katul et al. 1997). Clearly, novel tactics are needed to further the development of these constitutive laws for applications in the plant system at much finer scales (<10 μm), and of ways to properly scale them to coarser levels. Some answers may be found in microfluid dynamics, a field that is rapidly gaining attention (Squires & Quake 2005).

Formal applications of homogenization and averaging techniques (Torquato 2002) may guide the derivation of effective parameters for the current constitutive laws. Applications of such techniques to the set of equations described above are further complicated by three factors: (*a*) the inherent nonlinearities in the resistance to water flow, soil hydraulics, canopy turbulence, and plant responses to temperature and light, among others; (*b*) the fact that the external forcing can be highly intermittent (e.g.,

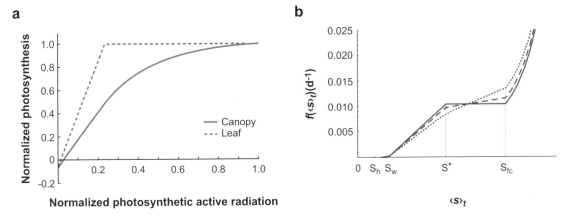

Figure 3

Examples of smoothing of nonlinearities appearing in macroscopic equations by averaging in space and time. (*a*) Space: differences in leaf-and-canopy-photosynthesis response to light when vertically averaging from leaf to entire canopy. A simplified version of the Farquhar photosynthesis model (Farquhar et al. 1980) is used for leaf photosynthesis, whereas the CANVEG model is used to scale it up for the canopy (Baldocchi 1992, Lai et al. 2000, Siqueira et al. 2002); the CANVEG model is a multilayer representation of the canopy that includes light attenuation, leaf physiology, leaf energy balance, and turbulent transport processes (assuming mean air temperature, relative humidity, wind speed, and CO_2 concentration above the canopy are unaltered). (*b*) Time: equivalent soil water loss function $f(\langle s \rangle_t)$ for the temporal soil moisture mean $\langle s \rangle_t$ (Laio et al. 2002). The continuous line is the nonlinear loss function at the daily level, whereas the dashed and dotted lines refer to long-term averages for different noise levels representing rainfall variability. S_h, the hygroscopic point; S_w, the wilting point; S^*, soil moisture at the onset of water stress; and S_{fc}, the field capacity.

rainfall, sunflecks, etc.), producing jumps that abruptly force the system toward different states; and (*c*) the rare presence of a clear scale separation. As a consequence, the up-scaling smoothes and reduces nonlinearities but never eliminates them (as illustrated spatially for photosynthesis and temporally for soil water in **Figure 3**). Furthermore, the classic simplifying techniques, such as "small-noise and system-size expansions," are not directly applicable in this context (Gardiner 2004, van Kampen 1992).

4.2. Origin of External Stochasticity

One of the main challenges in modeling the temporal dynamics of the soil-plant system is the noise structure of the hydroclimatic forcing. Depending on the temporal scale, some hydroclimatic components may be modeled as continuous Gaussian noise or processes driven by Gaussian noise (e.g., annual rainfall, daily temperature, and hourly wind velocity) and others may be modeled as intermittent jump processes (e.g., daily rainfall, subhourly photon flux density in the canopy understory). This strong intermittency alone suffices to prevent direct applications of traditional techniques used to simplify stochastic processes, such as small-noise expansions (Gardiner 2004, van Kampen 1992).

Next we show examples of the complex noise structure at different scales of temperature, light and rainfall, which are among the most important hydroclimatic drivers of the soil-plant system. We start in **Figure 4** with measurements of air temperature above the canopy at scales ranging from interannual to fractions of seconds. Note the large and organized turbulent fluctuations around the diurnal cycle in mean air temperature ($\sim5°C$). The figure shows that these excursions are intermittent and highly nonstationary during daytime as evidenced by progressive changes in the variance of air temperature. Even at sufficiently small timescales (~10 s), large ramp-like temperature excursions are often detected (**Figure 4**). The latter ones are connected with bursts of air having high velocity and low temperature (so-called sweep events) that penetrate the canopy from above, exchange heat with the vegetation, and then are ejected back to the atmosphere (ejection events) with lower velocity and higher temperature (e.g., Katul et al. 2006). Such rapid changes in air temperature can significantly affect temperature-dependent kinetic constants in plant processes.

Light above the canopy may also become intermittent at short timescales because of the passage of clouds (Knapp 1993). Furthermore, within the canopy the light regime can be highly stochastic with intermittent pulses (**Figure 5**) at scales of seconds to minutes owing to random shading by the overlying canopy. This means that photosynthesis of the understory, which is a nonlinear function of light intensity, is driven by random sunflecks (see Naumburg et al. 2001, Pearcy et al. 1997), as can be seen in **Figure 5**.

The effects of these intermittent and high-frequency variations in light levels on the spatial and temporal water movement within various plant organs have not been sufficiently studied (see Ewers et al. 2007, Oren et al. 1998). Intracrown shading can reduce transpiration of some leaves or branches but more water becomes readily available to support higher water fluxes of the better-illuminated branches. Some evidence of this compensation may be graphically demonstrated in **Figure 5**, where stem-level fluxes do not significantly vary in space and appear better behaved in following the overall diurnal cycle of light variations when compared to the branches downstream. The combination of recent advances in plant hydrodynamic models of water movement in trees and stochastic models of light attenuation within canopy has been employed to successfully represent such variability and its integrated effects on stomatal conductance and branch and tree transpiration (Ewers et al. 2007), and may be employed to a similar end in representing carbon uptake.

Finally, **Figure 6** shows rainfall variability from daily to century timescale. The annual rainfall can be decomposed into several embedded stochastic processes (D'Odorico et al. 2000, Porporato et al. 2006). Similar to temperature, at annual timescales the rainfall process is less intermittent (and near-Gaussian), while at the daily timescale it is highly intermittent. The propagation of such rainfall fluctuations to plant dynamics is gaining increasing usage for predicting an ecosystem's response to future climatic shifts or for explaining historical responses to climatic variations. It is now recognized that changes in total rainfall amounts are not sufficient to explain changes in net primary productivity (Knapp et al. 2002). Shifts in rainfall distributions,

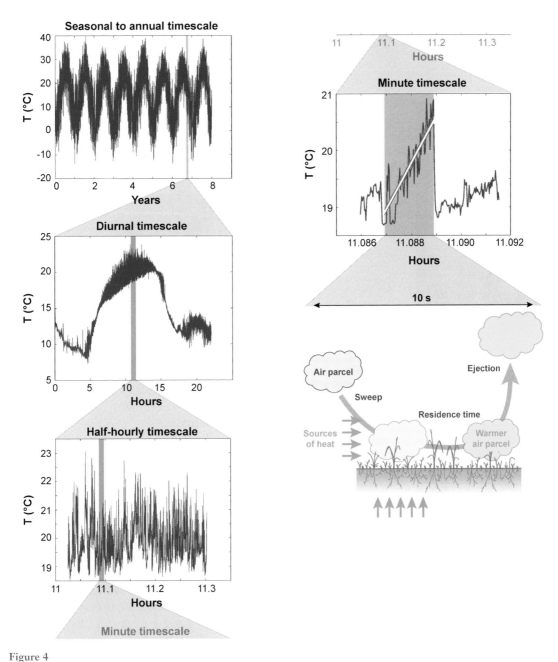

Figure 4

Illustration of the complexity in the hydroclimatic forcing using measured air temperature
fluctuations: (*top-left*) variations at seasonal-to-annual, (*middle-left*) daily, (*bottom-left*)
half-hourly, and (*top-right*) minute timescales. Note the increasing intermittency at shorter
timescales. Data were collected using sonic anemometry at the Duke Forest, near Durham,
North Carolina, from 1997–2005 (see Stoy et al. 2006).

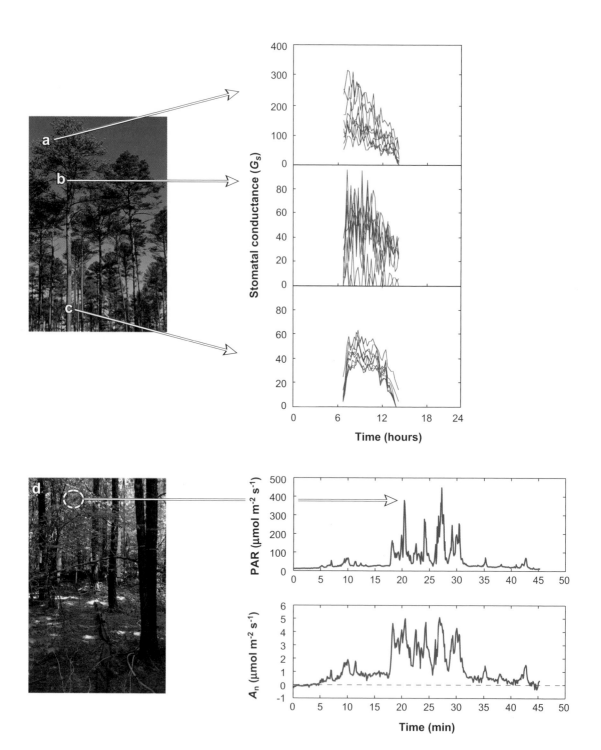

including extremes, cannot be ignored. These shifts remain unreliably predicted by climate models, and reconstructing the full distributional properties from mean projected scenarios continues to be problematic.

Models of rainfall that capture intermittency at daily timescales and the long-term variability at interannual timescale were embedded to describe interannual variability of plant productivity and its sensitivity to both rainfall amounts α and frequency λ (Porporato et al. 2006). Such hierarchical representation of rainfall variability can capture extreme events of storm intensity and drought duration. Interannual statistics of daily intensity and frequency of precipitation events also help explain the reconstructed annual net primary productivity (**Figure 6**).

Rainfall and, to a lesser extent, temperature, humidity, and radiation in turn control the statistics of water stress and carbon assimilation through soil moisture dynamics (Daly & Porporato 2006). Recently, the statistical properties of stochastic soil moisture models (Laio et al. 2001, Milly 1993, Porporato et al. 2004, Rodriguez-Iturbe et al. 1999) and plant water stress (Porporato et al. 2001) have been analytically derived assuming a nonlinear dependence between carbon assimilation and soil moisture at the daily level (Daly et al. 2004, Porporato et al. 2001). Ongoing research investigates how to link these soil moisture fluctuations to actual plant biomass growth and other biotic processes.

Competition for nutrients between plants and soil microbes is also mediated by stochastic dynamics of soil water availability (Kaye & Hart 1997). Characterization of small-scale water, root, and nutrient dynamics can be essential to quantifying plant-water interactions and strategies for carbon allocation and growth (Guswa et al. 2004, Jackson et al. 2000, Lai et al. 2002, Laio et al. 2006, Palmroth et al. 2006). In particular, fluctuating soil moisture conditions impact the cycling of nutrients and decomposition of organic matter in soils with important feedbacks on plants that can induce fluctuations at different timescales, possibly with self-sustained oscillations (D'Odorico et al. 2003, Ehrenfeld et al. 2005, Hogberg & Read 2006, Kaye & Hart 1997, Porporato et al. 2003, Yahdjian et al. 2006).

Figure 7 shows an example of modulations that may be triggered by variations in soil moisture using a simplified model of plant-soil microbial biomass competition (Manzoni & Porporato 2005, Porporato et al. 2003). This competition is quantified by the ratio between immobilization of mineral nitrogen by soil microbes (bacteria and fungi) and plant nitrogen uptake. Fluctuations of soil moisture create shifts in competitive advantage owing to the different responses of plants and soil microbial biomass to water stress. Rapid microbial responses to rewetting have also been

---

Figure 5

Sap flux-scaled stomatal conductance (G_s) sampled on a 15-minute time step from (*a*) nine upper branches, (*b*) lower branches, and (*c*) stems. These measurements, taken from Ewers et al. (2007), are for a *Pinus taeda* stand situated in the Sandhills of North Carolina. (*d*) High intermittency in photon flux density measured as photosynthetically active radiation (PAR) and leaf-level photosynthesis (A_n) for *Acer rubrum* in the Duke Forest, North Carolina (Naumburg et al. 2001). Here, $t = 0$ is 11:00 AM local time.

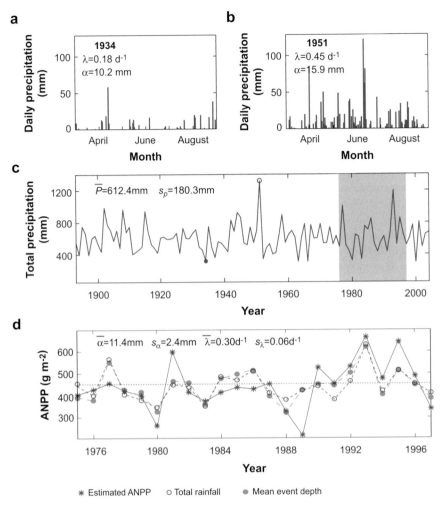

Figure 6

Growing season (April to September) rainfall regime at Manhattan, Kansas. Daily precipitation during (*a*) a very dry and (*b*) a very wet season; (*c*) total annual rainfall, *P*. (*d*) Time series of measured annual net primary productivity (ANPP) at the Konza Prairie Long Term Ecological Research (LTER) Program in Kansas (*asterisks*; **http://intranet.lternet.edu/cgi-bin/anpp.pl**). Also shown are the estimated ANPP obtained with a multiple linear regression model using only the total rainfall (*open circles*, $r^2 = 0.40$) versus the mean event depth, α, and the mean frequency of rainfall events, λ (*solid circles*, $r^2 = 0.47$). Overbars are interannual averages and s_λ is the standard deviation of λ. Redrawn after Porporato et al. (2006).

observed in arid ecosystems and trigger a pulse in nitrogen availability for plants (Austin et al. 2004, Schwinning & Sala 2004). However, prolonged dry conditions or frequent drying-rewetting cycles may damage the microbial community and increase its recovery time (Fierer & Schimel 2002, Schimel 2001).

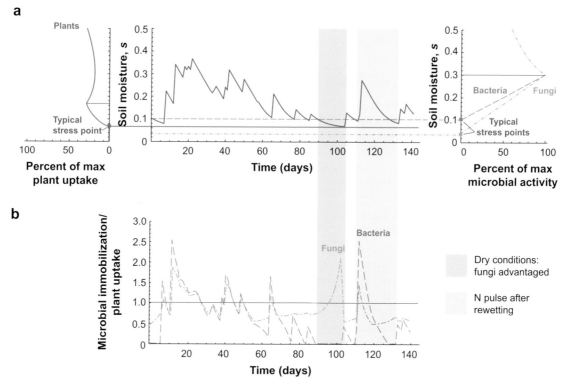

Figure 7

(*a*) Stochastic evolution of soil moisture *s* (*continuous line*) along with stress moisture levels for plants (*continuous line*) and microbes, bacteria (*dashed line*) and fungi (*dot-dashed line*). Left panel shows the plant nitrogen uptake and right panel shows microbial nitrogen (N) immobilization as a function of soil moisture. (*b*) Ratio between microbial immobilization and plant uptake for fungal-dominated (*dot-dashed line*) and bacterial-dominated (dashed line) microbial communities, illustrating the effect of soil moisture pulsing on this competition. Redrawn after Manzoni & Porporato (2005).

5. THE ISSUE OF SPATIAL DYNAMICS

Although so far our examples of external forcing have emphasized stochasticity in time, analogous problems exist in space. The random spatial variability in rainfall and soil moisture has been the subject of research within both hydrological and statistical sciences (Cox & Isham 1988, Rodriguez-Iturbe et al. 2006). However, recent analyses of vegetation spatial patterns arising from coupled carbon-water interaction seem to have neglected stochastic heterogeneities in space and time (Gilad et al. 2004, Klausmeier 1999, Lefever & Lejeune 1997, Meron et al. 2004, Rietkerk et al. 2004, Ursino 2005). Typically, variability in vegetation patterns caused by changes in annual rainfall regimes has been studied deterministically using reaction-diffusion (or activation-inhibition) equations that mirror the coupled carbon-water dynamics (see **Figure 1**), with one recent exception (D'Odorico et al. 2006a,b). An important extension of this line of work is to explore how spatial stochastic components may disrupt or enhance these vegetation patterns.

5.1. Plant-Water Interactions and Geomorphic Gradients

The up-scaling of local soil-plant dynamics to landscape and regional scales (e.g., 10–100 km) must account for topographic heterogeneities, hillslope and riparian processes, and river-network structures. Topographic position, slope, and aspect significantly vary in space within a watershed, influencing incident solar radiation, and hence, evapotranspiration and photosynthesis, which in turn impact water availability and vegetation distribution (Jones 1992, Ridolfi et al. 2003, Rodriguez-Iturbe & Porporato 2004, Western et al. 2002).

An example of a digital elevation model (DEM) analysis for the Konza prairie LTER site (7 km × 5.7 km) is reproduced in **Figure 8**. The stochastic dendritic structure of the landscape is clearly visible, notwithstanding the small elevation differences of the region. **Figures 8b** and **8c** provide examples of the frequency distributions of slope and aspect computed for the 30-m × 30-m grid elements of the DEM. The regularity of these histograms is suggestive of the possibility of a probabilistic description of the variations in vegetation properties across regional scales, building upon the framework of statistical geomorphology (e.g., Rodriguez-Iturbe & Rinaldo 1997, Western et al. 2002). Few analyses of this type have been conducted (e.g., Caylor et al. 2005, Istanbulluoglu & Bras 2006), and the field remains a fruitful area for research.

5.2. Plant-Water Interactions—Feedback to the Atmosphere

In previous examples (**Figures 6–8**), rainfall was considered entirely external to the soil-plant system. However, at sufficiently large spatial scales, ecosystems can modify their own precipitation regimes primarily because of two-way interactions between the soil-plant system and the atmosphere (Eagleson 1986). The strength of this two-way interaction can be quantified by examining how changes in land cover by human activities impact rainfall regimes (Chase et al. 2001, Kanae et al. 2002, Liston et al. 2002). Recent modeling efforts suggest that these impacts can be as large as those from other anthropogenic factors such as greenhouse gases and aerosols. For example, Roy et al. (2003) utilized a numerical model to show that land cover change in the United States over the past 300 years has significantly altered the local climate in July. Focusing on 1910 to 1990, they found that the increase in forested areas in the East Coast region of the United States resulted in lower surface temperature owing to higher evapotranspiration.

Higher evapotranspiration contributed moisture to the atmosphere thereby enhancing precipitation in the region (Roy et al. 2003). More broadly, precipitation recycling via evapotranspiration and the control that soil moisture exerts on the partitioning of sensible and latent heat fluxes can enhance the feedback between soil moisture and precipitation along a number of pathways (D'Odorico & Porporato

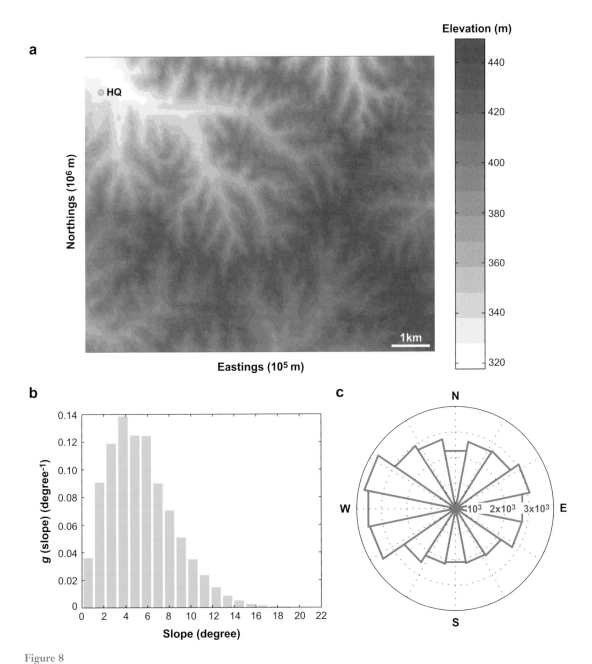

a

Northings (10^6 m)

HQ

Elevation (m)

440

420

400

380

360

340

320

1km

Eastings (10^5 m)

b

g (slope) (degree^{-1})

0.14

0.12

0.10

0.08

0.06

0.04

0.02

0

0 2 4 6 8 10 12 14 16 18 20 22

Slope (degree)

c

N

W

E

10^3 2×10^3 3×10^3

S

Figure 8

(*a*) Digital elevation map for an area of 7 km by 5.7 km in the Konza Prairie LTER site (HQ is the position of Konza headquarters); (*b*) probability distribution of slope *g* represented using histogram, and (*c*) of aspect represented using histogram in polar coordinates for the same landscape. The spatial probabilistic distribution of these geomorphologic properties affects solar radiation and soil moisture and thus plant growth.

2004; Entekhabi & Brubaker 1995; Findell & Eltahir 2003a,b; Freedman et al. 2001; Juang et al. 2007a,b; Wu & Dickinson 2005). In short, at sufficiently large scales, plant-water interactions can impact rainfall patterns, which thus can no longer be treated as mere external forcing.

6. CONCLUSIONS

The fundamental barrier to progress in plant-water interaction can be distilled to two main issues: (*a*) we do not know how to describe microscopic water movement in the soil-plant system, although microfluidics might offer a blueprint of how to proceed, and (*b*) we do not know how to scale up, spatially and temporally, these microscopic descriptions coherently, while preserving the effects of nonlinearity and stochasticity. The approaches reviewed here should be viewed as initial steps toward filling these knowledge gaps, with a bias toward the second knowledge gap. We discussed how existing equations for the soil-plant system, when averaged, are not scale-independent in time or space (Nykanen & Foufoula-Georgiou 2001) and remain subject to external stochasticity. With availability of long-term measurements and progress in nonlinear time series analysis, it is conceivable that low-dimensional nonlinear components of the dynamics of the soil-plant-atmosphere system can be extracted from data, thereby guiding the choice of model complexity.

These limitations notwithstanding, all representations must consider external stochasticity originating from the hydroclimatic forcing to the plant-soil system. This forcing is inherently intermittent with variability across all scales, which precludes the use of standard approximations often used in analysis of stochastic processes (e.g., small noise perturbations). Superposition of stochasticity at multiple scales (e.g., rainfall) can intensify extremes and must be taken into account for a reasonable description of ecosystem dynamics. Moreover, when propagated into the soil-plant system, such stochasticity may give rise to noise-induced oscillations and patterns, especially because the water-carbon-nitrogen cycles are tightly coupled.

DISCLOSURE STATEMENT

The authors are not aware of any biases that might be perceived as affecting the objectivity of this review.

ACKNOWLEDGMENTS

We thank S. Manzoni and G. Vico for their assistance in the preparation of the figures, references, and comments, and K. Novick, E. Daly, J.-C. Domec, and J. Rigby for their helpful suggestions. This work was funded, in part, by the National Science Foundation (NSF-EAR-0628432 & NSF-EAR-06-35787), and by the Office of Science (BER), U.S. Department of Energy, through the Terrestrial Carbon Processes Program (DE-FG02-00ER63015 and DEFG02-95ER62083).

LITERATURE CITED

Albertson J, Katul G, Wiberg P. 2001. Relative importance of local and regional controls on coupled water, carbon, and energy fluxes. *Adv. Water Resour.* 24:1103–18

Arnold L. 2003. *Random Dynamical Systems*. Berlin: Springer-Verlag. 586 pp.

Austin AT, Yahdjian L, Stark JM, Belnap J, Porporato A, et al. 2004. Water pulses and biogeochemical cycles in arid and semiarid ecosystems. *Oecologia* 141:221–35

Baldocchi D. 1992. A Lagrangian random-walk model for simulating water-vapor, CO_2 and sensible heat-flux densities and scalar profiles over and within a soybean canopy. *Bound.-Layer Meteorol.* 61:113–44

Bendat J, Piersol A. 1971. *Random Data: Analysis and Measurement Procedures*. New York: Wiley. 407 pp.

Boffetta G, Cencini M, Falcioni M, Vulpiani A. 2002. Predictability: A way to characterize complexity. *Phys. Rep.* 356:367–474

Bohrer G, Mourad H, Laursen TA, Drewry D, Avissar R, et al. 2005. Finite element tree crown hydrodynamics model (FETCH) using porous media flow within branching elements: A new representation of tree hydrodynamics. *Water Resour. Res.* 41:W11404

Bouchaud JP, Georges A. 1990. Anomalous diffusion in disordered media—statistical mechanisms, models and physical applications. *Phys. Rep.* 195:127–293

Brutsaert W. 1982. *Evaporation into the Atmosphere: Theory, History, and Applications.* Dordrecht: Reidel. 299 pp.

Buckley TN, Farquhar GD, Mott KA. 1997. Qualitative effects of patchy stomatal conductance distribution features on gas-exchange calculations. *Plant Cell Environ.* 20:867–80

Caylor KK, Manfreda S, Rodriguez-Iturbe I. 2005. On the coupled geomorphological and ecohydrological organization of river basins. *Adv. Water Resour.* 28:69–86

Chase TN, Pielke RA, Kittel TGF, Zhao M, Pitman AJ, et al. 2001. Relative climatic effects of landcover change and elevated carbon dioxide combined with aerosols: A comparison of model results and observations. *J. Geophys. Res. Atmos.* 106:31685–91

Chuang YL, Oren R, Bertozzi AL, Phillips N, Katul GG. 2006. The porous media model for the hydraulic system of a conifer tree: Linking sap flux data to transpiration rate. *Ecol. Model.* 191:447–68

Clark JS, Gelfand AE. 2006. A future for models and data in environmental science. *Trends Ecol. Evol.* 21:375–80

Cox DR, Isham V. 1988. A simple spatial-temporal model of rainfall. *Proc. R. Soc. London Ser. A* 415:317–28

Cox DR, Miller HD. 1977. *The Theory of Stochastic Processes*. London: Chapman & Hall. 408 pp.

Cross MC, Hohenberg PC. 1993. Pattern-formation outside of equilibrium. *Rev. Mod. Phys.* 65:851–1112

Daly E, Porporato A. 2006. Impact of hydroclimatic fluctuations on the soil water balance. *Water Resour. Res.* 42:W06401

Daly E, Porporato A, Rodriguez-Iturbe I. 2004. Coupled dynamics of photosynthesis, transpiration, and soil water balance. Part II: Stochastic analysis and ecohydrological significance. *J. Hydrometeorol.* 5:559–66

de Rooij GH. 2000. Modeling fingered flow of water in soils owing to wetting front instability: A review. *J. Hydrol.* 231:277–94

Dewar R. 2003. Information theory explanation of the fluctuation theorem, maximum entropy production and self-organized criticality in nonequilibrium stationary states. *J. Phys. A* 36:631–41

D'Odorico P, Laio F, Porporato A, Rodriguez-Iturbe I. 2003. Hydrologic controls on soil carbon and nitrogen cycles. Part I: A case study. *Adv. Water Resour.* 26:59–70

D'Odorico P, Laio F, Ridolfi L. 2005. Noise-induced stability in dryland plant ecosystems. *Proc. Natl. Acad. Sci. USA* 102:10819–22

D'Odorico P, Laio F, Ridolfi L. 2006a. Patterns as indicators of productivity enhancement by facilitation and competition in dryland vegetation. *J. Geophys. Res.* 111:G03010

D'Odorico P, Laio F, Ridolfi L. 2006b. Vegetation patterns induced by random climate fluctuations. *Geophys. Res. Lett.* 33:L19404

D'Odorico P, Porporato A. 2004. Preferential states in soil moisture and climate dynamics. *Proc. Natl. Acad. Sci. USA* 101:8848–51

D'Odorico P, Ridolfi L, Porporato A, Rodriguez-Iturbe I. 2000. Preferential states of seasonal soil moisture: The impact of climate fluctuations. *Water Resour. Res.* 36:2209–19

Durrett R, Levin SA. 1994. Stochastic spatial models—a user's guide to ecological applications. *Philos. Trans. R. Soc. London Ser. B* 343:329–50

Eagleson PS. 2002. *Ecohydrology: Darwinian Expression of Vegetation Form and Function.* New York: Cambridge Univ. Press. 496 pp.

Eagleson PS. 1986. The emergence of global-scale hydrology. *Water Resour. Res.* 22:S6–14

Eckmann JP, Ruelle D. 1985. Ergodic-theory of chaos and strange attractors. *Rev. Mod. Phys.* 57:617–56

Ehrenfeld JG, Ravit B, Elgersma K. 2005. Feedback in the plant-soil system. *Annu. Rev. Environ. Resour.* 30:75–115

Entekhabi D, Brubaker KL. 1995. An analytic approach to modeling land atmosphere interaction. 2. Stochastic formulation. *Water Resour. Res.* 31:633–43

Ewers BE, Mackay DS, Samanta S. 2007. Interannual consistency in canopy stomatal conductance control of leaf water potential across seven tree species. *Tree Physiol.* 27:11–24

Falkovich G, Gawedzki K, Vergassola M. 2001. Particles and fields in fluid turbulence. *Rev. Mod. Phys.* 73:913–75

Farquhar GD, Caemmerer SV, Berry JA. 1980. A biochemical model of photosynthetic CO_2 assimilation in leaves of C-3 species. *Planta* 149:78–90

Fierer N, Schimel JP. 2002. Effects of drying-rewetting frequency on soil carbon and nitrogen transformations. *Soil Biol. Biochem.* 34:777–87

Findell KL, Eltahir EAB. 2003a. Atmospheric controls on soil moisture-boundary layer interactions. Part I: Framework development. *J. Hydrometeorol.* 4:552–69

Findell KL, Eltahir EAB. 2003b. Atmospheric controls on soil moisture-boundary layer interactions. Part II: Feedbacks within the continental United States. *J. Hydrometeorol.* 4:570–83

Finnigan J. 2000. Turbulence in plant canopies. *Annu. Rev. Fluid Mech.* 32:519–71

Ford J. 1983. How random is a coin toss? *Phys. Today* 36:40–47

Freedman JM, Fitzjarrald DR, Moore KE, Sakai RK. 2001. Boundary layer clouds and vegetation-atmosphere feedbacks. *J. Clim.* 14:180–97

Gardiner CW. 2004. *Handbook of Stochastic Methods for Physics, Chemistry, and the Natural Sciences.* Berlin: Springer-Verlag. xvii+415 pp.

Gilad E, von Hardenberg J, Provenzale A, Shachak M, Meron E. 2004. Ecosystem engineers: From pattern formation to habitat creation. *Phys. Rev. Lett.* 93:098105

Grant BW, Vatnick I. 2004. Environmental correlates of leaf stomata. *Teach. Issues Exp. Ecol. (TIEE)*, Vol. 1:Jan. (Electronic)

Guswa AJ, Celia MA, Rodriguez-Iturbe I. 2004. Effect of vertical resolution on predictions of transpiration in water-limited ecosystems. *Adv. Water Resour.* 27:467–80

Hillel D. 2004. *Introduction to Environmental Soil Physics.* Amsterdam: Elsevier Acad. 494 pp.

Hogberg P, Read DJ. 2006. Towards a more plant physiological perspective on soil ecology. *Trends Ecol. Evol.* 21:548–54

Horsthemke W, Lefever R. 1984. *Noise-Induced Transitions: Theory and Applications in Physics, Chemistry, and Biology.* Berlin: Springer-Verlag. xv+318 pp.

Hsiao T. 1973. Plant responses to water stress. *Annu. Rev. Plant Physiol.* 24:519–70

Istanbulluoglu E, Bras RL. 2006. On the dynamics of soil moisture, vegetation, and erosion: Implications of climate variability and change. *Water Resour. Res.* 42:W06418

Jackson RB, Sperry JS, Dawson TE. 2000. Root water uptake and transport: Using physiological processes in global predictions. *Trends Plant Sci.* 5:482–88

Jaynes ET. 2003. *Probability Theory: The Logic of Science.* Cambridge, UK: Cambridge Univ. Press. 758 pp.

Jones HG. 1992. *Plants and Microclimate.* New York: Cambridge Univ. Press. 428 pp.

Juang JY, Katul GG, Porporato A, Stoy P, Siqueira M, et al. 2007a. Eco-hydrological controls on summertime convective rainfall triggers. *Glob. Change Biol.* 13:1–10, doi: 10.1111/j.1365-2486.2006.01315.x

Juang JY, Porporato A, Stoy PC, Siqueira M, Kim HY, Katul GG. 2007b. Hydrologic and atmospheric controls on initiation of convective precipitation events. *Water Resour. Res.* 43:W03421, doi:10.1029/2006WR004954.

Jury WA, Gardner WR, Gardner WH. 1991. *Soil Physics.* New York: Wiley. 328 pp.

Kanae S, Oki T, Musiake K. 2002. Principal condition for the earliest Asian summer monsoon onset. *Geophys. Res. Lett.* 29:1746

Kass RE, Raftery AE. 1995. Bayes factors. *J. Am. Stat. Assoc.* 90:773–95

Katul G, Leuning R, Oren R. 2003. Relationship between plant hydraulic and biochemical properties derived from a steady-state coupled water and carbon transport model. *Plant Cell Environ.* 26:339–50

Katul G, Porporato A, Cava D, Siqueira M. 2006. An analysis of intermittency, scaling, and surface renewal in atmospheric surface layer turbulence. *Physica D* 215:117–26

Katul G, Todd P, Pataki D, Kabala Z, Oren R. 1997. Soil water depletion by oak trees and the influence of root water uptake on the soil moisture content spatial statistics. *Water Resour. Res.* 33:611–23

Kaye JP, Hart SC. 1997. Competition for nitrogen between plants and soil microorganisms. *Trends Ecol. Evol.* 12:139–43

Klausmeier CA. 1999. Regular and irregular patterns in semiarid vegetation. *Science* 284:1826–28

Knapp AK. 1993. Gas-exchange dynamics in C-3 and C-4 grasses—consequences of differences in stomatal conductance. *Ecology* 74:113–23

Knapp AK, Fay PA, Blair JM, Collins SL, Smith MD, et al. 2002. Rainfall variability, carbon cycling, and plant species diversity in a mesic grassland. *Science* 298:2202–5

Kramer PJ, Boyer JS. 1995. *Water Relations of Plants and Soils.* San Diego: Academic. 495 pp.

Kull O, Jarvis PG. 1995. The role of nitrogen in a simple scheme to scale-up photosynthesis from leaf to canopy. *Plant Cell Environ.* 18:1174–82

Lai CT, Katul G, Butnor J, Siqueira M, Ellsworth D, et al. 2002. Modeling the limits on the response of net carbon exchange to fertilization in a southeastern pine forest. *Plant Cell Environ.* 25:1095–119

Lai CT, Katul G, Oren R, Ellsworth D, Schafer K. 2000. Modeling CO_2 and water vapor turbulent flux distributions within a forest canopy. *J. Geophys. Res. Atmos.* 105:26333–51

Laio F, D'Odorico P, Ridolfi L. 2006. An analytical model to relate the vertical root distribution to climate and soil properties. *Geophys. Res. Lett.* 33:L18401

Laio F, Porporato A, Fernandez-Illescas CP, Rodriguez-Iturbe I. 2001. Plants in water-controlled ecosystems: Active role in hydrologic processes and response to water stress—iv. Discussion of real cases. *Adv. Water Resour.* 24:745–62

Laio F, Porporato A, Ridolfi L, Rodriguez-Iturbe I. 2002. On the seasonal dynamics of mean soil moisture. *J. Geophys. Res. Atmos.* 107:4272

Larcher W. 1995. *Physiological Plant Ecology: Ecophysiology and Stress Physiology of Functional Groups.* Berlin: Springer-Verlag. xvi+506 pp.

Larson HJ, Shubert BO. 1979. *Probabilistic Models in Engineering Sciences.* New York: Wiley

Lathwell DJ, Grove TL. 1986. Soil-plant relationships in the tropics. *Annu. Rev. Ecol. Syst.* 17:1–16

Lefever R, Lejeune O. 1997. On the origin of tiger bush. *Bull. Math. Biol.* 59:263–94

Liston GE, McFadden JP, Sturm M, Pielke RA. 2002. Modelled changes in arctic tundra snow, energy and moisture fluxes due to increased shrubs. *Glob. Change Biol.* 8:17–32

Manzoni S, Porporato A. 2005. Role of model structure on the response of soil biogeochemistry to hydro-climatic fluctuations. *Eos Trans. AGU* 86. *Jt. Assem. Suppl.*, New Orleans

Martyushev LM, Seleznev VD. 2006. Maximum entropy production principle in physics, chemistry and biology. *Phys. Rep.* 426:1–45

Meron E, Gilad E, von Hardenberg J, Shachak M, Zarmi Y. 2004. Vegetation patterns along a rainfall gradient. *Chaos Solitons Fractals* 19:367–76

Milly PCD. 1993. An analytic solution of the stochastic storage problem applicable to soil water. *Water Resour. Res.* 29:3755–58

Naumburg E, Ellsworth DS, Katul GG. 2001. Modeling dynamic understory photosynthesis of contrasting species in ambient and elevated carbon dioxide. *Oecologia* 126:487–99

Nelson P. 2004. *Biological Physics: Energy, Information and Life.* New York: Freeman 598 pp.

Nobel PS. 2005. *Physicochemical and Environmental Plant Physiology.* Amsterdam: Elsevier Acad. 567 pp.

Norris JM. 1998. *Markov Chains.* Cambridge, UK: Cambridge Univ. Press. 237 pp.

Noy Meir I. 1973. Desert ecosystems: Environment and producers. *Annu. Rev. Ecol. Syst.* 4:25–44

Nykanen DK, Foufoula-Georgiou E. 2001. Soil moisture variability and scale-dependency of nonlinear parameterizations in coupled land-atmosphere models. *Adv. Water Resour.* 24:1143–57

Oren R, Phillips N, Katul G, Ewers BE, Pataki DE. 1998. Scaling xylem sap flux and soil water balance and calculating variance: a method for partitioning water flux in forests. *Ann. Sci. For.* 55:191–216

Oren R, Sperry JS, Katul GG, Pataki DE, Ewers BE, et al. 1999. Survey and synthesis of intra- and interspecific variation in stomatal sensitivity to vapour pressure deficit. *Plant Cell Environ.* 22:1515–26

Ozawa H, Ohmura A, Lorenz RD, Pujol T. 2003. The second law of thermodynamics and the global climate system: A review of the maximum entropy production principle. *Rev. Geophys.* 41:1018

Palmroth S, Oren R, McCarthy HR, Johnsen KH, Finzi AC, et al. 2006. Aboveground sink strength in forests controls the allocation of carbon below ground and its CO_2-induced enhancement. *Proc. Natl. Acad. Sci. USA* 103:19362–67

Papoulis A. 1991. *Probability, Random Variables, and Stochastic Processes.* Boston: McGraw-Hill. 666 pp.

Pearcy RW, Gross LJ, He D. 1997. An improved dynamic model of photosynthesis for estimation of carbon gain in sunfleck light regimes. *Plant Cell Environ.* 20:411–24

Phillips R, Quake SR. 2006. The biological frontier of physics. *Phys. Today* 59:38–43

Porporato A, Daly E, Rodriguez-Iturbe I. 2004. Soil water balance and ecosystem response to climate change. *Am. Nat.* 164:625–32

Porporato A, D'Odorico P. 2004. Phase transitions driven by state-dependent Poisson noise. *Phys. Rev. Lett.* 92:110601

Porporato A, D'Odorico P, Laio F, Ridolfi L, Rodriguez-Iturbe I. 2002. Ecohydrology of water-controlled ecosystems. *Adv. Water Resour.* 25:1335–48

Porporato A, D'Odorico P, Laio F, Rodriguez-Iturbe I. 2003. Hydrologic controls on soil carbon and nitrogen cycles. I. Modeling scheme. *Adv. Water Resour.* 26:45–58

Ridolfi L, D'Odorico P, Porporato A, Rodriguez-Iturbe I. 2003. Stochastic soil moisture dynamics along a hillslope. *J. Hydrol.* 272:264–75

Porporato A, Laio F, Ridolfi L, Rodriguez-Iturbe I. 2001. Plants in water-controlled ecosystems: Active role in hydrologic processes and response to water stress—iii. Vegetation water stress. *Adv. Water Resour.* 24:725–44

Porporato A, Rodriguez-Iturbe I. 2002. Ecohydrology—a challenging multidisciplinary research perspective. *J. Sci. Hydrol.* 47:811–21

Porporato A, Vico G, Fay PA. 2006. Super-statistics of hydro-climatic fluctuations and interannual ecosystem productivity. *Geophys. Res. Lett.* 33:L15402

Rietkerk M, Dekker SC, de Ruiter PC, van de Koppel J. 2004. Self-organized patchiness and catastrophic shifts in ecosystems. *Science* 305:1926–29

Roderick ML. 2001. On the use of thermodynamic methods to describe water relations in plants and soil. *Aust. J. Plant Physiol.* 28:729–42

Rodriguez-Iturbe I, Isham V, Cox DR, Manfreda S, Porporato A. 2006. Space-time modeling of soil moisture: Stochastic rainfall forcing with heterogeneous vegetation. *Water Resour. Res.* 42:W06D05

Rodriguez-Iturbe I, Porporato A. 2004. *Ecohydrology of Water-Controlled Ecosystems.* Cambridge, UK: Cambridge Univ. Press. 442 pp.

Rodriguez-Iturbe I, Porporato A, Ridolfi L, Isham V, Cox DR. 1999. Probabilistic modelling of water balance at a point: The role of climate, soil and vegetation. *Proc. R. Soc. A* 455:3789–805

Rodriguez-Iturbe I, Rinaldo A. 1997. *Fractal River Basins: Chance and Self-Organization.* New York: Cambridge Univ. Press. 547 pp.

Roy SB, Hurtt GC, Weaver CP, Pacala SW. 2003. Impact of historical land cover change on the July climate of the United States. *J. Geophys. Res. Atmos.* 108:4793–806

Sack L, Holbrook NM. 2006. Leaf hydraulics. *Annu. Rev. Plant Biol.* 57:361–81

Sagues F, Sancho JM. 2004. Patterns out of noise. *Contemp. Phys.* 45:503–14

Scheffer M, Carpenter S, Foley JA, Folke C, Walker B. 2001. Catastrophic shifts in ecosystems. *Nature* 413:591–96

Schimel JP. 2001. Biogeochemical models: Implicit vs explicit microbiology. In *Global Biogeochemical Cycles in the Climate System*, ed. ED Schulze, SP Harrison, M Heimann, EA Holland, JJ Lloyd, et al., pp. 177–84. New York: Academic

Schlesinger WH. 1997. *Biogeochemistry: An Analysis of Global Change.* San Diego: Academic. 588 pp.

Schwinning S, Sala OE. 2004. Hierarchy of responses to resource pulses in and semiarid ecosystems. *Oecologia* 141:211–20

Siqueira M, Katul G, Lai CT. 2002. Quantifying net ecosystem exchange by multilevel ecophysiological and turbulent transport models. *Adv. Water Resour.* 25:1357–66

Sivia DS, Skilling J. 2006. *Data Analysis: A Bayesian Tutorial.* Oxford: Oxford Univ. Press. 258 pp.

Sornette D. 2004. *Critical Phenomena in Natural Sciences.* Berlin: Springer-Verlag. 528 pp.

Sperry JS. 2000. Hydraulic constraints on plant gas exchange. *Agric. For. Meteorol.* 104:13–23

Sperry JS, Hacke UG. 2002. Desert shrub water relations with respect to soil characteristics and plant functional type. *Funct. Ecol.* 16:367–78

Sperry JS, Hacke UG, Oren R, Comstock JP. 2002. Water deficits and hydraulic limits to leaf water supply. *Plant Cell Environ.* 25:251–63

Squires TM, Quake SR. 2005. Microfluidics: Fluid physics at the nanoliter scale. *Rev. Mod. Phys.* 77:977–1026

Steffen WL, Denmead OT. 1988. *Flow and Transport in the Natural Environment: Advances and Applications.* Berlin: Springer-Verlag. 384 pp.

Stoy PC, Katul GG, Siqueira MBS, Juang JY, Novick KA, et al. 2006. Separating the effects of climate and vegetation on evapotranspiration along a successional chronosequence in the southeastern US. *Glob. Change Biol.* 12:2115–35

Strogatz SH. 1994. *Nonlinear Dynamics and Chaos: With Applications to Physics, Biology, Chemistry, and Engineering.* Cambridge, MA: Perseus Books. 498 pp.

Torquato S. 2002. *Random Heterogeneous Materials, Microstructure, and Macroscopic Properties.* New York: Springer-Verlag. 701 pp.

Ursino N. 2005. The influence of soil properties on the formation of unstable vegetation patterns on hillsides of semiarid catchments. *Adv. Water Resour.* 28:956–63

Vandenbroeck C, Parrondo JMR, Toral R. 1994. Noise-induced nonequilibrium phase transition. *Phys. Rev. Lett.* 73:3395–98

van Kampen NG. 1992. *Stochastic Processes in Physics and Chemistry.* Amsterdam: North Holland. xiv+465 pp.

Vanmarke E. 1983. *Random Fields: Analysis and Synthesis.* Cambridge, MA: MIT Press. xii+382 pp.

Vitousek PM. 1992. Global environmental change—An introduction. *Annu. Rev. Ecol. Syst.* 23:1–14

Western AW, Grayson RB, Bloschl G. 2002. Scaling of soil moisture: A hydrologic perspective. *Annu. Rev. Earth Planet. Sci.* 30:149–80

Whitfield J. 2005. Order out of chaos. *Nature* 436:905–7

Wu WR, Dickinson RE. 2005. Warm-season rainfall variability over the U.S. Great Plains and its correlation with evapotranspiration in a climate simulation. *Geophys. Res. Lett.* 32:L17402

Yahdjian L, Sala OE, Austin AT. 2006. Differential controls of water input on litter decomposition and nitrogen dynamics in the patagonian steppe. *Ecosystems* 9:128–41

Zimmerman MH. 1983. *Xylem Structure and the Ascent of Sap.* Berlin: Springer-Verlag. 143 pp.

Evolutionary Endocrinology: The Developing Synthesis between Endocrinology and Evolutionary Genetics

Anthony J. Zera,[1] Lawrence G. Harshman,[1] and Tony D. Williams[2]

[1]School of Biological Sciences, University of Nebraska, Lincoln, Nebraska 68588; email: azera1@unlnotes.unl.edu, lharsh@unlserve.unl.edu

[2]Department of Biological Sciences, Simon Fraser University, Burnaby, British Columbia, Canada V5A 1S6; email: tdwillia@sfu.ca

Annu. Rev. Ecol. Evol. Syst. 2007. 38:793–817

First published online as a Review in Advance on August 20, 2007

The *Annual Review of Ecology, Evolution, and Systematics* is online at http://ecolsys.annualreviews.org

This article's doi: 10.1146/annurev.ecolsys.38.091206.095615

Key Words

artificial selection, development, hormone, life history, polymorphism

Abstract

A productive synthesis of endocrinology and evolutionary genetics has occurred during the past two decades, resulting in the first direct documentation of genetic variation and correlation for endocrine regulators in nondomesticated animals. In a number of insect genetic polymorphisms (dispersal polymorphism in crickets, butterfly wing-pattern polymorphism), blood levels of ecdysteroids and juvenile hormone covary with morphology, development, and life history. Genetic variation in insulin signaling may underlie life history trade-offs in *Drosophila*. Vertebrate studies identified variation in brain neurohormones, bone-regulating hormones, and hormone receptor gene sequences that underlie ecologically important genetic polymorphisms. Most work to date has focused on genetically variable titers (concentrations) of circulating hormones and the activities of titer regulators. Continued progress will require greater integration among (*a*) traditional comparative endocrine approaches (e.g., titer measures); (*b*) molecular studies of hormone receptors and intracellular signaling pathways; and (*c*) fitness studies of genetically variable endocrine traits in ecologically appropriate conditions.

INTRODUCTION

The evolution of endocrine regulation has been a topic of interest to physiologists and evolutionary biologists for decades (reviewed in Adkins-Regan 2005, Matsuda 1987, Nijhout & Wheeler 1982, Norris 1997, West-Eberhard 2003, Zera 2004). Major issues in this topic include the nature of endocrine adaptations (i.e., evolutionary modifications of endocrine control mechanisms) and the role of these modifications in the evolution of morphology, behavior, development, and life history. Until fairly recently, the vast majority of evolutionary studies focused on comparative aspects of hormonal regulation among species or higher taxa. Typical examples include phylogenetic analyses of hormone structure/function (Colon & Larhammar 2005, Guilgur et al. 2006, Norris 1997) or comparative studies of various endocrine adaptations (e.g., differences in the hormonal control of ion balance among terrestrial, fresh-water, and salt-water vertebrates; reviewed in Norris 1997). Another important research focus has been the hormonal control of phenotypic trade-offs among life history traits in the field, such as testosterone-mediated trade-offs; (Ketterson et al. 2001, Marler 1988, Svensson et al. 2002). However, microevolutionary investigations of genetically variable hormonal regulators in nondomesticated animals were rare before 1990 (Kallman 1989, Zera & Teibel 1989).

This situation has changed dramatically, and the first substantial synthesis of endocrinology and evolutionary genetics, termed evolutionary endocrinology (Zera & Huang 1999), has begun. There are two main reasons for this development: First and foremost, during the past 20 years, evolutionary biology has focused increasingly on the microevolution and evolutionary genetics of complex organismal adaptations, such as life history traits (fecundity, longevity), life history trade-offs, and aspects of development (evolution of butterfly wing patterns, dispersal polymorphisms, body size). Hormones regulate most major components of development and life history, such as growth rate, body size, timing of metamorphosis, and sexual maturation. Furthermore, individual hormones typically affect numerous phenotypes (endocrine pleiotropy) and thus almost certainly give rise to genetic correlations that underlie constraints and trade-offs; these correlations strongly influence the evolution of development and life histories. A growing consensus has emerged that a detailed understanding of the microevolution of life histories and development requires a corresponding deep understanding of the microevolution of the endocrine mechanisms that control the expression of these traits (Brakefield et al. 2003, Finch & Rose 1995, Harshman & Zera 2007, West-Eberhard, 2003, Zera & Harshman 2001). Second, the recent development of evolutionary physiology (Feder et al. 1987, Garland & Carter 1994) highlighted the importance of investigating within-species genetic variation and covariation for physiological processes to understand the mechanisms by which physiology evolves. Thus, evolutionary-genetic analysis of within-population hormonal variation is increasingly viewed as an important research focus in its own right (Zera 2006, Zera & Huang 1999, Zera & Zhang 1995).

This review deals exclusively with animals, primarily insects and vertebrates, groups in which most evolutionary-endocrine studies have been conducted. We focus primarily on the following two topics: (*a*) the extent and nature of genetic variation and covariation for various endocrine traits and (*b*) the extent to which endocrine regulators (e.g., hormone levels, regulators of hormone titers, hormone receptor expression) covary genetically with whole-organism aspects of morphology, development, and life history, and the evolutionary implications of these associations. Our primary goal is to evaluate critically the most significant experimental findings in evolutionary endocrinology from the past two decades.

BACKGROUND IN EVOLUTIONARY ENDOCRINOLOGY

A Thumbnail Sketch of Endocrine Regulation

The mechanisms that comprise endocrine regulation are exceedingly complex, and we give only a brief outline of the main aspects here to aid the nonspecialist. For general reviews see, Adkins-Regan 2005, Gilbert et al. 2005, Kacsoh 2000, Nijhout 1994, and Norris 1997. The key components of endocrine signaling are (*a*) the systemic (e.g., blood) concentration or titer of a hormone (the signal), (*b*) the receptors that bind the hormone (the receiver; tissue sensitivity), and (*c*) the various intracellular mechanisms that convert hormone binding into production of effector molecules (e.g., enzymes; regulatory molecules). A hormone can be either the direct product of a gene (e.g., peptide hormones such as insulin and insulin-like growth factors), or a molecule that is the end product of a biosynthetic pathway (e.g., steroids such as testosterone and ecdysone). Hormones or prohormones are secreted into the circulatory system and may be modified (e.g., activated) in various tissues; hormones are transported either unbound (many peptide hormones) or bound to a carrier protein (many lipophilic hormones) to target tissues. The hormone can then interact with the outer cell membrane (e.g., polar peptide hormone binds to receptor), which initiates an often complex intracellular signaling cascade. Alternatively the hormone can enter the cell (e.g., steroid), where it often binds to a receptor protein and is transported to the nucleus where it modulates gene transcription. The hormone is then degraded and/or excreted.

A hallmark of hormonal regulation is the complex network of interactions that underlies such key phenomena as hormonal homeostasis, hormonal pleiotropy, and phenotypic integration. For example, the blood titers of many hormones (e.g., testosterone, estrogen, and ecdysone) are often maintained within a narrow range (homeostasis) by multiple negative feedback loops between the hormones and the neurohormonal regulators that control hormonal secretion. Similarly, the regulation of phenotypic expression by a specific hormone often requires the prior action of other hormones (priming effects) and modulation by additional hormones. Thus, the endocrine control of complex phenotypes, such as the timing of metamorphosis or egg production, typically requires a precisely orchestrated sequential production and elimination of numerous interacting hormones. Finally, a single hormone often influences the expression of multiple phenotypes (hormonal pleiotropy), which is a key aspect of the integrative expression of multiple components of a complex adaptation (phenotypic integration). These complex interactions can make the analysis of endocrine regulation exceedingly difficult.

Studied Endocrine Traits and Endocrine Methods

Of the three main components of endocrine regulation discussed above, thus far hormone titer has been studied most extensively from an evolutionary-endocrine perspective. This predominant focus on hormone titers is the consequence of a large body of evidence that implicates the regulation of phenotypic expression by changes in circulating hormone levels, such as the hormonal regulation of developmental and reproductive aspects of insect and vertebrate polymorphisms (Hartfelder & Emlen 2005, Nijhout 1994, Shi 2000). Moreover, specific and sensitive assays, such as radioimmunoassays, have long been available, which allow the measurement of hormone levels in the relatively large number of individuals often required in quantitative-genetic studies of endocrine variation (Zera & Cisper 2001, Zera & Zhang 1995, Zijlstra et al. 2004). Conversely, the effective titer available to tissue receptors may not necessarily correspond to the total blood hormone titer. For example, binding by plasma proteins can modulate the effective titer of a hormone (e.g., corticosterone; Breuner & Orchinik 2002), as can proximity to the site of hormone production/release (e.g., the concentration of testosterone, which is required for proper sperm production, is much higher in the testes than in the blood) (Kacsoh 2000). No evolutionary-endocrine studies (and few ecological/behavioral studies) have investigated these more complex aspects of hormone titers.

Some of the most detailed evolutionary-endocrine studies have focused on general regulators of the hormone titers, most notably, the activities of hormone-degrading enzymes (Zera 2006, Zera & Huang 1999). These endocrine traits are often easier to measure than hormone titers or receptor characteristics, often play an important role in hormone titer regulation, and can be measured in small individual organisms (e.g., insects), without severe damage, which is an important advantage in artificial selection studies (Zera & Zhang 1995).

Intraspecific variation in receptor attributes (e.g., tissue or stage-specific receptor-gene expression and hormone-binding affinity) may play an important role in endocrine adaptation. Receptor alterations can result in phenotypic changes that are restricted to the specific tissue or organ in which they occur, in contrast to the widespread phenotypic (i.e., pleiotropic) alterations expected to occur when the systemic titer of a hormone is changed. In some cases, such as for juvenile hormones, the receptor has not been unambiguously identified, thus precluding serious study of intraspecific variation in tissue sensitivity for this important group of hormones. Although researchers have investigated receptors extensively in comparative endocrine studies of vertebrates (Norris 1997), functional studies of receptor variation are rare in evolutionary-endocrine studies as well as in field-ecological studies (for an exception see Hoekstra et al. 2006, discussed in Vertebrate Studies, below).

The paucity of studies on receptors and components of intracellular signaling pathways testifies to the nascent state of evolutionary endocrinology. However, this situation almost certainly will change dramatically in the next decade. Researchers have undertaken extensive mutational/molecular analyses of signaling pathways and hormone receptors in a few model genetic organisms such as *Drosophila*, *C. elegans*, and mice (Harshman & Zera 2007, Henrich 2005, Partridge et al. 2005). These model organisms and molecular approaches are beginning to be incorporated into evolutionary-endocrine research at an increasing rate (Harshman & Zera 2007, Richard et al. 2005); a number of recent, notable examples are discussed throughout this review.

The identification of fitness effects of endocrine variation is also a key aspect of evolutionary endocrinology. This remains to be done for endocrine genotypes, although several detailed studies in vertebrates addressed this issue at the phenotypic level (discussed below in Vertebrate Studies).

The misuse of hormone manipulation. One widespread and growing problem in evolutionary endocrinology, which deserves special comment, is the inappropriate use of hormone manipulation (HM), which involves experimental manipulation of the in vivo titer of a hormone [see Zera (2007b) for a detailed discussion of this topic]. HM is very useful in evolutionary endocrinology, for example, to evaluate the functional significance of hormone-phenotype correlations identified in genetic analyses (Zijlstra et al. 2004; see Genetic Polymorphism for Wing Pattern and Life History in *Bicyclus*, below), or to produce phenotypes for experiments (e.g., Ketterson et al. 2001; see Vertebrate Studies, below). However, HM has been commonly misused; it should not be the sole or primary empirical method used to establish the role of a hormone in the regulation of phenotypic expression (e.g., the juvenile hormone studies of Emlen & Nijhout 1999, 2001; Meylan & Clobert 2005; for a detailed critique see Zera 2007b). By itself, HM can never provide more than weak support for the role of a particular hormone or its specific mechanisms of action in the regulation of phenotypic expression; applied hormones or analogs can have strong pharmacological (nonphysiological) effects, especially when applied in nonphysiological concentrations (which is often the case in insect studies) or at inappropriate times in the life cycle. Furthermore, introduced hormones can alter titers or receptors of other hormones. If variation in an endocrine trait is the focus of study, that endocrine trait should be quantified directly via the use of a well-validated assay, no matter how laborious.

Genetic Approaches in Evolutionary Endocrinology

Most evolutionary-endocrine studies to date focused on physiological aspects of hormonal variation such as hormone titers or activities of hormone-regulating enzymes, via the use of quantitative-genetic methods, the most common of which is artificial selection. For general reviews of artificial selection see Gibbs (1999), Harshman & Hoffmann (2000), Swallow & Garland (2005), and Falconer & MacKay (1996). The degree of response to artificial selection quantifies the genetic component of phenotypic variation for the selected trait (heritability) (Zera & Zhang 1995). Researchers identify endocrine factors that potentially regulate the expression of morphological/life history traits as indirect (correlated) responses when nonendocrine phenotypes (e.g., life history traits) are selected directly [Malisch et al. 2007, Zera 2006, Zera & Huang 1999, Zijlstra et al. 2004; see Discontinuous Variation (Genetic Polymorphism) in Insects, below].

A major advantage of artificial selection is the production of genetically differentiated populations for the selected and correlated phenotypes; these stocks can then be subjected to further analyses to identify the underlying causes of phenotypic divergence and correlations. For example, researchers used this approach to investigate the molecular and biochemical causes of line differences in the activity of the cricket juvenile hormone titer regulator, juvenile hormone esterase (JHE) (Zera 2006; discussed below in the section on Direct Selection on an Endocrine Regulator in *Gryllus assimilis*). Individual genes that contribute to physiological-genetic variation in an endocrine trait or a phenotype controlled by that regulator can also be identified in artificially selected lines via the use of standard quantitative-genetic methodologies, such as quantitative trait locus (QTL) mapping or the investigation of candidate genes. For example, QTL analysis of variation in the corticosterone-mediated stress axis in the rat identified a complex genetic architecture consisting of multiple interacting genetic factors with maternal and sex effects (e.g., Solberg et al. 2006). Until very recently, this approach was not often used in evolutionary-endocrine studies, and contributions of individual genes to a quantitative endocrine trait of interest are not well known [however see Crone et al. (2007) and Hoekstra et al. (2006); discussed below in the sections on Continuous Variation in Insects and on Vertebrate Studies]. We expect such analyses to become common in the next decade.

Quantitative-genetic studies of endocrine variation have been conducted with varying degrees of rigor. For example, some studies used full-sib breeding designs or regression of offspring on the female parent, which can significantly inflate heritabilities for a variety of reasons (e.g., maternal effects; King et al. 2004); many of these problems can be circumvented by half-sib analyses (Gu & Zera 1996; Roff et al. 1997). In some artificial selection studies (Suzuki & Nijhout 2006, Zera et al. 1989), researchers selected only one line in a particular direction (nonreplicated selection). Only suggestive genetic inferences, such as the existence of genetic differences between lines or genetic correlation, can be drawn from unreplicated studies (Falconer & Mackay 1996) without additional experiments such as line crosses (as in Zera & Teibel 1989).

Assessment of the functional significance of genetic correlation between a hormonal trait (e.g., hormone titer) and a phenotype (e.g., morphological trait) also requires care because correlation does not necessarily imply causation. Spurious, nonfunctional endocrine correlations may result from the extensive cross talk between hormonal regulators (see A Thumbnail Sketch of Endocrine Regulation, above). The identification of such correlations may be difficult but it is critically important and will typically require experimental manipulation; in this context, HM can be a powerful experimental technique (Zijlstra et al. 2004; see The Misuse of Hormone Manipulation, above). Researchers have substantially underestimated the difficulty involved in assessing the functional significance of endocrine correlations in many evolutionary-endocrine studies.

Mutational analysis is a more recent genetic approach in evolutionary endocrinology and is becoming more common. This approach uses laboratory-generated single-locus mutations, typically of large effect, to identify hormonal traits (e.g., components of signaling pathways) that control the expression of complex phenotypes. An example of this approach is the use of mutations in the insulin signaling pathway to identify the role of this pathway in life history variation and trade-offs in *Drosophila* (Flatt et al. 2005, Richard et al. 2005). Currently, mutational analysis is restricted almost exclusively to model genetic organisms (e.g., *C. elegans, Drosophila*). The great power of mutational analysis is its ability to dissect the mechanisms of hormone action on ecologically important traits by identifying the influence of often well-characterized individual endocrine loci on a phenotype of interest. The limitation of this approach is that mutations of large effect, which are typically used in mutational analysis, often do not segregate in natural populations. Thus, mechanisms that underlie the hormonal control of phenotypic variation identified by mutational analysis may be different from the mechanisms in natural or laboratory populations (Brakefield et al. 2003, Harshman & Zera 2007, Stern 2000). Relatively little is known currently about the nature of allelic variation in endocrine system genes that segregate in populations. A few notable physiological studies focused on spontaneous, single-locus endocrine mutations in the laboratory (Rountree & Nijhout, 1995a,b; Suzuki & Nijhout 2006); a spate of recent molecular studies investigated single-locus (or oligogenic) polymorphisms in natural populations of vertebrates (Colosimo et al. 2005, Hoekstra et al. 2006, Mundy et al. 2003, Nachman et al. 2003; all discussed below in Vertebrate Studies).

EMPIRICAL STUDIES

Discontinuous Variation (Genetic Polymorphism) in Insects

Wing (dispersal) polymorphism in crickets. During the past 15 years, investigators studied, from an endocrine-genetic perspective, a number of complex (multi-trait) polymorphisms that play an important role in the life cycle of the organism. Most notable are dispersal-polymorphism in crickets and wing-pattern polyphenism in butterflies (reviewed in Brakefield & Frankino 2007; Brakefield et al. 2003; Zera 2004, 2006). Polymorphism is defined here as discontinuous variation in some phenotype. If phenotypic variation results from variation in genotype, it is termed genetic polymorphism; if phenotypic variation is due to environmental variation, it is termed polyphenism.

Wing polymorphism in species of *Gryllus* (crickets) has been especially well-studied with respect to systemic variation in the titers of juvenile hormone (JH) and ecdysteroids, two key developmental and reproductive hormones, and the role of titer variation in regulating aspects of development (such as alternate morph production) and adult-life history (e.g., the trade-off between flight capability and reproduction). Wing polymorphism in *Gryllus* is both a genetic polymorphism (polygenic, with a threshold) (Roff, 1996) and an environmental polyphenism (Zera 2004, 2007a). However, virtually all endocrine work on this model focuses on genetic polymorphism, and employs artificial selection in a single laboratory environment. The polymorphism consists of a flight-capable morph that has long wings and large flight muscles, but delays egg production and has low fecundity. The alternate, flightless morph has the converse set of traits (Zera 2004, 2006, 2007a). In two *Gryllus* species, using field-collected individuals, researchers obtained selected lines that produced >90% of the dispersing or flightless/reproductive morph; these selected lines were subjected to endocrine analyses (reviewed in Zera 2004, 2006, 2007a).

In *G. rubens* and *G. firmus*, lines selected for the dispersing morph exhibited (*a*) much higher blood activity of the enzyme JHE, which degrades JH, (*b*) a reduced JH titer, and (*c*) an elevated ecdysteroid titer compared with lines selected for the flightless/reproductive morph. These differences occurred during a critical period in morph development. High JHE activity strongly cosegregated with the dispersing morph in crosses and backcrosses of selected populations of *G. rubens*, which is the first demonstration of genetic covariance (i.e., cosegregation) between a variable endocrine and developmental/life history trait in insects recently derived from the field. Because ecdysteroids promote growth and differentiation, whereas JH inhibits metamorphosis, the higher ecdysteroid titer and lower JH titer (and higher JHE activity) are consistent with a role for either or both of these hormones in the regulation of morph development (Zera 2006). In contrast, JH binding in the hemolymph (due to the JH binding protein) did not differ between morphs (Zera & Holtmeier, 1992), indicating that variation in JH binding does not likely play a major role in the endocrine regulation of morph development. Zera and coworkers have also undertaken one of the few direct comparisons between genetic polymorphism and environmental polyphenism for an endocrine trait, JHE activity (Zera 2006, Zera & Teibel 1989; also see Rountree & Nijhout 1995a,b and Suzuki & Nijhout 2006).

Endocrine studies of *Gryllus* have a number of limitations. For example, despite considerable study, the relative importance of JH versus ecdysteroids (or other hormones) in the regulation of aspects of morph expression still remains unresolved, and JH titer differences between morphs are not large (see extensive discussion in Zera 2004, 2006). Because a nuclear JH receptor has not been unequivocally identified in insects, the extent to which morph-specific traits result from variation in receptor characteristics, as opposed to variation in circulating hormone levels, cannot be assessed. Finally, the influence of hormones other than JH or ecdysteroids on morph development, or the endocrine mechanisms that coordinate morph development with morph-specific reproduction in adults, are unknown (Zera 2004, 2006).

Because JH plays an important role in insect reproduction (Gilbert et al. 2005, Nijhout 1994), endocrine-evolutionary models of complex polymorphism have long proposed that the blood JH titer should be higher in the morph with elevated fecundity. However, this hypothesis has been tested directly in dispersal-polymorphic species only during the past few years (Zera 2004, 2006). Surprisingly, in *G. firmus* the JH titer in the dispersing morph exhibits a high amplitude, circadian rhythm (50-fold titer change over a 6 hour period) that cycles above and below the relatively invariant titer in the flightless morph (Zhao & Zera 2004). These data suggest that morph-specific effects of JH may be determined not only by the titer, but also by the length of time during which the titer is elevated above a threshold (Zera et al. 2007). By contrast, the ecdysteroid titer is consistently elevated in flightless versus dispersing females and may contribute to the elevated egg production of the flightless morph in concert with or independent of JH. The unexpected morph-specific JH circadian rhythm illustrates the importance of measuring hormone titers directly over short timescales. A number of studies recently showed phenotypic variation in circadian or diurnal rhythms for endocrine traits in insects (Vafopoulou & Steel 2005; Zhao & Zera 2004). The failure to identify these endocrine titer rhythms, especially if the rhythms are morph-specific, can result in substantial errors in interpretation of the endocrine data (Zera 2006, 2007b).

Although field-endocrine studies are common in vertebrates (discussed in Vertebrate Studies, below), only a few limited studies have been undertaken in insects (reviewed in Zera et al. 2007). Thus, the correspondence between endocrine traits measured in selected lines in the laboratory and measured under field conditions is poorly understood. However, a recent multi-year field study found comparable results for blood JH and ecdysteroid titers in *G. firmus* in individuals sampled in the field or in selected lines raised under field conditions, as were found in laboratory studies described in the paragraph above (Zera et al. 2007; and A.J. Zera, unpublished observations). The morph-specific circadian rhythm for JH titer in field populations of *G. firmus* is a powerful model with which to investigate the microevolution of endocrine circadian rhythms, a poorly studied topic in evolutionary endocrinology.

Genetic polymorphism for wing pattern and life history in *Bicyclus*. Butterfly wing pattern components (e.g., size, color, shape) are exquisite adaptations for camouflage, thermoregulation, mate recognition, etc. Developmental, hormonal, biochemical, and more recently, molecular aspects of variation in wing pattern expression have been studied extensively, most often in the context of seasonal polyphenism (reviewed in Brakefield & Frankino 2007, Nijhout 1991). Various neuropeptides (Jones et al. 2006) and ecdysteroids are implicated as regulators of pigment biosynthesis and eyespot patterning.

The most detailed endocrine-genetic analyses of wing-pattern components are studies of ecdysteroid control of eyespot size and life history (rate of development) in the butterfly *Bicyclus anyana* (Brakefield & Frankino 2007, Koch et al. 1996, Zijlstra et al. 2004). Size of the ventral eyespot and rate of development are adaptations to a particular season (polyphenism) (Brakefield & Frankino 2007). Researchers conducted artificial selection in the laboratory in an environment intermediate between those producing the seasonal phenotypes, and on a base population that contained ventral eyespot sizes intermediate between those seen in wet and dry seasons. Selection on ventral eyespot size or rate of development demonstrated that these characteristics are correlated genetically with each other and with the hemolymph ecdysteroid titer during early pupal development. Results of ecdysteroid injection experiments implicated an earlier rise in the ecdysteroid titer as a cause of increased ventral eyespot size and faster rate of development in the wet-season morph, although a more complex picture emerged subsequently (Zijlstra et al. 2004).

Simultaneous two-trait selection on ventral eyespot size and rate of development produced four lines with the four possible trait combinations (fast-development rate and large eyespots, slow-development rate and large eyespots, etc.). Unexpectedly, investigators observed genetic covariance between the ecdysteroid titer and development rate, but not ventral eyespot size (Zijlstra et al. 2004). Nevertheless, injection of ecdysteroids increased eyespot size as well as rate of development, as observed previously (Koch et al. 1996). A confounding problem with hormone injection was the strong effect of injection per se on eyespot size (see Zijlstra et al. 2004). These authors concluded that the ecdysteroid titer primarily controls development rate, whereas ventral eyespot size can be modulated by ecdysteroid-dependent as well as ecdysteroid-independent mechanisms. These important studies illustrate that (*a*) endocrine-genetic control mechanisms can be more complex than they appear at first, and (*b*) the initial demonstration of hormone-phenotype associations should be viewed as the beginning, not the end point, of functional studies of endocrine control of phenotypic expression.

Evolution of a polyphenism by artificial selection. Although numerous studies investigated the mechanisms that underlie morph expression in polyphenisms, little is known about the evolutionary origins of polyphenism. In an innovative study, Suzuki & Nijhout (2006) addressed this issue with direct selection for plasticity in body color in *Manduca sexta*, a model organism in insect endocrine studies. The authors heat shocked the *Manduca sexta black* mutant to expose hidden genetic variation in coloration, which provided the means for selection. The authors produced a line (unreplicated) whose body color was always black irrespective of rearing temperature (monophenic line), as opposed to another line that was black or green depending on rearing temperature. Owing to the extensive endocrine database for this model species (Nijhout 1994), Suzuki & Nijhout (2006) could investigate the endocrine basis of the evolution of plasticity. These authors reported that the plastic line evolved by selection on genes that regulate temperature-dependent hemolymph juvenile hormone titer. One limitation of this study is that Suzuki & Nijhout (2006) used a bioassay to measure the JH titer; bioassays can be unreliable (see Baker 1990, Zera 2007b).

Other polymorphisms/polyphenisms. A number of other groups investigated JH regulation of complex polymorphism, such as horn size polymorphism in dung beetles and wing polymorphism in the soapberry bug (Dingle & Winchell 1997; Emlen & Nijhout 1999, 2001; Moczek & Nijhout 2002). Unfortunately, in these studies hormone manipulation (HM) was the only endocrine technique used by researchers to investigate JH regulation; they failed to measure any aspect of JH signaling directly, such as JH titer or activities of JH titer regulators [see Zera (2007b) for a detailed critique]. Conversely, preliminary data suggest ecdysteroid titer differences between nascent horn morphs (Emlen & Nijhout 1999). Rountree & Nijhout (1995a,b) investigated the ecdysteroid regulation of wing color pattern in the butterfly *Precis coenia* in natural polyphenic morphs and in a spontaneous laboratory mutant.

Continuous Variation in Insects

Direct selection on an endocrine regulator in *Gryllus assimilis*. The activity of the hormone-regulating enzyme JHE in the field cricket *Gryllus assimilis* was the first endocrine trait directly subjected to artificial selection in a laboratory population recently founded from field-collected individuals (Zera & Zhang 1995). These studies combined with other quantitative-genetic analyses (Gu & Zera 1996) and studies of

JHE activity in wing-polymorphic crickets (discussed above) make JHE the most intensively studied endocrine trait from an evolutionary-genetic perspective (reviewed in Zera 2006).

Researchers observed strong responses to replicated selection in either juvenile or adult stages; heritabilities were similar to those of nonendocrine enzymes (Zera 2006, Zera & Zhang 1995, Zera et al. 1998). Selection on JHE activity during the juvenile or adult stages resulted in no correlated responses on JHE activity during the alternate life cycle stage. These results demonstrate that some components of endocrine regulation can evolve independently in a stage-specific manner. The extent to which physiological mechanisms constrain various life cycle stages to evolve in concert is a key but poorly understood topic in evolution (Brakefield et al. 2003, West-Eberhard 2003, Zera 2006). Extensive studies of morphological, developmental, physiological, biochemical, and molecular correlates of JHE activity in selected lines of *G. assimilis* were reviewed recently in Zera (2006), and only more recent or salient findings are covered here.

Characterizations of selected lines of *G. assimilis* show that the response to divergent selection on blood JHE activity results equally from genes that contribute to: (*a*) whole-organism enzyme activity and (*b*) the degree of enzyme secretion into the blood, similar to *G. firmus* (Zera & Huang, 1999). Enzymatic properties do not differ between JHEs from high and low activity lines, nor are there nucleotide differences in the coding sequence of JHE genes within or between selected lines (Crone et al. 2007 and references therein). Cosegregation between line-specific JHE allele and blood JHE activity implicates DNA sequence variation at or near the JHE gene region as a contributor to line differences in blood JHE activity. The study by Crone and coworkers (2007) is one of the few that links molecular aspects of a candidate gene to physiological/phenotypic variation in an endocrine trait in an insect (analogous studies in vertebrates are discussed in the sections on Vertebrate Studies and on Single Locus Polymorphisms in the Field).

Comparisons among the results of artificial selection studies performed in *G. rubens*, *G. firmus*, and *G. assimilis* provide important insights into (*a*) the nature of the genetic factors responsible for microevolutionary changes in JHE activity, (*b*) the degree of change in JHE activity necessary to alter whole-organism JH metabolism (a prerequisite for affecting the expression of whole-organism traits), and (*c*) the role of modulation of JHE activity on the expression of specific phenotypes such as wing length and flight-muscle mass. Each of these issues is discussed in detail in Zera (2006).

Body size in insects. Body size is an important organismal trait (reviewed in Edgar 2006, Nijhout 2003). There has been a recent surge of molecular-endocrine studies on body size regulation in *Drosophila*, whereas physiological-hormonal investigations of body size regulation in *M. sexta* have been ongoing since the 1970s. These studies provided the impetus for the first evolutionary-endocrine studies of body size in insects.

Nijhout and colleagues (2003) developed a physiological model that relates attainment of size at metamorphosis (size when feeding ends in the last larval stage = adult size) to various whole-organism growth parameters, which are in turn related to underlying endocrine events (D'Amico et al. 2001, Edgar 2006, Nijhout 2003). Ninety percent of growth occurs during the last juvenile instar of *M. sexta*; when a genotype-specific critical weight is reached, a series of endocrine events are set in motion, leading ultimately to the cessation of feeding and onset of metamorphosis. The duration of time between attainment of critical weight and cessation of feeding [the interval to cessation of growth (ICG), when most weight gain occurs] is thought to be determined by the rate of decrease in the JH titer, subsequent secretion of the brain neurohormone (PTTH; secretion inhibited by JH), and release of ecdysteroids (release induced by PTTH); ecdysteroid release is a proximate initiator of molting (end of ICG) [see Nijhout (2003)].

Using this model, D'Amico and coworkers (2001) reported that a laboratory population of *M. sexta* evolved 50% higher body size during a 30-year period by altering growth rate, duration of ICG [i.e., purported timing of prothoracicotropic hormone (PTTH) release], and critical weight. Davidowitz & Nijhout (2004) identified how variation in growth rate and ICG interact to produce larger individuals at higher temperatures (thermal reaction norms). Preliminary studies using a full-sib design identified genetic variation for various size parameters (Davidowicz et al. 2003) and an extensive artificial selection study has been conducted (G. Davidowicz, personal communication).

The aforementioned *Manduca* studies are problematic regarding some endocrine aspects. For example, the timing of PTTH release and its functional relationship to the end of ICG is a major aspect of the model (e.g., see figure 3 of D'Amico et al. 2001 and Davidowicz et al. 2003). However, Nijhout and colleagues never measured PTTH release (i.e., PTTH titer) directly in any study that focused on the physiological control of body size; rather, they inferred PTTH release from the appearance of a morphological marker that is thought to be correlated with release of this neurohormone (e.g., see Methods section of D'Amico et al. 2001). However, the correlation has never been directly established, but instead was indirectly inferred in the 1970s on the basis of a best guess as to the timing of PTTH release during the last juvenile instar (Nijhout & Williams 1974 and references therein). PTTH release (measured directly in several moth species; Rybczynski 2005) is now known to be much more complex than suspected previously.

In *Drosophila*, recent studies have begun to unravel the detailed mechanisms by which various phylogentically conserved signaling pathways [most notably insulin/insulin-like growth factors and target of rapamycin (TOR)] influence body size [for a detailed review see Edgar (2006)]. Considerable discussion and empirical work is in progress on alterations in the signaling pathways described above in the context of the evolution of body size or size of organs (e.g., beetle horns) within and between species (De Jong & Bochdanovits 2003, Emlen et al. 2006, Shingleton et al. 2005).

Life history variation and trade-offs in *Drosophila*. The use of laboratory-generated mutants is a relatively new approach in evolutionary endocrinology that was pioneered in studies of the hormonal underpinnings of life history variation and trade-offs in *D. melanogaster*. The endocrine foci of these studies are insulin signaling, juvenile hormones, and ecdysteroids (Clancy et al. 2001, Richard et al. 2005, Tatar et al. 2001, Tu et al. 2005). A detailed discussion of these studies, some of which report contradictory findings, is useful to illustrate important issues regarding JH endocrinology in *Drosophila* and the use of mutational analysis. Laboratory-induced mutations of *D. melanogaster* that interfere with insulin signaling produce phenotypes that include small size, sterility, and extended life span (Bohni et al. 1999, Clancy et al. 2001, Tatar et al. 2001). Working with *Drosophila* insulin receptor (*DInR*) mutations, Tatar and coworkers (2001) proposed that these effects were caused by a defect in the cellular insulin receptor, which resulted in reduced systemic (e.g., circulating) levels of juvenile hormone. In other words, the phenotypic effects of insulin-signaling mutations were proposed to be nonautonomous, that is, not confined to cells with the insulin signaling defect. The key observation for this argument was that *DInR* mutations resulted in an approximately 75% decrease in biosynthesis of juvenile hormone in vitro, relative to wild type. Aerosol exposure of this mutant combination to methoprene, a juvenile hormone analog, rescued egg production, albeit at low levels, and the extended life span phenotype was lost. However, it remains unclear to what degree the rate of in vitro JH biosynthesis in *Drosophila* mutants and wild type corresponds to levels of these hormones in the body, especially levels of circulating hormones in the hemolymph. In fact, the circulating concentration of juvenile hormones remains unreported in the hemolymph. Preliminary analysis of the hemolymph of feeding last instar *D. melanogaster* larvae identified methyl farnesoate and bisepoxy JHIII as the two primary JH-like compounds (Jones & Jones 2007).

Another study in *D. melanogaster* that used a different mutation of insulin signaling (*chico¹*, an insulin substrate protein mutation) also found increased life span (Clancy et al. 2001). Tu and coworkers (2002, 2005) reported a reduced rate of JH biosynthesis and a reduced release of ecdysteroids from the ovary in vitro in *chico¹*, whereas Richard and coworkers (2005) found no reduction in either the rate of JH biosynthesis or the release of ecdysteroids from the ovary in this mutant in a different genetic background. In addition, the *DInR* mutation exhibited a reduced rate of ecdysteroid release by the ovaries in vitro (Tu et al. 2002). Most importantly, the blood level of ecdysteroids was not lower in the long-lived *chico¹* mutation compared with the other genotypes (Richard et al. 2005). This is the only case in which circulating levels of any hormone were measured directly in an insulin signaling mutation of *D. melanogaster*. Exposure to methoprene by topical or aerosol application across a range of doses failed to recover egg production in females homozygous for this mutation (Richard et al. 2005). These workers performed a critical experiment to determine whether female sterility is caused by a defect in insulin signaling that is autonomous to the egg or caused by a nonautonomous effect of the lowered level of circulating JH (or a change in some other systemic factor). They reciprocally transplanted immature ovaries between wild-type and *chico¹* mutants; wild-type ovaries produced mature eggs in *chico¹* females, whereas the *chico¹* ovaries did not mature in wild-type females. Therefore, Richard and coworkers (2005) determined that the effect of the *chico¹* mutation on egg maturation was ovary-autonomous [see also Drummond-Barbaosa & Spradling (2001)] in the presence of approximately wild-type levels of juvenile hormone biosynthesis and ecdsyteroids in the blood.

In conclusion, whereas genetic alteration of insulin signaling clearly affects lifespan in *D. melanogaster*, the influence of JH remains an open question. The answer to the question of whether variation in JH leads to variation in aging may depend on the development of a genetic lesion that can perturb levels of the hormone and development of the capability to measure JHs in the blood (Jones & Jones 2007). Major differences between studies (Tatar et al. 2001, and Tu et al. 2002, 2005 compared with Richard et al. 2005) remain unresolved and may be due to specific mutations in the insulin signaling pathway, genetic background, or differences in assays. Clearly there are problems with mutation analysis, such as different patterns of pleiotropic effects associated with different mutations or differences in genetic background, but in the long run this methodology will be powerful for the investigation of the effects of hormone variation on life histories and other complex traits.

Vertebrate Studies

Vertebrate research has provided especially important contributions to evolutionary endocrinology in two areas: (*a*) physiological and molecular aspects of single-locus endocrine polymorphisms that occur in the field and (*b*) fitness consequences of endocrine variation in the field. Conversely, artificial selection studies of physiological aspects of endocrine variation have not been conducted in as much detail as in insects.

Single locus polymorphisms in the field. In one of the first studies of a genetically variable endocrine trait in the field, Kallman (1989 and references therein) identified numerous alleles (five in one species) at the *P* (pituitary) locus that segregate in natural populations of *Xiphophorus* (platyfish). *P* alleles affect pituitary function strongly and male genotypes can differ substantially in adult size and the appearance of male secondary sexual characteristics (e.g., male caudal appendage) in the laboratory. Although it is termed the pituitary locus, the actual site of action of the *P* locus was traced to higher centers of the brain that involve the production, release, or fate of gonadotropin-releasing hormone (GnRH) (Halpern-Sebold et al. 1986, Kallman 1989). GnRH ultimately regulates the production of gonadal steroids (e.g., testosterone and estrogens) via effects on pituitary gonadotropins. Investigators conducted these genetic and endocrine studies during the 1970s and 1980s and unfortunately no additional endocrine work has been published on this system since then. Because both natural and sexual selection affect body size in *Xiphophorus* (Basolo & Wagner, 2004), the *P* locus may be an important factor in sexual selection and life history evolution in species of this genus in the field. The *P* locus–body size system shows great promise in the integration of endocrine-physiological and life history microevolutionary studies in the field.

A number of studies investigated adaptive variation at the melanocortin receptor gene *Mc1r* in vertebrates, which mediates melanocyte activity and body color. Researchers identified associations between *Mc1r* DNA sequence and body-color phenotype in some field populations of mice and lizards (Nachman et al. 2003, Rosenblum et al. 2004) and in several bird species (Mundy et al. 2003, Theron et al. 2001).

Nonsynonymous mutations at different sites of the *Mc1r* locus during the Pleistocene appear to have given rise to independently evolved melanic plumage polymorphisms in geese and skuas (Mundy et al. 2003). Results from molecular population genetic and association analyses support the hypothesis that selection has acted on *Mc1r* in the little striped whiptail (*Aspidoscelis inornata*) (Rosenblum et al. 2004). More recently, Hoekstra and coworkers (2006) reported that the *Mc1r* gene in extremely light-colored beach mouse populations of *Peromyscus polionotus* differs by a single, derived, charge-changing amino acid compared with *Mc1r* from more darkly colored mainland populations. Importantly, Hoekstra and coworkers (2006) reported that the amino-acid substitution changed the binding characteristics of the receptor. To our knowledge, this is the first direct demonstration in natural populations of adaptive genetic variation in receptor function.

Colosimo and coworkers (2005) investigated the role of DNA sequence variation in the *Ectodysplasin* (*Eda*) gene in the context of the evolution of reduced bony armor in the three-spined stickleback, *Gasterosteus aculeatus*, an extensively studied model in evolutionary and ecological genetics. The *Eda* gene encodes a secreted, locally acting signaling molecule (paracine regulator) that influences the development of dermal bones. Parallel evolution of the low-plated phenotype found in freshwater environments appears to have resulted from repeated selection on *Eda* alleles, derived from an ancestral haplotype, that are present in low frequencies in marine populations.

Artificial selection studies. Researchers performed several artificial selection studies of vertebrate endocrine traits in an evolutionary-behavioral context. In Japanese quail (*Coturnix coturnix japonica*), selection for low or high stress–induced corticosterone secretion led to changes in behavioral phenotype in the high stress line (e.g., greater avoidance, more fear-related behavior, and higher plasma corticosterone release in response to capture and restraint), but did not affect baseline corticosterone levels (Jones et al. 1994). A similar response occured for zebra finches selected directly on stress-induced corticosterone levels (Evans et al. 2006). These lines, together with testosterone titer manipulation, are in use to dissect the relative contributions of corticosterone and testosterone to the trade-off between immune function and the development of sexual signals (Roberts et al. 2007). Other studies involved selection on putative endocrine-mediated traits and investigated the correlated response of endocrine traits (mainly hormone titers). In a carefully controlled artificial selection study, Garland (Garland 2003) measured the titers of several hormones (e.g., corticosterone and leptin) in lines of the house mouse divergently selected for voluntary wheel-running (Malisch et al. 2007 and references therein). Ongoing research is aimed at quantifying levels of the corticosterone receptor and its binding protein in selected and control lines (T.G. Garland Jr., personal communication). In great tits (*Parus major*), divergent selection on personality produced lines of less aggressive, more cautious birds (slow explorers), and more aggressive fast explorers. Slow explorers showed a greater hypothalamic-pituitary-adrenal axis reactivity to social challenge compared with fast explorers (Carere et al. 2003, Groothuis & Carere 2005).

Hoeflich and coworkers (2004) found that high-growth lines of mice derived from several independent artificial selection experiments consistently exhibited elevated serum levels of insulin growth factor 1 (IGF1) but not growth factor binding proteins, nor elevated expression of IGF receptors in muscle. This is a unique study in that it compares the contribution of variable genes that encode plasma hormone levels versus hormone receptors to the response to selection in a key life history trait. The findings of Hoeflich and coworkers (2004) support the strong association between a specific IGF1 haplotype and size in dog breeds (Sutter et al. 2007), which suggests that IGF1 was also an endocrine target of selection on size during the early history of dog domestication.

Fitness correlates of endocrine variation in the field. Topics of central importance in evolutionary endocrinology are the extent to which endocrine genetic variation gives rise to variation in fitness in the field and the mechanisms involved (Ketterson et al. 1996). Several long-term field studies in vertebrates used hormonal and other physiological manipulations to address these topics. In a series of pioneering studies, Sinervo and colleagues (Sinervo 1999, Sinervo et al. 2006, Svensson & Sinervo 2004, and references therein) manipulated follicle-stimulating hormone (FSH) and follicle size to investigate the endocrine mechanisms that control the egg size/egg number trade-off. The same group used HM to study fitness effects of endocrine-mediated variation such as the trade-off between fecundity and offspring survival (Sinervo & Doughty 1996) and the cost of reproduction (Sinervo & DeNardo 1996; for a similar study in mammals see Oksanen et al. 2002). These authors proposed very specific endocrine regulatory mechanisms that underlie these trade-offs. For example, genetically based covariation between egg and clutch size is thought to arise from variation in synthesis, hormone titers, or receptor sensitivity for GnRH and FSH, and corticosterone may regulate female condition, immune function, and fecundity (Sinervo & Calsbeek 2003, Svensson et al. 2002). However, to date few components of hormonal regulation have been quantified directly (in some cases owing to a lack of appropriate hormone assays) and the extent to which analogous, natural hormonal variation regulates these trade-offs in unmanipulated lizards has yet to be established. In addition, the same manipulations that Sinervo (1999) used do not always generate consistent changes in egg and clutch size in other taxa (Christians & Williams 2002, Ji & Diong 2006).

In a very thorough study, Ketterson and colleagues (Ketterson et al. 1996, 2001) used HM to elevate plasma testosterone (T) in male dark-eyed juncos (*Junco hyemalis*) to determine the selection pressures that may have shaped the typical distribution of hormone titers in natural populations. Validation studies showed that T-implants cause a prolonged elevation of plasma T within the normal physiological range and this in turn affects many different phenotypic traits, presumably reflecting the pleiotropic actions of T. Compared with control males, high-T males have higher song rates, larger territory size, are more attractive to females, and gain more extrapair fertilizations, but they also show decreased parental behaviors (less nest defense, lower chick feeding rates) and have lower survival, perhaps related to lower body fat, higher plasma corticosterone, suppressed immune function, and delayed molt (Ketterson et al. 1996, 2001). In a recent paper, Reed and coworkers (2006) combined the long-term HM data from these studies with population modeling to test explicitly the prediction that experimentally manipulated high-T males (the extreme phenotype) have lower fitness than control males (where plasma T levels were shaped by selection). Surprisingly, they found the opposite: High-T males have higher fitness than control males owing to higher rates of extrapair copulation by T males. These authors suggested that testosterone levels may be constrained in natural populations, although there are other potential explanations such as selection via indirect effects of T on off-spring or females (Reed et al. 2006). A major unanswered question generated by these studies is the extent to which experimental manipulation mimics the pleiotropic effects of segregating genetic variation for endocrine regulators in these and other species.

SUMMARY AND CONCLUSIONS

During the past two decades, the first direct information on the extent, characteristics, and functional significance of genetic variation for endocrine traits in nondomesti-cated animals was obtained using well-validated endocrine techniques. This is the most important achievement of the nascent field of evolutionary endocrinology thus far. Insect and vertebrate studies contributed in complementary ways to our under-standing of evolutionary endocrinology: Insect studies provided the most detailed information on the endocrine-genetics of hormone titers, titer regulators, and in-tracellular signaling in the laboratory; vertebrate studies contributed the most to our understanding of the endocrine basis of single-locus polymorphisms and fitness effects of endocrine variation in the field.

Results to date clearly implicate variation in endocrine regulation as an im-portant and widespread aspect of organismal microevolution. Prominent examples include morphs of complex polymorphisms in insects (dispersal and wing-pattern polymorphisms in crickets and butterflies) and vertebrates (size and plate-morph polymorphisms in fish; color polymorphisms in various species). Genetic variation in endocrine regulation may play an important role in both developmental and repro-ductive aspects of these ecologically important polymorphisms. Similarly, mutational analysis implicates the modulation of insulin signaling in the microevolution of indi-vidual and suites of life history traits in *D. melanogaster*.

In spite of marked progress, evolutionary endocrinology is in its infancy, and studies of endocrine variation and microevolution have only scratched the surface of the mechanisms involved. Most empirical studies have focused on the physiological-genetic aspects of systemic hormone titers and titer regulators of only a few hormones (juvenile hormone and ecdysteroids in insects; testosterone and corticosterone in ver-tebrates), measured in a handful of species, under a limited number of environmental conditions. For example, only one study has investigated circadian aspects of en-docrine genetic variation, although such influences may be widespread, and may substantially complicate the interpretation of endocrine data if not identified (Zera & Cisper, 2001, Zhao & Zera 2004). Thus, a major goal of evolutionary endocrinology in the next decade should be to obtain direct and detailed measurements of titers and titer regulators on these and additional hormones. Unless this is done, it will not be possible to determine the extent to which results obtained to date represent general endocrine adaptations.

The next decade will see the considerable expansion of evolutionary endocrinology because key topics that have largely been ignored will almost certainly be the focus of intense study. Most notable will be studies of genetic variation in hormone receptors and signaling pathways and their role in organismal adaptation. Investigations of these topics have thus far been hampered by the lack of appropriate molecular tools; however, model organisms for molecular-genetic studies are being used increasingly to address these issues (Richard et al. 2005), and molecular techniques are becoming available for a wider range of organisms (Colosimo et al. 2005, Hoekstra et al. 2006). Data on hormone receptor variation will be critically important for assessment of the relative occurrence of endocrine alterations that give rise to localized, organ- or tissue-specific changes in phenotypic expression (Hoeflich et al. 2004). Indeed, a key topic for future research is the relative occurrence of endocrine adaptations that involve alterations in systemic (titers) versus localized (receptor) regulators, the factors that select for one type of adaptation or another, and functional constraints that limit the evolution of a particular type of endocrine adaptation.

Although a number of classic studies in vertebrates investigated fitness consequences of endocrine-mediated phenotypic variation (FSH-mediated egg size/number trade-offs; testosterone effects on life history trade-offs in juncos), these studies investigated almost exclusively variation produced by hormonal manipulation. Genetic variation has not been studied for these endocrine traits in unmanipulated individuals. Thus, the relevance of the manipulation studies to both the existence and fitness effects of natural endocrine genetic variation remains to be established. More integrative studies are needed in which detailed investigation of the endocrine-genetic correlates of phenotypic variation is combined with experimental manipulations performed in the appropriate ecological context.

In conclusion, new molecular techniques are beginning to allow investigation of important but previously unapproachable endocrine issues such as variation in receptor expression. However, as exciting as these developments are, studies which focus solely on the identification of candidate genes with overexpression studies or mutational analyses of single protein products will not fully describe the complexity and integrated nature of endocrine regulatory networks. Rather, molecular and biochemical studies of receptor expression, receptor function, and intracellular signaling must be integrated with classical approaches, which involve detailed measurement of hormone titers and titer regulators in quantitative-genetics experiments. Moreover, experimental manipulations must be incorporated to assess the functional significance of endocrine variation at both the whole organism and molecular levels, and ideally in some cases these studies will be integrated with fitness studies conducted under appropriate ecological conditions. These are exciting times for evolutionary endocrinology, which is beginning to contribute significantly to our understanding of the evolution of complex adaptations.

DISCLOSURE STATEMENT

The authors are not aware of any biases that might be perceived as affecting the objectivity of this review.

ACKNOWLEDGMENTS

A.J.Z. acknowledges support from NSF (most recently, IBN-9507388, IBN 0130665, and IBN-0212486), and L.G.H. acknowledges support from NIH (NIDDK, 1R01 DK074136–01) and NSF (EPS-0346476). T.D.W. was supported by a Discovery Grant from the Natural Sciences and Engineering Research Council of Canada. We thank J. Christian for comments on an earlier draft of the manuscript.

LITERATURE CITED

Adkins-Regan E. 2005. *Hormones and Animal Social Behavior*. Princeton: Princeton Univ. Press. 429 pp.

Baker FC. 1990. Techniques for identification and quantification of juvenile hormones and related compounds in arthropods. In *Morphogenetic Hormones of Arthropods*, Vol. 1, ed. AP Gupta, pp. 389–453. New Brunswick: Rutgers Univ. Press

Basolo A, Wagner W. 2004. Covariation between predation risk, body size and fin elaboration in the green swordtail, *Xiphophorus helleri*. *Biol. J. Linn. Soc.* 83:87–100

Bohni R, Riesgo-Escovar J, Oldham S, Brogiolo W, Stocker H, et al. 1999. Autonomous control of cell and organ size by CHICO, a *Drosophila* homolog of vertebrate IRS. *Cell* 97:865–75

Brakefield P, Frankino W. 2007. Polyphenisms in Lepidoptera: Multidisciplinary approaches to studies of evolution and development. In *Phenotypic Plasticity in Insects: Mechanisms and Consequences*, ed. T Ananthakrishnan, D Whitman. pp. 121–152. Science Publishers

Brakefield P, French V, Zwaan B. 2003. Development and the genetics of evolutionary change within insect species. *Annu. Rev. Ecol. Syst.* 34:633–60

Breuner CW, Orchinik M. 2002. Beyond carrier proteins: plasma binding proteins as mediators of corticosteriod action in vertebrates. *J. Endocrinol.* 175:99–112

Carere C, Groothuis T, Mostl E, Dann S, Koolhaus J. 2003. Fecal corticosteroids in a territorial bird selected for different personalities: daily rhythm and response to social stress. *Horm. Behav.* 43:540–48

Christians JK, Williams TD. 2002. Organ mass dynamics in relation to yolk production and egg formation in European starlings *Sturnus vulgaris*. *Physiol. Biochem. Zool.* 72:455–61

Clancy D, Gems D, Harshman LG, Oldham SH, Stocker H, et al. 2001. Extension of lifespan by loss of CHICO, a *Drosophila* insulin receptor substrate protein. *Science* 292:104–6

Colon J, Larhammar D. 2005. The evolution of neuroendocrine peptides. *Gen. Comp. Endocrinol.* 142:53–59

Colosimo P, Hosemann K, Balabhadra S, Villarreal G, Dickson M, et al. 2005. Widespread parallel evolution in sticklebacks by repeated fixation of ectodysplasin alleles. *Science* 307:1928–33

Crone E, Zera AJ, Anand A, Oakeshott J, Sutherland T, et al. 2007. JHE in *Gryllus assimilis*: Cloning, sequence-activity associations and phylogeny. *Insect Biochem. Molec. Biol.* In press

D'Amico L, Davidowicz G, Nijhout HF. 2001. The developmental and physiological basis of body size evolution in an insect. *Proc. R. Soc. London Ser.* B 268:1589–93

Davidowicz G, D'Amico L, Nijhout HF. 2003. Critical weight in the development of insect body size. *Evol. Dev.* 5:188–97

Davidowicz G, Nijhout HF. 2004. The physiological basis of reaction norms: The interaction among growth rate, the duration of growth and body size. *Integr. Comp. Biol.* 44:443–49

De Jong G, Bochdanovits Z. 2003. Latitudinal clines in *Drosophila melanogaster*: body size, allozyme frequencies, inversion frequencies, and the insulin signaling pathway. *J. Genet.* 82:207–23

Dingle H, Winchell R. 1997. Juvenile hormone as a mediator of plasticity in insect life histories. *Arch. Insect Biochem. Physiol.* 35:359–73

Drummond-Barbarosa D, Spradling A. 2001. Stem cells and their progeny respond to nutritional changes during *Drosophila* oogenesis. *Dev. Biol.* 231:265–78

Edgar B. 2006. How flies get their size: genetics meets physiology. *Nat. Rev. Genet.* 7:907–16

Emlen DJ, Nijhout HF. 1999. Hormonal control of male horn length dimorphism in the dung beetle *Onthophagus taurus* (Coleoptera: Scarabaeidae). *J. Insect Physiol.* 45:45–53

Emlen DJ, Nijhout HF. 2001. Hormonal control of male horn length dimorphism in *Onthophagus taurus* (Coleoptera: Scarabaeidae): a second critical period of sensitivity to juvenile hormone. *J. Insect Physiol.* 47:1045–54

Emlen DJ, Szafran Q, Corley L, Dworkin I. 2006. Insulin signaling and limb-patterning candidate pathways for the origin of evolutionary diversification. *Heredity* 97:179–91

Evans M, Roberts M, Buchanan K, Goldsmith A. 2006. Heritability of corticosterone responses and changes in life history traits during selection in the zebra finch. *J. Evol. Biol.* 19:343–52

Falconer DS, Mackay TF. 1996. *Introduction to Quantitative Genetics*, 4th ed. Essex: Longman. 464 pp.

Feder ME, Bennett AF, Burggren WW, Huey RB. 1987. *New Directions in Ecological Physiology*. Cambridge: Cambridge Univ. Press. 364 pp.

Finch CE, Rose MR. 1995. Hormones and the physiological architecture of life history evolution. *Q. Rev. Biol.* 70:1–51

Flatt T, Tu M-P, Tatar M. 2005. Hormonal pleiotropy and the juvenile hormone regulation of *Drosophila* development and life history. *BioEssays* 27:999–1010

Garland T Jr. 2003. Selection experiments: an underutilized tool in biomechnics and organismal biology. In *Vertebrate Biomechanics and Evolution*, ed. V Bels, J-P Gasc, A Casinos, pp. 23–65. Oxford: BIOS Scientific

Garland TJ Jr, Carter PA. 1994. Evolutionary physiology. *Annu. Rev. Physiol.* 56:579–621

Gibbs AG. 1999. Laboratory selection for the comparative physiologist. *J. Exp. Biol.* 202:2709–18

Gilbert LI, Iatrou K, Gill S, eds. 2005. *Comprehensive Molecular Insect Science*, Vol 3. *Endocrinology*. Amsterdam: Elsevier

Groothuis T, Carere C. 2005. Avain personalities: characterization and epigenesis. *Neurosci. Biobehav. Rev.* 29:137–50

Gu X, Zera AJ. 1996. Quantitative genetics of juvenile hormone esterase, juvenile hormone binding and general esterase activity in the cricket, *Gryllus assimilis*. *Heredity* 76:136–42

Guilgur L, Moncaut N, Canario A, Somoza G. 2006. Evolution of GnRH ligands and receptors in gnathostomata. *Comp. Biochem. Physiol.* 144A:272–83

Halpern-Sebold L, Schreibman M, Margolis-Nunno H. 1986. Differences between early- and late-maturing genotypes of the platyfish (*Xiphophorus maculatus*) in the morphology of their immunoreactive lutenizing hormone releasing hormone-containing cells: A developmental study. *J. Exp. Zool.* 240:245–57

Harshman LG, Hoffmann AA. 2000. Laboratory selection experiments using *Drosophila*: what do they really tell us? *Trends Ecol. Evol.* 15:32–36

Harshman LG, Zera A. 2007. The cost of reproduction: the devil in the details. *Trend. Ecol. Evol.* 22:80–86

Hartfelder K, Emlen DJ. 2005. Endocrine control of insect polyphenism. See Gilbert et al. 2005, pp. 651–703

Henrich VC. 2005. The ecdysteroid receptor. See Gilbert et al. 2005, pp. 243–85

Hoekstra HE, Hirschmann RJ, Bundey RA, Insel PA, Crossland JP. 2006. A single amino acid mutation contributes to adaptive beach mouse color pattern. *Science* 313:101–4

Hoeflich A, Bunger L, Nedbal S, Renne U, Elmlinger MW, et al. 2004. Growth in mice reveals conserved and redundant expression patterns of the insulin-like growth factor system. *Gen. Comp. Endocrinol.* 136:248–59

Ji X, Diong C-H. 2006. Does follicle excision always result in enlargement of offspring size in lizards? *J. Comp. Physiol.* 176:521–25

Jones G, Jones, D. 2007. Farnesoid secretions of dipteran ring glands: what we do know and what we can know. *Insect Biochem. Mol. Biol.* 37:771–98

Jones M, Rakes L, Yochum M, Dunn G, Wurster S, et al. 2006. The proximate control of pupal color in swallowtail butterflies: Implications for the evolution of environmentally cued pupal color in butterflies (Lepidoptera: Papilionidae). *J. Insect Physiol.* 53:40–46.

Jones R, Satterlee D, Ryder F. 1994. Fear of humans in Japanese quail selected for low and high adrenocortical response. *Physiol. Behav.* 56:379–83

Kacsoh B. 2000. *Endocrine Physiology*. New York: McGraw-Hill. 739 pp.

Kallman K. 1989. Genetic control of size at maturity in *Xiphophorus*. In *Ecology and Evolution of Livebearing Fishes (Peociliidae)*, ed. G Meffe, F Snelson Jr, pp. 163–84. Englewood Cliffs: Prentice-Hall

Ketterson E, Nolan V Jr, Castro J, Buerkle C, Clotfelter E, et al. 2001. Testosterone, phenotype, and fitness: a research program in evolutionary behavioral endocrinology. In *Avian Endocrinology*, ed. A Dawson, C Chaturvedi, pp. 19–40. New Delhi, India: Narosa Publishing House

Ketterson E, Nolan V Jr, Cawthorn M, Parker P, Ziegenfus C. 1996. Phenotypic engineering: using hormones to explore the mechanistic and functional bases of phenotypic variation in nature. *Ibis* 138:70–86

King R, Cline J, Hubbard C. 2004. Heritable variation in testosterone levels in male garter snakes (*Thamnophis sirtalis*). *J. Zool. Lond.* 264:143–47

Koch P, Brakefield P, Kesbeke F. 1996. Ecdysteroids control eyespot size and wing colour pattern in the polyphenic butterfly, *Bicyclus anynana* (Lepidoptera: Satyridae). *J. Insect Physiol.* 42:223–30

Malisch J, Salzman W, Gomes F, Rezende E, Jeske D, Garland T Jr. 2007. Baseline and stress-induced plasma corticosterone concentrations of mice selectively bred for high voluntary wheel running. *Physiol. Biochem. Zool.* 80:146–56

Marler CA. 1988. Evolutionary costs of aggression revealed by testosterone manipulations in free-living male lizards. *Behav. Ecol. Sociobiol.* 23:21–26

Matsuda R. 1987. *Animal Evolution in Changing Environments with Special Reference to Abnormal Metamorphosis*. New York: Wiley-Interscience. 355 pp.

Meylan S, Clobert J. 2005. Is corticosterone-mediated phenotype development adaptive? Maternal corticosterone treatment enhances survival in male lizards. *Horm. Behav.* 48:44–52

Moczek A, Nijhout HF. 2002. Developmental mechanisms of threshold evolution in a polyphenic beetle. *Evol. Dev.* 4:252–64

Mundy NI, Badcock NS, Hart T, Scribner K, Janssen K. 2003. Conserved genetic basis of a quantitative plumage trait involved in mate choice. *Science* 303:1870–73

Nachman M, Hoekstra H, D'Augustino S. 2003. The genetic basis of adaptive melanism in pocket mice. *Proc. Natl. Acad. Sci. USA* 100:5268–73

Nijhout HF. 1991. *The Development and Evolution of Butterfly Wing Patterns*. Washington and London: Smithsonian Institution Press. 336 pp.

Nijhout HF. 1994. *Insect Hormones*. Princeton: Princeton Univ. Press. 267 pp.

Nijhout HF. 2003. The control of body size in insects. *Dev. Biol.* 261:1–9

Nijhout HF, Williams C. 1974. Control of moulting and metamorphosis in the tobacco hornworm, *Manduca sexta* (L.): cessation of juvenile hormone secretion as a trigger for pupation. *J. Exp. Biol.* 61:493–501

Nijhout HF, Wheeler D. 1982. Juvenile hormone and the physiological basis of insect polymorphism. *Q. Rev. Biol.* 57:109–33

Norris DO. 1997. *Vertebrate Endocrinology*. San Diego: Academic. 634 pp.

Oksanen T, Koskela E, Mappes T. 2002. Hormonal manipulation of offspring number: maternal effort and reproductive costs. *Evolution* 56:1530–37

Partridge L, Gems D, Withers DJ. 2005. Sex and death: what is the connection? *Cell* 120:461–72

Reed WL, Clark ME, Parker PG, Raoulf SA, Arguedas N, et al. 2006. Physiological effects on demography: A long-term experimental study of testosterone's effects on fitness. *Am. Nat.* 167:667–83

Richard D, Rybczynski R, Wilson T, Wang Y, Wayne M, et al. 2005. Insulin signaling is necessary for vitellogenesis in *Drosophila melanogaster* independent of the roles of juvenile hormone and ecdysteroids: female sterility of the *chico¹* insulin signaling mutation is autonomous to the ovary. *J. Insect Physiol.* 51:455–64

Roberts M, Buchanan K, Hasselquist D, Evans M. 2007. Effects of testosterone and corticosterone on immunocompetence. *Horm. Behav.* 51:126–34

Roff DA. 1996. The evolution of threshold traits in animals. *Q. Rev. Biol.* 71:3–35

Roff DA, Sterling G, Fairbairn DJ. 1997. The evolution of threshold traits: a quantitative genetic analysis of the physiological and life-history correlates of wing dimorphism in the sand cricket. *Evolution* 51:1910–19

Rosenblum E, Hoekstra H, Nachman M. 2004. Adaptive reptile color variation and the evolution of the *Mc1r* gene. *Evolution* 58:1794–808

Rountree DB, Nijhout HF. 1995a. Genetic control of a seasonal morph in *Precis coenia* (Lepidoptera: Nymphalidae). *J. Insect Physiol.* 41:1141–45

Rountree DB, Nijhout HF. 1995b. Hormonal control of a seasonal polyphenism in *Precis coenia* (Lepidoptera: Nymphalidae). *J. Insect Physiol.* 41:987–92

Rybczynski R. 2005. Prothoracicotropic hormone. See Gilbert et al. 2005, pp. 61–123

Shi Y-B. 2000. *Amphibian metamorphosis. From Morphology to Molecular Biology.* Wiley-Liss, 288 pp.

Shingleton AW, Das J, Vinicius L, Stern DL. 2005. The temporal requirements for insulin signaling during development in *Drosophila. PLoS Biol.* 3:1607–17

Sinervo B. 1999. Mechanistic analysis of natural selection and a refinement of Lack's and Williams's principles. *Am. Nat.* 154:S26–42

Sinervo B, Calsbeek R. 2003. Physiological epistasis, ontogenetic conflict and natural selection on physiology and life history. *Integr. Comp. Biol.* 43:419–30

Sinervo B, DeNardo D. 1996. Costs of reproduction in the wild: path analysis of natural selection and experimental tests of causation. *Evolution* 50:1299–313

Sinervo B, Doughty P. 1996. Interactive effects of offspring size and timing of reproduction on offspring reproduction: experimental, maternal and quantiative genetics aspects. *Evolution* 50:1314–27

Sinervo B, Chaine A, Clobert J, Calsbeek R, Hazard L, et al. 2006. Self-recognition, color signals, and cycles of greenbeard mutualism and altruism. *Proc. Natl. Acad. Sci. USA* 103:7372–77

Solberg L, Baum A, Ahmadiyeh N, Shimomura K, Li R, et al. 2006. Genetic analysis of the stress-responsive adrenocortical axis. *Physiol. Genomics* 27:362–69

Stern DL. 2000. Perspective: Evolutionary developmental biology and the problem of variation. *Evolution* 54:1079–91

Sutter NB, Bustamante CD, Chase K, et al. 2007. A single *IGF1* allele is a major determinant of small size in dogs. *Science* 316:112–15

Suzuki Y, Nijhout HF. 2006. The evolution of a polyphenism by genetic accomodation. *Science* 311:650–52

Svensson E, Sinervo B. 2004. Spatial scales and temporal component of selection in side-blotched lizards. *Am. Nat.* 163:726–34

Svensson E, Sinervo B, Comendant T. 2002. Mechanistic and experimental analysis of condition and reproduction in a polymorphic lizard. *J. Evol. Biol.* 15:1034–47

Swallow J, Garland T Jr. 2005. Selection experiments as a tool in evolutionary and comparative physiology: insights into complex traits—an introduction to the symposium. *Integr. Comp. Biol.* 45:387–90

Tatar M, Kopelman A, Epstein D, Tu M-P, Yin C-M, Garofalo RS. 2001. A mutant *Drosophila* insulin receptor homolog that extends life-span and impairs neuroendocrine function. *Science* 292:107–10

Theron E, Hawkins K, Bermingham E, Ricklefs R, Mundy NI. 2001. The molecular basis of an avian plumage polymorphism in the wild. A melanocortin-1-receptor point mutation is perfectly associated with the melanic plumage morph of the bananaquit, *Coereba flaveola*. *Curr. Biol.* 11:550–57

Tu M, Yin C-M, Tatar M. 2002. Impared ovarian ecdysone synthesis of *Drosophila melanogaster* insulin receptor mutants. *Aging Cell.* 1:158–60

Tu M-P, Yin C-M, Tatar M. 2005. Mutations in insulin signaling pathway alter juvenile hormone synthesis in *Drosophila melanogaster*. *Gen. Comp. Endocrinol.* 142:347–56

Vafopoulou X, Steel CGH. 2005. Circadian organization of the endocrine system. See Gilbert et al. 2005, pp. 551–650

West-Eberhard M. 2003. *Developmental Plasticity and Evolution*. Oxford: Oxford Univ. Press. 816 pp.

Zera AJ. 2004. The endocrine regulation of wing polymorphism: State of the art, recent suprises, and future directions. *Integr. Comp. Biol.* 43:607–16

Zera AJ. 2006. Evolutionary genetics of juvenile hormone and ecdysteroid regulation in *Gryllus*: A case study in the microevolution of endocrine regulation. *Comp. Biochem. Physiol.* 144A:365–79

Zera AJ. 2007a. Wing polymorphism in *Gryllus* (Orthoptera:Gryllidae): Endocrine, energetic and biochemical bases of morph specializations for flight vs reproduction. In *Phenotypic Plasticity in Insects: Mechanisms and Consequences*, ed., T Ananthakrishnan, D Whitman, pp. 547–90. Plymouth, United Kingdom: Sci. Publ.

Zera AJ. 2007b. Endocrine analysis in evolutionary-developmental studies of insect polymorphism: Use and misuse of hormone manipulation. *Evol. Dev.* 9:499–513

Zera AJ, Cisper G. 2001. Genetic and diurnal variation in the juvenile hormone titer in a wing-polymorphic cricket: Implications for the evolution of life histories and dispersal. *Physiol. Biochem. Zool.* 74:293–306

Zera AJ, Harshman LG. 2001. Physiology of life history trade-offs in animals. *Annu. Rev. Ecol. Syst.* 32:95–126

Zera AJ, Holtmeier CL. 1992. In vivo and in vitro degradation of juvenile hormone-III in presumptive long-winged and short-winged *Gryllus rubens*. *J. Insect Physiol.* 38:61–74

Zera AJ, Huang Y. 1999. Evolutionary endocrinology of juvenile hormone esterase: Functional relationship with wing polymorphism in the cricket, *Gryllus firmus*. *Evolution* 53:837–47

Zera AJ, Sanger T, Cisper GL. 1998. Direct and correlated responses to selection on JHE activity in adult and juvenile *Gryllus assimilis*: implications for stage-specific evolution of insect endocrine traits. *Heredity* 80:300–9

Zera AJ, Strambi C, Tiebel KC, Strambi A, Rankin MA. 1989. Juvenile hormone and ecdysteroid titers during critical periods of wing morph determination in *Gryllus rubens*. *J. Insect Physiol.* 35:501–11

Zera AJ, Tiebel KC. 1989. Differences in juvenile hormone esterase activity between presumptive macropterous and brachypterous *Gryllus rubens*: Implications for the hormonal control of wing polymorphism. *J. Insect Physiol.* 35:7–17

Zera AJ, Zhang C. 1995. Direct and correlated responses to selection on hemolymph juvenile hormone esterase activity in *Gryllus assimilis*. *Genetics* 141:1125–34

Zera AJ, Zhao Z, Kaliseck K. 2007. Hormones in the field: Evolutionary endocrinology of juvenile hormone and ecdysteroids in field populations of the wing-dimorphic cricket *Gryllus firmus*. *Physiol. Biochem. Zool.* In press

Zhao Z, Zera AJ. 2004. The hemolymph JH titer exhibits a large-amplitude, morph-dependent, diurnal cycle in the wing-polymorphic cricket, *Gryllus firmus*. *J. Insect Physiol.* 50:93–102

Zijlstra W, Steigenga M, Koch P, Zwan B, Brakefield P. 2004. Butterfly selected lines explore the hormonal basis of interactions between life histories and morphology. *Am. Nat.* 163:E76–87

The Role of Behavior in the Evolution of Spiders, Silks, and Webs

Fritz Vollrath[1] and Paul Selden[2]

[1]Department of Zoology, Oxford University, Oxford OX1 3PS, United Kingdom;
email: fritz.vollrath@zoo.ox.ac.uk

[2]The Paleontological Institute, University of Kansas, Lawrence, Kansas 66045, and
Department of Palaeontology, Natural History Museum, London SW7 5BD,
United Kingdom; email: selden@ku.edu

Annu. Rev. Ecol. Evol. Syst. 2007. 38:819–46

First published online as a Review in Advance on
September 5, 2007

The *Annual Review of Ecology, Evolution, and
Systematics* is online at
http://ecolsys.annualreviews.org

This article's doi:
10.1146/annurev.ecolsys.37.091305.110221

Key Words

extended phenotype, fossil, morphology, silk

Abstract

Spiders' silks and webs have made it possible for this diverse taxon to
occupy a unique niche as the main predator for another, even more
diverse taxon, the insects. Indeed, it might well be that the spiders,
which are older, were a major force driving the insects into their
diversity in a coevolutionary arms race. The spiders' weapons were
their silks and here we explore the evidence for the evolution of silk
production and web building as traits in spider phylogeny.

INTRODUCTION

Spiders cover a large variety of morphological forms (Bristowe 1958, Comstock 1948, Foelix 1996) ranging from the huge, hairy mygalomorphs to pin-size, bald oonopids, from eight-eyed to two-eyed, from using lung to using trachea (or both or neither) to breathe, from having very long to very short legs. A spider's morphology (Bristowe 1958), anatomy (Snodgrass 1952), and nervous system (Barth 1985, 2002) typically reflect its general ecology (Main 1976, Nentwig 1986, Wise 1993) and behavior (Robinson 1975, Vollrath 1992). Most importantly, all spiders make and use silk throughout their lives, however diverse their morphology, ecology, and behavior. Thus, spider silk is not only an interesting material in its own right (Craig 1997, 2003; Vollrath & Knight 2001, Vollrath & Porter 2006), but it is an integral part of the behavior of all spiders, whether an individual uses it simply as a trailing safety line or integrates it into the often complex structures of the famous, and characteristic, spider's web (Shear 1986, Tilquin 1942, Witt et al. 1968).

Spiders tend to employ their different silks rather specifically. Hence, both silk type and silk deployment tend to be good indicators for a spider's specific lifestyle. Applied to spiders' webs, the character traits of web engineering and silk use are typically correlated with specific morphological and anatomical traits of the animal's body plan (Coddington 1986, Eberhard 1990, Shear 1986, Vollrath & Knight 2005, Witt et al. 1968). In addition, web geometry is a representation also of the spider's movements and thread manipulations (Vollrath 2000), which together comprise the spider's web-building behavior. Thus, the structure of the web represents an intimate interaction between morphology and behavior. Indeed, the web provides a rare example where a behavior pattern can be used analytically to provide quantitative character traits for large-scale cladistic analysis (Coddington & Levi 1991, Eberhard 1990, Griswold et al. 1998). It follows that a spider's web provides insights into not only spider taxonomy but also spider phylogeny by the analysis of present-day character traits, such as details of extant web architecture (Coddington & Levi 1991, Eberhard 1990, Griswold et al. 1998, Opell 2002), extant silk properties (Garb et al. 2006, Gatesy et al. 2001, Opell 2002, Opell & Bond 2001, Vollrath & Edmonds 1989; Vollrath & Knight 2003), and extant silk production systems (Coddington 1989; Glatz 1972, 1973; Shultz 1987; Vollrath & Knight 2001, 2005).

Of the tremendous diversity of spider web types (**Figure 1**), the orbicular webs of the araneid orb weavers are the most accessible analytically (Vollrath 1992; Zschokke & Vollrath 1995a,b) and to date these kinds of webs have provided the most important data sets for structure-function analysis. After all, two-dimensional geometries hanging freely in the air are more easily observed, drawn, photographed and filmed than three-dimensional structures that are often fully integrated into the vegetation. Still, both two-dimensional and three-dimensional web architectures are the outcome of dedicated building behavior patterns (Benjamin & Zschokke 2003; Eberhard 1986; Krink & Vollrath 1997, 1998; Opell 1996; Robinson & Lubin 1979; Zschokke & Vollrath 1995a,b), providing a wide range of traits for classification (Coddington & Levi 1991; Eberhard 1982, 1987, 1990; Gotts & Vollrath 1992; Griswold et al. 1998; Opell 2002).

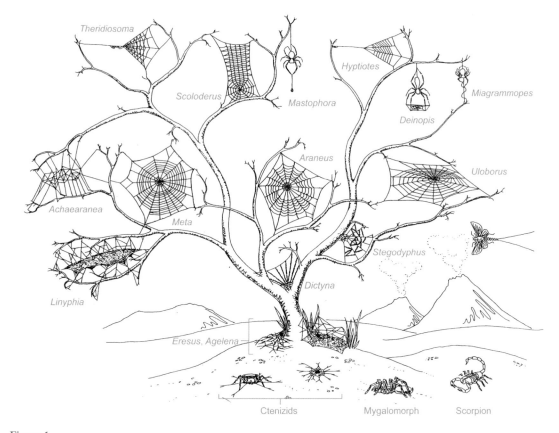

Figure 1

Schema of spider web evolution. A selection of orb web on a tree to demonstrate the various web types—this is certainly not a phylogenetic tree. Distant ancestors such as scorpions and more close ones such as mygalomorphs and ctenizids roam the ground, already using silk to line burrows and construct trip-lines. Further ancestral relatives (e.g., *Eresus* and *Agelena*) build their webs on the base of vegetation; a *Dictyna* web spans the fork of the tree. The right-hand branch contains (in order from its base) the webs of *Stegodyphus*, *Uloborus*, *Hypotiotes*, *Deinopis*, and *Miagrammopes*. The center branch holds a two-dimensional araneid orb web by *Araneus*. The left-hand branch holds a two-dimensional tetragnathid orb web by *Meta*. This branch also supports (*upper left to right*) derived orb webs by *Theridiosoma* and *Scoloderus*, and the minimalist *Mastophora* glue-drop web as well as (*below on the extreme left*) the highly derived three-dimensional webs by *Achaearanea* and *Linyphia* (adapted from Vollrath 1988).

Web and behavior are so closely linked that it is possible to deconstruct the web structure not only to provide a continuous record of the visible steps taken by the spider but also to infer from this visible record the underlying and hidden rules that are guiding these steps (Eberhard 1969; Gotts & Vollrath 1992; Krink & Vollrath 1997, 1998, 1999). A prerequisite for a successful behavioral dissection is a very good understanding of both, web engineering and spider activity (Eberhard 1981, 1986, 1988a,b; Heiling & Herberstein 1998; Herberstein & Heiling 1999; Vollrath 1987,

1988, 1992a; Vollrath et al. 1997). Fortunately, the modern techniques of filming and movement analysis are making this increasingly easy (Benjamin & Zschokke 2002, 2003; Zschokke & Vollrath 1995a) and accessible to modern simulation and modeling tools (Gotts & Vollrath 1992; Krink & Vollrath 1997, 1998, 1999, 2000). Integrating the analysis of thread manipulation and limb movements with the track of the spider's body, and using these data in combination with simulation and modeling of movement/track patterns, is thus beginning to elucidate the decision rules (algorithms) that govern the spider's web-building behavior. These kinds of analytical studies are taking Hans Peter's (1937), Bill Eberhard's (1969), Peter Witt's (1971) and Mike Robinson's (1975) seminal studies into the next phase of integrated modeling analysis (Gotts & Vollrath 1992; Krink & Vollrath 1997, 1999).

In summary, it appears that a detailed analysis of spider web structure gives access to a wide range of behavior patterns. One key to their analysis lies in the taxonomic position of the animals as well as specific morphological and anatomical characteristics.

Here we examine the question of whether (and how well) we can use a spider's morphology and anatomy (by studying overall body shape as well as specific organs) to deduce the animal's behavior (and specifically its use of silk) in the phylogenetic context. By this we hope to gain novel insights into the ecology and evolution of more ancestral spiders. Clearly, we are unlikely to find good evidence for many of the traits of interest in the fossils themselves. Hence we will have to infer ancestral traits from extant spiders and present-day web architecture and modern web-building characteristics.

LINKING MORPHOLOGY AND BEHAVIOR

Fossil spiders carry a surprising amount of detailed morphological and anatomical information. Such well-preserved structural data often allow us to infer specific behavior patterns. After all, the extant spiders provide many excellent examples of clear links between body structure and behavior. Thus, always considering likely differences in the ecological environment at the time of study, we can use preserved morphology to infer invisible behavior. In effect, we can use our understanding of the anatomical phenotype in combination with taxonomic status to deduce the architectural phenotype of a web—if a web was part of the extinct spider's hunting behavior. This should be possible because in spiders the trait "web-building behavior," i.e., its various constituent "fixed action" components, is a genetic trait not all that different from anatomical and morphological traits.

The evidence is strong that the web-building decision rules are inherited (Reed et al. 1970), although there is, of course, a large component of temporally adapted structural details that emerge from the composite action of these rules (Eberhard 1981, Krink & Vollrath 1998). Furthermore, there is also strong evidence that, in order to perform appropriately, the spider's actions and activites require very specific morphological adaptations in the body shape and leg dimensions (Bond & Opell 1998, Opell 1984, Vollrath 1987) and the spinning glands and spigots (Kovoor 1977, 1987; Opell 1989; Tillinghast & Townley 1987, 1994; Townley et al. 1993), as well as the sensory organs (Barth 1985, 2002; Vollrath 1995), and finally the claws, from which its walking type (whether on ground or in web, Comstock 1948) can be inferred. Many of these traits tend to be visible also in palaentological specimen (Selden 1990). In this review, we will examine the evidence of web evolution by combining morphological and ecological insights from extant species with palaentological data on extinct spiders and their main prey, insects.

Why would it be interesting to infer extinct webs from the morphology of extinct spiders? And why might such an exercise be important beyond its relevance for spider evolution? All evidence suggests that spiders got where they are today because of their silk and the way they use silk to make webs (**Figure 2**). Because those webs are the

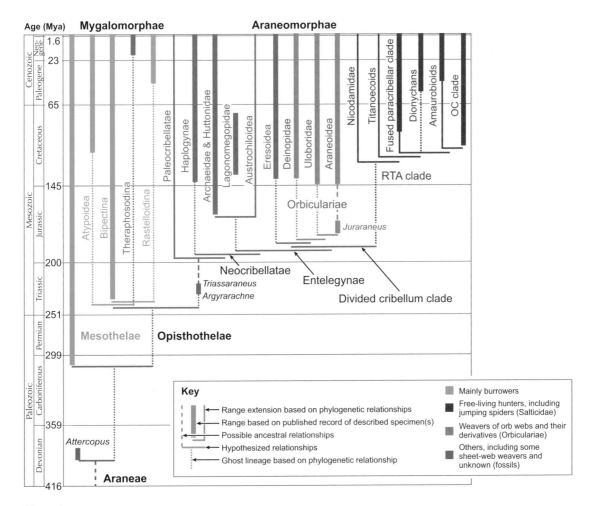

Figure 2

Phylogenetic tree of spiders. Data from the cladogram of Griswold et al. (2005) and published records of fossil spiders are combined to provide a view of spider evolution over geological time. Colors indicate major predation modes.

essence of spider behavior, we have in this entire group an excellent example where we can study the coevolution of behavior and morphology with implications far beyond the coevolution of spiders and insects.

After all, behavior is notoriously bad at being fossilized—except in the case of trace fossils (see below and Seilacher 1967). Having good evidence to infer behavior from morphology gives us a chance to study the evolution of a behavior pattern in the context of the overall climatic conditions at the time, which in turn could allow us to make inferences about the wider ecosystem that those animals inhabited (Robinson 1975). Being able to access the evolution of a complex behavior pattern by interpreting body morphology is of wider interest because typically it is behavior and not morphology that drives evolution, although it tends to be assumed otherwise both implicitly and explicitly. Behavior, with its great flexibility, allows the animal to take advantage of novel opportunities as they arise far quicker than morphology could. Highly variable circumstances as well as specific conditions that persist briefly provide behavior (with its rapid ability to adapt and habituate, even learn) with prospects far beyond the ability of morphology (with its long, averaging time frames) to exploit. Being able to take optimum advantage of environmental conditions has obvious rewards in terms of fitness, both physical and genetic. Thus behavior typically would lead the way in biasing reproductive success in the population. Morphological adaptations will follow as the paybacks of the changed behavior patterns accrue in genetic fitness.

Inferring fleeting behavior from morphological traits tends to be far from easy at the best of times; and it is rather tricky when the behavior is multifaceted. Spider's webs are a rare example to this rule of thumb, as discussed earlier. They provide a window into the evolution of a complex behavior because the extant species already provide an enormous variety of records (web types) with an ecological as well as structural diversity that can be firmly coupled to detailed insights into web-building behavior as well as foraging specialization (prey capture). Thus spiders allow us, virtually throughout their full range of ecotypes and without excessive effort yet with great detail, to study the structure-function relationship between an animal's body morphology and its behavior and ecology.

In summary, the evolution of the web, and especially the emergence of the orb web (and its occasional abandonment), has the potential to be a telling example for the evolution of behavior, if only we can begin to unravel the forces that drove the rise (and fall, if we want to call it that) of the orb-web spider tribe. Orb weavers are an important arachnid guild, with most of the extant species making orb webs or having evolved from orb-web builders. The guild of web makers per se is even bigger, of course, with a significant proportion of all spiders building a prey-capture web of some sort (Foelix 1996). Indeed, there is strong evidence that two important groups, the lycosid and pisaurid "wolf" spiders, derive from the builders of prey-capture webs, as do the salticid jumping spiders (Foelix 1996). In any case, with or without a web, spiders are among the most prominent of terrestrial predators. Indeed, the arachnids are the only major animal order that consists solely of predators. There is not a herbivore or detritivore among them, although the juveniles of some orb weavers seem to require air-borne pollen to get them through the first instars (Smith & Mommsen 1984).

This review touches on recent developments relevant to our core question: How might web-building behavior have affected the evolution of the spiders? Here we do not provide extended surveys of the literature on spider evolution and ecology, or on silk function or web-building behavior, because these topics have been superbly reviewed fairly recently by Coddington & Levi (1991) on spider evolution, Wise (1993) on spider ecology, Eberhard (1990) on web function and phylogeny, Craig (1997, 2003) on silk and silk evolution, Coddington (1989) on spinneret morphology, and Eberhard (1982) and Vollrath (1992) on behavior. However, a number of recent research studies have added significantly to our understanding of spider web evolution and these studies will form the focus of our review. More importantly however, we will use our vision of the evolution of the spider's web as an example to state the overriding importance of behavior in the evolution also of nonbehavior traits. Generally taxonomists and systematists ignore or underplay the role of behavior in the evolution of morphological traits because behavior can neither be easily measured nor genetically classified. Spiders and their webs are a rare, perhaps the only, exception to this generalization because a web structure can be analyzed just like a body morphology.

WHAT IS A WEB AND WHEN DID IT ORIGINATE?

A web can represent two types of character states. Web architecture is principally static with a semipermanent geometry, although its functional engineering is dynamic (Lin et al. 1995). However, web architecture is the outcome of web building, which is highly flexible, fleeting behavior. Thus, in analogy, web architecture compares with anatomy/morphology, while building behavior compares with embryology. Webs range from simple lines (laid down as the spider moves along) to complex structures (specifically assembled by the animal often over a considerable time span). If we include all spider structures made of silk then we must add to the tally the shroud of the prey-wrap, the tent or burrow of the retreat, and finally the cradle of the egg sac as well as the para-gliders of the ballooning spiderlings (Bell et al. 2005).

How did it all start? Sometime in the early Devonian (Selden et al. 1991, Shear et al. 1989) the first semiterrestrial spider-like arachnid—probably also carnivorous like all of today's spiders—shifted its prey-hunting behavior or its predator-escape behavior from the water to the land (**Figure 2**). This exposure to novel atmospheric conditions would have had grave implications not only for locomotion and reproduction but for breathing physiology (Selden & Edwards 1989) and (ecologically a tipping point) silk production mechanisms. Whatever the reasons for the first steps from water through intertidal (Churchill & Raven 1989, McQueen & McLay 1983) or freshwater swamps (Rovner 1987, 1989) to dry land, the spiders, once out of the water, quickly evolved silk producing organs that were fully functional (Rovner 1987). Silk is a biopolymer that, on the whole, functions best dry (Vollrath & Knight 2005), and we must assume that any underwater preadaptations for silk (Decae 1984, Rovner 1987) were released by this step onto land. This move thus quickly became a key step that allowed the adaptive radiation of the spiders into the taxon that is defined by its silk, both morphologically and ecologically.

There are two scenarios for the beginning of silk evolution. In one, the animal's eggs were at first covered by proteinaceous exudates from the coxal glands of the abdominal legs, which then evolved into a fibrous and sticky covering—perhaps to form a more effective glue or shield for deposited eggs as protection against the elements and predators (Shultz 1987). Scenario two assumes a protein mucus membrane to cover the whole abdomen, perhaps to shield and moisten the gills before they had time to evolve into the book lungs of the modern spiders (Damen et al. 2002, Strazny & Perry 1984). Either the membrane or the exudates could have evolved into a network of thin, individual thin filaments. Such a mesh would have been more effective as both a water-retaining sac when the land fell dry or an air-sac when the land was submerged. In either case, such a structure would have acted as a physical bubble or plastron lung (Messner & Adis 1995) allowing gas exchange between the inner and outer medium (be it air or water in- or outside). The presence of such a hypothetical silken lung would certainly have allowed the early spiders to invade drying-out flood-plains. Moreover, it would have given them an opportunity to explore their new environment while responding to its novel conditions by evolving their book lungs (which in effect are inverted gills housed inside a body cavity). Indeed, the highly derived *Argyroneta aquatica* today uses her silk to construct a diving-bell plastron lung allowing her to live fully underwater while other spiders, such as the sheet weaver *Desis marina* (Lamoral 1968a,b; McQueen & McLay 1983; Powell 1878) or the trapdoor spider *Idioctis* (Churchill & Raven 1989) and other purely terrestrial spiders living in areas prone to flooding (Rovner 1987), can survive extended periods of submersion in their silken sleeping sacs or cellars. The question of how the spiders first evolved (and used) silk is so far unresolved for lack of fossil data. Indeed, perhaps both selection pressures on proto-silk (to shield eggs as well as to provide a plastron) acted together. After all, *Argyroneta*'s air bubble not only prevents her eggs from drowning but the tough membraneous silk net also protects the eggs from predators.

Although the principal selective advantages leading to the evolution of silk fibers are still hypothetical, we have fairly good evidence on the morphological origin of the material. It is quite clear that spinnerets evolved from legs (Damen et al. 2002, Popadic et al. 1998) but the origin of silk glands remains somewhat shrouded. One hypothesis has it that they may be modified coxal glands (Bristowe 1958, Gertsch 1979, Kaston 1964, Marples 1967) assuming that silk evolved from an excretory product. However, while coxal glands are mesodermal in origin, silk glands are ectodermal (Craig 1997).

PRE-ADAPTATIONS FOR WEB BUILDING

Ancestral spiders, most likely, were freely roaming hunters that, after they had moved onto dry land, needed to seek shelter from the sun's rays or from tidal flooding for part of the day. As argued, silk initially may have evolved in response to the need to protect the animal's body as well as its eggs and young. After the uncoupling of silk production from reproduction and protection, single silk threads might have been extended beyond the shelter to provide guide lines for the spider as well as signal lines to detect nearby prey (Shear 1986). Eventually such single threads were interwoven to provide ever more efficient and effective traps that evolved to stop and retain potential prey.

However, building a trap requires not only the appropriate complement of silk producing organs but also the appropriate behavior patterns. In addition, constructing a silken trap, typically many times larger than the builder, requires spatial orientation. The evidence is strong that even the most ancestral spiders already had this capacity, as we outline now in a brief overview of the physiological ecology of extant spiders.

The Primitive *Liphistius* and the Mygalomorphs

Most of these spiders live in burrows, some rather complex with side chambers and several entrances (Main 1976) and many featuring concealed trap-door burrows. Some have silken lines radiating out from the opening to act as sensory trip-wires and/or home-finding devices on excursions. We know surprisingly little about the orientation mechanisms and survival mechanisms in the key taxon *Liphistius* (Foelix 1996, Main 1993), although recent work is beginning to provide new insights (Haupt 2003).

Hunting Spiders

Hunting spiders generally have good (and clear) vision in the principal eyes used for binocular, focal vision while the remaining eyes are for peripheral vision and movement detection (Barth 2002). The principal eyes may be used to identify prey or a mate, and may also help inform orientation. Some wolf spiders living at the edge of ponds can run away over the water; they return to firm ground by using visible landmarks or, if these are absent, astronomical cues such as the polarization pattern of the sky corrected by an internal clock (Barth 2002). In addition to such long-range orientation (employing direct landmarks) and navigation (with more indirect cues) hunting spiders also find their way much more locally (Vollrath 1992). For example, a hunting spider chased away from a prey or robbed of an egg sac tends to return in a straight line, even if the outward journey was along a circuitous route, indicating that the spider not only knows the direction of the shortcut but also its length (Görner & Claas 1985, Seyfarth et al. 1982). It seems that in these cases the spiders do not use any external cues but instead use an internal guidance system, often referred to as idiothetic memory (Barth 2002). The kinesthetic cues necessary for such a system may be provided by the lyriform slit sense organs in the cuticula of particular legs, as their immobilization affects the spider's performance (Barth 2002). As we shall see, orb weavers also seem to be able to use idiothetic orientation, perhaps also controlled by the lyriform organs. In any case, the anatomies of the eyes and of specific sensory organs on the legs and the chephalothorax are morphological features relevant to the spider's behavior.

Sheet-Web Spiders

The web of agelenid funnel spiders consists of a densely packed sheet of silk, often triangular and on ground level connected to a long, silken tube-retreat in one corner (Bristowe 1958). Prey falls onto the sheet and is grabbed in a dash by the spider running from its retreat on top of the sheet. Typically the prey is consumed inside the retreat after a return dash that is always straight in a bee-line, even if the outward journey was circuitous. It seems that the spider uses information about its outward path to constantly calculate the vector pointing homeward. Such navigation by path integration (Mittelstaedt 1985) could be done by using information about leg and body turns gathered with kinesthetic senses (Barth 2002) and stored in some form of idiothetic memory (Görner & Claas 1985, Seyfarth et al. 1982). Thus guided into the vicinity of the retreat (for such systems are inherently inaccurate), the spider then locates the mouth of its retreat using a variety of different mechanisms such as the shape and spatial position of the sheet as well as thread tensions and web elasticity, perhaps even smell (Barth 2002, Seyfarth et al. 1982). In addition the animal may use light or the polarization pattern of the sky to provide further reference cues (Barth 2002). Whether the necessary information is collected internally or externally, on site or enroute, the behavior of the funnel-web spiders clearly shows that they use a variety of cues and mechanisms to orient and navigate. The sensory organs on the legs and the claws as well as the spinnerets are relevant behavioral features.

Space-Web Spiders

The web of the linyphiid spiders consists of a bowl-shaped fabric in a three-dimensional network of tangled threads well integrated into vegetation. The bowl collects prey falling in from above, and the spider moves on the underside of the bowl, typically waiting at the lowest point to attack from below using web tensions to orient (Suter 1984). Conformation of the spinnerets is a key trait for use of silk. In addition, the theridiid spiders [which most probably are derived orb weavers (Griswold et al. 1998) although they exclusively build three-dimensional space webs] have little combs on their legs that are a key feature of this group and are probably connected to the way they have of throwing sticky silk at prey. Moreover, many theridiids, like some other spider groups (Witt & Rovner 1982), also have specific stridulation organs. These are used by theridiids to vibrate threads (or the surface, in the case of the lyosids) during courtship, and thus provide another trait closely linked to behavior and webs, albeit they have little to do with the predatory behavior mostly associated with webs and web building.

Orb-Web Spiders

The typical orb web consists of a flat wheel of stiff radial threads overlaid by a spiral of elastic and sticky threads suspended freely in vegetation from a few guy lines. Radials and spiral often show distinct asymmetries in shape and spacing typically associated with a vertical orientation of the web and constituting a fine-tuning to maximize prey capture (Vollrath et al. 1997). In order to orient in the web, the spider uses vibrations but also the direction of illumination, which, together with gravity provides a general compass direction (Vollrath 1992). In addition to hand-railing along existing threads and orienting by a set of rather simple decision rules (Krink & Vollrath 1999), some orb weavers also navigate using idiothetic path integration (Vollrath et al. 2000).

Orb-web construction thus requires the use of local as well as global cues. Because theridiid spiders derived from orb spiders (Griswold et al. 1998), it is not surprising that they share common orientation mechanisms (Benjamin & Zschokke 2002, 2003). These groups also share many silk related traits, which can be seen in the details of the spinnerets (Coddington 1989) as well as relevant sensory organs on the legs and body (Barth 1985, 2002, Hergenröder & Barth 1983).

Jumping Spiders

Modern jumping spiders are highly visual, webless hunters although the evidence is strong that they have evolved from web-based spiders (Jackson et al. 2001). The jumping spider's hunt consists of three stages: approach, stalk, and jump. Jumps are only over relatively short distances, whereas the spider can see prey over long distances. In a three-dimensional habitat this means that the prey insect may often have to be approached via a detour, which indeed is done by some jumping spiders with great skill, suggesting that an excellent sense of spatial orientation is coupled with a memory of the maze (Hill 1979, Tarsitano & Jackson 1997). Clearly, the spider acquires specific knowledge about its surroundings visually and it appears that the animal is thus able to calculate accurately the fly's position relative to its own from a combination of visual and idiothetic memory. In this group the main trait that links morphology to behavior is the highly specialist eyes (Land 1985).

SUMMARY

In summary, spiders employ a wide variety of sensory modes and orientation/navigation aides, which they use to either hunt without a web or locate a site for a web and inform its construction as well as find their way about in the web. Silk and the use of a retreat would have been one of the first apomorph traits setting all modern spiders aside from their sister groups such as scorpions, mites, and the only primarily aquatic chelicerate, the horseshoe crab *Limulus*. Behavior would have already been important for the first use of silk in wall papering a burrow, as outlined earlier, but the rules of laying down the silk must have been refined rapidly when silk started to be used for prey capture. Many of the behavioral traits have good morphological correlates, whereas others are more difficult to identify, often because of lack of data and insights.

FOSSIL EVIDENCE

Fossil evidence for spider silk and the use of silk in webs comes from a wide number of sources. In some cases there are not only excellent morphological data but also information on silk production or even fossil webs that provide indirect or even direct evidence for ancestral behavior patterns. However, most of our insights to date originate from analyzing the morphology of fossilized spiders and comparing it with that of modern spiders where we have good information on their silks and webs. Here we consider mainly the evidence for silk use inferred from spider morphology, although spider trace fossils, rare as they are, can illuminate important aspects of the behavior of the ancestors (Seilacher 1967). However, there are also a few examples of trace fossils produced by spiders, such as silk strands and bits of web (Bachofen-Echt 1949, Peñalver et al. 2006, Zschokke 2003), as well as traces of loco-motion (Repichnia) (Braddy 1995, Sadler 1993) and dwellings (Domichnia) (Gregory et al. 2006).

Locomotion traces recorded from the Permian Coconino Sandstone (275 million years old) of Arizona and New Mexico (Braddy 1995) were attributed to spiders by comparing the modern desert ecosystem of the southwest United States (in which large mygalomorph spiders are a common element of the fauna) with the Permian ichnofaunal association of these desert sandstones. A particular locomotion trackway, known as *Octopodichnus*, suggests that the behavior of that particular Permian spider walking across desert substrates was similar to the movement pattern of present-day descendants like Aphonopelma chalcodes or Brachypelma spp. (Sadler 1993).

It is rare that the maker of a trace fossil can be identified with certainty, but Pickford (2000) described fossils of the characteristic buck-spoor spider *Seothyra* from Miocene aeolianites from the ancient Namib Desert. *Seothyra* is widespread and common across southern African desert regions (Dippenaar-Schoeman 1990). It constructs a vertical burrow that opens out at the sand surface into a wide dish containing the horizontal web covered with sand; a pair of pits on either side of the web gives the impression at the desert surface of the footprint of a small antelope (Lubin & Henschel 1990). The Namib is an ancient desert in an area of south West Africa that may have experienced an arid or semiarid climate for some 80 million years (Cretaceous), whereas the dunes sands of the Kalahari Sequence date back 65 million years to the start of the Cenozoic era (Schneider 2004).

Evidence from mammalian fossils indicates that the distinctive Namib habitat dates back to at least the mid-Miocene (Pickford & Senut 1999). All in all, this particular and distinctive trace fossil is good evidence for the antiquity of the genus *Seothyra*, its behavior, and (indeed) its habitat.

Given the abundance of today's spider burrows in environments that are prone to flooding (and hence preservation in the fossil record) it is surprising that fos-silized spider burrows are rather rarely recorded. Perhaps this is a collection or rather identification artifact and they have been misidentified as being produced by other burowing arthropods such as Hymenoptera (Gregory et al. 2006).

Real fossil silks and webs, however, are not all that rare if we examine amber occurrences such as the Bachofen-Echt Cretaceous silk described by Zschokke (2003), or other samples from Canadian Cretaceous (P.A. Selden, unpublished observations). There are even records of flies caught in an amber web (Peñalver et al. 2006).

Before considering the body–fossil record of spiders, it is important to discuss ideas about the atmospheric composition of the mid-Palaeozoic Earth and whether this had any relevance to spider evolution and behavior at that time. For many years, it was assumed that the concentration of oxygen in the atmosphere in the mid-Palaeozoic was much lower than the present atmospheric level of 20% (Selden & Edwards 1989). Such a low level of oxygen would be insufficient to form an ozone layer to block lethal UV-B radiation and thus allow life to exist out of water. More recent work has shown that not only were today's oxygen levels finally reached in the Silurian period but that a possible drop in the level of atmospheric oxygen could explain a paucity of terrestrial fossils in early Carboniferous times, known as Romer's Gap (Ward et al. 2006). By the late Carboniferous, oxygen levels were somewhat higher than today, which could explain the greater incidence of charcoal (resulting from wildfires) in the fossil record (Scott & Glasspool 2006). Thus, atmospheric and climatic effects would have had major effects on the total terrestrial ecosystem at the time.

Atmospheric oxygen concentration could have had a more direct effect on spider behavior. Lowered oxygen concentration could have rendered respiratory systems ineffective (Ward et al. 2006), resulting in migration to different habitats or extinction as well as the evolution of a more efficient breathing apparatus such as the combination of book-lungs with trachea (Bromhall 1987a,b). UV-B radiation would have been a major problem for any organism moving from water (which effectively blocks this radiation) onto land, whenever this might have been (and some researchers have advocated, on biochemical rather than palaeontological evidence, that complex terrestrial organisms existed even as early as the Precambrian). Ways to avoid harmful UV-B radiation would be to venture onto land at night, and to retreat behind a barrier during the daytime: under a stone, in a burrow, under water, or perhaps beneath a silken canopy. Some spider silks are highly UV reflective (Craig 2003), and so would have been very useful to the earliest spiders were they around when UV-B radiation was still at a high level.

Devonian

The oldest known spider is *Attercopus fimbriunguis* (Shear et al. 1987) from the Middle Devonian of Brown Mountain, Gilboa, New York (Selden et al. 1991), first described by Shear et al. (1987) as a possible trigonotarbid (an extinct group of Palaeozoic arachnids related to spiders). However, compelling evidence of spider-like spinnerets in the sample (Shear et al. 1989) suggests that *Attercopus* was a real spider showing for many characters more plesiomorphic states than were found in the most primitive genera of all known extant spiders, i.e., *Liphistius* spp. Accordingly, *Attercopus* was placed within the Araneae as sister group to all other spiders (Selden et al. 1991). The animal material itself consists of small cuticle fragments recovered from the siltstone matrix by maceration with hydrofluoric acid (HF), which is a process that can yield astonishing results. For example, Selden et al. (1991) managed to isolate a specimen of *Attercopus* cuticle bearing 19–20 spigots. Each spigot consists of a bell-shaped base, about twice as long as wide at the base, supporting a narrow shaft about three times as long as the base and comparable to the simplest spigots of some extant spiders (**Figure 3**).

The mesothele spider *Liphistius* is typically considered a living fossil because of its segmented abdomen (opistosoma) (Bristowe 1975), but we recall that the single-articled median spinnerets of *Liphistius* typically bear only one spigot (or none), while its more complex lateral spinnerets are pseudosegmented (Haupt 2003). Curiously, the *Attercopus* spigot consists of a single, fusiform spinneret and thus has more in common with the median spinnerets of the more derived mygalomorph spiders, which are single-articled and bear many spigots. Hence, the Devonian spinneret seemed to be more comparable to mygalomorph spinnerets, at least in its superficial morphology, than to the bulk of mesothele spinning organs. It must be noted here

Figure 3

Macerated preparations of spigots of *Attercopus fimbriunguis* of Devonian (390 Mya) age, New York; and *Heptathela kimurai*, Recent (0 Mya), Japan.

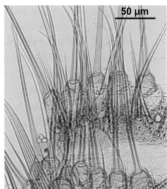

Attercopus fimbriunguis   *Heptathela kimurai*

that recent studies (Shear & Selden 2001) suggest that the original specimen is not a fusiform spinneret-like tube but a single sheet of cuticle folded over twice with the spigots arranged in two rows along one edge of the sheet. Other specimens also show spigots arranged in a double row along the edge of a piece of cuticle, while yet others could be pieces of spinnerets very much resembling flattened cylinders.

Thus, spiders were present among the earliest known terrestrial faunas and from the beginning were producing silk from fully formed spigots whether they were placed on spinnerets or directly on the body. Other arthropod fauna present in the *Attercopus* ecosystems included: scorpions, trigonotarbids, amblypygids, pseudoscorpions, mites, diplopods, chilopods, arthropleurids, and collembolans (Shear & Selden 2001). Many of these, e.g., collembolans, mites, and myriapods, would have been prey for *Attercopus* and other spiders, while other arachnids as well as chilopods would have been competitors for the diverse prey as well as predators on the early spiders. In any case, we may presume that the construction of a burrow (perhaps even with a trap door) would have been equally as beneficial to Devonian spiders as it is to burrowing spiders today.

Finally, we note that at that stage insects must have been in the infancy of their adaptive radiation. The only unequivocal evidence for true insects in the Devonian is a pair of jaws called *Rhyniognatha hirsti* from the Rhynie Chert of Scotland. These were first reported by Hirst & Maulik (1926); Tillyard (1928) described them and suggested that they were insect-like. The specimen was studied by many experts over the years, until Engel & Grimaldi (2004) confirmed that it belonged to a true insect. The next youngest insect is *Delitzschala bitterfeldensis* from the early Carboniferous of Germany (Brauckmann & Schneider 1996). There were certainly no flying animals at this time, which is important for any discussion of spider silk and web evolution, as well as the theory that spiders drove the evolution of the insects (Eisner et al. 1964).

Carboniferous

By Upper Carboniferous times the true Mesothelae had apparently become well established. However, although about 30 specimens of Carboniferous spiders have been reported, many are incorrectly identified as such, e.g., *Megarachne* (Selden et al. 2005). Nevertheless, one true mesothele has been identified: *Palaeothele montceauensis* from the Upper Carboniferous of France (Selden 1996a,b, 2000). In addition to the characteristic plesiomorphies of Mesothelae (dorsal opisthosomal tergites, two book-lung opercula, orthognath chelicerae, and fully-developed anterior median spinnerets), *Palaeothele* has a narrow sternum, which is an synapomorphy for mesothele spiders. At least five spinnerets can be seen in the holotype of *Palaeothele*: left anterior lateral (ALS), two anterior medians (AMS), and at least one posterior lateral (PLS). An additional spinneret adjacent to the PLS was interpreted as most likely to be the other PLS, implying absence of the posterior median spinneret (PMS) (Selden 1996a). A monograph on Carboniferous spiders is in preparation by P.A. Selden, but preliminary observations have already been published. Carboniferous spiders originally identified as araneomorphs (specifically the family Archaeometidae) are either not araneomorph spiders (e.g., *Archaeometa nephilina* (Selden et al. 1991) or *Eopholcus* (P.A. Selden, unpublished observations) or not even spiders at all (Penney & Selden 2006).

However, there were Carboniferous spiders families, e.g., the Arthrolycosidae and Arthromygalidae, that, as far as can be told from the specimens, were all Mesothelae (Penney & Selden 2006). By late Carboniferous times (c. 310 Mya) insects had become an important element of the predominantly forest fauna and insect flight had evolved (Grimaldi & Engel 2005). This was a critical time in the evolution of spider webs, but unfortunately direct evidence is lacking on the kinds of webs that spiders might have been producing at this time. Two scenarios could be envisioned: either that spider predation drove insects into the air with the spiders' webs coevolving with the insects wings or that insects took to the air for some other reason (such as dispersal or pollination) and that spiders followed them. At present, we have no fossil evidence for either scenario and cannot even say when spiders' webs first left the ground and ascended into the vegetation. Crucially, we have no good evidence that would allow us to pinpoint the timing of insect flight development. The abundant remains of the earliest true insects from the early Carboniferous of Germany (Brauckmann et al. 1985, 1996) show that a significant diversity of flying insect groups had already evolved by that time and in a later section we discuss in more detail the issue of insect radiation and flight (**Figure 4**).

In any case, we must consider that spiders were not the only Devonian predators of the early insects. For example, trigonotarbids were also sit-and-wait predators similar, and closely related, to spiders. But they, importantly, lacked both silk and venom, which we may assume to have been an ancestral (apomorph) trait for all spiders given that today only one spider taxon (Uloboridae) has secondarily lost venom glands. Nevertheless, the greater abundance in the late Palaeozoic of trigonotarbids compared with spiders would have meant that at that time trigonotarbids may have exerted a greater influence on insect evolutionary ecology. Other possible insect predators include other arachnids, such as scorpions, amblypygids and uropygids, and chilopods, as well as vertebrates, all of which (certainly the scorpions and vertebrates) would have left the water and become terrestrial by the late Carboniferous.

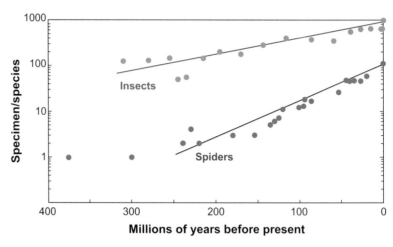

Figure 4

Graph showing relationship of spiders and insects over geological time. Note the similarity between the slopes of the two data sets. Further note the much greater age of the oldest specimen, the spider *Attercopus fimbriunguis*. Finally, note the logarithmic scale for the species distribution disguising the nonlinear increase in specimen of both taxa (data taken with permission from Penney 2004a).

Permian

Until recently, no arachnid fossils were known from the Permian period, in spite of abundant fossil insects from that period. The first Permian arachnid to be described was the trigonotarbid *Aphantomartus* (Rössler 1998); now we have the first spiders from this period, *Permarachne* and *Arthrolycosa* (Eskov & Selden 2005). *Permarachne*, especially, is an interesting specimen in that it is clearly a mesothele (showing plesiomorphies for spiders together with a narrow sternum—synapomorphic for mesotheles), but it also bears an elongate, flagelliform structure apparently emerging from the end of its abdomen. Eskov & Selden (2005) evaluated this flagelliform structure and considered it most likely to be an elongate spinneret. Elongate spinnerets are found in funnel-web spiders, such as the mygalomorph family Dipluridae and the araneomorph family Agelenidae, for example, but are not known in modern mesotheles. Hence, these researchers concluded that this was a new silk (and probably web) type for the Mesothelae. The argument went further (Eskov & Selden 2005), suggesting that there might have been a greater diversity of mesotheles in the late Palaeozoic than today. Finally, this kind of spinneret was considered to provide evidence for the late Palaeozoic development of a funnel web, which is primarily adapted to capture jumping insects.

If true, then the arms race between insects and spiders had begun in the Permian, i.e., c. 270 Mya (**Figure 4**). The hypothesis of a Permian period of rapid (co)adaptive radiation of both spiders and insects is intriguing. After all, this is a time with a rich fossil record of flying insects, at least from a few Fossil-Lagerstätten. Alas, this abundance of insects has yet to be matched by fossil spiders. As and when more spiders come to light from this period, we would hope to see more evidence of morphological features that would suggest the building of aerial webs. However, the end of the Permian, and the end of the Palaeozoic era, was marked by the greatest extinction event Earth has ever experienced. Its causes are, as yet, poorly understood (Erwin 2006), but its effects were profound: some 82% of genera, and more than half of all marine families, disappeared at this event.

How it might have affected the spiders and insects is not clear. For the insects, Labandeira & Sepkoski (1993) considered the skewing effect of Lagerstätten on their data (rich Permian deposits in Russia and Kansas but few in the Triassic), but concluded that the apparent drop in diversity at family level was real and due to the Permian-Triassic extinction. The effect on the spiders of the time is presently impossible to assess owing to lack of fossils.

Triassic

An extinction event is typically followed by a period of rapid adaptive radiation. Accordingly, among the insects there was a change at the end of the Palaeozoic as a large group of pterygote Palaeoptera, the Archaeoptera or palaeodictyopteroids, became extinct and were replaced by the pterygote Neoptera. The Palaeoptera were unable to fold their wings, whereas the Neoptera could do so. Interestingly, two groups of Palaeoptera survived the extinction and continue to the present day: the Odonata (dragon- and damselflies) and the Ephemeroptera (mayflies).

For spiders, the Triassic period saw the first mygalomorphs, such as *Rosamygale grauvogeli* in the modern family Hexathelidae (Selden & Gall 1992), as well as the first araneomorphs, i.e., modern web spiders (Selden et al. 1999). The hexathelids, as their name suggests, bear six spinnerets (ALS, PLS, PMS), and the PLS are elongated for the weaving of a funnel web (the infamous Sydney Funnelweb spider, *Atrax*, belongs in this family). *Rosamygale* was about 5-cm large and apparently lived in a semiarid environment on a delta, living presumably in a burrow or retreat from which a funnel web extended to catch jumping prey. Insects are abundant in the Voges region of France where it occured, mostly those with aquatic connections, e.g., aquatic larvae. Mygalomorphs include the tarantulas, funnel-web and bird-eating spiders. The importance of finding a mygalomorph spider in Triassic strata is enhanced by also finding species belonging to the sister group of that infraorder, the araneomorphs (Selden et al. 1999). *Argyrarachne* from Virginia is a juvenile, but *Triassaraneus* from South Africa is more likely to be an adult. These early araneomorphs already closely resemble modern orb weavers in general habitus. *Triassaraneus* has long, slender legs, a leg formula (longest to shortest) of 1243, no scopulae, and sparse bristles, all features suggestive of Araneomorphae. Furthermore, leg shape and arrangement, the lack of leg spines, the small tarsal claws and lack of scopulae, the paucity and arrangement of bristles, and the possible metatarsal trichobothrium seen on one leg are all suggestive of this specimen belonging to the extant group of Araneoidea.

Araneoids are primarily weavers of orb webs although some families and genera weave webs secondarily derived from orbs (Coddington 1986). Consequently it is possible, indeed likely, that the orb web dates from as early as the Triassic, i.e., is over 200 million years old. As we find during those times a great abundance of flying insects of modern aspect (neopteran pterygotes) that might have served as prey, we are beginning to see the predator-prey arms race in full swing.

Jurassic

The first Mesozoic spiders to be described date from the Jurassic period and are represented by modern-looking araneoids such as *Juraraneus rasnitsyni* (Eskov 1984) or even more contemporay araneomorphs such as *Jurarchaea zherikhini* (Eskov 1987). More recent finds of fossil spiders in the Jiulongshan Formation (Huang et al. 2006) include a wide range of mygalomorphs and araneoids as well as uloborids and palpimanoids (including arachaeid forms, see below) (P.A. Selden, D. Huang, D. Ren, in preparation). In conclusion, it appears that the Jurassic araneofauna contained a diversity of orb-web weavers (Orbiculariae), including both cribellate uloborids and ecribellate araneoids.

Because the orb web is considered to have originated among cribellate orbicularians (Coddington 1986), the presence of ecribellate araneoids is evidence of the split of ecribellate from cribellate orbicularians in at least late Jurassic times, and possibly earlier. By Jurassic times, holometabolous insects had originated and radiated, so that nearly all modern orders were present (Jarzembowski & Ross 1996). The diversity of spiders was obviously much higher than that presented by fossil evidence alone [Mesothelae; Mygalomorphae; Hexathelidae; Araneomorphae; Araneoidea (Juraraneidae, probably Tetragnathidae); Palpimanoidea (Archaeidae and others); Uloboridae]. By that time we had not only burrow dwelling spiders, but also funnel webs and orb webs. Most interestingly, modern Archaeidae are specialist spider hunters (araneophages) and the similarity in the morphology (elongate chelicerae) between the Jurassic and the modern species suggests that this mode of foraging had already evolved then.

Cretaceous

From early cretaceous rocks in Spain we have records of tetragnathid (including nephilinid) and uloborid spiders (Selden 1989, 1990; Selden & Penney 2003) that show the distinctive tarsal claw pattern of moderen orb-web weavers. Like the early Cretaceous Chinese Yixian formation (Zhou et al. 2003) yielding araneoids and uloborids (P.A. Selden, unpublished data), the Spanish deposits also represent lacustrine environments, some with volcanic ash falls. Today's tetragnathids are rather common along lakeshores, where they feed in the evening on the abundant insect life. Spiders do not fall into lake waters quite so readily as insects do (fossil insects in lacustrine deposits outnumber spiders by about 1000 to 1). But volcanic ash falls would certainly help to cause webs to collapse or spiders to lose their purchase and fall into the water. It is thus possible that we are already seeing a typical modern lake-shore spider fauna as far back as the early Cretaceous or Jurassic. It is only within the past dozen years that spiders have been described from Mesozoic ambers. Eskov & Wunderlich (1994) described the enigmatic new family Lagonomegopidae from two juvenile specimens in amber from Yantardakh, Taimyr, Siberia, and mentioned a further 50 undescribed specimens from the Upper Cretaceous of the region, some of which were placed in the superfamilies Araneoidea, Dysderoidea, and Thomisoidea. Eskov & Wunderlich also mentioned spiders from fossil resins from Azerbaijan and Armenia. The 47 amber spider specimens mentioned by Zherikhin & Sukatsheva (1973) from Yantardakh, Siberia may now be lost (Eskov & Wunderlich 1994). Other described Cretaceous amber spiders include the families Segestriidae, Oonopidae, Lagonomegopidae, Oecobiidae, Dictynidae, Araneidae, and Linyphiidae from New Jersey amber (Penney 2002, 2004b); Archaeidae, Pisauridae, Lagonomegopidae, and Ooniopidae in Burmese amber (Penney 2003a, 2004c, 2005, 2006); Linyphiidae and Deinopidae from Lebanese amber (Penney 2003b, Penney & Selden 2002); and Lagonomegopidae and Oonopidae from Canadian amber (Penney 2005, 2006). The first Mesozoic mygalomorphs were described by Eskov & Zonshtein (1990) from localities in the Lower Cretaceous of Siberia and Mongolia. These were placed in the modern families Mecicobothriidae, Antrodiaetidae, and Atypidae. The modern family Nemesiidae was reported from early Cretaceous amber from the Isle of Wight (Selden 2002), and Dipluridae from the early Cretaceous of Brazil (Selden et al. 2006).

Amber is an interesting fossil fixative as it not only conserves the spider but it also tends to preserve the silk and web. Hence amber spiders will take us closest to understanding the habits and habitat of fossil spiders (Zschokke 2003).

SUMMARY AND CONCLUSIONS

An analysis of spider and insect palaeontological data allowed Penney (2004a) to evaluate family richness through geological time. He concluded that both insect and spider fossil records show an exponential increase over time, which is the pattern typical of a radiating taxon. He further concluded that both taxa, insects and spiders, had comparable rates of diversification, which suggests that spiders and insects may have coradiated. The perceived main spurt of radiation would have happened at least 100 Mya before the origin of angiosperms (**Figure 4**). This suggests that insect evolution was driven less by flowering plants than by other factors, with spider predation being a strong possibility for a major selective force.

Present-day spiders, unlike their insect counterparts, are all carnivores, without exception (Foelix 1996). This suggests to us that the ancestral spiders were also carnivore insectivores. Alternatively, one would have to assume that an arachnid herbivore morphotype existed at some stage and perished without leaving a trace, fossil or otherwise. All evidence, from the mouthparts to the digestive tract and emzymes suggests specialist carnivory (Foelix 1996). Indeed, the whole body plan, including the silk glands and their position as well as the claws on the legs, suggests a deeply rooted predatory existence (Foelix 1996). A predatory life style would be greatly helped by the opportunistic flexibility offered by behavior patterns, which are exponentially increasing with increasing complexity of the rules as emergent properties add ever more flexibility of expression (Krink & Vollrath 1998). Such flexibility would be invaluable in an arms race between the predator and its prey. Here we must remember that spiders are not only predators but also prey; after all, the major predators of spiders are other spiders (Wise 1993). However this may be, a predatory life style with silk as a primary tool has served spiders very well, looking at their ecological diversity and importance.

DISCLOSURE STATEMENT

The authors are not aware of any biases that might be perceived as affecting the objectivity of this review.

LITERATURE CITED

Bachofen-Echt A. 1949. *Der Bernstein und seine Einschlüsse*. Vienna: Springer-Verlag. 204 pp.

Barth FG. 1985. *Neurobiology of Arachnids*. Berlin: Springer-Verlag. viii + 385 pp.

Barth FG. 2002. *A Spider's World: Senses and Behavior*. Berlin: Springer-Verlag. viii + 394 pp.

Bell JR, Bohan DA, Shaw EM, Weyman GS. 2005. Ballooning dispersal using silk: world fauna, phylogenies, genetics and models. *Bull. Entomol. Res.* 95:69–114

Benjamin SP, Zschokke S. 2002. Untangling the tangle-web: web construction behavior of the comb-footed spider *Steatoda triangulosa* and comments on phylogenetic implications (Araneae: Theridiidae). *J. Insect Behav.* 15:791–809

Benjamin SP, Zschokke S. 2003. Webs of theridiid spiders: Construction, structure and evolution. *Biol. J. Linn. Soc.* 78:93–105

Bond JE, Opell BD. 1998. Testing adaptive radiation and key innovation hypotheses in spiders. *Evolution* 52:403–14

Braddy SJ. 1995. The ichnotaxonomy of the invertebrate trackways of the Coconino Sandstone (Lower Permian), northern Arizona. *N. M. Mus. Nat. Hist. Sci. Bull.* 6:219–24

Brauckmann C, Brauckmann B, Gröning E. 1996. The stratigraphical position of the oldest known Pterygota (Insecta, Carboniferous, Namurian). *Ann. Soc. Géol. Belg.* 117:47–56

Brauckmann C, Koch L, Kemper M. 1985. Spinnentiere (Arachnida) und Insekten aus den Vorhalle-Schichten (Namurium B; Ober-Karbon) von Hagen-Vorhalle (West-Deutschland). *Geol. Paläontol. Westfal.* 3:1–131

Brauckmann C, Schneider J. 1996. Ein unter-karbonisches Insekt aus dem Raum Bitterfeld/Delitzsch (Pterygota, Arnsbergium, Deutschland). *Neues Jahrb. Geol. Paläontol. Monatsh.* 1996:17–30

Bristowe WS. 1958. *The World of Spiders*. New Nat. Ser. London: Collins. viii + 304 pp.

Bristowe WS. 1975. A family of living fossil spiders. *Endeavour* 34:115–17

Bromhall C. 1987a. Spider tracheal systems. *Tissue Cell* 19:793–807

Bromhall C. 1987b. Spider heart rates and locomotion. *J. Comp. Physiol.* B157:451–60

Churchill TB, Raven RJ. 1989. The circumtropical distribution and habits of the intertidal mygalomorph spider *Idioctis* (Barychelidae). *Rep. Dep. Biol. Univ. Turku* 15

Coddington JA. 1986. The monophyletic origin of the orb web. See Shear 1986, pp. 319–63

Coddington JA. 1989. Spinneret silk spigot morphology. Evidence for the monophyly of orb-weaving spiders, Cyrtophorinae (Araneidae) and the group Theridiidae-Nesticidae. *J. Arachnol.* 17:71–95

Coddington JA, Levi HW. 1991. Systematics and evolution of spiders (Araneae). *Annu. Rev. Ecol. Syst.* 22:565–92

Comstock JH. 1948. *The Spider Book*. New York: Comstock. viii + 729 pp.

Craig CL. 1997. Evolution of arthropod silks. *Annu. Rev. Entomol.* 42:231–67

Craig CL. 2003. *Spiderwebs and Silks: Tracing Evolution from Molecules to Genes to Phenotypes*. New York: Oxford Univ. Press. viii + 200 pp.

Damen WGM, Saridaki T, Averof M. 2002. Diverse adaptations of an ancestral gill: a common evolutionary origin for wings, breathing organs, and spinnerets. *Curr. Biol.* 12:1711–16

Decae AE. 1984. A theory on the origin of spiders and the primitive function of spider silk. *J. Arachnol.* 12:21–28

Dippenaar-Schoeman A. 1990. A revision of the African spider genus *Seothyra* Purcell (Araneae, Eresidae). *Cimbebasia* 12:135–60

Eberhard WG. 1969. Computer simulation of orb web construction. *Am. Zool.* 9:229–38

Eberhard WG. 1981. Construction behaviour and the distribution of tensions in orb webs. *Bull. Br. Arachnol. Soc.* 5: 189–204

Eberhard WG. 1982. Behavioural characters for the higher classification of orb-weaving spiders. *Evolution* 36:1067–95

Eberhard WG. 1986. Trail line manipulation as a character for higher level spider taxonomy. In *Proc. 9th Int. Congr. Arachnology Panamá, 1983*, ed. WG Eberhard, YD Lubin, BC Robinson, pp. 49–51. Washington, DC: Smithson. Inst. Press

Eberhard WG. 1987. The effect of gravity on temporary spiral construction by the spider *Leucauge mariana* (Araneae: Tetragnathidae). *J. Ethol.* 5:29–36.

Eberhard WG. 1988a. Memory of distances and directions moved as cues during temporary spiral construction in the Spider *Leucauge mariana* (Araneae : Araneidae). *J. Insect Behav.* 1:51–66

Eberhard WG. 1988b. Behavioural flexibility in orb web construction: effects of supplies in different silk glands and spider size and weight. *J. Arachnol.* 16:295–302

Eberhard WG. 1990. Function and phylogeny of spider webs. *Annu. Rev. Ecol. Syst.* 21:341–72

Eisner T, Alsop R, Ettershank G. 1964. Adhesiveness of spider silk. *Science* 146:1058–61

Engel MS, Grimaldi DA. 2004. New light shed on the oldest insect. *Nature* 427:627–30

Erwin DH. 2006. *Extinction. How Life on Earth Nearly Ended 250 Million Years Ago*. Princeton, NJ: Princeton Univ. Press. viii + 296 pp.

Eskov KY. 1984. A new fossil spider family from the Jurassic of Transbaikalia from (Araneae: Chelicerata). *Neues Jahrb. Geol. Paläontol. Monatsh.* 1984:645–53

Eskov KY. 1987. A new archaeid spider (Chelicerata: Araneae) from the Jurassic of Kazakhstan, with notes on the so-called "Gondwanan" ranges of recent taxa. *Neues Jahrb. Geol. Paläontol. Abh*. 175:81–106

Eskov KY, Selden PA. 2005. First record of spiders from the Permian period (Araneae: Mesothelae). *Bull. Br. Arachnol. Soc*. 13:111–16

Eskov KY, Wunderlich J. 1994. On the spiders of Taimyr ambers, Siberia, with the description of a new family and with general notes on the spiders from the Cretaceous resins (Arachnida: Araneae). *Beitr. Araneol*. 4:95–107

Eskov KY, Zonshtein S. 1990. First mesozoic mygalomorph spiders from the Lower Cretaceous of Siberia and Mongolia, with notes on the system and evolution of the infraorder Mygalomorphae (Chelicerata: Araneae). *Neues Jahrb. Geol. Paläontol. Abh*. 178:325–68

Foelix RF. 1996. *Biology of Spiders*. Oxford: Oxford Univ. Press. viii + 200 pp. 2nd ed.

Garb JE, DiMauro T, Vo V, Hayashi CY. 2006. Silk genes support the single origin of orb webs. *Science* 312:1762

Gatesy J, Hayashi C, Motriuk D, Woods J, Lewis R. 2001. Extreme diversity, conservation, and convergence of spider silk fibroin sequences. *Science* 291:2603–5

Gertsch WJ. 1979. *American Spiders*. New York: Van Nostrand Reinhold

Glatz L. 1972. Der Spinnapparat haplogyner Spinnen (Arachnida, Araneae). *Z. Morphol. Tiere* 72:1–26

Glatz L. 1973. Der Spinnapparat der Orthognatha (Arachnida, Araneae). *Z. Morphol. Tiere* 75:1–50

Görner P, Class B. 1985. Homing behaviour and orientation in the funnel-web spider, *Agelena labyrithica* Clerck. See Barth 1985, pp. 275–98

Gotts NM, Vollrath F. 1992. Physical and theoretical features in the simulation of animal behaviour, e.g., the spider's web. *Cybern. Syst*. 23:41–65

Gregory MR, Campbell KA, Zuraida R, Martin AJ. 2006. Plant traces resembling *Skolithos*. *Ichnos* 13:205–16

Grimaldi D, Engel MS. 2005. *Evolution of the Insects*. Cambridge/New York: Cambridge Univ. Press. xv + 755 pp.

Griswold CE, Coddington JA, Hormiga G, Scharff N. 1998. Phylogeny of the orb-web building spiders (Araneae, Orbiculariae: Deinopoidea, Araneoidea). *Zool. J. Linn. Soc*. 123:1–99

Griswold CE, Ramírez MJ, Coddington J, Platnick N. 2005. Atlas of phylogenetic data for entelegyne spiders (Araneae: Araneomorphae: Entelegynae) with comments on their phylogeny. *Proc. Calif. Acad. Sci., 4th Ser*. 56(Suppl. II):1–324

Haupt J. 2003. The Mesothelae—a monograph of an exceptional group of spiders (Araneae: Mesothelae). *Zoologica* 154:1–102

Heiling AM, Herberstein ME. 1998. The web of *Nuctenea sclopetaria* (Araneae, Araneidae): Relationship between body size and web design. *J. Arachnol*. 26:91–96

Herberstein ME, Heiling AM. 1999. Asymmetry in spider orb webs: A result of physical constraints? *Anim. Behav*. 58:1241–46

Hergenröder R, Barth FG. 1983. Vibratory signals and spider behavior: how do the sensory inputs from the eight legs interact in orientation. *J. Comp. Physiol. A* 152:361–71

Hill D. 1979. Orientation by jumping spiders of the genus *Phidippus* (Araneae: Salticidae) during the pursuit of prey. *Behav. Ecol. Sociobiol.* 5:301–22

Hirst S, Maulik S. 1926. On some arthropod remains from the Rhynie chert (Old Red Sandstone). *Geol. Mag.* 63:69–71

Huang D-Y, Nel A, Shen Y-B, Selden PA, Lin Q-B. 2006. Discussions on the age of Daohugou fauna—evidence from invertebrates. *Progr. Nat. Sci.* (Spec. Issue) 16:308–12

Jackson RR, Carter CM, Tarsitano MS. 2001. Trial-and-error solving of a confinement problem by a jumping spider, *Portia fimbriata. Behaviour* 138:1215–34

Jarzembowski EA, Ross AJ. 1996. Insect origination and extinction in the Phanerozoic. In *Biotic Recovery from Mass Extinction Events*, ed. MB Hart. *Geol. Soc. Spec. Publ.* 102:65–78

Kaston BJ. 1964. The evolution of spider webs. *Am. Zool.* 4:191–207

Kovoor J. 1977. La soie et les glandes sericigènes des arachnides. *Ann. Biol.* 16:97–171

Kovoor J. 1987. Comparative structure and histochemistry of silk-producing organs in Arachnids. In *Ecophysiology of Spiders*, ed. W Nentwig, pp. 160–86. Berlin/New York: Springer-Verlag

Krink T, Vollrath F. 1997. Analysing spider web-building behaviour with rule-based simulations and genetic algorithms. *J. Theor. Biol.* 185:321–31

Krink T, Vollrath F. 1998. Emergent properties in the behaviour of a virtual spider robot. *Proc. R. Soc. Ser. B* 265:2051–55

Krink T, Vollrath F. 1999. A virtual robot to model the use of regenerated legs in a web-building spider. *Anim. Behav.* 57:223–32

Krink T, Vollrath F. 2000. Optimal area use in orb webs of the spider Araneus diadematus. *Naturwissenschaften* 87:90–93

Labandeira CC, Sepkoski JJ. 1993. Insect diversity in the fossil record. *Science* 261:310–15

Lamoral BH. 1968a. On the species of the genus *Desis* Walckenaer, 1837 (Araneae: Amaurobiidae) found on the rocky shores of South Africa and south west Africa. *An. Natal Mus.* 20:139–50

Lamoral BH. 1968b. On the ecology and habitat adaptations of two intertidal spiders, *Desis formidabilis* (O. P.-Cambridge) and *Amaurobioides africanus* Hewitt, at "The Island" (Kommetjie, Cape Peninsula), with notes on the occurence of two other spiders. *An. Natal Mus.* 20:151–93

Land M. 1985. The morphology and optics of spider eyes. See Barth 1985, pp. 53–78

Lin LH, Edmonds DT, Vollrath F. 1995. Structural engineering of an orb-spider's web. *Nature* 373:146–48

Lubin YD, Henschel JR. 1990. Foraging at the thermal limit: burrowing spiders (*Seothyra*, Eresidae) in the Namib desert dunes. *Oecologia* 84:461–67

Main BY. 1976. *Spiders*. Sydney/London: Collins. viii + 296 pp.

Main BY. 1993. From flooding avoidance to foraging: adaptive shifts in trapdoor spider behaviour. *Mem. Qld. Mus.* 33:599–606

Marples BJ. 1967. The spinnerets and epiandrous glands of spiders. *J. Linn. Soc.* 46:209–22

McQueen DJ, McLay CL. 1983. How does the intertidal spider *Desis marina* (Hector) remain under water for such a long time? *NZ J. Zool.* 10:383–91

Messner B, Adis J. 1995. There is only facultative plastron respiration in diving web spiders (Araneae). *Dtsch. Entomol. Z.* 42:453–59

Mittelstaedt H. 1985. Analytical cybernetics of spider navigation. See Barth 1985, pp. 298–316

Nentwig W. 1986. *Ecophysiology of Spiders*. Berlin: Springer-Verlag

Opell BD. 1984. Comparison of carapace features in the family Uloboridae (Araneae). *J. Arachnol.* 12:105–14

Opell BD. 1989. Functional associations between the cribellum spinning plate and capture threads of *Miagrammopes animotus* (Araneida, Uloboridae). *Zoomorphology* 108:263–67

Opell BD. 1996. Functional similarities of spider webs with diverse architectures. *Am. Nat.* 148:632–48

Opell BD. 2002. How spider anatomy and thread configuration shape the stickiness of cribellar prey capture threads. *J. Arachnol.* 30:10–19

Opell BD, Bond JE. 2001. Changes in the mechanical properties of capture threads and the evolution of modern orb-weaving spiders. *Evol. Ecol. Res.* 3:567–81

Peñalver E, Grimaldi DA, Delclòs X. 2006. Early Cretaceous spider web with its prey. *Science* 312:7161

Penney D. 2002. Spiders in Upper Cretaceous amber from New Jersey (Arthropoda, Araneae). *Palaeontology* 45:709–24

Penney D. 2003a. *Afrarchaea grimaldii*, a new species of Archaeidae (Araneae) in Cretaceous Burmese amber. *J. Arachnol.* 31:122–30

Penney D. 2003b. A new deinopoid spider from Cretaceous Lebanese amber. *Acta Palaeontol. Polon.* 48:569–74

Penney D. 2004a. Does the fossil record of spiders track that of their principal prey, the insects? *Trans. R. Soc. Edinburgh: Earth Sci.* 94:275–81

Penney D. 2004b. New spiders in Upper Cretaceous amber from New Jersey in the American Museum of Natural History (Arthropoda, Araneae). *Palaeontology* 47:367–75

Penney D. 2004c. A new genus and species of Pisauridae (Araneae) in Cretaceous Burmese amber. *J. Syst. Palaeontol.* 2:141–45

Penney D. 2005. The fossil spider family Lagonomegopidae in Cretaceous ambers with description of a new genus and species from Myanmar. *J. Arachnol.* 33:439–44

Penney D. 2006. Fossil oonopid spiders in Cretaceous ambers from Canada and Myanmar. *Palaeontology* 49:229–35

Penney D, Selden PA. 2002. The oldest linyphiid spider, in Lower Cretaceous Lebanese amber (Araneae, Linyphiidae, Linyphiinae). *J. Arachnol.* 30:487–93

Penney D, Selden PA. 2006. Assembling the Tree of Life—Phylogeny of Spiders: a review of the strictly fossil spider families. In *European Arachnology 2005*, ed. C Deltshev, P Stoev. *Acta Zool. Bulg.* Suppl. 1:25–39

Peters HM. 1937. Studien am Netz der Kreuzspinne (*Aranea diadema* L.) I. Die Grundstruktur des Netzes und Beziehungen zum Bauplan des Spinnenkörpers. *Z. Morphol. Ökol. Tiere* 32:613–49

Pickford M. 2000. Fossil spider's webs from the Namib Desert and the antiquity of *Seothyra* (Araneae, Eresidae). *Ann. Paléontol.* 86:147–55

Pickford M, Senut B. 1999. Geology and paleobiology of the Namib Desert, southwestern Africa. *Geol. Surv. Namib. Mem.* 18, pp. 1–155

Popadic A, Panganiban G, Rusch D, Shear WA, Kaufman TC. 1998. Molecular evidence for the gnathobasic derivation of arthropod mandibles and for the appendicular origin of the labrum and other structures. *Dev. Genes Evol.* 208:142–50

Powell L. 1878. On *Desis robsoni*, a marine spider, from Cape Campbell. *Trans. NZ Inst.* 11:263–68

Reed CF, Witt PN, Scarboro MB, Peakall DB. 1970. Experience and the orb web. *Dev. Psychobiol.* 3:251–65

Robinson MH. 1975. The evolution of predatory behaviour in araneid spiders. In *Function and Evolution in Behaviour*, ed. G Baerends, C Beer, A Manning, pp. 292–312. Oxford, United Kingdom: Clarendon

Robinson MH, Lubin YD. 1979. Specialists and generalists: the ecology and behavior of some web-building spiders from Papua New Guinea. II *Psechrus argentatus* and *Fecinia* sp. (Araneae: Psechridae). *Pacific Insects* 21:133–64

Rössler R. 1998. Arachniden-Neufunde im mitteleuropäischen Unterkarbon bis Perm—Beitrag zur Revision der Familie Aphantomartidae Petrunkevitch 1945 (Arachnida, Trigonotarbida). *Paläontol. Z.* 72:67–88

Rovner JS. 1987. Nests of terrestrial spiders maintain a physical gill: flooding and evolution of silk constructions. *J. Arachnol.* 14:327–37

Rovner JS. 1989. Submersion survival in aerial web-weaving spiders from a tropical wet forest. *J. Arachnol.* 17:241–46

Sadler CJ. 1993. Arthropod trace fossils from the Permian De Chelly Sandstone, northeastern Arizona. *J. Paleontol.* 67:240–49

Schneider G. 2004. *The Roadside Geology of Namibia*. (*Samml. Geol. Führer* 97). Stuttgart: Gebrüder Borntraeger. 294 pp.

Scott AC, Glasspool IJ. 2006. The diversification of Paleozoic fire systems and fluctuations in atmospheric oxygen concentrations. *Proc. Natl. Acad. Sci. USA* 103:10861–65

Seilacher A. 1967. Fossil behaviour. *Sci. Am.* 217:72–80

Selden PA. 1989. Orb-web weaving spiders in the early Cretaceous. *Nature* 340:711–13

Selden PA. 1990. Lower Cretaceous spiders from the Sierra de Montsech, north-east Spain. *Palaeontology* 33:257–85

Selden PA. 1996a. First fossil mesothele spider, from the Carboniferous of France. *Rev. Suisse Zool.* 2:585–96

Selden PA. 1996b. Fossil mesothele spiders. *Nature* 379:498–99

Selden PA. 2000. *Palaeothele*, a replacement name for the fossil mesothele spider *Eothele* Selden *non* Rowell. *Bull. Br. Arachnol. Soc.* 11:292

Selden PA. 2002. First British Mesozoic spider, from Cretaceous amber of the Isle of Wight, southern England. *Palaeontology* 45:973–83

Selden PA, Anderson HM, Anderson JM, Fraser NC. 1999. The oldest araneomorph spiders, from the Triassic of South Africa and Virginia. *J. Arachnol.* 27:401–14

Selden PA, Casado F da C, Mesquita MV. 2006. Mygalomorph spiders (Araneae: Dipluridae) from the Lower Cretaceous Crato Lagerstätte, Araripe Basin, northeast Brazil. *Palaeontology* 49:817–26

Selden PA, Corronca JA, Hünicken MA. 2005. The true identity of the supposed giant fossil spider *Megarachne*. *Biol. Lett.* 1:44–48

Selden PA, Edwards D. 1989. Colonisation of the land. In *Evolution and the Fossil Record*, ed. KC Allen, DEG Briggs, pp. 122–52. London: Belhaven. xiii + 265 pp.

Selden PA, Gall J-C. 1992. A Triassic mygalomorph spider from the northern Vosges, France. *Palaeontology* 35:211–35

Selden PA, Penney D. 2003. Lower Cretaceous spiders (Arthropoda: Arachnida: Araneae) from Spain. *Neues Jahrb. Geol. Paläontol. Monatsh.* 2003:175–92

Selden PA, Shear WA, Bonamo PM. 1991. A spider and other arachnids from the Devonian of New York, and reinterpretations of Devonian Araneae. *Palaeontology* 34:241–81

Seyfarth E-A, Hergenröder R, Ebbes H, Barth FG. 1982. Idiothetic orientation of a wandering spider: compensation detours and estimates of goal distance. *Behav. Ecol. Sociobiol.* 11:139–48

Shear WA. 1986. *Spiders: Webs, Behavior, and Evolution*. Stanford: Stanford Univ. Press. xiii + 492 pp.

Shear WA, Palmer JM, Coddington JA, Bonamo PM. 1989. A Devonian spinneret: early evidence of spiders and silk use. *Science* 246:479–81

Shear WA, Selden PA. 2001. Rustling in the undergrowth: animals in early terrestrial ecosystems. In *Plants Invade the Land: Evolutionary and Environmental Perspective*, ed. PG Gensel, D Edwards, pp. 29–51. New York: Columbia Univ. Press. x + 304 pp.

Shear WA, Selden PA, Rolfe WDI, Bonamo PM, Grierson JD. 1987. New terrestrial arachnids from the Devonian of Gilboa, New York (Arachnida: Trigonotarbida). *Am. Mus. Novit.* 2901:1–74

Shultz JW. 1987. The origin of the spinning apparatus in spiders. *Biol. Rev.* 62:89–113

Smith RB, Mommsen TP. 1984. Pollen feeding in a orb-weaving spider. *Science* 226:1330–33

Snodgrass RE. 1952. *A Textbook of Arthropod Anatomy*. Ithaca, NY: Cornell Univ. Press

Strazny F, Perry SF. 1984. Morphometic diffusing capacity and functional anatomy of the book lungs in the spider *Tegenaria* spp. (Agelenidae). *J. Morphol.* 182:339–54

Suter RB. 1984. Web tension and gravity as cues in spider orientation. *Behav. Ecol. Sociobiol.* 16:31–36

Tarsitano MS, Jackson RR. 1997. Araneophagic jumping spiders discriminate between detour routes that do and do not lead to prey. *Anim. Behav.* 53:257–66

Tillinghast EK, Townley M. 1987. Chemistry, physical properties and synthesis of Araneidae orb webs. In *Ecophysiology of Spiders*, ed. W Nentwig, pp. 203–10. Berlin: Springer-Verlag

Tillinghast EK, Townley M. 1994. Silk glands of araneid spiders—selected morphological and physiological aspects. In *Silk Polymers: Materials Science and*

Biotechnology, ed. D Kaplan, W Adams, B Farmer, C Viney, pp. 29–44. Washington, DC: Am. Chem. Soc.

Tillyard RJ. 1928. Some remarks on the Devonian fossil insects from the Rhynie chert beds, Old Red Sandstone. *Trans. R. Entomol. Soc. London* 76:65–71

Tilquin A. 1942. *La Toile Géométrique des Araignées*. Paris: Presses Univ. France. viii + 536 pp.

Townley M, Tillinghast E, Cherim N. 1993. Moult-related changes in ampullate silk gland morphology and usage in the araneid spider *Araneus cavaticus*. *Philos. Trans. R. Soc. London Ser. B* 340:25–38

Vollrath F. 1987. Altered geometry of web in spiders with regenerated legs. *Nature* 328:247–48

Vollrath F. 1988. Untangling the spider's web. *Trend. Ecol. Evol.* 3:331–35

Vollrath F. 1992. Analysis and interpretation of orb spider exploration and web-building behaviour. *Adv. Stud. Behav.* 21:147–99

Vollrath F. 1995. Lyriform organs on regenerated spider legs. *Bull. Br. Arachnol. Soc.* 10:115–18

Vollrath F. 2000. Coevolution of behaviour and material in the spider's web. In *Biomechanics in Animal Behaviour*, ed. P Domenici, RW Blake, pp. 315–29. Oxford: BIOS Sci.

Vollrath F, Downes M, Krackow S. 1997. Design variability in web geometry of an orb-weaving spider. *Physiol. Behav.* 62:735–43

Vollrath F, Edmonds DT. 1989. Modulation of the mechanical properties of spider silk by coating with water. *Nature* 340:305–7

Vollrath F, Knight DP. 2001. Liquid crystalline spinning of spider silk. *Nature* 410:541–48

Vollrath F, Knight D. 2003. The nature of some spiders' silks. In *Elastomeric Proteins*, ed. PR Shewry, AS Tatham, AJ Bailey, pp. 152–74. New York: Cambridge Univ. Press

Vollrath F, Knight D. 2005. Biology and technology of silk production. In *Biotechnology of Biopolymers: From Synthesis to Patents*, ed. A Steinbuchel, Y Doi, 2:873–94. Weinheim: Wiley-VCH

Vollrath F, Norgaard T, Krieger M. 2000. Radius orientation in cross spider *Araneus diadematus*. In *European Arachnology 2000*, ed. S Toft, N Scharff, pp. 107–16. Aarhus, Den.: Aarhus Univ. Press

Vollrath F, Porter D. 2006. Spider silk as archetypal protein elastomer. *Softmatter* 2:377–85

Ward P, Labandeira C, Laurin M, Berner RA. 2006. Confirmation of Romer's Gap as a low oxygen interval constraining the timing of initial arthropod and vertebrate terrestrialization. *Proc. Natl. Acad. Sci. USA* 103:16818–22

Wise DH. 1993. *Spiders in Ecological Webs*. Cambridge, UK: Cambridge Univ. Press. viii + 328 pp.

Witt PN. 1971. Drugs alter web-building of spiders. *Behav. Sci.* 16:98–113

Witt PN, Reed CF, Peakall DB. 1968. *A Spider's Web: Problems in Regulatory Biology*. Berlin: Springer-Verlag. viii + 107 pp.

Witt PN, Rovner JS. 1982. *Spider Communication: Mechanisms and Ecological Significance*. 440 pp. Princeton, NJ: Princeton University Press

Zherikhin VV, Sukatsheva ID. 1973. On the Cretaceous insect-bearing "ambers" (re-tinites) of North Siberia. In *Voprosy paleontologii nasekomykh. Doklady na XXIV ezhegodnom chtenii pamyati N.A. Kholodkovskogo, 1–2 April, 1971.* Leningrad: Nauka Press. 3–48 [in Russian].

Zhou Z, Barrett PM, Hilton J. 2003. An exceptionally preserved Lower Cretaceous ecosystem. *Nature* 421:807–14

Zschokke S. 2003. Spider-web silk from the Early Cretaceous. *Nature* 424:636–37

Zschokke S, Vollrath F. 1995a. Unfreezing the behaviour of web spiders. *Behav. Physiol.* 58:1167–73

Zschokke S, Vollrath F. 1995b. Web construction patterns in a range of orb weaving spiders. *Europ. J. Entomol.* 92:523–41

Applications of Flow Cytometry to Evolutionary and Population Biology

Paul Kron,[1] Jan Suda,[2] and Brian C. Husband[1]

[1] Department of Integrative Biology, University of Guelph, Ontario, Canada N1G 2W1; email: bhusband@uoguelph.ca, pkron@uoguelph.ca

[2] Department of Botany, Faculty of Science, Charles University in Prague, CZ-128 01, Czech Republic and Institute of Botany, Academy of Sciences of the Czech Republic, CZ-252 43, Czech Republic; email: suda@natur.cuni.cz

Annu. Rev. Ecol. Evol. Syst. 2007. 38:847–76

The *Annual Review of Ecology, Evolution, and Systematics* is online at
http://ecolsys.annualreviews.org

This article's doi:
10.1146/annurev.ecolsys.38.091206.095504

Key Words

genome size, microorganism diversity, polyploidy, reproductive pathways, taxonomy

Abstract

Flow cytometry, a method of rapidly characterizing optical properties of cells and cell components within individuals, populations, and communities, is advancing research in several areas of ecology, systematics, and evolutionary biology. Measuring the light emitted or scattered from cells or cell components, often in combination with specific stains, allows a multitude of physical and genetic attributes to be evaluated simultaneously and the resulting information to be rapidly processed. As a result, the technique has enabled large-scale comparative analyses of genome-size evolution, taxonomic identification and delineation, and studies of polyploids, reproductive biology, and experimental evolution. It is also being used to characterize the structure and composition of microbial communities. Here, we outline the nature of these contributions, as well as future applications, and provide an online summary of protocols and sampling methods.

INTRODUCTION

Use of flow cytometry (FCM) in population biological research has expanded dramatically both in scope and frequency since its initial use in medical science (Robinson & Grégori 2007). FCM was originally used to identify and characterize cancerous cells through their DNA content, and now it has become increasingly useful in the context of ecology, evolutionary biology, and systematics, especially in the past 15 years. However, compared to its global use in scientific research, the technique is relatively new to these fields, and the methods and potential applications are unfamiliar to most population biologists.

In this review we summarize some of the most significant contributions of FCM to evolutionary and population biology. We summarize the basic principles of FCM and examine six conceptual areas advanced by this method. We also provide a central database of the most widely used FCM protocols and relevant citations (see the Supplemental Material link in the online version of this review or at **http://www.annualreviews.org/**).

GENERAL PRINCIPLES AND METHODS

Flow cytometry is a high-throughput analytical tool that simultaneously detects and quantifies multiple optical properties (fluorescence, light scatter) of single particles, usually cells or nuclei labeled with fluorescent probes, as they move in a narrow liquid stream through a powerful beam of light (**Figure 1**). The value of FCM lies both in the wide range of parameters that can be simultaneously recorded and the information that can be provided on how these parameters are distributed within the particle population. Basically, particles in suspension stained with one or more fluorescent dyes (fluorochromes) are hydrodynamically focused to form a precise single file, which intersects an illuminating beam at high speed. This intersection causes the particles to emit fluorescent light and to scatter light at various angles. The light signals are sent through a series of filters, photodetectors, converters, and multipliers to a computer, where the results are stored and typically displayed in the form of a distribution histogram (in 1-parameter studies) or two-dimensional or three-dimensional dot plots (in multiparameter studies) (**Figure 1**). These recorded values can be used to infer the physical and chemical structure of analyzed particles. The most common parameter measured in evolutionary, ecological, and biosystematics studies is relative fluorescence, which, in the case of DNA-selective fluorochromes, provides estimates of nuclear DNA amount with high precision.

The essential prerequisite of FCM analysis is the suspension of isolated particles. Whole cells may occasionally be measured but it is isolated nuclei that are predominantly used in current FCM protocols. The methodology typically involves selecting suitable tissue, isolating nuclei, staining with DNA-selective dyes, and running the sample on a flow cytometer (**Figure 1**). For details of methods and commonly used protocols, see the **Supplemental Appendix**; follow the Supplemental Material link from the Annual Reviews home page at **http://www.annualreviews.org/**.

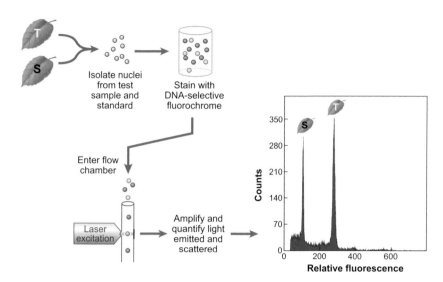

Figure 1

General flow cytometry procedure. Cells or nuclei are extracted from tissue of an individual to be tested (T) and from an individual to be used as a standard (S). Samples are often stained with, for example, a DNA-selective fluorochrome. Cells or nuclei in suspension are passed individually through a laser beam. Fluorescence and light scatter are measured, and data are output, typically as a histogram or scatterplot.

APPLICATIONS AND CONCEPTUAL DEVELOPMENTS

Since the 1980s, there have been numerous applications of FCM in ecology, evolution, and biosystematics. Most of these take advantage of its ability to estimate the DNA content of nuclei and to process large numbers of particles, providing information about the cellular composition of tissues and individuals, as well as the organismal composition of populations and communities. Below, we summarize six conceptual areas that are advancing because of this technique: genome size research, biosystematics and taxonomy, the study of polyploidy, reproductive pathway analysis, microorganism population and community structure, and experimental evolution. Some themes that recur throughout these discussions are (*a*) the way in which high sample throughput generates larger sample sizes than traditional methods, allowing for novel approaches to population studies; (*b*) the analytical power inherent in combining single particle measures (e.g., DNA content) with particle counts (e.g., nuclei number); and (*c*) the potential for addressing questions in unique ways by combining FCM with other, notably molecular, techniques.

1. CAUSES AND CONSEQUENCES OF GENOME SIZE VARIATION

1.1. Goals of Research on Genome Size

The term genome size is applied in two ways: It refers to both the DNA content of a single chromosome set (the base number) of an organism ("monoploid" or "basic" genome size; Cx-value), and to the DNA content of the unreplicated haploid genome ("holoploid genome size"; C-value) (Greilhuber et al. 2005, Leitch & Bennett 2004). More precisely, C-value refers to half the somatic DNA content (2C-value), regardless of the organism's ploidy, and Cx-value is the 2C-value divided by the ploidy level.

C-value and Cx-value are equivalent in diploids, but not in polyploids, in which the haploid state contains more than a single chromosome set. For this reason, careful and standardized use of terminology is essential to avoid confusion.

Across all living things, genome size is remarkably variable. Based on current estimates, C-values vary by a factor of $\sim$100,000x, from 0.007 pg in yeast to 700 pg in *Amoeba dubia* (Cavalier-Smith 1985, Gregory et al. 2007). There is wide heterogeneity within plants (0.065 to 127.4 pg; Leitch & Bennett 2007), animals (0.03 to 132.8 pg), and fungi ($\sim$0.007 to 0.81 pg) (Gregory et al. 2007). The observed variation (or lack thereof) has been a central focus of evolutionary biology, and now has increasing implications for cell and molecular biology, and for genomics research (Gregory 2005b).

Ever since measurement became feasible, evolutionists have tried to account for the magnitude of variation in genome size. Initially, the problem revolved around the observation that the size of the genome, which contains genes, is not related to the complexity of the organism (the C-value paradox). This paradox was later resolved by the discovery that much of the variation was composed of noncoding, repetitive DNA (reviewed in Gregory 2005a). Still, the extent of variation and its proximate and ultimate causes are largely unresolved, in what Gregory (2001) refers to as the C-value enigma.

A comprehensive understanding of genomes is still wanting, in part because of the relatively small number of species examined to date. Recent estimates indicate that genome size has been estimated for roughly 1.8% of angiosperms (Leitch & Bennett 2007), 2% of birds, 4% of reptiles, 9% of mammals, and 5% of teleost fish (Gregory 2007; these and other databases are reviewed in Gregory et al. 2007). In 2005, a special issue of *Annals of Botany* on plant genome size acknowledged the need for improved representation of the global flora, and similar needs are obvious for animals and fungi (Gregory et al. 2007).

1.2. Advances Enabled by Flow Cytometry

Many of the current questions regarding genome size variation will require additional data from a large taxonomic distribution. A variety of methods have been used to estimate genome size: Feulgen microdensitometry, reassociation kinetics, fluorometry, biochemical analysis, and more recently FCM. The most widely used method has been microdensitometry; however, the use of FCM has markedly increased, at least in plants (Bennett & Leitch 2005a). FCM offers obvious advantages to this kind of research: (*a*) sample preparation is relatively easy and rapid; (*b*) sample processing is rapid ($\sim$500–1000 nuclei/second, although rates about one order lower are common in most genome size projects); and (*c*) material does not need to be actively dividing and, in some cases (especially in animals), can be preserved. The main disadvantages relative to microdensitometry are a lack of direct visual observation of the nuclei being studied, and, to an ever-decreasing extent, cost of the instrument (Bennett & Leitch 2005a, Greilhuber et al. 2007). With the proper use of standards and attention to operating conditions (reviewed in Greilhuber et al. 2007), estimates are remarkably precise and sensitive to small variations. Thus FCM can provide rapid expansion not only of cross-species analysis but also of the capabilities of studying intraspecific and intraorganism variation in genome size.

1.3. Examples

1.3.1. Phenotypic correlates and fitness effects of genome size.

The evolutionary forces accounting for genome size variation can be reduced to two main arguments: It is a byproduct of the accumulation of noncoding DNA (Doolittle & Sapienze 1980, Ohno 1972) with no selective value, or it has phenotypic and fitness consequences and thus is adaptively significant. To address the selective argument, researchers have conducted large-scale comparative analyses and searched for, and found, associations between genome size and a wide range of morphological, cytological, developmental, physiological, reproductive, and ecological traits (reviewed by Bennett & Leitch 2005b, Gregory 2005c). For example, flow cytometry has been used to test for and demonstrate associations between genome size and metabolic rate (Vinogradov 1995, 1997), growth and development rate (Wakamiya et al. 1993), life strategy (Barow & Meister 2003), and environmental attributes (Wakamiya et al. 1993). The multitude of correlates suggests that genome size affects phenotype, likely through the physical consequences of increased nuclei size and volume, but the alternative, that variation in gene expression is a function of genome size, has only begun to be tested. Associations are likely to vary among taxonomic groups, depending on their biology and the environmental agents that limit fitness. Further development of a unified understanding of the adaptive significance of genome size will depend on expanding the taxonomic coverage and improving representation—arguably likely to be advanced more quickly with FCM than with any other method.

1.3.2. Genome downsizing.

The somatic DNA content (2C-value) of any given polyploid is expected to exceed its diploid progenitor(s) in proportion to its ploidy; however, its genome size (Cx-value) would be expected to be similar. Leitch & Bennett (2004) and coworkers examined this hypothesis using a database of angiosperm Cx-values (referred to as DNA content per basic genome) and found some striking patterns. In particular, while DNA content rose with increased ploidy, Cx-values declined. These results provide evidence for genome downsizing with increased ploidy across all major clades of flowering plants examined (Leitch & Bennett 2004). Similar patterns have been reported using FCM in more recent studies of taxonomic variation in plants (Jakob et al. 2004, Johnston et al. 2005, Price et al. 2005) and effects of radiation contamination on frogs (Vinogradov & Chubinishvili 1999). Artifacts of staining and ploidy identification may account for some of this pattern, although the ubiquity of downsizing suggests that a genuine loss of DNA accompanies genome duplication in many organisms. Several questions remain to be explored, including which genetic mechanisms and evolutionary processes may cause the loss of DNA. Potentially, DNA loss is selectively favored to minimize genetic instability or the phenotypic effects of increased nucleus and cell size. Additional genome size studies, particularly when integrated with molecular studies, will undoubtedly play an important role in elucidating this pattern.

1.3.3. Intraspecific variation in genome size.

The potential for intraspecific variation in genome size (Cx-value) is of longstanding interest but at odds with the initial notion of constancy in DNA content within individuals and species (Swift 1950). There are numerous reports of intraspecific variation in plants and animals, and this pattern is often attributed to chromosomal differences (aneuploidy, polyploidy, B-chromosomes, sex chromosomes), cryptic species (Greilhuber 1998), and variation in chromosome size (Gregory 2005a). However, evidence for intraspecific variation beyond chromosome polymorphism and cryptic taxonomic variation in genome size is rare and controversial. Previous studies, often based on densitometry or cytofluorometry, have observed such variation but these were not corroborated by subsequent FCM analyses and, in most cases, have been explained by taxonomic error or technical artifact (suboptimal staining, insufficient standardization; reviewed by Greilhuber 1998, 2005). It seems plausible that genome size may diverge in populations, even in the face of limited gene flow. However, conclusive evidence for this process will depend on increased attention not only to additional estimates of genome size, but also to the possible sources of error associated with FCM or other genome size measures (Loureiro et al. 2006, Noirot et al. 2003, Price et al. 2000, Walker et al. 2006).

2. SYSTEMATICS AND TAXONOMIC DELINEATION

2.1. Goals of Biosystematics

Investigations into the delineation and organization of biological diversity are older than biological science itself. In its current form, the discipline of biosystematics is evolving rapidly in concert with new criteria for assessing relatedness, methods of phylogenetic inference, and the expansion of traits (e.g., molecular, developmental, biochemical, and morphological) on which hypotheses can be evaluated. Biosystematics is increasingly relevant to other biological endeavors in providing a necessary framework for interpreting comparative data on trait evolution, community assembly (Pennington et al. 2006), and comparative genomics (Soltis & Soltis 2000a). By assisting in the enumeration and identification of biological diversity, it plays a fundamental role in global conservation initiatives and resource management (Krupnick & Kress 2005). The challenges facing this broad discipline are extensive and include detection and monitoring of new species, developing markers for increased resolution of relationships, and integrating processes of hybridization and genome duplication into mainstream analytical methods.

2.2. Impact of Flow Cytometry on Systematics and Taxonomic Research

A large and increasing number of cytological and cytogenetic characters such as chromosome number, morphology, and meiotic behavior, as well as nuclear DNA content, are used to circumscribe taxa and infer their relationships, and are therefore of central importance to the systematic community (Stace 2000). However, cytological techniques have historically been time-consuming and, in some cases, traits have been intractable to measure, limiting the number of accessions that can be processed and the extent to which cytological characters are used. FCM has altered the scope of biosystematics primarily by expanding the range of characters available to most practitioners, notably, nuclear DNA content (and thus, ploidy and genome size), and to a lesser extent, base pair composition (AT/GC ratio). More importantly, by making it possible to sample large numbers of individuals, cytotaxonomy has been transformed from an analysis of isolated individuals to the analysis of whole populations over large temporal and spatial scales.

The use of nuclear DNA content as a species trait expanded rapidly in the 1970s with the use of Feulgen microdensitometry. However, practical constraints usually limited examination to only a few individuals, which were then used as proxies for whole populations or species. DNA content estimates obtained using FCM frequently involve large sample sizes across multiple populations and are now commonplace in plant taxonomic studies (Suda et al. 2007). They have also been used in vertebrates (e.g., Birstein et al. 1993, MacCulloch et al. 1996), phytoplankton (Veldhuis et al. 1997), fungi (Eliam et al. 1994), and viruses (Brussaard et al. 2000).

With measures of DNA content and AT/GC ratios, FCM can distinguish not only between heteroploid taxa but also between homoploid species that have diverged in genome size or base pair composition, a task unachievable by other cytogenetic techniques at a comparable speed and cost. Aside from these basic taxonomic results, FCM, in combination with conventional karyological techniques and the use of genetic markers, can significantly contribute to the advancement of systematic research in animal and especially plant groups (the latter are usually more challenging owing to greater ploidy variation, recurrent origins, larger phenotypic plasticity, susceptibility to interspecific hybridization, and higher hybrid viability). Significant developments in systematics attributable to FCM include (*a*) improved detection and delineation of species, (*b*) improved data related to the contribution of chromosomal speciation to species diversity, and (*c*) increased availability of characters for the inference of phylogenetic relationships and polarity of character evolution.

2.3. Examples

2.3.1. Detection and delineation of species. Over the past decade, FCM has made a significant contribution in resolving complex low-level taxonomies, delimiting species boundaries, and revealing cryptic taxa, primarily by providing extensive data related to genome size and ploidy. Researchers have long been aware that polyploid complexes often pose serious taxonomic problems because an increase in ploidy level is commonly associated with blurring boundaries between taxa, even in alliances with easily recognizable diploids. Population-level studies using FCM have been used to clarify taxonomic boundaries in a number of vascular plant groups with ploidy variation, including dicots (e.g., Rosenbaumová et al. 2004), monocots (Pecinka et al. 2006), ferns (Bureš et al. 2003), and mosses (Melosik et al. 2005). The use of FCM to identify heteroploid animal taxa is relatively rare, either because of less ploidy variation in animal groups or preferences for noncytological (i.e., phenotypic or molecular) taxonomic criteria. Nevertheless, one such example concerns the diploid-tetraploid

Bufo viridis complex, in which all-triploid populations have been described as a separate taxon (Stöck et al. 2002).

FCM is also increasingly being used for resolving taxonomic complexities within homoploid groups when there are differences in genome size or AT/GC ratio. Variations in genome size (Cx-value) have been found in a number of plant and animal groups, either among species (Boulesteix et al. 2006, Mishiba et al. 2000) or within species (Dimitrova et al. 1999, Litvinchuk et al. 2005). Genome size data with potential taxonomic relevance for microorganisms have also started to appear in recent years, including data for phytoplankton (LaJeunesse et al. 2005, Veldhuis et al. 1997) and viruses (Brussaard et al. 2000). Similarly, AT/GC ratios are being used as taxonomic characters primarily to distinguish homoploid plant taxa (Vižintin et al. 2006). Additional within-taxon sampling may reveal previously undetected variation, as, for example, in *Pelobates fuscus*, where cryptic speciation was inferred from the presence of two groups with nonoverlapping nuclear DNA contents (Borkin et al. 2001a). Homoploid hybrids may also be detected, provided that the parental taxa differ sufficiently in genome size, opening up unprecedented possibilities for systematic research that have rarely been exploited to date in either plants (Jeschke et al. 2003, Mahelka et al. 2005) or animals (Ogielska et al. 2004).

Much valuable information of taxonomic significance is stored in the Plant DNA C-values database (Bennett & Leitch 2005c) and the Animal Genome Size database (Gregory 2007). In surveying the patterns of inter- and intraspecific genome size variation, we anticipate that interest in genome size as a taxonomically informative marker will increase, as will the utility of FCM for taxa circumscription and resolving taxonomies in homoploid groups.

2.3.2. Chromosomal speciation. Improved detection of polyploidy and genome size variation has raised the profile of chromosomal mechanisms of speciation, particularly through auto- and allo-polyploidization, but also through processes such as descending aneuploidy (e.g., Ramirez-Morillo & Brown 2001). An indirect outcome of more intensive population and broader geographical sampling is that many previously unknown cytotypes have been uncovered, particularly in plants (reviewed in Suda et al. 2007). MacCulloch et al. (1996) reported DNA content measures for 72 species of amphibians and reptiles, and concluded that, in some cases, cryptic species were present. Furthermore, use of FCM has revealed triploid cytotypes of putative hybrid origin, often for the first time, in a number of diploid-tetraploid plant groups (Baack 2004, Burton & Husband 1999, Husband & Schemske 1998, Petit et al. 1997). Hybrids of higher ploidy have also been observed (Suda & Lysák 2001).

In animal species, interspecific hybridization occasionally leads to the production of unisexual lineages, which may give rise to polyploid progeny. Eventually, complex diploid-polyploid groups may arise, with different cytotypes occurring in sympatry, as has been observed in about 70 vertebrate species (Vrijenhoek et al. 1989). Taxonomic treatment of these hybridogenetic complexes has long been inadequate, and only detailed insights into the sources of variation has allowed more satisfactory classifications that reflect evolutionary history. FCM has significantly contributed to this effort by providing rapid and reliable means of distinguishing between cytotypes

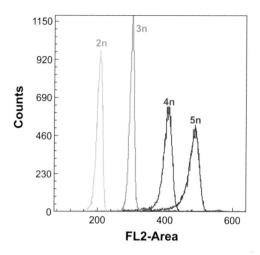

Figure 2

Integrated fluorescence histograms of nuclei from four different cytotypes of salamanders in the *Ambystoma* species complex, distinguished using FCM. Individuals were diploid (2n), triploid (3n), tetraploid (4n), and pentaploid (5n), representing various combinations of genomes in two sexual species and a variety of unisexual hybrids. Fluorescence was measured with a 585-nm (FL2) photodetector. Reproduced from figure 1 of Ramsden et al. 2006. *Mol. Ecol. Notes* 6:261–64.

(Doeringsfeld et al. 2004, Iguchi et al. 2003). For example, the FCM approach has proven competent in nonlethal discrimination between diploid sexual salamanders in the *Ambystoma laterale-jeffersonianum* group and their sympatric polyploid (3x, 4x, 5x) unisexuals (Ramsden et al. 2006) (**Figure 2**). In some cases, genome size measurements may be used to infer which parental genomes are maintained within lineages (Dawley et al. 1997).

2.3.3. Inferring phylogenetic relationships. Flow cytometry can also make powerful contributions to phylogenetic inference when used in combination with other techniques. There are a growing number of species for which DNA content (including ploidy classification) is coupled with genetic marker information, for example, using RAPDs (Spooner et al. 2001), AFLPs (Schönswetter et al. 2007), and DNA sequence (Vižintin et al. 2006). Potentially, the data on ploidy and, in some cases, genome size may be useful for corroborating insights into progenitor-derivative relationships and character evolution because of their clear polarity in many cases. The general trend is toward higher ploidy levels and larger genome sizes, although the evidence is mostly indirect (e.g., Leitch et al. 1998) and indices provided by FCM should always be evaluated in the context of more robust phylogenetic markers.

3. POLYPLOIDY

3.1. Goals of Polyploidy Research

Polyploidy (genome duplication) has played a significant role in the evolution and speciation of both plants and animals. Up to 80% of angiosperms and 95% of pteridophytes have experienced one or more episodes of polyploidization in their evolutionary history (Leitch & Bennett 1997). Although less attention has been paid to the incidence of polyploidy in animals, genome duplication has occurred in hundreds of invertebrate and vertebrate species, including insects, fishes, amphibians, reptiles, birds, and at least two mammalian species (Mable 2004, Otto & Whitton 2000).

Polyploidy research has been focused on the evolutionary and ecological dynamics of the process of polyploidization, particularly on the origin, establishment and maintenance of polyploids in plant and animal populations (Mable 2004, Soltis et al. 2003, Thompson & Lumaret 1992). In addressing these issues, key areas of interest have included (*a*) ploidy variation on different spatial and temporal scales; (*b*) the roles of allo- and auto-polyploidization, unreduced gamete production, and single vs multiple origins in the formation of polyploids; and (*c*) phenotypic and ecological traits of neo- and established polyploids, and the role these play in the ecology and evolution of cytotypes, populations, and species.

3.2. Advances Enabled by Flow Cytometry

The most significant contribution of FCM to polyploidy research has been through the improved screening of natural populations for ploidy variation, using a variety of tissues and large sample sizes. Although this application is more common for plants than animals, it has been used in studies of fishes (Doeringsfeld et al. 2004), insects with diploid-haploid mating systems (Aron et al. 2004), reptiles (Bickham et al. 1993), and amphibians (Stöck et al. 2002). Ploidy determination using FCM is also known in phytoplankton (Houdan et al. 2004) and even bacteria (Tobiason & Seifert 2006). The capacity of the technique to quickly process large numbers of samples means that, in some cases, FCM itself is the critical factor in allowing particular research questions to be asked (e.g., Thompson et al. 2004). Larger sample sizes can provide broader geographical information, more intensive sampling of populations and progeny, and measures of temporal variation.

Another significant advance in the study of polyploidy is the development of methods to quantify unreduced gamete production at early (prefertilization) stages. The use of FCM for this purpose in natural populations is still uncommon, but has tremendous potential. FCM has also been used to examine the origin of polyploid taxa, when DNA content estimates have provided information about constituent genomes. Finally, researchers are now able to address questions related to the occurrence and function of endopolyploidy (variation in ploidy among somatic nuclei of an individual) that were not possible with other techniques (Barow & Meister 2003, Galbraith et al. 1991, Joubès & Chevalier 2000, Korpelainen et al. 1997). A particular application of FCM for studying reproduction in endopolyploid (mosaic) animals is discussed in Section 4.3.4.

3.3. Examples

3.3.1. Ploidy variation on different spatial and temporal scales. A goal in polyploidy research is to better characterize the distribution of cytotypes (within species or

between closely related species) at various spatial and temporal scales. In recent years, a growing body of data has become available concerning the geographical distribution of cytotypes. Examples from the plant literature include studies of polyploid evolution (e.g., Baack 2004, Burton & Husband 1999, Ohi et al. 2003), as well as taxonomy and ecology (e.g., Bureš et al. 2003, Pecinka et al. 2006). In the animal literature, most studies have dealt with taxonomy, hybridization, and evolution of amphibians (e.g., Borkin et al. 2001b, Cavallo et al. 2002, Stöck et al. 2002).

Studies using FCM have also focused on the population level, using intensive sampling to describe smaller-scale patterns of cytotype distribution not readily obtainable with more cumbersome methods. This sampling has demonstrated distribution patterns within mixed populations with implications for intercytotype interaction, in both plants (e.g., Baack 2004, Husband & Schemske 1998) and animals (Martins et al. 1998). Intensive population sampling may also uncover rare and previously unrecognized cytotypes (Section 2.3.2), including ploidy mosaics (Section 4.3.4) and aneuploids. Cytotype surveys frequently ignore the presence of near-euploid aneuploids, grouping them with euploids, but careful application of FCM holds promise for the incorporation of this information into population surveys as well (e.g., Oshima et al. 2005, Roux et al. 2003; see also the **Supplemental Appendix**). Another potentially powerful application that is still underused is the monitoring of cytotype frequency fluctuations over time (but see Keeler 2004, Lampert et al. 2005).

3.3.2. Origins of polyploidy.
Three problems concerning the origins of polyploidy (i.e., production of neopolyploids) have been addressed with the help of FCM: (*a*) the identification of diploid progenitors, (*b*) the prevalence of multiple versus single origins of taxa, and (*c*) the role of unreduced gametes in polyploidy formation. FCM has been used to compare genome sizes of polyploids to those of their putative diploid progenitors, a comparison that in some cases has helped to identify probable parental taxa (e.g., Bennert et al. 2005, Dawley & Goddard 1988, Lysák et al. 1999). In studies of this kind, genome size data are helpful in generating or supporting hypotheses about parental taxa and, consequently, about allo- vs auto-polyploid origins. Their use is limited, however, when related diploid species vary little in DNA content, or when extinct taxa may be involved. For this reason, other molecular techniques generally play the central role in such studies, as they do in studies concerning the history and number of polyploidization events within a taxon (Soltis & Soltis 2000b). In this latter case, FCM has played a supportive role, through the screening of multiple populations and individuals to identify polyploids and to describe cytotype distribution patterns (e.g., Lampert et al. 2005, Ohi et al. 2003).

Unreduced gamete production is considered to be the most important mechanism in the generation of neopolyploids in animals (Mable 2004) and plants (Ramsey & Schemske 2002), although the primary data used to support this conclusion have mostly come from karyological studies (e.g., plant data reviewed by Bretagnolle & Thompson 1995). More recently, FCM has been used to evaluate rates of unreduced gamete production by screening for ploidy among progeny from controlled crosses (e.g., Burton & Husband 2001). In this particular study, the results revealed production of monoploid, diploid, and triploid gametes, and a significant role for triploid hybrids in the formation of tetraploids (triploid bridge; Husband 2004). FCM has greatly facilitated progeny screening in plants (Bretagnolle & Thompson 1995), and has had some limited application in animal populations (Oshima et al. 2005).

More important, FCM has emerged as a means to directly measure unreduced microgamete production in plants and thus to disentangle the rates of gamete production from differential fitness among cytotypes in progeny (Suda et al. 2007). In animals, this application is less developed, although FCM has been used to study variation in sperm DNA content in polyploids and hybrids artificially generated for aquaculture (Li et al. 2003), as well as in studies of reproduction in gynogenetic species complexes (Section 4.3.4). Laboratory studies of the effect of experimental treatments on unreduced gamete production (Akutsu et al. 2007) suggest an important potential use in natural populations: the comparison of variation in unreduced gamete production to variation in environmental factors.

3.3.3. Establishment, persistence, and distribution of polyploids.
Ecological studies of polyploidy have addressed questions related to phenotypic and fitness differences between polyploids and their diploid progenitors, ecological correlates of ploidy distribution, and the relationship between polyploidy and invasiveness (Thompson & Lumaret 1992). There are plentiful data indicating that cytotypes differ phenotypically in both plants (reviewed by Ramsey & Schemske 2002) and animals (Le Comber & Smith 2004, Otto & Whitton 2000), and clearly such differences may have ecological implications. Various ecological correlates to polyploidy have been described (Soltis et al. 2003), and cytotype surveys using FCM frequently include some environmental data (e.g., Baack 2004, Borkin et al. 2001b). Intensive sampling using FCM therefore has the potential to improve data sets correlating polyploidy with ecological factors. The extensive studies of a plant-insect interaction by Thompson and colleagues, relating plant ploidy to trophic interactions, depended on the large sample sizes made possible only with FCM (Thompson et al. 2004). Multipopulation surveys in native and introduced ranges have also demonstrated associations between cytotype and introduced status, as well as helping to identify the origin of introduced organisms (Amsellem et al. 2001, Lafuma et al. 2003, Mandák et al. 2003).

Comparison of neopolyploids to established polyploids can provide insights into the establishment process by distinguishing between phenotypic traits produced directly by genome duplication and those resulting from subsequent selection or genomic restructuring. Laboratory and greenhouse studies have shown that novel morphological and physiological traits may be present in neopolyploid plants (reviewed by Ramsey & Schemske 2002) and animals (Sakao et al. 2006), but experiments in the field are still rare (J. Ramsey, unpublished). FCM has facilitated this research, because extensive screening is required to identify neopolyploids, as well as to monitor their development (e.g., reversion to diploid state). Similarly, studies of reproductive isolation between polyploids, neopolyploids, and their diploid progenitors have used FCM to screen large numbers of parents and progeny (Husband & Sabara 2003; also H.A. Sabara & B.C. Husband, unpublished).

4. REPRODUCTIVE PATHWAYS

4.1. Goals of Research on Reproductive Pathways

Multiple reproductive pathways exist, even within single species, involving sexual or asexual reproduction, homosporous or heterosporous systems, reduced or unreduced gametes, and hybridization between taxa. These complex breeding and mating systems affect the ways in which genes are organized within organisms and within and among populations. Ultimately, evolution proceeds through the differential transmission of genes across generations, and so a goal of evolutionary biology has always been to understand the dynamics and variety of reproductive systems. A number of areas of research interest have been influenced by the use of FCM: (*a*) the study of gender ratios; (*b*) the dynamics of hybrid systems, including hetero- and homoploid hybridization and the potential for gene introgression; and (*c*) asexuality, including its prevalence and its relationship to hybridization and polyploidy.

4.2. Advances Enabled by Flow Cytometry

Major contributions of FCM in the study of reproductive systems have been: (*a*) rapid gender determination, based on the discrimination of individuals with heteromorphic chromosomes or with gender-linked ploidy; (*b*) improved screening of organisms to characterize variation and to detect rare types, such as polyploids, hybrids, or clonal lineages; and (*c*) discrimination of reproductive pathways in plants, using flow cytometry seed screening (FCSS). A recurrent theme in the application of FCM to reproductive systems is the ability to distinguish between the products of different reproductive pathways or gamete types at early stages (e.g., production of unreduced or gender-specific gametes, or apomictic vs sexual seeds), as well as later stages (e.g., progeny screening).

4.3. Examples

4.3.1. Gender determination. Gender can be difficult and time-consuming to determine in nonreproductive individuals, requiring techniques such as karyology or the use of genetic markers, and this has limited the study of sex ratios in plants (Stehlik et al. 2007). In organisms with chromosomal sex determination, DNA content has been used to distinguish genders, although this requires sufficient differences in DNA content of sex chromosomes to be detectable. In plants, FCM has been used to distinguish genders of *Melandrium album* (syn. *Silene latifolia*) (Doležel & Göhde 1995, Meagher & Costich 1994). In animals, gender determination using FCM appears to be restricted largely to insects with haploid/diploid systems (discussed below), and to birds (e.g., Canon et al. 2000). Aside from DNA content, sex chromosomes may also differ in AT/GC ratios, providing another means by which FCM might distinguish genders (Siljak-Yakovlev et al. 1996).

In organisms with heteromorphic sex chromosomes, counting and sorting of male- and female-determining gamete types is possible, provided there is sufficient DNA

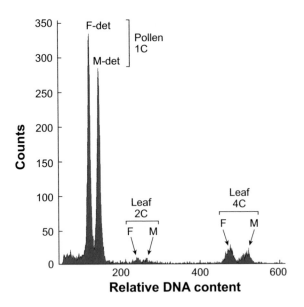

Figure 3

Integrated fluorescence histogram of nuclei from the dioecious plant *Rumex nivalis*. The
sample included 1C female- and male-determining pollen nuclei (F-det and M-det,
respectively), as well as 2C and 4C somatic nuclei from both female (F) and male (M) plants.
Female-determining pollen nuclei have 7 chromosomes, male-determining have 8, nuclei
from female somatic tissue have 2n = 14 and male somatic nuclei have 2n = 15. Reproduced
from figure 3c of Stehlik et al. 2007. *New Phytol.* 175:185–94.

content or AT/GC ratio difference between the two. In animals, the development
of techniques to count and sort sperm is directly attributed to FCM (Garner 2001).
Although this is now a well-developed technique in agriculture, applications in natural
populations are still being tested. Quantification of male- and female-determining
pollen in plants with XX/XYY systems has only recently been accomplished (Blocka-
Wandas et al. 2007, Stehlik et al. 2007) (**Figure 3**). These kinds of studies not only
provide information about determinants of sex ratios at the earliest stage (gamete
production), but also raise the possibility of larger-scale sampling to examine genetic
components, temporal change, and environmental correlates.

Butcher et al. (2000) used FCM to study a haplodiploid sex-determination system
in hymenopterans, in which males are typically haploid and females diploid, noting
that previous work on the evolution of this system had been hampered by the lack of a
simple and accurate way to distinguish ploidies. Aron et al. (2004) and Schrempf et al.
(2006) also used this method to screen eggs and larvae in haplodiploid hymenopterans,
noting the potential to use FCM to examine mortality-dependent changes in sex
ratios over time. Finally, in a conceptually similar application, discrimination between
unisexual and bisexual individuals may also be feasible in some plant species (e.g.,
Empetrum nigrum), owing to tight correlation between ploidy level and sex system
(Suda et al. 2004, Yeung et al. 2005).

4.3.2. Hybrid systems. Hybridization is important in evolutionary biology, systematics, and ecology, but its study may be limited when hybridization events are hard to detect, for example, when they are rare, when hybrids are phenotypically similar to parental taxa, or when available tissue is limited. FCM can provide data indicating hybrid states (e.g., intermediate DNA content and AT/GC ratios), and it can do so for large numbers of individuals in hybrid zones. In addition, it can be applied in some cases to early developmental stages, for example, in the seeds of *Malus* intercytotype hybrids (P. Kron & B.C. Husband, unpublished).

Hybrid detection using FCM has been employed to study invasiveness (Bleeker & Matthies 2005, Morgan-Richards et al. 2004), gene flow from cultivated to wild species (Herrera et al. 2002, Warwick et al. 2003), and the evolution of polyploidy (Petit et al. 1997), as well as in horticulture and aquaculture. It also has strong potential for future applications in conservation biology, with respect to the study of zones of overlap between rare species and closely related relatives. For example, Suda and coworkers (J. Suda, P. Vít & K. Seifertová, unpublished) were able to reliably discriminate among the endangered Czech serpentine endemic *Cerastium alsinifolium*, widespread *C. arvense*, and their homoploid hybrids in sympatric populations. In addition to plant groups, FCM also significantly contributed to the detection of complex hybridization patterns in animal systems, such as in sturgeons (Birstein et al. 1993, Ráb et al. 2004) and in unisexual species complexes (Section 4.3.4).

4.3.3. Asexual reproduction in plants. Understanding of facultative apomixis in plants was hampered until recently by limitations on methods for distinguishing between seeds produced sexually or through apomixis (Bicknell & Koltunow 2004). A major innovation allowing progress in this area has been the use of "flow cytometric seed screening" (FCSS) (Matzk et al. 2000). FCSS works by exploiting the nature of double fertilization, such that different modes of reproduction can be distinguished based on the relative ploidies of embryos and endosperm. The use of FCSS is extensively reviewed elsewhere (Matzk 2007). One recent example of its use is a study of 71 taxa within the genus *Hypericum* to show an association between polyploidy and apomixis in one section of the genus, and genome size change and apomixis in another (Matzk et al. 2003).

Although FCSS is a powerful new method for use in seed plants, it is not applicable in all organisms with sexual and asexual modes of reproduction. The (realized) relative frequencies of different modes of reproduction can also be estimated through the screening of large numbers of progeny from controlled crosses, if the ploidy of asexual offspring is known to be different from the ploidy of sexual crosses (e.g., Brutovská et al. 1998). Chapman et al. (2003) use this approach to demonstrate the unusual case of the origin of a 4x sexual lineage from a 5x asexual line in the plant genus *Hieracium*.

4.3.4. Gynogenetic and polyploid reproduction in vertebrates. In some vertebrate species complexes, a variety of clonal and hemiclonal reproductive modes exist that permit continuing reproduction when normal meiosis is disrupted (Dawley 1989). These reproductive modes are commonly associated with hybrid complexes (in which polyploidy may arise secondarily) and occasionally with autopolyploids. One example is gynogenesis, in which females produce unreduced eggs that require fertilization to initiate embryogenesis, but genetic material from the sperm is not normally incorporated. When it is incorporated, polyploid or mosaic (endopolyploid) progeny are produced, and populations may include a mix of cytotypes, including hybrid and nonhybrid individuals.

FCM has influenced the study of gynogenetic systems in two ways. First, it has been used to screen populations for cytotypes, either to generate a picture of their distribution (e.g., Doeringsfeld et al. 2004), or as a prelude to other experiments (e.g., Lampert et al. 2005). Distinctions have even been made between homopolyploids with different combinations of parental genomes, when these genomes vary in size (Goddard & Dawley 1990). Such population screening has increased the observations of ploidy mosaics, a group whose ecological and genetic roles in populations are not well understood, and whose prevalence and importance could be further revealed through increased use of FCM (Dawley & Goddard 1988, Tanaka et al. 2003).

Second, FCM has been used to identify and describe different reproductive pathways in hybrid and polyploid animals in an application conceptually similar to the detection of unreduced gametes in diploid plants (Section 3.3.2). In fish, FCM screening of parents and progeny has been used to verify gynogenesis (Goddard & Dawley 1990) and to infer gamete ploidy in autopolyploids (Oshima et al. 2005) and mosaics (Lamatsch et al. 2002). Also in fish, the direct measurement of DNA content in sperm and/or testicular tissue has been used to detect unreduced sperm production in a gynogenetic species complex (Alves et al. 1999) and sterility or reduced sperm production in autotriploids (Oshima et al. 2005, Zhang & Arai 1999). Bickham et al. (1993) measured DNA content of testicular tissue in autopolyploid and mosaic turtles and showed that this tissue is a ploidy mosaic, incorporating diploid tissue that produces normal haploid sperm. Finally, in an example from a hybridogenetic system, Vinogradov et al. (1991) detected production of two distinct sperm types in single individuals, corresponding to the two constituent genomes of the frog *Rana esculenta*.

5. POPULATION AND COMMUNITY DYNAMICS OF MICROORGANISMS

5.1. Goals of Research on Microorganism Populations and Communities

Microorganisms play a fundamental role at every level in global ecology (Gasol & del Giorgio 2000), but there are technical challenges to studying organisms that are remarkably diverse, highly variable in abundance, and difficult to observe. Studies of microbial populations are often motivated by the need to better understand organisms that directly affect human health, such as pathogens, parasites, and toxic plankton, and research goals range from developing better identification and monitoring tools to understanding the physiology of key organisms in various ecological contexts. Conversely, monitoring natural communities is increasingly a part of understanding the effects of human activity on the environment. Community-level studies have focused on expanding our understanding of critical components of the global ecology, such as aquatic food webs and soil communities.

5.2. Advances Enabled by Flow Cytometry

In many applications involving the detection, identification, and characterization of microorganisms, FCM in its current state cannot match the resolution provided by traditional techniques (e.g., microscopy, culturing), or more recently developed molecular techniques (e.g., sequencing). However, in population and community studies, traditional methods often provide only bulk measures (e.g., total chlorophyll content) or, when they do provide information about individual organisms, are limited by time and difficulty to low spatial or temporal coverage (Dubelaar et al. 2007). With FCM, whole populations and communities of microorganisms can be characterized in ways not possible with other techniques (Brussaard et al. 2000, Gasol & del Giorgio 2000, Gruden et al. 2004, Vives-Rego et al. 2000).

Advances in this area from the use of FCM include (*a*) rapid detection, classification, and simultaneous enumeration of a range of taxa, allowing for improved monitoring of populations of interest; (*b*) the associated ability to study whole communities at critical spatial and temporal scales, and therefore, to gain an increased understanding of population and community dynamics; and (*c*) improved understanding of the distribution and frequency of individuals in different sexual and asexual life stages, with demographic implications.

5.3. Examples

5.3.1. Detection, identification, and monitoring of populations. One of the most successful applications of FCM in natural populations has been in the study of microorganism food webs in aquatic environments (i.e., plankton communities) (Dubelaar et al. 2007, Gasol & del Giorgio 2000). This success is based on the relative ease with which some groups of organisms can be distinguished in mixed samples, including photosynthetic prokaryotes, picoeukaryotes, heterotrophic bacteria, and virioplankton (Marie et al. 1999a,b). Although marine viruses have only recently been subject to intensive study, detection of viruses as a group is now standard in marine surveys (e.g., Larsen et al. 2004, Marie et al. 1999a,b), and some species-level distinctions and DNA content measures have also been obtained using FCM (Brussaard et al. 2000). The use of FCM to distinguish components of plankton communities has sometimes led to the identification of new taxa, such as the prokaryote *Prochlorococcus*, the discovery of which changed the perception of how planktonic webs work (Chisholm et al. 1988).

Although aquatic systems dominate the application of FCM to natural populations, protocols are being developed for the study of bacterial populations in soils and other complex matrices (Gruden et al. 2004, Porter et al. 1997). In all environments and across all taxa, emerging protocols using species-specific fluorescent labeling, multiparameter measurements, and multivariate statistics, in combination with various molecular techniques, allow for the study of populations at increasingly fine scales (Dubelaar et al. 2007, Gruden et al. 2004, Larsen et al. 2004, Porter et al. 1997, Vives-Rego et al. 2000).

5.3.2. Population and community dynamics of plankton. The use of FCM has revised our understanding of marine food webs by facilitating the study of separate components of microorganism communities over "critical scales" (Dubelaar et al. 2007). For example, the discovery in recent years of new picoeukaryotes has focussed attention on their importance in food webs (Countway & Caron 2006, Guillou et al. 1999, Li 1994). Li (1994) used FCM to examine the partitioning of primary production by this group and other components of a plankton community, noting that the use of FCM advances the goal of understanding "bulk properties" of phytoplankton from "properties of the constituents." Similarly, autotrophic bacteria dominate primary production in some aquatic systems (Gasol & del Giorgio 2000), and major groups of these (*Prochlorococcus* and *Synechococcus*) can be readily distinguished with FCM; this has allowed for "advances in the knowledge of oceanic distributions" (Marie et al. 1997), across temporal, spatial, and resource gradients (e.g., Dandonneau et al. 2006, Veldhuis & Kraay 1993).

Flow cytometric examination of DNA-stained heterotrophic bacteria allows much faster and easier distinction and enumeration of types relative to microscopy, as well as providing cell cycle information (Marie et al. 1997). The ability to monitor changes in bacterial populations over time has provided insights into various ecological processes, including the study of carbon flux and predation in bacteria (Gasol & del Giorgio 2000, Vives-Rego et al. 2000). Grazing and predation among groups within plankton communities affect phytoplankton dynamics, and these processes may be quantified using FCM (Christaki et al. 1999, Vives-Rego et al. 2000). The use of FCM to measure physiological traits in plankton has raised awareness of variation in such traits, and has improved understanding of growth rates and reproductive cycles (Dubelaar et al. 2007, Vives-Rego et al. 2000).

The in situ study of marine viruses has taken place only in the past 15 years, aided by the advent of FCM, and the importance of these viruses in regulating bacterial populations is only now being appreciated (Brussaard 2004, Fuhrman 1999, Larsen et al. 2004). Marie et al. (1999a) and Larsen et al. (2004) detected distinct viral populations in seawater and, by relating their abundance to that of other taxa, generated hypotheses concerning infection patterns. Brussaard (2004) described FCM as one of the few techniques to provide data on infection rates of plankton in natural settings.

5.3.3. Life phase and demography. The difficulties inherent in studying microscopic organisms are further complicated by the presence of dormant resting stages that may be difficult to identify and count. The relative abundance of these life stages is a key part of microorganism demography and, as such, affects a variety of processes (e.g., phytoplankton blooms; Figueroa et al. 2006, Kremp & Parrow 2006). The use of FCM has improved the detection, identification, and enumeration of resting stage organisms. Under laboratory conditions, various spores and cysts have been identified using differential surface staining (Ochiai et al. 2006), antibody labeling (Ferrari et al. 2000), and autofluorescence measurement (Dreyer et al. 2006), all of which may ultimately be applied in natural populations. Beyond detection and identification, DNA content information obtained using FCM can clarify demographic pathways in populations. In dinoflagellates, Kremp & Parrow (2006) reported the first evidence for an asexual resting stage, and Figueroa et al. (2006) used FCM in a study of the relative importance of different sexual pathways.

6. EXPERIMENTAL EVOLUTION IN CULTURE

A recent application of FCM is in the study of evolutionary processes in bacterial and yeast cultures. Over multiple generations (hundreds to thousands), specific strains of microorganisms can be distinguished and monitored using FCM, on the basis of either T5 viral resistance (Lunzer et al. 2002, 2005; Zhu et al. 2005), DNA content (Gerstein et al. 2006), or fluorescent labeling (Desai et al. 2007, Thompson et al. 2006). Strains may be competed against one another (Desai et al. 2007, Lunzer et al. 2005, Thompson et al. 2006, Zhu et al. 2005) or monitored for some trait (e.g., genome size) over time (Gerstein et al. 2006). Studies have examined frequency-dependent selection (Lunzer et al. 2002), adaptive significance of a coenzyme type (Zhu et al. 2005), epistasis and adaptive landscapes (Lunzer et al. 2005), changes in genome size and ploidy over time (Gerstein et al. 2006), and mutation accumulation and evolution rates in asexual organisms (Desai et al. 2007, Thompson et al. 2006). Future work of this kind will undoubtedly advance our knowledge of a variety of evolutionary processes.

CONCLUSIONS AND FUTURE DIRECTIONS

Flow cytometry has had a profound influence on population and evolutionary biology. It combines the ability to collect qualitative information about biological particles with the capacity to quantify these measures for large numbers of particles. It expands the range of attributes that can be quantified beyond typical particle counters, and the rate of sampling beyond traditional microscope-based studies. It can be applied nondestructively to rare species, including endemics with extremely narrow ranges of distribution (e.g., Ramsden et al. 2006, Sgorbati et al. 2004). As a result, FCM is widening the diversity of attributes that can be explored on a population scale and the taxonomic, spatial, and temporal scope of these investigations. These advantages have led to insights regarding cryptic species and genome size variation, and are forging new research programs in microbial diversity, polyploid evolution, and complex reproductive systems.

Still, use of FCM to date has been limited relative to its full capabilities. Future applications will likely involve technical refinement of existing applications including protocol improvement to reduce stoichiometric error, advances in tissue preservation, expansion of staining and labeling approaches to enhance distinctions between particles, application of multiparametric analytical approaches, and less costly, more portable machines. As flow cytometers become more common in lab settings, current trends in population-level studies will continue, with expanded genome size and ploidy databases, including surveys of complete floras (e.g., Suda et al. 2005), multiparameter population surveys that incorporate changes in cytotype frequencies across spatial and temporal gradients, and dynamic surveys of microorganisms in aquatic and soil communities, with an ever-expanding capacity to distinguish between components of these communities. In the future, the reach of FCM will only increase as the range of attributes that can be measured is expanded using cell sorting and fluorescent probes specific to a variety of cellular attributes. This information alone, but perhaps more powerfully combined with other genetic markers, will open new avenues of investigation in evolutionary and population biology.

DISCLOSURE STATEMENT

The authors are not aware of any biases that might be perceived as affecting the objectivity of this review.

ACKNOWLEDGMENTS

We thank D. Schemske for helpful suggestions on the manuscript, P. Travnicek for assistance with supplemental material, the Czech Science Foundation, Ministry of Education, Youth, and Sports of the Czech Republic, and Academy of Sciences of the Czech Republic for support to J.S., and the Natural Science and Engineering Research Council of Canada, Canadian Foundation for Innovation, Ontario Innovation Trust, and Canada Research Chair Program for support to B.C.H.

LITERATURE CITED

Akutsu M, Kitamura S, Toda R, Miyajima I, Okazaki K. 2007. Production of $2n$ pollen of Asiatic hybrid lilies by nitrous oxide treatment. *Euphytica* 155:143–52

Alves MJ, Coelho MM, Próspero MI, Collares-Pereira MJ. 1999. Production of fertile unreduced sperm by hybrid males of the *Rutilus alburnoides* complex (Teleostei, Cyprinidae): an alternative route to genome tetraploidization in unisexuals. *Genetics* 151:277–83

Amsellem L, Chevallier M-H, Hossaert-McKey M. 2001. Ploidy level of the invasive weed *Rubus alceifolius* (Rosaceae) in its native range and in areas of introduction. *Plant Syst. Evol.* 228:171–79

Aron S, Passera L, Keller L. 2004. Evolution of miniaturisation in inquiline parasitic ants: timing of male elimination in *Plagiolepis pygmaea*, the host of *Plagiolepis xene*. *Insect. Soc.* 51:395–99

Baack EJ. 2004. Cytotype segregation on regional and microgeographic scales in snow buttercups (*Ranunculus adoneus*: Ranunculaceae). *Am. J. Bot.* 91:1783–88

Barow M, Meister A. 2003. Endopolyploidy in seed plants is differently correlated to systematics, organ, life strategy and genome size. *Plant Cell Environ.* 26:571–84

Bennert W, Lubienski M, Körner S, Steinberg M. 2005. Triplody in *Equisetum* subgenus *Hippochaete* (Equisetaceae, Pteridophyta). *Ann. Bot.* 95:807–15

Bennett MD, Leitch IJ. 2005a. Nuclear DNA amounts in angiosperms: progress, problems and prospects. *Ann. Bot.* 95:45–90

Bennett MD, Leitch IJ. 2005b. Genome size evolution in plants. In *The Evolution of the Genome*, ed. TR Gregory, 2:89–162. New York: Elsevier Acad.

Bennett MD, Leitch IJ. 2005c. *Plant DNA C-values Database* (release 4.0, Oct. 2005). **http://www.kew.org/cval/homepage.html**

Bickham JW, Hanks BG, Hale DW, Martin JE. 1993. Ploidy diversity and the production of balanced gametes in male twist-necked turtles (*Platemys platycephala*). *Copeia* 3:723–27

Bicknell RA, Koltunow AM. 2004. Understanding apomixis: recent advances and remaining conundrums. *Plant Cell* 16:S228–45

Birstein VJ, Poletaev AI, Goncharov BF. 1993. DNA content in Eurasian sturgeon species determined by flow cytometry. *Cytometry* 14:377–83

Bleeker W, Matthies A. 2005. Hybrid zones between invasive *Rorippa austriaca* and native *R. sylvestris* (Brassicaceae) in Germany: ploidy levels and patterns of fitness in the field. *Heredity* 94:664–70

Blocka-Wandas M, Sliwinska E, Grabowska-Joachimiak A, Musial K, Joachimiak AJ. 2007. Male gametophyte development and two different DNA classes of pollen grains in *Rumex acetosa* L., a plant with an XX/XY_1Y_2 sex chromosome system and a female-biased sex ratio. *Sex. Plant Reprod.* In press

Borkin LJ, Eremchenko VK, Helfenberger N, Panfilov AM, Rosanov JM. 2001a. On the distribution of diploid, triploid, and tetraploid green toads (*Bufo viridis* complex) in south-eastern Kazakhstan. *Russ. J. Herpetol.* 8:45–53

Borkin LJ, Litvinchuk SN, Rosanov JM, Milto KD. 2001b. Cryptic speciation in *Pelobates fuscus* (Anura, Pelobatidae): evidence from DNA flow cytometry. *Amphibia-Reptilia* 22:387–96

Boulesteix M, Weiss M, Biémont C. 2006. Differences in genome size between closely related species: the *Drosophila melanogaster* species subgroup. *Mol. Biol. Evol.* 23:162–67

Bretagnolle F, Thompson JD. 1995. Tansley Rev. No. 78. Gametes with the somatic chromosome number: mechanisms of their formation and role in the evolution of autopolypoid plants. *New Phytol.* 129:1–22

Brussaard CPD. 2004. Viral control of phytoplankton populations—a review. *J. Eukaryot. Microbiol.* 51:125–38

Brussaard CPD, Marie D, Bratbak G. 2000. Flow cytometric detection of viruses. *J. Virol. Methods* 85:175–82

Brutovská R, Čellárová E, Doležel J. 1998. Cytogenetic variability of in vitro regenerated *Hypericum perforatum* L. plants and their seed progenies. *Plant Sci.* 133:221–29

Bureš P, Tichý L, Wang YF, Bartoš J. 2003. Occurrence of *Polypodium xmantoniae* and new localities for *P. interjectum* in the Czech Republic confirmed using flow cytometry. *Preslia* 75:293–310

Burton TL, Husband BC. 1999. Population cytotype structure in the polyploid *Galax urceolata* (Diapensiaceae). *Heredity* 82:381–90

Burton TL, Husband BC. 2001. Fecundity and offspring ploidy in matings among diploid, triploid and tetraploid *Chamerion angustifolium* (Onagraceae): consequences for tetraploid establishment. *Heredity* 87:573–82

Butcher RDJ, Whitfield WGF, Hubbard SF. 2000. Complementary sex determination in the genus *Diadegma* (Hymenoptera: Ichneumonidae). *J. Evol. Biol.* 13:593–606

Canon NR, Tell LA, Needham ML, Gardner IA. 2000. Flow cytometric analysis of nuclear DNA for sex identification in three psittacine species. *Am. J. Vet. Res.* 61:847–50

Cavalier-Smith T. 1985. *The Evolution of Genome Size.* Chichester: Wiley

Cavallo D, De Vita R, Eleuteri P, Borkin L, Eremchenko V, et al. 2002. Karyological and flow cytometric evidence of triploid specimens in *Bufo viridis* (Amphibia Anura). *Eur. J. Histochem.* 46:159–64

Chapman H, Houliston GJ, Robson B, Iline I. 2003. A case of reversal: the evolution and maintenance of sexuals from parthenogenetic clones in *Hieracium pilosella*. *Int. J. Plant Sci.* 164:719–28

Chisholm SW, Olson RJ, Zettler ER, Goericke R, Waterbury JB, Welschmeyer NA. 1988. A novel free-living prochlorophyte abundant in an ocean euphotic zone. *Nature* 334:340–43

Christaki U, Jacquet S, Dolan JR, Vaulot D, Rassoulzadegan F. 1999. Growth and grazing on *Prochlorococcus* and *Synechococcus* by two marine ciliates. *Limnol. Oceanogr.* 44:52–61

Countway PD, Caron DA. 2006. Abundance and distribution of *Ostreococcus* sp. in the San Pedro Channel, California, as revealed by quantitative PCR. *Appl. Environ. Microbiol.* 72:2496–506

Dandonneau Y, Montel Y, Blanchot J, Giraudeau J, Neveux J. 2006. Temporal variability in phytoplankton pigments, picoplankton and coccolithophores along a transect through the North Atlantic and tropical southwestern Pacific. *Deep-Sea Res. Part 1. Oceanogr. Res. Pap.* 53:689–712

Dawley RM. 1989. An introduction to unisexual vertebrates. See Dawley & Bogart 1989, pp. 1–18

Dawley RM, Bogart JP, eds. 1989. *Evolution and Ecology of Unisexual Vertebrates.* Albany: NY State Mus.

Dawley RM, Goddard KA. 1988. Diploid-triploid mosaics among unisexual hybrids of the minnows *Phoxinus eos* and *Phoxinus neogaeus*. *Evolution* 42:649–59

Dawley RM, Rupprecht JD, Schultz RJ. 1997. Genome size of bisexual and unisexual *Poeciliopsis*. *J. Hered.* 88:249–52

Desai MM, Fisher DS, Murray AW. 2007. The speed of evolution and maintenance of variation in asexual populations. *Curr. Biol.* 17:385–94

Dimitrova D, Ebert I, Greilhuber J, Kozhuharov S. 1999. Karyotype constancy and genome size variation in Bulgarian *Crepis foetida* s. l. (Asteraceae). *Plant Syst. Evol.* 217:245–57

Doeringsfeld MR, Schlosser IJ, Elder JF, Evenson DP. 2004. Phenotypic consequences of genetic variation in a gynogenetic complex of *Phoxinus eos-neogaeus* clonal fish (Pisces: Cyprinidae) inhabiting a heterogeneous environment. *Evolution* 58:1261–73

Doležel J, Göhde W. 1995. Sex determination in dioecious plants *Melandrium album* and *M. rubrum* using high resolution flow cytometry. *Cytometry* 19:103–6

Doležel J, Greilhuber J, Suda J, eds. 2007. *Flow Cytometry with Plant Cells: Analysis of Genes, Chromosomes and Genomes*. Weinheim: Wiley-VCH

Doolittle WF, Sapienza C. 1980. Selfish genes, the phenotype paradigm and genome evolution. *Nature* 284:601–3

Dreyer B, Morte A, Pérez-Gilabert M, Honrubia M. 2006. Autofluorescence detection of arbuscular mycorrhizal fungal structures in palm roots: an underestimated experimental method. *Mycol. Res.* 110:887–97

Dubelaar GBJ, Casotti R, Tarran GA, Biegala IC. 2007. Phytoplankton and their analysis by flow cytometry. See Doležel et al. 2007, 13:287–322

Eliam T, Bushnell WR, Anikster Y. 1994. Relative nuclear DNA content of rust fungi estimated by flow cytometry of propidium iodide stained pycniospores. *Phytopathology* 84:728–35

Ferrari BC, Vesey G, Davis KA, Gauci M, Veal D. 2000. A novel two-color flow cytometric assay for the detection of *Cryptosporidium* in environmental water samples. *Cytometry* 41:216–22

Figueroa RI, Bravo I, Garcés E. 2006. Multiple routes of sexuality in *Alexandrium taylori* (Dinophyceae) in culture. *J. Phycol.* 42:1028–39

Fuhrman JA. 1999. Marine viruses and their biogeochemical and ecological effects. *Nature* 399:541–48

Galbraith DW, Harkins KR, Knapp S. 1991. Systemic endopolyploidy in *Arabidopsis thaliana*. *Plant Physiol.* 96:985–89

Garner DL. 2001. Sex-sorting mammalian sperm: concept to application in animals. *J. Androl.* 22:519–26

Gasol JM, del Giorgio PA. 2000. Using flow cytometry for counting natural planktonic bacteria and understanding the structure of planktonic bacterial communities. *Sci. Mar.* 64:197–224

Gerstein AC, Chun H-JE, Grant A, Otto SP. 2006. Genomic convergence toward diploidy in *Saccharomyces cerevisiae*. *PloS Genet.* 2:1396–401

Goddard KA, Dawley RM. 1990. Clonal inheritance of a diploid nuclear genome by a hybrid freshwater minnow (*Phoxinus eos-neogaeus*, Pisces: Cyprinidae). *Evolution* 44:1052–65

Gregory TR. 2001. Coincidence, coevolution, or causation? DNA content, cell size, and the C-value enigma. *Biol. Rev. Cambridge Philos. Soc.* 76:65–101

Gregory TR. 2005a. The C-value enigma in plants and animals: a review of parallels and an appeal for partnership. *Ann. Bot.* 95:133–46

Gregory TR. 2005b. Synergy between sequence and size in large-scale genomics. *Nat. Rev. Genet.* 6:699–708

Gregory TR. 2005c. Genome size evolution in animals. In *The Evolution of the Genome*, ed. TR Gregory, 1:3–87. New York: Elsevier Acad.

Gregory TR. 2007. *Animal Genome Size Database*. **http://www.genomesize.com**

Gregory TR, Nicol JA, Tamm H, Kullman B, Kullman K, et al. 2007. Eukaryotic genome size databases. *Nucleic Acids Res.* 35:D332–38

Greilhuber J. 1998. Intraspecific variation in genome size: a critical reassessment. *Ann. Bot.* 82(Suppl. A):27–35

Greilhuber J. 2005. Intraspecific variation in genome size in angiosperms: identifying its existence. *Ann. Bot.* 95:91–98

Greilhuber J, Doležel J, Lysák MA, Bennett MD. 2005. The origin, evolution and proposed stabilization of the terms 'genome size' and 'C-value' to describe nuclear DNA contents. *Ann. Bot.* 95:255–60

Greilhuber J, Temsch EM, Loureiro JCM. 2007. Nuclear DNA content measurement. See Doležel et al. 2007, 4:67–101

Gruden C, Skerlos S, Adriaens P. 2004. Flow cytometry for microbial sensing in environmental sustainability applications: current status and future prospects. *FEMS Microbiol. Ecol.* 49:37–49

Guillou L, Chrétiennot-Dinet M-J, Medlin LK, Claustre H, Loiseaux-de Goër S, et al. 1999. *Bolidomonas*: a new genus with two species belonging to a new algal class, the Bolidophyceae (Heterokonta). *J. Phycol.* 35:368–81

Herrera JC, Combes MC, Cortina H, Alvarado G, Lashermes P. 2002. Gene introgression into *Coffea arabica* by way of triploid hybrids (*C. arabica* x *C. canephora*). *Heredity* 89:488–94

Houdan A, Billard C, Marie D, Not F, Sáez AG, et al. 2004. Holococcolithophore-heterococcolithophore (Haptophyta) life cycles: flow cytometric analysis of relative ploidy levels. *Syst. Biodivers.* 1:453–65

Husband BC. 2004. The role of triploid hybrids in the evolutionary dynamics of mixed-ploidy populations. *Biol. J. Linn. Soc.* 82:537–46

Husband BC, Sabara HA. 2003. Reproductive isolation between autotetraploids and their diploid progenitors in fireweed, *Chamerion angustifolium* (Onagraceae). *New Phytol.* 161:703–13

Husband BC, Schemske DW. 1998. Cytotype distribution at a diploid-tetraploid contact zone in *Chamerion (Epilobium) angustifolium* (Onagraceae). *Am. J. Bot.* 85:1688–94

Iguchi K, Yamamoto G, Matsubara N, Nishida M. 2003. Morphological and genetic analysis of fish of a *Carassius* complex (Cyprinidae) in Lake Kasumigaura with reference to the taxonomic status of two all-female triploid morphs. *Biol. J. Linn. Soc.* 79:351–57

Jakob SS, Meister A, Blattner FR. 2004. The considerable genome size variation of *Hordeum* species (Poaceae) is linked to phylogeny, life form, ecology and speciation rates. *Mol. Biol. Evol.* 21:860–69

Jeschke MR, Tranel PJ, Rayburn AL. 2003. DNA content analysis of smooth pigweed (*Amaranthus hybridus*) and tall waterhemp (*A. tuberculatus*): implications for hybrid detection. *Weed Sci.* 51:1–3

Johnston JS, Pepper AE, Hall AE, Chen ZJ, Hodnett G, et al. 2005. Evolution of genome size in Brassicaceae. *Ann. Bot.* 95:229–35

Joubès J, Chevalier C. 2000. Endoreduplication in higher plants. *Plant Mol. Biol.* 43:735–45

Keeler KH. 2004. Impact of intraspecific polyploidy in *Andropogon gerardii* (Poaceae) populations. *Am. Midl. Nat.* 152:63–74

Korpelainen H, Ketola M, Hietala J. 1997. Somatic polyploidy examined by flow cytometry in *Daphnia*. *J. Plankton Res.* 19:2031–40

Kremp A, Parrow MW. 2006. Evidence for asexual resting cysts in the life cycle of the marine peridinoid dinoflagellate, *Scrippsiella hangoei*. *J. Phycol.* 42:400–9

Krupnick GA, Kress WJ, eds. 2005. *Plant Conservation: A Natural History Approach.* Chicago: Univ. Chicago Press

Lafuma L, Balkwill K, Imbert E, Verlaque R, Maurice S. 2003. Ploidy level and origin of the European invasive weed *Senecio inaequidens* (Asteraceae). *Plant Syst. Evol.* 243:59–72

LaJeunesse TC, Lambert G, Andersen RA, Coffroth MA, Galbraith DW. 2005. *Symbiodinium* (Pyrrhophyta) genome sizes (DNA content) are smallest among dinoflagellates. *J. Phycol.* 41:880–86

Lamatsch DK, Schmid M, Schartl M. 2002. A somatic mosaic of the gynogenetic Amazon molly. *J. Fish Biol.* 60:1417–22

Lampert KP, Lamatsch DK, Epplen JT, Schartl M. 2005. Evidence for a monophyletic origin of triploid clones of the Amazon molly, *Poecilia formosa. Evolution* 59:881–89

Larsen A, Flaten GAF, Sandaa R-A, Castberg T, Thyrhaug R, et al. 2004. Spring phytoplankton bloom dynamics in Norwegian coastal waters: microbial community succession and diversity. *Limnol. Oceanogr.* 49:180–90

Le Comber SC, Smith C. 2004. Polyploidy in fishes: patterns and processes. *Biol. J. Linn. Soc.* 82:431–42

Leitch IL, Bennett MD. 1997. Polyploidy in angiosperms. *Trends Plant Sci.* 2:470–76

Leitch IJ, Bennett MD. 2004. Genome downsizing in polyploid plants. *Biol. J. Linn. Soc.* 82:651–63

Leitch IJ, Bennett MD. 2007. Genome size and its uses: the impact of flow cytometry. See Doležel et al. 2007, 7:153–76

Leitch IJ, Chase MW, Bennett MD. 1998. Phylogenetic analysis of DNA C-values provides evidence for a small ancestral genome size in flowering plants. *Ann. Bot.* 82(Suppl. A):85–94

Li FH, Xiang JH, Zhang XJ, Zhou LH, Zhang CS, Wu CG. 2003. Gonad development characteristics and sex ratio in triploid Chinese shrimp (*Fenneropenaeus chinensis*). *Mar. Biotechnol.* 5:528–35

Li WKW. 1994. Primary production of prochlorophytes, cyanobacteria, and eucaryotic ultraphytoplankton: measurements from flow cytometric sorting. *Limnol. Oceanogr.* 39:169–75

Litvinchuk SN, Zuiderwijk A, Borkin LJ, Rosanov JM. 2005. Taxonomic status of *Triturus vittatus* (Amphibia: Salamandridae) in western Turkey: trunk vertebrae count, genome size and allozyme data. *Amphibia-Reptilia* 26:305–23

Loureiro J, Rodriguez E, Doležel J, Santos C. 2006. Flow cytometric and microscopic analysis of the effect of tannic acid on plant nuclei and estimation of DNA content. *Ann. Bot.* 98:515–27

Lunzer M, Miller SP, Felsheim R, Dean AM. 2005. The biochemical architecture of an ancient adaptive landscape. *Science* 310:499–501

Lunzer M, Natarajan A, Dykhuizen DE, Dean AM. 2002. Enzyme kinetics, substitutable resources and competition: from biochemistry to frequency-dependent selection in *lac. Genetics* 162:485–99

Lysák MA, Doleželová M, Horry JP, Swennen R, Doležel J. 1999. Flow cytometric analysis of nuclear DNA content in *Musa. Theor. Appl. Genet.* 98:1344–50

Mable BK. 2004. 'Why polyploidy is rarer in animals than in plants': myths and mechanisms. *Biol. J. Linn. Soc.* 82:453–66

MacCulloch RD, Upton DE, Murphy RW. 1996. Trends in nuclear DNA content among amphibians and reptiles. *Comp. Biochem. Physiol. B* 113:601–5

Mahelka V, Suda J, Jarolímová V, Trávníček P, Krahulec F. 2005. Genome size discriminates between closely related taxa *Elytrigia repens* and *E. intermedia* (Poaceae: Triticeae) and their hybrid. *Folia Geobot.* 40:367–84

Mandák B, Pyšek P, Lysák M, Suda J, Krahulcová A, Bímová K. 2003. Variation in DNA-ploidy levels of *Reynoutria* taxa in the Czech Republic. *Ann. Bot.* 92:265–72

Marie D, Brussaard CPD, Thyrhaug R, Bratbak G, Vaulot D. 1999a. Enumeration of marine viruses in culture and natural samples by flow cytometry. *Appl. Environ. Microbiol.* 65:45–52

Marie D, Partensky F, Jacquet S, Vaulot D. 1997. Enumeration and cell cycle analysis of natural populations of marine picoplankton by flow cytometry using the nucleic acid stain SYBR Green I. *Appl. Environ. Microbiol.* 63:186–93

Marie D, Partensky F, Vaulot D, Brussaard C. 1999b. Enumeration of phytoplankton, bacteria, and viruses in marine samples. In *Current Protocols in Cytometry*, ed. JP Robinson, Z Darzynkiewicz, PN Dean, A Orfao, P Rabinovitch, et al., pp. 11.11.1–11.11.15. New York: Wiley

Martins MJ, Collares-Pereira MJ, Cowx IG, Coelho MM. 1998. Diploids v. triploids of *Rutilus alburnoides*: spatial segregation and morphological differences. *J. Fish Biol.* 52:817–28

Matzk F. 2007. Reproduction mode screening. See Doležel et al. 2007, 6:131–52

Matzk F, Hammer K, Schubert I. 2003. Coevolution of apomixis and genome size within the genus *Hypericum*. *Sex. Plant Reprod.* 16:51–58

Matzk F, Meister A, Schubert I. 2000. An efficient screen for reproductive pathways using mature seeds of monocots and dicots. *Plant J.* 21:97–108

Meagher TR, Costich DE. 1994. Sexual dimorphism in nuclear DNA content and floral morphology in populations of *Silene latifolia* (Caryophyllaceae). *Am. J. Bot.* 81:1198–204

Melosik I, Odrzykoski II, Sliwinska E. 2005. Delimitation of taxa of *Sphagnum subsecundum* s.l. (Musci, Sphagnaceae) based on multienzyme phenotype and cytological characters. *Nova Hedwigia* 80:397–412

Mishiba KI, Ando T, Mii M, Watanabe H, Kokubun H, et al. 2000. Nuclear DNA content as an index character discriminating taxa in the genus *Petunia sensu* Jussieu (Solanaceae). *Ann. Bot.* 85:665–73

Morgan-Richards M, Trewick SA, Chapman HM, Krahulcová A. 2004. Interspecific hybridization among *Hieracium* species in New Zealand: evidence from flow cytometry. *Heredity* 93:34–42

Noirot M, Barre P, Duperray C, Louarn J, Hamon S. 2003. Effects of caffeine and chlorogenic acid on propidium iodide accessibility to DNA: consequences on genome size evaluation in coffee tree. *Ann. Bot.* 92:259–64

Ochiai N, Yarbrough LD, Parke JL. 2006. Two methods for distinguishing zoospores and cysts of *Phytophthora ramorum*. *Phytopathology* 96:S169–70

Ogielska M, Kierzkowski P, Rybacki M. 2004. DNA content and genome composition of diploid and triploid water frogs belonging to the *Rana esculenta* complex (Amphibia, Anura). *Can. J. Zool.* 82:1894–901

Ohi T, Kajita T, Murata J. 2003. Distinct geographic structure as evidenced by chloroplast DNA haplotypes and ploidy level in Japanese *Aucuba* (Aucubaceae). *Am. J. Bot.* 90:1645–52

Ohno S. 1972. So much 'junk' in our genomes. In *Evolution of Genetic Systems*, *Brookhaven Symp. Biol.*, ed. HH Smith, pp. 366–70. New York: Gordon & Breach

Oshima K, Morishima K, Yamaha E, Arai K. 2005. Reproductive capacity of triploid loaches obtained from Hokkaido Island, Japan. *Ichthyol. Res.* 52:1–8

Otto SP, Whitton J. 2000. Polyploid incidence and evolution. *Annu. Rev. Genet.* 34:401–37

Pecinka A, Suchánková P, Lysák MA, Trávníček B, Doležel J. 2006. Nuclear DNA content variation among central European *Koeleria* taxa. *Ann. Bot.* 98:117–22

Pennington RT, Richardson JE, Lavin M. 2006. Insights into the historical construction of species-rich biomes from dated plant phylogenies, neutral ecological theory and phylogenetic community structure. *New Phytol.* 172:605–16

Petit C, Lesbros P, Ge X, Thompson JD. 1997. Variation in flowering phenology and selfing rate across a contact zone between diploid and tetraploid *Arrhenatherum elatius* (Poaceae). *Heredity* 79:31–40

Porter J, Deere D, Hardman M, Edwards C, Pickup R. 1997. Go with the flow—use of flow cytometry in environmental microbiology. *FEMS Microbiol. Ecol.* 24:93–101

Price HJ, Dillon SL, Hodnett G, Rooney WL, Ross L, Johnston JS. 2005. Genome evolution in the genus *Sorghum* (Poaceae). *Ann. Bot.* 95:219–27

Price HJ, Hodnett G, Johnston JS. 2000. Sunflower (*Helianthus annuus*) leaves contain compounds that reduce nuclear propidium iodide fluorescence. *Ann. Bot.* 86:929–34

Ráb P, Flajšhans M, Ludwig A, Lieckfeldt D, Ene C, et al. 2004. The second highest chromosome count among vertebrates is associated with extreme ploidy diversity in hybrid sturgeons. *Cytogen. Genome Res.* 106:24

Ramírez-Morillo IM, Brown GK. 2001. The origin of the low chromosome number in *Cryptanthus* (Bromeliaceae). *Syst. Bot.* 26:722–26

Ramsden C, Bériault K, Bogart JP. 2006. A nonlethal method of identification of *Ambystoma laterale*, *A. jeffersonianum* and sympatric unisexuals. *Mol. Ecol. Notes* 6:261–64

Ramsey J, Schemske DW. 2002. Neopolyploidy in flowering plants. *Annu. Rev. Ecol. Syst.* 33:589–639

Robinson JP, Grégori G. 2007. Principles of flow cytometry. See Doležel et al. 2007, 2:19–40

Rosenbaumová R, Plačková I, Suda J. 2004. Variation in *Lamium* subg. *Galeobdolon* (Lamiaceae)—insights from ploidy levels, morphology and isozymes. *Plant Syst. Evol.* 244:219–44

Roux N, Toloza A, Radecki Z, Zapata-Arias FJ, Doležel J. 2003. Rapid detection of aneuploidy in *Musa* using flow cytometry. *Plant Cell Rep.* 21:483–90

Sakao S, Fujimoto T, Kimura S, Yamaha E, Arai K. 2006. Drastic mortality in tetraploid induction results from the elevation of ploidy in masu salmon *Oncorhynchus masou*. *Aquaculture* 252:147–60

Schönswetter P, Suda J, Popp M, Weiss-Schneeweiss H, Brochmann C. 2007. Circumpolar phylogeography of *Juncus biglumis* (Juncaceae) inferred from AFLP fingerprints, cpDNA sequences, nuclear DNA content and chromosome numbers. *Mol. Phylogenet. Evol.* 42:92–103

Schrempf A, Aron S, Heinze J. 2006. Sex determination and inbreeding depression in an ant with regular sib-mating. *Heredity* 97:75–80

Sgorbati S, Labra M, Grugni E, Barcaccia G, Galasso G, et al. 2004. A survey of genetic diversity and reproductive biology of *Puya raimondii* (Bromeliaceae), the endangered Queen of the Andes. *Plant Biol.* 6:222–30

Siljak-Yakovlev S, Benmalek S, Cerbah M, De la Peña TC, Bounaga N, et al. 1996. Chromosomal sex determination and heterochromatin structure in date palm. *Sex. Plant Reprod.* 9:127–32

Soltis DE, Soltis PS. 2000a. Contributions of plant molecular systematics to studies of molecular evolution. *Plant Mol. Biol.* 42:45–75

Soltis DE, Soltis PS, Tate JA. 2003. Advances in the study of polyploidy since *Plant speciation. New Phytol.* 161:173–91

Soltis PS, Soltis DE. 2000b. The role of genetic and genomic attributes in the success of polyploids. *Proc. Natl. Acad. Sci. USA* 97:7051–57

Spooner DM, Van den Berg RG, Rivera-Peña A, Velguth P, Del Rio A, Salas-López A. 2001. Taxonomy of Mexican and Central American members of *Solanum* series *Conicibaccata* (sect. *Petota*). *Syst. Bot.* 26:743–56

Stace CA. 2000. Cytology and cytogenetics as a fundamental taxonomic resource for the 20th and 21st centuries. *Taxon* 49:451–77

Stehlik I, Kron P, Barrett SCH, Husband BC. 2007. Sexing pollen reveals female bias in a dioecious plant. *New Phytol.* 175:185–94

Stöck M, Lamatsch DK, Steinlein C, Epplen JT, Grosse WR, et al. 2002. A bisexually reproducing all-triploid vertebrate. *Nat. Genet.* 30:325–28

Suda J, Kron P, Husband BC, Trávníček P. 2007. Flow cytometry and ploidy: applications in plant systematics, ecology and evolutionary biology. See Doležel et al. 2007, 5:103–30

Suda J, Kyncl T, Jarolímová V. 2005. Genome size variation in Macaronesian angiosperms: forty percent of the Canarian endemic flora completed. *Plant. Syst. Evol.* 252:215–38

Suda J, Lysák MA. 2001. A taxonomic study of the *Vaccinium* sect. *Oxycoccus* (Hill) W.D.J. Koch (Ericaceae) in the Czech Republic and adjacent territories. *Folia Geobot.* 36:303–20

Suda J, Malcová R, Abazid D, Banaš M, Procházka F, et al. 2004. Cytotype distribution in *Empetrum* (Ericaceae) at various spatial scales in the Czech Republic. *Folia Geobot.* 39:161–71

Swift H. 1950. The constancy of desoxyribose nucleic acid in plant nuclei. *Proc. Natl. Acad. Sci. USA* 36:643–54

Tanaka M, Kimura S, Fujimoto T, Sakao S, Yamaha E, Arai K. 2003. Spontaneous mosaicism occurred in normally fertilized and gynogenetically induced progeny of the kokanee salmon *Oncorhynchus nerka*. *Fish. Sci.* 69:176–80

Thompson DA, Desai MM, Murray AW. 2006. Ploidy controls the success of mutators and nature of mutations during budding yeast evolution. *Curr. Biol.* 16:1581–90

Thompson JD, Lumaret R. 1992. The evolutionary dynamics of polyploid plants: origins, establishment and persistence. *Trends Ecol. Evol.* 7:302–7

Thompson JN, Nuismer SL, Merg K. 2004. Plant polyploidy and the evolutionary ecology of plant/animal interactions. *Biol. J. Linn. Soc.* 82:511–19

Tobiason DM, Seifert HS. 2006. The obligate human pathogen, *Neisseria gonorrhoeae*, is polyploid. *PLoS Biol.* 4:1069–78

Veldhuis MJW, Cucci TL, Sieracki ME. 1997. Cellular DNA content of marine phytoplankton using two new fluorochromes: taxonomic and ecological implications. *J. Phycol.* 33:527–41

Veldhuis MJW, Kraay GW. 1993. Cell abundance and fluorescence of picoplankton in relation to growth irradiance and nitrogen availability in the Red Sea. *Neth. J. Sea Res.* 31:135–45

Vinogradov AE. 1995. Nucleotypic effect in homeotherms: body-mass-corrected basal metabolic rate of mammals is related to genome size. *Evolution* 49:1249–59

Vinogradov AE. 1997. Nucleotypic effect in homeotherms: body-mass independent resting metabolic rate of passerine birds is related to genome size. *Evolution* 51:220–25

Vinogradov AE, Borkin LJ, Günther R, Rosanov JM. 1991. Two germ cell lineages with genomes of different species in one and the same animal. *Hereditas* 114:245–51

Vinogradov AE, Chubinishvili AT. 1999. Genome reduction in a hemiclonal frog *Rana esculenta* from radioactively contaminated areas. *Genetics* 151:1123–25

Vives-Rego J, Lebaron P, Caron GN-v. 2000. Current and future applications of flow cytometry in aquatic microbiology. *FEMS Microbiol. Rev.* 24:429–48

Vižintin L, Javornik B, Bohanec B. 2006. Genetic characterization of selected *Trifolium* species as revealed by nuclear DNA content and ITS rDNA region analysis. *Plant Sci.* 170:859–66

Vrijenhoek RC, Dawley RM, Cole CJ, Bogart JP. 1989. A list of the known unisexual vertebrates. See Dawley & Bogart 1989, pp. 19–23

Wakamiya I, Newton RJ, Johnston JS, Price HJ. 1993. Genome size and environmental factors in the genus *Pinus*. *Am. J. Bot.* 80:1235–41

Walker DJ, Moñino I, Correal E. 2006. Genome size in *Bituminaria bituminosa* (L.) C.H. Stirton (Fabaceae) populations: separation of "true" differences from environmental effects on DNA determination. *Environ. Exp. Bot.* 55:258–65

Warwick SI, Simard M-J, Légère A, Beckie HJ, Braun L, et al. 2003. Hybridization between transgenic *Brassica napus* L. and its wild relatives: *Brassica rapa* L., *Raphanus raphanistrum* L., *Sinapis arvensis* L., and *Erucastrum gallicum* (Willd.) O.E. Schulz. *Theor. Appl. Genet.* 107:528–39

Yeung K, Miller JS, Savage AE, Husband BC, Igic B, Kohn JR. 2005. Association of ploidy and sexual system in *Lycium californicum* (Solanaceae). *Evolution* 59:2048–55

Zhang Q, Arai K. 1999. Distribution and reproductive capacity of natural triploid individuals and occurrence of unreduced eggs as a cause of polyploidization in the loach, *Misgurnus anguillicaudatus*. *Ichthyol. Res.* 46:153–61

Zhu G, Golding GB, Dean AM. 2005. The selective cause of an ancient adaptation. *Science* 307:1279–82

Cumulative Indexes

Contributing Authors, Volumes 34–38

Dudash MR, 35:375–403;
36:467–97
Dudley R, 38:179–201
Duffy JE, 38:739–66
Dukas R, 35:347–74
Duncan RP, 34:71–98
Dussault C, 35:113–47

E

Eckert CG, 36:47–79
Elena SF, 38:27–52
Elmqvist T, 35:557–81
Elser JJ, 36:219–42
Emmerson M, 36:419–44
Evans EW, 37:95–122
Eviner VT, 34:455–85
Ezenwa V, 34:517–47

F

Fahrig L, 34:487–515
Falkowski PG, 35:523–56
Finkel ZV, 35:523–56
Fitzpatrick BM,
38:459–87
Fletcher RJ Jr,
35:491–522
Foley WJ, 36:169–89
Folke C, 35:557–81
Fornoni J, 38:541–66
Ford SE, 35:31–54
Foster MS, 37:343–72
Frati F, 37:545–79
French V, 34:633–60
Funk DJ, 34:397–423

G

Gage DJ, 37:459–88
Gardner A, 38:53–77
Gerardo NM, 36:563–95
Gilbert GS, 35:675–700
Gioia P, 35:623–50
Gittleman JL, 34:517–47
Goodwillie C, 36:47–79
Goulson D, 34:1–26
Graham AL, 36:373–97
Graham CH, 36:519–39
Griffin AS, 38:53–77
Grosberg RK, 38:621–54

Grzebyk D, 35:523–56
Gunderson L, 35:557–81

H

Halanych KM, 35:229–56
Hallinan ZP, 35:175–97
Hampe A, 37:187–214
Hansen TF, 37:123–57
Harshman LG, 38:793–817
Harvell CD, 37:251–88
Hättenschwiler S,
36:191–218
Hauser MD, 36:499–518
Hawkins S, 37:373–404
Hay ME, 35:175–97
Hedrick PW, 37:67–93
Heil M, 34:425–53
Heiman KW, 36:643–89
Helmuth B, 37:373–404
Hill GE, 34:27–49
Hoffmann AA, 37:433–58
Holder MT, 37:19–42
Holling CS, 35:557–81
Holzapfel CM, 38:1–25
Hopper SD, 35:623–50
Huber SK, 35:55–87
Hungate BA, 37:611–36
Husband BC, 38:847–76

I

Irwin RE, 35:435–66
Islas S, 38:361–79

J

Jackson ST, 38:275–97
Jennions MD, 37:43–66
Johnston MO, 36:467–97
Jones KE, 34:517–47
Jones LE, 37:251–88
Jones MB, 37:519–44
Jordano P, 38:567–93
Joyce P, 36:445–66

K

Kabil H, 38:299–326
Kalisz S, 36:47–79
Katul G, 38:767–91

Katz ME, 35:523–56
Kaufman DM,
34:273–309
Kay MC, 36:643–89
Klingenberg CP, 36:1–21
Knight TM, 36:467–97
Knoll AH, 35:523–56
Knürr T, 38:595–619
Koch PL, 37:215–50
Koenig WD, 35:467–90
Kokko H, 37:43–66
Kron P, 38:847–76
Kronfeld-Schor N,
34:153–81
Kruckeberg AR, 36:243–66

L

Labandeira CC, 35:285–322
Lafferty KD, 35:31–54
Lartillot N, 36:541–62
Lazcano A, 38:361–79
Leamy LJ, 36:1–21
Lenihan HS, 36:643–89
Levin SA, 34:575–604
Levine JM, 34:549–74
Lewinsohn TM,
36:597–620
Lieberman BS, 34:51–69
Liebhold A, 35:467–90
Linder HP, 36:107–24
Liow LH, 35:323–45
Luo Y, 37:611–36;
38:683–712

M

Markow TA, 36:219–42
Marshall JC, 35:199–227
Mateos M, 36:219–42
Mazer SJ, 36:467–97
McDonald ME, 35:89–111
McGill BJ, 38:403–35
McGuire JA, 38:179–201
McKey D, 34:425–53
McLean S, 36:169–89
Meyer A, 34:311–38
Meyers LA, 38:203–30
Micheli F, 36:643–89
Mieszkowska N,
37:373–404

Milinski M, 37:159–86
Miller DJ, 37:489–517
Mitchell RJ, 36:467–97
Mooij WM, 36:147–68
Moore P, 37:373–404
Moriuchi KS, 38:437–57
Mueller UG, 36:563–95
Muller-Landau HC,
    34:575–604
Murrell DJ, 34:549–74
Mydlarz LD, 37:251–88
Myers JH, 34:239–72

N

Nathan R, 34:575–604
Nee S, 37:1–17
Niemi GJ, 35:89–111
Noor MAF, 34:339–64
Novotny V, 36:597–620
Nowak MD, 37:405–31
Núñez-Farfán J,
    38:541–66
Nunn CL, 34:517–47

O

Oakley CG, 38:437–57
Ohgushi T, 36:81–105
Olszewski TD, 35:285–322
Omland KE, 34:397–423
Oren R, 38:767–91

P

Pandolfi JM, 35:285–322
Parichy DM, 38:655–81
Parker IM, 35:675–700
Parker JD, 35:175–97
Parmesan C, 37:637–69
Partridge L, 38:299–326
Peck LS, 38:129–54
Pedersen AB, 34:517–47
Peichel CL, 38:655–81
Petit RJ, 37:187–214
Philippe H, 36:541–62
Pilson D, 35:149–74
Pletcher SD, 38:299–326
Podos J, 35:55–87
Porporato A, 38:767–91
Porter JW, 35:31–54

Poss M, 34:517–47
Prendeville HR, 35:149–74
Proches Ş, 38:275–97
Pulliam JRC, 34:517–47
Pyhäjärvi T, 38:595–619

R

Rand DM, 36:621–42
Raven JA, 38:255–73
Read AF, 36:373–97
Reich PB, 37:611–36
Reichman OJ, 37:519–44
Richardson DM,
    38:275–97
Ries L, 35:491–522
Ritchie MG, 38:79–102
Rodrigues ASL, 38:713–37
Ronce O, 38:231–53
Rooney TP, 35:113–47
Roopnarine PD, 34:605–32
Rose KA, 34:127–51
Rosenthal GG, 38:155–78
Rudall PJ, 36:107–24
Ruesink JL, 36:643—89
Rundel PW, 38:275–97

S

Sanjuán R, 38:27–52
Savolainen O, 38:595–619
Scheffer M, 35:557–81
Scheu S, 36:191–218
Schiel DR, 37:343–72
Schildhauer MP,
    37:519–44
Schlupp I, 36:399–417
Schultz TR, 36:563–95
Seibel BA, 38:129–54
Selden P, 38:819–46
Servedio MR, 34:339–64
Shine R, 36:23–46
Simmons LW,
    36:125–46
Simon C, 37:545–79
Sinervo B, 37:581–610
Sisk TD, 35:491–522
Sites JW Jr, 35:199–227
Six DL, 36:563–95
Smith JE, 38:327–59
Snyder WE, 37:95–122

Sodhi NS, 35:323–45
Sol D, 34:71–98
Srivastava DS, 36:267–94
Stachowicz JJ, 38:739–66
Stadler B, 36:345–72
St. Clair CC, 37:317–42
Steets JA, 36:467–97
Stevens JR, 36:499–518
Stevens RD, 34:273–309
Stewart IRK, 34:365–96
Stewart JB, 37:545–79
Strathmann RR, 38:621–54
Strauss SY, 35:435–66
Streelman JT, 38:655–81
Suda J, 38:847–76
Sullivan J, 36:445–66

T

Taft B, 35:55–87
Teskey RO, 38:275–97
Thatje S, 38:129–54
Thomson JD, 35:375–403
Thrall PH, 34:517–47
Tiffney BH, 35:1–29
Tiunov AV, 36:191–218
Tremblay J-P, 35:113–47
Trimble AC, 36:643–89
Turner MG, 36:319–44

U

Usher KM, 38:255–73

V

Valverde PL, 38:541–66
Vamosi JC, 36:467–97
Van Buskirk J, 37:433–58
van Oppen MJH,
    37:489–517
Vellend M, 36:267–94
Vollmer SV, 37:489–517
Vollrath F, 38:819–46

W

Wade MJ, 37:289–316
Wainwright PC, 38:381–401
Walker B, 35:557–81
Waller DM, 35:113–47

Weider LJ, 36:219–42
Welsh HH Jr, 35:405–34
Wertheim JO, 38:515–40
West SA, 38:53–77
Westneat DF, 34:365–96
Whigham DF, 35:583–621
Wiens JJ, 36:519–39
Wilga CD, 38:129–54
Willi Y, 37:433–58
Williams SL, 38:327–59

Williams TD, 38:793–817
Willig MR, 34:273–309
Willis BL, 37:489–517
Wilson AE, 35:175–97
Wilson P, 35:375–403
Wing SL, 35:285–322
Wingfield MJ, 38:275–97
Winn AA, 38:437–57
Wootton JT, 36:419–44
Worobey M, 38:515–40

Y

Yanoviak SP, 38:179–201
Yoder AD, 37:405–31

Z

Zardoya R, 34:311–38
Zera AJ, 38:793–817
Zwaan BJ, 34:633–60

Chapter Titles, Volumes 34–38

ANNUAL REVIEWS

Intelligent Synthesis of the Scientific Literature

Annual Reviews – Your Starting Point for Research Online
http://arjournals.annualreviews.org

- Over 1150 Annual Reviews volumes—more than 26,000 critical, authoritative review articles in 35 disciplines spanning the Biomedical, Physical, and Social sciences—available online, including all Annual Reviews back volumes, dating to 1932

- Current individual subscriptions include seamless online access to full-text articles, PDFs, Reviews in Advance (as much as 6 months ahead of print publication), bibliographies, and other supplementary material in the current volume and the prior 4 years' volumes

- All articles are fully supplemented, searchable, and downloadable — see http://ecolsys.annualreviews.org

- Access links to the reviewed references (when available online)

- Site features include customized alerting services, citation tracking, and saved searches

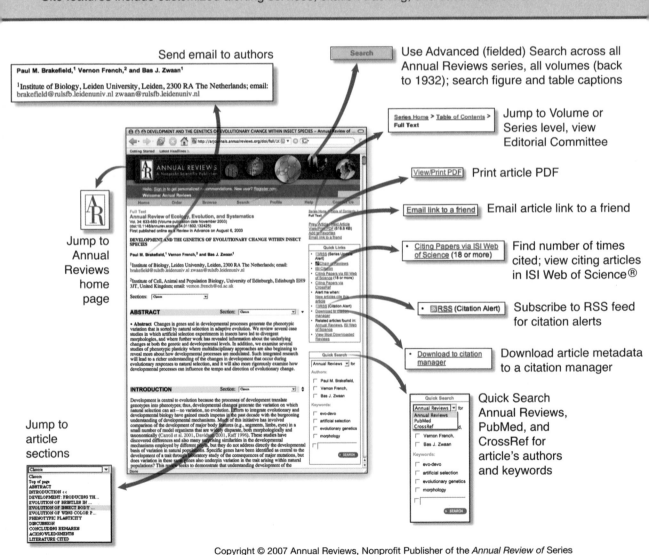

Send email to authors

Use Advanced (fielded) Search across all Annual Reviews series, all volumes (back to 1932); search figure and table captions

Jump to Volume or Series level, view Editorial Committee

Print article PDF

Email article link to a friend

Find number of times cited; view citing articles in ISI Web of Science®

Subscribe to RSS feed for citation alerts

Download article metadata to a citation manager

Quick Search Annual Reviews, PubMed, and CrossRef for article's authors and keywords

Jump to Annual Reviews home page

Jump to article sections